MECHANICAL BEHAVIOR
OF ANISOTROPIC SOLIDS

COMPORTEMENT MÉCANIQUE
DES SOLIDES ANISOTROPES

COLLOQUES INTERNATIONAUX
DU
CENTRE NATIONAL DE LA RECHERCHE SCIENTIFIQUE

——

N° 295

COMPORTEMENT MÉCANIQUE DES SOLIDES ANISOTROPES

Colloque Euromech 115
Villard-de-Lans, 19-22 juin 1979

Publié sous la direction de

Jean-Paul BOEHLER

Professeur à l'Université de Grenoble
Institut de Mécanique de Grenoble

Martinus Nijhoff Publishers
The Hague/Boston/London

Editions du CNRS
15, quai A. France - 75700 Paris

1982

COLLOQUES INTERNATIONAUX
DU
CENTRE NATIONAL DE LA RECHERCHE SCIENTIFIQUE

N° 295

MECHANICAL BEHAVIOR OF ANISOTROPIC SOLIDS

Proceedings of the Euromech Colloquium 115
Villard-de-Lans, June 19-22, 1979

Edited by

Jean-Paul BOEHLER

Professor, University of Grenoble, France
Institut de Mécanique de Grenoble

Martinus Nijhoff Publishers
The Hague/Boston/London

Editions du CNRS
15, quai A. France - 75700 Paris

1982

Distributors :
for the United States and Canada
Kluwer Boston, Inc.
190 Old Derby Street
Hingham, MA 02043
USA

Ventes en France :
Editions du CNRS
Service des ventes
295, rue Saint-Jacques
75005 PARIS

for all other countries
Kluwer Academic Publishers Group
Distribution Center
P.O. Box 322
3300 AH Dordrecht
The Netherlands

ISBN-13: 978-94-009-6829-5   e-ISBN-13: 978-94-009-6827-1
DOI: 10.1007/978-94-009-6827-1

Library of Congress Cataloging in Publication Data

Joint editions published by
MARTINUS NIJHOFF PUBLISHERS
P.O. Box 566, 2501 CN The Hague, The Netherlands
and CENTRE NATIONAL DE LA RECHERCHE SCIENTIFIQUE
15, quai Anatole-France - 75700 PARIS

Foreword

In 1978, the European Mechanics Committee and the French Centre National de la Recherche Scientifique agreed to the organization of an International Colloquium on the "Mechanical Behavior of Anisotropic Solids". The meeting was held at Villard-de-Lans (near Grenoble, France) from 19th to 22 nd June 1979.

The Colloquium considered mechanical aspects of the anisotropy of solids, both initial and induced by permanent deformation, anisotropic hardening and damage, oriented fissuration, etc. Topics concerned mathematical, experimental and engineering aspects of the anisotropy of metals, composites, soils and rocks. The aim of the Colloquium was to bring together experimentalists, theoretecians and engineers interested in various features of mechanical anisotropy, in order to permit an interdisciplinary exchange of understanding, experience and methods. A detailed description of the scope, aim and proposed topics is contained in the Preface.

The announcement of the Colloquium attracted a large number of submitted contributions. Conforming with the principles of Euromech Colloquia and of the Colloques Internationaux du CNRS, the accepted contributions were limited to 50 communications.

A general description of the scientific program is to be found in the Preface. Five general lectures gave state-of-the-art reports concerning some areas of the behavior of anisotropic solids ; the 50 communications were divided into 12 sessions dealing with specific topics (see "Contents"). In order to facilitate subsequent contact between the reader and the contributors, full addresses are given in the "List of Authors".

A conference room, meals and accommodation were provided at the "Grand Hôtel de Paris" in Villard-de-Lans, a small and pleasant holiday resort in the Alps. This arrangement encouraged personal contact between the participants after the sessions. Throughout the day, informal discussions were held in the various hotel suites.

We would like to express our thanks to the 84 participants representing 15 different countries. If the Colloquium was successful in achieving its aims, it is due to their common efforts. Special thanks are owed to Dr. BIGUENET for his constant and valuable assistance in resolving technical problems, and a word of appreciation to the two interpreters from Grenoble for their conscientious preparation and efficient services throughout the Colloquium.

A meeting such as this can only be organized if financial assistance is guaranteed. We are therefore indebted to our sponsors. Major support was provided by the CNRS. A grant from the DRET (Ministère de la Défense) financed the simultaneous translation services. Additional funds were granted by the French "Association Universitaire de Mécanique" and the "Conseil Général de l'Isère".

Finally, we are grateful to the "Editions Scientifiques du CNRS", which financed the publication of this volume of the Proceedings and to the "Imprimerie Louis-Jean" for its excellent job in printing the texts.

Jean-Paul BOEHLER

Avant-propos

En 1978, le Centre National de la Recherche Scientifique et le Comité Européen de Mécanique ont agréé l'organisation d'un Colloque International sur le "Comportement Mécanique des Solides Anisotropes". Cette rencontre s'est tenue à Villard-de-Lans du 19 au 22 juin 1979.

Le Colloque était spécialisé dans les aspects mécaniques de l'anisotropie des solides, aussi bien l'anisotropie initiale, que l'anisotropie induite par les déformations irréversibles, l'écrouissage et l'endommagement anisotropes, la fissuration orientée, etc. Les thèmes proposés concernaient les aspects mathématiques, expérimentaux et appliqués de l'anisotropie des métaux, des matériaux composites, des sols et des roches. Le but du Colloque était de réunir des expérimentateurs, des théoriciens et des ingénieurs s'intéressant aux différentes caractéristiques de l'anisotropie mécanique, afin de permettre un échange interdisciplinaire de conception, d'expérience et de méthode. Une description détaillée du domaine, du but et des thèmes proposés est incluse dans la Préface.

L'annonce du Colloque a attiré un grand nombre de propositions de contributions. Conformément aux principes des Colloques Internationaux du CNRS et des Colloques Euromech, les contributions acceptées ont été limitées à 50 communications.

Une description générale du programme scientifique est présentée dans la Préface. Cinq conférences générales ont permis de donner l'état actuel des connaissances dans certains domaines du comportement des solides anisotropes ; les 50 communications ont été regroupées en 12 sessions spécialisées, traitant de thèmes spécifiques (voir "Sommaire"). Afin de faciliter les contacts ultérieurs entre le lecteur et les auteurs, les adresses complètes sont données dans la "Liste des Auteurs".

La salle de conférences, les repas et les logements ont été fournis par le "Grand Hôtel de Paris" à Villard-de-Lans, une agréable petite station de vacances dans les Alpes. Cet arrangement a stimulé les contacts personnels entre les participants après les sessions. Tout au long des journées, des discussions libres se sont tenues dans les différents salons de l'hôtel.

Nous exprimons nos remerciements aux 84 participants, représentant 15 pays différents. Si le Colloque a réussi à atteindre ses buts, c'est grâce à leurs efforts communs. Nous remercions particulièrement M. BIGUENET pour son aide constante et appréciable dans la résolution des problèmes techniques, ainsi que les deux interprètes de Grenoble pour leur préparation consciencieuse et leurs services efficaces tout au long du Colloque.

Une telle rencontre ne peut être organisée que si une aide financière est assurée. C'est pourquoi nous sommes redevables à différents organismes pour avoir bien voulu parrainer ce projet. La contribution financière principale a été fournie par le CNRS. Un contrat de la DRET (Ministère de la Défense) a permis de financer les services de traduction simultanée. Des contributions complémentaires ont été accordées par l'Association Universitaire de Mécanique et le Conseil Général de l'Isère.

Finalement, nous sommes reconnaissants aux "Editions Scientifiques du CNRS", qui ont financé la publication de ce volume des Actes, et à l'"Imprimerie Louis-Jean" pour son excellent travail de composition et d'impression des textes.

Jean-Paul BOEHLER

Preface

On the Mechanical Behavior of Anisotropic Solids

The anisotropic properties of materials play an important role in numerous branches of Physics, Geology, Mechanics of Solids and Engineering. Situations commonly exist where oriented internal structures impart a directional character to the mechanical response of the material. Thus, the constitutive relations have to account for the fact that the material behavior is not invariant under arbitrary orthogonal transformations. A proper understanding, a rational description and adequate measurement of the anisotropic response of a material to mechanical, thermal, electrical or other agencies are vital in many areas of technology, engineering and biomechanics. For example, appropriate forming processes are needed in materials sciences to ensure suitable mechanical properties in composites or alloys ; proper methods must be defined to assess the carrying capacity of an oriented subsoil for an engineering structure ; suitable techniques must be proposed to evaluate the directional properties of alloys, composites, wood, rocks and living tissues.

This Colloquim at Villard-de-Lans in June 1979 focused specifically on the mechanical behavior of anisotropic solids, in particular on the variation of material properties, such as deformability and strength, according to the orientation of external agencies. It was therefore concerned with many problems that must be solved in engineering mechanics, in order to satisfy the multiple demands of our contemporary society. In an attempt to cover such a wide field of responsibilities, a suitable approach is to define a limited number of specific tasks.

In the first place, it is necessary to test, evaluate and describe in a mathematically objective manner the mechanical properties of anisotropic materials. The second task is to synthesize the material behavior into appropriate mathematical problems, which, as far as possible, must take into account the various couplings between the particular anisotropies, e.g. in the elastic, plastic, damage and failure domains, and their evolution during permanent deformation. To be useful, the model must be mathematically correct and sufficiently manageable to allow for applications. Once a model is proposed and justified, the question is then to develop suitable methods to solve boundary and initial valued problems from the usual systems of differential equations. The final task of the engineer is to solve specific problems concerning deformability, fissuration and failure, should he intend to put the devised technological processes into practice or to execute an engineering work.

Our Colloquium in such a scheme is focused primarily on aspects of model building within an objective framework. Mechanical anisotropy is the main concern, both initial anisotropy due to material formation or manufacturing, and induced anisotropy due to deformation, fissuration and polarisation of initially isotropic or anisotropic solids. This is considered, bearing in mind the significance of the directional properties of solids in metallurgy, engineering mechanics, naval engineering, aeronautics, geology, the development of composite or laminated materials, tunnel or storage zones in stratified rocks, the prevention of avalanches, seismology engineering, as well as in many other fields of materials sciences and engineering.

The evaluation of anisotropic properties and the devising of appropriate mathematical models of mechanical behavior require suitable objective methods to study general nonlinear and coupled phenomena in deformation and induced anisotropy. Considerable knowledge has already been accumulated in various branches of engineering, both in university centers and in specific industries. It appears, however, that theoretical developments are often based on mathematical models conceived ad hoc, and that the experimental work does not always account for essential features of anisotropic responses. The experimental techniques sometimes pass over some fundamental aspects, such as an objective evaluation of anisotropic plastic hardening or fissuration-induced softening, a proper interpretation of the "off-axis" tests for anisotropic solids, or a suitable correlation between the structural micro- and the phenomenological macro-anisotropy – to mention only a few essential theoretical and applied requirements. From both a technological and a scientific viewpoint, an assessment of the state-of-the-art was required along with the added possibility of defining a methodology soundly based on modern nonlinear mechanics and developed in the scope of a unified approach.

Taking into consideration, on the one hand, research aspects and, on the other, the need to furnish engineers with reliable theories and suggestions for suitable methods of approach to anisotropy in nonlinear cases, a specialized meeting was devised. The idea was to bring together theoreticians, experimentalists and engineers, interested in the fundamental aspects of mechanical anisotropy, for discussions and interdisciplinary exchanges on the problems of material modelling and the experimental evaluation of the directional properties of solids.

The idea was well received and approved by both the European Mèchanics Colloquia Committee and the French Centre National de la Recherche Scientifique. Their support enabled us to make the idea a reality and to organize this meeting on the Mechanical Behavior of Anisotropic Solids. The accepted papers are contained in this volume.

Our attention was focused primarily on the aspects of modelling and experiments. We felt it necessary, however, to include some studies concerning problem-solving methods, as well as certain specific solutions. This was in keeping with our principal line of thought, ranging from mathematical and experimental concepts to engineering applications, through the development of suitable mathematical aids, since good engineering requires mathematically sound and reliable theories.

This Colloquium was to enable mathematicians, physicists, metallurgists, specialists in mechanics of solids, soils and rocks, geologists and practising engineers, all interested in the understanding and application of the anisotropic properties of solids, to meet together for the first time.

Apart from its general purpose, any meeting, to be successful, has to define clear objectives and specific guidelines for discussions, in order, rather than simply accumulating a mass of knowledge, to arrive at a clear and ordered pattern. Thus, the objectives we hoped to attain were specified as follows: a) assess the actual state of knowledge in mechanical anisotropy of solids; b) permit an interdisciplinary exchange of understanding, methods and accumulated experience in describing various anisotropies; c) formulate or sketch a common basis and suggest suitable methods for mathematical modelling of anisotropic behavior; d) discuss and assess experimental techniques permitting an objective evaluation of the mechanical properties of anisotropic behavior, both for the anisotropy of formation and the anisotropy due to deformation and to the variation of the internal structure under the effect of external agencies; e) discuss the relations between micro- and macro- anisotropy, i.e. the connections between the oriented structure (inborn or induced) and mechanical anisotropy, including the evolution of such connections.

Besides these objectives, related to leading aspects of anisotropy, we had in mind that the meeting would help to specify the directions of future interdisciplinary research. The end result of the Colloquium was intended to determine the usefulness of such working sessions, both in accounting for the actual state-of-the-art, and in specifying particular domains, methods and techniques to be subsequently developed.

Five general lectures in well defined domains formed the framework of the Colloquium. Our first aim was to have a report on the present state of knowledge concerning the invariant formulation of constitutive equations for anisotropic behavior in general. Secondly, since deformation-induced anisotropy and its measurement are essential, a general lecture was devoted to plastic anisotropy. This domain appears to be largely studied experimentally. Hence, the relations between experimental facts and their actual theoretical explanation provide a possibility of revealing the failings of currently employed mathematical modelling. Micro- and macro-aspects of anisotropy were the topic of the third general lecture concerning the metallurgical features of anisotropy. It was intended to illustrate the states of research on the correlations between the directional characteristics of the internal structure and the bulk material behavior. Reports on the anisotropy of composites and the anisotropy of natural rocks close the group of general lectures, outlining practical aspects of the Mechanical Behavior of Anisotropic Solids.

Original contributions, chosen in accordance with the adopted guidelines, were grouped in special sessions, concerning e.g. the invariant formulation of constitutive equations for materials with directional characteristics, experimental and theoretical aspects of various oriented materials such as composites, soils, rocks and ice, and deformation-induced anisotropy in creep and continuous damage.

It seems that the Colloquium attained the goals proposed. After presentation and discussions, a clear picture was formed, concerning the most suitable formalism for the development of anisotropic constitutive equations and of relations between the micro-structural view and the phenomenological theories of mechanical behavior. This outlines a suitable approach for the study of the evolution of anisotropic properties during irreversible deformations, due to plasticity, fissuration and creep. Extensive theoretical work is needed on constitutive and evolution equations, presumably within the framework discussed here, to arrive at an objective and unified formalism.

We are confident that this volume contains pertinent and useful information on present-day knowledge and understanding concerning the mechanics of oriented solids, with particular emphasis laid on the nonlinear and inelastic aspects of anisotropy. This conviction is supported by the fact that the participants insisted on presenting the results available to a larger community and expressed their intention of maintaining the established contacts via appropriate interdisciplinary meetings on specific subjects in the near future. Among the promising subjects put forward for further studies were: appropriate accounting for materials internal structure, continuous oriented fissuration and damage under sustained or repeated external agencies, material hardening and softening, failure of oriented or stratified solids.

Mechanical anisotropies will no doubt continue to attract the attention of engineers in various fields, as well as that of material scientists and biologists in the years to come. We hope that this volume will be of help in their future studies.

J.P. BOEHLER
University of Grenoble

A. SAWCZUK
Polish Academy of Sciences

Grenoble and Warsaw, January 1981

Préface

Sur le Comportement Mécanique des Solides Anisotropes

Le comportement mécanique des solides présente souvent un caractère anisotrope, c'est-à-dire une variation de la réponse mécanique suivant l'orientation des sollicitations extérieures. L'origine de l'anisotropie mécanique réside dans le caractère orienté de la structure interne des matériaux, dû aux modes de formation ou aux procédés de fabrication (anisotropie initiale) : métaux laminés et filés, alliages et eutectiques orientés, composites fibreux ou lamellaires, roches stratifiées, argiles consolidées, sables compactés, bois, tissus biologiques (os), etc. Lorsqu'un matériau solide subit des déformations plastiques irréversibles, la structure interne évolue jusqu'à la rupture (réorientation des particules et des axes cristallographiques, glissements internes orientés, formation de fissures et de cavités orientées, etc.). Ces modifications de la structure microscopique provoquent une évolution de l'anisotropie des propriétés mécaniques macroscopiques (anistropie induite).

Les propriétés mécaniques des solides anisotropes jouent un rôle important dans la Métallurgie, la Science de l'Ingénieur, le Génie Civil et la Géologie. Une bonne compréhension, une description et une mesure appropriées de la réponse directionnelle des solides soumis à des sollicitations mécaniques, thermiques, électriques ou autres sont d'une importance vitale dans différentes branches de la technologie et concernent en général les problèmes de la Science de l'Ingénieur dans les objectifs de la société contemporaine. A titre d'exemples, on peut citer deux applications liées à des soucis de qualité des produits et d'économie de matières premières et d'énergie : optimisation des procédés de fabrication pour obtenir des matériaux présentant des caractéristiques mécaniques adaptées ; optimisation de la tenue des structures soumises à des conditions d'utilisation sévères. L'utilité pratique du développement des connaissances dans le domaine du comportement des solides anisotropes concerne des secteurs très diversifiés, comme la métallurgie, la mécanique de l'ingénieur, la construction navale, aéronautique et nucléaire, le développement des matériaux composites et renforcés, la géologie, la construction de tunnels et de zones de stockage sous-terrains, la prévention des avalanches, la sismologie, etc.

Les études du comportement plastique des solides anisotropes peuvent être regroupées suivant quatre axes : 1) tester, évaluer et décrire d'une manière objective les propriétés mécaniques des matériaux anisotropes ; 2) synthétiser le comportement matériel dans des modèles mathématiques adéquats, prenant en compte les effets des grandes déformations irréversibles ; 3) développer des méthodes appropriées pour la résolution des problèmes aux limites ; 4) résoudre des problèmes spécifiques concernant la déformabilité, la fissuration et la rupture, en vue de la mise en œuvre de procédés technologiques adaptés et de la construction de structures optimisées. Les perspectives à moyen terme concernent les trois premiers axes, dont le développement est indispensable pour pouvoir résoudre des problèmes spécifiques.

La détermination des propriétés anisotropes et le développement des modèles mathématiques du comportement mécanique dans le domaine non linéaire exigent des études théoriques et expérimentales adéquates. Une somme de connaissances et de résultats pour la description des différents aspects de l'anisotropie est déjà disponible, aussi bien dans les centres universitaires, que dans l'industrie. Cependant, les approches théoriques et expérimentales actuelles sont encore trop diversifiées dans les techniques utilisées et limitées dans les résultats obtenus. Des problèmes fondamentaux comme, par exemple, la description réaliste de l'écrouissage anisotrope, l'interprétation correcte des essais "hors-axe", la quantification des corrélations entre l'anisotropie de structure et l'anisotropie mécanique, sont encore mal formulés. D'autre part, la diversité des disciplines scientifiques et des domaines techniques dans lesquels l'anisotropie joue un rôle, a empêché une concertation entre les différentes approches des problèmes liés au comportement anisotrope des solides.

Devant une telle situation, il nous a semblé opportun de provoquer la réunion de théoriciens, d'expérimentateurs et d'ingénieurs s'intéressant aux diverses caractéristiques de l'anisotropie des solides. Le Comité Européen de Mécanique et le Centre National de la Recherche Scientifique ont bien voulu nous confier l'organisation d'un Colloque International sur le Comportement Mécanique des Solides Anisotropes.

Ce Colloque a réuni, pour la première fois semble-t-il, des Mathématiciens, des Physiciens, des Métallurgistes, des Mécaniciens des Solides, des Sols et des Roches, des Géologues et des Ingénieurs, spécialistes de l'anisotropie des solides dans leurs domaines de recherches spécifiques, pour essayer d'atteindre les objectifs suivants : a) évaluer l'état actuel des connaissances dans le domaine du comportement mécanique des solides anisotropes ; b) permettre un échange interdisciplinaire de conception, d'expérience et de méthode ; c) aider à développer ou améliorer les modélisations mathématiques des comportements anisotropes, les techniques expérimentales pour la mesure objective des propriétés mécanique et les études des corrélations entre l'anisotropie de la structure matérielle et l'anisotropie du comportement macroscopique ; d) définir les voies de recherches prioritaires ; e) favoriser la poursuite ultérieure de cette concertation interdisciplinaire.

Cinq conférences générales, sur des thèmes bien définies, ont formé le cadre général du Colloque. Notre premier but était d'avoir un rapport sur l'état actuel des connaissances concernant la formulation invariante des équations constitutives pour le comportement anisotrope en général. Ensuite, comme l'anisotropie induite par déformation et sa mesure sont essentielles, une conférence générale a été consacrée à l'anisotropie plastique des métaux. Les études expérimentales dans ce domaine sont assez nombreuses. La confrontation entre les réalités expérimentales et les tentatives actuelles pour leur explication théorique a fourni la possibilité de révéler les faiblesses des modèles mathématiques couramment utilisés. Les anisotropies microscopiques et macroscopiques ont constitué le thème de la troisième conférence générale, consacrée aux aspects métallurgiques de l'anisotropie. Le but était de présenter l'état actuel des recherches dans le domaine des corrélations entre les caractéristiques direction-nelles de la structure interne des métaux et l'anisotropie du comportement méca-nique global. Les matériaux composites jouent un rôle très important dans la technologie moderne. Une caractéristique essentielle de ces matériaux artificiels est l'anisotropie très prononcée de leur comportement mécanique. La quatrième conférence présente les différentes méthodes permettant de déterminer les pro-priétés mécaniques macroscopiques à partir des propriétés des consituants, ainsi que les résultats obtenus dans l'étude des déformations élastoplastiques de ces matériaux. La cinquième conférence générale a été consacrée aux connaissances actuelles dans le domaine du comportement mécanique des roches, en liaison avec les modes de formation de ces matériaux naturels et les déformations tectoniques subies au cours de leur histoire. Le cycle des conférences générales s'achève par ces deux derniers rapports, exposant les aspects pratiques du Comportement Mécanique des Solides Anisotropes.

Les contributions originales ont été regroupées en douze sessions traitant de thèmes spécifiques. Trois sessions spécialisées ont été consacrées à la formu-lation invariante des lois de comportement, aux propriétés physiques des maté-riaux anisotropes et aux problèmes de Génie Civil. Les neuf autres sessions ont été consacrées aux différents aspects théoriques, expérimentaux et appliqués des matériaux anisotropes, comme les métaux, les composites, les eutectiques, les cristaux liquides, les sols, les roches, le bois et la glace. Les comportements mécaniques étudiés concernent l'élasticité, la plasticité, la viscoplasticité, l'écrouissage, le fluage et l'endommagement anisotropes. Plusieurs communi-cations sont spécialisées dans l'étude de l'anisotropie induite par déformation irréversible, évolution de la structure interne et fissuration orientée.

Il semble que le Colloque a atteint les buts proposés. Après la présentation des contributions et les discussions, des idées claires se sont formées, concernant le formalisme le plus adéquat pour le développement des lois de comportement anisotrope et des relations entre l'approche au niveau de la microstructure et l'approche macroscopique phénoménologique du comportement mécanique des solides anisotropes. L'étude de l'évolution des propriétés anisotropes au cours des déformations irréversibles exige encore un travail théorique important, qui devrait être développé dans le cadre général défini par le Colloque, afin d'aboutir à un formalisme objectif et unifié.

Nous sommes persuadés que ce volume contient des renseignements pertinents et utiles sur les connaissances actuelles dans le domaine de la mécanique des solides à structure interne orientée et plus particulièrement sur les aspects non linéaires de l'anisotropie. Cette conviction est renforcée par le fait que les participants ont exprimé le souhait que les résultats présentés soient disponibles pour une plus grande communauté et que les contacts établis puissent être maintenus par des rencontres interdisciplinaires sur des sujets spécifiques dans un proche avenir. Parmi les voies de recherches prioritaires à développer, nous mentionnons la prise en compte appropriée des structures internes orientées, la croissance de la fissuration et de l'endommagement sous l'effet des solliciations externes, l'écrouissage anisotrope, la rupture des solides anisotropes.

Dans les années à venir, l'anisotropie mécanique continuera certainement d'attirer l'attention des ingénieurs dans les différents domaines d'applications, ainsi que celle des spécialistes de la science des matériaux et des biologistes. Nous espérons que ce volume apportera une aide dans leurs recherches futures.

J.P. BOEHLER
Université de Grenoble

A. SAWCZUK
Académie Polonaise des Sciences

Grenoble et Varsovie, Janvier 1981

Chairmen – *Organisateurs*

Prof. J.P. BOEHLER, Intitut de Mécanique de Grenoble, Université de Grenoble, B.P. 53X, 38041 Grenoble Cédex, France.

Prof. A. SAWCZUK, Polish Academy of Sciences, Institute of Fundamental Technological Research, Świętokrzyska 21, 00-049 Warsaw, Poland.

List of Authors – *Liste des Auteurs*

Numbers in parentheses indicate the pages on which the authors' contributions begin.

Les nombres entre parenthèses indiquent les premières pages des contributions des auteurs.

AMBARTSUMYAN S.A., (663), Yerevan State University, 375049 Yerevan, U.S.S.R.

ARULANANDAN K., (183), University of California, Department of Civil Engineering, Bainer Hall, Davis, California 95616, U.S.A.

AUBERT G., (107), Laboratoire Louis Néel associé à l'U.S.M.G., C.N.R.S., B.P. 166 X, 38042 Grenoble Cédex, France.

BACKHAUS G., (273), Herkulesstrasse 3, 8020 Desden, Germany-DDR.

BALTOV A., (243), Bulgarian Academy of Sciences, Inst. Mech. and Biomech., P.O. Box 373, 1090 Sofia, Bulgaria.

BAMBERGER Y., (715), Electricité de France, Direction des Etudes et Recherches, 1, avenue du Général de Gaulle, 92141 Clamart Cédex, France.

BERVEILLER M. (333), C.N.R.S.-Université Paris-Nord, Laboratoire PMTM, Avenue J.B. Clément, 93430 Villetaneuse, France.

BETTEN J., (65), RWTH Aachen, Templergraben 55, D-5100, Aachen, Germany-BRD.

BOEHLER J.P., (449), Institut de Mécanique de Grenoble, B.P. 53 X, 38041 Grenoble Cédex, France.

BONTCHEVA Nikolina, (243), Bulgarian Academy of Sciences, Inst. Mech. and Biomech., P.O. Box 373, 1090 Sofia, Bulgaria.

CANNARD G., (715), Laboratoire Régional des Ponts et Chaussées, 109, avenue Salvador Allende, B.P. 48, 69672 Bron Cédex, France.

CHABOCHE J.L., (737), O.N.E.R.A. (Office National d'Etudes et de Recherches Aérospatiales), 29, avenue de la Division Leclerc, 92320 Châtillon-sous-Bagneux, France.

CHAMBEROD A., (107), D.R.F., Physique du Solide, C.E.N.G., 85 X, 38041 Grenoble Cédex, France.

CHOU S.C., (471), U.S. Army Materials and Mechanics Research Center, Watertown, Massachusetts 02172, U.S.A.

CIVIDINI Annamaria, (581), Politecnico di Milano, Department of Structural Engineering, Piazza Leonardo da Vinci 32, 20133 Milano, Italy.

CLIFTON R.J., (365), Brown University, Division of Engineering, Providence, Rhode Island 02912, U.S.A.

CORDEBOIS J.P., (761), E.N.S.E.T. (Ecole Nationale Supérieure de l'Enseignement Technique), 61, avenue du Président Wilson, 94230 Cachan, France.

DAFALIAS Y.F., (183), University of California, Department of Civil Engineering, Bainer Hall, Davis, California 95616, U.S.A

DAHAN M., (913), Laboratoire de Mécanique Appliquée, Université de Franche-Compté, Route de Gray, La Boulois, 25030 Besançon Cédex, France.

DELAFIN M., (449), Institut de Mécanique de Grenoble, B.P. 53 X, 38041 Grenoble Cédex, France.

DIENES J.K., (901), University of California, Los Alamos Scientific Laboratory, P.O. Box 1663, Los Alamos, New Mexico 87545, U.S.A.

DONATH F.A., (819), CGS Inc., 104 W. University, P.O. Box 907, Urbana, Illinois 61801, U.S.A.

DUVAL P., (789), Laboratoire de Glaciologie-C.N.R.S., B.P. 53 X, 38041 Grenoble Cédex, France.

DVORAK G.J., (383), University of Utah, Civil Engineering Department, 3012 Merrill Engineering Building, Salt Lake City, Utah 84112, U.S.A.

EGGER P., (887), Ecole Polytechnique Fédérale de Lausanne, Laboratoire de Mécanique des Roches, 22, avenue de Provence, CH-1015 Lausanne, Switzerland.

FILIPINNI J.C., (149), Laboratoire d'Electrostatique, C.N.R.S., B.P. 166 X, 38042 Grenoble Cédex, France.

GATTI G., (581), Politecnico di Milano, Department of Structural Engineering, Piazza Leonarda da Vinci 32, 20133 Milano, Italy.

GIODA G., (581), Politechnico di Milano, Department of Structural Engineering, Piazza Leonardo da Vinci 32, 20133 Milano, Italy.

GOGUEL J., (533) B.R.G.M. (Bureau de Recherches Géologiques et Minières), 103, rue de Lille, 75007 Paris, France.

GUITARD D., (869), Institut National Polytechnique de Lorraine, E.N.S.E.M., 2, rue de la Citadelle, B.P. 850, 54011 Nancy Cédex, France.

GUZ' A.N., (675), Institute of Mechanics, The Ukrainian Academy of Sciences, 252057 Kiev, U.S.S.R.

HASHIN Z., (407), Tel-Aviv University, Materials and Structures, School of Engineering, Department of Solid Mechanics, Tel-Aviv, Israël.

HOMAND-ETIENNE Françoise, (565), Centre de Recherches en Mécanique et Hydraulique des Sols et des Roches de l'Ecole de Géologie, Rue du Doyen Marcel Roubault, Case Officielle n° 2, 54500 Vandœuvre, France.

HORGAN C.O., (853), Michigan State University, Department of Metallurgy, Mechanics and Materials Science, East Lansing, Michigan 48824, U.S.A.

HOUPERT R., (565), Centre de Recherches en Mécanique et Hydraulique des Sols et des Roches de l'Ecole de Géologie, Rue du Doyen Marcel Roubault, Case officielle n° 2, 54500 Vandœuvre, France.

IKEGAMI K., (201, 257), Tokyo Institute of Technology, Research Laboratory of Precision Machinery and Electronics, 4259 Nagatsuta, Midori-Ku, Yokohama 227, Japan.

JOHNSON A.F., (775), National Physical Laboratory, Division of Materials Applications, Teddington, Middlesex TW11 OLW, England.

KANEKO K., (257), Science University of Tokyo, 2-3, Kagurazaka, Shinjuku-ku, Tokyo 162, Japan.

LE GAC H., (789), Laboratoire de Glaciologie-C.N.R.S., B.P. 53 X, 38041 Grenoble Cédex; France.

LEHMANN Th., (285), Ruhr Universität Bochum, Lehrstuhl für Mechanik I, Postfach 10 21 48, D-4630 Bochum 1, Germany-BRD.

LICHARDUS S., (423), Institute of Construction and Architecture SAV, Dúbravska cestá, 885 46 Bratislava, Czechoslovakia.

LITEWKA A., (803), Technical University of Poznań, ul. Piotrowo 5, 60-965 Poznań, Poland.

MALRAISON B., (149), Laboratoire d'Electrostatique, C.N.R.S., B.P. 166 X, 38042 Grenoble Cédex, France.

MALMEISTERS A.K., (81), Institute of Polymer Mechanics, Latvian SSR Academy of Sciences, Aizkraukles Street 23, 226 006 Riga, U.S.S.R.

MARIGO J.J., (715), Laboratoire Central des Ponts et Chaussées, Département des Structures et Ouvrages d'Art, 58, boulevard Lefebvre, 75732 Paris Cédex 15, France.

MARKOV K.Z., (35), Centre for Mathematics and Mechanics, Bulgarian Academy of Sciences, P.O. Box 373, 1090 Sofia, Bulgaria.

MAUGIN G.A., (701), Université Paris VI, Laboratoire de Mécanique Théorique, Tour 66, 4, place Jussieu, 75230, Paris Cédex 05, France.

MEYSSONIER J., (789), Laboratoire de Glaciologie-C.N.R.S., B.P. 53 X, 38041 Grenoble Cédex, France.

MORLAND L.W., (553), University of East Anglia, School of Mathematics and Physics, Norwich NR4 7TJ, U.K.

MÜLLER I., (133), Fachbereich 9, HFI TU Berlin, Strasse des 17. Juni, 135, 1 Berlin 12, Germany-BRD.

MRÓZ Z., (597), Polish Academy of Sciences, Institute of Fundamental Technological Research, Świętokrzyska 21, 00-049 Warsaw, Poland.

NOVA R., (623), Politecnico di Milano, Istituto di Scienza e Tecnica delle Costruzioni, Piazza Leonardo da Vinci 32, 20133 Milano, Italy.

PARNIERE P., (303), I.R.S.I.D. (Institut de Recherche de la Sidérurgie Française), 185, rue Président Roosevelt, 78105 Saint-Germain-en-Laye Cédex, France.

PENELLE R., (347), Université Paris-Sud, Laboratoire de Métallurgie Physique, Bâtiment 413, 91405 Orsay Cédex, France.

PERNOT M., (347), Université Paris-Sud, Laboratoire de Métallurgie Physique, Bâtiment 413, 91405 Orsay Cédex, France.

PIETRUSZCZAK St., (597), Polish Academy of Sciences, Institute of Fundamental Technological Research, Świętokrzyska 21, 00-049 Varsaw, Poland.

PIAU Monique, (685), Institut de Mécnaique de Grenoble, B.P. 53 X, 38041 Grenoble Cédex, France.

POGGI Y., (149), Laboratoire d'Electrostatique, C.N.R.S., B.P. 166 X, 38042 Grenoble Cédex, France.

POUGET J., (701), Université Paris VI, Laboratoire de Mécanique Théorique, Tour 66, 4, place Jussieu, 75230 Paris Cédex 05, France.

RATHKJEN A., (47), Institute of Building Technology and Structural Engineering, Aalborg University Center, Danmarksgade 19, Postboks 159, DK-9000 Aalborg, Danmark.

RIVLIN R.S., (123), Center for the Application of Mathematics, Lehigh University, 203 E. Packer Ave., Bethlehem, Pa. 18015, U.S.A.

SAADA A.S., (833), Case Western Reserve University, Department of Civil Engineering, Cleveland, Ohio 44 106, U.S.A.

SACCHI G., (623), Politecnico di Milano, Istituto di Scienza e Tecnica delle Costruzioni, Piazza Leonardo da Vinci 32, 20133 Milano, Italy.

SAWCZUK A., (803), Polish Academy of Sciences, Institute of Fundamental Technological Research, Świętokrzyska 21, 00-049 Warsaw, Poland.

SAWICKI A., (409), Polish Academy of Sciences, Institute of Hydroengineering, ul. Cystersow 11, 80-953 Gdańsk-Oliwa, Poland.

SCHULER K.W., (819), Sandia Laboratories, Albuquerque, New-Mexico 87185, U.S.A.

SEICHEPINE J.L., (869), Institut National Polytechnique de Lorraine, E.N.S.E.M., 2, rue de la Citadelle, B.P. 850, 54011 Nancy Cédex, France.

SHIRATORI E., (257), Department of Mechanical Engineering, Faculty of Engineering, Saitama University, 255 Shimo-Okubo, Urawa-shi Saitama-ken, 338 Japan.

SHOOK Louise P., (833), Case Western Reserve University, Department of Civil Engineering, Cleveland, Ohio 44 106, U.S.A.

SIDOROFF F., (761), Ecole Centrale de Lyon, Laboratoire de Mécanique, 36, route de Dardilly, B.P. 17, 69130 Ecully, France.

SIRIEYS P.M., (481), I.N.S.A. (Institut National des Sciences Appliquées), Avenue de Rangueil, 31077 Toulouse Cédex, France.

SMITH G.F., (27), Center for the Application of Mathematics, Lehigh University, 203 E. Packer Ave., Bethlehem, Pennsylvania 18015, U.S.A.

SOBOTKA Z., (643), Ústav Teoretické a Aplikované Mechanicky, Vyšehradská 49, 128 49 Praha 2 − Nové Mĕsto, Czechoslovakia.

SPENCER A.J.M., (3), University of Nottingham, Department of Theoretical Mechanics, Nottingham NG7 2RD, England.

SUMEC J., (423), Institute of Construction and Architecture SAV, Dúbravská cesta, 885 46 Bratislava, Czechoslovakia.

TAMUZS V.P., (81), Institute of Polymer Mechanics, Latvian SSR Academy of Sciences, Aizkraukles Street 23, 226 006 Riga, U.S.S.R.

TILLERSON J.R., (819), Sandia Laboratories, Albuquerque, New Mexico 87185, U.S.A.

TSAI S.W., (435), U.S. Air Force Materials Laboratory, AFML/MBM, Wright Patterson Air Force Basis, Ohio 45433, U.S.A.

VAKULENKO A.A., (35), University of Leningrad, Faculty of Mathematics and Mechnics, Leningrad, U.S.S.R.

VAUTRIN A., (869), Institut National Polytechnique de Lorraine, E.N.S.E.M., 2, rue de la Citadelle, B.P. 850, 54011 Nancy Cédex, France.

VERCHERY G., (93), E.N.S.T.A. (Ecole Nationale Supérieure des Techniques Avancées), Centre de l'Yvette, 91120 Palaiseau, France.

WAINTAL A., (167), Laboratoire de Cristallographie et SNCI, C.N.R.S., B.P. 166 X, 38042 Grenoble Cédex, France.

WILMANSKI K., (133), Polish Academy of Sciences, Institute of Fundamental Technological Research, Świętokrzyska 21, 00-049 Warsaw, Poland.

ZAOUI A., (333), C.N.R.S.-Université Paris-Nord, Laboratoire PMTM, Avenue J.B. Clément, 93430 Villetaneuse, France.

SHIRATORI E. (237), Department of Mechanical Engineering, Faculty of Engineering, Saitama University, 255 Shimo-Okubo, Urawa-shi Saitama-ken 338 Japan.

SIDOR Linda P. (833), Case Western Reserve University, Department of Civil Engineering, Cleveland, Ohio 44106, U.S.A.

SIDOROFF F. (761), Ecole Centrale de Lyon et Laboratoire de Mécanique, 36, route de Dardilly, BP 163, 69130 Ecully, France.

SIRIEYS P.M. (681), I.N.S.A. (Institut National des Sciences Appliquées), Avenue de Rangueil, 31077 Toulouse Cedex, France.

SMITH E.P. (?), Center for the Application of Mathematics, Lehigh University, 601 E. Packer Ave., Bethlehem, Pennsylvania 18015 U.S.A.

SOBOTKA Z. (163), Ustav Teoretické a Aplikované Mechaniky, Vyšehradská 49, 126 49 Praha 2 – Nové Mesto, Czechoslovakia.

SPENCER A.J.M. (?3), University of Nottingham Department of Theoretical Mechanics Nottingham NG7 2RD, England.

SUMEC J. (825), Institute of Construction and Architecture, SAV, Dúbravská cesta, 885 64 Bratislava, Czechoslovakia.

TABUTEAU Y.P. (931), Institut de Polymer Mechanics Latvian SSR Academy of Science, Aizkraukles Street 23, 226 006 Riga, U.S.S.R.

TELKERSON J.R. (619), Sandia Laboratories, Albuquerque, New Mexico 87155, U.S.A.

TSAI S.W. (1451), US Air Force Materials Laboratory, AFML, MBM, Wright-Patterson Air Force Base, Ohio 45433, U.S.A.

VAKULENKO A.A. (1?), University of Leningrad Faculty of Mathematics and Mechanics, Leningrad, U.S.S.R.

VAUTRIN A. (809), Institut National Polytechnique de Lorraine, E.N.S.E.M., Rue de la Citadelle, B.P. 850, 54011 Nancy Cedex, France.

VERCHERY G. (911), E.N.S.T.A. (Ecole Nationale Supérieure des Techniques Avancées) Centre de l'Yvette, 91120 Palaiseau, France.

WALTER A.A. (161), Laboratoire de Cristallographie et S.M.C, C.N.R.S., B.P. 166 X, 38042 Grenoble Cedex, France.

WILMANSKI K. (133), vol. Z. Institute of Sciences, Institute of Fundamental Technological Research, Swietokrzyska 21, 00-049 Warsaw, Poland.

ZAOUI A. (101), C.N.R.S. Université Paris-Nord, Laboratoire P.M.M., Avenue J.B. Clément, 93430 Villetaneuse, France.

Contents – *Sommaire*

Session 1

Invariant Formulation of Constitutive Equations
Formulation Invariante des Lois de Comportement

Session 2

Physical Properties of Anisotropic Materials
Propriétés Physiques des Matériaux Anisotropes

Communications

Session 3

Changes of Macroscopic Anisotropy in Metals
Evolution de l'Anisotropie Macroscopique des Métaux

General Lecture ; Conférence Générale

Session 4

Anisotropy of Metallic Polycrystals
Anisotropie des Polycristaux Métalliques

Session 5

Analytical and Numerical Methods for the Determination of Mechanical Properties of Composites
Méthodes Analytiques et Numériques pour la Détermination des Propriétés Mécaniques des Composites

Session 6

Strength of Composites
Résistance des Composites

Session 7

Mechanics of Anisotropic Rocks
Mécanique des Roches Anisotropes

Session 8

Anisotropy of Consolidated Clays and of Materials
with Internal Friction
Anisotropie des Argiles Consolidées et des Matériaux à
Frottement Interne

Session 9

Vibrations, Waves' Propagation and Induced Anisotropy
Vibrations, Propagation des Ondes et Anisotropie Induite

Session 10

Damage and Creep
Endommagement et Fluage

Session 11

Experimental Investigations and Interpretation of Mechanical Tests
Recherches Expérimentales et Interprétation des Essais Mécaniques

Communications

XXX

Session 12

Problems of Civil Engineering
Problèmes de Génie Civil

Communications

General Lecture : *Conférence Générale*

The Formulation of Constitutive Equation for Anisotropic Solids

A.J.M. Spencer

The University of Nottingham, Nottingham, England.

1. Introduction

The plan of this lecture is that I shall first review some results in the theory of algebraic invariants of vectors and tensors in three dimensions, under the orthogonal group of transformations. Then I will show how, in some cases of interest, it is possible to use these results for invariants under orthogonal transformations to determine invariants of vectors and tensors under groups of transformations which are sub-groups of the orthogonal group. As examples, I shall consider (a) the case in which the group of transformations is the group of rotations about an axis, which is the symmetry group for a transversely isotropic material, and (b) the case in which the transformation group is generated by the set of reflections in three orthogonal planes, which is the symmetry group for an orthotropic material. Next I will show how these results can be applied to the problem of determining mechanical constitutive equations for anisotropic materials with various types of stress response; in particular I shall consider materials whose stress response is that of a linear or a non-linear elastic solid, or that of a plastic solid, and illustrate the results by considering transversely isotropic and orthotropic materials. Many fibre-reinforced and laminated materials are, on the macroscopic scale, either transversely isotropic or orthotropic, and I shall give particular consideration to materials of this kind, and especially to the case in which the preferred directions are not uniform, but vary with position in a body. Finally, I will discuss the effect on the constitutive equations of the kinematic constraints of incompressibility and inextensibility in specified directions.

Invariance problems arise frequently in continuum mechanics. Probably the simplest example is the strain-energy function of an elastic solid. Suppose that a body of an elastic material undergoes a deformation in which a typical particle which initially has position vector $\mathbf{X}$, with components X_R ($R = 1,2,3$) moves to the point with position vector $\mathbf{x}$ and components x_i ($i = 1,2,3$). Then the deformation is described by equations of the form

$$\mathbf{x} = \mathbf{x}(\mathbf{X}), \quad \text{or} \quad x_i = x_i(X_R). \tag{1.1}$$

Then it can be shown that the strain-energy function W can be expressed in the form

$$W = W(C_{RS}) , \tag{1.2}$$

where

$$C_{RS} = \frac{\partial x_i}{\partial X_R} \frac{\partial x_i}{\partial X_S} , \tag{1.3}$$

and C_{RS} are components of a second-order symmetric tensor $\mathbf{C}$. The repeated index summation convention is used in (1.3) and in future.

Now consider a second deformation in which the particle whose initial position is $\overline{\mathbf{X}} = \mathbf{Q} \cdot \mathbf{X}$ moves to the point with position vector x, thus

$$x = x(\overline{\mathbf{X}}) \quad \text{or} \quad x_i = x_i(\overline{X}_R) , \tag{1.4}$$

where $\mathbf{Q}$ is an orthogonal tensor with components Q_{RS} (all components are components in a fixed rectangular Cartesian coordinate system) so that

$$Q_{PR} Q_{PS} = Q_{RP} Q_{SP} = \delta_{RS} , \quad \det \mathbf{Q} = \pm 1 , \quad \overline{X}_R = Q_{RS} X_S , \tag{1.5}$$

where δ_{RS} represents the Kronecker delta. Then

$$\overline{C}_{RS} = \frac{\partial x_i}{\partial \overline{X}_R} \frac{\partial x_i}{\partial \overline{X}_S} = Q_{RP} Q_{SQ} C_{PQ} , \quad \text{or} \quad \overline{\mathbf{C}} = \mathbf{Q} \cdot \mathbf{C} \cdot \mathbf{Q}^T . \tag{1.6}$$

In general the deformations (1.1) and (1.4) will give rise to different values of W. However it may happen that, for certain orthogonal tensors $\mathbf{Q}$,

$$W(\overline{C}_{RS}) = W(C_{RS}) , \quad \text{or} \quad W(\mathbf{Q} \cdot \mathbf{C} \cdot \mathbf{Q}^T) = W(\mathbf{C}) , \tag{1.7}$$

where the C_{RS} are evaluated at the origin. It can be shown that the orthogonal tensors $\mathbf{Q}$ for which (1.7) is satisfied form a group, which is called the symmetry group of the material.

If (1.7) is satisfied for all orthogonal tensors $\mathbf{Q}$, then the symmetry group is the full orthogonal group in three dimensions, the material is isotropic, and W is an isotropic invariant of $\mathbf{C}$. If (1.7) is satisfied for some, but not all, orthogonal tensors $\mathbf{Q}$, then the symmetry group is a subgroup of the full orthogonal group. When the symmetry group is the proper orthogonal group, or rotation group, in three dimensions, then the material is hemitropic; in many cases, including those considered in this lecture, the distinction between isotropy and hemitropy is not important.

Let **n** be a unit vector with components n_i. Then a rotation through an angle α about an axis through the origin and in the direction of **n** is defined by the orthogonal tensor $\mathbf{Q}^{(n)}(\alpha)$, whose components are

$$Q_{RS}^{(n)}(\alpha) = \delta_{RS} \cos\alpha + e_{RSP} n_P \sin\alpha + (1 - \cos\alpha) n_R n_S , \qquad (1.8)$$

where e_{RSP} is the alternating symbol. The tensors $\mathbf{Q}^{(n)}(\alpha)$ $(0 \leqslant \alpha < 2\pi)$ form a group (the group of rotations about the axis **n**). A reflection in planes normal to **n** is characterized by the reflection tensor $\mathbf{R}^{(n)}$ whose components are

$$R_{ij}^{(n)} = \delta_{ij} - 2n_i n_j . \qquad (1.9)$$

If **a** is a unit vector normal to **n**, then $\mathbf{R}^{(a)}$ represents a reflection in planes normal to **a** (and parallel to **n**), and $\mathbf{Q}^{(a)}(\pi)$ represents a rotation through π about **a**. In all there are five continuous groups generated by $\mathbf{Q}^{(n)}(\alpha)$ and various combinations of $\mathbf{R}^{(n)}$, $\mathbf{R}^{(a)}$ and $\mathbf{Q}^{(a)}(\pi)$. The generators of these groups are

(i) $\mathbf{Q}^{(n)}(\alpha)$, (ii) $\mathbf{Q}^{(n)}(\alpha), \mathbf{R}^{(a)}$, (iii) $\mathbf{Q}^{(n)}(\alpha), \mathbf{R}^{(n)}$,

(iv) $\mathbf{Q}^{(n)}(\alpha), \mathbf{Q}^{(a)}(\pi)$, (v) $\mathbf{Q}^{(n)}(\alpha), \mathbf{Q}^{(a)}(\pi), \mathbf{R}^{(a)}, \mathbf{R}^{(n)}$. $\qquad (1.10)$

These are the symmetry groups for various forms of transverse isotropy. For our purposes the distinctions between these five cases are not important and we can characterize elastic transverse isotropy about the axis **n** by invariance of $W(\mathbf{C})$ under the rotations $\mathbf{Q}^{(n)}(\alpha)$.

The symmetry groups for the various kinds of crystal symmetry are finite subgroups of the full orthogonal group. For illustration we shall consider the case of orthotropy. Let **a**, **b** and **c** be three mutually orthogonal vectors. Then an orthotropic material has reflectional symmetry with respect to the planes normal to these three vectors, so that its symmetry group is generated by the three tensors $\mathbf{R}^{(a)}$, $\mathbf{R}^{(b)}$ and $\mathbf{R}^{(c)}$. The symmetry group is therefore comprised of the tensors

$$\mathbf{I}, \mathbf{R}^{(a)}, \mathbf{R}^{(b)}, \mathbf{R}^{(c)}, \quad \mathbf{Q}^{(a)}(\pi) = \mathbf{R}^{(b)} . \mathbf{R}^{(c)}, \quad \mathbf{Q}^{(b)}(\pi) = \mathbf{R}^{(c)} . \mathbf{R}^{(a)},$$

$$\mathbf{Q}^{(c)}(\pi) = \mathbf{R}^{(a)} . \mathbf{R}^{(b)}, \quad -\mathbf{I} = \mathbf{R}^{(a)} . \mathbf{R}^{(b)} . \mathbf{R}^{(c)} . \qquad (1.11)$$

For an orthotroqic elastic material, $W(\mathbf{C})$ is invariant under each of the transformations (1.11).

Another problem in continuum mechanics which gives rise to similar invariance problems is that of determining the form of the yield function for a plastic solid. The yield function is a function $F(T_{ij})$ of the stress components T_{ij} of the stress tensor $\mathbf{T}$. The yield function is invariant under any of the transformations represented by the tensors which form the symmetry group of the material. Thus,

$$F(\mathbf{T}) = F(\mathbf{Q} \cdot \mathbf{T} \cdot \mathbf{Q}^T), \tag{1.12}$$

for each $\mathbf{Q}$ which belongs to the symmetry group.

The above examples lead to special cases of the following algebraic problem. Let $\mathbf{A}, \mathbf{B}, \mathbf{C}, \ldots$ be a finite set of symmetric second-order tensors in three dimensions, and $\mathbf{a}, \mathbf{b}, \mathbf{c}, \ldots$ be a finite set of vectors in three dimensions (it is possible also to include a set of anti-symmetric tensors, but for brevity we omit these). Let

$$\bar{\mathbf{A}} = \mathbf{Q} \cdot \mathbf{A} \cdot \mathbf{Q}^T, \text{etc.}, \quad \bar{\mathbf{a}} = \mathbf{Q} \cdot \mathbf{a}, \text{etc.}, \tag{1.13}$$

where $\mathbf{Q}$ is an orthogonal tensor, and let

$$f(\mathbf{A}, \mathbf{B}, \mathbf{C}, \ldots, \mathbf{a}, \mathbf{b}, \mathbf{c}, \ldots)$$

be a polynomial in the components of the vectors and tensors. Then if

$$f(\bar{\mathbf{A}}, \bar{\mathbf{B}}, \bar{\mathbf{C}}, \ldots, \bar{\mathbf{a}}, \bar{\mathbf{b}}, \bar{\mathbf{c}}, \ldots) = f(\mathbf{A}, \mathbf{B}, \mathbf{C}, \ldots, \mathbf{a}, \mathbf{b}, \mathbf{c}, \ldots) \tag{1.14}$$

for all $\mathbf{Q}$ belonging to a group G, then f is an invariant under this group. The problem is to determine canonical forms for f; that is, to determine a set I_1, I_2, I_3, ... of invariants such that any invariant f can be expressed as a polynomial in I_1, I_2, I_3, ... The set I_1, I_2, I_3, ... is then called an integrity basis. A classical theorem due to Hilbert asserts the existence of a finite integrity basis. If the integrity basis is such that none of its elements can be expressed as a polynomial in the remainder, then the integrity basis is irreducible, and the aim is to determine an irreducible basis for a given set of vectors and tensors and a specified symmetry group.

An apparently more general problem is that of finding tensor polynomial functions of vectors and tensors which are form-invariant under a given symmetry group. Suppose, for example, that $\mathbf{T}$ is a symmetric second-order tensor function of $\mathbf{A}, \mathbf{B}, \mathbf{C}, \ldots, \mathbf{a}, \mathbf{b}, \mathbf{c}, \ldots$. Then $\mathbf{T}$ is said to be form-invariant under the transformation $\mathbf{Q}$ if

$$\mathbf{Q} \cdot \mathbf{T}(\mathbf{A}, \mathbf{B}, \mathbf{C}, \ldots, \mathbf{a}, \mathbf{b}, \mathbf{c}, \ldots) \cdot \mathbf{Q}^T = \mathbf{T}(\bar{\mathbf{A}}, \bar{\mathbf{B}}, \bar{\mathbf{C}}, \ldots, \bar{\mathbf{a}}, \bar{\mathbf{b}}, \bar{\mathbf{c}}, \ldots). \tag{1.15}$$

The problem is to determine canonical forms for **T**. This problem can be reduced to the invariance problem described above by a device due to Pipkin and Rivlin [5]. If **u** is an arbitrary vector, then

$$F(A,B,C,\ldots,a,b,c,\ldots;u) = u \cdot T(A,B,C,\ldots,a,b,c,\ldots) \cdot u \quad (1.16)$$

is an invariant of $A, B, C, \ldots, a, b, c, \ldots$, and **u**, and is of degree two in the components of **u**. Hence F can be expressed as a polynomial in the elements of the integrity basis for $A, B, C, \ldots, a, b, c, \ldots$, and **u**, of degree two in the components of **u**; therefore F is of the form

$$F = \sum_m \phi(I_1, I_2, \ldots, I_n) P_{ij}^{(m)} u_i u_j \, ,$$

where $I_1, I_2, \ldots I_n$ are elements of the integrity basis for $A, B, C, \ldots,$ $a, b, c, \ldots$, and $P_{ij}^{(m)} u_i u_j$ are invariants of $A, B, C, \ldots, a, b, c, \ldots$, and **u**, of degree two in **u**, which can be determined from the integrity basis for these tensors and vectors. It follows that **T** has the form

$$T = \sum_m \phi(I_1, I_2, \ldots, I_n) P^{(m)} \, . \quad (1.17)$$

Similar procedures can be applied to determine form-invariant tensors of any order.

A further problem is to relax the restriction that f be a polynomial in its arguments, and to seek a functional basis $I_1', I_2', I_3', \ldots$ of invariants of $A, B, C, \ldots, a, b, c, \ldots$ such that any invariant function $f'(A, B, C, \ldots, a, b, c, \ldots)$ can be expressed as a single-valued function of $I_1', I_2', I_3', \ldots$. An integrity basis is always a functional basis, but an irreducible integrity basis is not necessarily an irreducible functional basis.

2. Isotropic Integrity Bases

The problem of determining an irreducible integrity basis for an arbitrary number of vectors and tensors has been solved in a series of papers by Rivlin [6], Spencer and Rivlin [16-19] and Spencer [12,13]. A fairly complete account is given in Spencer [14]. Here we briefly outline some parts of the theory which will be used later in this lecture.

Consider a set of ν vectors $a^{(r)}$ $(r = 1, 2, \ldots, \nu)$ and λ second-order symmetric tensors $A^{(s)}$ $(s = 1, 2, \ldots, \lambda)$, and let I be a polynomial invariant of these vectors under the full or the proper orthogonal transformation group.

Without loss of generality it can be assumed that I is homogeneous in the components of the vectors and tensors. Hence I has the form

$$I = \beta_{i_1 i_2 \ldots i_n j_1 k_1 j_2 k_2 \ldots j_m k_m} \, a_{i_1}^{(r_1)} \, a_{i_2}^{(r_2)} \ldots a_{i_n}^{(r_n)} \, A_{j_1 k_1}^{(s_1)} \, A_{j_2 k_2}^{(s_2)} \ldots A_{j_m k_m}^{(s_m)},$$

(2.1)

where $r_1, r_2, \ldots, r_n$ are integers (not necessarily all different) chosen from $1, 2, \ldots, \nu$, and $s_1, s_2, \ldots, s_m$ are integers (not necessarily all different) chosen from $1, 2, \ldots, \lambda$, and $\beta_{i_1 i_2 \ldots i_n j_1 k_1 j_2 k_2 \ldots j_m k_m}$ are a set of numerical coefficients. Under an orthogonal transformation $\mathbf{Q}$, the components of $\mathbf{a}^{(r)}$ and $\mathbf{A}^{(s)}$ are transformed to

$$\bar{a}_p^{(r)} = Q_{pi} a_i^{(r)}, \quad \bar{A}_{qt}^{(s)} = Q_{qi} Q_{tj} A_{ij}^{(s)}.$$

(2.2)

Since I is unchanged if $\bar{\mathbf{a}}^{(p)}$ is substituted for $\mathbf{a}^p$ and $\bar{\mathbf{A}}^{(s)}$ is substituted for $\mathbf{A}^{(s)}$, it follows from (2.1) and (2.2) that

$$\beta_{i_1 i_2 \ldots i_n j_1 k_1 j_2 k_2 \ldots j_m k_m} = Q_{p_1 i_1} Q_{p_2 i_2} \cdots$$

$$\cdots Q_{p_n i_n} Q_{q_1 j_1} Q_{t_1 k_1} Q_{q_2 j_2} Q_{t_2 k_2} \cdots$$

$$\cdots Q_{q_m j_m} Q_{t_m k_m} \beta_{p_1 p_2 \ldots p_n q_1 t_1 q_2 t_2 \ldots q_m t_m} \quad (2.3)$$

A tensor β whose components have the property (2.3) for all orthogonal tensors $\mathbf{Q}$ is called an isotropic tensor; its components are the same in any rectangular Cartesian coordinate system.

To determine the isotropic tensors we consider the case in which I is an invariant only of the vectors $\mathbf{a}^{(r)}$, so that (2.1) reduces to

$$I = \beta_{i_1 i_2 \ldots i_n} \, a_{i_1}^{(r_1)} \, a_{i_2}^{(r_2)} \ldots a_{i_n}^{(r_n)}.$$

(2.4)

It was proved by Cauchy that an integrity basis for a set of vectors consists of the scalar products $\mathbf{a}^{(r)} \cdot \mathbf{a}^{(s)}$ $(r, s = 1, 2, \ldots, \nu)$ and the determinants $e_{ijk} a_i^{(r)} a_j^{(s)} a_k^{(t)}$ $(r, s, t = 1, 2, \ldots, \nu)$. Hence I can be expressed as follows:

$$I = \sum_n \beta_n \, (a_{i_1}^{(r_1)} a_{i_1}^{(r_2)}) \, (a_{i_2}^{(r_3)} a_{i_2}^{(r_4)}) \ldots (e_{jk\ell} \, a_j^{(s_1)} a_k^{(s_2)} a_\ell^{(s_3)}). \quad (2.5)$$

The determinant factor need occur at most once in any term because of the identity

$$
e_{ijk} \, e_{rst} = \begin{vmatrix} \delta_{ir} & \delta_{jr} & \delta_{kr} \\ \delta_{is} & \delta_{js} & \delta_{ks} \\ \delta_{it} & \delta_{jt} & \delta_{kt} \end{vmatrix}. \tag{2.6}
$$

By equating (2.4) and (2.5) it follows that the components of the isotropic tensor β are linear combinations of products of the form

$$
\delta_{i_1 j_1} \, \delta_{i_2 j_2} \cdots \delta_{i_p j_p} \tag{2.7}
$$

if β is of even order 2p, and of the form

$$
\delta_{i_1 j_1} \, \delta_{i_2 j_2} \cdots \delta_{i_p j_p} \, e_{k_1 k_2 k_3} \tag{2.8}
$$

if β is of odd order $2p + 3$. This method of determining the isotropic tensors is due to Smith and Rivlin [10]. For simplicity we consider only the case in which β is of even order, so that the invariant I in (2.1) is of even degree in the components of the vectors. This is also the only case which arises if the symmetry group is the full orthogonal group, because the determinant $e_{ijk} a_i^{(r_1)} a_j^{(r_2)} a_k^{(r_3)}$ changes sign under a transformation for which det $Q = -1$, and so is invariant only for proper orthogonal transformations.

By substituting for β in (2.1), it follows that I can be expressed as a polynomial in expressions of the forms

$$
\text{tr } \mathbf{P}^{(1)} = P_{ii}^{(1)}, \quad \mathbf{a}^{(r)} \cdot \mathbf{P}^{(2)} \cdot \mathbf{a}^{(s)} = a_i^{(r)} P_{ij}^{(2)} a_j^{(s)}, \tag{2.9}
$$

where $\mathbf{P}^{(1)}$ and $\mathbf{P}^{(2)}$ are tensor products formed by taking inner products of any number of the tensors $\mathbf{A}^{(s)}$ in any order. Hence (2.9) constitute an isotropic integrity basis for the vectors and tensors. However this integrity basis is not finite. The basis can be reduced to a finite basis by arguments based on a generalization of the Cayley-Hamilton theorem which show that tr $\mathbf{P}^{(1)}$ is reducible if $\mathbf{P}^{(1)}$ is of degree greater than six in $\mathbf{A}^{(s)}$, and $\mathbf{a}^{(r)} \cdot \mathbf{P}^{(2)} \cdot \mathbf{a}^{(s)}$ is reducible if $\mathbf{P}^{(2)}$ is of degree greater than four in $\mathbf{A}^{(s)}$ (an invariant is said to be reducible if it can be expressed as a polynomial in invariants of lower degree). It then remains to determine an irreducible integrity basis. This has been accomplished by detailed examination of all the possible cases. The process is rather laborious but has been carried out. The resulting irreducible integrity basis for an arbitrary number of vectors and second-order tensors is tabulated in Spencer [14]. The irreducibility of the basis has been established by Smith [7] using methods of the theory of group representations which yield the number of linearly independent invariants of each degree in the vectors and tensors. More recently, Smith [9] has shown how group-theoretic methods can be used to construct an integrity basis in a systematic way.

The problem of constructing a functional basis has been considered by a number of authors. Many of the published results are not correct ; the main source of error is failure to ensure single-valuedness of the representation of any invariant as a function of the elements of the basis. Complete functional bases have been given by Smith [8] and Boehler [1]. For the comparatively simple cases which arise in this lecture the irreducible integrity basis is also an irreducible functional basis.

3.   Transverse Isotropy

We shall consider the problem of determining an integrity basis for a single symmetric second-order tensor for the symmetry groups (1.10), but the method used may be extended to the more general problem defined by (1.14). The main results for a single tensor have probably been known for a long time ; they are given explicitly by Ericksen and Rivlin [3]. The usual approach is to refer the tensor to a coordinate system in which one of the axes coincides with the axis of transverse isotropy. However there is some advantage in developing the theory in a manner which does not depend on the introduction of a special coordinate system ; this becomes apparent when we consider applications to fibre-reinforced materials. If the fibres are suitably arranged, then a composite material which consists of an isotropic matrix reinforced by a single family of aligned fibres is, on the macroscopic scale, transversely isotropic with the fibre direction as the axis of transverse isotropy. However the fibres need not be arranged in straight lines, and so the direction of the axis of transverse isotropy may vary with position in a body. It is straightforward to adapt results which are based on a special choice of coordinates to allow for such variation, but it is also of interest to develop the theory in a coordinate-free manner from the outset, and we shall proceed to do this in two distinct ways.

The problem is now to determine canonical forms for a polynomial function $\phi(C_{RS})$ which satisfies

$$\phi(C) = \phi(Q \cdot C \cdot Q^T) \tag{3.1}$$

for all $Q = Q^{(a)}(\alpha)$, for $0 \leqslant \alpha < 2\pi$, where the components of $Q^{(a)}(\alpha)$ are given by (1.8) and a defines the axis of transverse isotropy. We solve this problem in two ways.

The first approach is to employ the anisotropic tensors which were introduced by Smith and Rivlin [10]. The invariant ϕ is of the form

$$\phi(C) = \alpha_{i_1 j_1 i_2 j_2 \ldots i_p j_p} \; C_{i_1 j_1} \; C_{i_2 j_2} \cdots C_{i_p j_p}. \tag{3.2}$$

It follows by arguments similar to those leading to (2.3) that

$$\alpha_{k_1 \ell_1 k_2 \ell_2 \ldots k_p \ell_p} = Q_{k_1 i_1} \, Q_{\ell_1 j_1} \, Q_{k_2 i_2} \cdots Q_{\ell_p j_p} \, \alpha_{i_1 j_1 j_2 \ldots i_p j_p} \tag{3.3}$$

for all $Q = Q^{(a)}(\alpha)$. A tensor α with the property (3.3) is said to be an anisotropic tensor for the symmetry group $\{Q^{(a)}(\alpha)\}$. The anisotropic tensors for a given symmetry group are readily calculated if an integrity basis for invariants, under that symmetry group, of an arbitrary number of vectors is known. For transverse isotropy this integrity basis is well-known ; it is essentially the integrity basis for two-dimensional vectors under orthogonal transformations in two dimensions, and consists of the scalar products $a^{(r)} . a^{(s)}$ together with the resolved components $a.a^{(r)}$ of the vectors $a^{(r)}$ in the direction of a. The corresponding anisotropic tensors then easily follow by a procedure analogous to that used in deriving the isotropic tensors (2.7) and (2.8). The result is that the components of α are of the form

$$\alpha_{i_1 j_1 i_2 j_2 \ldots i_p j_p} = \sum_n \gamma_N \, \delta_{k_1 \ell_1} \, \delta_{k_2 \ell_2} \cdots \delta_{k_q \ell_q} \, a_{m_1} \, a_{m_2} \cdots a_{m_r}, \tag{3.4}$$

where k_1, ℓ_1, k_2, ℓ_2, $\ldots$, k_q, ℓ_q, m_1, m_2, $\ldots$, m_r is a permutation of i_1, j_1, i_2, j_2, $\ldots i_p$, j_p and $2p = 2q + r$ (and hence r is even). Results equivalent to (3.4), with a chosen to coincide in direction with the x_3-axis, were given by Smith and Rivlin [10].

By substituting (3.4) into (3.2), it follows that $\phi(C)$ is a linear combination of polynomials in a_i and C_{ij}, with coefficients which are products of Kronecker deltas; thus $\phi(C)$ is an isotropic invariant of a and C. An isotropic integrity basis for a and C is a transversely isotropic integrity basis for C. It must be noted, however, that an irreducible isotropic integrity basis for a and C is not necessarily irreducible when it is regarded as a transversely isotropic basis for C.

The alternative derivation of this result is based more on physical argument. For definiteness, consider an elastic material reinforced with a single family of fibres, which are characterized by the unit vector a in the reference configuration. Then W must depend on the deformation gradients $\partial x_i / \partial X_R$ and on a, and by the usual arguments we obtain

$$W = W(C, a). \tag{3.5}$$

If the only anisotropic properties of the material are those which arise from the presence of the fibres, then W is unchanged if we choose a new reference configuration which is obtained by a rigid rotation of the undeformed material and the fibres, and in which the particles are initially at $\overline{\mathbf{X}} = \mathbf{Q} . \mathbf{X}$ and the fibre direction is $\mathbf{Q} . \mathbf{a}$ where $\mathbf{Q}$ is any proper orthogonal tensor. Hence

$$W(\mathbf{C}, \mathbf{a}) = W(\mathbf{Q} . \mathbf{C} . \mathbf{Q}^{\mathrm{T}}, \mathbf{Q} . \mathbf{a}), \tag{3.6}$$

and thus W is an isotropic invariant of $\mathbf{C}$ and $\mathbf{a}$.

To determine the transversely isotropic integrity basis for $\mathbf{C}$ explicitly, we first note that ϕ ($\mathbf{C}$) must be even in $\mathbf{a}$. Hence it is sufficient to construct an isotropic integrity basis for $\mathbf{C}$ and $\mathbf{a} \otimes \mathbf{a}$, where $\otimes$ denotes the dyadic product ($\mathbf{a} \otimes \mathbf{a}$ is the symmetric second-order tensor whose components are $a_i a_j$). This integrity basis can be read off from tables; it consists of the traces of the following tensor products:

$$\mathbf{C}, \mathbf{C}^2, \mathbf{C}^3, \mathbf{a} \otimes \mathbf{a}, (\mathbf{a} \otimes \mathbf{a})^2, (\mathbf{a} \otimes \mathbf{a})^3, \mathbf{C} . \mathbf{a} \otimes \mathbf{a}, \mathbf{C} . (\mathbf{a} \otimes \mathbf{a})^2,$$
$$\mathbf{C}^2 . \mathbf{a} \otimes \mathbf{a}, \mathbf{C}^2 . (\mathbf{a} \otimes \mathbf{a})^2. \tag{3.7}$$

However, since $\mathbf{a}$ is a unit vector

$$\mathbf{a} \otimes \mathbf{a} = (\mathbf{a} \otimes \mathbf{a})^2 = (\mathbf{a} \otimes \mathbf{a})^3 = \dots. \tag{3.8}$$

Also

$$\operatorname{tr} \mathbf{a} \otimes \mathbf{a} = 1, \quad \operatorname{tr} \mathbf{C} . \mathbf{a} \otimes \mathbf{a} = \mathbf{a} . \mathbf{C} . \mathbf{a}, \quad \operatorname{tr} \mathbf{C}^2 . \mathbf{a} \otimes \mathbf{a} = \mathbf{a} . \mathbf{C}^2 . \mathbf{a}. \tag{3.9}$$

With (3.8) and (3.9), the set (3.7) reduces to

$$\operatorname{tr} \mathbf{C}, \quad \operatorname{tr} \mathbf{C}^2, \quad \operatorname{tr} \mathbf{C}^3, \quad \mathbf{a} . \mathbf{C} . \mathbf{a}, \quad \mathbf{a} . \mathbf{C}^2 . \mathbf{a}, \tag{3.10}$$

and this is the irreducible transversely isotropic integrity basis for $\mathbf{C}$. It agrees with the results of Ericksen and Rivlin [3]. An equivalent and, for our purposes, more convenient set is

$$\operatorname{tr} \mathbf{C}, \quad 1/2 \{(\operatorname{tr} \mathbf{C})^2 - \operatorname{tr} \mathbf{C}^2\}, \quad \det \mathbf{C}, \quad \mathbf{a} . \mathbf{C} . \mathbf{a}, \quad \mathbf{a} . \mathbf{C}^2 . \mathbf{a}. \tag{3.11}$$

4. Orthotropic Symmetry

The symmetry group for orthotropic symmetry with respect to three orthogonal planes normal to the unit vectors $\mathbf{a}$, $\mathbf{b}$ and $\mathbf{c}$ is given by (1.11).

The anisotropic tensors for all of the finite symmetry groups which characterize the crystal classes have been obtained by Smith and Rivlin [11]. For orthotropic symmetry these tensors, expressed in coordinate-free form, are linear combinations of outer products formed from

$$\mathbf{a} \otimes \mathbf{a}, \quad \mathbf{b} \otimes \mathbf{b}, \quad \mathbf{c} \otimes \mathbf{c} \tag{4.1}$$

Since **a**, **b** and **c** are mutually orthogonal, we have

$$\mathbf{a} \otimes \mathbf{a} + \mathbf{b} \otimes \mathbf{b} + \mathbf{c} \otimes \mathbf{c} = \mathbf{I}, \tag{4.2}$$

and we may discard one of the set (4.1), say **c** ⊗ **c**, in favour of the unit tensor **I**, and take, as a basic set of anisotropic tensors, the tensors

$$\mathbf{a} \otimes \mathbf{a}, \quad \mathbf{b} \otimes \mathbf{b}, \quad \mathbf{I}. \tag{4.3}$$

It follows by arguments analogous to those used in Section 3 that any polynomial invariant of **C**, under the transformation group (1.11), is a polynomial in the components C_{RS} of **C**, the components a_i of **a**, and the components b_i of **b**, which is of even degree in the components a_i and of even degree in the components b_i, and is invariant under all orthogonal transformations. Thus in this case also the tables of isotropic invariants can be used to determine an integrity basis for **C** in the case of orthotropic symmetry; specifically, an isotropic integrity basis for **a** ⊗ **a**, **b** ⊗ **b** and **C** will be an integrity basis for **C** for orthotropic symmetry with respect to planes normal to **a**, **b** and **c** = **a** × **b**.

This result also can be obtained by more direct physical arguments. One method of constructing an orthotropic material is by building up layers of thin sheets of unidirectionally reinforced fibre-reinforced materials in some regular sequence, with the reinforcement in each layer lying initially in the direction either of **a** or of **b**, where **a** and **b** are orthogonal. Then, in the case of an elastic material, W depends on **C**, **a** and **b**, and since the sense of the fibres is not important, W has to be an even function of **a** and an even function of **b**. As in the case of Section 3, W is unchanged if we choose a new reference configuration in which the particles are at $\overline{\mathbf{X}} = \mathbf{Q} \cdot \mathbf{X}$ and the fibre directions are **Q**.**a** and **Q**.**b**, and it follows that W is an isotropic invariant of **C**, **a** ⊗ **a** and **b** ⊗ **b**.

From tables of isotropic invariants, an isotropic integrity basis for **C**, **a** ⊗ **a** and **b** ⊗ **b** is (omitting invariants which can be eliminated by relations analogous to (3.8)) the traces of the following tensor products:

$$\mathbf{C}, \; \mathbf{C}^2, \; \mathbf{C}^3, \; \mathbf{a} \otimes \mathbf{a}, \; \mathbf{b} \otimes \mathbf{b}, \; \mathbf{C}.\mathbf{a} \otimes \mathbf{a}, \; \mathbf{C}^2.\mathbf{a} \otimes \mathbf{a}, \; \mathbf{C}.\mathbf{b} \otimes \mathbf{b},$$
$$\mathbf{C}^2.\mathbf{b} \otimes \mathbf{b}, \; \mathbf{a} \otimes \mathbf{a}.\mathbf{b} \otimes \mathbf{b}, \; \mathbf{a} \otimes \mathbf{a}.\mathbf{b} \otimes \mathbf{b}.\mathbf{C}, \; \mathbf{a} \otimes \mathbf{a}.\mathbf{b} \otimes \mathbf{b}.\mathbf{C}^2 \tag{4.4}$$

Since $\mathbf{a}$ and $\mathbf{b}$ are orthogonal unit vectors

$$\text{tr } \mathbf{a} \otimes \mathbf{a} = \text{tr } \mathbf{b} \otimes \mathbf{b} = 1, \quad \mathbf{a} \otimes \mathbf{a}.\mathbf{b} \otimes \mathbf{b} = (\mathbf{a}.\mathbf{b}) \mathbf{a} \otimes \mathbf{b} = 0, \tag{4.5}$$

and so the integrity basis reduces (after some rearrangement) to

$$I_1 = \text{tr } \mathbf{C}, \quad I_2 = 1/2 \{(\text{tr } \mathbf{C})^2 - \text{tr } \mathbf{C}^2\}, \quad I_3 = \det \mathbf{C},$$

$$\mathbf{a}.\mathbf{C}.\mathbf{a}, \quad \mathbf{a}.\mathbf{C}^2.\mathbf{a}, \quad \mathbf{b}.\mathbf{C}.\mathbf{b}, \quad \mathbf{b}.\mathbf{C}^2.\mathbf{b}. \tag{4.6}$$

Another way to construct an orthotropic material is by building up thin sheets of unidirectionally reinforced fibre-reinforced material according to a regular sequence, but with the fibres initially aligned in the directions of two unit vectors $\mathbf{d}$ and $\mathbf{e}$ which are not necessarily orthogonal, but are inclined at an angle 2Φ. In the general case, the symmetry transformations for such a material are

$$\mathbf{I}, \quad -\mathbf{I}, \quad \mathbf{R}^{(c)}, \quad \mathbf{Q}^{(c)}(\pi), \tag{4.7}$$

where c is orthogonal to $\mathbf{d}$ and $\mathbf{e}$. If the material is elastic, then its strain-energy function W is an isotropic invariant of

$$\mathbf{C}, \quad \mathbf{d} \otimes \mathbf{d}, \quad \mathbf{e} \otimes \mathbf{e}, \tag{4.8}$$

and an integrity basis is

$$I_1, \quad I_2, \quad I_3, \quad \mathbf{d}.\mathbf{C}.\mathbf{d}, \quad \mathbf{d}.\mathbf{C}^2.\mathbf{d}, \quad \mathbf{e}.\mathbf{C}.\mathbf{e}, \quad \mathbf{e}.\mathbf{C}^2.\mathbf{e},$$

$$\cos 2\Phi \, \mathbf{d}.\mathbf{C}.\mathbf{e}, \quad \cos 2\Phi \, \mathbf{d}.\mathbf{C}^2.\mathbf{e}, \quad \text{and} \quad \cos^2 2\Phi. \tag{4.9}$$

However it can be shown that $\cos 2\Phi \, \mathbf{d}.\mathbf{C}^2.\mathbf{e}$ may be expressed in terms of the other invariants, and so this invariant may be omitted from the integrity basis.

The symmetry group (4.7) does not describe an orthotropic material. However, if the two families of fibres are mechanically equivalent (as, for example, when the successive sheets of unidirectionally reinforced material are alternately aligned in the directions of $\mathbf{d}$ and $\mathbf{e}$, but are otherwise identical) then the bisectors of $\mathbf{d}$ and $\mathbf{e}$ are also planes of symmetry, and the material is orthotropic. In this case W has to be a symmetric function of $\mathbf{d}$ and $\mathbf{e}$, and so (4.9) (with $\cos 2\Phi \, \mathbf{d}.\mathbf{C}^2.\mathbf{e}$ eliminated) may be replaced by the following set of invariants:

$$I_1, \quad I_2, \quad I_3, \quad \mathbf{d}.\mathbf{C}.\mathbf{d} + \mathbf{e}.\mathbf{C}.\mathbf{e}, \quad (\mathbf{d}.\mathbf{C}.\mathbf{d})(\mathbf{e}.\mathbf{C}.\mathbf{e}),$$

$$\mathbf{d}.\mathbf{C}^2.\mathbf{d} + \mathbf{e}.\mathbf{C}^2.\mathbf{e}, \quad (\mathbf{d}.\mathbf{C}^2.\mathbf{d})(\mathbf{e}.\mathbf{C}^2.\mathbf{e}), \quad \cos 2\Phi \, \mathbf{d}.\mathbf{C}.\mathbf{e}, \quad \cos^2 2\Phi.$$

However $(d.C^2.d)(e.C^2.e)$ can be expressed in terms of the other invariants, and omitted, and so W can be expressed as a function of

$$I_1, \; I_2, \; I_3, \; d.C.d + e.C.e, \; (d.C.d)(e.C.e),$$

$$d.C^2.d + e.C^2.e, \; d.C.e \cos 2\Phi, \; \cos^2 2\Phi. \tag{4.10}$$

The two sets of invariants (4.6) and (4.10) are both integrity bases for C for the case of orthotropic symmetry, and so they must be equivalent. This can be shown directly. The unit vectors a and b which bisect e and d are

$$a = (d + e)/2 \cos \Phi, \quad b = (d - e)/2 \sin \Phi. \tag{4.11}$$

so that

$$d = a \cos \Phi + b \sin \Phi, \quad e = a \cos \Phi - b \sin \Phi. \tag{4.12}$$

If (4.12) are substituted into (4.10), we obtain a set of invariants which can be shown to be equivalent to the set (4.6), with the addition of $\cos^2 2\Phi$.

It is apparent that this approach can be extended to the determination of integrity bases for systems of vectors and tensors, and can be applied to other symmetry groups besides those considered in Sections 3 and 4. A similar method has been applied to a variety of problems by Boehler [2].

5. Linear Elasticity

For a linear elastic solid the strain-energy function is a quadratic function of the components E_{ij} of the infinitesimal strain tensor **E**.

For a transversely isotropic material the most general quadratic function which can be formed from the invariants (3.11) (with C replaced by E) is of the form

$$W = 1/2 \, \lambda (\operatorname{tr} E)^2 + \mu_T \operatorname{tr} E^2 + \alpha(a.E.a) \operatorname{tr} E + 2(\mu_L - \mu_T) \, a.E^2.a +$$

$$+ 1/2 \, \beta(a.E.a)^2, \quad (5.1)$$

where $\lambda, \mu_T, \mu_L, \alpha$ and β are elastic constants. The stress is given by

$$T_{ij} = \partial W / \partial E_{ij}, \tag{5.2}$$

from which it follows from (5.1) that

$$T_{ij} = \lambda E_{kk} \delta_{ij} + 2\mu_T E_{ij} + \alpha(a_k a_m E_{km} \delta_{ij} + a_i a_j E_{kk}) \tag{5.3}$$

$$+ 2(\mu_L - \mu_T)(a_i a_k E_{kj} + a_j a_k E_{ki}) + \beta a_i a_j a_k a_m E_{km}.$$

This is in agreement with the well-known expression for the stress in a transversely isotropic linearly elastic material. The constants μ_L and μ_T represent shear moduli. The other constants λ, α and β can be related to elastic constants which have more direct physical interpretations, such as extension moduli and Poisson's ratios.

For an orthotropic material the most general quadratic form for W is, from (4.6),

$$W = 1/2 \, \lambda (\text{tr } E)^2 + \mu \, \text{tr } E^2 + \alpha_1 (a.E.a) \, \text{tr } E$$
$$+ \alpha_2 (b.E.b) \, \text{tr } E + 2\mu_1 \, a.E^2.a + 2\mu_2 \, b.E^2.b$$
$$+ 1/2 \, \beta_1 (a.E.a)^2 + 1/2 \, \beta_2 (b.E.b)^2 + \beta_3 (a.E.a)(b.E.b) \tag{5.4}$$

where λ, μ, α_1, α_2, μ_1, μ_2, β_1, β_2 and β_3 are elastic constants. The corresponding expression for the stress is

$$T_{ij} = (\lambda E_{rr} + \alpha_1 a_r a_s E_{rs} + \alpha_2 b_r b_s E_{rs}) \delta_{ij}$$
$$+ (\alpha_1 E_{rr} + \beta_1 a_r a_s E_{rs} + \beta_3 b_r b_s E_{rs}) a_i a_j$$
$$+ (\alpha_2 E_{rr} + \beta_3 a_r a_s E_{rs} + \beta_2 b_r b_s E_{rs}) b_i b_j$$
$$+ 2\mu E_{ij} + 2\mu_1 (a_i a_k E_{kj} + a_j a_k E_{ki})$$
$$+ 2\mu_2 (b_i b_k E_{kj} + b_j b_k E_{ki}) . \tag{5.5}$$

This expression also is in agreement with known results.

There is nothing in this formulation which requires a to be constant in the case of transverse isotropy or a and b to be constant in the case of orthotropy. Therefore the constitutive equations (5.3) and (5.5) can be used if the directions which characterize the anisotropy vary from point to point, as would be the case, for example, for a fibre-reinforced material in which the fibres are arranged in families of curves which are not straight lines. Similar remarks apply to the constitutive equations which will be formulated in the next two sections.

6. Finite Elasticity

The constitutive equation for a finite elastic solid with strain-energy function $W(C_{RS})$ is

$$T_{ij} = \frac{\rho}{\rho_0} \frac{\partial x_i}{\partial X_R} \frac{\partial x_j}{\partial X_S} \left[\frac{\partial W}{\partial C_{RS}} + \frac{\partial W}{\partial C_{SR}} \right], \tag{6.1}$$

where x_i, X_R and C_{RS} were defined in Section 1 and ρ_0 and ρ denote the densities in the reference and deformed configurations respectively. Consider first a transversely isotropic material in which the axis of transverse isotropy is in the direction of a vector $\mathbf{a}_0$ in the reference configuration. Then, from (3.11), W is a function of

$$I_1 = \mathrm{tr}\,\mathbf{C}\,, \quad I_2 = 1/2\,\{(\mathrm{tr}\,\mathbf{C})^2 - \mathrm{tr}\,\mathbf{C}^2\}\,, \quad I_3 = \det \mathbf{C} = (\rho_0/\rho)^2\,,$$

$$I_4 = \mathbf{a}_0.\mathbf{C}.\mathbf{a}_0\,, \quad I_5 = \mathbf{a}_0.\mathbf{C}^2.\mathbf{a}_0\,. \tag{6.2}$$

We note also that a material line element with direction $\mathbf{a}_0$ in the reference configuration has the direction of a unit vector $\mathbf{a}$ in the deformed configuration, where

$$\lambda_a a_i = \frac{\partial x_i}{\partial X_R}\,a_R^{(0)} \tag{6.3}$$

and λ_a is the stretch of the line element, given by

$$\lambda_a^2 = I_4\,. \tag{6.4}$$

It is also convenient to introduce the tensor $\mathbf{B}$, whose components B_{ij} are given by

$$B_{ij} = \frac{\partial x_i}{\partial X_R}\,\frac{\partial x_j}{\partial X_R}\,. \tag{6.5}$$

From (6.1) and (6.2) it follows that T_{ij} can be expressed in the form

$$T_{ij} = \frac{\rho}{\rho_0}\,\frac{\partial x_i}{\partial X_R}\,\frac{\partial x_j}{\partial X_S}\,\sum_{\alpha=1}^{5}\,\frac{\partial W}{\partial I_\alpha}\left[\frac{\partial I_\alpha}{\partial C_{RS}} + \frac{\partial I_\alpha}{\partial C_{SR}}\right]\,. \tag{6.6}$$

By calculating the derivatives $\partial I_\alpha/\partial C_{RS}$, substituting these derivatives into (6.6), and using the relations (6.3), (6.4) and (6.5), and the Cayley-Hamilton theorem for $\mathbf{B}$, we find eventually that (6.6) can be expressed in the form

$$\mathbf{T} = 2I_3^{-1/2}\,\{(I_2 W_2 + I_3 W_3)\,\mathbf{I} + W_1\,\mathbf{B} - I_3 W_2 \mathbf{B}^{-1} + I_4 W_4\,\mathbf{a}\otimes\mathbf{a}$$

$$+ I_4 W_5\,(\mathbf{a}\otimes\mathbf{B}.\mathbf{a} + \mathbf{a}.\mathbf{B}\otimes\mathbf{a})\}\,, \tag{6.7}$$

where W_α denotes $\partial W/\partial I_\alpha$.

The constitutive equation for an orthotropic elastic body can be obtained in a similar way. In this case W is a function of

$$I_1, \ I_2, \ I_3, \ I_4, \ I_5, \ I_6 = b_0.C.b_0, \quad I_7 = b_0.C^2.b_0 \,, \tag{6.8}$$

where a_0 and b_0 are unit vectors normal to two planes of reflectional symmetry in the reference configuration. If a and b denote unit vectors in the deformed configuration in the directions of line elements which had the directions of a_0 and b_0 in the reference configuration, then (6.1) gives

$$T = 2I_3^{-1/2} \{(I_2 W_2 + I_3 W_3) I + W_1 B - I_3 W_2 B^{-1} + I_4 W_4 \, a \otimes a$$

$$+ I_6 W_6 \, b \otimes b + I_4 W_5 (a \otimes B.a + a.B \otimes a)$$

$$+ I_6 W_7 (b \otimes B.b + b.B \otimes b)\} . \tag{6.9}$$

The constitutive equations (6.7) and (6.9) are equivalent to those given by Ericksen and Rivlin [3] and by Green and Adkins [4] for transversely isotropic and orthotropic elastic materials, but they are here expressed in a coordinate-free form and the method of derivation is rather different.

7.   Plasticity

Most theories of plasticity assume the existence of a yield condition. We consider yield conditions of the form

$$F(T_{ij}) \leqslant k^2 \,, \tag{7.1}$$

where k depends on the deformation history, and $F(T_{ij})$ is the yield function, and is homogeneous of degree two in the stress components T_{ij}. $F(T_{ij})$ is invariant under the transformations of the symmetry group. Hence, from (3.10), for a transversely isotropic material, $F(T_{ij})$ can be expressed as a function of

$$J_1 = \text{tr} \, T, \quad J_2 = 1/2 \, \text{tr} \, T^2, \quad J_3 = 1/3 \, \text{tr} \, T^3, \quad J_4 = a.T.a,$$

$$J_5 = 1/2 \, a.T^2.a, \tag{7.2}$$

where a is the axis of transverse isotropy. For an orthotropic material, $F(T_{ij})$ can, from (4.6) (with a minor modification), be expressed as a function of

$$J_1, \ J_2, \ J_3, \ J_4, \ J_5, \ J_6 = b.T.b, \quad J_7 = 1/2 \, b.T^2.b, \tag{7.3}$$

where **a** and **b** are unit vectors normal to two of the planes of reflectional symmetry.

In plasticity theories it is often assumed that $F(T_{ij})$ is a plastic potential, such that the components D_{ij}^p of the plastic strain-rate $\mathbf{D}^p$ are given by

$$D_{ij}^p = \dot{\Lambda}\, \frac{\partial F}{\partial T_{ij}}, \tag{7.4}$$

where $\dot{\Lambda}$ is a scalar multiplier. From (7.2) this gives, for the case of transverse isotropy,

$$\mathbf{D}^p = \dot{\Lambda}\, \{F_1 \mathbf{I} + F_2 \mathbf{T} + F_3 \mathbf{T}^2 + F_4\, \mathbf{a} \otimes \mathbf{a} + 1/2\, F_5\, (\mathbf{a} \otimes \mathbf{T}.\mathbf{a} + \mathbf{a}.\mathbf{T} \otimes \mathbf{a})\}, \tag{7.5}$$

and, for the case of orthotropy,

$$\mathbf{D}^p = \dot{\Lambda}\, \{F_1 \mathbf{I} + F_2 \mathbf{T} + F_3 \mathbf{T}^2 + F_4\, \mathbf{a} \otimes \mathbf{a} + F_6\, (\mathbf{b} \otimes \mathbf{b})$$
$$+ 1/2\, F_5\, (\mathbf{a} \otimes \mathbf{T}.\mathbf{a} + \mathbf{a}.\mathbf{T} \otimes \mathbf{a}) + 1/2\, F_7\, (\mathbf{b} \otimes \mathbf{T}.\mathbf{b}.\mathbf{T} \otimes \mathbf{a})\}, \tag{7.6}$$

where $F_\alpha = \partial F/\partial J_\alpha$.

8.   Kinematic constraints. Elasticity

It is well-known that many problems in continuum mechanics are greatly simplified if the material concerned is regarded as incompressible, and that for some real materials the volumetric strain is small compared to the shear strain under loading conditions which are normally encountered. As a first approximation such materials may be regarded as incompressible.

For an incompressible linearly elastic solid, $\operatorname{tr} \mathbf{E} = 0$. Hence the strain energy function (5.1) for an incompressible transversely isotropic material may be written as

$$W = \mu_T \operatorname{tr} \mathbf{E}^2 + 2(\mu_L - \mu_T)\, \mathbf{a}.\mathbf{E}^2.\mathbf{a} + 1/2\, \beta(\mathbf{a}.\mathbf{E}.\mathbf{a})^2 - p \operatorname{tr} \mathbf{E}, \tag{8.1}$$

where p is a Lagrangian multiplier. From (5.2), the stress is

$$\mathbf{T} = 2\mu_T \mathbf{E} + 2(\mu_L - \mu_T)\, (\mathbf{a} \otimes \mathbf{E}.\mathbf{a} + \mathbf{a}.\mathbf{E} \otimes \mathbf{a}) + \beta(\mathbf{a}.\mathbf{E}.\mathbf{a})\, \mathbf{a} \otimes \mathbf{a} - p\mathbf{I}. \tag{8.2}$$

The mechanical effect of the kinematic constraint of incompressibility is to produce a reaction in the form of the arbitrary pressure p.

In a fibre-reinforced material which is constructed by reinforcing a relatively soft matrix material by strong stiff fibres, the resistance to deformation by extension in a fibre-direction may greatly exceed the resistance to other deformation modes. As a first approximation such a material may be regarded as inextensible in the fibre-direction. For small deformations, the condition for inextensibility in the direction of a is $a.E.a = 0$. In this case the strain-energy function (5.1) reduces to

$$W = 1/2 \,\lambda(\mathrm{tr}\,E)^2 + \mu_T \,\mathrm{tr}\,E^2 + 2(\mu_L - \mu_T)\,a.E^2.a + Ta.E.a \,, \qquad (8.3)$$

where T is a Lagrangian multiplier. Then, from (5.2),

$$T = \lambda I\,\mathrm{tr}\,E + 2\mu_T E + 2(\mu_L - \mu_T)(a \otimes E.a + a.E \otimes a) + Ta \otimes a. \qquad (8.4)$$

The mechanical effect of the inextensibility constraint is a reaction in the form of an arbitrary tension T in the direction of inextensibility. If the material is incompressible and also inextensible in the direction a, the constitutive equation (5.3) reduces to

$$T = 2\mu_T E + 2(\mu_L - \mu_T)(a \otimes E.a + a.E \otimes a) - pI + Ta \otimes a\,. \qquad (8.5)$$

Corresponding results are readily obtained for an orthotropic material. For an incompressible material, (5.5) reduces to

$$T = (\beta_1 a.E.a + \beta_3 b.E.b)\,a \otimes a + (\beta_3 a.E.a + \beta_2 b.E.b)\,b \otimes b$$
$$+ 2\mu E + 2\mu_1(a \otimes E.a + a.E \otimes a) + 2\mu_2(b \otimes E.b + b.E \otimes b) - pI\,. \qquad (8.6)$$

For a material which is inextensible in the orthogonal directions a and b, (5.5) reduces to

$$T = \lambda I\,\mathrm{tr}\,E + 2\mu E + 2\mu_1(a \otimes E.a + a.E \otimes a)$$
$$+ 2\mu_2(b \otimes E.b + b.E \otimes b) + T_a\,a \otimes a + T_b\,b \otimes b\,, \qquad (8.7)$$

where T_a and T_b are arbitrary tensions in the directions a and b. For a material which in incompressible and also inextensible in the directions a and b, (5.5) becomes

$$T = 2\mu E + 2\mu_1(a \otimes E.a + a.E \otimes a) + 2\mu_2(b \otimes E.b + b.E \otimes b)$$
$$- pI + T_a\,a \otimes a + T_b\,b \otimes b\,. \qquad (8.8)$$

Analogous results hold for finite elastic deformations. The condition for incompressibility is $I_3 = \det C = 1$; the condition for inextensibility in the direction a_0 in the reference configuration is $I_4 = a_0.C.a_0 = 1$. In the case of transverse isotropy, (6.7) is modified as follows:

(a) Incompressible material ($I_3 = 1$)

$$T = 2\{W_1 B - W_2 B^{-1} + I_4 W_4\, a \otimes a + I_4 W_5\,(a \otimes B.a + a.B \otimes a)\}$$
$$- pI. \quad (8.9)$$

(b) Material which is inextensible in the direction a ($I_4 = 1$)

$$T = 2I_3^{-1/2}\,\{(I_2 W_2 + I_3 W_3)\,I + W_1 B - I_3 W_2 B^{-1}$$
$$+ W_5\,(a \otimes B.a + a.B \otimes a)\} + Ta \otimes a. \quad (8.10)$$

(c) Material which is incompressible and inextensible in the direction a ($I_3 = 1$, $I_4 = 1$)

$$T = 2\{W_1 B - W_2 B^{-1} + W_5\,(a \otimes B.a + a.B \otimes a)\} - pI + Ta \otimes a. (8.11)$$

For the case of an orthotropic material, (6.9) is modified as follows:

(d) Incompressible material ($I_3 = 1$)

$$T = 2\{W_1 B - W_2 B^{-1} + I_4 W_4\, a \otimes a + I_5 W_6\, b \otimes b$$
$$+ I_4 W_5\,(a \otimes B.a + a.B \otimes a) + I_5 W_7\,(b \otimes B.b + b.B \otimes b)\} - pI. \quad (8.12)$$

(e) Material which is inextensible in the directions a and b ($I_4 = 1$, $I_6 = 1$)

$$T = 2I_3^{-1/2}\,\{(I_2 W_2 + I_3 W_3)\,I + W_1 B - I_3 W_2 B^{-1}$$
$$+ W_5\,(a \otimes B.a + a.B \otimes a) + W_7\,(b \otimes B.b + B \otimes b)\} + T_a\, a \otimes a + T_b\, b \otimes b. \quad (8.13)$$

(f) Material which is incompressible and inextensible in the directions a and b ($I_3 = 1$, $I_4 = 1$, $I_6 = 1$)

$$T = 2\{W_1 B - W_2 B^{-1} + W_5\,(a \otimes B.a + B \otimes a)$$
$$+ W_7\,(b \otimes B.b + b \otimes b)\} - pI + T_a\, a \otimes a + T_b\, b \otimes b. \quad (8.14)$$

For an orthotropic material which is inextensible in two non-orthogonal directions **d** and **e**, it is more convenient to express W as a function of the invariants (4.10), and note that the inextensibility conditions take the forms **d.C.d** = 1 and **e.C.e** = 1. This theory is described in Spencer [15], but the constitutive equations derived in [15] contain some redundant terms.

9. Kinematic constraints. Plasticity

The stress in a material subject to kinematic constraints can be divided into two parts. Thus

$$\mathbf{T} = \mathbf{S} + \mathbf{R}, \quad T_{ij} = S_{ij} + R_{ij}, \tag{9.1}$$

where **R** is the reaction stress, and **S** is the extra-stress. The extra-stress is determined by constitutive equations; the reaction stress is arbitrary and is determined by equations of motion or equilibrium and boundary conditions. For an incompressible material $\mathbf{R} = -p\mathbf{I}$; without loss of generality the hydrostatic part of **S** may be absorbed into $p\mathbf{I}$, and we may assume tr **S** = 0, so that in this case **S** becomes the deviatoric stress. Similarly, for a material which is incompressible and inextensible in the direction **a**, we have

$$\mathbf{R} = -p\mathbf{I} + T\mathbf{a} \otimes \mathbf{a}, \tag{9.2}$$

and without loss of generality it may be assumed that

$$\text{tr } \mathbf{S} = 0, \quad \mathbf{a.S.a} = 0. \tag{9.3}$$

It follows from (9.1), (9.2) and (9.3) that in this case

$$\mathbf{S} = \mathbf{T} - \frac{1}{2}\left(\text{tr } \mathbf{T} - \mathbf{a.T.a}\right)\mathbf{I} - \frac{1}{2}\left(\text{tr } \mathbf{T} - 3\mathbf{a.T.a}\right)\mathbf{a} \otimes \mathbf{a}. \tag{9.4}$$

For a material which is incompressible and inextensible in the two orthogonal directions **a** and **b**, we have

$$\mathbf{R} = -p\mathbf{I} + T_a\,\mathbf{a} \otimes \mathbf{a} + T_b\,\mathbf{b} \otimes \mathbf{b}, \tag{9.5}$$

and it may be assumed that

$$\text{tr } \mathbf{S} = 0, \quad \mathbf{a.S.a} = 0, \quad \mathbf{b.S.b} = 0. \tag{9.6}$$

In this case S takes the form

$$\mathbf{S} = \mathbf{T} - (\text{tr } \mathbf{T} - \mathbf{a.T.a} - \mathbf{b.T.b}) \mathbf{I} + (\text{tr } \mathbf{T} - 2\mathbf{a.T.a} - \mathbf{b.T.b}) \mathbf{a} \otimes \mathbf{a}$$

$$+ (\text{tr } \mathbf{T} - \mathbf{a.T.a} - 2\mathbf{b.T.b}) \mathbf{b} \otimes \mathbf{b}. \quad (9.7)$$

For most metals it is observed that plastic yielding is, to a good approximation, independent of the hydrostatic pressure. The yield function $F(T_{ij})$ can then be expressed as a function of the components of the deviatoric stress $\mathbf{T} - 1/3 \mathbf{I} \text{ tr } \mathbf{T}$. In the case of a transversely isotropic material, the set of invariants (7.2) may then be replaced by

$$J_2' = 1/2 \text{ tr } \mathbf{S}^2, \quad J_3' = 1/3 \text{ tr } \mathbf{S}^3, \quad J_4' = \mathbf{a.S.a}, \quad J_5' = 1/2 \, \mathbf{a.S}^2 \mathbf{.a}, \quad (9.8)$$

and in the case of an orthotropic material, (7.3) may be replaced by

$$J_2', J_3', J_4', J_5', J_6' = \mathbf{b.S.b}, \quad J_7' = \mathbf{b.S}^2 \mathbf{.b} \quad (9.9)$$

where $\mathbf{S} = \mathbf{T} - 1/3 \mathbf{I} \text{ tr } \mathbf{T}$.

If a plastic material is inextensible in one or more directions, it is to be expected that yielding will not be affected by arbitrary tensions in those directions. Hence the yield function may be expressed as a function of the extra-stress, rather than of the total stress. Thus in the case of a transversely isotropic material which is inextensible in the direction $\mathbf{a}$, the set (9.8) may be replaced by

$$J_2'' = 1/2 \text{ tr } \mathbf{S}^2, \quad J_3'' = 1/3 \text{ tr } \mathbf{S}^3, \quad J_5'' = 1/2 \, \mathbf{a.S}^2 \mathbf{.a}, \quad (9.10)$$

where now S is given by (9.4). In the case of an orthotropic material which is inextensible in the directions $\mathbf{a}$ and $\mathbf{b}$, then S is given by (9.7), and the set (9.9) may be replaced by

$$J_2''' = 1/2 \text{ tr } \mathbf{S}^2, \quad J_3''' = 1/3 \text{ tr } \mathbf{S}^3, \quad J_5''' = 1/2 \, \mathbf{a.S}^2 \mathbf{.a}, \quad J_7''' = 1/2 \, \mathbf{b.S}^2 \mathbf{.b}. \quad (9.11)$$

Similar arguments can be applied in the case of an orthotropic plastic material which is inextensible in two non-orthogonal directions $\mathbf{d}$ and $\mathbf{e}$. This case was discussed in [15].

If the associated flow rule (7.4) is adopted, then the plastic strain-rate $\mathbf{D}^p$ is obtained from the appropriate form for $F(T_{ij})$.

The current yield stress k which was introduced in (7.1) depends on the deformation history. It is sometimes assumed that k depends on the plastic work W_p, where

$$\dot{W}_p = T_{ij} D_{ij}^p. \quad (9.12)$$

If we denote $\Phi(T_{ij}) = \{F(T_{ij})\}^{1/2}$, then $\Phi(T_{ij})$ is homogeneous of degree one, and (9.4) gives

$$D_{ij}^p = \dot{\epsilon}\,\partial\Phi/\partial T_{ij}, \tag{9.13}$$

where $\dot{\epsilon}$ is a factor of proportionality. Hence

$$\dot{W}_p = \dot{\epsilon}T_{ij}\,\partial\Phi/\partial T_{ij} = \dot{\epsilon}\Phi(T_{ij}) = \dot{\epsilon}k(W_p). \tag{9.14}$$

If we choose $\epsilon = 0$ when $W_p = 0$, and k is an increasing function of W_p, then (9.14) establishes a correspondence between ϵ and W_p, and k may equally be regarded as a function of ϵ. In effect, ϵ is an equivalent plastic strain. To obtain an explicit expression for $\dot{\epsilon}$, it is necessary to solve (9.13) for T_{ij} in terms of D_{ij}^p and $\dot{\epsilon}$, and then substituting for T_{ij} in the yield condition gives $\dot{\epsilon}$ as a function of D_{ij}^p. As a simple example, a plausible generalization of von Mises' yield condition for a transversely isotropic material which is inextensible in the direction **a** is

$$J_2'' + cJ_5'' = k^2, \tag{9.15}$$

where c is constant. Then the flow rule gives

$$D_{ij}^p = \frac{\dot{\epsilon}}{2k}\left(\frac{\partial J_2''}{\partial T_{ij}} + c\,\frac{\partial J_5''}{\partial T_{ij}}\right) = \frac{\dot{\epsilon}}{2k}\{S_{ij} + 1/2\,c\,(a_i a_p S_{pj} + a_j a_p S_{pi})\},$$

and hence it follows that

$$D_{ij}^p D_{ij}^p = \frac{\dot{\epsilon}^2}{4k^2}\{2J_2'' + c(c+4)\,J_5''\},\qquad a_i a_j D_{ik}^p D_{jk}^p = \frac{\dot{\epsilon}^2}{2k^2}(1+1/2c)^2\,J_5''. \tag{9.16}$$

Therefore, from (9.15) and (9.16),

$$\dot{\epsilon}^2 = 2D_{ij}^p D_{ij}^p - \frac{4c}{c+2}\,a_i a_j D_{ik}^p D_{jk}^p. \tag{9.17}$$

and this determines the 'equivalent strain-rate' $\dot{\epsilon}$ for any given deformation. To assume that k is a function of ϵ is equivalent to assuming that k is a function of W_p.

REFERENCES

[1] BOEHLER J.P. – "On irreductible representations for isotropic scalar functions", *ZAMM*, 57 (1977): 323-327.

[2] BOEHLER J.P. – "A simple derivation of representations for non-polynomial constitutive equations in some cases of anisotropy". *ZAMM*, 59 (1979): 157-167.

[3] ERICKSEN J.E. and R.S. RIVLIN. – "Large elastic deformations of homogeneous anisotropic materials". *J. Rat. Mech. Anal.*, 3 (1954): 281-301.

[4] GREEN A.E. and J.E. ADKINS. – *"Large elastic deformations"*. Oxford University Press, 1960.

[5] PIPKIN A.C. and R.S. RIVLIN. – "The formulation of constitutive equations in Continuum Physics, I". *Arch. Rat. Mech. Anal.*, 4 (1959): 129-144.

[6] RIVLIN R.S. – "Further remarks on the stress-deformation relations for isotropic materials". *J. Rat. Mech. Anal.*, 4 (1955): 681-701.

[7] SMITH G.F. – "On isotropic integrity bases". *Arch. Rat. Mech. Anal.*, 18 (1965): 282-292.

[8] SMITH G.F. – "On isotropic functions of symmetric tensors, skew-symmetric tensors and vectors". *Int. J. Engng. Sci.*, 9 (1971): 899-916.

[9] SMITH G.F. – "On the generation of integrity bases". *Atti Accad. Naz. Lincei*, Ser. 8, 9 (1968): 51-101.

[10] SMITH G.F. and R.S. RIVLIN. – "The anisotropic tensors". *Q. Appl. Math.*, 15 (1957): 309-314.

[11] SMITH G.F. and R.S. RIVLIN. – "Integrity Bases for Vectors – the Crystal Classes". *Arch. Rat. Mech. Anal.*, 15 (1964): 169-221.

[12] SPENCER A.J.M. – "The Invariants of Six Symmetric 3 × 3 Matrices". *Arch. Rat. Mech. Anal.*, 7 (1961): 64-77.

[13] SPENCER A.J.M. – "Isotropic Integrity Bases for Vectors and Second-Order Tensors, Part II". *Arch. Rat. Mech. Anal.*, 18 (1965): 51-82.

[14] SPENCER A.J.M. – "Theory of Invariants". In *Continuum Physics*, 1. Academic Press (1971): 239-353.

[15] SPENCER A.J.M. – *"Deformations of Fibre-Reinforced Materials"* Oxford University Press, 1972.

[16] SPENCER A.J.M. and R.S. RIVLIN. – "The theory of matrix polynomials and its application to the mechanics of isotropic continua", *Arch. Rat. Mech. Anal.*, 2 (1959) : 309-336.

[17] SPENCER A.J.M. and R.S. RIVLIN. – "Finite integrity bases for five or fewer symmetric 3 × 3 matrices". *Arch. Rat. Mech. Anal.*, 2 (1959): 435-446.

[18] SPENCER A.J.M. and R.S. RIVLIN. – "Further results in the theory of matrix polynomials". *Arch. Rat. Mech. Anal.*, 4 (1960): 214-230.

[19] SPENCER A.J.M. and R.S. RIVLIN. – "Isotropic Integrity Bases for Vectors and Second-Order Tensors, Part I". *Arch. Rat. Mech. Anal.*, 9 (1962): 45-63.

RESUME

(Sur la formulation des équations constitutives pour les solides anisotropes)

Nous passons d'abord en revue les problèmes algébriques de la détermination des invariants, dans un groupe de transformations orthogonales, d'un

certain nombre de vecteurs et de tenseurs cartésiens du second ordre. On montre
ensuite comment la solution de ce problème algébrique peut être appliquée pour
la formulation des équations constitutives des matériaux anisotropes, et en
particulier des matériaux orthotropes et orthotropes de révolution. Cette
approche est illustrée par la formulation des équations constitutives en élasticité
linéaire, en élasticité finie et en plasticité. Une attention particulière est réservée
pour les matériaux renforcés par des fibres. L'effet des restrictions cinématiques
de l'incompressibilité et de l'inextensibilité dans certaines directions est
également discuté.

Anisotropic Constitutive Expressions

G. F. Smith

Lehigh University, Bethlehem, Pennsylvania, U.S.A.

1. Introduction

We are concerned with the problem of determining the general form of the tensor-valued polynomial expression

$$T_{i_1 \ldots i_n} = \Phi_{i_1 \ldots i_n} (B_{i_1 \ldots i_p}, C_{i_1 \ldots i_q}, \ldots) \tag{1.1}$$

which is invariant under a group of transformations $\{A\} = \{A_1, A_2, \ldots\}$. Thus the function $\Phi_{i_1 \ldots i_n}$ appearing in (1.1) must satisfy

$$A_{i_1 j_1} \ldots A_{i_n j_n} \Phi_{j_1 \ldots j_n} (B_{i_1 \ldots i_p}, C_{i_1 \ldots i_q}, \ldots) =$$

$$\Phi_{i_1 \ldots i_n} (A_{i_1 j_1} \ldots A_{i_p j_p} B_{j_1 \ldots j_p}, A_{i_1 j_1} \ldots A_{i_q j_q} C_{j_1 \ldots j_q}, \ldots) \tag{1.2}$$

for all $A = \|A_{ij}\|$ belonging to $\{A\}$. We are also concerned with determining the general form of expressions such as

$$T_{i_1 i_2} = C_{i_1 \ldots i_5} B_{i_3 i_4 i_5} + C_{i_1 \ldots i_8} B_{i_3 i_4 i_5} B_{i_6 i_7 i_8} \tag{1.3}$$

which are invariant under a group of transformations $\{A\}$. This is of course a special case of (1.1). The restrictions imposed on the property tensors $C_{i_1 \ldots i_5}$ and $C_{i_1 \ldots i_8}$ by the requirement that (1.3) be invariant under $\{A\}$ are that

$$A_{i_1 j_1} \ldots A_{i_5 j_5} C_{j_1 \ldots j_5} = C_{i_1 \ldots i_5}, \quad A_{i_1 j_1} \ldots A_{i_8 j_8} C_{j_1 \ldots j_8} = C_{i_1 \ldots i_8} \tag{1.4}$$

must hold for all A belonging to $\{A\}$. Tensors $C_{i_1 \ldots i_5}$ and $C_{i_1 \ldots i_8}$ which satisfy (1.4) for all A belonging to $\{A\}$ are said to be invariant under $\{A\}$. There is an extensive literature devoted to the determination of the general expressions for tensors which are invariant under the various crystallographic groups. See, for example, references [1], …, [3]. In principle, the problem of determining the form of (1.3) is solved once the general forms of the tensors

$C_{i_1 \dots i_5, \dots}$ are known. In practice, difficulties frequently arise. For example, let us consider the problem of determining the general form of

$$T_{ij} = C_{ijk \dots q} B_{k\ell m} B_{npq}, \quad T_{ij} = T_{ji}, \quad B_{ijk} = B_{ikj} \qquad (1.5)$$

which is invariant under the three-dimensional orthogonal group O_3. The general eighth order tensor which is invariant under O_3 is expressible as

$$C_{i_1 \dots i_8} = \alpha_1 \delta_{i_1 i_2} \delta_{i_3 i_4} \delta_{i_5 i_6} \delta_{i_7 i_8} + \alpha_2 \delta_{i_1 i_2} \delta_{i_3 i_4} \delta_{i_5 i_7} \delta_{i_6 i_8} + \dots (1.6)$$

where the right hand side of (1.6) denotes a linear combination of the 105 distinct isomers of the tensor $\delta_{i_1 i_2} \delta_{i_3 i_4} \delta_{i_5 i_6} \delta_{i_7 i_8}$. Only 91 of these isomers are linearly independent. Explicit expressions for the 91 linearly independent 8th order tensors invariant under O_3 are given by Kearsley and Fong [4]. If we employ the results of [4] and substitute the general expression for $C_{i_1 \dots i_8}$ in (1.5), we obtain 91 terms. However, only 15 of these terms are linearly independent and one must solve a tedious algebraic problem in order to obtain the appropriate expression. It is preferable to proceed as follows. We set

$$T_{ij} = T_{ij}^{(1)} + T_{ij}^{(2)}, \quad T_{ij}^{(1)} = \frac{1}{3} T_{kk} \delta_{ij}, \quad T_{ij}^{(2)} = T_{ij} - T_{ij}^{(1)},$$

$$B_{ijk} = B_{ijk}^{(1)} + B_{ijk}^{(2)} + B_{ijk}^{(3)} + B_{ijk}^{(4)},$$

$$B_{ijk}^{(1)} = \frac{1}{15} (B_i \delta_{jk} + B_j \delta_{ik} + B_k \delta_{ij}), \quad B_i = B_{ipp} + B_{pip} + B_{ppi},$$

$$B_{ijk}^{(2)} = \frac{1}{6} (2C_i \delta_{jk} - C_j \delta_{ik} - C_k \delta_{ij}), \quad C_i = B_{ipp} - B_{ppi}, \qquad (1.7)$$

$$B_{ijk}^{(3)} = \frac{1}{3} (2B_{ijk} - B_{jki} - B_{kij}) - B_{ijk}^{(2)},$$

$$B_{ijk}^{(4)} = \frac{1}{6} (B_{ijk} + B_{jik} + B_{kji} + B_{ikj} + B_{jki} + B_{kij}) - B_{ijk}^{(1)}.$$

The tensors $B_{ijk}^{(1)}, \dots, B_{ijk}^{(4)}$ have 3, 3, 5 and 7 independent components respectively. We may then write (1.5) as

$$T_{ij}^{(1)} + T_{ij}^{(2)} = c_{ijk \dots q} (B_{k\ell m}^{(1)} + \dots + B_{k\ell m}^{(4)}) (B_{npq}^{(1)} + \dots + B_{npq}^{(4)}) \quad (1.8)$$

and consider the 20 separate problems

$$T_{ij}^{(\alpha)} = c_{ijk \dots q} B_{k\ell m}^{(\beta)} B_{npq}^{(\gamma)}, \quad (\alpha = 1,2 \; ; \; \beta, \gamma = 1,2,3,4 \; ; \; \beta \leqslant \gamma). \quad (1.9)$$

We may compute the number of linearly independent terms in the expressions obtained from (1.9) by setting $(\alpha, \beta, \gamma) = (1,1,1), (1,1,2), \ldots, (2,4,4)$ and we find that there is just one linearly independent term in 15 cases and none in the remaining 5 cases. Thus, we have reduced the complicated algebraic problem of determining the general form of (1.5) which is invariant under the group O_3 to 15 essentially trivial problems. The validity of this procedure is based on the fact that the independent components of the six tensors $T_{ij}^{(1)}, \ldots, B_{ijk}^{(4)}$ form carrier spaces for irreducible representations of the group O_3. We next show that a variant of this procedure may be effectively employed to establish the general form of expressions such as (1.3) which are invariant under any given crystallographic group.

2. Decomposition Procedure

We now consider the problem of determining the form of

$$T_{i_1 \ldots i_n} = C_{i_1 \ldots i_n j_1 \ldots j_m} B_{j_1 \ldots j_m} \tag{2.1}$$

which is invariant under a group $\{A\} = \{A_1, \ldots, A_N\}$. Let $T_1, \ldots, T_p$ and $B_1, \ldots, B_q$ denote the independent components of $T_{i_1 \ldots i_n}$ and $B_{j_1 \ldots j_m}$ respectively. Thus, (2.1) may be written as

$$\mathbf{T} = \mathbf{CB}, \quad \mathbf{T} = \left\| \begin{matrix} T_1 \\ \vdots \\ T_p \end{matrix} \right\|, \quad \mathbf{B} = \left\| \begin{matrix} B_1 \\ \vdots \\ B_q \end{matrix} \right\|. \tag{2.2}$$

The requirement of the form (1.2) that

$$A_{i_1 p_1}^{(k)} \ldots A_{i_n p_n}^{(k)} T_{p_1 \ldots p_n} = c_{i_1 \ldots i_n j_1 \ldots j_m} A_{j_1 q_1}^{(k)} \ldots A_{j_m q_m}^{(k)} B_{q_1 \ldots q_m} \tag{2.3}$$

must hold for all $\mathbf{A}_k = \| A_{ij}^{(k)} \|$ belonging to $\{A\}$ may be written as

$$S(\mathbf{A}_k) \mathbf{T} = \mathbf{C} R(\mathbf{A}_k) \mathbf{B} . \tag{2.4}$$

The sets of $Np \times p$ matrices $S(\mathbf{A}_k)$ and $Nq \times q$ matrices $R(\mathbf{A}_k)$ which define the transformation properties of $\mathbf{T}$ and $\mathbf{B}$ respectively are said to form matrix representations of the group $\{A\}$. With (2.2) and (2.4), we see that the $p \times q$ matrix $\mathbf{C}$ is subject to the restrictions that

$$S(\mathbf{A}_k) \mathbf{C} = \mathbf{C} R(\mathbf{A}_k) \tag{2.5}$$

must hold for $k = 1, \ldots, N$. We may determine matrices Q and P such that the matrix representations $QS(A_k)Q^{-1}$ and $PR(A_k)P^{-1}$, which are said to be equivalent to the representations $S(A_k)$ and $R(A_k)$, are decomposed into the direct sum of irreducible representations of $\{A\}$. Thus, we have

$$QS(A_k)Q^{-1} = n_1 \Gamma_1(A_k) \dotplus \cdots \dotplus n_r \Gamma_r(A_k),$$

$$PR(A_k)P^{-1} = m_1 \Gamma_1(A_k) \dotplus \cdots \dotplus m_r \Gamma_r(A_k) \tag{2.6}$$

where the right hand side of $(2.6)_1$ denotes a block diagonal matrix which contains n_1 $p_1 \times p_1$ matrices $\Gamma_1(A_k), \ldots,$ n_r $p_r \times p_r$ matrices $\Gamma_r(A_k)$ along the diagonal. We note that there is a finite number r of inequivalent irreducible matrix representations associated with a given crystallographic group and we denote these by $\Gamma_1(A_k), \ldots, \Gamma_r(A_k)$. The representations $\Gamma_1(A_k), \ldots, \Gamma_r(A_k)$ are said to be of degrees $p_1, \ldots, p_r$ respectively. Let

$$QT = \tau_1 \dotplus \cdots \dotplus \tau_r, \quad PB = \beta_1 \dotplus \cdots \dotplus \beta_r,$$

$$\tau_i = \tau_{i1} \dotplus \cdots \dotplus \tau_{in_i}, \quad \beta_i = \beta_{i1} \dotplus \cdots \dotplus \beta_{im_i}, \tag{2.7}$$

$$\tau_{ij} = \left\| \begin{matrix} \tau_{j_1}^{(i)} \\ \vdots \\ \tau_{jp_i}^{(i)} \end{matrix} \right\|, \quad \beta_{ij} = \left\| \begin{matrix} \beta_{j1}^{(i)} \\ \vdots \\ \beta_{jp_i}^{(i)} \end{matrix} \right\|$$

where the matrices Q and P are those appearing in (2.6) and where the $\tau_{jk}^{(i)}$, $\beta_{jk}^{(i)}$ are of course linear combinations of the components of T and B respectively. With (2.2) and (2.4), we have

$$QT = QCB = QCP^{-1}PB = DPB, \quad D = QCP^{-1},$$

$$QS(A_k)Q^{-1}D = DPR(A_k)P^{-1} \tag{2.8}$$

where $(2.8)_3$ must hold for $k = 1, \ldots, N$. With (2.7), we see that $(2.8)_1$ may be written as

$$\left\| \begin{matrix} \tau_1 \\ \vdots \\ \tau_r \end{matrix} \right\| = \left\| \begin{matrix} D^{11}, \ldots, D^{1r} \\ \vdots \qquad \vdots \\ D^{r1}, \ldots, D^{rr} \end{matrix} \right\| \left\| \begin{matrix} \beta_1 \\ \vdots \\ \beta_r \end{matrix} \right\|, \quad D^{ij} = \left\| \begin{matrix} D_{11}^{ij}, \ldots, D_{m_j}^{ij} \\ \vdots \qquad \vdots \\ D_{n_i 1}^{ij}, \ldots, D_{n_i m_j}^{ij} \end{matrix} \right\|. \tag{2.9}$$

The matrices $D_{11}^{ij}, \ldots$ are $p_i \times p_j$ matrices and the matrix D^{ij} is a $p_i n_i \times p_j m_j$ matrix. With (2.6), (2.8)$_3$ and (2.9)$_1$, we see that the matrices $D_{\alpha\beta}^{ij}$ appearing in (2.9)$_2$ are subject to the restrictions that

$$\mathbf{\Gamma}_i(A_k) \, D_{\alpha\beta}^{ij} = D_{\alpha\beta}^{ij} \, \mathbf{\Gamma}_j(A_k) \tag{2.10}$$

must hold for $k = 1, \ldots, N$. We may then employ Schur's lemma [5] which tells us that the matrices $D_{\alpha\beta}^{ij}$ are zero matrices if $i \neq j$ and are multiples of the $p_i \times p_i$ identity matrix if $i = j$. For example, if $n_1 = 2, n_2 = 1, m_1 = 2, m_2 = 1, p_1 = 1, p_2 = 2$, we have

$$
\begin{Vmatrix} \tau_{11}^{(1)} \\ \tau_{21}^{(1)} \\ \tau_{11}^{(2)} \\ \tau_{12}^{(2)} \end{Vmatrix}
=
\left\Vert
\begin{array}{cc|cc}
a_1, & a_2 & & \\
 & & \multicolumn{2}{c}{0} \\
a_3, & a_4 & & \\
\hline
 & & a_5, & 0 \\
\multicolumn{2}{c|}{0} & 0, & a_5
\end{array}
\right\Vert
\begin{Vmatrix} \beta_{11}^{(1)} \\ \beta_{21}^{(1)} \\ \beta_{11}^{(2)} \\ \beta_{12}^{(2)} \end{Vmatrix} . \tag{2.11}
$$

Thus, the problem of determining the form of $T = CB$ which is invariant under $\{A\}$ may be replaced by the equivalent problem of determining the form of $QT = DPB$ which is invariant under $\{A\}$. However, appropriate choice of Q and P renders this problem trivial and essentially reduces the problem of determining the form of $T = CB$ to the problem of determining the decomposition of T and B into the sum of quantities whose transformation properties under $\{A\}$ are defined by the irreducible representations $\mathbf{\Gamma}_1(A_k), \ldots, \mathbf{\Gamma}_r(A_k)$. We plan to employ this technique as the basis of a computer program which will generate explicit expressions for scalar-valued and tensor-valued functions of a number of vectors and second-order tensors which are invariant under any given crystallographic group. This work is in progress and we foresee no essential difficulties. For a more detailed discussion of the above notions, see references $[6], \ldots, [10]$.

3. The General Case

The decomposition procedure employed in § 2 may be employed effectively when we seek to determine the general form of the polynomial expression

$$\Phi = \Phi(B_{i_1 \ldots i_p}, \; C_{i_1 \ldots i_q}, \ldots) \tag{3.1}$$

which is invariant under a crystallographic group $\{A\}$. We note that the problem of determining the form of the function

$$T_{i_1 \ldots i_n} = \Phi_{i_1 \ldots i_n}(B_{i_1 \ldots i_p}, \; C_{i_1 \ldots i_q}, \ldots)$$

which is invariant under $\{A\}$ may be reduced to the problem of determining the form of a scalar-valued function $\Phi^*(B_{i_1 \ldots i_p}, C_{i_1 \ldots i_q}, \ldots, T_{i_1 \ldots i_n})$ which is invariant under $\{A\}$ where Φ^* is linear in $T_{i_1 \ldots i_n}$. There is consequently no loss in generality in restricting consideration to the case (3.1). Let $S(A_k)$, $R(A_k)$, ... denote the representations of $\{A\}$ which define the transformation properties under $\{A\}$ of the column matrices $B, C, \ldots$ whose entries are the independent component $B_1, \ldots, B_p$, $C_1, \ldots, C_q, \ldots$ of the tensors $B_{i_1 \ldots i_p}, C_{i_1 \ldots i_q}, \ldots$. We may then express (3.1) as

$$\Phi = \bar{\Phi}(B, C, \ldots) \tag{3.2}$$

where the polynomial function $\bar{\Phi}$ is subject to the restriction that

$$\bar{\Phi}(B, C, \ldots) = \bar{\Phi}(S(A_k) B, R(A_k) C, \ldots) \tag{3.3}$$

for all A_k belonging to $\{A\}$. Let us choose matrices Q, P so that

$$QS(A_k) Q^{-1} = n_1 \Gamma_1(A_k) \dotplus \cdots \dotplus n_r \Gamma_r(A_k),$$
$$PR(A_k) P^{-1} = m_1 \Gamma_1(A_k) \dotplus \cdots \dotplus m_r \Gamma_r(A_k), \ldots,$$
$$QB = \beta_{11} \dotplus \cdots \dotplus \beta_{1n_1} \dotplus \cdots \dotplus \beta_{r1} \dotplus \cdots \dotplus \beta_{rn_r}, \tag{3.4}$$
$$PC = \beta_{1,n_1+1} \dotplus \cdots \dotplus \beta_{1,n_1+m_1} \dotplus \cdots \dotplus \beta_{r,n_r+1} \dotplus \cdots \dotplus \beta_{r,n_r+m_r}, \ldots$$

where the transformation properties under $\{A\}$ of the $\beta_{1i}, \ldots, \beta_{ri}$ are defined by the irreducible representations $\Gamma_1(A_k), \ldots, \Gamma_r(A_k)$ respectively. With (3.4), we see that (3.2) is expressible in the form

$$\Phi = \Phi^*(\beta_{1i}, \ldots, \beta_{rj}) \quad (i = 1, 2, \ldots; \ldots; j = 1, 2, \ldots) \tag{3.5}$$

where the function Φ^* is subject to the restrictions that

$$\Phi^*(\beta_{1i}, \ldots, \beta_{rj}) = \Phi^*(\Gamma_1(A_k) \beta_{1i}, \ldots, \Gamma_r(A_k) \beta_{rj}) \tag{3.6}$$

must hold for all A_k belonging to $\{A\}$. We observe that the problem of determining the form of $\Phi(B, C)$ and $\Psi(D, E, F)$ which are invariant under $\{A\}$, when translated into the form (3.6), will differ only in the number of quantities $\beta_{1i}, \ldots, \beta_{rj}$ which appear as arguments of the function Φ^*. Thus, if we solve (3.6) for the case where $i = 1, \ldots, k_1, \ldots, j = 1, \ldots, k_r$ where $k_1, \ldots, k_r$ are arbitrary, we have then solved the most general problem which may arise. In [11], [12], Kiral, Smith and Smith have obtained results of this generality for 27 of the 32 crystal classes. Work is in progress to extend these results to the remaining crystal classes (the 5 cubic crystal classes).

4. Syzygies

We observe that the problem of determining the general for of $\Phi(\mathbf{B}, \mathbf{C}, \ldots)$ where Φ is a scalar-valued polynomial function of the components of $\mathbf{B}, \mathbf{C}, \ldots$ which is invariant under $\{A\}$ includes the problems considered in §2 such as (2.1) as a special case. Let us consider the problem of determining the general expression for a polynomial scalar-valued function of a single tensor $\mathbf{B} = \|B_{ij}\|$ which is invariant under $\{A\}$. Let I_1, I_2, I_3 be polynomial functions which are of degrees 1, 2, 3 respectively in $\mathbf{B}$ and which are invariant under $\{A\}$ such that any polynomial function of $\mathbf{B}$ which is invariant under $\{A\}$ is expressible as a polynomial in the I_1, I_2, I_3. The invariants I_1, I_2, I_3 are referred to as the elements of an integrity basis for functions of $\mathbf{B}$ which are invariant under $\{A\}$. Thus, we have

$$\Phi(\mathbf{B}) = c_0 + c_1 I_1 + c_2 I_1^2 + d_2 I_2 + c_3 I_1^3 + d_3 I_1 I_2 + e_3 I_3$$
$$+ c_4 I_1^4 + d_4 I_1^2 I_2 + e_4 I_2^2 + f_4 I_1 I_3 + \ldots . \tag{4.1}$$

If we consider the problem of determining the general form of the functions

$$c_{ij} B_{ij}, \; c_{ijk\ell} B_{ij} B_{k\ell}, \; c_{ijk\ell mn} B_{ij} B_{k\ell} B_{mn}, \ldots \tag{4.2}$$

which are invariant under $\{A\}$, we may of course proceed by setting

$$c_{ij} B_{ij} = c_1 I_1, \ldots ,$$
$$c_{ijk\ell mnpq} B_{ij} B_{k\ell} B_{mn} B_{pq} = c_4 I_1^4 + d_4 I_1^2 I_2 + e_4 I_2^2 + f_4 I_1 I_3, \ldots \tag{4.3}$$

provided that all of the terms of degree n in $\mathbf{B}$ appearing on the right hand side of (4.1) are linearly independent. In general, we are not sure that this is the case. For example, it might be the case that $I_1 I_3 = I_2^2$. Such a relationship is referred to as a syzygy. The problem of determining all of the syzygies which relate the elements $I_1, I_2, I_3, \ldots$ of an integrity basis is one of the main problems of the theory of invariants. Solution of this problem would enable us to eliminate all of the redundant terms appearing in an expression such as (4.1) and we would then be able to read off from (4.1) the general expression for

$$c_{i_1 \ldots i_{2n}} B_{i_1 i_2} \ldots B_{i_{2n-1} i_{2n}}$$

for any n. In references [13], [14], [15], we have considered the problem of eliminating all redundant terms from the general expressions for scalar-valued and tensor-valued polynomial functions which are invariant under a given group $\{A\}$.

Acknowledgement

The work described in this paper was supported by a grant from the National Science Foundation.

REFERENCES

[1] BIRSS R.R. − *Symmetry and Magnetism*. North Holland Publishing Company, 1964.

[2] FIESCHI R. and F.G. FUMI. − *Nuovo Cimento*, 10 (1953): 865.

[3] SMITH G.F. − *An. N.Y.A.S.*, 172 (1970): 57.

[4] KEARSLEY E.A. and J.F. FONG. − *J. Res. Nat. Bur. Stds.*, 79B (1975): 49.

[5] LOMONT J.S. − *Applications of Finite Groups*, New York: Academic Press, 1959.

[6] SMITH G.F. and E. KIRAL. − *Int. J. Engng. Sci.*, 16 (1978): 773.

[7] SMITH G.F. − *Proceedings of the Symposium on Group Theoretical Methods in Mechanics*, Novosibirsk, USSR, 1978.

[8] CALLEN H. − *Am. J. Physics*, 36 (1968): 735.

[9] CALLEN H., E. CALLEN and J. KALVA. − *Am. J. Physics*, 38 (1970): 1278.

[10] DUFFEY G.H. − *18th Polish Solid Mechanics Conference*, 1975.

[11] KIRAL E. and G.F. SMITH. − *Int. J. Engng. Sci.*, 12 (1974): 471.

[12] KIRAL E., M.M. SMITH and G.F. SMITH. − *Int. J. Engng. Sci.*, 18, (1980): 569.

[13] SMITH G.F. − *Arch. Rat'l Mech. Anal.*, 10 (1962): 108.

[14] SMITH G.F. − *Quart. Applied Math.*, 31 (1973): 373.

[15] SMITH M.M. and G.F. SMITH. − Quart. Applied Math., 34 (1982) : 509.

RESUME

(Sur les expressions constitutives anisotropes)

Nous discutons du problème de la détermination de la forme générale d'une fonction tensorielle polynomiale $T = \Phi(B, C, ...)$ d'un certain nombre de tenseurs $B, C, ...$, lorsqu'elle est invariante dans un groupe $\{A\}$.

On the Structure of Anisotropic Tensor Functions in Continua

K. Z. Markov
Centre for Mathematics and Mechanics, Bulgarian Academy of Sciences, Sofia, Bulgaria.

A. A. Vakulenko
Faculty of Mathematics and Mechanics, University of Leningrad, Leningrad, USSR.

1. Introduction

In the analysis of response of anisotropic media there exists a general problem about the material symmetry restrictions imposed on the constitutive equations. Numerous studies on this question have been carried out (cf. the survey [18]). However, these studies have been based on the assumption that the tensor relations involved are polynomial (or that they could be approximated by polynomials). According to the apt remark of Pipkin and Wineman [8], this assumption 'is a matter of mathematical convenience; it seems somewhat foreign to a discussion of material symmetry'.

The present work is devoted to an analysis of the symmetry restrictions imposed on the tensor function $\mathcal{F}$ of arbitrary type. No proofs are given here, for brevity of this note. We put no requirements on regularity properties of $\mathcal{F}$. But unlike [8], [21], we do not rely on the known polynomial results in order to extend these to a more general class of functions.

We begin our study by introducing the sets of anisotropic tensors $\mathcal{C}(G)$ and tensor functions $\mathcal{S}(G)$ the symmetry groups of which contain an orthogonal subgroup G (Sec. 2). The orthogonal subgroups G are described by the so-called determining tensors [4], [5], i.e. by tensors with a symmetry group coinciding with G. The fact that we deal only with the tensor representations of the orthogonal group automatically restricts the class of subgroups which we are interested in when describing the sets $\mathcal{C}(G)$ and $\mathcal{S}(G)$: this is the class consisting of those groups which can be determined by tensors (Sec. 3). If a subgroup G is determined by a tensor X, it turns out that the sets of anisotropic tensors $\mathcal{C}(G)$ consist of tensors which could be constructed from X, using the invariant tensor operations; the values $\mathcal{F}(T)$ of the anisotropic functions $\mathcal{F} \in \mathcal{S}(G)$ are constructed in the same manner from the tensor-argument T and the determining tensor X (Sec. 4). The latter result leads to the so-called Canonical Form of the tensor functions with a given group of symmetry (Theorem 4.2). The Canonical Form reduces the investigation of the anisotropic tensor functions to two basic problems: 1) the construction of a tensor basis (i.e. a basic system of form-invariant tensors) and 2) the construction of a complete system of scalar invariants. Some facts concerning the tensor bases are discussed in Sec. 5. The concept of a complete system of scalar invariants is considered in Sec. 6; it is emphasized there that the completeness depends only on the requirements regarding the type of polynomiality, smoothness, etc., put on the tensor functions. The relationship between the construction of a tensor basis and that of a complete system of scalar invariants is discussed in Sec. 7.

2. Symmetry group of tensors and tensor functions

Let E be a three-dimensional Euclidean space over the field R of the real numbers and let

$$R, \ E, \ E \otimes E, \dots, \underset{p}{\otimes} E, \dots \tag{2.1}$$

be the tensor powers of E, so that $\mathfrak{G}(E) = \overset{\infty}{\underset{p=0}{\cup}} \otimes E$ is the set of all the Cartesian tensors over E (R is treated as the 'zeroth' tensor power of E). The space $\underset{p}{\otimes} E$ of p-th order tensors is generated by the polyads $x_1 \otimes x_2 \dots \otimes x_p$, i.e. by the tensor products of the vectors $x_1, \dots, x_p \in E$ (hereafter we omit the sign of tensor product of vectors) [1]. Viewing the second-order tensors as linear transformations of E, we can associate with any $A \in E \otimes E$ a linear transformation $A^p \in \underset{2p}{\otimes} E$ of the space $\underset{p}{\otimes} E$, which acts on the polyads according to the formula

$$A^p: \ x_1 x_2 \dots x_p \longrightarrow (A.x_1) \ (A.x_2) \dots (A.x_p),$$

where $x_i \in E$, $i = 1, 2, \dots, p$, and the dot denotes the 'semi-scalar' product, e.g. $(A.x)_n = A_{nm} x^m$. The tensor $A^p = \underset{p}{\otimes} A$ is the p-th tensor power of the tensor $A \in E \otimes E$.

As usual, we denote by $O(E) \subset E \otimes E$ the orthogonal group of the space E, so $O(E) = \{U \in E \otimes E \,|\, U.U^* = I\}$, with U^* being the tensor conjugated to U and I the unit tensor.

We call $\mathfrak{F}: V_1 \longrightarrow V_2$ a tensor function over E if V_1 and V_2 are certain Cartesian products of a finite number of tensor powers (2.1). For simplicity we restrict our attention to the tensor functions of the particular type

$$\mathscr{F} : \underset{p}{\otimes} E \longrightarrow \underset{q}{\otimes} E, \quad p, q > 0, \tag{2.2}$$

though all the considerations and results given below are, mutatis mutandis, valid in the general case as well.

We connect with every tensor function (2.2) its symmetry group $G_{\mathscr{F}} \subset O(E)$. This is the group which consists of all $U \in O(E)$ that 'commute' with $\mathscr{F}$, i.e.

$$G_{\mathscr{F}} = \{ U \in O(E) \mid \mathscr{F}(U^p(T)) = U^q(\mathscr{F}(T)), \ \forall T \in \underset{p}{\otimes} E \} \tag{2.3}$$

The tensor function is called isotropic, if $G = O(E)$. If G is a subgroup of $O(E)$ then $\mathscr{F}$ is called anisotropic.

In the case of a linear function $\mathscr{F}$ we have $\mathscr{F}(T) = F \cdot T$, with a tensor $F \in \underset{s}{\otimes} E$, $s = p + q$ (the boldface dot means contraction with respect to p pairs of indices). Then (2.3) reduces to the familiar definition of the symmetry group

$$G = \{ U \in O(E) \mid U^s(F) = F \} \tag{2.4}$$

for the tensor $F \in \underset{s}{\otimes} E$.

As a simple consequence of (2.3) and (2.4) we obtain the useful relation

$$G_T \cap G_{\mathscr{F}} \subset G_{\mathscr{F}(T)}, \quad \forall T \in \underset{p}{\otimes} E, \tag{2.5}$$

between the symmetry groups of the argument and the value for a tensor function $\mathscr{F}$.

Let $G \subset O(E)$ be an orthogonal subgroup. In continuum mechanics applications, the basic problem concerns the description of the set

$$\mathscr{S}(G) = \{ \mathscr{F} \mid G \subset G_{\mathscr{F}} \} \tag{2.6}$$

consisting of tensor functions $\mathscr{F}$, each of which has the transformations $U \in G$ as symmetry elements. In the terminology adopted, the elements of $\mathscr{S}(G)$ are called form-invariant tensors for the group G [18], or tensor concomitants for G [2], [3]. For brevity we shall call them G-invariants or invariants, if the group G is fixed.

If we assume that the tensor functions $\mathscr{F}$ are linear, then the set (2.6) reduces to the set

$$\mathscr{C}(G) = \{ T \in \mathscr{C}(E) \mid G \subset G_T \} = \{ T \in \mathscr{C}(E) \mid U^p(T) = T, \ \forall U \in G \} \tag{2.7}$$

of all the Cartesian tensors over E which do not change under the transformation U^p, $U \in G$. We have the decomposition $\mathscr{C}(G) = \overset{\infty}{\underset{n=0}{\cup}} I_n(G)$, where $I_n(G)$ is the linear space of all G-invariant tensors of order n, $n = 0, 1, \ldots$.

3. Determining tensors

We introduce the class $\mathscr{G}$ of orthogonal subgroups each of which is a symmetry group for a certain tensor over E, i.e.

$$\mathscr{G} = \{G \subset O(E) \mid \exists X \in \mathscr{C}(E), \ G = G_X\}. \tag{3.1}$$

LEMMA 3.1.:   The class $\mathscr{G}$ consists of all finite and texture[1] orthogonal subgroups.

LEMMA 3.2.:   Let G_0 be an orthogonal subgroup which does not belong to the class $\mathscr{G}$, then

$$\mathscr{S}(G_0) = \mathscr{S}(G_*), \quad \mathscr{C}(G_0) = \mathscr{C}(G_*), \tag{3.2}$$

where $G_* = \underset{G_0 \subset G}{\cap}$, $G \in \mathscr{G}$, is the smallest group in $\mathscr{G}$ which contains G_0 as a subgroup.

Hence, analyzing the structure of the sets $\mathscr{S}(G)$ of G-invariants, we can restrict our attention to the subgroups $G \in \mathscr{G}$.

Let $G \in \mathscr{G}$. Following Lokhin and Sedov [5], we call $X \in \mathscr{C}(E)$ the determining tensor for G, if $G = G_X$. Equivalently, the subgroups $G \in \mathscr{G}$ could also be determined by a number of tensors $X_1, \ldots, X_n$ so that $G = G_{X_1} \cap \ldots \cap G_{X_n}$. The point is that, within the accuracy of an inversion, $G_X = G_{X_1} \cap \ldots \cap G_{X_n}$, where the tensor X is the polyad $X = X_1 \otimes \ldots \otimes X_n$; the inversion can be eliminated by adding the alternating tensor as a tensor multiplier into the polyad $X_1 \otimes \ldots \otimes X_n$.

Obviously, the class $\mathscr{G}$ is wide enough for continuum mechanics applications, because it includes all 32 symmetry groups of crystals and the seven texture subgroups; the table of the determining tensors for these 39 subgroups is given in [5].

It is important to note that the groups $G \in \mathscr{G}$ are compact. Moreover, the assumption that the group G is determined by a tensor presents a set of polynomial conditions; consequently, the groups $G \in \mathscr{G}$ are algebraic and

---

[1] An orthogonal subgroup is a texture, if it contains all rotations about a fixed axis.

therefore, according to a known result, they are Lie groups. But in our case this is clear from their description given by the Lemma 3.1.

4. Description of the sets of anisotropic tensors and tensor functions

Let $G \in \mathcal{G}$ be a certain orthogonal subgroup with the determining tensor X, so that $G = G_X$. We consider the sets

$$I_p(X) \subset \underset{p}{\otimes} E, \quad I(X) = \underset{p=0}{\overset{\infty}{\cup}} I_p(X) \subset \mathscr{C}(E), \tag{4.1}$$

consisting, respectively, of the p-th order tensors and of the tensors over E, which can be constructed from X using the invariant tensor operations, i.e. forming linear combinations, tensor products, contractions and isomers [1].

The invariant tensor operations are polynomial isotropic tensor functions, so, according to (2.5), $G_X \subset G_T$ for each $T \in I(X)$. Hence, in the notation (2.7) and (4.1), we have that $I(X) \subset \mathscr{C}(G)$. It turns out, however, that these two sets coincide:

THEOREM 4.1 : If $G \in \mathcal{G}$ and $G = G_X$, then

$$I(X) = \mathscr{C}(G), \quad I_p(X) = I_p(G), \quad p = 0, 1, \ldots . \tag{4.2}$$

In other words, every tensor T with a symmetry group $G_T \supset G = G_X$ can be constructed from X, using only the invariant tensor operations.

Theorem 4.1 has been announced by Lokhin [4] for the class of symmetry groups of crystals and textures.

If $G = O(E)$, Theorem 4.1 is well-known (see, for example, [3]); in this case it states that every isotropic tensor is a linear combination of isomers of the tensor $I \otimes I \otimes \ldots \otimes I$, because the unit second-order tensor I determines the full orthogonal group, i.e. $G_I = O(E)$.

THEOREM 4.2 : If $G \in \mathcal{G}$, $G = G_X$, then every tensor function $\mathscr{F} \in \mathscr{S}(G)$ of the type (2.2) can be represented in the so-called Canonical Form:

$$\mathscr{F}(T) = \sum_{i=1}^{M} f_i(T) P_i(T), \quad T \in \underset{p}{\otimes} E, \tag{4.3}$$

---

[1] A is an isomer of B, if there exists a permutation s such that

$$a_{i_1 i_2 \cdots i_p} = b_{i_{s(1)} i_{s(2)} \cdots i_{s(p)}} ; \quad A = \| a_{i_1 i_2 \cdots i_p} \|, \quad B = \| b_{i_1 i_2 \cdots i_p} \| \in \underset{p}{\otimes} E.$$

where $P_i \in \mathcal{S}(G)$ are polynomial tensor functions such that for every $T \in \underset{p}{\otimes} E$ the set of their values $P_1(T), \ldots, P_M(T)$ contains a basis for the space $I_q(X, T)$, generated by the tensors X and T by means of the invariant tensor operations, and where f_i are scalar-valued G-invariants.

A stronger result was proposed in [7], where it was shown that the tensor functions P_i could be chosen so that the scalar-valued functions f_i in (4.3) are polynomial, provided the function $\mathcal{F}$ is polynomial. In this case (4.3) is called the Polynomial Canonical Form. It is clear that every Polynomial Canonical Form is also a Canonical Form [8], [21], but the converse is not generally valid.

It turns out that the Polynomial Canonical Forms are smooth in the sense of the following theorem.

THEOREM 4.3 : Let (4.3) be the Polynomial Canonical Form for the tensor functions $\mathcal{F} \in \mathcal{S}(G)$. Then a tensor function $\mathcal{F} \in \mathcal{S}(G)$ is smooth, i.e. $\mathcal{F} \in C^\infty$, if and only if the scalar-valued G-invariants f_1 to f_M in the representation (4.3) for $\mathcal{F}$ are also smooth.

However, in the case of a tensor function $\mathcal{F} \in C^m$, $m < \infty$, the coefficients f_i are in C^n, and $m \neq n$ in general. The problem of relating m to n seems to be difficult. To the best of our knowledge the only result in this direction has been established in [13] (see also [19]), where it has been shown that an isotropic symmetric second-order tensor function is continuous if and only if the coefficients f_1, f_2 and f_3 (there are three of them in the case as given below; see (5.2)) possess three continuous derivatives, i.e. $n = m + 3$ in the notation just introduced.

The following consequence of Theorems 4.1 and 4.2 is often used in applications:

THEOREM 4.4 : Every anisotropic tensor function $\mathcal{F} \in \mathcal{S}(G)$ can be represented in the form

$$\mathcal{F}(T) = \mathcal{H}(T, X),$$ (4.4)

where $\mathcal{H}$ is a certain isotropic tensor function of two tensor arguments T and X, with X denoting the determining tensor for the group G.

5.   Tensor bases

Following Smith and Rivlin [14], we call a tensor basis for the group G every basis in the linear space $I_p(G)$ of anisotropic p-th order tensors with the given symmetry group G.

Theorem 4.1 presents the full description of the space $I_p(G)$ of anisotropic tensors. However, it gives non information about the concrete construction of the tensor bases. Note that a general method and examples of finding such bases have been proposed in [14], [16].

According to Theorem 4.1, the construction of a tensor basis means the construction of a basis in the linear space $I_p(X) \subset \underset{p}{\otimes} E$ consisting of the p-th order tensors which are obtained by applying the invariant tensor operations to the tensor X, $G = G_X$. The same problem occurs when analyzing the G-invariant tensor functions, because the values of the functions $P_i(T)$, i.e. of the basic form-invariant tensors in (4.3), form a generating system for the space $I_q(X, T)$, $T \in \underset{q}{\otimes} E$. That is why we shall take the liberty to call the polynomial tensor functions $P_i(T)$ in the Canonical Form (4.3) also the tensor basis for the group G.

The general representation (4.3) reduces the investigation of the anisotropic tensor functions to two basic problems:

A. Construction of a tensor basis for the group G;

B. Construction of a complete system of scalar-valued G-invariants allowing a representation of the scalar-valued G-invariants f_i by the elements of this system.

Let $E \underset{s}{\otimes} E = \{T \in E \otimes E \mid T = T^*\}$ be the space of second-order symmetric tensors. The problem of tensor basis construction has been solved, first of all, for the spaces of symmetric second-order tensors

$$I_2^s(x_1, x_2, \ldots, x_{r_1}, A_1, A_2, \ldots, A_{r_2}) \subset E \underset{s}{\otimes} E,$$

generated by certain vectors $x_i \in E$ and second-order symmetric and/or skew-symmetric tensors; the basic results in this field have been surveyed in [18].

For example, for every tensor $T \in E \underset{s}{\otimes} E$, the space $I_2^s(T) \in E \underset{s}{\otimes} E$ is generated by the tensors I, T, and $T.T$. As a matter of fact, this is the Hamilton-Cayley theorem. Obviously, $I_2^s(I, T) = I_2^s(T)$, so that Theorem 4.2 then leads to the general representation of the isotropic second-order tensor functions of the type

$$\mathcal{T} : E \underset{s}{\otimes} E \longrightarrow E \underset{s}{\otimes} E \tag{5.1}$$

in the form

$$\mathcal{T}(T) = f_1(T)I + f_2(T)T + f_3(T) T.T, \quad T \in E \underset{s}{\otimes} E, \tag{5.2}$$

with the isotropic scalar-valued functions f_i, $i = 1, 2, 3$.

For two symmetric tensors $T_1, T_2 \in E \underset{s}{\otimes} E$, the space $I_2^s(T_1, T_2)$ $\subset E \underset{s}{\otimes} E$ is generated by the eight Rivlin-Ericksen tensors [10]

$$I, T_1, T_2, T_1.T_1 . T_2.T_2, T_1 \circ T_2, T_1 \circ (T_2.T_2), T_2 \circ (T_1.T_1), \quad (5.3)$$

where $T_1 \circ T_2 = \frac{1}{2}(T_1.T_2 + T_2.T_1)$, etc. The basis (5.3) allows us to construct the general representations (4.3) for the tensor functions of the type (5.1) which are rotationally invariant or possess a rhombic symmetry group; this is so because these are the only cases in which the symmetry groups are determined by symmetric second-order tensors. The determining tensors for the rest of the crystal point groups are of order higher than two [5]. Therefore, to get the concrete form of the general representation (4.3) for the functions (5.1) with the crystal symmetry groups which do not belong to the rhombic syngony, we need tensor basis constructions in the spaces $I_2^s(X, T)$, formed by the tensors $T \in E \underset{s}{\otimes} E$, $X \in \underset{p}{\otimes} E$, $p > 2$. This is a problem much more complicated than the tensor basis construction by using two symmetric second-order tensors. One method, considered in [6], is to obtain all vectors and/or second-order tensors which can be found invariantly by means of the tensors T and X given, and then to apply, for instance, the basis (5.3). Using this method, a tensor basis in the space $I_2^s(O_h, T) \subset E \underset{s}{\otimes} E$ has been constructed in [6];

$$T \in E \underset{s}{\otimes} E, \quad O_h = e_2^4 + e_3^4 \in \underset{s}{\otimes} E$$

with e_1, e_2, e_3 being an orthonormal basis in E. The tensor basis consists of the following 14 tensors

$$I, T, T.T, T_1, T_1.T_1, T_2, T_2.T_2, T \circ T_1, T \circ T_2, T_1 \circ T_2, \quad (5.4)$$
$$T \circ (T_1.T_1), T \circ (T_2.T_2), T_1 \circ (T.T), T_2 \circ (T.T),$$

where $T_1 = O_h : T, T_2 = O_h : (T.T)$ are the two symmetric second-order tensors which can be constructed by using T and O_h (the colon means a contraction with respect to two pairs of indices). The tensor basis (5.4) leads to the general representation of the tensor functions (5.1) which possess the cubic gyroidal crystal point group as a symmetry group, because just this point group is determined by the tensor O_h.

Here we propose a more general method for constructing tensor bases. The method is based on the essential relationship between the tensor bases and the completeness of the systems of scalar-valued invariants (Theorems 7.1 and 7.2 given below). In a particular case, the method has been proposed in [20].

6. Complete systems of scalar invariants

First of all, we will discuss the concept of completeness for a system of scalar-valued invariants.

Let $G \subset O(E)$ be an orthogonal subgroup and let

$$J_p(G) = \{f : \bigotimes_p E \longrightarrow R \mid G \subset G_f\} \qquad (6.1)$$

be the set of all scalar-valued invariants of the p-th order tensors under the group G. By $J_p^r(G) \subset J_p(G)$ we denote the set of scalar-valued G-invariants which possess r continuous derivatives, $p, r = 0, 1, \ldots$.

DEFINITION 6.1 : The system of scalar-valued G-invariants $g_1, \ldots, g_n$ is called complete if every G-invariant $f \in J_p(G)$ can be represented by g_1 to g_n, i.e.

$$f = \varphi(g_1, g_2, \ldots, g_n). \qquad (6.2)$$

It is to be noted that we put no requirements on the continuity or differentiability of representation (6.2) of the invariants by the basic ones; thus, the system g_1 to g_n, complete in the sense of the Definition 6.1, is the so-called functional basis [8], [21].

In continuum mechanics applications, however, we not only need the possibility to represent an arbitrary invariant through the basic ones, but the differentiability of the representation as well. That is why we introduce one more definition:

DEFINITION 6.2 : The system of scalar-valued G-invariants $g_1, \ldots, g_n$ $\in J_p^r(G)$ is called C^r-complete, if every G-invariant $f \in J_p^r(G)$ can be represented by g_1 to g_n in the form (6.2) with a function $\varphi \in C^r$. If $r = \infty$, the system is called smoothly-complete.

However, it is always traditionally assumed that the invariants under consideration are polynomial. As it is well known, such a tradition is connected with the essential results in the theory of invariants, obtained by the methods of algebra (see, for example [2], [3]). The polynomial requirement leads to the introduction of the polynomial completeness, in the sense that the function φ in (6.2) must be polynomial, if the invariant f is polynomial; this is the familiar definition of an integrity basis for the group G considered.

From the point of view of continuum mechanics, the polynomial requirement is unnecessary. The only thing we need there is the possibility to differentiate (at least once) the function φ in (6.2) for a differentiable f.

The simplest example we could mention here is the representation of the elastic potential $W = W(T_\sigma)$ for an anisotropic elastic solid by the basic system of invariants for the stress tensor T_σ; in order to obtain the deformation tensor we have to differentiate W once with respect to T_σ. Besides, there exist useful invariants of the stress or strain tensors which are not polynomial.

The following results give the basic known relations between the various definitions of the completeness.

LEMMA 6.1 :   ([8], [21]). Every integrity basis is a functional basis.

Moreover, the same methods, used in [8], [21] to prove Lemma 6.1, show that the following stronger result is valid as well:

LEMMA 6.2 :   Every C^r-complete system is a functional basis, $r = 0, 1, \ldots$

LEMMA 6.3 :   ([12]). Every integrity basis is smoothly-complete.

There is a hypothesis that the integrity basis is also C^r-complete, $r < \infty$ ([9], p. 63).

7.   Relations between tensor bases and complete systems of invariants

As stated in Sec. 5, finding the Canonical Form (4.3) of the anisotropic tensor functions $\mathscr{T} \in \mathscr{S}(G)$ requires the solution of two problems A and B, namely, firstly, the construction of a tensor basis for G and, secondly, the construction of a complete system of scalar-valued G-invariants. It is to be noted that the definition of the completeness depends on the restrictions imposed on the tensor function $\mathscr{T}$. If $\mathscr{S}$ is smooth, then the completeness we use is the smooth one (in the sense of Def. 6.2). If $\mathscr{T}$ is polynomial, then the completeness is polynomial as well, i.e. we have to use an integrity basis, etc.

As it follows from the following two theorems, Problems A and B are closely connected.

THEOREM 7.1 :   Let $G \in \mathscr{G}$, $G = G_X$, and g_i, $i = 1, \ldots, n$, be a system of polynomial scalar-valued G-invariants of p-th order tensors. If the gradients $\nabla_T g_i \in I_p(X, T)$, $T \in \bigotimes_p E$, form a tensor basis for the functions (2.2) with $p = q$, then the system g_1 to g_n is a functional basis for G.

THEOREM 7.2 :   Let $G \in \mathscr{G}$, $G = G_X$, be a finite orthogonal subgroup and g_i, $i = 1, \ldots, n$, be a system of polynomial scalar-valued G-invariants of p-th order tensors. If the system g_1 to g_n is an integrity basis for G, then the gradients $\nabla_T g_i \in I_p(X, T)$, $T \in \underset{p}{\otimes} E$, form a tensor basis for the functions (2.2) with $p = q$.

It is important to note that, under the conditions of Theorem 7.2, the tensor basis $\nabla_T g_i$ may not be a polynomial tensor basis in the sense that the scalar-valued functions f_i in the Canonical Form (4.3) may not be polynomial, if the tensor function $\mathscr{T}$ is polynomial. Such an example has been presented by Smith [17].

The integrity bases for the symmetric second-order tensors have been constructed in [15] for all the crystal point groups. According to Theorems 4.2 and 7.2, we need only differentiate these integrity bases in order to obtain the corresponding tensor bases and the general representations (4.3) for the tensor functions (5.1) with crystal symmetry groups.

Theorem 7.2 generalizes the well-known fact that the tensor basis I, T, $T.T$ in the space $I_2^s(T)$, $T \in E \otimes E$, can be found by differentiating the complete system $\mathrm{tr}(T)$, $\mathrm{tr}(T.T)$, $\mathrm{tr}(T.T.T)$ of isotropic invariants for the symmetric second-order tensor T. But if we reject the assumption that the tensors are symmetric, i.e. if we consider the isotropic tensor functions $\mathscr{T} : E \otimes E \rightarrow E \otimes E$, then the gradients of the full system of isotropic invariants for the tensors $T \in E \otimes E$ are not sufficient to form a tensor basis [11]; hence, the condition that the group G is finite is essential for the validity of Theorem 7.2.

Acknowledgment

One of the authors (K.M.) had a very useful discussion with Prof. G.F. Smith, whose criticism led to improvements in the precision of some formulations in the first version of the paper. The authors greatly appreciate Prof. Smith's interest in the paper.

REFERENCES

[1] BOURBAKI N. *Eléments de mathématique. Algèbre.* Paris : Hermann, 1970, Ch. III.

[2] DIEUDONNE J.A. and J.B. CARRELL. *Invariant Theory, Old and New.* N.Y. – London : Academic Press, 1971.

[3] GOUREVICH G.B. *Foundations of the Theory of Algebraic Invariants.* Sijthoff and Noordhoff, 1964.

[4] LOKHIN V.V. "A system of determining parameters which characterize geometrical properties of anisotropic media" (in Russian). *Dokladi AN SSSR (DAN SSSR),* 149 (1963): 295-297.

[5] LOKHIN V.V. and L.I. SEDOV. – "Nonlinear tensor functions of several tensor arguments" (in Russian). *Prikl. matematika i mekhanika (PMM)* 27 (1963): 393-417.

[6] MARKOV K.Z. – Thesis. Leningrad: University of Leningrad, 1972.

[7] PIPKIN A.C. and R.S. RIVLIN. – "The formulation of constitutive equations in continuum physics". Part I. *Archive Rat. Mech. Anal.*, 4 (1959): 129-144.

[8] PIPKIN A.C. and A.S. WINEMAN. – "Material symmetry restrictions on non-polynomial constitutive equations". *Ibid.* 12 (1963): 420-426.

[9] POENARU V. – *"Singularités C$^\infty$ en présence de symétrie"*. Lecture Notes in Mathematics, 510. Springer-Verlag, 1976.

[10] RIVLIN R.S. and J.L. ERICKSEN. – "Stress-deformation relation for isotropic materials". *J. Rat. Mech. Anal.* 4 (1955): 323-425.

[11] RYCHLEWSKI J. – "The general form of isotropic transformation of tensor spaces". *Bull. Acad. Polon. Sci., ser. sci. techn.* 18 (1970): 533-538.

[12] SCHWARZ G. – "Smooth functions invariant under the action of a compact Lie group". *Topology* 14 (1975): 63-68.

[13] SERRIN J. – "The derivation of stress-deformation relations for a Stokesian Fluid". *J. Math. Mech.*, 8 (1959): 459-470.

[14] SMITH G.F. and R.S. RIVLIN. – "The anisotropic tensors". *Quart. Appl. Math.*, 15 (1957): 308-314.

[15] SMITH G.F. and R.S. RIVLIN. – "The strain-energy function for anisotropic elastic materials". *Trans. Amer. Math. Soc.*, 88 (1958): 175-193.

[16] SMITH G.F. and R.S. RIVLIN. – "Integrity bases for vectors – the crystal classes". *Archive Rat. Mech. Anal.*, 15 (1964): 169-221.

[17] SMITH G.F. – Private communication.

[18] SPENCER A.J.M. – *Theory of Invariants*. In A.C. Eringen, ed., *Continuum Physics*. 1,239-353.New-York: Academic Press, 1971.

[19] VAKULENKO A.A. – *Polylinear Algebra and Tensor Analysis in Mechanics* (in Russian). University of Leningrad, Leningrad, 1972.

[20] VAKULENKO A.A. and K.Z. MARKOV. – "On the bases in the space of second-order tensors". *Bull. Acad. Polon. Sci. ser. sci. techn.* 24 (1976): 283-292.

[21] WINEMAN A.S. and A.C. PIPKIN. – "Material symmetry restrictions on constitutive equations". *Archive Rat. Mech. Anal.*, 17 (1964): 184-214.

RESUME

(Sur la structure des fonctions tensorielles anisotropes pour les milieux continus)

Dans ce travail, le problème général des restrictions imposées par les symétries matérielles sur les fonctions tensorielles est discuté. On montre que la représentation générale des fonctions tensorielles anisotropes est uniquement une conséquence des propriétés de symétrie. Cette représentation réduit l'investigation de ces fonctions à deux problèmes fondamentaux : la construction d'une base tensorielle et d'une base d'intégrité, respectivement. L'équivalence de ces deux problèmes est formulée pour la classe des groupes discrets.

Symmetry Relations
for Anisotropic Materials

A. Rathkjen

Instituttet for Bygningsteknik, Aalborg Universitetscenter, Danmark.

1. Introduction

In continuum mechanics the description of physical properties of materials takes place in constitutive equations expressing relations between tensors which otherwise appear in some field equations. The possible symmetries that an anisotropic material may have will be reflected one way or another in the constitutive equations. The relations may be linear or non-linear tensor functions or functionals of tensor arguments. In Sections 2-5 the case of linear relations will be discussed while non-linear relations are briefly mentioned in Section 6.

2. Symmetry relations

In Cartesian coordinates a linear relation between two tensors t and e may be given as

$$t^{k\ldots}_{\ell\ldots} = a^{k\ldots n}_{\ell\ldots m} \; e^{m\ldots}_{n\ldots} \tag{2.1}$$

where the material property in question is represented by the tensor a. When the tensor t is of order P, the tensor e of order Q the tensor a will be of order $P + Q$. In a linear theory the components of a are constants, or probably functions of position.

Let the constitutive equation (2.1) be valid in a coordinate system with coordinates x^k. In an−other system X^ℓ the corresponding equation is

$$T^{k\ldots}_{\ell\ldots} = A^{k\ldots n}_{\ell\ldots m} \; E^{m\ldots}_{n\ldots} \tag{2.2}$$

With the coordinate transformations

$$\partial x^k/\partial X^\ell = \beta^k_\ell; \; \partial X^k/\partial x^\ell = \gamma^k_\ell \tag{2.3}$$

where

$$\beta^k_\ell \gamma^\ell_m = \beta^\ell_m \gamma^k_\ell = \delta^k_m \tag{2.4}$$

the tensor transformations are

$$A_{\varrho\;\cdots}^{k\;\cdots} = \beta_{\varrho}^{s} \ldots \gamma_{r}^{k} \ldots a_{s\;\cdots}^{r\;\cdots} \tag{2.5}$$

For an arbitrary tensor the components $A_{\varrho\;\cdots}^{k\cdots}$ are different from $a_{\varrho\;\cdots}^{k\cdots}$ unless the transformation is the identity transformation $\beta_{\varrho}^{k} = \gamma_{\varrho}^{k} = \delta_{\varrho}^{k}$. If the components $A_{\varrho\;\cdots}^{k\cdots}$ happen to be equal to $a_{\varrho\;\cdots}^{k\cdots}$ for some transformation the tensor posseses a kind of symmetry, and the transformation is a symmetry transformation.

When a physical property of a material is represented by a tensor and the material possesses some kind of symmetry, the components of the tensor satisfy the symmetry relations

$$a_{\varrho\;\cdots}^{k\cdots} = \beta_{\varrho}^{s} \ldots \gamma_{r}^{k} \ldots a_{s\;\cdots}^{r\cdots} \tag{2.6}$$

with

$$\beta_{\varrho}^{k} = \beta_{\varrho}^{k(1)}, \beta_{\varrho}^{k(2)\cdots}, \gamma_{\varrho}^{k} = \gamma_{\varrho}^{k(1)}, \gamma_{\varrho}^{k(2)} \ldots \tag{2.7}$$

where $\beta_{\varrho}^{k(\nu)}, \gamma_{\varrho}^{k(\nu)}, \nu = 1, 2, \ldots$ constitute a group of transformations s known as the *symmetry group* for the material property.

If both $\beta_{\varrho}^{k(M)}$ and $\beta_{\varrho}^{k(N)}$ are elements of the symmetry group then also

$$\beta_{\varrho}^{k} = \beta_{m}^{k(M)} \beta_{\varrho}^{m(N)} \tag{2.8}$$

is an element of the symmetry group. This indicates that the symmetry group may possibly be represented by a relatively small number of generating symmetry transformations.

For a given symmetry transformation the relations (2.6) express some linear relations between the components of **a** and thus impose some restrictions on these components. The restriction placed on a component is one of the following:

1) a component must be equal to zero,
2) a component is a linear combination of some of the others, or
3) no restriction at all is placed on a component. Such components are called independent and they can have any value.

3.   Symmetry rotations

A symmetry transformation can be a rotation, i.e. a proper orthogonal transformation characterized by an axis of rotation and an angle of rotation. Since the identity transformation is an element of the symmetry group the angle of rotation χ may be expressed as

$$\chi = 2\pi/n \tag{3.1}$$

where n is an integer, or χ may have any value in which case the material is *transversally isotropic.* An axis corresponding to a particular value of n is termed an *n-fold symmetry axis.*

Choosing the x^3-axis as a symmetry axis the matrices corresponding to the transformation are

$$[\beta_\varrho^k] = \begin{bmatrix} \cos\chi & \sin\chi & 0 \\ -\sin\chi & \cos\chi & 0 \\ 0 & 0 & 1 \end{bmatrix}, \quad [\gamma_\varrho^k] = \begin{bmatrix} \cos\chi & -\sin\chi & 0 \\ \sin\chi & \cos\chi & 0 \\ 0 & 0 & 1 \end{bmatrix} \tag{3.2}$$

and the symmetry relations (2.6) are

$$a_\varrho^{k\cdots}{}_{\cdots} = \beta_\varrho^s \ldots \gamma_r^k \ldots a_s^{r\cdots}{}_{\cdots} \tag{3.3}$$

In another coordinate system the symmetry relations are

$$A_\varrho^{k\cdots}{}_{\cdots} = B_\varrho^s \ldots G_r^k \ldots A_s^{r\cdots}{}_{\cdots} \tag{3.4}$$

where

$$A_\varrho^{k\cdots}{}_{\cdots} = b_\varrho^s \ldots g_r^k \ldots a_s^{r\cdots}{}_{\cdots} \tag{3.5}$$

$$B_\varrho^s = b_\varrho^m g_n^s \beta_m^n \tag{3.6}$$

$$G_r^k = b_r^m g_n^k \gamma_m^n \tag{3.7}$$

Following Hermann [1] a coordinate transformation is chosen in such a way that the matrices corresponding to B_ϱ^s and G_r^k become diagonal. This is performed by choosing

$$[b_\varrho^k] = \begin{bmatrix} 1/\sqrt{2} & -i/\sqrt{2} & 0 \\ -i/\sqrt{2} & 1/\sqrt{2} & 0 \\ 0 & 0 & 1 \end{bmatrix}, \quad [g_\varrho^k] = \begin{bmatrix} 1/\sqrt{2} & i/\sqrt{2} & 0 \\ i/\sqrt{2} & 1/\sqrt{2} & 0 \\ 0 & 0 & 1 \end{bmatrix} \tag{3.8}$$

The matrices corresponding to B_ϱ^s and G_r^k thus become

$$[B_\varrho^k] = \begin{bmatrix} \exp(i\chi) & 0 & 0 \\ 0 & \exp(-i\chi) & 0 \\ 0 & 0 & 1 \end{bmatrix}, [G_\varrho^k] = \begin{bmatrix} \exp(-i\chi) & 0 & 0 \\ 0 & \exp(i\chi) & 0 \\ 0 & 0 & 1 \end{bmatrix}$$

$$(3.9)$$

In indicial notation this may be written

$$B_\varrho^k = \lambda_\varrho \delta_\varrho^k ; \quad G_k^\varrho = \lambda^\varrho \delta_k^\varrho ; \quad \text{no sum on } \varrho \tag{3.10}$$

with

$$\lambda_1 = \lambda^2 = \exp(i\chi); \quad \lambda_2 = \lambda^1 = \exp(-i\chi); \quad \lambda_3 = \lambda^3 = 1 \tag{3.11}$$

Introducing (3.10) in (3.4) the symmetry relations now read

$$A_\varrho^{k\cdots}_{\cdots} = \lambda_\varrho \delta_\varrho^s \ldots \lambda^k \delta_r^k \ldots A_s^{r\cdots}_{\cdots} = \lambda_\varrho \ldots \lambda^k \ldots A_\varrho^{k\cdots}_{\cdots},$$

$$\text{no sum on } k, \varrho \ldots \tag{3.12}$$

Since no sums occur in (3.12) the restriction imposed on each single component $A_\varrho^{k\cdots}_{\cdots}$ by the symmetry relation can be read off directly. For a combination of indices $k, \ldots, \varrho, \ldots$ making the coefficient $\lambda_\varrho \ldots \lambda^k \ldots$ equal to 1 the component $A_\varrho^{k\cdots}_{\cdots}$ can have any value. Among other combinations this is seen to be the case for all $k = \cdots = \varrho = \cdots = 3$ following from the choice of identifying the symmetry axis with the x^3-axis. For a combination of indices $k, \ldots, \varrho, \ldots$ not making the coefficient $\lambda_\varrho \ldots \lambda^k \ldots$ equal to 1 the component $A_\varrho^{k\cdots}_{\cdots}$ has to be zero in order to satisfy the symmetry relation.

It is seen from the transformation matrices (3.8) that the components $A_\varrho^{k\cdots}_{\cdots}$ are complex and furthermore it is the components $a_\varrho^{k\cdots}_{\cdots}$ that are used in the constitutive equation (2.1). To determine the components $a_\varrho^{k\cdots}_{\cdots}$ from the components $A_\varrho^{k\cdots}_{\cdots}$ the transformation

$$a_\varrho^{k\cdots}_{\cdots} = b_m^k \ldots g_\varrho^n \ldots A_n^{m\cdots}_{\cdots} \tag{3.13}$$

has to be used. The restrictions imposed on the components $a_\varrho^{k\cdots}_{\cdots}$ may also be determined from the symmetry relations

$$\beta_k^n \ldots a_\varrho^{k\cdots}_{\cdots} = \beta_\varrho^s \ldots a_s^{n\cdots}_{\cdots} \tag{3.14}$$

which follows from the symmetry relations (3.3) and the orthogonality conditions (2.4).

The great value of the form (3.12) of the symmetry relations lies in the fact that it makes it possible to determine the greatest number of symmetry rotations about a particular axis to which a tensor of given order is sensitive. If an axis is an n-fold symmetry axis with $n > R$, where R is the order of the tensor under consideration then it is a symmetry axis for rotations through any angle χ.

To show this, note that the symmetry relations (3.12) may be written as

$$A_\varrho^{k\cdots}{}_{\cdots} = \exp(iq\chi) \, A_\varrho^{k\cdots}{}_{\cdots} \tag{3.15}$$

where

$$\exp(iq\chi) = \lambda_\varrho \ldots \lambda^k \ldots \tag{3.16}$$

and q is an integer

$$- R \leqslant q \leqslant R \tag{3.17}$$

R being the order of the tensor $\mathbf{A}$.

When $q = 0$ the coefficient $\exp(iq\chi) = 1$ and the corresponding component $A_\varrho^{k\cdots}{}_{\cdots}$ may have any value. Other values of q making $\exp(iq\chi) = 1$ are determined from

$$q\chi = m2\pi \; ; \quad m = 1, 2, \ldots \tag{3.18}$$

or, using (3.1)

$$q/n = m \tag{3.19}$$

For $n > R$ which implies $n > q$ it is seen that (3.19) cannot be satisfied.

If follows that if $\exp(iq\chi) \neq 1$ for one value of $n > R$ then $\exp(iq\chi) \neq 1$ for any value of $n > R$ and the corresponding component $A_\varrho^{k\cdots}{}_{\cdots}$ has to be zero.

As a consequence only rotations with $n = 1, 2, \ldots R$ are distinguishable from transverse isotropy.

4. Other symmetry transformations

Apart from rotations the central inversion, reflections and translations may happen to be symmetry transformations.

The transformation corresponding to the central inversion is

$$\beta_\varrho^k = \gamma_\varrho^k = - \delta_\varrho^k \tag{4.1}$$

or
$$\beta_\ell^k = \lambda_\ell \delta_\ell^k \; ; \quad \gamma_k^\ell = \lambda^\ell \delta_k^\ell \; ; \quad \text{no sum on } \ell \tag{4.2}$$
with
$$\lambda_\ell = \lambda^k = -1 \tag{4.3}$$

From (3.12) it follows that for tensors of odd order all components have to be zero while for tensors of even order the central inversion imposes no restrictions.

The transformation corresponding to a reflection in the $x^1 x^2$-plane is given by
$$\beta_\ell^k = \lambda_\ell \delta_\ell^k \; ; \quad \gamma_k^\ell = \lambda^\ell \delta_k^\ell \; ; \quad \text{no sum on } \ell \tag{4.4}$$
with
$$\lambda_1 = \lambda^1 = 1 \; ; \quad \lambda_2 = \lambda^2 = 1 \; ; \quad \lambda_3 = \lambda^3 = -1 \tag{4.5}$$

This means that components with an odd number of indices 3 have to be zero while all other components can have any value.

From the fact that the transformation corresponding to a 2-fold symmetry rotation about the x^3-axis is given by (4.4) with
$$\lambda_1 = \lambda^1 = -1 \; ; \quad \lambda_2 = \lambda^2 = -1 \; ; \quad \lambda_3 = \lambda^3 = 1 \tag{4.6}$$

it is seen that for tensors of even order a reflection in the $x^1 x^2$-plane and a 2-fold rotation about the x^3-axis are equivalent transformations as far as symmetry restrictions are concerned.

In general a reflection in the $x^1 x^2$-plane is equivalent to a 2-fold rotation about the x^3-axis followed by a central inversion or vice versa.

As mentioned above the components of the tensor **a** may be functions of position. If any translation is a symmetry transformation this functional dependence vanishes and the material is homogeneous. If not isotropic the material is said to be *rectilinear anisotropic*. For some materials a translation has to be accompanied by a certain rotation in order to satisfy the symmetry relation. In such cases the material is called *curvilinear anisotropic*.

It should be noticed that while rotations and translations are operations which can be performed on any real material, no real material can be subjected to an inversion or a reflection. Two experimenters, however, using coordinate systems of different handedness, may *observe* that an inversion or a reflection is a symmetry transformation.

5. Results for tensors of order four

In section 3 it was shown that only rotations with $n = 1, 2, \ldots R$ can be distinguished from transverse isotropy as far as tensors of order R are concerned. For tensors of fourth order this means that an axis of symmetry can be 1-fold, 2-fold, 3-fold, 4-fold or it can be an axis of transverse isotropy.

The symmetry group of a tensor will be indicated by the symbol $\{m\ n\ p\}$ containing the information that the first axis is an m-fold symmetry axis, the second an n-fold and the third a p-fold. The three axes are orthogonal and numbered arbitrarily. Transverse isotropy is indicated as an ∞-fold axis of symmetry.

It is immediately seen that the symmetry of a tensor of order four having only one axis of symmetry is given by

$$\{1\ 1\ 1\},\quad \{1\ 1\ 2\},\quad \{1\ 1\ 3\},\quad \{1\ 1\ 4\}\quad \text{or}\quad \{1\ 1\ \infty\} \tag{5.1}$$

In the appendix the independent components, the zero-components and the relations between the components are given for the symmetry rotations (5.1). The results are displayed in 9×9 matrices, the numbering of the components may be taken from the matrix corresponding to $\{1\ 1\ 1\}$ which is actually no symmetry at all.

When a tensor has more than one axis of symmetry the results for the basic symmetry rotations (5.1) have to be combined. If all symmetry axes are orthogonal it is found that only rotations given by

$$\{1\ 2\ 2\},\quad \{1\ 2\ 3\},\quad \{1\ 2\ 4\},\quad \{1\ 4\ 4\},\quad \{1\ 2\ \infty\}\quad \text{and}\quad \{\infty\ \infty\ \infty\} \tag{5.2}$$

lead to different restrictions.

Finally, a tensor may have symmetry axes which are not orthogonal. The only combination of rotations found to impose new restrictions may be written as

$$\{1\ 1\ 2/3\} \tag{5.3}$$

where 1 1 2 indicates that one of three orthogonal axes is a 2-fold axis of symmetry and /3 indicates that an axis making equal angles with these axes is a 3-fold axis of symmetry.

The results for the symmetry rotations (5.2) and (5.3) too, are given in the appendix. Similar results, but in indicial notation and for all crystal classes are found in a paper by Lokhin and Sedov [2].

Since the central inversion does not place any restrictions on a tensor of even order and since a reflection is equivalent to a 2-fold rotation for such tensors the symmetry of a fourth order tensor is one of the symmetries in (5.1), (5.2) and (5.3).

A set of generating transformations for the symmetry groups (5.1), (5.2) and (5.3) is given in the table below together with some of the corresponding crystallographic symbols.

Symmetry group	Generating transformations	Hermann-Maugin	Shubnikov	Schoenflies
$\{1\,1\,1\}$	I	1	1	C_1
$\{1\,1\,2\}$	R_3^2	2	2	C_2
$\{1\,1\,3\}$	R_3^3	3	3	C_3
$\{1\,1\,4\}$	R_3^4	4	4	C_4
$\{1\,1\,\infty\}$	R_3^∞	6	6	C_6
$\{1\,2\,2\}$	R_2^2, R_3^2	222	2:2	V
$\{1\,2\,3\}$	R_2^2, R_3^3	32	3:2	D_3
$\{1\,2\,4\}$	R_2^2, R_3^4	422	4:2	D_4
$\{1\,2\,\infty\}$	R_2^2, R_3^∞	622	6:2	D_6
$\{1\,4\,4\}$	R_2^4, R_3^4	432	3/4	O
$\{1\,1\,2/3\}$	R_3^2, R_d^3	23	3/2	T
$\{\infty\,\infty\,\infty\}$	$R_1^\infty, R_2^\infty, R_3^\infty$	.... Isotropy		

In the table

$$I = \begin{bmatrix} 1 & 0 & 0 \\ 0 & 1 & 0 \\ 0 & 0 & 1 \end{bmatrix}$$

is the identity transformation, and

$$R_1^n = \begin{bmatrix} 1 & 0 & 0 \\ 0 & \cos\chi & \sin\chi \\ 0 & -\sin\chi & \cos\chi \end{bmatrix}$$

$$R_2^n = \begin{bmatrix} \cos\chi & 0 & -\sin\chi \\ 0 & 1 & 0 \\ \sin\chi & 0 & \cos\chi \end{bmatrix}$$

$$R_3^n = \begin{bmatrix} \cos\chi & \sin\chi & 0 \\ -\sin\chi & \cos\chi & 0 \\ 0 & 0 & 1 \end{bmatrix}$$

are rotations about orthogonal axes (1 0 0), (0 1 0) and (0 0 1) with $\chi = 2\pi/n$, $n = \infty$ indicating any angle of rotation. Finally

$$R_d^3 = \begin{bmatrix} 0 & 0 & 1 \\ 1 & 0 & 0 \\ 0 & 1 & 0 \end{bmatrix}$$

is a 3-fold rotation about the axis $(1\ 1\ 1)/\sqrt{3}$.

6. Non-linear relations

So far only linear relations between tensors have been considered. When the relations are non-linear the influence of material symmetry becomes more involved. It is not the subject of this paper to discuss the problem of determining the restrictions imposed on non-linear constitutive equations by material symmetry so only one of the results will be given. It can be shown that when a tensor **t** is a function of a number of tensor arguments then each component can be expressed as

$$t^{k\ldots m} = \sum_{\alpha=1}^{N} C_\alpha \, \partial K^\alpha/\partial \psi_{k\ldots m} \tag{6.1}$$

where the C's are functions of some scalar invariants of the argument tensors depending on the symmetry group, ψ is an arbitrary tensor of the same order as **t**, and the K's are N scalar invariants of the argument tensors and ψ which are linear in ψ. Details and further references are found in papers by Rivlin [3] and Spencer [4].

While the results from the preceding sections cannot be used directly in connection with non-linear relations some information about the type of material symmetry can be obtained if the increments of the tensors are used instead of the tensors proper.

Taking as an example a second order tensor **t** to be a function of another second order tensor **e** and the symmetry group **s**, it follows that

$$dt^{k\ell} = (\partial t^{k\ell}/\partial e^{mn}) \, de^{mn} \tag{6.2}$$

when **s** is assumed not to change as **e** and **t** change. With the notation

$$a_{mn}^{k\ell} = \partial t^{k\ell}/\partial e^{mn} \tag{6.3}$$

it is evident that the relation

$$dt^{k\ell} = a_{mn}^{k\ell} \, de^{mn} \tag{6.4}$$

is linear. In contrast to the situation in sections 2-5 the components $a^{k\ell}_{mn}$ are not constants but functions of the invariants mentioned above. The symmetry of **a** however, has to be one of the symmetries discussed in section 5.

APPENDIX

Identifying superscripts with row numbers and subscripts with column numbers, the components of a tensor of fourth order $a^{k\ell}_{mn}$ can be arranged in a 9 × 9 matrix. For each of the 12 rotation groups in section 5 this is done below.

In the place belonging to a particular component, as an example a^{12}_{21} is chosen, one may find one of the following symbols:

$\begin{smallmatrix}12\\21\end{smallmatrix}$ — a^{12}_{21} is an independent component

$\textcircled{\begin{smallmatrix}12\\12\end{smallmatrix}}$ — a^{12}_{21} is a dependent component equal to a^{12}_{12}

$\textcircled{\begin{smallmatrix}12\\21\end{smallmatrix}}$ — a^{12}_{21} is a dependent component, the dependence given to the right of the matrix.

0 indicates a component equal to zero.

Tensors may have symmetries such as

$$a^{k\ell}_{mn} = a^{\ell k}_{mn} = a^{k\ell}_{nm} = a^{mn}_{k\ell}$$

which do not arise from material symmetry. In the table concluding this appendix the number of independent components is given when such symmetries are present.

$$\{1\ 1\ 1\}\begin{bmatrix}
\begin{smallmatrix}11\\11\end{smallmatrix} & \begin{smallmatrix}11\\12\end{smallmatrix} & \begin{smallmatrix}11\\13\end{smallmatrix} & \begin{smallmatrix}11\\21\end{smallmatrix} & \begin{smallmatrix}11\\22\end{smallmatrix} & \begin{smallmatrix}11\\23\end{smallmatrix} & \begin{smallmatrix}11\\31\end{smallmatrix} & \begin{smallmatrix}11\\32\end{smallmatrix} & \begin{smallmatrix}11\\33\end{smallmatrix} \\
\begin{smallmatrix}12\\11\end{smallmatrix} & \begin{smallmatrix}12\\12\end{smallmatrix} & \begin{smallmatrix}12\\13\end{smallmatrix} & \begin{smallmatrix}12\\21\end{smallmatrix} & \begin{smallmatrix}12\\22\end{smallmatrix} & \begin{smallmatrix}12\\23\end{smallmatrix} & \begin{smallmatrix}12\\31\end{smallmatrix} & \begin{smallmatrix}12\\32\end{smallmatrix} & \begin{smallmatrix}12\\33\end{smallmatrix} \\
\begin{smallmatrix}13\\11\end{smallmatrix} & \begin{smallmatrix}13\\12\end{smallmatrix} & \begin{smallmatrix}13\\13\end{smallmatrix} & \begin{smallmatrix}13\\21\end{smallmatrix} & \begin{smallmatrix}13\\22\end{smallmatrix} & \begin{smallmatrix}13\\23\end{smallmatrix} & \begin{smallmatrix}13\\31\end{smallmatrix} & \begin{smallmatrix}13\\32\end{smallmatrix} & \begin{smallmatrix}13\\33\end{smallmatrix} \\
\begin{smallmatrix}21\\11\end{smallmatrix} & \begin{smallmatrix}21\\12\end{smallmatrix} & \begin{smallmatrix}21\\13\end{smallmatrix} & \begin{smallmatrix}21\\21\end{smallmatrix} & \begin{smallmatrix}21\\22\end{smallmatrix} & \begin{smallmatrix}21\\23\end{smallmatrix} & \begin{smallmatrix}21\\31\end{smallmatrix} & \begin{smallmatrix}21\\32\end{smallmatrix} & \begin{smallmatrix}21\\33\end{smallmatrix} \\
\begin{smallmatrix}22\\11\end{smallmatrix} & \begin{smallmatrix}22\\12\end{smallmatrix} & \begin{smallmatrix}22\\13\end{smallmatrix} & \begin{smallmatrix}22\\21\end{smallmatrix} & \begin{smallmatrix}22\\22\end{smallmatrix} & \begin{smallmatrix}22\\23\end{smallmatrix} & \begin{smallmatrix}22\\31\end{smallmatrix} & \begin{smallmatrix}22\\32\end{smallmatrix} & \begin{smallmatrix}22\\33\end{smallmatrix} \\
\begin{smallmatrix}23\\11\end{smallmatrix} & \begin{smallmatrix}23\\12\end{smallmatrix} & \begin{smallmatrix}23\\13\end{smallmatrix} & \begin{smallmatrix}23\\21\end{smallmatrix} & \begin{smallmatrix}23\\22\end{smallmatrix} & \begin{smallmatrix}23\\23\end{smallmatrix} & \begin{smallmatrix}23\\31\end{smallmatrix} & \begin{smallmatrix}23\\32\end{smallmatrix} & \begin{smallmatrix}23\\33\end{smallmatrix} \\
\begin{smallmatrix}31\\11\end{smallmatrix} & \begin{smallmatrix}31\\12\end{smallmatrix} & \begin{smallmatrix}31\\13\end{smallmatrix} & \begin{smallmatrix}31\\21\end{smallmatrix} & \begin{smallmatrix}31\\22\end{smallmatrix} & \begin{smallmatrix}31\\23\end{smallmatrix} & \begin{smallmatrix}31\\31\end{smallmatrix} & \begin{smallmatrix}31\\32\end{smallmatrix} & \begin{smallmatrix}31\\33\end{smallmatrix} \\
\begin{smallmatrix}32\\11\end{smallmatrix} & \begin{smallmatrix}32\\12\end{smallmatrix} & \begin{smallmatrix}32\\13\end{smallmatrix} & \begin{smallmatrix}32\\21\end{smallmatrix} & \begin{smallmatrix}32\\22\end{smallmatrix} & \begin{smallmatrix}32\\23\end{smallmatrix} & \begin{smallmatrix}32\\31\end{smallmatrix} & \begin{smallmatrix}32\\32\end{smallmatrix} & \begin{smallmatrix}32\\33\end{smallmatrix} \\
\begin{smallmatrix}33\\11\end{smallmatrix} & \begin{smallmatrix}33\\12\end{smallmatrix} & \begin{smallmatrix}33\\13\end{smallmatrix} & \begin{smallmatrix}33\\21\end{smallmatrix} & \begin{smallmatrix}33\\22\end{smallmatrix} & \begin{smallmatrix}33\\23\end{smallmatrix} & \begin{smallmatrix}33\\31\end{smallmatrix} & \begin{smallmatrix}33\\32\end{smallmatrix} & \begin{smallmatrix}33\\33\end{smallmatrix}
\end{bmatrix}$$

{1 1 2}

11/11	11/12	0	11/21	11/22	0	0	0	11/33
12/11	12/12	0	12/21	12/22	0	0	0	12/33
0	0	13/13	0	0	13/23	13/31	13/32	0
21/11	21/12	0	21/21	21/22	0	0	0	21/33
22/11	22/12	0	22/21	22/22	0	0	0	22/33
0	0	23/13	0	0	23/23	23/31	23/32	0
0	0	31/13	0	0	31/23	31/31	31/32	0
0	0	32/13	0	0	32/23	32/31	32/32	0
33/11	33/12	0	33/21	33/22	0	0	0	33/33

{1 1 3} (circled entries shown in parentheses)

11/11	11/12	11/13	11/21	11/22	(12/13)	11/31	11/32	11/33
(12/11)	12/12	12/13	(12/21)	12/22	(-11/13)	(11/32)	(-11/31)	12/33
(-13/22)	(13/21)	13/13	13/21	13/22	13/23	13/31	13/32	0
(-12/22)	(12/21)	(12/13)	(12/12)	(-12/11)	(-11/13)	(11/32)	(-11/31)	(-12/33)
(11/22)	(-11/21)	(-11/13)	(11/12)	(11/11)	(-12/13)	(-11/31)	(-11/32)	(11/33)
(13/21)	(13/22)	(13/-23)	(13/22)	(13/-21)	(13/13)	(13/-32)	(13/31)	0
31/11	(32/11)	31/13	(32/11)	(-31/11)	(-32/13)	31/31	31/32	0
32/11	(-31/11)	32/13	(-31/11)	(-32/11)	(31/13)	(-31/32)	(31/31)	0
33/11	33/12	0	(-32/12)	(33/11)	0	0	0	33/33

$$\binom{12}{11} = \frac{12}{22} \cdot \frac{11}{21} \cdot \frac{11}{12}$$

$$\binom{12}{21} = \frac{11}{11} \cdot \frac{11}{22} \cdot \frac{12}{12}$$

$$\{1\ 1\ 4\}\begin{bmatrix}
\frac{11}{11} & \frac{11}{12} & 0 & \frac{11}{21} & \frac{11}{22} & 0 & 0 & 0 & \frac{11}{33} \\[4pt]
\frac{12}{11} & \frac{12}{12} & 0 & \frac{12}{21} & \frac{12}{22} & 0 & 0 & 0 & \frac{12}{33} \\[4pt]
0 & 0 & \frac{13}{13} & 0 & 0 & \frac{13}{23} & \frac{13}{31} & \frac{13}{32} & 0 \\[4pt]
\left(-\tfrac{12}{22}\right) & \left(\tfrac{12}{21}\right) & 0 & \left(\tfrac{12}{12}\right) & \left(-\tfrac{12}{11}\right) & 0 & 0 & 0 & \left(-\tfrac{12}{33}\right) \\[4pt]
\left(\tfrac{11}{22}\right) & \left(-\tfrac{11}{21}\right) & 0 & \left(-\tfrac{11}{12}\right) & \left(\tfrac{11}{11}\right) & 0 & 0 & 0 & \left(\tfrac{11}{33}\right) \\[4pt]
0 & 0 & \left(-\tfrac{13}{23}\right) & 0 & 0 & \left(\tfrac{13}{13}\right) & \left(-\tfrac{13}{32}\right) & \left(\tfrac{13}{31}\right) & 0 \\[4pt]
0 & 0 & \frac{31}{13} & 0 & 0 & \left(-\tfrac{32}{13}\right) & \frac{31}{31} & \frac{31}{32} & 0 \\[4pt]
0 & 0 & \frac{32}{13} & 0 & 0 & \left(\tfrac{31}{13}\right) & \left(-\tfrac{31}{32}\right) & \left(\tfrac{31}{31}\right) & 0 \\[4pt]
\frac{33}{11} & \frac{33}{12} & 0 & \left(-\tfrac{33}{12}\right) & \left(\tfrac{33}{11}\right) & 0 & 0 & 0 & \frac{33}{33}
\end{bmatrix}$$

$$\{1\ 1\ \infty\}\begin{bmatrix}
\frac{11}{11} & \frac{11}{12} & 0 & \frac{11}{21} & \frac{11}{22} & 0 & 0 & 0 & \frac{11}{33} \\[4pt]
\left(\tfrac{12}{11}\right) & \frac{12}{12} & 0 & \left(\tfrac{12}{21}\right) & \frac{12}{22} & 0 & 0 & 0 & \frac{12}{33} \\[4pt]
0 & 0 & \frac{13}{13} & 0 & 0 & \frac{13}{23} & \frac{13}{31} & \frac{13}{32} & 0 \\[4pt]
\left(-\tfrac{12}{22}\right) & \left(\tfrac{12}{21}\right) & 0 & \left(\tfrac{12}{12}\right) & \left(-\tfrac{12}{11}\right) & 0 & 0 & 0 & \left(-\tfrac{12}{33}\right) \\[4pt]
\left(\tfrac{11}{22}\right) & \left(-\tfrac{11}{21}\right) & 0 & \left(-\tfrac{11}{12}\right) & \left(\tfrac{11}{11}\right) & 0 & 0 & 0 & \left(\tfrac{11}{33}\right) \\[4pt]
0 & 0 & \left(-\tfrac{13}{23}\right) & 0 & 0 & \left(\tfrac{13}{13}\right) & \left(-\tfrac{13}{32}\right) & \left(\tfrac{13}{31}\right) & 0 \\[4pt]
0 & 0 & \frac{31}{13} & 0 & 0 & \left(-\tfrac{32}{13}\right) & \frac{31}{31} & \frac{31}{32} & 0 \\[4pt]
0 & 0 & \frac{32}{13} & 0 & 0 & \left(\tfrac{31}{13}\right) & \left(-\tfrac{31}{32}\right) & \left(\tfrac{31}{31}\right) & 0 \\[4pt]
\frac{33}{11} & \frac{33}{12} & 0 & \left(-\tfrac{33}{12}\right) & \left(\tfrac{33}{11}\right) & 0 & 0 & 0 & \frac{33}{33}
\end{bmatrix}$$

$$\left(\tfrac{12}{11}\right) = \tfrac{12}{22} \cdot \tfrac{11}{21} \cdot \tfrac{11}{12}$$

$$\left(\tfrac{12}{21}\right) = \tfrac{11}{11} \cdot \tfrac{11}{22} \cdot \tfrac{12}{12}$$

$\{1\,2\,2\}$

$\frac{11}{11}$	0	0	0	$\frac{11}{22}$	0	0	0	$\frac{11}{33}$
0	$\frac{12}{12}$	0	$\frac{12}{21}$	0	0	0	0	0
0	0	$\frac{13}{13}$	0	0	0	$\frac{13}{31}$	0	0
0	$\frac{21}{12}$	0	$\frac{21}{21}$	0	0	0	0	0
$\frac{22}{11}$	0	0	0	$\frac{22}{22}$	0	0	0	$\frac{22}{33}$
0	0	0	0	0	$\frac{23}{23}$	0	$\frac{23}{32}$	0
0	0	$\frac{31}{13}$	0	0	0	$\frac{31}{31}$	0	0
0	0	0	0	0	$\frac{32}{23}$	0	$\frac{32}{32}$	0
$\frac{33}{11}$	0	0	0	$\frac{33}{22}$	0	0	0	$\frac{33}{33}$

$\{1\,2\,3\}$

$\frac{11}{11}$	0	$\frac{11}{13}$	0	$\frac{11}{22}$	0	$\frac{11}{31}$	0	$\frac{11}{33}$
0	$\frac{12}{12}$	0	$\left(\frac{12}{21}\right)$	0	$\left(-\frac{11}{13}\right)$	0	$\left(-\frac{11}{31}\right)$	0
$\left(-\frac{13}{22}\right)$	0	$\frac{13}{13}$	0	$\frac{13}{22}$	0	$\frac{13}{31}$	0	0
0	$\left(\frac{12}{21}\right)$	0	$\left(\frac{12}{12}\right)$	0	$\left(-\frac{11}{13}\right)$	0	$\left(-\frac{11}{31}\right)$	0
$\left(\frac{11}{22}\right)$	0	$\left(-\frac{11}{13}\right)$	0	$\left(\frac{11}{11}\right)$	0	$\left(-\frac{11}{31}\right)$	0	$\left(\frac{11}{33}\right)$
0	$\left(\frac{13}{22}\right)$	0	$\left(\frac{13}{22}\right)$	0	$\left(\frac{13}{13}\right)$	0	$\left(\frac{13}{31}\right)$	0
$\frac{31}{11}$	0	$\frac{31}{13}$	0	$\left(-\frac{31}{11}\right)$	0	$\frac{31}{31}$	0	0
0	$\left(-\frac{31}{11}\right)$	0	$\left(-\frac{31}{11}\right)$	0	$\left(\frac{31}{13}\right)$	0	$\left(\frac{31}{31}\right)$	0
$\frac{33}{11}$	0	0	0	$\left(\frac{33}{11}\right)$	0	0	0	$\frac{33}{33}$

$$\left(\frac{12}{21}\right) = \frac{11}{11} \cdot \frac{11}{22} \cdot \frac{12}{12}$$

$$\{1\ 2\ 4\}\begin{bmatrix}
\frac{11}{11} & 0 & 0 & 0 & \frac{11}{22} & 0 & 0 & 0 & \frac{11}{33} \\[4pt]
0 & \frac{12}{12} & 0 & \frac{12}{21} & 0 & 0 & 0 & 0 & 0 \\[4pt]
0 & 0 & \frac{13}{13} & 0 & 0 & 0 & \frac{13}{31} & 0 & 0 \\[4pt]
0 & \left(\frac{12}{21}\right) & 0 & \left(\frac{12}{12}\right) & 0 & 0 & 0 & 0 & 0 \\[4pt]
\left(\frac{11}{22}\right) & 0 & 0 & 0 & \left(\frac{11}{11}\right) & 0 & 0 & 0 & \left(\frac{11}{33}\right) \\[4pt]
0 & 0 & 0 & 0 & 0 & \left(\frac{13}{13}\right) & 0 & \left(\frac{13}{31}\right) & 0 \\[4pt]
0 & 0 & \frac{31}{13} & 0 & 0 & 0 & \frac{31}{31} & 0 & 0 \\[4pt]
0 & 0 & 0 & 0 & 0 & \left(\frac{31}{13}\right) & 0 & \left(\frac{31}{31}\right) & 0 \\[4pt]
\frac{33}{11} & 0 & 0 & 0 & \left(\frac{33}{11}\right) & 0 & 0 & 0 & \frac{33}{33}
\end{bmatrix}$$

$$\{1\ 2\ \overset{.}{\infty}\}\begin{bmatrix}
\frac{11}{11} & 0 & 0 & 0 & \frac{11}{22} & 0 & 0 & 0 & \frac{11}{33} \\[4pt]
0 & \frac{12}{12} & 0 & \left(\frac{12}{21}\right) & 0 & 0 & 0 & 0 & 0 \\[4pt]
0 & 0 & \frac{13}{13} & 0 & 0 & 0 & \frac{13}{31} & 0 & 0 \\[4pt]
0 & \left(\frac{12}{21}\right) & 0 & \left(\frac{12}{12}\right) & 0 & 0 & 0 & 0 & 0 \\[4pt]
\left(\frac{11}{22}\right) & 0 & 0 & 0 & \left(\frac{11}{11}\right) & 0 & 0 & 0 & \left(\frac{11}{33}\right) \\[4pt]
0 & 0 & 0 & 0 & 0 & \left(\frac{13}{13}\right) & 0 & \left(\frac{13}{31}\right) & 0 \\[4pt]
0 & 0 & \frac{31}{13} & 0 & 0 & 0 & \frac{31}{31} & 0 & 0 \\[4pt]
0 & 0 & 0 & 0 & 0 & \left(\frac{31}{13}\right) & 0 & \left(\frac{31}{31}\right) & 0 \\[4pt]
\frac{33}{11} & 0 & 0 & 0 & \left(\frac{33}{11}\right) & 0 & 0 & 0 & \frac{33}{33}
\end{bmatrix}$$

$$\left(\frac{12}{21}\right) = \frac{11}{11} \cdot \frac{11}{22} \cdot \frac{12}{12}$$

{1 4 4}

Circled entries are shown in (parentheses).

11/11	0	0	0	11/22	0	0	0	(11/22)
0	12/12	0	12/21	0	0	0	0	0
0	0	(12/12)	0	0	0	(12/21)	0	0
0	(12/21)	0	(12/12)	0	0	0	0	0
(11/22)	0	0	0	11/11	0	0	0	(11/22)
0	0	0	0	0	(12/12)	0	(12/21)	0
0	0	(12/21)	0	0	0	(12/12)	0	0
0	0	0	0	0	(12/21)	0	(12/12)	0
(11/22)	0	0	0	11/22	0	0	0	(11/11)

{1 1 2/3}

11/11	0	0	0	11/22	0	0	0	11/33
0	12/12	0	12/21	0	0	0	0	0
0	0	13/13	0	0	0	13/31	0	0
0	(13/31)	0	(13/13)	0	0	0	0	0
(11/33)	0	0	0	(11/11)	0	0	0	(11/22)
0	0	0	0	0	(12/12)	0	(12/21)	0
0	0	(12/21)	0	0	0	(12/12)	0	0
0	0	0	0	0	(13/31)	0	(13/13)	0
(11/22)	0	0	0	(11/33)	0	0	0	(11/11)

$$\{\infty\ .\infty\ \ \infty\}\begin{bmatrix}
\substack{11\\11} & 0 & 0 & 0 & \substack{11\\22} & 0 & 0 & 0 & \left(\substack{11\\22}\right)\\
0 & \substack{12\\12} & 0 & \left(\substack{12\\21}\right) & 0 & 0 & 0 & 0 & 0\\
0 & 0 & \left(\substack{12\\12}\right) & 0 & 0 & 0 & \left(\substack{12\\21}\right) & 0 & 0\\
0 & \left(\substack{12\\21}\right) & 0 & \left(\substack{12\\12}\right) & 0 & 0 & 0 & 0 & 0\\
\left(\substack{11\\22}\right) & 0 & 0 & 0 & \substack{11\\11} & 0 & 0 & 0 & \left(\substack{11\\22}\right)\\
0 & 0 & 0 & 0 & 0 & \left(\substack{12\\12}\right) & 0 & \left(\substack{12\\21}\right) & 0\\
0 & 0 & \left(\substack{12\\21}\right) & 0 & 0 & 0 & \left(\substack{12\\12}\right) & 0 & 0\\
0 & 0 & 0 & 0 & 0 & \left(\substack{12\\21}\right) & 0 & \left(\substack{12\\12}\right) & 0\\
\left(\substack{11\\22}\right) & 0 & 0 & 0 & \left(\substack{11\\22}\right) & 0 & 0 & 0 & \left(\substack{11\\11}\right)
\end{bmatrix}$$

$$\left(\substack{12\\21}\right) = \substack{11\\11}\cdot\substack{11\\22}\cdot\substack{12\\12}$$

Symmetry group	$a^{k\ell}_{mn}$	$a^{k\ell}_{mn}=a^{\ell k}_{mn}$	$a^{k\ell}_{mn}=a^{\ell k}_{mn}$ $=a^{k\ell}_{nm}$	$a^{k\ell}_{mn}=a^{\ell k}_{mn}$ $=a^{k\ell}_{nm}=a^{mn}_{k\ell}$
{1 1 1}	81	54	36	21
{1 1 2}	41	28	20	13
{1 1 3}	27	18	12	7
{1 1 4}	21	14	10	7
{1 1 ∞}	19	12	8	5
{1 2 2}	21	15	12	9
{1 2 3}	14	10	8	6
{1 2 4}	11	8	7	6
{1 2 ∞}	10	7	6	5
{1 4 4}	4	3	3	3
{1 1 2/3}	7	5	4	3
{∞ ∞ ∞}	3	2	2	2

Number of independent components for various symmetries.

REFERENCES

[1] HERMANN C. – "Tensoren und Kristallsymmetrie". *Zeitschrift für Kristallographie,* 89 (1934) : 32-48.
[2] LOKHIN V.V. and L.I. SEDOV. – "Nonlinear tensor functions of several tensor arguments", *Journal of Applied Mathematics and Mechanics (PMM),* 27 (1963) : 597-629.
[3] RIVLIN R.S. – "The application of the theory of invariants to the study of constitutive equations". in G. Fichera, ed. *Trends in Applications of Pure Mathematics to Mechanics.* London, Pitman Publishing, (1976), pp. 299-310.
[4] SPENCER A.J.M. – "Theory of Invariants". in A.C. Eringen, ed., *Continuum Physics Vol. I,* New York and London, Academic Press, (1971), pp. 239-353.

RESUME

(Relations symétriques pour les matériaux anisotropes)

En mécanique des milieux continus, on suppose généralement que les relations symétriques pour les propriétés anisotropes peuvent être décrites par 13 types de symétries. Il y a les 11 types de symétries cristallographiques, l'orthotropie de révolution et l'isotropie. On développe une méthode qui permet de limiter le nombre des transformations symétriques à examiner en relation avec une propriété particulière du matériau. Comme conséquence, le groupe entier des symétries pour une propriété du matériau peut être trouvé parmi un nombre limité de combinaisons de transformations symétriques. Les propriétés matérielles qui sont représentées par des tenseurs d'ordre quatre ont été choisies comme exemple. On montre qu'il n'existe aucun autre type de symétrie que les 13 mentionnés ci-dessus. Les transformations symétriques, les composantes indépendantes et les relations entre les composantes des tenseurs d'ordre quatre sont données pour chaque type de symétrie.

Theory of Invariants in Creep Mechanics of Anisotropic Materials

J. Betten

Technische Hochschule Aachen, Germany-BRD.

1. Introduction

Many mathematicians have studied the theory of algebraic invariants in detail. Many results can be found, for instance, in [13, 14, 29]. Very extensive accounts of algebraic invariant theory from the point of view of its application to modern continuum mechanics are presented, for example, in [17, 26, 28].

It is convenient to employ the theory of invariants in creep mechanics of anisotropic materials in a manner similar to that used in the theory of plasticity [4].

Creep deformations of the 'secondary' stage are large and of a similar character as 'pure' plastic deformations. For instance, creep deformations of metals will usually[1] be uninfluenced by a superimposed hydrostatic pressure.

In [2] the author bases a generalized theory of invariants in creep mechanics on the following hypotheses: incompressible and isotropic material, creep rate independence of superimposed hydrostatic pressure, existence of a flow potential, and the Norton-Bailey's power law [1, 19] valid for the special case of uni-axial stress. Therefore, the flow potential is expressed in a general form as the second-order and third-order invariant of the stress deviator. It is also assumed that the equivalent stress is a function of dissipation. In the special case of the Mises potential the generalized theory leads to Odqvist's theory.

In this paper a theory of invariants in creep mechanics of anisotropic materials is investigated. For this purpose the mentioned 'isotropic concept' [2] will be used by substituting a mapped stress tensor.

2. Invariants of the Mapped Stress Tensor and its Deviator

To describe the isotropic creep-behaviour it is expedient to start from a creep potential $F = F(\sigma_{ij})$, which is, in the isotropic case, a scalar-valued tensor

---

[1]The usual assumption of plastic incompressibility can be considered as a special case [3, 6].

function of only the stress tensor σ_{ij}. This function is said to be isotropic if

$$F(a_{ip} a_{jq} \sigma_{pq}) = F(\sigma_{ij}) \tag{2.1}$$

under any orthonormal transformation a_{ij}. Such a scalar function is termed an isotropic invariant [11]. The representation of the function (2.1) is furnished by an irreducible functional basis, i.e. by an irreducible set of isotropic invariants such that any isotropic function (2.1) can be expressed as a single-valued function of the basic invariants of the stress tensor

$$S_1(\sigma) \equiv \delta_{ij}\sigma_{ji}, \quad S_2(\sigma) \equiv \sigma_{ij}\sigma_{ji}, \quad S_3(\sigma) \equiv \sigma_{ij}\sigma_{jk}\sigma_{ki}. \tag{2.2a,b,c}$$

Often the representation

$$F = F[I_1(\sigma), I_2(\sigma), I_3(\sigma)] \tag{2.3}$$

is used with the principal invariants :

$$I_1(\sigma) \equiv \delta_{ij}\sigma_{ji}, \quad I_2(\sigma) \equiv (\sigma_{ij}\sigma_{ji} - \sigma_{ii}\sigma_{jj})/2, \tag{2.4a,b}$$

$$I_3(\sigma) \equiv (2\sigma_{ij}\sigma_{jk}\sigma_{ki} - 3\sigma_{ij}\sigma_{ji}\sigma_{kk} + \sigma_{ii}\sigma_{jj}\sigma_{kk})/6 \tag{2.4c}$$

Assuming incompressibility, it is practical to use the invariants

$$I_2(\sigma') \equiv \sigma'_{ij}\sigma'_{ji}/2, \quad I_3(\sigma') \equiv \sigma'_{ij}\sigma'_{jk}\sigma'_{ki}/3 \tag{2.5a,b}$$

of the stress deviator $\sigma'_{ij} = \sigma_{ij} - \sigma_{kk}\delta_{ij}/3$, so that the creep potential (2.3) takes the form :

$$F = F[I_2(\sigma'), I_3(\sigma')]. \tag{2.6}$$

In anisotropic solids, equal stresses σ_{ij} cause various deformations ϵ_{ij} in different directions. Therefore, anisotropic behaviour can be described by substituting the mapped stress tensor

$$\tau_{ij} = \alpha_{ij} + \beta_{ijk\ell}\sigma_{k\ell} + \gamma_{ijk\ell mn}\sigma_{k\ell}\sigma_{mn} + \cdots \tag{2.7}$$

for the actual stress tensor σ_{ij} in the 'isotropic concept' (2.3). The anisotropic behaviour is expressed by the tensors $\alpha_{ij}, \beta_{ijk\ell}, \gamma_{ijk\ell mn}, \cdots$ of rank 2, 4, 6,..., respectively, the components of which are to be determined from experiments. By analogy with (2.4a, b, c), we form the corresponding invariants of the symmetric tensor (2.7) :

$$I_1(\tau) \equiv \delta_{ij}\tau_{ji}, \quad I_2(\tau) \equiv (\tau_{ij}\tau_{ji} - \tau_{ii}\tau_{jj})/2, \tag{2.8a,b}$$

$$I_3(\tau) \equiv (2\tau_{ij}\tau_{jk}\tau_{ki} - 3\tau_{ij}\tau_{ji}\tau_{kk} + \tau_{ii}\tau_{jj}\tau_{kk})/6, \tag{2.8c}$$

Because of (2.7) they can be represented in the following manner:

$$I_1(\tau) = I_1(\alpha) + \{I_1(\beta)\}_{pq}\, \sigma_{pq} + \{I_1(\gamma)\}_{pqrs}\, \sigma_{pq}\, \sigma_{rs}, \tag{2.9a}$$

$$\begin{aligned}
I_2(\tau) = {} & \{I_2(\alpha\alpha)\} + 2\{I_2(\alpha\beta)\}_{pq}\, \sigma_{pq} + \\
& + [2\{I_2(\alpha\gamma)\}_{pqrs} + \{I_2(\beta\beta)\}_{pqrs}]\, \sigma_{pq}\, \sigma_{rs} + \\
& + \{I_2(\beta\gamma)\}_{p\ldots u}\, \sigma_{pq}\, \sigma_{rs}\, \sigma_{tu} + \\
& + \{I_2(\gamma\gamma)\}_{p\ldots w}\, \sigma_{pq}\, \sigma_{rs}\, \sigma_{tu}\, \sigma_{vw},
\end{aligned} \tag{2.9b}$$

$$\begin{aligned}
I_3(\tau) = {} & \{I_3(\alpha\alpha\alpha)\} + [\{I_3(\alpha\alpha\beta)\}_{pq} + 2\{I_3(\alpha\beta\alpha)\}_{pq}]\, \sigma_{pq} + \\
& + [\ldots + \{I_3(\beta\beta\alpha)\}_{pqrs} + \ldots]\, \sigma_{pq}\, \sigma_{rs} + \\
& + [2\{I_3(\alpha\beta\gamma)\}_{p\ldots u} + \ldots + \{I_3(\beta\beta\beta)\}_{p\ldots u}]\, \sigma_{pq}\, \sigma_{rs}\, \sigma_{tu} + \ldots \\
& + \ldots + \{I_3(\gamma\gamma\gamma)\}_{p\ldots za}\, \sigma_{pq}\, \sigma_{rs}\, \sigma_{tu}\, \sigma_{vw}\, \sigma_{xy}\, \sigma_{za}.
\end{aligned} \tag{2.9c}$$

In (2.9 a,b,c) the symbols $\{I_1(\beta)\}_{pq}$, $\{I_2(\alpha\gamma)\}_{pqrs}$, $\{I_3(\alpha\beta\alpha)\}_{pq}$, etc. are tensors of rank 2, 4, 2, etc., respectively. These tensors are formed by using the operations from (2.8 a, b, c) with respect to the first index pairs of the tensors α, β and γ. The remaining free indices are appended to the curved brackets, for example:

$$\{I_1(\beta)\}_{pq} \equiv \delta_{ij}\, \beta_{jipq}, \tag{2.10a}$$

$$\{I_2(\alpha\gamma)\}_{pqrs} \equiv (\alpha_{ij}\, \gamma_{jipqrs} - \alpha_{ii}\, \gamma_{jjpqrs})/2, \tag{2.10b}$$

$$\begin{aligned}
\{I_3(\alpha\beta\alpha)\}_{pq} \equiv {} & (2\,\alpha_{ij}\,\beta_{jkpq}\,\alpha_{ki} - 3\alpha_{ij}\,\beta_{jipq}\,\alpha_{kk} + \\
& + \alpha_{ii}\,\beta_{jjpq}\,\alpha_{kk})/6.
\end{aligned} \tag{2.10c}$$

In the special case of vanishing tensors α_{ij} and $\gamma_{ijk\ell mn}$ the mapping (2.7) leads to the transformed stresses

$$\tau_{ij} = \beta_{ijk\ell}\, \sigma_{k\ell}. \tag{2.11}$$

Then the image tensor τ_{ij} is a linear transformation, which is used in the theory of plasticity of anisotropic materials [4, 9, 10, 25]. For (2.11) the invariants are simplified:

$$I_1(\tau) = A_{pq}\, \sigma_{pq}, \quad I_2(\tau) = A_{pqrs}\, \sigma_{pq}\, \sigma_{rs}/2, \tag{2.12a,b}$$

$$I_3(\tau) = A_{pqrstu}\, \sigma_{pq}\, \sigma_{rs}\, \sigma_{tu}/3, \tag{2.12c}$$

if we define:

$$A_{pq} \equiv \{I_1(\beta)\}_{pq} \equiv \beta_{iipq}, \tag{2.13a}$$

$$A_{pqrs} \equiv 2\{I_2(\beta\beta)\}_{pqrs} \equiv \beta_{ijpq}\,\beta_{jirs} - \beta_{iipq}\,\beta_{jjrs}, \tag{2.13b}$$

$$\left.\begin{array}{l} A_{pqrstu} \equiv 3\{I_3(\beta\beta\beta)\}_{pqrstu} \equiv \\[4pt] \equiv \beta_{ijpq}\,\beta_{jkrs}\,\beta_{kitu} - 3\,\beta_{ijpq}\,\beta_{jirs}\,\beta_{kktu}/2 + \\[4pt] \qquad\qquad + \beta_{iipq}\,\beta_{jjrs}\,\beta_{kktu}/2. \end{array}\right\} \tag{2.13c}$$

With (2.12a,b,c) the function (2.3) takes a form which is often used as plastic potential of anisotropic materials [5, 7, 12, 24]. Analogous to the stress deviator σ'_{ij} we form the deviator $\tau'_{ij} = \tau_{ij} - \tau_{kk}\,\delta_{ij}/3$, which can be expressed by

$$\tau'_{ij} = \beta'_{\{ij\}pq}\,\sigma'_{pq} + \beta'_{\{ij\}rr}\,\sigma_{ss}/3 \tag{2.14}$$

after considering (2.11) and defining [4] the tensor

$$\beta'_{\{ij\}pq} \equiv \beta_{ijpq} - \beta_{kkpq}\,\delta_{ij}/3, \tag{2.15}$$

which is deviatoric with respect to the free indices $\{ij\}$. From the decomposition (2.14) we see that the deviator τ'_{ij} of the image tensor (2.11) is influenced by the hydrostatic pressure $p = -\sigma_{kk}/3$. Thus, incompressibility requires $\beta'_{\{ij\}rr} \overset{!}{=} 0$. If the actual stress tensor σ_{ij} in (2.11) is split into its deviator and spherical tensor,

$$\tau_{ij} = \beta_{ijpq}\,\sigma'_{pq} + \beta_{ijpq}\,\delta_{pq}\,\sigma_{ss}/3, \tag{2.16}$$

we see that the trace $\beta_{ijpq}\,\delta_{pq}$ must be zero in the incompressible case. Together, we have the following equivalent conditions of incompressibility:

$$\beta_{ijrr} \overset{!}{=} 0, \quad \beta'_{\{ij\}rr} \overset{!}{=} 0, \quad \beta'_{ij\{pq\}} \overset{!}{=} \beta_{ijpq}. \tag{2.17a,b,c}$$

The requirement (2.17c) results from the definition

$$\beta'_{ij\{pq\}} \equiv \beta_{ijpq} - \beta_{ijrr}\,\delta_{pq}/3 \tag{2.18}$$

combined with (2.17a). Contrary to (2.15), the tensor (2.18) is deviatoric corresponding to the free indices $\{pq\}$.

Because of (2.17a,b,c) in connection with (2.11), (2.14), and (2.16), incompressibility results, if we use one of the following linear transformations:

$$\tau_{ij} = \beta_{ij\,pq}\,\sigma'_{pq}\,, \quad \tau'_{ij} = \beta'_{\{ij\}\,pq}\,\sigma'_{pq}\,, \quad \tau_{ij} = \beta_{ij\,\{pq\}}\,\sigma_{pq} \qquad (2.19a,b,c)$$

instead of (2.11).

As seen from (2.17a,c), the transformations (2.19a,c) are equivalent, while (2.19b) is the deviator of (2.19a, c). Its invariants are given, as in (2.5a,b), by

$$I_2\,(\tau') \equiv \frac{1}{2}\,\tau'_{ij}\,\tau'_{ji} = \frac{1}{2}\,\beta'_{\{ij\}\,pq}\,\beta'_{\{ji\}\,rs}\,\sigma'_{pq}\,\sigma'_{rs}\,, \qquad (2.20)$$

$$I_3\,(\tau') \equiv \frac{1}{3}\,\tau'_{ij}\,\tau'_{jk}\,\tau'_{ki} = \frac{1}{3}\,\beta'_{\{ij\}\,pq}\,\beta'_{\{jk\}\,rs}\,\beta'_{\{ki\}\,tu}\,\sigma'_{pq}\,\sigma'_{rs}\,\sigma'_{tu}\,, \qquad (2.21)$$

from which we form the gradient dyadics

$$\partial I_2\,(\tau')/\partial\sigma_{ij} = \beta''_{\{pq\}\,\{ij\}}\,\beta'_{\{qp\}\,k\ell}\,\sigma'_{k\ell}\,, \qquad (2.22)$$

$$\frac{\partial I_3\,(\tau')}{\partial\sigma_{ij}} = \beta''_{\{pq\}\,\{ij\}}\,\beta'_{\{qr\}\,k\ell}\,\beta'_{\{rp\}\,mn}\,\sigma'_{k\ell}\,\sigma'_{mn}\,. \qquad (2.23)$$

In (2.22), (2.23) the tensor $\beta''_{\{pq\}\{ij\}}$ is deviatoric with respect to the index pair $\{ij\}$ just as to the index pair $\{pq\}$:

$$\beta''_{\{pq\}\{ij\}} \equiv \beta'_{\{pq\}\,ij} - \beta'_{\{pq\}\,rr}\,\delta_{ij}/3\,, \qquad (2.24a)$$

$$\beta''_{\{pq\}\{ij\}} \equiv \beta'_{pq\,\{ij\}} - \beta'_{rr\,\{ij\}}\,\delta_{pq}/3\,. \qquad (2.24b)$$

Because of (2.17b) and (2.24a) the double deviator $\beta''_{\{pq\}\{ij\}}$ in (2.22), (2.23) can be replaced by $\beta'_{\{pq\}\,ij}$, as can also be concluded from (2.17c) together with (2.15). However, the symbol $\beta''_{\{pq\}\{ij\}}$ indicates the vanishing traces of the dyadics (2.22), (2.23) immediately.

The isotropic special case is given by $\beta_{ijk\ell} = \delta_{ik}\,\delta_{j\ell}$ and

$$\beta'_{\{ij\}\,k\ell} = \delta_{ik}\,\delta_{j\ell} - \delta_{ij}\,\delta_{k\ell}/3\,.$$

Then the gradient (2.22) agrees with the stress deviator, and (2.23) leads to the deviator of the square of the reduced stress:

$$\frac{\partial I_3\,(\sigma')}{\partial\sigma_{ij}} = \sigma'_{jk}\,\sigma'_{ki} - \frac{2}{3}\,I_2\,(\sigma')\,\delta_{ji} \equiv (\sigma'^{(2)}_{ji})' \equiv \sigma''_{ji}\,. \qquad (2.25)$$

Finally, from (2.7) we find the generalized form of (2.14):

$$\tau'_{ij} = \alpha'_{ij} + \beta'_{\{ij\}\, pq}\, \sigma'_{pq} + \gamma'_{\{ij\}\, pq\, rs}\, \sigma'_{pq}\, \sigma'_{rs} +$$

$$+ \frac{1}{3}\, [\beta'_{\{ij\}\, rr} + (\gamma'_{\{ij\}\, pq\, rr} + \gamma'_{\{ij\}\, rr\, pq})\, \sigma'_{pq} +$$

$$\left.+ \frac{1}{3}\, \gamma'_{\{ij\}\, pp\, qq}\, \sigma_{rr}\,]\, \sigma_{ss}, \right\} \qquad (2.26)$$

in which the last term must be zero in the incompressible case. To this end we require:

$$\alpha_{kk} = \beta'_{\{ij\}\, kk} = \gamma'_{\{ij\}\, kk\, \ell\ell} \overset{!}{=} 0\,, \qquad (2.27)$$

so that the deviator takes the form

$$\tau'_{ij} = \alpha'_{ij} + \beta'_{\{ij\}\, pq}\, \sigma'_{pq} + \gamma'_{\{ij\}\, pq\, rs}\, \sigma'_{pq}\, \sigma'_{rs} \qquad (2.28)$$

instead of (2.19b).

3.   Creep Potential and Constitutive Equations

In this section the constitutive equations for the secondary creep stage of anisotropic materials will be formulated, and the material constants involved in the equations will be related to experimental data of calibration tests under uniaxial tension or compression. The strain rate-stress relations for creep given below are based on the assumption of the existence of a creep potential. It is evident from the theory of isotropic tensor functions that in an isotropic medium the creep potential (2.3) or (2.6) can depend only on invariants of the stress tensor or the stress deviator, respectively. Certain considerations that favour the creep potential hypothesis are presented, for instance, in [22].

As mentioned in Section 2, the anisotropic behaviour is described by substituting the invariants (2.8a,b,c) or (2.20), (2.21) of the mapped stress tensor (2.7) or its deviator (2.28) for the corresponding invariants (2.4a,b,c) or (2.5a,b,c) in the isotropic creep potential (2.3) or (2.6), respectively. This idea is schematically shown in Figure 1.

The actual creep state of an anisotropic solid is mapped on to a fictitious isotropic state with equivalent creep rate $\dot{\epsilon} = \dot{\gamma}$ by a suitable transformation $\tau_{ij} = \tau_{ij}(\sigma_{k\ell})$. The 'limiting creep stresses' σ_{cx}, σ_{cy} etc. in Figure 1 for specimens cut along the x-direction, y-direction etc., and the transformed fictitious isotropic 'limiting creep stress' τ_c produce a defined creep strain ϵ_c in a defined

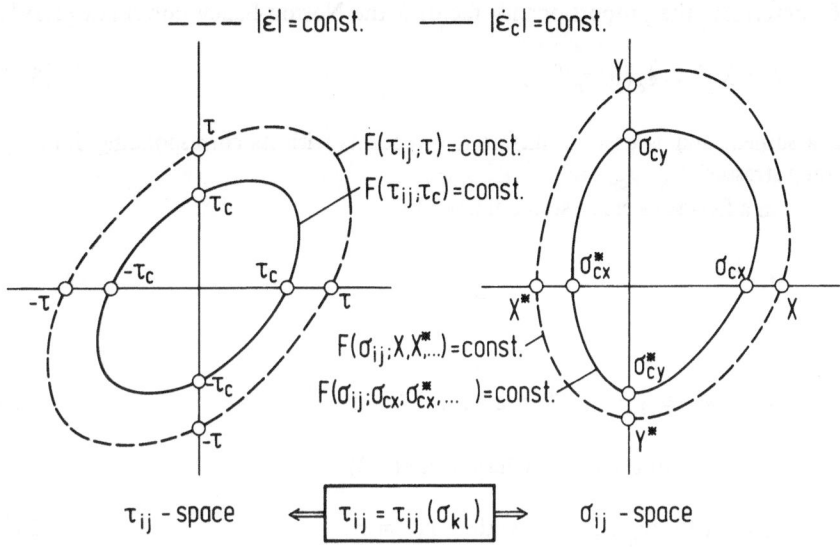

Figure 1. – *Creep potentials.*

time, e.g. 1 per cent creep strain in 10^5 hours. Therefore, $\dot{\epsilon}_c$ and σ_c are materials constants [20, 21].

The theory of the creep potential, like the theory of the plastic potential [16, 18], is based on the principle of maximum dissipation rate. Following Lagrange's method in connection with a creep condition $F(\sigma_{ij}) = $ const. as a secondary condition, the flow rule is obtained from the principle of maximum dissipation rate as follows:

$$\dot{\epsilon}_{ij} = \dot{\lambda}[\partial F(\sigma)/\partial \sigma_{ij}] . \tag{3.1a}$$

or:

$$\dot{\epsilon}_{ij} = \dot{\lambda}[\partial F(\tau)/\partial \tau_{pq}] (\partial \tau_{pq}/\partial \sigma_{ij}) \equiv (\partial \tau_{pq}/\partial \sigma_{ij}) \dot{\gamma}_{pq} . \tag{3.1b}$$

In (3.1a,b) the factor $\dot{\lambda}$ is Lagrange's multiplier. As it is known, the surfaces $F(\sigma_{ij}) = $ const. or $F(\tau_{ij}) = $ const. must be convex in the σ_{ij}-space or τ_{ij}-space, respectively. The Jacobian matrix $I_{pqij} \equiv \partial \tau_{pq}/\partial \sigma_{ij}$ in (3.1b) is given by the selected transformation $\tau_{pq} = \tau_{pq}(\sigma_{ij})$. When using (2.7) the Jacobian matrix depends on the stress state, while the linear transformation (2.11) leads to the Jacobian matrix $\beta_{ijk\ell}$. Then the flow rule (3.1b) is therefore given by:

$$\dot{\epsilon}_{ij} = \beta_{pqij}[\partial F(\tau)/\partial \tau_{pq}] \dot{\lambda} \equiv \beta_{pqij} \dot{\gamma}_{pq} . \tag{3.2}$$

To determine the proportionality factor λ the Norton-Bailey power law [1,19],

$$\dot{e} = K\sigma^n \equiv \dot{e}_c (\sigma/\sigma_c)^n \, , \tag{3.3}$$

is assumed, and is used in different directions with its corresponding 'limiting creep stresses' σ_{cx}, σ_{cy} etc.

In a fictitious creep state, defined by

$$\dot{\gamma}_c \overset{!}{=} \dot{e}_c \quad \text{or} \quad \dot{\gamma} \overset{!}{=} \dot{e} \, , \tag{3.4a,b}$$

we have by analogy of (3.3):

$$\dot{\gamma} = L\tau^m \equiv \dot{\gamma}_c (\tau/\tau_c)^m = \dot{e}_c (\tau/\tau_c)^m = \dot{e} \, , \tag{3.5}$$

so that, because of $\dot{e}_{11} \equiv \dot{e}$, we have from (3.2):

$$\dot{\lambda} = \dot{e}/(\partial F/\partial \tau_{ij})_{i=j=1} \quad \text{with} \quad \tau_{11} = \tau \, . \tag{3.6}$$

In (3.6) the fictitious 'isotropic' creep stress τ appears, which can be determined by the hypothesis of the equivalent dissipation rate. Thus, in connection with (3.4b) we require

$$\tau\dot{e} \overset{!}{=} \sigma_{ij} \dot{e}_{ij} \equiv \dot{D} \, , \tag{3.7}$$

so that, using (3.2), (3.6) and the inverse transformation $\sigma_{ij} = \beta_{ijk\varrho}^{(-1)}\tau_{k\varrho}$ of (2.21), the relation

$$\tau(\partial F/\partial \tau_{ij})_{i=j=1} = \tau_{ij}(\partial F/\partial \tau_{ij}) \tag{3.8}$$

results, from which we can determine the fictitious stress τ. The rate of dissipation of creep energy is obtained from (3.7), combined with (2.11), (3.2) and (3.6):

$$\dot{D} = \dot{\lambda} \, \tau_{pq} \, [\partial F(\tau)/\partial \tau_{pq}] \, . \tag{3.9}$$

Considering homogeneous creep potentials $F(\tau_{ij})$ of degree r, we use Euler's theorem on homogeneous functions,

$$F(S\tau_{ij}) = S^r F(\tau_{ij}) \implies \tau_{ij} [\partial F(\tau)/\partial \tau_{ij}] = r F(\tau_{ij}) \, , \tag{3.10}$$

and find from (3.9) the dissipation rate $\dot{D} = \varphi r\tau^r \dot{\lambda}$, if we assume a creep condition $F(\tau_{ij}) = \varphi\tau^r$ of degree r. For instance, the square creep condition

$F = \tau'_{ij} \tau'_{ij}/2 = \tau^2/3$ with $r = 2$ and $\varphi = 1/3$ leads to the dissipation rate $\dot{D} = 2\lambda\tau^2/3$ which agrees with (3.7), because of (3.6), i.e. $\dot{\lambda} = 3\dot{\epsilon}/2\tau$.

In the following, incompressibility is adoqted, i.e., by analogy of (2.6) a creep potential of the form

$$F = F[I_2(\tau'), I_3(\tau')] \tag{3.11}$$

is based on the anisotropic case. The invariants in (3.11) of the deviator (2.19b) are given by (2.20) and (2.21). From the flow rule (3.2), combined with relations (2.17c), (3.5), (3.6), and (3.8), we finally obtain the constitutive equations:

$$\dot{\epsilon}_{ij} = \Phi \beta'_{pq\,\{ij\}} \left(\frac{\partial F}{\partial I_2} \tau'_{qp} + \frac{\partial F}{\partial I_3} \tau''_{qp} \right), \tag{3.12}$$

in which the function Φ is defined by:

$$
\left.
\begin{aligned}
\Phi &\equiv \frac{1}{2} L \left\{ 3 \Big/ \left[\left(\frac{\partial F}{\partial I_2} \right)_V + \frac{1}{3}\tau \left(\frac{\partial F}{\partial I_3} \right)_V \right] \right\}^{\frac{m+1}{2}} \times \\
&\quad \times \left[\frac{\partial F}{\partial I_2} I_2(\tau') + \frac{3}{2} \frac{\partial F}{\partial I_3} I_3(\tau') \right]^{\frac{m-1}{2}}.
\end{aligned}
\right\} \tag{3.13}
$$

The index V at the brackets in (3.13) indicates the uniaxial equivalent fictitious stress strate $(\tau_{ij})_V \equiv \mathrm{diag}\,\{\tau_{11} \equiv \tau, 0, 0\}$. The symbol τ''_{pq} in (3.13) is, like (2.25), the deviator of the square of the reduced tensor (2.19b).

Inserting (3.12), together with (3.13), into (3.7), we obtain the rate of dissipation of the creep energy

$$D = [2(\partial F/\partial I_2) I_2(\tau') + 3(\partial F/\partial I_3) I_3(\tau')] \Phi \tag{3.14}$$

in considering (2.19c) and because of

$$\tau'_{qp} \tau_{pq} = 2I_2(\tau'), \quad \text{and} \quad \tau''_{pq} \tau_{qp} = 3I_3(\tau').$$

In the isotropic special case, given by $\beta_{pq\,ij} = \delta_{pi}\delta_{qj}$, $L \to K$, $m \to n$, $\tau'_{ij} \to \sigma'_{ij}$ and $\tau''_{ij} \to \sigma''_{ij}$, the constitutive equations (3.12), together with (3.13), immediately lead to the corresponding relations derived in [2].

4. Orthotropic behaviour

The parameters L, m in (3.5), (3.13) are determined by experimental data. For instance, we consider an orthotropic material and use the creep law (3.3) in tests on specimens cut along the mutually perpendicular directions x, y, z. Then, with the notations from Figure 1, we have :

$$\dot{\epsilon} = K_x X^{n_x}, \quad \dot{\epsilon} = K_y Y^{n_y}, \quad \dot{\epsilon} = K_z Z^{n_z}, \tag{4.1a,b,c}$$

from which we find in the 'limiting creep stress' state :

$$\dot{\epsilon}_c = (K_x K_y K_z)^{1/3} (\sigma_{cx}^{n_x} \sigma_{cy}^{n_y} \sigma_{cz}^{n_z})^{1/3} \tag{4.2}$$

According to the idea in Figure 1 and using the mapping (2.11), the 'limiting creep stresses' can be expressed by the fictitious isotropic 'limiting creep stress' τ_c :

$$\tau_c \equiv \tau_{xx} \equiv \beta_{xxxx} \sigma_{cx} \equiv \ell_x \sigma_{cx}, \quad \tau_c \equiv \tau_{yy} \text{ etc.,} \tag{4.3}$$

so that then relation (4.2) according to (3.5) takes the form $\dot{\epsilon}_c = L\tau_c^m$, if the fictitious creep factor L is the geometrical mean value

$$L \equiv [(K_x/\ell_x^{n_x})(K_y/\ell_y^{n_y})(K_z/\ell_z^{n_z})]^{1/3} \tag{4.4}$$

and the fictitious creep exponent m is the arithmetical mean value

$$m \equiv (n_x + n_y + n_z)/3 \tag{4.5}$$

from corresponding experimental data. In the case of an existing Bauschinger effect, the compression test data $\sigma_{cx}^*, \sigma_{cy}^*$, etc. appear in equations (4.3), (4.4) and (4.5) as well.

In the orthotropic case, the transformation (2.11) can be specified according to [4] :

$$\beta_{ijpq} \equiv \omega_{ip} \omega_{jq} \Rightarrow \tau_{ij} = \omega_{ip} \omega_{jq} \sigma_{pq} . \tag{4.6}$$

If the second order tensor ω_{ij} is real and symmetric, then its principal values $\omega_I, \omega_{II}, \omega_{III}$ are all real. For isotropic materials the tensor ω_{ij} is identical to Kronecker's tensor δ_{ij}. Using the notations from Figure 1 and considering (4.6), the diagonal form of the orthotropic tensor ω_{ij} is given by[1]

$$\omega_{ij} = \text{diag} \{\sqrt{\tau/X}, \quad \sqrt{\tau/Y}, \quad \sqrt{\tau/Z}\}. \tag{4.7}$$

---

[1] If ω_{ij} has the diagonal form, then, in accordance with (4.6), the tensors τ_{ij} and σ_{ij} are coaxial.

For example, the assumptions of incompressibility (2.19b) and orthotropic material (4.6) with $\beta'_{\{ij\}pq} = \omega_{ip}\,\omega_{jq} - \omega^{(2)}_{pq}\,\delta_{ij}/3$ immediately lead from the quadratic creep condition $I_2(\tau') = \tau^2/3$ to the Hill-condition [16].

The following considerations are based upon the creep potential

$$F = I_2(\tau') + \alpha I_3(\tau')/\tau \quad \text{where} \quad -3 \leq \alpha \leq 3/2, \tag{4.8}$$

which is suitable to describe 'second-order effects' as shown in [2, 5] for the isotropic case, for instance. The range $-3 \leq \alpha \leq 3/2$ in (4.8) results from the convexity of the potential surface $F = \text{const.}$ Using (4.8), the constitutive equations (3.12) together with (3.13) become

$$\dot{\epsilon}_{ij} = \frac{1}{2}L\left(\frac{3}{1+\alpha/3}\right)^{\frac{m+1}{2}}\left[I_2(\tau') + \frac{3}{2}\alpha\frac{I_3(\tau')}{\tau}\right]^{\frac{m-1}{2}} \times \left.\begin{array}{c} \\ \\ \times\ \beta'_{pq\{ij\}}(\tau'_{qp} + \alpha\,\tau''_{qp}/\tau), \end{array}\right\} \tag{4.9}$$

which is applied in the next section.

4.1. *The Poynting Effect*

Considering a pure shear stress state

$$\sigma_{ij} = \sigma'_{ij} = \sigma_{12}\begin{pmatrix} 0 & 1 & 0 \\ 1 & 0 & 0 \\ 0 & 0 & 0 \end{pmatrix}, \tag{4.10}$$

then the deviator (2.19b) and the deviator of its square are given by

$$\tau'_{ij} = \omega_I\,\omega_{II}\,\sigma'_{ij} \quad \text{and} \quad \tau''_{ij} = \frac{1}{3}\,\omega_I^2\,\omega_{II}^2\,\sigma_{12}^2\,\text{diag}\,\{1,1,-2\}, \tag{4.11a,b}$$

if the stress tensor σ_{ij} and its image τ_{ij} are coaxial. This coaxiality exists in the orthotropic case (4.6), if the orthotropic tensor ω_{ij} has the diagonal form (4.7). Because of (4.10), the invariants of the deviator (4.11a) are given by $I_2(\tau') = \omega_I^2\,\omega_{II}^2\,\sigma_{12}^2$ and $I_3(\tau') = 0$, so that the constitutive equations (4.9) are reduced to

$$\epsilon_{ij} = \frac{1}{6}L\left(\frac{3\omega_I^2\,\omega_{II}^2}{1+\alpha/3}\right)^{\frac{m}{2}}\sigma_{12}^m\begin{pmatrix} \alpha\varsigma_I & 3\dfrac{\tau}{\sigma_{12}} & 0 \\[2mm] 3\dfrac{\tau}{\sigma_{12}} & \alpha\varsigma_{II} & 0 \\[2mm] 0 & 0 & -2\alpha\varsigma_{III} \end{pmatrix} \tag{4.12}$$

in using (2.18), (4.6), (4.10) and (4.11a,b). In the matrix (4.12) the abbreviations

$$\varsigma_I \equiv (2\,\omega_I^2 - \omega_{II}^2 + 2\,\omega_{III}^2)/3, \tag{4.13a}$$

$$\varsigma_{II} \equiv (-\omega_I^2 + 2\,\omega_{II}^2 + 2\,\omega_{III}^2)/3, \tag{4.13b}$$

$$\varsigma_{III} \equiv (\omega_I^2 + \omega_{II}^2 + 4\,\omega_{III}^2)/6 \tag{4.13c}$$

are used, while the quotient

$$\tau/\sigma_{12} = \omega_I\,\omega_{II}\,[3/(1+\alpha/3)]^{1/2} \tag{4.14}$$

is determined from (3.8), considering (4.8), (4.10) and (4.11a).

From (4.12) and (4.13c) we obtain the quotient of the longitudinal strain rate $\dot\epsilon_{33}$ and the shearing strain rate $\dot\epsilon_{12}$,

$$\frac{\dot\epsilon_{33}}{\dot\epsilon_{12}} = -\frac{\omega_I^2 + \omega_{II}^2 + 4\,\omega_{III}^2}{6\,\omega_I\,\omega_{II}} \cdot \frac{2\,\alpha}{3\sqrt{3}}\left(1 + \frac{\alpha}{3}\right)^{1/2}, \tag{4.15}$$

which can be considered as a suitable measure of the Poynting effect (Fig. 2).

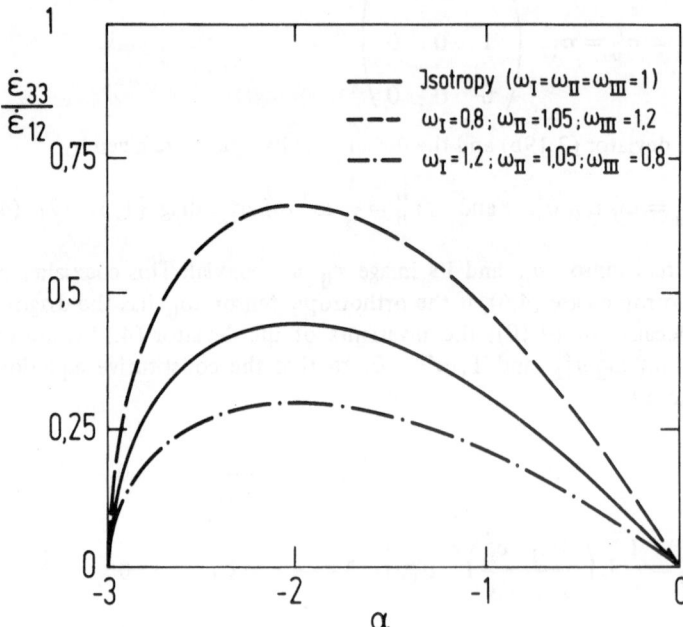

Figure 2. – *Poynting effect.*

Regarding experimental investigations [15, 27], the relation (4.15) is numerically evaluated only for negative α-values, $-3 \leq \alpha \leq 0$, as shown in Figure 2. The limit $\alpha = -3$ is given by the convexity of the creep potential (4.8) and is compatible with equation (4.15), in which the square root must be real. In the isotropic case, $\omega_I = \omega_{II} = \omega_{III} = 1$, the results from Figure 2 are identical with those calculated in [2].

The anisotropic influence is contained in the factor

$$(\omega_I^2 + \omega_{II}^2 + 4\omega_{III}^2)/(6\,\omega_I\,\omega_{II}),$$

which is equal to one for isotropic materials, while the function

$$-2\alpha(1 + \alpha/3)^{1/2}/(3\sqrt{3})$$

of equation (4.15) expresses the 'second order effect'. This possibility of a 'polar' decomposition is based on the fact that, because of (4.7), the tensors σ_{ij} and τ_{ij} are coaxial.

5. Comparison with the representation theory for anisotropic tensor functions

If the isotropic behaviour $\dot{\epsilon}_{ij} = \epsilon_{ij}(\sigma_{pq})$ can be approximated by a tensor power series in one variable,

$$\epsilon_{ij} = \epsilon_{ij}(\sigma_{pq}) = \sum_{\nu=0}^{N} A_\nu\,\sigma_{ij}^{(\nu)}, \tag{5.1}$$

then, by using Hamilton-Cayley's theorem,

$$\sigma_{ij}^{(\nu)} = P_{\nu-2}\,\sigma_{ij}^{(2)} + Q_{\nu-1}\,\sigma_{ij} + R_\nu\,\delta_{ij}, \tag{5.2}$$

where $P_{\nu-2}$, $Q_{\nu-1}$, and R_ν are scalar-valued functions of the principal invariants (2.4a,b,c), and of the orders $\nu-2$, $\nu-1$, ν, respectively, in the stresses σ_{ij} [8], the isotropic tensor-valued tensor function (5.1) is represented by [28]:

$$\dot{\epsilon}_{ij} = \eta_0\,\delta_{ij} + \eta_1\,\sigma_{ij} + \eta_2\,\sigma_{ik}\,\sigma_{kj} \equiv \sum_{\nu=0}^{2} \eta_\nu\,\sigma_{ij}^{(\nu)}. \tag{5.3}$$

An inverse form of (5.3) is used, for example, in [23]. The η_ν in (5.3) are invariants of σ and hence can be expressed as functions $\eta_\nu = \eta_\nu(I_1, I_2, I_3)$ of the invariants (2.4a,b,c). Comparing (5.3) with the consitutive equations obtained from the flow rule (3.1a) and a creep potential of the form (2.3), we have

$$\eta_0 \equiv \dot{\lambda}(\partial F/\partial I_1 - I_1 \, \partial F/\partial I_2 - I_2 \, \partial F/\partial I_3), \tag{5.4a}$$

$$\eta_1 \equiv \dot{\lambda}(\partial F/\partial I_2 - I_1 \, \partial F/\partial I_3), \quad \eta_3 \equiv \dot{\lambda} \, \partial F/\partial I_3. \tag{5.4b,c}$$

By analogy with (5.3) we obtain for the incompressible case with (2.5a,b) and (2.6):

$$\dot{\epsilon}_{ij} = \vartheta_0 \delta_{ij} + \vartheta_1 \sigma'_{ij} + \vartheta_2 \sigma'_{ik} \sigma'_{kj}, \tag{5.5}$$

and as in (5.4a,b,c) we have :

$$\vartheta_0 \equiv -\frac{2}{3} I_2 (\sigma') \, \vartheta_2, \quad \vartheta_1 \equiv \dot{\lambda} \frac{\partial F}{\partial I_2}, \quad \vartheta_2 \equiv \dot{\lambda} \frac{\partial F}{\partial I_3}. \tag{5.6a,b,c}$$

Because of the incompressibility not all the values $\vartheta_0, \vartheta_1, \vartheta_2$ are independent, as we see from (5.6a). Equation (5.6a) immediately results from (5.5), too, if $\dot{\epsilon}_{ii} = 0$.

As in (5.1) and (5.3) the anisotropic behaviour can be expressed by

$$\dot{\gamma}_{ij} = \dot{\gamma}_{ij}(\tau_{pq}) = \sum_{\nu=0}^{N} B_\nu \, \tau_{ij}^{(\nu)} \quad \text{where} \quad \tau_{ij} = \beta_{ijk\ell} \sigma_{k\ell} \tag{5.7}$$

and

$$\left. \begin{array}{l} \gamma_{ij} = \varphi_0 \delta_{ij} + \varphi_1 \tau_{ij} + \varphi_2 \tau_{ip} \tau_{pj} \equiv \sum_{\nu=0}^{2} \varphi_\nu \, \tau_{ij}^{(\nu)} \\[12pt] \text{where} \quad \varphi_\nu = \varphi_\nu [I_1(\tau), \, I_2(\tau), \, I_3(\tau)] , \end{array} \right\} \tag{5.8}$$

respectively, so that, because of (3.2), we find the representation

$$\dot{\epsilon}_{ij} = \beta_{pqij}(\varphi_0 \delta_{qp} + \varphi_1 \tau_{qp} + \varphi_2 \tau_{qr} \tau_{rp}) . \tag{5.9}$$

In view of the requirement of incompressibility (2.17c), we have an 'anisotropic' representation instead of (5.9)

$$\dot{\epsilon}_{ij} = \beta'_{pq\{ij\}} (\psi_0 \delta_{qp} + \psi_1 \tau'_{qp} + \psi_2 \tau'_{qr} \tau'_{rp}) , \tag{5.10}$$

in which, by comparing (5.10) with the constitutive equations (3.12), (3.13), the scalar-valued functions ψ_0, ψ_1, ψ_2 are:

$$\psi_0 \equiv -\frac{2}{3} I_2(\tau') \, \psi_2, \quad \psi_1 \equiv \Phi \frac{\partial F}{\partial I_2(\tau')}, \quad \psi_2 \equiv \Phi \frac{\partial F}{\partial I_3(\tau')} \cdot \tag{5.11a,b,c}$$

As in the expressions (5.6a,b,c), not all functions (5.11a,b,c) are independent. Finally, the dissipation function (3.7) is obtained from (5.10) when considering (2.19c) and (5.11a):

$$\dot{D} = 2 \psi_1 I_2(\tau') + 3 \psi_2 I_3(\tau') , \tag{5.12}$$

which is, because of (5.11b,c), equivalent to (3.14), and which agrees with the corresponding one of isotropic materials ($\tau' \longrightarrow \sigma'$) given in [23], if in [23] isochoric behaviour ($\epsilon_{kk} = 0$) is required.

REFERENCE

[1] BAILEY R.W. – "The utilization of creep test data in Engineering Design". *Proceedings, Inst. Mech. Eng.*, 131 (1935) : 131-349.

[2] BETTEN J. – "Zur Verallgemeinerung der Invariantentheorie in der Kriechmechanik", *Rheol. Acta*, 14 (1965): 715-720.

[3] BETTEN J. – "Beitrag zum isotroqen kompressiblen plastischen Fliessen", *Arch. Eisenhüttenwes.*, 46 (1965) : 317-323.

[4] BETTEN J. – "Ein Beitrag zur Invariantentheorie in der Plastomechenik anisotroper Werkstoffe", *Z. angew. Math. Mech. (ZAMM)*. 56 (1976): 557-559.

[5] BETTEN J. – "Plastische Anisotropie und Bauschinger-Effekt; allgemeine Formulierung und Vergleich mit experimentell ermittelten Fliessortkurven", *Acta Mechanica*, 25 (1976): 79-94.

[6] BETTEN J. – "Zur Modifikation des Spannungsdeviators", *Acta Mechanica*, 27 (1977) : 173-184.

[7] BETTEN J. – "Plastische Stoffgleichungen inkompressibler anisotroper Werkstoffe", *Z. angew. Math. Mech. (ZAMM)*, 57 (1977) : 671-673.

[8] BETTEN J. – *Elementare Tensorrechnung für Ingenieure*, Vieweg Verlag, Braunschweig, 1977.

[9] BOEHLER J.P. and A. SAWCZUK – "Equilibre limite des sols anisotropes", *Journal de Mécanique*, 9 (1970): 5-33.

[10] BOEHLER J.P. and A. SAWCZUK – "On yielding of oriented solids", *Acta Mechanica*, 27 (1977) : 185-206.

[11] BOEHLER J.P. – "On irreducible representations for isotropic scalar functions". *Z. angew. Math. Mech. (ZAMM)* 57 (1977): 323-327.

[12] DUBEY R.N. and M.J. HILLER – "Yield criteria and the Bauschinger effect for a plastic solid" *Journal of Basic Engineering*, Transactions of the ASME, March 1972, pp. 228-280.

[13] GRACE J.H. and A. YOUNG – *The Algebra of Invariants*. London and New York : Cambridge Univers. Press, 1903.

[14] GUREVICH G.B. – *Foundations of the Theory of Algebraic Invariants*, (1964) Noorhoff, Groningend.

[15] HECKER F.W. – "Die Wirkung des Bauschinger-Effekts bei grossen Torsions-Formänderungen", Dissertation, Techn. Hochschule Hannover, 1967.

[16] HILL R. – *The Mathematical Theory of Plasticity*. Oxford, 1950.

[17] LEIGH D.C. – *Nonlinear Continuum Mechanics*. New York: Mc Graw-Hill Book Company, 1968.

[18] V. MISES R. – "Mechanik der plastischen Formänderung von Kristallen", *Z. angew. Math. Mech. (ZAMM).* 8 (1928) : 161-185.

[19] NORTON F.N. – *Creep of High Temperatures.* New-York: Mc Graw Hill, 1929.

[20] ODQUIST F.K.G. and J. HULT. – *Kriechfestigkeit metallischer Werkstoffe.* Berlin, 1962.

[21] ODQUIST F.K.G. – *Mathematical Theory of Creep and Creep Rupture.* Oxford, 1966.

[22] RABOTNOV Yu. N. – *Creep Problems in Structurel Members.* Amsterdam/London, 1969.

[23] SAWCZUK A. and P. STUTZ. – "On formulation of stress-strain relations for solids at failure". *Z. angew. Math. Phys. (ZAMP),* 19 (1968): 770-778.

[24] SAYIR R. – "Zur Fliessbedingung der Plastizitätstheorie", *Ing.-Archiv,* 39 (1970): 414-432.

[25] SOBOTKA Z. – "Theorie des plastischen Fliessens von anisotropen Körpern". *Z. angew. Math. Mech. (ZAMM),* 49 (1969): 25-32.

[26] SPENCER A.J.M. – "Theory of Invariants" in A.C. Eringen, ed., *Continuum Physics, Vol. 1 (Mathematics),* New York/London, 1971.

[27] SWIFT H.W. – "Length changes in metals under torsional overstrain". *Engineering,* 163 (1947): 253-257.

[28] TRUESDELL C. and W. NOLL. – "The non-linear field theories of mechanics" in S. Flügge, ed., *Handbuch der Physik Vol. III/3,* Berlin/Heidelberg/New York, 1965.

[29] WEITZENBÖCK R. – *Invariantentheorie,* Groningen : Noordhoff, 1923.

RESUME

(La théorie des invariants dans la mécanique du fluage des matériaux anisotropes)

Dans ce travail, les équations constitutives pour le fluage secondaire des matériaux anisotropes sont formulées, et les constantes matérielles intervenant dans ces équations sont reliées à des données expérimentales. La théorie est basée sur l'hypothèse de l'existence d'un potentiel pour le fluage, qui, lorsque le matériau est isotrope, ne peut dépendre que des invariants du tenseur des contraintes ou du déviateur des contraintes. Le comportement anisotrope est décrit par la substitution des invariants d'un tenseur contrainte transformé ou de son déviateur avec les invariants correspondants du tenseur des contraintes réelles dans le potentiel pour le fluage isotrope. L'effet Poynting, par exemple, est pris en compte. Finalement, l'approche par le tenseur contrainte transformé est comparée avec la théorie des représentations des fonctions tensorielles anisotropes.

Description of Deformation and Fracture of Anisotropic Solids

A. K. Malmeisters and V. P. Tamuzs

Institute of Polymer Mechanics, Latvian SSR Academy of Sciences, Riga, USSR.

1. Strength and elasticity tensors of different ranks

To describe the curve of deformation or the ultimate strength surface in complex stress state, the tensorial polynomial series may be useful. In the case of deformation of nonlinearly elastic material, the relationship $\sigma_{ij}(\epsilon_{ij})$ may be written in the form

$$\sigma_{ij} = a_{ijk\ell}\,\epsilon_{k\ell} + a_{ijk\ell mn}\,\epsilon_{k\ell}\epsilon_{mn} + \dots , \tag{1}$$

where σ_{ij} is the stress, ϵ_{ij} the strain, and $a_{ij\dots mn}$ the elasticity tensors of different ranks.

To describe the strength surface in the space of stresses σ_{ij}, a similar relationship is helpful [8]:

$$p_{ij}\,\sigma_{ij} + p_{ijk\ell}\,\sigma_{ij}\sigma_{k\ell} + \dots = 1 . \tag{2}$$

Here $p_{ijk\ell}\dots$ are the strength tensors of different ranks.

Anisotropy of the medium imposes certain restrictions on the structure of the tensors of elasticity $a_{ij\dots k\ell}$ and strength $p_{ij\dots k\ell}$. The simplest way for arriving at Eqs. (1) and (2) is by forming various linearly independent combinations of different orders of the invariants J_i of the tensor σ_{ij}. In Eq. (2) such combinations of invariants with coefficients directly define the structure of the strength tensors. Similarily, one may determine the number of the elasticity tensor components in Eq. (1), whereas the structure of $a_{ij\dots k\ell}$ can be obtained by forming the polynomial potential $W(J_i)$ and deriving the function $\sigma_{ij}(\epsilon_{ij})$ from the relationships:

$$\sigma_{ij} = \frac{\partial W}{\partial J_k}\,\frac{\partial J_k}{\partial \epsilon_{ij}}$$

The structure and the number of elasticity tensor components of order up to the twelfth inclusive for the most common classes of anisotropy have been obtained in [3]. The number of the components of elasticity (strength tensors) for the two-dimensional stress state is given in [10]. Table 1 lists the number of elasticity (strength tensor) components of higher order in the two and three-dimensional case for the following symmetries of the medium: complete anisotropy (triclinic class), orthotropy (rhombic class), cubic symmetry, transverse isotropy and isotropy.

It is clear from Table 1 that the number of independent components of the tensors increases rapidly with the increase of the tensor order. This means that in practice the series (1) and (2) should be terminated at a rather small number of terms. Retaining the linear and square summands of Eq. (2), one obtains the following relationship

$$p_{11}\,\sigma_{11} + p_{22}\,\sigma_{22} + p_{1111}\,\sigma_{11}^2 + p_{1122}\,\sigma_{11}\,\sigma_{22} + p_{2222}\,\sigma_{22}^2 + \tag{3}$$
$$+ p_{1212}\,\sigma_{12}^2 = 1$$

for an orthotropic body in two-dimensional stress state. The coefficients of Eq. (3) are provided from strength tests of simple stress states. Whenever the number of tests corresponds to the number of indefinite coefficients in formula (3), the necessary values are defined by quite simple equations

TABLE 1. Numerators show the number of components for the plane stress state; denominators, in the three-dimensional case.

Symmetry class	Designation of symmetry	Number of tensor components rank					
		2	4	6	8	10	12
triclinic		$\frac{3}{6}$	$\frac{6}{21}$	$\frac{10}{56}$	$\frac{15}{126}$	$\frac{21}{252}$	$\frac{28}{462}$
rhombic	$\dfrac{2 \cdot m}{m \cdot 2 : m}$	$\frac{2}{3}$	$\frac{4}{9}$	$\frac{6}{20}$	$\frac{9}{42}$	$\frac{12}{78}$	$\frac{16}{138}$
cubic	$\dfrac{4 \cdot m}{\tau/4}$	$\frac{1}{1}$	$\frac{3}{3}$	$\frac{3}{6}$	$\frac{6}{11}$	$\frac{6}{18}$	$\frac{10}{32}$
hexagonal	$\dfrac{6 \cdot m}{m \cdot 6 : m}$	$\frac{1}{2}$	$\frac{2}{5}$	$\frac{3}{10}$	$\frac{4}{23}$	$\frac{5}{33}$	$\frac{7}{57}$
transverse isotropy	$\dfrac{m \cdot \infty}{m \cdot \infty : m}$	$\frac{1}{2}$	$\frac{2}{5}$	$\frac{2}{9}$	$\frac{3}{16}$	$\frac{3}{25}$	$\frac{4}{39}$
isotropy	$\dfrac{m \cdot \infty}{\infty/\infty \cdot m}$	$\frac{1}{1}$	$\frac{2}{2}$	$\frac{2}{3}$	$\frac{3}{4}$	$\frac{3}{5}$	$\frac{4}{7}$

[8]. However, if the number of tests exceeds the number of coefficients, it is advisable to use the method of least squares and to determine the coefficients $p_{ijk\ell}$ so that deviation of the approximation surface (3) from the experimental data be minimal. Figures 1 and 2 show the ellipsoids approximating the strength surface of a Kevlar type reinforced fiber (with symmetry class 2.m) at various test temperatures [6]. These figures show that the test data are well approximated by the surfaces of rank two.

To conclude, we note that the invariants of tensor σ_{ij} for different medium symmetries can be found in [1]. The same invariants may be obtained by complete contractions of the tensor σ_{ij} and the anisotropy tensors which are found in [5] for all the classes of symmetry. The tensors of anisotropy $\mathbf{T}^{(n)}$ for the symmetry group Q are determined as follows:

a) the intersection of symmetry groups Q_n of these tensors coincides with $Q = Q_1 \cap Q_2 \cap \ldots \cap Q_n$;

b) each tensor whose symmetry group contains Q may be derived from the tensor $\mathbf{T}^{(n)}$ using linear combinations, multiplication and contraction of the tensors;

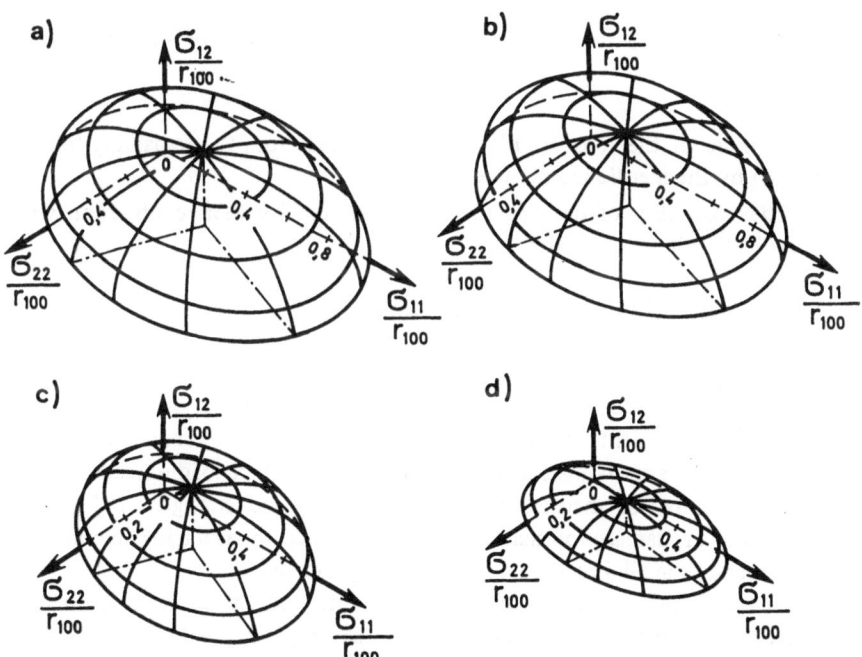

Figure 1 – *Strength surfaces of kevlar type reinforced plastic (rhombic class symmetry) at different temperatures:* $T = 20°C(a),\ 50°(b),\ 100°(c),\ 150°(d)$

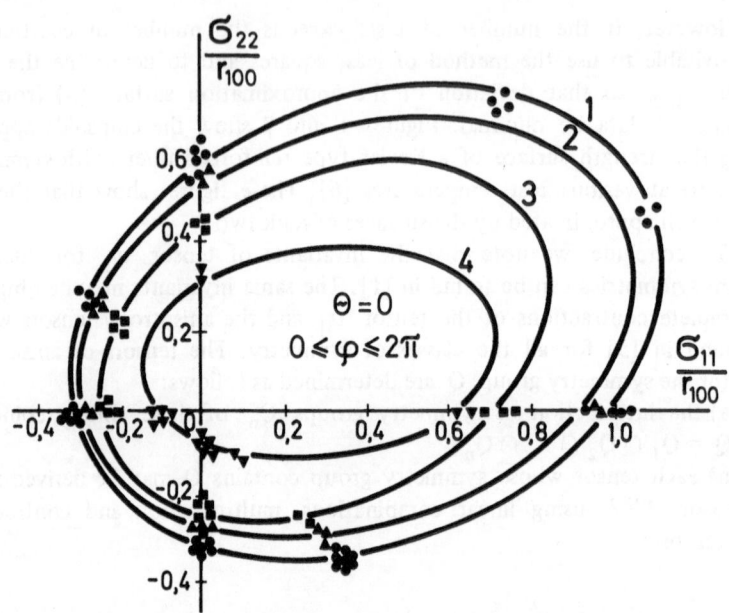

Figure 2. – *Sections of strength surfaces (Fig. 1) in the plane $\sigma_{12} = 0$. Experimental points marked.*

c) none of the tensors $\mathbf{T}^{(n)}$ can be derived from the remaining tensors of this set by the mentioned method.

The possibility of deriving the invariants σ_{ij} by their contraction with tensors $\mathbf{T}^{(n)}$ will be used later.

2. Main equations of the local strain theory

In describing deformation and strength, Eqs. (1) and (2) are useful for the case of simple loading or when the mechanical properties in the nonlinear region are independent of the loading history (which is permissible in practice as a rough approximation). The effects of complex loading can be considered by statistical theories of plasticity [7, 9] and strength [11], where macrostrains or damage of material are determined by averaging microshears and microcracks in all directions in the material volume.

In the statistical theory of plasticity, namely in the local strain theory, the deformation tensor ϵ_{ij} in the principal directions $i, j = 1, 2, 3$ is derived by averaging the strains $\epsilon_{i'j'}$ calculated in the arbitrary mobile coordinate system $i', j' = x, y, z$. The mutual arrangement of the axes ij and $i'j'$ is defined by the direction cosines $\ell_{ix}, \ell_{iy} \dots$. The direction of the z axis is defined either by the angles θ, ϕ or the vector $e_{3'}$ (Fig. 3). The values $\epsilon_{i'j'} = \epsilon_{xz}, \epsilon_{yz}, \epsilon_{zz}$ are arbitrary functions of the direction in space, besides $\epsilon_{zz}(\theta, \phi)$ is invariant relative to rotations of the $i'j'$ axes about the z axis,

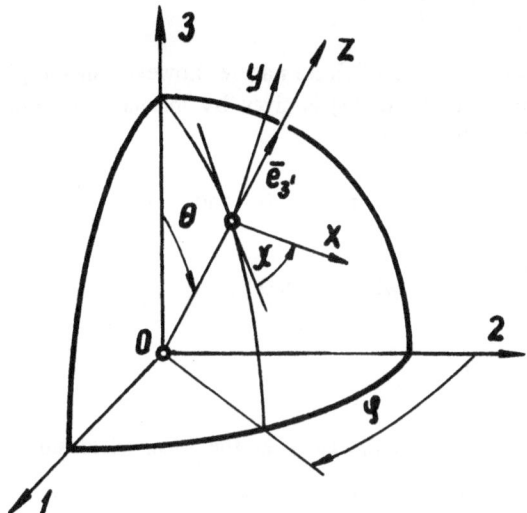

Figure 3. – *Main coordinate system $i, j = 1, 2, 3$ and local (mobile) system $i', j' = x, y, z$. The position of the Z axis is defined by the coordinates θ, ϕ or the vector e'_3.*

while $\epsilon_{xz}(\theta, \phi)$ and $\epsilon_{yz}(\theta, \phi)$ change as the vector components with the notation of the same axes. Consequently, $\epsilon_{zz}(\theta, \phi)$ is a scalar function on the sphere, and $\epsilon_{xz}, \epsilon_{yz}$ are the components of the vectorial function $\underset{\sim}{\gamma} = \epsilon_{xz} e'_1 + \epsilon_{yz} e'_2$ on the sphere. The values $\epsilon_{zz}, \epsilon_{xz}, \epsilon_{yz}$ are called the local strain components, to emphasize that they change arbitrarily from point to point on the sphere S, i.e., they are not components of some single (constant on S) tensor. The principal formula of the theory of local strains is actually that of averaging:

$$\epsilon_{ij} = \frac{3}{S} \int_S \epsilon_{zz} \ell_{iz} \ell_{jz} \, dS$$

$$+ \frac{3}{2S} \int_S [\epsilon_{xz}(\ell_{ix}\ell_{jz} + \ell_{jx}\ell_{iz}) + \epsilon_{yz}(\ell_{iy}\ell_{jz} + \ell_{jy}\ell_{iz})] \, dS \qquad (4)$$

It is easy to verify that integration of Eq. (4) over every

$$\epsilon_{i'j'}(\theta, \phi) \quad (i', j' = x, y, z)$$

leads to a second order tensor in ij axes. The construction of the theory
is completed by setting a law which would express the connection of local
strains at an arbitrary point on the sphere with the stress tensor:

$$\epsilon_{i'j'} = \epsilon_{i'j'} \{ \sigma, \mathbf{e}_3' \} \tag{5}$$

The connection (5) may be of finite nature, however, in the plasticity theories
it is more common that Eq. (5) is given by means of increments and taking
into account the loading history:

$$d\epsilon_{i'j'} = \begin{cases} \gamma_{i'j'} \, df, & \text{when } f = 0 \text{ and } df = \dfrac{\partial f}{\partial \sigma_{k'\varrho'}} \, d\sigma_{k'\varrho'} > 0, \quad (6) \\[4pt] 0, & \text{when } f < 0, \end{cases}$$

$$\epsilon_{i'j'} = \int_{L'} d\epsilon_{i'j'} . \tag{7}$$

Here $\gamma_{i'j'} = h \, \dfrac{\partial \Phi}{\partial \sigma_{i'j'}}$; $f(\sigma_{i'j'}, \epsilon_{i'j'})$ is the local yield surface in the space
of the $\sigma_{i'j'}$ at a point θ, ϕ on the sphere; the scalar function on the sphere
$h(\sigma, \epsilon, \theta, \phi)$ is a local function of strain hardening; $\Phi(\sigma, \epsilon, \theta, \phi)$ is a
local plastic potential and also a scalar function on the sphere (in particular,
Φ may be equal to f, which in the local system of coordinates i'j' leads
of the associated flow rule); $L'(\sigma_{i'j'}, \theta, \phi)$ is the loading path at the point
θ, ϕ on the sphere.

It is easy to see that Eqs. (6) and (7) are just a generalization of the
classical incremental plasticity theory, involving a single smooth load surface f = 0, to a theory with an infinite number of load surfaces. The classical
theory is obtained when the given functions f, h, Φ are independent of
the position on the sphere.

3.   Spherical invariants of the stress tensor

Anisotropy of the medium imposes certain limitations on the relations
f, h, $\Phi(\sigma, \theta, \phi)$. The problem we face is to exhibit these limitations for
different symmetries. Let us examine the anisotropy of the medium which
is characterized by the rotation group Q. We shall designate the arbitrary
position of the coordinate axes by the letter u, a transformation of the coordinates compatible with the medium symmetry Q will be denoted by g,
and the axes obtained by rotating the coordinate system x, y, z by an arbitrary
angle about the z axis will be designated by χ.

Let us introduce the following definition:

By a spherical invariant of the tensor σ relative to the group Q we call a scalar function on the sphere $f(\sigma, u)$ which is invariant under rotations about the z axis,

$$f(\sigma, u) = f(\sigma, u\chi), \tag{8}$$

and such that

$$f(\sigma, ug) = f(\sigma, u), \quad g \in Q. \tag{9}$$

Since the tensor σ is defined in a unique way by the components $\sigma_{i'j'}$ in an arbitrary coordinate system, the conditions (8) and (9) can be rewritten 'componentwise' in the following form:

$$f[\sigma_{i'j'}(u)] \equiv f[\sigma_{i'j'}(u\chi)] \tag{10}$$

and

$$f[\sigma_{i'j'}(u), u] \equiv f[\sigma_{i''j''}(ug), ug], \quad \text{if} \quad \sigma_{i'j'} = \sigma_{i''j''}. \tag{11}$$

The condition (10) signifies that f does not depend on a rotation about the z axis, while condition (11) shows that at two positions of the coordinate system which are mutually identical from the viewpoint of the medium symmetry, the scalar invariants have an identical structure in the local coordinate system.

In [4] there was given an algorithm for calculating the full base of spherical polynomial invariants S_i of the second order symmetric tensor σ. The polynomially independent spherical invariants of tensor σ, relative to the group of coordinate transformations Q, can be obtained by deriving polynomially independent joint invariants of the tensors σ and $T^{(n)}$ defining the group Q, and replacing their even number indexes by z.

Obviously, the usual scalar invariants are included in a set of spherical invariants as a particular case, whenever the invariant is independent of the point on the sphere.

As an example, let us consider a construction of the spherical invariants for an orthotropic medium (rhombic class). The symmetry tensors $T^{(n)}$ in the principal axes assume the form

$$A_{ij} = \delta_{i1}\delta_{j1}, \quad B_{ij} = \delta_{i2}\delta_{j2}, \quad C_{ij} = \delta_{i3}\delta_{j3}. \tag{12}$$

The scalar invariants of orthotropy may be derived by contractions of the A_{ij}, B_{ij}, C_{ij}, and σ_{ij}:

$$\sigma_{ij}A_{ij}, \quad \sigma_{ij}B_{ij}, \quad \sigma_{ij}C_{ij}, \quad \sigma_{ij}\sigma_{k\ell}A_{ik}B_{j\ell}, \quad \sigma_{ij}\sigma_{k\ell}B_{ik}C_{j\ell},$$
$$\sigma_{ij}\sigma_{k\ell}A_{ik}C_{j\ell}, \quad \sigma_{ij}\sigma_{k\ell}\sigma_{mn}A_{ik}B_{jm}C_{\ell n}. \tag{13}$$

In the principal axes, Eq. (13) assumes the form

$$\sigma_{11}, \sigma_{22}, \sigma_{33}, \sigma_{12}^2, \sigma_{23}^2, \sigma_{31}^2, \sigma_{12}\sigma_{13}\sigma_{23}. \tag{14}$$

The spherical polynomial invariants used for the construction of the function on the sphere are obtained according to the above formulated algorithm. The invariants are grouped according to their orders with respect to $\sigma_{i'j'}$.

a) Geometrical invariants:

$$A_{zz} = A_{ij}\ell_{iz}\ell_{jz} = \ell_{1z}^z; \quad B_{zz} = \ell_{2z}^z; \quad C_{zz} = \ell_{3z}^z; \tag{15}$$

$$(\ell_{1z}^z + \ell_{2z}^z + \ell_{3z}^z = 1).$$

b) Linear invariants.

Besides the scalar invariants there are:

$$\sigma_{i'z}A_{i'z} = \sigma_{i'z}\delta_{i1}\delta_{j1}\ell_{ii'}\ell_{jz} = \sigma_{i'z}\ell_{1i'}\ell_{1z}, \quad \sigma_{i'z}B_{i'z}, \quad \sigma_{i'z}C_{i'z},$$

$$\sigma_{k'\varrho'}A_{zk'}B_{z\varrho'}, \quad \sigma_{k'\varrho'}A_{zk'}C_{z\varrho'}, \quad \sigma_{k'\varrho'}B_{zk'}C_{z\varrho'}.$$

Since not all the above given invariants are linearly independent, the selection of just the independent invariants leads to the following remaining relationships (in the principal axes):

$$\sigma_{11}, \sigma_{22}, \sigma_{33}, \sigma_{12}\ell_{1z}\ell_{2z}, \sigma_{13}\ell_{1z}\ell_{3z}, \sigma_{23}\ell_{2z}\ell_{3z}, \tag{16}$$

or abbreviated

$$\sigma_{\alpha\alpha}, \sigma_{\alpha\alpha+1}\ell_{z\alpha+1}\ell_{z\alpha},$$

(no summation over α).

c) The square invariants are derived in the form:

$$\sigma_{\alpha\alpha+1}^2, \sigma_{\alpha\alpha+1}\sigma_{\alpha\alpha+2}\ell_{z\alpha+1}\ell_{z\alpha+2};$$

(assume $\alpha = 1, 2, 3$ and $3 + 1 = 1$).

d) The cubic invariant is a follows:

$$\sigma_{12}\sigma_{13}\sigma_{23}.$$

In total, there are fifteen polynomial spherical invariants (two geometric, six linear, six quadratic and one cubic).

For a plane stress state when $\sigma_{11} \neq 0, \sigma_{22} \neq 0, \sigma_{12} \neq 0$ and other σ_{ij} vanish (the axes 1,2 coincide with symmetry axes), the following spherical invariants S_i remain:

$$\ell_{1z}^2, \ell_{2z}^2, \sigma_{11}, \sigma_{22}; \sigma_{12}^2, \sigma_{12}\ell_{1z}\ell_{2z}, \sigma_{ij}\sigma_{jk}\sigma_{ik}. \tag{19}$$

The following equivalent set of S_i can also be used:

$$\ell_{1z}^2, \ell_{2z}^2, \sigma_{zz}, \sigma_{xz}^2 + \sigma_{yz}^2, \sigma_{11}, \sigma_{22}, \sigma_{12}^2, \sigma_{ij}\sigma_{jk}\sigma_{ik}. \tag{20}$$

For an isotropic material (with the symmetry tensor δ_{ij}) the base comprises the five polynomially independent spherical invariants

$$\sigma_{ii}, \sigma_{ij}\sigma_{ij}, \sigma_{ij}\sigma_{ik}\sigma_{jk}, \sigma_{zz}, \sigma_{zi'}\sigma_{zi'},$$

i.e., three scalar invariants, and the normal and shear stress in the local coordinate system.

4. Application of spherical invariants to strength and failure theory.

The main principles of the theory of dispersed failure using functions on the sphere are expounded in [11]. It is assumed that any damage of the vicinity of any point 0 on of the medium can be entirely characterized by a centrally symmetric scalar function Π_z on a unit sphere with the center at 0. Each value of this function $\Pi_z(\theta, \phi)$ can be interpreted as damage of the vicinity of the point 0 in the θ, ϕ direction. Local failure occurs if some invariant characteristic of the function Π_z reaches the critical value. A kinetic equation, connecting the function Π_z with the stress tensor should be established. As a first approach it can be assumed, that the damage rate is a function of the instantaneous stress state only:

$$\dot{\Pi}_z = f[S_i(\sigma_{i'j'})]. \tag{22}$$

As the failure criterion, the natural condition

$$\max \Pi_z = 1 \tag{23}$$

can be used.

Expressions (22) and (23) offer a possibility of estimating failure time for isotropic as well as for anisotropic material under complex loading. Of course the formula (22) should be chosen in more concrete form and the indefinite coefficients at the spherical invariants in (22) should be determined from the basic tension, compression and shear tests.

Expressions (22), (23) can be also used to describe the strength surface when the loads increase proportionally (simple loadings). The equation of the strength surface according to (22) and (23) has the following form:

$$\max f[S_i(\sigma_{i'j'})] = 1 . \tag{24}$$

Expressions (19) or (20) should be used together with formula (24) in the case of orthotropic materials and the plane stress state. In particular, when in (24) the usual scalar invariants are used as the S_i, the function f does not depend on the direction on sphere and condition (2) is obtained. Using in the strength criterion (24) all the spherical invariants, we obtain strength surfaces generalizing the above mentioned strength criterion. Then, if linear and quadratic terms are used, a strength surface in the main stress plane has a form of intersected elliptic arcs corresponding to different failure directions, i.e., directions in which the damage Π_z first reaches the value 1.

REFERENCES

[1] GREEN A. and J. ADKINS. — *Large elastic deformations and nonlinear continuum mechanics.* Oxford: Clarendon Press, 1960.

[2] LAGZDIN A. — "Integral representations of second rank tensors and field theory of plasticity." *Mekhanika Polimerov*, 3 (1979): 424-433 (in Russian).

[3] LAGZDIN A. and V. TAMUZS. — "High-elasticity tensors". *Mekhanika Polimerov*, 6 (1965): 40-48 (in Russian).

[4] LAGZDIN A. and V. TAMUZS. — "On the construction of phenomonological theory of fracture in anisotropic solids". *Mekhanika Polimerov*, 4 (1971): 634-644 (in Russian).

[5] LOKHIN V. and L. SEDOV. — "Nonlinear tensorial functions of several tensorial arguments". *Prikladnaya matematika i mekhanika*, 27, 3 (1963): 393-417 (in Russian).

[6] MAKSIMOV R., E. PLUME and E. SOKOLOV. — "Investigation of temperature dependence of fabric composite strength under plain stress state". *Mekhanika Polimerov*, 3 (1978): 452-457 (in Russian).

[7] MALMEISTERS A. — "Fundamentals of the theory of local character of deformations". (review 1) *Mekhanika Polimerov*, 4 (1965): 12-27 (in Russian).

[8] MALMEISTERS A. — "Geometry of theories of strength". *Mekhanika Polimerov*, 4 (1966): 519-534 (in Russian).

[9] MALMEISTERS A., V. TAMUZS and G. TETERS. — *Mechanik der Polymerwerkstoffe.* Berlin: Academie Verlag, 1977.

[10] RIKARDS R. — "High order elasticity tensors and strength surfaces for the two-dimensional case". *Mekhanika Polimerov*, 2 (1974): 372-374 (in Russian).

[11] V. KUKSENKO and TAMUZS V. — *Fracture micromechanics of polymer materials.* Martinus Nijhoff Publishers. The Hague, Boston, London 1981.

RESUME

(Une description de la déformation et de la rupture des solides anisotropes)

Une description de la déformation des solides anisotropes élastiques non-linéaires et la surface des résistances des solides anisotropes sont données en utilisant une série polynomiale tensorielle. Le nombre des composantes des tenseurs d'élasticité d'ordres supérieurs est analysé. Des corrélations entre la méthode proposée et les données expérimentales pour des matériaux composites sont présentées. La théorie statistique de la plasticité des solides anisotropes a été développée. Pour établir les équations constitutives, le concept nouveau des "invariants sphériques des contraintes" pour les solides anisotropes a été introduit. Les invariants sphériques des contraintes sont utilisés pour la description de la rupture dispersée des solides anisotropes.

Les Invariants des Tenseurs d'Ordre 4 du Type de l'Élasticité

G. Verchery

Ecole Nationale Supérieure de Techniques Avancées, Palaiseau, France.

1. Introduction

On étudie dans ce travail les invariants des tenseurs $T_{ijk\ell}$ d'ordre 4 ayant les symétries des tenseurs de comportement de l'élasticité linéaire :

$$T_{ijk\ell} = T_{jik\ell} = T_{ij\ell k} = T_{k\ell ij} \tag{1}$$

et possédant donc, pour le cas tridimensionnel, 81 composantes, dont 21 distinctes, ou, pour le cas bidimensionnel, 16 composantes, dont 6 distinctes.

Le problème de la formation de ces invariants, qui semble n'avoir reçu d'attention que récemment, ne constitue pas un problème complètement original. Tout d'abord, d'un point de vue fondamental, la théorie des groupes expose des méthodes systématiques permettant d'établir les invariants des représentations linéaires d'un groupe (ici le groupe orthogonal); toutefois ces méthodes sont lourdes de mise en œuvre et les applications publiées en ont été faites à d'autres quantités que les tenseurs du type de l'élasticité. Par ailleurs, par des méthodes algébriques empiriques, divers auteurs s'intéressant aux matériaux composites (Tsai et Pagano [3], Wu [5], Hahn [2]) ont donné des invariants des tenseurs du type de l'élasticité dans le cas bidimensionnel.

Dans le présent travail, pour le cas bidimensionnel, on complète les résultats cités ci-dessus en utilisant une méthode de variables complexes, qui simplifie considérablement le problème. On forme alors sans difficulté six invariants polynomiaux irréductibles liés par une relation (base d'intégrité minimale avec une relation de syzygie). On peut également former facilement les conditions imposées par la nature définie positive de l'énergie de déformation et par les symétries matérielles.

On présente ensuite une méthode qui tente une généralisation de l'équation caractéristique classique des tenseurs d'ordre 2. Il existe en effet deux tenseurs du type de l'élasticité qui sont isotropes, soit $\mathbf{I}$ et $\mathbf{E}$. On cherche alors à déterminer un tenseur symétrique du second ordre $\mathbf{P}$ et deux scalaires m et k, tels que :

$$T_{ijk\ell} P^{k\ell} = (2m\, I_{ijk\ell} + k\, E_{ijk\ell})\, P^{k\ell} \tag{2}$$

Ceci conduit à annuler le déterminant Δ de ce système linéaire homogène par rapport aux composantes indépendantes de **P**. Les coefficients de l'équation $\Delta(m, k) = 0$ sont des invariants du tenseur **T**. Dans le cas bidimensionnel, cette méthode fournit bien cinq invariants polynomiaux indépendants ; de plus, l'équation $\Delta = 0$ est une cubique du plan (m, k), sur laquelle apparaissent les symétries matérielles. Dans le cas tridimensionnel, cette méthode ne résout pas entièrement la question : elle ne fournit que 11 invariants polynomiaux ; une extension semble toutefois fournir des résultats supplémentaires.

Enfin, on donne quelques applications des méthodes et résultats exposés.

2. La méthode de variables complexes

La méthode classique de la variable complexe en élasticité introduite par Kolosov et Muskhelishvili, a été interprétée par Green et Zerna [1] comme un changement de repère complexe. On propose ici un autre changement de repère complexe, qui s'y rattache simplement mais qui est unitaire :

$$\left. \begin{array}{l} X^1 = \dfrac{\overline{kx + ky}}{\sqrt{2}} = \dfrac{\overline{kz}}{\sqrt{2}} \\[3mm] X^2 = \dfrac{kx + \overline{ky}}{\sqrt{2}} = \dfrac{kz}{\sqrt{2}} \end{array} \right\} \tag{3}$$

avec $z = x + iy$ et $k = e^{i\pi/4}$

Sous forme matricielle, ceci s'écrit :

$$\overline{X} = m_1 \, \overline{x} \tag{4}$$

en désignant par $\overline{X}$ le vecteur des composantes contravariantes. On a alors, avec des notations évidentes :

$$\left\{ \begin{array}{l} \underline{X} = \underline{\underline{g}}\,\overline{X} \quad ; \quad \overline{X} = \overline{\underline{\underline{g}}}\,\underline{X} \\[2mm] \underline{X} = M_1 \underline{x} \quad ; \quad \underline{x} = \overline{x} = M_1\,\overline{X} = m_1 \underline{X} \end{array} \right. \tag{5}$$

où $\overline{\underline{\underline{g}}} = \underline{\underline{g}} = \begin{pmatrix} 0 & 1 \\ 1 & 0 \end{pmatrix}$ est la matrice du tenseur métrique en coordonnées complexes, et où M_1 est la matrice complexe conjuguée et inverse de $m_1 : M_1 = m_1^{-1} = m_1^*$.

Une transformation linéaire T d'un vecteur, de matrice t_1 en coordonnées réelles :

$$\overline{x}' = t_1 \, \overline{x} \tag{6}$$

se traduit en coordonnées complexes par une matrice $\mathbf{T_1} = \mathbf{m_1}\,\mathbf{t_1}\,\mathbf{M_1}$:

$$\overline{\mathbf{X}}' = \mathbf{T_1}\,\overline{\mathbf{X}} \tag{7}$$

Plus généralement pour un tenseur A d'ordre n de composantes complètement contravariantes rangées en vecteurs $\overline{a}$ et $\overline{A}$, et de composantes complètement covariantes $\underline{a} = \overline{\underline{a}}$ et $\underline{A}$, on a :

$$\begin{cases} \overline{A} = m_n\,\overline{a} & ; \quad \overline{a} = M_n\,\overline{A} \\ \underline{A} = M_n\,\underline{a} & ; \quad \underline{a} = m_n\,\underline{A} \end{cases} \tag{8}$$

avec $M_n = m_n^{-1} = m_n^*$, ces matrices M_n et m_n se déduisant de M_1 et m_1 de façon classique.

C'est ainsi que la relation $\overline{a} = M_2\,\overline{A}$ s'explicite en :

$$\begin{bmatrix} A_{xx} \\ A_{xy} \\ A_{yx} \\ A_{yy} \end{bmatrix} = \frac{1}{2} \begin{bmatrix} i & 1 & 1 & -i \\ 1 & i & -i & 1 \\ 1 & -i & i & 1 \\ -i & 1 & 1 & i \end{bmatrix} \begin{bmatrix} A^{11} \\ A^{12} \\ A^{21} \\ A^{22} \end{bmatrix} \tag{9}$$

et que la restriction $\overline{t} = \widetilde{M}_4\,\overline{T}$ aux tenseurs du type de l'élasticité, de la relation $\overline{a} = M_4\,\overline{A}$ s'écrit :

$$\begin{bmatrix} T_{xxxx} \\ T_{xxxy} \\ T_{xxyy} \\ T_{xyxy} \\ T_{xyyy} \\ T_{yyyy} \end{bmatrix} = \frac{1}{4} \begin{bmatrix} -1 & 4i & 2 & 4 & -4i & -1 \\ i & 2 & 0 & 0 & 2 & -i \\ 1 & 0 & -2 & 4 & 0 & 1 \\ 1 & 0 & 2 & 0 & 0 & 1 \\ -i & 2 & 0 & 0 & 2 & i \\ -1 & -4i & 2 & 4 & 4i & -1 \end{bmatrix} \begin{bmatrix} T^{1111} \\ T^{1112} \\ T^{1122} \\ T^{1212} \\ T^{1222} \\ T^{2222} \end{bmatrix} \tag{10}$$

tandis que l'on a $\overline{T} = \widetilde{m}_4\,\overline{t}$, avec $\widetilde{m}_4 = \widetilde{M}_4^*$.

La transformation T opérant sur A donne A' et on a :

$$\overline{a}' = t_n\,\overline{a} \quad ; \quad \overline{A}' = T_n\,\overline{A} \tag{11}$$

avec $T_n = m_n\,t_n\,M_n$.

Le résultat essentiel est que les matrices $\mathbf{T}_n$ ont une forme particulièrement simple pour les rotations $R(\theta)$ d'un angle θ autour de l'origine et les
symétries $S(\alpha)$ par rapport à la droite $y = x \, tg \, \alpha$. On ne donnera ici en détail
que le cas de rotation pour lequel les matrices $\mathbf{R}_n(\theta)$ sont diagonales. Tout
calcul fait, on a donc pour les premiers ordres de tenseurs, avec $r = e^{i\theta}$:

$$\begin{cases} A^{1'} = r\,A^1 \\ A^{2'} = \bar{r}\,A^2 \end{cases} \tag{12}$$

$$\begin{cases} A^{11'} = r^2\,A^{11} \\ A^{12'} = A^{12} \\ A^{21'} = A^{21} \\ A^{22'} = \bar{r}^2\,A^{22} \end{cases} \tag{13}$$

$$\begin{cases} A^{111'} = r^3\,A^{111} & A^{211'} = r\,A^{211} \\ A^{112'} = r\,A^{112} & A^{212'} = \bar{r}\,A^{212} \\ A^{121'} = r\,A^{121} & A^{221'} = \bar{r}\,A^{221} \\ A^{122'} = \bar{r}\,A^{122} & A^{222'} = \bar{r}^3\,A^{222} \end{cases} \tag{14}$$

On forme donc très facilement la loi de transformation en coordonnées
complexes. Pour les tenseurs du type de l'élasticité, on a, compte tenu des
symétries :

$$\begin{cases} T^{1111'} = r^4\,T^{1111} \\ T^{1112'} = r^2\,T^{1112} \\ T^{1122'} = T^{1122} \\ T^{1212'} = T^{1212} \\ T^{1222'} = \bar{r}^2\,T^{1222} \\ T^{2222'} = \bar{r}^4\,T^{2222} \end{cases} \tag{15}$$

De ces formules, on peut tirer immédiatement tous les invariants polynomiaux. Dans le cas des tenseurs du type de l'élasticité, on obtient six invariants polynomiaux irréductibles réels liés par une syzygie :

$$
\begin{cases}
L_1 = T^{1122} \\
L_2 = T^{1212} \\
Q_1 = T^{1111}\,T^{2222} \\
Q_2 = T^{1112}\,T^{1222} \\
C_1 + iC_2 = T^{1111}\,(T^{1222})^2
\end{cases}
\quad
\begin{array}{l}
\text{invariants linéaires} \\[1em]
\text{invariants quadratiques} \\[1em]
\text{invariants cubiques}
\end{array}
\qquad (16)
$$

avec la relation :

$$
C_1^2 + C_2^2 = Q_1 Q_2^2 \qquad (17)
$$

On peut expliciter ces quantités à l'aide des composantes réelles du tenseur **T** :

$$
L_1 = \frac{1}{4}\,(T_{xxxx} + T_{yyyy} - 2T_{xxyy} + 4T_{xyxy}) \qquad (18)
$$

$$
L_2 = \frac{1}{4}\,(T_{xxxx} + T_{yyyy} + 2T_{xxyy}) \qquad (19)
$$

$$
Q_1 = \frac{1}{16}\,(T_{xxxx} + T_{yyyy} - 2T_{xxyy} - 4T_{xyxy})^2 + (T_{xxxy} - T_{xyyy})^2 \qquad (20)
$$

$$
Q_2 = \frac{1}{16}\,(T_{xxxx} - T_{yyyy})^2 + \frac{1}{4}\,(T_{xxxy} + T_{xyyy})^2 \qquad (21)
$$

$$
\begin{aligned}
C_1 = {} & \frac{1}{64}\,(T_{xxxx} + T_{yyyy} - 2T_{xxyy} - 4T_{xyxy})\,\cdot \\
& \cdot\,[(T_{xxxx} - T_{yyyy})^2 - 4(T_{xxxy} + T_{xyyy})^2) \\
& + \frac{1}{4}\,(T_{xxxy}^2 - T_{xyyy}^2)\,(T_{xxxx} - T_{yyyy})
\end{aligned}
\qquad (22)
$$

$$
\begin{aligned}
C_2 = {} & \frac{1}{16}\,(T_{xxxy} - T_{xyyy})\,[(T_{xxxx} - T_{yyyy})^2 - 4(T_{xxxy} + T_{xyyy})^2] \\
& - \frac{1}{16}\,(T_{xxxx} + T_{yyyy} - 2T_{xxyy} - 4T_{xyxy})\,\cdot \\
& \cdot\,(T_{xxxx} - T_{yyyy})\,(T_{xxxy} + T_{xyyy})
\end{aligned}
\qquad (23)
$$

Rappelons que ces quantités sont des invariants dans le groupe des rotations autour de l'origine. Une symétrie par rapport à l'axe $y = x \, tg \, \alpha$ se traduit par les relations :

$$\begin{cases} T^{1111''} = s^8 \, T^{2222} \\[1ex] T^{1112''} = s^4 \, T^{1222} \\[1ex] T^{1122''} = T^{1122} \\[1ex] T^{1212''} = T^{1212} \\[1ex] T^{1222''} = \bar{s}^4 \, T^{1222} \\[1ex] T^{2222''} = \bar{s}^8 \, T^{1111} \end{cases} \tag{24}$$

avec $s = e^{i\alpha}$, d'où on déduit immédiatement que les symétries-miroirs laissent invariants L_1, L_2, Q_1, Q_2 et C_1, mais transforment C_2 en $-C_2$.

Les invariants linéaires et quadratiques, à des facteurs multiplicatifs près, ont été donnés dans [3] et [5]. La référence [2] donne en outre un invariant cubique s'exprimant à l'aide de L_1, L_2, Q_1, Q_2 et C_1.

Les symétries éventuelles du matériau introduisent des conditions supplémentaires sur ces 6 invariants irréductibles. Pour trouver l'effet d'un miroir faisant un angle α avec l'axe Ox, il est commode d'effectuer un changement de repère en prenant la direction du miroir suivant la bissectrice des axes ; le miroir échange alors les nouvelles coordonnées réelles ou complexes $X^I = k \bar{s} X^1$, $X^{II} = \bar{k} s X^2$; les composantes dans ce nouveau repère du tenseur T doivent être insensibles à cet échange. On en tire que $-e^{-4i\alpha} \, T^{1111}$ et $i e^{-2i\alpha} \, T^{1112}$ sont des quantités réelles, et qu'en conséquence C_2 est nul.

D'où les résultats :

– *pour le cas orthotrope* (2 miroirs à $\pi/2$)
il existe 4 invariants indépendants non nuls L_1, L_2, Q_2, C_1, avec $Q_1 = (C_1/Q_2)^2$ et $C_2 = 0$.
les angles α et $\alpha + \pi/2$ des miroirs sont déterminés par :

$$tg \, 2\alpha = -\frac{Re(T^{1112})}{Im(T^{1112})}$$

– *pour le cas à symétrie carrée* (4 miroirs à $\pi/4$)
il existe 3 invariants indépendants non nuls L_1, L_2, Q_1, avec $C_1 = C_2 = Q_2 = 0$ (et $T^{1112} = T^{1222} = 0$).
les angles $\alpha, \alpha + \pi/4, \alpha + \pi/2$ et $\alpha + 3\pi/4$ des miroirs sont déterminés par :

$$tg \, 4\alpha = -\frac{Im(T^{1111})}{Re(T^{1111})}$$

— *pour le cas isotrope*

il existe deux invariants indépendants non nuls L_1 et L_2, avec $C_1 = C_2 = Q_1 = Q_2 = 0$ (et $T^{1111} = T^{1112} = T^{1222} = T^{2222} = 0$).

Enfin la nature définie positive de l'énergie de déformation impose des inégalités aux invariants. On obtient les conditions nécessaires et suffisantes :

$$\begin{cases} L_1 > \sqrt{Q_1} \\ L_2 > 0 \\ L_1^2 L_2 + 2C_1 - 2L_2 Q_1 - 2L_1 Q_2 > 0 \end{cases} \tag{25}$$

Les conditions suivantes, redondantes avec les précédentes sont aussi vérifiées :

$$\begin{cases} L_1 L_2 > Q_2 \\ L_1^4 L_2^2 > C_1^2 + C_2^2 \end{cases} \tag{26}$$

Les diverses propriétés données ci-dessus mettent en évidence la simplicité et l'efficacité de la méthode des variables complexes dans le cas bidimensionnel.

3. Méthode des équations caractéristiques généralisées

On sait que, pour les tenseurs du 2e ordre symétriques, l'équation caractéristique fournit les invariants par ses coefficients, cette méthode ayant l'avantage d'être valable aussi bien dans le cas bidimensionnel que dans le cas tridimensionnel. La méthode du paragraphe précédent étant limitée au cas bidimensionnel, il serait intéressant de disposer d'un équivalent de cette équation caractéristique, pour les tenseurs du type de l'élasticité.

Pour cette généralisation, on utilise le fait qu'il existe deux tenseurs du type de l'élasticité isotropes et indépendants I et E :

$$I_{ijk\ell} = -\frac{1}{n} g_{ij} g_{k\ell} + \frac{1}{2}(g_{ik} g_{j\ell} + g_{i\ell} g_{jk}) \tag{27}$$

$$E_{ijk\ell} = g_{ij} g_{k\ell} \tag{28}$$

où $n = 2$ ou 3 est la dimension de l'espace considérée et g le tenseur métrique.

Le tenseur unité en est une combinaison :

$$U = I + \frac{1}{n} E \tag{29}$$

Introduisons en outre, pour les tenseurs d'ordre 4 $A, T, \ldots$, opérant sur les tenseurs symétriques d'ordre 2 X, T, $\ldots$, les notations suivantes :

$$A : X) = A^{ijk\ell} X_{k\ell}$$
$$(X : A : Y) = A^{ijk\ell} X_{ij} Y_{k\ell} \tag{30}$$

avec comme cas particulier :

$$(X : Y) = (X : U : Y) = X_{ij} Y^{ij}$$

et, pour un tenseur T symétrique :

$$(X : T : Y) = (Y : T : X)$$

Une propriété essentielle des tenseurs E et I est d'opérer séparément, respectivement sur les parties sphériques X_S, Y_S et déviatoires X_D, Y_D des tenseurs symétriques du 2e ordre. Plus précisément on a :

$$\left. \begin{array}{l} (X : E : Y) = (X_S : E : Y_S) \\[2mm] (X : I : Y) = (X_D : E : Y_D) \end{array} \right\} \tag{31}$$

Ceci permet, on le voit, de fixer indépendamment les valeurs de $(X : E : X)$ et de $(X : I : X)$ sans pour autant fixer complètement X.

Si alors $Y = T : X)$ est le transformé de X par le tenseur T, on peut poser pour X divers problèmes d'extremum, à $(X : E : X)$ et $(X : I : X)$ fixés :

a) déterminer X pour que $(X : Y)$ soit extremum.
b) déterminer X pour que $(Y : Y)$ soit extremum.

En représentant les tenseurs par leurs matrices dans un repérage donné, ceci conduit aux équations suivantes :

$$\det (T - kE - 2mI) = 0 \quad \text{pour a)} \tag{32}$$

$$\det (T^2 - KE - 2MI) = 0 \quad \text{pour b)} \tag{33}$$

où k, K, $2m$ et $2M$ sont les multiplicateurs de Lagrange associés aux conditions imposées sur X.

Ces équations constituent la généralisation proposée des équations caractéristiques. A priori, elles ne se réduisent pas l'une à l'autre (alors que dans le cas classique des tenseurs du 2e ordre symétriques, les deux problèmes analogues constituent en fait un problème unique). On n'étudie cependant dans ce travail que la première des expressions, (32), correspondant au problème suivant : trouver un tenseur du 2e ordre symétrique X_{ij} et deux scalaires k et m tels que :

$$T_{ijk\ell} X^{k\ell} = (2m\, I_{ijk\ell} + k\, E_{ijk\ell})\, X^{k\ell} \tag{34}$$

Dans le cas bidimensionnel, le déterminant $\Delta(m, k)$ du système (34) (donné par (32)) dont l'annulation fournit l'équation caractéristique, s'écrit, tout calcul fait, à l'aide de 5 invariants indépendants de **T** :

$$\Delta(m, k) = -4m^2 k + 4m^2 L_2 + 4km\, L_1 - 4m(L_1 L_2 - Q_2)$$
$$- k(L_1^2 - Q_1) + L_1^2 L_2 - 2L_1 Q_2 - L_2 Q_1 + 2C_1 \tag{35}$$

On remarquera que les combinaisons constituant les coefficients de cette expression sont tous positifs d'après les inégalités découlant de la nature définie positive de l'énergie de déformation.

La forme de la cubique $\Delta(m, k) = 0$ dans le plan (m, k) est caractérisée par les symétries matérielles :

— dans le cas anisotrope général, on peut écrire l'équation sous la forme :

$$k = L_2 + \frac{Q_2 + C_1/\sqrt{Q_1}}{2m - L_1 - \sqrt{Q_1}} + \frac{Q_2 - C_1/\sqrt{Q_1}}{2m - L_1 + \sqrt{Q_1}}$$

où les deux numérateurs sont positifs, d'où l'allure de la figure 1.

— dans le cas orthotrope, la cubique se décompose en une hyperbole et une droite (Fig. 2) :

$$(2m - L_1 - \sqrt{Q_1})\, [2Q_2 + (L_2 - k)\, (2m - L_1 - C_1/Q_2)] = 0$$

— dans le cas de symétrie carrée, la cubique se décompose en trois droites distinctes (Fig. 3) :

$$(L_2 - k)\, [(L_1 - 2m)^2 - Q_1] = 0$$

— enfin dans le cas isotrope, on a une droite et une droite double (Fig. 4) :

$$(L_2 - k)\, (L_1 - 2m)^2 = 0$$

Dans le cas tridimensionnel, l'expression $\Delta(m, k)$ est de degré six en m et 2 en k. On montre en effet facilement, par la règle de dérivation des déterminants que (comme dans le cas bidimensionnel) :

$$\frac{\partial^2 \Delta}{\partial k^2} = 0 \tag{36}$$

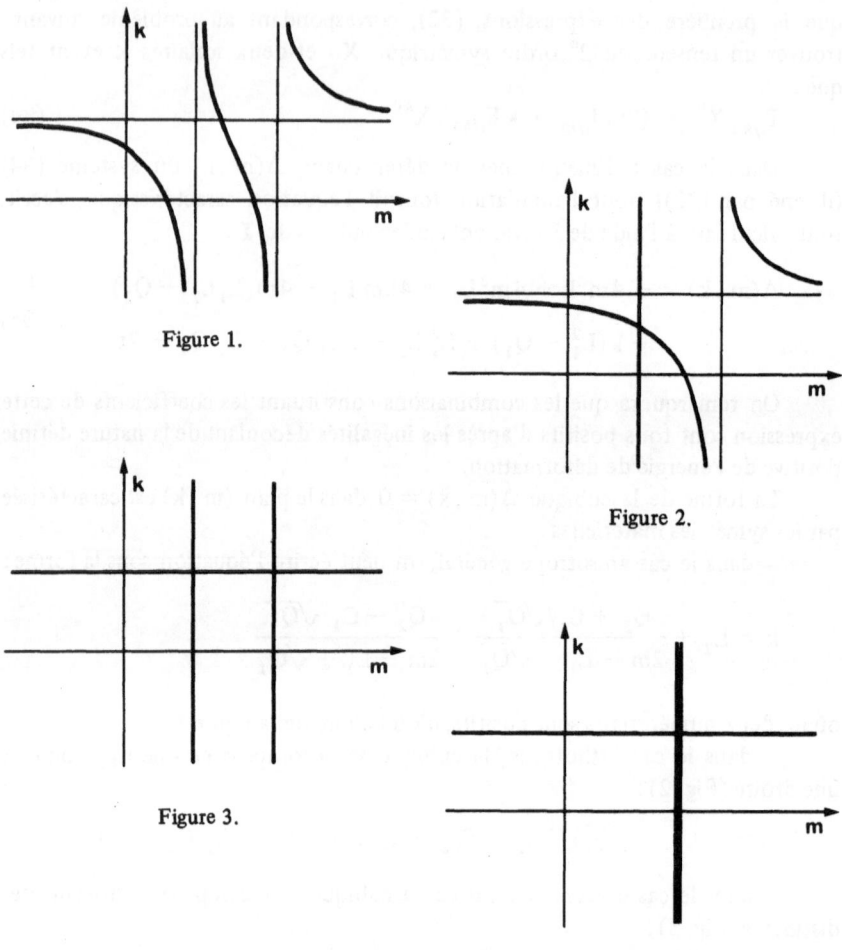

Figure 1.

Figure 2.

Figure 3.

Figure 4.

On obtient donc ainsi 11 invariants : 2 linéaires, 2 quadratiques, 2 cubiques, 2 de degré 4, 2 de degré 5 et un $(\Delta(0,0))$ de degré 6. La deuxième équation caractéristique généralisée (33), également linéaire en K et de degré 6 en M, possède aussi 11 coefficients (de degré double par rapport aux précédents) qui sont invariants. Il n'a toutefois pas été possible dans ce travail d'établir les relations qui les lient éventuellement aux 11 invariants provenant de la première équation caractéristique. Cette étude ne règle donc pas la question de l'établissement des 18 invariants possibles pour les 21 composantes indépendantes des tenseurs du type de l'élasticité.

4. Applications

Ces méthodes et ces résultats donnent lieu à diverses applications pour les matériaux anisotropes et en particulier les matériaux composites stratifiés. Leur exposition complète exigeant des développements algébriques longs mais sans difficulté (figurant dans un rapport, à paraître, de l'auteur [4]), on se borne ici à les citer :

— formation de 20 invariants indépendants (au lieu de 15 seulement donnés par [2]) pour un tenseur d'élasticité dans les rotations autour d'un axe.

— formation des invariants indépendants de la loi classique de comportement des plaques stratifiées : 17 dans le cas négligeant les effets de cisaillement transversal (loi (37)) ; 20 dans le cas avec cisaillement transversal (lois (37) et (38)) :

$$\begin{bmatrix} N \\ \hline M \end{bmatrix} = \begin{bmatrix} A & B \\ \hline B & D \end{bmatrix} \begin{bmatrix} \epsilon^0 \\ \hline K \end{bmatrix} \tag{37}$$

$$(Q) = [d]\,(\Gamma) \tag{38}$$

— établissement des lois de comportement à partir de résultats expérimentaux, en étendant la méthode due à Wu *et al.* [6].

— reconnaissance des symétries, exactes ou "approchées", pour ces divers comportements : classes de symétrie et position des axes privilégiés : les résultats expérimentaux donnent lieu en général à des réalisations approximatives des symétries matérielles.

— extension de la méthode de variables complexes à d'autres types de tenseurs.

5. Conclusion

Ce travail a étudié les invariants des tenseurs d'ordre 4 du type de l'élasticité par deux méthodes originales : la méthode des variables complexes et la méthode des équations caractéristiques généralisées. Bien que les résultats qu'elles fournissent ne soient ni entièrement nouveaux, ni complets, elles se révèlent bien adaptées au problème.

REFERENCES

[1] GREEN A.E. and W. ZERNA. — *Theoretical Elasticity*. 2nd édition, Oxford : Clarendon Press, 1968.
[2] HAHN H.T. — "A derivation of invariants of fourth rank tensors", *J. Composite Materials*, 8, 1, Jan. 1974, pp. 1-14.

[3] TSAI S.W. and N.J. PAGANO. – "Invariant properties of composite materials". In *Composite Materials Workshop*, S.W. Tsai. J.C. Halpin and N.J. Pagano; Eds., Technomic, Stamford (Conn.), 1968, pp. 233-253.
[4] VERCHERY G. – Rapport ENSTA, Paris, à paraître.
[5] WU E.M. – "Fourth order tensor invariants and geometric representation". *Office of Naval Research/Advanced Research Projects Agency, Report HPC* 1970, 70-123.
[6] WU E.M., K.L. JERINA and R.E. LAVENGOOD. – "Data averaging of anisotropic composite material constants" *Analysis of Test Methods for High Modulus Fibers and Composites*, STP 521, ASTM, 1973, pp. 229-252.

ABSTRACT

In this paper, the invariants of fourth rank tensors with symmetries of the linear elasticity tensors are studied by two new methods. In the two dimensional case, a complex variables method gives very simple transformation formulas for the rotation of tensor components. From this, the six irreducible polynomial invariants related by one relation are quite easily established. Conditions due to material symmetries and the positive definiteness of the strain energy are also easily deduced. The second method use a generalization of the concept of characteristic equation. In the two-dimensional case, it gives a cubic equation $\Delta(m, k) = 0$, the coefficients of which are 5 independant invariants. Its geometric representation in the m, k plane provides the material symmetries. In the three dimensionnal case, it has not been possible to determine all the properties of the 11 or 22 invariants given by the method. Finally some applications of the results are indicated.

Session 2

Physical Properties of Anisotropic Materials
Propriétés Physiques des Matériaux Anisotropes

Communications

G. AUBERT et A. CHAMBEROD
L'Anisotropie Magnétique : un Moyen de Caractérisation des Matériaux.

R.S. RIVLIN
Birefringence in Finitely-Deformed Materials.

I. MÜLLER and K. WILMANSKI
State Functions for a Pseudoelastic Body.

Y. POGGI, J.C. FILIPPINI and B. MALRAISON
Anisotropies and Elasticity in Liquid Crystals.

A. WAINTAL
Aspects Théoriques et Expérimentaux des Equations d'Etat d'Origine Elastique de Solides Anisotropes.

Y.F. DAFALIAS and K. ARULANANDAN
The Formation Factor Tensor in Relation to Structural Characteristics of Anisotropic Granular Soils

L'Anisotropie Magnétique :
un Moyen de Caractérisation des Matériaux

G. Aubert
Laboratoire Louis Néel, C.N.R.S., Grenoble, France.

A. Chamberod
D.R.F., Physique du Solide, C.E.N.-Grenoble, France.

1. Introduction

L'étude des propriétés magnétiques des matériaux occupe une place importante dans la physique de la matière condensée. L'interprétation de ces propriétés nécessite de faire appel aux modèles et méthodes les plus élaborés de la physique fondamentale. Il s'agit en effet d'un problème à grand nombre de particules en interactions (problème à N corps), ces interactions ayant une origine purement quantique (le théorème de Bohr-Miss van Leewen, 1919, démontre de manière tout à fait générale que la mécanique classique conduit à un moment magnétique moyen nul pour toute assemblée de particules chargées). D'autre part, le champ des applications est extrêmement vaste et il est difficile d'imaginer une civilisation industrielle sans matériaux magnétiques. Même si la plus ancienne des applications, la boussole, a été nettement supplantée par des système gyroscopiques, les gyroscopes les plus sophistiqués en cours de développement font tout de même appel à la suspension magnétique d'une sphère supraconductrice !

Enfin la très grande sensibilité de certaines mesures magnétiques permet d'utiliser le magnétisme comme un moyen de caractérisation des matériaux dont les propriétés magnétiques ont pu être modifiées même très faiblement par des influences diverses. Par exemple, les renversements du dipôle magnétique terrestre ont laissé leurs traces dans les matériaux issus des dorsales médio-océaniques et les mesures magnétiques effectuées sur ceux-ci ont apporté une preuve indiscutable de l'expansion de la croûte terrestre au niveau de ces dorsales. Les contraintes mécaniques auxquelles est ou a été soumis un échantillon modifient également ses propriétés magnétiques qui peuvent donc être utilisées comme un moyen indirect d'investigation dans des problèmes au départ essentiellement mécaniques. Il est clair que dans ce domaine, ce sont les propriétés magnétiques anisotropes qui devront être envisagées.

2. Description des propriétés magnétiques anisotropes des matériaux dans l'approximation du milieu continu

2.1. L'aimantation

Un échantillon de matière condensée est un ensemble de particules possédant un moment magnétique orbital et un moment magnétique de spin. Pour une particule de charge q et de masse m, le moment cinétique orbital est proportionnel au moment cinétique $\vec{\ell}$, $\vec{\mu_\ell} = \dfrac{q}{2m} \vec{\ell}$, et donc directement lié au mouvement de la particule. $\vec{\mu_\ell}$ et $\vec{\ell}$ sont des opérateurs de la mécanique quantique mais pour ce qui nous concerne ici on pourra faire comme s'il s'agissait de vecteurs quantifiés spatialement (description dite semi-classique), la composante ℓ_z étant un multiple entier de $\hbar$, $\ell_z = m\hbar$. Pour un électron $\mu_{\ell_z} = -\dfrac{e\hbar}{2m_e} m = -m\,\mu_B$ où $\mu_B = \dfrac{e\hbar}{2m_e} = 0{,}9274.10^{-23}$ J/T est le magneton de Bohr qui donne l'ordre de grandeur des moments magnétiques microscopiques. L'ordre de grandeur des moments magnétiques nucléaires est le magnéton nucléaire $\mu_N = \dfrac{e\hbar}{2m_p} = 5{,}051.10^{-27}$ J/T beaucoup plus petit que μ_B, aussi les propriétés magnétiques des matériaux sont-elles essentiellement électroniques.

Le moment magnétique de spin est proportionnel au moment cinétique intrinsèque dit de spin $\vec{s}$ et pour un électron $\vec{\mu_s} = -g_s\,\mu_B \vec{s}$ où $g_s = 2{,}0023$ est le rapport gyromagnétique de l'électron et $s_z = \pm 1/2$.

Nous ne nous intéresserons pas ici au problème du calcul ab initio du moment magnétique résultant des électrons constituant un atome ou un ion isolé ou en interaction avec d'autres. Dans l'approximation du milieu continu, nous introduisons un champ vectoriel $\vec{M}$ appelé aimantation du matériau dont la valeur en un point P est la moyenne volumique des moments magnétiques microscopiques $\vec{\mu_i}$ de toute origine des particules contenues dans un élément de volume $d\tau$ entourant P, soit $M = \dfrac{1}{d\tau} \displaystyle\sum_{d\tau} \vec{\mu_i}$. Les dimensions de $d\tau$ tont de l'ordre de quelques centaines d'Å c'est-à-dire que la moyenne porte sur quelques milliers de particules, ce qui assure la continuité du champ $\vec{M}$ à l'exception des surfaces de séparation.

En l'absence d'interaction (paramagnétisme) et sans champ magnétique appliqué on a $\vec{M} \equiv \vec{O}$. L'application d'un champ donne alors lieu à deux effets concurrents :

– un effet purement électromagnétique de même origine que les phénomènes d'induction fait apparaître une aimantation en sens inverse du champ (diamagnétisme)

— un effet d'orientation des moments microscopiques existants parallèlement au champ qui produit une aimantation de même sens que le champ et d'autant plus faible que la température est plus élevée à cause de la compétition avec l'agitation thermique (paramagnétisme de Langevin-Brillouin). Ces deux effets sont toujours très faibles (sauf à très basse température et pour des champs élevés pour le dernier) si on les compare aux effets obtenus dans des matériaux où, par suite d'interactions qui ne peuvent être clairement comprises que dans le cadre de la mécanique quantique, les moments magnétiques microscopiques ont tendance à s'ordonner parallèlement entre eux (ferromagnétisme), antiparallèlement (antiferro et ferrimagnétisme) ou suivant des structures non colinéaires plus complexes (hélimagnétisme par exemple). Cette tendance spontanée conduit à l'existence d'une aimantation dite spontanée, $\|\vec{M}\| \neq 0$, en l'absence de champ. L'aimantation spontanée est évidemment fonction de la température avec une température critique de changement de phase (généralement du second ordre) entre la phase ordonnée et la phase désordonnée appelée par extension phase paramagnétique. La température critique d'ordre est connue sous le nombre de température de Curie pour les ferromagnétiques et de température de Néel dans les autres cas.

Pour les matériaux magnétiques ordonnés et les ferromagnétiques en particulier qui nous occuperont essentiellement par la suite, une complication intervient dans la description macroscopique des matériaux du fait de l'existence d'une structure à une plus grande échelle que l'échelle macroscopique (dimensions du $d\tau$ précédent) et que nous appelerons mégascopique. Les moments magnétiques spontanément ordonnés créent en effet un champ magnétique non négligeable dans lequel ils se trouvent eux-mêmes plongés. Cette interaction est l'interaction dipolaire magnétique bien connue en électromagnétisme et tout le problème vient du fait qu'elle ne décroît qu'en $1/r^3$ avec la distance et il est donc impossible de s'en affranchir quelle que soit la taille de l'échantillon. Le champ créé par l'échantillon lui-même est d'une manière générale en sens inverse des moments magnétiques d'où son nom de champ démagnétisant. Il apparaît que la meilleure solution énergétique pour l'échantillon n'est pas de diminuer son aimantation spontanée mais de se subdiviser en domaines (domaines de Weiss pour les ferromagnétiques) dont les dimensions sont mégascopiques et qui peuvent dans certains cas être du même ordre que les dimensions de l'échantillon. Les aimantations des domaines ont des directions différentes et le champ démagnétisant est ainsi réduit sans que soit réduite dans chaque domaine l'aimantation spontanée conséquence des interactions autres que dipolaires. Il conviendrait donc de bien distinguer dans le langage l'aimantation $\vec{M}$ qui est la propriété macroscopique, donc définie à l'intérieur même de chaque domaine, et l'aimantation magascopique $\vec{\mathfrak{M}}/V$ où $\mathfrak{M}$ est le moment magnétique total de l'échantillon et V son volume. Ces deux quantités ont la même dimension et l'on dit souvent quand on mesure $\mathcal{M}/V$ que l'on fait une mesure d'aimantation alors que l'on ne fait qu'une mesure de moment. $\mathcal{M}/V$ ne correspond à $\vec{M}$ que dans le cas où l'échantillon est monodomaine, ont dit aussi saturé.

2.2. *Les champs magnétiques*

On décrit en tout point de l'espace vide le champ électromagnétique au moyen de deux champs $\vec{E}$ et $\vec{B}$ qui interviennent dans la formule de Lorentz $\vec{F} = q(\vec{E} + \vec{v} \times \vec{B})$. En présence de matière, la situation reste la même dans le vide mais se complique à l'intérieur de la matière elle-même.

Considérons un système de courants créant en l'absence d'échantillon un champ dans le vide $\vec{B}_0$. Les moments magnétiques de l'échantillon créent eux-mêmes un champ magnétique $\vec{b}'$, somme des champs magnétiques créés par cette assemblée de dipôles dans le vide. A l'extérieur de l'échantillon, on est suffisamment loin de tous les dipôles et la variation spatiale de $\vec{b}'$, varie spatialement à l'échelle des distances interatomiques. Nous définissons dans l'approximation du milieu continu un champ $\vec{B}_i'$ identique à $\vec{b}'$ à l'extérieur de l'échantillon et égal à la moyenne spatiale de $\vec{b}'$ sur un volume de l'ordre de l'ordre de grandeur du $d\tau$ défini plus haut pour l'intérieur de l'échantillon. Nous définissons ainsi en tout point de l'espace un champ $\vec{B} = \vec{B}_0 + \vec{B}'$ qui comme $\vec{B}_0$ et $\vec{b}'$ est à divergence nulle, c'est-à-dire à flux conservatif.

On peut alors démontrer sur la base des équations de l'électromagnétisme que le champ $\vec{H} = \dfrac{\vec{B}}{\mu_0} - \vec{M}$ obéit au théorème d'Ampère $\overrightarrow{\mathrm{rot}}\,\vec{H} = \vec{j}$ où $\vec{j}$ est la densité des courants de conduction (nous utilisons ici le système d'unités MKSA rationalisé où $\mu_0 = 4\pi.10^{-7}$, $\vec{B}$ se mesurant en Tesla, $\vec{H}$ et $\vec{M}$ en A/m et les moments en A/m^2 ou J/T). Dans le vide $\vec{H} = \vec{B}/\mu_0$ et n'est donc pas distinct de $\vec{B}$ sinon par les unités utilisées.

Le champ démagnétisant dont nous avons parlé plus haut est le champ $\vec{H}'$ tel que $\vec{H} = \vec{H} + \vec{H}'$ ou $\vec{H}_0 = \dfrac{\vec{B}_0}{\mu_0}$. Par exemple dans le cas d'un échantillon sphérique monodomaine d'aimantation uniforme $\vec{M}$, on démontre que $\vec{H}' = -\dfrac{1}{3}\vec{M}$ à l'intérieur de l'échantillon. D'une manière plus générale, pour un échantillon limité par une quadrique et uniformément aimanté $\vec{H}' = -\overrightarrow{\overrightarrow{D}}.\vec{M}$ à l'intérieur de l'échantillon où $\overrightarrow{\overrightarrow{D}}$ est le tenseur de champ démagnétisant, tenseur symétrique de rang 2 et de trace unité. Seules les quadriques conduisent à un champ démagnétisant uniforme à l'intérieur de l'échantillon pour une aimantation pour une aimantation uniforme, aussi ces formes d'échantillon sont-elles à préférer pour les mesures magnétiques. Pratiquement, la sphère est la forme la plus commode à réaliser. Des disques très aplatis ou des cylindres très allongés (fils) constituent des approximations acceptables dans certaines conditions.

2.3. Les énergies magnétiques [1] [8]

Les équations de Maxwell conduisent pour l'expression du travail fourni par le milieu extérieur (opérateur et sources de champ) dans une transformation quasi statique d'un échantillon à :

$$\delta W = \iiint \vec{B}_0 . \, d\vec{M} \, d\tau$$

où $\vec{B}_0$ est le champ créé par les sources de champ dans le vide en l'absence d'échantillon. Une transformation quasi statique d'un échantillon magnétique ordonné n'est pas obligatoirement réversible à cause des phénomènes d'hystérésis qui sont liés à la structure en domaines. Ainsi les fonctions d'état thermodynamique ne sont définies de façon unique, c'est-à-dire caractéristiques du matériau, que dans le domaine des champs appliqués suffisamment élevés pour que l'échantillon soit saturé.

Un échantillon limité par une surface quadrique placée dans un champ magnétique extérieur $\vec{B}_0$, uniforme sur son volume et suffisamment grand, est mono-domaine avec une aimantation uniforme $\vec{M}$. On peut alors définir une énergie interne par unité de volume U telle que :

$$dU = TdS + \vec{B}_0 . \, d\vec{M}$$

et une énergie libre :

$$F = U - TS$$

telle que :

$$dF = - SdT + \vec{B}_0 . \, d\vec{M}$$

avec :

$$(dF)_T = (\delta W_{rev})_T = \vec{B}_0 . \, d\vec{M}$$

travail réversible isotherme par unité de volume.

Dans un souci de simplification, nous supposons ici le réseau invariable et n'introduisons pas les variables volume et pression hydrostatique ni les tenseurs de contrainte et de déformation. L'énergie libre F contient l'énergie d'interaction dipolaire dont l'électromagnétisme fournit l'expression $-\dfrac{1}{2} \mu_0 \vec{M}.\vec{H}'$ par unité de volume. On posera donc :

$$F = E_a - \frac{1}{2} \mu_0 \vec{M}.\vec{H}'$$

où E_a est une densité d'énergie caractéristique du matériau appelée, pour un monocristal, énergie d'anisotropie magnétocristalline car elle dépend évidemment en principe de la direction de $\vec{M}$ par rapport aux axes cristallographiques. L'énergie dipolaire :

$$- \frac{1}{2} \mu_0 \vec{M} . \vec{H}' = \frac{1}{2} \mu_0 \vec{M} . \vec{D} . \vec{M}$$

dépend de la forme de l'échantillon d'où le nom anisotropie de forme qui lui est souvent donné. Pour une sphère, ce terme se réduit à $\frac{1}{6} \mu_0 \vec{M}^2$.

Quel est maintenant le potentiel thermodynamique dont le minimum correspond à l'équilibre de l'échantillon placé dans un champ extérieur $\vec{B_0}$ fixé et à une température fixée T ? On voit facilement qu'il s'agit de $G = U - TS - \mu_0 \vec{M} . \vec{B_0}$ car $dG = - SdT - \mu_0 \vec{M} . d\vec{B_0}$ (on utilise ici la même notation pour le potentiel thermodynamique et la fonction d'état correspondante égale au minimum d'équilibre du potentiel).

Désignons par $\vec{u}$ le vecteur unitaire de $\vec{M}$ et par $\vec{u_\theta}$ et $\vec{u_\phi}$ les deux autres vecteurs unitaires du trièdre local associé à la direction $\vec{u}(\theta, \phi)$ de $\vec{M}$ par rapport à des axes liés à l'échantillon (Fig. 1). Introduisons la fonction d'état $E'_a = E_a - \mu_0 \vec{M} . \vec{H}$ où $\vec{H} = \vec{H_0} + \vec{H}'$ est appelé champ intérieur ou champ appliqué "corrigé" du champ démagnétisant. En posant $H_M = \vec{H} . \vec{u}$, on trouve facilement que :

$$M = - \frac{1}{\mu_0} \left(\frac{\partial E'_a}{\partial H_M} \right)_{T, \theta, \phi} \tag{1}$$

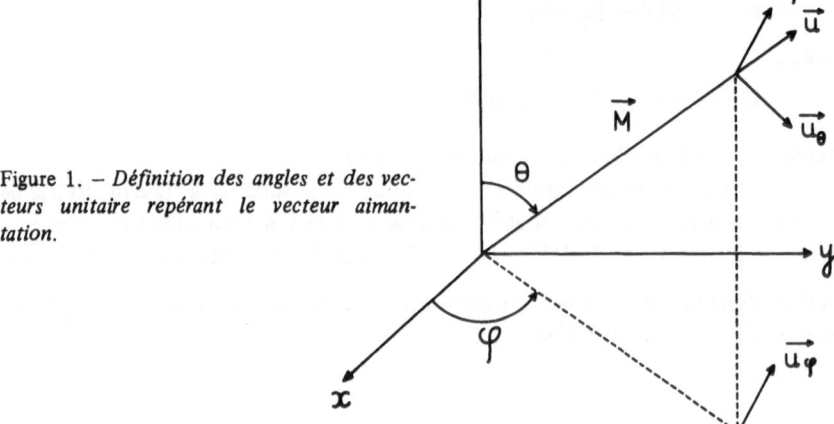

Figure 1. – *Définition des angles et des vecteurs unitaire repérant le vecteur aimantation.*

et $\quad dG = \left(\dfrac{\partial E_a'}{\partial \theta} - \mu_0 M \vec{H}.\vec{u}_\theta \right) d\theta + \left(\dfrac{\partial E_a'}{\partial \phi} - M \sin\theta\, H.\vec{u}_\phi \right) d\phi$

d'où la condition de minimum de G :

$$\mu_0 \vec{M} \times \vec{H}_0 = - \mu_0 \vec{M} \times \vec{H}' + \frac{1}{\sin\theta} \left(\frac{\partial E_a'}{\partial \phi} \right)_{T, H_M, \theta} \vec{u}_\theta + \left(\frac{\partial E_a'}{\partial \theta} \right)_{T, H_M, \phi} \vec{u}_\phi$$

$$\mu_0 \vec{M} \times \vec{H}_0 = \vec{M} \times \vec{B}_0 = \vec{\Gamma} \tag{2}$$

n'est autre que le couple par unité de volume exercé par l'échantillon sur l'appareil qui le maintient fixe dans le champ extérieur $\vec{B}_0$.

Les relations (1) et (2) montrent que des mesures d'aimantation et de couple permettent d'accéder aux dérivés de la fonction E_a' d'où l'on pourra remonter à E_a par :

$$E_a = E_a' + \mu_0 \vec{M}.\vec{H} \tag{3}$$

3.   L'anisotropie magnétocristalline et l'anistropie induite

3.1. *Phénoménologie*

Ainsi qu'elle a été introduite plus haut, la fonction $E_a' a$ est une fonction des variables T, H_M, θ, ϕ et caractéristique du matériau. On peut la développer sous la forme :

$$E_a' = \sum_n K_n'(T, H_M)\, f_n(\theta, \phi)$$

où les fonctions $f_n(\theta, \phi)$ sont entièrement déterminées par des considérations de symétrie et les K_n' des coefficients d'anisotropie. D'après (1) :

$$M = \sum_n M_n(T, H_M)\, f_n(\theta, \phi)$$

avec :

$$M_n(T, H_M) = - \frac{1}{\mu_0} \left(\frac{\partial K_n'}{\partial H_M} \right)_T$$

L'aimantation peut donc présenter elle-aussi une anisotropie de son module bien que celle-ci soit en général très faible et que la partie principale de M soit isotrope. On tire de (3) :

$$E_a = \sum_n (K_n' + \mu_0 M_n H_M)\, f_n(\theta, \phi) = \sum_n K_n\, f_n(\theta, \phi)$$

où les $K_n(T, H_M)$ sont appelées constantes d'anisotropie bien qu'elles dépendent de T et de H_M. On a par ailleurs :

$$\left(\frac{\partial K_n}{\partial H_M}\right)_T = \mu_0 H_M \left(\frac{\partial M_n}{\partial H_M}\right)_T.$$

Si comme le montre souvent l'expérience K_n et M_n ne dépendent pratiquement pas de H_M dans le domaine des champs usuels $K'_n = K_n - \mu_0 M_n H_M$ est une fonction linéaire de H_M à une température déterminée dont la mesure en fonction de celui-ci donne immédiatement K_n et M_n.

Donnons quelques exemples de développements de E_a (ou $E'_a a$). La méthode la plus générale consiste à prendre pour $f_n(\theta, \phi)$ les combinaisons linéaires d'harmoniques sphériques d'ordre n, soit :

$$f_n(\theta, \phi) = \sum_{m=-n}^{+n} C_n^m Y_n^m(\theta, \phi),$$

répondant aux conditions de symétrie imposée. Ainsi pour un cristal de symétrie cubique, on construit des "harmoniques cubiques" [2], mais l'usage a généralement consacré d'autres méthodes de développement reflétant plus directement la symétrie comme des polynômes symétriques [2] des cosinus directeurs $\alpha_1, \alpha_2, \alpha_3$ de la direction de l'aimantation par rapport aux axes quaternaires du cristal cubique. On voit facilement que seules puissances paires des α_i peuvent intervenir (l'énergie ne dépend pas du sens de l'aimantation ou est invariante par rapport au renversement du temps) et comme le polynôme d'ordre 2 est invariant :

$$(\alpha_1^2 + \alpha_2^2 + \alpha_3^2 = 1),$$

on a :

$$E_a = K_0 + K_1 (\alpha_1^2 \alpha_2^2 + \alpha_2^2 \alpha_3^2 + \alpha_3^2 \alpha_1^2) + K_2 \alpha_1^2 \alpha_2^2 \alpha_3^2 + \cdots$$

Pour un cristal hexagonal, en prenant pour axe z de la figure 1, l'axe c de la structure, on a :

$$E_a = K_0 + K_1 \sin^2\theta + K_2 \sin^4\theta + K_3 \sin^6\theta \ \cos^6\phi + \cdots$$

l'anisotropie dans le plan de base n'intervenant qu'à l'ordre 6 en $\sin\theta$.

Supposons maintenant que le cristal ait subi des influences extérieures diverses possédant des caractères de symétrie différents de ceux qui caractérisent la structure cristalline, par exemple une contrainte mécanique ou un traitement thermique pendant lequel l'aimantation a été fixée dans une direction (θ', ϕ') ou $\vec{\alpha}'$ déterminée par un champ extérieur appliqué. Il apparaît alors une anisotropie dite induite E_u dont l'expression phénoménologique reflêtera les propriétés de symétrie de l'ensemble cristal-influence extérieure. Ainsi pour un cristal cubique ayant subi le traitement thermomagnétique précédent :

$$E_u(\vec{\alpha}, \vec{\alpha}') = - F \sum_i \alpha_i^2 \alpha_1'^2 - G \sum_{i \neq j} \alpha_i \alpha_j \alpha_i' \alpha_j'$$

qui contient maintenant des termes du deuxième ordre en α_i, c'est-à-dire une anisotropie induite uniaxiale (N.B. : les coefficients F et G correspondent à la notation consacrée par l'usage et n'ont rien à voir avec les fonctions F et G utilisées plus haut).

Dans le cas d'un échantillon polycristallin dont les cristallites sont de dimensions petites par rapport aux dimensions de l'échantillon et orientées au hasard, l'énergie magnétocristalline a une moyenne nulle sur l'ensemble de l'échantillon aux fluctuations statistiques près alors que l'énergie uniaxiale induite est en moyenne :

$$E_u = - K_u \cos^2(\vec{u}, \vec{u}')$$

où $(\vec{u}, \vec{u}')$ est l'angle entre l'aimantation de l'échantillon de direction $\vec{u}$ et la direction $\vec{u}'$ qu'elle avait pendant le traitement thermomagnétique ou bien la direction $\vec{u}'$ d'une contrainte uniaxiale appliquée pendant ou avant la mesure. K_u est appelée constante d'anisotropie uniaxiale induite.

3.2. Méthodes de mesure

Il existe plusieurs méthodes de mesure de l'anisotropie magnétique faisant appel à des expériences statiques comme des mesures d'aimantation en fonction de la direction du champ appliqué ou dynamiques comme la résonance ferromagnétique par laquelle on accède aux dérivées secondes de E_a' par rapport à θ et ϕ. Nous ne décrirons ici que la méthode de loin la plus performante et permettant d'atteindre des précisions suffisantes pour déceler en particulier des anisotropies induites très faibles superposées à des anisotropies magnétocristallines importantes. Il s'agit de la méthode basée directement sur la relation (2) et qui consiste à mesurer le couple nécessaire pour maintenir un échantillon dans un champ extérieur $\vec{B}_0$ uniforme suffisant pour le rendre monodomaine. Le terme $- \mu_0 \vec{M} \times \vec{H}'$ dépend de la forme de l'échantillon (anisotropie de forme) et les conditions expérimentales sont en général choisies pour qu'il soit nul (échantillon sphérique par exemple).

On a alors :

$$\vec{\Gamma} = \vec{M} \times \vec{B}_0 = \frac{1}{\sin\theta} \left(\frac{\partial E'_a}{\partial\phi}\right)_{T,H_M,\theta} \vec{u}_\theta + \left(\frac{\partial E'_a}{\partial\theta}\right)_{T,H_M,\phi} \vec{u}_\phi$$

Sans entrer dans le détail des arrangements expérimentaux souvent fort sophistiqués et allant jusqu'à des appareillages entièrement automatiques et pilotés par un ordinateur [6], le schéma le plus courant actuellement consiste à fixer l'échantillon sur un système libre de tourner autour d'un axe z vertical ("balance de torsion") dont on asservit la position en compensant le couple exercé sur l'échantillon par le couple créé par le courant d'asservissement qui parcourt une bobine convenablement placée dans le champ d'un aimant permanent. La détection de la position est en général optique et permet d'obtenir des précisions angulaires dans l'asservissement de la position de quelques secondes d'arc [8]. Le champ extérieur $\vec{B}_0$ est créé par un électroaimant qui peut tourner autour de l'axe z. $\vec{B}_0$ prend ainsi diverses orientations dans un plan horizontal, orientations connues par rapport à l'échantillon avec une précision analogue par des dispositifs le plus souvent optiques. L'anisotropie magnétique dépendant très fortement de la température, il importe de réguler la température de l'échantillon avec une très grande précision de l'ordre de quelques mK dans le domaine 2-300 K par exemple ou quelques centièmes de Kelvin au-dessus de la température ambiante. L'échantillon est orienté par les méthodes habituelles de la radiocristallographie de telle façon que le plan xoy perpendiculaire à Oz vertical soit un plan de symétrie. $\vec{B}_0$ et M pour l'échantillon saturé sont alors dans ce plan et repérés chacun par une seule variable (Fig. 2).

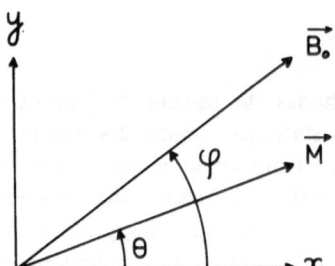

Figure 2. – *Définition des angles pour les mesures de couples.*

(N.B. : On ne confondra pas les angles θ et ϕ de cette figure avec ceux de la figure 1 et de l'expression 2).

On a alors : $\Gamma_z = MB_0 \sin(\theta - \phi)$ et l'on .mesure à chaque température Γ_z en fonction de ϕ. Pour chaque valeur de ϕ connue expérimentalement, θ n'est pas connu directement mais connaissant M mesurable par ailleurs, on a immédiatement :

$$\theta - \phi = \text{Arc sin} \; \frac{\Gamma_z}{MB_0}$$

et l'on sait donc passer de $\Gamma_z(\phi)$ à $\Gamma_z(\theta)$ lié plus directement par la relation (2) aux coefficients d'anisotropie.

N.B. : $\theta - \phi$ est en général très faible (quelques minutes d'arc) et de ce fait fut longtemps négligé dans les interprétations comme s'il s'agissait d'une erreur anglaire. A la différence d'une erreur, cet écart n'est pas aléatoire mais lié directement à la propriété mesurée et le fait de le négliger a été à l'origine de nombreuses conclusions erronées.

La grande finesse des mesures d'anisotropie résulte de la possibilité de réaliser une analyse harmonique de Γ_z fonction de ϕ ou de θ [9]. Prenons par exemple le cas d'un cristal cubique tel que xoy soit un plan (100).

Avec pour axe origine [100], $\Gamma_z = \dfrac{K_1'}{2} \sin 4\theta +$ (termes d'ordre supérieur en $8\theta, 12\theta, \ldots$). Une anisotropie uniaxiale induite dans la direction θ_0 du plan, introduira un couple supplémentaire $K_u \sin 2(\theta - \theta_0)$ facilement séparable par décomposition en série de Fourier de la courbe de couple, même si K_u est mille fois plus faible que K_1' [10]. L'étude de $K_1' = K_1 - \mu_0 M_1 H$ en fonction du champ $\|\vec{B}\|$ appliqué permet de déterminer K_1 et M_1. Dans le cas des métaux ferromagnétiques courants (Fe, Ni, Co ....) et de leurs alliages, M_1 est de l'ordre de $10^{-4} M_0$ où M_0 est la partie isotrope de M. La grande sensibilité et la grande précision atteintes par les mesures de couple (on peut par exemple détecter 1 dyne.cm avec un appareillage permettant d'en mesurer 10^6) a permis néanmoins de mesurer M_1 à 1 % près [1], [8]. On imagine sans peine que les mesures d'anisotropie magnétique puissent également être utilisées comme des tests de la qualité des cristaux.

3.3. *Origine microscopique de l'anisotropie magnétique*

L'origine microscopique de l'anisotropie magnétocristalline est essentiellement quantique. Considérons le cas simple dans lequel les moments magnétiques microscopiques sont bien localisés, par exemple des ions de transition dans un oxyde magnétique. L'interaction responsable de l'ordre magnétique est l'interaction d'échange qui fait intervenir les opérateurs de spin des ions. Pour deux ions i et j, elle s'écrit : $-2 J_{ij} \vec{S}_i \cdot \vec{S}_j$ où J_{ij} est l'intégrale d'échange (positive pour le ferro-magnétisme) et $\vec{S}_i$ l'opérateur de spin de l'ion i. Cette interaction ne dépend que de l'orientation mutuelle de $\vec{S}_i$ et $\vec{S}_j$ mais est invariante dans une rotation d'ensemble par rapport au cristal. Les moments orbitaux des ions sont liés aux densités de charges électroniques et donc sensibles aux champs électriques créés par les ions environnants (effet Stark cristallin). Enfin les moments de spin et orbital de chaque ion sont couplés entre eux par l'interaction spin-orbite qui provient du fait que le moment magnétique intrinsèque de chaque électron (moment de spin) est placé dans le champ magnétique créé par le propre mouvement de l'électron, mouvement qui implique l'existence d'un moment orbital. Ainsi l'énergie du système des moments magnétiques couplés par l'interaction d'échange va dépendre de leur orientation par rapport aux axes cristallins, la symétrie du cristal fixant l'environnement de chaque ion donc la symétrie du champ cristallin.

Les origines des énergies d'anisotropie induites sont fondamentalement les mêmes mais se prêtent à des interprétations semi-phénoménologiques variées qui mettent mieux en évidence leurs relations avec leurs causes comme le montreront les quelques exemples qui suivent.

N.B. : Pour une revue bibliographique récente sur l'anisotropie magnéto-cristalline, on pourra consulter la référence [7].

4.   Exemple : matériaux magnétiques texturés

Nous nous proposons de montrer au moyen d'un exemple simple comment l'énergie d'anisotropie magnétique peut traduire les propriétés d'anisotropie "mécanique" de certains matériaux.

Rappelons tout d'abord brièvement que l'on dira d'un matériau polycristallin qu'il est "texturé" lorsque la distribution des orientations des axes cristallins des différents grains n'est pas au hasard. Une texture peut résulter de différentes opérations : solidification, déformation, recuit sous contrainte ou après déformation, recristallisation sous champ magnétique. On la caractérise habituellement par rapport aux dimensions extérieures du matériau : axe d'un fil ou plan et bords d'une feuille.

Nous considérerons dans ce qui suit, la texture obtenue sur un échantillon plan par laminage. On peut obtenir dans ce cas un ensemble de grains ayant tous un plan cristallographique de même symétrie (par exemple : (110) ou (101) ou (011) etc.) parallèle au plan de l'échantillon. On caractérise alors la texture en donnant le plan cristallographique correspondant au plan de l'échantillon, et ensuite la direction cristallographique parallèle à la direction de laminage dans ce plan : ex. {110} ⟨100⟩.

Au point de vue magnétique, on pourra admettre, si les grains sont assez gros, que l'énergie d'anisotropie mesurée dans des champs assez élevés est la somme algébrique des énergies de chacun des grains. Or chaque grain, qui est un monocristal, possède une énergie d'anisotropie magnétocristalline de la forme :

$$E_a = K_0 + K_1 (\alpha_1^2 \alpha_2^2 + \alpha_2^2 \alpha_3^2 + \alpha_3^2 \alpha_1^2) + K_2 (\alpha_1^2 \alpha_2^2 \alpha_3^2)$$

comme cela a été décrit précédemment (§ 3). Si les grains sont assez nombreux et répartis au hasard, la somme de leurs énergies ne dépendra plus de la direction de mesure, par effet de moyenne. Par contre on conçoit facilement que si les grains ne sont pas répartis au hasard, on mesurera une énergie d'anisotropie magnétique résultante.

A titre d'exemple, considérons un polycristal texturé $\{100\}\langle 100\rangle$ avec des écarts à une texture parfaite que l'on peut décrire comme des rotations dans le plan autour de la direction de laminage. L'aimantation dans chaque grain tourne dans un plan $\{100\}$ lors de la mesure du couple, et le couple dû à un grain s'écrit :

$$L = -\frac{K_1}{2} [\sin (4 (\theta + \alpha))] v$$

où K_1 est la première constante d'anisotropie du cristal, θ est l'angle de l'aimantation par rapport à la direction de laminage, α l'écart entre la direction $\{100\}$ du grain et la direction de laminage, et v le volume du grain.

Le couple mesuré par unité de volume de l'échantillon total est la somme des couples :

$$\frac{L}{V} = \sum_i \frac{\left(-\dfrac{K_1}{2} \sin (4 (\theta + \alpha)) v_i\right)}{\sum_i v_i}$$

$$= \int_{-\alpha}^{+\alpha} \frac{-\dfrac{K_1}{2} \sin (4 (\theta + \alpha)) \, d\alpha}{\int d\alpha} = \frac{K_1}{2} \sin 4\theta \, \frac{\sin 4\alpha}{4\alpha} .$$

On vérifie bien ainsi que l'échantillon polycristallin texturé présente une forme de couple tout à fait analogue à celle d'un disque monocristallin dans un plan $\{100\}$, mais l'amplitude de l'anisotropie est diminuée d'un facteur $\dfrac{\sin 4\alpha}{4\alpha} < 1$. Connaissant la valeur de K_1 du monocristal, on peut déduire la valeur de α caractérisant la texture.

La figure 3 compare les courbes de couple mesurées sur un monocristal (100) $\langle 100\rangle$ et sur un échantillon texturé (110) $\langle 100\rangle$. On utilise

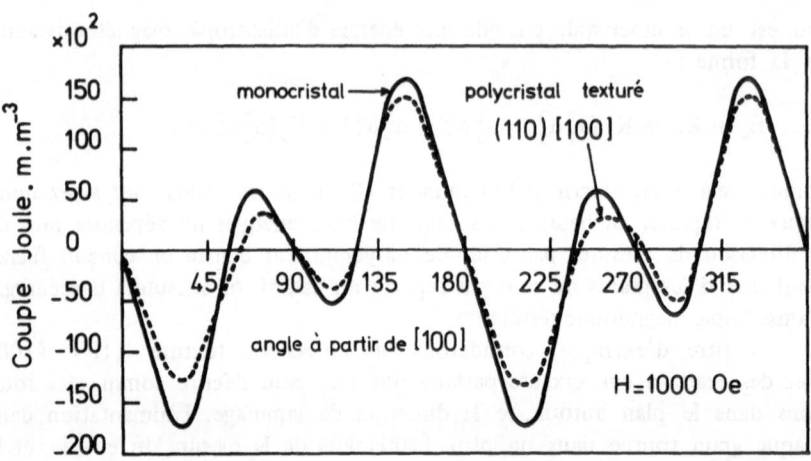

Figure 3. – *Comparaison des courbes de couples mesurés dans un monocristal (110) [100] ou un polycristal texturé (110) [100].*

beaucoup ce genre de résultats dans l'alliage FeSi [4], servant à de nombreuses applications, en particulier pour les tôles de transformateur, dans lequel on cherche à développer des textures de laminage de façon à obtenir des directions de facile aimantation. Des résultats analogues pourraient être donnés concernant les alliages Fer-Nickel, les Alnico etc. Pour tous ces alliages, la méthode de mesure de l'énergie d'anisotropie est très utilisée [3] : elle est en effet plus pratique et plus rapide à faire, pour déterminer la texture, qu'une mesure de rayons X par exemple.

5. Conclusion

Nous avons vu sur un exemple comment l'anisotropie magnétique peut refléter une anisotropie "mécanique" et, d'une manière plus générale, permet d'accéder à certaines propriétés métallurgiques d'un alliage. On pourra trouver dans différents ouvrages [3], [5] une revue de telles études.

Certes, la limitation de cette méthode d'investigation apparaît vite puisqu'on ne peut faire pratiquement des mesures d'anisotropie magnétique que dans des matériaux ferromagnétiques. Par contre, lorsque ces mesures sont possibles, elles permettent d'obtenir très facilement des résultats rapides avec une précision souvent supérieure à celle de bien d'autres méthodes plus longues à mettre en œuvre.

REFERENCES

[1] AUBERT G. – "Etude de l'anisotropie magnetocristalline. Application au Nickel Thèse, Grenoble, 1966.

[2] AUBERT G., Y. AYANT, E. BELORIZKY and R. CASALEGNO. – "Various methods for analyzing data on anisotropic scalar properties in cubic symmetry: Application to magnetic anisotropy energy of nickel." *Phys. Rev. B*, **14**, 12 (1976): 5314-5326.

[3] BERKOWITZ A.E. and E. KNELLER. Eds. – *Magnetism and Metallurgy, Vol. 2*, New York and London: Academic Press, 1969, chap. XII and XV.

[4] BOZORTH R.M. – "Orientation of crystals in silicon-iron." *Trans. Am. Soc. Metals*, 23 (1935): 1107-1111.

[5] BOZORTH R.M. – *Ferromagnetism*. New York: Van Norstrand Company Inc., 1951.

[6] CHAMBRON W. and A. CAPLAIN. – "Etude des lacunes en très faible concentration dans l'alliage fer-nickel à 70 % de nickel par la méthode de l'anisotropie magnétique". *Acta Metallurgica*, **22** (1974): 357-366.

[7] DARBY M.I. and E.D. ISAAC. – "Magnetocrystalline anisotropy of ferro and ferrimagnetics." *I.E.E.E. Trans. on Magnetics*, 10, 2 (1974): 259-304.

[8] ESCUDIER P. – "L'anisotropie de l'aimantation : un paramètre important de l'anisotrpie magnétocristalline." *Ann. Phys.*, 9 (1975): 125-173.

[9] GERSDORF R. and G. AUBERT. – "The fitting of several magnetic anisotropy constants to measured torque curves with an application to nickel." *Physica*, 95B (1978): 135-146.

[10] PECHART R. – "Contribution à l'étude de l'ordre directionnel induit par traitement thermomagnétique dans les monocristaux d'alliages binaires ferromagnétiques." Thèse, Grenoble, 1972.

ABSTRACT

We first recall the description of the magnetic properties of materials in the continuous medium approximation: magnetization, shape anisotropy, magneto-crystalline anisotropy (M.C.A.). By various means (heat treatment, strains, irradiation,...), one can create a magnetic anisotropy called induced anisotropy (I.A.) superimposed on the M.C.A. which is an intrinsic property of the material. Inversely, measurement of this I.A. can give us information about the treatments to which the sample was previously submitted.

After studying the principal method of measurement of magnetic anisotropy and evoking the microscopic mechanisms which are responsible for the M.C.A., we give a simple example in which this method of characterization is used.

DISCUSSION

QUESTION POSEE PAR M. PERNOT : Il semble que les conclusions sur les relations entre les courbes de couple d'anisotropie magnétocristalline
et la texture cristallographique, obtenues sur une texture simple, ne
soient pas généralisables à des textures plus complexes. En effet, dans
le cas des alliages Fe-Si, les textures cubiques ou de Goss développées
sont très accusées ; cependant il a été montré [1] dans le cas général
que les renseignements que donnent les courbes de couple sur la texture sont trop incomplets pour être utilisés, par exemple, dans le cas
des relations entre les courbes de couple et l'aptitude à l'emboutissage
des aciers extra-doux. Ceci est dû au fait que le couple est la dérivée
de l'énergie ; le terme d'isotropie plane est donc perdu, alors qu'il est
prépondérant dans les textures des aciers extra-doux. Existe-t-il une
méthode simple et rapide pour remonter directement à l'énergie d'anisotropie magnétocristalline et non à sa dérivée ?

[1] M. PERNOT. Thèse de Doctorat d'Etat : Université Paris XI, 1977.

RÉPONSE DE A. CHAMBEROD : La mesure de l'énergie d'anisotropie fournit
dans de nombreux cas des renseignements qualitatifs et souvent quantitatifs sur les textures. Le cas des fer-silicium est particulièrement significatif. Cela dit, on ne saurait conclure que ce type de mesure peut
donner à coup sûr des renseignements tout à fait satisfaisants sur les
textures, notamment lorsque celles-ci sont complexes ou orientées de
façon défavorable pour permettre une mesure efficace, ou tout simplement lorsque l'énergie d'anisotropie résultante est trop faible.

Il n'existe pas en tout cas de mesure simple et rapide pour obtenir
directement l'énergie d'anisotropie autrement que par la mesure du couple,
c'est-à-dire de la dérivée de cette énergie.

Birefringence in Finitely-Deformed Materials

R. S. Rivlin

Lehigh University, Bethlehem, Pennsylvania, U.S.A.

1. Introduction

In this paper we develop the equations for the propagation of electromagnetic waves in a material which is subjected to a finite static deformation. The theory developed follows the lines of that presented earlier by Smith and Rivlin [2].

The theory proceeds from the assumption that the electric displacement field is linearly related to the instantaneous value of the electric field and the magnetic intensity field is linearly related to the magnetic induction field, the coefficient matrices being functions of the instantaneous values of the displacement gradient matrix. For semantic simplicity we shall refer to such materials as elastic materials.

The practically important cases when the material is isotropic and non-magnetic are derived as special cases.

Two formulations of the theory, which we have called the Lagrangian and Eulerian formulations, are presented. These correspond roughly to the Piola-Kirchhoff and Cauchy formulations of the equations of continuum mechanics.

The application of the equations which are derived to materials in which the coefficient matrices in the constitutive equations depend on the history of the deformation gradient matrix are also discussed.

2. Electromagnetic constitutive equations for a deformed elastic material.

The constitutive equations governing the propagation of an electromagnetic wave in a deformed isotropic elastic material can be obtained by the now standard procedures of constitutive equation theory.

We denote by $\bar{e}, \bar{h}, \bar{d}, \bar{b}$, the electric field, the magnetic intensity field, the electric displacement field and the magnetic induction field respectively. We consider the elastic material to be subjected to a deformation in which a particle initially at vector position $\mathbf{X}$ with respect to a fixed origin moves to vector position $\mathbf{x}$ as a result of the deformation. We denote by x_i and X_A (i, $A = 1, 2, 3$) the components of $\mathbf{x}$ and $\mathbf{X}$ respectively in a rectangular

Cartesian coordinate system x. Similarly, we denote the components in the system x of $\overline{e}, \overline{h}, \overline{d}, \overline{b}$ by $\overline{e}_i, \overline{h}_i, \overline{d}_i, \overline{b}_i$ respectively.

We assume that $\overline{h}$ and $\overline{d}$ at the instant of time t are linearly related to the values of $\overline{e}$ and $\overline{b}$ at time t and are also functions of the instantaneous value of the deformation gradient matrix g defined by

$$g = \|g_{iA}\| = \|x_{i,A}\|, \tag{2.1}$$

where the notation $_{,A}$ is used to denote the operator $\partial/\partial X_A$. Accordingly, we have

$$\overline{d} = k\overline{e} + \overline{k}\overline{b}, \quad \overline{h} = \overline{\omega}\overline{e} + \omega\overline{b}, \tag{2.2}$$

where $k, \overline{k}, \omega$ and $\overline{\omega}$ are matrix functions of g. It can easily be shown that if the material has a center of symmetry

$$\overline{k} = \overline{\omega} = 0, \tag{2.3}$$

and we shall restrict our discussion to this case, so that

$$\overline{d} = k(g)\overline{e}, \quad \overline{h} = \omega(g)\overline{b}. \tag{2.4}$$

If an electromagnetic potential exists for the material, then k and ω are symmetric matrices. However, in the following analysis we shall not make explicit use of this fact.

We superpose on the assumed deformation a rigid rotation and simultaneously subject the electromagnetic fields to an identical rigid rotation. Then, in (2.4) $g, \overline{e}, \overline{b}, \overline{d}, \overline{h}$ are replaced by $ag, a\overline{e}, a\overline{b}, a\overline{d}$ and $a\overline{h}$ respectively, where a is a proper orthogonal (i.e. a rotation) matrix. Equations (2.4) become

$$\overline{d} = a^{-1}k(ag)a\overline{e}, \quad \overline{h} = a^{-1}\omega(ag)a\overline{b}, \tag{2.5}$$

and these equations must be valid for all proper orthogonal a. It can easily be shown that the necessary and sufficient condition for this to be the case is that the relations (2.4) be expressible in the form

$$\overline{D} = K(C)\overline{E}, \quad \overline{H} = \Omega(C)\overline{B}, \tag{2.6}$$

where K and Ω are matrix functions of the Cauchy strain matrix C defined by[1]

$$C = g^\dagger g, \tag{2.7}$$

---

[1] The dagger denotes the transpose

and $\bar{\mathbf{E}}, \bar{\mathbf{B}}, \bar{\mathbf{H}}$ and $\bar{\mathbf{D}}$ are defined by

$$\bar{\mathbf{E}} = \mathbf{g}^\dagger \bar{\mathbf{e}}, \quad \bar{\mathbf{H}} = \mathbf{g}^\dagger \bar{\mathbf{h}},$$
$$\bar{\mathbf{B}} = (\det \mathbf{g}) \, \mathbf{g}^{-1} \bar{\mathbf{b}}, \quad \bar{\mathbf{D}} = (\det \mathbf{g}) \, \mathbf{g}^{-1} \bar{\mathbf{d}}. \tag{2.8}$$

We can also write the constitutive equations in the forms (2.6) with many other definitions for $\bar{\mathbf{E}}, \bar{\mathbf{H}}, \bar{\mathbf{B}}, \bar{\mathbf{D}}$ than (2.8). The particular choice in (2.8) is made because it leads to felicitous expressions for Maxwell's equations in terms of $\bar{\mathbf{E}}, \dots, \bar{\mathbf{D}}$ (see §4 below).

Equations (2.6) provide a Lagrangian formulation of the constitutive equations, whether or not the material is isotropic. If the material is isotropic, $\mathbf{K}$ and $\boldsymbol{\Omega}$ must be isotropic matrix functions of $\mathbf{C}$ and must therefore be expressible in the forms

$$\mathbf{K} = K_0 \, \boldsymbol{\delta} + K_1 \mathbf{C} + K_2 \mathbf{C}^2,$$
$$\boldsymbol{\Omega} = \Omega_0 \, \boldsymbol{\delta} + \Omega_1 \mathbf{C} + \Omega_2 \mathbf{C}^2, \tag{2.9}$$

where $K_0, K_1, \dots, \Omega_2$ are scalar functions of the basic orthogonal invariants of $\mathbf{C}$,

$$I_1 = \operatorname{tr} \mathbf{C}, \quad I_2 = \frac{1}{2}\{(\operatorname{tr} \mathbf{C})^2 - \operatorname{tr} \mathbf{C}^2\}, \quad I_3 = \det \mathbf{C}. \tag{2.10}$$

$\boldsymbol{\delta}$ is the unit matrix. We note that $I_3 = 1$ if the deformation is isochoric, as it must be if the material is incompressible.

From (2.6) and (2.8), we can obtain relations of the form (2.4), where

$$\mathbf{k} = \frac{1}{\det \mathbf{g}} \, \mathbf{g} \mathbf{K}(\mathbf{C}) \, \mathbf{g}^\dagger, \quad \boldsymbol{\omega} = (\det \mathbf{g}) \, (\mathbf{g}^\dagger)^{-1} \boldsymbol{\Omega}(\mathbf{C}) \, \mathbf{g}^{-1}. \tag{2.11}$$

With (2.11), equations (2.4) provide an Eulerian formulation of the constitutive equations, whether or not the material considered is isotropic. If the material is isotropic, we obtain the appropriate expressions for $\mathbf{k}$ and $\boldsymbol{\omega}$ by substituting from (2.9) and (2.7) in (2.11). We thus obtain

$$\mathbf{k} = k_0 \boldsymbol{\delta} + k_1 \mathbf{c} + k_2 \mathbf{c}^2,$$
$$\boldsymbol{\omega} = \omega_0 \boldsymbol{\delta} + \omega_1 \mathbf{c} + \omega_2 \mathbf{c}^2, \tag{2.12}$$

where $\mathbf{c}$ is the Finger strain matrix defined by

$$\mathbf{c} = \mathbf{g}\mathbf{g}^\dagger, \tag{2.13}$$

and $k_0, k_1, \ldots, \omega_2$ are given by

$$k_0 = I_3^{1/2} K_2, \quad k_1 = I_3^{-1/2} (K_0 - I_2 K_2), \quad k_2 = I_3^{-1/2} (K_1 + I_1 K_2),$$

$$\omega_0 = I_3^{1/2} \Omega_1 + I_2 I_3^{-1/2} \Omega_0, \quad \omega_1 = I_3^{1/2} \Omega_2 - I_1 I_3^{-1/2} \Omega_0, \quad \omega_2 = I_3^{-1/2} \Omega_0. \tag{2.14}$$

In deriving these relations, we have used the Cayley-Hamilton theorem for
c and we note that I_1, I_2, I_3 defined by (2.10), may also be expressed by

$$I_1 = \text{tr } \mathbf{c}, \quad I_2 = \frac{1}{2} \{(\text{tr } \mathbf{c})^2 - \text{tr } \mathbf{c}^2\}, \quad I_3 = \det \mathbf{c}. \tag{2.15}$$

We also note that $\det \mathbf{g} = I_3^{1/2}$.

We now consider that the deformation is homogeneous and the electromagnetic fields are associated with a plane electromagnetic wave propagating in
the direction of the unit vector n. Then,

$$(\bar{\mathbf{e}}, \bar{\mathbf{h}}, \bar{\mathbf{d}}, \bar{\mathbf{b}}) = (\mathbf{e}, \mathbf{h}, \mathbf{d}, \mathbf{b}) \exp i\omega (s\mathbf{x} \cdot \mathbf{n} - t), \tag{2.16}$$

where s is the slowness of the wave and $\mathbf{e}, \mathbf{h}, \mathbf{d}, \mathbf{b}$ are constant vectors. The
relations (2.4) then yield

$$\mathbf{d} = k\mathbf{e}, \quad \mathbf{h} = \omega\mathbf{b}. \tag{2.17}$$

We may also write

$$(\bar{\mathbf{E}}, \bar{\mathbf{H}}, \bar{\mathbf{D}}, \bar{\mathbf{B}}) = (\mathbf{E}, \mathbf{H}, \mathbf{D}, \mathbf{B}) \exp i\omega (S\mathbf{X} \cdot \mathbf{N} - t), \tag{2.18}$$

where (cf. (2.8))

$$\mathbf{E} = \mathbf{g}^\dagger \mathbf{e}, \quad \mathbf{H} = \mathbf{g}^\dagger \mathbf{h},$$

$$\mathbf{B} = (\det \mathbf{g}) \mathbf{g}^{-1} \mathbf{b}, \quad \mathbf{D} = (\det \mathbf{g}) \mathbf{g}^{-1} \mathbf{d} \tag{2.19}$$

and

$$\mathbf{N} = \frac{s}{S} (\mathbf{g}^\dagger \mathbf{n}), \quad \frac{S}{s} = \{(\mathbf{g}^\dagger \mathbf{n}) \cdot (\mathbf{g}^\dagger \mathbf{n})\}^{1/2} = (\mathbf{n}^\dagger \mathbf{c}\mathbf{n})^{1/2}. \tag{2.20}$$

Then, we obtain from (2.6)

$$\mathbf{D} = \mathbf{K}(\mathbf{C}) \mathbf{E}, \quad \mathbf{H} = \mathbf{\Omega}(\mathbf{C}) \mathbf{B}. \tag{2.21}$$

3. The secular equation — Eulerian formulation

Taking the velocity of electromagnetic waves in free-space as unity[1], Maxwell's equations yield

$$\text{s} \mathbf{n} \times \mathbf{e} = \mathbf{b}, \quad \text{s} \mathbf{n} \times \mathbf{h} = -\mathbf{d}. \tag{3.1}$$

Using (2.17) to eliminate $\mathbf{e}, \mathbf{h}$ and $\mathbf{b}$ from (3.1), we obtain

$$\text{s}^2 \mathbf{n} \times \{\omega[\mathbf{n} \times (\mathbf{k}^{-1}\, \mathbf{d})]\} + \mathbf{d} = 0. \tag{3.2}$$

Alternatively using (2.17) to eliminate $\mathbf{e}, \mathbf{h}$ and $\mathbf{d}$ from (3.1), we obtain

$$\text{s}^2 \mathbf{n} \times \{\mathbf{k}^{-1}\, (\mathbf{n} \times \omega \mathbf{b})\} + \mathbf{b} = 0. \tag{3.3}$$

Equations (3.2) and (3.3) determine the polarizations of the $\mathbf{d}$ and $\mathbf{b}$ vectors. Of course, they both lie in the plane normal to $\mathbf{n}$. If the material considered is a non-magnetic dielectric, so that $\omega = \mathbf{\delta}$, then from $(3.1)_2$ we see that $\mathbf{d}$ is also perpendicular to $\mathbf{b}$.

The necessary and sufficient condition that (3.2) and (3.3) have non-trivial solutions for $\mathbf{d}$ and $\mathbf{b}$ is

$$|k_{ij} + \text{s}^2\, \epsilon_{ipq}\, \epsilon_{jsr}\, n_p\, n_r\, \omega_{qs}| = 0. \tag{3.4}$$

With some algebraic manipulation this can be re-written as

$$\phi \text{s}^4 - \psi \text{s}^2 + \theta = 0, \tag{3.5}$$

where

$$\phi = (\mathbf{n}^\dagger \mathbf{k} \mathbf{n})\,(\mathbf{n}^\dagger \omega^{-1} \mathbf{n})\, \det \omega,$$
$$\psi = \mathbf{n}^\dagger \{(\text{tr}\, \mathbf{k}\omega^\dagger)\, \mathbf{k} - \mathbf{k}\omega^\dagger \mathbf{k}\}\, \mathbf{n}, \tag{3.6}$$
$$\theta = \det \mathbf{k}.$$

Equation (3.5) is the secular equation for the slowness s. We note that in deriving (3.4)–(3.6), we have made no use of the fact that the material considered is isotropic.

Equation (3.5) yields

$$\text{s}^2 = \frac{1}{2\phi}\{\psi \pm (\psi^2 - 4\theta\phi)^{1/2}\}. \tag{3.7}$$

---

[1] The refractive index is then $1/\text{s}$.

The dielectric constant matrix **k** and the magnetic permeability matrix $\boldsymbol{\omega}^{-1}$ are necessarily positive definite. Accordingly θ and ϕ are necessarily positive and the necessary and sufficient conditions that all the values of s given by (3.7) be real are

$$\theta > 0, \quad \phi > 0, \quad \psi > 0, \quad \psi^2 - 4\theta\phi \geq 0. \tag{3.8}$$

In practically interesting cases, the material will generally be non-magnetic so that

$$\boldsymbol{\omega} = \boldsymbol{\delta}. \tag{3.9}$$

Then

$$\phi = \mathbf{n}^\dagger \mathbf{kn}, \quad \psi = \mathbf{n}^\dagger [(\text{tr } \mathbf{k}) \, \mathbf{k} - \mathbf{k}^2] \, \mathbf{n}, \quad \theta = \det \mathbf{k}. \tag{3.10}$$

Also, equations (3.2) and (3.3) become

$$s^2 \, \{(\mathbf{nk}^{-1}\mathbf{d}) \, \mathbf{n} - \mathbf{k}^{-1}\mathbf{d}\} + \mathbf{d} = \mathbf{0}, \tag{3.11}$$

and

$$s^2 \, \mathbf{n} \times \mathbf{k}^{-1} \, (\mathbf{n} \times \mathbf{b}) + \mathbf{b} = \mathbf{0}. \tag{3.12}$$

4.   The secular equation – Lagrangian form

It can be shown fairly easily [3] that in terms of the Lagrangian electromagnetic fields defined by (2.8) Maxwell's equations may be written as

$$\text{curl } \overline{\mathbf{E}} = -\partial \overline{\mathbf{B}}/\partial t, \quad \text{curl } \overline{\mathbf{H}} = \partial \overline{\mathbf{D}}/\partial t. \tag{4.1}$$

With (2.18) we obtain

$$S\mathbf{N} \times \mathbf{E} = \mathbf{B}, \quad S\mathbf{N} \times \mathbf{H} = -\mathbf{D}. \tag{4.2}$$

We note that both **B** and **D** are perpendicular to **N**. Then, with (2.21) we obtain

$$S^2\mathbf{N} \times \{\boldsymbol{\Omega}(\mathbf{N} \times \mathbf{K}^{-1}\mathbf{D})\} + \mathbf{D} = 0 \tag{4.3}$$

and

$$S^2\mathbf{N} \times \{\mathbf{K}^{-1} \, (\mathbf{N} \times \boldsymbol{\Omega}\mathbf{B})\} + \mathbf{B} = \mathbf{0}. \tag{4.4}$$

From either of these equations we obtain the secular equation

$$\Phi S^4 - \Psi S^2 + \Theta = 0, \tag{4.5}$$

where

$$\Phi = (N^\dagger \, KN) \, (N^\dagger \Omega^{-1} \, N) \, \det \Omega,$$

$$\Psi = N^\dagger \, \{(\text{tr } K\Omega^\dagger) \, K - K\Omega^\dagger \, K\} \, N, \tag{4.6}$$

$$\Theta = \det K.$$

When the material is non-magnetic, i.e. when $\omega = \delta$, we have, from $(2.11)_2$,

$$\Omega = (\det g)^{-1} \, C. \tag{4.7}$$

5. Viscoelastic materials

The procedure used in arriving at the constitutive equations (2.6) with (2.9), or (2.4) with (2.12), can also be used to obtain corresponding constitutive equations of the Lagrangian or Eulerian type for an isotropic rate-dependent material.

We can envisage two types of rate-dependence. In a particular material either or both of these may be present. If both are present and the material is electromagnetically linear, then the constitutive assumption (2.4) is replaced by the assumption that $\overline{d}$ and $\overline{h}$ at time t are vector functionals of the histories of the deformation gradient matrix $g(\tau)$, up to and including time t, and linear functionals of the histories $\overline{e}(\tau)$ and $\overline{b}(\tau)$ respectively.

However, here we will consider that the material is non-magnetic and make the constitutive assumption that $\overline{d}$ is a linear function of the instantaneous value of $\overline{e}$ and a functional of the history $g(\tau)$ of the deformation gradient matrix, with support $(-\infty, t]$. Thus, we write

$$\overline{d} = k \, [g(\tau)] \overline{e}, \quad \overline{h} = \overline{b}. \tag{5.1}$$

It can then be shown by considering the effect of a superposed time-dependent rigid rotation that $\overline{D}$ and $\overline{H}$ are related to $\overline{E}$ and $\overline{B}$ by

$$\overline{D} = K \, [C(\tau)] \, \overline{E}, \quad \overline{H} = (\det g)^{-1} \, C\overline{B}, \tag{5.2}$$

where K is a matrix functional of $C(\tau)$ with support $(-\infty, t]$.

The Eulerian constitutive equations corresponding to (5.2) have the forms

$$\overline{d} = gk \, [C(\tau)] \, g^\dagger \overline{e}, \quad \overline{h} = \overline{b}. \tag{5.3}$$

The matrix functional k in (5.3) is, in general, different from that in (5.1).

If the electromagnetic fields are those associated with an electromagnetic wave, then from (5.2) and (2.18) we have

$$D = K[C(\tau)]\,E, \quad H = \frac{1}{\det g}\,CB, \tag{5.4}$$

and from (5.3) and (2.16) we have

$$d = gk\,[C(\tau)]\,g^\dagger e, \quad h = b. \tag{5.5}$$

If the material is isotropic then K and k are isotropic matrix functionals.

Provided that the period of the electromagnetic wave is small compared with the times involved in carrying out the deformation and in establishing equilibrium following a deformation, we can neglect time derivatives of K, k, g in comparison with those of the electromagnetic fields. Then, the Lagrangian secular equation still takes the form (4.5), with Θ, Φ, Ψ given by (4.6) in which K is replaced by K and Ω is replaced by $(\det g)^{-1} C$. Also, the Eulerian secular equation still takes the form (3.5) with θ, ϕ, ψ given by (3.6) in which k is replaced by k and ω by δ.

If the functional K in (5.4) is of the fading memory type, and at some instant of time, say $t = 0$, the material is subjected to a deformation with is thereafter held constant, then (cf. [1]) if t is not too small, K may be replaced by an ordinary matrix function, K say, of the Cauchy strain C corresponding to the constant deformation, which also depends on t. Thus, equation $(5.4)_1$ may be replaced by

$$\overline{D} = K(C, t)\,\overline{E}. \tag{5.6}$$

If the material is isotropic then K is an isotropic matrix function of C and can therefore be expressed in the form $(2.9)_1$ where K_0, K_1, K_2 are now functions of t as well as of I_1, I_2, I_3 defined by (2.10). Analogous consideration apply to equation $(5.5)_1$, which can be replaced by

$$d = ke, \tag{5.7}$$

where k is expressible in the form $(2.12)_1$, in which k_0, k_1 and k_2 are functions of t as well as of I_1, I_2, I_3.

Acknowledgement

This paper was written with the support of a grant from the National Science Foundation to Lehigh University.

REFERENCE

[1] RIVLIN R.S. – "Stress relaxation in incompressible elastic materials at constant deformation". *Q. Appl. Mech.* **14** (1956): 83-89.
[2] SMITH G.F. and R.S. RIVLIN. – "Photoelasticity with finite deformations". *J. App. Math. Phys.*, **21** (1970): 101-115.
[3] WALKER J.B. and A.C. PIPKIN, R.S. RIVLIN. – "Maxwell's equations in a deformed body". *Rend. Accad. Naz. d. Lincei*, **38** (1965): 674-676.

RESUME

(Biréfringence des matériaux sous déformation finie)

Nous discutons de la propagation des ondes électromagnétiques dans un matériau isotrope à symétrie centrale, soumis à une déformation finie. Deux formulations de la théorie, Lagrangienne et Eulérienne, sont présentées.

State Functions for a Pseudoelastic Body

I. Müller
Technical University, Berlin, Germany – BRD.

K. Wilmanski
Institute of Fundamental Technical Research, Warsaw, Poland.

1. Introduction

The shape of the load-deformation curves of a pseudo-elastic body depends strongly on the temperature. While at low temperatures the body exhibits properties akin to those of a plastic body, it behaves like a true (non-linear) elastic body at high temperatures. At intermediate temperatures the body has a yield load, at which the deformation increases markedly, but upon unloading this increase of deformation is recovered and there is no residual deformation.

This peculiar behaviour is due to a phase transition in the body and to twinformation. At high temperatures the body is in an austenitic phase while at low temperatures it exists in a martensitic phase. This latter phase may occur in different modifications, called twins, because they are energetically identical.

Reviews of the macroscopic properties of pseudo-elastic bodies and of their lattice structure may be found in the paper [1] by Delaey, Krishnan, Tas and Warlimont and in the book [5], edited by Perkins.

This paper presents a structural model which ignores the complexities of the molecular structure of a pseudo-elastic body and yet it simulates most of the properties of such bodies.

2. Phenomenology

The form of the load-deformation curve of a pseudo-elastic body is illustrated by Figures 1a through 1f which refer to increasing temperatures.

At low temperatures the possible load-deformation curves are reminiscent of those of an elastic-ideally plastic body. In particular, one recognises in Figures 1a and 1b an elastic range at small loads and small deformations and a plastic range, once P_Y is exceeded. In contrast to the behaviour of a plastic body, there is a second elastic branch of the load-deformation curve which occurs at large deformations and which enables the body to support loads beyond P_Y. Once the body has been subjected to a stress greater than P_Y, it will show a residual deformation D_P upon unloading. Figure 1b shows that P_Y decreases when the temperature grows.

At higher temperatures the load-deformation curves have the shapes shown in Figures 1c and 1d. Upon loading there is still an elastic range with $|P| < P_Y$, as well as a plastic range at $P = P_Y$ and a second elastic range at $|P| > P_Y$. However, upon unloading from $P > P_Y$ the body does not contract to a residual deformation, rather it returns to the original configuration with $D = 0$ along a path that contains the recovery range at $P = P_R$. This is the phenomenon that has been termed pseudo-elasticity. Both P_Y and P_R increase with increasing temperature and the slope of the initial elastic branch also increases.

As the temperature becomes still higher, the difference between P_Y and P_R diminishes and eventually vanishes. Thus we observe curves as those in Figure 1e, and, at still higher temperature, in Figure 1f. Each branch of those curves can be traversed in both directions, as indicated by the arrows.

The most dramatic property of a pseudo-elastic body is the shape memory which is implied by the Figures 1: If, at a low temperature we deform the body plastically and leave it unloaded at D_P, heating to a high temperature will let the body creep back to $D = 0$, since that is the only deformation which the body can sustain at high temperature when unloaded.

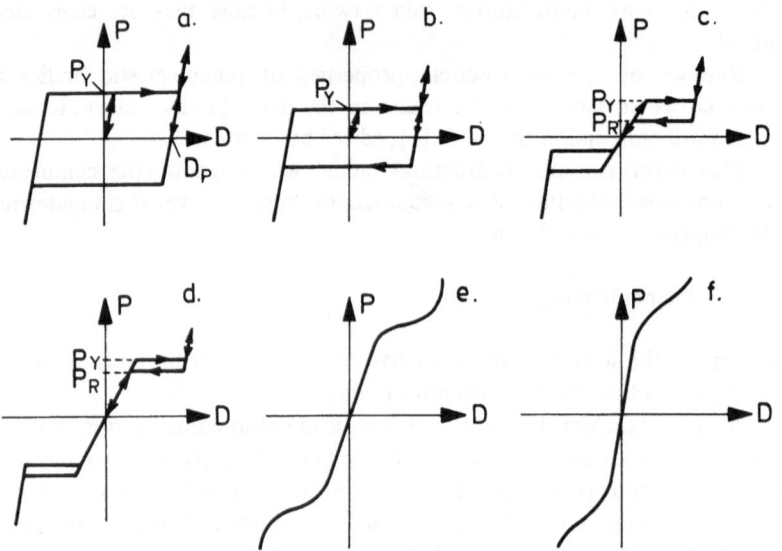

Figure 1. – *Stress-strain temperature behaviour of a pseudoelastic body.*

The object of the following chapters is the description of the observed behaviour of a pseudo-elastic body by a model which we proceed to describe now.

3. The Model

The basic element of the model is a lattice-particle which has two stable modifications and a metastable one. Those three modifications are schematically drawn in Figure 2 which also shows the potential energy of the lattice-particle as a function of the displacement Δ. We see that the three equilibria are separated by energetic barriers and that the minima at $\Delta = \pm J$ are deeper than the minimum at $\Delta = 0$.

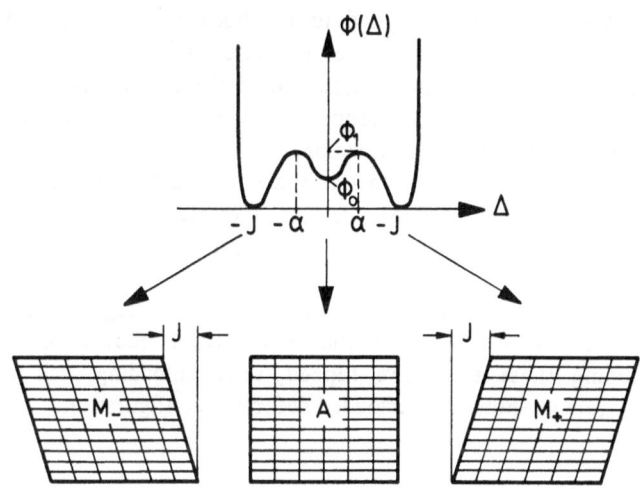

Figure 2. — *Configurations of a lattice-particle.*

We refer to the three modifications as types $M_\pm$ and type A, respectively, as indicated in the figure. M stands for martensite and A stands for austenite and these names refer to the phases which the body may assume. The indices $\pm$ on M identify the twins of the martensitic phase.

The body itself is modelled as a superposition of many lattice-particles as shown schematically in Figure 3, such that the lattice particles are arranged in long layers at an angle of $\pi/4$ to the boundaries.

Figures 3a and 3b show two realisations of the reference configuration, in which we take the deformation D to be equal to zero. In Figure 3b this configuration is realised by a non-uniform distribution of lattice particles, because adjacent layers are alternately of types M_+ and M_-.

These two realizations of the reference configuration may seem to be different in the schematic representation of Figures 3a and 3b, but macroscopically that difference cannot be detected, because a body is made up of very many layers — rather than the ten layers of the figures — and their dimensions are microscopic so that to the naked eye the body shown in Figure 3b would have a smooth surface just as the body of Figure 3a. Also there are many other non-uniform realizations of the reference configuration, because the types of lattice particles need not alternate between different layers.

We shall come back to a discussion of non-uniform distributions of lattice-particles at the end of this paper. To begin, however, we consider uniform distributions and show how a deformation may come about under the action of a load on this model.

When a tensile load is applied to the body in the state of Figure 3a, the lattice particles come under the influence of a shear stress τ which tends to tilt them in the manner indicated in Figure 4a, where a small part of the body is shown. As a consequence, there is a small overall deformation which may be recovered by unloading.

Once the load exceeds a critical value, the lattice particles of type A flip over the potential barrier and assume a displacement somewhat smaller than $\Delta = -J$, as depicted in Figure 4b. A considerable deformation is the result.

Some of this deformation is recovered by unloading, but not all of it. The body is left with a large residual deformation, because all lattice particles, including those that have flipped, after unloading. will become of type M_ . Figure 4c shows this unloaded deformed configuration.

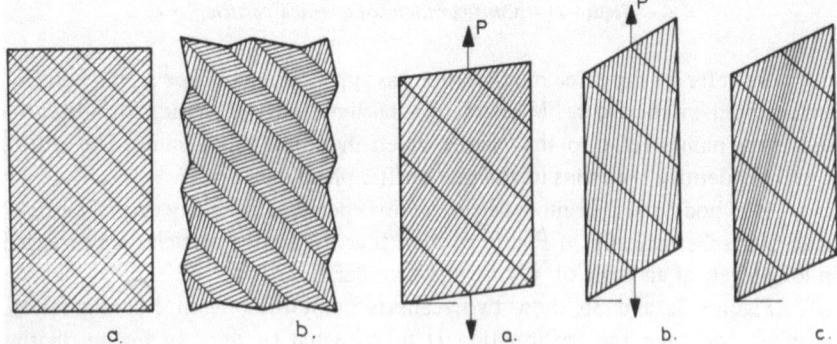

Figure 3. — *Uniform realizations of the reference configuration.*

Figure 4. — *Deformation of the model under loading and unloading.*

It is clear from this description that the overall deformation of the body is given by the relation

$$D = \frac{1}{\sqrt{2}} \sum_{i=1}^{N} \Delta_i \, , \qquad (3.1)$$

where the summation extends over all lattice particles of which there are N. The factor $1/\sqrt{2}$ results from the arrangement of the particles at 45° to the original boundaries.

Figure 4 shows that during the deformation the body changes its shape rather drastically. This must lead to internal stresses, if the body is clamped or, indeed, if it forms part of a bigger body that is subject to a load. Such internal stresses and their effects are ignored here.

The above arguments are supposed to make the reader familiar with the comportment of the model under a load. But these arguments ignore the fact that the lattice particles participate in the thermal motion. We must take the thermal motion into account, however, because shape memory and pseudo-elasticity are consequences of an adjustment of the body to the thermal motion.

4. Free Energy of the Body

4.1. *Probability Distribution* $P_\Delta = N_\Delta/N$

If there is appreciable thermal motion, i.e. at high temperatures, it is no longer useful to ask for the displacement of a particular lattice particle. We may instead ask for the probability P_Δ that a lattice particle has the displacement Δ. For big N this probability is equal to the ratio N_Δ/N, where N_Δ is the number of particles with displacement Δ.

For the calculation of P_Δ we assume that, at high temperatures, the lattice particles do not feel the influence of the bottom of the energy well in which they move. Therefore, for the purpose of calculating P_Δ we replace the true potential (Fig. 5a) by a box potential (Fig. 5b).

Thus P_Δ is constrained by only two conditions

$$\sum_\Delta P_\Delta = 1 \quad \text{and} \quad \frac{1}{\sqrt{2}} \sum_\Delta \Delta P_\Delta = \frac{D}{N} \qquad (4.1)$$

of which the first is the usual normalization of probabilities and the second follows from (3.1).

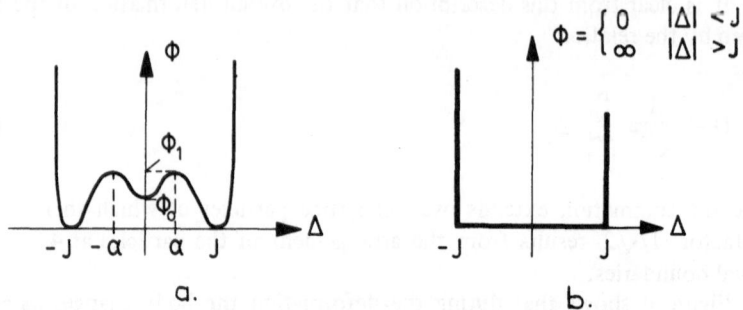

Figure 5. – *The potential and its simplification as a box potential.*

4.2. Entropy

The probability is actually determined from the requirement, that the entropy

$$H = - Nk \sum_\Delta P_\Delta \ln P_\Delta \qquad (4.2)$$

be maximal under the constraints (4.1). Hence, by a simple calculation,

$$P_\Delta = \frac{e^{-\beta \Delta}}{\sum_\Delta e^{-\beta \Delta}} \qquad (4.3)$$

where β is the solution of $\mathscr{L}(\beta J) = - \dfrac{D}{NJ}$, and $\mathscr{L}(x) = \cotg x - \dfrac{1}{x}$ is the Langevin function. The equation for β follows from $(4.1)_2$ when the summation is replaced by integration. Substitution of P_Δ from (4.3) into formula (4.2) leads to the following expression for the entropy

$$H = Nk \left(\ln \frac{\sinh \beta J}{\beta J} - \beta J \, \mathscr{L}(\beta J) \right), \qquad (4.4)$$

where again summation was replaced by integration, and an unimportant constant was deleted. According to this formula, H is a function of D, but unfortunately this function cannot be written in analytic form, because the Langevin function cannot be explicitly inverted to give β in terms of D.

4.3. Energy

The above replacement of the true potential by a box potential is obviously not good enough for the calculation of the energy, because energy would vanish in this approximation. We obtain, however, a reasonably good expression for the energy, if we take the probability distribution P_Δ from (4.3) and write

$$E = N \sum_\Delta \Phi(\Delta) P_\Delta .^1 \qquad (4.5)$$

To be sure, this is the expectation value of the *potential* energy only, but we may ignore the *kinetic* part of the energy, since we are interested here in the D-dependence of the energy.

The energy E depends on the form of the potential $\Phi(\Delta)$ and we can simplify calculations by replacing the true potential (Fig. 5a) by that of Figure 6, thus obtaining, after a little calculation in which again the sum over Δ is replaced by an integral, the formula

$$E = N\Phi_0 \frac{\sinh \beta\alpha}{\sinh \beta J} \left\{ 1 + \frac{\Phi_1}{\Phi_0} \left(1 - 2 \frac{\mathcal{L}(\beta\alpha)}{\beta\alpha} \right) \right\} . \qquad (4.6)$$

Here again β is given in terms of the deformation D, as the solution of the equation $\mathcal{L}(\beta J) = - \dfrac{D}{NJ}$.

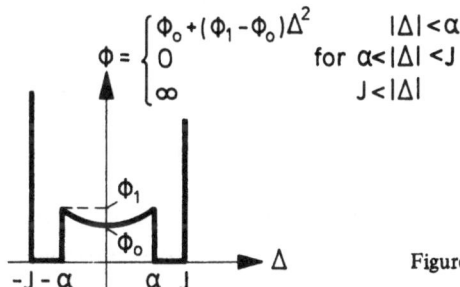

$$\Phi = \begin{cases} \Phi_0 + (\Phi_1 - \Phi_0)\Delta^2 & |\Delta| < \alpha \\ 0 & \text{for } \alpha < |\Delta| < J \\ \infty & J < |\Delta| \end{cases}$$

Figure 6. – *An approximation to the potential.*

4.4. *Graphical Representation of the free energy* $\Psi = E - TH$

From (4.4) and (4.6) we conclude that the free energy of the body is of the form

$$\Psi = N \left[\Phi_0 \frac{\sinh \beta\alpha}{\sinh \beta J} \left\{ 1 + \frac{\Phi_1}{\Phi_0} \left(1 - 2 \frac{\mathcal{L}(\beta\alpha)}{\beta\alpha} \right) \right\} \right.$$
$$\left. - kT \left\{ \ln \frac{\sinh \beta J}{\beta J} - \beta J \, \mathcal{L}(\beta J) \right\} \right] . \qquad (4.7)$$

---

[1] A justification of this formula will be given by the authors in a forthcoming paper.

Since $\mathcal{L}(\beta J) = -\dfrac{D}{NJ}$ cannot be solved analytically for β, it is impossible to find an analytic expression for $\Psi(D, T)$. However, there is no difficulty to numerically evaluate equation (4.7), and Figure 7 shows the function $\Psi(D, T)$ for the special choices $\dfrac{\Phi_1}{\Phi_0} = 3$, $\dfrac{\alpha}{J} = 0{,}6$ and for an interesting range of temperatures.

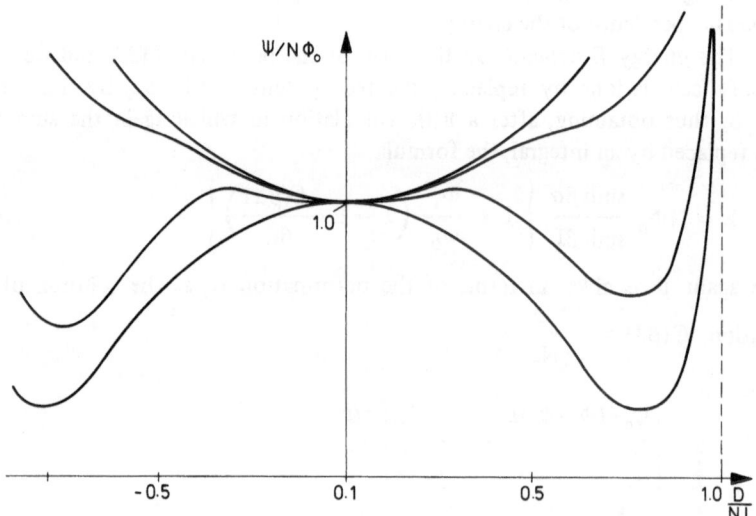

Figure 7. – *Free energy as a function of deformation.*

All curves $\Psi(D, T)$ coincide at $D = 0$ and all tend to infinity when D tends to $\pm NJ$. Apart from those common characteristics, the curves differ widely: At low temperatures, there are two lateral minima and a maximum in-between. At higher temperatures, a third minimum develops at $D = 0$ which is first shallower than the lateral minima and then deeper. At still higher temperatures only the central minimum is left and the ascending branches of the curves show two points of inflection. Eventually, at higher temperatures yet, the inflection points vanish and we are left with curves of positive curvature everywhere and a minimum at $D = 0$.

5. Load-Deformation curves and their interpretation

5.1. *Graphical representation*

As is well-known, the load can be obtained from the free energy by differentiation with respect to deformation,

$$P = \frac{\partial \Psi(D, T)}{\partial D}. \tag{5.1}$$

It follows that the shapes of load-deformation curves differ considerably for different temperatures due to the differences in shape of the functions $\Psi(D, T)$. Figure 8 illustrates some typical examples. Curves like these, except that they were graphs of polynomials, were abstracted by Falk [2] from the experimental data about pseudo-elastic bodies.

The dashed parts of these load-deformation curves represent unstable equilibria and we shall never be able to observe these.

To discuss the load-deformation curves of Figure 8 we recall that they refer to uniform bodies. Within the theory this is put in evidence by the fact that we have the same probability P_Δ for all lattice-particles and correspondingly we may define one free energy Ψ/N per particle.

5.2. Interpretation at low temperatures

Figures 8a and 8b refer to a low temperature and we have stable stress-free configurations only at $D \approx \pm 0.8\,NJ$ where all lattice-particles are of type $M_\pm$, i.e. their displacements are close to the sites of the deep minima of the potential energy $\Phi(\Delta)$. In particular, there is no stable configuration at $D = 0$, even though the lattice-particles could realize such a configuration by all attaining the metastable displacement $\Delta = 0$ as shown Figure 3a. However, this configuration cannot occur at low temperatures, because the lattice particles prefer to assume the minimal energy at $\Delta = + J$ or $\Delta = - J$, and the low temperature does not allow them to leave this position.

If, starting at $D \approx - 0.8\,NJ$, where all particles are of type M_-, we apply a tensile load to the body, the deformation will increase with a growing load until D_C is reached; then all particles are ready to flip over to type M_+ and upon a further slight increase of load the body will yield along the dotted line in the load-deformation curve until it reaches the right solid branch of that curve. On this branch we may move up upon further loading or down upon unloading. Complete unloading leads to a stress-free configuration at $D \approx + 0.8\,NJ$ and from there we may move back to the left solid branch by appropriate compression loading.

The so-described loading path corresponds to the observed behaviour of a pseudo-elastic body at a low temperature as exhibited in Figures 1a and 1b. P_Y corresponds there to the height of the dotted line and this height decreases with increasing temperature (see Fig. 8b), according to observations.

The Figures 1a and 1b also contain a virginal curve that starts in the stress-free state at $D = 0$. According to Figures 8a and 8b, the states on that curve cannot be stable in a uniform body; therefore, since the curve is

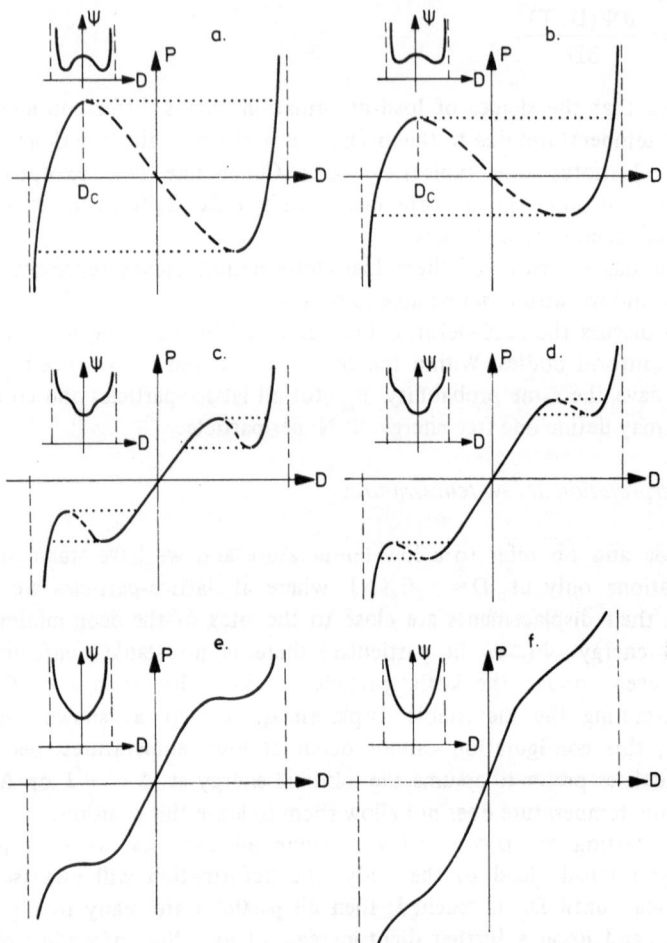

Figure 8. – *Load-deformation-temperature behaviour of the model.*

observed, it must be realized in a non-uniform body. We proceed to discuss this.

5.3. *The virginal curve*

While a uniform body, by the above theory, cannot attain any state within the plastic region enclosed by the solid and dotted lines of Figures 8a and 8b, a non-uniform body can have stable states there. Indeed, Figure 3b shows a non-uniform body in a stress-free configuration at $D = 0$, where all lattice-particles are in stable positions at either $\Delta = J$ or $\Delta = - J$.[1] Obviously this non-uniform body can be deformed elastically by tilting the lattice-particles unless the load exceeds P_Y, when the body yields, because the lattice-particles of type M_- flip to become of type M_+.

A detailed discussion of the deformation of a non-uniform body is presented by Müller & Villaggio [3] who formulated a 'snap-spring model' for an elastic-plastic body.

In a body which has been cooled from the melt and which has not been deformed yet we expect non-uniformity to be the rule, because crystallization of the lattice starts at many places and the two different lattice types M_+ and M_- will form at random, roughly in equal numbers.

5.4. *Interpretation at intermediate temperatures*

The Figures 8c and 8d refer to an intermediate temperature and they permit only one stress-free configuration, namely at $D = 0$. This means that the temperature is now high enough for the lattice-particles to be lifted from the stable minima of the potential energy at $\Delta = \pm J$. They have thus a tendency, assisted by the metastable minimum at $\Delta = 0$, to assemble in the middle of the potential well and form particles of type A, i.e. austenitic particles.

As a consequence a body which has been deformed at low-temperature from the virginal stress-free state to the stress-free state at $D \approx 0.8 \, NJ$, will creep back to $D = 0$ upon heating, i.e. the body will exhibit shape memory. Under a small load the body cannot quite make it back to $D = 0$, but it will still creep to a value $D > 0$ which decreases as the temperature gets higher. Therefore the initial part of the load-deformation curve is steeper for higher temperatures, as observed in Figures 1c and 1d and confirmed by Figures 8c and 8d.

If the load increases beyond the maximum in the first quadrant of Figures 8c and 8d, the body yields along the upper dotted line until it reaches the right solid branch of the load-deformation curve and upon unloading it recovers along the lower dotted line. This shows that our model exhibits the phenomenon of pseudo-elasticity, as observed in Figures 1c and 1d. Figures 8c and 8d also confirm that the yield load P_Y and the recovery load P_R increase with increasing temperature.

Along the dotted yield lines in Figures 8c and 8d there occurs a transition from the austenitic phase A to the martensitic phase M_+. This phase

---

[1] In Figure 3a the lattice-particles are in metastable positions and the body cannot assume this state, because there is a competing state with lower energy.

transition is reversed along the dotted recovery line. This is so because under a big load the lattice-particles find in type M_+ a lower energy than in type A.

Here again the points between the dotted lines cannot be attained by a uniform body, but they can be attained, if the distribution of lattice-particles is non-uniform.

5.5. *Interpretation at high temperature*

The discussion of Figures 8e and 8f is short: Figure 8e represents the critical limiting case in which the maximum and the minimum of the previous figures grow together to form a point of inflection with a horizontal tangent. Figure 8f is a monotone curve whose character does not change even if temperature is raised, except that the slope gets steeper. Both these curves represent a fully elastic behaviour and they are in agreement with the observations shown in Figures 1e and 1f. We may understand this behaviour, if we consider that at high temperature the thermal energy of the lattice particles is so great that the particles may ignore the details of the potential energy at the bottom of the energy well. They behave as they would in a simple box-potential.

Curves like the one in Figure 8f were derived for a 'snap-spring model' of a body with shape memory at a high temperature by Müller in [4].

5.6. *Phase Equilibrium*

To those interested in thermodynamic phase transitions, the free energy-deformation curves of Figure 7 and the load deformation curves of Figures 8c through f may be reminiscent of the isotherms of van der Waals fluids which permit a phase transition. There is one essential difference though: In a van der Waals isotherm there is no hysteresis of the type shown in Figures 8c and 8d. This is so, because in a van der Waals fluid, the liquid and the vapour are always in phase-equilibrium, whereas the austenite and martensite of a pseudo-elastic body are not in phase equilibrium, due to the difficulties of the geometrical readjustment that must occur with the phase change.

REFERENCES

[1] DELAEY L., R.O. KRISHNAN, H. TAS and H. WARLIMONT. – Review: "Thermo-elasticity, pseudoelasticity and the memory effects associated with martensitic trans-formations", *Journal of Materials Science*, 9 (1974): 1521-1555.

[2] FALK F. – "Wie hängt die Freie Energie einer Memory Legierung von der Verzerrung und der Temperatur ab"? *ZAMM Sonderband über GAMM Tagung 1979 in Wiesbaden* (in press).

[3] MÜLLER I. and P. VILLAGGIO. – "A model for an elastic-plastic body." *Arch. Rat. Mech. Anal.*, 65 (1977): 25-46.

[4] MÜLLER I. – "A model for a body with shape memory." *Arch. Rat. Mech. Anal.*, 70 (1979): 61-77.

[5] PERKINS J. ed. – *Shape memory effects in alloys.* New York and London: Plenum Press, 1975.

RESUME

(Fonctions d'état pour un corps pseudo-élastique)

Nous donnons un bref exposé de la phénoménologie du comportement pseudo-élastique des alliages à mémoire. On procède à la construction d'un modèle capable de simuler qualitativement ce comportement. Plus spécifiquement, le traitement du modèle par la mécanique statistique permet la compréhension du comportement en charge – déformation – température des alliages à mémoire comme une conséquence d'une transition de phase martensite – austénite.

DISCUSSION

COMMENT BY R.S. RIVLIN: It may be worth mentioning some papers which were published about 1946 by W.A. Wooster and collaborators. These were concerned with the stress induced phase transitions between the electrical twins of crystalline quartz. The twins are mirror images of each other and accordingly have, in their undeformed states, the same free energies. However, if forces are applied, their free energies are, in general, different and a phase transition from one twin to another may take place. From a structural point-of-view, this transition involves the motion of a simple oxygen atom in each cell. The possibility of a phase change taking place when specified forces are applied arises only if the Gibbs free energy would be decreased by such a change. Even when the free-energy balance is favorable to a phase transition, this does not occur unless the temperature is raised sufficiently (by a few hundred degrees) in order to provide the oxygen atoms with sufficient mobility to overcome the barriers which exist between their possible equilibrium positions.

The free-energy difference in the presence of applied forces may be easily calculated from the known elastic constants of quartz in the case when the competing systems are both homogeneous. Of course, even if we have initially a homogeneous body, consisting entirely of a single phase, and we deform it homogeneously, the transition to the other phase need not take place homogeneously since, as soon as a change of phase has taken place at one point of the test-piece, the stress field will no longer be homogeneous.

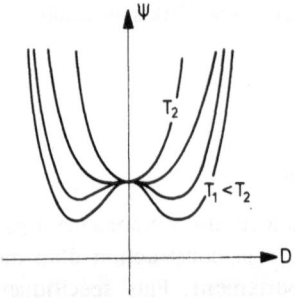

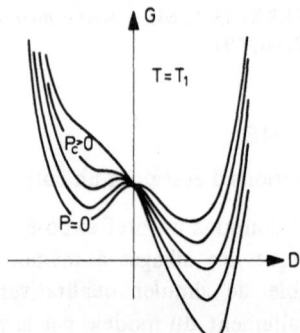

Figure A. – *Free energy versus deformation for different temperatures.*

Figure B. – *GIBBS' free energy versus deformation for different loads.*

I mention this work since somewhat similar considerations may apply to the phase transitions discussed by Professor Müller, particularly to transitions between the martensitic phases. Of course, in the case of martensitic-austenitic transitions a complication arises from the fact that the free-energies of the two phases are different, even if the material is undeformed. Account would presumably be taken of this by introducing the latent heat associated with such a transition.

ANSWER BY I. MÜLLER: If a lattice allows twin formation, and no more than that, the potential energy has two lateral minima like those shown in Figure 2 but no minimum at the center. In such a case, to the best of my knowledge, the free-energy as a function of deformation and for different temperatures has the form shown in Figure A.

In order to obtain the Gibbs free-energy G, appropriate to the load P, one must subtract PD from the free-energy Ψ and thus one obtains the curves of Figure B which refer to one temperature T_1 and different (positive) loads. Thus Professor Rivlin is right when he says that the load favors the formation of one of the twins at the expense of the other one. Indeed, for P = 0 and a low temperature T_1, the curve $G(D, T_1)$ has two equal minima, so that both twins are equally likely to occur, while for P > 0 the left minimum becomes shallower and eventually, for P greater than a critical value P, it vanishes altogether. At a high temperature T_2, where $\Psi(D, T_2)$ has only one minimum, the effect of the load is less drastic: Even at P = 0 both twins and all intermediate states are present, and $P > 0$ favors one twin gradually, rather than suddenly at reaching a critical load.

The form of the free-energy function $\Psi(D, T)$ in my paper differs a little from the above, because the lattice allows twin formation and an additional stable configuration at $\Delta = 0$ (see Fig. 2). This leads to a free energy which may have three minima, or one minimum and four points of inflection (see Fig. 7). Nevertheless, here too subtraction of PD from Ψ leads to the Gibbs free-energy in the same way as above. Due to the greater complexity of the curves $\Psi(D, T)$, however, we may still have a sudden formation of one of the twins at temperatures where $\Psi(D, T)$ has only one minimum. This may occur as long as the curves $\Psi(D, T)$ have points of inflection (see Fig. 8c and 8d).

Anisotropies and Elasticity in Liquid Crystals

Y. Poggi, J. C. Filippini and B. Malraison

Laboratoire d'Electrostatique, C.N.R.S., Grenoble, France.

1. Introduction

The state of matter, discovered by Reinitzer [20] termed liquid crystalline, or mesomorphic, has structural properties intermédiate between those of a solid crystal and a liquid. They possess many of the mechanical properties of a liquid e.g. high fluidity, inability to support shear. At the same time, they are similar to crystals, in that they exhibit anisotropy in their optical, electrical and magnetic properties. So, the quintessential property of a liquid crystal is its anisotropy. In this paper, we will list some of its manifestations and introduce an order parameter which characterizes and quantifies it.

Liquid crystals are found among organic compounds. The molecules are elongated. Liquid crystallinity is more likely to occur if the molecules have flat segments i.e. benzene rings. Friedel [8] has proposed a classification of the states of matter based on their structural differences. He has distinguished two types of liquid crystals:

i) the nematic state is characterized by a long range orientational order, the long axes of the molecules tend to align, locally along a preferred direction which may vary throughout the unstrained medium. Many of the interesting phenomena are described by the fluctuations of this preferred axis, and it is useful to define a unit field vector n(r), called the director, giving its local orientation.

ii) the smectic phase is characterized by a layer structure. The centers of gravity of the elongated molecules are arranged in equidistant planes: the long molecular axis being perpendicular to this planes (smectic A) or tilted (smectic B, C...). In practice the layers are rather thick (20 to 30 Å) and they give rise to typical Bragg reflections in small angle X-ray experiments. The layers can slide freely over one another.

We note that there is still another type of liquid crystals, the cholesteric state, classified by Friedel [8] as a variety of the nematics. Their structure is the same as of the nematics, but it has an additional twist about an axis perpendicular to the long axes of molecules. So, a nematic is a cholesteric of infinite pitch.

For all these mesomorphic phases, such an arrangement takes place only below a transition temperature T_{NI}. Above T_{NI}, the compounds present an isotropic phase characterized by a short range order responsible for the anomalous physical properties of this phase, such a structural order, characterized by $n(r)$, is easily distorted and can be aligned, thanks to the elasticity of these compounds, by magnetic and electric fields, or by surfaces which have been properly prepared.

We can see that this is a very broad subject, and so in this paper, we shall limit ourselves to discussing the nematic phase and its transitions to isotropic and smectic A phases.

The behaviour of the system under varying temperature can be well defined, if it is characterized by an order parameter which is non − zero in the liquid crystalline phase and vanishes in the isotropic one. Let us suppose, for simplicity, that the orientation of the molecule is described by a unit vector ν_i which should not be confused with the director n giving the average prefered direction of the molecules: the order parameter is commonly defined by a numerical parameter

$$S = \frac{1}{2} < 3 \cos^2 \theta - 1 > \tag{1}$$

where θ is the angle between ν_i and n, and the bracket $<>$ denotes a thermal average. An instantaneous picture at the molecular level should look like the picture of Figure 1: Figure 1a refers to a temperature close to the first order transition to the isotroqic phase at a temperature T_{NI}; Figure 1b is obtained at a lower temperature; Figure 1c shows the picture of the isotropic phase.

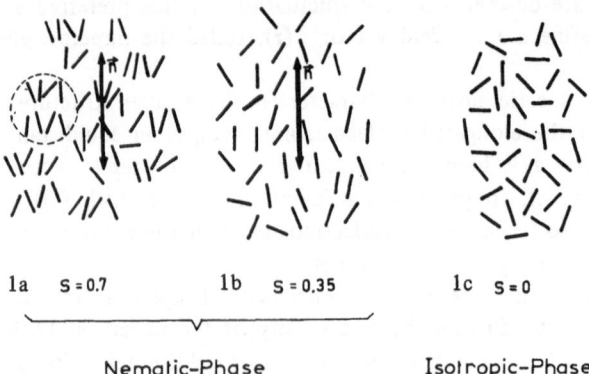

1a   S = 0.7   1b   S = 0.35   1c   S = 0

Nematic-Phase   Isotropic-Phase

Figure 1. − *Instantaneous picture of a nematic phase at the molecular level.*

2. Anisotropies and order parameter: the nematic-isotropic transition

In real liquid crystals, the molecules may be flexible in contrast to the rigid model considered above. So, as suggested by de Gennes [1], it is convenient to define the amount of order in terms of a macroscopic property which is independent of any assumption about the rigidity of the molecules. Thus, we have chosen to represent the amount of order by the diamagnetic susceptibility anisotropy χ_a, because the relation between χ_a and molecular properties is well understood. Any other anisotropy could have been chosen as well. Figure 2 shows for MBBA (p-methoxybenzylidene p-n-butylaniline) the behaviour, with varying temperature, of the diamagnetic susceptibility. In the isotropic phase, far from the transition, we have a disorganized state, and the diamagnetic susceptibility average $\overline{\chi}$ is measured. In the nematic phase, the long molecular axis remains parallel to the measuring magnetic field direction, and the susceptibility parallel to the magnetic field is measured ($\chi_\parallel^N$). From these measurements we derive the magnetic anisotropy in the nematic phase [9],

$$\chi_a = \frac{3}{2} (\chi_\parallel - \overline{\chi}),\qquad(2)$$

and its behaviour with varying temperature as shown in Figure 3. The same curves are obtained for different liquid crystals.

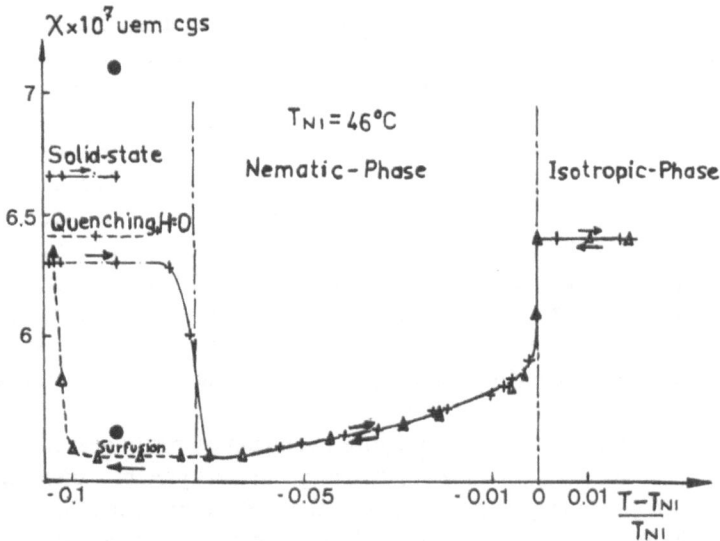

Figure 2. — *Temperature behaviour of the diamagnetic susceptibility of MBBA.*

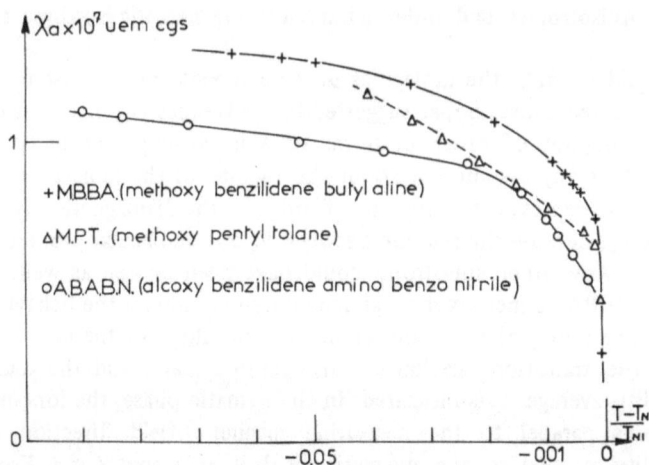

Figure 3. – *Temperature behaviour of the diamagnetic anisotropy for different nematic liquid crystals.*

Concerning the dielectric anisotropy, it is determined by a capacity measurement. The liquid crystal is sandwiched between two sealed flat ceramic plates. The faces in contact with the liquid crystal are coated with chromium and gold. The cell is thermostated to an accuracy better than 0.05°C and placed in a 10 kOe magnetic field. The measuring electric field can be parallel or perpendicular to that of the orienting magnetic field, and it allows to measure the dielectric anisotropy $\Delta\epsilon = \epsilon_{//} - \epsilon_{\perp}$, where $\epsilon_{//}$ is the dielectric constant in a direction parallel to the magnetic field, and $\epsilon_{\perp}$ the dielectric constant in an orthogonal direction. Figure 4 shows for different nematic liquid crystals the variations of $\Delta\epsilon$ with temperature.

On the other hand, the refractive indices have been measured with a prism method. The prism is made of two flat electrodes inclined at a small angle. The electrodes are rubbed parallel to the prism ridge, in order to induce a wall orientation effect. The incident beam is orthogonal to the entering direction and then refracted. It gives two deflected directions connected with the ordinary and extraordinary refractive indices. To this configuration, Descartes' laws are applied. The prism is in an over regulated to an accuracy better than 0.05°C and placed on a goniometer plate. Figure 5 shows the variations, with varying temperature, of the refractive indices anisotropy Δn.

From Figures 2, 3, 5 we can derive a common property: near the nematic isotropic transition T_{NI}, all these physical parameters present a discontinuity. So, the nematic-isotropic transition is of the first order. Hence, it should be worthwhile to analyze the behaviour of the system near such a transition, and to study the temperature dependence of the order parameter S defined as the

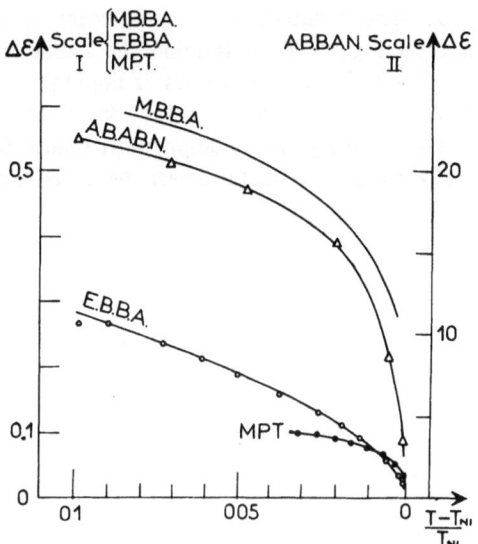

Figure 4. – *Temperature behaviour of the dielectric anisotropy for different nematic liquid crystals.*

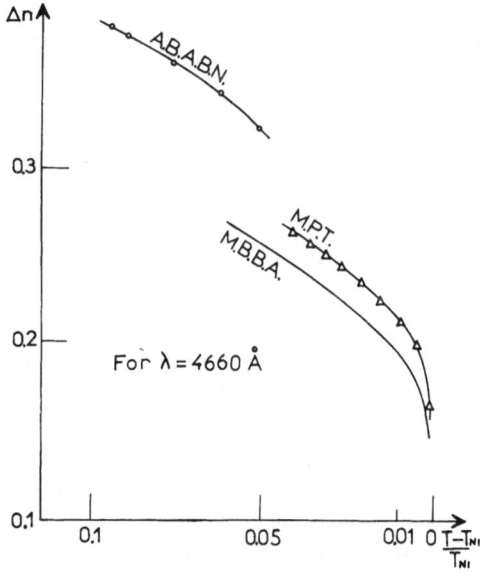

Figure 5. – *Temperature behaviour of the refractive indices anisotropy for different nematic liquid crystals.*

ratio [5] of the diamagnetic susceptibility anisotropy in the nematic phase χ_a and in a monocrystal $\chi_{a,0}$. To determine the behaviour of $S = \chi_a/\chi_{a,0}$ close to T_{NI} [18] we have chosen a new class of liquid crystals, the biphenyls, having a very good chemical stability, and most of our experiments were performed on 7 CB (4'-n-heptyl-4-cyanobiphenyl). Figure 6 gives the temperature dependence of the order parameter S near the nematic-isotropic transition

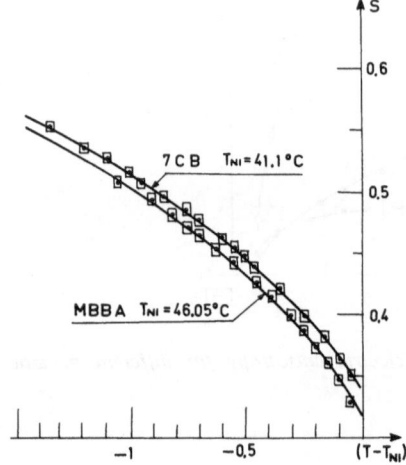

Figure 6. — *Temperature behaviour of the order parameter near the nematic-isotropic transition.*

The first order character of the nematic-isotropic transition has been indicated by various statistical theories [5, 17] which give only a qualitative description of the nematic phase. The mean field theory of Maier and Saupe [21] also predicts a first-order transition, and leads to a better evaluation of S, but the universal function $S(T/T_{NI})$ departs from the experimental results, especially in the vicinity of T_{NI}. All the approaches are unable to predict the important pretransitional phenomena observed [2, 3, 13] in the isotropic phase. This is not surprising, because these statistical theories are very simple, with essentially one kind of forces being retained : steric repulsions in Onsager's theory [17], and the Van-der-Waals forces in the Maïer and Saupe approach [21]. A more complete treatment should take into account both energetic attractions and steric repulsions, which appear to have comparable importance. In the absence of such a theory, the Landau approximation [19] (expansion of the free energy in terms of an order parameter) [12] yields a phenomenological approach of great interest. Contrary to the magnetic case, the presence of a non-vanishing term of order S^3 in such an expansion implies the phase transition to be of the first order :

$$F - F_0 = \alpha(T - T^*) S^2 - \frac{1}{3} \beta S^3 + \frac{1}{4} \gamma S^4 + \ldots \tag{3}$$

where α, β, γ are temperature independent, and T^* is a temperature slightly below T_{NI}, representing a second order phase transition temperature if such a phenomenon was not shifted by a first order transition occuring at T_{NI}.

In zero field there is no term linear in S in the previous expansion: this ensures the state of minimum F to be a state of zero S, i.e. an isotropic one. Thus for T_{NI}, F and $\partial F/\partial S$ must be zero. By delating the isotropic solution $S = S_C$, we get the following law of behaviour of the order parameter:

$$\frac{3}{2}\frac{S}{Sc} - \left(\frac{S}{Sc}\right)^2 = \frac{1}{2}\frac{T - T^*}{T_{NI} - T^*}. \tag{4}$$

As shown in Ref. [19], our experimental results satisfactorily follow this law for $T^* = 40.2 \pm 0.1\,^\circ$C. The theoretical curve is represented in Figure 6 by the solid line. Contrarily to the Maïer and Saupe theory, Landau's theory accounts very well for the experimental results and describes the behaviour of the system in the vicinity of the nematic isotropic transition. Moreover, the difference $T_{NI} - T^* = 0.9 \pm 0.1\,^\circ$C agrees very well with direct measurements we have performed in the isotropic phase as we are going to show it in the next section.

3. Pretransitional Phenomena in the isotropic phase. Electric and magnetic birefringences

Since the nematic-isotropic transition is only weakly of the first order, we expect that short range order effects will be prominent at temperatures just above the transition point T_{NI}; particularly, the coherence length $\xi(t)$ will be rather large, typically a few hundred angströms, ten times the molecular length (a few thousand angstroms in the nematic phase). So, if in the isotropic phase liquid crystals look like ordinary liquids, this short-range order influences the behaviour of many physical properties which are quite different from those of ordinary liquids. Two of the various methods of investigating these pretransitional phenomena consist in measuring the Cotton-Mouton and Kerr effect of the substance. The short range effects near the nematic-isotropic transition have been discussed by de Gennes [1]. For the magnetic birefringence, the theory gives

$$n_{\parallel} - n_{\perp} = (T - T^*)^{-\gamma} H^2 = C_M \lambda H^2, \tag{5}$$

where $n_{\parallel}$ and $n_{\perp}$ are the refractive indices measured for polarization respectively parallel and normal to the magnetic field H, and $\gamma = 1$ in a mean field theory. For the electric birefringence, one may expect a similar formula expressing the dependence on the electric field. The experimental devices are shown in Figures 7 and 8. Detailed information is given in references [2] and [3]. In Figures 9 and 10 we have reported, for 7 CB, the inverse of the Cotton-Mouton and Kerr coefficients C and B, respectively, versus temperature. These plots exhibit a linear part. By extrapolating it to the temperature axis, one can get the temperature T^*. Then, C and B are proportional to $(T - T^*)^{-1}$.

The temperature difference $T_{NI} - T^* = 1.1 \pm 0.1$ °C is in good agreement with that we have calculated from experiments performed in the nematic phase and described in Section 2. We notice that in the vicinity of the transition, a systematic deviation from the mean field theory is observed as in the case of the nematic phase. This experiment indicates a cross-over between a classical mean field behaviour sufficiently above T_{NI} and a critical one, in close analogy with the cross-over between the classical mean field dependence of the magnetic susceptibility and the critical one, as transition temperature is approached.

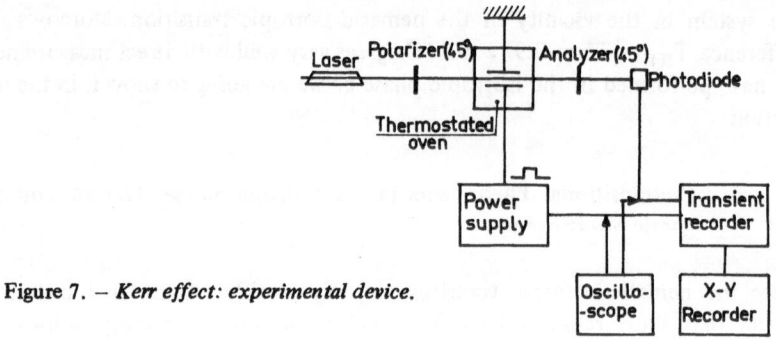

Figure 7. – *Kerr effect: experimental device.*

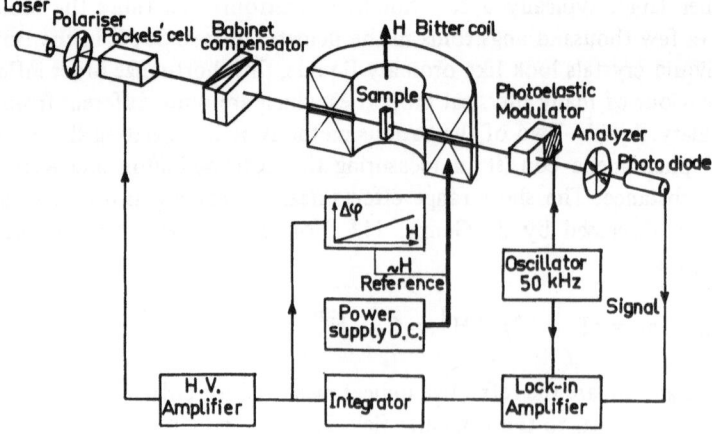

Figure 8. – *Cotton-Mouton effect: experimental device.*

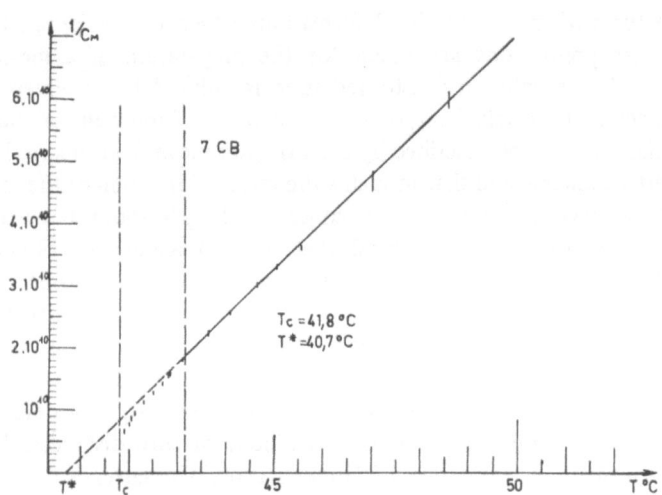

Figure 9. – *Temperature behaviour of the inverse of the Cotton-Mouton coefficient.*

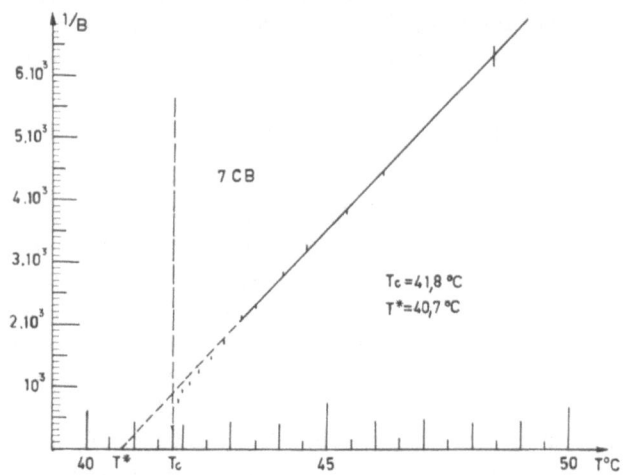

Figure 10. – *Temperature behaviour of the inverse of the Kerr coefficient.*

4. Elasticity and orientation fluctuations in nematics: smectic A – nematic transition

The subject of this section is the particular elasticity of the mesophases, that means the energetic effects associated to a static deformation of a monocrystal. Thus, we will describe the continuous or discontinuous deformations of the ideal structures of monocrystals: deformations which can be produced when no particular precautions are taken for the preparation of a mesomorphic sample, or when a voluntary distorted state is induced by an external stress, i.e. a magnetic or electric field, or some treatment of the walls of the sample. Such an elasticity can be described by a continuous mean field theory. Nevertheless it must be emphazised that in such a theory, the 'direction of the molecules' means the average orientation on a coherence length about a few thousand angströms. Keeping this fact in mind, there is no fundamental difference with the elastic theory of solids: same phenomenological theory, same expansion of the energy for weak deformations. The originality of the elasticity of the mesomorphic compounds lies in the fact that the elastic energy is not a function of the local motions of the molecules, but of local rotations of the orientation of the molecules as it is expected by the Frank theory [6]. For deformations, the wave length of which is greater than the intermolecular distances, the elastic energy is given by an expansion of Hooke's type. The latter is valid only for weak deformations: it is a quadratic form of the local curvatures expressed in terms of the director $n(r)$ and its gradient. For an uniaxial system as the nematic one, there are only three fundamental distortions: the splay (Fig. 11a) expressed by $\operatorname{div} n$, the twist (Fig. 11b) expressed by $n \cdot \operatorname{rot} n$, and the bend (Fig. 11c) expressed by $n \times \operatorname{rot} n$. So, the free energy is given by the expression

$$ F = \frac{1}{2} \int [K_1(\operatorname{div} n)^2 + K_2(n \cdot \operatorname{rot} n)^2 + K_3(n \times \operatorname{rot} n)^2]\, dv\,, \qquad (6) $$

K_1, K_2, K_3 being the elastic constants related respectively to the previous quoted deformations. These elastic deformations have been the subject of many studies [10, 11, 16]. The elastic constants are measured using the Freedericksz transition [7] which describes the distortion of a nematic film, initially aligned by a wall effect, by a magnetic field H. The molecules align their long axes parallel to the magnetic field direction, and in Figures 12a, 12b, 12c, we can see the three geometries allowing the determination of the elastic constant (respectively twist, splay, bend). It is experimentally ascertained that a rocking motion of the molecules takes place if the magnetic field exceeds

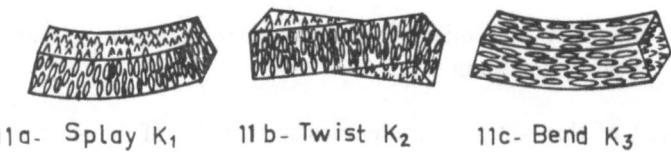

11a- Splay K₁   11b- Twist K₂   11c- Bend K₃

Figure 11. − *Fundamental distortions in a nematic liquid crystal.*

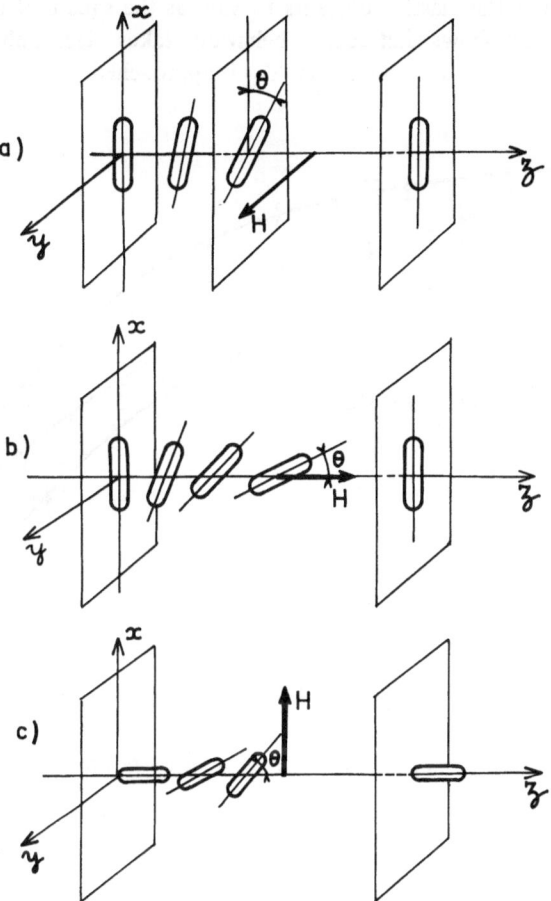

Figure 12. — *Geometries allowing the determination of the elastic constants.*

the critical value H_C of the applied magnetic field directly related to the elastic constant, the thickness d of the sample and the diamagnetic anisotropy χ_a, and given by

$$H_{c,i} = \frac{\pi}{d} \sqrt{\frac{K_i}{\chi_a}}, \quad \text{where} \quad i = 1, 2, 3. \tag{7}$$

Thus, if χ_a and d are known, the measurements of H_C in each of the three previous geometries leads to the determination of the elastic constants, whose dependence on temperature is given in Figures 13, 15. The mean field theory predicted the elastic constants to vary as the square of the order parameter. Figure 14 shows that such a behaviour takes place only far from the transition temperature T_{NI}, but fails if T_{NI} is approached.

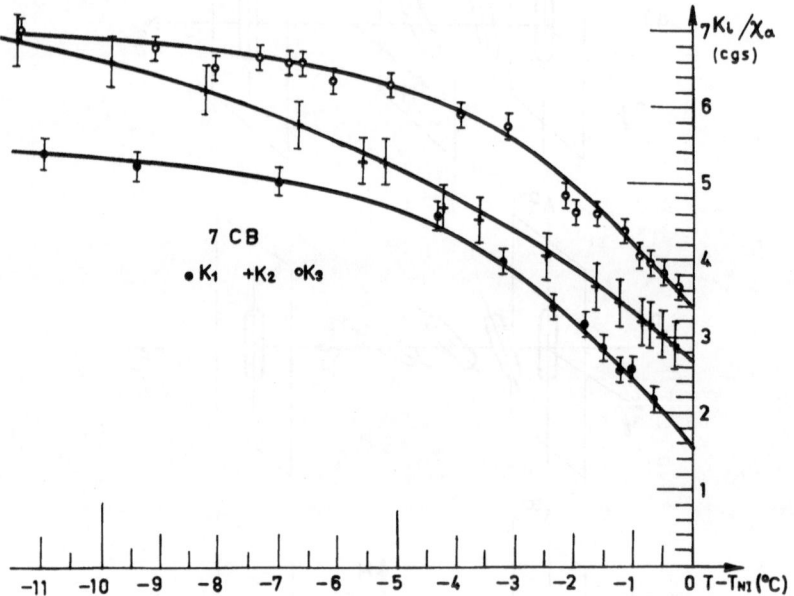

Figure 13. — *Temperature dependence of the elastic constants of 7 CB.*

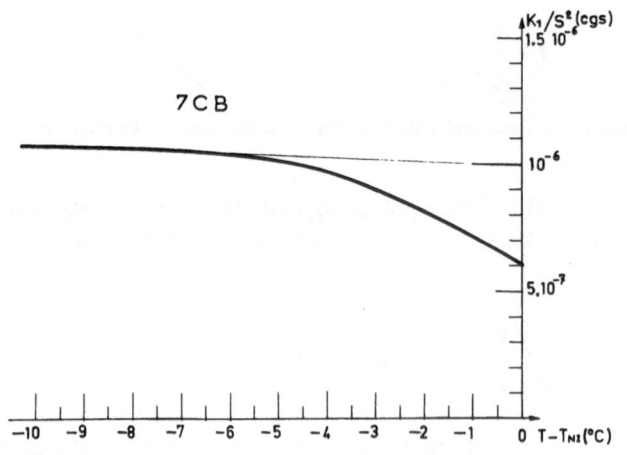

Figure 14. — *Departure from the Maïer and Saupe theory of the temperature dependence of the elastic constants.*

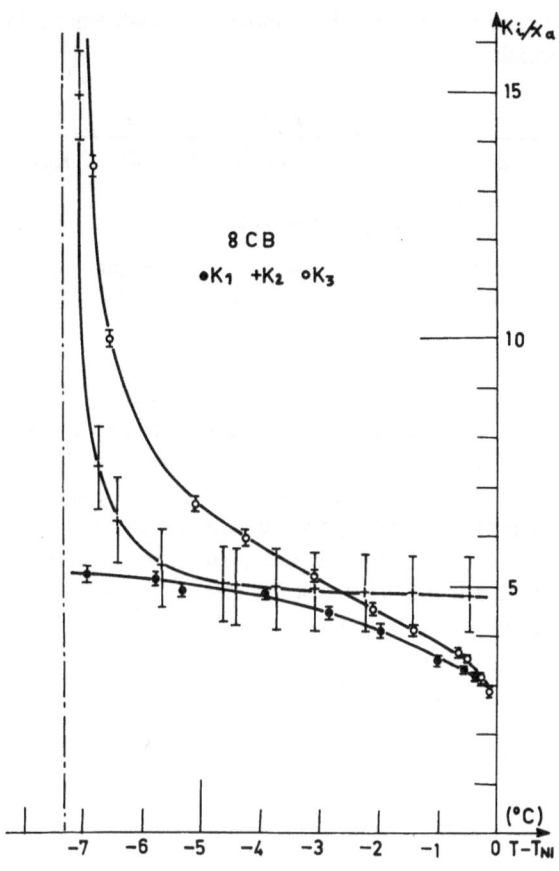

Figure 15. – *Temperature dependence of the elastic constants of 8 CB which has a smectic A phase.*

These elastic deformations, strongly dependent on temperature, are induced in the nematic medium by fluctuations of the orientation of the director $n(r)$. It is interesting to study them, and their quenching by an external stress such as a high magnetic field. De Gennes [1] has considered the effect of a magnetic field parallel to the optical axis of a uniformly aligned nematic film. We generalized the calculation, which was performed with only one elastic constant K, to include the three elastic constants K_1, K_2, K_3. Consequently, in formula (6) we must add a term of the magnetic energy $\chi_a \, n(r) \, H^2$. The effect of a magnetic field parallel to the optical axis of a nematic single crystal is to decrease the effect of the fluctuations and thus to increase the birefringence of the sample. The mathematical treatment [15] leads to the formula

$$\frac{\Delta n(H, T) - \Delta n(H = 0, T)}{\Delta n_0} = \frac{k_B T}{4\pi} \sqrt{\frac{\chi_a}{K_3}} \left(\frac{1}{K_1} + \frac{1}{K_2}\right) |H| . \qquad (8)$$

At this stage, we should make several remarks:

i) The predicted effect is very small. Using $\chi_a \sim 10^{-7}$ C.G.S., $K_i \sim 10^{-6}$ C.G.S., one calculates $\delta(\Delta n)/\Delta n_0 \sim 10^{-4}$ for $H = 10^5$ Oe. Hence, very sensitive optical techniques as well as unusually large fields are required.

ii) Near a nematic-smectic transition, the twist K_2 and bend K_3 elastic constants diverge; whereas the splay constant remains regular, as shown in Figure 15 for 8 CB (octylcyanobiphenyl), a compound which presents a nematic smectic A transition for $T = T_{NA} = 32.5\,°C$. In the smectic phase, the presence of incompressive layers prevents the existence of twist and bend distorsions. In the pretransitional state, one can understand the singular behaviour as coming from the increasing role of fluctuating smectic domains. The critical behaviour is reflected in the continuous freezing of $\delta(\Delta n)$ as T_{NA} is approached from above.

iii) On the other hand, the decrease of the elastic constants near the isotropic nematic transition ($T = T_{NI}$) is faster. So, one expects an increase of $\delta(\Delta n)$ as T_{NI} is approached from below.

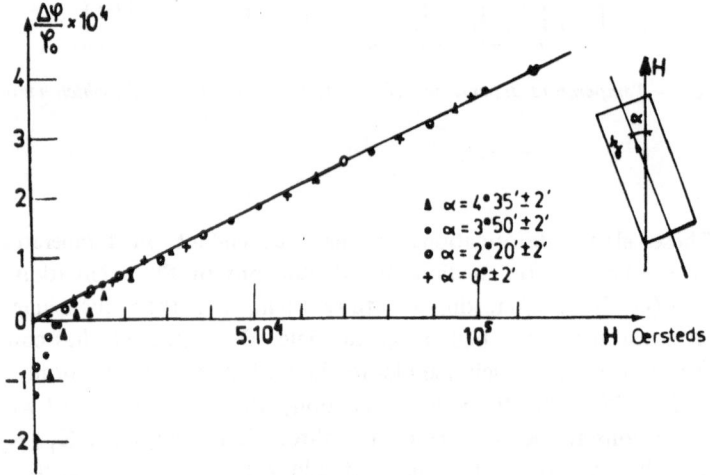

Figure 16. – *Relative phase shift versus magnetic field for different values of the angle* α *between the optical axis of the sample and the direction of the magnetic field.*

Thus we see in Figure 16 that, in agreement with the theory, a characteristic feature is the non-analytic dependence such as that of |H|. Indeed, we verify that by changing the direction of the magnetic field by 180° we get the same value of the birefringence. This result can be easily understood as coming from the symetry $n(r) \rightarrow -n(r)$ of the nematic phase. Figure 17 gives the temperature dependence of the increase of the slope of the induced excess birefringence (I.E.B.) $\delta(\Delta n)$ normalized to $k_B T H$. We see that, for a given field, $\delta(\Delta n)$ increases drastically near the nematic-isotropic transition $(T = T_{NI})$, whereas it decreases continuously to zero at the smectic A − nematic transition $(T = T_{NA})$. We note that experimental results agree with the theoretical predictions we have previously made.

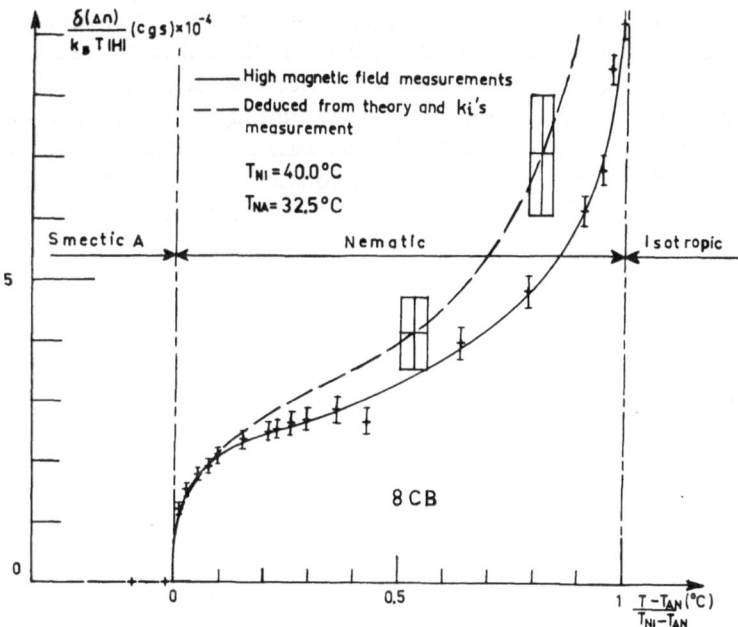

Figure 17. − *Temperature dependence of the I.E.B. in the nematic phase of 8 CB.*

5. Conclusion

The anisotropic nature of liquid crystals and the elastic properties of these materials have led to important applications during the last five years. Most of them are display devices in which optical effects are induced by an electric field: the liquid crystal material is contained between two closely spaced plates, both coated on the inside with a transparent conductive layer. Deformation of the initial structure occurs above a critical voltage, due to the competition between the orientation action of the electric field, because of the dielectric anisotropy, and the elastic forces. Various types of display devices have been developed: digital or alphanuperic displays for calculators, electronic clocks or watches, multimeters, data output equipment, signal displays in traffic and industrial control systems, advertising displays, large-area picture screens, TV pannels, image converters, page composers for hologram storage, etc.

The anisotropic nature of liquid crystals also appears in the isotropic phase, in the vicinity of the nematic-isotropic transition temperature where large values of the Kerr constants, up to 100 times larger than for nitrobenzene, are reached. Recently synthetized nematogenic liquids may be used in Kerr cells applications in place of nitrobenzene: they have two important advantages over this liquid: (a) the driving voltages are 4 to 5 times lower than for nitrobenzene, (b) the resistivity is high enough for the cell to be sealed in every case, whereas a dynamic deionization was necessary for certain applications of nitrobenzene. Various electrooptical devices using liquid crystals in the isotropic phase, e.g., d.c. biased Kerr cell light modulators, a.c. transverse light modulators, a.c. transverse light modulators, rotating field Kerr cell modulators, light deflectors — have been recently developed in the laboratory for Electrostatics of the CNRS in Grenoble [4].

REFERENCES

[1] de GENNES P.G. – ed. *The Physics of Liquid Crystals*. Oxford: Clarendon Press, 1974.
[2] FILIPPINI J.C. and Y. POGGI. – *Journal de Physique Lettres* 35 (1974): L99.
[3] FILIPPINI J.C., Y. POGGI and G. MARET. – Physique sous champs magnétiques intenses. *Colloques Internationaux du C.N.R.S.* 242 (1974): 67-69.
[4] FILIPPINI J.C. and Y. POGGI. – Submitted to publication in Optics.
[5] FLORY P.J. – *Proceedings of the Royal Society* A234 (1956): 73.
[6] FRANK F.C. – *Discussions of the Faraday Society* 25I (1968).
[7] FREEDERICKSZ V. and V. ZOLINA. – *Translations of the Faraday Society* 29 (1933): 919.
[8] FRIEDEL G. – *Annales de Physique*, 19 (1922): 273.
[9] GASPAROUX H., B. REGAYA and J. PROST. – *Comptes Rendus de l'Académie des Sciences Paris*. 276B (1973): 643.
[10] GRULER H. – *Zeitung für Naturforschung*. 28A (1973): 474.
[11] JAHNIG E. and H. SCHMIDT. – *Annales de Physique*. 71 (1972): 129.
[12] LANDAU L.D. – ed. *D. Ter Haar (Gordon and Breach)*. collected papers N.Y. 1974.
[13] Mac COLL J.R. and C.S. SMITH. – *Physical Revue Letters* 29 (1972: 85.
[14] MALRAISON B., Y. POGGI and J.C. FILIPPINI. – *Solid State Communications* 31 (1979): 843-845.
[15] MALRAISON B., Y. POGGI and E. GUYON. – *Physical Review A* 21: 1012-1024.
[16] NEHRING J. and A. SAUPE. – *Journal of Chemical Physic* 54 (1971): 337.

[17] ONSAGER L. – *Annales New-York Accademy Sciences* 51 (1949): 627.
[18] POGGI Y., R. ALEONARD and J. ROBERT. – *Physics Letters* 54A (1975): 393.
[19] POGGI Y., P. ATTEN and R. ALEONARD. – *Physical Review A* 14A (1976): 466.
[20] REINITZER F. – *Wiener Monatsch Chemie* 9 (1888): 421.
[21] SAUPE A. and W. MAIER. – *Zeitung für Naturforschung* 14a (1959): 882, 15a
(1960): 287, 16a (1961): 816.

RESUME

(Anisotropies et élasticité dans les cristaux liquides).

Après une courte présentation des différentes phases des cristaux liquides on examine l'ordre moléculaire dans la phase nématique : on présente les aniso-tropies de susceptibilité magnétique, de permittivité, d'indice de réfraction et on étudie l'évolution du paramètre d'ordre – représenté par l'anisotropie de susceptibilité diamagnétique – en fonction de la température, plus parti-culièrement près de la transition nématique-isotrope. On examine ensuite les phénomènes prétransitionnels au voisinage de la transition nématique-isotrope et on montre comment le modèle de Landau rend compte des résultats expé-rimentaux. Une grande partie de la communication est consacrée à l'élasticité dans les nématiques et les smectiques : on présente la théorie de Frank et on montre comment il est possible de déterminer expérimentalement les trois constantes élastiques à partir de la transition de Freedericksz. L'importance de l'élasticité dans les cristaux liquides est illustrée par la description d'une expérience récente sur la réduction des fluctuations thermiques d'un mono-cristal nématique par un champ magnétique. La conclusion fait apparaître le rôle de l'anisotropie et de l'élasticité dans les applications des cristaux liquides nématiques.

Aspects Théoriques et Expérimentaux des Équations d'État d'Origine Élastique de Solides Anisotropes

A. Waintal

Laboratoire Louis Néel, CNRS, Grenoble, France.

1. Introduction

Le but de cet article est de donner un aperçu général des expériences de propagation des ondes ultrasonores que le physicien utilise pour l'étude des milieux anisotropes soumis à des contraintes et plus spécialement à une pression hydrostatique. La théorie de la thermoélasticité non linéaire est l'instrument privilégié pour de telles études, car elle permet, entre autres, par l'interprétation de ces expériences, le calcul des coefficients élastiques d'ordre deux et trois et la détermination des équations d'état d'origine élastique.

Après quelques rappels sur les équations de base de la thermoélasticité nécessaires à l'expérimentateur pour l'interprétation des expériences de propagation d'ondes ultrasonores, on décrit quelques ensembles expérimentaux classiques (appareillages de pression hydrostatique avec leurs électroniques de mesure). Finalement, comme application, on déduit une équation d'état d'origine mécanique pour les cristaux cubiques que l'on illustre sur le silicium par comparaison avec l'équation d'état obtenue expérimentalement par diffraction de RX jusqu'à 150 kbar.

2. Rappel sur les équations de la thermoélasticité

La théorie de la thermoélasticité, appliquée à l'étude de la propagation des ondes ultrasonores dans les solides sous contrainte, a fait l'objet de nombreux travaux [1, 2, 3, 10, 17]..., mais le point de vue adopté dans cette présentation est celui de Wallace [24], car il présente l'avantage d'être directement utilisable par l'expérimentateur.

Quand on étudie les propriétés des ondes ultrasonores dans les solides, on décrit les phénomènes en terme de mouvement de particules matérielles autour de leur configuration d'équilibre, c'est-à-dire en terme de déformations mesurées à partir d'une configuration initiale. Si on considère des ondes élastiques de grande longueur d'onde, la déformation est considérée comme homogène et infinitésimale. Par contre, quand le solide est soumis à des contraintes extérieures importantes, la déformation, bien que petite, ne peut plus être considérée comme infinitésimale.

Ainsi, trois sortes de configurations vont être définies :

— une configuration initiale sans déformation $(\overline{X})$;
— une configuration déduite de la précédente par application d'une contrainte, (X). Cette configuration peut être considérée comme initiale quand on effectue des expériences d'ondes ultrasonores ;
— une configuration finale déduite des précédentes par une déformation infinitésimale (x).

On désignera par X et x la position d'un petit élément de masse dans les configurations initiale et finale, par $\rho(X)$ et $\rho(x)$ les densités.

On appellera D_{ij} le tenseur symétrique de Green-Lagrange :

$$D_{ij} = \frac{1}{2}\,(\alpha_{ki}\alpha_{kj} - \delta_{ij}),\quad \alpha_{ij} = \frac{\partial x_i}{\partial X_j}$$

2.1. *Fonctions thermodynamiques*

L'énergie libre dépend seulement des positions relatives des particules matérielles :

$$F(x, T) = F(X, D_{ij}, T)$$

Puisque la déformation est petite, bien que non infinitésimale, il est possible d'utiliser un développement limité à partir de l'état initial X.

$$\rho(X)\,F(X, D_{ij}, T) = \rho(X)\,F(X, O, T) + C_{ij}\,D_{ij} + \frac{1}{2}\,C_{ijk\ell}\,D_{ij}\,D_{k\ell}$$
$$\frac{1}{3!}\,C_{ijk\ell mn}\,D_{ij}\,D_{k\ell}\,D_{mn}\,.$$

Par définition, les coefficients $C_{ijk\ell}$... sont les dérivées d'ordre 2 et 3 de $\rho(X)F$ par rapport aux D_{ij} et sont appelés les constantes élastiques d'ordre 2, 3 ...

C_{ij} est le tenseur des contraintes de Cauchy

$$C_{ij} = T_{ij} = \rho(X)\,\frac{\partial F}{\partial D_{ij}}\quad \text{dans la configuration d'origine}$$

Le nombre des composantes d'un tenseur de rang 6 est égal à 729 et se réduit par la symétrie du tenseur D_{ij}. La symétrie du cristal réduira encore le nombre des composantes. Ainsi ce nombre tombe-t-il à 8 pour les classes m3 et 23 et à 6 pour les classes $\overline{4}3$m, 432, m3m des cristaux cubiques et à 3 pour un corps isotrope.

2.2. *Equations de base*

a) Les principes de la thermodynamique, le principe d'Alembert sur les travaux virtuels permettent de calculer les composantes des contraintes, à partir de l'énergie, dans la configuration finale.

$$T_{ij} = \rho(x)\, \alpha_{ik}\, \alpha_{j\ell}\ \frac{\partial F}{\partial D_{k\ell}}$$

b) L'équation de continuité, basée sur la géométrie des déformations donne :

$$\frac{\rho(X)}{\rho(x)} = I = \det\,[\alpha_{ij}]$$

c) La loi fondamentale de la mécanique donne l'équation du mouvement :

$$\rho(x)\ \frac{\partial^2 x_i}{\partial t^2} = \frac{\partial T_{ji}}{\partial x_j}$$

2.3. *Propagation des ondes élastiques de faibles amplitudes*

On considère maintenant les petits mouvements à partir de la configuration d'origine et on linéarise les équations du mouvement en X.

On montre que : [24]

$$\rho(X)\ \frac{\partial^2 x_i}{\partial t^2} = A_{ijk\ell}\ \frac{\partial^2 x_k}{\partial X_j\, \partial X_\ell}$$

avec

$$A_{ijk\ell} = T_{j\ell}\,\delta_{ik} + C_{ijk\ell}$$

$A_{ijk\ell}$ sont appelés coefficients de propagation d'ondes.

Il faut maintenant construire la matrice de propagation L. La solution de l'équation différentielle est de la forme :

$$x - X = W \exp\,(kX - \omega t)$$

$$\rho(X)\, V^2 W_i = A_{ijk\ell}\, K_j\, K_\ell\, W_k$$

$$V = \frac{\omega}{k} \quad K = \frac{k}{|k|} \qquad\qquad L_{ik} = A_{ijk\ell}\, K_j\, K_\ell\,.$$

La discussion des modes propres de la matrice L est un problème classique d'algèbre linéaire. On retrouve les formules habituelles relatives à la configuration initiale sans contrainte en faisant $T_{ij} = 0$. Les coefficients de propagation se confondent alors avec les constantes élastiques usuelles. La grandeur expérimentale à laquelle on a accès est la vitesse V par l'intermédiaire de la mesure des temps t de propagation des ondes.

2.4. *Calcul des coefficients élastiques du 3e ordre dans une configuration initiale sans contrainte*

On a considéré, jusqu'à présent, une configuration initiale dans laquelle le solide était sous contrainte et on a vu que la propagation des ondes de faible amplitude se calculait à partir des composantes initiales des contraintes et des constantes élastiques du deuxième ordre. Cependant, pour la plupart des expériences, la contrainte initiale est faible, comparée aux constantes élastiques et il est alors intéressant de formuler les constantes élastiques dans la configuration sous contrainte au moyen d'un développement limité des constantes élastiques de la configuration sans contrainte.

Les données surlignées représentent les paramètres spécifiant la déformation de la configuration $\overline{X}$ (contrainte nulle) à celle X (sous contrainte).

On a alors :

$$\rho_0 F(\overline{D_{ij}}, T) = \rho_0 F(O, T) + \frac{1}{2} \, \overline{C_{ijk\ell}} \, \overline{D_{ij}} \, \overline{D_{k\ell}}$$

et après de longs calculs, on trouve l'expression de $C_{ijk\ell}$ en fonction des paramètres de la déformation et des constantes élastiques d'ordre 2 et 3 de la configuration $\overline{X}$.

Les constantes élastiques du second ordre dans la configuration sous contrainte se calculent à partir des constantes d'ordre supérieur de la configuration sans contrainte et des paramètres de la déformation. Pour interpréter les expériences de propagation sous contrainte, il est pratique d'introduire [22] la notion de "vitesse naturelle" W. $W = \ell_0/t$, ℓ_0 étant le chemin acoustique dans la configuration sans contrainte, t le temps de parcours de l'onde acoustique dans la configuration sous contrainte.

Alors que la vitesse de propagation est V : $V = \ell_1/t$, ℓ_1 étant le vrai chemin acoustique, on a alors :

$$W = \ell_0/\ell_1(V), \quad \rho_0 W^2 = \left(\frac{\rho_0}{\rho(X)} \right) \left(\frac{\ell_0}{\ell_1} \right)^2 (\rho(X) \, V^2)$$

On définit de même les coefficients de la matrice de propagation relativement à la "vitesse naturelle".

$$M_{ik} = \left(\frac{\rho_0}{\rho(X)}\right) \left(\frac{\ell_0}{\ell_1}\right)^2 A_{ijk\ell} K_j K_\ell \, .$$

Et en effectuant un développement limité au premier ordre à partir de la configuration à contrainte nulle,

$$M_{ik} = \overline{M}_{ik} + \delta M_{ik}$$

on trouve après une longue algèbre l'expression de $\dfrac{\partial(\rho_0 W^2)}{\partial T_{pq}}$, en fonction des paramètres de la propagation et des constantes élastiques d'ordre 2 et 3 de la configuration $\overline{X}$.

Cette formule est à la base des interpétations expérimentales, car elle permet, par simple mesure de la pente à l'origine de $\rho_0 W^2$, le calcul de combinaison de constantes élastiques. De plus, dans les domaines usuels de contraintes, $1/t^2$ peut être considéré comme variant linéairement avec T_{pq}, ce qui permet une bonne précision dans la mesure de ces constantes.

Cette formule a été évaluée dans [24], pour les classes de symétrie les plus hautes, pour différents types d'ondes se propageant selon les directions cristallographiques parallèles aux axes de symétrie en fonction de contraintes uniaxiales ou de la pression hydrostatique appliquée.

3. Aspects expérimentaux de la propagation des ondes ultrasonores dans les solides soumis à une pression hydrostatique

3.1. *Principe*

A l'aide d'un transducteur piézoélectrique (généralement en quartz), on envoie dans l'échantillon à étudier des impulsions répétées de fréquence de quelques MHz. Les ondes se propagent à l'intérieur de l'échantillon, se réfléchissent à son extrêmité libre et le traversent à nouveau. A chaque réflexion sur l'interface non libre de l'échantillon, une partie de l'énergie est transmise vers le quartz qui le transforme en signal électrique. Ce dernier est amplifié à l'aide d'un dispositif électronique et envoyé sur un oscilloscope ; on observe donc sur ce dernier toute une série d'échos. L'intervalle de temps θ entre deux échos successifs est lié à la durée t du parcours aller et retour 2ℓ de l'onde dans l'échantillon. Si V est la vitesse de propagation du mode propagée, on a :

$$C = \rho V^2, \quad t = \frac{2\ell}{V}, \quad \theta = t + \frac{\gamma}{2\pi f}$$

γ étant l'argument du coefficient de réflexion Γ à l'interface non libre de l'échantillon pour la radio fréquence f, C la constante élastique, ρ la densité. La mesure expérimentale à laquelle on a accès est celle de θ et doit se faire de manière très précise. Les méthodes ultrasoniques qui permettent ces mesures sont classiques ([11, 12, 13, 14, 18]) mais en fait peuvent se réduire à deux grandes familles selon l'appareillage haute pression envisagé. Dans le cas où la pression est purement hydrostatique, le transducteur peut être directement collé sur le cristal et il est alors sous pression. Dans le cas où la pression est quasi-hydrostatique, il n'est plus possible de travailler avec des transducteurs internes, et la génération des ondes élastiques dans le cristal doit se faire à partir d'un transducteur via un piston.

Dans le premier cas, le volume de la chambre de travail peut être suffisamment important et l'on peut travailler avec un cristal de taille normale (quelques mm) relativement aux expériences ultrasonores, mais les pressions atteintes ne dépassent pas alors 30 kbar. Il est aussi possible de travailler dans ce cas à basse température, le milieu transmetteur de pression étant alors l'hélium.

Dans le deuxième cas, cas des générateurs de pression où le milieu transmetteur est un milieu à faible coefficient de cisaillement, les pressions atteintes sont plus élevées, jusqu'à 100 kbar, mais le volume utile de la cellule restreint les dimensions de l'échantillon à quelques dixièmes de mm. Par contre, il est possible de travailler à haute température. Ce type de manipulation est approprié aux mesures ultrasonores des matériaux polycristallins (roches, minéraux,...). Pour les raisons mentionnées ci-dessus, on utilisera généralement dans le premier cas la méthode de "superposition des échos" et dans le deuxième cas les méthodes qui font appel à des comparaisons de phase. Ces méthodes permettent des mesures très précises des temps de parcours.

3.2. *Méthodes ultrasonores*

3.2.1. *Superposition des échos (Fig. 1)*
On choisit deux échos dans le train d'onde à l'aide de la ligne à retard et on module leur brillance sur l'axe Z de l'oscilloscope. La fréquence de répétition du pulse est ajustée à l'aide du générateur de fréquence de façon à correspondre à un multiple entier du temps d'aller et retour du pulse acoustique dans le solide.

Il est ainsi possible de superposer cycle par cycle deux échos et : $\theta = 1/F_0$
F_0 : fréquence de répétition.

Les temps de parcours que l'on peut mesurer sont de l'ordre de quelques μs.

3.2.2. *Méthode de la "mesure de phase" (Fig. 2)*
Un synthétiseur de fréquence est la pièce principale du système électronique. Son signal est modulé par une porte qui génère des pulses radio-fréquence

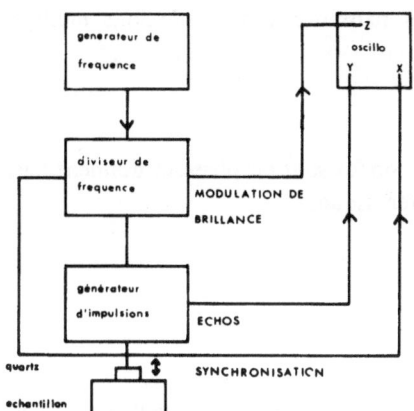

Figure 1. – *Méthode de superposition des échos.*

de largeurs variables et à des taux de répétition différents. Ces pulses sont stables quant à leurs phases, par rapport à la fréquence de la porteuse. Le circuit de porte permet de synchroniser l'oscilloscope par rapport à cette même phase. Un train d'ondes répété (long si on le compare au temps de parcours dans l'échantillon, mais plus court que le temps de parcours dans le piston) est principalement réfléchi à l'interface échantillon-piston mais partiellement

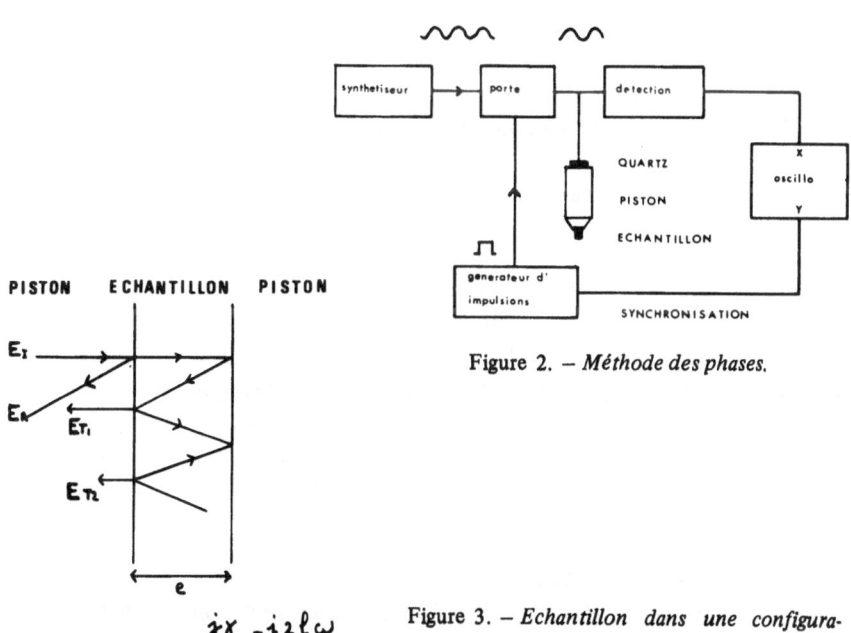

Figure 2. – *Méthode des phases.*

Figure 3. – *Echantillon dans une configuration quasi hydrostatique.*

$$E_{T_2} = E_{T_1} \, A \, e^{j\delta} \, e^{-j\frac{2\ell\omega}{V}}$$

transmis dans l'échantillon pour donner naissance à des réflexions multipes (Fig. 3) :

$$E_{T_2} = E_{T_1} \, A \, e^{j\gamma} \, e^{-j2\ell\omega/V}$$

A des fréquences critiques f_n, ces ondes sont en phase et donnent naissance sur l'oscilloscope à une figure caractéristique.

$$\gamma - \frac{2\ell \, \omega_n}{V} = 2\pi n$$

$$V = \frac{2\ell f_n}{n + \gamma/2\pi}$$

n étant le nombre de longueur d'ondes dans deux fois la longueur de l'échantillon et peut être mesuré sans ambiguité.

$$n \sim \frac{f_n}{\Delta f}$$

Δf mesurant la séparation entre des valeurs successives de f_n. Il est donc possible de mesurer V connaissant f_n, n, γ. Les temps de transit sont inférieurs à la μs.

3.3. *Aspects expérimentaux*

3.3.1. *Influence de* γ

Une analyse mathématique de γ est possible et montre que ce paramètre peut devenir important si la fréquence de résonance du transducteur varie avec la pression (cas des transducteurs internes). Il n'est toutefois pas possible de connaître le comportement de γ avec la pression et l'hypothèse généralement admise est de le négliger car dans le domaine des 10 kbar, il a été montré que la fréquence de résonance du quartz variait très peu avec la pression [16, 20]. γ dépend aussi de la nature et de l'épaisseur de la colle qui assure la liaison entre l'échantillon et le transducteur, et doit être le plus petit possible. Il n'existe cependant pas de recette universelle pour trouver la colle la mieux adaptée au type d'onde transmis et à la nature de l'échantillon. En effet, il faut que le système transducteur-colle-échantillon soit le plus homogène possible sous pression afin d'éviter les ruptures dues aux différences de comportement.

3.3.2. *Préparation des échantillons*

Les échantillons doivent posséder des faces polies et parallèles et les garder sous pression. Ce besoin peut nécessiter un compactage préalable quand on travaille sur des matériaux polycristallins. Le polissage et l'usinage s'effectuent, alors, après que l'échantillon ait été soumis à la pression.

3.4. *Passages électriques – Cellules ultra-son*

De nombreux montages sont possibles [9, 23], mais dans tous les cas, l'échantillon est placé dans une enceinte contenant un fluide qui comprimé transmet une pression hydrostatique.

Sur la figure 4 est donné le principe d'une cellule ultra-son. La liaison électrique qui amène les ondes radiofréquences s'effectue au moyen d'un câble coaxial à impédance itérative adapté afin d'éviter les réflexions parasites. On utilise généralement des thermocoaxiaux qui sont brasés sur l'obturateur de la chambre de pression, afin d'obtenir une bonne étanchéité. L'isolation électrique, ainsi que la tenue mécanique de l'âme du thermocoaxial sont assurées par un petit manchon en téflon rempli de colle isolante.

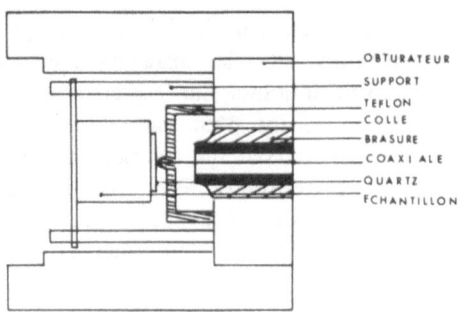

Figure 4. – *Cellule ultra-son.*

3.5. *Appareillage haute pression*

Tous les appareillages délivrant de la pression peuvent théoriquement s'adapter aux mesures ultrasonores [4, 6, 8, 9, 16, 23]... Il est hors de question de donner une liste exhaustive des expériences décrites dans la littérature spécialisée, mais en fait, ces expériences peuvent être classées en deux familles : celles qui sont effectuées sous pression hydrostatique et celles effectuées sous pression quasi hydrostatique. Dans le premier cas, le domaine des pressions est de l'ordre de 20 kbar, alors qu'il est beaucoup plus élevé pour le deuxième type d'expérience, mais il est alors obligatoire d'utiliser un transducteur externe à cause de la non hydrostaticité de la pression, ce qui impose l'emploi de la "méthode de comparaison des phases" comme méthode de mesure.

3.5.1. Appareillage délivrant une pression hydrostatique

Sur la figure 5 est donné le principe d'un appareillage délivrant une pression hydrostatique, le milieu transmetteur de pression étant du pétrole.

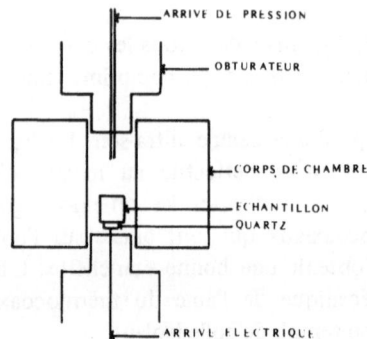

Figure 5. — Cellule haute pression (pétrole).

La chambre de pression est constituée par un corps de chambre et de deux obturateurs, généralement en acier maraging. L'étanchéité de ces obturateurs est assurée au moyen de joints. Sur le premier est brasé un tube capillaire qui amène la pression venant d'un multiplicateur de pression, sur le deuxième est brasé le thermocoaxial qui permet l'arrivée des ondes hautes fréquences. La mesure de la résistance électrique de jauges de manganine permet la mesure des pressions. Ces appareillages peuvent être utilisés à basse température au moyen d'une cryogénie appropriée et par l'emploi d'hélium comme transmetteur de pression.

3.5.2. Appareillage délivrant une pression quasi hydrostatique

a) Enclumes de Brigman (Fig. 6).

Schématiquement, l'échantillon placé dans un anneau de pyrophilitte, milieu transmetteur de pression, est pressé entre deux enclumes généralement en carbure de tungstène, au moyen des pistons et est ainsi soumis à une pression quasi-hydrostatique. Les transducteurs sont positionnés dans les échancrures des enclumes afin qu'ils ne soient pas écrasés quand l'ensemble est mis sous presse hydraulique. Un manchon extérieur assure l'alignement des différentes parties.

L'aspect le plus critique de ce genre de manipulation consiste dans la figure compliquée des échos qui proviennent des réflexions multiples. De plus, l'alignement des pièces en présence doit être de l'ordre de quelques fractions de longueur d'onde si l'on veut recevoir un signal exploitable. Par contre, il est possible d'étendre les performances de cet appareil en insérant un four autour de l'échantillon.

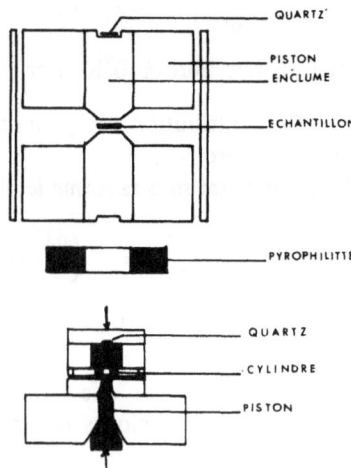

Figure 6. – *Enclume de Brigdman et pis-ton cylindre.*

b) Piston-cylindre (Fig. 6).

La configuration piston-cylindre est une des méthodes les plus simples d'obtention de hautes pressions. Très schématiquement, le piston soumet à une pression quasi-hydrostatique l'échantillon placé dans un cylindre. Un système de joints, placé entre le piston et le cylindre, minimise l'extrusion de l'échantillon au niveau du piston. De la même manière que pour les enclumes de Bridgman, les ondes haute fréquence sont amenées au moyen de transducteurs externes, mais l'épaisseur de l'échantillon pouvant être plus importante, la figure d'échos est plus facilement interprétable.

En conclusion, les appareils du deuxième type sont difficiles à mettre en œuvre quand ils sont couplés à des mesures de vitesse du son, d'autant plus qu'il faut superposer à l'état hydrostatique une contrainte uniaxiale de compression qui intervient dans le calcul des constantes élastiques du troisième ordre.

Les appareillages qui permettent l'application de contraintes uniaxiales ne sont pas présentés ; la simplicité du principe ne doit pourtant pas masquer à l'expérimentateur les difficultés d'obtenir des contraintes homogènes dans l'échantillon.

4. Equation d'état d'origine mécanique, application au silicium

4.1. *Introduction*

L'équation de base nécessaire au calcul des équations d'état des solides est l'énergie libre de Helmotz [21]

$$F(V,T) = U_0(V) + F^*(V,T)$$

$U_0(V)$ énergie libre à $0°K$, d'origine purement mécanique.

$F^*(V,T)$ contribution des phonons à l'énergie libre qui tend vers zéro quand T tend vers zéro.

L'équation d'état sous sa forme le plus générale s'écrit :

$$P(V,T) = \left(-\frac{\partial F}{\partial V}\right)_T = -\left(\frac{\partial U_0}{\partial V}\right)_T - \left(\frac{\partial F}{\partial V}\right)_T$$

$$P(V,T) = P_0(V) + P^*(V,T).$$

P_0 fait intervenir des effets mécaniques tandis que P^* fait intervenir des termes d'origines thermiques qui ne dépendent pas de la même manière du volume.

On se place dans le cas où $P \gg P^*$, on néglige alors les termes d'origines thermiques et on peut alors déterminer une équation d'état isotherme tirée de la thermo-élasticité non linéaire.

4.2. *Equation d'état d'origine mécanique*

L'équation d'état d'origine mécanique se déduit immédiatement du tenseur des contraintes en faisant $T_{ij} = -P\delta_{ij}$

$$-P\delta_{ii} = \frac{1}{V}\, \alpha_{ik}\, \alpha_{i\ell}\, \frac{\partial F}{\partial D_{k\ell}}$$

avec :
$$\delta_{ii} = 3\,, \quad \alpha_{ik}\,\alpha_{i\ell} = \delta_{k\ell} + 2\,D_{k\ell}$$
d'où :

$$P = -\frac{1}{3V}\, (\delta_{k\ell} + 2\,D_{k\ell})\, \frac{\partial F}{\partial D_{k\ell}}$$

On pose $U_0 = F/V_0$; U_0 est l'énergie de déformation rapportée à l'unité de volume de la configuration initiale.

$$P = -\frac{V_0}{3V}\, (\delta_{k\ell} + 2\,D_{k\ell})\, \frac{\partial U_0}{\partial D_{k\ell}}.$$

Cette équation est valable pour tous les types de solides anisotropes [5].

4.3. *Application au silicium*

Le silicium cristallise dans une structure cubique du type diamant de classe de symétrie m3m. Trois constantes élastiques du deuxième ordre et six du troisième ordre caractérisent cette classe.

Elles s'écrivent dans la notation à deux indices.

$$C_{11} \quad C_{12} \quad C_{44}$$
$$C_{111} \quad C_{112} \quad C_{123} \quad C_{144} \quad C_{155} \quad C_{166}$$

avec

$$C_{111} = C_{222} = C_{333}$$
$$C_{112} = C_{113} = C_{233} \cdots.$$
$$\cdots\cdots\cdots\cdots\cdots\cdots\cdots$$

L'ellipsoïde des déformations, associé à la compression hydrostatique, est sphérique car il doit posséder, d'après le principe de Curie-Newman, au moins la symétrie du groupe ponctuel m3m.

L'énergie de déformation s'écrit alors :

$$U_0 = \frac{1}{2} C_{11} (D_1^2 + D_2^2 + D_3^2) + C_{12}(D_1 D_2 + D_2 D_3 + D_3 D_1)$$

$$+ \frac{1}{6} C_{111}(D_1^3 + D_2^3 + D_3^3)$$

$$+ \frac{1}{2} C_{112}(D_1^2 D_2 + D_2^2 D_3 + D_3^2 D_1 + D_2^2 D_1 + D_3^2 D_2 + D_1^2 D_3)$$

$$+ C_{123} D_1 D_2 D_3$$

avec $D_1 = D_2 = D_3$ et l'équation d'état prend la forme :

$$P = -\frac{V_0}{3V} (1 + 2 D_i) \frac{\partial U_0}{\partial D_i}.$$

Après dérivation, et en faisant $D_1 = D_2 = D_3$, on trouve :

$$P = -\frac{V_0}{V} \left[D_1(C_{11} + 2C_{12}) \right.$$
$$+ D_1^2 \left(2C_{11} + 4C_{12} + \frac{C_{111}}{3} + 3C_{112} + C_{123} \right)$$
$$\left. + D_1^3 (C_{111} + 6C_{112} + 2C_{123}) \right].$$

L'équation de continuité donne :

$$\left(\frac{V}{V_0}\right) = \det\left[\alpha_{ij}\right]$$

$$\left(\frac{V}{V_0}\right)^2 = (1 + 2D_1)^3$$

d'où :

$$D_1 = \left[\frac{1}{2}\left(\frac{V}{V_0}\right)^{2/3} - 1\right]$$

L'équation d'état s'écrit finalement au troisième ordre :

$$P = -\frac{V_0}{V}\left[\frac{1}{2}\left(\left(\frac{V}{V_0}\right)^{2/3} - 1\right)(C_{11} + 2C_{12})\right.$$

$$+ \frac{1}{4}\left(\left(\frac{V}{V_0}\right)^{2/3} - 1\right)^2\left(2C_{11} + 4C_{12} + \frac{C_{111}}{2} + 3C_{112} + C_{123}\right)$$

$$\left. + \frac{1}{8}\left(\left(\frac{V}{V_0}\right)^{2/3} - 1\right)^3 (C_{111} + 6C_{112} + C_{123})\right].$$

Application numérique.

Les valeurs expérimentales des constantes élastiques du silicium ont été mesurées par Mc Skimin [15] sous pression.

$$C_{11} + 2C_{12} = 29{,}364 \; 10^{11} \text{ dynes/cm}^2$$

$$2C_{11} + 4C_{12} + \frac{C_{111}}{2} + 3C_{112} + C_{123} = -310{,}054 \; 10^{11}$$

$$C_{111} + 6C_{112} + 2C_{123} = -363{,}782 \; 10^{11}$$

Nous avons tracé la courbe de compressibilité théorique pour les valeurs de V/V_0 variant de 1 à 0,90 (courbe silicium) que nous avons comparés à la courbe expérimentale tirée des expériences de diffraction de RX [19] (Fig. 7).

Compte-tenu des erreurs expérimentales, on peut considérer que l'accord entre les deux courbes est satisfaisant, mais semble quand même meilleur à basse pression, ce qui pourrait signifier qu'à haute pression (> 50 kbar) il faudrait peut être tenir compte des constantes élastiques d'ordre supérieur à trois.

Il est à noter que des mesures de données ultrasonores effectuées à faible pression (< 10 kbar) permettent de déduire de façon satisfaisante des

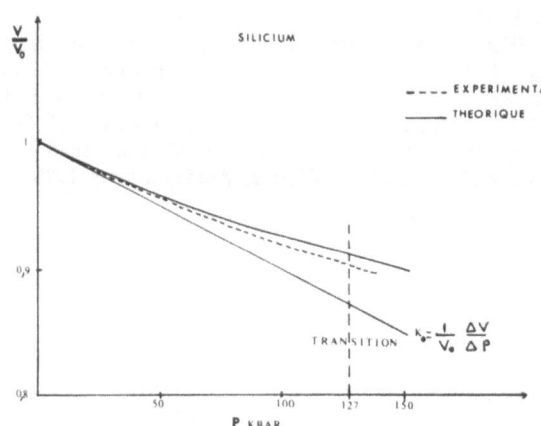

Figure 7

compressions à très haute pression qu'il est quelquefois impossible d'atteindre expérimentalement — ceci montre, entre autre, l'intérêt de telles expériences pour de nombreux laboratoires.

REFERENCES

[1] BIRCH F. – *Phys. Rev.*, 71 (1947): 809.
[2] BORN M. and K. KUANG. – *Dynamical theory of crystal lattices.* London: Oxford University Press, 1954.
[3] BRUGGER. – *Phys. Rev.*, 133 (1964): A 1611.
[4] DANIELS W.B. and C.S. SMITH. – *The physics and chemistry of high pressure.* London: Society of Chemical Industry, 1963, 50.
[5] DELANNOY and PERRIN. – *Le Journal de Physique*, 36 (1975): 1123.
[6] HEYDERMANN P. – *Rev. Sci. Instr.*, 41 (1970): 1896.
[7] HIDEYUKI FUJISAWA – *4ᵉ Conférence Internationale des hautes pressions*, Kyoto, 1974.
[8] HITOSHI MIZUTANI. – *4ᵉ Conférence Internationale des hautes pressions*, Kyoto, 1974.
[9] LAZARUS D. – *Phys. Rev.*, 76, 4 (1949).
[10] LEIBFRIED, LUDWIG. – *Solid State Phys.*, 12 (1961): 275.
[11] Mc SKIMMIN H.J. – *J. Acoust. Soc. Am.*, 33 (1961): 12.
[12] Mc SKIMMIN H.J. – *J. Acoust. Soc. Am.*, 33 (1961): 606.
[13] Mc SKIMMIN H.J. and P. ANDREATCH. – *J. Acoust. Soc. Am.*, 34 (1962): 609.
[14] Mc SKIMMIN H.J. and W.L. BOND. – *Phys. Rev.*, 105 (1957): 116.
[15] Mc SKIMMIN H.J. and P. ANDREATCH. – *J. of Applied Physics*, 35 (1964): 2161.
[16] MORRIS C.E. – *J. of Applied Physics*, 47 (1976): 3979.
[17] MURNAGHAN. – *Finite deformation of an elastic solid.* New York: Wiley, 1951.
[18] PAPADAKIS E.P. – *J. Acoust. Soc. Am.*, 42 (1967): 1045.
[19] PAUREAU J.J., A. WAINTAI, R. FRUCHART and J.P. SENATEUR. – *European high pressure Research Group, 16th Annual Conference*, Reading 11-13 April 1978.

[20] SUSSE C. − *J. Phys. Radium*, 15 (1955): 348.
[21] SWENSON C.A. − *International Study Institute on Physics and Chemistry of Solid Under high pressure*. Vol. 1, Delft, the Netherlands, 2-14 August 1970.
[22] THURSTON R.N. and K. BRUGGER. − *Phys. Rev.*, 133 (1964): A 1604.
[23] WAINTAL A., J. ROUCHY and A. DRAPERI. − *5ᵉ Conférence Internationale de haute pression et technologie*, Moscou, 26-31 mai 1975.
[24] WALLACE. − *Solid state Physics*, Academic Press, 1970.

RESUME

On décrit les appareillages de pression avec leurs électroniques de mesure qui permettent la mesure des constantes élastiques d'ordre deux sous pression. La théorie de la thermoélasticité non linéaire permet d'en déduire les constantes élastiques d'ordre trois .qui sont nécessaires, entre autres, à la dérivation des équations d'état d'origine mécanique. On utilise cette équation d'état au troisième ordre pour calculer la courbe de compressibilité du silicium et on la compare à la courbe expérimentale.

The Formation Factor Tensor in Relation to Structural Characteristics of Anisotropic Granular Soils

Y. F. Dafalias and K. Arulanandan
University of California, Davis, California, U.S.A.

1. Introduction

The mechanical properties of granular soils depend on their grain and aggregate structural characteristics which are particles' shape (elongation, flakiness, roundness/angularity, surface roughness), gradation, porosity, particles' orientation and contacts' orientation (measured in terms of the normals to contact planes). Anisotropic properties develop due to particles and contact orientations [1, 3, 17, 18, 19]. In view of the difficulty to obtain undisturbed samples from the field for laboratory tests, the mechanical properties can be correlated with proper indices measured in situ, which must depend on the above structural characteristics. Scalar indices like relative density and penetration resistance cannot possibly account for anisotropy which is tensorial in character.

The formation factor tensor F_{ij} is defined as the ratio ρ_{ij}/ρ_e, where ρ_{ij} is the electric resistivity tensor of an aggregate of non-conductive, non-polarizable particles saturated by a solution with resistivity ρ_e. It will be shown theoretically and experimentally that F_{ij} is independent of ρ_e and depends only on the structural characteristics including orientation. Thus, F_{ij} is a proper index for correlation with anisotropic properties as will be briefly discussed.

Historically, the scalar equivalent F of F_{ij} for isotropic aggregates has been used primarily for the determination of porosity, beginning with the pioneering work of Maxwell [14], and an extensive review can be found in [16]. Its tensorial character, however, has not been fully investigated. The systematic use of F for soil characterization can be found in a series of papers by Arulanandan et al. [2, 3, 4, 5, 6, 10, 11].

2. Basic equations

In the following the summation convention applies over repeated latin indices which vary over the range 1, 2, 3 and are associated with the axes of a cartesian coordinate system. Electric current is passing through the solution of a saturated unit volume aggregate of non-conductive, non-polarizable particles with porosity n. If E_i, J_i denote the electric field and current density respectively, the superscripts $+$ and $-$ indicate the space outside and inside the particles respectively and a bar over E_i, J_i denotes their mean value over the corresponding space (total, outside, inside), the following average equations can be written

$$\bar{E}_i = \rho_{ij} \bar{J}_j \tag{1}$$

$$\bar{E}_i = (1 - n) \bar{E}_i^- + n \bar{E}_i^+ \tag{2}$$

$$\bar{J}_j = n \bar{J}_j^+ \tag{3}$$

$$\bar{E}_j^+ = \rho_e \bar{J}_j^+ \tag{4}$$

Although the particles and solution are isotroqic, the ρ_{ij} has different 'normal' and 'shear' components due to particles' and contacts' orientation. The form of Eq. (1) in terms of the conductivity tensor has been stated in [13]. In addition, one can write

$$\bar{E}_i^- = \bar{f}_{ij} \bar{E}_j^+ \tag{5}$$

where $\bar{f}_{ij}$ is the form factor tensor which depends on all structural characteristics. Since the $\bar{E}_j^+$ can vary independently, a combination of Eqs. (1) through (5) yields

$$F_{ij} = \delta_{ij} + \frac{1 - n}{n} \bar{f}_{ij} \tag{6}$$

where $F_{ij} = \rho_{ij}/\rho_e$ and δ_{ij} is the Kronecker delta. As $n \to 1$, $F_{11}, F_{22}, F_{33} \to 1$ $F_{ij} \to 0$ for $i \neq j$. Observe that F_{ij} is independent of ρ_e because ρ_{ij} increases proportionally with ρ_e for non-conductive particles. Analytical expressions for the dependence of $\bar{f}_{ij}$ on the structural characteristics will be derived first for very dilute dispersions, and subsequently the effect of concentration will be investigated.

3. Dilute dispersions

For very dilute dispersions ($.85 \leqslant n \leqslant 1$) the form factor $\bar{f}_{ij}$ will be henceforth denoted by f_{ij}. Expressions for f_{ij} will be obtained based on the assumption that the electric fields surrounding each particle do not perturb each other to any appreciable extent. Therefore, the f_{ij} will not depend on these structural characteristics which are pertinent, with respect to their effect on field interaction, only when the particles are close together, such as porosity (measure of 'closeness'), gradation and contact orientation (measure of relative particles' position). But, the f_{ij} will depend on particles' shape and orientation which are pertinent for dilute dispersions. Even then, analytical solutions can be obtained only for the case of ellipsoidal particles.

3.1. *Ellipsoidal Particles*

Modeling a grain as a non-conductive, non-polarizable ellipsoid of average axes $a > b > c$, its orientation with respect to the cartesian system 1, 2, 3 is defined in terms of the direction cosines $\ell_{\alpha i} = \cos(\alpha, i)$ where $\alpha = a, b, c$.

Assume first that all particles are identically oriented. Without field interaction, the effective field acting upon a particle is equal to the average field $\bar{E}_i^+$ in the solution [12]. The field $\bar{E}_i^-$ inside the particles can be obtained by standard methods requiring the solution of the Laplace equation for the electric potential in confocal elliptic coordinates subject to proper continuity boundary conditions [12]. In the process, $\bar{E}_i^+$ is projected along α and, the internal field $\bar{E}_\alpha^-$ along α is found and projected back along i by means of $\ell_{\alpha i}$. Finally, the following set of equations is obtained

$$\bar{E}_i^- = f_{ij}^\theta \bar{E}_j^+ \tag{7a}$$

$$f_{ij}^\theta = \sum_\alpha \ell_{\alpha i} \ell_{\alpha j} f_\alpha \tag{7b}$$

$$f_\alpha = (1 - S_\alpha)^{-1} \tag{8a}$$

$$S_\alpha = \frac{1}{2} abc \int_0^\infty (\alpha^2 + x)^{-1} ((a^2 + x)(b^2 + x)(c^2 + x))^{-1/2} dx \tag{8b}$$

$$S_a + S_b + S_c = 1 \tag{8c}$$

The values of the S_α depend only on the ratios among a, b, c, hence the S_α will be called the shape factors. The superscript θ of f_{ij}^θ indicates the identical orientation of the ellipsoids.

Not all particles, however, are identically oriented. A statistical orientation will be introduced by properly defined probability density functions associated with the $\ell_{\alpha i}$. Since $\ell_{\alpha i} \ell_{\alpha j} = \delta_{ij}$ the $\ell_{\alpha i}$ are note independent but can be expressed in terms of three independent random variables θ_i, i = 1, 2, 3. Letting the axes a, b, c be initially parallel to axes 3, 2, 1 respectively, Figure 1 shows clearly how the orientation of an ellipsoid can be fully defined in terms

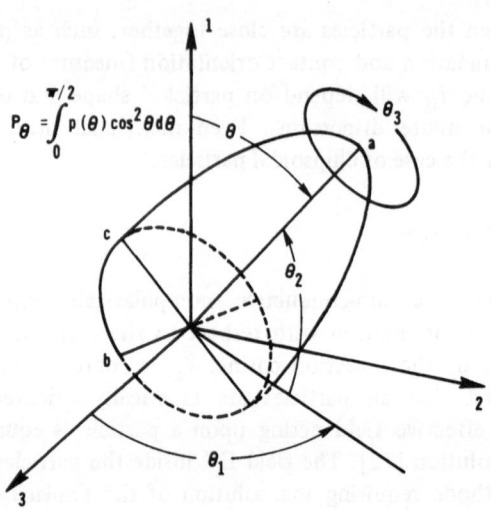

Figure 1. – *Orientation of an ellipsoid defined in terms of three rotations* $\theta_1, \theta_2, \theta_3$.

of three angles of rotation $\theta_1, \theta_2, \theta_3$, such that $0 \leqslant \theta_1 \leqslant 2\pi, 0 \leqslant \theta_2 \leqslant \pi/2$, $0 \leqslant \theta_3 \leqslant 2\pi$. The $\ell_{\alpha i}$ are related to θ_i according to

$$
[\ell_{\alpha i}] = \begin{bmatrix} s\theta_2 & -s\theta_1 c\theta_2 & c\theta_1 c\theta_2 \\ -c\theta_2 s\theta_3 & c\theta_1 c\theta_3 - s\theta_1 s\theta_2 s\theta_3 & s\theta_1 c\theta_3 + c\theta_1 s\theta_2 s\theta_3 \\ c\theta_2 c\theta_3 & c\theta_1 s\theta_3 + s\theta_1 s\theta_2 c\theta_3 & s\theta_1 s\theta_3 - c\theta_1 s\theta_2 c\theta_3 \end{bmatrix} \quad (9)
$$

where $s\theta_i, c\theta_i$ stand for $\sin\theta_i, \cos\theta_i$ respectively.

Let $p_i(\theta_i)$ now denote the probability density function of the random variable θ_i, where $\displaystyle\int_0^{2\pi} p_i d\theta_i = 1$ for $i = 1,3$ and $\displaystyle\int_0^{\pi/2} p_2 \cos\theta_2 d\theta_2 = 1$.

Since θ_i are independent, the joint probability density of the θ_i's is given by $p(\theta_1, \theta_2, \theta_3) = p_1 p_2 p_3 \cos\theta_2$. Knowledge of $p_i(\theta_i)$ determines statistically the particles' orientation. In order to abbreviate the following expressions, the operator P on $X(\theta_1, \theta_2, \theta_3)$ is defined as

$$
P[X] = \int_0^{2\pi} \int_0^{\pi/2} \int_0^{2\pi} p_1(\theta_1) p_2(\theta_2) p_3(\theta_3) \cos\theta_2 \, X \, d\theta_1 d\theta_2 d\theta_3 \quad (10)
$$

with $P[1] = 1$. The expected value of F_{ij} for a given $p(\theta_1, \theta_2, \theta_3)$ is $P[F_{ij}^\theta]$ which in combination with Eqs. (6), (7b) and (10) yields the expected value of f_{ij}

$$f_{ij} = P[f_{ij}^\theta] = \sum_\alpha P[\ell_{\alpha i} \ell_{\alpha j}] f_\alpha \tag{11}$$

For a fixed α the $\ell_{\alpha i} \ell_{\alpha j}$ is a second order symmetric tensor. Eq. (11) is the sought closed form expression of the form factor tensor f_{ij} in terms of the shape factors S_α (through f_α and Eq. (8a)) and the tensorial quantities $P[\ell_{\alpha i} \ell_{\alpha j}]$ which will be called the orientation factors.

Of particular interest are the normal components F_{11}, F_{22}, F_{33} and f_{11}, f_{22}, f_{33} which are obtained in terms of $P[\ell_{\alpha i}^2]$ when $i = j$ in Eqs. (6) and (11). Using the notations

$$P_{\theta_i^2} = \int_0^{\hat\theta_j} p_i(\theta_i) \cos^2\theta_i \, d\theta_i \tag{12a}$$

$$P_{\theta_i^3} = \int_0^{\hat\theta_i} p_i(\theta_i) \cos^3\theta_i \, d\theta_i \tag{12b}$$

$$P_{2\theta_i} = \int_0^{\hat\theta_i} p_i(\theta_i) \sin 2\theta_i \, d\theta_i \tag{12c}$$

where $\hat\theta_i$ denotes the proper limit of integration for θ_i, application of the operator P on $\ell_{\alpha i}^2$ obtained from Eq. (9) yields the following explicit expressions for the three columns of the matrix $P[\ell_{\alpha i}^2]$:

$$P[\ell_{\alpha 1}^2] = \left\{ \begin{array}{l} 1 - P_{\theta_2^3} \\ P_{\theta_2^3}(1 - P_{\theta_3^2}) \\ P_{\theta_2^3} P_{\theta_3^2} \end{array} \right\} \tag{13a}$$

$$P[\ell_{\alpha 2}^2] = \left\{ \begin{array}{l} P_{\theta_2^3}(1 - P_{\theta_1^2}) \\ P_{\theta_1^2} P_{\theta_3^2} + (1 - P_{\theta_2^2})(1 - P_{\theta_3^3})(1 - P_{\theta_2^2}) - (1/8) P_{2\theta_1} P_{2\theta_2} P_{2\theta_3} \\ P_{\theta_1^2}(1 - P_{\theta_3^2}) + P_{\theta_3^2}(1 - P_{\theta_1^2})(1 - P_{\theta_3^3}) + (1/8) P_{2\theta_1} P_{2\theta_2} P_{2\theta_3} \end{array} \right\} \tag{13b}$$

$$P[\ell_{\alpha 3}^2] = \left\{ \begin{array}{l} P_{\theta_1^2} P_{\theta_2^3} \\ P_{\theta_3^2}(1 - P_{\theta_1^2}) + P_{\theta_1^2}(1 - P_{\theta_3^3})(1 - P_{\theta_2^2}) + (1/8) P_{2\theta_1} P_{2\theta_2} P_{2\theta_3} \\ (1 - P_{\theta_1^2})(1 - P_{\theta_3^2}) + P_{\theta_1^2} P_{\theta_3^2}(1 - P_{\theta_3^3}) - (1/8) P_{2\theta_1} P_{2\theta_2} P_{2\theta_3} \end{array} \right\} \tag{13c}$$

3.2. Invariant Properties

Of particular interest are the first invariants f_{ii} and F_{ii}. Recalling that $\sum\limits_i \ell^2_{\alpha i} = 1$ and $P[1] = 1$, from Eq. (11) follows

$$f_{ii} = \sum_i \sum_\alpha P[\ell^2_{\alpha i}]\, f_\alpha = \sum_\alpha f_\alpha P\left[\sum_i \ell^2_{\alpha i}\right] = \sum_\alpha f_\alpha \qquad (14)$$

which shows that f_{ii} is independent of particles' orientation and depends only on particles' shape, a property not shared by the other two invariants $f_{ij} f_{ji}$ and $f_{ij} f_{jk} f_{ki}$. Defining now the average values $F = (1/3)F_{ii}$ and $f = (1/3)\,f_{ii} = (f_a + f_b + f_c)/3$, from Eqs. (6) and (14) follows

$$F = 1 + \frac{1-n}{n}\, f \qquad (15)$$

Therefore, F depends only on porosity and particles' shape through f, but not orientation. This can provide a unique correlation of F with the n of dilute dispersions for a given shape, independent of anisotropy, if electrical measurements are performed along 3 perpendicular but otherwise arbitrary directions since particles' orientation leaves F invariant.

3.3. Isotropy

A dilute dispersion is isotropic if the particles are randomly oriented. According to the definition of the $p_i(\theta_i)$ isotropy implies $p_1 = p_3 = 1/2\ \pi$ and $p_2 = 1$. A substitution of these values and use of Eqs. (6), (9), (10), (11), (12) and (13), yields $f_{ij} = 0$, $F_{ij} = 0$, for $i \neq j$, and Eq. (15) with $F_{11} = F_{22} = F_{33} = F$ and $f_{11} = f_{22} = f_{33} = f$. In other words, the average form and formation factors of an anisotropic dilute aggregate are related in exactly the same way (Eq. (15)) as the form and formation factors of the corresponding isotropic aggregate consisting of the same particles and having the same porosity as the anisotropic one.

3.4 Transverse Isotropy

In most natural deposits or laboratory prepared samples, the soil particles' orientation (and contact orientation for dense aggregates) is rotationally symmetric with respect to a vertical axis called the axis of transverse isotropy. Mechanical properties exhibit transversely isotropic symmetries as a direct reflection of the structural arrangement. In terms of the present formulation structural transverse isotropy is characterized by the values $p_1 = p_3 = 1/2\pi$ while θ_2 has a variable probability density $p_2(\theta_2)$ measuring the degree of transverse isotropy. If $\theta = (\pi/2) - \theta_2$ is the azimuthal angle of the particles long axis with the vertical axis, Figure 1, its probability density $\overline{p}(\theta)$ can be defined from $\overline{p}(\theta) = p_2[(\pi/2) - \theta]$. For further reference, the frequency function $p(\theta) = \overline{p}(\theta) \sin \theta$ is introduced and the orientation factor P_θ is defined as

$$P_\theta = \int_0^{\pi/2} \overline{p}(\theta) \sin \theta \cos^2 \theta \, d\theta = \int_0^{\pi/2} p(\theta) \cos^2 \theta \, d\theta \qquad (16)$$

where it follows from Eq. (12b) that $P_\theta = 1 - P_{\theta 3}$. The $p(\theta)$ measures the number of particles at an angle θ over a variation $d\theta$, and P_θ, which varies from 0 to 1, can be interpreted in terms of

$$\overline{\theta} = \cos^{-1} \sqrt{P_\theta} \qquad (17)$$

$\overline{\theta}$ being the average azimuthal angle of particles' orientation. Using the above values of the p_i's from Eqs. (9), (10), and (11) follows that $f_{ij} = 0$ for $i \neq j$ (because $P[\ell_{\alpha i} \ell_{\alpha j}] = 0$), and Eqs. (12), (13), and (16) yield

$$P[\ell_{\alpha i}^2] = \begin{bmatrix} P_\theta & \dfrac{1}{2}(1 - P_\theta) & \dfrac{1}{2}(1 - P_\theta) \\[2ex] \dfrac{1}{2}(1 - P_\theta) & \dfrac{1}{4}(1 + P_\theta) & \dfrac{1}{4}(1 + P_\theta) \\[2ex] \dfrac{1}{2}(1 - P_\theta) & \dfrac{1}{4}(1 + P_\theta) & \dfrac{1}{4}(1 + P_\theta) \end{bmatrix} \qquad (18)$$

which in conjunction with Eq. (11) gives the principal values of the vertical $f_v = f_{11}$ and horizontal $f_h = f_{22} = f_{33}$ form factors as follows

$$f_v = P_\theta f_a + \frac{1}{2}(1 - P_\theta)(f_b + f_c) \qquad (19a)$$

$$f_h = \frac{1}{2}[(1 - P_\theta) f_a + \frac{1}{2}(1 + P_\theta)(f_b + f_c)] \qquad (19b)$$

3.5. *Spheroidal Particles*

For the case of prolate spheroids $b = c < a$ or oblate spheroids $b = c > a$, $S_b = S_c = S$ and $S_a = 1 - 2S$ from Eq. (8c). In addition, S can be found in terms of the axial ratio $R = b/a$ since it is possible to integrate now the elliptic integrals of (8b). Carrying out these computations, a combined use of Eqs. (6), (8a), and (19) yields

$$F_v = 1 + \frac{1 - n}{n} f_v, \quad f_v = \frac{2S - (3S - 1) P_\theta}{2S(1 - S)} \tag{20a}$$

$$F_h = 1 + \frac{1 - n}{n} f_h, \quad f_h = \frac{S + 1 + (3S - 1) P_\theta}{4S(1 - S)} \tag{20b}$$

where

$$S = \frac{1}{2(1 - R^2)} \left[\frac{R^2}{2\sqrt{1 - R^2}} \ln\left[\frac{1 - \sqrt{1 - R^2}}{1 + \sqrt{1 - R^2}}\right] + 1 \right], \quad 0 \leqslant R \leqslant 1 \tag{21a}$$

$$S = \frac{1}{2(R^2 - 1)} \left[\frac{R^2}{\sqrt{R^2 - 1}} \tan^{-1}(\sqrt{R^2 - 1}) - 1 \right], \quad 1 \leqslant R \tag{21b}$$

As R changes from 0 (needles) to ∞ (laminae) and P_θ from 0 ($\bar{\theta} = \pi/2$, all particles horizontal) to 1 ($\bar{\theta} = 0$, all particles vertical), the variation of f_v and f_h from Eqs. (20), (21) is eloquently shown in Figure 2, [11], where also some experimentally calculated values of f_v are presented. It follows immediately from Eqs. (19), (20) that

$$f = \frac{1}{3} (f_v + 2f_h) = \frac{1}{3} (f_a + 2f_b) = \frac{3S + 1}{6S(1 - S)} \tag{22}$$

which is independent of P_θ as expected according to Eq. (14), and F is now obtained from Eqs. (15) and (22). For isotropy $\bar{p}(\theta) = 1$, Eq. (16) yields $P_\theta = 1/3$ and Eq. (15) is resumed from Eqs. (19) or (20).

4. The effect of concentration

As the field interaction increases with concentration, the $\bar{f}_{ij}$ will depend on porosity, gradation, and contacts' orientation in addition to particles' shape and orientation. An exact solution to the problem does not exist even for the simple case of spherical particles, and different approximate methods for isotropic aggregates are summarized in [16]. The most widely accepted is

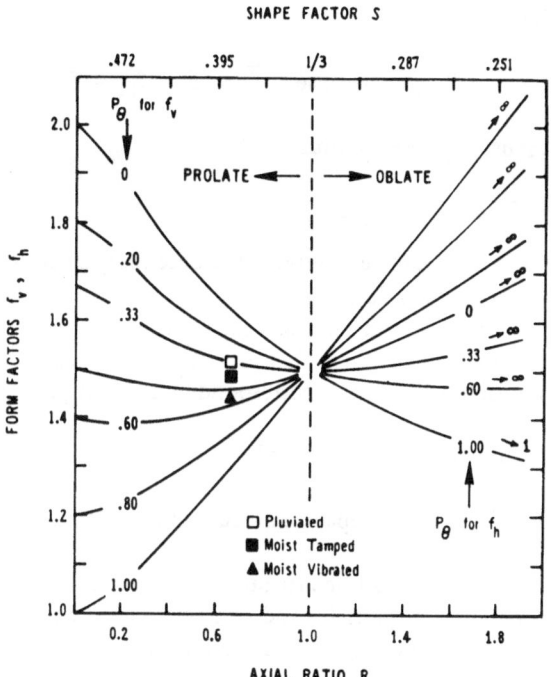

Figure 2. – *Variation of the vertical and horizontal form factors with axial ratio and orientation factor for transversely isotropic aggregates of spheroids.*

Bruggeman's integration technique [9], which was applied to transversely isotropic sand aggregates [10, 11] yielding $F_v = n^{-f_v}$, $F_h = n^{-f_h}$. It was found that while the previously derived relations for dilute dispersions under-estimate the measured values of the formation factor in spherical aggregates and sand samples, Bruggeman's equations overestimate it and, in addition, do not preserve the tensorial character of F_{ij}.

4.1. The Concentration Tensor

A novel approach is presented here. For any dense dispersion charac-terized by f_{ij}, a corresponding fictitious dilute dispersion with a statistically identical particles' orientation and a form factor f_{ij} can by uniquely defined (but not vice-versa). The concept of a fourth order concentration tensor $C_{ijk\ell}$ is introduced, function of porosity, gradation, shape, contacts' and particles' orientation, such that

$$\overline{f}_{ij} = C_{ijk\ell} f_{k\ell} \tag{23}$$

with the properties

$$C_{ijk\ell} = C_{jik\ell} = C_{ij\ell k} \quad \text{because} \quad \bar{f}_{ij} = \bar{f}_{ji}, \; f_{k\ell} = f_{\ell k} \tag{24a}$$

As $n \longrightarrow 1$ it must $\bar{f}_{ij} \longrightarrow f_{ij}$ thus

$$C_{ijk\ell} \longrightarrow \frac{1}{2} \left(\delta_{ik}\delta_{j\ell} + \delta_{i\ell}\delta_{jk} \right) \tag{24b}$$

The concentration tensor can be interpreted as describing an electrical resistivity hardening process as porosity decreases.

Of particular interest will be a so-called uncoupled process with respect to a coordinate system i, j, k, such that each $\bar{f}_{ij}$ is obtained by a different (in general) scalar multiplication of the corresponding f_{ij}. Analytically this requires

$$C_{ijk\ell} = 0 \quad \text{if} \quad i = j, \; k = \ell, \; i \neq k \tag{25a}$$

$$C_{ijk\ell} = 0 \quad \text{if any index appears an odd number of times} \tag{25b}$$

The cause of such a process can be attributed to the effect of a proper combination of the orientation of particles and contacts on the field interaction. For the same reason, a process may be characterized by special symmetries. It follows that for isotropic concentration symmetries

$$C_{ijk\ell} = I_{ijk\ell} = p\,\delta_{ij}\delta_{k\ell} + q(\delta_{ik}\delta_{j\ell} + \delta_{i\ell}\delta_{jk}) \tag{26}$$

and for transversely isotropic concentration symmetries [20], with 1 being the axis of transverse isotropy,

$$\begin{aligned}
C_{ijk\ell} = A_{ijk\ell} &= p\,\delta_{ij}\delta_{k\ell} + q(\delta_{ik}\delta_{j\ell} + \delta_{i\ell}\delta_{jk}) \\
&+ r(\delta_{ij}\delta_{k1}\delta_{\ell 1} + \delta_{k\ell}\delta_{i1}\delta_{j1}) + s\,\delta_{i1}\delta_{j1}\delta_{k1}\delta_{\ell 1} \\
&+ t(\delta_{ik}\delta_{j1}\delta_{\ell 1} + \delta_{jk}\delta_{i1}\delta_{\ell 1} + \delta_{j\ell}\delta_{i1}\delta_{k1} + \delta_{i\ell}\delta_{j1}\delta_{k1})
\end{aligned} \tag{27}$$

The p, q, f, s, t are scalar functions of porosity, gradation, shape and proper invariants of the particles' and contacts' orientation. According to Eq. (24b) as $n \longrightarrow 1$, p, r, s, t $\longrightarrow 0$ and q $\longrightarrow 1/2$. For uncoupled processes Eq. (25a) requires p = r = 0 [8] while Eq. (25b) is identically satisfied.

4.2. *Particular Cases*

In the following the effect of Eq. (23), with $C_{ijk\ell}$ given by Eqs. (26) and (27), on initially isotropic and transversely isotropic dilute dispersions will be considered. The axes of transverse isotropy (when applicable) for both the dilute dispersion and the process will be identical. The abbreviations $\alpha = p + 2r + 2q + s + 4t$, $\beta = p + r$ and $\gamma = p + q$ will be used, and in order to avoid repetitions, it can be proved for all cases that $\overline{f}_{ij} = 0$ when $i \neq j$.

A. Isotropic dilute dispersion $(f_{11} = f_{22} = f_{33} = f_v = f_h = f)$

  a) Isotropic process (Eq. (26))

$$\overline{f}_v = \overline{f}_h = (3p + 2q) f \qquad (28a)$$

  Uncoupled $(p = 0)$

$$\overline{f}_v = \overline{f}_h = 2qf \qquad (28b)$$

  b) Transversely isotroqic process (Eq. (27))

$$\overline{f}_v = (\alpha + 2\beta) f, \quad \overline{f}_h = (\beta + 2\gamma) f \qquad (29a)$$

  Uncoupled $(p = r = 0 \longrightarrow \beta = 0)$

$$\overline{f}_v = \alpha f, \quad \overline{f}_h = 2\gamma f \qquad (29b)$$

B. Transversely isotropic dilute dispersion $(f_{11} = f_v, f_{22} = f_{33} = f_h)$

  a) Isotropic process (Eq. (26))

$$\overline{f}_v = (p + 2q) f_v + 2pf_h, \quad \overline{f}_h = pf_v + 2(p + q) f_h \qquad (30a)$$

  Observe that $\overline{f}_v + 2\overline{f}_h = (3p + 2q) (f_v + 2f_h) = 3(3p + 2q) f$

  Uncoupled $(p = 0)$

$$\overline{f}_v = 2qf_v, \quad \overline{f}_h = 2qf_h \qquad (30b)$$

  b) Transversely isotropic process (Eq. (27))

$$\overline{f}_v = \alpha f_v + 2\beta f_h, \quad \overline{f}_h = \beta f_v + 2\gamma f_h \qquad (31a)$$

  Uncoupled $(p = r = 0 \longrightarrow \beta = 0)$

$$\overline{f}_v = \alpha f_v, \quad \overline{f}_h = 2\gamma f_h \qquad (31b)$$

The invariants of particles' or contacts' orientation, associated with transversely isotropic symmetries of the dilute dispersion and/or the process, can be expressed by means of average azimuthal angles on which the scalar quantities of the above equations depend. Of particular interest are Eqs. (29) if viewed in relation to the aggregates of spherical particles which, being isotropic in the dilute state, develop transversely isotropic symmetries (observe $\overline{f}_v \neq \overline{f}_h$) due to a transversely isotroqic contacts' orientation distribution (orientation of particles has no meaning). This is corroborated by electrical measurements which yield $F_v \neq F_h$ for packing procedures causing anisotropy (plunging, tapping, etc.).

5. Comparison with experiments

Measured values of F_v and F_h for saturated sand aggregates prepared by different methods at different porosities are shown in Figure 3 (F_v only) and Table 1 [13]. Values of F_h along perpendicular horizontal directions were found equal within 1 % difference, confirming the transversely isotropic character of the aggregates. The different values of F_v, Figure 3, for the same porosity and sand according to the method of preparation, indicate different orientation.

To study this in detail, the expression $q = (1/2)\, n^{-c}$ is proposed for Eq. (30b) in accordance with the general requirements, which together with Eqs. (15) and (20) yields

$$F_g = 1 + \frac{1-n}{n^{1+c}}\, f_g \tag{32}$$

where g stands for v, h or nothing and c is an exponential concentration factor function of shape, gradation, and orientation. If R is known, S and f can be

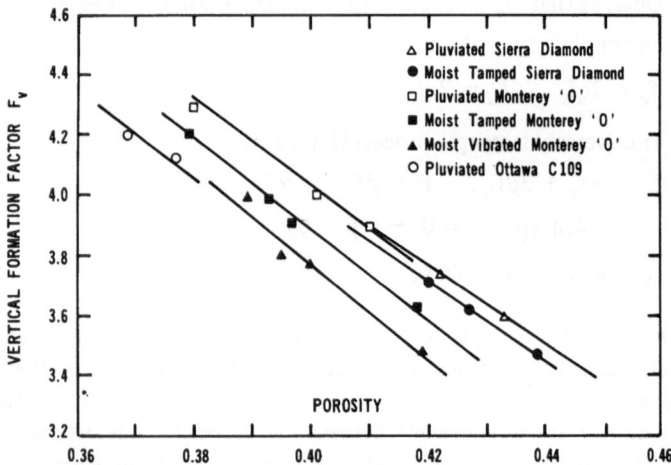

Figure 3. – *Change of the vertical formation factor with porosity for various sands and methods of placement.*

computed from Eqs. (21) and (22), and c can be determined from Eq. (32) and measured values of $F = (F_v + 2F_h)/3$. Subsequently f_v, P_θ and $\bar\theta$ can be calculated from Eqs. (32), (20a), and (17). This is shown in Table 1 for Monterey 0 ($R = .650$, $S = .386$, $f = 1.517$) and Sierra Diamond ($R = .562$, $S = .402$, $f = 1.529$) sands. The increasing 'horizontality' shown by $\bar\theta$ in the order moist vibrated, moist tamped, pluviated agrees with thin sections studies [17, 18, 19]. In [11] histograms of such studies on similar samples [17] were used to obtain independently an estimate of P_θ by the discretized version of Eq. (16)

$$P_\theta = \sum_{\theta=5°}^{\theta=85°} (p_\theta + p_{\pi-\theta}) \cos^2 \theta \tag{33}$$

where p_θ is the percent frequency of the long axis orientation. The corresponding $\bar\theta$ agree qualitatively with the calculations in Table 1.

TABLE 1. Measured values of F_v, F_h and calculation of the average azimuthal angle $\bar\theta$ for transversely isotropic Pluviated (P1), Moist Tamped (MT) and Moist Vibrated (MV) samples of Monterey 0($f = 1.517$) and Sierra Diamond ($f = 1.529$) sands.

Sand/ Method of preparation	Measured			Calculated				
					Eq.(32)	Eq.(32)	Eq.(20a)	Eq.(17)
	n	F_v	F_h	F	c	f_v	P_θ	$\bar\theta$
Monterey 0								
P1	.380	4.28	4.04	4.12	.240	1.594	.105	71.1
MT	.397	3.87	3.95	3.92	.258	1.489	.419	49.6
MV	.419	3.50	3.74	3.66	.270	1.426	.608	38.7
Sierra								
Diamond	.439	3.48	3.61	3.57	.331	1.478	.453	47.7
MT								

Finally, Eq. (32) is used to calculate F for isotropic aggregates (Eq. (28b)) of spherical particles ($R = 1$, $S = 1/3$, $f = 3/2$) using an average $c = .128$ (independent of orientation) obtained by curve fitting measured values of F [21]. The small error, as shown in Table 2, shows the pertinence of the assumption $q = (1/2) n^{-c}$ in relation to concentration effects due to porosity decrease. The corresponding formulas of Maxwell $F = (3 - n)/2n$ [14] and Bruggeman $F = n^{-1.5}$ [9] give a percent error as high as -28 and 58 respectively.

TABLE 2. Comparison of measured and calculated values of F for aggregates of spherical particles using c = .128 in Eq. (32)

Porosity n	Measured F	Calculated F	Error %
.40	3.55	3.53	− 0.5
.35	4.16	4.19	0.6
.30	5.00	5.08	1.6
.20	8.30	8.37	0.9
.10	20.00	19.13	− 4.3

6.   Discussion on correlation with mechanical anisotropy

For any correlation of F_{ij} with mechanical properties, the tacit assumption is made that the initial structural characteristics on which F_{ij} depends, determine, to a large extent, the mechanical property of interest.

There are two ways to use F_{ij} for considering anisotropy. The first is to achieve a correlation with a mechanical property that involves a definite direction in the anisotropic granular mass. For example, the cyclic stress ratio required to cause initial liquefaction in 10 cycles involves cycling of the vertical principal stress and has been found to be affected by as much as 250 % from the particles' and contacts' orientation with respect to the vertical for the same sand at the same initial porosity [17]. In [3] a definite correlation was found between this ratio and the structural index $I = D_r A^{-7}$ where D_r is the initial relative density, and $A = (F_v/F_h)^{1/2}$ is the anisotropy index.

The second way requires first a rigorous mathematical determination of all the necessary anisotropic parameters and their variations. For example, in [7] it is shown that a generalized Coulomb criterion for transversely isotropic sands can be characterized by 3 parameters. For a proper correlation between electrical measurements and anisotropic Coulomb failure, F_v, F_h, or any function of them must be correlated with each one of these parameters.

Acknowledgment

The support of this work by the National Science Foundation under Grant $N°$: ENG 76-13146 is acknowledged.

REFERENCES

[1] ARTHUR J.R.F. and B.K. MENZIES. − "Inherent anisotropy in sand', *Geotechnique*, 22, 1 (1972): 115-128.

[2] ARULANANDAN K. and Y.F. DAFALIAS. – 'Significance of formation factor in sand structure characterization', *Letters in Applied and Engineering Sciences,* 17 (1979): 109-112.

[3] ARULANANDAN K. and B. KUTTER. – 'A directional structure index related to sand liquefaction', *Proceedings,* A.S.C.E., Geotechnical Engineering Division Specialty Conference, Earthquake Engineering and Soil Dynamics, Pasadena, CA, 1 (1978): 213-230.

[4] ARULANANDAN K. and J.K. MITCHELL. – 'Low frequency dielectric dispersion of clay water electrolyte systems', *Clay and Clay Minerals,* 16 (1968): 337-351.

[5] ARULANANDAN K. and S. MITRA. – "Soil characterization by use of electrical Network", *Proceedings, 4th Asilomar Conference on Circuits and Systems,* Nov. (1970): 480-485.

[6] ARULANANDAN K. and S.S. SMITH. – "Electrical dispersion in relation to soil structure, *J. Soil Mechanics and Foundation Division, ASCE,* SM12 (1978): 1113-1133.

[7] BOEHLER J.P. – "Contributions théoriques et expérimentales a l'étude des milieux plastiques anisotrope". Thèse, Grenoble, 1975.

[8] BOEHLER J.P. and A. SAWCZUK. – "Equilibre limite des sols anisotropes", *J. de Mechanique,* 9, 1 (1970) : 5-33.

[9] BRUGGEMAN D.A.G. – "Berechnung Verschiedenez Physikalischer Konstanten von Heterogenensubstanzen, *Ann. Phys. Lpz.* 5, 24 (1935): 636.

[10] DAFALIAS Y.F. and K. ARULANANDAN. – "The structure of anisotropic sands in relation to electrical measurements", *Mechanics Research Communications,* 5, 6(1978): 325-330.

[11] DAFALIAS Y.F. and K. ARULANANDAN. – "Electrical characterization of transversely isotropic sands. *Archives of Mechanics,* 31, 5 (1979): 723-739.

[12] FRICKE H. – "A mathematical treatment of the Electric Conductivity and Capacity of Disperse Systems", *Phys. Rev.* 24 (1924) : 575-587.

[13] KUTTER B.L. – "Electrical properties in relation to Structure of Cohesionless Soils", Master thesis, Department of Civil Engineering, University of California, Davis, 1978.

[14] MAXWELL J.C. – *A treatise on electricity and magnetism,* London: Oxford University Press, 1892.

[15] MEIROVITCH L. – *Methods of Analytical Dynamics,* New York: McGraw-Hill Advanced Engineering series, 1970.

[16] MEREDITH R.E. and C.W. TOBIAS. – *Conduction in Heterogeneous Systems. Advances in Electrochemistry and Electrochemical Engineering,* 2. New York: 1962, Ch. 2, John Wiley and Sons, Inc.

[17] MITCHELL J.K., J.M. CHATOIAN, G.C. CARPENTER. – *"The influence of fabric on the liquefaction behavior of sand".* Report to U.S. Army Engineering Waterways Experiment Station, Vicksburg, Mississippi, contract No. DACA 39-75-MO260, University of California, Berkeley, 1976.

[18] ODA M. – "Initial fabrics and their relation to mechanical properties of granular materials" *Soils and Foundations,* 12, 1(1972) : 17-37.

[19] ODA M., I. KOISHIKAWA and T. HIGUCHI. – "Experimental study of anisotropic shear strength of sand by plane strain test" *Soils and Foundations,* 18, 1 (1978): 25-38.

[20] ROGERS T.C. and A.C. PIPKIN. – "Asymmetric relaxation and compliance matrices in linear viscoelasticity" *Z.A.M.P.* 14 (1963): 334-343.

[21] WYLLIE M.R.J. and GREGORY A.R. – "Formation factors of unconsolidated porous media: Influence of particle shape and effect of cementation" *Petroleum Transactions, A.I.M.E.,* 198 (1953): 103-110.

RESUME

(Le tenseur du facteur de formation en relation avec les caractéristiques de structure des sols granulaires anisotropes)

Le facteur de formation est défini comme la résistance électrique d'une dispersion de particules non conductrices et non polarisables, normalisée par la résistance de la solution. On montre que ce facteur est une quantité tensorielle reliée à la porosité et le tenseur du facteur de forme, lui-même fonction de la porosité, de la grosseur et de la forme des particules, ainsi que de l'orientation statistique des particules et de leurs contacts. Cette dépendance vis-à-vis de l'orientation peut être utilisée pour caractériser l'anisotropie de structure des sols granulaires par des mesures électriques. Les tenseurs du facteur de forme pour une dispersion dense et la dispersion diluée correspondante sont reliés par un tenseur de concentration. Des comparaisons avec des résultats expérimentaux sont présentés et les corrélations avec les propriétés mécaniques anisotropes sont discutées brièvement.

Changes of Macroscopic Anisotropy in Metals

Evolution de l'Anisotropie Macroscopique des Métaux

General Lecture : *Conférence Générale*

Experimental Plasticity
on the Anisotropy of Metals

K. Ikegami

Tokyo Institute of Technology, Yokohama, Japan.

1. Introduction

The mechanical behavior of metals has been investigated by many researchers since long ago. A large number of experimental works on the plastic behavior of metals are found from earlier times. In the initial stage of those studies [1], uniaxial behavior of various metals is examined, as related to yield points and the Bauschinger effect. From the beginning of the twentieth century [10], systematic experiments are conducted on the mechanical behavior of metals under combined stress state. These studies have started by the determination of the initial yielding of metals and were followed by the verification of the validity of the flow theory and the deformation theory [15, 68] for the case of non-proportional loading. A detailed chronological table of research on experimental plasticity is given in the article [11].

In 1928, Mises [181] generalized the flow theory by the use of a plastic potential. After that, the thermodynamic concept of the plastic potential and the associated flow rule was proposed by Drucker [185]. These results open up a possibility for the completion of the theory of plasticity including the anisotropy or the strain history. Many assumptions on the behavior of yield surfaces caused by preloadings are proposed [182, 194]. The necessity of the verification of those assumptions gives an impetus to experimental works on the subsequent yield surfaces after various preloadings. Experimental plasticity is strongly affected by the slip theory [193]. The slip theory suggests the existence of a corner in the preloading direction of the subsequent yield surface. Various experimental confirming methods have been proposed and tried out [157, 166]. From these historical facts, it is found that the experimental plasticity has two objectives. The first objective is to elaborate the explicit form of the plastic stress-strain relations or the yield surfaces including the anisotropy or the strain history. The second one is to verify the assumption of the law of plasticity.

The purpose of this article is to synthesize the phenomenological research results on the plastic behavior of metals under combined stress state. The emphasis is especially placed on the behavior of yield surfaces caused by preloading or textures anisotropy. The effect of pressure and creep on plastic deformations is not treated. Some other articles [2, 14] are reported on the recent research progress in the field of the theory of plasticity.

2.   Experimental methods

There are many experimental methods to obtain a combined stress state. The specimens used in combined stress tests are classified into three groups according to their shapes: thin-wall specimens, plate specimens, and block specimens. The method of subjecting thin-wall specimens to combined axial load, torsion and internal pressure (or external pressure) is most favourable as regards the uniformity of a combined stress state. In this method, however, it is difficult to test plates under combined stress state. Furthermore the thin-wall specimen is not easy to produce to a clear tolerance and the cost is high. For these reasons, the plate specimens and the block specimens are used for practical combined stress testing methods. Figure 1 shows the method of obtaining the combined stress state by using plate and block specimens. Figure 1(a), [69, 75], indicates the method of applying tension to a notched or grooved strip. The various combined stress states are produced in the groove according to their orientations. Figure 1(b), [76, 78], shows the method of loading rhomboidal plates of different shapes transversely at the corners of the plate. The combined stress state is produced in the central part of the plate by bending. Figure 1(c), [86, 98], illustrates the method of using strips cut from a prestressed plate or a plate having an anisotropic texture. Uniaxial tensile tests are conducted with those strips. The tensile stress component of a strip specimen is interpreted as a plane stress state by considering the uniaxial tensile component as related to the initial direction of prestrain. Figure 1(d), [85], gives an example of the method of obtaining a combined stress state from biaxial tensions in a cross-shaped specimen. The block type specimens are used because of the difficulty of the compression test on plate specimens. Figures 1(e) and (f), [79, 84], indicate the methods of using a block of thick plate and a block of cemented thin plates, respectively. The combined stress state is obtained by subjecting those blocks to compression and tension. Figure 1(g), [99, 103], is the method of determining the yield surfaces by using the Knoop hardness. The Knoop hardness numbers, representing six orientations of the indentor (the marks a to f in the figure) with respect to the principal directions in Figure 1(g), are proportional to the stress deviator necessary to cause plastic flow in metals. The yield surface in the deviatoric plane can be constructed from the Knoop numbers.

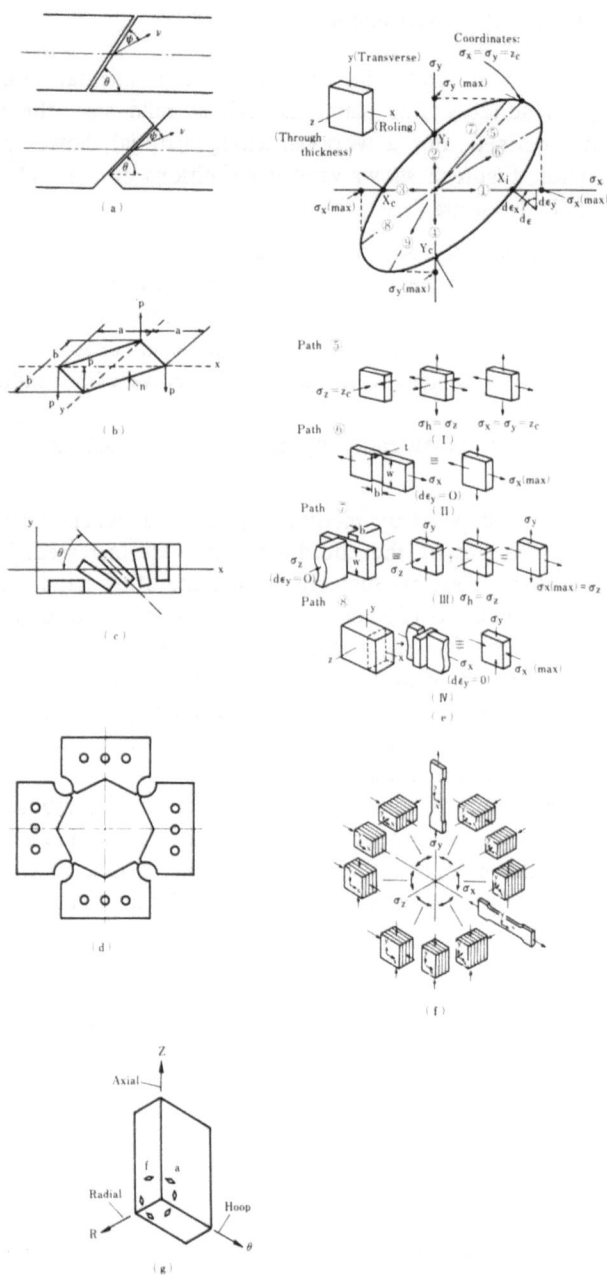

Figure 1. — *Testing methods for combined stress state by using plates and blocks (a)-(g).*

3.   Definition of the yield-point

The definition of the yield-point affects the experimental results on the subsequent yield surfaces. For materials such as mild steel the lower yield stress is clearly defined. For a work-hardening material, however, the yield-point is not clear. Figure 2 shows various definitions of the yield-point used for such materials. These are :

(1) Deviation from linearity, A (proportional limit, LP) ;
(2) Extrapolation of a tangent to the stress-strain curve to the elastic slope, B_1, or to the ordinate of stress, B_2 (extrapolation method, EP) ;
(3) Stress point by a small permanent set, C (proof strain or stress) ;
(4) Contact point of a tangent to the stress-strain curve with a multiple elastic slope, D ;
(5) Deviation from proportionality by introducing a backward linear extrapolation, E.

The value of the yield point by method (1) decreases when the precision of measurement or the size of the coordinate scale are increased. The value is greatly influenced by personal subjectivity. On the other hand, to determine the yield-point by methods (2) and (4), the specimen has to be deformed by a considerable amount of plastic strain. The preloading condition of specimens is changed by the process of defining the yield-point. Therefore only one point on the yield surface is determined by one specimen. Many specimens of the same dimensions and material are necessary. In many experimental works on the subsequent yield surfaces, method (3) is adopted.

The determination of the yield-point in a state of combined stresses is carried out on the effective stress-strain curve or the stress-strain curve exhibiting the maximum strain component. Strictly speaking, this is a wrong method, because an anisotropic effective stress-strain relation should be used to determine a yield point of anisotropic materials. But nobody knows the

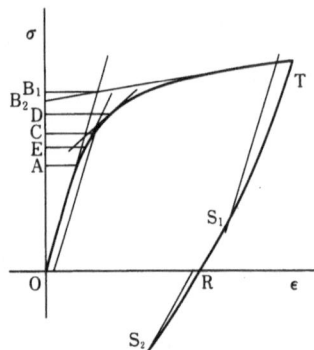

Figure 2. – *Definition of yield points.*

anisotropic condition of a materials before the experiment. Therefore one is forced to adopt such approximate methods.

Two cases have to be distinguished in the determination of the yield-point at the opposite side from the preloading direction. Referring to Figure 2, one method is that in which the yield-point is determined from the completely unloaded state to zero stress (the point R in Fig. 2) after preloading. In the other method, the yield-point (the point S_1 in Fig. 2) is determined by the deviation from the linearity of the stress-strain curve in subsequent unloading after preloading. The yield points in the opposite direction are determined by reloading from the partially unloaded state, such as at the point T in Figure 2.

The yield surfaces in Figure 3 [146] have been determined for the same material by using two different definitions of yield points at the opposite side to preloading. The results shown in Figure 3(a) were obtained after complete unloading always contains the origin of the stress space. But for the yield surface determined after partially unloading, the origin of the stress space is outside the yield surfaces. Only these yield surfaces which after preloading contain the complete unloading process, are obtained by the methods in Figures 1(a), (c), (e), (f) and (g). The difficulty of precise determination of yield points is discussed in other articles [10, 37]. In Tables 1-3 are summarized the test materials, the shape of specimens, the loading type and the definition of yield point used in the experiments on the subsequent yield surfaces.

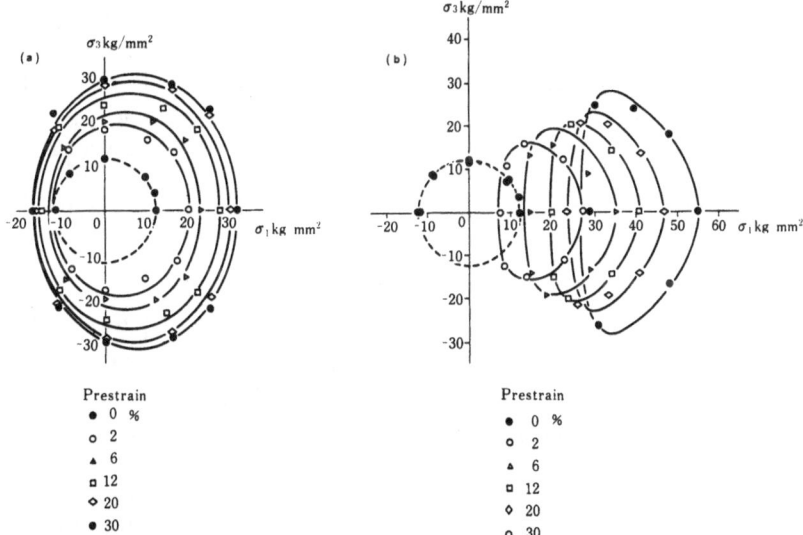

Figure 3. – *Subsequent yield surfaces determined from complete unloading state and from partial unloading state.*
(a) after complete unloading.   (b) after partial unloading.

TABLE 1. Experiments on the effect of preloading and anisotropy on yield surfaces (for complete unloading)

Authors	Ref.	Material	Specimen	Load	Def. yield
Yoshimura, Takenaka, Abe	114	mild steel	OD = 15, ID = 12	AL. TR. AL. TR.	$\times 10^{-6}$ EP
Iagn, Shishmarev	115	nickel		AL. TR.	180
Naghdi, Essenberg, Koff	116	alumi. alloy (24S-T-4)	ID = 0.75″ WT = 0.075″	AL. TR.	PL
Hu, Bratt	117	alumi. alloy (2S-F)	OD = 15/8″ ID = 1″	AL. IP.	
McComb	118	alumi. alloy (2014-T61)	OD = 4.5″ WT = 0.156″	AL. TR.	PL
Mair, Pugh	119	copper	ID = 1″ WT = 0.04″	AL. TR.	EP 10^3
Parker, Bassett	120	brass	OD = 1.125″ WT = 0.0625″ 0.0530″	IP. TR.	EP, PL
Duong	121	alumi., steel		AL. TR.	20
Jenkins	122	zinc alloy	OD = 0.778″ ID = 0.6875″	AL. TR. IP.	PL
Miastkowski, Szczepinski	123	brass	ID = 30 WT = 1	AL. IP.	PL, 100 200×10^3, 5×10^3
Miastkowski	125				
Turski	128	brass	ID = 30 WT = 1	AL. IP.	PL, 100 200×10^3, 5×10^3
Marjanovic, Szczepinski	134 135	brass	ID = 30 WT = 1	AL. IP.	PL, 100 200×10^3, 5×10^3
Theocaris, Hazell	77	alumi. (6061-T651)	plate	Bending	proof strain
Duong	124	alumi., copper		AL. TR.	20
Shiratori, Ikegami	85	brass	plate	Ten. Ten.	200
Rogan, Shelton	126	steel alloy (En 25)	ID = 0.7564″ WT = 0.027″	AL. TR.	EP
Hayashi, Kawaguchi, Fukuda	124	mild steel	ID = 30 WT = 1.5	IP. AL.	500, 10^3 1.5×10^3, 2×10^3

Frederking, Sidebottom	129	steel, copper alloy alumi. alloy	ID = 1.790" WT = 0.050"	AL. TR. IP.	25
Tozawa, Nakamura	83	alumi. alloy brass, steel	block	Ten. Cop.	20, 100, 500, 10³
Hecker	130	alumi., copper	ID = 0.9853" WT = 0.04"	AL. IP.	100, 200, 500, 10³ 2 × 10³
Hecker	131	alumi., copper	ID = 0.9853" WT = 0.04"	AL. IP.	5, 2 × 10³
Azuma, Sugimoto Saito	132	brass	OD = 22 ID = 19	AL. IP. TR.	750
Michino, Findley	133	steel (SAE1017)	OD = 1" WT = 0.060"	AL. IP.	10 ~ 50
Mehan	153	Zircaloy	OD = 0.630 ID = 0.550	AL. IP.	PL, 250, 500, 750 10³, 2×10³, 10⁶
Lee, Backofen	79	Ti., Ti. alloy	block	Ten. Cop.	20
Lee, Backofen	80	Ti., Ti. alloy	block	Ten. Cop.	20
Kelly, Hosford	82	Mg alloy, Mg.	block	Ten. Cop.	1% 5% 10% 0.5%
Dillamore, Hazel, Watson, Hadden	154	mild steel, stainless steel magnesium	OD = 0.75" WT = 0.04" OD = 0.5" WT = 0.015"	AL. IP.	Plastic work
Althoff, Wincierz	155	copper. alumi.	OD = 19 WT=0.5±0.01	AL. IP.	500, 10, 2 × 10³
Sawada, Saito	156	Zn-22Al	OD = 33 ID = 30	AL. IP.	2 × 10³
Shih, Lee	82	Zircaroy	block	Ten. Cop.	2 × 10³
Zieb, Kühn, Ledworuski	75	Ti, brass alumi. alloy	grooved plate	Ten.	2 × 10³

OD: outer diameter   AL: axial load
ID: inner diameter   TR: torsion
WT: wall thickness   IP, EP: internal (external) pressure

TABLE 2. Experiments on the effect of preloading on yield surfaces (for partial unloading).

Authors	Ref.	Material	Specimen	Load	Def. yield
Ivey	136	alumi. alloy (198)	OD = 1.080″ ID = 1.060″	AL. TR.	PL. x 10^{-6}
Bertsch, Findley	137	alumi. alloy (6061-T6)	OD = 1.000″ WT = 0.03″	AL. TR.	10
Shishmarev	138	mild steel	OD = 12 WT = 0.5	AL. TR.	100, 200, 10^3, 75 × 10 10^4
William, Svensson	139	alumi. (1100-F)	OD = 1.100″ ID = 1.000″	AL. TR.	0, 10, 20, 40, 50, 60. 80, 100, 200, 300
Smith, Almroth	140	alumi. alloy (6061-T6)	OD = 1.06″ ID = 1.005″	AL. TR.	5
William, Svensson	141	alumi. (1100-F)	OD = 1.100″ ID = 1.000″	AL. TR.	0, 10, 20, 40, 50, 60, 80, 100, 200, 300
Shiratori, Ikegami, Kaneko	142	brass	ID = 20 WT = 1.5	AL. TR. IP. EP.	200, 500, 10^3, 2×10^3 3×10^3, 10^4
Michino, Findley	143	stainless st. (AISI 304L)	OD = 1″ WT = 0.06″	AL. TR.	10
Shiratori, Ikegami, Kaneko, Yoshida, Koike	144	alumi. alloy (17S)	OD = 23 WT = 1.5	AL. TR.	200
Shiratori, Ikegami, Kaneko	145	brass	ID = 20 WT = 1.5	AL. TR. IP.	200
Shiratori, Ikegami, Kaneko, Sugibayashi	146	brass	ID = 23 WT = 1.5	AL. TR.	200
Shiratori, Ikegami, Kaneko, Takada	148	brass	ID = 20 WT = 1.5	Al. TR.	200
Phillips, Moon	149	alumi. (1100-0)	ID = 21/16″ WT = 0.050″	AL. TR.	PL
Moreton, Moffat, Hornby	150	alloy steel	ID = 1.3″ WT = 0.040″	AL. IP. EP	PL

OD: outer diameter   AL: axial load
ID: inner diameter   TR: torsion
WT: wall thickness   IP. EP: internal (external) pressure

TABLE 3. Experiments on the effects of temperature, time and neutron irradiation on yield surfaces

Authors	Ref.	Material	Specimen	Load	Def. yield
Phillips Phillips, Tang Phillips, Liu, Justusson Phillips, Kasper Phillips, Ricciuti	157 158 159 160 171	alumi (1100-0)	ID = 21/16'' WT = 0.050''	AL. TR.	PL ($<3\times10^{-6}$)
Brown	162	alumi. (2024-T81)	OD = 1.000'' ID = 0.880''	AL. TR.	10^2, 10^3 (strain rate)
Tanaka, Ishizaki, Inoue	163	alumi. alloy	OD = 21 ID = 19	AL. TR.	30
Ishizaki, Okabe, Tanaka, Inoue	164	steel (STB 35)	OD = 20 ID = 19	AL. TR.	200
Lebedyev Novikov	165	steel, iron alumi. alloy		AL. IP.	
Daneshi, Hawkyard	166	alumi. copper	OD = 0.725'' ID = 0.625''	AL. TR.	100
Talypov	167	steel	OD = 40 ID = 38.5	AL. IP.	2000
Kumakura, Takeda, Konuki	168	steel	OD = 10 ID = 8		500, 1000 EP
Michino, Findley	133	stainless st. (AI SI 304L)	OD = 1'' WT = 0.06''	AL. TR.	10
Lindholm, Yeakley	169	mild steel	ID = 0.750'' WT = 0.025''	AL. TR.	upper yield stress
Dudderar, Dufty	170	copper	ID = 1.000'' WT = 0.035'' 0.050''	AL. TR.	7.5, 15

4. Factors which influence the yield surface

On the assumption that the plastic deformation is independent of hydrostatic pressure and that the material is incompressible, the behaviour of the subsequent yield surface is experimentally investigated as concerns the following factors,

(1) Magnitude and direction of preloading — proportional preloading, preloading along the path with a corner and cyclic preloading.

(2) Anisotropy of textures.
(3) Temperature effect.
(4) Time effects: strain ageing, creep strain and strain rate.
(5) Irradiation by neutrons.

As to the above factors, the discussions are directed to the following points.

(1) Behavior of yield surfaces: change of the shapes, translation and rotation.
(2) Features of the shapes of yield surfaces: Bauschinger effect, cross effect, existence of the corner and congruency among yield surfaces.
(3) Normality of the plastic strain incremental vectors to the yield surfaces. This is an experimental method to verify the flow rule of plastic strain incremental vectors.

5.   Stress spaces used to represent yield surfaces

These are three stress spaces used to represent yield surfaces.

(1) Coordinates representing the two or three variable stresses are used in the test.
(2) The deviatoric stress space [104].
(3) The deviatoric plane, called the π plane, in the principal stress space [105, 106, 110].

In the stress space (1), the subsequent yield surfaces show a complicated change in their shapes depending on their preloading direction and magnitude. It is impossible to establish an exact formula relating the preloading and the subsequent yield surface by using such complicated results. The stress space for which the change of the subsequent yield surfaces is simplified would by more convenient for formulating the subsequent yield surfaces. In the case of a material for which an equivalent stress-strain relation in various proportional loadings can be uniquely represented by the stress-strain relation of Mises type there are available for this purpose the deviatoric stress space and the deviatoric plane, the so called π plane. In such stress space, the subsequent yield surfaces after proportional loading the materials of Mises type are symmetrical with respect to the preloading direction, and the various subsequent yield surfaces are congruent after the same amount of preloading in different directions. An example is shown in Figure 4, [142]. Several subsequent yield surfaces after different proportional loadings are represented in a deviatoric plane. In this figure, the various different preloading directions which are given by combined stress are adjusted to the direction $\zeta = 0$. The symmetry and congruency properties of those yield surfaces are observed. The stress space in which such symmetry and congruency occur is called an "isotropic stress space". In an

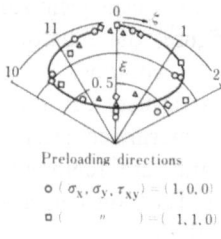

Preloading directions

○ (σ_x , σ_y , τ_{xy}) = (1, 0, 0)
□ ("   ) = (1, 1, 0)
◇ ("   ) = (2, 1, 0)
△ ("   ) = (1, 1, 1)

Figure 4. – *Subsequent yield surfaces re-presented in the deviatoric plane.*

isotropic stress space, the shapes of subsequent yield surfaces after proprtional preloading depend only on the magnitude of preloading, though they depend on both the magnitude and direction of preloading in the ordinary stress space.

When metals are loaded along a path with a corner, for example, along the strain path of the inserted Figure in Figure 5, [145], transient phenomena arise in their deformations, [104, 109, 111, 112]. This means that the stress increment vector does not follow the change of the strain increment vector. The lag of coincidence between them occurs. With the increase of plastic deformation past a corner, this lag is recovered and stationary plastic hardening takes place. During the transient period, just after the corner of a strain path, the subsequent yield surface shows complicated changes in its shape. Figure 5 shows are example of the change of the subsequent yield surface for the preloading along a strain path with a corner of 90 degrees. The subsequent yield surfaces which are represented in the deviatoric stress plane indicate complicated changes including rotations and translations. The symmetry is difficult to detect. But, after some duration of the transient period, the subsequent yield surface, for example indicated by circle points, recovers its symmetry. The congruency

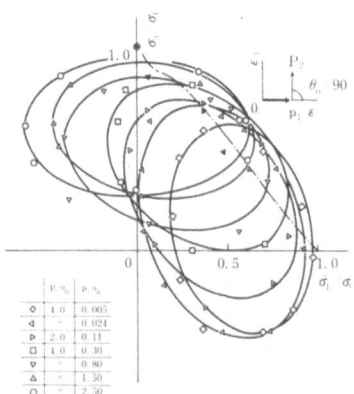

Figure 5. – *Subsequent yield surfaces after preloading along a path with a corner.*

between subsequent yield surfaces is also observed, as shown in Figure 6, [145]. In this figure, the subsequent yield surface represented by triangular, square and circular contours is determined for the preloading along a strain path without a corner. On the other hand, the yield surfaces marhed by black signs are determined after the transient period for the path with a corner. These results are represented in the deviatoric stress plane and their final preloading direction coincides with the vertical axis of the stress plane. Such recovery of the symmetry and congruency among subsequent yield surfaces is considered to be the fading phenomenon of strain history [94, 135].

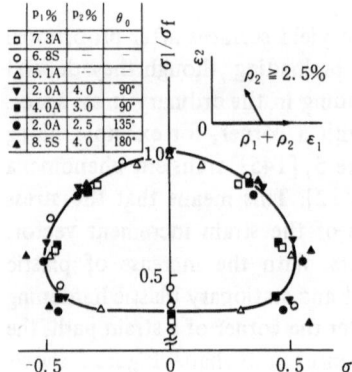

Figure 6. – *Congruency among subsequent yield surfaces.*

In the case of matherials for which plastic behavior is affected by both the second and third invariant of stress, the deviatoric stress plane, or the π plane, is not available to observe such simplified plastic behavior as the symmetry or congruency of subsequent yield surfaces. Some modifications are necessary for the deviatoric stress space or the deviatoric plane. Methods of such modifications are proposed in other articles, [107, 108, 111, 113].

6. Experimentally determined subsequent yield surfaces

6.1. *Effect of preloading by the path of a constant stress ratio*

6.1.1. *Subsequent yield surfaces by complete unloading to zero stress state after preloading*

Figures 7(a) and (b), [119, 121], show the results under combined axial stress and shear stress. The preloadings are given by torsion in Figure 7(a) and tension in Figure 7(b). The negative cross[1] effect is observed in Figure 7(a),

---

[1] The negative cross effect is the effect of increasing yield stress perpendicular to the preloading direction. The positive cross effect is the opposite one.

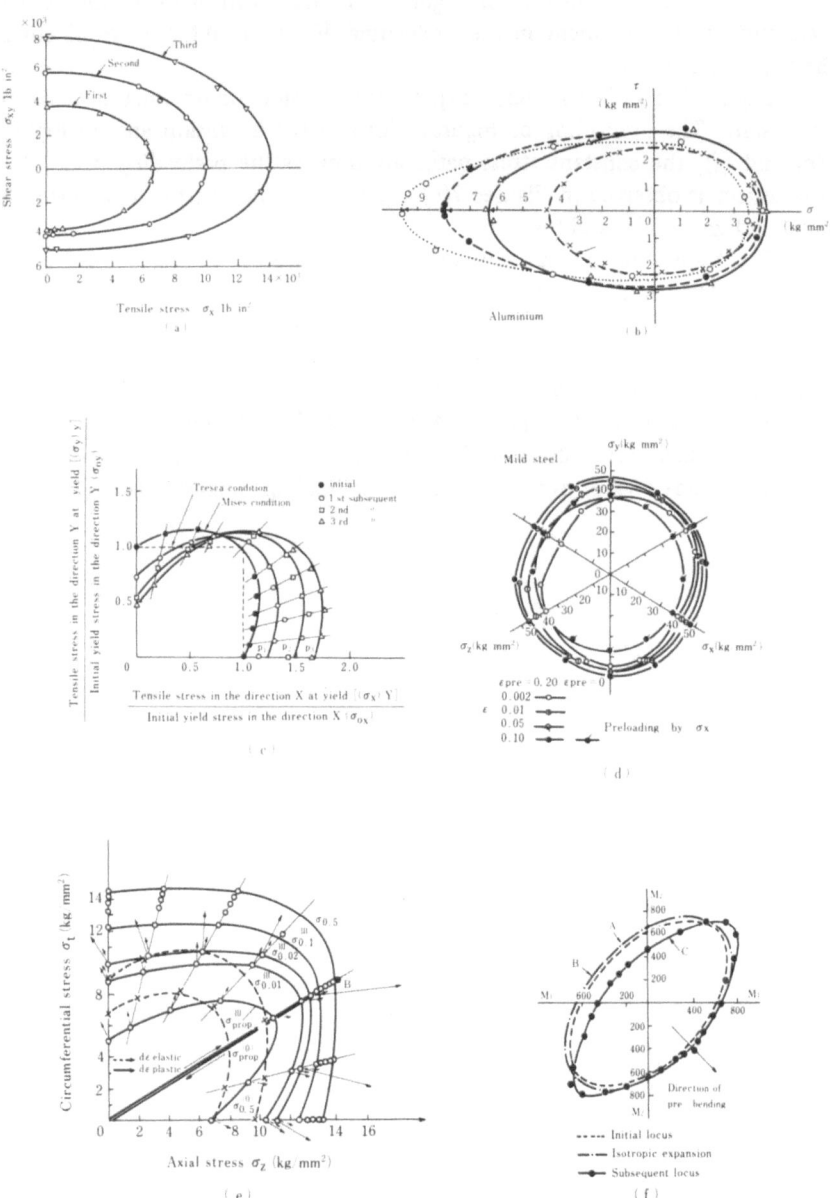

Figure 7. – *Subsequent yield surfaces after proportional preloading (a)-(f) (the case of complete unloading after preloading).*

but the cross effect is not clear in Figure 7(b). The distortion of the subsequent yield surfaces is prominent in the preloading direction, but not prominent in the opposite direction.

Figures 7(c), (f), [85, 83, 123], show the results under combined biaxial stress state. The preloading of Figures 7(c) and (d) is tension and, in Figures 7(e) and (f), the constant stress paths are used as the preloading paths. The cross effect is observed in Figures 7(c) and (e), and the negative cross effect is found in Figures 7(d) and (f).

The translation of the subsequent yield surfaces is observed in Figures 7(a), (f). The normality of the plastic strain incremental vector to the subsequent yield surfaces holds good as shown in Figure 7(e).

6.1.2. *Subsequent yield surfaces by partially unloading after preloading*

Figures 8(a) and (b), [139, 142], depict the subsequent yield surfaces under combined axial stress and shear stress, i.e., Figure 8(a) for shear preloading and Figure 8(b) for tensile preloading. Figure 8(c), [142], is

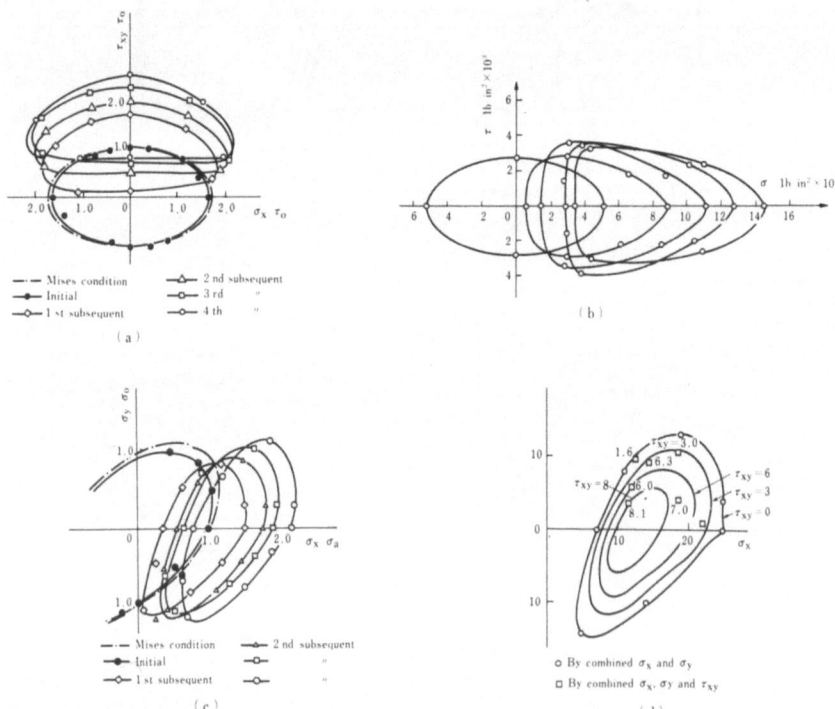

Figure 8. – *Subsequent yield surfaces after proportional preloading (a)-(d) (the case of partial unloading after preloading).*

the subsequent yield surface after tensile preloading, which is determined by the biaxial stress state. The subsequent yield surface in Figure 8(d), [142], shows the result determined by combined biaxial stresses and shear stress after tensile preloading.

The subsequent yield surfaces bulge in the preloading direction and are flattened opposite to the preloading direction. After tensile or shear preloading, the subsequent yield surfaces contract with a small amount of preloading and then they expand with increase of preloading. The cross effect varies depending on the magnitude of preloading. The translations of the subsequent yield surfaces are observed in the preloading direction.

6.2. *Subsequent yield surfaces after preloading by the path of variable stress ratios*

Figure 9(a), [124], shows the subsequent yield surfaces after preloading by increasing the shear stress under constant axial stress from a point in the initial yield surface. The results are determined in the stress plane of combined axial stress and shear stress. Figure 9(b), [85], shows the result in the biaxial stress state. The preloading is given by increasing one stress component of the biaxial stresses while keeping the other stress component constant. Those results were obtained after complete unloading preceding preloading. The subsequent yield surface translates with rotation by preloadings.

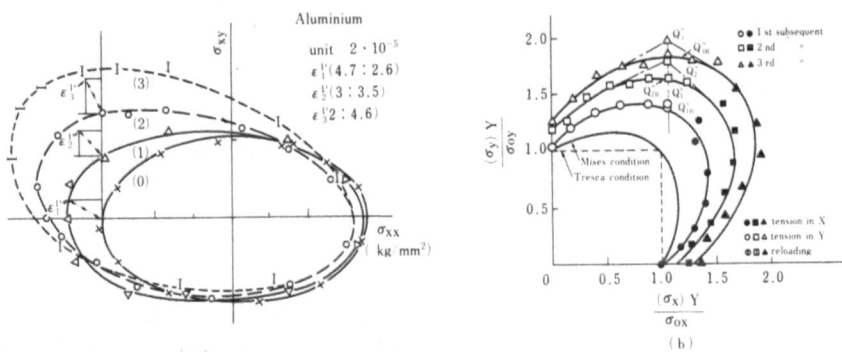

Figure 9. – *Subsequent yield surfaces after preloading along a path of variable stress ratios (the case of complete unloading after preloading) (a)-(b).*

6.3. *Subsequent yield surfaces after cyclic preloadings*

Figures 10(a) and (b), [148], show the effect of the amplitude of cyclic plastic strain and of the cycle numbers on the subsequent yield surfaces. The cyclic

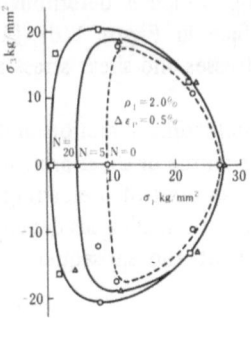 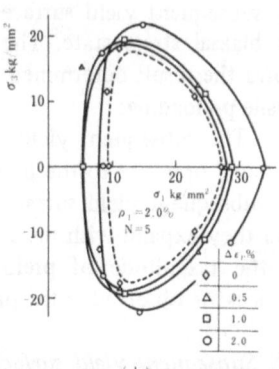

Figure 10. – *Subsequent yield surfaces after cyclic preloading (a)-(b)*.

preloading is given after tensile prestrain of magnitude 2.0 %. For large plastic strain amplitudes, the negative cross effect and translation of the yield surfaces are observed as shown in Figure 10(a). With the increase of cycle numbers, the subsequent yield surface expands to the side opposite of the preloading and saturates into one shape. The correlated effect of plastic strain amplitude, cyclic numbers and magnitude of prestrain on the subsequent yield surfaces is complicated, [148]. The fading phenomena during cyclic loading are investigated experimentally in other articles [94, 134, 135, 147].

6.4. *Effect of the anisotropy of textures*

The yield surfaces are determined for various initially anisotropic materials, i.e., zircaloy, [82, 153], titanium and its alloy, [79, 80], magnesium and its alloy, [81, 154], cold rolled brass, [85], copper and its alloy, [129, 155], aluminium and its alloy, [83, 129], steel alloy, [154], stainless steel, [154], and zinc alloy, [124, 158]. The experimental procedures for determining the yield surfaces for those materials are summarized in Table 1. Figures 11(a), (d), [153, 78, 82, 83] show some examples of the yield surface for those anisotropic materials.

6.5. *Effect of proof strains*

Figures 12(a) and (b), [141, 142], show the subsequent yield surfaces after tensile and shear preloadings, respectively. Those results are determined from the state unloaded partially after preloading. The examples determined from the completely unloaded state are shown in Figure 7(a) and (e). The effect of proof

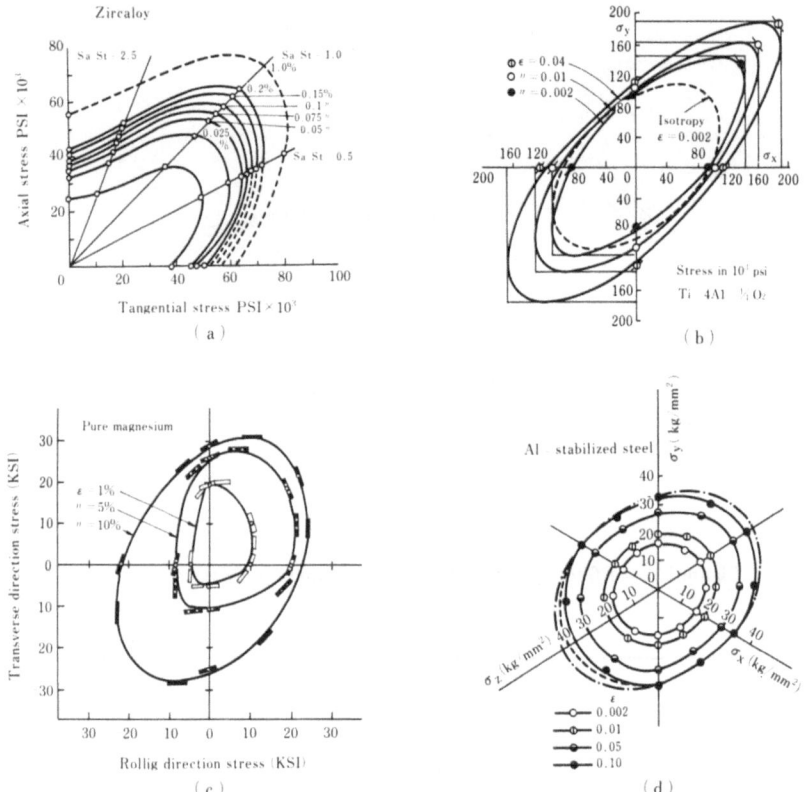

Figure 11. – *Yield surfaces of initially anisotropic materials (a)-(d).*

strain is more pronounced at the side opposite to the preloading direction than in the preloading direction.

6.6. Temperature effect

Examples of the initial yield surface for high temperature conditions are shown in Figures 13(a) and (b), [158,163], and in Figures 4(g) and (h), [166], the corresponding examples are shown for low temperature conditions. The shapes are altered isotropically by change of temperature. Figures 13(c) to (h), [158, 163, 166], exhibit the effects of both preloading and temperature. When the temperature rises, the subsequent yield surfaces after preloading are contracted isotropically in comparison with those at room temperature. As shown in Figures 13(c) to (h), the yield surfaces are translated in the preloading direction

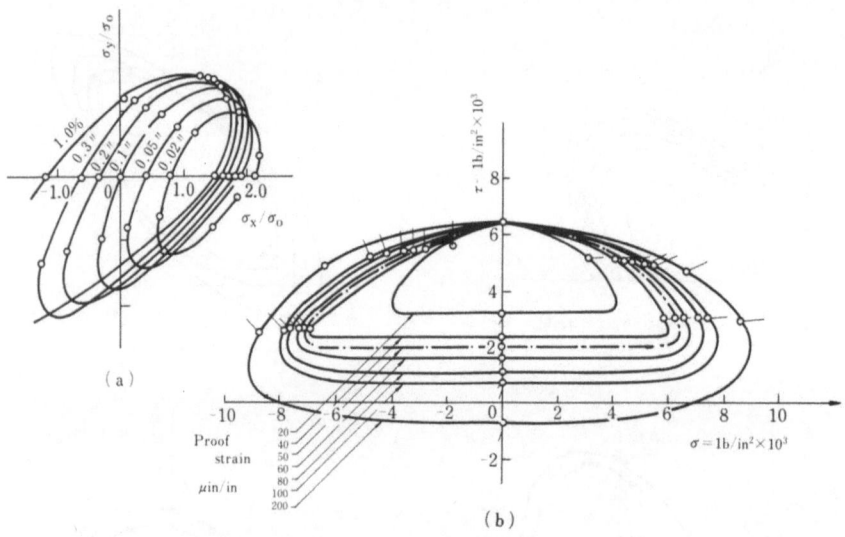

Figure 12. – *Effect of proof strains on subsequent yield surfaces (a)-(b).*

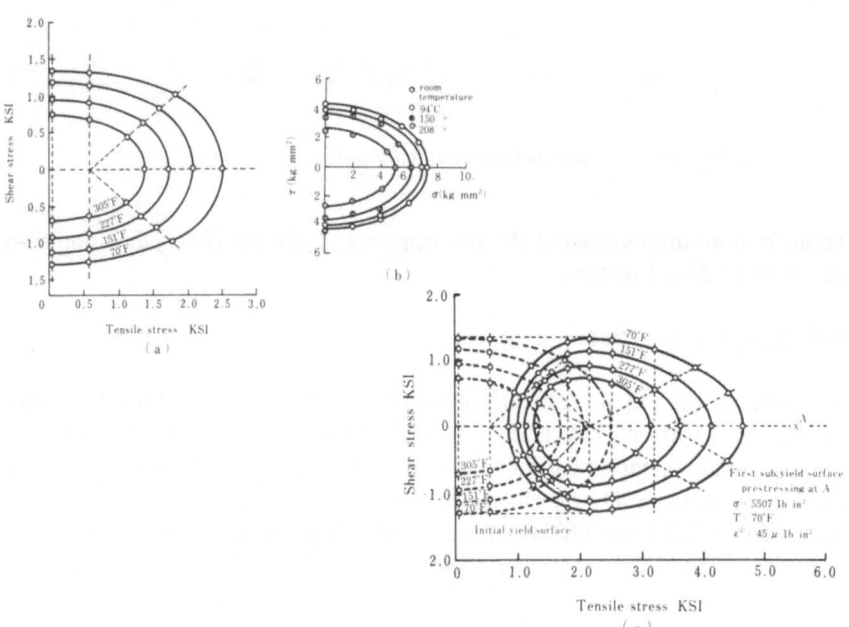

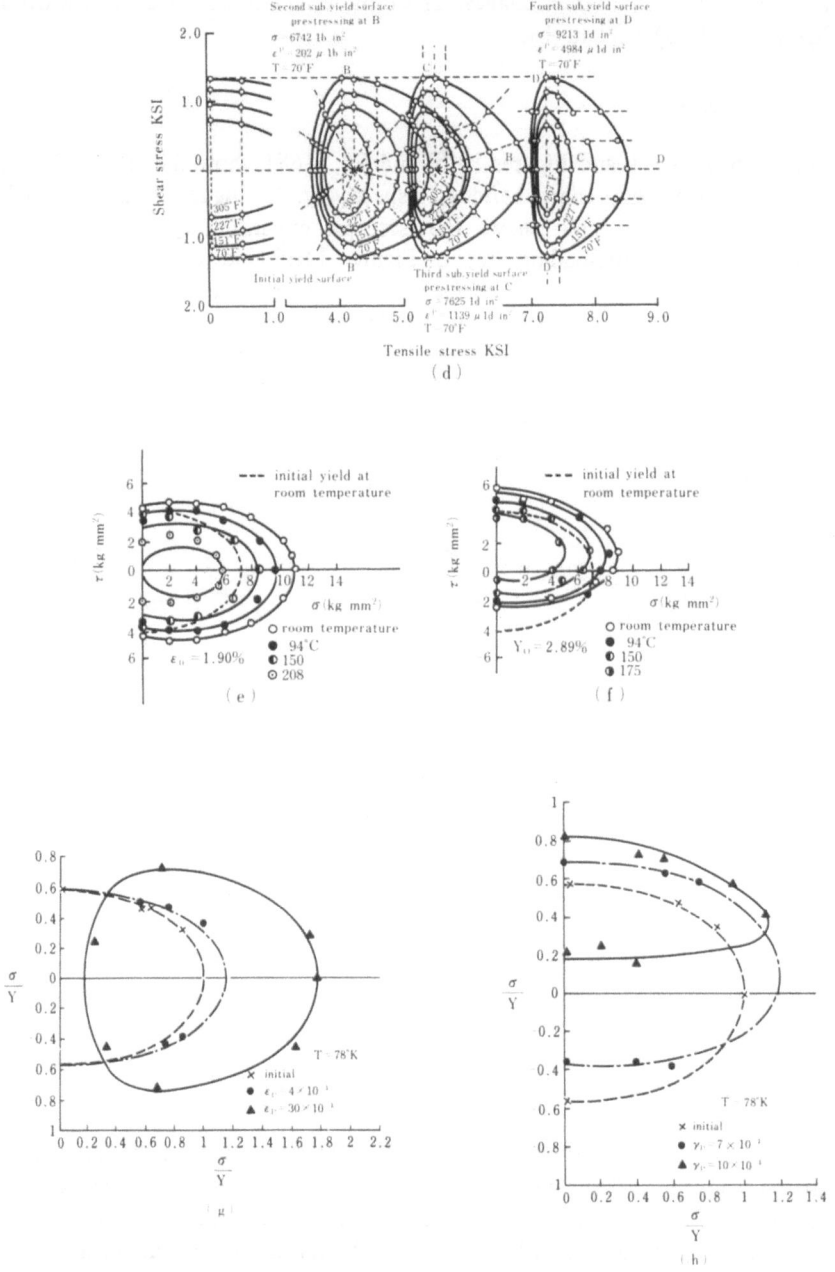

Figure 13. – *Effect of temperature on subsequent yield surfaces (a)-(h).*

with the shrinkage of their shapes. The cross effect is not clear, as shown in Figures 13(c), (d), (g) and (h).

6.7. *Time effect*

Figure 14(a), [167], and Figures 14(b) and (c), [168], show the effect of strain aging over long and short periods, respectively. In Figure 14(a), the yield surface after strain aging shows the isotropic expansion of the subsequent yield surface after preloading, but, in Figures 14(b) and (c), this expansion is not observed. The change of shape is more pronounced in the preloading direction than in the direction perpendicular to the preloading. In a long period of strain aging, the recovery of the shape of the yield surface is found to be as shown in Figure 14(a).

Besides these results, there are the reports on the effect of the creep strain, [162, 164], and strain rate, [169], on the yield surface.

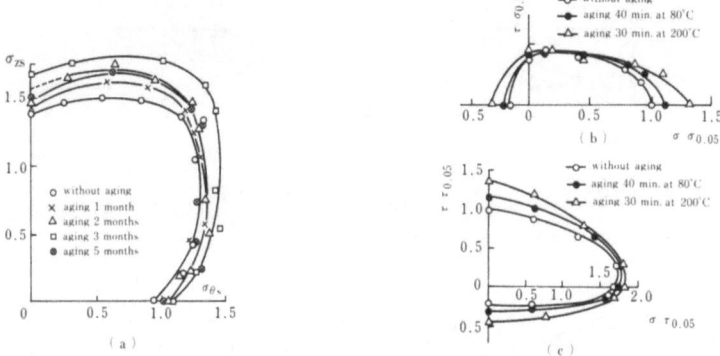

Figure 14. – *Effect of aging on subsequent yield surfaces (a)-(c).*

6.8. *Effect of neutron irradiation*

Figures 15(a) and (b), [170], show the effect of irradiation on the initial yield surface and the subsequent yield surface after preloading, respectively. In Figure 15(a), the yield surfaces before and after neutron irradiation are denoted by CU 1 and CU 2. Figure 15(b) exhibits the subsequent yield surface after neutron irradiation and tensile preloading. The subsequent yield surface contracts and translates in the preloading direction.

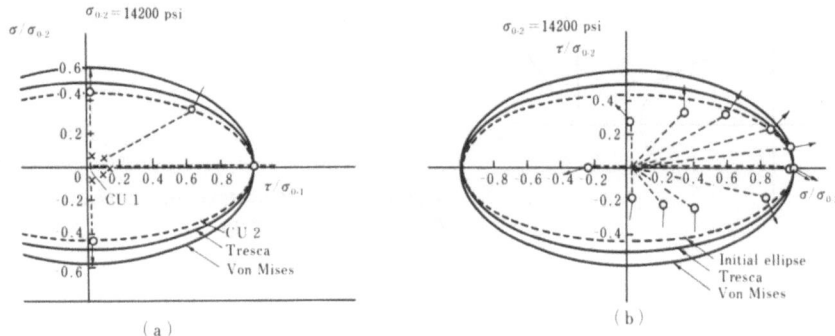

Figure 15. – *Effect of neutron irradiation on subsequent yield surfaces (a)-(b).*

7. Experimental methods for the verification of the corner[1]

The slip theory has suggested the existence of the corner of the subsequent yield surface in the preloading direction. The corner is experimentally verified by the following methods.

(1) Determination of yield surfaces

The existence of the corner is examined by determining the subsequent yield surface in some quadrants of the stress space, e.g. [116], or by determining the yield surface in the vicinity of a preloading point with a large number of yield points, e.g. [137].

(2) Neutral loading

The elastic or plastic change in strains is examined by carrying out a neutral loading test [60] in which the stress state lies in the yield surface predicted by the slip theory.

(3) Measurement of the initial shear coefficients

In this test, a specimen is deformed to a certain degree by axial load as shown in Figure 16. Then the specimen is subjected by combined axial load and shear to proportional loading. The initial shear coefficients for such loading paths are compared with the values predicted by the slip theory [179].

(4) Zig-zag loading (wiggle loading)

Let us consider the zig-zag loading path A-B-C as shown in Figure 17(a). If the yield surface is smooth, the plastic strain increment vector is normal to it and the plastic strain increments AB and BC do not oscillate appreciably.

---

[1] The 'corner' means the convexly distorted part of a yield surface. In this report it is used in a wider sense than in the slip theory.

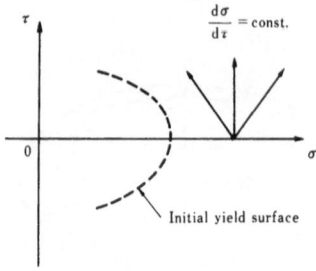

Figure 16. – *Investigation of corners by initial shear coefficients.*

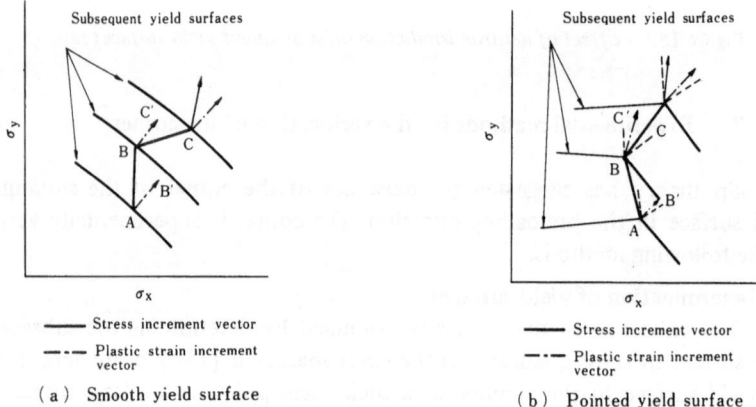

(a) Smooth yield surface   (b) Pointed yield surface

Figure 17. – *Investigation of corners by zig-zag loadings (a)-(b).*

On the other hand, if the yield surface has a corner, the plastic strain increment vector is not necessarily fixed in direction and it lies between the normals on each side of the point, [171]. The methods (1) to (3) prove that the yield surface may or may not have a protruded part. It is found by method (4) that the corner may be rounded or sharp. Table 4 shows the summary of the experimental observations of the corner by both methods (3) and (4).

8. Hardening rule for yield surfaces

As to the behavior of yield surfaces by preloading, the following assumptions are proposed.

(1) Perfect plasticity
    The yield surface does not change at all by preloading.

TABLE 4. Experiments on the existence of corners in yield surfaces.

(a)

Authors	Ref.	Material	Specimen	Loading type	Loading type	Corner
Naghdi, Rowley, Beadle	171	alumi. alloy (24S-T4)	ID = 0.75" WT = 0.075"	AL TR	ZG	
Drucker, Stockton	172	alumi. alloy (24S-T4)	OD = 2.200" WT = 0.100"	AL IP	ZG	
Phillips	173 175	alumi. (2S-0)	OD = 0.850" WT = 0.05"	AL TR	ZG	
Bertsch, Findley	137	alumi. alloy (6061-T6)	OD = 1.000" WT = 0.03"	AL TR	ZG	
Paul, Chen, Lee	175	alumi. alloy (24S-T4)	OD = 0.90" ID = 0.75"	AL TR	ZG	
Mair	176	copper alumi. alloy alumi.	OD = 1.00" WT = 0.04"	AL TR IP	ZG	
Shiratori, Ikegami	85	brass	plate	Ten Ten	ZG	
Hecker	177	alumi. copper	ID = 0.9853" WT = 0.040"		ZG	
Michno, Findley	133	stainless st. (AISL 304L)	OD = 1" WT = 0.06"	AL TR	ZG	

ZG: zig-zag loading path

(b)

Authors	Ref.	Material	Specimen	Loading type	Loading type	Corner
Peters, Dow, Batdorf	178	alumi. alloy (14S-T4)	OD = 3.75" ID = 3.50	AL TR	VS	
Budiansky, Dow, Peters, Shepherd	179	alumi. alloy (14S-T4)	OD = 4.5" WT = 0.156"	AL TR	VS	
Naghdi, Rowley	180	alumi. alloy (24S-T4)	ID = 0.75" WT = 0.076"	AL TR	VS	
Naghdi, Essenberg, Koff	116	alumi. alloy (24S-T4)	ID = 0.75 WT = 0.075"	AL TR	VS	

VS: torsion after axial loading

Existence of corner

positive	OD: outer diameter	AL: axial loading
negative	ID: inner diameter	TR: torsion
occasionally positive	WT: wall thickness	IP: internal pressure

(2) Isotropic hardening, [182, 184]

The yield surface expands isotropically by preloading. This assumption is used in the flow theory and the deformation theory.

(3) Kinematic hardening, [211, 218]

The yield surface is translated in the preloading direction. Many kinematic hardening rules are proposed.

(4) Formation of corners (Piecewise linear hardening), [193, 210]

The yield surface is assumed to consist of several straight lines, such as in the Tresca condition, and in the yield condition predicted by the slip theory.

The composite model valid for the combination of the assumptions (1)-(4) are described in [119, 126]. On the plastic potential surface, in cyclic loading or in unstable behavior, concepts of the loading surface are proposed in [186, 192].

Figure 18, [263], shows the comparison of stress-strain relations in loading along stress paths with a corner. The stress paths are indicated in Figure 18(a). Those experiments were conducted under combined axial stress and shear stress by using thin-wall specimens of brass. Figures 18(b) and (c) show the equivalent stress-strain relations and the strain trajectories, respectively. The solid lines in those figures indicate the experimental results. The calculated results, by using a kinematic model of the Prager-Ziegler type, are shown by dotted lines. This model gives good results in the case of a smaller corner angle. But, for a large corner angle, the discrepancy between experimental results and calculated results becomes pronounced. This fact suggests that this kinematic model is suitable only for the case of the loading paths whose directions do not greatly change.

9. Anisotropic yield conditions

To obtain the explicit form of subsequent yield surfaces, one needs to take into consideration many yield conditions including the anisotropy, [243, 251], or the strain history, [227, 242]. The following difficulties arise in adapting those tranditional yield conditions to the calculation of stress-strain relations in the plastic range.

(1) The shape of the subsequent yield surfaces

Most of the traditional yield conditions are derived from the quadratic equation of stress components. However, as shown in Figure 7, the subsequent yield surfaces experimentally determined after proportional preloading have the convex part in the preloading direction and the flat part in the opposite direction. Their shapes are similar to an egg-shape, when they are represented in the stress space. Such characteristic shape of subsequent yield surfaces cannot be represented by the yield conditions derived from the quadratic equations of stress components.

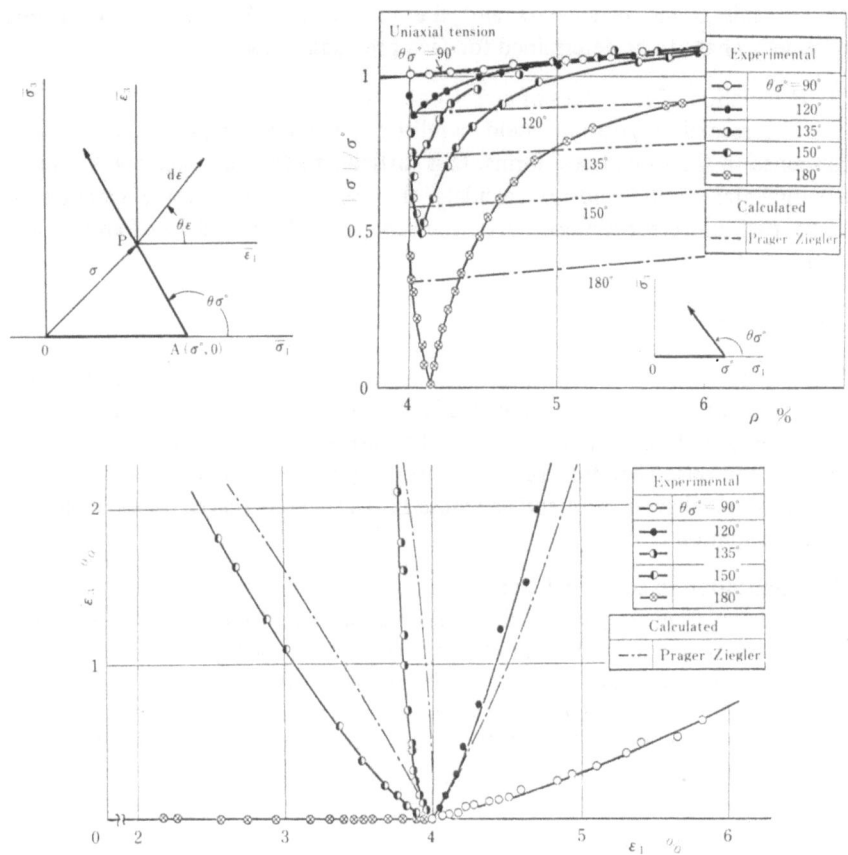

Figure 18. – *Comparison of stress-strain relations between calculated results by Prager-Ziegler's model and experimental results.*
(a) Stress paths with a corner
(b) Equivalent stress-strain relations
(c) Strain trajectories.

(2) Dependency of anisotropic parameters on preloading

In the ordinary stress space, the subsequent yield surfaces show a complicated change in their shape depending on both the magnitude and direction of preloading. When the anisotropic parameters contained in anisotropic yield conditions are determined by using such complicated changed results of the subsequent yield surfaces, the values of anisotropic parameters depend also in a complex manner on both the magnitude and direction of the preloading. It is impossible to establish a general relation between the values of the anisotropic parameters and the preloadings. The set of anisotropic parameters have to be determined for each preloading case.

(3) Calculation of the values of anisotropic parameters

In several anisotropic yield conditions, the anisotropic parameters are contained in their quadratic forms. It is difficult to fit such yield conditions to experimental results and to calculate the values of anisotropic parameters. Some trial and error methods are necessary to settle the values of anisotropic parameters.

(4) Initial yielding

Most of the traditional anisotropic yield conditions are reduced to the Mises condition by substituting certain specific values for the anisotropic parameters or by putting the values of prestrain equal to zero. This means that the Mises condition is used as the initial yielding in most traditional anisotropic yield conditions. Therefore such anisotropic yield conditions cannot apply to the yielding behavior of materials whose initial yielding obeys, for example, the Tresca condition.

(5) Application to complex loading

The subsequent yield surfaces show complicated changes in their shapes in the case of such complex loading as cyclic loading or loading along the path with a corner. Examples can be seen in Figures 5 and 10. Those results imply that the yield condition and its hardening rules vary in a complicated manner for such complex loadings [252, 257]. It is difficult to correlate the change of yield conditions with the complex loading process.

10.   Multi-loading surfaces

To reduce the difficulties in applying the traditional yield conditions, as mentioned in the previous section, the method of using multi-loading surfaces has been proposed for the calculation of the stress-strain relation, [258, 265]. The multi-loading surfaces are defined by the stress locus in which materials have the same work-hardening variables. The magnitude of the plastic strain values or the tangent modulus of the stress-strain curves are used as the work-hardening variables. One method, by using the multi-loading surfaces, [263, 265], is briefly explained as follows.

(1) Derivation of the isotropic stress and strain spaces

In the ordinary stress space, the shapes of loading surfaces show also complicated changes in a similar manner as subsequent yield surfaces. To simplify such complicated behavior, one needs an isotropic stress space with symmetry and congruency among the loading surfaces.

(2) Formulation of the multi-loading surfaces

An equation in the fourth powers of the stress components is used to represent the loading surfaces. The loading surfaces of Mises type materials are represented by a set of surfaces whose shapes vary from egg-shape to circle with the increase of subsequent preloading.

(3) Calculation of the plastic strain increment vector or the stress vector

The plastic strain increment vector for a given stress path can be calculated by applying the flow rule to the loading surface at each intersection between a given stress path and a loading surface. The stress increment vector for a given strain path can be obtained by finding the stress point of each loading surface at which the plastic strain increment vector has a given direction.

Though a loading path changes its direction during the deformation process, if the change does not contain an unloading process, only a slight change will occurr in the shape of the loading surface. On the other hand, the subsequent yield surfaces are sensitive to changes of the loading path. (see e.g. Figure 7 in the article [264]). This simple property of the multi-loading surfaces reduces the complicated procedure to the calculation of the stress-strain relations.

Figure 19, [265], shows an example of the multi-loading surfaces, denoted by f, in mechanical ratcheting tests of a thin-wall tube. This expe-

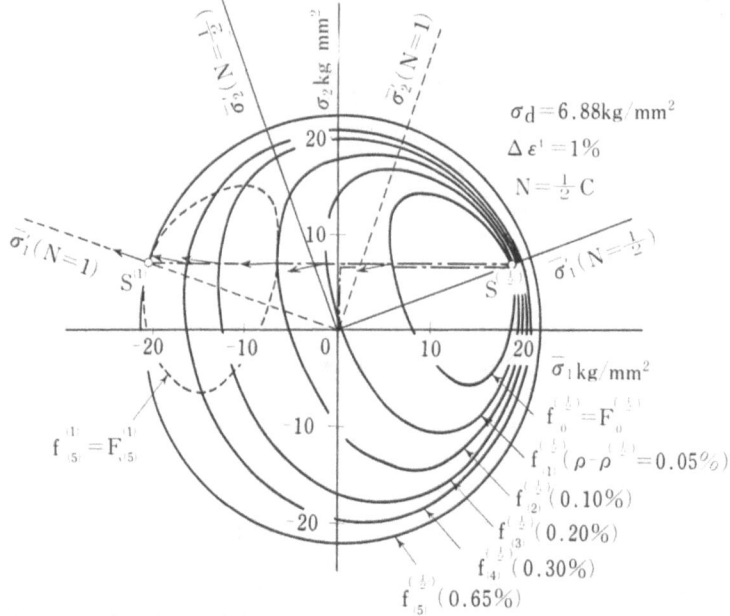

Figure 19. – *Multi-loading surfaces during ratcheting loading.*

riment is conducted under combined steady internal pressure and cyclic axial
load. The results are represented in the isotropic stress space. The chain line
in Figure 19, [265], indicates the stress path during a half cycle of ratcheting.
The arrow points in the direction of the strain increment vectors which arise
at the intersections between the multi-loading surfaces and the stress path.

Figure 20, [265], shows an example of ratcheting strain behavior. In
this figure, two different experimental results are represented in an isotropic

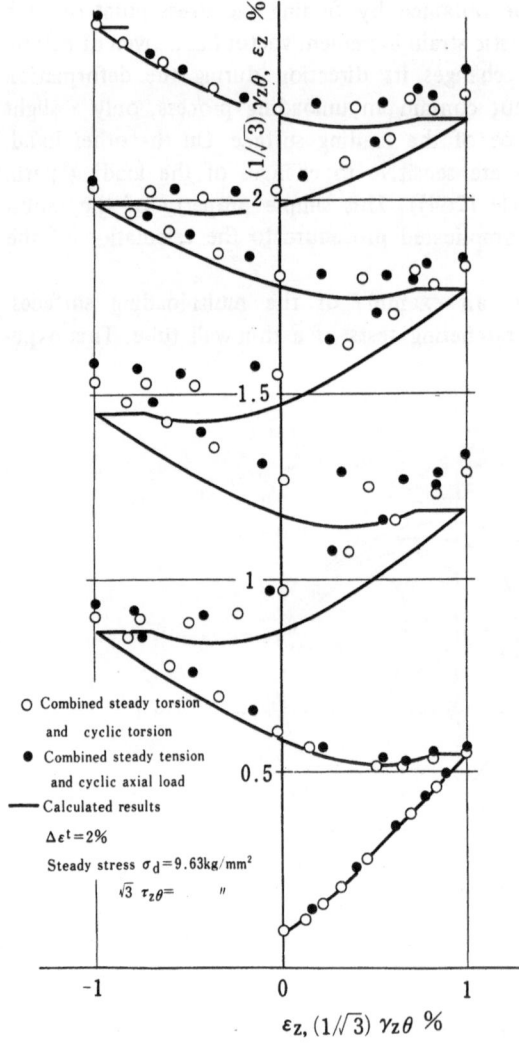

O Combined steady torsion
and cyclic torsion
● Combined steady tension
and cyclic axial load
——Calculated results

$\Delta \varepsilon^t = 2\%$

Steady stress $\sigma_d = 9.63 \text{kg/mm}^2$

$\sqrt{3}\ \tau_{z\theta}=$ "

$(1/\sqrt{3})\ \gamma_{z\theta},\ \varepsilon_z\ \%$

$\varepsilon_z,\ (1/\sqrt{3})\ \gamma_{z\theta}\ \%$

Figure 20. − *Comparison of strain
trajectories during ratcheting
loading between calculated results
by the multi-loading surfaces and
experimental results.*

strain plane. One is the result of combined steady axial load and cyclic torsion. The other is the result of combined steady torsion and cyclic axial load. The vertical and horizontal lines of Figure 20 correspond to the ratcheting and steady strain, respectively. The solid line indicates the results calculated by using the multi-loading surfaces. The calculated results predict correctly the ratcheting behavior. Moreover, two different ratcheting behaviors can be uniquely represented in the isotropic strain plane.

11. Conclusions

The experimental and theoretical investigations of plastic behavior under combined stress state are reviewed from the phenomenological point of view. The emphasis is placed on the research results on yield surfaces. Much research has been done in this field, but, to complete the theory of plasticity, more extensive and systematic studies are needed as concerns cyclic plasticity, and viscoplasticity including time and temperature effects.

Recently, some theories of plasticity without yield surfaces were proposed, using the tensor theory, [104], the functional theory, [266], the hypoelasticity theory, [267], and the endochronic theory, [268]. One of the most interesting research projects in experimental plasticity is to examine the validity of those new theories.

Acknowledgements

The author wishes to thank Centre National de la Recherche Scientifique for the financial support in presenting this article, and also wishes to thank Dr J.P. Boehler of Université Scientifique et Médicale de Grenoble and Prof. A. Sawczuk of the Polish Academy of Sciences for their discussions to this article.

REFERENCES

[1] BELL J.F. – "Festkörpermechanik". *Handbuch der Physik*, vol. 4a/1, Springer Verlag, 1973.
[2] NADAI A. – "Theories of strength". *Trans. ASME*, 55 (1933): A111-A129.
[3] NAGHDI P.M. – "Stress-strain relations in plasticity and thermoplasticity". Plasticity, *Proc. 2nd Symp. Naval Structure Mech.*, Pergamon Press, (1960): 122-166.
[4] DRUCKER D.C. – "On the role of experiment in development of theory". *Proc. 4th U.S. Natl. Cong. Appl. Mech.* (1962): 15-33.
[5] ZYCZKOWSKI M. – "Combined loadings in the theory of plasticity". *Int. J. Non-Linear Mech.*, 2 (1967): 173-205.

[6] PAUL B. – "Macroscopic criteria for plastic flow and brittle fracture". *Fracture*, vol. 2, Academic Press (1968): 313-496.

[7] SINDO A. – "Recent trend in the theory of plasticity". *Science of Machine*, 20 1968): 295-302.

[8] OHASHI Y. – "A view on elastoplastic analysis". *J. Japan Soc. Mech. Engrs.*, 72 (1969): 11-20.

[9] MICHNO M.J. and W.N. FINDLEY. – "An historical perspective of yield surface investigations for metals". *Int. J. Non-Linear Mech.*, 11 (1976): 59-82.

[10] PHILLIPS A. – "Experimental plasticity". *Mech. Plastic Solids, Proc. Int. Symp. Foundations of Plasticity*, 2 Noordhoff Int. Pub. (1974): 193-233.

[11a] IKEGAMI K. – "An historical perspective of experimental study on subsequent yield surfaces for metals – Parts 1 and 2". *J. Soc. Mat. Sci. Japan*, 24 (1975): 491-504, 709-719.

[11b] IKEGAMI K. – "An historical perspective of experimental study on subsequent yield surfaces for metals – Parts 1 and 2". *BISI*, 14420 (1976), British Ind. Sci. Int. Translation Service (The Metal Society).

[12] IKEGAMI K. – "A survey of the plastic potential theory for metals". *J. Japan Soc. Tech. Plasticity*, 18 (1977): 715-724.

[13] HECKER S.S. – "Experimental studies of yield phenomena in biaxially loaded metals". *Constitutive Equations in Viscoplasticity*, AMS-Vol. 20 (1976): 1-33.

[14] GOTOH M. – "On a trend of the plasticity theory and its future". *J. Japan Soc. Tech. Plasticity*, 18 (1977): 664-670.

[15] GUEST J.J. – "On the strength of ductile materials under combined stress". *Philo. Mag.*, 50 (1900): 69-132.

[16] HANCOCK E.L. – "Results of tests on materials subjected to combined stresses". *ibid.*, 11 (1906): 276-282. 12 (1906): 418-425. 15 (1908): 214-222. 16 (1908): 720-725.

[17] SCOBLE W.A. – "The strength and behavior of ductile materials under combined stress". *Philo. Mag.*, 12 (1906): 553-547.

[18] HANCOCK E.L. – "The effect of combined stresses on the elastic properties of iron and steel". *ibid.*, 12 (1906): 418-425.

[19] HANCOCK E.L. – "The effect of combined stresses in the elastic properties of steel". *Proc. ASTM*, 6 (1906): 295-307.

[20] MASON W. – "Mild steel tubes in compression and under combined stresses". *Proc. Inst. Mech. Engr.*, 4 (1909): 1205-1236.

[21] SMITH C.A.M. – "Compound stress experiments". *ibid.*, 3 (1909): 1237-1277.

[22] TURNER L.B. – "The elastic breakdown of materials submitted to compound stresses". *Engineering*, 87 (1909): 169-170, 203-209. 92 (1911): 115-117, 183-185, 246-250, 305-307.

[23] WESTERGAARD H.M. – "On the resistance of ductile materials to combined stresses in two or three directions perpendicular to one another". *J. Franklin Inst.*, 189 (1920): 627-640.

[24] HAIGH B.R. – "The strain-energy function and the elastic limit". *Engineering*, 109 (1920): 158-160.

[25] LODE W. – "Versuche über den Einfluss der Mittleren Hauptspannung auf das fliessen der Metalle Eisen, Kupfer and Nickel". *Z. f. Physik*, 36 (1926): 913-939.

[26] TAYLOR G.L. and H. QUINNEY. – "The plastic distortion of metals". *Philo. Trans. Roy. Soc., Ser. A.*, 230 (1931): 323-362.

[27] MORRISON J.L.M. – "The yield of mild steel with particular reference to the effect of size of specimen". *Proc. Inst. Mech. Engr.*, 142 (1940): 193-323.

[28] LESSELLS J.M. and C.W. MacGREGOR. – "Combined stress experiments on a nickel chrome-molybdenum steel". *J. Franklin Inst.*, **230** (1940): 163-181.

[29] DAVIS E.A. – "Increase of stress with permanent strain and stress-strain relations in the plastic state for copper under combined stresses". *J. Appl. Mech.*, **10** (1943): A187-A196.

[30] DAVIS E.A. – "Yielding and fracture of medium-carbon steel under combined stresses". *ibid.*, **12** (1945): A13-A24.

[31] OSGOOD W.R. – "Combined stress tests on 24S-T aluminum alloy tubes". *ibid.*, **14** (1947): A147-A153.

[32] FRAENKEL S.J. – "Experimental studies of biaxially stressed mild steel in the plastic range". *ibid.*, **15** (1948): 193-200.

[33] DAVIS H.E. and E.R. PARKER. – "Behavior of steel under biaxial stresses determined by tests on tubes". *J. Appl. Mech.*, **15** (1948): A201-A215.

[34] MORRISON J.L.M. – "The criterion on yield gun steels". *Proc. Inst. Mech. Engrs.*, **159** (1948): 81-94.

[35] SHEPHERD W.M. – "Plastic stress-strain relations". *ibid.*, **159** (1948): 95-114.

[36] MARIN J. – "Stress-strain relations in the plastic range for biaxial stresses". *J. Franklin Inst.*, **248** (1949): 231-249.

[37] DAVIS E.A. – "Combined tension-torsion tests with fixed principal directions". *J. Appl. Mech.*, **22** (1935): 411-415.

[38] ZUKOV A.M. – "On the plastic deformation of isotropic material in combined stresses". *Izv. Akad. Nauk. CCCP*, **12** (1956): 72-87.

[39] BAIRD B.L. – "Biaxial stress-strain properties of welds in high strength alloys". *Welding Research Supplement*, **42** (1963): 571S-576S.

[40] JOHNSON K.R. and O.M. SIDEBOTTOM. – "Strain-history effect on isotropic and anisotropic plastic behavior". *Exp. Mech.*, **12** (1972): 264-271.

[41] SIDEBOTTOM O.M. and R.K. BHATTACHANYYA. – "Evaluation of finite-plasticity theories for torsion-tension members made of Tresca materials". *Exp. Mech.*, **13** (1972): 238-245.

[42] HOHENEMSER K. – "Fliessversuche an Rohren aus Stahl bei kombinierter Zugund Torsionsbeanspruchung". *ZAMM*, **11** (1931): 15-19.

[43] HOHENEMSER K. and W. PRAGER. – "Beitrag zur Mechanik des Bildsame Verhaltens von Flusstahl". *ibid.*, **12** (1932): 1-14.

[44] CUNNINGHAM D.E., E.G. THOMSEN and J.E. DORN. – "Plastic flow of magnesium alloy under biaxial stresses". *Proc. ASTM*, **47** (1947): 546-553.

[45] MORRISON J.L. and W.M. SHEPHERD. – "An experimental investigation of plastic stress-strain relations". *Proc. Inst. Mech. Engrs.*, **163** (1950): 1-19.

[46] FEIGN M. – "Inelastic behavior under combined tension and torsion". *Proc. 2nd U.S. Natl. Congr. Appl. Mech.* (1955): 469-476.

[47] SAITO K. – "Anisotropy produced by plastic deformation – Part 2". *Trans. Japan Soc. Mech. Engrs.*, **20** (1954): 777-783.

[48] SAITO K. and H. IGAKI. – "Anisotropy produced by plastic deformation – Part 3". *ibid.*, **21** (1955): 468-473.

[49] SAITO K. and H. IGAKI. – "Anisotropy produced by plastic deformation – Part 4". *ibid.*, **23** (1957): 893-899.

[50] SAITO K., H. SHOTA and H. IGAKI. – "Anisotropy produced by plastic deformation – Part 5." *ibid.*, **23** (1957): 899-904.

[51] PHILLIPS A. and L. KAECHELE. – "Combined stress tests in plasticity". *J. Appl. Mech.*, **23** (1956): 43-48.

[52] PHILLIPS A. – "An experimental investigation on plastic stress-strain relations". *Proc. 9th Int. Congr. Appl. Mech.*, 8 (1957): 23-33.

[53] SHAMMANY M.R. and O.M. SIDEBOTTOM. – "Incremental versus total-strain theories for proportionate and nonproportionate loading of torsion-tension". *Exp. Mech.*, 7 (1967): 497-505.

[54] SCHLAFER J.L. and O.M. SIDEBOTTOM. – "Experimental evaluation of incremental theories for nonproportionate loading of thin-walled cylinders". *ibid.*, 9 (1969): 500-506.

[55] SIDEBOTTOM O.M. – "Evaluation of finite-plasticity theories for non-proportionate loading of torsion-tension members". *ibid.*, 12 (1972): 8-24.

[56] CHU S.C. and J.D. VASILAKIS. – "Inelastic behavior of thick-walled cylinders subjected to nonproportionate loading". *ibid.*, 13 (1973): 113-119.

[57] MARIN J. and L.W. HU. – "Plastic stress-strain relations for biaxial tension and variable stress ratio". *Trans. ASME*, 52 (1952): 1095-1125.

[58] MARIN J. and L.W. HU. – "On the validity of assumptions made in theories of plastic flow for metals". *Trans. ASME*, 75 (1953): 1181-1190.

[59] MARIN J. and L.W. HU. – "Biaxial plastic stress-strain relations of a mild steel for variable stress ratios". *Trans. ASME*, 78 (1956): 499-509.

[60] HU L.W. and J. MARIN. – "Anisotropic loading functions for combined stresses in the plastic range". *J. Appl. Mech.*, 22 (1955): 77-85.

[61] GILL S.S. – "Three neutral loading tests". *J. Appl. Mech.*, 23 (1956): 497-502.

[62] GILL S.S. and J. PARKER. – "Plastic stress-strain relationships – some experiments on the effect of loading path and loading history". *J. Appl. Mech.*, 26 (1959): 77-87.

[63] PARKER J. and J. KETTLEWELL. Jr. – "Plastic stress-strain relationships – further experiments on the effect of loading history". *J. Appl. Mech.* 28 (1961): 439-446.

[64] ROGAN J. and A. SHELTON. – "Yield and subsequent flow behavior of some annealed steels under combined stress". *J. Strain Anal.* 4 (1969): 127-137.

[65] MITTAL R.K. – "Biaxial loading of aluminium and a generalization of the parabolic law". *J. Materials* 9 (1971): 67-81.

[66] SHAHABI S.N. and A. SHELTON. – "The yield, flow and creep behavior of annealed En24 steel under combined stress". *J. Mech. Engng. Sci.* 17 (1975): 93-104.

[67] DANESHI G.H. and J.B. HAWKYARD. – "A tension-torsion machine for testing yield criteria and stress-strain relations at $T = 78°K$". *Int. J. Mech. Sci.* 18 (1976): 57-62.

[68] MOON H. – "An experimental study of the outer yield surface for annealed polycrystalline aluminium". *Acta Mech.* 24 (1976): 191-208.

[69] HILL R. – "A new method for determining the yield criterion and plastic potential of ductile metals". *J. Mech. Phys. Solids* 1 (1953): 271-276.

[70] HUNDY B.B. and A.P. GREEN. – "A determination of plastic stress-strain relations". *ibid.*, 3 (1954): 16-21.

[71] LIANIS G. and H. FORD. – "An experimental investigation of the yield criterion and the stress-strain law". *ibid.*, 5 (1957): 215-222.

[72] ELLINGTON J.P. – "An investigation of plastic stress-strain relationships using grooved tensile specimens". *ibid.*, 6 (1958): 270-281.

[73] CORRINGAN D.A., R.E. TRAVIS, V.P. ARDITO and C.M. ADAMS Jr. – "Axial strength of welds in heat-treated sheet steel". *Welding Research Supplement*, 41 (1962): 123S-128S.

[74] BARAYA G.L. and J. PARKER. – "Determination of yield surfaces by notched strip specimens". *Int. J. Mech. Sci.*, 5 (1963): 353-363.

[75] ZIEBS J., H.D. KUHN and S. LEDWORUSKI. – "Fliessbedingung und Spannung-Formänderung-Beziehung mit ausgekehlten Flachproben." *Z. Metallkde*, 66 (1975): 58-66.

[76] LERNER S. and W. PRAGER. – "On the flexure of plastic plates". *J. Appl. Mech.*, 27 (1960): 353-354.

[77] THEOCARIS P.S. and C.R. HAZELL. – "Experimental investigation of subsequent yield surfaces using the Moiré method". *J. Mech. Phys. Solids*, 13(1965): 281-294.

[78] HAZELL C.R. and J. MARIN. – "A possible specimen for the study of biaxial yielding of materials". *Int. J. Mech. Sci.*, 9 (1967): 57-63.

[79] LEE B. and W.A. BACKOFEN. – "An experimental determination of the yield locus for titanium and titanium-alloy sheet". *Trans. Metallurg. Soc. AIME*, 236 (1916): 1077-1084.

[80] LEE B. and W.A. BACKOFEN. – "Yielding and plastic deformation in textured sheet of titanium and its alloys". *ibid.*, 236 (1966): 1696-1704.

[81] KELLY E.W. and W.F. HOSFORD. – "The determination characteristics of textured magnesium". *ibid.*, 242 (1968): 654-661.

[82] SHIH C.F. and D. LEE. – "Further developments in anisotropic plasticity". *Trans. ASME Ser. H.*, 100 (1978): 294-302.

[83] TOZAWA Y. and M. NAKAMURA. – "Yield locus for anisotropic materials". *J. Japan Soc. Mech. Engrs.*, 75 (1972): 541-546.

[84] TOZAWA Y. and M. NAKAMURA. – "Compressive testing method of thin plates in plane state". *J. Japan Soc. Tech. Plasticity*, 8 (1967): 444-448.

[85] SHIRATORI E. and K. IKEGAMI. – "Experimental study of the subsequent yield surface by using cross-shaped specimen". *J. Mech. Phys. Solids*, 16 (1968): 373-394.

[86] KLINGLER L.J. and G. SACHS. – "Dependence of the stress-strain curves of cold-worked metals upon the testing method". *J. Aero. Sci.*, 15 (1948): 151-154.

[87] KLINGLER L.J. and G. SACHS. – "Plastic flow characteristics of aluminum-alloy plate". *J. Aero. Sci.*, 15 (1948): 599-604.

[88] FORD H. – "Researches into the deformation of metals by cold rolling". *Inst. Mech. Engrs.*, 159 (1945): 115-143.

[89] SAITO K. – "Anisotropy produced by plastic deformation". *Trans. Japan Soc. Mech. Engrs.*, 20 (1954): 771-777.

[90] GAROFALO F. and J.R. LOW. – "The effect of prestraining in simple tension and biaxial tension on flow and fracture behavior of a low carbon deep-drawing steel sheet". *J. Mech. Phys. Solids*, 3 (1955): 275-294.

[91] SAITO K., H. IGAKI and M. UNO. – "On the anisotropy parameters in the Hill's anisotropic yield criterion". *J. Japan Soc. Tech. Plasticity*, 4 (1963): 508-514.

[92] SZCZEPINSKI W. – "On the effect of plastic deformation on yield criterion". *Arch. Mech. Stos.*, 15 (1963): 275-296.

[93] SVENSSON N.L. – "Anisotropy and the Bauschinger effect in cold rolled aluminium". *J. Mech. Engng. Sci.*, 8 (1966) : 162-172.

[94] SZCZEPINSKI W. and J. MIASTOWSKI. – "An experimental study of the effect of the prestraining history in the yield surfaces of an aluminum alloy" *J. Mech. Phys. Solids*, 16 (1968): 153-162.

[95] ROLFE S.T., R.P. HAAK and J.H. GROSS. – "Effect of state-of-stress and yield criterion on the Bauschinger effect". *J. Basic Engng.*, 90 (1968) : 403-408.

[96] PASCOE K.J. — "Directional effects of prestrain in steel". *J. Strain Anal.*, 6 (1971): 181-184.

[97] KISHI T. and T. TANABE. — "The Bauschinger effect and its role in mechanical anisotropy". *J. Mech. Phys. Solids*, 21 (1973): 303-315.

[98] TANABE T., T. OKUBO and I. GOKYU. — "A study of the equivalent stress-equivalent strain relation of titanium sheets". *J. Japan Inst. Metals*, 38 (1974): 511-518.

[99] WHEELER R.G. and D.R. IRELAND. — "Multiaxial plastic flow of Zircaloy-2 determined from hardness data". *Elect. Tech.*, 4 (1966): 313-317.

[100] LEE D.S., F.S. JABARA and W.A. BACKOFEN. — "The Knoop-hardness yield loci for two titanium alloys". *Trans. Metallurg. Soc.*, 239 (1967): 1476-1478.

[101] WONSIEWICZ B.C. and W.W. WILKENING. — "A comparison of conventional and Knoop-hardness yield loci for magnesium and magnesium alloys". *ibid.*, 245 (1969): 1313-1319.

[102] GRESIK D. — "Bestimmung von Fliessortkurven an Knupfer- und Aluminum-Kristallen mit Hilfe der Knoophärte". *Z. Metallkde.*, 63 (1972): 618-622.

[103] AMATEAU M.F. and W.D. HANNA. — "Comparison of first quadrant yield loci for Ti-6A1-4V with those predicted by Knoop hardness measurements". *Metal. Trans. Ser. A*, 6A (1975): 417-419.

[104a] ILYUSHIN A.A. — "On the relation between stress and strain increment in the continuum mechanics". *PMM*, 18 (1954): 641-666.

[104b] LENSKY V.S. — "Analysis of plastic behavior of metals under complex loading". Plasticity, *Proc. 2nd Symp. Naval Structure Mech.*, Pergamon Press (1960): 259-278.

[105] HSU T.C. — "The effect of the rotation of the stress axes on the yield criterion of prestrained materials". *Trans. ASME Ser. D*, 88 (1966): 61-70.

[106] SHIRATORI E. and K. IKEGAMI. — "Studies of the anisotropic yield condition". *J. Mech. Phys. Solids*, 17 (1975): 473-491.

[107] SHIRATORI E., K. IKEGAMI and F. YOSHIDA. — "The subsequent yield surface and stress-strain relation of the material of Tresca-type". *Trans. Japan Soc. Mech. Engrs.*, 16 (1975): 1073-1080.

[108] SHIRATORI E., K. IKEGAMI and F. YOSHIDA. — "The subsequent yield surfaces after proportional preloading of the Tresca-type materials". *Trans. Japan Soc. Mech. Engrs.*, 42 (1976): 406-411.

[109] SHIRATORI E., K. IKEGAMI and K. KANEKO. — "Stress and plastic strain increment after corners on strain paths". *J. Mech. Phys. Solids*, 23 (1975): 325-334.

[110] TOZAWA Y. and H. SHIRAI. — "A method of systematical representation of plastic deformation behavior — Plastic deformation behavior of prestrained materials 1". *J. Japan Soc. Tech. Plasticity*, 16 (1975): 550-558.

[111] OHASHI Y., K. KAWASHIMA and T. YOKOCHI. — "Anisotropy due to plastic deformation of initially isotropic mild steel and its analytical formulation". *J. Mech. Phys. Solids*, 23 (1975): 277-294.

[112] OHASHI Y., M. TOKUDA and H. YAMASHITA. — "Effect of third invariant of stress deviator on plastic deformation of mild steel". *ibid.*, 23 (1975): 295-323.

[113] OHASHI Y.M. TOKUDA and Y. YAMASHITA. — "Effect of the curvature of strain trajectory on plastic deformation". *Trans. Japan Soc. Mech. Engrs.*, 40 (1974): 2784-2796.

[114] YOSHIMURA Y., Y. TAKENAKA and S. ABE. — "Yield condition and rate of work-hardening proper to a metal and their dependence on its strain history of extension and twist (2nd report)". *Trans. Japan Soc. Mech. Engrs.*, 25 (1958): 140-147.

[115] LAGN I.I. and O.A. SHISHMAREV. – "Some results of an investigation on the elastic limit of plastically extended nickel sample". *Dokl. Acad. Nauk. SSSR*, 119 (1958): 431-433.

[116] NAGHDI P.M., F. ESSENBERG and W. KOFF. – "An experimental study of initial and subsequent yield surfaces in plasticity". *J. Appl. Mech.*, 25 (1958): 201-209.

[117] HU L.W. and J.F. BRATT. – "Effect of tensile plastic deformation on yield condition". *ibid.*, 25 (1958): 411.

[118] McCOMB H.G. – "Some experiments concerning subsequent yield surfaces in plasticity". *NASA TN*, D-396 (1960).

[119] MAIR W.M. and H. LI, D. PUGH. – "The effect of prestrain on yield surfaces in copper". *J. Mech. Engng. Sci.*, 6 (1964): 150-163.

[120] PARKER J. and M.B. BASSETT. – "Plastic stress-strain relationships – some experiments to devise a subsequent yield surface". *J. Appl. Mech.*, 31 (1964): 676-682.

[121] DUONG B.H. – *C.R. Acad. Sc. Paris*, 259 (1964): 4509-4512.

[122] JENKINS D.R. – "Kinematic hardening in zinc alloy tubes". *J. Appl. Mech.*, 32 (1965): 849-858.

[123] MIASTKOWSKI J. and W. SZCZEPINSKI. – "An experimental study of yield surfaces of pre-strained brass". *Int. J. Solids Structures*, 1 (1965): 189-194.

[124] DUONG B.H. – *C.R. Acad. Sc. Paris*, 262 (1966): 401-404.

[125] MIASTOKOWSKI J. – "Analysis of the memory effect of plastically prestrained material". *Arch. Mech. Stos.*, 20 (1968): 261-276.

[126] ROGAN J. and A. SHELTON. – "Effect of pre-stress on the yield and flow of En25 steel". *J. Strain Anal.*, 4 (1969): 138-161.

[127] HAYASHI I., K. KAWAGUCHI and H. FUKUDA. – "An experimental study on the subsequent yield surfaces of mild steel – A study on the yielding mechanism of prestrained mild steel, 1st report". *J. Japan Soc. Tech. Plasticity*, 11 (1970): 17-23.

[128] TURSKI K. – "Effect of plastic deformation on subsequent behavior of metals in a different loading program". *Bull. Acad. Polonaise Sci.*, 19 (1971): 89-96.

[129] FREDERKING R.M.W. and O.M. SIDEBOTTOM. – "An experimental evaluation of plasticity theory for anisotropic metals". *J. Appl. Mech.*, 38 (1971): 15-22.

[130] HECKER S.S. – "Yield surfaces in prestained aluminum and copper". *Metallurg. Trans.*, 2 (1971): 2077-2086.

[131] HECKER S.S. – "Influence of deformation history on the yield locus and stress-strain behavior of aluminum and copper". *Metallurg. Trans.*, 4 (1973): 985-989.

[132] AZUMA Y., M. SUGIMOTO and K. SAITO. – "Yield condition of anisotropic materials prestrained by combined stress state". *Preprint of Japan Soc. Tech. Plasticity*, (1973): 282-292.

[133] MICHNO M.J. and W.N. FINDLEY. – "Subsequent yield surfaces for annealed mild steel under dead-weight loading: aging, normality, convexity, corners, Bauschinger, and cross effects". *J. Engng. Mat. Tech.*, 96 (1974): 56-64.

[134] MARJANOVIC R. and W. SZCZEPINSKI. – "Yield surfaces of the M-63 brass prestrained by cyclic biaxial loading". *Arch. Mech.*, 26 (1974): 311-320.

[135] MARJANOVIC R. and W. SZCZEPINSKI. – "On the effect of biaxial cyclic loading on the yield surface of M-63 brass". *Acta Mech.*, 23 (1975): 65-74.

[136] IVEY H.J. – "Plastic stress-strain relations and yield surfaces for aluminum alloys". *J. Mech. Engng. Sci.*, 3 (1961): 15-31.

[137] BERTSCH P.K. and W.N. FINDLEY. – "An experimental study of subsequent yield surfaces: corners, normality, Bauschinger, and allied effects". *Proc. 4th U.S. Natl. Congr. Appl. Mech.* (1962): 893-907.

[138] SISHMAREV O.A. – "Study of a portion of the yield curve opposite to the point of loading". *Akad. Nauk. SSSR Izvestia, Otd. Tekh. Nauk. Mekh. i Mash.*, 4 (1962): 159-164.

[139] WILLIAMS J.F. and N.L. SVENSSON. – "Effect of tensile prestrain on the yield locus of 1 100-F aluminum" *J. Strain Anal.*, 5 (1970): 128-139.

[140] SMITH S. and B. O. ALMROTH. – "An experimental investigation of plastic flow under biaxial stress". *Exp. Mech.*, 10 (1970): 217-224.

[141] WILLIAMS J.F. and N.L. SVENSSON. – Effect of torsional prestrain on the yield locus of 1 100-F aluminum". *J. Strain Anal.*, 6 (1971): 263-272.

[142] SHIRATORI E., K. IKEGAMI and K. KANEKO. – "The influence of the Bauschinger effect on the subsequent yield condition". *Trans. Japan Soc. Mech. Engrs.* 39 (1973): 458-471.

[143] MICHNO M.J. and W.N. FINDLEY. – "Experiments to determine small offset yield surfaces for 304L stainless steel under combined tension and torsion". 18 (1973): 163-179.

[144] SHIRATORI E., K. IKEGAMI, K. KANEKO, F. YOSHIDA and S. KOIKE. – "The subsequent yield surfaces after preloading under combined axial load and torsion". *Trans. Japan Soc. Mech. Engrs.*, 41 (1975): 3430-3437.

[145] SHIRATORI E., K. IKEGAMI and K. KANEKO. – "The stress vector and the subsequent yield surface in loading along the strain path with a corner". *Trans. Japan Soc. Mech. Engrs.*, 40 (1974): 671-679.

[146] SHIRATORI E., K. IKEGAMI, K. KANEKO and T. SUGIBAYASHI. – "Subsequent yield surfaces after large tensile or torsional prestrain". *Preprint of Japan Soc. Mech. Engrs.*, 75-7-2 (1975): 41-44.

[147] SIERAKOWSKI R.L. and A. PHILLIPS. – "The effect of repeated loading on the yield surface". *Acta Mech.*, 6 (1968): 217-231.

[148] SHIRATORI E., K. IKEGAMI, K. KANEKO and J. TAKADA. – "Effect of cyclic preloading on subsequent yield surfaces". *Trans. Japan Soc. Mech. Engrs.*, 42 (1976): 3338-3347.

[149] PHILLIPS A. and H. MOON. – "An experimental investigation concerning yield surfaces and loading surfaces". *Acta Mech.*, 27 (1977): 91-102.

[150] MORETON D.M., D.G. MOFFAT and R.P. HORNBY. – "Techniques for investigating the yield surface behavior of pressure-vessel materials". *J. Strain Anal.*, 13 (1978): 185-191.

[151] LAMBA H.S. and O.M. SIDEBOTTOM. – "Cyclic plasticity for nonproportional paths: Part 1 – Cyclic hardening, erasure of memory, and subsequent strain hardening experiments". *Trans. ASME Ser. H*, 100 (1978): 96-103.

[152] LAMBA H.S. and O.M. SIDEBOTTOM. – "Cyclic plasticity for nonproportional paths: Part 2 – Comparison with predictions of three incremental plasticity models". *Trans. ASME Ser. H*, 100 (1978): 104-111.

[153] MEHAN R.L. – "Effect of combined stress on yield and fracture behavior of Zircaloy-2". *J. Basic Engng.*, 83 (1961): 499-512.

[154] DILLAMORE I.L., R.J. HAZEL, T.W. WATSON and P. HADDEN. – "An experimental study of the mechanical anisotropy of some common metals". *Int. J. Mech. Sci.*, 13 (1971): 1049-1061.

[155] ALTHOFF J. and P. WINCIERZ. – "The influence of texture on the yield loci of copper and aluminum". *Z. Metallkde.*, 63 (1972): 623-633.

[156] SAWADA T. and S. SAITO. – "The mechanical behavior of Zn-22Al superplastic alloy under the combined stress field at room temperature". *J. Japan Soc. Tech. Plasticity*, 16 (1975): 285-290.

[157] PHILLIPS A. – "Yield surfaces of pure aluminum at elevated temperatures". Thermoinelasticity, *Proc. IUTAM*, Symp. Thermal Inealasticity, Pergamon Press (1968): 241-258.

[158] PHILLIPS A. and J.L. TANG. – "The effect of loading path on the yield surface at elevated temperature". *Int. J. Solids Structures*, 8 (1972): 463-474.

[159] PHILLIPS A., C.S. LIU and J.W. JUSTUSSON. – "An experimental investigation of yield surfaces at elevated temperature". *Acta Mech.*, 14 (1972): 119-146.

[160] Phillips A. and R. KASPER. – "On the foundations of thermoplasticity – experimental investigation". *J. Appl. Mech.*, 40 (1973): 891-896.

[161] PHILLIPS A. and M. RICCIUTI. – "Fundamental experiments in plasticity and creep and aluminum – extension of previous results". *Int. J. Solids Structures*, 12 (1976): 159-171.

[162] BROWN G.M. – "Inelastic deformation of an aluminum alloy under combined stress at elevated temperature". *J. Mech. Phys. Solids*, 18 (1970): 383-396.

[163] TANAKA K., Y. ISHIZAKI and T. INOUE. – "Study on yield surface at elevated temperature". *Trans. Japan Soc. Mech. Engrs.*, 38 (1972): 2439-2445.

[164] ISHIZAKI Y., T. OKABE, K. TANAKA and T. INOUE. – "On the effect of plastic strain and creep strain on the yield surface at elevated temperature". *ibid.*, 39 (1973): 3010-3015.

[165] LEBEDYEV A.A., and N.V. NOVIKOV. – "Deformation and fracture of structural metals under complex stress at low temperature". Mech. Plastic Solids, *Proc. Int. Sym. Foundations of Plasticity*, 2 (1974) Noordhoff Int. Pub.: 463-475.

[166] Daneshi G.H. and J.B. HAWKYARD. – "An investigation into yield surfaces and plastic flow for F.C.C. metals at room temperature and a low temperature". *Int. J. Mech. Sci.*, 18 (1976): 195-200.

[167] TALYPOV G.P. – "Investigation of influence of preliminary plastic strain and natural aging on behavior of low-carbon steel". *Izvest. Akad. Nauk SSSR*, 6 (1961): 125-130.

[168] KUMAKURA S., T. TAKEDA and A. AKIYOSHI. – "Change of yield surfaces of mild steels by strain aging". *Preprint of the 17th Japan Congress on Materials Research* (1973): 13-14.

[169] LINDHOLM U.S. and L.M. YEAKLEY. – "A dynamic biaxial testing machine". *Exp. Mech.*, 7 (1967): 1-7.

[170] DUDDRAR T.D. and J. DUFTY. – "Neutron irradiation and the yield surfaces of copper". *J. Appl. Mech.*, 34 (1967): 200-206.

[171] NAGHDI P.M., J.C. ROWLEY and C.W. BEADLE. – "Experiments concerning the yield surface and the assumption of linearity in the plastic stress-strain relation". *J. Appl. Mech.*, 22 (1955): 416-420.

[172] DRUCKER D.C. and F.D. STOCKTON. – "Instrumentation and fundamental experiments in plasticity". *Proc. SESA*, 10 (1953): 127-142.

[173] PHILLIPS A. – "Pointed vertices in plasticity". Plasticity, *Proc. 2nd Symp. Naval Structure Mech.*, Pergamon Press (1960): 202-214.

[174] PHILLIPS A. and G.A. GRAY. – "Experimental investigation of corners in the yield surface". *J. Basic Engng.*, 83 (1961): 275-288.

[175] PAUL B., W. CHEN and L. LEE. – "An experimental study of plastic flow under stepwise increments of tension and torsion". *Proc. 4th U.S. Natl. Congr. Appl. Mech.*, 2 (1962): 1031-1038.

[176] MAIR W.M. – "An investigation into the existence of corners on the yield surface". *J. Strain Anal.*, 2 (1967): 188-195.

[177] HECKER S.S. – "Experimental investigation of corners in the yield surface". *Acta Mech.*, 13 (1972): 69-86.

[178] PETERS R.W., N.F. DOW and S.B. BATDORF. – "Preliminary experiments for testing basic assumptions of plasticity theories". *Proc. SESA*, 7 (1950): 127-140.

[179] BUDIANSKY B., N.F. DOW, R.W. PETERS and R.P. SHEPHERD. – "Experimental studies of polyaxial stress-strain laws of plasticity". *Proc. 1st U.S. Natl. Congr. Appl. Mech.* (1951): 503-512.

[180] NAGHDI P.M. and J.C. ROWLEY. – "An experimental study of biaxial stress-strain relations in plasticity". *J. Mech. Phys. Solids*, 3 (1954): 63-80.

[181] MISES R. – "Mechanik der plastischen Formänderung von Kristallen". *ZAMM*, 8 (1928): 161-185.

[182] PRAGER W. – "Exploring stress-strain relations of isotropic solids". *J. Appl. Phys.*, 15 (1944): 65-71.

[183] PRAGER W. – "Strain hardening under combined stresses". *J. Appl. Phys.*, 16 (1945): 837-840.

[184] PRAGER W. – "The stress-strain relation laws of the mathematical theory of plasticity – a survey of recent progress". *J. Appl. Mech.*, 15 (1948): 226-233.

[185] DRUCKER D.C. – "A more fundamental approach to plastic stress-strain relations". *Proc. 1st U.S. Natl. Congr. Appl. Mech.* (1951): 487-491.

[186] PALMER A.C., G. MAIER and D.C. DRUCKER. – "Normality relations and convexity of yield surfaces for unstable materials or structural elements". *Trans. ASME Ser. E.*, 34 (1967): 464-470.

[187] PHILLIPS A. and R.L. SIERAKOWSKI. – "On the concept of the yield surface". *Acta Mech.*, 1 (1965): 29-35.

[188] JUSTUSSON J.W. and A. PHILLIPS. – "Stability and convexity in plasticity" *Acta Mech.*, 2 (1966): 251-267.

[189] EISENBERG M.A. and A. PHILLIPS. – "A theory of plasticity with non-coincident yield and loading surfaces". *ibid.*, 11 (1971): 247-260.

[190] McLAUGHLIN P.V. – Properties of workhardening materials with a limit surface", *Trans. ASME, Ser. E*, 40 (1973): 803-807.

[191] NAGHDI P.M. and J.A. TRAPP. – "The significance of formulating plasticity theory with reference to loading surfaces in strain space". *Int. J. Engng. Sci.*, 13 (1975): 785-797.

[192] NAGHDI P.M. and J.A. TRAPP. – "On the nature of normality of plastic strain rate and convexity of yield surfaces in plasticity". *Trans. ASME Ser. E*, 42 (1975): 61-66.

[193] BATDORF S.B. and V. BUDIANSKY. – "A mathematical theory of plasticity based on the concept of slip". *NASA*, TN 1871 (1949).

[194] KOITER W. – Stress-strain relations, uniqueness and variational theorems for elastic-plastic materials with singular yield surface". *Quart. Appl. Math.*, 11 (1953): 350-354.

[195] SANDERS J.L. – "Plastic stress-strain relations based on linear loading functions". *Proc. 2nd U.S. Natl. Congr. Appl. Mech.* (1954): 455-460.

[196] LIN T.H. – "A proposed theory of plasticity based on slips". *Proc. 2nd U.S. Natl. Congr. Appl. Mech.* (1954): 461-468.

[197] LIN T.H. – "On stress-strain relations based on slips". *Proc. 3rd U.S. Natl. Congr. Appl. Mech.* (1958): 581-587.

[198] WARNER W.H. and G.H. HANDELMAN. – "A modified incremental strain law for workhardening materials". *Quart. J. Mech. Appl. Math.*, 9 (1956): 279-293.

[199] BLAND D.R. – "The associated flow rule of plasticity". *J. Mech. Phys. Solids*, 6 (1957): 71-78.

[200] HODGE P.G. – "A general theory of piecewise linear plasticity based on maximum shear". *J. Mech. Phys. Solids*, 5 (1957): 242-260.

[201] BERMAN I. and P.G. HODGE. – "A gneral theory of piecewise linear plasticity for initially anisotropic materials". *Arch. Mech. Stos.*, 11 (1959): 513-540.

[202] SAITO K., H. IGAKI and M. SUGIMOTO. – "A study on the quivalent stress and the equivalent strain rate, (Theory of anisotropy plasticity based on the maximum shear stress hypothesis)". *Trans. Japan Soc. Mech. Engrs.*, 37 (1970): 883-888.

[203] SUGIMOTO M., H. IGAKI and K. SAITO. – "Equivalent stress and equivalent plastic strain rate for work-hardening materials, (Theory of anisotropic plasticity based on the maximum shear hypothesis)". *ibid.*, 39 (1973): 1164-1174.

[204] BUDIANSKY B. – "A reassessment of deformation theories of plasticity". *Trans. ASME, Ser. E*, 26 (1959): 259-264.

[205] KLIUSHNIKOV V.D. – "On a possible manner of establishing the plasticity relations". *PMM*, 23(1959): 282-291.

[206] KLIUSHNIKOV V.D. – "New concept in plasticity and deformation theory". *ibid.*, 23 (1959): 722-731.

[207] RABOTNOV Iu. N. – "Model illustrating some properties of a hardening plastic body". *ibid.*, 23 (1959): 164-169.

[208] MAIER G. – "Linear flow-laws of elastoplasticity: a unified general approach". *Rendiconti*, 47 (1969): 266-276.

[209] SEWELL M.J. – "A plastic flow rule at a yield vertex". *J. Mech. Phys. Solids*, 22 (1974): 469-490.

[210] SVENSSON N.L. and R. METCALFE. – "Corners in the yield locus". *J. Strain Anal.*, 11 (1976): 46-55.

[211] PRAGER W. – "The theory of plasticity: a survey of recent achievement". *Proc. Inst. Mech. Engrs.*, 169 (1955): 41-57.

[212] PRAGER W. – "A new method of analysing stresses and strains in workhardening plastic solids". *Trans. ASME, Ser. E*, 23 (1956): 493-496.

[213] HODGE P.G. – "Piecewise linear plasticity". *Proc. 9th Int. Congr. Appl. Mech.*, 8 (1956): 65-72.

[214] PERRONE N. and P.G. HODGE. – "Strain hardening solutions with generalized kinematic models". *3rd U.S. Natl. Congr. Appl. Mech.* (1958): 641-648.

[215] SHIELD R.T. and H. ZIEGLER. – "On Prager's hardening rule". *ZAMP*, 9 (1958): 260-276.

[216] CLAVUOT C. and H. ZIEGLER. – "Uber einige Verbestigungsregeln". *Ing. Arch.*, 28 (1959): 13-26.

[217] ZIEGLER H. – "A modification of Prager's hardening rule". *Quart. Appl. Math.*, 17 (1959): 55-56.

[218] ZIEGLER H. – "Bemerkung zu einem Hauptachsenproblem in der Plastizitätstheorie". *ZAMP*, 9 (1960): 157-163.

[219] HODGE P.G. – "Discussion of the paper (212)". *Trans. ASME*, 24 (1957): 482.

[220] KADASHEVICH Iu. I. and V.V. NOVOZHILOV. – "The theory of plasticity which takes into account residual microstresses". *PMM*, 22 (1958): 78-79.

[221] EISENBERG M.A. and A. PHILLIPS. – "On nonlinear kinematic hardening". *Acta Mech.*, 5 (1968): 1-13.

[222] KECHZHI K. – "On work-hardening of plastic solids". *PMM*, 22 (1958): 544-546.

[223] TANAKA M. and Y. MIYAGAWA. – "On the generalized kinematic hardening theory of plasticity". *Ing. Arch.*, 44 (1975): 255-268.

[224] MAIER G. – "A matrix structural theory of piecewise linear elastoplasticity". *Meccanica*, 3 (1970): 54-66.

[225] YAMADA K. and K. TAKATSUKA. – "Finite element analysis of nonlinear problem". *J. Japan Soc. Tech. Plasticity*, 14 (1973): 758-765.

[226] TATERETSU T. and K. SAITO. – "A study on relation of stress and strain for numerical analysis of plastic deformation". *ibid.*, 15 (1974): 180-187.

[227] EDELMAN F. and D.C. DRUCKER. – "Some extensions of elementary plasticity theory". *J. Franklin Inst.*, 251 (1951): 581-605.

[228] YOSHIMURA Y. – "Theory of plasticity for small and finite deformations based on legitimate concept of strain". *Aero. Res. Inst. Univ. Tokyo*, 349 (1959).

[229] BALTOV A. and A. SAWCZUK. – "A rule of anisotropic hardening". *Acta. Mech.*, 2 (1965): 81-92.

[230] SVENSSON N.L. – "Anisotropy and the Bauschinger effect in cold rolled aluminum". *J. Mech. Engng. Sci.*, 8 (1966): 162-172.

[231] BACKHAUS G. – "Zur Fiessgrenze bei allgemeiner Verfestigung". *ZAMM*, 48 (1968): 99-108.

[232] WILLIAMS J.F. and N.L. SVENSSON. – "A rationally based yield criterion for work hardening materials". *Meccanica*, 4 (1971): 104-114.

[233] LEHMANN Th. – "Einige Bemerkungen zu einer allgemeiner Klasse von Stoffgesetzen für grosse elasto-plastische Formänderungen". *Ing. Arch.*, 41 (1972): 297-310.

[234] TURSKI K. – "Yield criterion for a prestrained material". *Bull. Acad. Polonaise Sci.*, 20 (1972): 425-430.

[235] PHILLIPS A. and G.J. WENG. – "An analytical study of an experimentally verified hardening law". *Trans. ASME Ser. E*, 42 (1976): 375-378.

[236] SHRIVASTAVA H.P., Z. MROZ and R.N. DUBEY. – "Yield criterion and the hardening rule for a plastic solids". *ZAMM*, 53 (1973): 625-633.

[237] SHIRAI H. and T. TOZAWA. – "Equation of yield condition for prestrained initial isotropic materials". *J. Japan Soc. Tech. Plasticity*, 18 (1977): 100-105.

[238] KANEKO K., K. IKEGAMI and E. SHIRATORI. – "The yield condition and flow rule of a metal for the various pre-strain paths". *Trans. Japan Soc. Mech. Engrs.*, 41 (1975): 2793-2802.

[239] BACKHAUS G. – "Zur analytischen Darstellung des Materialverhaltens". *ZAMM*, 51 (1971): 471-477.

[240] BACKHAUS G. – "Zur analytischen Erfassung des allgemeinen Bauschingereffektes". *Acta Mech.*, 14 (1972): 31-42.

[241] BACKHAUS G. – "Zum Stoffgesetz des elastisch-plastischen Körpers". *ZAMM*, 52 (1972): T293-T305.

[242] INOUE T. and K. TANAKA. – "Subsequent yield conditions of metals under cyclic loading at elevated temperature". *Ing. Arch.*, 44 (1975): 53-62.

[243] HILL R. – "A theory of the yielding and plastic flow of anisotropic metals". *Proc. Roy. Soc. London, Ser. A*, 193 (1948): 281-297.

[244] DORN J.E. – "Stress-strain rate relations for anisotropic plastic flow". *J. Appl. Phys.*, 20 (1949): 15-20.

[245] HSU T.C. – "A theory of the yield locus and flow rule of anisotropic materials". *J. Strain Anal.*, 1 (1966): 204-215.

[246] SAYIR M. – "Zur Fliessbedingung der Plastizitätstheorie". *Ing. Arch.*, 39 (1970): 414-432.

[247] SOBOTKA Z. - "Theorie des plastischen Fliessens von anisotropen Körpern". *ZAMM,* 49 (1969): 25-32.

[248] SETH B.R. - "Yield conditions in plasticity". *Arch. Mech.,* 24 (1972): 769-776.

[249] DUBEY R.N. and M.J. HILLIER. - "Yield criteria and the Bauschinger effect for a plastic solid". *Trans. ASME, Ser. D,* 94 (1972): 228-230.

[250] BETTEN J. - "Plastische Anisotropie und Bauschinger-Effekt; allgemeine Formulierung und Vergleich mit experimentell ermittelten Fliessortkurven". *Acta Mech.,* 25 (1976): 79-94.

[251] GOTOH M. - "An admissible form of an inelastic constitutive equation - II Examples of elastoplastic constitutive equation". *Int. J. Non-linear Mech.,* 12 (1977): 175-187.

[252] BACKHAUS G. - "Fliessspannungen und Fliessbedingung bei zyklischen Verformungen". *ZAMM,* 56 (1976): 337-348.

[253] SHRIVASTAVA H.P., Z. MROZ and R.N. DUBEY. - "Yield criterion and second order effects in plane-stress". *Acta Mech.,* 17 (1973): 137-143.

[254] BRUHNS O. - "On the description of cyclic deformation processes using a more general elasto-plastic constitutive law". *Arch. Mech.,* 25 (1973): 535-546.

[255] DAFALIAS Y.F. and E.P. POPOV. - "A model of nonlinearly hardening materials for complex loading". *Acta Mech.,* 21 (1975): 173-192.

[256] NICHOLSON D.W. - "On the hardening and yield surface motion in plastic plasticity". *ibid.,* 21 (1975): 193-207.

[257] EISENBERG M.A. - "A generalization of plastic flow theory with application to cyclic hardening and softening phenomena". *Trans. ASME, Ser. H,* 98 (1976): 221-228.

[258] MROZ Z. - "On the description of anisotropic workhardening". *J. Mech. Phys. Solids,* 15 (1967): 163-175.

[259] MROZ Z. - "An attempt to describe the behavior of metals under cyclic loads using a more general work hardening model". *Acta Mech.,* 7 (1969): 199-212.

[260] IWAN W.D. - "On a class of models for the yielding behavior of continuous and composite system". *Trans. ASME, Ser. E,* 34 (1967): 612-617.

[261] MORITOKI H. - "On the field of workhardening moduli - 1st report. A method of determining the field". *Trans. Japan Soc. Mech. Engrs.,* 42 (1976): 3398-3409.

[262] KANEKO K., K. IKEGAMI and E. SHIRATORI. - "Plastic deformation behavior of metals for complex loading paths". *Trans. Japan Soc. Mech. Engrs.,* 41 (1975): 750-767.

[263] YOSHIDA F., T. MURAKAMI, K. KANEKO, K. IKEGAMI and E. SHIRATORI. - "Plastic behavior in the combined loading along straight stress paths with a bend". *ibid.,* 43 (1977): 1231-1241.

[264] YOSHIDA F., K. IKEGAMI and E. SHIRATORI. - "Stress-strain relation of the material completely unloaded after plastic deformation in combined reloading". *J. Japan Soc. Tech. Plasticity,* 18 (1977): 525-532.

[265] YOSHIDA F., N. TAJIMA, K. IKEGAMI and E. SHIRATORI. - "Plastic theory of the mechanical ratcheting". *Trans. Japan Soc. Mech. Engrs.,* 43 (1977): 2500-2510.

[266] HOLSAPPLE K.A. - "Elastic-plastic materials as simple materials". *ZAMM* 53 (1973): 261-270.

[267] TOKUOKA K. - "Yield conditions and flow rules derived from hypoelasticity". *Arch. Rat. Mech. Anal.,* 42 (1972): 239-252.

[268] VALANIS K.C. - "A theory of viscoplasticity without a yield surface". *Arch. Mech. Stos.,* 23 (1971): 517-533.

RESUME

(Sur l'anisotropie des métaux en plasticité expérimentale).

Les études phénoménologiques sur la plasticité des métaux sont passées en revue. L'accent est mis sur le comportement plastique sous des états de contraintes combinées. Différentes méthodes d'essais permettant d'obtenir des conditions de contraintes combinées sont présentées. La définition des points d'écoulement est discutée en relation avec la détermination des surfaces d'écoulement. On donne quelques résultats expérimentaux pour la détermination des surfaces d'écoulement, en liaison avec les effets dûs à la précharge, l'anisotropie de texture, la température, le temps et l'irradiation par neutrons. Les différentes méthodes expérimentales pour la vérification de l'existence d'un coin sur les surfaces d'écoulement sont examinées. Les difficultés qui apparaissent pour l'ajustage des conditions d'écoulement traditionnels aux résultats expérimentaux sur les surfaces d'écoulement successives sont soulignées. Pour réduire ces difficultés, une méthode utilisant un ensemble de surfaces de charges multiples est introduite. En conclusion, les objectifs futurs de la plasticité expérimentale sont évoqués.

A Rule of Anisotropic Hardening with Taking into Account Preliminary and Actual Plastic Deformations and Strain Rates

A. Baltov and N. Bontcheva
Institute of Mechanics and Biomechanics, Sofia, Bulgaria.

1. Statement of the problem

The aim of this paper is to propose an extension of the Baltov-Sawczuk anisotropic hardening rule [2] to the case of nonlinear hardening and strain rate sensitivity. Strain rates of the order 10^{-5} s^{-1} to 10^{-1} s^{-1} are considered. The processes are quasi-dynamic at such strain rates and the wave and inertia effects may be neglected. The considered metals change their plastic behaviour according to the strain rate but their visco-plastic properties bring negligible effects. The proposed hardening rule may have application in the cases when elements which had undergone metal forming plastic processes are subjected during their exploatation to loadings with different strain rates.

The preliminary plastic forming process causes internal structural changes which are rate dependent. During the actual deformation process, the internal structure changes again, depending on the actual strain rate and plastic deformation as well as on the preliminary plastic deformation and strain rate. These structural changes manifest in two effects: in the volume average change and in appearance of micrononhomogeneous stresses and strains. On the macrolevel the first effect leads to isotropic hardening, the second one to the deformational anisotropy of the plastic properties. Residual plastic and elastic microstrains are obtained after unloading and the crystals tend to reach their origin state. This causes a rapid microstress relaxation. On the macrolevel the same effect leads to a difference in the mechanical behaviour of the material at instantaneous and durable processes (see sec. 3). Similar problems are considered in different aspects in [1, 3, 5, 8, 9, 12, 14, etc.].

2. Basic assumptions

The model of a rate sensitive quasidynamically deformable plastic material is based on the following assumptions:

a) The strain tensor and its rate consist of an elastic and a plastic part,

$$\epsilon_{ij} = \epsilon^e_{ij} + \epsilon^p_{ij}, \quad \dot{\epsilon}_{ij} = \dot{\epsilon}^e_{ij} + \dot{\epsilon}^p_{ij}, \quad (i, j = 1, 2, 3) \tag{1}$$

where a dot denotes differentiation with respect to time.

b) The Hooke law is valid for the elastic part of the strain tensor,

$$\sigma_{ij} = E_{ijk\ell} \, \epsilon^e_{k\ell}, \quad (i, j, k, \ell = 1, 2, 3) \tag{2}$$

where $E_{ijk\ell}$ is the elastic moduli tensor and σ_{ij} is the stress tensor.

c) The material is plastically incompressible,

$$\epsilon^p_{kk} = 0, \quad \dot{\epsilon}^p_{kk} = 0. \tag{3}$$

d) A yield condition $F = 0$ exists, and it depends on the stress tensor σ_{ij}, the microstress tensor σ^μ_{ij}, the average intensity of the preliminary strain rate β_h, the intensity of the preliminary plastic deformation γ_h, the intensity of the actual strain rate β and the intensity of the actual plastic deformation γ,

$$F(\sigma_{ij}, \sigma^\mu_{ij}, \beta_h, \gamma_h, \beta, \gamma) = 0 \tag{4}$$

where

$$\beta_h = \frac{1}{t_f - t_0} \int_{t_0}^{t_f} \sqrt{\frac{2}{3} \dot{e}_{ij} \dot{e}_{ij}} \, dt, \quad e_{ij} = \epsilon_{ij} - \frac{1}{3} \delta_{ij} \epsilon_{kk},$$

$$\gamma_h = \int_{t_0}^{t_f} \sqrt{\frac{1}{2} \dot{e}^p_{ij} \dot{e}^p_{ij}} \, dt, \quad \beta = \sqrt{\frac{2}{3} \dot{e}_{ij} \dot{e}_{ij}}, \quad \gamma = \sqrt{\frac{1}{2} e^p_{ij} e^p_{ij}}. \tag{5}$$

δ_{ij} is the Kroneker delta and (t_0, t_f) is the time interval in which the preliminary deformation process took place.

The yield condition (4) is specified in the form

$$F \equiv \frac{1}{2} N_{ijk\ell} \, \bar{s}_{ij} \bar{s}_{k\ell} - (\tau^*_p)^2 = 0 \tag{6}$$

where

$$N_{ijk\ell} = M_{ijk\ell} + A s^\mu_{ij} s^\mu_{k\ell}, \quad A = A(\beta_h, \gamma_h, \beta, \gamma)$$

$$\bar{s}_{ij} = s_{ij} - s^\mu_{ij}, \quad s_{ij} = \sigma_{ij} - \frac{1}{3} \delta_{ij} \sigma_{kk}, \quad \bar{\sigma}_{ij} = \sigma_{ij} - \sigma^\mu_{ij}, \tag{7}$$

$$\tau^*_p = \tau^*_p(\beta_h, \gamma_h, \beta, \gamma), \quad s^\mu_{ij} = \sigma^\mu_{ij} - \frac{1}{3} \delta_{ij} \sigma^\mu_{kk}.$$

$M_{ijk\ell}$ may describe initial anisotropic plastic properties of the material, according to Olszak and Urbanowski [11]. In the case of initially plastically isotropic material one has

$$M_{ijk\ell} = I_{ijk\ell} = \frac{1}{2} (\delta_{i\ell} \delta_{jk} + \delta_{j\ell} \delta_{ik}). \tag{8}$$

In (7), τ_p^* is the isotropic yield point which is to be determined in two-dimensional experiments (see sec. 4). $\bar{\sigma}_{ij}$ is the active stress tensor [4].

The separate consideration of the influence of β_h, γ_h and β, γ enables one to take into account the influence of the history of plastic deformation and strain rate, as well as the effect of this influence on the actual deformation process. In the particular case when $A = \text{const}$, $\tau_p^* = \text{const}$ and $s_{ij}^\mu = C e_{ij}^p$, $C = \text{const}$, the condition (6) yields the Baltov-Sawczuk anisotropic hardening rule [2].

e) The microstress deviator s_{ij}^μ is treated as an internal state variable, connected with the micrononhomogeneous stresses and strains, caused by the plastic deformation. The following equation of evolution is assumed:

$$\dot{s}_{ij}^\mu + L\gamma \, s_{ij}^\mu = B_{ijk\ell} \, \dot{\beta} e_{k\ell}^p + R_{ijk\ell} \, \dot{\gamma} e_{k\ell}^p + Q \dot{e}_{ij}^p + T_{ijk\ell} \, e_{k\ell}^p . \tag{9}$$

Here $L, B_{ijk\ell}, R_{ijk\ell}, Q, T_{ijk\ell}$ depend on β_h, γ_h, β and γ. If the strain rate is constant, $\dot{\beta} = 0$. As indicated by experimental results, it may be often assumed that $B_{ijk\ell}, R_{ijk\ell}$ and $T_{ijk\ell}$ are isotropic tensors, i.e. $B_{ijk\ell} = BI_{ijk\ell}$, $R_{ijk\ell} = RI_{ijk\ell}$, $T_{ijk\ell} = TI_{ijk\ell}$. In the case of an instanteneous deformation, viscous behaviour cannot take place because of short duration. Hence the terms in (9) containing L and $T_{ijk\ell}$ may be neglected. Then

$$\dot{s}_{ij}^\mu = R_{ijk\ell} \, e_{k\ell}^p \, \dot{\gamma} + Q \dot{e}_{ij}^p . \tag{10}$$

If no realoading takes place after unloading, Eq. (9) assumes the form:

$$\dot{s}_{ij}^\mu + L\gamma \, s_{ij}^\mu = T_{ijk\ell} \, e_{k\ell}^p . \tag{11}$$

f) The plastic strain rate is given by the flow law associated with the yield condition (6):

$$\dot{e}_{ij}^p = \dot{\lambda} \, \frac{\partial F}{\partial \sigma_{ij}}, \quad \dot{\lambda} = \begin{cases} 0, \text{ if } F < 0 \text{ or } F = 0, \dfrac{\partial F}{\partial \sigma_{k\ell}} \, \dot{\sigma}_{k\ell} \leqq 0, \\[3mm] > 0, \text{ if } F = 0, \dfrac{\partial F}{\partial \sigma_{k\ell}} \, \dot{\sigma}_{k\ell} > 0. \end{cases} \tag{12}$$

$\dot{\lambda} > 0$ results from the condition that $\dot{F} = 0$.

g) The plastic strain causes on the microlevel a nonhomogeneous stress and strain state. Consider a small characteristic volume V in the neighbourhood of the body point x. At each point $x' \in V$ there exist a stress $\sigma'_{ij}(x, x', t)$ and a strain $\epsilon'_{ij}(x, x', t)$. We define the tensors of average stress and strain $\overline{\sigma}_{ij}(x, t)$ and $\overline{\epsilon}_{ij}(x, t)$, as well as the deviations from them $\sigma^*_{ij}(x, x', t)$ and $\epsilon^*_{ij}(x, x', t)$, by

$$\sigma'_{ij} = \overline{\sigma}_{ij} + \sigma^*_{ij}, \quad \overline{\sigma}_{ij} = \frac{1}{V} \int_V \sigma'_{ij} \, dV, \quad \int_V \sigma^*_{ij} \, dV = 0,$$

$$\epsilon'_{ij} = \overline{\epsilon}_{ij} + \epsilon^*_{ij}, \quad \overline{\epsilon}_{ij} = \frac{1}{V} \int_V \epsilon'_{ij} \, dV, \quad \int_V \epsilon^*_{ij} \, dV = 0. \tag{13}$$

We consider the metal as a composite consisting of two components, namely a phase (a), the crystal grains, and the phase (b), the boundaries [4]. The average phase tensors $\overline{\sigma}_{ij(\alpha)}(x, t)$ and $\overline{\epsilon}_{ij(\alpha)}(x, t)$, as well as their deviations $\sigma_{ij(\alpha)}(x, t)$ and $\epsilon_{ij(\alpha)}(x, t)$, $(\alpha = a, b)$ are

$$\overline{\sigma}_{ij(\alpha)} = \frac{1}{V_\alpha} \int_{V_\alpha} \sigma'_{ij} \, dV, \quad \overline{\sigma}_{ij(\alpha)} = \overline{\sigma}_{ij} + \sigma_{ij(\alpha)}, \quad \sigma_{ij(\alpha)} = \frac{1}{V_\alpha} \int_{V_\alpha} \sigma^*_{ij} \, dV,$$

$$\overline{\epsilon}_{ij(\alpha)} = \frac{1}{V_\alpha} \int_{V_\alpha} \epsilon'_{ij} \, dV, \quad \overline{\epsilon}_{ij(\alpha)} = \overline{\epsilon}_{ij} + \epsilon_{ij(\alpha)}, \quad \epsilon_{ij(\alpha)} = \frac{1}{V_\alpha} \int_{V_\alpha} \epsilon^*_{ij} \, dV, \tag{14}$$

where V_α is the volume of the phase $\alpha(\alpha = a, b)$. Then $\vartheta_\alpha = \dfrac{V_\alpha}{V}$, $\vartheta_a + \vartheta_b = 1$.

Moreover we define the second order deviation tensors $\sigma^*_{ij(\alpha)}(x, x', t)$ and $\epsilon^*_{ij(\alpha)}(x, x', t)$ as follows:

$$\sigma^*_{ij} = \sigma_{ij(\alpha)} + \sigma^*_{ij(\alpha)}, \quad \int_{V_\alpha} \sigma^*_{ij(\alpha)} \, dV = 0,$$

$$\epsilon^*_{ij} = \epsilon_{ij(\alpha)} + \epsilon^*_{ij(\alpha)}, \quad \int_{V_\alpha} \epsilon^*_{ij(\alpha)} \, dV = 0. \tag{15}$$

We introduce further the microstress tensor σ^μ_{ij} on the macrolevel and the phase microstress tensor $\sigma^\mu_{ij(\alpha)}$ on the microlevel, using the following interpretation resulting from energy considerations:

$$\dot{W} = \frac{1}{V} \int_V \sigma'_{ij} \dot{\epsilon}'_{ij} \, dV = \overline{\sigma}_{ij} \dot{\overline{\epsilon}}_{ij} + \dot{W}^*$$

$$\dot{W}^* = \vartheta_a \sigma_{ij(a)} \dot{\epsilon}_{ij(a)} + \vartheta_b \sigma_{ij(b)} \dot{\epsilon}_{ij(b)} + \dot{W}^{**} \tag{16}$$

$$\dot{W}^* = \sigma^\mu_{ij} \dot{\overline{\epsilon}}_{ij}, \quad \dot{W}^{**} = \vartheta_a \sigma^\mu_{ij(a)} \dot{\epsilon}_{ij(a)} + \vartheta_b \sigma^\mu_{ij(b)} \dot{\epsilon}_{ij(b)}.$$

We assume that

$$\sigma_{ij} = \bar{\sigma}_{ij} + \sigma_{ij}^{\mu}, \quad \epsilon_{ij} = \bar{\epsilon}_{ij}. \tag{17}$$

$\dot{W}^{**}$ in (16) may be often neglected.

We assume that during a preliminary loading process without plastic deformation, i.e. when $\epsilon_{ij}^{p} = 0$, $s_{ij}^{\mu} = 0$, $\epsilon_{ij} = \epsilon_{ij}^{e}$, we have

$$\sigma_{ij} = \bar{\sigma}_{ij} = \bar{\sigma}_{ij(\alpha)}, \quad \sigma_{ij(\alpha)} = 0. \tag{18}$$

If both components have equal elastic properties, then $\epsilon_{ij(\alpha)} = 0$. If not, the Reuss model is assumed:

$$s_{ij} = \mu e_{ij}^{e}, \quad \frac{1}{\mu} = \frac{\vartheta_a}{\mu_a} + \frac{\vartheta_b}{\mu_b}, \quad \mu = \text{const},$$

$$\sigma_{kk} = 3K\epsilon_{kk}^{e}, \quad \frac{1}{K} = \frac{\vartheta_a}{K_a} + \frac{\vartheta_b}{K_b}, \quad K = \text{const},$$

$$\bar{\epsilon}_{ij(\alpha)} = \bar{\epsilon}_{ij(\alpha)}^{(e)0}, \quad \epsilon_{ij(\alpha)} = \epsilon_{ij(\alpha)}^{(e)0}, \quad (\alpha = a, b) \tag{19}$$

$$\bar{\epsilon}_{ij(\alpha)}^{(e)0} = \epsilon_{ij}^{e} + \epsilon_{ij(\alpha)}^{(e)0},$$

$$e_{ij(\alpha)}^{(e)0} = \left(\frac{\mu}{\mu_\alpha} - 1\right) e_{ij}^{e}, \quad \epsilon_{kk(\alpha)}^{(e)0} = \left(\frac{K}{K_\alpha} - 1\right) \epsilon_{kk}^{e}$$

In the case of unloading taking place after a preliminary loading process with plastic deformation i.e. when $\epsilon_{ij}^{e} = 0$, $\sigma_{ij} = 0$, $\epsilon_{ij}^{p} = \epsilon_{ij} \neq 0$, $s_{ij}^{\mu} \neq 0$, we have

$$\bar{\epsilon}_{ij(\alpha)} = \bar{\epsilon}_{ij(\alpha)}^{r} = \bar{\epsilon}_{ij(\alpha)}^{(e)r} + \bar{\epsilon}_{ij(\alpha)}^{p}, \quad \eta_a = \frac{\vartheta_a}{\vartheta_b}$$

$$\epsilon_{ij(\alpha)} = \epsilon_{ij(\alpha)}^{r} = \epsilon_{ij(\alpha)}^{(e)r} + \epsilon_{ij(\alpha)}^{p},$$

$$e_{ij}^{p} = \vartheta_a \bar{e}_{ij(a)}^{r} + \vartheta_b \bar{e}_{ij(b)}^{r}, \quad (\alpha = a, b)$$

$$-\sigma_{ij}^{\mu} = \bar{\sigma}_{ij}^{r} = \vartheta_a \bar{\sigma}_{ij(a)}^{r} + \vartheta_b \bar{\sigma}_{ij(b)}^{r}, \quad \sigma_{ij(\alpha)} = \sigma_{ij(\alpha)}^{r} = \bar{\sigma}_{ij(\alpha)}^{r} - \bar{\sigma}_{ij}^{r}, \tag{20}$$

$$\bar{s}_{ij(\alpha)} = \bar{s}_{ij(\alpha)}^{r} = \mu_\alpha \bar{e}_{ij(\alpha)}^{(e)r}, \quad \bar{\sigma}_{kk(\alpha)} = \bar{\sigma}_{kk(\alpha)}^{r} = 3K_\alpha \bar{\epsilon}_{kk}^{(e)r},$$

$$e_{ij(a)}^{(e)r} = \frac{1}{\mu_a} [s_{ij(a)}^{r} + \vartheta_b(1 - \eta_a) \bar{s}_{ij(b)}^{r}],$$

$$e_{ij(b)}^{(e)r} = \frac{1}{\mu_b} \left[s_{ij(b)}^{r} + \vartheta_a \frac{1}{1 - \eta_a} \bar{s}_{ij(a)}^{r} \right].$$

During a preliminary loading process with plastic deformation when $\sigma_{ij} \neq 0$, $\epsilon_{ij}^e \neq 0$, $\epsilon_{ij}^p \neq 0$, $s_{ij}^\mu \neq 0$, we have

$$\bar{\epsilon}_{ij(\alpha)} = \bar{\epsilon}_{ij(\alpha)}^{(e)0} + \bar{\epsilon}_{ij(\alpha)}^r, \quad \epsilon_{ij(\alpha)} = \epsilon_{ij(\alpha)}^{(e)0} + \epsilon_{ij(\alpha)}^r,$$

$$\bar{\sigma}_{ij} = \bar{\sigma}_{ij}^0 + \bar{\sigma}_{ij}^r, \quad \bar{\sigma}_{ij(\alpha)} = \bar{\sigma}_{ij(\alpha)}^0 + \bar{\sigma}_{ij(\alpha)}^r,$$

$$\sigma_{ij(\alpha)} = \sigma_{ij(\alpha)}^0 + \sigma_{ij(\alpha)}^r, \quad e_{ij}^e = \frac{\vartheta_a}{\mu_a} \bar{s}_{ij(a)}^0 + \frac{\vartheta_b}{\mu_b} \bar{s}_{ij(b)}^0, \tag{21}$$

$$\bar{e}_{ij(\alpha)}^{(e)r} = \frac{1}{\mu_\alpha} \bar{s}_{ij(\alpha)}^r, \quad \bar{e}_{ij(\alpha)}^{(e)0} = \frac{1}{\mu_\alpha} \bar{s}_{ij(\alpha)}^0,$$

$$\bar{\epsilon}_{kk}^{(e)r} = \frac{1}{3K_\alpha} \bar{\sigma}_{kk(\alpha)}^r, \quad \bar{\epsilon}_{kk(\alpha)}^{(e)0} = \frac{1}{3K_\alpha} \bar{\sigma}_{kk(\alpha)}^0.$$

The relation between the volumes of both phases generally changes during the plastic deformation, i.e. $\eta_a = \eta_a(\epsilon_{ij}^p)$.

Considering a preliminary loading process with plastic deformation and unloading taking place in the time interval (t_0, t_f), we integrate the expression for $\dot{W}^*$ in (16) and obtain:

$$\int_{t_0}^{t_f} s_{ij}^\mu \dot{e}_{ij}^p \, dt = \int_{t_0}^{t_f} \eta_a \sigma_{ij(a)} (\dot{\epsilon}_{ij(a)}^{(e)0} + \dot{\epsilon}_{ij(a)}^r) \, dt. \tag{22}$$

Using this expression one can devise experiments giving a relation between the micro and macro effects [4, 6].

3. One-dimensional stress state

In order to discuss the proposed model on the basis of some experimental results, the one-dimensional stress state will be considered.

In the case of uniaxial tension, when

$$\sigma_{11} > 0, \; \sigma_{22} = \sigma_{33} = 0, \; \epsilon_{11}^p = e_{11}^p > 0, \; \epsilon_{22}^p = \epsilon_{33}^p = -\frac{1}{2} \epsilon_{11}^p$$

the yield condition (6) takes the form

$$\frac{3}{4} \left(\frac{2}{3} \sigma_{11} - s_{11}^\mu \right)^2 \left[1 + \frac{3}{2} A(s_{11}^\mu)^2 \right] = (\tau_p^*)^2. \tag{23}$$

In the case of isotropic tensors $B_{ijk\ell}$, $R_{ijk\ell}$ and $T_{ijk\ell}$, the evolution equation (9) becomes:

$$\dot{s}_{11}^\mu + L\gamma s_{11}^\mu = B\dot{\beta} e_{11}^p + R\dot{\gamma} e_{11}^p + Q\dot{e}_{11}^p + T e_{11}^p. \tag{24}$$

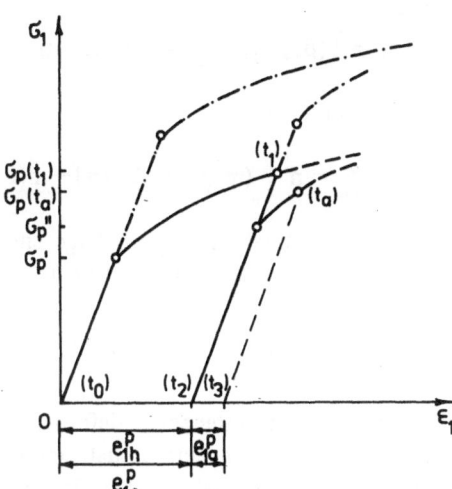

Figure 1. – *Several stages deformation process.*

We shall consider a deformation process consisting of several stages, Fig. 1:

a) Preliminary instanteneous deformation with plastic deformation and constant strain rate. Then $\beta_h = 0$, $\gamma_h = 0$, $\beta = \beta_c = $ const, $\dot{\beta} = 0$, $e^P_{11}(t_0) = 0$, $s^\mu_{11}(t_0) = 0$ in the time interval (t_0, t_1). Eq. (10) yields

$$\dot{s}^\mu_{11} = (R\gamma + Q)\,\dot{e}^P_{11}\,, \quad \gamma = \frac{\sqrt{3}}{2}\,e^P_{11}\,, \tag{25}$$

where $R = \bar{R}(\beta_c, \gamma)$, $Q = \bar{Q}(\beta_c, \gamma)$.

After integrating (25) over the interval (t_0, t_1) we obtain

$$s^\mu_{11}(t_1) = s^\mu_{1h} = C_1(0, 0, \beta_c, \gamma_{1h})\,e^P_{1h}\,, \tag{26}$$

where $\gamma_{1h} = \gamma(t_1)$, $e^P_{1h} = e^P_{11}(t_1)$.

b) Elastic unloading in the time interval (t_1, t_2). In that case

$$e^P_{11}(t_2) = e^P_{1h}\,, \quad s^\mu_{11}(t_2) = s^\mu_{1h}\,, \quad \gamma(t_2) = \gamma_{1h}\,.$$

c) No realoading exists in the time interval (t_2, t_3). Then $\sigma_{11} = 0$, $e^P_{11} = e^P_{1h} = $ const, $\dot{e}^P_{11} = 0$, $\gamma = \gamma_h = \gamma_{1h} = $ const, $\dot{\gamma} = 0$, $\beta_h = \beta_c = $ const, $\dot{\beta} = 0$, and Eq. (11) takes the form:

$$\dot{s}^\mu_{11} + m s^\mu_{11} = n\,, \tag{27}$$

where
$$n = \overline{T}(\beta_c, \gamma_{1h}) e^P_{1h} = \text{const}, \quad m = \overline{L}(\beta_c, \gamma_{1h}) \gamma_{1h} = \text{const}.$$

The solution of this equation is

$$ms^\mu_{11} = n + (ms^\mu_{1h} - n) \exp[-m(t - t_2)] \tag{28}$$

If the time interval (t_2, t_3) is large enough, we might assume that the relaxation process has attenuated and then

$$s^\mu_{11}(t_3) = s^\mu_{1(3)} = \varphi(\beta_c, \gamma_{1h}) e^P_{1h}, \quad \varphi = \frac{\overline{T}}{\overline{L}\gamma_{1h}}. \tag{29}$$

d) Secondary instanteneous deformation process with additional plastic deformation and constant actual strain rate $\beta = \beta_a = \text{const}$, $\dot\beta = 0$, takes place in the time interval (t_3, t_a). In this case Eq. (25) is valid but

$$R = R(\beta_c, \gamma_{1h}, \beta_a, \gamma) \quad \text{and} \quad Q = Q(\beta_c, \gamma_{1h}, \beta_a, \gamma).$$

Integration of Eq. (25) over the interval (t_3, t_a) yields

$$s^\mu_{11}(t_a) = s^\mu_{1a} = C_1(\beta_c, \gamma_{1h}, \beta_a, \gamma_a) e^P_{1c} + D(\beta_c, \gamma_{1h}, \beta_a) e^P_{1h}, \tag{30}$$

where

$$D = \varphi - C_1(\beta_c, \gamma_{1h}, \beta_a, \gamma_{1h}), \quad \gamma_a = \gamma(t_a), \quad e^P_{1c} = e^P_{11}(t_a) = e^P_{1h} + e^P_{1a}.$$

If the additional plastic strain is small as compared with the preliminary one, i.e. $e^P_{1c} \approx e^P_{1h}$, then Eq. (30) takes the form

$$s^\mu_{1a} = S(\beta_c, \gamma_{1h}, \beta_a, \gamma_a) e^P_{1h}, \quad S = C_1 + D. \tag{31}$$

A comparison of the results obtained for the process under consideration and the yield condition (23) leads to the following conclusions:

If $\beta_h = \beta_a$, the yield point $\sigma_p(t_1)$ is higher than the yield point $\sigma^*_p = \sqrt{3}\,\tau^*_p(t_1)$, associated with an isotropic hardening. This is due to $s^\mu_{1h} \neq 0$. The yield point of the secondary process σ^{II}_p (Fig. 1) is lower than $\sigma_p(t_1)$, because after the relaxation one has $s^\mu_{1(3)} < s^\mu_{1h}$. The hardening curve from σ^{II}_p to $\sigma_p(t_a)$ has a different shape as compared to the hardening curve belonging to the preliminary process. This is due to the fact that s^μ_{1a} depends on $\beta_h, \gamma_{1h}, \beta_a$ and γ_a. Figure 2 shows same experimental results. It is seen that the behaviour of the investigated metals is well described by the proposed model.

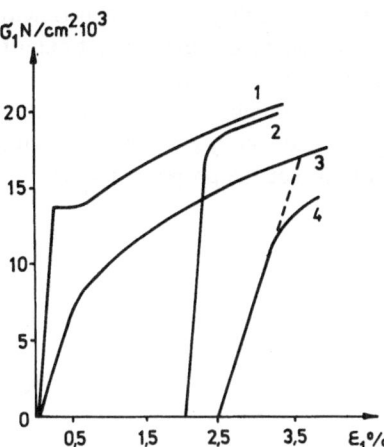

Figure 2. – *Stress-strain curves before and after material prestraining*
  1. *preliminary process, alluminium alloy*
  2. *secondary process,* $e_{1h}^p = 1,92\,\%$, *alluminium alloy*
  3. *preliminary process, electrolytic copper*
  4. *secondary process,* $e_{1h}^p = 2,5\,\%$, *electrolytic copper*

If $\beta_h < \beta_a$, the stress-strain curve of the secondary process is higher than the curve mentioned above (the dotted line in Fig. 1). This curve tends to the curve corresponding to a preliminary process with $\beta_h = \beta_a$. A similar effect occurs in the case of torsion or compression. The experimental results confirm the theoretical conclusions [10, 15].

4.   Two-dimensional stress state

Let us consider the two-dimensional stress state

$$\sigma_{11} \neq 0, \quad \sigma_{22} \neq 0, \quad \sigma_{12} \neq 0, \quad \sigma_{33} = \sigma_{23} = \sigma_{13} = 0$$

in order to discuss the developed theoretical model on the basis of experimental results. We shall apply the method with plates, according to Szczepinski's procedure, developed in the quasidynamical case in [3, 5]. If a preliminary plastic deformation e_{1h}^p is realized in the plate in the direction Ox_1, at a constant strain rate $\beta_h = const$, the yield condition (6) takes the form:

$$F \equiv R_1(\sigma_{11} - \sigma_{11}^\mu)^2 + \frac{1}{4}(1 + R_1)\,\sigma_{22}^2 - R_1\sigma_{22}(\sigma_{11} - \sigma_{11}^\mu)$$

$$+ \sigma_{12}^2 - (\tau_p^*)^2 = 0 \quad (32)$$

where

$$R_1 = \frac{1}{3} + \frac{1}{2}\,A(s_{11}^\mu)^2, \quad \sigma_{11}^\mu = \frac{3}{2}\,s_{11}^\mu, \quad \tau_p^* = \tau_p^*(\beta_h, \gamma_h, \beta, \gamma)$$

$$A = A(\beta_h, \gamma_h, \beta, \gamma), \quad \gamma_h = \frac{\sqrt{3}}{2}\,e_{1h}^p.$$

Since, according to the experimental method, a one-dimensional stress state is realized in different directions of the plate, the s_{11}^{μ} is determined by the relations given in Sec. 3.

A number of experiments, following Szczepinski's procedure [13], show that the yield surface is successfully approximated by the yield condition (32) of the proposed model.

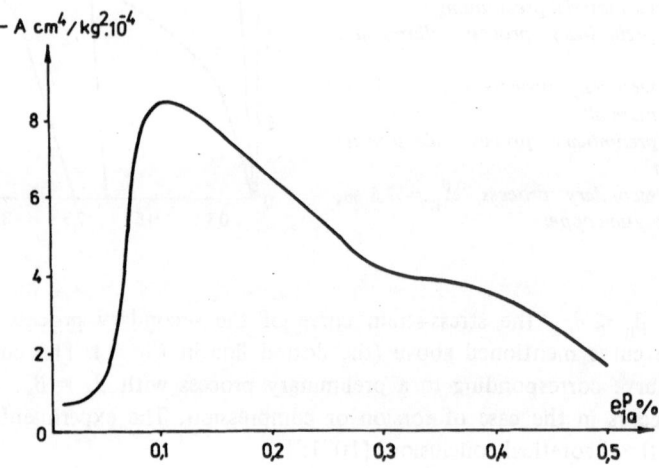

Figure 3. – *Material function A versus plastic strain.*

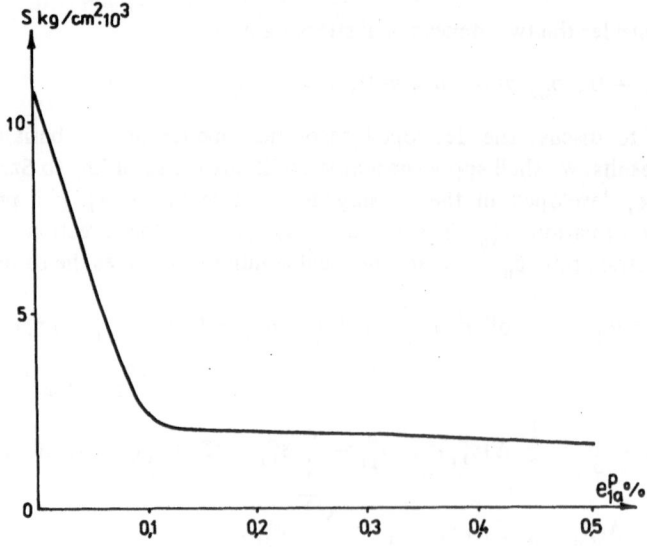

Figure 4. – *Material function S versus plastic strain.*

The functions A, S and τ_p^* are plotted in Figures 3, 4 and 5, as obtained by experiments with aluminium alloy at $\beta_h = \beta_a$ and $e_{1h}^p = 1{,}92\,\%$. An approximation of the experimental curves by the method of least square, using

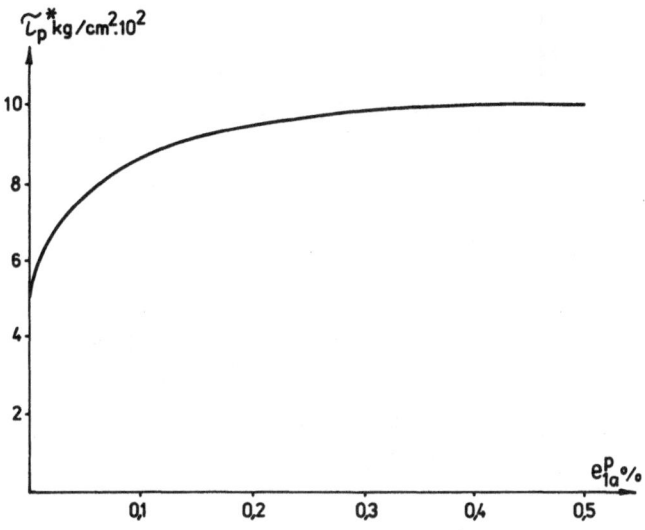

Figure 5. – *Material function τ_p^* versus plastic strain.*

a computer leads to the following expressions for the functions mentioned above [7]:

$$A = e_{1a}^p[-0{,}208.10^6(e_{1a}^p)^{2,5} + 0{,}268(e_{1a}^p)^{1,25} - 0{,}180.10^4]^{-1},$$
$$[cm^4/kg^2]$$

$$S = 1025(e_{1a}^p)^{-0,507}, \ [kg/cm^2].$$

$$\tau_p^* = 760(e_{1a}^p)^{0,342}, \ [kg/cm^2].$$

These approximations describe satisfactorily the experimentally obtained fact that at low values of the actual plastic strain, the kinematic hardening effect dominates. When e_{1a}^p increases, the anisotropic hardening effect becomes dominating. A further increase of e_{1a}^p leads to isotropic hardening.

Figure 6 shows the influence of the parameters β_h and β_a on the function τ_p^* at $e_{1a}^p = 0{,}2\,\%$ and at different values of e_{1h}^p. The experiments were made on electrolytic copper. The figure shows the influence of the strain rate history.

Figure 7 shows the function S for $\beta_a = \beta_h = 1{,}0.10^{-5}\,s^{-1}$ and $e_{1a}^p = 0{,}02\,\%$ in the case of electrolytic copper. It is seen from the figure that with increasing e_{1h}^p, S tends to a constant value.

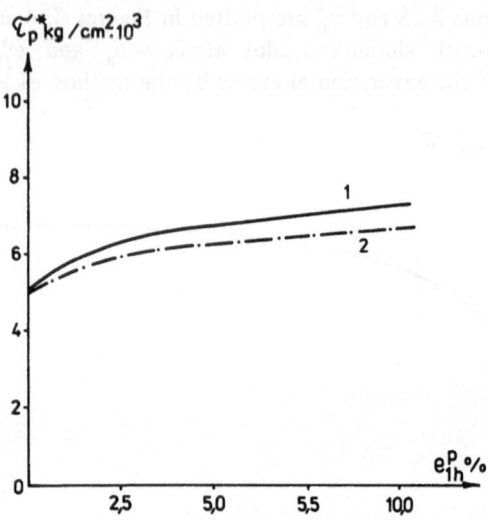

Figure 6. – *Influence of rate of straining on the yield stress*
  1. *secondary process,* $\beta_h = \beta_a = 1,62.10^{-2} \text{ s}^{-1}$
  2. *secondary process,* $\beta_h = 0,83.10^{-5} \text{ s}^{-1}$ *and* $\beta_a = 1,62.10^{-2} \text{ s}^{-1}$:

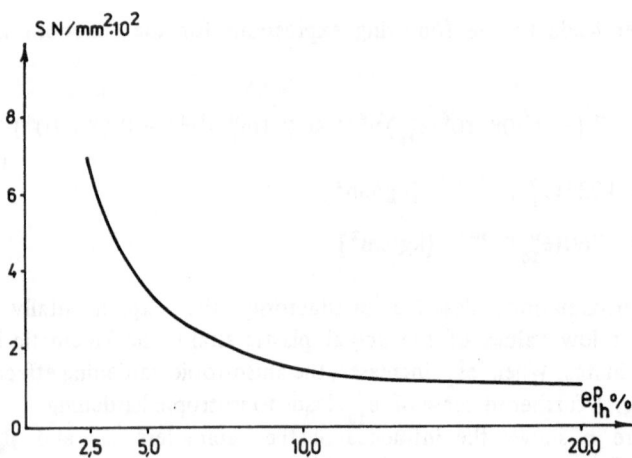

Figure 7. – *Material function S versus preliminary plastic strain for electrolytic copper.*

The experimental method mentioned here may be successfully combined with X-ray measurements of the residual elastic microstrains [4, 6]. On the basis of Eq. (22), using an X-ray defractometer, we may obtain the residual elastic microstrains $e^{(e)r}_{11(a)}$ and $e^{(e)r}_{22(a)}$ as functions of β_h, γ_h and e^p_{1h}. This enables to get information about η_a as a function of the plastic deformation.

REFERENCES

[1] BACKHAUS G. – "Zur Fliessgrenze bei allgemeiner Verfestigung". *ZAMM*, **48** (1968): 99-108.

[2] BALTOV A. and A. SAWCZUK. – "A rule of anisotropic hardening." *Acta Mechanica*, I/2 (1965): 81-92.

[3] BALTOV A. and S. VODENICHAROV. – "Deformation anisotropy of rate sensitive metals." *Bull. Acad. Pol. Sci., Ser. sci. tech.* **XXIII** (1973): 105-110.

[4] BALTOV A., N. BONTCHEVA, R. KAZANDJIEV, I. RADOVANOV and ST. VODENICHAROV. – "On the theoretical and experimental description of the mechanical behaviour of metals during metal forming processes." *Metal Forming Plasticity, IUTAM Symposium Tutzing/Germany*. Berlin, Heidelberg, New York: Springer-Verlag, 1979, pp. 14-26.

[5] BALTOV A. and S. VODENICHAROV. – "A method for analysing the influence of the strain velocity history over metal plastic properties". (in Bulg.), *Theoret. and Appl. Mech.*, 1 (1979): 63-70.

[6] BALTOV A., N. BONTCHEVA, R. KAZANDJIEV, I. RADOVANOV and ST. VODENICHAROV. – "A method of investigating the microstresses in rate sensitive plastic materials". *Theoret. and Appl. Mech.*, (in print).

[7] BALTOV A. and T. DUCHEVA. – "On the identification of the material behaviour functions in the anisotropic yield condition". *Theoret. and Appl. Mech.*, (in print).

[8] KADASHEVITCH JU. and V.V. NOVOZHILOV. – "Creep of micrononhomogeneous continua", (in Russ.). *Issledovanija po uprugosti i plastitchnosti*, 12, Leningrad, 1978.

[9] KAFKA V. – "The general theory of isothermal elastic-plastic deformation of polycrystals based on an analysis of the microscopic state of stress". *ZAMM*, 48 (1968): 265-282.

[10] KLEPACZKO J. – "Strain rate history effects for polycrystalline aluminium and theory of intersections." *J. Mech. Phys. Sol.*, 16 (1968): 255-266.

[11] OLSZAK W. and W. URBANOWSKI. – "The plastic potential and the generalized distortion energy". *Arch. Mech. Stos.*, 8 (1965): 671-694.

[12] OTTOSEN N.S. – "A non-linear kinematic hardening function", *Report Risö M-1938*. Roskilde, Denmark.

[13] SZCZEPINSKI W. – "On the effect of plastic deformation on yield condition". *Arch. Mech. Stos.*, 15 (1963): 275-296.

[14] UGODCHIKOV A.G. and YU.G. KOROTKIKH. – "Constitutive equations of nonisothermal elastic-plastic deformation and methods of their practical application." *Foundation of Plasticity*. A. Sawczuk, ed. Groningen, Noordhoff, 1972.

[15] USUI E. and T. SHIRAKASHI. – "Effect of temperature and strain rate upon flow stress of metals in compression." *Bull. Japan. Soc. of Proc. Engg.*, 4 (1971): 91-100.

RESUME

(Une loi d'écrouissage anisotrope, prenant en compte les déformations plastiques et les vitesses de déformation initiales et actuelles)

Un modèle pour les matériaux plastiques écrouissables à comportement non linéaire et dépendant de la vitesse de déformation est proposé. Le critère d'écoulement de Baltov-Sawczuk est développé de façon à pouvoir décrire le comportement de tels matériaux. L'influence de la déformation plastique et de la vitesse de déformation initiales et celle de la déformation plastique et de la vitesse de déformation actuelles sont considérées séparément. Le modèle est applicable pour la description des processus technologiques de formage plastique des éléments structuraux, suivis de processus d'exploitation secondaire. Les possibilités du modèle dans les cas particuliers des états de contrainte mono- et bi- dimensionnels sont discutées. Les résultats d'une série d'essais sont analysés.

Plastic Behaviours
of Initially Anisotropic Metals
after Multi Prestrainings

E. Shiratori, K. Ikegami
Tokyo Institute of Technology, Yokohama, Japan.

K. Kaneko
Science University of Tokyo, Japan.

1. Introduction

Commercially obtained industrial metallic materials are generally subjected to many processes of workings and heat treatments, and they show various initial anisotropies. The anisotroqic theories [1, 2, 4, 5, 7] for initially isotropic materials which have succeeded, in some degree, in describing the subsequent plastic behaviours of the materials after various prestrainings cannot easily be applied to those initially anisotropic materials, as the required information on the type and amount of the prestrain previously given to the materials during their manufacturing processes is generally not given to the users. Hence, the authors try to derive an anisotropic theory of plasticity for an initially anisotroqic material by experimentally studying stress-strain relations of commercially obtained metals in the reloading stages after various pre-loadings. The reloading stage is assumed to be divided into three regions, namely, elastic region, transitional hardening region and steady hardening region. Strain hardening characteristic is formulated for each of these regions.

2. Experiment

Three kinds of material as received are used without any additional heat treatment.

(1) Steel tube   JIS[1]   G3454   STP G 42
(2) Brass rod   JIS   H3422   BsBM 2
(3) Al-alloy rod   JIS   H4163   A2B2 H(5056).

---

[1] Japanese Industrial Standard.

Tubular specimens are made by machining. Dimensions of the specimens are as follows:

(1) for mild steel,

d_i = inner diameter of the tube = 27.0 ± 0.02 mm,
t = thickness of the tube = 1.5 ± 0.02 mm,
ℓ_p = length of the parallel part of tube = 50 mm.

(2) for brass and Al-alloy,

d_i = 18.0 ± 0.02 mm,
t = 1.2 mm   (for internal pressure),
    = 2.0 mm   (for axial load and torsion),
ℓ_p = 50 mm.

A multi axial combined stress testing machine [6] is used for the experiment. Stress-strain relations were measured for various prestraining and reloading paths. They are represented by means of the effective stress and effective strain of Mises type which are given by the equations

$$\left.\begin{aligned}
\sigma_{eq} &= (\sigma_x^2 - \sigma_x \sigma_y + \sigma_y^2 + 3\sigma_{xy}^2)^{1/2} , \\
\epsilon_{eq} &= \frac{2}{\sqrt{3}} (\epsilon_x^2 + \epsilon_x \epsilon_y + \epsilon_y^2 + \epsilon_{xy}^2)^{1/2} ,
\end{aligned}\right\} \tag{1}$$

where $\sigma_x (= \sigma_1)$, σ_y, $\sigma_{xy} (= \sigma_3)$ and $\epsilon_x (= \epsilon_1)$, ϵ_y, $\epsilon_{xy} (= \epsilon_3)$ are longitudinal, circumferential and torsional components of true stress and true (logarithmic) plastic strain, respectively. Some stress-strain curves in proportional loadings of combined axial load (tension or compression), internal pressure and torsion are shown in Figure 1 and Figure 2, respectively, for brass, Al-alloy, and steel as received. It is found from these figures that the materials have some initial anisotropies. Initial stress-strain curves before prestraining can be represented well by the equation

$$\sigma_{eq,i} = \alpha_i (\beta_i + \epsilon_{eq,i})^{n_i} \quad (i = 1, 3, \theta) \tag{2}$$

where the subscript i designates the loading direction. Values of material constants α, β, n are shown in Table 1. The values for an arbitrary θ can be related to those for tension and torsion by the following equation:

$$M_\theta = M_1 - \frac{\theta}{90} (M_1 - M_3) \quad (M = \alpha, \beta, n) . \tag{3}$$

Stress-strain curves calculated by means of Eq. (2) are shown in Figures 1 and 2.

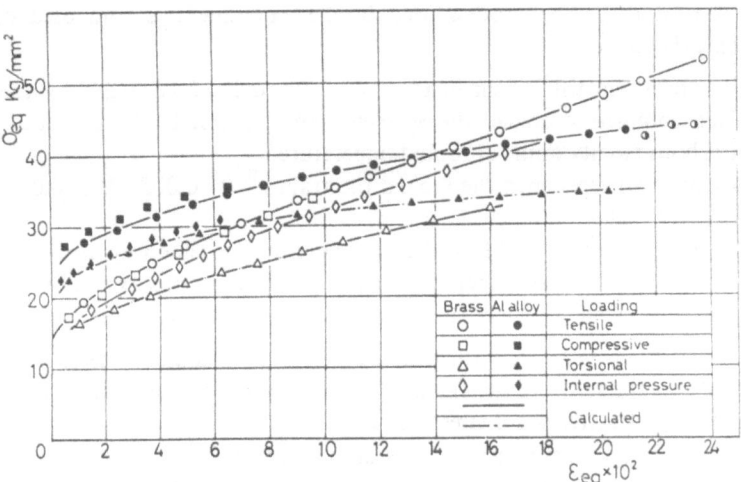

Figure 1. – *Initial equivalent stress-strain curves for brass and Al-alloy.*

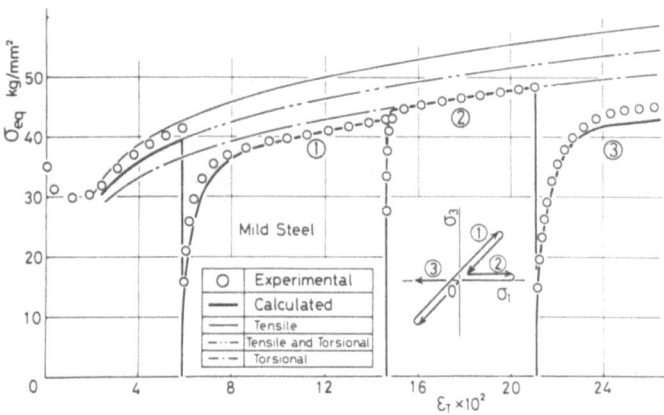

Figure 2. – *Stress-strain curves of mild steel in loading along the stress path illustrated in the figure.*

TABLE 1. Coefficients in the approximate equation of the initial stress-strain curve.

	Mild Steel			Brass			Al alloy		
	Tension	Torsion	Int.pres.	Tension	Torsion	Int.pres.	Tension	Torsion	Int.pres.
α	75.4	65.9	51.3	121.6	73.3	103.9	59.2	40.3	
β	-0.0131	-0.0103	-0.0212	0.0753	0.0469	0.0524	0.0063	-0.0300	
n	0.184	0.195	0.121	0.720	0.520	0.623	0.205	0.090	

A typical loading process used in the experiment is composed of the following five stages:

(1) First prestrain the specimen to ϵ^I (= 0.04, 0.08, 0.16, 0.24) by tension.
(2) Then completely unload the specimen and leave it in this unloaded state for about twenty hours at room temperature.
(3) Apply to the specimen the second prestrain ϵ^{II} (= 0.005, 0.02, 0.04, 0.08, 0.16) by torsion.
(4) Completely unload and leave the specimen in the same way as in stage (2).
(5) Finally reload the specimen in four directions in the stress space by tension, compression and torsion (in positive or negative direction).

Stress-strain relations during these stages were measured. The yield surface which defines the fully elastic region at reloading stage was determined as a locus of the 0.02 % effective proof stress points.

Some of the experimental results are shown in Figures 2 and 3. Figure 2 shows effective stress-effective strain curves of mild steel in loading along the stress path illustrated in the figure. The abscissa is the total length ϵ_T of the prestrain path. Thin lines are initial stress-strain curves. It is found from this figure that the stress-strain curve during reloading in a certain loading direction after various prestrainings lies beneath the initial stress-strain curve in the same loading direction, and that the former becomes parallel to the latter with the increase of plastic deformation. Figure 3 shows some yield loci after prestraining along prestress paths illustrated in the figure. It is found from this figure that the yield locus after prestraining of a complex loading path moves along the

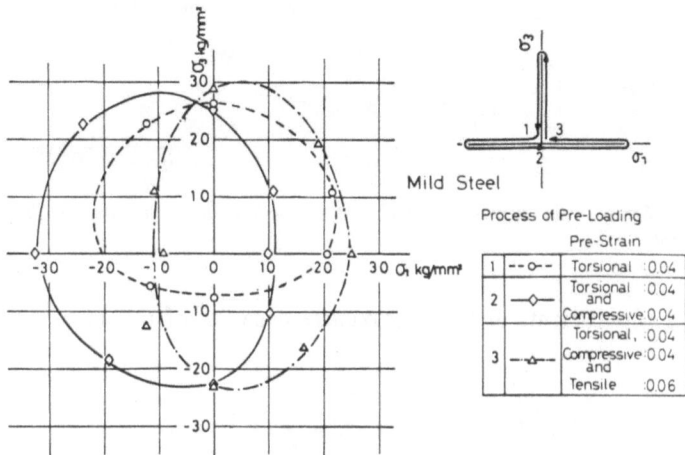

Figure 3. — *Subsequent yield loci after prestraining along prestress paths illustrated in the figure.*

latest preloading direction and has no nose, unlike in the case of simple pres-training of the initially isotropic material.

3. Formulation of the strain hardening characteristic

3.1. *Division of the reloading stage*

The reloading stage after a complex prestraining can be divided into three regions, namely, elastic region, transitional hardening region and steady hardening region. Figure 4 illustrates this division in the case where the specimen is reloaded in the direction θ in the stress-plane (σ_I, σ_{II}) after it was first prestrained in the direction σ_I. The curves $\sigma_t^\theta (\epsilon^{II})$ and $\sigma_\theta(\epsilon)$ in Figure 4(a) represent the stress-strain curve in this case and the transitional hardening part of the stress-strain curve in the case where the first prestraining direction coincides with the reloading direction, respectively. The yield point $\sigma_{p_0}^\theta$ in Figure 4(a) corresponds to the intersection of the reloading path and the actual yield locus F, after prestraining in the direction σ_I, as shown in Figure 4(b). Point ST, the point of contact of reloading stress-strain curve $\sigma_t^\theta (\epsilon^{II})$ with a parallel line to the stress-strain curve $\sigma_\theta(\epsilon)$, corresponds to the intersection of the reloading path and the boundary curve H between transitional and steady hardening regions. Point $\sigma_{p_1}^\theta$, the intersection of the ordinate and the tangent line to $\sigma_t^\theta (\epsilon^{II})$ which is parallel to $\sigma_\theta(\epsilon)$, corresponds to the intersection of the reloading path and the rigid-plastic yield locus G after prestraining. Assuming that the reloading stress-strain curve is parallel to $\sigma_\theta (\epsilon)$ in the steady hardening region, the reloading stress-strain relation for any θ is completely determined by three closed curves F, G, H and the characteristic of the transional hardening

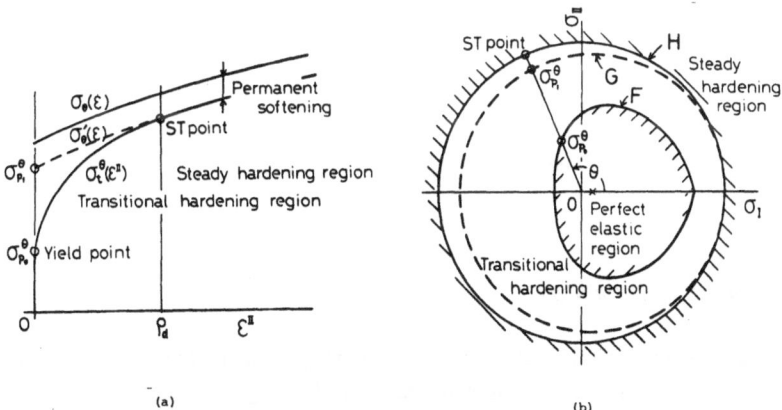

(a)　　　　　　　　　　　　(b)

Figure 4. − *Division of the reloading stage.*

part $\sigma_t^\theta (\epsilon^{II})$. As the boundary curve H can be determined by the magnitude of the transitional strain ρ_d, ρ_d is formulated instead of H.

3.2. Elastic limit curve (yield locus F)

Figure 5 shows some yield loci after multi-prestraining to nearly equal effective strain value, which are obtained by rotating them so that their latest prestraining directions coincide with the direction of σ_1 in the figure. The observed values of the elastic limit can approximately be shown by a single closed curve, namely, the full line or dotted line in the figure, independently of the prestraining path. The full line is an elliptic approximation, and the dotted line is a modified elliptic approximation [3]. The former is sufficient for the usual applications.

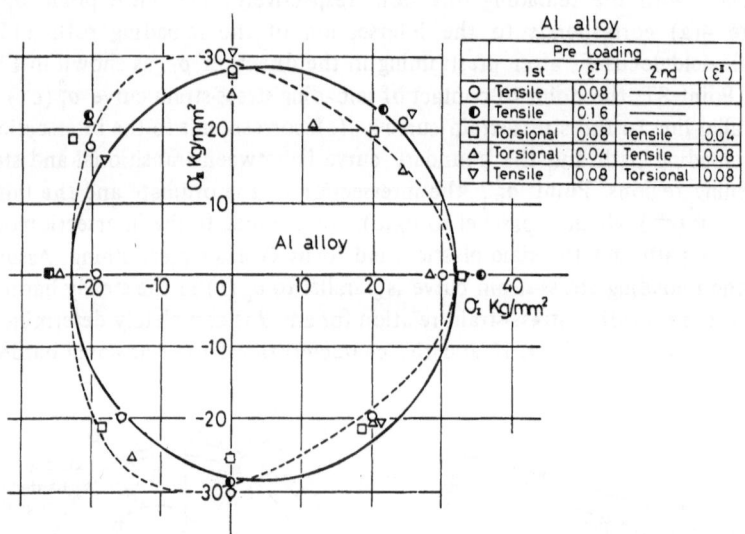

Figure 5. – Elastic region for Al-alloy.

3.3. Transitional strain ρ_d

Our experimental results show that the magnitude of the transitional strain ρ_d of a given material depends mainly on the angle θ_d between the latest preloading and reloading directions, and is nearly independent of the prestraining path. Figure 6 shows the relation between ρ_d and θ_d for mild-steel

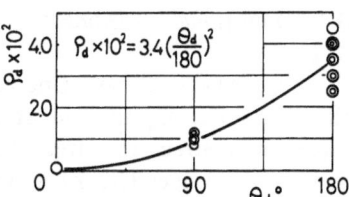

Figure 6. – *Magnitude of transitional strain.*

specimens. Table 2 shows the prestraining paths used. The relation can be approximately expressed by the equation

$$\rho_d \times 10^2 = \xi(\theta_d/180)^m,$$
(4)[1]

where

$\quad$ m = 2, $\quad \xi = 3.4 \quad$ for mild steel,

$\quad$ m = 1, $\quad \xi = 4.4 \quad$ for brass and Al-alloy.

TABLE 2. Preloading path in Figures 6 and 7.

	Loading Pattern	ε	θ_d	ρ_d
□	CO after TE	0.04	180°	0.045
⊡	do.	0.08	180°	0.040
◪	do.	0.12	180°	0.030
◇	RE after TO	0.04	180°	0.025
◈	TE after CO	0.04	180°	0.040
▲	RE after TE – TO(σ_1/σ_3= 1/1)	0.08	180°	0.030
▽	RE after CO – TO(σ_1/σ_3=-1/-1)	0.06	180°	0.025
○	CO after TO	0.04	90°	0.012
⊙	and TE after CO	0.04	180°	0.040
◖	and TO after TE	0.06	90°	0.012
△	TO after TE	0.04	90°	0.008
▽	do.	0.08	90°	0.010

$\quad$ TE ; Tensile Loading $\qquad$ TO ; Torsional Loading

$\quad$ CO ; Compressive Loading $\quad$ RE ; Reverse Loading

$\quad$ – ; Combined Loading

---

[1] For θ_d = 0, Eq. (4) gives ρ_d = 0 which is different from the observed value ρ_d = 0.1 × 10^{-2}. The difference may be neglected.

3.4. Transitional hardening part

A transitional hardening part of the reloading stress-strain curve starts at the yield point σ_{p0}^{θ} and ends at the point ST. Let X and Y denote the quantities representing strain and stress, respectively, on the transitional hardening part, and let the values (X, Y) at the start and end points be (0, 0) and (1, 1), respectively.

We have

$$X = \epsilon^{j+1}/\rho_d^{\theta d}, \tag{5}$$

$$Y = 1 - \{\sigma_{\theta}'(\epsilon) - \sigma_t^{\theta}(\epsilon^{j+1})\}/(\sigma_{p1}^{\theta d} - \sigma_{p0}^{\theta d}), \tag{6}$$

where ϵ^{j+1} denotes the strain occurred during reloading after the j-th prestraining.

The experimental X-Y relation for a given material can be approximately represented by a single curve, as shown in Figure 7, irrespective of the prestraining path (Table 2). The curve is given by the following equation:

$$X^{0.5} + (1 - Y)^{0.43} = 1 \tag{7}$$

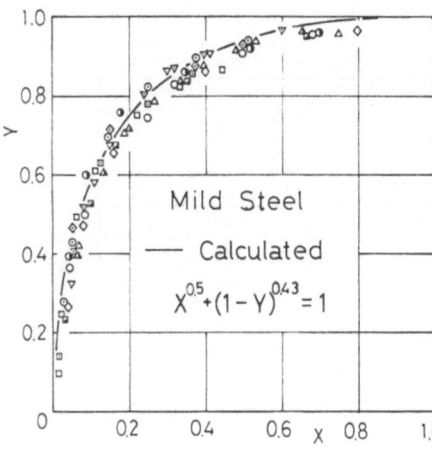

Figure 7. – Transitional hardening curve for mild steel.

3.5. Rigid-plastic yield locus after prestraining (Curve G)

The curve G can be determined experimentally by extrapolating steady hardening parts of the reloading stress-strain curves to the ordinate axis, as shown in Figure 4(a). The shape of the curve G is found to be approximately represented by an ellipse whose axes in the axial and torsional loading directions are denoted by 2A and 2B, respectively, and the distance between the centre of the ellipse and the origin of the co-ordinates is indicated as C.

Observed values of A and B are plotted for the magnitude of total prestrain ϵ_T, as shown in Figure 8. Thick full and chain lines in the figure are the relations for initial tensile and torsional loadings (without prestraining), respectively.

The experimental relations for a simple (mono) prestraining path are very similar in shape to the relation for initial loadings. This implies that the following equations of a similar type to Eq. (2) can be derived for the values A and B in the case of simple prestraining:

$$S_i = \alpha_i^S \, (\beta_i^S + \epsilon_i)^{n_i^S} ,$$
(8)

$$M_\theta^S = M_1^S - \frac{\theta}{90} (M_1^S - M_3^S).$$
(9)

Here
$$S = A, B, \quad i = 1, 3, \theta, \quad M = \alpha, \beta, n.$$

Table 3 gives the values of α, β and n for tensile and torsional preloadings. Relations calculated from Eqs. (8) and (9) are shown in Figure 8 by thin lines with small marks inserted.

Experimental relations between C and ϵ_T are given in Figure 9. The following equations are approximately assumed:

$$C_i = \sigma_i \, (\epsilon_i) - S_i(\epsilon_i)$$
(10)

$$C_\theta = C_1 - \frac{\theta}{90} \, (C_1 - C_3)$$
(11)

TABLE 3. Coefficients in the approximate equation (8).

		Mild Steel		Brass		Al alloy	
		Tension	Torsion	Tension	Torsion	Tension	Torsion
A	α	70.6	71.1	112.0	94.9	59.2	50.6
	β	-0.0116	-0.0118	0.0702	0.0606	0.0063	-0.0097
	n	0.190	0.189	0.684	0.616	0.205	0.154
B	α	65.9	61.3	93.0	67.1	52.8	40.3
	β	-0.0103	-0.00089	0.0595	0.0426	-0.0053	-0.030
	n	0.195	0.200	0.608	0.490	0.168	0.090

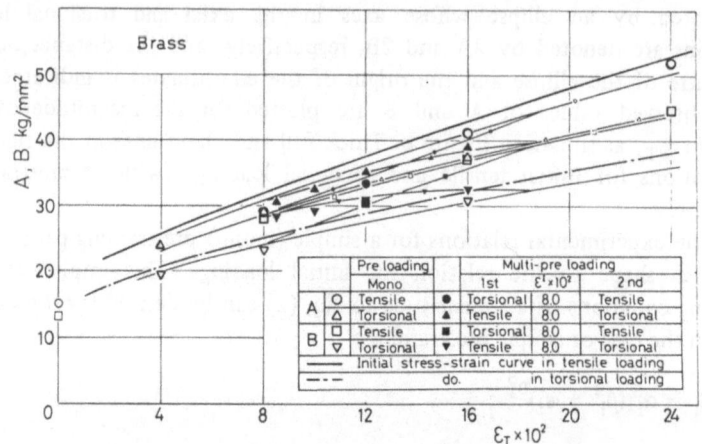

Figure 8. − *Experimental values of the major and minor axes of elliptic G curve.*

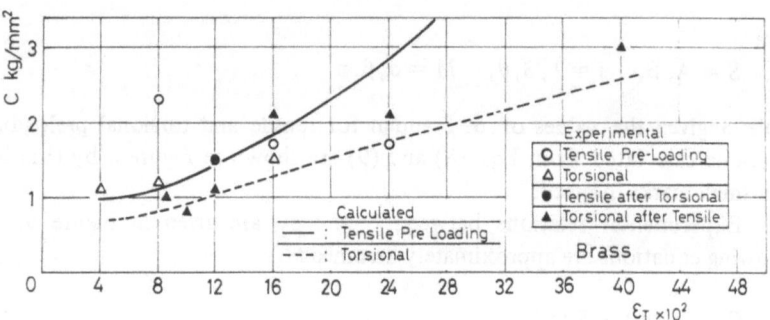

Figure 9. − *Center of the G-ellipse.*

where $S_i = A$ for $i = 1$, and $S_i = B$ for $i = 3$. Full and dotted lines in Figure 9 show the relations calculated by means of Eq. (10) for tensile and torsional preloadings, respectively.

Now examine the changing of A and B during the progress of the second prestrain in the case of double-prestraining. Blackened triangular marks in Figure 8 correspond to the case, where the second prestrain by torsion is applied after tensile prestraining. The values of A and B are found to gradually approach, as the torsional strain increases, the chain line which corresponds to the torsional simple prestraining.

Assume that $S_{\theta 1}(\epsilon)$ and $S_{\theta 2}(\epsilon)$ in Figure 10 are the relations between $S(= A, B)$ and ϵ_T for simple prestrainings in direction θ_1 and θ_2, respectively. Then the $S - \epsilon_T$ relation for the second prestrain ϵ^{II} in the direction θ_2 after

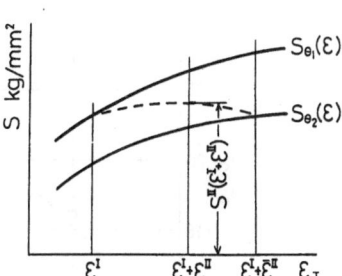

Figure 10. – *Values of S (= A,B) after multi-prestraining.*

the first prestrain ϵ^I in the direction θ_1 will schematically be represented by the dotted line in the Figure. The equation of this dotted line is assumed to be of the form

$$S^{II}(\epsilon) = \alpha^{S2} (\beta^{S2} + \epsilon)^{nS2} \qquad (12)$$

with variable material constants α^{S2}, β^{S2} and n^{S2}, depending on the value ϵ^{II}. In order to satisfy the conditions

$$S^{II}(\epsilon) = \epsilon_{\theta 1}(\epsilon) \quad \text{at} \quad \epsilon = \epsilon^I$$

and

$$S^{II}(\epsilon) = \epsilon_{\theta 2}(\epsilon) \quad \text{at} \quad \epsilon = \epsilon^I + \bar{\epsilon}^{II},$$

these material constants were assumed to be given by the relation

$$M^{S2} = M_{\theta_1}^{S1} + \frac{\epsilon^{II}}{\bar{\epsilon}^{II}} (M_{\theta_2}^{S1} - M_{\theta_1}^{S1}), \quad (M = \alpha, \beta, n), \qquad (13)$$

where

$$S_{\theta_1} = \alpha_{\theta_1}^{S1} (\beta_{\theta_1}^{S1} + \epsilon)^{n_{\theta_1}^{S1}}, \qquad (14)$$

$$S_{\theta_2} = \alpha_{\theta_2}^{S1} (\beta_{\theta_2}^{S1} + \epsilon)^{n_{\theta_2}^{S1}}, \qquad (15)$$

and $M_{\theta_i}^{Sj}$ denotes the coefficients α, β, n in the equation for S after j-th prestraining in the direction θ_i. Hence simple prestraining corresponds to the case j = 1. $\bar{\epsilon}^{II}$ denotes the strain increment in the second prestraining beyond which plastic anisotropy due to the first prestraining vanishes. The dependence of $\bar{\epsilon}^{II}$ on the value of ϵ^I was derived from Eqs. (12) and (13) with the experimental data of the B value, and the result is shown in Figure 11. The result is approximately expressed by the equation

$$\bar{\epsilon}^{II} = \delta \sqrt{\epsilon^I} \qquad (16)$$

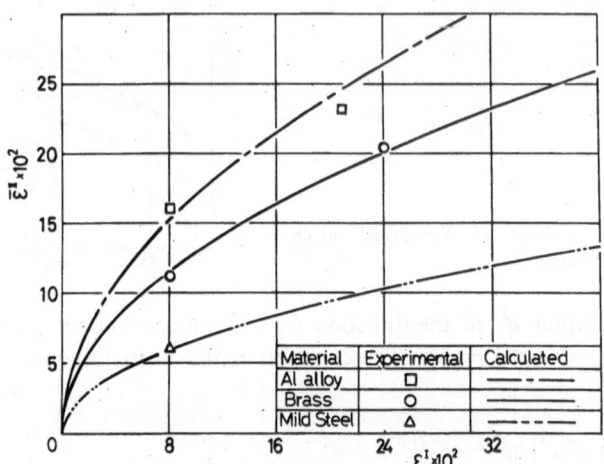

Figure 11. – *Relation between* ϵ^{I} *and* $\bar{\epsilon}^{II}$.

where the value of δ is 0.21 for mild steel, 0.41 for brass, and 0.54 for Al-Alloy. Results calculated from Eqs. (12) to (16) are shown in Figure 8, by the thin lines marked with small black marks. A pretty good agreement is observed between the calculation and the experiment.

The center of the G-ellipse lies approximately in the latest preloading direction for a range of not too small prestrain. Hence, the value of C for double-prestraining may be assumed to be equal to the value of C for simple prestraining in the same direction as that of the latest preloading and of the same total prestrain as that in the double-prestraining. Applicability of this assumption is confirmed, in Figure 9, by comparing the values of C (black mark) for double prestraining with those for simple prestraining.

Finally, the G-ellipse determined by the above-mentioned method must be magnified or reduced, so that the latest prestress point will lie on the ellipse.

This method of determining the G-ellipse can be extended easily beyond the double-prestraining. The thick line in Figure 2 shows the calculated stress-strain relation for the multi-preloading path represented in the insert, which gives a good approximation to the actual relation.

4. Determination of the material constants necessary for the calculation

The material constants necessary for calculating the stress-strain relation of an initially anisotropic material in reloading after a certain multi-preloading are as follows:

(1) Constants α_i, β_i and n_i in Eq. (2) for tensile and torsional stress-strain relations;

(2) Constants A_e, B_e and C_e in the following equation of an F-curve:

$$\frac{(\sigma_I - C_e)^2}{A_e^2} + \frac{\sigma_{II}^2}{B_e^2} = 1 \, ;$$

(3) Constants ξ and m in Eq. (4) for the transitional strain ρ_d;

(4) Constants α_i^S, β_i^S and n_i^S in Eq. (8) for the major and minor axes of the G-ellipse for tensile and torsional prestrainings;

(5) Constant δ in Eq. (16) for $\bar{\epsilon}^{II}$.

At least the following six tests on tubular specimens are necessary for detemination of these constants:

(1) Tensile loading test (to strain as large as possible);

(2) Torsional loading test (to strain as large as possible);

(3) Tensile loading to the strain of practical concern. After completely unloading, torsional and tensile loadings (to $\epsilon_3 \doteqdot \epsilon_1 \doteqdot 2.0 \times 10^{-4}$). Then, compressive loading (to $\epsilon_1 < -5 \times 10^{-2}$).

(4) Torsional loading (to $\epsilon_3 \doteqdot \epsilon_1$ in test (3)). After completely unloading, tensile and torsional loadings (to $\epsilon_1 \doteqdot \epsilon_3 \doteqdot 2.0 \times 10^{-4}$). Then, torsional loading in the opposite direction (to $\epsilon_3 < -5 \times 10^{-2}$).

(5) Tensile loading (to nearly the same strain ϵ_1 as in test (3)). After completely unloading, torsional loading (to $\epsilon_3 \doteqdot \epsilon_1/2$). Then, torsional loading in the opposite direction (to $\epsilon_3 < -5 \times 10^{-2}$).

(6) Torsional loading (to $\epsilon_3 \doteqdot \epsilon_1$ in test (3)). After completely unloading, tensile loading (to $\epsilon_1 \doteqdot \epsilon_3/2$). Then, compressive loading (to $\epsilon_1 < -5 \times 10^{-2}$).

5. Conclusion

Stress-strain curves for initially anisotropic metals in reloading after multi-prestrainings under combined axial load and torsion were investigated on the basis of a hardening model composed of elastic, transitional hardening and steady hardening regions. The following results were obtained:

(1) The elastic limit curve can be approximately represented by a single ellipse, idependently of the prestraining path.

(2) The magnitude of the transitional strain ρ_d is mainly dependent on the angle θ_d between the latest preloading and reloading directions, and approximately independent of the prestraining path.

(3) The dimensionless stress-strain curve of the transitional hardening region can be approximately represented by a single curve irrespective of prestraining path.

(4) The boundary between the transitional hardening and steady hardening regions can be approximately given by an ellipse. Equations of the major and minor axes and the central position of this ellipse are experimentally derived.

(5) It becomes possible to estimate approximately the plastic behaviour of an initially anisotropic metal after multi-prestrainings under combined tension and torsion, by the use of the six specimens' data.

Acknowledgements

The authors would like to thank Messrs. H. Suzumura and Y. Suzuki for their help in the experiments.

REFERENCES

[1] BALTOV A. and A. SAWCZUK. – "A rule of anisotropic hardening". *Acta Mechanica*, 1 (1965): 81-92.

[2] EISENBERG M.A. and A. PHILLIPS. – "A theory of plasticity with non-coincident yield and loading surfaces". *Acta Mechanica*, 11 (1971): 247-260.

[3] KANEKO K., K. IKEGAMI and E. SHIRATORI. – "The yield condition and flow rule of a metal for the various pre-strain paths". *Bull. Japan Soc. Mech. Engrs.*, 19 (1976): 577-583.

[4] MROZ Z. – "On the description of anisotropic work hardening". *J. Mech. Phys. Solids*, 15 (1967): 163-175.

[5] PRAGER W. – "The theory of plasticity: A survey of recent achievements". *Proceedings of the Institution of Mechanical Engineers*, 169 (1955): 41-57.

[6] SHIRATORI E., K. IKEGAMI and K. KANEKO. – "The influence of the Bauschinger effect on the subsequent yield condition". *Bull. Japan Soc. Mech. Engrs.*, 16 (1973): 1482-1493.

[7] WILLIAMS J.F. and N.L. SVENSSON. – "Effect of torsional pre-strain on the yield locus of 1100-F Aluminium". *J. of Strain Analysis*, 6 (1971): 263-272.

RESUME

(Comportements plastiques des métaux initialement anisotropes après prédéformations multiples)

Les comportements plastiques de trois types de métaux initialement anisotropes, à savoir de l'acier doux et des alliages de cuivre et d'aluminium, après avoir été soumis à des prédéformations multiples sous des charges axiales et des torsions combinées, ont fait l'objet d'une étude expérimentale. Les courbes contraintes — déformations en recharge après différentes prédéformations ont été calculées sur la base d'un modèle d'écrouissage comportant des régions élastiques, à écrouissage transitoire et à écrouissage constant. Les courbes calculées sont en assez bon accord avec les courbes expérimentales.

DISCUSSION

QUESTIONS BY J.J. ENGEL: 1) Which definition of the elastic limit did you use for experimental determination of elastic domain (proportional limit, residual plastic strain Δe^p, . . .) and with which precision?

2) The influence of the precision of the determination is very important for concluding on the existence of corners of the frontier. Did you try to verify the reproductivity of your determination?

REPLY BY E. SHIRATORI: 1) The yield point is defined by the stress point corresponding to the plastic strain 200×10^{-6}. We mesured the strain by strain-gauges, so the precision is expected within $\pm 10 \times 10^{-6}$.

2) The effect of determining a sequence of yield points is discussed in the authors' article [1]. The effect is noticeable at the opposite side to preloading. Therefore, when we obtain several yield points by using one specimen, we must determine the yield points in the direction opposite to the preloading.

For verifying the existence of corners, the determining method of yield surface is not preferable. The zig-zag loading method or initial shear measuring method should be used instead.

[1] E. SHIRATORI, K. IKEGAMI and K. KANEKO. — "Subsequent yield surface in consideration of the Bauschinger Effect", in A. Sawczuk ed., *Foundations of Plasticity* (Alphen aan den Rijn: Sijthoff and Noordhoff, 1972).

Anisotropic Behaviour
at Cyclic Plastic Deformation

G. BACKHAUS
Technische Universität, Dresden, Germany-DDR.

1. Introduction

Proceeding from a previously proposed constitutive equation for plastic materials the anisotropy produced by permanent deformation is considered. Then the case of cyclic deformation is studied. Experimental results show that the superposition of equal strain amplitudes and a constant plastic strain produces a mean stress relaxing very slowly to zero with increasing number of cycles. An analytical representation employing a relaxation function dependent on the history of cyclic deformation is developed and yields good agreement with experimental data.

Starting from the Huber-Mises yield condition and neglecting shape changes of the yield surface, the following constitutive equation for plastic materials of any deformation history with continuous change of the deformation direction $d\epsilon_{ij}^p/d\epsilon_v$ is presented by the author, among others in [1, 2]:

$$S_{ij} = \frac{2}{3} k_f(\epsilon_v) \frac{d\epsilon_{ij}^p}{d\epsilon_v}(\epsilon_v)$$

$$- \frac{2}{3} \int_0^{\cdot\,\epsilon_v} z(\bar{\epsilon}_v) k_f(\bar{\epsilon}_v) \varphi(\epsilon_v - \bar{\epsilon}_v) \frac{d^2\epsilon_{ij}^p}{d\epsilon_v^2}(\bar{\epsilon}_v) d\bar{\epsilon}_v \tag{1}$$

where (see Fig. 1)

2z is the Bauschinger characteristic $= \dfrac{\sigma_F^+ + \sigma_F^-}{\sigma_F^+}$,

$k_f(\epsilon_v)$ represents the flow curve at monotonic loading,

$\epsilon_v = \int d\epsilon_v = \int \sqrt{2/3\, d\epsilon_{ij}^p\, d\epsilon_{ij}^p}$ is the equivalent plastic strain,

$\varphi(\epsilon_v - \bar{\epsilon}_v)$ takes into account the strain-induced relaxation of the internal stress states causing the Bauschinger effect.

According to test results, φ may be written in the form

$$\varphi(\epsilon_v - \bar{\epsilon}_v) = \exp\left[-\kappa_1(\epsilon_v - \bar{\epsilon}_v)^\rho\right]. \tag{2}$$

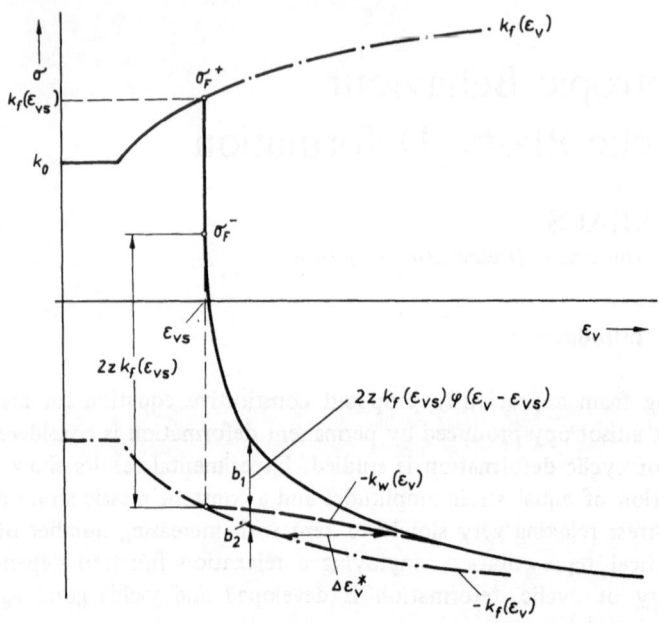

Figure 1. – *Flow stress at loading reversal.*

For any sudden direction change $\Delta \dfrac{d\epsilon_{ij}^p}{d\epsilon_v}$ at $\epsilon_v = \epsilon_{vs}$ Eq. (1) becomes

$$S_{ijF} = \frac{2}{3} \cdot k_f(\epsilon_{vs}) \left[\frac{d\epsilon_{ij}^p}{d\epsilon_v}(\epsilon_{vs}) + \Delta \frac{d\epsilon_{ij}^p}{d\epsilon_v} \right] \qquad \longrightarrow t_{ij}$$

$$-\frac{2}{3} \int_0^{\epsilon_{vs}} z(\bar{\epsilon}_v) \, k_f(\bar{\epsilon}_v) \, \varphi(\epsilon_v - \bar{\epsilon}_v) \frac{d^2 \epsilon_{ij}^p}{d\epsilon_v^2}(\bar{\epsilon}_v) \, d\bar{\epsilon}_v \quad \longrightarrow \Delta t_{ij} \qquad (3)$$

$$-\frac{2}{3} z(\epsilon_{vs}) \, k_f(\epsilon_{vs}) \, \varphi(0) \, \Delta \frac{d\epsilon_{ij}^p}{d\epsilon_v}. \qquad\qquad \longrightarrow \Delta S_{ijB}$$

In the deviatoric stress space t_{ij} represents a hypersphere with the radius $R_0 = \sqrt{2/3}\, k_f$ corresponding to isotropic strain hardening (dot-dash circle) (Fig. 2). Δs_{ijB} representing the Bauschinger effect yields a hypersphere with the radius $R_E = R_0(1 - z)$ (dashed circle), whereas Δt_{ij} taking into account the influence of the preceding deformation history determines the position of the yield surface s_{ijF} and the degree of anisotropy.

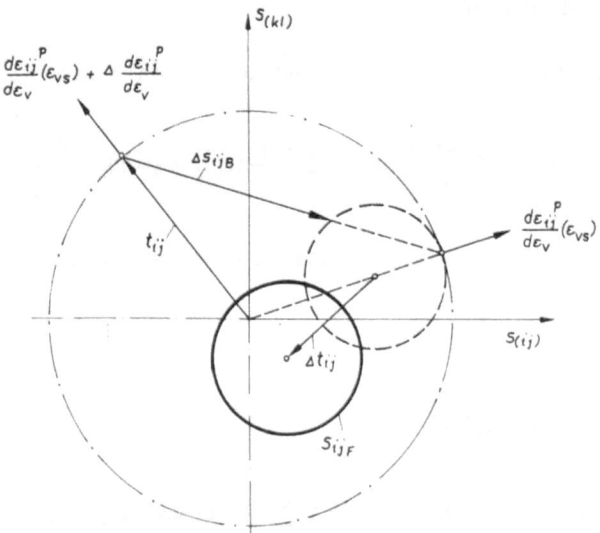

Figure 2. – *Yield surface $F(s_{ijF}) = 0$ in the deviatoric stress space.*

For a deformation process with consecutive sudden direction changes $\Delta de_{ij}^p/de_v$ at $\epsilon_v = \epsilon_{vs}$ Eq. (1) is obtained in the form

$$s_{ij}(\epsilon_v) = \frac{2}{3} k_f(\epsilon_w) \frac{de_{ij}^p}{de_v}(\epsilon_v)$$

$$- \frac{2}{3} \sum_{s=1}^{n} z\, k_f(\epsilon_{ws})\, \varphi_s(\epsilon_v - \epsilon_{vs})\, \Delta \frac{de_{ij}^p}{de_v}(\epsilon_{vs}),\qquad (4)$$

where $k_f(\epsilon_v)$ is replaced by $k_f(\epsilon_w)$ due to a softening effect appearing in the case of a sudden direction change. In [2] the softening effect is described by a reduction of the equivalent strain ϵ_v by $\Delta\epsilon_v^*$ (see Fig. 1) :

$$\epsilon_w = \epsilon_v - \Delta\epsilon_v^*.\qquad (5)$$

The value z in (4) can be assumed to be a constant.

The present paper demonstrates that the Bauschinger relaxation function $\varphi(\epsilon_v - \bar{\epsilon}_v)$ in the considered case, in general, no longer possesses the above presented simple form (2). As a result of consecutive sudden direction changes the influence of preceding plastic deformations on the stresses can be strongly prolonged. Thus, the function φ_s in Eq. (4) will be dependent on the history of direction changes.

2. Cyclic deformation

In the case of cyclic deformation, the following relations exist :

$$\frac{de_{ij}^p}{de_v}(\epsilon_v) = (-1)^n \frac{de_{ij}^p}{de_v}(0) ; \quad \Delta \frac{de_{ij}^p}{de_v}(\epsilon_{vs}) = (-1)^s 2 \frac{de_{ij}^p}{de_v}(0). \tag{6}$$

Taking these relations into account and introducing the equivalent stress

$$\sigma_v = \sqrt{\frac{3}{2} s_{ij} s_{ij}} \tag{7}$$

Eq. (4) is obtained as follows:

$$\sigma_v(\epsilon_v > \epsilon_{vn}) = k_f(\epsilon_w) - (-1)^n \sum_{s=1}^{n} (-1)^s 2z \, k_f(\epsilon_{ws}) \, \varphi_s(\epsilon_v - \epsilon_{vs}) . \tag{8}$$

Cyclic deformation tests with equal strain amplitudes after relatively large plastic prestraining, besides the above-mentioned cyclic softening, show decreasing differences of consecutive stress amplitudes, i.e., a relaxation of the mean stress. In Figure 3 and 4 test results are given for a mild steel and for brass [4]. Here, $\Delta\epsilon_{v1}$ means the width of the first cycle, $\Delta\epsilon_v$ that of the following cycles. For comparison, the results of symmetrical deformation cycles are also plotted in the figures. The measured decrease of the stress amplitude differences can be used for checking the applicability of Eq. (8).

Considering the case of equal strain amplitudes and introducing

$$\epsilon_{vn+1} - \epsilon_{vs} = (n + 1 - s) \Delta\epsilon_v$$

$$k_f(\epsilon_{ws}) = k_{ws} \tag{9}$$

from Eq. (8) we obtain the differences of consecutive stress amplitudes as follows

$$\Delta\sigma_A(n) = \sigma_v(\epsilon_{vn+1}) - \sigma_v(\epsilon_{vn})$$

$$= k_{wn+1} - k_{wn} + (-1)^n 2z k_{w1} \varphi(n \Delta\epsilon_v)$$

$$- (-1)^n \sum_{s=2}^{n} (-1)^s 2z(k_{ws} - k_{ws-1}) \varphi((n + 1 - s) \Delta\epsilon_v) . \tag{10}$$

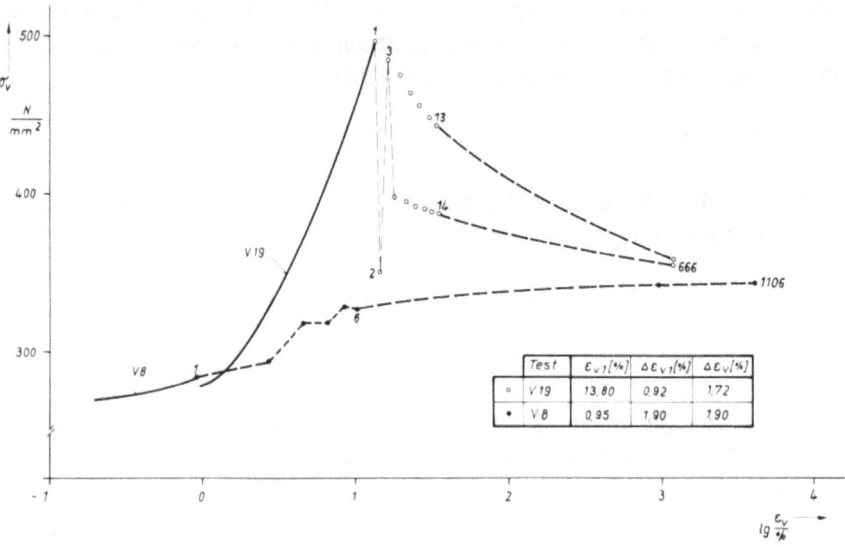

Figure 3. — *Stress amplitudes from tests with cyclic plastic deformation (steel).*

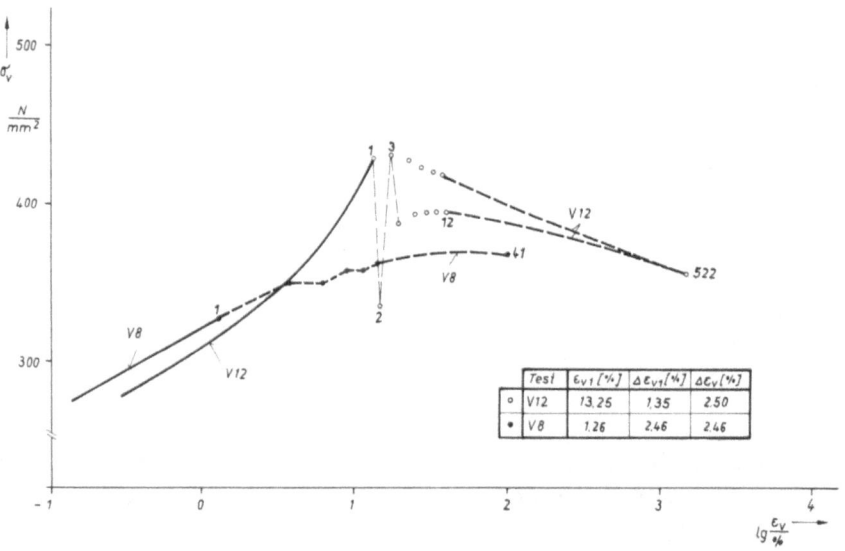

Figure 4. — *Stress amplitudes from tests with cyclic plastic deformation (brass).*

The terms within the sum expression possess alternating signs. Therefore and with respect to the fact that for large prestraining the differences $k_{ws} - k_{ws-1}$ are very small the sum expression can be neglected:

$$\Delta\sigma_A(n) \approx k_{wn+1} - k_{wn} + (-1)^n 2z\,k_{w1}\,\varphi(n\,\Delta\epsilon_v)\,. \tag{11}$$

Neglecting also the small differences $k_{wn+1} - k_{wn}$ and using Eq. (2) we obtain with regard to $k_{w1} = k_f(\epsilon_{v1})$

$$|\Delta\sigma_A(n)| \approx 2z\,k_f(\epsilon_{v1})\exp\left[-\,\kappa_1\,n^\rho\,\Delta\epsilon_v^\rho\right]\,. \tag{12}$$

For instance, Figure 5 shows the result of a calculation[1] corresponding to Test 20 in Table 1 and using the values κ_1, ρ from Table 2. For comparison, the experimentally obtained values $|\Delta\sigma_A(n)|$ of Test 20 are presented also in Figure 5. It is seen that there is a large difference regarding the decrease of $|\Delta\sigma_A(n)|$.

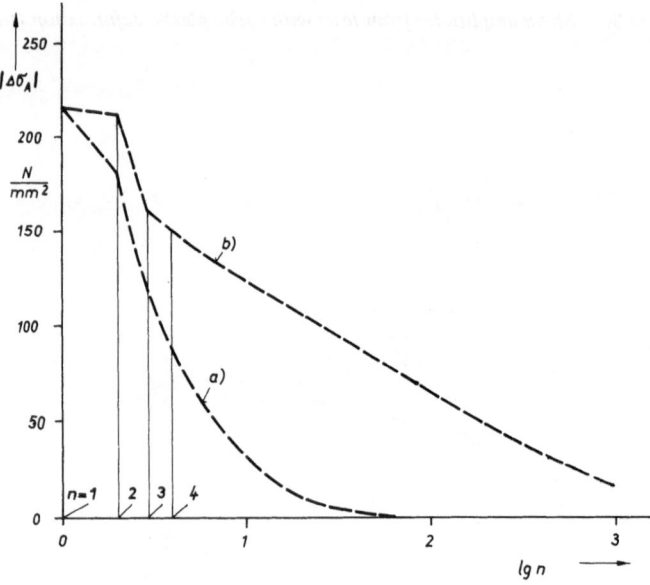

Figure 5. — *Stress amplitude differences against number of loading reversals*
a) *calculated for* κ = *const*
b) *experimental.*

---

[1] With regard to $\Delta\epsilon_{v1} \neq \Delta\epsilon_v$ a modified equation (12) is employed for calculation.

The considerable delay of the measured decrease compared to the calculated decrease can be interpreted in such a way that the function $\varphi_s(\epsilon_v - \epsilon_{vs})$ attached to the loading reversal at $\epsilon_v = \epsilon_{vs}$ is influenced by the following reversals, i.e. φ_s is dependent on the cyclic deformation history.

3. History dependence of the function φ_s

The behaviour discussed above regarding the relaxation of the mean stress can be described analytically by introducing a variable exponent κ dependent on the number of passed loading reversals. Thus, the function $\varphi(\epsilon_v - \epsilon_{v1})$ associated with the first loading reversal at $\epsilon_v = \epsilon_{v1}$ takes the following forms valid in the intervals:

$$\epsilon_{v1} < \epsilon_v \leqslant \epsilon_{v2}: \overline{\varphi}_1 = \exp\left[-\kappa_1 (\epsilon_v - \epsilon_{v1})^\rho\right],$$

$$\epsilon_{v2} < \epsilon_v \leqslant \epsilon_{v3}: \overline{\varphi}_2 = \overline{\varphi}_1(\epsilon_{v2}) \frac{\exp\left[-\kappa_2 (\epsilon_v - \epsilon_{v1})^\rho\right]}{\exp\left[-\kappa_2 (\epsilon_{v2} - \epsilon_{v1})^\rho\right]}, \tag{13}$$

$$\epsilon_{v3} < \epsilon_v \leqslant \epsilon_{v4}: \overline{\varphi}_3 = \overline{\varphi}_2(\epsilon_{v3}) \frac{\exp\left[-\kappa_3 (\epsilon_v - \epsilon_{v1})^\rho\right]}{\exp\left[-\kappa_3 (\epsilon_{v3} - \epsilon_{v1})^\rho\right]}$$

and so on, where $\kappa_1 > \kappa_2 > \kappa_3$, $\kappa_i > \kappa_{i+1}$.
From the above relations it follows for the interval

$$\epsilon_v > \epsilon_{vn}: \overline{\varphi}_n = \exp\left[-\sum_{i=1}^{n-1} (\kappa_i - \kappa_{i+1})(\epsilon_{vi+1} - \epsilon_{v1})^\rho - \kappa_n(\epsilon_v - \epsilon_{v1})^\rho\right] \tag{14}$$

The functions $\overline{\varphi}_n$ are shown in Figure 6 for

$$\epsilon_{v2} - \epsilon_{v1} = \epsilon_{v3} - \epsilon_{v2} = 0{,}003$$

and using the relation (18) given below for κ_i and the values $\kappa_1, \kappa_\infty, \rho$ for steel from Table 2. It is seen that the relaxation is retarded due to the occurence of loading reversal.

For the relaxation function attached to the loading reversal with the number $s < n$ and effective in the interval $\epsilon_v > \epsilon_{vn}$ from Eq. (14) we obtain when employing the notation from Eq. (4):

$$\varphi_S(\epsilon_v > \epsilon_{vn}) = \exp\left[-\sum_{i=1}^{n-s} (\kappa_i - \kappa_{i+1})(\epsilon_{vi+s} - \epsilon_{vs})^\rho\right.$$

$$\left. - \kappa_{n-s+1}(\epsilon_v - \epsilon_{vs})^\rho\right]. \tag{15}$$

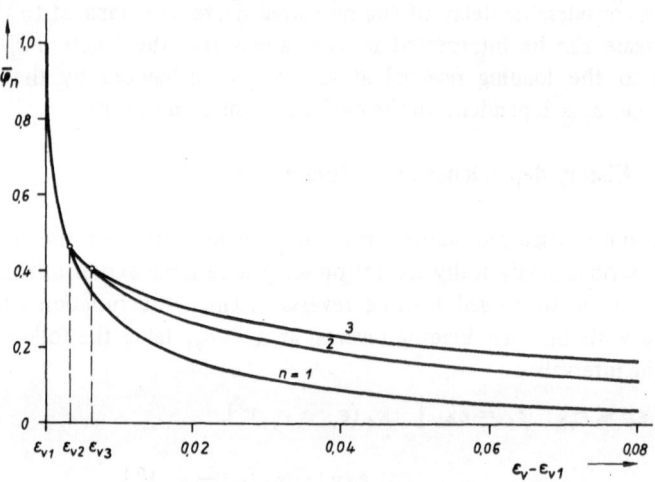

Figure 6. – *Functions* $\bar{\varphi}_n$.

For s = n, φ_n must be identical to the relaxation caused by a single reversal :

$$\varphi_n \ (\epsilon_v > \epsilon_{vn}) = \exp\left[-\kappa_1 \ (\epsilon_v - \epsilon_{vn})^\rho\right] \equiv \bar{\varphi}_1. \tag{16}$$

4. Relation for κ_i

The application of the relation developed above for the function φ_S to Eq. (12) yields:

$$|\Delta\sigma_A (n)| \approx 2 \, zk_f \, (\epsilon_{v1}) \, \varphi_1 \, (n \Delta\epsilon_v)$$

where

$$\varphi_1 \ (n\Delta\epsilon_v) = \exp\left[-\sum_{i=1}^{n-1} (\kappa_i - \kappa_{i+1})(i\Delta\epsilon_v)^\rho - \kappa_n \ (n\Delta\epsilon_v)^\rho\right]. \tag{17}$$

Eq. (17) can serve in the successive determination of the exponents κ_i from experimental values $|\Delta\sigma_A(n)|$.

Tests are carried out [4] by subjecting thin-walled tubular specimens to torsion. The specimens are of mild steel (St38u-2) and brass (Ms58), Table 1 gives the test data.

TABLE 1

Material	Test	ϵ_{v1} [%]	$\Delta\epsilon_{v1}$ [%]	$\Delta\epsilon_v$ [%]	n to failure
St 38u-2	V18	13,62	2,05	4,03	152
	V19	13,80	0,92	1,72	700
	V20	13,85	0,32	0,56	> 1 250
Ms 58	V11	13,40	3.16	6,26	49
	V12	13,25	1,35	2,50	604
	V13	12,86	0,79	1,42	1 252

In Figures 7 and 8 the experimental values of $|\Delta\sigma_A(n)|$ are plotted. For determining the exponents κ_i from these results, a more general relation instead of Eq. (17) is required because of the difference between $\Delta\epsilon_{v1}$ (first cycle) and $\Delta\epsilon_v$ (for the following cycles). The evaluation yields

$$\kappa_i \approx \kappa_\infty + \frac{\kappa_1 - \kappa_\infty}{i} . \tag{18}$$

Table 2 gives the values κ_1, κ_∞ together with 2z and ρ^1.

TABLE 2

Material	κ_1	κ_∞	2z	ρ
St 38 u-2	10,4	1	1	0,45
Ms 58	19	0,3	1	0,59

The dashed curves in Figures 7 and 8 represent the calculated results employing Eq. (18) and using the values of Table 2. The zig-zag-course of the experimental points (Figure 7) is caused by the strong decrease of the stress amplitudes (Figure 3). The calculation yields the mean values.

To illustrate the influence of different Bauschinger relaxation functions φ according to Eq. (2) or (15), rsp., on the course of the stress amplitudes, the results of calculation employing Eq. (8) are plotted in Figure 9 in the form of relative stresses $\mathring{\sigma}_v(n + 1)/\sigma_v(1)$. It is supposed for simplification that, in Eq. (8), $k_f(\epsilon_w) = k_f(\epsilon_{v1})$.

The analysis of the test results shows that the history dependence of the function φ_S disappears for symmetrical deformation cycles: $\kappa_\infty = \kappa_1$. From

---

[1] The values 2z and ρ are determined from the difference between the flow curve and the curve with loading reversal taking into account the softening effect [4] and neglecting the small reversible plastic strain at load zero.

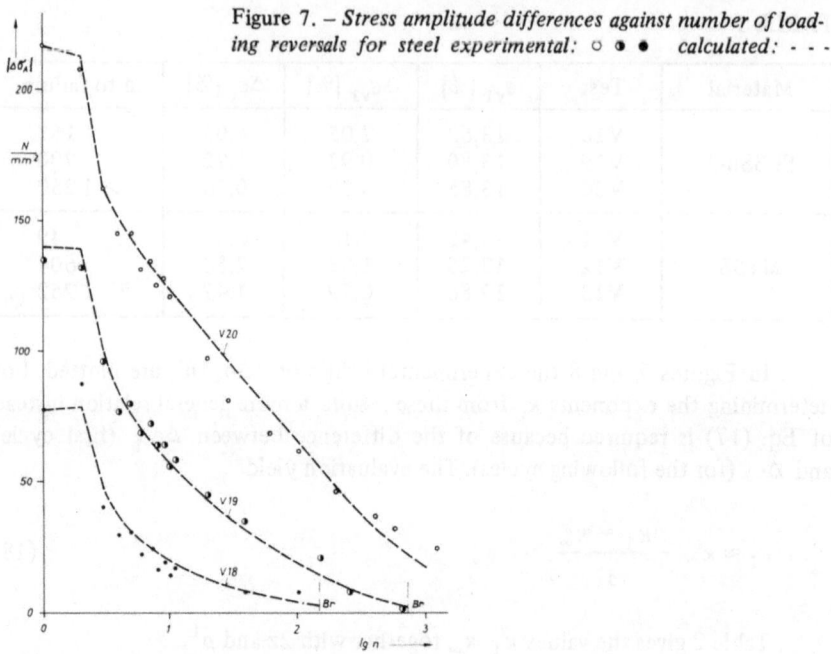

Figure 7. – *Stress amplitude differences against number of load-ing reversals for steel experimental:* ○ ◑ ●   *calculated:* - - -

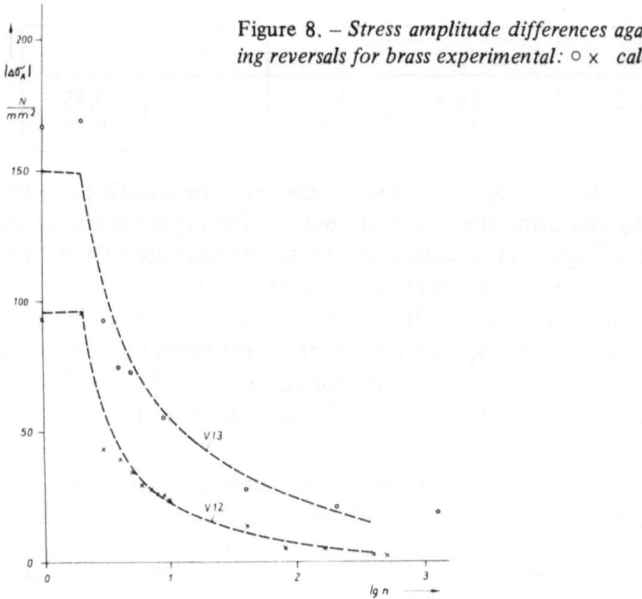

Figure 8. – *Stress amplitude differences against number of load-ing reversals for brass experimental:* ○ ×   *calculated:* - - -

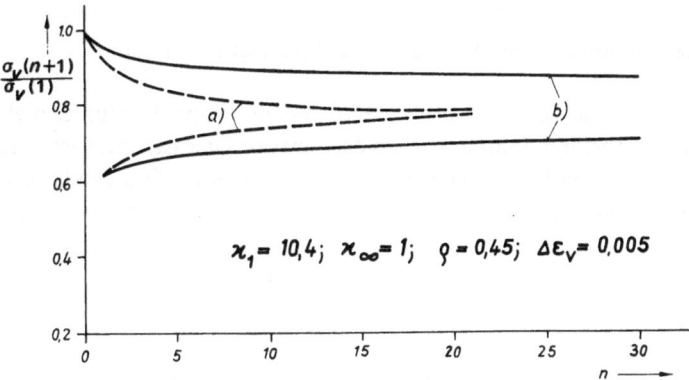

Figure 9. – *Stress amplitudes for strain-controlled cyclic loading calculated using a) Eq. (2) b) Eq. (15).*

this an influence of the mean strain ϵ_{vm} to κ_∞ follows. Furthermore, it can be presumed that there is a dependence on the amount of the direction change $\Delta \dfrac{d\epsilon_{ij}^p}{d\epsilon_v}$. Thus, a dependence of κ_∞ may be supposed in the following manner:

$$\kappa_\infty = \kappa_\infty \left(\epsilon_{vm} ; \ \Delta \frac{d\epsilon_{ij}^p}{d\epsilon_v} \ \Delta \frac{d\epsilon_{ij}^p}{d\epsilon_v} \right) \tag{19}$$

so that $\kappa_\infty (0; 0) = \kappa_1$.

The latter $(\kappa_\infty = \kappa_1)$ also holds for deformation processes having continuous direction changes [3].

REFERENCES

[1] BACKHAUS G. – "Zur analytischen Erfassung des allgemeinen Bauschingereffekts." *Acta Mechanica*, **14** (1972): 31-42.

[2] BACKHAUS G. – "Fliessspannungen und Fliessbedingung bei zyklischen Verformungen". *ZAMM*, **56** (1976): 337-348.

[3] BACKHAUS G. – "Plastic deformation in form of strain trajectories of constant curvature – theory and comparison with experimental results". *Acta Mechanica*, **34** (1979): 193-204.

[4] RICHTER K. – "Experimentelle und theoretische Untersuchungen zum Spannungs-Verformungs-Verhalten von St 38 und Ms 58 bei Zug-Torsions-Belastung im plastischen Bereich". Dissertationsschrift der Techn. Universität Dresden, 1978.

RESUME

(Comportement anisotrope sous déformation plastique cyclique)

A partir d'une loi constitutive précédemment proposée pour les matériaux plastiques, l'anisotropie induite par des déformations permanentes est examinée. Le cas des déformations cycliques est ensuite étudié. Des résultats expérimentaux montrent que la superposition de déformations à amplitudes égales et d'une déformation plastique constante produit une contrainte moyenne se relaxant très lentement vers zéro lorsque le nombre de cycles augmente. Une représentation analytique utilisant une fonction de relaxation dépendant de l'histoire des déformations cycliques est développée ; elle présente une bonne concordance avec les données expérimentales.

On Changes of Anisotropy during Thermo-Mechanical Processes

Th. Lehmann

Institut für Mechanik der Ruhr-Universität Bochum, Bochum, Germany-BRD.

1. Introduction

Many materials, such as polycrystalline metals, can be considered as classical continua. In many cases we may assume that the material is isotropic in the initial state of a thermo-mechanical process. During inelastic deformations, however, the material, generally, becomes anisotropic. In other cases the material is already anisotropic in the initial state, but the shape of the anisotropy changes during the thermo-mechanical process.

The aim of this paper is to give a frame for the description of such processes (particularly in polycrystalline materials) based on the foundations of classical continuum mechanics and thermodynamics. The frame shall embrace large non-isothermic deformations as well as solid phase transformations, recrystallization, recovery, and aging phenomena. Processes at the microscale, i.e. inside the crystall lattice, are not considered in detail; they may be regarded as the physical background of the phenomenological considerations. In a brief supplement the possible extension to non-classical continua is discussed.

2. Kinematics of large deformations in classical continua

The kinematics of large deformations of classical continuum are treated extensively in some recent papers [11, 16, 20]. For our purpose it is most convenient to base the description of the deformations on the metric changes of a body-fixed coordinate system ξ^i. This leads to the definition of the Almansi strain tensor

$$
\epsilon_{ik} = \frac{1}{2} \{g_{ik}(\xi^r, t) - g_{ik}(\xi^r, \overset{\circ}{t})\} = \frac{1}{2} \{g_{ik} - \overset{\circ}{g}_{ik}\}
$$

$$
\epsilon^i_k = \frac{1}{2} \{\delta^i_k - g^{ir} \overset{\circ}{g}_{rk}\} = \frac{1}{2} \{\delta^i_k - (q^{-1})^i_k\},
$$

(1)

where the superscript o denotes the initial state. $e^i_k k$ as well as all other quantities introduced in the following are related to the base of the body-fixed coordinate system in the deformed state. The strain rate is given by

$$d^i_k = \frac{1}{2} g^{ir} \dot{g}_{rk} = g^{ir} \dot{\epsilon}_{ik} = (\dot{\epsilon})^i_{.r} q^r_k \quad \text{(with } q^i_k = \overset{o}{g}{}^{ir} g_{rk}). \tag{2}$$

The dot denotes the substantial derivation with respect to time, which coincides with the partial derivation while ξ^i is held constant. Introducing an imaginary intermediate configuration [9, 11, 16, 20] marked by a superscript * we can decompose the total strain into its elastic and inelastic parts according to

$$q^i_k = \overset{o}{g}{}^{im} \overset{*}{g}_{mr} \overset{*rs}{g} g_{sk} = \underset{(i)}{q^i_{.r}} \underset{(e)}{q^r_k}$$

$$(q^{-1})^i_k = g^{im} \overset{*}{g}_{mr} \overset{*rs}{g} \overset{o}{g}_{sk} = \underset{(e)}{(q^{-1})^i_r} \underset{(i)}{(q^{-1})^r_{.k}}. \tag{3}$$

The corresponding splitting of the strain rate yields [9]

$$d^i_k = \underset{(e)}{d^i_k} + \underset{(i)}{d^i_k}. \tag{4}$$

In some cases it may be useful to decompose the strain into volume changes and distortions. This can be done by introducing the Hencky strain tensor [11]

$$\tilde{\epsilon}^i_k = \frac{1}{2} (\ln q)^i_k = \frac{1}{3} \epsilon \delta^i_k + \tilde{\gamma}^i_k, \tag{5}$$

where $\epsilon = \tilde{\epsilon}^r_r$ denotes the volume changes and $\tilde{\gamma}^i_k$ the distortions. The covariant derivation with respect to time (Zaremba-Jaumann derivation [11]) leads to

$$\tilde{\epsilon}^i_k|_0 = (\dot{\tilde{\epsilon}})^i_{.k} + d^i_r \tilde{\epsilon}^r_k - d^r_k \tilde{\epsilon}^i_r = \frac{1}{3} \dot{\epsilon} \delta^i_k + \tilde{\gamma}^i_k|_0. \tag{6}$$

In general, however, is

$$\tilde{\epsilon}^i_k|_0 \neq d^i_k. \tag{7}$$

Therefore some difficulties arise in the use of the Hencky strain tensor unless the elastic behaviour (including total volume changes) remains isotropic under all inelastic deformations. For small volume changes we can put

$$\epsilon = \tilde{\epsilon}^r_r \approx \epsilon^r_r = g^{ri} \epsilon_{ir} \tag{8}$$

and decompose

$$\dot{\epsilon}^i_k = \frac{1}{3}\, \dot{\epsilon}^r_r \delta^i_k + (\dot{\epsilon}^i_k - \frac{1}{3}\, \dot{\epsilon}^r_r \delta^i_k) \approx \frac{1}{3}\, \dot{\epsilon}\, \delta^i_k + \dot{\gamma}^i_k\, . \tag{9}$$

3. Thermodynamics of thermo-mechanical processes in classical continua

The rate of specific mechanical work is given by

$$\dot{w} = \frac{1}{\rho}\, \sigma^i_k\, d^k_i = \frac{1}{\overset{o}{\rho}}\, s^i_k\, d^k_i = \frac{1}{\overset{o}{\rho}}\, s^i_k\, q^r_j(\dot{\epsilon})^k_{.r}\, , \tag{10}$$

where σ^i_k means the Cauchy stress tensor and

$$s^i_k = \frac{\overset{o}{\rho}}{\rho}\, \sigma^i_k \tag{11}$$

denotes the weighted Cauchy stress tensor. The total work can be decomposed again according to the splitting of the strain rate. This leads to

$$\dot{w} = \frac{1}{\overset{o}{\rho}}\, s^i_k\, \underset{(e)}{d^k_i} + \frac{1}{\overset{o}{\rho}}\, s^i_k\, \underset{(i)}{d^k_i} = \underset{(e)}{\dot{w}} + \underset{(i)}{\dot{w}}\, . \tag{12}$$

The inelastic part of work rate must be split once more into one part $\underset{(d)}{\dot{w}}$ which is dissipated immediately and into another part $\underset{(h)}{\dot{w}}$ which is connected with the changes of the internal structure of the material. Therefore we have to write

$$\underset{(i)}{\dot{w}} = \underset{(h)}{\dot{w}} + \underset{(d)}{\dot{w}}\, , \tag{13}$$

and we obtain finally

$$\dot{w} = \underset{(e)}{\dot{w}} + \underset{(h)}{\dot{w}} + \underset{(d)}{\dot{w}}\, . \tag{14}$$

The first law of thermodynamics (energy balance) states

$$\dot{u} = \dot{w} - \frac{1}{\rho}\, (q^i + h^i)|_i + r\, . \tag{15}$$

In this formula mean: u the specific internal energy, q^i the heat flux, h^i the other energy fluxes (e.g. due to the migration of lattice defects without macroscopic deformations), r the energy sources.

Adopting the usual assumption of classical thermodynamics that each material element has the properties of a local thermodynamic system whose state is uniquely defined by a set of state variables, u must be expressible as a function of the state variables, i.e.

$$u = u \{ \underset{(e)}{\epsilon_k^i}, s, p_{(\nu)}, b, \beta_k^i, B_{ks}^{ir}, \ldots \}. \tag{16}$$

Herein s denotes the specific entropy, $p_{(\nu)}$ the mass fractions of the n solid phases ($\nu = 1$ to $n - 1$), b, β_k^i, B_{ks}^{ir} further internal state variables.

In contradiction to many other authors (e.g. [4, 6, 7, 15, 17, 22, 23]) the total strain ϵ_k^i cannot be used as a thermodynamic state variable. This can be shown by consideration of the processes at the microscale. The state of a material element depends on the average of the elastic deformation of the crystall lattice, the temperature, and the distribution of the lattice defects (e.g. dislocations and disclinations), but not on macroscopic inelastic deformations produced by slip processes which have passed through the crystall grains.

Introducing the specific free energy φ by a Legendre transformation we obtain

$$\varphi = u - Ts = \varphi \{ \underset{(e)}{\epsilon_k^i}, T, p_{(\nu)}, b, \beta_k^i, B_{ks}^{ir} \ldots \} \tag{17}$$

and

$$\dot{\varphi} = \dot{u} - T\dot{s} - s\dot{T} = \underset{(e)}{\dot{w}} + \underset{(h)}{\dot{w}} + \underset{(d)}{\dot{w}} - \frac{1}{\rho} (q^i + h^i)|_i + r - T\dot{s} - s\dot{T} \tag{18a}$$

$$= \frac{\partial \varphi}{\partial \underset{(e)}{\epsilon_k^i}} \underset{(e)}{(\dot{\epsilon})_{.k}^i} + \frac{\partial \varphi}{\partial T} \dot{T} + \sum_{\nu=1}^{n-1} \frac{\partial \varphi}{\partial p_{(\nu)}} \dot{p}_{(\nu)} + \frac{\partial \varphi}{\partial b} \dot{b} + \frac{\partial \varphi}{\partial \beta_k^i} \beta_k^i|_0$$

$$+ \frac{\partial \varphi}{\partial B_{ks}^{ir}} B_{ks}^{ir}|_0 + \ldots \tag{18b}$$

From the Legendre transformation and the comparison of the Eqs. (18a) and (18b) we can conclude

$$s = - \frac{\partial \varphi}{\partial T} \qquad \text{(caloric state eq.),} \tag{19a}$$

$$\frac{1}{\overset{\circ}{\rho}} \underset{(e)}{s_k^r} \underset{(e)}{q_r^i} = \frac{\partial \varphi}{\partial \underset{(e)}{\epsilon_i^k}} \qquad \text{(thermic state eq.),} \tag{19b}$$

and

$$T\dot{s} = \underset{(h)}{\dot{w}} + \underset{(d)}{\dot{w}} - \frac{1}{\rho} (q^i + h^i)|_i + r - \sum_{\nu=1}^{n-1} \frac{\partial \varphi}{\partial p_{(\nu)}} \dot{p}_{(\nu)} - \frac{\partial \varphi}{\partial b} \dot{b}$$

$$- \frac{\partial \varphi}{\partial \beta_k^i} \beta_k^i|_0 - \frac{\partial \varphi}{\partial B_{ks}^{ir}} B_{ks}^{ir}|_0 - \ldots \tag{20a}$$

$$= -T\left\{\frac{\partial^2\varphi}{\partial T^2}\,\dot{T} + \frac{\partial^2\varphi}{\partial\epsilon_k^i\,\partial T}\,\underset{(e)}{(\dot{\epsilon})_{;k}^i} + \sum_{\nu=1}^{n-1}\frac{\partial^2\varphi}{\partial p_{(\nu)}\,\partial T}\,\dot{p}_{(\nu)} + \frac{\partial^2\varphi}{\partial b\,\partial T}\,\dot{b}\right.$$

$$\left. + \frac{\partial^2\varphi}{\partial\beta_k^i\,\partial T}\,\beta_k^i|_0 + \frac{\partial^2\varphi}{\partial B_{ks}^{ir}\,\partial T}\,B_{ks}^{ir}|_0 + \cdots\right\}. \qquad (20b)$$

The second law of thermodynamics (the entropy balance) demands

$$T\dot{s} = \underset{(h)}{\dot{w}} + \frac{T}{\rho}\left(\frac{q^i}{T}\right)|_i + \frac{1}{\rho}\,h^i|_i - r + \dot{\eta}$$

$$+ \sum_{\nu=1}^{n-1}\frac{\partial\varphi}{\partial p_{(\nu)}}\,\dot{p}_{(\nu)} + \frac{\partial\varphi}{\partial b}\,\dot{b} + \frac{\partial\varphi}{\partial\beta_k^i}\,\beta_k^i|_0 + \frac{\partial\varphi}{\partial B_{ks}^{ir}}\,B_{ks}^{ir}|_0 + \cdots$$

$$= -\underset{(h)}{\dot{w}} + \frac{T}{\rho}\left(\frac{q^i}{T}\right)|_i + \frac{1}{\rho}\,h^i|_i - r + \dot{\eta} - T\frac{\partial^2\varphi}{\partial T^2}\,\dot{T} - T\frac{\partial^2\varphi}{\partial\epsilon_k^i\,\partial T}\,\underset{(e)}{(\dot{\epsilon})_{;k}^i}$$

$$+ \sum_{\nu=1}^{n-1}\left\{\frac{\partial\varphi}{\partial p_{(\nu)}} - T\frac{\partial^2\varphi}{\partial p_{(\nu)}\,\partial T}\right\}\dot{p}_{(\nu)} + \left\{\frac{\partial\varphi}{\partial b} - T\frac{\partial^2\varphi}{\partial b\,\partial T}\right\}\dot{b}$$

$$+ \left\{\frac{\partial\varphi}{\partial\beta_k^i} - T\frac{\partial^2 T}{\partial\beta_k^i\,\partial T}\right\}\beta_k^i|_0 + \left\{\frac{\partial\varphi}{\partial B_{ks}^{ir}} - T\frac{\partial^2\varphi}{\partial B_{ks}^{ir}\,\partial T}\right\}B_{ks}^{ir}|_0 + \cdots$$

$$= \underset{(d)}{\dot{w}} + \dot{\eta} - \frac{1}{\rho T}\,q^i T|_i \geqslant 0. \qquad (21)$$

$\dot{\eta}$ characterizes the dissipated energy connected with internal irreversible processes. It also covers that part of the energy applied by the sources r and the divergence of h^i which is dissipated immediately.

Eqs. (19), (20) and (21) give the general frame and the restrictions for the formulation of the constitutive law. In order to fill this frame we have to determine

a) the free energy φ as a function of the state variables,
b) the evolution laws for the dependent external process variables,
c) the evolution laws for the internal state variables,
d) the evolution laws for the dissipated energy $\underset{(d)}{w}$ and η, and
e) the laws for the energy fluxes q^i and h^i.

The balance equations for the free energy and the entropy are field equations. Therefore it is suggested that the evolution laws of the internal variables $p_{(\nu)}$, b, β_k^i, B_{ks}^{ir} etc. are of the same kind. This means we should expect that the evolution of the internal variables is governed by partial differential equations representing sources and the divergence of respective fluxes. In solid bodies, however, the fluxes are certainly small under normal conditions, since the migration of lattice defects etc. is hampered by grain bounderies and other obstacles. Therefore the source terms prevail in most cases. In long time processes or in other particular cases, however, this may change.

4. Elementary thermo-mechanical processes in classical continua

For the sake of simplicity we focus our considerations in the following to elementary thermo-mechanical processes which can be assumed to be homogeneous throughout the body. In such cases all fluxes vanish. No field equations enter the description of the processes. The (reversible) applied heat can be summarized to $\dot{q}$. Eqs. (20a) and (20b) can be condensed to

$$\underset{(h)}{\dot{w}} + \underset{(d)}{\dot{w}} + \dot{q} = - \underbrace{T \frac{\partial^2 \varphi}{\partial T^2}}_{c_v} \dot{T} - T \frac{\partial^2 \varphi}{\partial \epsilon_k^i \partial T} \underset{(e)}{(\dot{\epsilon})^i_{.k}} \tag{22}$$

$$+ \sum_{\nu=1}^{n-1} \left\{ \frac{\partial \varphi}{\partial p_{(\nu)}} - T \frac{\partial^2 \varphi}{\partial p_{(\nu)} \partial T} \right\} \dot{p}_{(\nu)} + \left\{ \frac{\partial \varphi}{\partial b} - T \frac{\partial^2 \varphi}{\partial b \partial T} \right\} \dot{b}$$

$$+ \left\{ \frac{\partial \varphi}{\partial \beta_k^i} - T \frac{\partial^2 T}{\partial \beta_k^i \partial T} \right\} \beta_k^i|_0 + \left\{ \frac{\partial \varphi}{\partial B_{ks}^{ir}} - T \frac{\partial^2 \varphi}{\partial B_{ks}^{ir} \partial T} \right\} B_{ks}^{ir}|_0 + \ldots$$

where c_v means the heat capacity at constant strain. The second law reduces to

$$\underset{(d)}{\dot{w}} + \dot{\eta} \geqslant 0 . \tag{23}$$

The stresses s_k^i and the temperature T, or their conjugated process variables, the total strain ϵ_k^i and the applied heat q, can act as independent (external) process variables of a thermo-mechanical process. ϵ_k^i and q are not thermodynamical state variables, therefore we prefer, in the following, s_k^i and T as independent process variables, for simplicity. But no difficulties arise when ϵ_k^i (or a combination of ϵ_k^i and s_k^i) and q must be choosen as independent process variables in accordance with the given process. The introduction of s_k^i instead of $\underset{(e)}{\epsilon_k^i}$ as state variable into the free energy may be achieved by a corresponding Legendre transformation. Details of this procedure are omitted here.

Dependent process variables are in our case: the conjugated external process variables ϵ_k^i, q ; the internal process variables (at the same time state variables) $p_{(\nu)}, b, \beta_k^i, B_{ks}^{ir}, \ldots$; the dissipated energy $\underset{(d)}{w}, \eta$, and others, if interesting.

The initial state of the material elements is determined by the initial configuration $\overset{\circ}{g}_{ik}$, the initial values of the external state variables $\overset{\circ}{s}{}^{i}_{k}$, $\overset{\circ}{T}$, and the initial values of the internal variables $\overset{\circ}{p}_{(\nu)}, \overset{\circ}{b}, \overset{\circ}{\beta}{}^{i}_{k}, \overset{\circ}{B}{}^{ir}_{ks}, \ldots$. The next step consists in the formulation of the constitutive law. It contains

a) the expression for the free energy

$$\varphi = \varphi \underset{(e)}{\{\epsilon^{i}_{k} \, T, \, p_{(\nu)}, \, b, \, \beta^{i}_{k}, \, B^{ir}_{ks}, \ldots\}},$$

b) the evolution laws for the conjugated external process variables

$$d^{i}_{k} = \underset{(e)}{d^{i}_{k}} \{\cdots\cdots\} + \underset{(e)}{d^{i}_{k}} \{\cdots\cdots\}$$

$$\dot{q} = \dot{q} \{\cdots\cdots\},$$

c) the evolution laws for the internal variables

$$\dot{p}_{(\nu)} = \dot{p}_{(\nu)} \{\cdots\cdots\cdots\}$$

$$\cdots\cdots\cdots\cdots\cdots\cdots,$$

d) the evolution laws for the dissipated energy

$$\underset{(d)}{\dot{w}} = \underset{(d)}{\dot{w}} \{\cdots\cdots\}$$

$$\dot{\eta} = \dot{\eta} \{\cdots\cdots\}.$$

The evolution laws must be compatible with Eq. (22) and the condition (23). They form a set of ordinary first order differential equations, which may be supplemented by auxiliary conditions, such as a yield condition. With respect to the structure of the evolution laws we can distinguish three different types:

A) The equilibrium type (incremental type, rate insensitive), which corresponds to a quasistatic sequence of thermodynamic equilibrium states. It is characterized by the fact, that the evolution of the considered quantity is correlated to the increments of the independent process variables (examples: elastic and elastic-plastic deformations without solid phase transformations or recrystallization etc.)

B) The non-equilibrium type (flow type, rate sensitive), which corresponds essentially to thermodynamic non-equilibrium states. This means that the evolution of the considered quantity is governed only by the actual state (examples: viscoplastic deformations, recrystallization). The non-equilibrium type may also depend explicitly on time. This is, for instance, the case in aging processes [21].

C) The mixed type (examples: elastic-viscoplastic deformations without solid phase transformations, elastic-plastic deformations with recrystallization and recovery).

In the following we shall restrict ourselves to a more particular but nevertheless fairly general approach for a constitutive law representing an elastic-viscoplastic body.

5. An approach to the constitutive law for elementary processes of a classical elastic-viscoplastic body

We assume that the free energy of a classical elastic-viscoplastic body can be written in the form

$$\varphi = \underset{(e)}{\varphi} \{\epsilon_k^i, T, p_{(\nu)}, b, \beta_k^i, B_{ks}^{ir}, \ldots\} = \underset{(e)}{\varphi} \underset{(e)}{\{\epsilon_k^i, T, p_{(\nu)}\}}$$

$$+ \underset{(h)}{\varphi} \{T, p_{(\nu)}, b, \beta_k^i, B_{ks}^{ir}\}. \qquad (24)$$

This approach implies that the relations between the elastic deformations of the crystall lattice and the stresses are not influenced by the internal variables b, β_k^i and B_{ks}^{ir}, which represent the distribution of the lattice defects and determine the hardening state of the material (together with $p_{(\nu)}$ and T). It may be emphasized, however, that this assumption is not essential for the following considerations. It only simplifies the discussion on plastic anisotropy which, for polycrystalline materials, is usually the dominant phenomenon.

Concerning the reversible volume changes inherent in solid phase transformations, we have two possibilities of treatment: a) we may consider these volume changes as pseudo-elastic (as usually done in the theory of so-called memory-alloys, see e.g. [3]): in this case the reversible work done during solid phase transformations belongs to $\underset{(e)}{w}$. b) we may also consider these volume changes as reversible inelastic deformations, the respective work forming a part of $\underset{(h)}{w}$. The first interpretation is sometimes more convenient, since the reference configuration for the determination of ϵ_k^i remains unchanged by solid phase transformations.

$$\frac{1}{\overset{\circ}{\rho}} s_k^r \underset{(e)}{q_r^i} = \frac{\partial \varphi}{\partial \underset{(e)}{\epsilon_i^k}} \qquad (25)$$

holds true and represents the thermic state equation which correlates s_k^i, $\underset{(e)}{\epsilon_k^i}$ T and $p_{(\nu)}$. We may assume in our case that this relation is an isotropic function; otherwise we cannot expect the thermic state equation to be independent of the inelastic deformations. We therefore focus our considerations on the evolution of anisotropy to the second term of the expression for the free energy. Once more we emphasize that this is not an essential assumption of our considerations.

For simplification we restrict ourselves in the next step to processes without phase transformations and assume that the free energy can be specified as

$$\varphi = \underset{(e)}{\varphi} \{\epsilon_k^i, T\} + \underset{(h)}{\varphi^{\bullet}}\{T, b, \beta_k^i, B_{ks}^{ir}\} = \underset{(e)}{\varphi} \{\epsilon_k^i, T\} + \tag{26}$$

$$+ f(T) + g(b) + h(B) \quad \text{with} \quad B = B_{ks}^{ir} \beta_i^k \beta_r^s \quad \text{and} \quad B_{ks}^{ir} = B_{sk}^{ri}.$$

This approach implies that the free energy stored in the lattice defects does not depend on the temperature T. This assumption may be justified as a first approximation, remembering that also in thermoelasticity the coupling term between $\underset{(e)}{\epsilon_k^i}$ and T has no important relevance for many problems. Substituting (26) into (22) we obtain

$$\underset{(h)}{\dot{w}} + \underset{(d)}{\dot{w}} + \dot{q} = c_v \dot{T} - T \frac{\partial^2 \underset{(e)}{\varphi}}{\partial \underset{(e)}{\epsilon_k^i} \partial T} (\dot{\epsilon})_{.k}^i \tag{27}$$

$$+ g'(b) \dot{b} + h'(B) \underbrace{\{2B_{ks}^{ir} \beta_i^k \beta_r^s|_0 + \beta_i^k \beta_r^s B_{ks}^{ir}|_0\}}_{\dot{B}}.$$

Assuming that $\dot{b}, \beta_k^i|_0$ and $B_{ks}^{ir}|_0$ do not depend on $\dot{T}$ and $(\dot{\epsilon})_{.k}^i$, this equation can be split into

$$\underset{(h)}{\dot{w}} - \dot{\eta} = g'(b) \dot{b} + h'(B) \{2B_{ks}^{ir} \beta_i^k \beta_r^s|_0 + \beta_i^k \beta_r^s B_{ks}^{ir}|_0\} \tag{28a}$$

$$\dot{q} + \underset{(d)}{\dot{w}} + \dot{\eta} = c_v \dot{T} - T \frac{\partial^2 \underset{(e)}{\varphi}}{\partial \underset{(e)}{\epsilon_k^i} \partial T} (\dot{\epsilon})_{.k}^i. \tag{28b}$$

The next step concerns the definition of the evolution laws. Taking, as proposed, s_k^i and T as independent process variables, the increments of the conjugated external process variables $(d_k^i, \dot{q})$, of the internal variables $(\dot{b}, \beta_k^i|_0, B_{ks}^{ir}|_0)$, and of the dissipated energy $(\underset{(d)}{\dot{w}}, \dot{\eta})$ must be expressible as functions of the actual state (where $\underset{(e)}{\epsilon_k^i}$ can be replaced by s_k^i) and (for evolution laws of the rate insensitive equilibrium type) of the increments of the independent process variables $(s_k^i|_0, \dot{T})$. The evolution law for the elastic deformation can be derived from Eq. (26) using the relation (25). Differentiation with respect to time leads to an evolution law of the equilibrium type

$$\underset{(e)}{d_k^i} = \underset{(e)}{d_k^i}\{s_k^i, T, d_k^i; s_k^i|_0, \dot{T}\}. \tag{29}$$

For elastic isotropic materials this can be replaced in many cases by an hypo-elastic linear approximation [8]

$$\underset{(e)}{d_k^i} = \left\{\frac{1}{9K}(\dot{s}_r^r) + \alpha\dot{T}\right\}\delta_k^i + \frac{1}{2G}t_k^i|_0, \tag{30}$$

where

$$t_k^i = s_k^i - \frac{1}{3}s_r^r\delta_k^i \tag{31}$$

denotes the deviator of s_k^i.

Concerning the inelastic deformations we assume that the stress can be decomposed according to the different internal mechanisms into a plastic stress $\overline{s}_k^i$ (sometimes called: a-thermal stress) which operates independently of strain rate, and a viscous stress (vanishing for elastic-plastic materials) [13, 14, 18] putting

$$s_k^i = \overline{s}_k^i + (s_k^i - \overline{s}_k^i). \tag{32}$$

Furthermore we adopt the assumption that inelastic deformations occure only if a certain yield condition is fulfilled. In accordance with the expression (26) for the free energy it may be assumed that this yield condition takes the form

$$F(s_k^i, T, b, \beta_k^i, B_{ks}^{ir}) = B_{ks}^{ir}(t_i^k - c\beta_i^k)(t_r^s - c\beta_r^s) - \overline{k}^2(b, T)$$
$$= f^2 - \overline{k}^2 > 0 \tag{33a}$$

$$\overline{F}(\overline{s}_k^i, T, b, \beta_k^i, B_{ks}^{ir}) = B_{ks}^{ir}(\overline{t}_i^k - c\beta_j^k)(\overline{t}_r^s - c\beta_r^s) - \overline{k}^2(b, T)$$
$$= \overline{f}^2 - \overline{k}^2 = 0 \tag{33b}$$

This form of the yield condition represents a rather wide frame for the description of inelastic anisotropy [2, 8]. Following the usual assumptions, we may finally derive the evolution law for the inelastic deformations from the general approach

$$\underset{(i)}{d_k^i} = \dot{\lambda}\frac{\partial\overline{F}}{\partial\overline{s}_i^k} \tag{34a}$$

$$= \frac{\dot{\gamma}_0}{\sqrt{\overline{f}^2}}\Phi\left(\frac{f^2}{\overline{k}^2} - 1\right)\frac{\partial F}{\partial s_i^k}, \tag{34b}$$

where Φ denotes a suitable function of the given argument. The plastic stress $\overline{s}_k^i$ occuring in (33b) and (34b) does not represent an internal state variable; it has only the meaning of a dependent auxiliary process variable and can be eliminated by equating the expressions (34a) and (34b) [10]. With (29) or (30), respectively and (34a) the evolution law for the total strain is defined. The evolution law for the applied heat, the other conjugated external process variable, follows from (28b) as soon as the evolution law for the dissipated energy is formulated. Together with the definition of the evolution laws for the internal variables, this is the remaining task.

It can be verified that a complete compatible set of evolution laws for the internal variables and the dissipated energy can be formulated by the approaches

$$\dot{b} = \frac{\xi}{\overset{\circ}{\rho}\, g'(b)}\, \overline{s}_k^i\, \underset{(i)}{d_i^k} - \vartheta_{(1)}(T - T_R, \dots)\, b\,, \tag{35a}$$

$$B_{ks}^{ir}\, \beta_r^s\big|_0 = \frac{\zeta(1-\xi)}{2\overset{\circ}{\rho}\, h'(B)}\, \underset{(i)}{d_k^i} - \vartheta_{(2)}(T - T_R, \dots)\, \frac{1}{2}\, B_{ks}^{ir}\, \beta_r^s\,, \tag{35b}$$

$$\beta_r^s\, B_{ks}^{ir}\big|_0 = \frac{(1-\zeta)(1-\xi)}{\overset{\circ}{\rho}\, h'(B)}\, \underset{(i)}{d_k^i} - \vartheta_{(3)}(T - T_R, \dots)\, B_{ks}^{ir}\, \beta_r^s\,, \tag{35c}$$

$$\underset{(d)}{\dot{w}} = \frac{1-\xi}{\overset{\circ}{\rho}}\, (\overline{s}_k^i - c\beta_k^i)\, \underset{(i)}{d_i^k} + \frac{1}{\overset{\circ}{\rho}}\, (s_k^i - \overline{s}_k^i)\, \underset{(i)}{d_i^k}\,, \tag{35d}$$

$$\dot{\eta} = g'(b)\, b\, \vartheta_{(1)}(T - T_R, \dots) \tag{35e}$$
$$+ h'(B)\, B\, \{\vartheta_{(2)}(T - T_R, \dots) + \vartheta_{(3)}(T - T_R, \dots)\}.$$

$$\text{with}\quad \vartheta_{(r)}(T - T_R, \dots) \begin{cases} = 0 & \text{for } T \leqslant T_R \\ > 0 & \text{for } T > T_R\,. \end{cases}$$

When the material is isotropic in the original state then the initial conditions for the evolution of the internal parameters are

$$b(\overset{\circ}{t}) = \overset{\circ}{b} \qquad \text{(can be choosen equal to zero)} \tag{36a}$$

$$\beta_k^i(\overset{\circ}{t}) = \overset{\circ}{\beta}_k^i = 0 \tag{36b}$$

$$B_{ks}^{ir}(\overset{\circ}{t}) = \overset{\circ}{B}_{ks}^{ir} = \delta_s^i\, \delta_k^r\,. \tag{36c}$$

But we can also take into account a given anisotropy in the initial state using initial conditions derivating from (36b) and (36c). When B_{ks}^{ir} remains constant (equal to $\overset{\circ ir}{B}_{ks}$ according to (36c)) during the whole process, we obtain the known superposition of isotropic and kinematic hardening, but now in an extended sense, taking the coupling of different thermo-mechanical processes into account as well.

The approach (35a, e) characterizes the interaction between isotropic and anisotropic hardening due to inelastic deformations, on the one hand, and annealing by recrystallization and recovery at temperature above T_R, on the other hand. In this approach the parameters ξ, ζ, and the functions $\vartheta_{(\nu)}(T - T_R, \ldots)$ can still depend on the whole set of internal variables. Therefore it remains a wide frame for fitting this approach to the real material behaviour. But of course, the approach (35a, e) does not represent the only possibility for the formulation of the evolution laws. Even when we keep fixed the expressions (26) for the free energy and (33) for the yield condition, other formulations of the evolution laws are possible. Therefore the above approach (35a, e) serves rather as an example in order to show how a consistent formulation of the evolution laws can be found. It demonstrates at the same time that thermodynamics represents only a frame for the definition of the constitutive law; the constitutive law itself cannot be derived only from thermodynamic considerations.

From the equations (35a, e) we derive the evolution law for the work used for the change of internal structure.

$$\underset{(h)}{\overset{\bullet}{w}} = \frac{\xi}{\overset{\circ}{\rho}} \, \overline{s}^i_k \underset{(i)}{d^k_i} + \frac{1 - \xi}{\overset{\circ}{\rho}} \, c\beta^i_k \underset{(i)}{d^k_i} \, , \tag{37}$$

and the evolution law for the applied heat

$$\overset{\bullet}{q} = c_v \, \overset{\bullet}{T} - T \, \underset{(e)}{\frac{\partial^2 \varphi}{\partial \epsilon^i_k \, \partial T}} \, \underset{(e)}{(\overset{\bullet}{\epsilon})^i_{.k}} - \underset{(d)}{\overset{\bullet}{w}} - \overset{\bullet}{\eta} \, . \tag{38}$$

The transition to elastic-plastic bodies involves no difficulties. In identifying s^i_k with $\overline{s}^i_k$, we must, on the one hand, cancel the equations (33a) and (34a) and on the other hand, supplement the yield condition (33b) by a loading condition, which can be derived from (33b) and the evolution laws (35a, c) for the internal parameters.

The extention of our considerations to material in which the elastic behaviour is influenced by the inelastic deformations leads to no essential difficulties as already stated. In such cases it becomes, however, impossible to decompose the free energy into separated terms as assumed in Eqs. (24) and (26). The consequence is that the evolution laws become more complicated, but the general scheme of the considerations remains unaltered.

Larger difficulties arise when we want to take into account solid phase transformations. In this case we have to assume that the different solid phases have different states. Therefore we have to treat the intermediate states during solid phase transformations as a mixture of differently structured media. The situation becomes simplier only when all phases behave isotropically.

6. Some remarks on general thermo-mechanical processes in classical continua

General thermo-mechanical processes in classical solid bodies are governed by (for more details see [11])

(A) the general field equations, i.e. the balance equation for mass (implicitly fulfilled by the introduction of a body-fixed coordinate system), the balance equations for linear and angular momentum, the balance equations for internal energy and entropy;

(B) the constitutive law of the material consisting of the state function for free energy, the evolution laws for the internal state variables and the evolution laws for inelastic strain (in both cases including statements about the initial state), the laws for the flux of heat and other internal energy, the laws for entropy production;

(C) the history of the independent process variables, namely of the thermo-mechanical boundary conditions, the body forces and the energy sources (which both in some cases can also belong to the dependent process variables).

When we disregard the energy fluxes different from heat and corresponding the flux terms in the evolution laws for the internal variables, the evolution laws for the internal variables may be taken from the elementary processes. But even in this case, additional interaction phenomena occur concerning the evolution of anisotropy during inelastic deformations. This is due to the heat flux, which requires going back from Eqs. (22) to (21). Therefore the evolution of anisotropy becomes inhomogeneous, even when the mechanical boundary conditions would allow for homogeneous processes.

7. Some remarks on possible generalizations to non-classical continua

Until now we have assumed that the evolution of the anisotropy during inelastic deformations depends besides thermic influences uniquely on these deformations. But this cannot be true in every case; when, for instance, an anisotropically structured medium is deformed by pure slip processes, the orientation of the anisotropy remains unchanged although the body undergoes macroscopic inelastic shear strains. In such cases we have to introduce generalized (oriented) continua.

A first step in this direction consists in the introduction of a Cosserat-continuum [5, 12, 19]. The generalized strains and stresses in a Cosserat-continuum remain second order tensors. Therefore we can extend our therm-dynamical considerations to a Cosserat-continuum without essential difficulties. We have only to observe that the number of variables which determine the state of stress and strain increases correspondingly.

It may happen, however, that the freedom given by the introduction of a Cosserat-continuum is not sufficient to describe the evolution of anisotropy during inelastic deformation. We must then proceed to a generalized continuum as, for instance, used in [1]. The theory of such a continuum is characterized by the fact that, besides second order tensors, also third order tensors enter the description of strains and stresses. This complicates the theory of such a continuum seriously. Nevertheless we must face the fact that the descrip-tion of the evolution of anisotropy in a continuum may require the introduc-tion of such generalized continua.

REFERENCES

[1] ANTHONY K.H. – a) "Die Reduktion von nichteuklidischen Objekten in eine euklidische Form und physikalische Deutung der Reduktion durch Eigenspannung-stände in Kristallen." *Arch. Rat. Mech. Anal.*, 37 (1970): 161-180; b) "Die Theorie der Disklinationen." *Arch. Rat. Mech. Anal.*, 39 (1970): 43-88; c) "Die Theorie der nichtmetrischen Spannungen in Kristallen." *Arch. Rat. Mech. Anal.*, 40 (1971): 50-78.

[2] BALTOV A. and A. SAWCZUK. – "A rule of anisotropic hardening." *Acta Mech.*, 1 (1965): 81-92.

[3] DELAY L., R.V. KRISHNAN, H. TAS and H. WARLIMONT. – "Thermoplasticity, pseudoelasticity and the memory effects associated with martensitic transforma-tions." *Journ. Mat. Sci.*, 9 (1974): 1521-1555.

[4] GREEN A.E., P.M. NAGHDI. – "A general theory of an elastic-plastic continuum." *Arch. Rat. Mech. Anal.*, 18 (1965): 251-281.

[5] GUENTHER W. – "Zur Statik un Kinematik des Cosseratschen Kontinuums." *Abh. Braunschweig. Wiss. Ges.*, 10 (1958): 195-213.

[6] KESTIN J. – "On the application of the principles of thermodynamics to strained solid materials." *IUTAM Symp., Wien, 1966*, eds. H. Parkus and L.I. Sedov, Berlin: Springer, 1968, p. 177-212.

[7] KLUITENBERG G.A. – "Thermodynamical theory of elasticity and plasticity." *Physica*, 28 (1962): 217-232.

[8] LEHMANN Th. – "Einige Bemerkungen zu einer allgemeinen Klasse von Stoff-gesetzen für grosse elasto-plastische Formänderungen." *Ing. Arch.*, 41 (1972): 297-310.

[9] LEHMANN Th. – "On large elastic-plastic deformations." Paper contribution, Intern. Symp. on *Foundations of Plasticity, (Warsaw 1972)*, e. A. Sawczuk, Noordhoff Int. Publ., Leyden, 1973.

[10] LEHMANN Th. – "On the theory of large non-isothermic elastic-plastic and elastic-visco-plastic deformations." *Arch. Mech.*, 29 (1977): 393-409.

[11] LEHMANN Th. – "Some aspects of non-isothermic large inelastic deformations." *SM Archives*, 3, 2 (1978): 261-317.

[12] LIPPMANN H. – "Eine Cosserat-Theorie des plastischen Fliessens." *Acta Mech.*, 8 (1969): 255-284.

[13] NABARRO F.R.N. – *Theory of crystal dislocations.* Oxford: Clarendon press, 1967.

[14] PERZYNA P. – "Thermodynamic theory of viscoplasticity." *Advances in Appl. Mech.*, 11 (1971): 313-354.

[15] PHILLIPS A. – "Yield surfaces of pure aluminium at elevated temperatures." *Proc. of the IUTAM symposium on thermo-inelasticity*, East Kilbridge, 1968, Berlin: springer-verlag, 1970, pp. 241-258.

[16] RANIECKI B. and K. THERMANN. – "Infinitesimal thermoplasticity and kinematics of finite elastic-plastic deformations." *Mitt. Inst. Mech. RUB Bochum*, 2 1978.

[17] RICE J.R. – "Inelastic constitutive relation for solids: an internal-variable theory and its application to metal plasticity." *J. Mech. Phys. Sol.*, 19 (1971): 433-455.

[18] ROSENFIELD A. and G. HAHN. – "Numerical descriptions of the ambient low-temperature and high-strain rate flow and fracture of plain carbon steel." *Trans. Amer. Soc. Metals*, 59 (1966): 962-980.

[19] SAWCZUK A. – "On the yielding of cosserat continua." *Arch. Mech. Stosow.*, 19 (1967): 471-480.

[20] SIDOROFF F. – "The geometrical concept of intermediate configuration and elastic-plastic finite strain." *Arch. Mech.*, 25 (1973): 299-308.

[21] STOUFFER D.C. and A.M. STRAUSS. – "Principle of age-temperature shifting: a phenomenological theory of annealing". *Acta Mech.*, 27 (1977): 55-67.

[22] TING E.C. – "A thermodynamic theory for finite elastic-plastic deformations." *Z. Angew. Math. Phys., (ZAMP)*, 22 (1971): 702-713.

[23] ZIEGLER H. – "Plastizität ohne Thermodynamik ?" *Z. Angew. Math. Phys., (ZAMP)*, 21 (1970): 798-805.

RESUME

(Sur l'évolution de l'anisotropie au cours des processus thermodynamiques)

Basé sur la mécanique des milieux continus et la thermodynamique classiques, un cadre général est développé pour la description de l'évolution de l'anisotropie au cours des processus thermodynamiques associés aux grandes déformations inélastiques. Les considérations sont centrées sur l'interaction entre l'écrouissage dû aux déformations plastiques ou viscoplastiques d'une part, et le recuit par recristallisation et les phénomènes de recouvrance, etc. d'autre part. Quelques remarques sur des généralisations possibles sont incluses.

Session 4

Anisotropy of Metallic Polycrystals

Anisotropie des Polycristaux Métalliques

General Lecture : *Conférence Générale*

Relations entre Textures et Comportement Mécanique Anisotrope des Métaux

P. Parnière

I.R.S.I.D., Saint-Germain en Laye, France.

1. Introduction

De par certains procédés de transformation utilisés lors de leur fabrication (laminage, étirage, ... [2, 19]) la plupart des matériaux métalliques ont une structure anisotrope. Cette anisotropie structurale a, entre autres, comme conséquence une anisotropie des propriétés et caractéristiques mécaniques. Il peut donc être nécessaire d'en tenir compte dans les modèles rhéologiques et en particulier dans les modèles élasto-plastiques utilisés pour en décrire le comportement.

De plus parallèlement à cette anisotropie initiale (ou "anisotropie de formation" [9] la mise en œuvre du matériau et en particulier les déformations lors d'opérations de mise en forme (emboutissage, filage, ...) entraînent des modifications de structure et la création d'une anisotropie de déformation. Il peut aussi être nécessaire d'introduire cette évolution anisotrope dans les modèles de comportement des métaux.

Dans ce texte nous présentons, illustrés de nombreux exemples, quelques aspects de nos connaissances sur les anisotropies structurales et leurs relations avec le comportement anisotrope des métaux.

2. Divers types d'anisotropie structurale dans les métaux. Description

Dans les matériaux métalliques les causes d'anisotropie sont très nombreuses. Suivant l'échelle à laquelle on se place on peut distinguer entre :

a) les anisotropies morphologiques :

— anisotropie de la répartition des diverses phases constituant le métal. Un exemple en est la structure en bandes des aciers au carbone (Fig. 1).

— anisotropie de la forme des constituants des diverses phases, par exemple grains allongés dans les tôles minces (Fig. 2) ou inclusions de sulfure de manganèse dans les aciers (Fig. 3).

Figure 1. – *Structure en bande dans un acier 18 M 5 Nb brut de laminage à chaud. Photo prise dans la direction transverse DT (la ferrite est en blanc et la perlite en noir).*

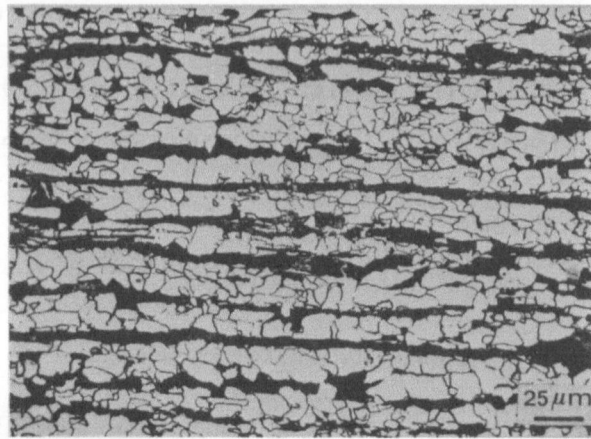

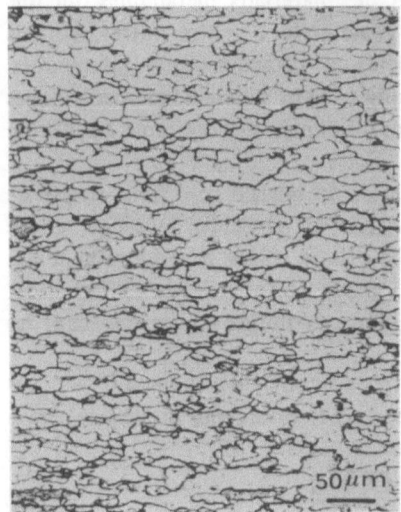

Figure 2. – *Grains allongés dans une tôle mince d'acier extra doux calmé à l'aluminium laminée à froid et recuite (photo prise dans la direction transverse DT).*

Figure 3. – *Sulfures de manganèse MnS dans une tôle d'acier A 52 brut de laminage à chaud.*

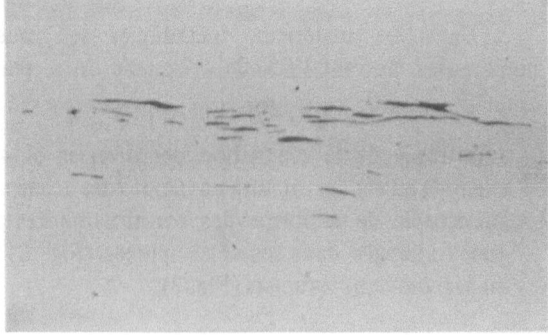

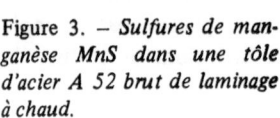

b) les anisotropies cristallographiques :
— présence d'orientations cristallographiques préférentielles (ou texture cristallographique) des grains (Fig. 4).
— présence de colonies de grains ayant des orientations cristallographiques voisines, par exemple bandes d'orientations dans des tôles d'acier inoxydable ferritique à 17 % de chrome (Fig. 5).

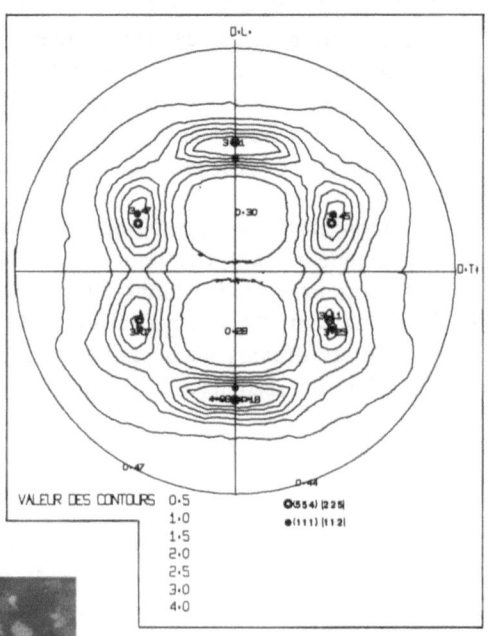

Figure 4. — *Figure de pôles {200} montrant la texture cristallographique d'une tôle d'acier extra doux contenant du niobium. Orientation préférentielle {554}⟨225⟩*.*

* Dans une tôle mince les orientations des grains sont généralement repérées par les indices de Miller {hkℓ} du plan cristallin parallèle au plan de laminage et par les indices ⟨uvw⟩ de la direction cristalline parallèle à la direction de laminage [43].

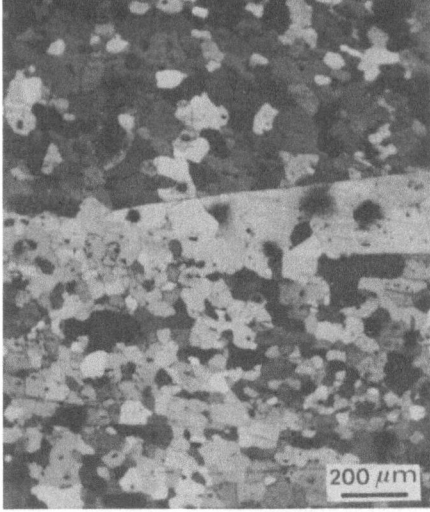

Figure 5. — *Bandes de grains ayant approximativement la même orientation cristallographique dans une tôle d'acier inoxydable ferritique à 17 % Cr. (Document communiqué par MM. Baroux et Kraemer — Centre de Recherches Métallurgiques d'Ugine-Aciers).*

2.1. *Description des anisotropies morphologiques.*

Ces anisotropies peuvent être décrites

— soit par l'évolution en fonction de l'orientation d'un ou des paramètres morphologiques caractéristiques de la structure,
— soit par un paramètre spécifique faisant plus ou moins référence à une forme géométrique simple (rectangle, ellipse, ...).

La définition de ces paramètres relève de la morphologie mathématique [54, 55, 26], leur détermination expérimentale de la métallographie quantitative [26]. Ces dernières années sous l'impulsion du Centre de Morphologie Mathématique de l'Ecole Nationale Supérieure des Mines de Paris[1] et de l'IRSID les progrès de ces méthodes et techniques ont été très importants.

2.1.1. *Description de l'anisotropie de la répartition des phases*

La grandeur essentielle pour décrire la répartition des phases X_i d'un solide est l'ensemble des fonctions covariances $C_{ij}(\vec{h})$ [54, 55, 26, 22, 16, 29]. Ces fonctions sont définies par la probabilité pour que l'élément structurant $\vec{h}$ (deux points distants de $\vec{h}$) soit tel que son premier point $\vec{x}$ appartienne à la phase X_i et son second $(\vec{x} + \vec{h})$ appartienne à la phase X_j :

$$C_{ij}(\vec{h}) = \Pr[\vec{x} \in X_i \ \ et \ \ (\vec{x} + \vec{h}) \in X_j] \tag{1}$$

Elles dépendent de l'orientation de l'élément structurant $\vec{h}$ dans le solide, et permettent donc de décrire l'anisotropie de la répartition des diverses phases le constituant. Par exemple sur la Figure 6 sont représentées les fonctions covariance de l'acier dont la structure micrographique est montrée sur la Figure 1 (structure en bandes) mesurées dans la direction de laminage DL et dans la direction normale DN. L'anisotropie de la structure est clairement mise en évidence par les différences entre les deux courbes : oscillations amorties dans la direction normale, croissance monotone dans la direction de laminage. Dans la direction de laminage les oscillations sont caractéristiques de la périodicité de bandes parallèles de ferrite et de perlite, et l'amortissement de ces oscillations est caractéristique d'écarts à cette périodicité [14]. Dans la direction de laminage la croissance monotone et lente vers l'asymptote montre toute absence de macrostructure (dans la limite des distances étudiées $h \lesssim 140 \ \mu m$).

Une "quantification" de l'anisotropie peut être obtenue en déterminant l'évolution en fonction de l'orientation de divers paramètres morphologiques déduits de ces courbes : pente à l'origine, portée, ... [22, 16]. Représentée

---

[1] 35, rue Saint Honoré – 77305 Fontainebleau (France).

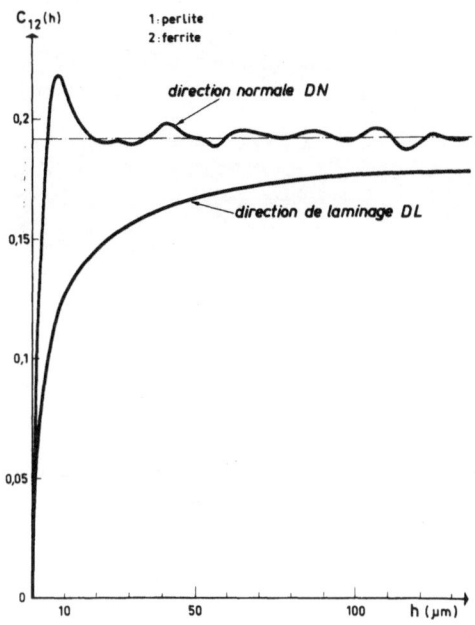

Figure 6. — *Graphes des fonctions covariances perlite-ferrite d'une tôle d'acier au carbone dans les directions normales et de laminage. Mise en évidence de la structure en bandes. D'après T. Hersant* [24].

en coordonnées polaires cette évolution peut être dans certains cas approximée par une courbe simple comme une ellipse dont l'excentricité e peut être prise comme paramètre quantifiant l'anisotropie [56]. Par exemple en utilisant comme paramètre morphologique la pente à l'origine des covariances, Chermant *et coll.* ont, pour des fontes blanches lamellaires (Fig. 7) ainsi déterminé une relation quantitative entre la vitesse de solidification, l'anisotropie de la structure (Fig. 8) [14] et les propriétés mécaniques [15].

2.1.2. *Description de l'anisotropie de forme des divers constituants d'une phase*

L'anisotropie des constituants individualisables est dans de nombreux cas décrite par un "facteur de forme" faisant référence explicitement ou implicitement à une forme géométrique anisotrope bien définie (ellipse, rectangle,...) [59, 41, 37]. Ces facteurs de forme sont des combinaisons des grandeurs spécifiques de base : surface A, périmètre L, courbure M,... [26]. Par exemple pour analyser l'influence d'additions de cérium sur la forme des inclusions de sulfures dans l'acier Mathy *et coll.* ont utilisé des facteurs de forme tels que $\dfrac{A}{L^2}$ en faisant référence explicitement à une ellipse et ont ainsi

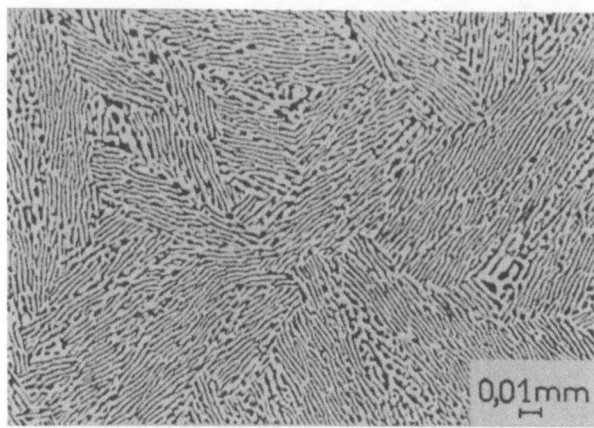

Figure 7. – *Structure d'une fonte blanche lamellaire dans la direction perpendiculaire au front de solidification. D'après Chermant et coll. [14].*

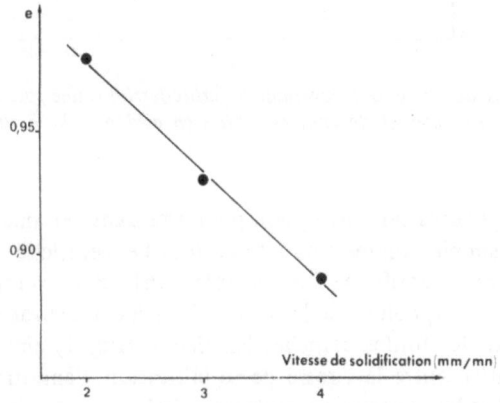

Figure 8. – *Relation entre la vitesse de solidification et l'anisotropie e de la structure d'une fonte blanche lamellaire. D'après [14].*

montré que l'addition de cérium "globulise" bien les sulfures [37] (Fig. 9). Mais il est important de noter que si pour une forme géométrique donnée (par exemple une ellipse), il existe bien une relation biunivoque entre le facteur de forme et l'anisotropie (Fig. 10), un même facteur de forme peut correspondre à des formes très différentes, certaines isotropes, d'autres anisotropes (Fig. 11). Il est donc préférable d'utiliser des paramètres d'anisotropie indépendants de la forme. Par exemple dans le cas des grains ferritiques tels que ceux représentés sur la figure 2 l'anisotropie de forme des grains (élongation

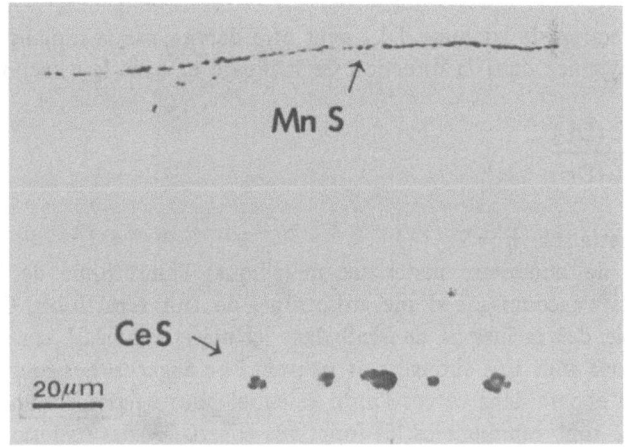

Figure 9. – *Inclusions de MnS et CeS dans un acier A 52 laminé. D'après Bernard et coll.* [*4*].

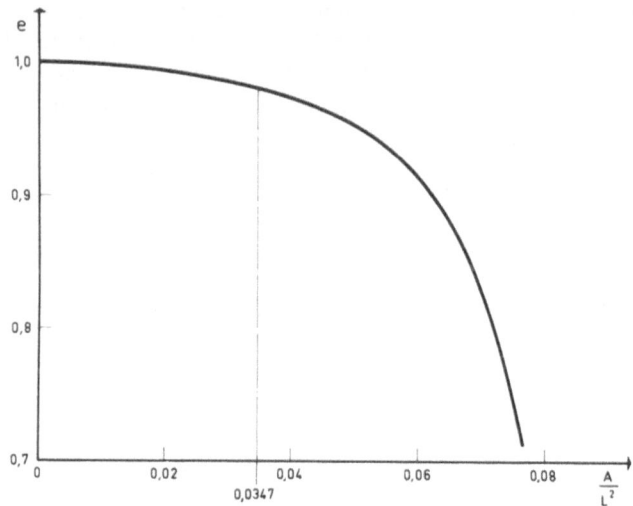

Figure 10. – *Relation entre le facteur de forme A/L^2 et l'excentricité e pour les ellipses.* *D'après C. Lafond* [*33*].

Figure 11. – *Exemples de formes dif-* *férentes ayant le même facteur de* *forme A/L^2.*

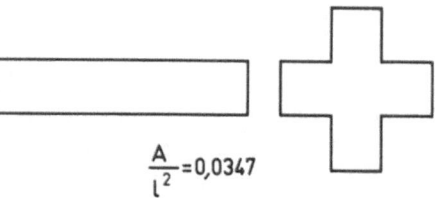

dans la direction de laminage DL) peut être décrite par le rapport η des traversées moyennes dans la direction de laminage et dans la direction normale

$$\eta = \frac{\overline{L}(DL)}{\overline{L}(DN)} \tag{2}$$

avec pour cette tôle $\eta = 2$.

Dans de nombreux matériaux métalliques l'anisotropie de forme des constituants s'accompagne d'une anisotropie de leur répartition. C'est le cas par exemple des inclusions de MnS dans les aciers laminés à chaud (Fig. 3). Les inclusions sont très allongées et peuvent être alignées. Les fonctions covariance sont encore dans ce cas l'outil essentiel pour décrire l'anisotropie. Sur la figure 12 sont représentées les fonctions covariance des inclusions de MnS mesurées dans la direction de laminage DL et dans la direction normale DN. On constate que dans la direction normale la portée de l'ordre de 7 μm correspond bien à l'épaisseur moyenne des inclusions de MnS. Par contre dans la direction de laminage la portée serait très grande, bien supérieure à la longueur moyenne des inclusions dans cette direction, ce qui traduit l'existence d'alignements d'inclusions. Hersant a montré que la longueur moyenne de ces alignements est la caractéristique morphologique principale de la structure inclusionnaire dans cette direction [23, 27].

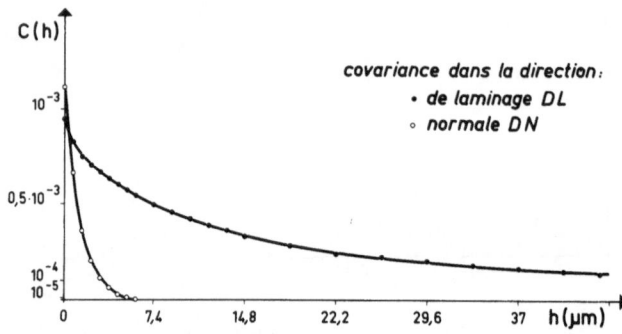

Figure 12. – *Graphes des fonctions covariances des inclusions de MnS d'une tôle d'acier A 52 mesurées dans les directions normale et de laminage. D'après T. Hersant [25] (mesures faites sur l'analyseur de texture TAS).*

Notons que le développement et la généralisation de l'utilisation de ces fonctions pour décrire les anisotropies morphologiques [36, 57, 30] est liée à la mise au point de l'analyseur de textures TAS qui permet de les mesurer automatiquement et correctement [32, 40, 34].

2.2. *Description des textures cristallographiques*

Dans un solide polycristallin la fraction volumique dV/V des grains ayant l'orientation cristallographique Ω (à dΩ) près[1] est donnée par la fonction de répartition des orientations $F(\Omega)$ [60, 10]

$$\frac{dV}{V} = F(\Omega)\, d\Omega \qquad (3)$$

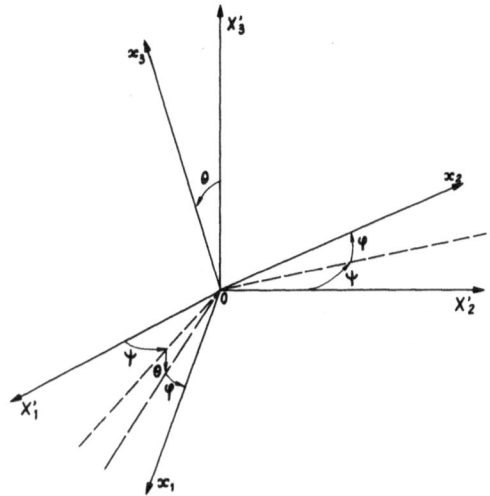

Figure 13. – *Définition des angles d'Euler (ψ, θ, Φ) précisant l'orientation du trièdre $Ox_1 x_2 x_3$ lié au cristal par rapport au trièdre $OX_1 X_2 X_3$ lié au solide.*

Cette fonction peut être calculée à partir de la détermination par diffraction des rayons X ou des neutrons de figures de pôles [43, 10, 52]. Elle peut être représentée :

– soit par son développement en série sur la base des fonctions harmoniques sphériques généralisées $T_\ell^{mn}(\Omega)$ [10, 52] :

$$F(\Omega) = \sum_{\ell=0}^{\infty} \sum_{m=-\ell}^{+\ell} \sum_{n=-\ell}^{+\ell} C_\ell^{mn}\, T_\ell^{mn}(\Omega) \qquad (4)$$

---

[1] L'orientation cristallographique Ω d'un grain est définie comme étant la rotation qui amène le trièdre de référence lié au solide $(OX_1X_2X_3)$ en coïncidence avec le repère lié au grain $(Ox_1 x_2 x_3)$. De nombreuses descriptions de cette rotation sont possibles [43]. On utilise généralement les angles d'Euler (ψ, θ, Φ) dont la définition est donnée sur la figure 13.

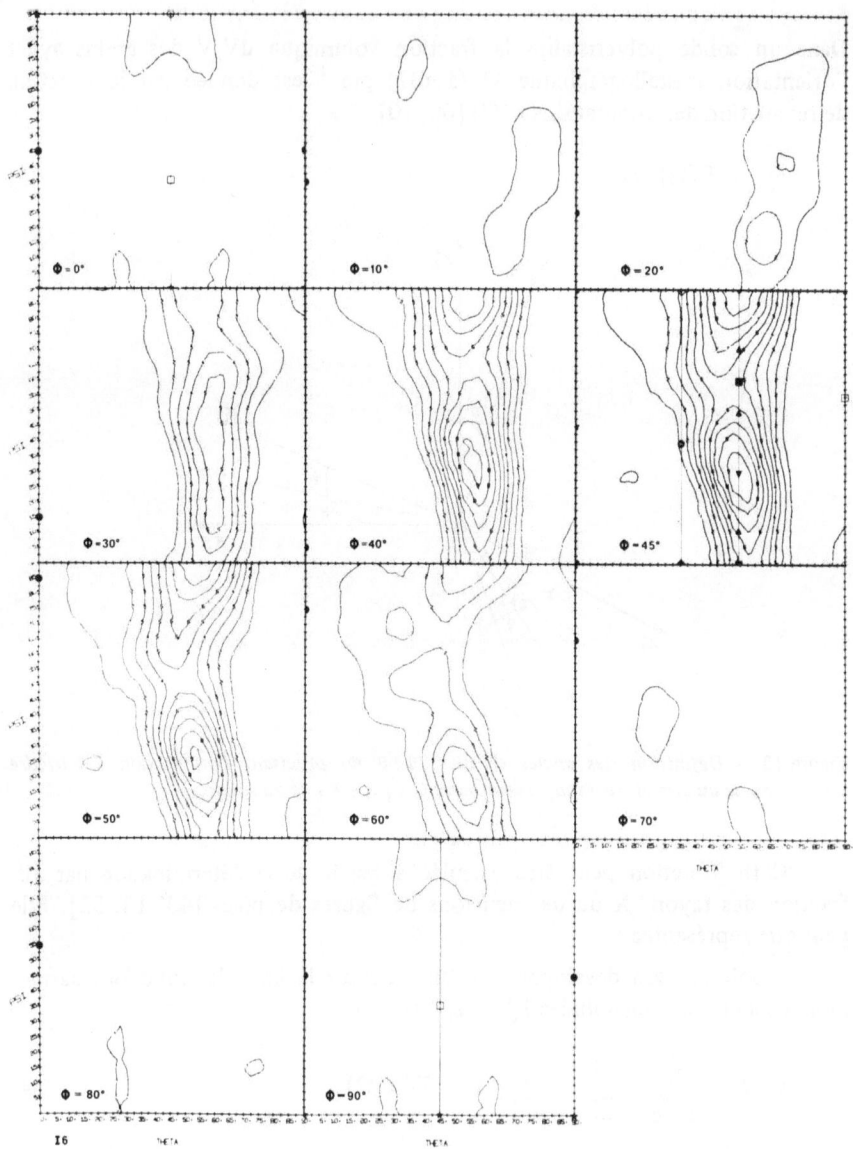

Figure 14. – *Diagramme* $F (\psi, \theta, \Phi = \Phi_i) = Cte$ *montrant la fonction de répartition des orientations d'une tôle mince d'acier extra doux. L'orientation cristallographique préférentielle principale est* $\{111\} \langle 110 \rangle$ ▼.

— soit graphiquement par l'intermédiaire de sections planes dans l'espace des orientations (Fig. 14).

— soit d'une manière plus synthétique sous forme de figures de pôles (Fig. 4) (une figure de pôles représente la distribution dans la solide des normales à un plan cristallin de type {abc}. Ces figures présentent l'avantage de pouvoir être mesurées assez facilement au moyen de la diffraction des rayons X ou des neutrons) [43].

La fonction de répartition des orientations $F(\Omega)$ ne permet de décrire que la seule texture cristallographique indépendamment de la position des grains, de leur taille,... Lorsque la prise en compte de ces paramètres est nécessaire, par exemple pour décrire les structures telles que celles représentées sur la figure 5 (bandes d'orientation) il est nécessaire d'utiliser les fonctions de corrélation des orientations [28].

La fonction de corrélation des orientations d'ordre n notée $C_n(\vec{h}_i, \Omega_i)$ est définie par la relation :

$$\frac{dV}{V} = C_n(\vec{h}_1 \ldots \vec{h}_i \ldots \vec{h}_{n-1}, \Omega_1 \ldots \Omega_i \ldots \Omega_n)\, d\vec{h}_1 \ldots d\vec{h}_{n-1}$$

$$d\Omega_1 \ldots d\Omega_n \quad (5)$$

avec Ω_i orientation cristallographique au point de coordonnées $\vec{X}_i$

$$\vec{h}_i = \vec{X}_{i+1} - \vec{X}_1 \quad \text{(Fig. 15)}$$

et dV/V fraction volumique du polycristal corrélée suivant C_n.

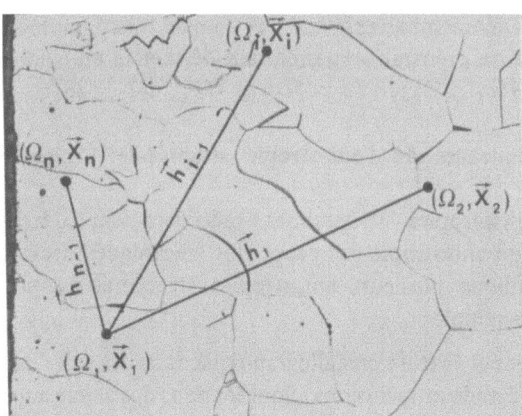

Figure 15. — *Corrélations des orientations en divers points d'une structure polycristalline. Définition des paramètres $\vec{h}_i$ et Ω_i.*

Les fonctions de corrélation des orientations d'ordres 1, 2 et 3 sont les seules mesurables facilement et sont d'ailleurs les seules pratiquement utiles [28].

La fonction de corrélation d'ordre 1 est définie par la relation :

$$\frac{dV}{V} = C_1(\Omega_1)\, d\Omega_1 \tag{6}$$

dV/V est la fraction volumique du matériau ayant l'orientation Ω (à $d\Omega$ près). Elle est donc identique à la fonction de répartition des orientations $F(\Omega)$ précédemment définie.

La fonction de corrélation des orientations d'ordre 2 est définie par la relation :

$$\frac{dV}{V} = C_2(\vec{h}_1, \Omega_1\,\Omega_2)\, d\vec{h}_1\, d\Omega_1\, d\Omega_2 \tag{7}$$

Cette relation met en évidence que cette fonction est identique à la fonction covariance $C_{ij}(\vec{h})$ introduite précédemment (§ 2.1.1) dans laquelle le rôle des deux phases i et j et joué par les orientations Ω_1 et Ω_2 (ou plus exactement les ensembles d'orientations $\{\Omega_1, d\Omega_1\}$, $\{\Omega_2, d\Omega_2\}$).

La fonction de corrélation des orientations d'ordre 3 est définie par la relation

$$\frac{dV}{V} = C_3(\vec{h}_1, \vec{h}_2, \Omega_1, \Omega_2, \Omega_3)\, d\vec{h}_1\, d\vec{h}_2\, d\Omega_2\, d\Omega_3 \tag{8}$$

Elle est encore mesurable sur des coupes micrographiques si on dispose d'une méthode de préparation permettant de mettre en évidence les ensembles d'orientations $\{\Omega_1, d\Omega_1\}$, $\{\Omega_2, d\Omega_2\}$ et $\{\Omega_3, d\Omega_3\}$ [28].

3. Conséquences de l'anisotropie structurale. Anisotropie mécanique.

Toutes les anisotropies structurales précédemment décrites ont comme conséquence une anisotropie des propriétés mécaniques. Généralement dans un matériau métallique plusieurs anisotropies structurales sont présentes simultanément ; par exemple :

— grains allongés et texture cristallographique dans des tôles minces,
— structure en bande et inclusions allongées dans des aciers au carbone.

Leurs effets sur l'anisotropie des propriétés mécaniques se combinent, ce qui en rend l'analyse très complexe. Seuls quelques cas relativement simples ont fait l'objet d'études approfondies et seront présentés ici :

1) l'influence de la texture cristallographique sur l'anisotropie mécanique des matériaux polycristallins monophasés,

2) l'influence de la structure en bandes et de l'anisotropie des inclusions sur l'anisotropie de la ductilité des tôles épaisses.

3.1. *Relations texture cristallographique – anisotropie mécanique*

3.1.1. *Caractérisation de l'anisotropie mécanique.*

La quasi totalité des études ont été faites dans le cas des tôles minces (épaisseur faible devant les autres dimensions, symétrie orthotropique). L'anisotropie des propriétés mécaniques d'une tôle mince est caractérisée par :

a) la variation dans le plan de la tôle des caractéristiques mécaniques : module d'élasticité, limite d'élasticité, résistance, coefficient d'écrouissage n. Des exemples sont donnés sur les figures 16 et 17.

b) l'anisotropie de la géométrie de déformation d'une éprouvette de traction, anisotropie décrite par le coefficient d'anisotropie $r(\epsilon_1 ; \alpha)$ (Fig. 18 et 19).

De nombreuses études empiriques (cf. [45 et 46]) ont permis de relier qualitativement cette anisotropie à la nature des orientations préférentielles présentes dans la tôle. La forme la plus sophistiquée de ces relations qualitatives a été l'établissement de corrélations entre les orientations préférentielles et les propriétés mécaniques. Un exemple d'une telle corrélation est montré sur la figure 20. Dans l'ensemble ces relations et corrélations sont très insuffisantes. La mise au point de méthodes quantitatives de description des textures au moyen de la fonction de répartition des orientations $F(\Omega)$ ont permis de développer des analyses plus approfondies.

3.1.2. *Analyse théorique des relations texture-anisotropie mécanique*

Cette analyse est en fait une généralisation des modèles de déformation des polycristaux isotropes aux polycristaux anisotropes dont la texture est connue par l'intermédiaire de la fonction de répartition des orientations $F(\Omega)$.

Dans un premier temps Bunge a généralisé le modèle de Taylor [58] aux tôles polycristallines anisotropes de matériaux de réseau de structure c.f.c. se déformant par glissement sur des systèmes $\{111\} < 110 >$ et de réseau c.c. se déformant par glissement sur des systèmes $\{110\} < 111 >$ [11, 13].

Les hypothèses de base du modèle de Taylor sont :

– les déformations élastiques sont nulles $E_{ij}^e = \epsilon_{ij}^e = 0$.

– chaque grain subit la même déformation plastique que le solide polycristallin, déformation supposée homogène : $\epsilon_{ij}^p = E_{ij}^p$.

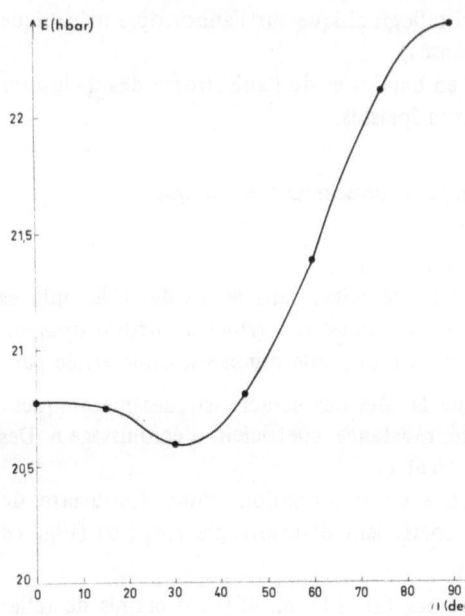

Figure 16. – *Variation du module d'élasticité d'une tôle mince d'acier extra doux en fonction de l'angle α entre la direction d'observation et la direction de laminage de la tôle.*

Figure 17. – *Evolution du coefficient d'écrouissage n (défini à partir de la loi de comportement σ = k εⁿ) en fonction de l'angle entre la direction de traction et la direction de laminage pour trois tôles d'acier extra doux.*

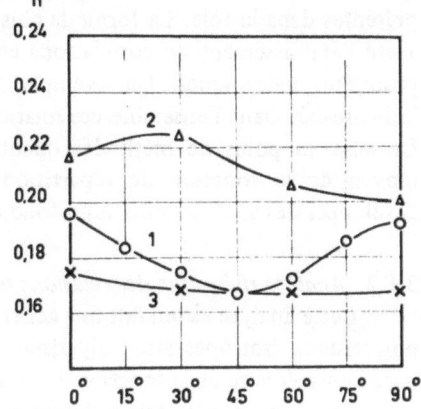

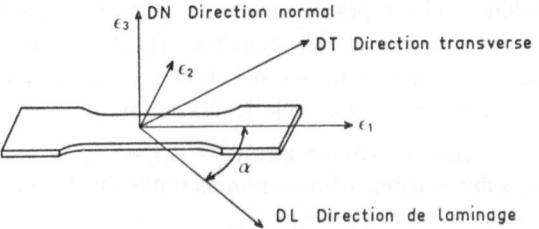

Figure 18. – *Définition du coefficient d'anisotropie d'une tôle mince :* $r(\epsilon_1 ; \alpha) = \epsilon_2/\epsilon_3$.

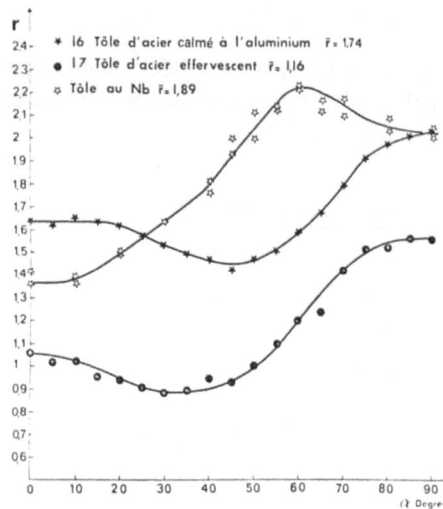

Figure 19. – *Evolution du coefficient d'anisotropie* $r (\epsilon = 0,2 ; \alpha)$ *pour trois tôles d'acier extra doux de composition chimique différente.*

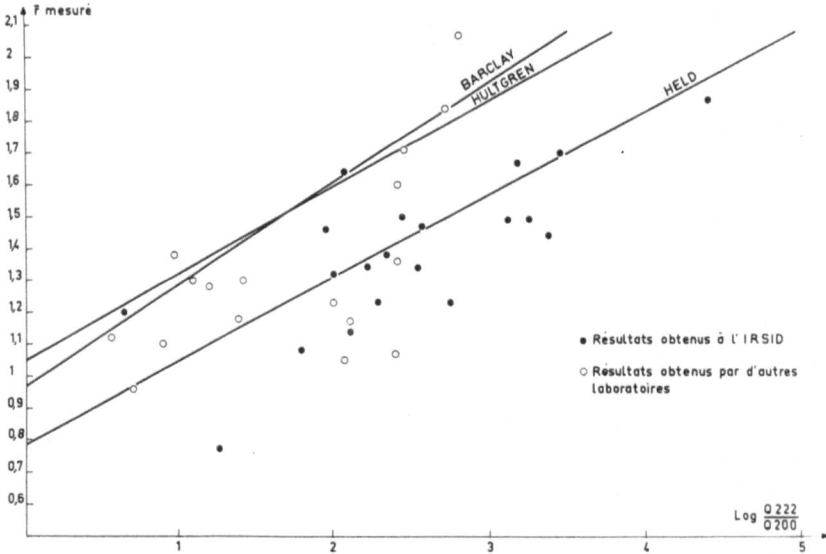

Figure 20. – *Corrélation entre le coefficient d'anisotropie* $\bar{r}$ *et les intensités relatives des orientations* $\{111\}$ *et* $\{100\}$ *dans les tôles minces d'acier extra doux. Comparaison des résultats obtenus à l'IRSID et dans divers autres laboratoires.*

Pour une déformation décrite par le paramètre $q = -\dfrac{E_2}{E_1}$ (E_i déformations principales du solide)[1] la déformation d'un grain d'orientation cristallographique Ω est caractérisée par le facteur de Taylor $M(q; \Omega)$ (proportionnel à la limite d'élasticité) :

$$M(q; \Omega) = \sum_p \frac{|s_p|}{E_1}$$

(s_p cisaillement sur les systèmes de glissement actifs).

Pour le polycristal ayant une texture décrite par la fonction de répartition des orientations le facteur de Taylor est donné par la relation

$$\overline{M}(q) = \int F(\Omega) \, M(q; \Omega) \, d\Omega$$

Le paramètre q^* caractérisant la déformation du polycristal est donné par la relation de minimisation

$$\left[\frac{d\overline{M}(q)}{dq} \right]_{q=q^*} = 0$$

La figure 21 montre l'anisotropie de la limite d'écoulement (pour $\epsilon = 0{,}001$) d'une tôle d'acier extra doux calmé à l'aluminium ainsi calculée comparée aux valeurs mesurées (les deux ensembles de valeurs sont mis en coïncidence pour $\alpha = 45°$).

Parnière et Roesch ont appliqué ce modèle au cas de matériaux de structure cubique centré se déformant par glissement non cristallographique multiple sur des plans de la zone $<111>$ [51]. La figure 22 montre l'évolution du coefficient d'anisotropie $r(\alpha)$ d'une tôle d'acier extra doux ayant une forte texture de composante principale $\{111\} <110>$ calculée avec cette hypothèse. On constate que la courbe calculée est nettement plus proche de la courbe expérimentale que la courbe calculée avec les hypothèses de Bunge, ce qui est normal compte tenu du plus grand réalisme de l'hypothèse du glissement non cristallographique multiple pour le fer et les aciers extra doux [47, 50]. Cependant malgré cette amélioration les valeurs calculées dans le cadre du modèle de Taylor sont encore relativement éloignées des valeurs expérimentales. Des constatations analogues ont été faites par tous les chercheurs ayant utilisé ce modèle pour

---

[1] Ce paramètre est relié au coefficient d'anisotropie r au moyen de la relation

$$r = \frac{q}{1-q}$$

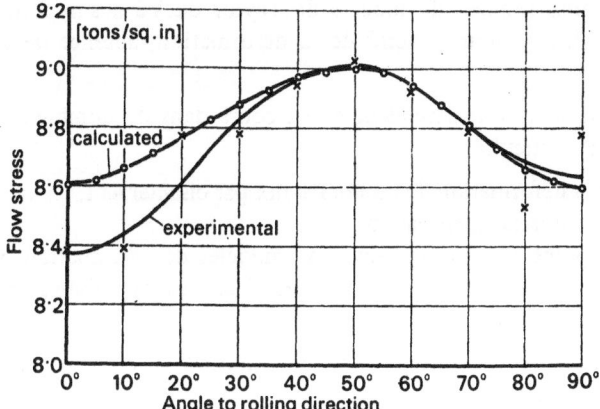

Figure 21. – *Courbes montrant l'évolution de la limite d'écoulement ($\epsilon_1 = 0{,}001$) d'une tôle mince d'acier extra doux calmé à l'aluminium, en fonction de l'angle entre la direction de traction et la direction de laminage. Comparaison des valeurs mesurées et des valeurs calculées au moyen de la théorie de Taylor. D'après Bunge [13].*

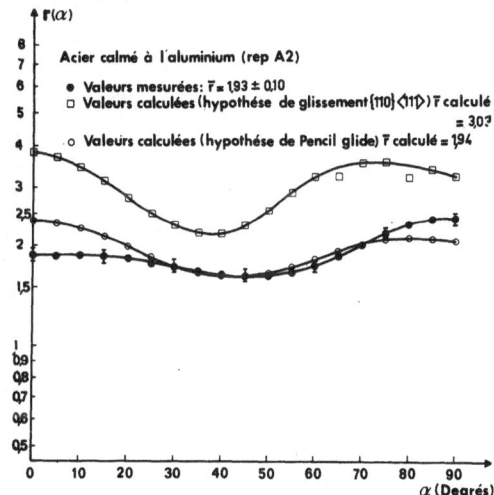

Figure 22. – *Courbes r(α) expérimentales et calculées par application de la théorie de Taylor. Tôle d'acier extra doux calmé à l'aluminium. D'après [49].*

l'aluminium [20], le cuivre [31], le laiton [31] et le nickel [12]. Les écarts sont encore plus grands dans le cas des matériaux de structure hexagonale [17]. Les causes de ces écarts sont multiples [48]. Les principales sont :

a) une mauvaise connaissance des mécanismes de déformation des grains (nature des systèmes de glissement possibles, valeurs des cissions critiques, durcissement) et une mauvaise prise en compte dans les modèles.

b) les insuffisances du modèle de Taylor dues à des hypothèses simplifi-
catrices irréalistes : homogénéité de la déformation, absence de déformations
élastiques.

Des travaux sont actuellement en cours dans de nombreux laboratoires
pour lever ces difficultés :

— étude de la déformation de monocristaux par cisaillement.
— mesures de durcissement latent.
— mise au point et mise en œuvre de modèles de déformation plastique des
  polycristaux permettant de lever certaines des hypothèses restrictives du
  modèle de Taylor [8, 7].

Au cours de ce colloque une présentation de ces travaux est faite par
Berveiller et Zaoui,

D'importants progrès sont donc à prévoir dans l'analyse théorique des
relations entre la texture et l'anisotropie mécanique au cours des prochaines
années.

3.2. *Influence de la structure en bandes et de l'anisotropie des inclusions sur l'anisotropie de la ductilité des tôles épaisses*

La figure 23 montre pour une tôle d'acier E 36 l'anisotropie croissante
de la striction Z^1 dans les directions transverse et normale avec l'augmentation
de l'anisotropie de la structure inclusionnaire. La présence d'une structure en
bandes a le même effet. Bien que ces deux causes d'anisotropie soient toujours
dans la pratique simultanément présentes, il semble que l'anisotropie de la
structure inclusionnaire soit la cause essentielle de l'anisotropie de la ductilité
entre les directions normale et transverse [21].

L'analyse théorique des relations entre ces anisotropies est très peu
avancée car les modèles décrivant l'influence de précipités ou d'inclusions sur
les propriétés de ductilité des métaux ont été établis pour des précipités et
inclusions isotropes et ne sont pas applicables au cas des inclusions allongées
[53, 35, 1]. Moussy et coll. ont montré que pour des tôles épaisses en acier au
carbone-manganèse ayant une structure inclusionnaire très anisotrope les
processus de croissance des défauts et donc les processus de déformation
plastique et de rupture sont très différents suivant la position de la direction de
traction dans la tôle [6].

---

[1] La striction Z est la réduction de section d'une éprouvette de traction déformée
jusqu'à la rupture.

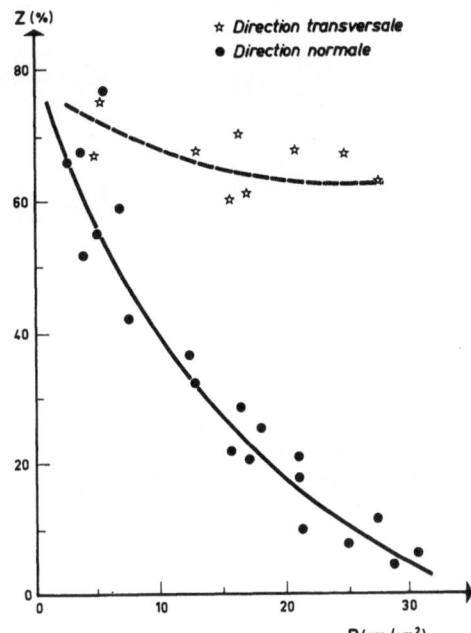

Figure 23. – *Relation entre l'aniso-tropie de la ductilité et l'anisotropie de la structure inclusionnaire. Acier E 36. D'après Bernard* et coll. [4].

Quelque soit la direction de traction une décohésion métal-inclusion apparaît dès le début de la déformation plastique autour de certaines inclusions de MnS. Mais :

— pour une traction dans la direction normale dès l'amorçage des trous la plasti-fication du métal se développe de maniére hétérogène (Fig. 24a). Cette structure hétérogène se conserve lorsque la déformation augmente. La zone B, la plus déformée se rompt et un premier stade de coalescence se poursuit entre inclusions coplanaires (Fig. 24b), créant des macrofissures. Ces macrofissures coalescent lors du stade de rupture finale, par cisaillement sur des plans orientés entre 0 et 45° par rapport à la direction de traction.

— pour une traction dans la direction transverse (ou dans la direction de laminage), dès l'amorçage des trous ceux-ci croissent perpendiculairement à la direction de traction (Fig. 25a) ; la coalescence se produisant suivant le schéma représenté sur la figure 25 b. Autour des inclusions la déformation est homogène.

Ces différences se retrouvent sur les faciès de rupture [6].

Ces résultats montrent que l'anisotropie de la structure inclusionnaire est un paramètre essentiel qui joue un rôle spécifique dans le mécanisme de la rupture ductile, rôle qui devra être pris en compte par les modèles. Leur interpré-

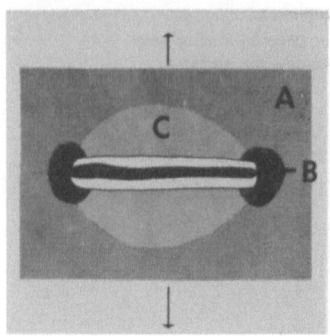
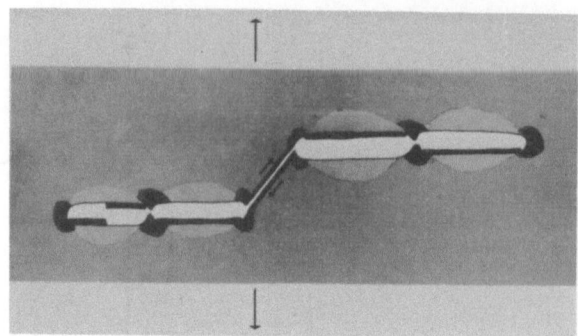

Figure 24. – *Schéma illustrant la déformation autour d'inclusions allongées. Traction dans la direction normale.*
Zone A : déformation uniforme de la matrice.
Zone B : déformation plus importante que celle de la matrice.
Zone C : de part et d'autre de l'inclusion déformation moins importante que celle de la matrice. D'après Moussy et coll. [6].

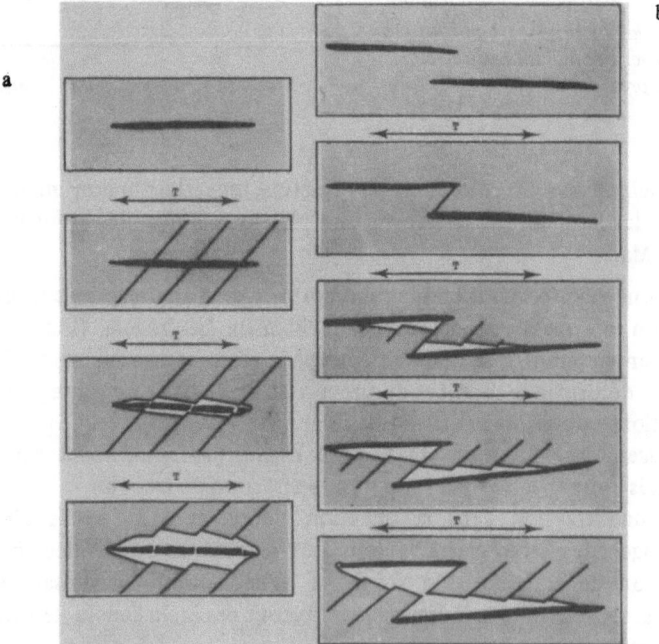

Figure 25a. – *Schéma illustrant la déformation autour d'une inclusion allongée. Traction dans la direction transverse.*
Figure 25b. – *Schéma illustrant la création de fissures par coalescence de cavités formées autour d'inclusions voisines. Traction dans la direction transverse. D'après Moussy et coll. [6].*

tation par Hersant et Moussy à l'aide d'une description quantitative de la morphologie de la structure inclusionnaire en est une première étape [23, 6].

4.   Evolution de l'anisotropie avec la déformation.

Toute déformation plastique d'un matériau métallique entraîne :

— un changement de forme des divers constituants du matériau (grains, inclusions,...),
— des rotations de ces constituants et en particulier des rotations du réseau cristallin,

ce qui se traduit par la création d'une anisotropie structurale ou la modification de l'anisotropie initiale.

4.1. *Exemples de l'influence de la déformation plastique sur l'anisotropie.*

4.1.1. *Changement de forme des grains au cours du laminage à froid de l'acier extra doux polycristallin* (Fig. 26).

Au cours du laminage les grains s'allongent dans la direction de laminage. Dans l'exemple présenté on avait initialement :

$$\eta = \frac{\overline{L}(DL)}{\overline{L}(DN)} = 1,1$$

après un laminage avec un taux de réduction de 33 % ($\epsilon_{DL} = -\epsilon_{DN} = 0,4$). L'anisotropie des grains a considérablement augmenté :

$$\eta = \frac{\overline{L}(DL)}{\overline{L}(DN)} = 2,4$$

Montjoie a montré que la déformation moyenne des grains caractérisée par ce changement de forme est égale à la déformation macroscopique du polycristal [39] (Fig. 27) (les écarts observés pour les grandes déformations sont essentiellement imputables aux difficultés qu'il y a à mettre en évidence micrographiquement les grains).

4.1.2. *Déformation des inclusions dans l'acier*

Suivant leur composition chimique, leur forme, la température à laquelle est faite la déformation, les inclusions se déforment plus ou moins lorsque l'acier est déformé plastiquement [38]. La figure 28 montre pour un acier de construction l'évolution de la longueur des inclusions de MnS en fonction du taux de

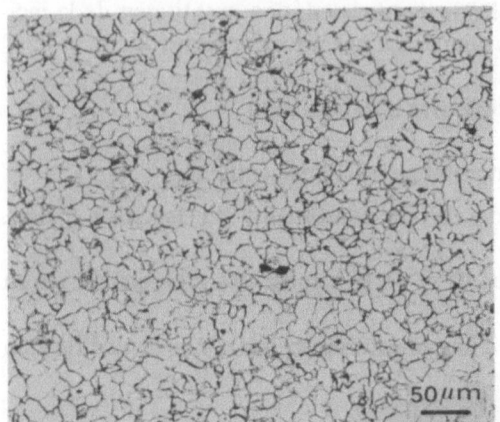

Figure 26. – *Evolution de la forme des grains au cours du laminage à froid d'acier extra doux polycristallin. D'après [39].*
a) avant laminage à froid, η = 1,1.
b) après laminage à froid – taux de réduction 33 %. η = 2,4.

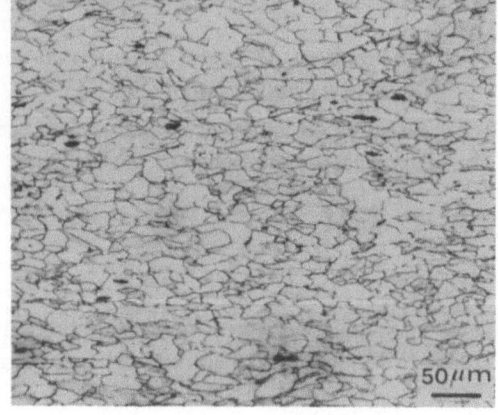

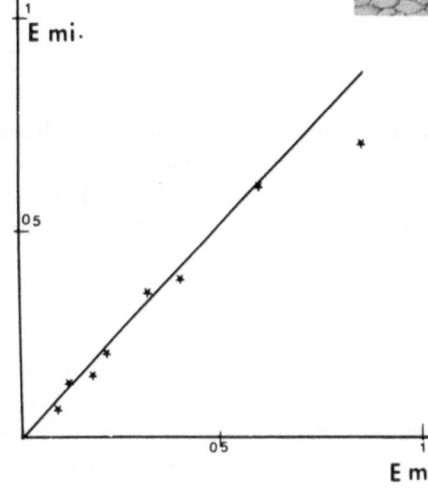

Figure 27. – *Relation entre la déformation du polycristal E_{ma} et la déformation moyenne des grains $E_{mi} = \frac{1}{2} Log \frac{\eta}{\eta_{init.}}$ au cours du laminage à froid d'acier extra doux. D'après [39].*

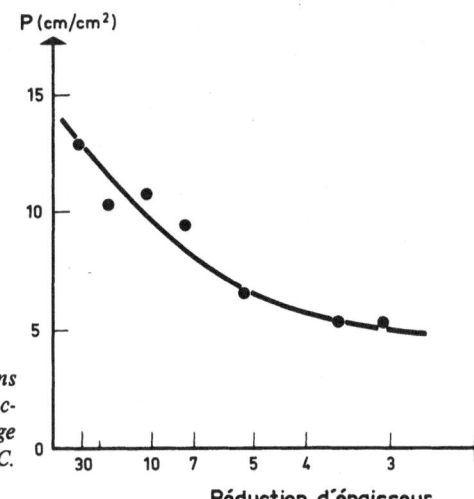

Figure 28. – *Longueur des inclusions dans la direction de laminage en fonction du taux de réduction. Laminage d'un acier de construction à 1150°C. D'après [5].*

réduction au cours d'un laminage à 1 150°C [5]. Compte tenu de son importance pratique ce problème de la déformation des inclusions dans l'acier a fait l'objet de nombreuses études et on dispose de nombreux résultats expérimentaux [38] (influence de la composition des inclusions, de la nature de la déformation, de la température, ...). En particulier, à la température ambiante, quelque soit leur nature les inclusions se déforment peu (cf. § 2.2), la morphologie et donc l'anisotropie de la structure inclusionnaire ne sont donc pas fondamentalement modifiées.

4.1.3. *Modifications de la texture cristallographique au cours d'une déformation plastique*

Les figures 30 et 31 montrent les textures cristallographiques d'une tôle polycristalline d'acier extra doux (dont la texture initiale est représentée sur la figure 29) après une déformation par traction dans la direction de laminage DL($\epsilon_1 = 0,26$, Fig. 29), et après une déformation par expansion biaxiale symétrique dans le plan de la tôle (Fig. 30) [3]. Ces exemples montrent que même des déformations plastiques relativement faibles conduisent à des textures très marquées et peuvent donc entraîner des modifications de texture importantes. Les textures de déformation dépendent :

— de la texture initiale,
— du type et de l'amplitude de la déformation.

Dans le cas des métaux du réseau cubique centré le modèle de déformation plastique des polycristaux de Taylor a permis une assez bonne analyse de la

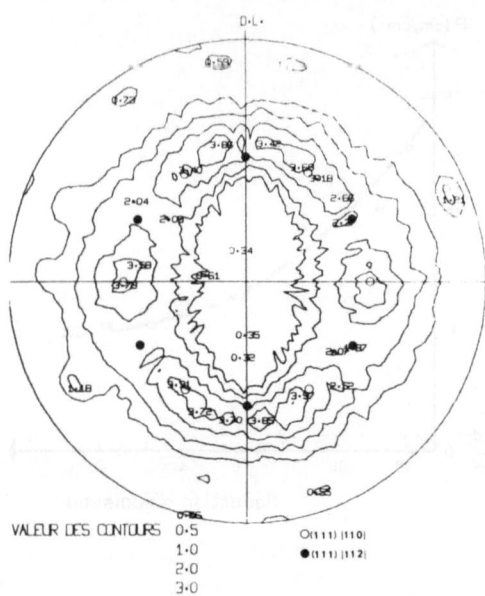

VALEUR DES CONTOURS   0·5
                       1·0
                       2·0
                       3·0
                       4·0

○(111) [110]
●(111) [112]

Figure 29. − *Figure de pôles {200} montrant la texture d'une tôle d'acier extra doux calmé à l'aluminium.*

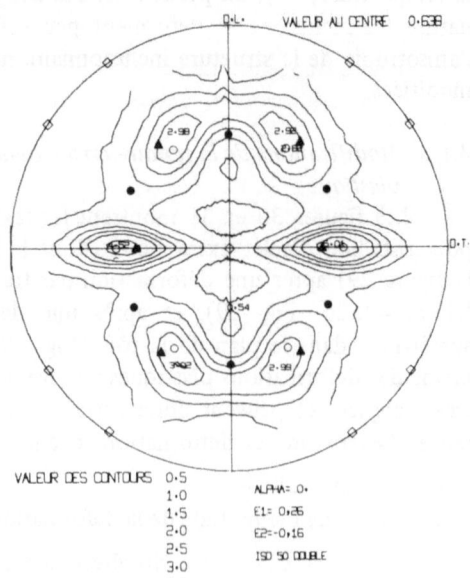

VALEUR DES CONTOURS   0·5
                       1·0
                       1·5
                       2·0
                       2·5
                       3·0
                       4·0

ALPHA= 0·
E1= 0·26
E2=-0·16
ISO 50 DOUBLE

Figure 30. − *Texture de la tôle 16 (Fig. 28) après un allongement en traction de 30 % dans la direction de laminage.*

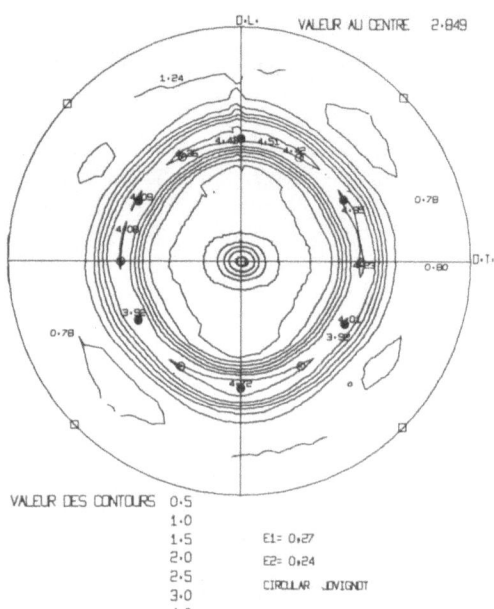

Figure 31. – *Texture de la tôle 16 après une déformation par expansion biaxiale symétrique.*

formation des textures de déformation [44]. A titre d'exemple sur la figure 32 sont représentées :

— la fonction de répartition des orientations d'une tôle polycristalline d'acier extra doux laminée à froid (taux de réduction 70 %) [11].

— les orientations finales de 648 grains, dont les orientations initiales étaient distribuées de manière homogène dans l'espace des orientations, calculées par Dillamore et Katoh pour une déformation identique (modèle de Taylor — glissement non cristallographique multiple) [18].

On constate qu'il y a assez bon accord entre la texture réelle et la texture calculée. Il n'en est pas de même pour les matériaux ayant un réseau cristallin cubique à faces centrées ou hexagonal. Là encore le recours à des modèles de déformation plastique plus réalistes est nécessaire.

5. Conclusions

Dans les matériaux métalliques les causes d'anisotropie sont très nombreuses. Ces anisotropies structurales et leurs conséquences sur l'anisopie mécanique ont fait l'objet de nombreuses études. Ces études ont été faites indépendamment les unes des autres et il se pose un difficile problème de synthèse

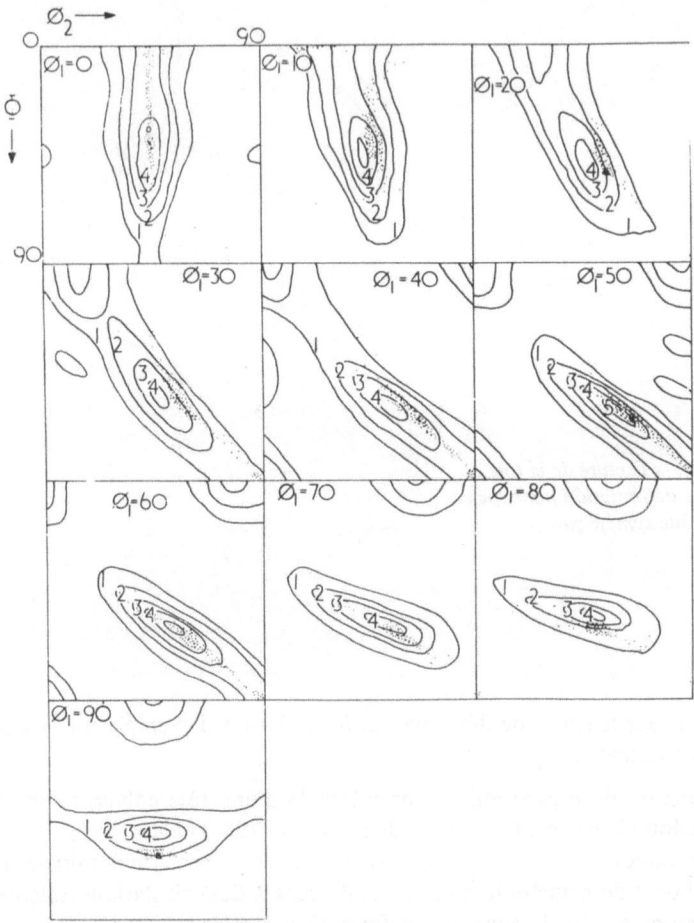

Figure 32. – *Diagrammes F (φ_1 = Cte, Φ, φ_2) = Cte, montrant la fonction de répartition des orientations d'une tôle d'acier doux. Comparaison avec la texture calculée (modèle de Taylor (hypothèse du glissement non cristallographique multiple). D'après Dillamore et Katoh [18].*

de ces travaux et résultats afin d'établir une théorie ou un modèle simple des relations entre anisotropie structurale et anisotropie mécanique.

Le développement d'une telle théorie nécessite une rationalisation des résultats expérimentaux. Une telle rationalisation devient maintenant possible avec le développement des méthodes de description quantitative des structures (morphologie mathématique – analyse quantitative des textures cristallographiques).

Mais un rôle essentiel revient aux théories de la plasticité des milieux continus anisotropes. Malgré leurs limites ces théories peuvent et doivent fournir le formalisme général permettant décrire la diversité des comportements anisotropes observés. Il est donc essentiel que s'établisse un lien solide entre les mécaniciens travaillant sur ces questions et les métallurgistes.

Sur un plan pratique ce travail de synthèse est très important car les anisotropies structurales et leurs conséquences sur l'anisotropie mécanique ont des conséquences très importantes sur les propriétés d'emploi des matériaux métalliques. Par exemple dans les tôles minces l'anisotropie mécanique induite par la texture cristallographique peut être à l'origine de deux effets contradictoires vis-à-vis de l'emboutissabilité des tôles :

— certaines textures entraînant une augmentation du coefficient d'anisotropie $\bar{r}$ conduisent à une amélioration très importante de l'emboutissabilité (profondeur d'emboutissage sans rupture) (Fig. 33) [45, 46].
— certaines textures sont à l'origine de la présence dans les emboutis de cornes d'emboutissage (Fig. 34), défaut qu'il est nécessaire d'éliminer par une opération d'usinage supplémentaire [45, 46].

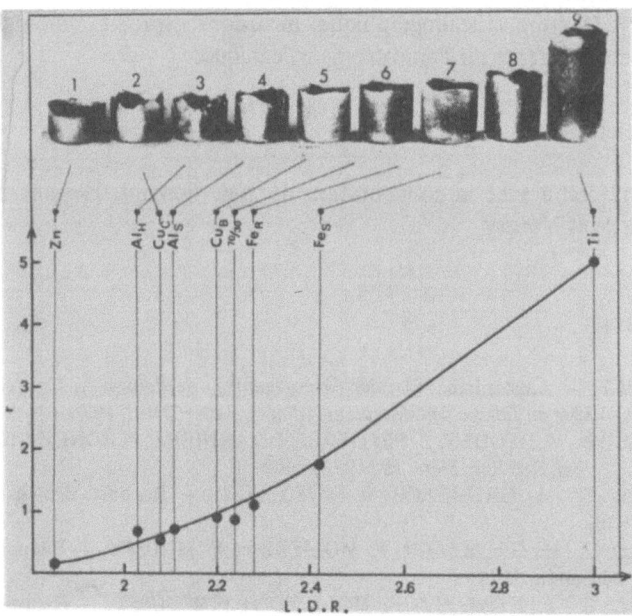

Figure 33. – *Evolution de l'emboutissabilité (profondeur des emboutis : LDR), en fonction du coefficient d'anisotropie moyen $\bar{r}$* [61].

Figure 34. – *Emboutis en coupelles montrant l'existence de cornes d'emboutissage (acier extra doux)* [46].

Il est donc nécessaire que les théories de la déformation plastique des milieux continus anisotropes puissent rendre compte de ces divers effets, en particulier lorsque plusieurs causes d'anisotropie structurale sont présentes simultanément (texture cristallographique, inclusions, structure en bandes, ...) ayant des effets différents sur l'anisotropie mécanique.

Remerciements

Ce texte a été établi avec la collaboration de MM. Bernard, Hersant, Lafond, Le Bon, Moliexe et Moussy.

REFERENCES

[1] ASHBY M.F. – Amsterdam : Elsevier *Strengthening Mechanism in Crystals,* 1970.
[2] BAQUE P. – *Mise en Forme des Métaux et Alliages.* Paris, CNRS, 1976.
[3] BAUDELET B., M. DEGUEN, L. FELGERES, P. PARNIERE, F. RONDE-OUSTAU et G. SANZ. – *Mem. Sci. Rev. Mét.,* 75 (1978) : 409.
[4] BERNARD G., M. GRUMBACH et F. MOLIEXE. – Rapport IRSID RE 220, octobre 1976.
[5] BERNARD G., M. GRUMBACH, F. MOLIEXE et F. MOUSSY. – Rapport IRSID RFP 185, décembre, 1978.
[6] BERNARD G., T. HERSANT, F. MOLIEXE et F. MOUSSY. – Rapport IRSID RE 595, janvier, 1979.
[7] BERVEILLER M. – Thèse. Université de Paris-Villetaneuse, 1979.
[8] BERVEILLER M. et A. ZAOUI. – *Proceedings of the 4th ICSMA Nancy* – France, 1 (30-08/03-09 1976) : 196.

[9] BOEHLER J.P. – Thèse. Université Scientifique et Médicale de Grenoble, 1975.

[10] BUNGE H.J. – Berlin (RDA): Akademie Verlag, *Mathematische Methoden der Texturanalyse*, 2 1969.

[11] BUNGE H.J. – *Kristall und Technik*, 5 (1970): 145.

[12] BUNGE H.J. – *Proceedings of the International Seminar*. Académie Cracovie (Pologne) : Polonaise des Sciences, "Quantitative analysis of textures". 1971.

[13] BUNGE H.J. et W.T. ROBERT. – *J. Applied Cryst.*, 9 (1969) : 116.

[14] CAMARD P., J.L. CHERMANT et M. COSTER. – *Mém. Sci. Rev. Mét.* (1978): 671.

[15] CAMARD P., J.L. CHERMANT et A. DESCHANVRES. – *Fonderie* (1978) : 384.

[16] CAUWE Ph. et J. SERRA. – *Fascicule de Morphologie Mathématique Appliquée*, 3. C.M.M. Fontainebleau, 1975.

[17] DERVIN P. – Thèse de Docteur-Ingénieur Faculté des Sciences d Orsay, 1978.

[18] DILLAMORE I.L. et H. KATOH. – *Metal Science*, 8 (1974) : 21.

[19] DOUSSAN G. – Mise en forme des métaux et alliages, Paris, CNRS, 1976.

[20] FERREIRA O. – Thèse de Docteur-Ingénieur. Faculté des Sciences d'Orsay, 1978.

[21] GRANGE R.A. – *Met. Trans.*, 2 (1971): 417.

[22] HAAS A., G. MATHERON et J. SERRA. – *Annales des Mines* (1976): 39.

[23] HERSANT T. – "Analyse quantitative des microstructures en sciences des matériaux, biologie et médecine". Dr. Riederer Verlag Gmbh, Stuttgart (RFA), 1978.

[24] HERSANT T. – *IRSID* St Germain en Laye. Communication personnelle. Résultats non publiés.

[25] HERSANT T. et D. JEULIN. – Rapport *IRSID P*, 241, 1975.

[26] HERSANT T., D. JEULIN et P. PARNIERE. – *Rapport IRSID RE*, 322, 1976.

[27] HERSANT T., C. LAFOND, F. MOLIEXE et P. PARNIERE. – *Rapport IRSID RE*, 574, 1978.

[28] HERSANT T. et P. PARNIERE. – *Rapport IRSID RE*, 557, 1978.

[29] JEULIN D. – Rapport final des travaux effectués dans le cadre de la convention CECA-IRSID 6210 AA/3/304. *Rapport IRSID RE*, 578, 1978.

[30] JEULIN D. – *Proceedings of the International Symposium on Quantitative Metallography*, Milan: AIM, 1978.

[31] KALLEND J.S. et G.J. DAVIES. – *J.I.M.*, 98 (1970): 242.

[32] KLEIN J.C. et J. SERRA. – *J. Microscopy*, 45 (1972): 349.

[33] LAFOND C., G. LECLERCQ, F. MOLIEXE, R. NAMDAR-IRANI, L. ROESCH et G. SANZ. – *Rapport IRSID RE*, 346, 1976.

[34] LEITZ T.A.S. – Document Leitz réf. 521-43 a.

[35] McCLINTOCK F. – *Ductility*, Metals Park, Ohio (USA): ASM, 1968.

[36] MARIAUX A., O. PERAY et J. SERRA. – *N.B.S. Special Publication*, 431. *Proceedings of the 4th International Congress for Stereology*, 1976.

[37] MATHY H., N. LAMBERT et T. GREDAY. – *AIM Ed. Milan (Italie)*. Proceedings of the International Symposium on Quantitative Metallography, 1978.

[38] MOLIEXE F. et M. WANIN. – *Rapport IRSID RFP*, 92, 1974.

[39] MONTJOIE M. – (IRSID). Résultats non publiés.

[40] MULLER W. et SERRA J. – *Leitz Scientific and Technical Information*. Suppl. I, 4 (1974): 101.

[41] ONDRACEK G. – *NBS Special Publication*, 431. Proceedings of the 4th International Congress for Stereology, 1976.

[42] PARNIERE P. – *Rapport IRSID RFP*, 245, 1978.

[43] PARNIERE P. – *Rapport IRSID RFP*, 86 bis, 1978.

[44] PARNIERE P. – *Mém. Sci. Rev. Mét.* (1978): 713.

[45] PARNIERE P. et G. POMEY. – *C.I.T. du C.D.S.*, 2 (1974): 423.

[46] PARNIERE P. et G. POMEY. – *Mécanique, Matériaux, Electricité* (1974): 1.

[47] PARNIERE P. et L. ROESCH. – *Rapport IRSID P,* **223,** 1973.

[48] PARNIERE P. et L. ROESCH. – *C.I.T. du C.D.S.* (1974): 2848.

[49] PARNIERE P. et L. ROESCH. – Compte rendu du contrat DGRST-IRSID, n° 72-7-0349. *Rapport IRSID RE,* **424** (1974).

[50] PARNIERE P. et L. ROESCH. – *Mém. Sci. Rev. Mét.* (1975): 221.

[51] PARNIERE P., L. ROESCH, M. GRUMBACH et C. SAUZAY. – *Mém. Sci. Rev. Mét.* (1975): 241.

[52] ROE R.J. – *J. Appl. Phys.,* **36** (1965): 2024.

[53] ROESCH L., G. HENRY, M. EUDIER et J. PLATEAU. – *Mém. Sci. Rev. Mét.,* **63** (1966): 927.

[54] SERRA J. – *Masson Ed. Paris.* "Traité d'Informatique Géologique", 1972.

[55] SERRA J. – "Lectures on image analysis by mathematical morphology", Fontainebleau: Publ. C.M.M., N 475, 1976.

[56] SERRA J. – *Fascicule de Morphologie Mathématique Appliquée,* 8, Fontainebleau: C.M.M., 1975.

[57] SERRA J. – *Image Analysis and Mathematical Morphology.* Academic Press, à paraître.

[58] TAYLOR G.I. – *J. Inst. Metals,* **62** (1938): 307.

[59] UNDERWOOD E.E. – *N.B.S. Special Publication,* **431,** Proceedings of the 4th International Congress for Stereology, 1976.

[60] VIGLIN A.S. – *Fiz. Tver. Tela,* **2** (1960): 2463.

[61] WILSON D.V. – *J.I.M.,* **94** (1966): 84.

ABSTRACT

Due to their fabrication processing, most metallic materials have an anisotropic structure. This paper shows, illustrated with many examples, some aspects of our knowledge on the structural anisotropies and their relations with anisotropic mechanical behaviour of metals.

It is divided into three parts :

– various types of structural anisotropies in metals. Description.
– consequences of the structural anisotropy. Mechanical anisotropy.
– influence of deformation on anisotropy.

The great variety of the examples shown put forward the necessity of a rationalization of the experimental results and the interest of a theory of the relations between structural and mechanical anisotropies. This can be made by a collaboration between metallurgists and specialists in the field of theoretical mechanics.

Etude de l'Anisotropie Élastique, Plastique et Géométrique dans les Polycristaux Métalliques

M. Berveiller et A. Zaoui
Laboratoire PMTM, Université Paris-Nord, Villetaneuse, France.

1. Introduction

Le présent article étudie, à partir d'une solution généralisée du problème de l'inclusion et de son utilisation dans le cadre d'un schéma self-consistent, certains aspects du problème général suivant : déduire et caractériser l'anisotropie des propriétés physiques d'un agrégat homogène à partir de celle de ses constituants et de la structure de l'agrégat. Nous limitant aux propriétés mécaniques des polycristaux en élastoplasticité, nous envisagerons comme sources d'anisotropie à l'échelle du grain celles de l'élasticité, supposée linéaire, et de la plasticité : les grains monocristallins sont astreints à ne se déformer plastiquement que par glissements sur des systèmes cristallographiquement déterminés, les glissements aux joints de grain étant exclus. La structure de l'agrégat sera caractérisée par une anisotropie "géométrique" (liée à l'écart à la sphéricité des grains, assimilés à des ellipsoïdes) et par une anisotropie cristallographique, attachée à la présence d'une texture.

Nous appliquerons la solution générale du problème de l'inclusion à l'explicitation quantitative des propriétés suivantes :

— Un état isotrope de contraintes appliquées (pression hydrostatique) peut provoquer, dans les grains à symétrie hexagonale d'un polycristal isotrope, des cissions telles que la limite élastique soit atteinte.

— Si, à elle seule, la déformation élastique d'un polycristal n'induit qu'un changement négligeable de texture, il en va autrement de l'écoulement plastique. Pour des grains sphériques et à élasticité isotrope, c'est la rotation plastique seule qui est la cause directe de modification de la texture (mesurée par les rotations élastiques) ; en cas d'anisotropie élastique ou géométrique (grains non sphériques) celle-ci dépend également de la déformation plastique.

— Cette évolution de texture est responsable du développement de l'anisotropie plastique. L'utilisation d'une approximation isotrope de l'accommodation plastique intergranulaire, à la place du modèle de Kröner d'accommodation élastique, permet, dans la prévision des textures et de l'anisotropie plastique, une meilleure prise en compte des caractéristiques intracristallines, notamment de l'anisotropie de l'écrouissage.

2. Problème de l'inclusion et applications

Le problème classique de l'inclusion ellipsoïdale, à élasticité isotrope ou non, déformée plastiquement de façon uniforme ainsi que la matrice infinie dans laquelle elle est située, peut être formulé de façon telle qu'il couvre également le cas d'une matrice à élasticité anisotrope et permette la prise en compte explicite des rotations élastiques et plastiques, en vue d'application aux problèmes de formation des textures de déformation dans les polycristaux. On peut notamment, à partir de là, mettre en évidence la possibilité de plastification sous pression hydrostatique d'un polycristal isotrope constitué de grains à symétrie hexagonale, ainsi que les différentes sources de rotation élastique (et donc de modification de texture) dans les grains d'un polycristal en écoulement plastique.

2.1. *Principe de calcul et résultats*

L'inclusion (V), de constantes élastiques C^I, est située dans une matrice infinie de constantes C^o. Nous désignons par β^{pI} (resp. β^{po}) la partie plastique du gradient du déplacement total $\beta_{ji}^T = u_{i,j}$ de l'inclusion (resp. de la matrice).
    A l'aide de la distribution d'Heaviside $\delta_V^o(\vec{r})$ définie par

$$\delta_V^o(\vec{r}) = 0 \quad \text{si} \quad \vec{r} \notin (V)$$
$$\delta_V^o(\vec{r}) = 1 \quad \text{si} \quad \vec{r} \in (V)$$

on peut poser en tout point du milieu :

$$C(\vec{r}) = C^o + (C^I - C^o)\,\delta_V^o(\vec{r}) = C^o + \Delta C\,\delta_V^o(\vec{r}) \tag{1}$$

$$\beta^p(\vec{r}) = \beta^{po} + (\beta^{pI} - \beta^{po})\,\delta_V^o(\vec{r}) = \beta^{po} + \Delta\beta^p\,\delta_V^o(\vec{r}) \tag{2}$$

Le gradient du déplacement total β^T est la somme de la partie élastique β et de la partie plastique β^p et le champ de contraintes est lié par la loi de Hooke à β (en transformation infinitésimale). Les équations d'équilibre ($\sigma_{ij,i} = 0$) permettent alors de mettre les équations différentielles du problème sous la forme (avec $\delta_i(S) = -\delta_{,i}^o(\vec{r})$) :

$$C_{ijk\ell}^o\,u_{\ell,ki}' = f_{ij}\,\delta_i(S) \tag{3}$$

avec :

$$f_{ij} = -\Delta C_{ijk\ell} u'_{\ell,k} - \Delta C_{ijk\ell} \beta^o_{k\ell} + C^o_{ijk\ell} \Delta \beta^p_{k\ell} + \Delta C_{ijk\ell} \Delta \beta^p_{k\ell} \tag{4}$$

et

$$u'_{i,j} = u_{i,j} - \beta^o_{ji} - \beta^{po}_{ji} \tag{5}$$

où β^o_{ji} désigne la partie élastique du gradient du déplacement total à l'infini. La solution de (3) peut être obtenue par l'intermédiaire du tenseur de Green G_{ij} pour le milieu infini de constantes C^o sous la forme (dans l'inclusion):

$$\beta^I_{k\ell} = P^I_{k\ell mn} (\beta^o_{mn} + (R^{oI}_{mn pq} - \delta_{mp} \delta_{nq}) \Delta \beta^p_{pq}) \tag{6}$$

avec

$$\delta_{ij} = \text{symbole de Kronecker}$$

$$R^{oI}_{mnk\ell} = -\int_V G_{jn,im} (r - r') \cdot C^o_{ijk\ell} \, dV'$$

$$P^I = (E + R)^{-1} \tag{7}$$

$$E_{mnk\ell} = \delta_{mk} \delta_{n\ell}$$

$$R_{mnk\ell} = -\int_V G_{jn,im} (r - r') \Delta C_{ijk\ell} \, dV'$$

β^I est uniforme dans l'inclusion car il en est ainsi de l'intégrale

$$\int_V G_{jn,im} (r - r') \, dV' \, [I].$$

Dans le cas de l'élasticité homogène et isotrope (μ, ν) et pour une inclusion sphérique, (6) permet de calculer les contraintes σ_{ij} dans l'inclusion. On obtient :

$$\sigma_{ij} = \Sigma_{ij} + 2\mu (1 - \beta) (E^p_{ij} - \epsilon^p_{ij}) \tag{8}$$

où Σ_{ij} est le champ de contraintes à l'infini

$$E^p_{ij} = \frac{1}{2} (\beta^{po}_{ij} + \beta^{po}_{ji})$$

$$\epsilon^p_{ij} = \frac{1}{2} (\beta^{pI}_{ij} + \beta^{pI}_{ji})$$

$$(1 - \beta) = \frac{7 - 5\nu}{15 (1 - \nu)} \cong \frac{1}{2}$$

Il faut noter que la formule (8) est précisément la loi d'interaction intergranulaire adoptée dans le modèle de Kröner (où E_{ij}^p est la moyenne des ϵ_{ij}^p), dont nous reparlerons plus loin. De même, (6) permet de calculer les rotations élastiques dans l'inclusion, rotations qui sont la mesure des rotations du réseau à l'origine de la formation des textures de déformation.

2.2. *Influence de l'anisotropie élastique sur la limite élastique de polycristaux isotropes constitués de grains à symétrie hexagonale.*

Certains essais [2] ont montré que la pression hydrostatique p (premier invariant scalaire de σ) avait une influence sensible sur l'écoulement plastique des polycristaux constitués de grains à symétrie hexagonale. Cette influence peut être reliée aux interactions élastiques entre les grains se développant du fait de leur anisotropie élastique. En partant de (6) et en supposant $\Delta\beta^p = 0$, on a

$$\beta_{k\ell}^I = P_{k\ell mn}^I \beta_{mn}^{\circ} \tag{9}$$

Si β_{mn}° est de la forme $\epsilon^{\circ}\delta_{mn}$, il n'en est pas forcément ainsi de β^I, de sorte que des cissions peuvent se développer dans l'inclusion et y activer du glissement plastique.

La déformation élastique ϵ_{ij}^I dans l'inclusion, supposée sphérique, dont l'anisotropie élastique a une symétrie hexagonale et située dans une matrice isotrope s'obtient à partir de (9) et (7) sous la forme

$$\epsilon_{33}^I = \epsilon^{\circ} \frac{1 + R_{11}^I + R_{21}^I - 2R_{31}^I}{(1 + R_{33}^I)(1 + R_{11}^I + R_{21}^I) - R_{13}^I(R_{31}^I + R_{11}^I)}$$

$$\epsilon_{22}^I = \epsilon_{11}^I = \epsilon^{\circ} \frac{1 - R_{13}\,\epsilon_{33}^I}{1 + R_{11} + R_{21}}$$

où R_{ij}^I et R_{ij} sont les composantes de $R_{ijk\ell}^I$ et $R_{ijk\ell}$ en notation matricielle de Voigt.

La cission maximale τ s'obtient alors à l'aide de

$$\tau = \frac{1}{2}\{\epsilon_{11}^I(C_{11}^I + C_{12}^I - 2C_{13}^I) + \epsilon_{33}^I(C_{13}^I - C_{33}^I)\} \tag{10}$$

où la notation de Voigt est aussi utilisée pour les constantes élastiques.

Dans le cas d'un polycristal de zinc, on trouve ainsi que :

$$\tau \simeq 0{,}0426\,P$$

où P est la pression hydrostatique appliquée.

Pour des valeurs habituelles de P, la valeur de τ ainsi calculée n'est donc pas négligeable par rapport à la cission critique qui est de l'ordre de quelques $10^{-2} \times$ daN/mm^2 et peut être suffisante pour déclencher le glissement plastique.

2.3. *Influence de l'anisotropie géométrique sur les textures de déformation plastique*

L'équation (6), dont la partie antisymétrique représente la rotation élastique ω^I, mesure de la rotation du réseau, montre qu'en général trois facteurs interviennent sur la formation des textures :

– la rotation plastique ω^{pI} (partie anitsymétrique de β^{pI})
– l'anisotropie et l'hétérogénéité élastiques (par l'intermédiaire de P^I et de R°)
– l'anisotropie de forme par l'intermédiaire du tenseur

$$- \int_V G_{jn,im}\, dV$$

Pour préciser l'influence de l'anisotropie de forme, nous supposons l'élasticité isotrope et homogène et l'inclusion de forme ellipsoïdale (axes principaux a, b = a, c). De plus nous supposons que $\beta^{po} \equiv \beta^\circ \equiv 0$.

La rotation élastique ω^I dans l'inclusion s'écrit alors d'après (6)

$$\omega^I_{ij} = - \omega^{pI}_{ij} + \Omega_{ij}$$

ω^{pI} étant la partie antisymétrique de β^{pI} et Ω_{ij} représentant l'influence de l'anisotropie géométrique ($\Omega_{ij} = 0$ dans le cas d'une inclusion sphérique).

A partir de (6), on obtient alors la dépendance de ω^I avec ϵ^p :

$$\Omega_{mn} = - R^{\circ I}_{\}mn} \{_{pq}\ \epsilon^{pI}_{pq} \tag{11}$$

En utilisant les notations de Bunge [3] qui repère les axes cristallographiques du cristal par rapport aux axes macroscopiques par les angles (ϕ_1, Φ, ϕ_2), on trouve pour $\Delta\phi_2$ les "corrections" d'anisotropie géométrique qui sont données sur la figure 1 dans le cas où il y a uniquement glissement sur le système primaire. Il apparait donc que cette "correction" est souvent du même ordre de grandeur que la valeur à laquelle elle s'applique, propriété pourtant négligée dans les prévisions courantes de textures, notamment par l'utilisation du modèle de Taylor qui, dans le cas présent, s'accompagnerait de la relation $\omega_{ij} = - \omega^p_{ij}$.

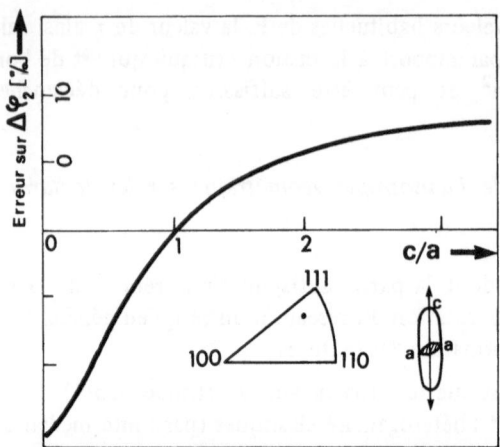

Figure 1. – *Correction d'anisotropie de forme : erreur relative sur* $\Delta\Phi_2$.

3.   Plasticité des polycristaux

Dans cette partie, nous reprenons la solution du problème de l'inclusion comme base d'un modèle self consistent pour la prédiction de la plasticité des polycristaux. Ce schéma self-consistent est celui de Kröner [4] : grâce à la façon dont nous avons retrouvé la formule de Kröner (formule 8), nous pouvons établir en quoi et pourquoi il est mal adapté à la prévision de l'écoulement plastique des polycristaux. Il repose en effet sur une hypothèse implicite d'accommodation purement élastique entre les grains. Pour introduire une part plastique dans l'accommodation entre les grains et, du même coup, limiter l'amplitude des contraintes internes telles qu'elles sont prévues par le modèle de Kröner, nous étudions ensuite le problème de l'inclusion dans une matrice en écoulement élastoplastique supposé obéir, pour simplifier, aux lois de Hencky-Mises. On arrive ainsi à introduire une fonction d'accommodation plastique directement mesurable à partir d'essais mécaniques simples et utilisable dans le cadre du schéma self-consistent. On applique le modèle obtenu à l'étude de la texture de déformation en traction des métaux C.F.C. et on indique une méthode pour calculer le coefficient de Lankford de polycristaux possédant une texture.

3.1. *Modèle self-consistent à accommodation élastique*

En plus de la loi d'interaction entre les grains qui est celle déduite en 2 (formule 8), l'élaboration d'un modèle self-consistent nécessite la représentation du comportement plastique du monocristal et de la structure du polycristal (texture, taille et forme des grains).

Nous limitant à la plasticité à froid et aux métaux C.F.C., nous repré-senterons le comportement plastique du monocristal par une matrice d'écrouis-sage H^{mn} reliant l'accroissement du glissement $d\gamma^n$ sur le système de glis-sement (n) à l'accroissement de la cission critique réduite $d\tau^m$ sur le système (m). En appelant R_{ij}^m les éléments de la matrice d'orientation reliant l'état de contrainte σ_{ij} à la cission réduite sur le système (m), on a si (n) est actif :

$$d\tau^m = H^{mn}\, d\gamma^n \quad \text{(écrouissage du système m)}$$

$$d\tau^m = R_{ij}^m\, d\sigma_{ij}$$

$$(12)$$

soit en négligeant l'effet de la rotation élastique sur la variation de la cission réduite, responsable d'un terme spécifique d'écrouissage "géométrique" :

$$H^{mn}\, d\gamma^n = R_{ij}^m\, (d\Sigma_{ij} + \mu(dE_{ij}^P - d\epsilon_{ij}^P)) \qquad (13)$$

où l'on a posé : $2(1 - \beta) \cong 1$ dans la formule (8).

L'accroissement de déformation plastique $d\epsilon_{ij}^P$ du grain considéré est relié à l'accroissement du glissement $d\gamma^n$ sur les systèmes de glissements actifs (n) par $d\epsilon_{ij}^P = R_{ij}^n\, d\gamma^n$, et on a donc pour (13) :

$$(H^{mn} + \mu R_{ij}^m\, R_{ij}^n)\, d\gamma^n = R_{ij}^m\, (d\Sigma_{ij} + \mu dE_{ij}^P) \qquad (14)$$

La relation (14) fait apparaitre les différentes sources d'écrouissage pour un système de glissement dans un grain du polycristal :

— l'écrouissage propre intracristallin, de matrice H^{mm} dont les termes sont typiquement de l'ordre de $\mu/200$.
— l'écrouissage intergranulaire de matrice $\mu R_{ij}^m\, R_{ij}^n$ provenant des contraintes internes de grain à grain et dont les termes sont de l'ordre de quelques fractions de μ ($\mu/2$, $\mu/6 \ldots$)

Cette décomposition montre que pour le modèle de Kröner le second terme est largement prépondérant par rapport au premier, les contraintes internes masquant alors les caractéristiques plastiques propres du monocristal.

L'utilisation d'un tel modèle [5] montre en effet qu'il faut augmenter artificiellement les termes H^{mn} jusqu'à des valeurs de l'ordre de $\mu/10$ pour obtenir des courbes de traction calculées compatibles avec les résultats expéri-mentaux. Cette insuffisance du modèle de Kröner provient essentiellement de l'hypothèse implicite, dans le problème de l'inclusion, d'une déformation plastique uniforme de la matrice, conduisant à un schéma d'accommodation purement élastique entre les grains du polycristal et donc à une surévaluation des contraintes internes. Pour limiter l'amplitude de ces contraintes internes en autorisant une part plastique dans l'accommodation des incompatibilités entre les grains, nous formulons un nouveau problème d'inclusion conduisant à un modèle self-consistent d'accommodation élastoplastique.

3.2. Modèle self-consistent à accommodation elastoplastique

Nous supposons que l'écoulement élastoplastique de la matrice peut se représenter par des relations analogues à celles de Prandtl et Reuss

$$d\epsilon_{ij}^T = C_{ijk\ell}^{0-1} d\sigma_{k\ell} + A_{ijk\ell} s_{k\ell} d\lambda \quad (\text{si } f = 0 \quad \text{et} \quad f_{hk} d\sigma_{hk} > 0) \quad (15)$$

f étant le potentiel plastique défini par

$$f = A_{ijk\ell} s_{ij} s_{k\ell} - k^2$$

et $\quad d\lambda = g f_{hk} d\sigma_{hk} \left(f_{hk} = \dfrac{\partial f}{\partial \sigma_{hk}} \right)$

$A_{ijk\ell}$ et k^2 représentent les paramètres de l'écrouissage et de l'écoulement plastique et s_{ij} le déviateur de σ_{ij}.

En définissant une contrainte équivalente $s_E = \sqrt{A_{ijk\ell} s_{ij} s_{k\ell}}$ et en supposant le chargement radial en tout point de la matrice et l'état d'écrouissage (supposé uniforme) représenté par un paramètre $\overline{h} = \dfrac{1}{s_E} \displaystyle\int_0^{s_E} g s^2 \, ds$, on peut écrire (15) sous la forme

$$\epsilon_{ij}^T = (C_{ijk\ell}^{0-1} + \overline{h} A_{ijk\ell}) \, \sigma_{k\ell} \equiv L_{ijk\ell} \, \sigma_{k\ell} \quad (16)$$

(16) se présente alors, pour un état plastique donné, comme une relation de comportement d'élasticité linéaire, anisotrope pour la déformation totale ϵ_{ij}^T de la matrice et les coefficients $L_{ijk\ell}$ sont homogènes dans toute la matrice et fixées par les conditions d'écrouissage à l'infini. On est alors ramené au problème non homogène de l'inclusion plastique dans une matrice "élastique" de compliance $L_{ijk\ell}$ et soumise aux contraintes homogènes Σ_{ij} à l'infini. On obtient alors à partir de (6)

$$\sigma = \Sigma - L^{-1}(E + R^I)^{-1} (R^{0I} - E)(E^p - \epsilon^p) \quad (17)$$

Dans la suite, nous nous limitons à une approximation isotrope des interactions entre inclusion (sphérique) et matrice, le comportement élastoplastique de cette dernière, caractérisé par le tenseur L, étant, dans l'établissement de la loi d'interaction, supposé isotrope. On trouve alors [6] :

$$\sigma_{ij} = \Sigma_{ij} + 2\mu\alpha(1 - \beta) (E_{ij}^p - \epsilon_{ij}^p) \quad (18)$$

avec

$$
\alpha = \frac{1 + 6\mu\bar{h}\ \dfrac{1 + \nu}{7 - 5\nu}}{1 + 2\mu\bar{h}\ \dfrac{13 - 5\nu}{15(1 - \nu)} + 8\mu^2\,\bar{h}^2\ \dfrac{1 + \nu}{15(1 - \nu)}} \simeq \frac{1}{1 + \mu\bar{h}} \tag{19}
$$

Dans le cas d'un essai de traction (Σ, E^p), on a pour α

$$
\alpha \simeq \frac{1}{1 + \dfrac{3}{2}\ \mu\ \dfrac{E^p}{\Sigma}} \tag{20}
$$

Dans (18), la forme du modèle self-consistent est sauvegardée mais l'introduction de la fonction d'accommodation plastique α limite les contraintes internes puisque, dans le cas d'un essai de traction par exemple, α est égal à 1 au début de l'écoulement plastique $(E^p \simeq 0)$ mais tend très rapidement vers des valeurs inférieures d'un ordre de grandeur au moins lorsque E^p croît.

L'équation (14) devient alors

$$
(H^{mn} + \alpha\mu R_{ij}^m R_{ij}^n)\, d\gamma^n = R_{ij}^m\, (d\Sigma_{ij} + \alpha\mu\, dE_{ij}^p) = R_{ij}^m\, dQ_{ij} \tag{21}
$$

et on voit qu'il n'est plus nécessaire d'augmenter artificiellement les termes H^{mn} puisque les coefficients $\mu R_{ij}^m R_{ij}^n$ sont maintenant divisés par un facteur de l'ordre de 10.

De plus, les résultats qu'on attend d'un tel modèle seront sensibles à l'anisotropie de la matrice H telle qu'elle peut apparaître lors d'essais de durcissement latent sur des monocristaux de cuivre ou d'aluminium [7].

3.3. *Application à la prévision du comportement plastique et des textures de déformation des polycristaux C.F.C.*

3.3.1. *Prévision des courbes de traction*

Dans le cas où, pour simplifier, on suppose α constant dans (18), on montre que, si l'on connaît la réponse du modèle de Kröner $(\alpha = 1)$ – soit $\bar{\Sigma}$, $\bar{E}^p$ pour $\bar{H}^{hg}$ –, celle du présent modèle s'en déduit par : $\Sigma = \bar{\Sigma}$, $E^p = \bar{E}^p/\alpha$ pour $H^{hg} = \alpha\bar{H}^{hg}$. Le résultat en est bien, globalement, une plus grande facilité d'écoulement plastique en même temps qu'une meilleure sensibilité aux caractéristiques de l'écrouissage intracristallin. On retrouve ainsi, beaucoup plus simplement, des prévisions conformes à celles du modèle de Hill et Hutchinson [8] qui résultent d'une méthode itérative complexe à

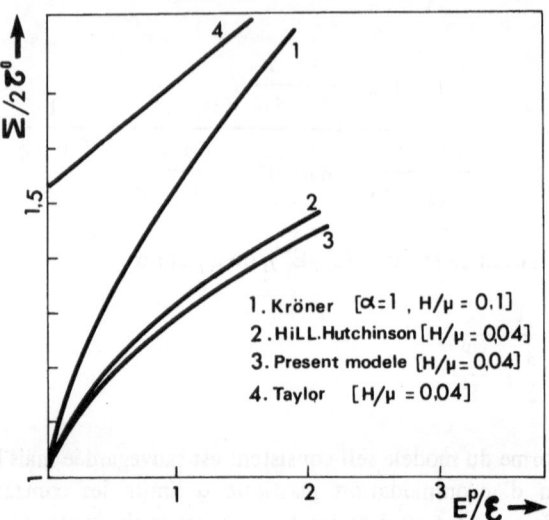

Figure 2. – *Courbes de traction pour H isotrope. Comparaison entre différentes prédictions théoriques.*

renouveler sur chaque nouveau problème. C'est ce qu'illustre la figure 2, dans le cas d'un écrouissage intracristallin isotrope, où sont comparées les prévisions de différents modèles.

3.3.2. *Prévision des textures de déformation en traction*

La figure 3 représente les changements d'orientation de l'axe de traction (par rapport aux axes $\langle 100 \rangle$, $\langle 111 \rangle$ et $\langle 110 \rangle$) déduits de (21) pour plusieurs valeurs de $\alpha\mu/H$, en prenant une matrice H isotrope. De même, la figure 4 indique les changements d'orientation lorsque H est anisotrope, (H_1 désignant le module d'auto-écrouissage et $H_2 > H_1$ celui de l'écrouissage latent), le rôle de l'anisotropie de H étant d'autant plus important que α est plus faible. Ces résultats, plus conformes aux observations expérimentales que ceux déduits du modèle plus classique de Taylor, confirme la réalité des mécanismes de relaxation d'accommodation plastique, malgré la simplicité de la représentation qui vient d'en être faite.

3.3.3. *Calcul du coefficient de Lankford en traction simple (principe)*

Lorsque le polycristal possède une texture initiale, la résolution des équations (21) se complique puisque la forme du tenseur $dQ_{ij} = d\Sigma_{ij} + \alpha\mu dE^p_{ij}$ n'est pas connue a priori.

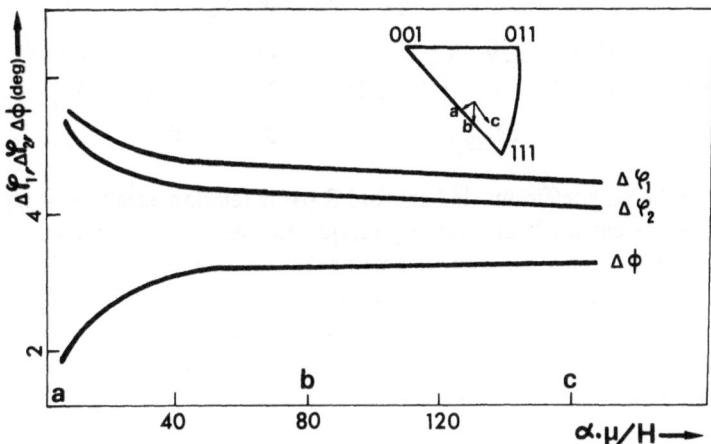

Figure 3. – *Changement d'orientation, pour un grain donné, en fonction de αμ/H (traction).*

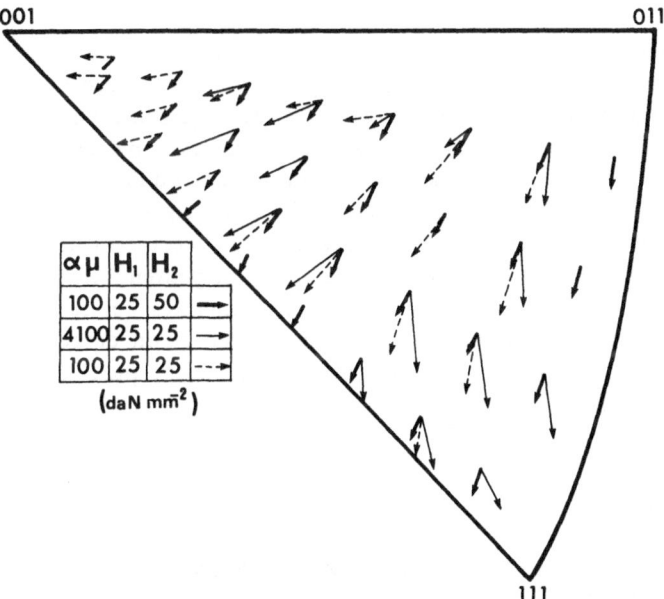

Figure 4. – *Changement d'orientation pour différents grains et différentes valeurs de* αμ, H_1 *et* H_2 *(traction).*

En effet, on a en traction simple selon l'axe Ox_3 :

$$dQ_{ij} = \begin{pmatrix} 0 & 0 & 0 \\ 0 & 0 & 0 \\ 0 & 0 & d\Sigma_{33} \end{pmatrix} + \alpha\mu E_{33}^p \begin{pmatrix} -m & 0 & 0 \\ 0 & -(1-m) & 0 \\ 0 & 0 & 1 \end{pmatrix}$$

m étant relié au coefficient de Lankford R par la relation usuelle $R = m/1 - m$.

Dans le cas où l'anisotropie plastique est négligée, m = 1/2 et le tenseur dQ ne dépend que d'un seul paramètre. Lorsqu'une texture initiale est présente, m est inconnu et le tenseur dQ dépend de deux paramètres. On est alors obligé de recourir à une méthode itérative en partant d'une valeur initiale arbitraire m_0 pour m, à corriger cette valeur initiale, en vérifiant à chaque étape la "self-consistence" du modèle puisqu'on doit avoir

$$E_{ij}^p = \int_{(\Omega)} \epsilon_{ij}^p(\Omega)\, f(\Omega)\, d\Omega$$

$f(\Omega)$ étant la fonction de répartition des orientations cristallines par rapport au repère macroscopique.

4. Conclusion

Les exemples qui viennent d'être traités illustrent l'intérêt de l'utilisation d'un schéma self-consistent fondé, suivant le cas, sur tel ou tel problème d'inclusion, dans la prévision des propriétés d'un agrégat dont les constituants ont des propriétés anisotropes — agrégat dont la structure peut elle-même être anisotrope. Les limites d'un tel schéma tiennent cependant au fait qu'il ne prend en compte qu'assez grossièrement le caractère granulaire de l'agrégat, en évaluant les interactions entre les grains à partir de celles entre une inclusion plastifiée de façon uniforme et une matrice homogène. Ces limites ne peuvent être franchies qu'en envisageant des "motifs" d'interaction plus complexes, tel celui d'une paire d'inclusions [9], imageant, selon le point de vue qu'on désire adopter, soit des grains contigus, soit des sous-grains différemment plastifiés à l'intérieur d'un même grain.

REFERENCES

[1] ESHELBY J.D. — "Elastic inclusions and inhomogeneities". *Prog. in Sol. Mech.* (1961): 87-140.
[2] YOSHIDA S., A. OGUEHI and M. NOBUKI. — "Influence of high hydrostatic pressure on the flow-stress of zinc and zirconium polycrystals". *Trans. J. Inst. Met.,* **13** (1972): 69-75.

[3] BUNGE H.J. – *"Mathematische Methoden der Texturanalyse"*. Berlin: Akademic Verlag, 1969.

[4] KRONER E. – "Zur plastichen Verformung des Vielkristalls". *Acta Met.*, 9 (1961): 155-162.

[5] HUTCHINSON J.W. – "Plastic stress-strain relations of FCC polycrystalline metals hardening according to Taylors's rule". *J. Mech. Phys. Sol.*, 12 (1964): 11-24.

[6] BERVEILLER M. and A. ZAOUI. – "An extension of the self-consistent scheme to plastically flowing polycrystals". *J. Mech. Phys. Sol.* 26 (1979): 326-344.

[7] FRANCIOSI P., M. BERVEILLER and A. ZAOUI. – "Latent hardening in Copper and Aluminium single crystals". *Acta Met.*, 28 (1980): 273-283.

[8] HUTCHINSON J.W. – "Elastic-plastic behaviour of polycrystalline metals and composites". *Proc. Roy. Soc.*, A 319 (1970): 247-272.

[9] BERVEILLER M. – "Contribution à l'étude du comportement plastique et des textures de déformation des polycristaux métalliques". Thèse de Doctorat d'Etat, Université Paris XIII, 1978.

ABSTRACT

The solution of the generalized problem of an ellipsoïdal inclusion in an infinite matrix is used in order to point out the influence of the elastic anisotropy on the plastic flow of a polycrystal submitted to a hydrostatic pressure, as well as the one of a geometrical anisotropy on the elastic rotations and the texture development. Starting from a new inclusion problem, a polycrystalline self-consistent scheme with a plastic accommodation is then proposed: this model allows a better integration of the anisotropic properties of single crystal strain-hardening and a more realistic prediction of deformation textures and the subsequent plastic anisotropy.

Anisotropie de Comportement Plastique de Tôles d'Aluminium et de Titane Possédant une Texture Cristallographique

M. Pernot et R. Penelle
Laboratoire de Métallurgie Physique,
Université Paris-Sud, Orsay, France.

1. Introduction

Le comportement en déformation plastique d'un agrégat polycristallin est généralement anisotrope, car d'une part le monocristal présente un comportement anisotrope de par la cristallographie de la déformation, et d'autre part car un polycristal possède presque toujours une texture cristallographique et, s'il n'en possède pas avant la mise en charge, il va s'en créer une au cours de la déformation.

Suivant les cas, cette anisotropie est néfaste ou bénéfique ; cependant, la tendance actuelle est de développer des textures qui utilisent au mieux les propriétés directionnelles afin d'atteindre un but fixé avec une meilleure rentabilité qui comporte souvent une économie de matières premières.

Il est donc très important de prévoir l'anisotropie à partir des propriétés du monocristal, du comportement d'un grain dans une matrice polycristalline et des paramètres décrivant la texture.

Le modèle de comportement d'un grain dans un polycristal retenu dans ce travail est le modèle simple de Taylor [19], généralisé pour tenir compte de la texture cristallographique, par l'intermédiaire de la Fonction de Distribution des Orientations des Cristallites (FDOC), associé à un comportement du monocristal en glissement ou en maclage régi par une loi de cission critique.

La FDOC, qui ne tient compte que des paramètres d'orientations (la taille, forme des grains, etc. sont négligés), est calculée à partir des figures de pôles mesurées, au moyen de développements en séries sur des bases d'harmoniques sphériques [8].

Le modèle, déjà utilisé par d'autres auteurs [4, 7, 17], permet de prévoir la position et la hauteur relative des cornes d'emboutissage, qui sont, rappelons-le, des défauts qui apparaissent lors d'essais d'emboutissage dans une tôle mince, d'un godet cylindrique, sous forme de variations de la hauteur de la paroi du godet.

La prédiction des cornes est faite en calculant théoriquement les variations dans le plan de la tôle du rapport $R(\alpha) = E_{22}/E_{33}$ dans un essai de traction simple, α étant l'angle entre la direction de mesure et la direction de laminage.

Ceci a été appliqué à des tôles d'aluminium de pureté commerciale pour des déformations à différents taux de laminage suivies ou non d'un recuit standard, en utilisant le glissement $\{111\}\langle 110 \rangle$, et à une tôle de titane non allié avec plusieurs systèmes de glissement et en introduisant le maclage.

2. Rappel des hypothèses du modèle de Taylor anisotrope

Les hypothèses du modèle de Taylor sont les suivantes :

1) La déformation plastique est homogène, soit $[E] = [\epsilon]$, si $[E]$ et $[\epsilon]$ représentent les tenseurs de déformation, respectivement, macroscopique et microscopique.

La déformation élastique est négligée.

2) La déformation a lieu par glissement. Cinq systèmes de glissement au moins sont nécessaires pour accommoder une déformation quelconque. La déformation peut également avoir lieu par maclage, ce mode de déformation étant considéré comme statistiquement homogène.

3) Parmi toutes les combinaisons de systèmes de déformation qui sont compatibles géométriquement avec la déformation, celle(s) qui minimise(nt) l'énergie de déformation est (sont) retenue(s). Il s'agit du Principe de Travail Minimum.

4) Le glissement est régi par une loi de cission critique, ainsi que le maclage. Si l'on veut prendre en compte le durcissement, il doit être isotrope.

Dans un essai de traction simple, on suppose que le tenseur de déformation macroscopique s'écrit :

$$[E] = \begin{bmatrix} E_{11} & 0 & 0 \\ 0 & E_{22} & 0 \\ 0 & 0 & E_{33} \end{bmatrix} \tag{1}$$

avec $E_{11} + E_{22} + E_{33} = 0$ \hfill (2)

Si l'on pose : $R = E_{22}/E_{33}$, il vient :

$$[E] = E_{11} \begin{bmatrix} 1 & 0 & 0 \\ 0 & \dfrac{-R}{1+R} & 0 \\ 0 & 0 & \dfrac{-1}{1+R} \end{bmatrix} \qquad (3)$$

Soient ψ, θ, ϕ les angles d'Euler repérant un référentiel lié à un cristal de l'agrégat par rapport au référentiel macroscopique des axes principaux de déformation de l'échantillon. Pour un cristal d'orientation ψ, θ, ϕ et se déformant avec un rapport R, il est possible de calculer le facteur de Taylor pour un incrément de déformation suivant la direction 1 :

$$M(\psi, \theta, \phi, R) = \frac{\delta w}{\tau_c \, \delta E_{11}} \qquad (4)$$

où τ_c est la cission critique du système de glissement s'il n'y a qu'une famille de considérée, ou bien la cission du système de déformation pris en référence s'il y a plusieurs familles. δw est le travail de déformation :

$$\delta w = \sum_h \tau_c^h \sum_k |\delta \gamma_k^h| \qquad (5)$$

la sommation sur h est faite sur les différentes familles de systèmes de déformation chacune ayant une cission critique τ_c^h, $\delta \gamma_k^h$ est l'incrément de cisaillement sur le k-ième système de la famille h ; ces termes sont reliés à [E] par une matrice d'orientation.

Le facteur de Taylor est calculé par minimisation par rapport aux quantités de glissement géométriquement compatibles avec la déformation macroscopique. Il peut également être calculé en utilisant l'analyse de Bishop et Hill [3], fondée sur le Principe du Travail Maximum ; il s'agit alors de construire la surface limite d'écoulement plastique du cristal, et de maximiser le travail de déformation sur les états de contrainte correspondant aux sommets de la surface de charge. Nous avons utilisé cette seconde méthode qui est plus rapide.

Le Principe du Travail Minimum a été généralisé au polycristal anisotrope par Hosford et Backofen [14]. Le facteur de Taylor moyen du polycristal est considéré comme étant la somme des facteurs de Taylor de tous les cristaux pondérée par la fraction volumique de cristaux de chaque orientation, soit la valeur de la FDOC, $F(\psi, \theta, \phi)$:

$$\overline{M}(R) = \iiint M(\psi, \theta, \phi, R) \cdot F(\psi, \theta, \phi) \sin\theta \, d\theta \, d\psi \, d\phi \qquad (6)$$

On obtient ainsi une courbe $\overline{M} = f(R)$ qui généralement possède un seul minimum. Dans cette direction de mesure, on suppose que le comportement du polycristal est celui qui minimise l'énergie, on retient donc comme valeur de R celle qui correspond au minimum de la courbe.

Par changement de base, on fait varier la direction de mesure dans le plan de laminage, on obtient alors une prédiction des courbes $R = f(\alpha)$.

Pour faciliter les calculs, le facteur de Taylor $M(\psi, \theta, \phi, R)$ est développé sur la même base que la FDOC, l'intégrale de l'équation (6) se ramène alors à une somme de produit des coefficients des deux développements.

3. Application à l'aluminium

Nous avons, dans une première partie, étudié, sur une tôle d'aluminium de pureté commerciale d'épaisseur 1 mm à l'état recristallisé, théoriquement et expérimentalement, l'évolution du rapport R au cours de déformations par traction de 5, 10 et 20 % dans la direction de laminage avec une vitesse de déformation $\overset{\circ}{\varepsilon} \simeq 10^{-4}\,\mathrm{s}^{-1}$.

Pour chaque état, nous avons déterminé les figures de pôles {111}, {200} et {220} par diffraction des rayons X avec la méthode en reflexion et transmission. Les FDOC ont été calculées en développant les séries jusqu'à l'ordre 34 [11] pour ajuster correctement la texture la plus accusée (erreur de troncature inférieure à 10 %).

La FDOC de la tôle avant déformation est présentée sur la figure 1, le maximum de la fonction est situé près de l'orientation idéale {123} <111> où {hkℓ} est le plan parallèle au plan de laminage et <uvw> la direction parallèle à la direction de laminage. L'évolution du maximum de la FDOC en fonction de la déformation ainsi que la dispersion en θ sont présentées sur la figure 2 ; le maximum ne se déplace pratiquement pas mais il augmente en valeur en même temps que la déformation.

Le calcul de R a donc été fait jusqu'à 20 % de déformation avec les quatre FDOC correspondantes pour des développements à l'ordre 16 qui est suffisant pour atteindre une bonne convergence. Les valeurs calculées sont comparées aux valeurs mesurées de R sur la figure 3 ; on remarque que les résultats théoriques et expérimentaux sont en assez bon accord et que, malgré de grandes variations du maximum de la FDOC, R ne varie pas de façon significative avec la déformation dans le domaine 0-20 %.

On peut se poser la question de savoir s'il est possible de raisonner sur le monocristal correspondant à la composante majeure de la texture, sur la figure 4 sont tracées les courbes $M = f(R)$ pour l'orientation {123} <111> ($\psi = 20°$, $\theta = 35°$, $\phi = 30°$) et pour l'orientation {112} <111> ($\psi = 0°$, $\theta = 35°$, $\phi = 45°$) ; ces orientations proches correspondent respectivement au maximum de la FDOC et à un point de la dispersion ayant une valeur encore élevée.

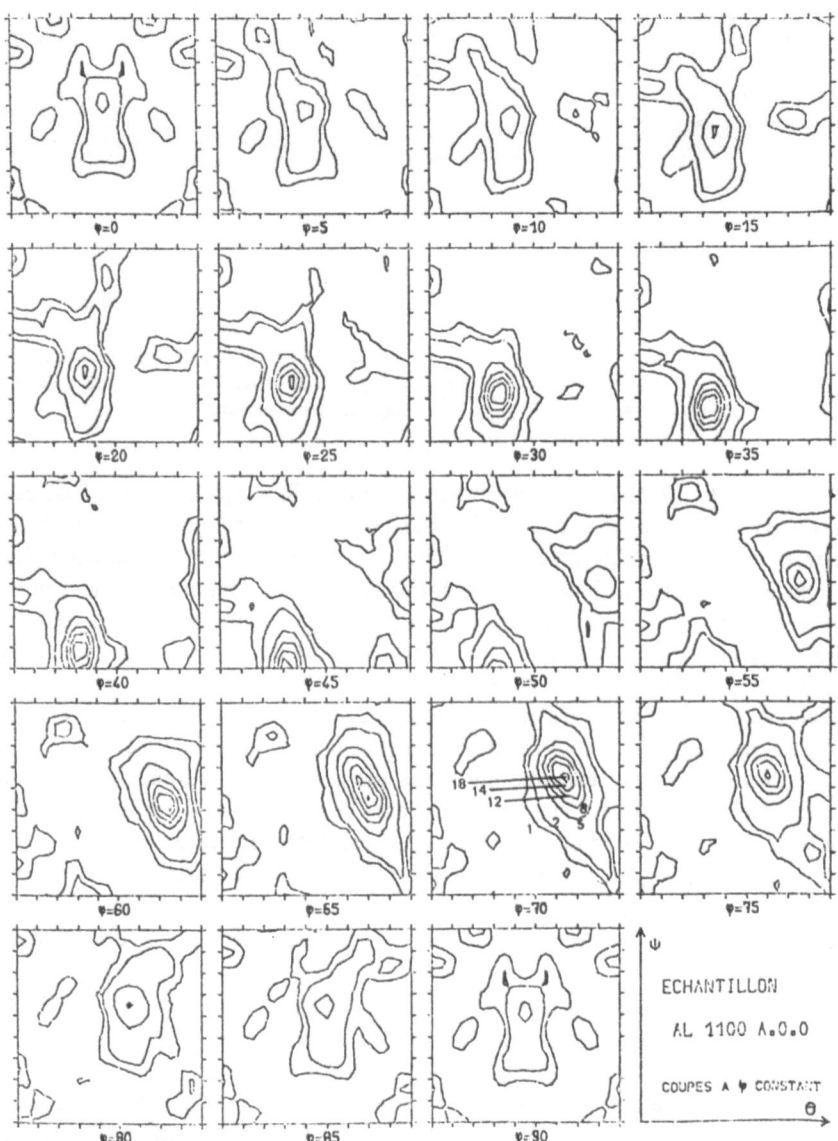

Figure 1. — *Courbes de niveaux représentant les variations de la FDOC dans l'espace d'Euler pour le matériau avant déformation.*

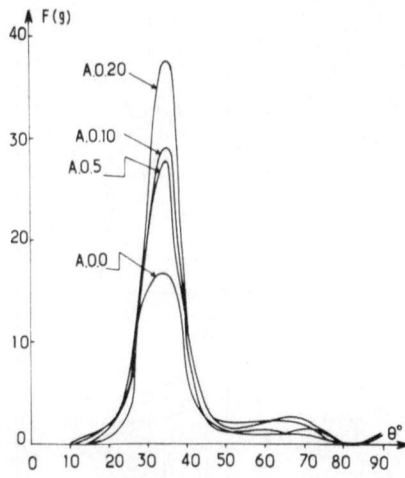

Figure 2. — *Evolution de la dispersion de la FDOC suivant θ pour ψ = 20° et φ = 30°, en fonction de la déformation. Le chiffre suivant A.O. représente le pourcentage de déformation.*

Dans le premier cas, le minimum est situé à R = 5,7, dans le second cas, les variations de M sont très faibles mais le minimum est situé à R = 0. Nous constatons donc qu'il suffit de s'écarter légèrement d'une orientation idéale donnée pour que la valeur de R correspondant au minimum de la courbe M = f(R) varie de 0 à l'infini. Ceci explique que le minimum de la courbe du polycristal puisse se trouver aux environs de 0,6 ; en conséquence, il semble qu'il ne soit pas possible d'obtenir des résultats même qualitatifs en utilisant seulement quelques orientations idéales, car la dispersion a une influence décisive sur la valeur moyenne de R. Cette valeur, qui reste sensiblement constante avec la déformation, paraît être plus influencée par la dispersion autour du maximum (Fig. 2) que par la valeur maximale croissante de la fonction ; en effet dans le cas présent, la "largeur à mi-hauteur" de la fonction reste constante.

Dans une seconde partie, nous avons étudié, sur de l'aluminium de même pureté que précédemment, l'évolution des cornes d'emboutissage pour différents taux de laminage suivis ou non d'un recuit de recristallisation à 350°C pendant 5 heures.

L'état de départ est une tôle de 10 mm homogénéisée, qui a été laminée à froid à huit épaisseurs comprises entre 5 et 0.1 mm.

Les pourcentages de cornes sont portés sur la figure 5 en fonction de la déformation. H étant la hauteur de la paroi mesurée après emboutissage d'un godet cylindrique à partir d'un flan de 60 mm avec un poinçon de 32 mm, le pourcentage de corne est donné par :

$$\frac{2\,(H_{max} - H_{min})}{H_{max} + H_{min}} \times 100$$

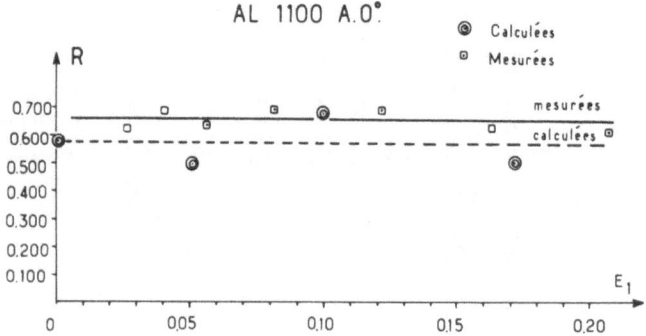

Figure 3. – *Valeurs de R mesurées et calculées en fonction de la déformation.*

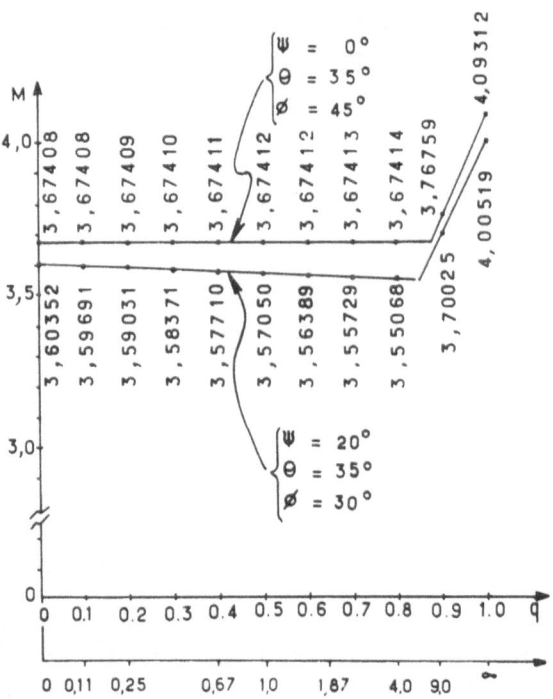

Figure 4. – *Valeurs des facteurs de Taylor en fonction de R ou de $q = \dfrac{R}{1 + R}$ pour deux orientations idéales.*

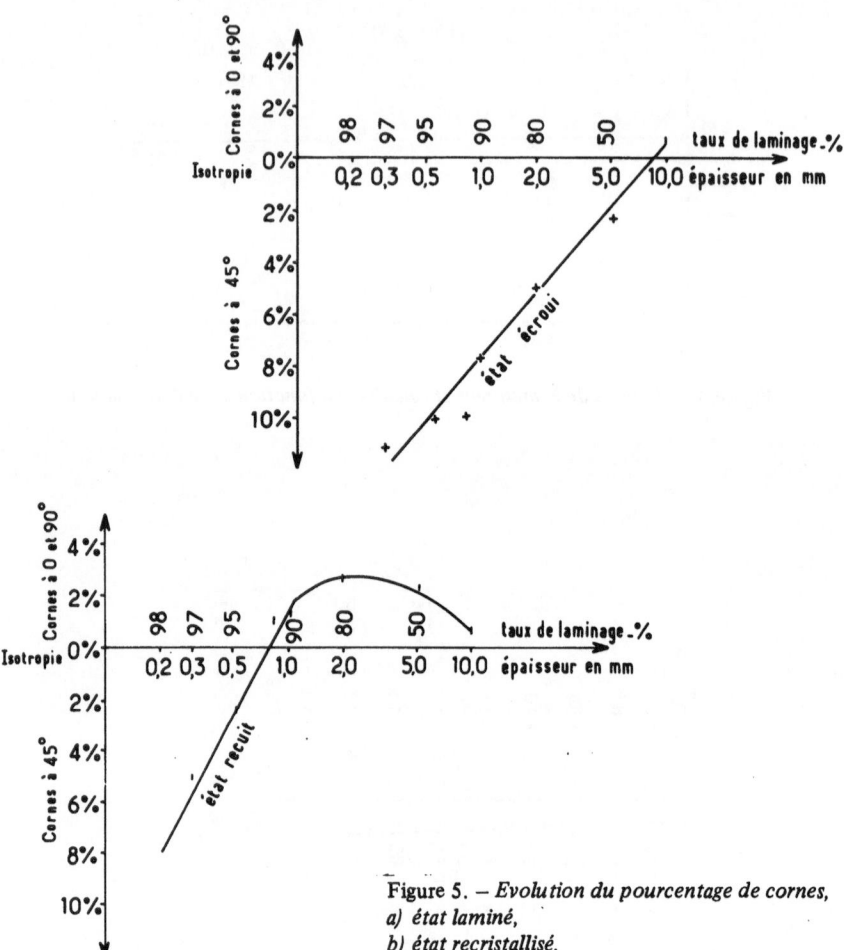

Figure 5. – *Evolution du pourcentage de cornes,*
a) état laminé,
b) état recristallisé.

Ces courbes présentent un domaine sensiblement linéaire pour lequel on observe une inversion de quatre cornes à 0° et 90° de la direction de laminage à quatre cornes à 45° de la direction de laminage, avec passage par un état sans corne d'un intérêt industriel évident.

Afin de calculer R, les textures globales ont été déterminées par diffraction des neutrons [16] sur les tôles de 10 et 5 mm, et par diffraction des rayons X pour les autres. Les FDOC ont été calculées pour des troncatures à l'ordre 24, les évolutions des composantes les plus importantes sont tracées en fonction de la déformation sur la figure 6. L'état initial comporte trois composantes avec de larges dispersions, la première entre {123} <111> et {112} <111>, la

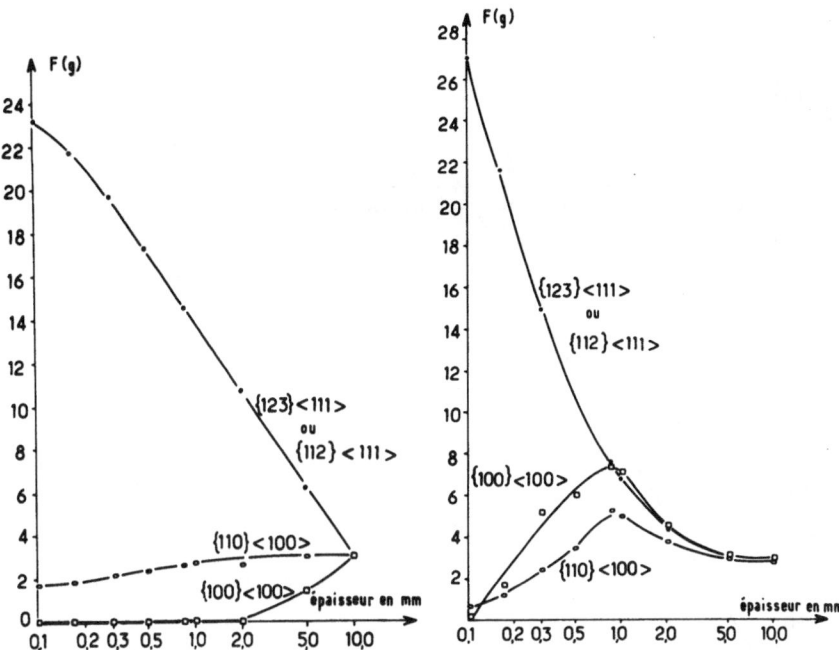

Figure 6. – *Evolution des composantes importantes de la texture,*
a) état laminé,
b) état recristallisé.

seconde $\{110\} <001>$ et la troisième $\{100\} <001>$. Au cours du laminage, la première composante est stable en position mais augmente très fortement en fraction volumique alors que les deux autres décroissent. Après recristallisation, la première composante reste stable alors que les deux autres se développent pour passer par un maximum à 1 mm d'épaisseur ; en outre, il a été montré qu'il n'existe pas de gradient de texture dans l'épaisseur des tôles.

Pour tous les états métallurgiques, nous avons calculé les courbes $R = f(\alpha)$. Il a été montré que, dans l'aluminium particulièrement, les courbes $R = f(\alpha)$ calculées ont la même allure que les courbes mesurées mais que le modèle de Taylor surestimait l'anisotropie [18]. Nous n'avons donc pas cherché à prédire quantitativement la hauteur des cornes, comme avait tenté de le faire Tucker [22] pour des monocristaux d'aluminium et Grumbach *et coll.* [13] pour des polycristaux d'acier doux. Nous nous sommes attachés à calculer l'évolution de hauteur des cornes en fonction du taux de déformation à l'aide de l'estimation de $\Delta R = 1/2 \; (R_0 - 2R_{45} + R_{90})$, ce paramètre ayant été depuis longtemps relié à la formation des cornes [12].

Les valeurs de ΔR calculées sont portées sur la figure 7 en fonction de la déformation. La comparaison avec la figure 5 montre que l'on obtient les

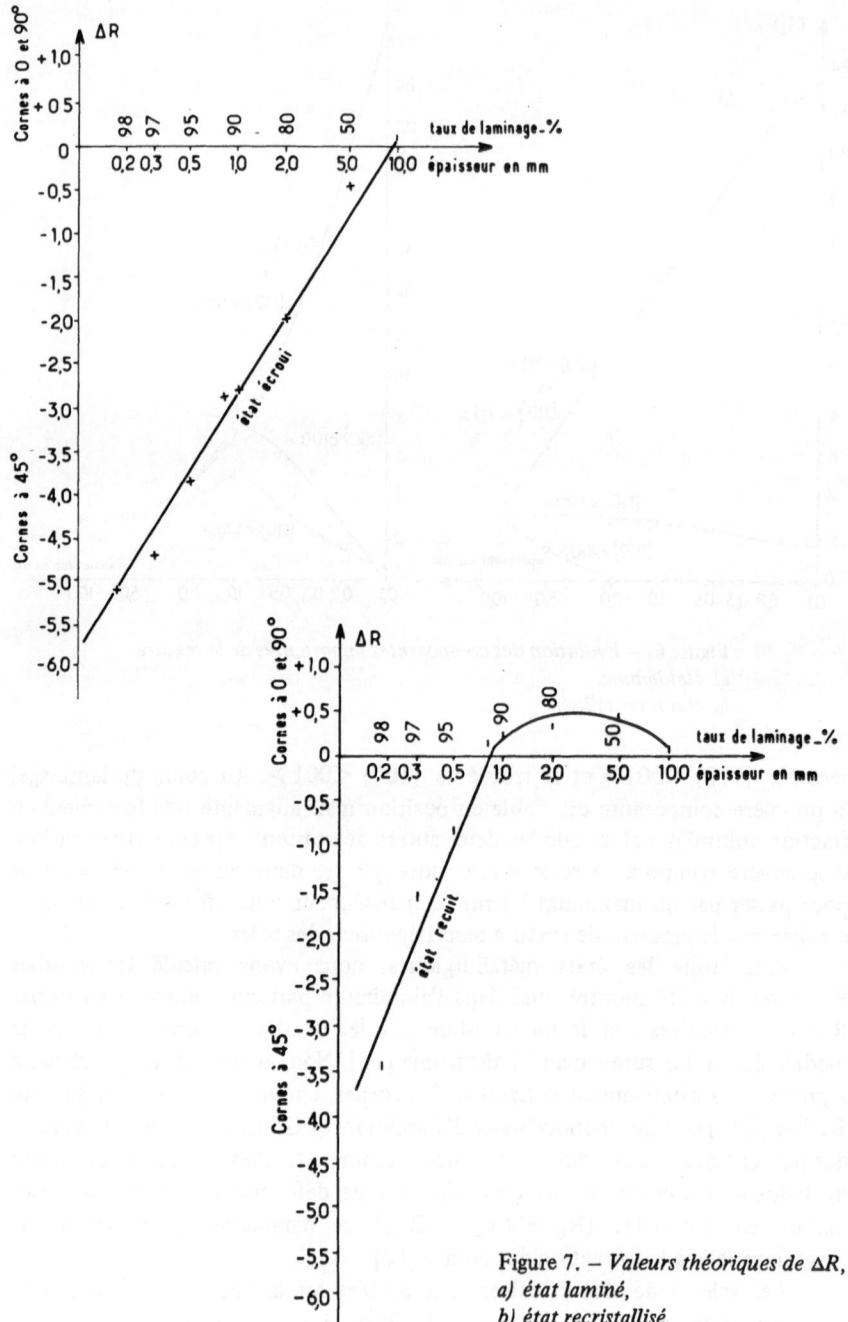

Figure 7. — *Valeurs théoriques de* ΔR,
a) *état laminé*,
b) *état recristallisé*.

mêmes allures pour les courbes théoriques et expérimentales, et que les états d'isotropie plane sont prévus pour des valeurs de déformations très proches de celles observées expérimentalement. De plus, les rapports entre les échelles de pourcentage de corne et de ΔR pour l'état laminé et pour l'état recristallisé sont sensiblement les mêmes et respectivement égales à 2,8 et 2,9.

Si l'on se réfère à la figure 6, on remarque que l'état d'isotropie plane est obtenu pour un mélange des trois composantes de la texture avec des valeurs sensiblement égales des maximums, mais ceci ne tient pas compte des dispersions autour des composantes.

Rappelons que, parmi ces composantes de la texture, les composantes $\{100\}<001>$ et $\{110\}<001>$ donnent des cornes à $0°$ et $90°$ de la direction de laminage, alors que la composante du type $\{123\}<111>$ donne des cornes à $45°$ de la direction de laminage.

Le modèle utilisé a donc permis de prévoir la position des cornes, l'évolution de leur hauteur et aussi le traitement ne donnant pas de corne.

4. Application au titane

Les métaux de structure hexagonale possèdent une anisotropie de déformation plastique plus accusée que celle des métaux cubiques, en particulier les valeurs de R du titane α sont très élevées. Malgré l'intérêt industriel qu'ils présentent, ces matériaux ont été peu étudiés en raison de la complexité des modes de déformation, glissement ou maclage et de l'absence de données sérieuses pour les valeurs de cissions critiques surtout dans le cas des nuances industrielles.

Nous avons calculé la variation de R dans le plan de la tôle en nous fondant sur les travaux de Thornburg et Piehler [20] qui ont été les premiers à prédire les textures de laminage du Titane à l'aide du modèle de Taylor-Bishop-Hill.

Le matériau utilisé dans la présente étude est une tôle de titane, de 1 mm d'épaisseur avec une taille de grains de l'ordre de 30 μm, contenant environ 1 200 ppm d'oxygène ; la texture cristallographique de cette tôle, déterminée comme précédemment, possède une composante majeure $(\overline{2}115)[0\overline{1}10]$ avec une forte dispersion en direction de $(\overline{1}013)[1\overline{2}10]$.

Les systèmes de glissement prismatique $\{10\overline{1}0\}$, basal $\{0001\}$ et pyramidal $\{10\overline{1}1\}$ en zone avec une direction $<11\overline{2}0>$ contenue dans le plan basal ne permettent pas d'accommoder une déformation selon l'axe $[0001]$. A température ambiante, cette impossibilité peut être palliée par intervention du maclage à l'aide des systèmes $\{11\overline{2}2\}<11\overline{2}3>$ et $\{10\overline{1}2\}<\overline{1}011>$ [1] qui permettent respectivement une contraction et un allongement parallèlement à cette direction. Outre l'incertitude sur les valeurs de cission critique de glissement due à la présence d'éléments interstitiels, l'existence même d'une loi de cission critique pour le maclage est actuellement très controversée, cependant en dépit de ces réserves, nous avons postulée, en accord avec le modèle de Taylor, la validité d'une telle loi.

Pour le glissement, nous avons retenu les valeurs de cission critique de Churchman [6]

$$\tau(1\,0\overline{1}0) = 9,19 \text{ kg/mm}^2$$

$$\tau(0001) = 10,90 \text{ kg/mm}^2$$

$$\tau(1\,0\overline{1}1) = 9,90 \text{ kg/mm}^2$$

qui utilisa un titane de pureté similaire au nôtre ; par contre, en l'absence de données pour le maclage, nous avons admis que ce mode de déformation intervenait lorsque le glissement ne permettait pas d'accommoder une déformation quelconque. Ceci implique donc que les cissions critiques de mâclage possèdent des valeurs supérieures ou égales à celles de glissement.

La compétition entre les différents modes de glissement basal, prismatique et pyramidal ayant une direction de glissement $<1\,1\overline{2}0>$ a été analysée par Chin et Mammel [5] en fonction des rapports de cission critique, les cinq régions

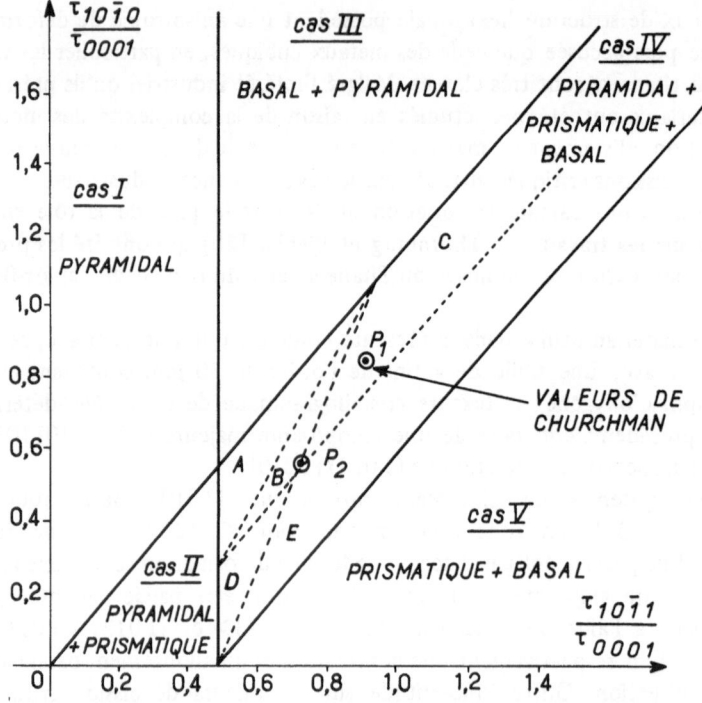

Figure 8. – *Domaines d'existence des différents systèmes de glissement en fonction des rapports de cissions critiques.*

correspondant aux différentes possibilités de glissement sont donnés, figure 8, selon les rapports $^\tau(1\,0\bar{1}1)/^\tau(0001)$ et $^\tau(1\,0\bar{1}0)/^\tau(0001)$.

Dervin [9] adoptant une analyse similaire à celle de Thornburg et Piehler [21] fondée sur le formalisme de Bishop et Hill [3] calcula les états de contraintes physiquement possibles faisant intervenir glissement et maclage afin de déterminer les systèmes de déformation actifs. Cette méthode ne sera pas détaillée ici, elle est fondée sur l'hypothèse que la déformation se produit sur quatre systèmes de glissement et un ou plusieurs systèmes de maclage. Il est ainsi possible, dans un premier temps, d'explorer la surface de charge définie pour le glissement seulement et de déterminer les états de contraintes permettant le déclenchement de toutes les combinaisons de quatre systèmes de glissement. Afin de déterminer totalement le tenseur des contraintes, un système de maclage est alors sélectionné.

Les états de contraintes varient selon la région considérée ; dans le cadre de la présente étude et en fonction des systèmes retenus compte tenu de la pureté du matériau utilisé, le domaine à retenir correspond au sous cas C de la région IV, ainsi existe-t-il 216 possibilités de 5 systèmes de glissement et maclage.

Sur le plan expérimental, le coefficient R extrapolé à 0 % de déformation a été déterminé pour des angles α de 15 en 15° à partir de la direction de laminage sur éprouvettes normalisées à une vitesse de déformation $\overset{\circ}{\epsilon} \simeq 10^{-4}\,s^{-1}$.

La figure 9 montre que R ne varie pas avec le taux de déformation, cependant si l'on compare la courbe $R_{exp} = f(\alpha)$ et la courbe $R_{th} = f(\alpha)$ calculée à partir du modèle de Taylor en introduisant la FDOC (Fig. 10), nous

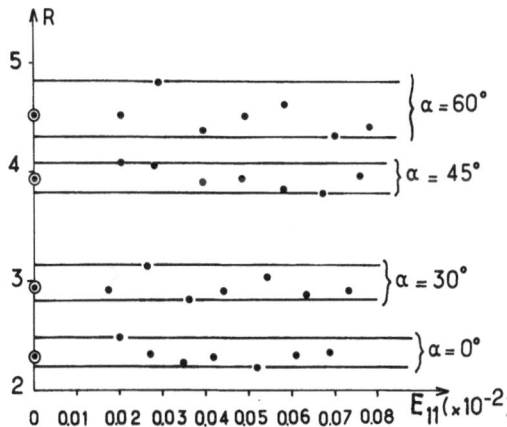

Figure 9. — *Variations de R en fonction de la déformation. Sur l'axe des ordonnées sont portées les valeurs de R extrapolées à 0 % de déformation.*

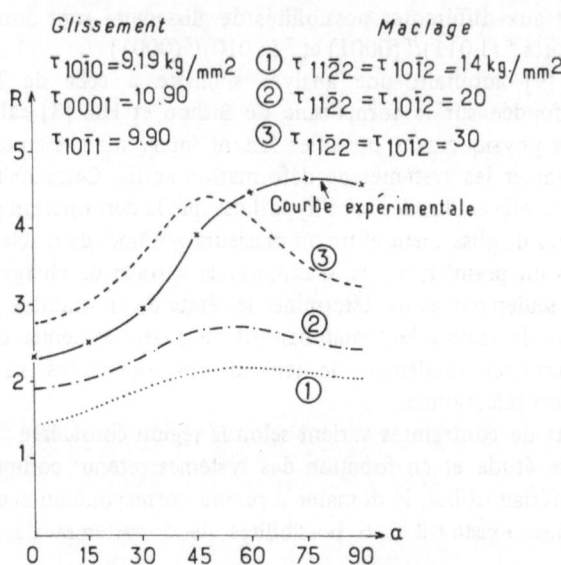

Figure 10. – *Comparaison entre les courbes expérimentales et théoriques pour le point P₁ (Fig. 8).*

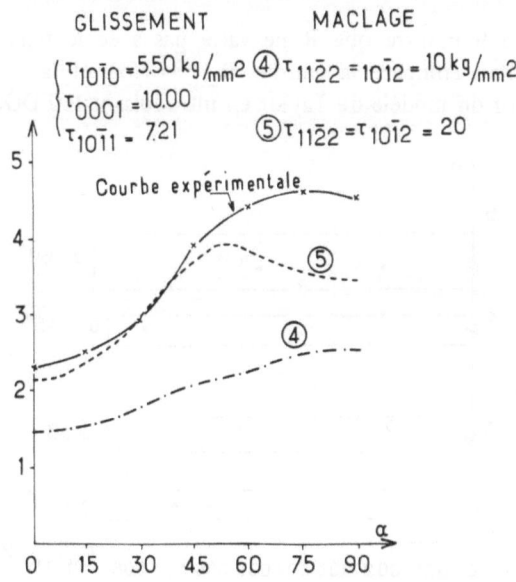

Figure 11. – *Comparaison entre les courbes expérimentales et théoriques pour le point P₂ (Fig. 8).*

pouvons constater que l'accord reste médiocre, cet écart est dû en partie au fait que les valeurs de cission critique déterminées par Churchman sont fortement entâchées d'erreurs, compte tenu de la précision des mesures. En effet, les cissions pour l'ensemble des systèmes sont équivalentes, ce qui ne correspond pas à la réalité, le matériau se comporterait alors comme un solide isotrope si l'on admet de plus que les cissions de maclage sont identiques à celles de glissement.

En favorisant le glissement prismatique (point P_2, Fig. 8), conformément aux observations expérimentales et en accroissant les cissions de maclage, nous retrouvons un accord raisonnable entre théorie et expérience (Fig. 11).

5. Conclusions

Le modèle de Taylor généralisé aux polycristaux possédant une texture cristallographique, permet de prévoir semi-quantitativement le comportement anisotrope en déformation plastique.

Il semble que, dans tous les calculs théoriques de courbes $R(\alpha)$ pour des matériaux de réseau cubique, tant cubique-centré comme les aciers doux avec le glissement du type "pencil glide", que cubique à faces centrées, on obtient l'allure des courbes déterminées expérimentalement, avec cependant une surestimation de l'anisotropie.

Pour un matériau de réseau hexagonal comme le titane α, les conclusions sont beaucoup plus délicates car, de par le nombre de mécanismes de déformation qui interviennent et les incertitudes sur les cissions critiques, il existe plus de variables.

Les critiques que l'on peut faire au modèle de Taylor sont les suivantes :

1) Expérimentalement, il s'avère que la déformation n'est pas homogène.
2) La continuité des contraintes n'est pas respectée d'un grain à l'autre.
3) Le modèle surestime les contraintes internes aux joints de grain puisqu'il n'autorise pas de relaxation.
4) Le modèle ne tient pas compte de la taille des grains, un des paramètres les plus importants après la texture cristallographique.
5) Dans le calcul de $R(\alpha)$, lorsque α est tel que la direction de mesure n'est pas un des axes de symétrie de la texture, il est peu probable que les axes principaux de déformation coïncident avec les axes principaux de contrainte.

Le modèle permet, néanmoins, une approche de la compréhension de la formation des cornes d'emboutissage du point de vue texture cristallographique, ainsi que la prévision de leur évolution au cours de traitements thermomécaniques. Ceci avec un volume de calculs numériques déjà important.

Nous ne pensons pas que le modèle de Taylor puisse donner des résultats autres que semi-quantitatifs, mais qu'il faut s'orienter vers des modèles plus élaborés, comme celui de Kröner [15] qui accepte une accommodation élastique aux joints, ou son adaptation par Berveiller et Zaoui [2] qui admettent une accommodation élasto-plastique, pour le comportement d'un grain dans une matrice polycristalline, et celui de Zarka [23] pour le comportement du monocristal.

Actuellement, nous développons un programme de calcul, en collaboration avec M. Engel du Groupe de M. Zarka (Ecole Polytechnique), pour prévoir quantitativement l'évolution des textures de déformation de matériaux c.f.c., au moyen des modèles de Kröner et de Zarka [10].

Il s'agit d'une première étape afin de tester les modèles et leur programmation numérique avant de passer aux calculs de comportement plastique anisotrope.

REFERENCES

[1] AKHTAR A. – "Basal slip and twinning in α-titanium single crystals". *Met. Trans.*, 6 A (1975): 1105-1113.

[2] BERVEILLER M. and A. ZAOUI. – "Comments on deformation texture development and polycristalline modelling". In Gottstein and Lucke (ed.), *ICOTOM 5*, Springer Verlag., 1 (1978): 319-327.

[3] BISHOP J.F.W. and R. HILL. – "A theory of the plastic distortion of a polycristalline aggregate under combined stresses" *Phil. Mag.*, 42 (1951): 414-427.

[4] BUNGE H.J. and W.T. ROBERTS. – "Orientation distribution, elastic and plastic anisotropy in stabilized steel sheet". *J. Appl. Cryst.*, 2 (1969): 116-128.

[5] CHIN G.Y. and W.L. MAMMEL. – "Competition among basal, prim, and pyramidal slip modes in Hcp metals". *Met. Trans.* 1 (1970): 357-361.

[6] CHURMAN A.T. – "The slip modes of titanium and the effect of purity on their occurence during tensile deformation of single crystals". *Proc. Roy. Soc.*, A 226 (1954): 216-226. ·

[7] DAVIES G.J., D.J. GOODWILL, J.S. KALLEND and T. RUBERG. – "The correlation of structure and texture with formability" *J. Inst. Metals*, 101 (1973): 270-273.

[8] DERVIN P., J.P. MARDON, M. PERNOT, R. PENELLE and P. LACOMBE. – "Application de la représentation tridimensionnelle des textures à l'étude de l'évolution des textures de laminage et de recristallisation du titane". *J. Less Common Metals*, 55 (1977): 25-43.

[9] DERVIN P. – "Analyse quantitative des textures cristallographiques de matériaux de système cubique ou hexagonal. Applications à l'aluminium et au titane. Relation avec l'anisotropie de déformation plastique dans le cas de tôles minces de titane". Thèse Docteur Ingénieur, Paris XI, 1978.

[10] ENGEL J.J., M. PERNOT and R. PENELLE. – "Quantitative prediction of texture evolution of aluminium during tensile strain. Comparison with experimental orientation distribution function". In *Strengh of Metals and Alloys*. Haasen *et al.* (eds.), 763-768, Pergamon Press, 1979.

[11] FERREIRA O., P. DERVIN, M. PERNOT and R. PENELLE. – "Influence of a rolling or a tensile deformation on the evolution of texture – symmetrical or not – of aluminium sheets. Relation with the strain ratio R". In Gottstein and Lucke (eds.), *ICOTOM 5*, 337-346. Springer Verlag, 1978.

[12] GREWEN J. – "Textures and deep drawing process". In *Quantitative analysis of textures*. Proc. of the International Seminar, Cracow (1971): 195-218.

[13] GRUMBACH M., P. PARNIERE, L. ROESCH and C. SAUZAY. – "Etude des relations quantitatives entre le coefficient d'anisotropie, les cornes d'emboutissage et la texture des tôles minces d'aciers extra-doux". *Mém. Sci. Rev. Mét.*, 72 (1965): 241-253.

[14] HOSFORD W.F. and W.A. BACKOFEN. – "Strength and plasticity of textured metals". In *Fundamentals of Deformation Processing*, 259-292, Syracuse Univ. Press, 1964.

[15] KRONER E. – "Zur plastichen Verformung des Vielkristalls". *Acta Met.*, 9 (1961): 155-161.

[16] MADRON J.P., M. PERNOT, P. DERVIN, R. PENELLE and M. ENGLANDER. – "Contribution de la diffraction neutronique à l'étude de la fonction texture du titane recristallisé. Comparaison avec la diffraction des rayons X". *J. Appl. Cryst.*, 10 (1977): 372-375.

[17] PARNIERE P. and L. ROESCH. – "Analyse quantitative des relations entre le coefficient d'anisotropie et la texture des tôles minces d'aciers extra-doux". *Proc. du 3ᵉ Colloque Européen sur les textures*, Pont-à-Mousson, 1973, pp. 585-617.

[18] PERNOT M. – "Determination quantitative des textures cristallographiques de tôles minces: relations avec l'anisotropie magnétocristalline d'aciers extra-doux, et l'anisotropie de comportement plastique de divers métaux et alliages de réseau cubique". Thèse d'Etat, Paris XI, 1977.

[19] TAYLOR G.I. – "Plastic strain in metals". *J. Inst. Metals*, 62 (1938): 307-324.

[20] THORNBURG D.R. and H.R. PIEHLER. – "Cold rolling texture development in titanium and titanium-aluminium alloys". *Titanium Science and Technology* 2 (1973): 1187-1197.

[21] THORNBURG D.R. and H.R. PIEHLER – "An analysis of constrained deformation by slip and twinning in hexagonal close-packed metals and alloys". *Met. Trans.*, 6 (1975): 1511-1523.

[22] TUCKER G.E.G. – "Texture and earing in deep drawing of aluminium". *Acta Met.*, 9 (1961): 275-286.

[23] ZARKA J. – "Etude du comportement des monocristaux métalliques. Application à la traction du monocristal c.f.c." *J. de Mécanique*, 12 (1973): 276-318.

ABSTRACT

The Taylor model generalized to plastic behaviour of anisotropic polycristals, has been used to predict the position and relative height of drawing ears, by mean of the evolution of the ratio $R = E_{22}/E_{33}$ measured in a tensile test, and by using the Crystallite Orientation Distribution Function.

In a commercial purity aluminium with cubic lattice, the evolution of ears as a function of the amount of reduction by rolling with or without annealing, has been predicted. A metallurgical state without ears, of a large industrial interst, has been predicted and observed.

In a second part, the model has been applied with introduction of twinning, for titanium, a metal with hexagonal lattice. Results are in good agreement with experiments when prismatic slip is favoured.

Dynamic Plastic Deformation
for Non-Proportional Loading Paths

R. J. Clifton

Brown University, Presidence, Rhode Island, U.S.A.

1. Introduction

Improved understanding of the plastic response of metals subjected to loading histories which include marked changes in direction of the stress trajectories (in a hyperspace of components of the stress tensor) is of fundamental importance in plasticity theory and its applications. In particular, the plastic response following abrupt changes in direction of the stress trajectory provides, for rate independent models of plastic response, a critical distinction between models incorporating smooth yield surfaces and those incorporating yield surfaces with vertices. Differences between predictions obtained using these two classes of models appear to be important in a number of applications including plastic buckling, stability of finite plastic deformations, strain localization, and plastic waves.

Experiments involving non-proportional loading paths (i.e. changes in direction of the stress trajectories) are normally conducted quasi-statically. Under these conditions, nominally homogeneous states of deformation can be produced. Both stress and deformation can be monitored. Consequently, the plastic response can be obtained for a variety of stress histories. Quasi-static experiments are most satisfactory for smooth stress trajectories at moderate strain-rates, say less than $1 \sec^{-1}$. For more abrupt changes in stress trajectories and for higher strain rates the measurements of stress become less reliable due to response time limitations of the load transducers.

An alternative means for investigating plastic response for non-proportional loading paths is to use plastic waves of combined stress. The theory of such waves has been considered by several investigators (e.g. [1, 3, 5, 7, 11, 16, 17]). So far, there have been relatively few experimental investigations (e.g. [10, 13, 14]). However, the number of experimental investigations of plastic waves of combined stress can be expected to increase considerably in the near future as more laboratories develop the capability for plate impact experiments in which both normal and shear tractions are imposed uniformly over one face of a plate specimen.

Such experiments have a number of advantages for investigating plastic response for non-proportional loading paths. First, since two types of waves are generated, the stress trajectories obtained in these experiments include sharp changes of direction during the transition from one type of wave to the other. Second, the wave profile associated with predominantly shearing deformation is a sensitive indicator of the shear resistance characteristics of the material. In particular, the speed of propagation of the front of this quasi-shear wave is related directly to the instantaneous response of the material for stress increments in the new direction of the stress trajectory; furthermore, the continuous wave profile near the quasi-shear wave front is indicative of the plastic response shortly after the change in direction of the stress trajectory occurs. Because the continuous quasi-shear wave spreads with distance of propagation, high resolution of the near wavefront features can be obtained by monitoring the wave profile at sufficient distances from the impact face.

In this paper, the theory of plastic waves of combined stress is reviewed. A technique for conducting pressure-shear wave experiments is described briefly and experimental results are compared with theoretical predictions. Discrepancies between theory and experiment are viewed as indicative of the need for futher examination of the adequacy of smooth yield surface models for loading paths with sharp changes in stress directions. Previous work on combined longitudinal and torsional plastic waves is cited as another case in which self-consistent slip models give better agreement with experiment for the early part of the wave profile following a change in direction of the stress trajectory.

2. Theory

Partial differential equations governing the propagation of plastic waves of combined stress in the x-direction can, for most constitutive models, be written in the form [6]

$$Aw_t + Bw_x + d = 0 \tag{1}$$

In Equation (1), $w(x, t)$ is an n-dimensional vector that can be partitioned in the form

$$w(x, t) = \begin{bmatrix} v(x, t) \\ \sigma(x, t) \end{bmatrix} \quad \begin{array}{l} v(x, t) \in V^{(m)} \\ \sigma(x, t) \in V^{(n-m)} \end{array} \tag{2}$$

where $v(x, t)$ and $\sigma(x, t)$ are vectors whose components are, respectively, the particle velocities and stresses which are of interest in the problem; A and **B** are matrices that can be partitioned in the form

$$\mathbf{A} = \begin{bmatrix} \mathbf{M} & 0 \\ 0 & \mathbf{S} \end{bmatrix}, \quad \mathbf{B} = \begin{bmatrix} 0 & -\mathbf{N} \\ -\mathbf{N}^{\mathrm{T}} & 0 \end{bmatrix} \tag{3}$$

where **M**, **S** are, for usual material models, symmetric positive-definite matrices of dimension m an n-m respectively and **N** is a matrix of m rows and n-m columns. The matrix **M** is a mass matrix in that the kinetic energy is given by $\frac{1}{2}$ **v**·**Mv**. For pressure-shear waves $\mathbf{M} = \rho\mathbf{I}$ where ρ is the mass density and **I** is the 2×2 identity matrix; the matrix **S** is an instantaneous compliance matrix in that $\mathbf{S}(\delta\sigma)$ gives the instantaneous strain increment associated with an instantaneous infinitesimal change $\delta\sigma$ in the stress vector; the matrix **N** has the property that **N**σ is the traction on a plane with its normal in the positive x-direction. The vector **d** in (1) accounts for time dependent relaxation processes in the material. The first m equations of (1) express conservation of momentum conditions and the last n-m equations are compatibility conditions in which constitutive equations have been used to eliminate the total strain rates.

For most material models **A**, **B** and **d** are functions of σ, α, x where α is a vector of parameters that characterize the plastically deformed state of the material. With the assumption that spatial gradients of σ and α are unimportant in characterizing the plastically deformed state, the dependence of α on the loading history can be expressed in the form

$$\alpha(x, t) = \underset{\tau = -\infty}{\overset{t}{\mathbf{H}}}\ (\sigma(x(\tau), \tau)) \tag{4}$$

where **H** is a vector valued functional of the loading history $\sigma(x(\tau), \tau)$ following the motion $x(\tau)$ of the particle that is at x at time t. The parameters $\alpha(x, t)$ can include physical quantities such as plastic strains and measures of dislocation structures as well as empirical quantities used to characterize loading surfaces or flow potentials during plastic flow.

Three essentially different types of constitutive equations are of interest in characterizing the plastic response of materials subjected to non-proportional loading histories. In two of these types the plastic response is modelled as rate independent whereas in the third the plastic response is rate dependent. The rate independent models are subdivided further into those that use smooth loading surfaces and those that include vertices in the loading surfaces. For both types of rate independent models the vector **d** in (1) is identically zero. For the case of smooth loading surfaces the matrix **S** depends on the current values of σ and $\varkappa$ only whereas for the case of loading surfaces with vertices at the current stress state the matrix S depends also on the stress rate σ_t; however, the dependence of S on σ_t is homogeneous of degree zero in order to retain the rate independence of the model. The latter behaviour is characteristic of slip models in which the plastic compliance depends on the direction of loading because the collection of slip systems that will be active during an incremental loading is dependent on the loading direction. Mathematically, the slip models can be viewed as cases for which the dependence of the vector $\varkappa$ on the current value of the stress rate σ_t, is homogeneous of degree zero.

If jumps in σ_t do not cause jumps in α, then the usual investigation of characteristics for systems of equations of the form (1) leads to

$$(-cA + B)[w_t] = 0 \tag{5}$$

where $c = \dfrac{dx}{dt}$ is the characteristic velocity of the front accross which w is continuous but discontinuities in the derivatives of w occur. The jump $[w_t]$ is the difference $(w_t^+ - w_t^-)$ between the values of w_t in front of and behind the wavefront. Equation (5) can be written in the expanded form

$$cM[v_t] + N[\sigma_t] = 0 \tag{6a}$$

$$N^T[v_t] + cS[\sigma_t] = 0 \tag{6b}$$

Use of (6b) to eliminate $[\sigma_t]$ in (6a) gives

$$C[v_t] = 0 \tag{7}$$

where $C = (NS^{-1}N^T - c^2M)$. Equation (7) has non-trivial solutions $[v_t]$ only if the determinant of the matrix is zero. The symmetry of $NS^{-1}N^T$ and the positive definiteness of M insure the existence of real wave velocities c that satisfy $\det. C = 0$. The jumps $[v_t]$ associated with distinct wave speeds are mutually orthogonal. Furthermore, from (6a) the corresponding jumps in traction rates are mutually orthogonal.

This orthogonality of jumps in traction rates associated with different types of waves suggests that non-proportional loading paths with orthogonal changes in loading direction are to be expected when more than one type of combined stress plastic wave is generated. Such non-proportional loading paths can be illustrated most readily for the case of rate independent materials with smooth loading surfaces. In this case, and for step loading conditions, there is no characteristic time or length in the problem so that centered simple wave solutions of the form $w = w(x, t)$ are expected. Substituting this expression for w in (1), one obtains (for $d = 0$ and $A = A(\sigma, \alpha)$)

$$(-cA(\sigma, \alpha) + B) w' = 0 \qquad (8)$$

where c is the velocity of propagation x/t and w' is the derivative of w with respect to c. From the similarity of (8) and (5), an expanded form of (8) analogous to (6) can be used to write

$$(NS^{-1} N^T - c^2 M) v' = 0 \qquad (9)$$

which is analogous to (7). Thus, for non-trivial simple wave solutions (i.e. $v' \neq 0$), the speeds of propagation of a particular state of stress σ and deformation α are the same as the characteristic speeds for the propagation of discontinuities $[v_t]$ into a region in the state (σ, α). The orthogonality of jumps in derivatives of traction vectors that was indicated previously becomes, for simple waves, orthogonality in the traction trajectories for different types of simple waves which transmit the state (σ, α) at different speeds.

The combination of different types of simple waves to obtain a solution for pressure-shear loading of a half space is illustrated in Figure 1. The fast simple wave is purely longitudinal whereas the slow simple wave involves both longitudinal and transverse motion. The constant state region at b links the fast and slow simple waves. The orthogonal change in direction of the traction trajectory at b illustrates the orthogonality properties deduced from (9).

Simple wave solutions also exist for the case of rate independent descriptions of plastic response based on slip models. In these cases, $S = S(\sigma, \alpha, \sigma_t)$ where the dependence on σ_t is homogeneous of degree zero so that

$$S(\sigma, \alpha, \sigma_t) = S(\sigma, \alpha, \beta\sigma_t) \qquad (10)$$

where β is an arbitrary scalar. If one again assumes a centered simple wave solution $w = w(x/t)$ then, using (10), one obtains

$$(-cA(\sigma, \alpha, \sigma') + B) w' = 0 \qquad (11)$$

which is analogous to (8) except that in (11) the matrix A depends also on the derivative σ'. Decomposition of (11) in the form analogous to (6) and elimination of v' gives

$$(-N^T M^{-1}N + c^2 S(\sigma, \alpha, \sigma'))\sigma' = 0 \qquad (12)$$

Because S in (12) depends on σ' the problem of finding speeds c and stress trajectory directions $\sigma'/|\sigma'|$ ($|\sigma'|$ denotes the magnitude of the vector σ')

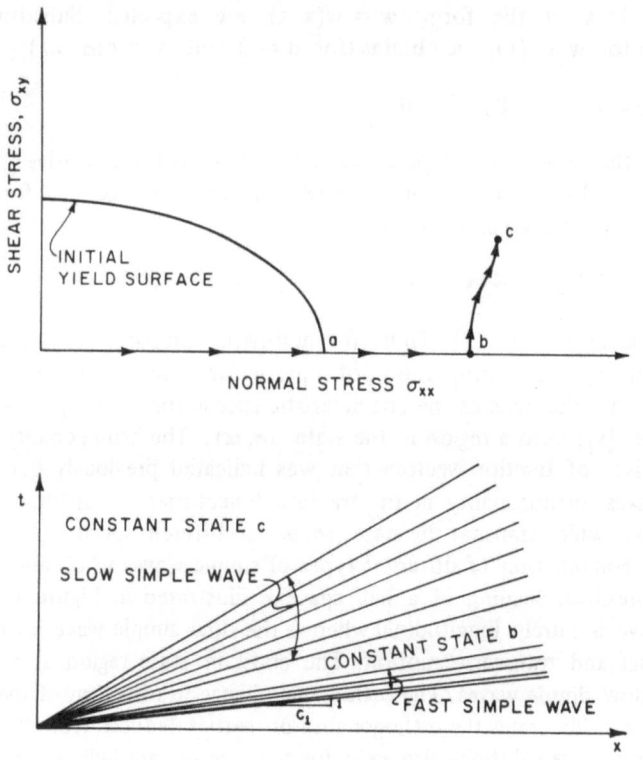

Figure 1. – *Schematic of simple wave solution for pressure-shear waves.*

that satisfy (12) is considerably more complicated than for the case of smooth loading surfaces for which σ' does not appear in the characteristic matrix (cf. (8)). For (12) the usual results on the existence of real wave speeds and the orthogonality of the vectors σ' associated with distinct wave speeds do not apply. Conversely, removal of the limitations on the number of wave speeds and the orthogonality of trajectories associated with distinct wave speeds increases the variety of wave phenomena that can be described by such equations.

A good example of new features of combined stress plastic waves that are predicted by slip models occurs in the theory of combined longitudinal and torsional plastic waves in pre-stressed, thin-walled tubes. In this case there are two non-zero components of stress: the longitudinal stress σ and the shear stress τ. A simple wave solution [9] for a tube subjected to a static torque followed by longitudinal impact is shown in Figure 2 for a case in which the plastic response of the tube is modelled by means of a rate independent self-consistent slip model [4]. The stress trajectory along ab corresponds to a fast

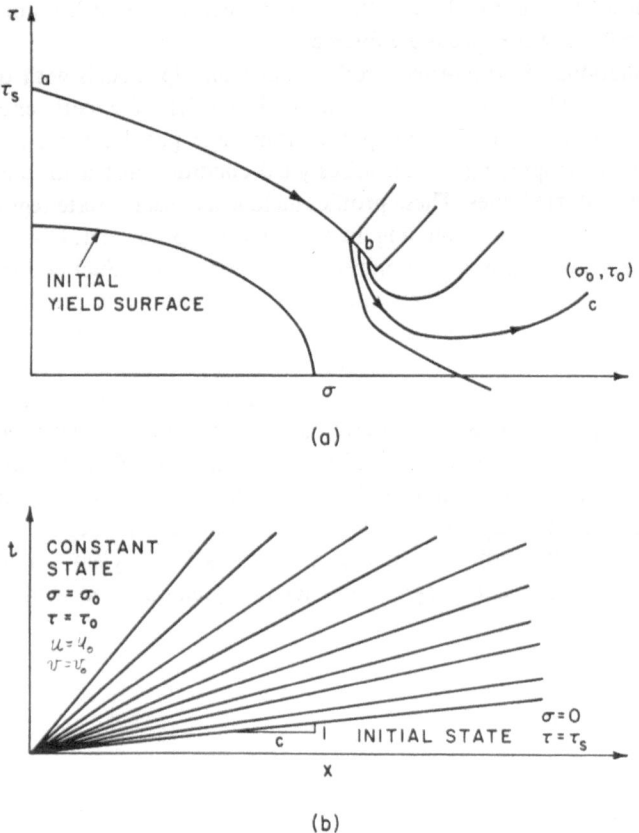

Figure 2. – *Schematic of simple wave solutions for longitudinal impact of a pre-torqued tube with its plastic response modelled by means of a self-consistent slip model.*

simple wave in which the speeds of propagation of the various states of stress are greater than the elastic shear wave speed. At b there are three possible ways to continue the simple wave solution. One is to continue along the path ab with the speed of propagation decreasing monotonically. Another is to turn counterclockwise at right angles and have a discontinuous drop in wave speed that requires the insertion of a constant state region between two simple waves as in Figure 1. The third possibility is the path bc that involves a discontinuous change in loading direction with essentially a continuous decrease in wave speed. The existence of the third possible path is a feature that distinguishes predictions based on a slip model from those based on smooth yield surface models. The latter models admit only two wave speeds at a given point

on a path analogous to ab and the stress trajectories at such a point must be
orthogonal if the wave speeds are distinct.

Predictions of strain-time profiles based on slip models with those based
on smooth yield surface models and with results of experiments [14] are
shown in Figure 3 [9]. The computed strain-time profiles for a smooth yield
surface model employing a Von Mises yield condition and isotropic hardening
are shown as dotted lines. These profiles include a constant state region between
a fast simple wave corresponding to the largest root of det. C = 0 (See (7))
and a slow simple wave corresponding to the other root. Such a constant state
region is not observed in the experiments. It is also not predicted by self-
consistent slip models. Results are shown for two different hardening models:
(i) isotropic hardening in which slip on one slip system hardens all slip systems
equally and (ii) independent hardening in which slip on one slip system hardens
only that slip system. Slightly better agreement between theory and experiment has been obtained with a self-consistent slip model that employs latent
hardening in which slip on one slip system hardens intersecting slip systems
more than parallel slip systems [8]. Overall, agreement between thery and
experiment is better for slip models than for smooth yield surface models.
However, fully satisfactory agreement has not been obtained with any models
used so far.

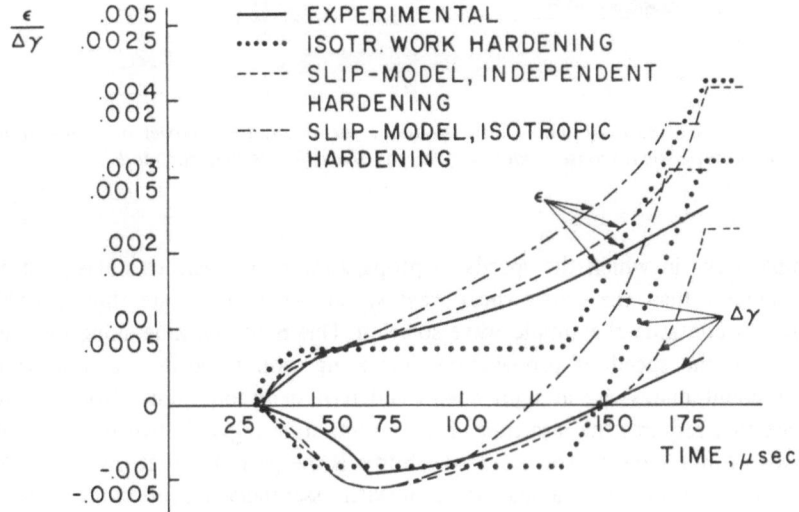

Figure 3. – *Comparison of computed and measured strain-time profiles at x = 15.9 cm
in a tube of 3003 Aluminum ($\tau_s = 22.6$ MPa, $u_0 = 660$ cm/s, $v_0 = 43$ cm/s).*

3. Pressure-shear experiments

Pressure-shear experiments [13] offer definite advantages over combined longitudinal and torsional wave experiments for examining dynamic plastic response under non-proportional loading paths. Figure 4 shows that in pressure-shear experiments two parallel flat plates are impacted with the direction of approach not aligned with the normal to the impact plane. Before reflected waves arrive from lateral boundaries the wave propagation in the flyer and target is one dimensional. Thus, pressure-shear waves are governed essentially exactly by one-dimensional wave equations such as (1). Consequently, interpretation of experimental results is significantly more straightforward than for combined longitudinal and torsional plastic waves which become nominally one dimensional only at distances of several diameters from the impact end where lateral inertia effects are negligible. In the configuration shown in Figure 4 the rotation of the projectile is prevented by a key in the back of the projectile gliding in a keyway in the barrel. The velocity of the projectile at impact is obtained by recording the times at which a series of wires are shorted out. The inclination of the flyer at impact is monitored by means of a series of voltage-biased tilt pins that are positioned in the impact plane of the target and shorted out by contact with the flyer. Impart occurs in a vacuum chamber in order to minimize an air cushion that could lubricate the impact faces and allow slip to occur.

The time profiles of the waves propagating in the target are monitored by means of an interferometer, Figure 5, that is used to monitor, simulta-

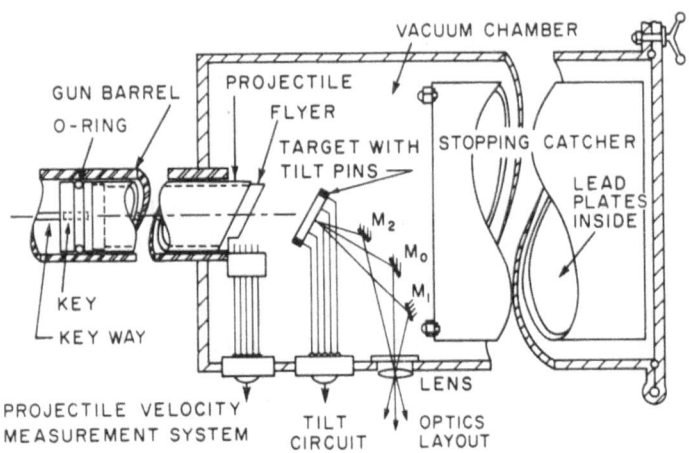

Figure 4. – *Schematic of pressure-shear experiment.*

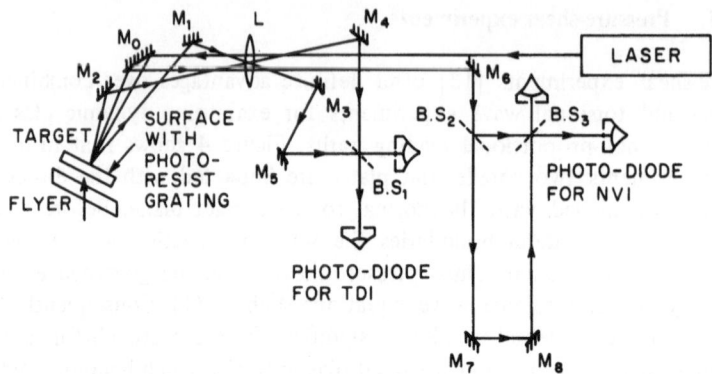

Figure 5. – *Schematic of transverse displacement interferometer (TDI) with normal velocity interferometer (NVI).*

neously and independently, the normal and in-plane components of the motion of the rear surface of the target. The normal velocity interferometer (NVI in Fig. 5) [2] operates by the combination of two light beams that have been at the target at times that differ by the time required for light to go around the delay leg $B.S_2$, M_7, M_8, $B.S_3$. One peak-to-peak variation in intensity at the photo-diode detector corresponds to a change in normal velocity of $\lambda/2\tau$ where λ is the wavelength of the laser light and τ is the time required for the light to go around the delay leg. The transverse displacement interferometer (TDI in Fig. 5) [12] makes use of two beams diffracted symmetrically from a diffraction grating on the rear surface of the target. One peak-to-peak variation in intensity at the photodiode detector corresponds to a transverse displacement of $d/2n$ where d is the pitch of the grating and n is the order of the diffracted beams.

Experimental results for pressure-shear impact of 6061-T6 aluminium plates at a projectile velocity of 0.215 mm/μsec and an inclination of 26.6° are shown in Figures 6 and 7. The corresponding values of normal and transverse components of particle velocity imposed at the impact face of the target are 0.096 mm/μsec and 0,048 mm/μsec, respectively. The oscilloscope trace in Figure 6 shows the output of the NVI. When the wavefront arrives there is a burst of fringes (peak-to-peak variations in intensity) that is not recorded because the frequency of the fringes exceeds the response capability of the detection system (approx. 400 MHz). Thus, the initial jump in normal velocity is indeterminate to an integer multiple of the NVI constant $\lambda/2\tau$. Knowledge of the final velocity (i.e. 2 x 0.096 mm/μsec) enables one to determine this integer with certainty. The normal velocity increases behind the wavefront and then becomes nearly constant until approximately 0.35 μsec after the arrival of the wavefront. At this time the longitudinal wavefront

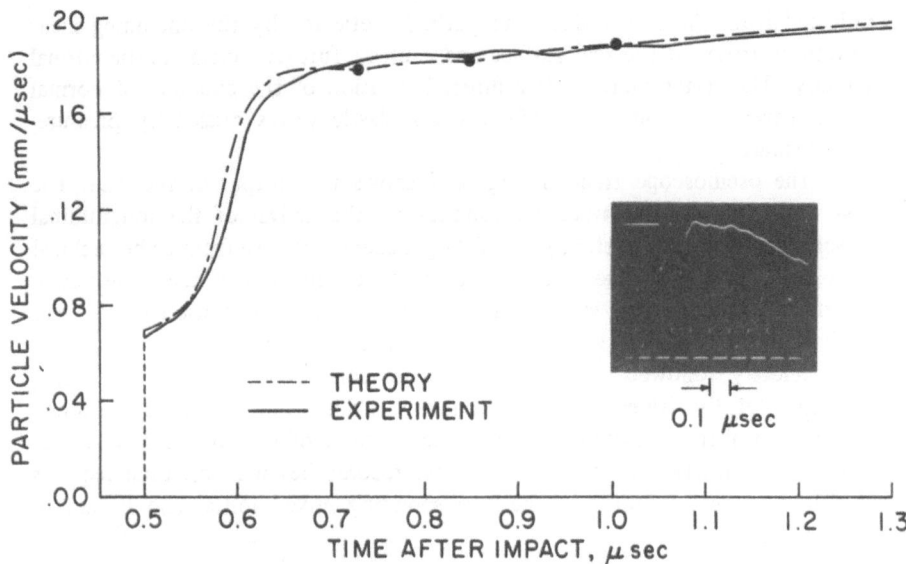

Figure 6. – *Normal velocity-time profile at the rear surface for symmetric impact of 3.2 mm thick plates of 6061-T6 aluminum; NVI sensitivity* $\lambda/2\tau$ *is 0.0442 mm/μsec./ fringe.*

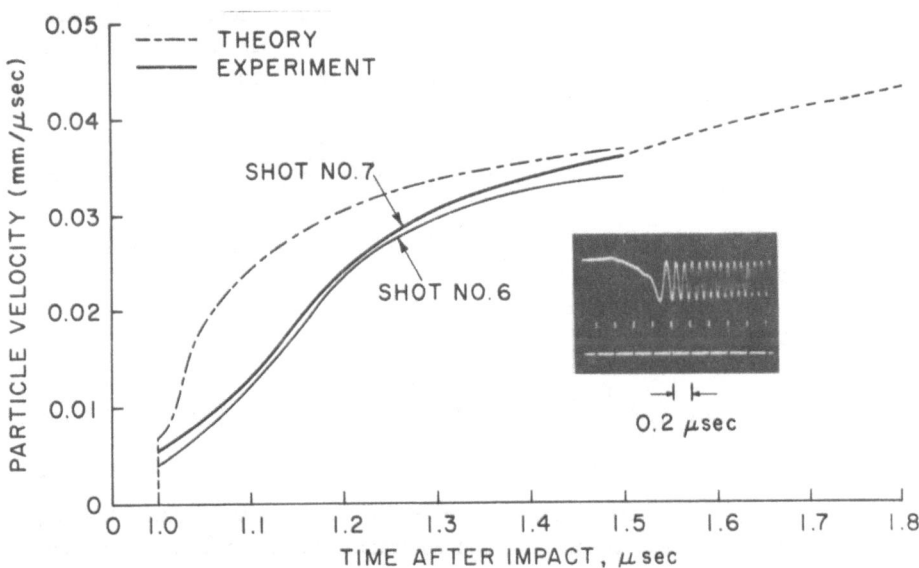

Figure 7. – *Transverse velocity-time profile at the rear surface for symmetric impact of 3.2 mm thick plates of 6061-T6 aluminum; TDI sensitivity* $d/2n$ *is 2.5 μm/fringe.*

reflected from the rear surface and partially reflected by the oncoming shear wavefront arrives at the rear surface and causes a further increase in the normal velocity. The latter increase is a direct indication of the coupling of normal and transverse motion in combined stress plastic waves caused by pressure-shear loading.

The oscilloscope trace in Figure 7 shows the output of the TDI. The first deflection of the trace corresponds to the arrival of the longitudinal wavefront. Tilt between the flyer and target causes the wavefront to be inclined with respect to the target rear surface so that reflection of the longitudinal wavefront from the rear surface causes a small transverse motion of the rear surface. When the shear wavefront arrives there is an abrupt increase in transverse velocity followed by a gradual increase corresponding to the increasing frequency of the fringes. At approximately 1.5 μsec. after impact the longitudinal wavefront reflected from the back surface of the flyer arrives at the target. Interpretation of the experimental records beyond this time requires as full numerical solution of the corresponding mixed initial and boundary value problem.

4.   Comparison of theory and experiment

Figures 6 and 7 also include computed velocity-time profiles based on an elastic/visco-plastic model for 6061-T6 aluminum that is described fully elsewhere [1]. The main features of the model, which is similar in form to the one proposed by Perzyna [15], is that the constitutive equations for the infinitesimal strain rates $\dot{\epsilon}_{ij}$ have the form

$$\dot{\epsilon}_{ij} = \frac{1 + \nu}{E} \dot{\sigma}_{ij} - \frac{\nu}{E} \delta_{ij} \dot{\sigma}_{kk} + \langle \phi(\bar{\tau}, \bar{\gamma}^p) \rangle \frac{\partial f}{\partial \sigma_{ij}} \tag{13}$$

where ν, E, δ_{ij} are Poisson's ratio, Young's modulus, and the Kronecker delta, respectively; $f = (S_{ij} S_{ij}/2)^{1/2}$ is the Huber-Mises yield function in which S_{ij} is the stress deviator. The symbol $\langle \phi(\bar{\tau}, \bar{\gamma}^p) \rangle$ is a plastic strain-rate function defined by

$$\langle \phi(\bar{\tau}, \bar{\gamma}^p) \rangle = \begin{cases} 0 & \text{for } \bar{\tau} < \tau_p(\bar{\gamma}^p) \\ \\ \phi(\bar{\tau}, \bar{\gamma}^p) & \text{for } \bar{\tau} > \tau_p(\bar{\gamma}^p) \end{cases} \tag{14}$$

where $\bar{\tau}, \phi, \bar{\gamma}^p$ are, respectively, an effective shear stress, a corresponding equivalent plastic shear strain rate, and equivalent plastic shear strain defined by

$$\bar{\tau} = (S_{ij} S_{ij}/2)^{1/2} \tag{15a}$$

$$\phi = (2\,\dot{\epsilon}_{ij}^{vp}\,\dot{\epsilon}_{ij}^{vp})^{1/2} \tag{15b}$$

$$\overline{\gamma}^p = \int_0^t \phi\,dt$$

where $\dot{\epsilon}_{ij}^{vp}$ is the visco-plastic strain rate given by the last term in (13). The quantities $\overline{\tau}$ and ϕ are defined in such a way that $\overline{\tau}\phi$ is the rate of plastic working. The function $\tau_p(\overline{\gamma}^p)$ in (14) is the quasi-static stress-strain curve in a pure shear test.

The important feature of this model relative to understanding the experimental results is that, for constant $\overline{\gamma}^p$, ellipsoids f = constant are surfaces of constant equivalent plastic strain-rate. The plastic strain-rate vector is normal to these ellipsoids. Thus, for the stress trajectory that occurs in pressure-shear experiments the plastic shear strain-rate after the shear wave-front arrives is only a small fraction of the equivalent plastic strain-rate since the normal to the surface f = constant is nearly parallel to the axis of the normal stress σ_{xx}. A theory that underestimates the plastic shear strain-rate in the region behind the shear wavefront tends to underestimate the decay of the transverse velocity in this region. Such an effect is a possible explanation for the discrepancy between theory and experiment shown in Figure 7. This discrepancy would be reduced if the normal to the surface f = constant had a larger component along the axis of the shear stress σ_{xy}. Consequently, it appears that a flow potential with a smaller radius of curvature at the final stress state reached along the σ_{xx} axis during the longitudinal wave loading would predict wave profiles that agree better with those observed in experiments. Flow potentials that develop relatively sharp curvature near the current stress state after continued loading in one direction are to be expected for slip models of plastic response. Thus, as in the case of combined longitudinal and torsional plastic waves, it appears that the pressure-shear experiments indicate that slip models have promise for predicting the dynamic plastic response of metals subjected to non-proportional loading paths.

Acknowledgement

This work has been supported by the National Science Foundation, including support provided through the Materials Research Laboratory at Brown University.

REFERENCES

[1] ABOU-SAYED A.S. and R.J. CLIFTON. – "Analysis of combined pressure-shear waves in an elastic-visco-plastic material." *Journal of Applied Mechanics*, **44** (1977): 79-84.

[2] BARKER L.M. – "Fine structure of compressive and release wave shapes in aluminum measured by the velocity interferometer technique." In *Behavior of Dense Media Under High Dynamic Pressures*. New York: Gordon and Breach, 1968, pp. 483-505.

[3] BLEICH N.N. and I. NELSON. – "Plane waves in an elastic-plastic half space due to combined surface pressure and shear." *Journal of Applied Mechanics*, 33 (1969): 149-158.

[4] BUDIANSKY B. and T.T. WU. – "Theoretical prediction of plastic strains of polycrystals." *Proceedings of the Fourth U.S. National Congress of Applied Mechanics* (1962): 1175-1185.

[5] CLIFTON R.J. – "An Analysis of combined longitudinal and torsional plastic waves in a thin-walled tube." *Proceedings of the Fifth U.S. National Congress of Applied Mechanics* (1966): 465-480.

[6] CLIFTON R.J. – "Plastic waves: theory and experiment." *Mechanics Today*, 1, (1974): 102-167.

[7] CRISTESCU N. – "On the propagation of elastic-plastic waves for combined stresses." *Journal of Applied Mathematics and Mechanics*, 23 (1959): 1605-1612.

[8] GULDENPFENNIG J. – "Anwendung eines Modells der Vielkristallplastizität auf ein Problem, gekoppelter elasto-plastischer Wellen." Mitteilungen aus dem Institut fur Mechanik, Ruhr-Universität, Bochum, 1977.

[9] GULDENPFENNIG J. and R.J. CLIFTON. – "Plastic waves of combined stress based on self-consistent slip models." *Journal of Mechanics and Physics of Solids*, 28 (1980): 201-219.

[10] HSU J.C.C. and R.J. CLIFTON. – "Plastic waves in a rate-sensitive material, II: waves of combined stress." *Journal of Mechanics and Physics of Solids*, 22 (1974): 255-266.

[11] KALISKY S., W.K. NOWACKI and E. WLODARCZYK. – "Propagation of plane loading and unloading biwaves in an elastic-viscoplastic semi-infinite body, Part I: Theory and Part II: Numerical analysis." *Proceedings of Vibrations Problems*, 8 (1967).

[12] KIM K.S., R.J. CLIFTON and P. KUMAR. – "A combined normal and transverse displacement interferometer with an application to impact of Y-cut quartz." *Journal of Applied Physics*, 48 (1977): 4132-4139.

[13] KIM K.S. and R.J. CLIFTON. – "Pressure-shear waves in 6061-T6 Aluminum." *Journal of Applied Mechanics*, 47 (1980): 11-16.

[14] LIPKIN J. and R.J. CLIFTON. – "Plastic waves of combined stresses due to longitudinal impact of a pretorqued tube, Part 1: Experimental results; Part 2: Comparison of theory with experiment." *Journal of Applied Mechanics*, 37 (1970): 1107-1120.

[15] PERZYNA P. – "Fundamental problems in viscoplasticity." *Advances in Applied Mechanics*, 9, New York: Academic Press, 1966, pp. 243-377.

[16] RAKHMATULIN Kh. A. – "On the propagation of elastic-plastic waves owing to combined loadings." *Journal of Applied Mathematics and Mechanics*, 22 (1958): 1079-1088.

[17] TING T.C.T. and N. NAN. – "Plane waves due to combined compressive and shear stresses in a half space." *Journal of Applied Mechanics*, 36 (1969): 1389-1397.

RESUME

(Déformations plastiques dynamiques pour des chemins de charge non proportionnels)

Les ondes plastiques sous contraintes combinées sont discutées comme moyen d'investigation pour les déformations plastiques dynamiques pour des chemins de charge non proportionnels. La comparaison entre la théorie et les expériences pour les ondes longitudinales et torsionnelles dans les tubes minces et pour les ondes en pression-cisaillement dans des plaques, suggère qu'un meilleur accord pourrait être obtenu si la réponse plastique était modelée au moyen de modèles de glissements au lieu de modèles basés sur l'existence d'une surface lisse d'écoulement.

Analytical and Numerical Methods for the Determination
of Mechanical Properties of Composites

Méthodes Analytiques et Numériques pour la Détermination
des Propriétés Mécaniques des Composites

General Lecture/Conférence Générale

G.J. DVORAK
  Mechanical Properties of Composites.

Communications

Z. HASHIN
  Analysis of the Effects of Fiber Anisotropy on the Properties of Carbon
  and Graphite Fiber Composites.

A. SAWICKI
  Elasto-Plastic Theory of Composites with Regular Internal Structure.

S. LICHARDUS and J. SUMEC
  Finite Strip Method for the Analysis of Visco-Elastic Two-Dimensional
  Structures.

General Lecture : *Conférence Générale*

Mechanical Properties of Composites

G.J. Dvorak

The University of Utah, Salt Lake City, Utah, U.S.A.

1. Introduction

Determination of macroscopic mechanical properties of heterogeneous media from the properties of their constituents is a problem of long standing in the mechanics of solids. Current activity in the field includes many areas of applications, such as particulate and fibrous composites, polycrystals, concrete, and geological materials. This lecture first presents a survey of constitutive theories of elastic, anisotropic heterogeneous media, which have been useful in these applications. The second part reviews our recent results in the area of elastic-plastic deformation of laminated metal matrix composite plates. Finally, fracture and fatigue in fibrous composites are briefly discussed.

2. Constitutive relations for anisotropic heterogeneous media

2.1. Overall mechanical properties

The general approach to the evaluation of overall properties of an aggregate from the properties of its constituents is outlined in Figures 1 and 2. The notation and procedure used is formally similar to that of Hill [1] and Laws [2]. Second-order tensors are denoted by boldface lower-case letters, e.g. σ, l, m; fourth order tensors are denoted by boldface upper case letters, e.g., A, L, M; and scalars are denoted by lower case letters, e.g., c, θ. The inner product of second order tensors is written as ab, the second order inner product of tensor A and b as Ab. Similarly AB stands for the fourth-order inner product of A and B. The second order tensors are all symmetric, unless otherwise noted. The fourth-order tensors have symmetries $A_{ijk\ell} = A_{jik\ell} = A_{ij\ell k}$, but $A_{ijk\ell} \neq A_{k\ell ij}$. The transpose $A^T_{ijk\ell} = A_{k\ell ij}$. The inverse, when it exists is denoted as A^{-1} so that $AA^{-1} = I = A^{-1}A$, where I is defined in Figure 1 in terms of Kronecker's delta δ_{ij}.

Suppose that the composite consists of n phases bonded together. The information about the microstructure is limited, only phase volume fractions $c_r (\Sigma c_r = 1)$, phase properties L_r, M_r, l_r, m_r, and the general shape of each phase are known. We consider a representative volume R, which is typical of the composite on average and contains a sufficient number of constituents. First, the relationship between local stress rate averages $\dot{\sigma}_r$, and overall stress rate average $\dot{\bar{\sigma}}$ is written in terms of the constituent volume fractions c_r. The same is done for strain rate averages $\dot{\varepsilon}_r$ and $\dot{\bar{\varepsilon}}$. Second, known instantaneous constituent properties, moduli L_r, compliances M_r, thermal stress vectors l_r, and thermal expansion vectors m_r are used to write local constitutive relations. Third, relations between local and overall rate averages are written in terms of concentration factors A_r, a_r, and B_r, b_r. This permits the derivation of the equations containing only the concentration factors and volume ratios.

The procedure is continued in Figure 2, where one finds the final result, i.e., expressions for overall instantaneous properties of the composite, L, M, l, m, in terms of constituent properties, volume ratios, and the concentration factors. Specialized results are obtained for binary mixtures, where it is possible to find the overall instantaneous moduli L, and compliances M in terms of a single concentration factor, and the overall thermal vector l without an explicit knowledge of concentration factors.

The procedure described in Figures 1 and 2 was developed by Hill [1] and extended to thermostatics by Laws [2]. In principle, the procedure can

---

OVERALL MECHANICAL PROPERTIES OF HETEROGENEOUS MEDIA

Overall Averages in R_x

$$\dot{\bar{\sigma}} = c_1 \dot{\sigma}_1 + c_2 \dot{\sigma}_2 + \ldots \qquad \dot{\bar{\varepsilon}} = c_1 \dot{\varepsilon}_1 + c_2 \dot{\varepsilon}_2 + \ldots$$

Constituent Properties ($r = 1, 2, \ldots n$)

$$\dot{\sigma}_r = L_r \dot{\varepsilon}_r - \dot{\theta} \ell_r \qquad \dot{\varepsilon}_r = M_r \dot{\sigma}_r + \dot{\theta} m_r , M_r = L_r^{-1}$$

$$\ell_r = L_r m_r \qquad m_r = M_r \ell_r$$

Relations Between Averages in R_x

$$\dot{\sigma}_r = B_r \dot{\bar{\sigma}} + \dot{\theta} b_r \qquad \dot{\varepsilon}_r = A_r \dot{\bar{\varepsilon}} - \dot{\theta} a_r$$

$$c_r B_1 + c_2 B_2 + \ldots = I \qquad c_1 A_1 + c_2 A_2 + \ldots = I$$

$$c_1 b_1 + c_2 b_2 + \ldots = 0 \qquad c_1 a_1 + c_2 a_2 + \ldots = 0$$

$$I = \frac{1}{2} \left(\delta_{ik} \delta_{j\ell} + \delta_{i\ell} \delta_{jk} \right)$$

Figure 1. –

OVERALL MECHANICAL PROPERTIES OF HETEROGENEOUS MEDIA

Relations Between Averages in $\underset{x}{R}$

$$\dot{\bar{\sigma}} = c_1 \dot{\sigma}_1 + c_2 \dot{\sigma}_2 + \ldots \qquad \dot{\bar{\varepsilon}} = c_1 \dot{\varepsilon}_1 + c_2 \dot{\varepsilon}_2 + \ldots$$

$$\dot{\sigma}_r = B_r \dot{\bar{\sigma}} + \dot{\theta} b_r \qquad \dot{\varepsilon}_r = A_r \dot{\bar{\varepsilon}} - \dot{\theta} a_r$$

$$\dot{\bar{\sigma}} = L \dot{\bar{\varepsilon}} - \dot{\theta} \ell \qquad \dot{\bar{\varepsilon}} = M \dot{\bar{\sigma}} + \dot{\theta} m$$

$$L = c_1 L_1 A_1 + \ldots \qquad M = c_1 M_1 B_1 + c_2 M_2 B_2 + \ldots = L^{-1}$$

$$\ell = \Sigma c_r (\ell_r + L_r a_r) \qquad m = \Sigma c_r (m_r + M_r b_r)$$

Binary Mixtures $(r = 1, 2)$

$$L - L_2 = c_1 (L_1 - L_2) A_1 \qquad M - M_2 = c_1 (M_1 - M_2) B_1$$

$$\ell = (L - L_2)(L_1 - L_2)^{-1} \ell_1 + (L - L_1)(L_2 - L_1)^{-1} \ell_2$$

Figure 2. –

be used for anisotropic inclusions, as well as anisotropic mixtures, providing that the appropriate concentration factors are known. Typical examples of applications are isotropic mixtures consisting of isotropic or anisotropic phases modelled as spherical, ellipsoidal, or thin disc inclusions, e.g., polycrystals [3, 6], geological materials [7]; and fibrous composites, which are transversely isotropic in the overall sense, but may consist of isotropic phases [8, 9]. It is advantageous to establish the overall degree of anisotropy of the mixture by assumption, and calculate the concentration factors accordingly. The formulation in terms of stress and strain rates is useful in applications involving inelastic deformation and can be readily converted to total values in the elastic case.

2.2. *The self-consistent method*

This is a frequently used approach to calculation of the concentration factors in linear elasticity problems, which rests on the assumption that each of the constituents can be represented by an ellipsoidal inclusion embedded in the homogeneous composite medium. Then, as shown by Eshelby [10], the strain in the inclusion is uniform. This facilitates the evaluation of the concentration factors for many practically useful inclusion shapes [11]. A similar approach can be used in axisymmetric plasticity problems in fibrous composites [12], but not in nonsymmetric plasticity problems because the local field in the inclusion is not uniform. Applications to viscoelastic composites have been made as well [13]. To fix ideas, Figure 3 shows schematically a specific geometry of the self-consistent model. The composite consists here of two phases ($r = 1, 2$) modelled as cylindrical inclusions in the medium with unknown overall properties L, l, M, m. In this particular case one of the inclusions ($r = 1$) is a composite cylinder containing a fiber surrounded by a layer of matrix. Let c_f denote the volume fraction of the fiber in the composite cylinder. Then, for $c_f = 1$ the matrix layer shall vanish and only the fiber will remain in the first inclusion. The second inclusion ($r = 2$) represents the continuous matrix, but it is regarded also as a cylinder in the self-consistent approximation. When one denotes the inclusion volume fractions as c_1 and c_2, the actual fiber and matrix volume fractions as v_f and v_m, it follows that

$$v_f + v_m = c_1 + c_2 = 1, \quad c_1 = v_f/c_f, \tag{1}$$

and

$$v_f \leqslant c_f \leqslant 1 \quad \text{with} \quad c_2 = 0 \quad \text{for} \quad c_f = v_f.$$

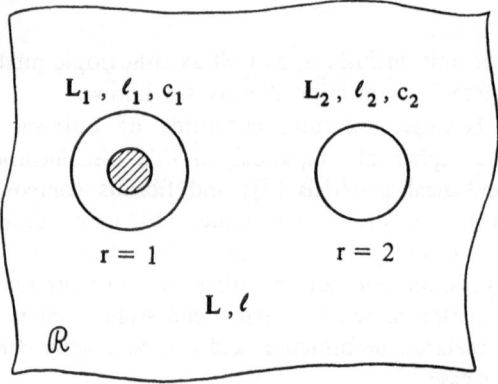

Figure 3. – *The Modified Self-Consistent Model.*

The calculation of the concentration factors for the self-consistent model is illustrated in Figure 4 for the case of n inclusions. This particular procedure, developed by Hill [14], rests on the evaluation of "constraint tensors" L^* and M^* which describe the response of an ellipsoidal cavity in the composite medium to a uniform traction rate $\dot{\sigma}^*$ applied at the cavity boundary. A superposition of the uniform fields $\dot{\bar{\sigma}}$, $\dot{\bar{\varepsilon}}$, with $\dot{\sigma}^*$, $\dot{\varepsilon}^*$ permits an evaluation of the actual local fields in the inclusions which reside in the

THE SELF-CONSISTENT METHOD

Ellipsoidal cavity in medium L loaded by uniform traction rate $\dot{\sigma}^*$:

$$\dot{\sigma}^* = -L^* \dot{\varepsilon}^* \qquad\qquad \dot{\varepsilon}^* = -M^* \dot{\sigma}^*$$

Superpositon of uniform fields $\dot{\bar{\sigma}}$ and $\dot{\bar{\varepsilon}}$ with $\dot{\sigma}_r$ and $\dot{\varepsilon}_r$

$$\dot{\sigma}^* = \dot{\sigma}_r - \dot{\bar{\sigma}} \qquad\qquad \dot{\varepsilon}^* = \dot{\varepsilon}_r - \dot{\bar{\varepsilon}}$$

$$\dot{\sigma}_r - \dot{\bar{\sigma}} = L^* (\dot{\bar{\varepsilon}} - \dot{\varepsilon}_r) \qquad\qquad \dot{\varepsilon}_r - \dot{\bar{\varepsilon}} = M^* (\dot{\bar{\sigma}} - \dot{\sigma}_r)$$

Recall :

$$\dot{\bar{\sigma}} = L \dot{\bar{\varepsilon}} \qquad\qquad \dot{\bar{\varepsilon}} = M \dot{\bar{\sigma}}$$

$$\dot{\sigma}_r = L_r \dot{\varepsilon}_r \qquad\qquad \dot{\varepsilon}_r = M_r \dot{\sigma}_r$$

$$\dot{\sigma}_r = B_r \dot{\bar{\sigma}} \qquad\qquad \dot{\varepsilon}_r = A_r \dot{\bar{\varepsilon}}$$

Evaluation of concentration factors A_r, B_r, a_r, b_r

$$(L^* + L_r) \dot{\varepsilon}_r = (L^* + L) \dot{\bar{\varepsilon}}, \qquad (M^* + M_r) \dot{\sigma}_r = (M^* + M) \dot{\bar{\sigma}}$$

$$A_r = (L^* + L)(L^* + L_r)^{-1}, \qquad B_r = (M^* + M)(M^* + M_r)^{-1}$$

$$a_r = (\ell - \ell_r)(L^* + L_r)^{-1}, \qquad b_r = (m - m_r)(M^* + M_r)^{-1}$$

Figure 4. –

cavities, and leads eventually to the expressions for the concentration factors, as shown in Figure 4.

It is clear that the self-consistent model makes no distinction between the phases, e.g., between fiber and matrix in a composite. Also, the results found with this approach may become unreliable when the constituent properties and their volume fractions are extremely different. These and other limitations of the method have been discussed in the literature [6, 14] and will be recalled in the sequel.

2.3. *The differential scheme*

A variant of the self-consistent method, due to Roscoe [15], can be developed from the result of Einstein [16] concerning a dilute concentration of rigid spheres in an incompressible Newtonian fluid. In principle, the overall properties are evaluated in an incremental fashion; a small amount of inclusions is first added to the matrix, new overall properties of the composite are calculated; then an additional amount of inclusions is added to the current composite. The process is repeated until the added inclusions occupy the desired volume fraction c_1. Figure 5 describes some aspects of the procedure, as presented by McLaughlin [17].

THE DIFFERENTIAL SCHEME FOR BINARY MIXTURES

Recall (for $r = 1, 2$)

$$L - L_2 = c_1 (L_1 - L_2) A_1$$

Denote:

L_1 — inclusion moduli

c_1 — current inclusion concentration

$L(c_1)$ — current composite moduli

$\bar{L} = L(c_1 + \delta c_1)$ — new composite moduli

Obtain:

$$\bar{L} - L = \frac{\delta c}{(1 + \delta c)} (L_1 - L) E_1 \;\; , \; E_1 = [I + P(L_1 - L)]^{-1}$$

$$\frac{dL}{dc_1} = \frac{1}{(1 - c_1)} (L_1 - L) E_1,$$

$$L = L_2 \quad \text{at} \quad c_1 = 0$$

Figure 5. –

2.4. *Bounding methods*

Appealing as the methods of § 2.2 and § 2.3 are, they lack a rigorous foundation which would guarantee the correctness of the results. Such assurance can be obtained from classical energy principles which have been used to develop bounds on elastic moduli, primarily for binary composite aggregates.

Elementary bounds can be obtained using either the Voigt assumption $A_1 = A_2 = I$, or the Reuss assumption $B_1 = B_2 = I$. The elastic moduli follow from the equations of Figures 1 and 2, with $M_R \leqslant M \leqslant M_V$ [1].

Hashin and Shtrikman [18, 19] derived new variational principles which led to tighter bounds than the Voigt and Reuss averages. These principles have been rederived by Hill [20] who aslo showed their relation to the classical principles of potential and complementary energy. A schematic sequence of equations used in the derivation of the bounds is shown in Figure 6. Complete expositions can be found in original references [21 to 24].

HASHIN-SHTRIKMAN BOUNDS*

(Best bounds in terms of volume fractions and phase moduli)

Reinforced material	Comparison material
$\mu, \varepsilon, \sigma, L$	$\mu^*, \varepsilon^*, L^*$
$U = W V$	$U^* = W^* V$

Let $\mu' = \mu - \mu^*$

         $\varepsilon' = \varepsilon - \varepsilon^*$

Def $\tau = \sigma - L^*\varepsilon = (L - L^*)\varepsilon$

$2(U^* - U) \geqslant \text{OR} \leqslant \int \tau[(L-L^*)^{-1}\tau - \varepsilon' - 2\varepsilon^*]\, dV$

max U when $(L - L^*)$ is positive definite
min U when $(L - L^*)$ is negative definite

div $(L^*\varepsilon' + \tau) = 0$, $\mu' = 0$ on S

$(L^*\varepsilon' + \tau)$ n, and μ' are continuous on S_r

* The derivation presented here was made by Hill [20].

Figure 6. –

The relationship between the self-consistent results and the Hashin-Shtrikman bounds has been established for a number of composite systems, including the binary fibrous composites [21]. The conclusion is that the self-consistent results fall between the bounds in these systems, subject to certain restrictions in specific instances of limited practical significance.

More accurate boundings methods for elastic heterogeneous media, which require more information about the geometry of the microstructure, have been developed. Such approaches to the evaluation of elastic moduli have been based on statistical treatment [25 to 27] and on quantum mechanical scattering theory [28].

2.5. *Results for fibrous composites*

A specific form of the overall constitutive law for a unidirectional lamina is shown in Figure 7 [8]. Here, k is the plane strain bulk modulus for lateral dilatation without longitudinal extension, n is the bulk modulus for longitudinal axial straining, and ℓ is the associated cross modulus; p and m are the longitudinal and transverse shear moduli, respectively. The conventional Young's modulus E and Poisson's ratio ν under longitudinal load (in the fiber direction x_3) are given as well. Although the composite is regarded as a homogeneous, transversely isotropic solid in the overall sense, it is possible to use microstructural considerations related to the fiber reinforcement to establish universal connections between the elastic moduli k, ℓ, and n [8], and also between these moduli and the overall thermal expansion coefficients α, β [12]. These connections have been extended to certain elastic-plastic composites subjected to axisymmetric mechanical and uniform thermal loads [12].

ELASTIC MODULI OF FIBROUS COMPOSITES

(Transverse isotropy)

$$\frac{1}{2}(\sigma_{11} + \sigma_{22}) = k(\epsilon_{11} + \epsilon_{22}) + \ell\epsilon_{33}$$

$$\sigma_{33} = \ell(\epsilon_{11} + \epsilon_{22}) + n\epsilon_{33}$$

$$(\sigma_{11} - \sigma_{22}) = 2m(\epsilon_{11} - \epsilon_{22})$$

$$\sigma_{12} = 2m\epsilon_{12}$$

$$\sigma_{13} = 2p\epsilon_{13}, \quad \sigma_{32} = 2p\epsilon_{32}$$

$$E = n - \ell^2/k, \quad \nu = \ell/2k$$

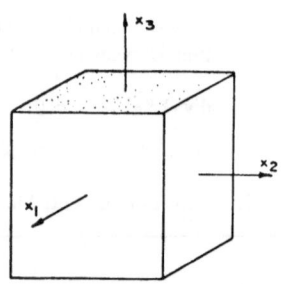

Universal Connections (Hill 1964, Dvorak and Bahei-El-Din 1979).

$$\frac{k - k_1}{\ell - \ell_1} = \frac{k - k_2}{\ell - \ell_2} = \frac{\ell - c_1\ell_1 - c_2\ell_2}{n - c_1 n_1 - c_2 n_2} =$$

$$= \frac{k\alpha + \ell\beta - c_1(k_1\alpha_1 + \ell_1\beta_1) - c_2(k_2\alpha_2 + \ell_2\beta_2)}{\ell\alpha + n\beta - c_1(\ell_1\alpha_1 + n_1\beta_1) - c_2(\ell_2\alpha_2 + n_2\beta_2)}$$

Figure 7. −

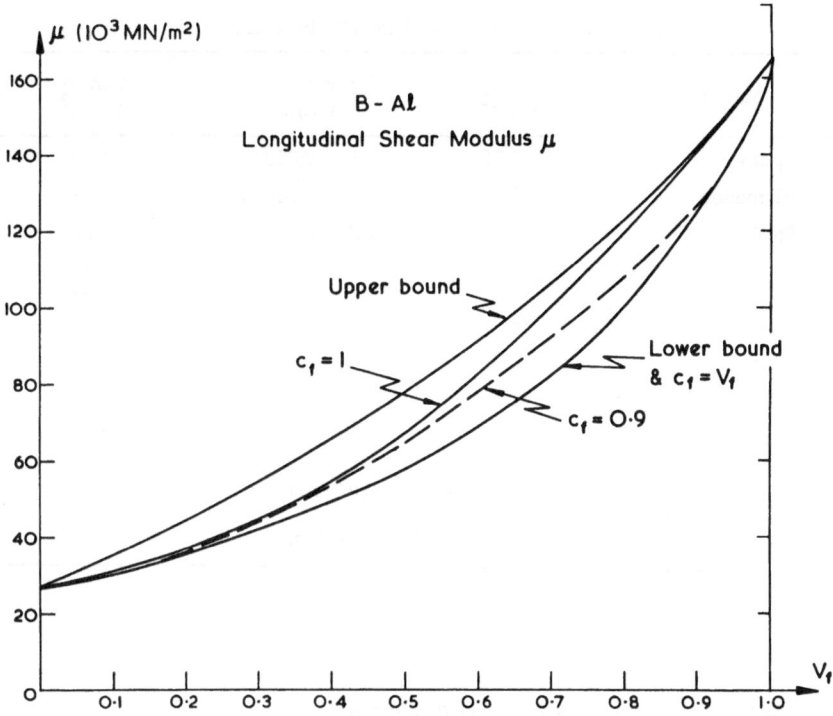

Figure 8. — -*Longitudinal Shear Modulus of B-Al.*

It follows that only one of the moduli k, ℓ, n is independent, so that in general the composite properties can be described by three moduli, such as k, m, and p.

An illustration of the relationship between H-S bounds and SCM values of the composite moduli in fibrous composites is given in Figure 8. The longitudinal shear modulus p (denoted as μ in Fig. 8) of a B-Al composite is plotted as function of the fiber volume fraction v_f. The results were obtained for the modified model of Figure 3; $c_f = 1$ denotes the standard self-consistent estimate, $c_f = 0.9$ shows the effect of a thin matrix layer at the fiber on the result. The Hashin-Shtrikman upper and lower bounds are shown as well, one can find in this particular case that for $c_f = v_f$ the SCM result coincides with the lower bound.

Finally, Figure 9 shows a list of elastic moduli and strength values (with Y indicating yield strength) of typical matrix and fiber materials. Values for steel are added only for comparison. Even elementary considerations show that fibrous composites have high strength/density and stiffness/density ratios, which are desirable in applications.

FIBER AND MATRIX PROPERTIES			
Material	E GN/m^2	Tensile strength GN/m^2	Density kN/m^3
Epoxy	3.38	0.029	11.9
Aluminum	73	0.069 (Y)	26.3
Steel	207	0.2 – 4.0 (Y)	76.6
FP	350	1.4	39.0
Graphite (T-50)	386	3.4	13.8
Boron	400	3.4	25.2
E-Glass	72	3.4	25.0
S-Glass	86	4.8	24.4

$$1GN/m^2 = 145 * 10^3 lb/in^2$$

$$1 * 10^6 lb/in^2 = 6.9 \, GN/m^2$$

Figure 9. –

3. Elastic-plastic behavior of fibrous composites

3.1. *Introduction*

Almost all practically used metal matrix composites may experience significant inelastic deformation in service. In fact, the elastic range of most systems is small in comparison with their ultimate strength. Therefore, to take advantage of the high strength of fibrous composite materials, it is necessary to admit working loads which exceed their elastic limit.

The inelastic component of the overall deformation usually originates in the matrix, most fibers remain elastic until failure, and thus help to retain high stiffness of the material in the inelastic range, c.f., Figure 9. The composite matrices fall into two broad categories. One consists of polymer materials, such as epoxy resins which are time-dependent; another of metals, such as aluminum, magnesium, and titanium, which are elastic-plastic. The time-dependent behavior of fibrous composites has been extensively studied in the litterature and approximate linear and nonlinear viscoelasticity theories have been developed [13, 30, 31]. However, a detailed discussion of these results is beyond the scope of this presentation. Instead, attention will be given to our recent results pertaining to elastic-plastic deformation of metal matrix composites.

In addition to the elastic and inelastic deformation modes, in which the constituents deform without failure, there exists a variety of deformation modes involving progressive cracking in the matrix, at fiber-matrix interfaces, and also in the fibers. However, such cracking appears to be more pronounced in polymer-matrix rather than metal-matrix systems.

3.2. *Choice of material model*

Metal-matrix composites reinforced by continuous elastic fibers may exhibit an appreciable amount of elastic-plastic deformation depending on the state of stress and temperature. The inelastic component of the overall deformation is caused by plastic flow of the matrix. Although the fibers strengthen the matrix substantially and are the principal source of composite strength, they have a relatively small effect on the overall stress level which causes the onset of plastic flow. Indeed, the presence of the reinforcing fibers may be the very cause of plastic deformation, as in the case of heat treatment of unidirectional materials [32].

In choosing an appropriate model for construction of elastic-plastic constitutive relations, one is tempted to examine the utility of material models which have been used with success in formulations of elastic constitutive relations, as described in Chapter 2. However, both the evaluation of bounds on moduli and the self-consistent estimates depend on the solution of an inclusion problem in which the stress field is found within a cylindrical inclusion as well as in the adjoining composite. The inclusion problem can be solved for linear elastic and viscoelastic constituents, where the strains in the inclusions are uniform. However, the solution of the elastic-plastic counterpart requires the use of a numerical scheme [33] which precludes the development of tractable constitutive relations that could be useful in solving geometrically complex problems.

Although it is possible to simplify the elastic-plastic inclusion problem, and such simplifications have been made in self-consistent estimates of instantaneous elastic-plastic properties of polycrystals [34, 35], our recent results suggest that the approach is unsatisfactory in the case of fibrous composites [12]. Figue 10 indicates the essence of the difficulty: The self-consistent estimate of the initial yield stress in longitudinal shear exceeds the upper bound on limit load of the composite lamina. The modified model of Figure 3 appears to alleviate this difficulty (c.f., results for $c_f = 0.3$, 0.9 in Figure 10), but the elastic-plastic inclusion problem can be solved in simple form only for axisymmetric mechanical and uniform thermal loads.

COMPARISON OF INITIAL YIELD AND LIMIT LOADS IN LONGITUDINAL SHEAR	
Al-B $v_f = 0.3$	$\bar{\tau}_{13}^Y/k^*$
Self consistent method-Initial yielding	
$c_f = v_f = 0.3$	0.7077
$c_f = 0.9$	0.6485
$c_f = 1.0$	1.3245
Finite element method*-Initial yielding	
$v_f = 0.3$	0.7270
Limit analysis – Upper bound**	
$v_f = 0.3$	1.0820

* Dvorak, Rao and Tarn (1974)
** Majumdar and McLaughlin (1973)

Figure 10. –

Under such circumstances, it is necessary to use rather simple material models which represent only the essential aspects of the elastic-plastic behavior. Figure 11 shows a schematic drawing of such a model of a lamina, which is used in the sequel. It consists of a matrix unidirectionally reinforced by continuous elastic fibers. Each of the fibers is assumed to be of very small diameter, so that although the fibers occupy a finite volume fraction of the composite, they do not interfere with matrix deformation in the transverse plane. As a result, the transverse tension and shear, as well as longitudinal shear response of the composite are derived from that of the matrix, except when there is an axial prestrain, and coupling of axial and transverse plastic strain components is encountered. The model can be represented by parallel fiber and matrix bars or plates with axial coupling.

Figure 11 presents schematically the equations derived from the model, more complete description can be found in References [29, 35]. A consequence of the elastic constraint imposed on the matrix by the fiber is the existence of an axial residual normal stress in the plastically deforming matrix. When this residual stress is accounted for in the lamina yield condition, it appears

ELASTIC-PLASTIC LAMINA

Stresses

$$\sigma_m = A_m \bar{\sigma}$$

$$\sigma_f = A_f \bar{\sigma}$$

Yield surface

$$f(\sigma_m) = 0 \quad , \quad f(A_m \bar{\sigma}) = 0$$

With Mises matrix

$$f = \bar{\sigma}^T A_m^T c A_m \bar{\sigma} - Y^2 = 0$$

$$\bar{\sigma} \equiv [\bar{\sigma}_{11} \ \bar{\sigma}_{22} \ \bar{\sigma}_{33} - \bar{\alpha}_{33} \ \bar{\sigma}_{12} \ \bar{\sigma}_{23} \ \bar{\sigma}_{31}]$$

Hardening and flow rules

$$d\bar{\alpha} = df/(\partial f/\partial \bar{\sigma}_{33}) = (I - A_{me}^{-1} A_m) \, d\bar{\sigma}$$

$$d\bar{\epsilon} = c_f \, d\epsilon_f + c_m \, d\epsilon_m, \quad d\bar{\epsilon}_{33} = d\epsilon_{33}^f = d\epsilon_{33}^m$$

$$d_f = M_{fe} A_f \, d\bar{\sigma}, \quad d\epsilon_m = M_{me} A_m \, d\bar{\sigma} + d\lambda_m \frac{\partial f}{\partial \sigma_m}$$

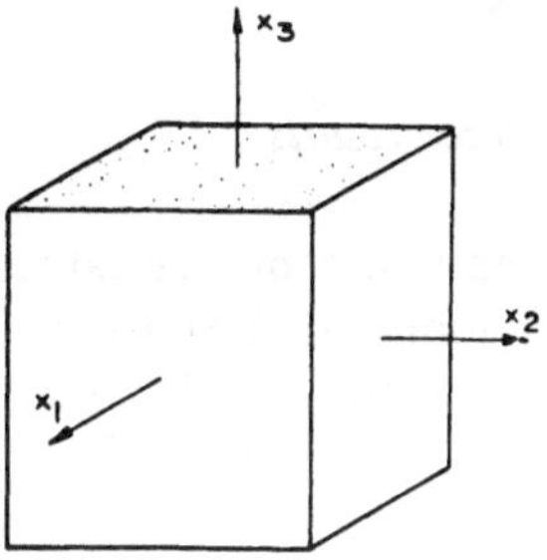

Figure 11. –

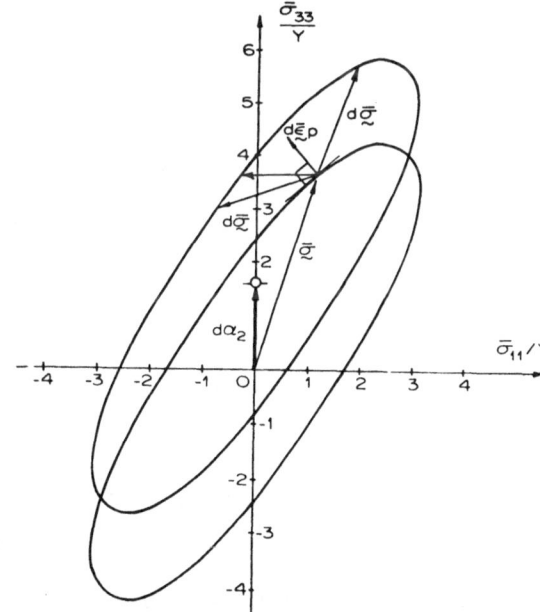

Figure 12. – *Kinematic Motion of Lamina Yield Surface.*

there as the translation factor $\bar{\alpha}_{33}$ causing motion of the original yield surface in the direction $\bar{\sigma}_{33}$. This is illustrated in Figure 12 (where $d\alpha_2 = d\bar{\alpha}_{33}$). It follows that similar translations will appear in analysis of laminated composite plates.

3.3. *Selected results*

We present now some typical results which have been obtained with the model in applications to B-Al laminated plates. Figure 13 shows the results of an evaluation of the overall initial yield surface for biaxial in-plane loading, and of the failure envelope. The initial yield surface, which bounds the region of purely elastic material behavior, is an inner envelope of the local yield surfaces of individual laminae which are plotted as two intersecting ellipses in the overall stress space of Figure 13. The failure envelope was constructed for a family of proportional loading directions, at overall stress levels which

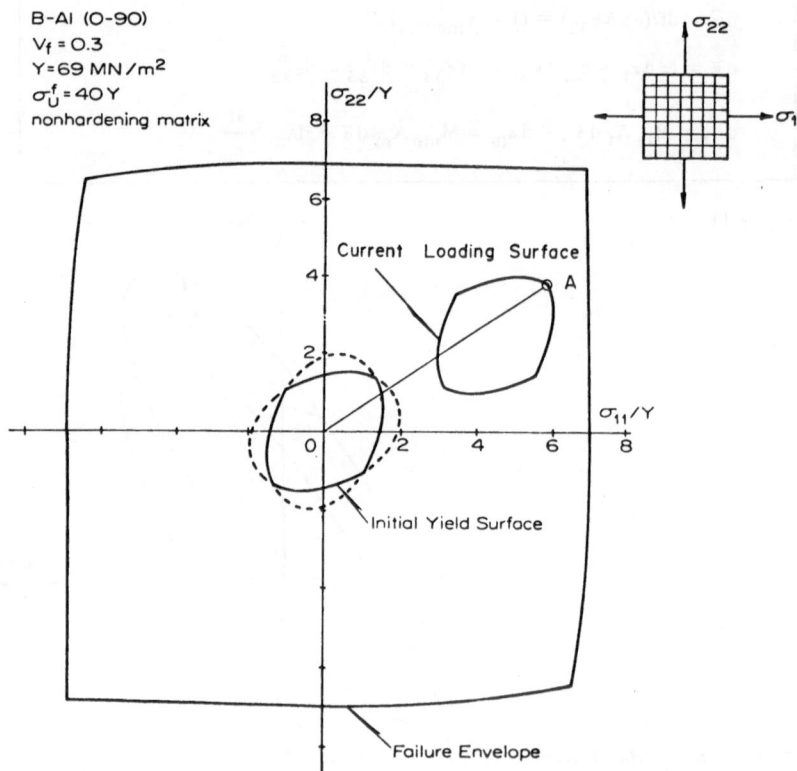

Figure 13. — *The Elastic-Plastic Deformation Range of B-Al Plates.*

would cause onset of local fiber failure in the plastically deforming plate without buckling, assuming equal tensile and compression fiber strengths.

To evaluate the accuracy of the material model, we have made calculations which replicated experimental loading conditions so that the results could be compared with actual measurements. Dr. Byron Pipes has very kindly given us the experimental records which he obtained for a group of laminated plates of different layup, which were tested in simple tension.

To define the properties of the aluminum matrix, we have approximated the experimental curves by Ramberg-Osgood type relation, Figure 14. However, the curve that fits best the behavior of unreinforced matrix provides a poor approximation for the subsequent experimental data. It was found that another matrix stress-strain law, with $n = 5.5$, $\kappa = 0.05$ in Figure 14, gave better results. The difference between the properties of unreinforced and reinforced matrix is not surprising. The fit which we made to the reinforced matrix was made empirically, this might not have been necessary if transverse tension test data were available for a unidirectional material.

Using the two fits indicated in Figure 14, calculations were made for several laminated plates. The theory described earlier was extended to the case of a hardening matrix which follows the Ziegler modification of the Prager's kinematic hardening rule.

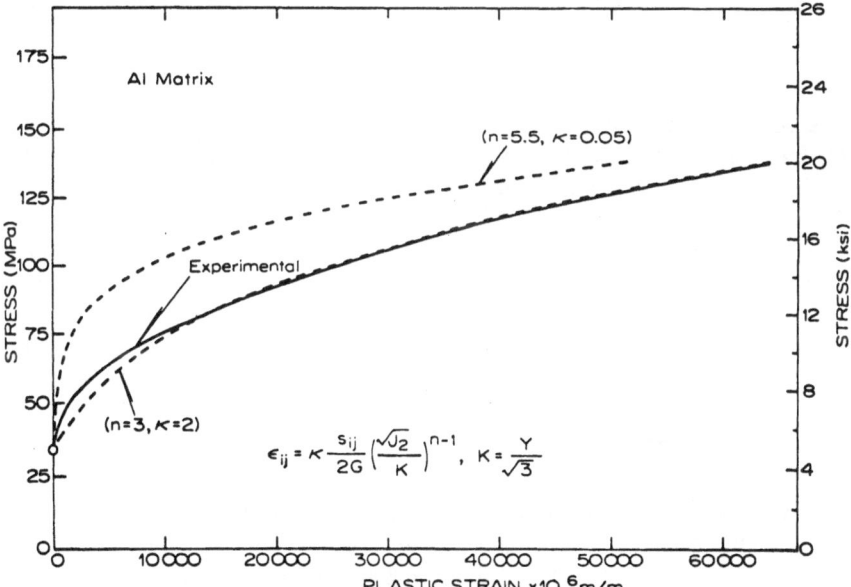

Figure 14. — *Ramberg-Osgood Approximations of Matrix Stress-Strain Curve.*

Figures 15 to 18 indicate how the calculated and experimental results compare. The figures are self-explanatory, and refer to the behavior of two types of plates. One drawing, e.g., 15, or 17, shows the response in simple tension. Another, e.g., 16, or 18, the comparison of the axial and transverse strains. Other types of laminated plates were considered as well, with similar results.

Finally, Figures 19 to 22 show the development of plastic zones at a round hole in $(0/90)_s$ plate made of FP-Al. A finite element program was developed using the constitutive relations described in Figure 11 and used in this calculation [35]. Figure 19 indicates the geometry and loading of the plate. First yielding is observed in the 90-degree ply, at a relatively low load. Figures 20 to 22 show the plastic zones in the two layers (with the zone in the 0-degree layer in darker shade). We note that the 90-degree layer has yielded almost completely at about 10 ksi of applied load (Fig. 22), which corresponds to about 25 % of the expected strength of the plate.

These selected results indicate that the simple material model provides a reasonably accurate description of the observed elastic-plastic behavior of laminated plates. The model can be incorporated in a finite element computer program for solution of geometrically complex problems. Plastic yielding in the matrix must be expected to be the dominant deformation mode at most loading conditions in laminated plates. Elastic behavior is of limited significance in applications which hope to utilize the high strength of the composite.

4. Closing remarks

The scope of this survey does not permit a more detailed written discussion of fracture and fatigue of fibrous composites. The reader may consult Wu [36] as an introduction into recent work on phenomenological failure criteria, Reifsnider [37] who investigated failure mechanisms and damage states in composite laminates, Whitney and Nuismer [38], and Slepetz [39], among many others, on fracture of fibrous laminates. Our own work on shakedown and fatigue of metal matrix composites and its recent extension to laminates is described in [40, 41]. The work by Pipes on the stress distribution at the edges of laminated plates and its relation to fracture are discussed in [42].

Acknowledgement

Drs. R. Byron Pipes, Kenneth L. Reifsnider, John M. Slepetz, and Edward M. Wu have kindly supplied their original results for the oral presentation of this lecture. This work was supported by a grant from the U.S. Army Research Office.

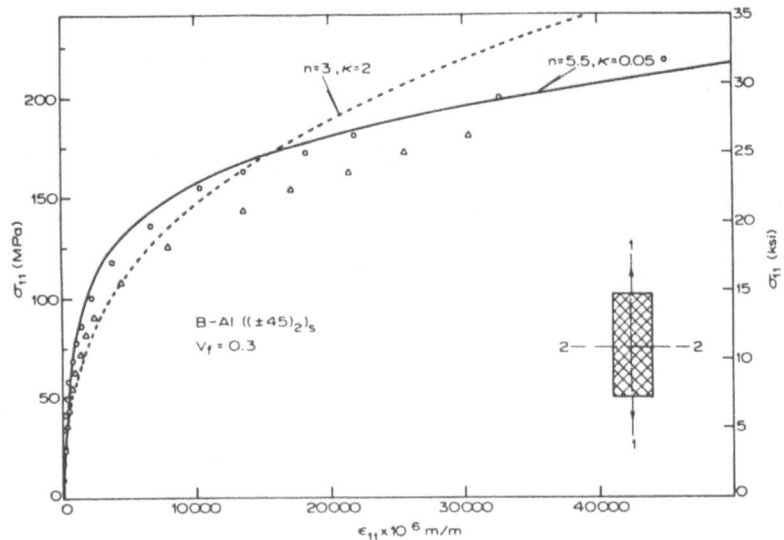

Figure 15. – *Comparison of Calculated and Measured Stress-Strain Curves for a Crossply B-Al Plate.*

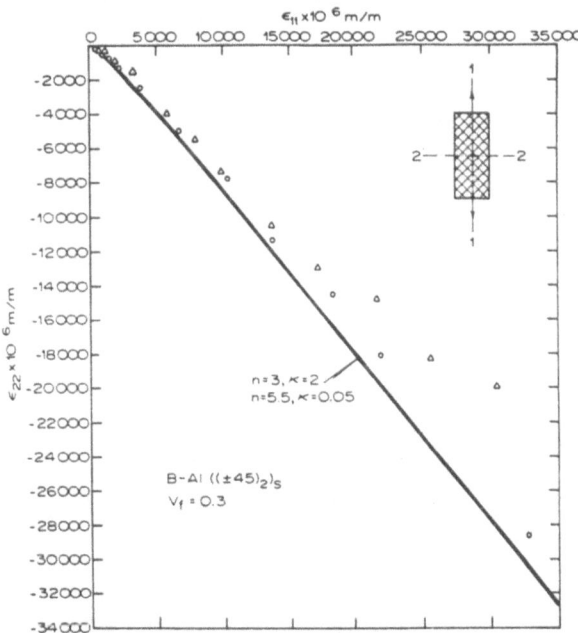

Figure 16. – *Comparison of Calculated and Measured Strains in a Crossply B-Al Plate.*

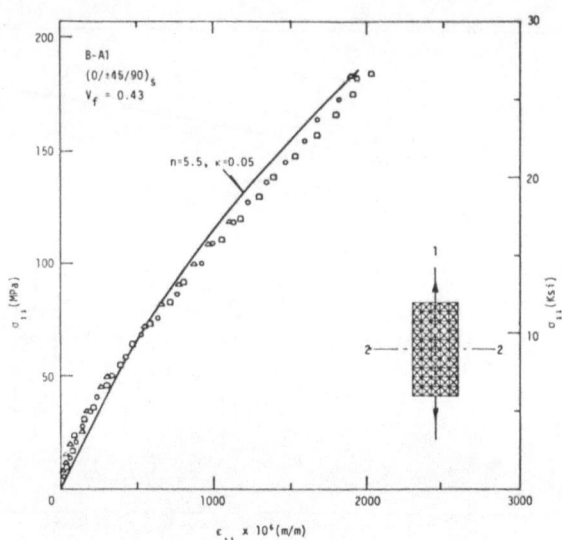

Figure 17. – *Comparison of Calculated and Measured Stress-Strain Curves for a Laminated B-Al Plate.*

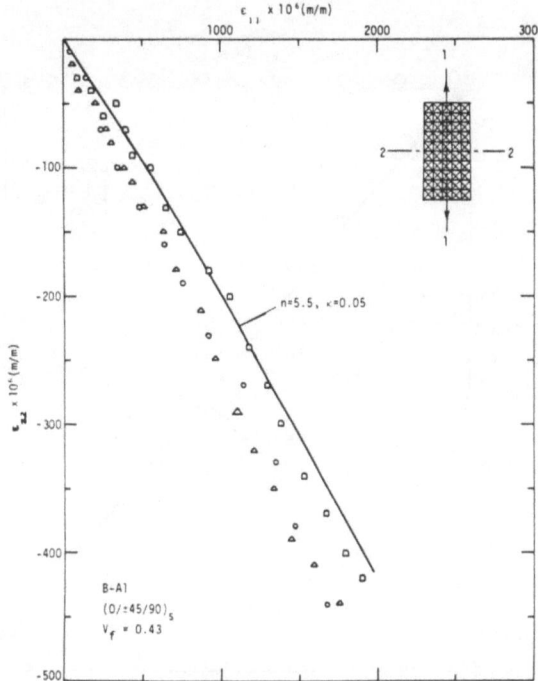

Figure 18. – *Comparison of Calculated and Measured Strains in a Laminated B-Al Plate.*

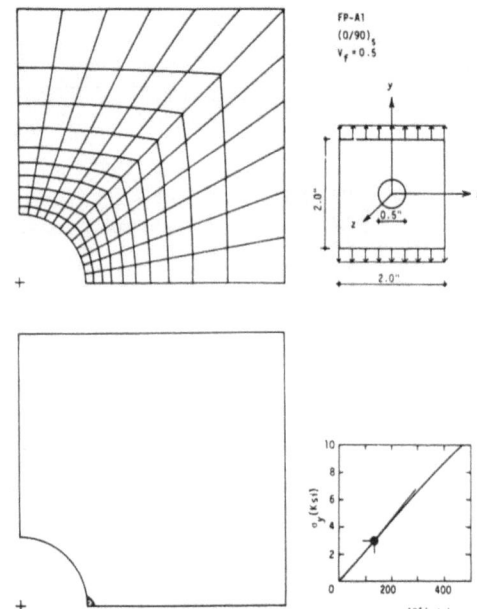

Figure 19. – *Plastic Zones at a Hole in a (0/90)$_s$ FP-Al Plate.*

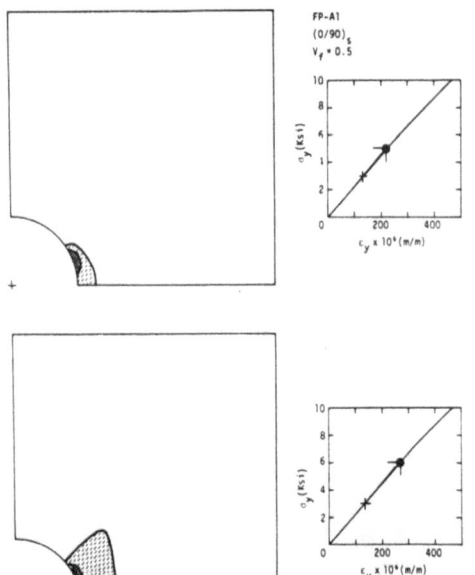

Figure 20. – *Plastic Zones at a Hole in a (0/90)$_s$ FP-Al Plate.*

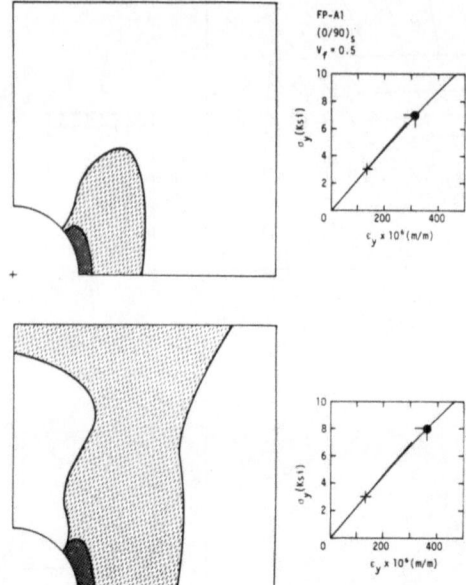

Figure 21. – *Plastic Zones at a Hole in a (0/90)ₛ FP-Al Plate.*

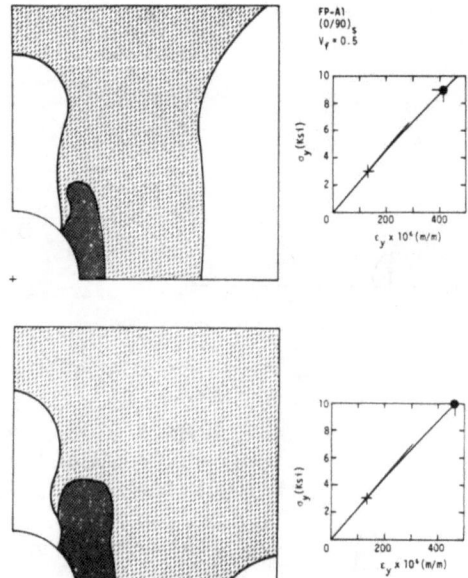

Figure 22. – *Plastic Zones at a Hole in a (0/90)ₛ FP-Al Plate.*

REFERENCES

[1] HILL R. – "Elastic properties of reinforced solids: some theoretical principles". *Journal of Mechanics and Physics of Solids*, 11 (1963): 357.

[2] LAWS N. – "On the thermostatics of composite materials". *Journal of Mechanics and Physics of Solids*, 21 (1973): 9.

[3] HILL R. – "Continuum micro-mechanics of elastoplastic polycristals". *Journal of Mechanics and Physics of Solids*, 13 (1965): 89.

[4] HERSHEY A.V. – "The elasticity of an isotropic aggregate of anisotropic cubic crystals". *Journal of Applied Mechanics*, 21 (1954): 236.

[5] KRÖNER E. – "Berechnung der Elastichen Konstanten des Vielkristalls aus den Konstanten des Einkristalls". *Zeit. Phys.*, 151 (1958): 504.

[6] BUDIANSKY B. – "On the elastic moduli of some heterogeneous materials". *Journal of Mechanics and Physics of Solids*, 13 (1965): 223.

[7] WATT J.P., G.F. DAVIES and J.O. O'CONNELL. – "The elastic properties of composite materials". *Rev. Geophys. and Space Phys.*, 14 (1976): 541.

[8] HILL R. – "Theory of mechanical properties of fibre-strengthened materials: I. Elastic behavior". *Journal of Mechanics and Physics of Solids*, 12 (1964): 199.

[9] HILL R. – "Theory of mechanical properties of fibre-strengthened materials: III. Self-consistent model". *Journal of Mechanics and Physics of Solids*, 13 (1965): 189.

[10] ESHELBY J.D. – "The determination of the elastic field of an ellipsoidal inclusion and related problems". *Proceedings of the Royal Society (London)*, A241 (1957): 376.

[11] WALPOLE L.J. – "On the overall elastic moduli of composite materials". *Journal of Mechanics and Physics of Solids*, 17 (1969): 235.

[12] DVORAK G.J. and Y.A. BAHEI-EL-DIN. – "Elastic-plastic behavior of fibrous composites". *Journal of Mechanics and Physics of Solids*, 27 (1979): 51.

[13] LAWS N. and R. McLAUGHLIN. – "Self-consistent estimates of the viscoelastic creep compliances of composite materials". *Proceedings of the Royal Society (London)*, A359 (1978): 251.

[14] HILL R. – "A self-consistent mechanics of composite materials". *Journal of Mechanics and Physics of Solids*, 13 (1965): 213.

[15] ROSCOE R. – *British Journal of Applied Physics*, 3 (1952): 267.

[16] EINSTEIN A. – "Eine neue Bestimmung der Moleküldimensionen". *Ann. Phys.*, 19 (1906): 289.

[17] McLAUGHLIN R. – "A study of the differential scheme for composite materials". *International Journal of Engineering Science*, 14 (1977): 237.

[18] HASHIN Z. and S. SHTRIKMAN. – "On some variational principles in anisotropic and nonhomogeneous elasticity". *Journal of Mechanics and Physics of Solids*, 10 (1962): 335.

[19] HASHIN Z. and S. SHTRIKMAN. – "A variational approach to the theory of the elastic behavior of multiphase materials". *Journal of Mechanics and Physics of Solids*, 11 (1963): 127.

[20] HILL R. – "New derivations of some elastic extremum principles". In *Progress in Applied Mechanics, The Prager Anniversary Volume*. New York: MacMillan, 1963, p. 99.

[21] HASHIN Z. and W. ROSEN. – "The elastic moduli of fiber-reinforced materials". *Journal of Applied Mechanics* (1964): 223.

[22] HASHIN Z. – "On elastic behavior of fiber reinforced materials of arbitrary transverse phase geometry". *Journal of Mechanics and Physics of Solids*, 13 (1965): 119.

[23] WALPOLE L.J. – "On bounds for the overall elastic moduli of inhomogeneous systems". I. *Journal of Mechanics and Physics of Solids*, 14 (1966): 151. II. *Ibid.*, 14 (1966): 289.

[24] WILLIS J.R. – "Bounds and self-consistent estimates for the overall properties of anisotropic composites". *Journal of Mechanics and Physics of Solids*, 25 (1977): 185.

[25] BERAN M.J. – "*Statistical Continuum Theories*". *Monogr. on Statist. Phys.*, Vol. 9. Interscience: New York, 1968.

[26] KRÖNER E. – "*Statistical Continuum Mechanics*". New York: Springer 1972.

[27] BERAN M.J. – "Statistical continuum theories". *Trans. Soc. Rheol.*, 9 (1965): 338.

[28] GUBERNATIS J.E. and J.A. KRUMHANSL. – "Macroscopic engineering Properties of polycristalline materials: elastic properties". *Journal of Applied Physics*, 46 (1975): 1875.

[29] DVORAK G.J. and Y.A. BAHEI-EL-DIN. – "Plasticity of Composite Laminates". *Research Workshop on Mechanics of Composite Materials*. Duke University (1978): 32. also: Bahei-El-Din Y.A. and G.J. Dvorak. "Plastic yielding at a circular hole in a laminated FP-Aℓ plate". *Modern Developments in Composite Materials and Structures*. J.R. Vinson, ed., ASME, New York: 1979, p. 123.

[30] HASHIN Z. – "Theory of fiber reinforced materials". NASA-CR-1974 (1974): 705.

[31] LOU Y.C. and R.A. SCHAPERY. – "Viscoelastic characterization of nonlinear fiber-reinforced plastic". *Journal of Composite Materials, 5 (1971): 208.*

[32] DVORAK G.J. and M.S.M. RAO. – "Thermal stresses in heat-treated fibrous composites". *Journal of Applied Mechanics*, 98 (1976): 619.

[33] HUANG W.C. – "Plastic behavior of some composite materials". *Journal of Composite Materials*, 5 (1971): 320.

[34] HUTCHINSON J.W. – "Elastic-plastic behavior of polycristalline metals and composites". *Proceedings of the Royal Society (London)*, A319 (1970): 247.

[35] BAHEI-EL-DIN Y.A. – Ph.D. Dissertation. Duke University, 1979.

[36] WU E.M. – "Phenomenological anisotropic failure criteria". Mechanics of *Composite Materials*, 2. G.P. Sendeckyj, ed., New York: Academic Press, 1974, pp. 353-431.

[37] REIFSNIDER K.L. and A. TALUG. – "Characteristic Damage States in Composite Laminates". In *Research Workshop on Mechanics of Composite Materials*. Duke University, 1978, p. 147.

[38] WHITNEY J.M. and R.J. NUISMER. – "Stress fracture criteria for laminated composites containing stress concentrations". *Journal of Composite Materials*, 8 (1974): 253.

[39] SLEPETZ J.M. and L. CARLSON. – "Fracture of composite compact tension specimens". *ASTM-STP593* (1975): 143.

[40] DVORAK G.J. and J.Q. TARN. – "Fatigue and shakedown in metal matrix composites". *ASTM-STP569* (1965): 145.

[41] JOHNSON W.S. – Ph. D. Dissertation. Duke University, 1979.

[42] PIPES R.B. – "Boundary layer effects in composite laminates". In *Research Workshop on Mechanics of Composite Materials*. Duke University, 1978, p. 78.

RESUME

(Propriétés mécaniques des matériaux composites)

Une revue des théories du comportement des milieux anisotropes hétérogènes est présentée. L'évaluation des propriétés mécaniques globales de l'aggrégat à partir des propriétés des composants et de la géométrie de la microstructure est discutée. La méthode "self-consistent" et l'évaluation de bornes rigoureuses pour les modules sont exposées. Une attention spéciale est portée sur le comportement des composites fibreux élasto-plastiques et sur les plaques en composites lamellaires présentant des imperfections géométriques.

Analysis of the Effects of Fiber Anisotropy on the Properties of Carbon and Graphite Fiber Composites

Z. Hashin

Tel-Aviv University, Tel-Aviv, Israël.

ABSTRACT

Graphite and Carbon fibers are highly anisotropic materials which can be adequately modeled as transversely isotropic, with symmetry axis in fiber direction. An analytical method is developed to determine elastic, viscoelastic, thermal expansion, heat conduction and electrical properties of composites containing such fibers on the basis of known results for isotropic fibers.

It is shown how the results can be used to determine experimentally the thermomechanical properties of Graphite and Carbon fibers on the basis of measured properties of the composite. It should be noted that the small diameters (.01 mm) of such fibers precludes direct measurement of their properties.

The important implications of fiber anisotropy for metal matrix composites are discussed. It is shown that Graphite and Carbon fiber reinforcement results in transverse (to fiber direction) elastic moduli which are substantially lower than the metal matrix elastic moduli.

Details are given in Ref. [1].

RESUME

(Influence de l'anisotropie des fibres sur les propriétés des composites à fibres de carbone et de graphite).

Une analogie entre les équations de l'élasticité pour un corps isotrope et celles pour un corps orthotrope de révolution est utilisée pour le développement des expressions et des bornes pour les cinq constantes d'élasticité efficaces d'un composite fibreux à matrice orthotrope de révolution. L'application de ces résultats pour le calcul des cinq constantes d'élasticité efficaces des fibres de carbone et de graphite est discutée. Les coefficients de dilatation thermique sont obtenus à partir d'un théorème général. Les conductivités thermiques, les constantes diélectriques et les perméabilités magnétiques sont obtenues également par l'utilisation de certaines analogies mathématiques. Des détails sont donnés dans la Réf. [1].

REFERENCE

[1] HASHIN Z. – "Analysis of properties of fiber composites with anisotropic constituents". *Journal of Applied Mechanics* **46** (1979): 543-550.

Elasto-Plastic Theory of Composites with Regular Internal Structure

A. Sawicki

Institute of Hydroengineering, Gdańsk-Oliwa, Poland.

1. Introduction

The material considered is the two-component medium $\mathcal{B} = \{\mathcal{B}_i\}$, $i = 1, 2$ such that :

a) $\forall$ i $\mathcal{B}_i$ is a continuous, elastic-ideal plastic medium.

b) At every point of the body $\mathcal{B}$ there exist simultaneously the two components : $X = \{X_i\}$, $X_i \in \mathcal{B}_i$.

c) $\forall V \in E_3$, $V_1, V_2 \rightarrow V_1 + V_2 = V$ or $\eta_1 + \eta_2 = 1$, where $\eta_i = \dfrac{V_i}{V}$.

The volume V is called the representative elementary volume. Following the above assumptions there are defined, at every point of the composite, the three stress (strain) tensors, such that:

$$\sigma = \eta_1 \sigma^{(1)} + \eta_2 \sigma^{(2)} , \tag{1.1}$$

$$\varepsilon = \eta_1 \varepsilon^{(1)} + \eta_2 \varepsilon^{(2)} . \tag{1.2}$$

Here, $\sigma^{(i)}(\varepsilon^{(i)})$ is related to the i-th component and is called the microstress (-strain) tensor. $\sigma(\varepsilon)$ is called the macrostress (-strain) tensor and may be interpreted as the average over the elementary volume. It is clear that the introduced quantities present a simplification of the real physical situation. In such an attempt the composite is treated as a macroscopically homogeneous material, generally anisotropic. Macroscopic anisotropy is induced by the internal structure of the composite. The following linear relations are accepted after Hill [1], for the elastic range of the composite behaviour:

$$\varepsilon^{(i)} = A^{(i)} \varepsilon , \tag{1.3}$$

$$\sigma^{(i)} = B^{(i)} \sigma . \tag{1.4}$$

Further, the matrix notation will be used, such that:

$$\sigma = [\sigma_{11}, \sigma_{22}, \sigma_{33}, \sigma_{12}, \sigma_{13}, \sigma_{23}]^T , \text{etc.} \tag{1.5}$$

Then the quantities $\mathbf{A}^{(i)}$ and $\mathbf{B}^{(i)}$ will be called the structural matrices. These matrices are assumed to be nonsingular. For the elastic range of composite the constituents are assumed to be the linear Hookean bodies:

$$\sigma^{(i)} = \mathbf{L}^{(i)} \, \varepsilon^{(i)} \, , \tag{1.6}$$

where $\mathbf{L}^{(i)}$ is the matrix of elastic moduli for the i-th component. From relations (1.1), (1.3) and (1.6) it follows that:

$$\sigma = (\eta_1 \, \mathbf{L}^{(1)} \, \mathbf{A}^{(1)} + \eta_2 \, \mathbf{L}^{(2)} \, \mathbf{A}^{(2)}) \, \varepsilon = \mathbf{L}\varepsilon \, , \tag{1.7}$$

Where $\mathbf{L}$ is the matrix of "effective elastic moduli" for the composite. This matrix depends on the internal geometry of the material and on the elastic moduli of the constituents. In order to determine the structural matrices and the matrix $\mathbf{L}$, the following energy postulate is accepted:

$$\int_V \sigma^\mathbf{T} \, \varepsilon \, dV = \int_{V_1} \sigma^{(1)\mathbf{T}} \, \varepsilon^{(1)} dV + \int_{V_2} \sigma^{(2)\mathbf{T}} \, \varepsilon^{(2)} \, dV \, . \tag{1.8}$$

This means that potential energy expressed by macrostrains and macrostresses equals to the energy expressed by microstrains and microstresses. The detailed derivation of respective matrices for a layered composite and a unidirectionally reinforced one are presented in [2, 3]. In general, the matrix $\mathbf{L}$ is like that for anisotropic material. There is no substantial difficulty in the determination of both the structural and constitutive matrices, but in the case of not simple internal structure of the composite, mathematical calculations would be onerous. A clear physical situation exists in some cases of practical interest (for example orthotropy), where the internal geometry of the material is regular. The introduced notions form the basis for the elasto-plastic theory of composites. These basic concepts will be developed in the subsequent sections.

2. Basic concepts

Relation (1.4) may be interpreted as an affine transformation from the space of macrostresses into space of microstresses. If $f^{(i)} = 0$ denotes the yield condition for the i-th component, then:

$$\mathbf{B}^{(i)} : f^{(i)} \longrightarrow \overline{f}^{(i)} \, , \tag{2.1}$$

where the transformation $\mathbf{B}^{(i)}$ is represented by a nonsingular matrix $\mathbf{B}^{(i)}$. It is known that under an affine transformation, a hypersurface is transformed into

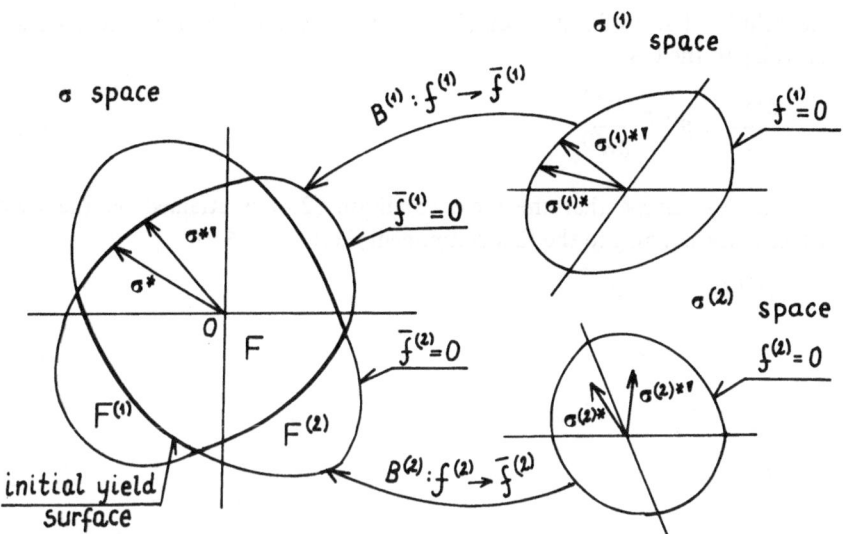

Figure 1. – *Transformation of yield surfaces.*

another hypersurface. The hypersurfaces $\overline{f}^{(i)} = 0$ and $f^{(i)} = 0$ are affinely equivalent. Let $\overline{F}^{(i)}$ denote the interior of the space bounded by the surface $\overline{f}^{(i)} = 0$. The common part of both regions is as follows:

$$\overline{F} = \overline{F}^{(1)} \cap \overline{F}^{(2)}. \tag{2.2}$$

The region $\overline{F}$ is bounded by the hypersurface $\overline{f} = 0$. All the points satisfying the relation

$$\overline{f}(\sigma) < 0 \tag{2.3}$$

represent the elastic behaviour of the composite and there the structural relations (1.3) and (1.4) are valid. When the condition

$$\overline{f}(\sigma) = 0 \tag{2.4}$$

is attained, one of the components remains plastic (alternatively both of them remain plastic if the corner of the surface (2.4) is attained) (Fig. 1). The surface (2.4) will be called the initial yield surface. If the process of further loading continues, i.e. the macrostress vector passes through the surface (2.4), the structural relations (1.3) and (1.4) do not hold. Henceforth, the associated flow rule will be assumed valid for both components:

$$d\varepsilon^{(i)} = d\lambda^{(i)} \, \frac{\partial f^{(i)}}{\partial \sigma^{(i)}} \tag{2.5}$$

The right-hand side of expression (2.5) transforms into the macrostresses space according to the law:

$$\frac{\partial \overline{f}^{(i)}}{\partial \sigma} = B^{(i)T} \frac{\partial f^{(i)}}{\partial \sigma^{(i)}}. \tag{2.6}$$

Let us assume that the yield condition (2.4) is attained on the yield surface corresponding to the first component, i.e.:

$$f^{(1)} (\sigma^{(1)}*) = 0, \tag{2.7}$$

$$f^{(2)} (\sigma^{(2)}*) < 0. \tag{2.8}$$

Eq. (2.7) in the space of macrostresses has the form:

$$\overline{f}^{(1)} (\sigma^*) = 0, \tag{2.9}$$

where* distinguishes the set of stress vectors satisfying the relation (2.4).

Let $d\sigma$ denote the macrostress increment, such that

$$\sigma = \sigma^* + d\sigma \tag{2.10}$$

If $d\sigma$ is directed into the interior of the initial yield surface, the process of unloading takes place. If $d\sigma$ is tangent to the initial yield surface the regrouping of elastic states in both components may appear. The component being in an elastic state imposes constraints on the plastic flow of the first component. However, in some cases there are no constraints and the plastic flow of the first component is possible, as for example for the Reuss composite discussed in Sec. 4.

The stress increment $d\sigma$ outward to the initial yield surface cannot be directed arbitrarily. The microstresses caused by $d\sigma$ should satisfy the equilibrium conditions, and the respective microstrains should satisfy the compatibility conditions.

Let us consider the particular forms of $d\sigma$, which prevent the vector σ^* from gliding on the initial yield surface. Such forms of the macrostress increment $d\sigma$ will be discussed in the next sections.

Because the structural relations (1.4) do not hold, a substitutional hypothesis should be adopted. Such a natural hypothesis is that the macrostress increment $d\sigma$ is taken over by the constituent being in an elastic state, i.e.:

$$d\sigma^{(2)} = \frac{1}{\eta_2} d\sigma \tag{2.11}$$

The microstress increment (2.11) produces in the second component the following increment of the microstrain vector:

$$d\varepsilon^{(2)\text{e}\ell} = M^{(2)} \, d\sigma^{(2)} = \frac{1}{\eta_2} M^{(2)} \, d\sigma, \tag{2.12}$$

where $M^{(2)}$ denotes the matrix of the elastic compliances for the second constituent.

The increment of plastic microstrains in the first constituent is given by

$$d\varepsilon^{(1)\text{p}\ell} = d\lambda^{(1)} \, (B^{(1)\text{T}})^{-1} \, \frac{\partial \overline{f}^{(1)}}{\partial \sigma}. \tag{2.13}$$

The internal structure of the composite imposes additional kinematical restrictions which should be indirectly satisfied by the macrostress increment $d\sigma$. For example, for the composite reinforced by fibers in the x_3 direction one must have

$$d\varepsilon^{(2)\text{e}\ell}_{33} = d\varepsilon^{(1)\text{p}\ell}_{33} \tag{2.14}$$

Let $\Delta\sigma$ denote the finite increment of the macrostress vector. According to the law (2.11), this is taken over by the 2-nd component.

The macrostress vector

$$\sigma = \sigma^* + \Delta\sigma \tag{2.15}$$

produces the following microstress state:

$$\sigma^{(2)} = B^{(2)} \, \sigma^* + \frac{1}{\eta_2} \Delta\sigma, \tag{2.16}$$

or

$$\sigma^{(2)} = \left(B^{(2)} - \frac{1}{\eta_2} 1\right) \sigma^* + \frac{1}{\eta_2} \sigma. \tag{2.17}$$

Substitution of relation (2.17) into the yield condition

$$f^{(2)} \, (\sigma^{(2)}) = 0 \tag{2.18}$$

gives the yield surface for the 2-nd component:

$$\overline{\overline{f}}^{(2)} \, (\sigma^*, \sigma) = 0, \tag{2.19}$$

where σ^* satisfies Eq. (2.9). The second component remains plastic at the points belonging to the surface (2.19) which moves along the direction of $\Delta\sigma$. The set of these points, if it is non-empty, forms the global yield surface for the composite.

The above results have been obtained on the assumption that the stress increment $d\sigma$ may be taken over by the unplastic component. The internal geometry of the material is such that the unplastic component imposes constraints on the plastic flow of the second component. Constraints influence the flow rule of the composite. It is difficult to define the general law at the present stage of investigation. The discussion will be performed on four simple examples.

3. The Voigt composite

The definition of the Voigt composite may be written as follows:

$$A^{(i)} = 1, \quad i = 1, 2. \tag{3.1}$$

The relations (3.1) mean that microstrains and macrostrains are equal in both components. The structural matrices for stresses are of the form:

$$B^{(i)} = \begin{bmatrix} b_1^{(i)}, & b_2^{(i)}, & b_2^{(i)}, & 0, & 0, & 0 \\ b_2^{(i)}, & b_1^{(i)}, & b_2^{(i)}, & 0, & 0, & 0 \\ b_2^{(i)}, & b_2^{(i)}, & b_1^{(i)}, & 0, & 0, & 0 \\ 0, & 0, & 0, & b_3^{(i)}, & 0, & 0 \\ 0, & 0, & 0, & 0, & b_3^{(i)}, & 0 \\ 0, & 0, & 0, & 0, & 0, & b_3^{(i)} \end{bmatrix} \tag{3.2}$$

where:

$$b_1^{(i)} = \frac{2 G_i (\lambda + G) + G\lambda_i}{G (3\lambda + 2 G)}, \quad b_2^{(i)} = \frac{\lambda_i G - \lambda G_i}{G(3\lambda + 2 G)},$$

$$\lambda = \eta_1 \lambda_1 + \eta_2 \lambda_2, \quad G = \eta_1 G_1 + \eta_2 G_2, \quad b_3^{(i)} = b_1^{(i)} - b_2^{(i)}.$$

Here, λ_i and G_i are the elastic moduli for the i-th component. It follows from (3.2) that the Voigt composite behaves macroscopically like an isotropic one. The Huber-Mises yield condition:

$$f^{(i)} = (\sigma_1^{(i)} - \sigma_2^{(i)})^2 + (\sigma_1^{(i)} - \sigma_3^{(i)})^2 + (\sigma_2^{(i)} - \sigma_3^{(i)})^2 +$$

$$- 2(\sigma_0^{(i)})^2 = 0 \tag{3.3}$$

is transformed into the space of macrostresses as:

$$\overline{f}^{(i)} = (\sigma_1 - \sigma_2)^2 + (\sigma_1 - \sigma_3)^2 + (\sigma_2 - \sigma_3)^2 - 2\left(\frac{\sigma_0^{(i)}}{b_3^{(i)}}\right)^2 = 0 \qquad (3.4)$$

Let

$$\min\left(\frac{\sigma_0^{(i)}}{b_3^{(i)}}\right) = \frac{\sigma_0^{(1)}}{b_3^{(1)}}. \qquad (3.5)$$

Then

$$\overline{f}^{(1)} = (\sigma_1^* - \sigma_2^*)^2 + (\sigma_1^* - \sigma_3^*)^2 + (\sigma_2^* - \sigma_3^*)^2 - 2\left(\frac{\sigma_0^{(1)}}{b_3^{(1)}}\right)^2 = 0 \quad (3.6)$$

From the definition of the Voigt composite it follows that:

$$d\varepsilon^{(1)p\ell} = d\varepsilon^{(2)e\ell} \qquad (3.7)$$

where:

$$d\varepsilon^{(2)e\ell} = M^{(2)} \, d\sigma^{(2)} = \frac{1}{\eta_2} \, M^{(2)} \, d\sigma \qquad (3.8)$$

Relations (2.5), (2.6) and (3.2) imply that :

$$d\varepsilon^{(1)p\ell} = d\lambda^{(1)} \frac{\partial f^{(1)}}{\partial \sigma^{(1)}} = d\lambda^{(1)} \, (B^{(1)})^{-1} \frac{\partial f^{(1)}}{\partial \sigma}. \qquad (3.9)$$

One has

$$d\sigma = \frac{d\sigma}{\left|\frac{\partial \overline{f}^{(1)}}{\partial \sigma}\right|} \frac{\partial \overline{f}^{(1)}}{\partial \sigma}. \qquad (3.10)$$

From Eqs. (3.7)-(3.10) one obtains

$$\left(d\lambda^{(1)} \left(B^{(1)}\right)^{-1} - \frac{d\sigma}{\eta_2 \left|\frac{\partial \overline{f}^{(1)}}{\partial \sigma}\right|} M^{(2)}\right) \frac{\partial \overline{f}^{(1)}}{\partial \sigma} = 0 \qquad (3.11)$$

Thus, the problem of $d\lambda^{(1)}$ determination is reduced to the eigenvalue problem for Eq. (3.11). In the presented example, the unplastic material imposes isotropic constraints on the plastic flow of the 1-st component. Eq. (2.19) takes now the form

$$\overline{\overline{f}}^{(2)} = [(\sigma_1 - \sigma_2) - \eta_1 b_3^{(1)} (\sigma_1^* - \sigma_2^*)]^2 +$$
$$+ [(\sigma_1 - \sigma_3) - \eta_1 b_3^{(1)} (\sigma_1^* - \sigma_3^*)]^2 +$$
$$+ [(\sigma_2 - \sigma_3) - \eta_1 b_3^{(1)} (\sigma_2^* - \sigma_3^*)]^2 - 2 (\eta_2 \sigma_0^{(2)})^2 = 0, \qquad (3.12)$$

representing a family of cylinders with the axes lying on the cylinder

$$(\sigma_1 - \sigma_2)^2 + (\sigma_1 - \sigma_3)^2 + (\sigma_2 - \sigma_3)^2 - 2 (\eta_1 \sigma_0^{(1)})^2 = 0. \qquad (3.13)$$

The global yield condition for the composite is the envelope of the surfaces (3.12)

$$\mathscr{F} = (\sigma_1 - \sigma_2)^2 + (\sigma_1 - \sigma_3)^2 + (\sigma_2 - \sigma_3)^2 - 2 (\eta_1 \sigma_0^{(1)} + \eta_2 \sigma_0^{(2)})^2 = 0 \tag{3.14}$$

A representation of the above results is given in Figure 2.

4. The Reuss composite

A simple illustration of the Voigt composite in the uniaxial state of stress and strain is presented in Figure 3a. A similar illustration for the Reuss composite

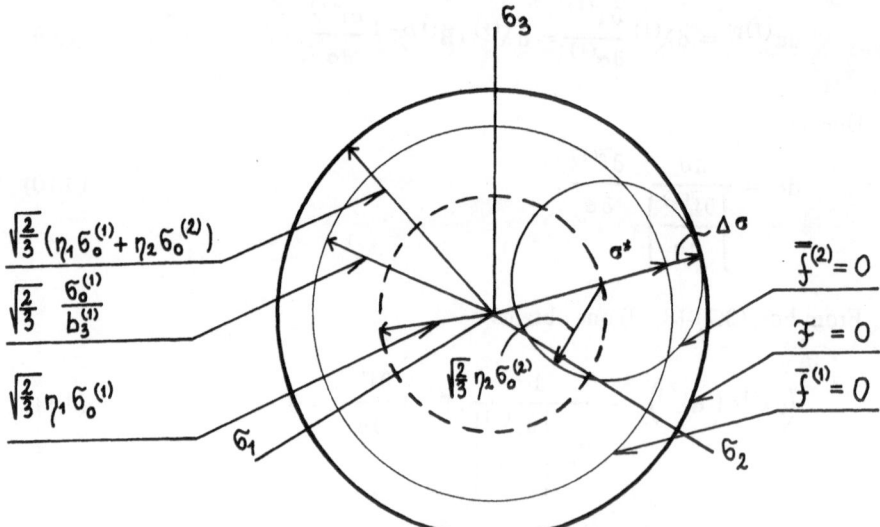

Figure 2. – *Yield surfaces for the Voigt composite.*

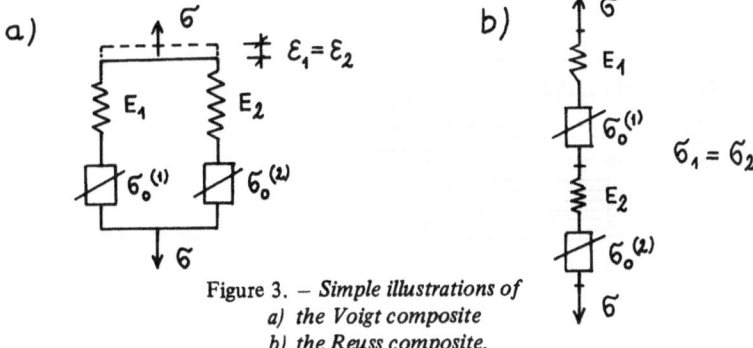

Figure 3. — *Simple illustrations of*
a) the Voigt composite
b) the Reuss composite.

is shown in Figure 3b. Generalization of this scheme leads to the following definition of the Reuss composite:

$$B^{(i)} = 1, \quad i = 1,2. \tag{4.1}$$

Relations (4.1) mean that microstresses, in both components, are equal to macrostresses. The Huber-Mises yield condition for the i-th component is of the form:

$$f^{(i)} = \overline{f}^{(i)} = (\sigma_1 - \sigma_2)^2 + (\sigma_1 - \sigma_3)^2 + (\sigma_2 - \sigma_3)^2 - 2(\sigma_0^{(i)})^2 = 0 \tag{4.2}$$

It follows from the definition of the material and from the equilibrium condition (1.1) that no increment dσ may be taken over by the unplastic material. Then the plastic flow of the composite is possible when one of the constituents remains unplastic. There are no constraints imposed by the unplastic component on the plastic flow of the first component.

The above discussed examples are rather simple. Somewhat more complicated examples will be discussed in the next chapters.

5. Layered composite

The scheme of the considered medium is presented in Figure 4. The structural matrices for stresses have the shape

$$B^{(i)} = \begin{bmatrix} b_1^{(i)}, & b_2^{(i)}, & b_3^{(i)}, & 0, & 0, 0 \\ b_2^{(i)}, & b_1^{(i)}, & b_3^{(i)}, & 0, & 0, 0 \\ 0, & 0, & 1, & 0, & 0, 0 \\ 0, & 0, & 0, & b_4^{(i)}, & 0, 0 \\ 0, & 0, & 0, & 0, & 1, 0 \\ 0, & 0, & 0, & 0, & 0, 1 \end{bmatrix}, \tag{5.1}$$

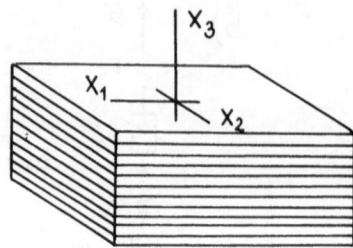

Figure 4. – *Layered composite.*

where the coefficients $b_k^{(i)}$ depend on the elastic moduli of the components and their fractional concentrations. The detailed derivation of the matrix (5.1) is presented in [2]. The Huber-Mises yield condition for the i-th component:

$$f^{(i)} = (\sigma_{11}^{(i)} - \sigma_{22}^{(i)})^2 + (\sigma_{11}^{(i)} - \sigma_{33}^{(i)})^2 + (\sigma_{22}^{(i)} - \sigma_{33}^{(i)})^2 +$$
$$+ 6(\sigma_{12}^{(i)2} + \sigma_{13}^{(i)2} + \sigma_{23}^{(i)2}) - 2(\sigma_0^{(i)})^2 = 0 \qquad (5.2)$$

takes the following form in the macrostresses space:

$$\overline{f}^{(i)} = (b_1^{(i)2} + b_2^{(i)2} - b_1^{(i)} b_2^{(i)})(\sigma_{11}^2 + \sigma_{22}^2) + (b_3^{(i)} - 1)^2 \sigma_{33}^2 +$$
$$+ (2b_1^{(i)} b_2^{(i)} - (b_1^{(i)} - b_2^{(i)})^2) \sigma_{11} \sigma_{22} +$$
$$+ (b_3^{(i)} - 1)(b_1^{(i)} + b_2^{(i)})(\sigma_{11} \sigma_{33} + \sigma_{22} \sigma_{33}) +$$
$$+ 3(b_4^{(i)2} \sigma_{12}^2 + \sigma_{13}^2 + \sigma_{23}^2) - (\sigma_0^{(i)})^2 = 0 . \qquad (5.3)$$

The yield condition (5.3) is similar to that given by Sobotka [5], for the anisotropic materials. A more detailed discussion will be given for the axisymmetrical stress and strain states, (Fig. 5). The stress structure is described by the matrices

$$B^{(i)} = \begin{bmatrix} b_1^{(i)} + b_2^{(i)} , & b_3^{(i)} \\ 0 & , & 1 \end{bmatrix} , \quad i = 1, 2 . \qquad (5.4)$$

The yield conditions (5.3) take the form:

$$\overline{f}^{(i)} = [(b_1^{(i)} + b_2^{(i)}) \sigma_1 + (b_3^{(i)} - 1) \sigma_3]^2 - (\sigma_0^{(i)})^2 = 0 , \quad i = 1, 2 . \quad (5.5)$$

Eqs. (5.5) represent the pair of two parallel lines in the space of macrostresses. The dotted region shown in Figure 5 corresponds to the elastic behaviour of

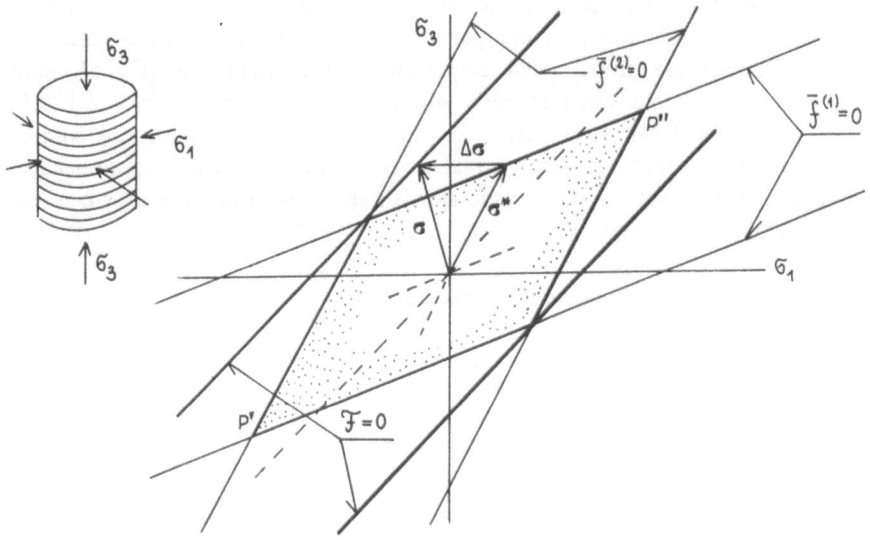

Figure 5. – *Yield conditions for the layered composite in the axisymmetrical case.*

both components. Let the yield condition be attained in the first component, i.e. $\bar{f}^{(1)} = 0$. The statically admissible macrostress increment has the form:

$$d\boldsymbol{\sigma} = d\sigma \begin{bmatrix} 1 \\ 0 \end{bmatrix} . \tag{5.6}$$

The compatibility condition:

$$d\varepsilon_1^{(1)\,pl} = d\varepsilon_1^{(2)\,el} . \tag{5.7}$$

allows to obtain the value of the parameter $d\lambda^{(1)}$:

$$d\lambda^{(1)} = \frac{d\sigma(1 - \nu_2)}{2\eta_2\,E_2\,\sigma_0^{(1)}} \tag{5.8}$$

It is easy to prove that the 2-nd material becomes plastic when the macrostress vector attaints the condition

$$F = (\sigma_1 - \sigma_3)^2 - (\eta_1\,\sigma_0^{(1)} + \eta_2\,\sigma_0^{(2)})^2 = 0 . \tag{5.9}$$

The presented procedure leads to exactly the same result when applied to the case when first the 2-nd component becomes plastic. Then the functional (5.9) presents the global yield condition for the composite, (Fig. 5). Both components become simultaneously plastic also at the points P' and P''. In the case of particular stress states (σ_{13} or $\sigma_{23} \neq 0$, and the remaining components of the macrostress vector equal to zero) the layered composite behaves like the Reuss material, i.e. plastic flow of the composite appears when one of the components remains unplastic.

6. Unidirectionally reinforced composite

Consider the fibre-reinforced composite, (Fig. 6). The structural matrices for stresses are as follows:

$$
\mathbf{B}^{(i)} =
\begin{bmatrix}
1, & 0, & 0, & 0, & 0, & 0 \\
0, & 1, & 0, & 0, & 0, & 0 \\
b_1^{(i)}, & b_1^{(i)}, & b_2^{(i)}, & 0, & 0, & 0 \\
0, & 0, & 0, & 1, & 0, & 0 \\
0, & 0, & 0, & 0, & b_3^{(i)}, & 0 \\
0, & 0, & 0, & 0, & 0, & b_3^{(i)}
\end{bmatrix}
, \quad i = 1, 2 . \tag{6.1}
$$

For the axisymmetrical state ($\sigma_1 = \sigma_2$) the matrices (6.1) reduce to the form:

$$
\mathbf{B}^{(i)} =
\begin{bmatrix}
1, & 0 \\
2b_1^{(i)}, & b_2^{(i)}
\end{bmatrix}
, \quad i = 1, 2 , \tag{6.2}
$$

where: $b_1^{(1)} = \eta_2 \alpha$, $b_1^{(2)} = -\eta_1 \alpha$, $b_2^{(1)} = E_1/E$, $b_2^{(2)} = E_2/E$,

$\alpha = (\nu_1 E_2 - \nu_2 E_1)/E$, $E = \eta_1 E_1 + \eta_2 E_2$.

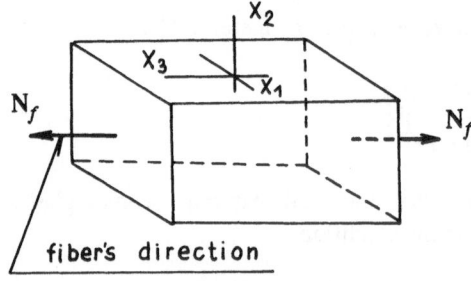

Figure 6. – *Fiber-reinforced composite.*

The Huber-Mises yield condition

$$f^{(i)} = (\sigma_1^{(i)} - \sigma_3^{(i)})^2 - (\sigma_0^{(i)})^2 = 0 \tag{6.3}$$

takes the following form:

$$\overline{f}^{(i)} = [(1 - 2b_1^{(i)})\,\sigma_1 - b_2^{(i)}\,\sigma_3]^2 - (\sigma_0^{(i)})^2 = 0 \ . \tag{6.4}$$

The statically admissible macrostress increment has the following form:

$$d\boldsymbol{\sigma} = d\sigma \begin{bmatrix} 0 \\ 1 \end{bmatrix} \ . \tag{6.5}$$

The compatibility condition imposes that

$$d\varepsilon_3^{(1)\,pl} = d\varepsilon_3^{(2)\,el} \ . \tag{6.6}$$

Applying the known procedure one obtains:

$$d\lambda^{(1)} = \frac{d\sigma\,b_2^{(2)}}{2\eta_2\,E_2\,\sigma_0^{(1)}} \ . \tag{6.7}$$

The global yield condition has the form (5.9).

7. Conclusions

The presented theory is a proposal of an efficient engineering method for the elasto-plastic analysis of composites. Although the presented examples are rather simple, there is no substantial difficulty in analysing more complicated stress states and different yield conditions. More attention should be payed to the composites' mechanisms of failure.

REFERENCES

[1] HILL R. — "Elastic properties of reinforced solids. Some theoretical principles". *J. Mech. Phys., Solids*, 11 (1963): 357-372.

[2] SAWICKI A. — "On application of effective moduli theory to layered soil". *Hydrot. Trans.*, 39 (1978): 3-13.

[3] SAWICKI A. — "Continuum elastic theory of reinforced earth". *Arch. Hydrot.*, XXV, 4 (1978): 477-489.

[4] SAWICKI A. – "Plasticity of the Voigt composite". *Bull. Acad. Polon. Sci.*, sér. Sci. Techn., (submitted).

[5] SOBOTKA Z. – "Energetic yield condition for anisotropic materials". *Stav. Časopis*, XIV, 2 (1966): 118-120.

RESUME

(Théorie élastoplastique des composites à structure interne régulière).

On considère une théorie élastoplastique pour les matériaux composites. Connaissant les propriétés mécaniques des composants et la géométrie interne du composite, on essaie de décrire les propriétés mécaniques macroscopiques du matériau composite. On suppose que les deux composants sont élastiques – parfaitement plastiques et qu'ils suivent la condition de plasticité de Huber-Misès, ainsi que la loi d'écoulement associée. Les relations fondamentales de la théorie sont présentées. La discussion concerne des composites d'intérêt pratique (composites à renforcement unidirectionnel, composites lamellaires) et des états de contrainte particuliers.

Finite Strip Method for the Analysis of Visco-Elastic Two-Dimensional Structures

S. Lichardus and J. Sumec

Institute of Construction and Architecture
of the Slovak Academy of Sciences, Bratislava, Czechoslovakia.

1. Introduction

Despite of the fact that in the title the analysis of two-dimensional structures is indicated, we shall be concerned with three-dimensional structures made from composite non-homogeneus anisotropic linear visco-elastic materials. The problem considered is a quasi-static one in which inertia forces due to deformation could be neglected. The constitutive equation of an arbitrary linear visco-elastic material can be written in the form

$$\sigma^{ij} = E^{ijk\ell} \epsilon_{k\ell} , \tag{1.1}$$

where σ^{ij} is the stress, $E^{ijk\ell}$ is the tensor operator of moduli of elasticity and $\epsilon_{k\ell}$ are the strains. All the terms are functions of space and time variables.

The division of the structure into Layered Finite Strips (Fig. 1) simplifies the form of $E^{ijk\ell}$. The material properties may be different for each layer finite strip, bu they are assumed constant within a particular strip. The material of an arbitrary strip can be isotropic, orthotropic, or anisotropic, also with reinforcement or weakened by cracks.

In the analysis the correspondence principle is used for calculating, the time-dependent visco-elastic response by using the solution of an associated elastic problem. This method departs from the principle deduced by E.H. Lee [12] for isotropic media and by M.B. Biot [1] for materials with a so-called homogeneus relaxation spectrum. In the general case of visco-elastic anisotropy this analogy is not applicable. J. Brilla [3, 8] has proved, however, that also in this case, elastic problems can be associated in the sense of the Laplace transform to the visco-elastic ones. The solution of the associated boundary value problem is equivalent to the problem of minimization of the general potential energy. The inverse transform is obtained numerically in the form of a Dirichlet series. The particular material of the structure is characterized by the visco-elastic coefficients which are derived on the basis of the constitutive equations applied, rheological models and the theory of reinforcing [2, 16].

As appropriate method for the solution the authors chose the Finite Strip Method (FSM) and developed the Layer Finite Strip Method (LFSM) [11, 13, 14].

2. Finite Strip Method

Because each strip can be subjected to membrane stresses and to bending stresses, it is convenient to write the total value of the potentional energy of the structure analysed as a sum of the potential of normal stresses and bending stresses. In this case the displacements $\widetilde{u}_\alpha\,(\alpha = 1, 2)$ are best written in the form [11]

$$\widetilde{u}_1(x,y) = \sum_{n=1}^{r} [C_i(x)\,\widetilde{u}_{1in} + C_j(x)\,\widetilde{u}_{1jn}]\sin kny\,,$$

$$\widetilde{u}_2(x,y) = \sum_{n=1}^{r} [C_i(x)\,\widetilde{u}_{2in} + C_j(x)\,\widetilde{u}_{2jn}]\cos kny\,, \qquad (2.1)$$

where

$$k_n = \frac{n\pi}{\ell}$$

and the displacement function for $\widetilde{w}(x,y)$ is

$$\widetilde{w}(x,y) = \sum_{n=1}^{r} [B_i(x)\,\widetilde{w}_{in} + B_j(x)\,\widetilde{w}_{jn} + $$
$$+ \; E_i(x)\,\widetilde{\theta}_{in} + E_j(x)\,\widetilde{\theta}_{jn}]\sin kny. \quad (2.2)$$

The polynomial expressions for $\widetilde{u}_\alpha(x, y)$ are linear and for $\widetilde{w}(x, y)$ they are of third or fifth degree.

3. Layer Finite Strip Method (LFSM).

We shall analyse a layered viscoelastic rectangular plate of bridge type (Fig. 1). Our task is to find the set of vector functions $\{\delta(x, y, z, t)\}$ and the corresponding stress $\{\sigma(x, y, z, t)\}$ which satisfies the equilibrium conditions and also the given boundary conditions. The method applied is based on the well-known idea of minimizing the energetical potential of the associated problem (tildas denote the Laplace transform also in Section 2) over a finite space of functions $V_N \subset V$, where V is an infinite dimensional space. V will be the

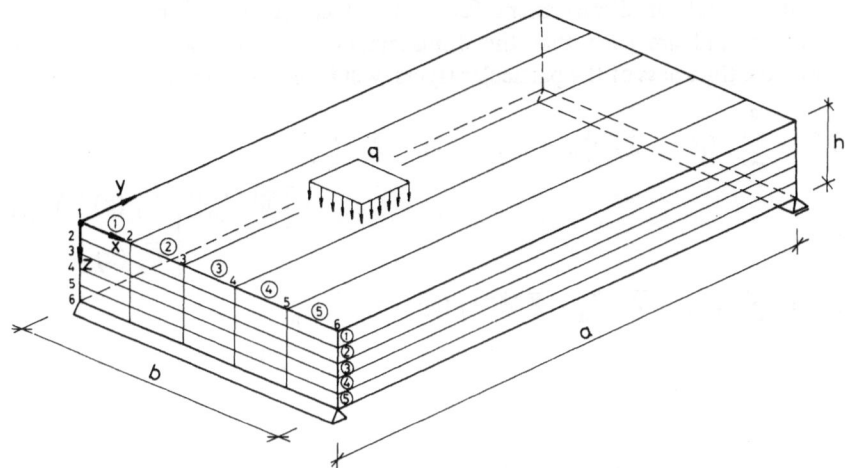

Figure 1. – *Plate divided into layers and strips.*

subspace of a Sobolev space $W_2^{(2)}(\Omega)$ generated by all functions in $W_2^{(2)}(\Omega)$ which satisfy the main (also stable) boundary conditions for the associate problem (similar conditions apply also for the Finite Element Method for viscoelastic structures [9]).

A necessary condition for the existence of the minimum of the functional $\tilde{\phi}$ of the given associated problem is

$$\frac{\partial \tilde{\phi}}{\partial \tilde{\delta}} = 0, \quad \delta = (\tilde{\delta}_1, \tilde{\delta}_2, \ldots, \tilde{\delta}_N)^T, \tag{3.1}$$

where

$$\tilde{\phi} = \frac{1}{2} \tilde{\delta}^T \tilde{K} \tilde{\delta} - \tilde{\delta}^T \tilde{F} = \tilde{W} + \tilde{U} \tag{3.2}$$

and $\tilde{\phi}$ = total potential energy,

$\tilde{K}$ = stiffness matrix,

$\tilde{F}$ = load vector,

$\tilde{W}$ = potential energy due to external forces,

$\tilde{U}$ = strain energy.

A suitable set of displacement functions which satisfy the given boundary condition and are such that the displacements u, v and w change linearly across the thickness of the particular layers, will be chosen in the form

$$\widetilde{u}^k = \sum_{n=1}^{s} [[(1-\overline{Z})\widetilde{u}_{i,n}^k + \overline{Z}\widetilde{u}_{i,n}^{k+1}][1-\overline{Z})\widetilde{u}_{i+1,n}^k$$

$$+ \overline{Z}\widetilde{u}_{i+1,n}^{k+1}]] \begin{bmatrix} \overline{C}_1'(x) \\ \overline{C}_2'(x) \end{bmatrix} Y_{n(y)},$$

$$\widetilde{v}^k = \sum_{n=1}^{s} [[(1-\overline{Z})\widetilde{v}_{i,n}^k + \overline{Z}\widetilde{v}_{i,n}^{k+1}][(1-\overline{Z})\widetilde{v}_{i+1,n}^k$$

$$+ \overline{Z}\widetilde{v}_{i+1,n}^{k+1}] \begin{bmatrix} \overline{C}_1(x) \\ \overline{C}_2(x) \end{bmatrix} Y_{n(y)}',$$

$$\widetilde{w}^k = \sum_{n=1}^{s} [[(1-\overline{Z})\widetilde{w}_{i,n}^k + \overline{Z}\widetilde{w}_{i,n}^{k+1}][(1-\overline{Z})\widetilde{w}_{i+1,n}^k$$

$$+ \overline{Z}\widetilde{w}_{i+1,n}^{k+1}]] \begin{bmatrix} \overline{C}_1(x) \\ \overline{C}_2(x) \end{bmatrix} Y_{n(y)},$$

$$(3.3)$$

where $\widetilde{u}_{i,n}^k$, $\widetilde{u}_{i,n}^{k+1}$, $\widetilde{v}_{i,n}^k$, $\widetilde{v}_{i,n}^{k+1}$, ... are the displacement parameters, whose location is indicated on Figure 2. $y_{n(y)}$, $y_{n(y)}'$ $(n=1,2,...,s)$ are the basis functions and their first derivatives, and these must a priori satisfy the given boundary conditions for y = 0 and y = a. In our case,

$$Y_n(y) = \sin \frac{n\pi y}{a}.$$

$C_1(x), C_2(x), \overline{C}_1(x), \overline{C}_2(x)$, are the polynomial parts of the displacements [11].

The system of linear algebraic equations, derived on the above mentioned assumptions, whose solution is the set of vector functions $\{\widetilde{\delta}\}$, has formaly the same form for FSM and also for LFSM,

$$[\widetilde{K}] \{\widetilde{\delta}\} = \{\widetilde{F}\}, \tag{3.4}$$

but the stiffness matrix for FSM has the form

$$[\widetilde{K}] = \int_{\Omega} [B]^* [\widetilde{D}] [B] \, d\Omega. \tag{3.5}$$

and for LFSM

$$[K] = \sum_i \sum_k [K_{i\,k}] = \sum_i \sum_k \left(\sum_n [K_{i\,k\,n}] \right), \tag{3.6}$$

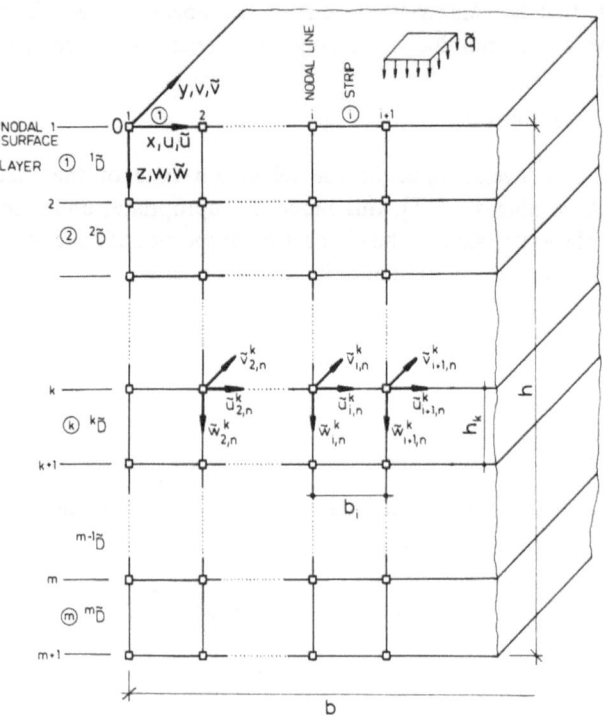

Figure 2. – *Coordinate system, indication of layers, nodal lines and nodal points in cross section of the plate.*

where

$$[K_{i\ k\ n}] = \int_0^a \int_0^{b_i} \int_0^{h_k} [B_{i\ k\ n}]^* [D_{i\ k\ n}] [B_{i\ k\ n}]\, dx\, dy\, dz \qquad (3.7)$$

and $[B]$ is the strain matrix.

In both cases the determination of the property matrix $[\widetilde{D}]$ is an important task requiring adequate attention.

4. Derivation of the property matrix.

As a special case, we consider a standard linear viscoelastic material where

$E_{(\alpha)}^{ijk\ell}$ is the tensor of moduli of elasticity,

$\eta_{(\alpha)}^{ijk\ell}$ is the tensor of moduli of viscosity.

We assume, that the Maxwell element of the linear viscoelastic material, also called Zener's material has a homogeneous relaxation spectrum, and hence one has

$$\eta_{(\alpha)}^{ijk\ell} = K\, E_{(\alpha)}^{ijk\ell} \tag{4.1}$$

where K is the inverse value of the relaxation time of the material of the structure. According to [15], this model is appropriate, under some simplifications, for the expression of the viscoelastic properties of concrete.

The differential equation describing this model is

$$K\,(E_{(1)}^{\alpha\beta\gamma\delta} + E_{(2)}^{\alpha\rho\gamma\delta})\,\dot\epsilon_{\gamma\delta} + E_{(1)}^{\alpha\rho\gamma\delta}\,\epsilon_{\gamma\delta} = K\dot\sigma^{\alpha\beta} + \sigma^{\alpha\beta}. \tag{4.2}$$

Applying the Laplace transformation to equation (4.2), where

$$p_i = 1/\gamma_i$$

is the parameter of the Laplace transform and tildas denote the result of the transformation [10], we obtain

$$\left(p + \frac{1}{K}\right)\tilde\sigma^{ij} = \left[p(E_{(1)}^{ijk\ell} + E_{(2)}^{ijk\ell}) + \frac{1}{K}\,E_{(1)}^{ijk\ell}\right]\tilde\epsilon_{k\ell}. \tag{4.3}$$

In contracted form, if we denote

$$E^{ij} = E_{(1)}^{ij} + E_{(2)}^{ij},$$

we obtain

$$\left(p + \frac{1}{K}\right)\tilde\sigma^{11} = \left(pE^{11} + \frac{1}{K}\,E_{(1)}^{11}\right)\tilde\epsilon_{11} + \left(pE^{12} + \frac{1}{K}\,E_{(1)}^{12}\right)\tilde\epsilon_{22},$$

$$\left(p + \frac{1}{K}\right)\tilde\sigma^{22} = \left(pE^{12} + \frac{1}{K}\,E_{(1)}^{12}\right)\tilde\epsilon_{11} + \left(pE^{22} + \frac{1}{K}\,E_{(1)}^{22}\right)\tilde\epsilon_{22}, \tag{4.4}$$

$$\left(p + \frac{1}{K}\right)\tilde\sigma^{12} = \left(pE^{66} + \frac{1}{K}\,E_{(1)}^{66}\right)\tilde\gamma_{12}.$$

In the case of orthotropy, introducing a contracted form E^{ij}, we can rewrite equation (4.3) as

$$[\tilde\sigma^{ij}] = \left(p + \frac{1}{K}\right)^{-1}[d^{ij}]\,[\tilde\epsilon_{k\ell}], \tag{4.5}$$

where the matrix of the viscoelastic coefficients is given by

$$
d^{ij} = \begin{bmatrix} pE^{11} + \dfrac{1}{K} E^{11}_{(1)} & pE^{12} + \dfrac{1}{K} E^{12}_{(1)} & 0 \\[2ex] pE^{12} + \dfrac{1}{K} E^{12}_{(1)} & pE^{22} + \dfrac{1}{K} E^{22}_{(1)} & 0 \\[2ex] 0 & 0 & pE^{66} + \dfrac{1}{K} E^{66}_{(1)} \end{bmatrix} \tag{4.6}
$$

The stresses in the elastic orthotropic two-dimensional structure are obtained from the known equation

$$
[\sigma^{ij}] = [c^{ij}] [\epsilon_{k\ell}], \tag{4.7}
$$

where the matrix $[C^{ij}]$ depends on the arrangement of reinforcing bars or fibres. If the reinforcing bars are arranged in two perpendicular directions, parallel to the directions of the coordinate axes X and Y, then the matrix $[C^{ij}]$ is in the form (4.8). We later suppose that the standard linear viscoelastic model has isotropic material coefficients $E^{ijk\ell}_{(2)}$ and $\eta^{ijk\ell}_{(2)}$ (Maxwell element) and that the coefficients $E^{ijk\ell}_{(1)}$ have orthotropic properties (elastic spring element).

$$
C^{ij} = \begin{bmatrix} \bar{E}^{11} A^{-1} & \bar{E}^{11} \mu_{21} A^{-1} & 0 \\[1ex] \bar{E}^{22} \mu_{21} A^{-1} & \bar{E}^{22} A^{-1} & 0 \\[1ex] 0 & 0 & G'_{12} \end{bmatrix} ; A = 1 - \mu_{12} \mu_{21} \tag{4.8}
$$

If we introduce

$$
E^{ij}_{(1)} = \frac{\bar{E}^{ij}_{(1)}}{1 - \mu_{12} \mu_{21}}, \tag{4.9}
$$

where $\bar{E}^{ij}_{(1)}$ are the moduli of elasticity calculated according to the theory of reinforcing [2, 16] for the ideal structure made from the composite material, then we can easily determine the matrix $[d^{ij}]$ which fully describes the viscoelastic properties of the material considered.

REFERENCES

[1] BIOT M.A. – "Variational and Lagrangian methods in visco-elasticity deformation and flow of solids". In: *IUTAM Coll., Madrid 1955*, Berlin: Springer Verlag, 1956.

[2] BOLOTIN V.V. – "Vibration of layered elastic plates". In: *Proc. Vibration Probl.*, 4, 1963.

[3] BRILLA J. – "Linear visco-elastic bending analysis of anisotropic plates". In: *Proc. XIth Int. Congr. Appl. Mechanics, Munich 1964*, Springer 1966.

[4] BRILLA J. – "Convolutional variational principles and methods in linear visco-elasticity of anisotropic bodies". In: *Proc. Int. Conf. Variational Methods in Engineering* Southampton, 1972.

[5] BRILLA J. – "Convolutional variational principles and methods in linear visco-elasticity". In: *ZAMM-Sonderheft* 54 (1974).

[6] BRILLA J. – "Finite element method in linear visco-elasticity". In: *ZAMM-Sonderheft*, 54 (1974).

[7] BRILLA J. – "Finite element analysis of a system of quasi-parabolic partial differential equations". In: *Acta Univ. Carolinae-Mathematica et Physica* 1974, Nr. 1-2, 5-9.

[8] BRILLA J. – "Generalized variational methods in linear visco-elasticity". In: *IUTAM Symp. Mech. Viscoelastic Media and Bodies, Gothenburg, 1974*.

[9] BRILLA J., S. LICHARDUS, A. NEMETHY. – "The Generalization of the finite element method for the solution of viscoelastic two-dimensional problems". In: *IUTAM Symp. Mech. viscoelastic Media and Bodies, Gothenburg 1974*.

[10] CARSLAW H.S. and J.C. JAEGER. – *"Operational Methods in Applied Mathematics"*. Oxford University Press, 1948.

[11] CHEUNG Y.K. – *"Finite Strip Method in Structural Analysis"*. Oxford: Pergamon Press, 1976.

[12] LEE E.H. – "Stress analysis in visco-elastic bodies". *Quart. Appl. Math.*, 13, 2 (1955).

[13] LICHARDUS S. and A. NEMETHY. – "Generalization of finite strip method for analysis of visco-elastic shallow shells". *Stavebnicky casopis*, 24, 5 (1976) (In English). VEDA, Bratislava.

[14] LICHARDUS S. and J. SUMEC. – "Analysis of structural elements of various dimensions made from composite visco-elastic materials". (In Slovak: "Analýza konštrukčných prvkov rôznych dimenzií z kompozitných väzkopružných materiálov"). *Internal Research Report of Inst. of Const. and Archit. of SAV*, Bratislava, USTARCH-SAV, 1978.

[15] MALMEISTER A.K. – *"Elasticity and Inelasticity of Concrete"*. (In Russian). Riga, 1957.

[16] MALMEISTER A.K., V.P. TAMUZ and G.A. TETERS. – "Strength of Polymers". (In Russian: Soprotivlenyje zhostkych polimernych materialov"). Riga, Izdat. ZINATNE, 1972.

RESUME

(Méthode des bandes finies pour l'analyse des structures viscoélastiques bidimensionnelles)

Nous présentons une analyse des structures viscoélastiques bidimensionnelles pour un matériau anisotrope non-homogène, en utilisant la méthode des couches et bandes finies. L'analyse part du potentiel minimum de l'énergie de la structure étudiée et utilise les principes variationnels de convolution. Le matériau constitutif de la structure est de première importance dans le calcul numérique pour la détermination de la matrice des coefficients viscoélastiques. Il s'agit d'un matériau composite, pour lequel on suppose que la théorie des matériaux armés peut être appliquée et que les propriétés viscoélastiques peuvent être simulées à l'aide de modèles rhéologiques. La solution associée, au sens de la transformation de Laplace, est inversée par la méthode de colocation et exprimée sous la forme d'une série de Dirichlet pour les lignes nodales discrètes.

Session 6

Strength of Composites

Résistance des Composites

Strength Ratios of Orthotropic Materials

S.W. Tsai

U.S. Air Force Materials Laboratory, Ohio, U.S.A.

1. Introduction

Orthotropic materials such as unidirectional composites have a high degree of directionality in their strength. For example, most modern composite materials have longitudinal strength ten to twenty times that of the transverse and shear strengths. Under the imposed combined stresses or strains, the failure criteria must be more exact than that required for isotropic materials. One strength parameter such as the uniaxial tensile strength will suffice for the description of the behavior of the isotropic material under combined stresses such as the von Mises criterion.

In this paper, the tensor polynomial criterion will be specialized so that the magnitude of the interaction term is assumed to be equal to the von Mises criterion for the isotropic material. All composite materials can then be shown on the same generalized von Mises ellipse. In addition, the strength ratio can also be introduced to the failure criterion so that its utility can be enhanced for the design and testing of composites. Finally, the analytic predictions will be compared with available data.

2. Strength ratios

The standard form for the tensor polynomial failure criterion in stress space is given as follows:

$$F_{ij} \sigma_i \sigma_j + F_i \sigma_i = 1 \tag{1}$$

where F_{ij} and F_i are strength parameters associated with this failure criterion and σ_i are the stress components. For orthotropic materials under the state of plane stress, there are six material parameters defined as follows:

$$F_{11} = \frac{1}{X_L X_L^-} \; , \qquad F_1 = \frac{1}{X_L} - \frac{1}{X_L^-}$$

$$F_{22} = \frac{1}{X_T X_T^-} \; , \qquad F_2 = \frac{1}{X_T} - \frac{1}{X_T^-} \qquad\qquad (2)$$

$$F_{66} = \frac{1}{X_{LT}} \; , \qquad F_{16} = F_{26} = F_6 = 0$$

where

X_L, X_L^- = tensile and compressive strength in the longitudinal direction,
X_T, X_T^- = tensile and compressive strength in the transverse direction,
X_{LT} = shear strength in the longitudinal direction.

The only remaining term in Equation 1 is the interaction term F_{12}. The determination of this term requires truly combined or biaxial stresses. This is different from all the quantities listed in Equation 2 above where the parameters can be determined from simple tensile, compressive, or shear tests. More will be said on the interaction term later.

For a given material subjected to a given state of stress, the left-hand side of Equation 1 can be computed. This failure criterion states that when the left-hand side of Equation 1 reaches a numerical value of 1, failure occurs. If this value is less than 1, failure has not occurred. The left-hand side of Equation 1 cannot have value greater than 1. That is not admissable under the present concept of failure criterion. Because of the existence of the linear term in Equation 1, the physical interpretation of the numerical value of the left-hand side of this equation conveys only a go or no-go situation. That is whether or not the numerical value is equal to or less than unity.

This situation, however, can be improved if the following strength ratio is defined as follows:

$$S = \frac{\sigma_{i(a)}}{\sigma_i} = \frac{\epsilon_{i(a)}}{\epsilon_i} \qquad\qquad (3)$$

where

S = strength ratio (between the strength and stress vectors as in Ref. 1[1]).
$\sigma_{i(a)}$ = allowable or maximum stress components.
σ_i = applied or imposed stress components.

---

[1] E.M. Wu, "Strength and Fracture of Composites", (L.J. Brautman, Ed., *Composite Materials* Vol. 5, *Fracture and Fatigue*, Academic Press, 1974, pp. 191-247.

The definition in Equation 3 requires additional assumptions.

1) Loading is proportional in that the three ratios of the stress components among the allowed and the imposed remain constant and are equal to S, the strength ratio.

2) We further assume that our material is linearly elastic up to failure and there is one-to-one correspondence between the stress and the strain components. The second equality of Equation 3 implies the existence of this linearity. The stress components in Equation 1 at the time of failure are equal to the allowed stress components, and Equation 1 can be written as follows:

$$F_{ij}\, \sigma_{i(a)}\, \sigma_{j(a)} + F_i\, \sigma_{i(a)} = 1 \qquad (4)$$

In place of the allowed stress components in Equation 4, we can substitute the relations in Equation 3 into Equation 4, we have

$$[F_{ij}\, \sigma_i\, \sigma_j]\, S^2 + [F_i\, \sigma_i]\, S = 1 \qquad (5)$$

In this equation the stress components are those imposed or applied. The strength ratios S in this equation can be obtained by solving the quadratic equation for a given material under a given state of imposed stress. The value of S so obtained contained the following significance:

1) When $S = 1$, we have failure.

2) When S is less than 1, this is not an admissable situation, which means that the imposed stresses have already exceeded the failure combinations.

3) When S is greater than 1, this is a value that can be applied to the imposed stress in order to induce failure. For example, if $S = 2$, the imposed stresses can be doubled before failure is reached.

It is believed that the strength ratio so defined provides a practical means of determining not only if failure has or will occur but will also give numerical description as to how far or the degree or margin of safety that a given state of stress can sustain. This is an improvement over the original failure criterion as stated in Equation 1, where only the go or no-go judgment can be made.

Since Equation 5 is a quadratic equation, there exists two roots. The one strength ratio S is the solution of the imposed stresses. The conjugate root defined S^- can be shown to be the solution of the equation if all the stress components change signs. For example, if all the stress components are positive and the resulting strength ratio is S, then the conjugate root S^- is the strength ratio when all the imposed stress components become negative.

3. Generalized von mises criterion

Equation 5 can be shown to be a generalized von Mises criterion for orthotropic materials if we can make the following substitutions:

$$x = \sqrt{F_{11}}\,\sigma_1, \quad y = \sqrt{F_{22}}\,\sigma_2, \quad z = \sqrt{F_{66}}\,\sigma_6 \tag{6}$$

Equation 5 then becomes

$$[x^2 + 2F_{12}^*\,xy + y^2 + z^2]\,S^2 + [F_1^* x + F_2^* y]\,S = 1 \tag{7}$$

where $F_{12}^* = F_{12}/\sqrt{F_{11}F_{22}}$, $F_1^* = F_1/\sqrt{F_{11}}$, $F_2^* = F_2/\sqrt{F_{22}}$.
The non-dimensionalized strength ratio in Equation 7 can be further simplified by a coordinate translation so that the linear terms can be eliminated. This can be accomplished by substituting the following:

$$x' = x + \Delta x$$
$$y' = y + \Delta y \tag{8}$$

Into Equation 7 we then have

$$x'^2 + 2F_{12}^*\,x'y' + y'^2 = k^2 \tag{9}$$

where

$$k^2 = \frac{1}{S^2} + [\Delta x^2 + 2F_{12}^*\,\Delta x \Delta y + \Delta y^2]$$

$$\Delta x = \frac{1}{2S}\,[F_1^* - F_{12}^*\,F_2^*]/[1 - F_{12}^{*2}]$$

$$\Delta y = \frac{1}{2S}\,[F_2 - F_{12}^*\,F_1^*]/[1 - F_{12}^{*2}]$$

In Equation 9 the shear stress is assumed to be zero since it does not play a significant part with the discussion of the interaction term which is in the x-y plane only. The normalized von Mises conditions using the same parameters as those in Equation 9 is

$$x'^2 - x'y' + y'^2 = k^2 \tag{10}$$

It is therefore convenient to equate Equation 9 as a generalized von Mises criterion. With this assumption

$$F_{12}^* = -1/2 \tag{11}$$

In addition to a reasonable agreement between the prediction based on Equation 9 and available data, which will be shown later, the concept of a fixed F_{12}^* value has an additional advantage. All orthotropic materials can now be mapped on the same generalized von Mises ellipse. Only one combined stress-space failure curve is needed for all orthotropic materials. To recap the various transformation and mapping operations that enable us to go from Equation 1 to Equation 9, this will be shown in a series of figures in Figure 1.

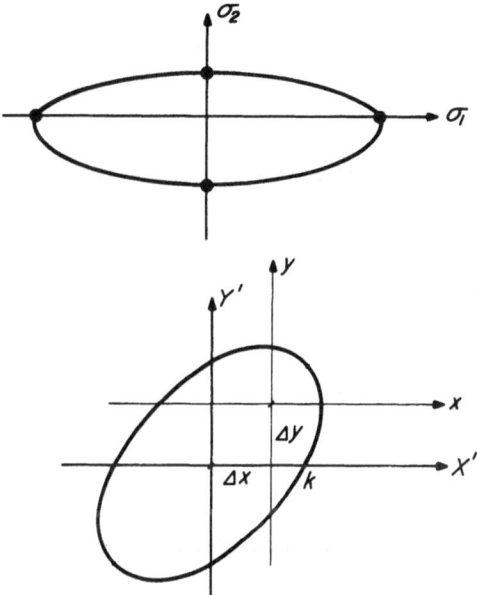

Figure 1. – *Transformation of failure surface in stress-space from the physical stress plane in the top figure to non-dimensional stress-space in the bottom figure as represented by Equation 7. The coordinate axis can be centralized by displacement shown in Equations 8 and 9. The classical von Mises criterion showing Equation 10 can be recovered from Equation 9 if the normalized interaction term is to be – 1/2.*

With a fixed F_{12}^* as stated in Equation 11, all orthotropic materials can be mapped onto the unit ellipse shown in the last figure of Figure 1. What distinguish from one orthotropic material to another are the displacements of the origin of the reference coordinates, i.e., x and y as specified in Equation 9 and the normalizing factor stated in Equation 6. For laminated composite materials, it is a standard practice to assume the state of strain either constant or varying linearly across the thickness. Since strain is so specified, it is more covenient to use the strength ratio equations in strain-space rather than stress-space. With the assumed linearity up to failure, it is simple to convert the strength ratio equation in stress-space such as that in Equation 5 to strain-space by making the following substitutions of the stress-strain relations.

$$\sigma_i = Q_{ij} \epsilon_j \tag{12}$$

where

Q_{ij} = the plane stress elastic moduli and

ϵ_j = the strain components

Substituting Equation 12 into Equation 5, we have

$$[F_{ij} Q_{ik} Q_{j\ell} \epsilon_k \epsilon_\ell] S^2 + [F_i Q_{ij} \epsilon_j] S = 1 \tag{13}$$

If we define the strength parameter in strain-space in terms as G_{ij}, we have

$$[G_{k\ell} \epsilon_k \epsilon_\ell] S^2 + [G_j \epsilon_j] S = 1 \tag{14}$$

where $G_{k\ell} = F_{ij} Q_{ik} Q_{j\ell}$, $G_j = F_i Q_{ij}$

When a laminate is subjected to bending and twisting movements, it is usually assumed that the strain distribution across the laminate is linear. The strength distribution is defined as

$$\epsilon_i = zk_i \tag{15}$$

where k_i = curvature, i = 1, 2, 6
      z = the coordinate in the thickness direction

If we substitute Equation 15 into Equation 14, we have

$$[G_{ij} k_i k_j] [zS]^2 + [G_i k_i] [zS] = 1 \tag{16}$$

Note that the product of the strength ratio and the z coordinate appear together in this equation. Thus, for the same value of this product the highest value of z will result in the lowest value of the strength ratio S. When a laminate is subjected to bending or twisting, within each ply assembly it is the highest z value or at the outer surface away from the neutral axis where failure is most likely to occur.

4.   Initial strains

Composite laminates are normally cured at elevated temperature. Curing strains are induced from shrinkage during curing as well as subsequent cooling to room temperature. In addition, the moisture absorption of the epoxy matrix will induce expansion strains in the laminate. A first order approximation of the induced strains can be described by the expansion coefficients in

$$e_i = \alpha_i^T \, \Delta T + \alpha_i^H \, c, \; i = L, T \tag{17}$$

where
α_i^T = longitudinal and transverse thermal expansion coefficients.

α_i^H = longitudinal and transverse moisture expansion coefficients.

ΔT = temperature change (normally negative).

c = moisture content.

The equivalent or nonmechanical stresses that would induce the same expansional strain as those in Equation 17 can be obtained using the following plane stress-strain relations.

$$\sigma_i^N = Q_{ij} e_j \tag{18}$$

where σ_i^N are the nonmechanical stress components which are the stress required to induce the expansional strains. The stress resultants in a laminate that would accommodate and self-equilibrate all the free expansions of the plies as called for in Equation 17 are as follows:

$$N_i^N = \int_{-h/2}^{h/2} \sigma_i^N \, dz \tag{19}$$

where N_i^N are the nonmechanical stress resultants. In the case of unsymmetrical laminates, we shall also have nonmechanical moments defined as follows:

$$M_i^N = \int_{-h/2}^{h/2} \sigma_i^N \, z dz \tag{20}$$

where M_i^N are the nonmechanical moments due to differential expansions of an unsymmetrical laminate. The total stress resultants and moments that can be applied to a laminate are the sum of the nonmechanical and the mechanical or applied resultants and moments, as follows:

$$\begin{aligned} N_i^{Total} &= N_i^N + N_i^M \\ M_i^{Total} &= M_i^N + M_i^M \end{aligned} \tag{21}$$

The strength ratio relation in Equation 14 must be modified in order to include the initial strains. It is assumed that for a given laminate the curing process has already taken place so that the temperature difference ΔT in Equation 17 is fixed. We can further assume that the moisture content is also fixed either at zero for the newly cured laminate or some saturation level after a long period of time. In either event the moisture level for our present purpose can also be assumed to be fixed. Then the strength ratio that appears in Equation 14 sould be specialized to include only the effect of the mechanical component of the loading. In other words, the nonmechanical strains are assumed to be predetermined. Then the definition of strength ratio instead of Equation 3 should have superscript M added to it, that

$$S^M = \frac{\sigma_{i(a)}^M}{\sigma_i^M} = \frac{\epsilon_{i(a)}^M}{\epsilon_i^M} \tag{22}$$

The mechanical strength ratio in this equation will specify the ratio of the increase of the applied or mechanical stress or strain that a given laminate (with fixed nonmechanical strains) can sustain. Figure 2 shows the relationship between the initially uncured laminate as in (a) and the laminate subjected to free expansions such as the e_i in Equation 17. Then, finally, the non-

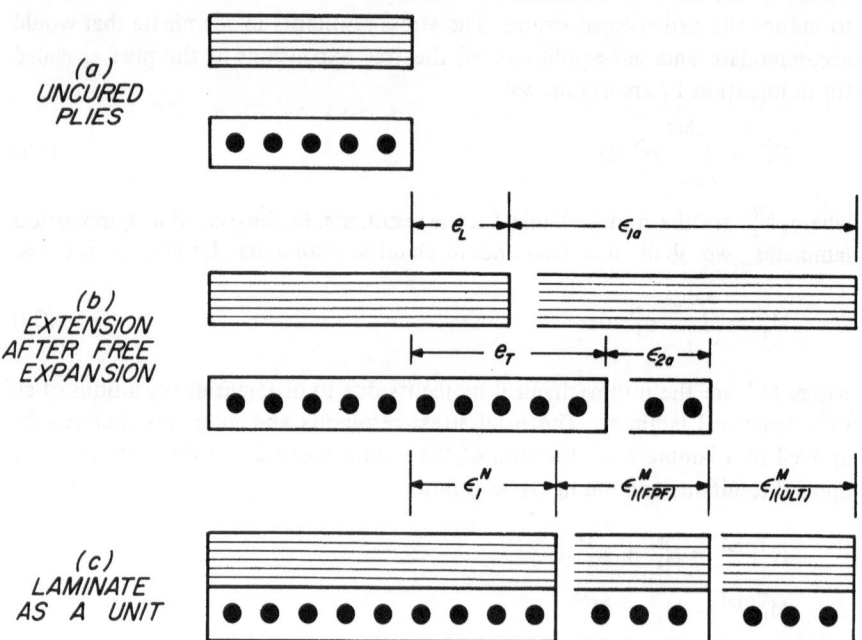

Figure 2. — *Relationship between uncured ply in (a) and the free-expansion strain in (b). In (c) the laminate as a unit is subjected to nonmechanical as well as mechanical stress resultants.*

mechanical and mechanical stress resultants that can be applied to the laminate as a unit. This is shown in (c). The strain components that must be substituted in the strength ratio equation of Equation 14 must take into account the fact that the ultimate strains of laminas are normally measured at room temperature after free expansions. That is the (b) in Figure 2. Thus the following strains must be used for the determination of strength ratio in Equation 14.

$$\epsilon_i = S^M \epsilon_i^M + \epsilon_i^N - e_i \tag{23}$$

where

$$\epsilon_i^M = a_{ij} N_j^M, e_i = \{e_L, e_T, 0\}$$

$$\epsilon_i^N = a_{ij} N_j^N, a_{ij} = \text{in-plane compliance}$$

Substituting Equation 23 into Equation 14, we have

$$G_{ij} [S^M \epsilon_i^M + \epsilon_i^N - e_i] [S^M \epsilon_j^M + \epsilon_j^N - e_j]$$
$$+ G_i [S^M \epsilon_i^M + \epsilon_i^N - e_i] = 1 \tag{24}$$

where the S value in Equation 14 is assumed to be unity. Rearranging Equation 24 in terms of the quadratic equation of S^M, we have

$$[G_{ij} \epsilon_i^M \epsilon_j^M] [S^M]^2 + \{G_{ij} [\epsilon_i^M (\epsilon_j^N - e_j) + \epsilon_j^M (\epsilon_i^N - e_i)] + G_i \epsilon_i^M\} S^M$$
$$+ [G_{ij} (\epsilon_i^N - e_i) (\epsilon_j^N - e_j) + G_i (\epsilon_i^N - e_i)] = 1 \tag{25}$$

or we have

$$a [S^M]^2 + b [S^M] + c = 0 \tag{26}$$

This solution of this equation will yield two conjugate roots, S^M and S^{-M}. One corresponds to the given state of strains and the conjugate corresponds to the solution when all the signs of the strain components are reversed. For multidirectional laminates, there is a strength ratio for each ply orientation. The lowest of the strength ratios is referred to as the first-ply failure; the highest, the ultimate.

5. Numerical example

The strength ratios of AS/3501 graphite/epoxy have been predicted using the theory developed in this work with the following properties of the unidirectional composites.

$$E_L = 20.01 \times 10^6 \text{ psi} \qquad E_T = 1.3 \times 10^6 \text{ psi}$$

$$G_{LT} = 1.03 \times 10^6 \text{ psi} \qquad \nu_{LT} = .3$$

$$X_L = 210 \times 10^3 \text{ psi} \qquad X_L^- = 210 \times 10^3 \text{ psi} \qquad (27)$$

$$X_T = 7.5 \times 10^3 \text{ psi} \qquad X_T^- = 30 \times 10^3 \text{ psi}$$

$$X_{LT} = 13.5 \times 10^3 \text{ psi}$$

$$F_{11} = 22.676 \times 10^{-12} \quad F_{22} = 4.444 \times 10^{-9} \quad F_{66} = 5.487 \times 10^{-9}$$
$$\tag{28}$$
$$F_1 = 0 \quad F_2 = 100 \times 10^{-6} \quad F_{12}^* = -1/2 \quad F_{12} = -.159 \times 10^{-12}$$

From the stiffness and strength data above, we can compute the following using Equations 13 and 14,

$$G_{11} = 7363, \quad G_{22} = 7440, \quad G_{12} = -1743$$
$$\tag{29}$$
$$G_{66} = 5821, \quad G_1 = 39.22, \quad G_2 = 130.76$$

Using the data above, we can calculate the off-axis tensile and compressive strength of unidirectional composites. The results are shown in Figure 3. The unpublished experimental results were furnished by R.Y. Kim of the University of Dayton Research Institute in Dayton, Ohio. Using the property data above, plus the following expansion coefficients, we can calculate the strength of angle-ply laminates.

$$\alpha_L^T = 5.55 \times 10^{-9}/°F$$
$$\tag{30}$$
$$\alpha_T^T = 6.94 \times 10^{-6}/°F$$

The temperature difference is assumed to be $-200°F$. It is further assumed that moisutre expansion is zero because the laminates were tested soon after fabrication. Using Equation 24, the mechanical strength ratios for these angle-ply laminates were computed and the results are shown in Figure 4 as solid lines. Also shown in this figure as dashed lines are the prediction based on an interaction term $F_{12}^* = 0$. Although the solid line ($F_{12}^* = -1/2$) makes better prediction of strength than the dashed line, there is still a gap between the theory and the data observed as shown as dots. This discrepancy is being investigated at this time. One may be able to attribute the discrepancy due to the length to width ratio of the tensile coupon. The data in this figure were also furnished by Dr. Kim.

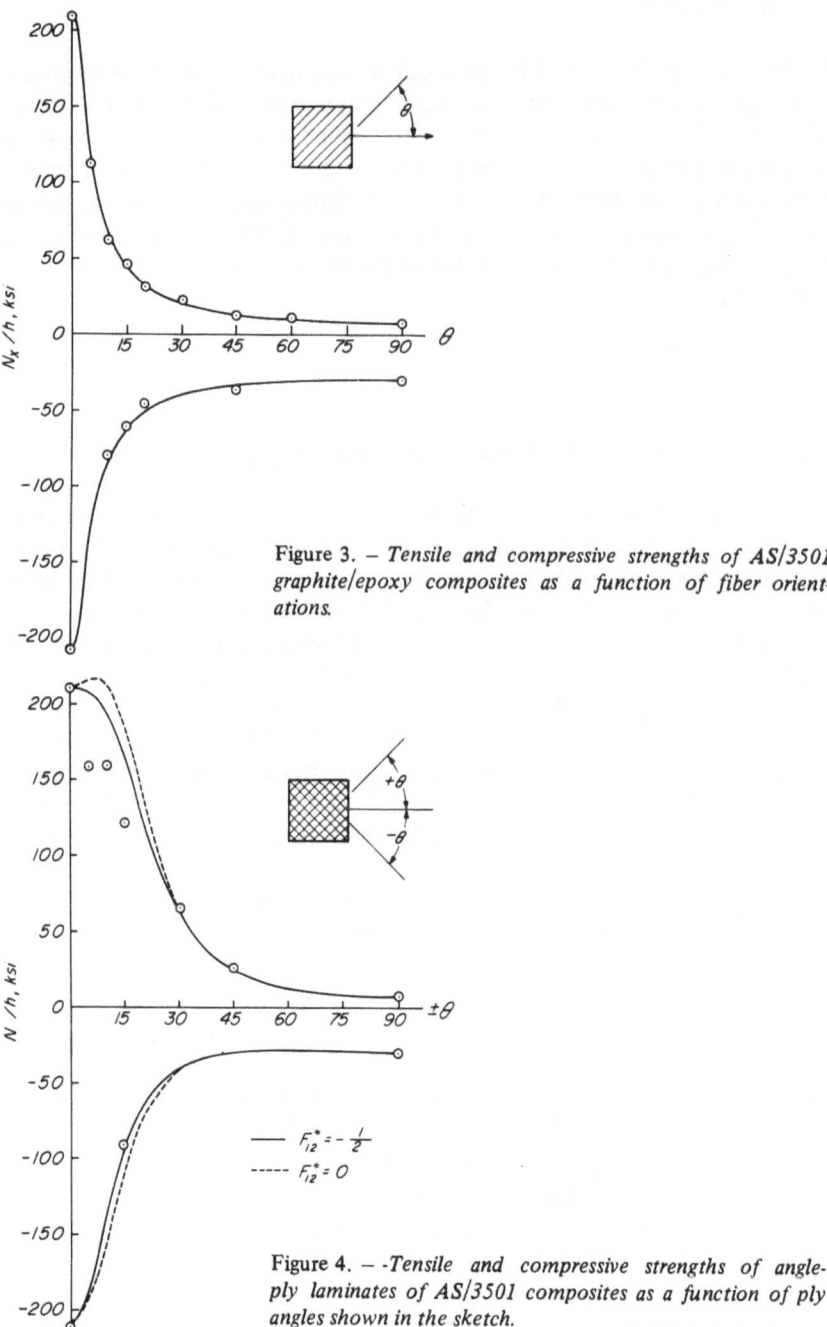

Figure 3. – *Tensile and compressive strengths of AS/3501 graphite/epoxy composites as a function of fiber orientations.*

Figure 4. – *Tensile and compressive strengths of angle-ply laminates of AS/3501 composites as a function of ply angles shown in the sketch.*

6.   Conclusion

We have shown that strength ratios can be calculated from the conventional failure criteria. In particular, the tensor polynomial criterion is easy to use and provides reasonable agreement with available data. If the value of the interaction term in this failure criteria is based on a generalization of the von Mises criterion for isotropic materials, this simplification appears to provide reasonable prediction, and it also provides the opportunity of mapping the strength properties of all orthotropic materials onto the same generalized von Mises ellipse.

RESUME

(Rapports de résistances pour les matériaux orthotropes)

Les rapports de résistances sont des mesures commodes des capacités portantes des matériaux orthotropes soumis à des contraintes combinées. Le critère tensoriel polynomial peut être considéré comme un critère de Von Misès généralisé si une hypothèse est faite sur la valeur du terme d'interaction F_{12}. Avec cette simplification, tout matériau orthotrope peut avoir comme surface de rupture la même ellipse généralisée de Von Misès. Seules l'échelle de l'ellipse et la localisation de son origine dans les axes de coordonnées des deux contraintes normales permettent de distinguer un matériau orthotrope d'un autre. L'analyse des résistances des composites lamellaires peut également inclure la déformation initiale due aux effets de température et d'humidité. Pour les composites lamellaires multidirectionnels, il existe un rapport de résistance pour chaque couche et pour une charge donnée. On peut aisément calculer les charges qui provoqueraient une rupture dans une première couche, des ruptures intermédiaires et la rupture ultime.

DISCUSSION

COMMENT BY Z. HASHIN: In the writer's opinion the F_{12} difficulty is due to the author's choice of modeling the failure criterion of a unidirectional fiber composite by *one* quadratic polynomial. In a previous publication [1] the failure criterion in plane stress has been modeled in terms of two different failure modes-fiber rupture or matrix failure (crack along fibers) as is observed in testing.

In this approach the F_{12} coefficient does not appear at all, the failure criterion is piecewise smooth and agree well with experimental data. The approach has been recently generalized [2] to three dimensional states of stress.

It should be borne in mind that all of the approaches mentioned above are merely curve fittings by quadratic polynomials. The writer believes that since a unidirectional fiber composite has several distinct failure modes, it is more advantageous to model each mode separately, by a quadratic, thus obtaining a piecewise smooth failure criterion instead of one which is entirely smooth.

[1] HASHIN Z. and A. ROTEM. – "A Fatigue Failure Criterion for Fiber Reinforced Materials". *J. Composite Materials* 7 (1973): 448.
[2] Z. HASHIN. – "Failure Criteria for Unidirectional Fiber Composites". *J. of Applied Mechanics* (to appear).

COMMENT BY E.M. WU: I would appreciate your comment on Professor Hashin's concern on the physics of failure. He believes that failure of composites can be classified into two modes; i.e. the fiber failure and the matrix failure. Experimental observations (fractography and acoustic emission) suggest that failure is a continuous and concurrent process. Is Hashin's concern and proposed approach valid on useful ?

REPLY BY S.W. TSAI: There is no physics in my phenomenological approach. The approach you cited was initially proposed by Puck which can be viewed as a special case of the tensor polynomial criterion.

If the transverse and shear strengths are infinite, only fiber failure can occur. Then we have for this assumption:

$$F_{22} = F_2 = F_{66} = F_{12}^* = 0 .$$ (1)

The resulting failure criterion is:

$$F_{11} \sigma_1^2 + F_1 \sigma_1 = 1.$$ (2)

Now, if the longitudinal strengths are infinite, only "matrix" failure can occur. We have:

$$F_{11} = F_1 = F_{12}^* = 0 .$$ (3)

The resulting failure criterion is:

$$F_{22} \sigma_2^2 + F_2 \sigma_2 + F_{66} \sigma_6^2 = 1 .$$ (4)

The interaction term F_{12} is lost in the uncoupling process above. Puck's approach can be viewed as a combined maximum stress criterion in Equation 2 and a shortened tensor polynomial criterion in Equation 4. One equation is easier to use than two simultaneous equations. While both approaches are valid mathematically, neither has much physics.

COMMENT BY Y.F. DAFALIAS: I would like to thank Dr. Tsai for the very informative lecture on failure of orthotropic materials. I would like also to make a comment that may be proved usefull in the future. It is based on a work of mine titled "Anisotropic Hardening of Initially Orthotropic Materials", ZAMM, in press. The failure criterion examined by Dr. Tsai had the usual 6 constants of orthotropy. Since failure was the primary objective, it was not necessary to consider hardening. If, however, one assumes that the orthotropic fourth order tensor (which gives the 6 constants) is not constant, but a properly invariant tensor function of the plastic strain, it can be shown, using the works of Adkins and Smith on integrity bases, that additional terms develope in the yield criterion involving scalar functions of the plastic strain on the second order. For small strains, these terms can be omitted, resuming the 6 initial constants. For larger strains, which may occur up to failure, it is possible that consideration of these additional terms can give a better agreement of the experimental results with the distorted shape of the yield surface.

REPLY BY S.W. TSAI: Thank you for pointing out relevant papers which I am not familiar with. I will try to study these papers and see if my present work can be extended to a more general case following your suggestion. Your constructive criticism is much appreciated.

Failure Criteria for Unidirectional Fiber-Reinforced Composites under Confining Pressure

J. P. Boehler and M. Delafin
Institut de Mécanique de Grenoble, Grenoble, France.

1. Introduction

For the past fifteen years, theoretical and experimental research in composite material mechanics has been stimulated by several applications in modern technology, such as aeronautics, the aerospace industry, etc. More recently, structured solids are widely employed in current structural applications, such as in the car industry, the building industry, the construction of sports equipement, etc.

The principal advantage of composite materials is their high strength-to-weight ratio, which allows lighter structural elements. In order not to lose this advantage by over-dimensioning, but to arrive at a rational application of composites when designing structures, an appropriate knowledge regarding the material behavior under complex loading is required. An essential feature of composites is their pronounced anisotropy of mechanical properties. In this paper, we concentrate our attention on developing a failure criterion for structured solids under complex stress.

The presently proposed criteria for composites are developed either in the micromechanics or the macromechanics approach. The micromechanics approach consists in deriving the criterion, considering the mechanical properties of the different constituents of the composite and making use of limit analysis methods. This approach, however, has been used only under particular assumptions regarding the material composition ([17]-[19]) or produces reliable criteria only for restricted types of composites [15].

In the macromechanics approach, which is purely phenomenological, the structured solid is considered as a homogeneous anisotropic material. Various isotropic failure theories have been generalized for anisotropic composite materials, such as the maximum stress theory [22], the maximum strain theory [24], and the Tresca condition ([13], [14]); a discussion of these adaptations is given in [12].

The Von Mises distorsional energy theory has been generalized by Hill [11] for orthotropic metals and applied by Azzi and Tsai ([3], [4]) to unidirectional fiber-reinforced composites. In this theory, it is assumed that the strengths under compression and tension are the same and that the material is insensitive to hydrostatic stress. These assumptions, however, are not suitable for composites [16].

In order to avoid such limitations, Tsai and Wu [23] formulated a failure condition by proposing the general quadratic in stress

$$F_i \, \sigma_i + F_{ij} \, \sigma_i \sigma_j = 1 \,, \tag{1}$$

where $i, j = 1, 2, \ldots, 6$. The coefficients F_i and F_{ij} are material constants. The linear terms in (1) take stress sign sensitivity of the material into account. Expressed in terms of the actual stress components, the criterion is not restricted to the stress deviator as is the case for Hill's condition; thus, hydrostatic stress effects are involved.

The criterion of Tsai and Wu is widely used in engineering, but it was tested solely in rather simple loading conditions. The current tendency for applications of composites requires a reliable criterion for failure under complex stress. With this objective in mind, we have performed a number of tests consisting of compressions under confining pressure on an epoxy resin reinforced by unidirectional glass-fibers [8]. We observed that the criterion of Tsai and Wu is not confirmed by our experimental results.

In this paper we develop a criterion of failure for fiber-reinforced composites in an objective and rational manner. A consistent and uniform formulation of the mechanical behavior of structured solids can be developed within the theory of representations of anisotropic tensor functions ([1], [6], [7], [20]). In such an approach, the derived constitutive relations satisfy automatically the material symmetries of the structured solid considered. Moreover, the minimal number and the type of variables involved in anisotropic constitutive relations are specified. Thus, the essential variables to be observed in experiments are well defined.

An example of application of the tensor functions representation approach is given in ([5], [6], [9]) by the development of a general theory for the plastic behavior of anisotropic solids. In Section 2, we present the basic theoretical setting with a specific application to transversely isotropic media. The most general forms of both the flow law and the yield condition are thus derived. The stress and mixed stress-anisotropy invariants entering the general form of the criteria for transversely isotropic solids are specified. After specialization to triaxial tests consisting of oriented compressions or tractions under confining pressure, the so reduced general form of the yield conditions is visualized in the space of the basic invariants. Some properties of the invariants' space, as well as restrictions concerning the shape and location of the yield surface are presented and discussed in Section 3. The yield surface is divided into three zones, each attainable by two different specific tests.

In Section 4, the general form of the yield condition is specified for unidirectional fiber-reinforced composites. The proposed criterion is an appropriate generalization of the Von Mises condition for transversely isotropic solids. The new generalization is expressed in a homogeneous form of order two in stress and it takes stress sign sensitivity of the material into account, as well as hydrostatic stress effects. Specializations of the proposed criterion to oriented triaxial and simple compressive or tensile tests are derived.

After having presented, in Section 5, our experimental results concerning the behavior of a fiber-reinforced composite under confining pressure, we compare the proposed criterion and the Tsai and Wu criterion with the tests' data in Section 6. We first specify the number of independent material constants involved in criterion (1), making use of the hypotheses of Tsai and Wu concerning the shear stress sign insensitivity of the material and introducing the restrictions due to the material symmetries of a transversely isotropic solid. We then establish the invariant form of condition (1) and derive the specializations for triaxial and simple compression tests. Finally, the material constants are computed from seven experimental data. The same calculation for the proposed criterion shows that nine experimental data are necessary in order to determine the values of the different constants involved. Comparisons of the two criteria with our experimental results show that the proposed criterion fits much better the tests' data.

Some concluding remarks concerning the suitability and objectivity of the employed approach of tensor functions representations to the description of yielding and failure of structured solids are given in Section 7. Problems related to different modes of failure are also stressed.

2. Theoretical setting

2.1. *General form of the plastic constitutive law*

Consider a homogeneous transversely isotropic solid, where the unit vector $\mathbf{v}$ specifies the material privileged direction. For a unidirectional fiber-reinforced composite, $\mathbf{v}$ denotes the fiber direction. The material structure is thus defined by the following tensor:

$$\mathbf{M} = \mathbf{v} \otimes \mathbf{v} = \begin{vmatrix} 0 & 0 & 0 \\ 0 & 0 & 0 \\ 0 & 0 & 1 \end{vmatrix} (\cdot, \cdot, \mathbf{v}). \tag{2}$$

For futher investigations of the plastic behavior, we restrict our attention to the general form of the constitutive relation for a rate type material. In the case of a transversely isotropic solid, such a relation is expressed in the form of a tensor function

$$T = F(D, M),\tag{3}$$

where D denotes the rate of deformation tensor and T is the stress tensor. The relation (3) can be viewed as a constitutive equation provided it satisfies the principle of isotropy of space [21]. This principle requires that an arbitrary transformation Q of the orthogonal group $\mathcal{O}$, applied to both the medium and the kinematic variables, results in the same transformation of the material response:

$$\forall Q \in \mathcal{O}: \quad F(QDQ^T, QMQ^T) = QF(D, M)Q^T.\tag{4}$$

This property implies that the function F is isotropic with respect to its two arguments D and M, thus anisotropic with respect to D, the type of anisotropy being transverse isotropy ([6], [7]).

The form of the relation (3) is further specified by the representation theorems for tensor functions. In the case of a general (not necessarily polynomial) function F, the irreducible representation is given by ([6], [7]):

$$\begin{aligned}
T &= \alpha_0 I + \alpha_1 M + \alpha_2 D + \alpha_3 (MD^2 + D^2 M) \\
\alpha_i &= \alpha_i (\mathrm{tr}\, D, \mathrm{tr}\, D^2, \mathrm{tr}\, D^3, \mathrm{tr}\, MD, \mathrm{tr}\, MD^2).
\end{aligned}\tag{5}$$

In (5), the stress tensor T is expressed as a linear combination of six symmetric second order tensor generators. The coefficients α_i are scalar-valued functions of five basic invariants. Representation (5) constitutes the most general form of the constitutive relation (3) for a transversely isotropic rate type material.

Plastic behavior may be introduced in (5) by the homogeneity condition of order zero with respect to time or, which is equivalent, with respect to the rate of deformation:

$$\frac{\partial T}{\partial D} D = 0 \quad \text{if} \quad \frac{\partial T}{\partial D} \neq 0.\tag{6}$$

Introducing condition (6) in the representation (5), one obtains the general form of both the flow law,

$$\frac{D}{\mathrm{tr}^{1/2} D^2} = \psi_0 I + \psi_1 M + \psi_2 T + \psi_3 (MT + TM) + \psi_4 T^2 + \\
+ \psi_5 (MT^2 + T^2 M)\tag{7}$$

$$\psi_i = \psi_i \left(\frac{\mathrm{tr}\, D}{\mathrm{tr}^{1/2}\, D^2} , \frac{\mathrm{tr}^{1/3}\, D^3}{\mathrm{tr}^{1/2}\, D^2} , \frac{\mathrm{tr}\, MD}{\mathrm{tr}^{1/2}\, D^2} , \frac{\mathrm{tr}^{1/2}\, MD^2}{\mathrm{tr}^{1/2}\, D^2} \right),$$

and the yield condition,

$$f(\mathrm{tr}\, T , \mathrm{tr}\, T^2 , \mathrm{tr}\, T^3 , \mathrm{tr}\, MT , \mathrm{tr}\, MT^2) = 0. \tag{8}$$

Details regarding the derivation of relations (7) and (8) are given in ([5], [6], [9]). In order to make the paper self-contained, we give here below the essential characteristics of the presented theory.

The flow law (7) specifies the rate of deformation tensor only to within a scalar multiplier. The coefficients ψ_i are homogeneous scalar-valued functions of order zero with respect to D. The anisotropic character of the relation between the tensors T and D derives from the presence of the tensor-generators involving the structural tensor M. A characteristic feature of the flow law is that if the principal directions of tensors T and M do not coincide (for example in off-axis tests), then the principal directions of D differ from those of T. It is also worthwhile to point out that the flow law (7) is not necessarily associated with the yield condition (8) by the restrictive hypothesis of the plastic potential.

In the general form (8) of the yield condition, the independent variables involve in addition to the isotropic invariants $\mathrm{tr}\, T$, $\mathrm{tr}\, T^2$, $\mathrm{tr}\, T^3$, two mixed invariants $\mathrm{tr}MT$ and $\mathrm{tr}MT^2$, which account for the anisotropic character of the material behavior. They specify the orientation of the stress tensor with respect to the material attached structural tensor M.

The relation (8) is of importance in the sense that it specifies the minimal number and the type of independent variables to be observed in experiments. An equivalent form of the yield condition (8) is obtained by introducing the stress deviator S:

$$S = T - \left(\frac{1}{3}\, \mathrm{tr}\, T \right) I. \tag{9}$$

Four of the five basic invariants of T and M can be expressed in terms of the basic invariants of S and M and the invariant $\mathrm{tr}\, T$.

$$\mathrm{tr}\, T^2 = \mathrm{tr}\, S^2 + \frac{1}{3}\, \mathrm{tr}^2\, T ,$$

$$\mathrm{tr}\, T^3 = \mathrm{tr}\, S^3 + \mathrm{tr}\, S^2\, \mathrm{tr}\, T + \frac{1}{9}\, \mathrm{tr}^3\, T ,$$

$$\mathrm{tr}\, MT = \mathrm{tr}\, MS + \frac{1}{3}\, \mathrm{tr}\, T ,$$

$$\mathrm{tr}\, MT^2 = \mathrm{tr}\, MS^2 + \frac{2}{3}\, \mathrm{tr}\, MS\, \mathrm{tr}\, T + \frac{1}{9}\, \mathrm{tr}^2\, T .$$

$$(10)$$

In view of the relations (10), the general form (8) can be written equivalently as

$$g(\text{tr}\,T,\ \text{tr}\,S^2,\ \text{tr}\,S^3,\ \text{tr}\,MS,\ \text{tr}\,MS^2) = 0. \tag{11}$$

2.2. Application to tests under confining pressure

Consider triaxial tests consisting of compressions or tractions of oriented specimens under confining pressure. We denote by p the confining pressure, by σ_n the axial stress and by θ the angle between the axis of the specimen and the privileged direction **v** of the transversely isotropic solid (Fig. 1).

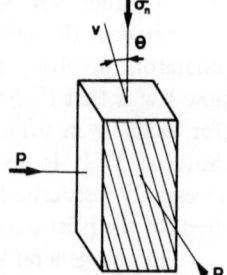

Figure 1. – *Fiber-reinforced material in compression tests under confining pressure.*

The arguments of criterion (11) can be expressed in terms of the three variables p, σ_n and θ:

$$\text{tr}\,T \quad = 2p + \sigma_n,$$

$$\text{tr}\,S^2 \quad = \frac{2}{3}\,(\sigma_n - p)^2,$$

$$\text{tr}\,S^3 \quad = \frac{2}{9}\,(\sigma_n - p)^3, \tag{12}$$

$$\text{tr}\,MS \quad = \frac{1}{3}\,(\sigma_n - p)\,(3\cos^2\theta - 1),$$

$$\text{tr}\,MS^2 = \frac{1}{9}\,(\sigma_n - p)^2\,(3\cos^2\theta + 1).$$

Thus, if restricted to tests under confining pressure, the five arguments of the general form (11) are no longer independent; they are related by the equations

$$\text{tr}\,MS^2 = \frac{1}{3}\,\text{tr}\,S^2 + \frac{\epsilon}{\sqrt{6}}\,\text{tr}\,MS\,\text{tr}^{1/2}\,S^2,$$

$$\text{tr}\,S^3 \quad = \epsilon\,\frac{2}{3\sqrt{6}}\,\text{tr}^{3/2}\,S^2$$

$$\text{where}\ \begin{cases} \epsilon = +1\ \text{if}\ \sigma_n - p > 0, \\[4pt] \epsilon = -1\ \text{if}\ \sigma_n - p < 0. \end{cases} \tag{13}$$

Only three invariants remain independent and therefore the general form of the yield condition for triaxial tests is reduced to

$$h(\text{tr}\,T, \text{tr}\,S^2, \text{tr}\,MS) = 0 . \tag{14}$$

3. Visualization of the yield condition in the space of basic invariants

Before developing a specific criterion of failure for fiber-reinforced composites, we first show how the restriction (14) to triaxial tests of the general yield condition (11) is represented in the space of the basic invariants $\text{tr}\,T$, $\text{tr}\,S^2$, $\text{tr}\,MS$. For this purpose, we introduce an alternative form of (14). Considering that the invariant $\text{tr}\,S^2$ is non-negative and that an expression for the orientation angle θ in terms of the basic invariants can be derived from the relations (12),

$$\cos^2\theta = \frac{1}{3} + \epsilon\,\sqrt{\frac{2}{3}\,\frac{\text{tr}\,MS}{\text{tr}^{1/2}S^2}}\,, \qquad \text{where} \qquad \begin{cases} \epsilon = +1 \ \ \text{if} \ \ \sigma_n - p > 0 , \\[2mm] \epsilon = -1 \ \ \text{if} \ \ \sigma_n - p < 0 , \end{cases} \tag{15}$$

an alternative form of the criterion (14) is given by:

$$\ell\left(\text{tr}\,T, \ \sqrt{\tfrac{3}{2}\,\text{tr}\,S^2}, \theta\right) = 0 . \tag{16}$$

Thus, the critical states of stress can be represented in the plane of the invariants $\left(\text{tr}\,T, \sqrt{\tfrac{3}{2}\,\text{tr}\,S^2}\right)$ by a family of curves parametrized by the orientation angle θ.

We shall now analyse the visualization of the compressive and tensile strengths under confining pressure. We adopt the convention of positive compression.

3.1. *Compressive strengths under confining pressure*

Consider tests consisting of axial compressions σ_n under positive confining pressures p and suppose that the axial stress is greater than p :

$$\sigma_n > 0, \ \ p > 0, \ \ \sigma_n - p > 0 . \tag{17}$$

An application of relations (12) results in

$$\text{tr}\,T = 2p + \sigma_n, \ \ \sqrt{\tfrac{3}{2}\,\text{tr}\,S^2} = \sigma_n - p > 0. \tag{18}$$

Conditions (18) show that for a given value of p, the invariants $\mathrm{tr}\,T$ and $\sqrt{\frac{3}{2}\,\mathrm{tr}\,S^2}$ are related by

$$\sqrt{\frac{3}{2}\,\mathrm{tr}\,S^2} = \mathrm{tr}\,T - 3\,p.\tag{19}$$

Thus, for fixed confining pressure p, the critical states of stress, corresponding to the different values of the orientation angle θ, are represented by points lying on the same straight line (19), which is parallel to the first bisectrix of the plane $\left(\mathrm{tr}\,T,\ \sqrt{\frac{3}{2}\,\mathrm{tr}\,S^2}\right)$. The region corresponding to compression tests under confining pressure subjected to the condition (17) is visualized by zone 1 of Figure 2.

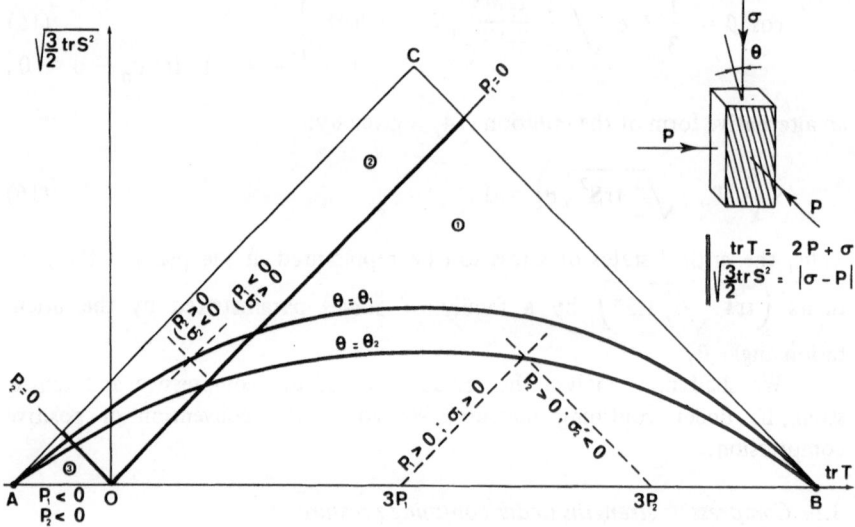

Figure 2. – *Visualization of the yield criteria in the invariants' plane.*

3.2. *Tensile and pseudo-tensile strengths under confining pressure*

Suppose now that the axial stress σ_n is lower than the confining pressure p. The axial deviatoric stress σ_n' is then negative:

$$\frac{3}{2}\,\sigma_n' = \sigma_n - p < 0.\tag{20}$$

According to the sign of the axial stress σ_n, two cases can be distinguished:

$$p > \sigma_n > 0: \text{ pseudo-tensile tests,}$$

$$p > 0 > \sigma_n: \text{ tensile tests.} \tag{21}$$

For both cases, the relations (12) result in

$$\text{tr}\,T = 2p + \sigma_n, \quad \sqrt{\tfrac{3}{2}\,\text{tr}S^2} = p - \sigma_n > 0. \tag{22}$$

Thus, for a given value of p, the invariants $\text{tr}T$ and $\sqrt{\tfrac{3}{2}\,\text{tr}S^2}$ are related by

$$\sqrt{\tfrac{3}{2}\,\text{tr}S^2} = 3p - \text{tr}T. \tag{23}$$

The critical states of stress, corresponding to the different values of θ and to a fixed confining pressure p, are represented by points lying on the same straight line (23), which is parallel to the second bisectrix of the plane $\left(\text{tr}T, \sqrt{\tfrac{3}{2}\,\text{tr}S^2}\right)$. The region scanned by tensile and pseudo-tensile tests under confining pressure consists of the two zones 1 and 2 of Figure 2.

3.3. *Scanning of the plane* $\left(\text{tr}\,T, \sqrt{\tfrac{3}{2}\,\text{tr}S^2}\right)$

From the preceding analysis, we can conclude that each point in zone 1 corresponds to two different strengths. One of the strengths is obtained by a test with positive axial deviatoric stress:

$$p_1 > 0, \quad \tfrac{3}{2}\sigma_1' = \sigma_1 - p_1 > 0 \tag{24}$$

The other strength corresponds to a negative axial deviatoric stress:

$$p_2 > 0, \quad \tfrac{3}{2}\sigma_2' = \sigma_2 - p_2 < 0. \tag{25}$$

The introduced quantities are related by

$$p_2 = \tfrac{1}{3}(2\sigma_1 + p_1), \quad \sigma_2 = \tfrac{1}{3}(4p_1 - \sigma_1), \quad \sigma_2' = -\sigma_1'. \tag{26}$$

In zone 2, each point corresponds also to two strengths obtained by two different tests. For one of the tests $(p_2 > 0, \sigma_2)$, the axial deviatoric stress is negative (cf. 3.2). Zone 2 being characterized by

$$\sqrt{\frac{3}{2} \, \mathrm{tr} S^2} > \mathrm{tr} T, \tag{27}$$

the following inequality can be derived:

$$p_2 + 2\sigma_2 < 0. \tag{28}$$

Since p_2 is assumed to be positive, the axial stress σ_2 is necessarily negative and corresponds to a tensile test under confining pressure:

$$p_2 > 0, \quad \sigma_2 < 0. \tag{29}$$

The other test corresponds to a positive axial stress (axial compression) and a negative confining pressure (confining traction):

$$p_1 < 0, \quad \sigma_1 > 0. \tag{30}$$

The introduced quantities are related by

$$p_1 = \frac{1}{3} \, (p_2 + 2\sigma_2) < 0, \quad \sigma_1 = \frac{1}{3} \, (4p_2 - \sigma_2) > 0, \quad \sigma_1' = -\sigma_2' > 0. \tag{31}$$

It is possible to show in a similar manner that each point in zone 3 of Figure 2 corresponds also to two different tests; both tests have to be performed under negative confining pressure (confining traction).

3.4. Concluding remarks.

Compression tests under confining pressure allow to cover only zone 1 of the plane $\left(\mathrm{tr} T, \sqrt{\frac{3}{2} \mathrm{tr} S^2} \right)$. In order to have access to zone 2, it is necessary to perform axial tensile tests under confining pressure. Zone 3 is unattainable by tests under positive confining pressure. Since negative confining pressures are difficult to realize practically, it is necessary to perform biaxial tensile tests so as to come into zone 3; biaxial strengths can also be represented in the invariant plane $\left(\mathrm{tr} T, \sqrt{\frac{3}{2} \, \mathrm{tr} S^2} \right)$.

Each straight line (19) for $p_1 = \mathrm{const.}$ and (23) for $p_2 = \mathrm{const.}$ can intersect a given curve $\theta = \mathrm{const.}$ at only a single point. This condition leads to restrictions for the shape of the curves $\theta = \mathrm{const.}$ In another connection, it is reasonable to assume that for an anisotropic solid failure will occur under high isotropic compression. Experimental evidence of this effect for a stratified rock is given in [2]. Thus, all the curves $\theta = $ const. intersect the positive semi-axis $\text{tr}\,T$ at a single point. From the preceding remarks, it can be concluded that the curves $\theta = $ const. are located within a boundary triangle (ABC), as shown in Figure 2.

4. Proposed criterion

The general form (8) of the yield condition requires specifications for each classs of transversely isotropic materials by further hypotheses. As was stated in Section 1, most of the presently proposed criteria for anisotropic solids are appropriate generalizations of the available criteria for isotropic media. It is worthwhile pointing out that the generally accepted criteria for isotropic materials are homogeneous in stress. This is in particular the case of the Von Mises condition. Thus, we propose a new generalization of the Von Mises criterion to unidirectional fiber-reinforced composites in developing a homogeneous form of order two in stress and employing the simplest combinations of the basic invariants involved in (8) and their square roots. The proposed criterion is expressed by:

$$\text{tr}\,T^2 + a_1\,\text{tr}^2 T + a_2\,\text{tr}\,MT^2 + a_3\,\text{tr}^2 MT + a_4\,\text{tr}\,MT\,\text{tr}\,T + a_5\,\sqrt{\text{tr}\,T^2}\,\text{tr}\,T +$$

$$+ a_6\,\sqrt{\text{tr}\,T^2}\,\text{tr}\,MT + a_7\,\sqrt{\text{tr}\,MT^2}\,\text{tr}\,MT + a_8\,\sqrt{\text{tr}\,MT^2}\,\text{tr}\,T +$$

$$+ a_9\,\sqrt{\text{tr}\,MT^2}\,\sqrt{\text{tr}\,T^2} = K^2, \tag{32}$$

where the $a_i\,(i = 1, 2, \ldots, 9)$ are the material constants and K the constant of the criterion.

In expression (32), terms involving square roots of the basic invariants take stress sign sensitivity of the material into account. It is therefore no longer necessary to add linear terms as is the case for the Tsai and Wu criterion.

The proposed criterion (32), when applied to axial compression tests σ_n in direction θ and under the confining pressure p, results in

$$\sigma_n^2\,[1 + a_1 + (a_2 + a_4)\cos^2\theta + a_3\cos^4\theta] +$$

$$+ p[(2 + 4a_1 + a_2 + a_3 + 2a_4) - (a_2 + 2a_3 + 2a_4)\cos^2\theta + a_3\cos^4\theta] +$$

$$+ \sigma_n p[4a_1 + a_4 + (2a_3 + a_4)\cos^2\theta - 2a_3\cos^4\theta] +$$

$$+ \sqrt{2p^2 + \sigma_n^2}\,[(2a_5 + a_6 - a_6\cos^2\theta)\,p + (a_5 + a_6\cos^2\theta)\,\sigma_n] +$$

$$+ \sqrt{p^2 + (\sigma_n^2 - p^2)\cos^2\theta}\,[(2a_8 + a_7 - a_7\cos^2\theta)\,p +$$

$$+ (a_8 + a_7\cos^2\theta)\,\sigma_n] +$$

$$+ a_9\,\sqrt{p^2 + (\sigma_n^2 - p^2)\cos^2\theta}\,\sqrt{2p^2 + \sigma_n^2} = K^2. \tag{33}$$

For simple compression tests, the ratio $\left(\dfrac{\sigma_\theta}{\sigma_{90}}\right)^2$, where σ_θ and σ_{90} are the strengths, respectively, in the direction θ and direction $\theta = \dfrac{\pi}{2}$, can be calculated from the criterion (32). One obtains

$$\left(\frac{\sigma_\theta}{\sigma_{90}}\right)^2 = \frac{1}{1 + A\cos\theta + B\cos^2\theta + C\cos^3\theta + D\cos^4\theta} \tag{34}$$

where A, B, C and D are related to the material constants a_i by

$$A = \frac{a_g + a_9}{1 + a_1 + a_5}, \quad B = \frac{a_2 + a_4 + a_6}{1 + a_1 + a_5}, \quad C = \frac{a_7}{1 + a_1 + a_5},$$

$$D = \frac{a_8}{1 + a_1 + a_5}. \tag{35}$$

Moreover, the criterion constant K is related to the strength in direction $\theta = \dfrac{\pi}{2}$ by

$$K^2 = (1 + a_1 + a_5)\,\sigma_{90}^2. \tag{36}$$

From the relations (34) and (35), we conclude that oriented simple compression tests determine the values of the four material constants a_i. The five other constants specify the evolution of the directional strengths with the confining pressure.

5. Experimental behavior of a fiber-reinforced composite under confining pressure

In order to test the reliability of the proposed criterion and of the Tsai and Wu criterion for complex loading states, we performed a number of experiments consisting of directional axial compressions of a glass fiber reinforced epoxy resin under confining pressure.

5.1. *Testing procedure*

The specimens were prepared in seven different orientations: $\theta = 0$, $15°$, $30°$, $45°$, $60°$, $75°$ and $90°$ (Fig. 1). They were cut to the shape of parallelepipeds with dimensions $12 \times 12 \times 24$ mm. We applied five different confining pressures: $p = 0$, 250, 500, 750 and 1 000 kg/cm^2. Thus, we obtained 35 different test conditions.

For the orientations $\theta = 0$ and $\theta = 15°$, some problems regarding the modes of failure arose. Because of the parallelepiped shape, the phenomenon of "brooming" appeared at the ends of the specimens. In order to prevent this effect, we first employed longer specimens and fitted the ends in steel rings, but failure started at the level of the rings. Finally, we employed for the orientations $\theta = 0$ and $\theta = 15°$ bone shaped specimens. We obtained then failure by kink in the central part of all specimens, except for the specimen ($\theta = 0$, $p = 0$), where failure occured by longitudinal splitting. This failure mode and the corresponding limit stress being quite different from those of the other orientations and confining pressures, we excluded the results for ($\theta = 0$, $p = 0$). We shall come back to the problem of failure modes in Section 7. Other details regarding the testing procedure are given in [8].

5.2. *Experimental results*

An evolution of the material behavior under confining pressure and given material orientation was observed. The respective documentation regarding the obtained stress-strain curves and the observed phenomena are presented in [8].

Characteristic stress-strain curves for some orientations θ and the confining pressure $p = 250 \, \text{kg/cm}^2$ are shown in Figure 3. For the orientations $\theta = 0$ and $\theta = 15°$, we obtained curves with very marked peaks. For the other orientations, the peaks disappear. For each test, the chosen limit stress corresponds to the peak or to the maximum stress observed during the compression.

The obtained values for the limit stresses are presented in Table 1. The value given for ($\theta = 0$, $p = 0$) is not an experimental result, but a fictive value obtained by continuous extrapolation of the data for $\theta = 0$ and the other four confining pressures. The results of Table 1 will now be employed in order to test the proposed criterion and the Tsai and Wu criterion.

TABLE 1. Compressive strengths under confining pressure:
experimental data (kg/cm^2).

P (kg/cm^2)	$\theta = 0$	$\theta = 15°$	$\theta = 30°$	$\theta = 45°$	$\theta = 60°$	$\theta = 75°$	$\theta = 90°$
0	7300	2662	1359	1254	1236	1264	1311
250	7850	3003	1407	1290	1369	1378	1483
500	8300	3091	1549	1396	1412	1456	1579
750	8650	3518	1774	1728	1925	1657	1644
1000	8900	3617	2314	1841	2016	1660	1854

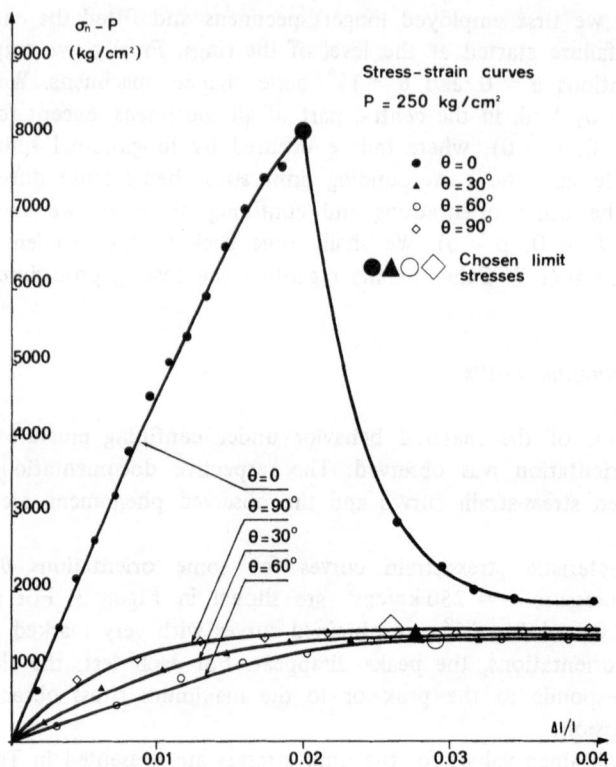

Figure 3. — *Fiber-reinforced epoxy resin: stress-strain curves of oriented compression tests under 250 kg/cm² confining pressure.*

6. Comparison of the criteria with experimental results

6.1. *Criterion of Tsai and Wu*

We first specify the non-vanishing material constants F_i and F_{ij} of criterion (1). The hypotheses of Tsai and Wu imply that the material is sensitive to the sign of normal stress, but intensitive to the sign of the shear stress. Thus, the following material constants vanish:

$$F_i = 0 \quad \text{if} \quad i > 3 ,$$

$$F_{ij} = 0 \quad \text{if} \quad \begin{cases} i \neq i \\ \text{and} \\ i, j > 3. \end{cases} \tag{37}$$

For an orthotropic solid, only twelve non-vanishing material constants remain:

$$F_1, F_2, F_3, F_{11}, F_{22}, F_{33}, F_{44}, F_{55}, F_{66}, F_{23}, F_{31}, F_{12}. \tag{38}$$

Consider now a transversely isotropic solid with an attached reference system (v_1, v_2, v_3), v_3 being the privileged axis v of the material. The rotational symmetry of the medium about the v_3 axis results in the following restrictions:

$$F_1 = F_2, \quad F_{11} = F_{22}, \quad F_{44} = F_{55}, \quad F_{66} = F_{11} - F_{12},$$
$$F_{31} = F_{23}. \tag{39}$$

Finally, for transversely isotropic materials, the criterion of Tsai and Wu involves only seven independent material constants:

$$F_1, F_3, F_{11}, F_{33}, F_{44}, F_{31}, F_{12}. \tag{40}$$

The criterion is then expressed in the (v_1, v_2, v_3) reference system by:

$$F_1(\sigma_{11} + \sigma_{22}) + F_3\sigma_{33} + F_{11}(\sigma_{11}^2 + \sigma_{22}^2 + 2\sigma_{12}^2) + F_{33}\sigma_{33}^2 +$$
$$+ F_{44}(\sigma_{31}^2 + \sigma_{23}^2) + 2F_{12}(\sigma_{11}\sigma_{22} - \sigma_{12}^2) +$$
$$+ 2F_{31}\sigma_{33}(\sigma_{11} + \sigma_{22}) = 1. \tag{41}$$

Let us establish now the invariant form of the criterion (41). Of course, any criterion for transversely isotropic solids is a specific form of the general criterion (8). The basic invariants which appear in (8) are expressed in terms of the stress components in the (v_1, v_2, v_3) frame by

$$\begin{cases} \text{tr}\,T = \sigma_{11} + \sigma_{22} + \sigma_{33}, \\ \text{tr}\,T^2 = \sigma_{11}^2 + \sigma_{22}^2 + \sigma_{33}^2 + 2(\sigma_{23}^2 + \sigma_{31}^2 + \sigma_{12}^2), \\ \text{tr}\,MT = \sigma_{33}, \\ \text{tr}\,MT^2 = \sigma_{33}^2 + \sigma_{23}^2 + \sigma_{31}^2. \end{cases} \tag{42}$$

The fifth basic invariant $\text{tr}\,T^3$ is not involved in the criterion (41). Making use of the relations (42) in (41), we obtain the invariant form of the Tsai and Wu criterion:

$$b_1\,\text{tr}\,T + b_2\,\text{tr}\,MT + b_3\,\text{tr}^2\,T + b_4\,\text{tr}^2\,MT + b_5\,\text{tr}\,MT\,\text{tr}\,T + b_6\,\text{tr}\,T^2 +$$
$$+ b_7\,\text{tr}\,MT^2 = 1, \tag{43}$$

which is, indeed, a particular form of (8). The seven coefficients b_i are new material constants which are related to the constants (40) by:

$$b_1 = F_1, \quad b_2 = F_3 - F_1, \quad b_3 = F_{12},$$
$$b_4 = F_{11} + F_{33} - F_{44} - 2F_{31}, \quad b_5 = 2(F_{31} - F_{12}),$$
$$b_6 = F_{11} - F_{12}, \quad b_7 = F_{44} + 2(F_{12} - F_{11}). \tag{44}$$

When applied to tests consisting of axial compressions σ_n in various directions θ and under confining pressures p, the yield condition (41) yields

$$\sigma_n[F_1 \sin^2 \theta + F_3 \cos^2 \theta] + p[F_1(1 + \cos^2 \theta) + F_3 \sin^2 \theta] +$$
$$+ \sigma_n^2[F_{11} \sin^4 \theta + F_{33} \cos^4 \theta + (F_{44} + 2F_{31}) \sin^2 \theta \cos^2 \theta] +$$
$$+ p^2[F_{11}(1 + \cos^4 \theta) + F_{33} \sin^4 \theta + (F_{44} + 2F_{31}) \sin^2 \theta \cos^2 \theta +$$
$$+ 2F_{12} \cos^2 \theta + 2F_{31} \sin^2 \theta] + 2\sigma_n p[(F_{11} + F_{33} - F_{44}) \sin^2 \theta \cos^2 \theta +$$
$$+ F_{31}(\sin^2 \theta + \sin^4 \theta + \cos^4 \theta) + F_{12} \sin^2 \theta] = 1. \tag{45}$$

For simple compression tests, we obtain the expression:

$$F_1 \sigma_n \sin^2 \theta + F_3 \sigma_n \cos^2 \theta + F_{11} \sigma_n^2 \sin^4 \theta + F_{33} \sigma_n^2 \cos^4 \theta +$$
$$+ (F_{44} + 2F_{31}) \sigma_n^2 \cos^2 \theta \sin^2 \theta = 1. \tag{46}$$

Relation (46) involves five independent material constants. Thus, in the criterion of Tsai and Wu, five constants are determined by oriented simple compression tests. The two remaining constants specify the evolution of the directional strengths with the confining pressure.

In Table 2, the values of the seven material constants b_i obtained from our experimental results and the seven tests chosen for the calculation are presented. The chosen tests are those which resulted in the best theoretical representation for the experimental data.

6.2. *Proposed criterion*

The invariant form of the proposed criterion is given by equation (32). The applications to triaxial and simple compression tests result, respectively, in the relations (33) and (34). Criterion (32) involves nine material constants a_i and the criterion constant K.

TABLE 2. Criterion of Tsai and Wu.

- General form: $F_i \sigma_i + F_{ij} \sigma_i \sigma_j = 1$ $i, j = 1, 2, \ldots, 6$.

- Invariant form for transversely isotropic solids:

$$b_1 \, \text{tr} \, T + b_2 \, \text{tr} \, MT + b_3 \, \text{tr}^2 \, T + b_4 \, \text{tr}^2 \, MT + b_5 \, \text{tr} \, T \, \text{tr} \, MT +$$
$$+ b_6 \, \text{tr} \, T^2 + b_7 \, \text{tr} \, MT^2 = 1 .$$

- Experimental data chosen for the calculation of the material constants b_i:

$p = 0$ $\qquad\qquad$: $\quad \theta = 0°, 15°, 30°, 60°, 90°$.

$p = 1000 \, \text{kg/cm}^2$: $\quad \theta = 0°, 90°$.

- Numerical values of the material constants b_i:

$b_1 = 0{,}668 \, 10^{-2}$ $\qquad$ $b_2 = -0{,}510 \, 10^{-2}$ $\qquad$ $b_3 = 0{,}283 \, 10^{-5}$

$b_4 = 0{,}152 \, 10^{-5}$ $\qquad$ $b_5 = -0{,}593 \, 10^{-5}$ $\qquad$ $b_6 = -0{,}737 \, 10^{-5}$

$b_7 = 0{,}873 \, 10^{-5}$

As a first step, we established by means of the relation (34) a theoretical representation of the experimental data in simple compression. Calculation of the four constants (35), A, B, C and D, showed that for the tested glass fiber reinforced composite the constant A can be taken equal to zero. Substituting the value $A = 0$ in (35), we obtain the relation

$$a_9 = - a_8 \tag{47}$$

Thus, for glass fiber reinforced composites, criterion (32) takes the reduced form:

$$\text{tr} \, T^2 + a_1 \, \text{tr}^2 \, T + a_2 \, \text{tr} \, MT^2 + a_3 \, \text{tr}^2 \, MT + a_4 \, \text{tr} \, MT \, \text{tr} \, T + a_5 \sqrt{\text{tr} \, T^2} \, \text{tr} \, T +$$
$$+ a_6 \sqrt{\text{tr} \, T^2} \, \text{tr} \, MT + a_7 \sqrt{\text{tr} \, MT^2} \, \text{tr} \, MT +$$
$$+ a_8 \sqrt{\text{tr} \, MT^2} \, (\text{tr} \, T - \sqrt{\text{tr} \, T^2}) = K^2 . \tag{48}$$

and the application of (34) to simple compression tests gives

$$\left(\frac{\sigma_\theta}{\sigma_{90}}\right)^2 = \frac{1}{1 + B \cos^2\theta + C \cos^2\theta + D \cos^4\theta} \cdot \tag{49}$$

From relations (35), (36) and (49), we can conclude that the values of the criterion constant K and of the three material constants a_i are determined from four experimental data of oriented simple compression tests. The values of the five other material constants are obtained from the data of five compression tests under non-zero confining pressures. The results of the calculation and the nine chosen tests are given in Table 3. The chosen tests are those which resulted in the best theoretical representation for the experimental data.

TABLE 3. Proposed criterion.

- Invariant form :

$$\mathrm{tr}\ T^2 + a_1\ \mathrm{tr}^2\ T + a_2\ \mathrm{tr}\ MT^2 + a_3\ \mathrm{tr}^2\ MT + a_4\ \mathrm{tr}\ T\ \mathrm{tr}\ MT +$$
$$+ a_5\ \sqrt{\mathrm{tr}\ T^2}\ \mathrm{tr}\ T + a_6\ \sqrt{\mathrm{tr}\ T^2}\ \mathrm{tr}\ MT + a_7\ \sqrt{\mathrm{tr}\ MT^2}\ \mathrm{tr}\ MT +$$
$$+ a_8\ \sqrt{\mathrm{tr}\ MT^2}\ \mathrm{tr}\ T + a_9\ \sqrt{\mathrm{tr}\ MT^2}\ \sqrt{\mathrm{tr}\ T^2} = K^2$$

- Monoaxial compressive strengths :

$$\left(\frac{\sigma_\theta}{\sigma_{90}}\right)^2 = \frac{1}{1 + A \cos\theta + B \cos^2\theta + C \cos^3\theta + D \cos^4\theta}$$

- Experimental data chosen for the calculation of the materials constants a_i :

p = 0   : $\theta = 0°, 30°, 75°, 90°$.

p = 250 kg/cm^2 : $\theta = 0°, 60°, 90°$.

p = 1 000 kg/cm^2 : $\theta = 0°, 90°$.

- Numerical values of the material constants a_i :

$a_1 = -0,077$ $a_2 = 0,492$ $a_3 = -1,668$ $a_4 = 1,789$

$a_5 = -0,512$ $a_6 = -1,946$ $a_7 = 0,936$ $a_8 = -1,168$

$K = 840,36$ kg/cm^2

6.3. *Comparisons*

In Figures 4 and 5, the experimental data and the theoretical curves obtained by application of the proposed criterion and the Tsai and Wu criterion are presented in the plane of the invariants $\operatorname{tr}\mathbf{T}$ and $\sqrt{\dfrac{3}{2}\operatorname{tr}\mathbf{S}^2}$. For p = 0, i.e., in simple compression, both criteria fit well the experimental data. But when the confining pressure increases, great divergencies appear for the Tsai and Wu criterion, which in general overestimates the strengths of the material under non-zero confining pressures.

The proposed criterion fits better the experimental data for all values of the confining pressure p and the orientation θ. In a few cases, in particular for $\theta = 15°$, some differences appear, but the theoretical curves are always located on the safe side.

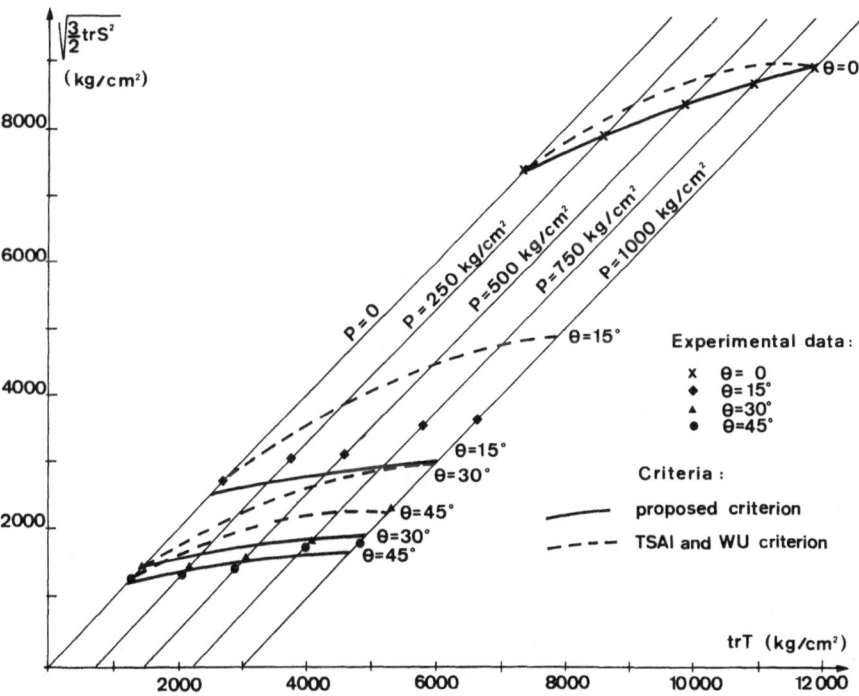

Figure 4. – *Fiber-reinforced epoxy resin: compressive strengths under confining pressures; orientations* $\theta = 0°, 15°, 30°$ *and* $45°$.

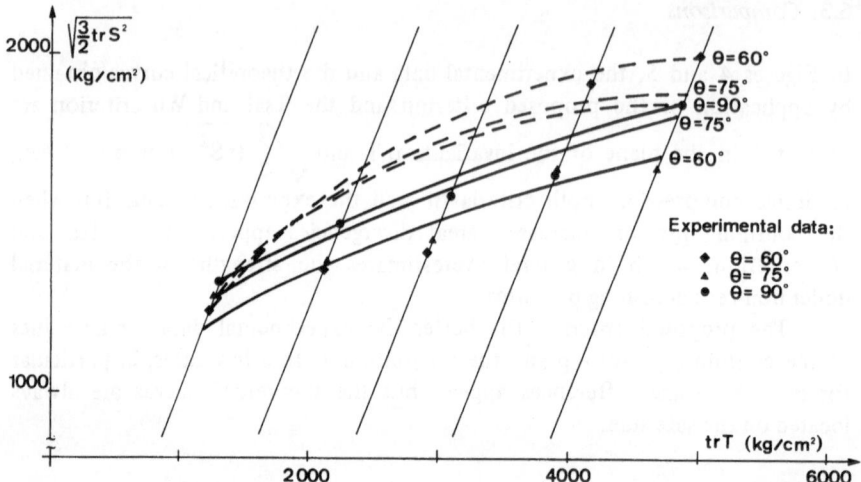

Figure 5. – *Fiber-reinforced epoxy resin: compressive strengths under confining pressures: orientations* θ = 60°, 75° *and* 90°.

7. Conclusions

Application of theorems about representations of tensor functions to the description of the plastic behavior of structured solids allows to derive the most general form of the yield conditions for anisotropic materials. Such an approach indicates the minimal number and the type of basic stress and mixed stress-anisotropy invariants involved in the failure criteria. The essential variables to be observed in experiments are thus specified.

An illustration of the approach was given here on the example of a transversely isotropic fiber-reinforced composite. The general form of the yield condition was specialized to oriented compressive and tensile tests under confining pressure and visualized in the space of basic invariants. Restrictions regarding the shape and the location of the yield surface, as well as the type of experiments allowing to cover entirely the yield surface were specified.

From the general form of the yield conditions for transversely isotropic solids, a new generalization of the Von Mises criterion to fiber-reinforced composites was proposed. The new criterion is expressed in terms of the basic stress and mixed stress-anisotropy invariants by a homogeneous form of order two in stress. Despite the fact that no linear terms are involved, stress-sign sensitivity of the material is taken into account. Hydrostatic stress effects are also involved. Comparisons of the proposed criterion and the Tsai and Wu criterion with our experimental results, obtained in compression tests of a glass fiber reinforced resin under confining pressure, show that for non-zero confining pressures the proposed criterion fits much better the tests data and is located on the safe side.

Experiments consisting of compression tests under confining pressure allow to cover the yield surface only in a limited region of the invariants' space, more precisely only in zone 1 of Figure 2. In order to come also to zone 2, we are presently performing oriented tensile tests under confining pressure. The modes of failure observed in such tests are quite different from those occuring in compression tests under confining pressure. Thus, the criterion developed here and the calculated material constants will most probably not be consistent with the data in tensile tests. In fact, as it was recently pointed out by Hashin [10], all of the different failure modes cannot be represented by a single smooth function. We are presently developing a specific failure condition for fiber-reinforced composites subjected to tensile stress under confining pressure.

REFERENCES

[1] ADKINS J.E. – "Symmetry relations for orthotropic and transversely isotropic materials". *Arch. Rat. Mech. Anal.*, 4 (1960): 193-213.

[2] ALLIROT D., J.P. BOEHLER and A. SAWCZUK. – "Irreversible deformations of an anisotropic rock under hydrostatic pressure". *Int. J. Rock Mech. Min. Sc. and Geomech. Abstr.*, 14 (1977): 77-83.

[3] AZZI V.D. and S.W. TSAI. – "Anisotropic strength of composites". *Experimental Mechanics*, Vol. 5, n° 9 (1965): 283-288.

[4] AZZI V.D. and S.W. TSAI. – "Strength of laminated composite materials". *A.I.A.A. Journal*, Vol. 4, n° 2 (1966): 296-301.

[5] BOEHLER J.P. – "Contributions théoriques et expérimentales à l'étude des milieux plastiques anisotropes". Thèse de Doctorat ès Sciences, Grenoble: Université Scientifique et Médicale, 1975.

[6] BOEHLER J.P. – "Lois de comportement anisotrope des milieux continus". *J. Méc.*, Vol. 17, n° 2 (1978): 153-190.

[7] BOEHLER J.P. – "A simple derivation of representations for non-polynomial constitutive equations in some cases of anisotropy". *Z.A.M.M.* 59 (1979) : 157-167.

[8] BOEHLER J.P. and M. DELAFIN. – "Experimental behavior of a fiber-reinforced composite under confining pressure". (in preparation).

[9] BOEHLER J.P. and A. SAWCZUK. – "On yielding of oriented solids". *Acta Mechanica* 27 (1977): 185-204.

[10] HASHIN Z. – "Failure criteria for unidirectional fiber composites". *J. of Applied Mech.*, 102, n° 2 (1980): 329-334.

[11] HILL R. – *The Mathematical Theory of Plasticity.* Oxford: Clarendon Press, 1950.

[12] KAMINSKY B.E. and LANTZ R.B. – "Strengths theories of failure for anisotropic materials". In *Composite Materials: Testing and Design.* Philadelphia: A.S.T.M., STP 460 (1969): 160-169.

[13] KELLY A. and G.J. DAVIES. – "The principle of fiber reinforcement of metals". *Metallurgical Reviews,* Vol. 10, n° 37 (1965): 1-77.

[14] LANCE R.H. and D.N. ROBINSON. – "A maximum shear stress theory of plastic failure of fiber-reinforced materials". *J. Mech. Phys. Solids*, 19 (1971): 49-60.

[15] LE NIZERHY D. – "Comportement plastique des matériaux composites multi-couches". Thèse de Docteur-Ingénieur, Paris : Université Pierre et Marie Curie, 1976.

[16] LIN T.H., SALINAS D. and Y.M. ITO. – "Effects of hydrostatic stress on the yielding of cold-rolled metals and fiber-reinforced composites". *J. Composite Materials*, 6 (1972): 409-413.

[17] Mc. LAUGHLIN P.V. – "Plastic limit behaviour and failure of filament reinforced materials". *Int. J. Solids Structures*, 8 (1972): 1299-1318.

[18] Mc. LAUGHLIN P.V. and S.C. BATTERMAN. – "Limit behaviour of fibrous materials". *Int. J. Solids Structures*, 6 (1970): 1357-1376.

[19] Mc. LAUGHLIN P.V. and S.C. BATTERMAN. – "Limit surfaces for fibrous composite plates". *A.I.A.A. Journal*, Vol. 8, n° 12 (1970): 2136-2140.

[20] SPENCER A.J.M. – "Theory of Invariants". In C. ERINGEN, ed., *Continuum Physics*. Academic Press, 1971, pp. 239-353.

[21] TRUESDELL C. and NOLL W. – *"The Non-Linear Field Theory of Mechanics"*, *Handbuch der Physik III, 3*. Berlin: Springer-Verlag, 1965.

[22] TSAI S.W. – "Strength theories of filamentary structures". In R.T. SCHWARTZ and H.S. SCHWARTZ, eds., *Fundamental Aspects of Fiber Reinforced Plastic Composites*. New-York: Wiley Interscience, 1968, pp. 3-11.

[23] TSAI S.W. and E.M. WU. – "A general theory of strength for anisotropic materials". *J. Composite Materials*, 5 (1971): 58-80.

[24] WADDOUPS M.E. – "Advanced composite material mechanics for the design and stress analyst". *General Dynamics, Fort Worth Division Report FZM-4763*, 1967.

RESUME

(Critères de rupture pour les composites à fibres unidirectionnelles sous pression de confinement).

Les tendances actuelles dans les applications des matériaux composites exigent un critère de rupture fiable pour les états de contraintes complexes. Dans ce but, nous avons réalisé une série d'essais de compressions sous pression de confinement sur des échantillons orientés d'une résine epoxy renforcée par des fibres de verre unidirectionnelles. Le critère de Tsai et Wu, généralement utilisé dans les bureaux d'étude, n'est pas confirmé par nos résultats expérimentaux. Nous proposons un nouveau critère, formulé dans le cadre d'une théorie générale du comportement plastique des solides anisotropes, développée par l'application des théories de représentation des fonctions tensorielles. Le critère proposé s'exprime par une forme homogène du second ordre par rapport aux invariants mixtes du tenseur des contraintes et d'un tenseur structurel lié au matériau. Il prend en compte les différences entre les résistances en traction et en compression, les effets dûs à la pression de confinement et concorde bien avec les résultats des essais.

Delamination and Interlaminar Strength of Graphite/Epoxy Laminates

S. C. Chou

Army Materials and Mechanics Research Center, Watertown, Massachusetts, U.S.A.

1. Introduction

Strength prediction of composite laminates has been studied by many researchers. To predict the strength of a composite one must have the knowledge of failure modes and the corresponding failure criteria. There are many failure modes in a laminate composite such as fiber breaking, matrix cracking, crazing, delamination, fiber pull-through, etc. This study deals primarily with the delamination failure mode.

In their study, Pagano and Pipes [2] assume that the delamination is solely due to interlaminar normal stress exceeding a critical value. Based on the results [3] of an analysis of a $[0/90]_s$ laminate subjected to uniform axial extension, Pagano demonstrates the existence of maximum interlaminar tension σ_z at the midplane of the laminate. To design delamination specimens, they further assume that the distribution of σ_z versus the width of the specimen for any composite laminate with a configuration such as $[(\pm \theta)_2/90]_s$ will be the same as that for the $[0/90]_s$ laminate. Based on this assumption, the maximum amplitude of interlaminar normal stress, σ_z^{max} can be determined from the equilibrium conditions for a ply in the laminate:

$$\sigma_z^{max}/\epsilon_x = \frac{45}{28} \frac{(\bar{\nu}_{xy}^{(1)} - \bar{\nu}_{xy}^{(2)}) \bar{Q}_{yy}^{(1)} \bar{Q}_{yy}^{(2)} \eta}{(1 + \eta) (\bar{Q}_{yy}^{(1)} \eta + \bar{Q}_{yy}^{(2)})} \tag{1}$$

where ϵ_x is the applied axial strain, $\bar{\nu}_{xy}$ is the ply Poisson's ratio and $\bar{Q}_{yy}$ the transverse plane stress stiffness coefficient, with superscripts (1) and (2) referring to the angle plies and the 90° plies, η is the ratio of the total thickness of two angle-ply units, h_1/h_2. In equation (1), it is clear that a large Poisson's ratio mismatch between the two angle-ply units tends to magnify the stress, σ_z^{max}. Therefore, the interior plies should be at 90°, so that the value of $\bar{\nu}_{xy}^{(2)}$ will be a minimum. The test specimen may now be designed, according to Pagano and Pipes, by maximizing equation (1) with respect to η while holding θ as a constant. This leads to the expression:

$$\eta = [\bar{Q}_{yy}^{(2)}/\bar{Q}_{yy}^{(1)}]^{1/2} \tag{2}$$

from which the relative number of angle and $90°$ plies can be determined. The value of θ has been chosen by maximizing $\bar{\nu}_{xy}^{(1)} - \bar{\nu}_{xy}^{(2)}$ with respect to θ for a given material. In [1] the HTS/ERLA 2256 graphite/epoxy prepreg tape was used, the resulting design point was $\theta = 25°$ and $\eta = 3.1$ which was rounded to 4.0 for practical purposes. The specimens were then fabricated as $[(\pm 25)_2/90]_s$ and tested. Test results showed that delaminations were located at the midplane. Harris and Orringer [1] used the same approach as Pagano and Pipes [2] but with Hercules Magnamite AS/3501-6 graphite/ epoxy prepreg tapes; the resulting design point was $\theta = 26°$ and $\eta = 3.05$ (rounded to 4.0). Their experimental results showed that delaminations took place at the interface between the angle ply and $90°$ ply rather than at the midplane of the specimen. Their conclusion was that delamination failure was due to combined effects of interlaminar normal and shear stresses rather than the normal stress σ_z alone. Harris and Orringer further suggested that the designed laminate configuration might not have the largest interlaminar normal stress for all possible θ values in the configuration such as $[(\pm \theta)_2/90]_s$. This is due to the manner of maximizing equation (1) suggested in [2]. Strictly speaking, equation (1) is a continuous function of two independent variables θ and η, and maxima should therefore be sought from the behavior of the partial derivatives. Since this approach involves complicated trigonometric expression in θ, Harris and Orringer use the graphic method to determine the maxima of $\sigma_z^{max}/\epsilon_x$ by plotting a family of curves of $\sigma_z^{max}/\epsilon_x$ versus θ with integer values of η as a parameter. They find that the true maxima of $\sigma_z^{max}/\epsilon_x$ for AS/3501-6 $[(\pm \theta)_2/90]_s$ laminate tend to occur at θ between $30°$ to $40°$, instead of near $26°$ as originally predicted. In summary, their findings are that (a) delamination failure may be due to combined-stress effects, rather than due to the interlaminar normal stress alone; (b) if the delamination in a laminate is solely due to σ_z^{max} and occurs at the midplane where the shear stresses vanish because of symmetry, the value of θ shall be between $30°$ and $40°$.

This paper is intended to explore these two points further by mechanically testing various configurations of angle-ply laminates, and examining the experimental results. The analytical capability required to formulate the delamination criterion is only briefly discussed.

2.   Test Specimen Configurations

Material used to fabricate composite laminates used in this study is the T300/5208 graphite/epoxy prepreg tape with the following ply properties:

$E_{11} = 22 \times 10^6$ psi $\nu_{12} = 0.28$

$E_{22} = 1.54 \times 10^6$ psi $G_{12} = 0.81 \times 10^6$ psi.

These properties are very close to those of HTS/ERLA graphite/epoxy used in [2] and AS/3501-6 in [1]. It is anticipated that conclusions drawn from the comparison of these two studies with the current study will not be affected by the slight difference in material properties.

Five $12'' \times 12''$ plates were fabricated from T300/5208 graphite/epoxy prepreg tapes, following the supplier's recommendations for tooling, layup, and cure cycle. The stacking sequence of these plates is $[(\pm \theta)_2/90]_s$, and the θ values for the five plates fabricated are $5°, 15°, 25°, 35°$ and $45°$. The average thickness for the ten-ply laminates is 0.055 inch, and the volume fraction is about 55 %. Test coupons of $1'' \times 8''$ were taken from each plate and each specimen was bonded with $1'' \times 2'' \times 0.125''$ glass/epoxy tabs at both ends leaving a $1'' \times 4''$ gage section. Each specimen was also instrumented with an FAE-12-35PL strain gage, and stored in a desiccator under vacuum before it was tested at the ambient condition.

Design parameters determined according to the procedure suggested in [2] are listed in Table 1; the values of η are rounded to the next integer for practical purposes. However, other factors are also taken into consideration; for instance, the case of $\theta = 35°$, $\eta = 3.0$ will be an unbalanced laminate and is undesirable for the purpose of this study; therefore, $\eta = 4$ is used. For the case of $\theta = 45°$, from equation (2) $\eta = 2$; but in the fabrication of the plate $\eta = 4$ is used, so that all specimens have the same geometric configuration and delamination load may be compared. Table 1 also lists the values of $\overline{v}_{12}^{(1)}$ for the five configuration of interest. Since the value of $\overline{v}_{12}^{(2)}$ for the 90° ply is the same for all cases, the laminate which has the highest value of $\overline{v}_{12}^{(1)}$ will have the largest value of σ_z^{max}. From Table 1, it is noticed that the laminate with $\theta = 25°$ has the largest value of $\overline{v}_{12}^{(1)}$, therefore, the laminate $[(\pm 25)_2/90]_s$ is expected to have delamination at the midplane under the smallest load of the five cases tested. However, test results discussed in the next section are contradictory to the expectation.

TABLE 1. Delamination specimen configurations and test results.

$[(\pm\theta)_2/90]_s$		$\theta = 5°$	$15°$	$25°$	$35°$	$45°$
$\eta = h_1/h_2$		$3.77\approx4$	$3.6\approx4$	$3.3\approx4$	$2.4\approx3$	$1.8\approx2$
$\overline{v}_{12}^{(1)}$		0.37	0.97	1.61	1.19	0.77
$\overline{v}_{12}^{(2)}$		0.02	0.02	0.02	0.02	0.02
Delamination Test	Stress (ksi)	99.3 (69.7)	46.7	30.8	23.9	18.2
	Strain (%)	0.697 (0.493)	0.383	0.336	0.406	0.62

3. Test Results

The specimens were tested individually in an electrohydraulic, servocontrolled closed-loop testing machine which was run at the load-control mode with the load rate of one pound per second. The specimen under test was constantly monitored with a traveling microscope along the edge of a specimen. Load and strain were recorded simultaneously and tests were terminated whenever delamination occurred in the specimen. Since the load rate was so slow, the measured delamination load could be considered as the threshold of delamination. The strain and nominal stress at the threshold of delamination are listed in Table 1 for the five cases investigated. The delamination stress for the case of $\theta = 35°$ is less than that for $\theta = 25°$, which indicates that the design procedure for delamination specimens suggested in [2] does not lead to the optimized configuration. The stress and strain in parentheses for the case of $\theta = 5°$ were recorded when the matrix in 90° ply failed, which occurred, obviously, prior to delamination. The strain of 0,493 % at which the matrix failed seems to be larger than the maximum strain for 90° unidirectional laminar; this may be due to the effect of interlaminar stresses.

The typical edge view of the delaminated specimens for the five configurations are shown in Figure 1. First, it is observed that none of the specimens tested delaminates at the midplane of specimen; all specimens have delamination at the interface between the angle ply and 90° ply. As mentioned previously, specimens with $\theta = 5°, 15°, 25°, 35°$ had matrix failure prior to the delamination; this phenomenon was detected not only visually by using the travelling microscope, but also auditorily. The matrix failure in 90° ply creates very little acoustic activity, being just a few clicks, while the delamination generates a loud cracking noise. For the laminate with $\theta = 45°$, the matrix failure in the 90° ply takes place simultaneously with the delamination at the interface of angle ply and 90° ply. It is evident from this series of experiments that the interlaminar shear stresses may play a significant role in the process of delamination, because the interlaminar shear stresses are maxima at the interface of angle ply and 90° ply, although they may not have the same order of magnitude as that of interlaminar normal stress. The interlaminar shear stresses vanish at the midplane of a symmetric laminate, while the normal stress reaches its maximum.

4. Theoretical Approach

Formulation of delamination criterion requires a reliable numerical scheme to calculate the interlaminar stresses. This problem has been of interest to many researches recently, however the difficulties encountered in calculating the stresses right at the free edge of a coupon specimen have not been resolved

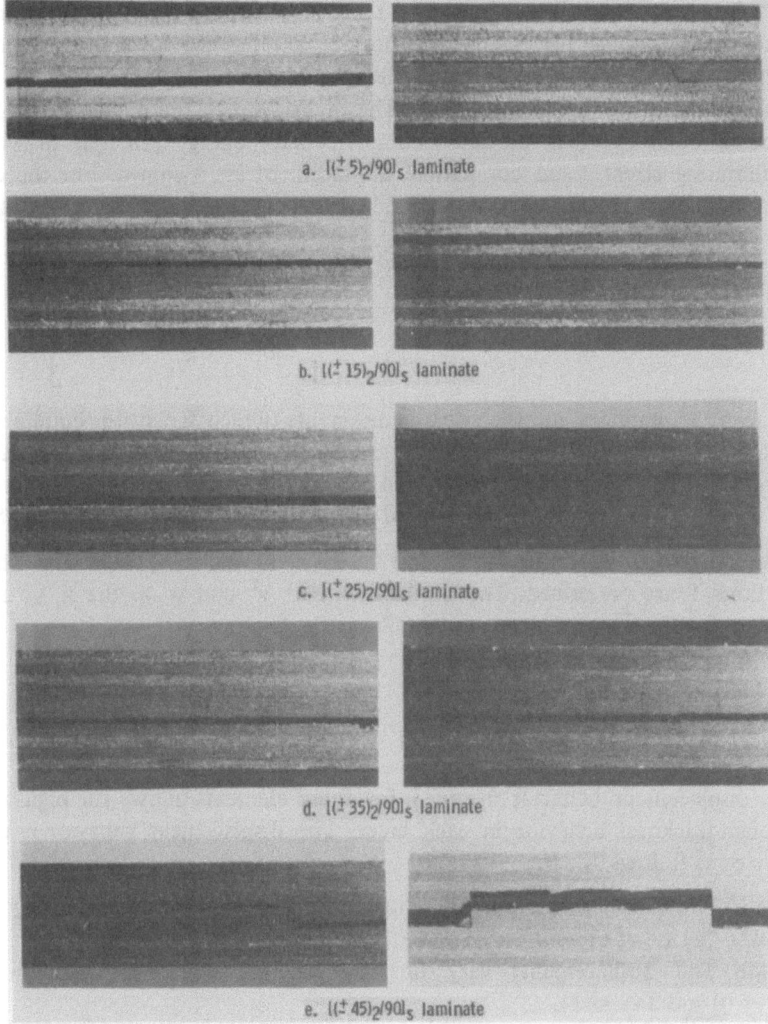

Figure 1. – *Typical delamination of T300-5 208 graphite/epoxy laminates.*

(see [4]). Spilker and Chou [4] show that the hybrid-stress finite element model seems to be able to resolve some of the difficulties. Their approach has, so far, been developed for two-dimensional linear analysis. However, due to the material anisotropy and specimen geometry, the two-dimensional analysis is not adequate for this purpose. The delamination criterion cannot be rigorously formulated without the three-dimensional analysis for a thick laminate. In the following paragraph, a summarized theoretical approach for the two-dimensional analysis from [4] is given to provide a reference for those who may wish to use the hybrid-stress finite element model to explore further in the area of modelling failure criterion. The hybrid-stress finite element model is based on a two-field modified complementary energy principle in which equilibrating stresses and compatible displacement are assumed. For the case of no applied external forces, the hybrid-stress functional π_{mc} is given by [4, 5]

$$\pi_{mc} = \sum_{n_e} \left\{ \sum_N \left[1/2 \int_{V_{n_i}} \sigma^{iT} S^i \sigma^i dv - \int_{V_{n_i}} \sigma^{iT} \hat{\varepsilon}^i dv \right. \right.$$
$$\left. \left. + \int_{S_{\sigma_{n_i}}} \bar{T}^T u^i ds \right] \right\} \quad (3)$$

where $\sigma^i, \varepsilon^i, u^i$ and s^i are the stress components (which satisfy the equilibrium equations), strain components (which are calculated from the assumed displacement field), displacement vector, and material property matrix, respectively, for the ith layer. V_{ni} is the volume of the ith layer for the nth element, $S_{\sigma_{n_i}}$ is the portion of the boundary of the ith layer of the nth element on which tractions $\bar{T}$ are prescribed. The displacement u^i, v^i, and w^i in the x, y, z directions for the ith layer are assumed to be in the form

$$u^i(x,y,z) = x\varepsilon_x + u^i(y,z)$$
$$v^i(x,y,z) = v^i(y,z)$$
$$w^i(x,y,z) = w^i(y,z)$$

The displacement behavior assumed for these elements allows for high-order through-thickness behavior in each layer. The displacement behavior is summarized as follows (see Fig. 2):

a) The through-thickness distributions in the ith layer $u^i(y, z)$, $v^i(y,z)$, and $w^i(y,z)$ are of order z^3, z^3, and z^2, respectively.

b) The displacements vary linearly with y between element nodes 1 (y = 0) and 2 (y = 1).

c) Continuity of displacements v and w along interlayer boundaries is exactly satisfied.

d) An element composed of N layers will have (10N + 4) degree of freedom. The assumed stress behavior is summarized as follows:

a) The through-thickness distributions for σ_y^i, σ_z^i, σ_{yz}^i, σ_{xy}^i, and σ_{xy}^i within each layer are of order z^3, z^5, z^4, z^3, and z^4, respectively.

b) The in-plane distributions for σ_y^i, σ_z^i, σ_{yz}^i, σ_{xy}^i, and σ_{xz}^i are of order y^3, y, y^2, y^3, and y^2, respectively.

c) The stress equilibrium equations are exactly satisfied for all locations in the element.

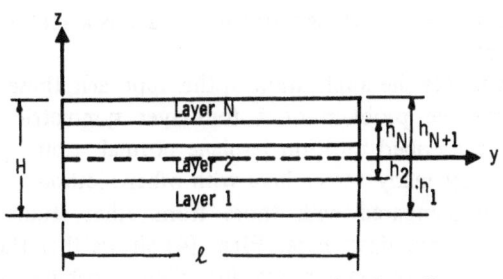

a. Element Geometry, Layer Numbering Convention

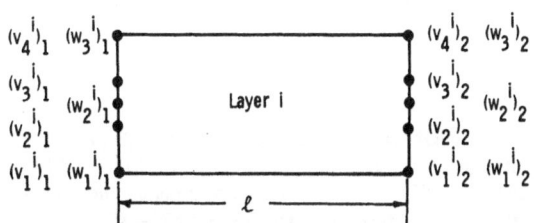

$(v_j^i)_k$: jth in-plane displacement (in y direction) for the ith layer at element node k.

$(w_j^i)_k$: jth transverse displacement (in z direction) for the ith layer at element node k.

b. Element Degrees of Freedom for a Typical Layer

Figure 2. – *Plane strain multilayer element conventions and degrees of freedom.*

d) Continuity of tractions (in the case of stresses σ_x, σ_{yz} and σ_{xz}) along interlayer boundaries is satisfied exactly.

e) The traction-free conditions at the upper and lower surfaces of the laminate are satisfied exactly (i.e., $\sigma_z = \sigma_{yz} = \sigma_{xz} = 0$ at $z = h_1$ and h_{N+1}). At the traction-free edge a special element is required and the assumed stresses are the same as above but with additional constraint, i.e.

f) the left edge of the element ($y = 0$) is taken as the traction-free edge and the traction-free conditions $\sigma_{yz}^i = \sigma_y^i = \sigma_{xy}^i = 0$ are satisfied exactly for all z in each layer.

The expression for and formulation of the element stiffness matrix K can be found in [4] and [5]; here, an equivalent element load vector Q, corresponding to prescribed strain $\bar{\epsilon}_x$ is defined. These relations are

$$K = G^T H^{-1} G$$

$$Q = \epsilon_x G^T H^{-1} f$$

where G represents the work done by assumed stresses (tractions) and displacements on the element boundary, H represents the complementary strain energy corresponding to assumed stresses, and f is a vector resulting from the elimination of σ_x^i.

To demonstrate the application of this approach, these two elements were used to analyze the problem of a four-layer symmetric cross-ply laminate subjected to a prescribed uniform in-plane normal strain $\bar{\epsilon}_x$ [4]. This problem has been solved by many researchers with other methods. Results obtained in this study were compared with those from other studies. Two significant findings can be summarized here. First, [4] shows that the analysis leads to convergence of all observed stresses, including distributions at the free edge; this had not been accomplished previously. Secondly, the converged peak values calculated by the present approach are lower than those of other investigators.

REFERENCES

[1] HARRIS A. and O. ORRINGER. – "Investigation of angle-ply delamination specimen for interlaminar Strength test". *J. Comp. Mater.*, 12, (1978): 285-299.
[2] PAGANO N.J. and R.B. PIPES. – "Some Observations on the interlaminar strength of composite laminates". *Int. J. Mech. Sci.*, 15, (1973): 679-688.
[3] PAGANO N.J. – "On the calculation of interlaminar normal stress in composite laminate". *J. Comp. Mater.*, 8 (1974): 65-82.
[4] SPILKER R.L. and S.C. CHOU. – "Edge effects in symmetric composite laminates: Importance of satisfying the traction-free-edge condition". *J. Comp. Mater.*, 14 (1980): 2-20.
[5] SPILKER R.L. – "A hybrid-stress finite-element formulation for thick multilayer laminates". To appear in *Computers and Structures*.

RESUME

(Délaminage et résistance interlamellaire)

Des composites multicouches à fibres croisées $[(\pm \theta)_2/90]_s$ en graphite/époxy T 300/5 208 ont été fabriqués pour $\theta = 5°$, $15°$, $25°$, $35°$ et $45°$. Des échantillons prélevés sur ces matériaux ont été testés jusqu'à l'initialisation d'une délamination. Les résultats expérimentaux montrent que les contraintes de cisaillement interlamellaires jouent un rôle significatif dans le processus de délaminage. Par conséquent, on peut conclure que les composites lamellaires $[(\pm \theta)_2/90]_s$ ne peuvent pas être utilisés pour la détermination de la résistance interlamellaire. Une analyse, plus précise que celles qui sont couramment disponibles, est nécessaire pour améliorer le procédé de dimensionnement de tels échantillons. Une méthode par éléments finis, basée sur des contraintes hybrides et brièvement évoquée ici, semble être une approche prometteuse, mais un plus grand développement est nécessaire avant que le modèle puisse être utilisé pour analyser l'état de contrainte dans un échantillon en cours de délaminage ou pour améliorer les procédés de dimensionnement pour de tels échantillons.

Session 7

Mechanics of Anisotropic Rocks

Mécanique des Roches Anisotropes

Colloques internationaux du CNRS
N° 295 — COMPORTEMENT MÉCANIQUE DES SOLIDES ANISOTROPES

General Lecture : *Conférence Générale*

Anisotropie Mécanique des Roches

P. M. Sirieys

Institut National des Sciences Appliquées, Toulouse, France.

Sommaire

---

* Figures 1, 2, 3, 4b, 9 : By permission from *Structural Analysis of Metamorphic Tectonites*, by Turner/Weiss. Copyright, 1963. McGraw-Hill Book Company.

Figure 4a : Extrait de Maurice Mattauer: *Les déformations des matériaux de l'écorce terrestre*, Hermann, Paris, collection Méthodes, 2e édition revue et corrigée, 1979.

Symboles principaux

S_i, S_f	Structure initiale, Structure finale
S_0	Stratification
S	Schistosité
S_1, S_2, S_3	Surfaces représentatives des modules élastiques
ε, ϵ_{ik}	Tenseur petite déformation
H, h_{ik}	Tenseur carré de la déformation finie
σ, σ_{ik}, $\boldsymbol{\Sigma}$	Tenseur contrainte
s, $s_{ijk\ell}$	Tenseur élasticité
c, $c_{ijk\ell}$	Tenseur rigidité
h_1, h_2, h_3	Composantes principales de H
$\bar{e}_1$, $\bar{e}_2$, $\bar{e}_3$	Déformations principales logarithmiques $\left(\bar{e}_i = \dfrac{1}{2} \log h_i \right)$
σ_1, σ_2, σ_3	Composantes principales de $\underset{\sim}{\sigma}$
X, Y, Z	Directions principales

$1 \text{ bar} = 14 \text{ psi} = 10 \text{ N/cm}^2 = 10^2 \text{ KPa}$

1. Introduction

L'étude de l'anisotropie mécanique des roches se présente comme un vaste sujet, compte tenu de la grande variété des milieux rocheux : variété dans le mode de formation, dans les compositions minéralogiques, dans les transformations naturelles. Les structures anisotropes observées sur le terrain résultent souvent d'une succession de déformations tectoniques ; l'étude mécanique expérimentale d'un échantillon en laboratoire se présente alors comme une nouvelle déformation qui se superpose aux précédentes, y compris celle lors de son extraction.

L'échelle est en outre un facteur essentiel dans l'étude des mécanismes anisotropes mis en jeu ; les résultats expérimentaux sur le comportement anisotrope des roches et massifs sont analysés ici aux échelles macroscopiques et mésoscopiques. Cet examen débute par une analyse sommaire des structures rocheuses, de l'influence de leur histoire, génétique et tectonique et de la caractérisation de leur anisotropie structurale ; ensuite est abordée l'étude de l'anisotropie de comportement dans les domaines des petites déformations (dilatation thermique, élasticité, inélasticité, dilatance), des déformations finies (plasticité) et des déformations discontinues (rupture) ; enfin sont indiquées des corrélations entre anisotropies. Il n'a pas été effectué d'étude systématique de roches particulières mais les réponses à certains types de sollicitations font intervenir des groupements de roches, généralement fonction de leur passé tectonique.

2. Structures initiales

Les structures actuelles des roches et des massifs, observables sur échantillon ou sur le terrain, sont le résultat de toute leur histoire, décomposée habituellement en deux parties : celle concernant le mode de formation des roches (structures initiales) et celle concernant la déformation naturelle (structures tectoniques). La superposition des mécanismes de mise en place, de ceux de déformation naturelle et des transformations métamorphiques induit une anisotropie dans le milieu actuel. On distingue deux types fondamentaux (Fig. 1) de structures initiales :

La structure planaire est constituée par une famille de plans parallèles, un plan de la structure est caractérisé par 2 directions d'axes géométriques a et b orthogonales. Exemples : litage, orientation préférentielle des grains dans un plan, orientation préférentielle des frontières de grains (Fig. 1b).

La structure linéaire est constituée par des éléments, ou amas de grains, tous allongés dans la même direction et équants normalement à cette direction appelée axe géométrique b (Fig. 1c). Exemples : intersection de litages, orientation préférentielle de fossiles ou de minéraux prismatiques allongés ayant une direction constante, de grains plats parallèles à une direction constante (Fig. 1d).

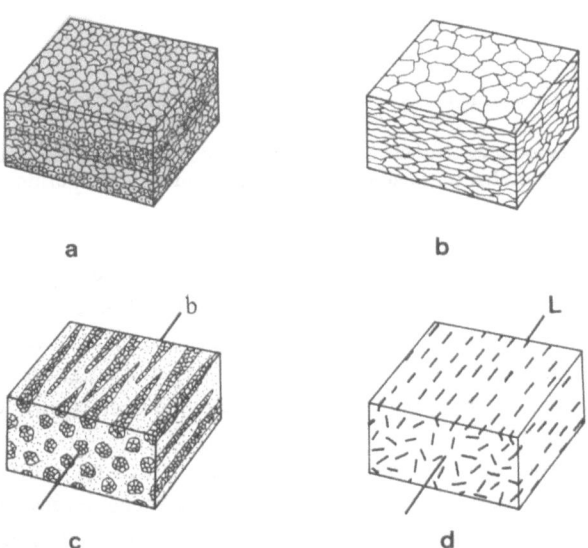

Figure 1. -- *Structures initiales* (d'après Turner et Weiss [45]).

a et b : Structures planaires. (a : litage, b : orientation planaire des frontières de grains). c et d : Structures linéaires. (c : amas de grains allongés, d : éléments plats parallèles à L).

Les éléments d'une structure sont ainsi définis par les trois axes a b c
(c étant normal au plan a b). La structure se compose parfois de plusieurs sous-
structures $a_1 b_1 c_1$, $a_2 b_2 c_2$ qui ont leur propre incidence sur le comportement
mécanique. On distingue (Fig. 2) des éléments structuraux de localisation
(litage) et des éléments - structuraux d'orientation, éventuellement non
confondus, entraînant alors des sous-structures de localisation et d'orientation.

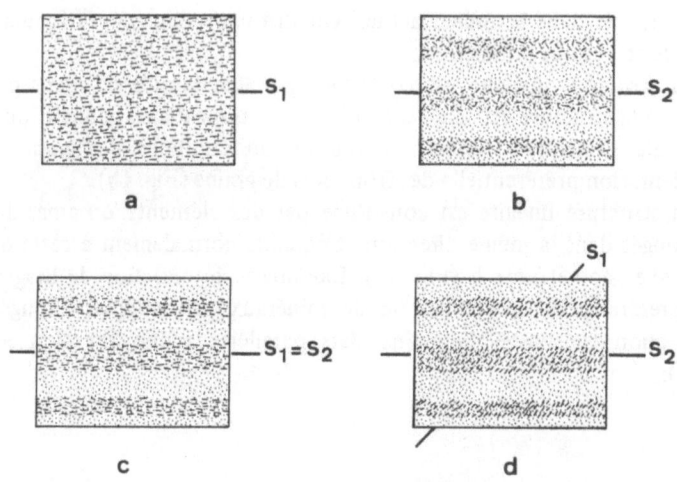

Figure 2. – *Eléments structuraux d'orientation et de localisation* (d'après Turner et Weiss
[45]). -
     a : orientation préférentielle S_1, b : localisation préférentielle S_2, c : sous-structures
d'orientation et de localisation identiques, d : sous-structures S_1 et S_2 distinctes.

La structure initiale des roches sédimentaires est de type planaire, caractérisée par la stratification (ou litage) horizontale par suite du dépôt dans le
champ gravitaire ; celle des roches ignées est de type planaire (cas de litage ou
strates composites, d'écoulement stratifié ou d'orientation planaire des
minéraux) ou de type linéaire (cas de linéation d'écoulement, d'orientation
linéaire des grains).
     Cinq classes de symétrie caractérisent les structures initiales (Fig. 3) :
     S1 – Symétrie sphérique – (isotropie) – La roche se présente comme un
milieu pseudo-isotrope (absence d'axes a b c). C'est le cas, pour les roches sédimentaires, d'agrégats de grains équants (par exemple un grès massif avec des
grains sensiblement sphériques), pour les roches ignées, d'agrégats avec une
orientation sensiblement aléatoire des grains (par exemple des granites ou des
basaltes).

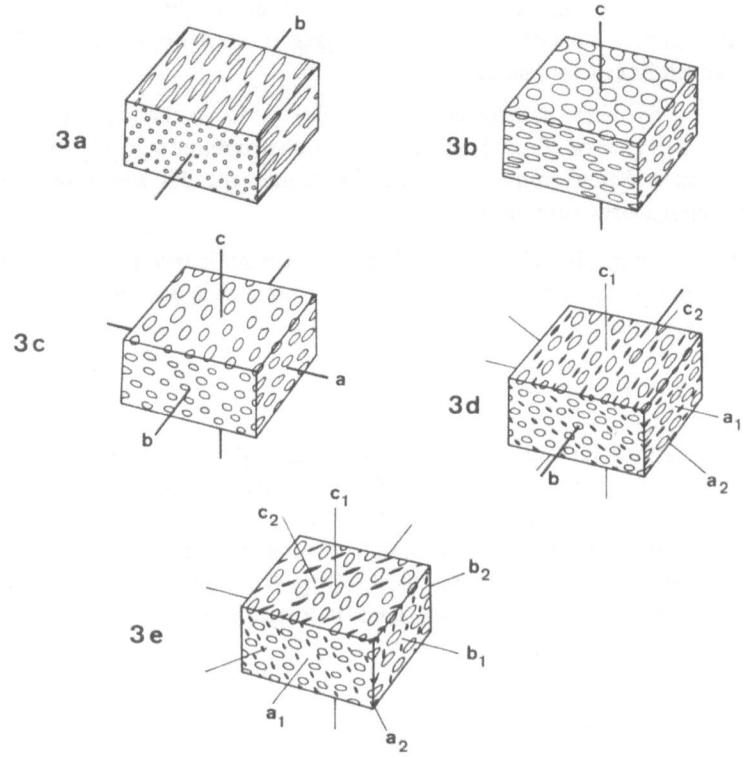

Figure 3. — *Symétrie des structures initiales* (d'après Turner et Weiss [45]).

a : symétrie axiale, linéation b ; b : symétrie axiale, structure planaire normale à c ; c : symétrie orthorhombique, axes abc ; d : symétrie monoclinique, sous-structures $a_1 b c_1$ et $a_2 b c_2$; e : symétrie triclinique, sous-structures $a_1 b_1 c_1$ et $a_2 b_2 c_2$.

S2 — Symétrie axiale — (symétrie de révolution, isotropie transverse) — (Fig. 3a – 3b) — La roche présente un plan de symétrie (appelé plan isotrope) et un axe de révolution normal à ce plan, qui est soit l'axe c d'une structure planaire (cas de sédiments stratifiés, calcaires par exemple), soit l'axe b d'une structure linéaire, par exemple les laves.

S3 — Symétrie orthorhombique — (Fig. 3c) — La roche présente trois plans de symétrie tri-orthogonaux, d'intersections a, b, c : les trois axes de la structure. C'est le cas d'une structure planaire avec une linéation b dans le plan a b (par exemple des sédiments lités avec linéation ou des agrégats cristallins avec strates et linéation).

S4 — Symétrie monoclinique — (Fig. 3d) — La roche présente un plan de symétrie. C'est le cas de deux sous-structures planaires $a_1 b c_1$ et $a_2 b c_2$

ayant un axe b commun, par exemple des sédiments lités comportant une structure linéaire d'orientation ou des agrégats comportant des structures d'écoulement linéaire et laminaire.

S5 – Symétrie triclinique – (Fig. 3e) – La roche ne présente pas d'élément de symétrie. C'est le cas de deux sous-structures $a_1 b_1 c_1$ et $a_2 b_2 c_2$ non coaxiales. Par exemple sédiments lités contenant des structures d'écoulement non linéaires ou d'agrégats.

Ces structures initiales sont parfois observées sur le terrain, dans la mesure où elles n'ont pas subi de déformation postérieure à celle de leur formation lors de leur mise en place.

3. Structures tectoniques

Les déformations tectoniques modifient l'anisotropie des structures initiales. La nature et les mécanismes de ces déformations dépendent des conditions thermodynamiques de pression et température. On définit [26] trois étages structuraux (Fig. 4a) pour lesquels la déformation est de type fragile, ductile ou visqueux. Des expériences sur un calcaire (Fig. 4b) montrent, en outre, que la transition fragile-ductile n'est pas, dans le diagramme (p, T), la même en compression qu'en traction.

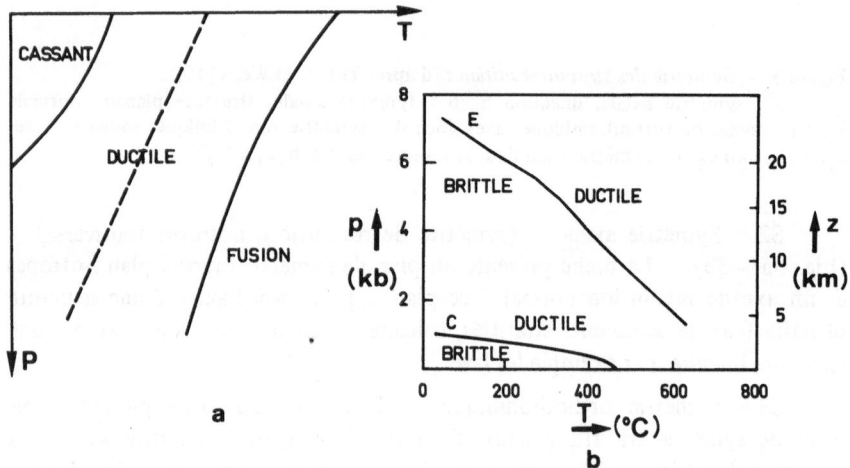

Figure 4. – *Etages structuraux.*
    a : les trois domaines d'après Mattauer [26]). (T = température, p = pression) ;
b : Transition fragile-ductile (d'après Heard in Turner et Weiss [45]) du calcaire de Solenhofen, distincte en compression (C) et en traction (E) (T = température, p = pression de confinement, z = profondeur).

3.1. *Déformation continue finie (Tectonique souple)*

Les déformations naturelles finies de type continu se sont effectuées lors des orogénèses pour lesquelles les conditions pression-température permettaient un écoulement continu. Chronologiquement, on distingue essentiellement, en Europe, les orogénèses antécambriennes, calédonienne, hercynienne, alpine. Ces déformations viscoplastiques ont pu atteindre des valeurs très élevées à des vitesses estimées de l'ordre de $10^{-14}\,\mathrm{sec}^{-1}$. Par réorientation de grains et minéraux, déformation de grains, création de nouveaux minéraux dans des directions préférentielles elles ont modifié les caractères et les symétries des structures initiales.

Une déformation homogène finie résulte d'une transformation **T**, reliant dans un référentiel (xi) les coordonnées actuelles ξ aux coordonnées initiales **X** d'un point, qui s'écrit, en écriture symbolique, sous la forme :

$$\xi = \mathbf{T\,X} \tag{1}$$

Cette transformation se décompose [16] de deux façons (décomposition polaire droite et gauche) en un produit d'une déformation par une rotation :

$$\mathbf{T} = \mathbf{RD} = \mathbf{D'\,R} \tag{2}$$

La déformation est caractérisée par son carré :

$$\mathbf{H} = \mathbf{D}^2 = {}^t\mathbf{T\,T} \tag{3a}$$

$$\mathbf{H} = \mathbf{D'}^2 = \mathbf{T\,{}^t T} \tag{3b}$$

D est la déformation avant rotation, **D'** la déformation après rotation, ${}^t\mathbf{T}$ la transposée de **T**. Pour une transformation homogène **T**, **R**, **D** et **D'** sont constants dans le volume considéré ; dans le cas d'une transformation hétérogène, on utilise la transformation linéaire tangente définie par $T_{ik} = \xi_{i,k}$.

Une déformation pure $(\mathbf{D} = \mathbf{D'})$ est caractérisée par 3 paramètres d'orientation dans (x_i) (on utilise généralement pour (x_i) le référentiel géographique) et 3 paramètres d'intensité qui se réduisent à 2 car la déformation tectonique peut être considérée comme isovolume. En adoptant la déformation logarithmique, la relation d'invariance de volume s'écrit :

$$\bar{e}_1 + \bar{e}_2 + \bar{e}_3 = 0$$

avec $\bar{e} = \mathrm{Log}\,h$, h = composante principale de H, et :

$$\bar{e}_1 \geqslant \bar{e}_2 \geqslant \bar{e}_3.$$

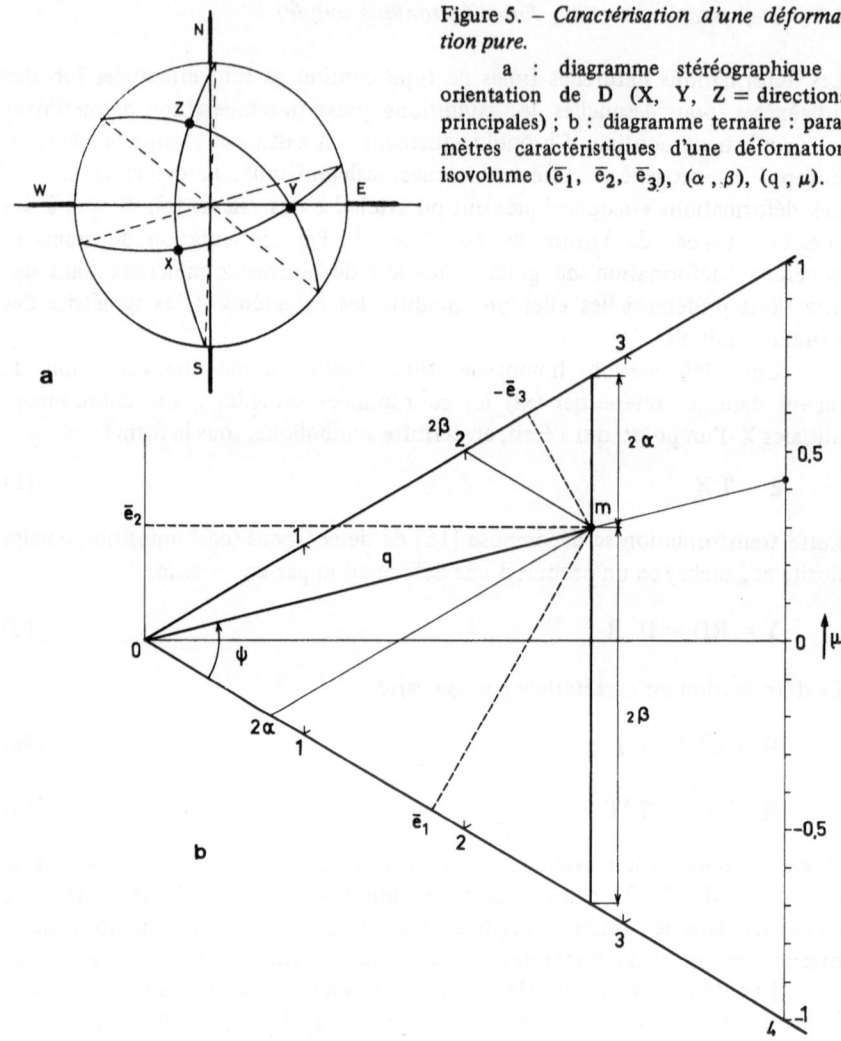

Figure 5. – *Caractérisation d'une déformation pure.*
    a : diagramme stéréographique :
orientation de D (X, Y, Z = directions
principales) ; b : diagramme ternaire : paramètres caractéristiques d'une déformation
isovolume $(\bar{e}_1, \bar{e}_2, \bar{e}_3)$, (α, β), (q, μ).

La déformation finie isovolume est donc imagée à l'aide de deux
diagrammes : le diagramme stéréographique (Fig. 5a) qui fournit son orientation dans le référentiel géographique, et le diagramme ternaire (Fig. 5b) qui
permet de caractériser sa grandeur par deux paramètres liés aux second et
troisième invariants du déviateur de déformation, l'intensité q et le coefficient
μ (paramètre de Lode).

$$q = \overline{om} = 2(\alpha^2 + \beta^2 + \alpha\beta)^{1/2} \tag{4a}$$

$$\mu = \frac{\beta - \alpha}{\beta + \alpha} \tag{4b}$$

avec $3\alpha = \overline{e}_1 - \overline{e}_2$

$3\beta = \overline{e}_2 - \overline{e}_3$

On utilise parfois ψ défini par $\tan \psi = \mu/\sqrt{3}$, appelé phase. Une déformation est dite monophasée lorsque ψ, c'est-à-dire μ, est resté constant au cours de son évolution, elle est schisteuse pour $\mu > 0$ et fibreuse pour $\mu < 0$.

Dans un milieu initial isotrope une déformation schisteuse induit une structure planaire, une déformation fibreuse induit une structure linéaire. La symétrie de ces structures est orthorhombique, sauf dans le cas $\mu = \pm 1$ où elle est de révolution.

Dans une transformation rotationnelle, les axes principaux de la déformation tournent au cours de la transformation. L'axe de rotation étant en général quelconque par rapport au trièdre principal de **D**, la transformation tectonique dépend de 8 paramètres : 3 paramètres d'orientation de **D**, 2 d'intensité de **D**, 2 d'orientation de **R**, 1 d'intensité de **R**. Lorsque l'axe de rotation coïncide constamment avec un axe principal de **D** le nombre de paramètres de **T** s'abaisse. Un exemple de ce cas, fréquemment observé sur le terrain, est le glissement simple (Fig. 6) caractérisé en déformation plane par la transformation **G** qui s'exprime dans (x_i) tel que x_1 est la direction de glissement par :

$$\begin{pmatrix} 1 & \gamma \\ 0 & 1 \end{pmatrix} \tag{5}$$

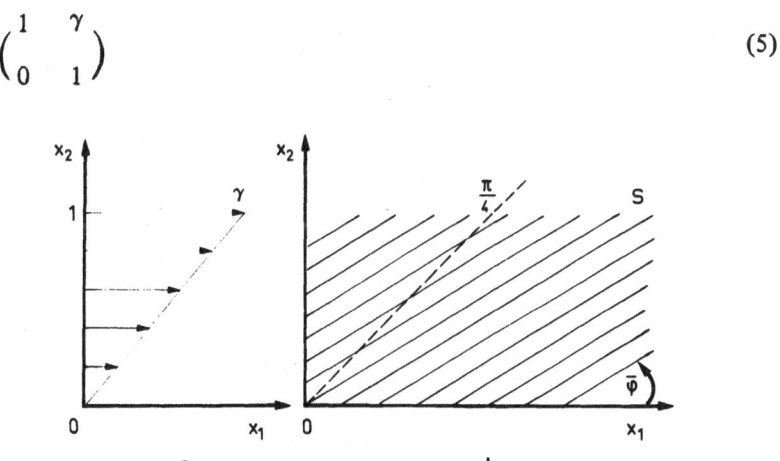

Figure 6. – *Le glissement simple.*
a : champ des déplacements ; b : la schistosité S.

(γ étant le taux de glissement). Les éléments caractéristiques de **G** sont : l'orientation $\bar{\varphi}$ par rapport à x_1 de la déformation **D**' après rotation, matérialisée sur le terrain par la schistosité de flux, la rotation Φ des directions principales, la valeur $\bar{e}_1$ de la déformation principale logarithmique majeure, ils dépendent du seul paramètre γ par :

$$\tan 2\,\bar{\varphi} = \frac{2}{\gamma} \tag{6a}$$

$$\tan \Phi = -\frac{\gamma}{2} \tag{6b}$$

$$\mathrm{Sh}\,\bar{e}_1 = \frac{\gamma}{2} \tag{6c}$$

La structure résultante ne dépend pas seulement de la déformation finie, mais également du chemin suivi par cette déformation, c'est-à-dire de la déformation infinitésimale. Pour une déformation monophasée la déformation infinitésimale a constamment le même paramètre μ, celui de **D** dont l'évolution est alors représentée par un trajet linéaire dans le diagramme ternaire. Lorsque μ varie au cours de **D** elle est dite polyphasée. Les structures résultantes et l'anisotropie de la roche sont étroitement liées à ce chemin.

Ces résultats concernent une déformation homogène, ils peuvent s'appliquer aussi à une déformation moyenne (ou déformation image, ou déformation régionale) caractérisant une déformation hétérogène à petites fluctuations, dans le cas du plissement par exemple.

3.2. *Déformation discontinue (Tectonique cassante)*

3.2.1. *Failles*

A pression et température moyennes, la déformation plastique se localise sur des surfaces, planes en déformation homogène. Il en résulte des discontinuités de déplacements tangentiels, la formation de plans d'anisotropie et de structures planaires. Ce mécanisme est celui de la genèse des failles qui sont susceptibles d'être ultérieurement réactivées et d'entraîner de nouvelles discontinuités de déplacements, comme dans le cas des séismes.

Dans un milieu initial isotrope (et homogène) une déformation homogène induit un système discret de plans de glissement, dont l'orientation dépend de celle des tenseurs tectoniques limites, c'est-à-dire vérifiant le critère d'écoulement du milieu envisagé. Dans le cas d'une loi élémentaire linéaire (loi de Coulomb) ce système de plans comporte deux familles passant par l'axe principal relatif à la contrainte principale intermédiaire, d'angle dièdre égal à $2\mu = \pi/2 - \Phi$ (Φ étant l'angle de frottement interne). Le milieu

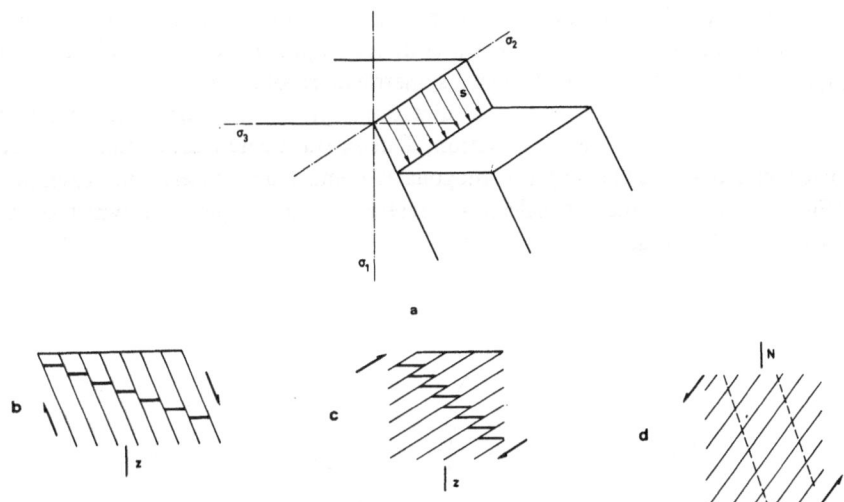

Figure 7. — *Structure planaire faillée.*

a : stries s sur un plan de faille ; b : failles normales, σ_1 vertical ; c : chevauchements, σ_3 vertical ; d : décrochements senestres, σ_2 vertical.

se déforme alors selon un schéma de blocs. Lors d'une seule famille de plans mobilisée la déformation continue image, ou déformation moyenne, est rotationnelle, du type de glissement simple **G**. Sur ces surfaces planes, la direction de glissement est caractérisée par la strie de glissement (Fig. 7a). Les discontinuités des déplacements sont caractérisées par les rejets, vecteurs colinéaires à la strie de glissement. Les structures (ou sous-structures) ainsi créées sont de type planaire avec une direction préférentielle, celle de la strie, donc à symétrie orthorhombique. Des exemples géologiques fréquents [37] sont ceux des tenseurs tectoniques ayant une direction principale verticale, susceptibles d'engendrer trois types de structures courantes : les failles normales (à σ_1 vertical et extension horizontale) (Fig. 7b), les chevauchements (à σ_3 vertical et contraction horizontale) (Fig. 7c) et les décrochements (à σ_2 vertical et rejets horizontaux) (Fig. 7d).

3.2.2. *Diaclases*

A basse pression et température, les roches ont un comportement fragile, leur déformation s'effectue par discontinuité de la composante normale du vecteur déplacement, et non de la composante tangentielle comme dans le cas des failles. Ce mécanisme s'accompagne d'une variation de volume positive (expansion) contrairement à celui de jeu ou rejeu des failles qui est isovolume. Le milieu initialement isotrope présente alors une famille discrète

de surfaces de discontinuité, planes en champ homogène, les diaclases d'extension, orientées selon YZ (normales à la direction principale majeure d'extension
X), sa structure est de type planaire à symétrie de révolution.

Des tenseurs tectoniques à directions principales constantes peuvent
conduire, sous l'effet de la relaxation, à deux ou même trois familles [22] de
diaclases alors sensiblement triorthogonales comme on l'observe, par exemple,
(Fig. 8) dans des massifs calcaires, gréseux ou granitiques. La symétrie est
alors orthorhombique.

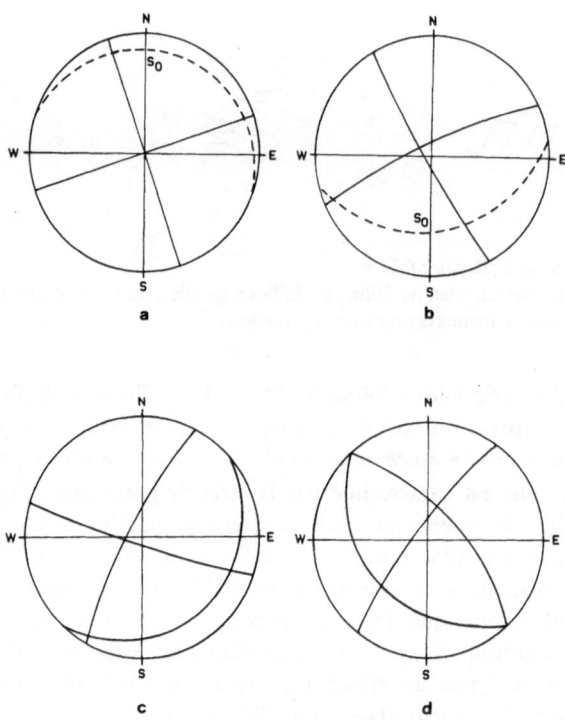

Figure 8. — *Familles de diaclases* (d'après Laurent et Allard [22]).
a : calcaires de Pagny (Meuse) ; b : Grès du Cap Frehel (Côtes du Nord) ; c : Granite
de la Bresse (Vosges) ; d : granite de Ligron (Massif Armoricain). (S_0 = stratification).

4. Structures actuelles (Superposition des déformations tectoniques)

Les études de terrains montrent que leur histoire tectonique est une superposition de ces divers mécanismes. La structure du milieu actuel est donc le
résultat, sur les structures initiales, des déformations tectoniques successives,

souples et/ou cassantes ; sa symétrie dépend du caractère passif ou actif de la structure initiale. Turner et Weiss [45] énoncent trois principes fondamentaux :

1) Lorsque les éléments d'une structure initiale sont cinématiquement passifs, la symétrie mécanique de cette structure est sphérique, le milieu initial est considéré comme isotrope.

2) Lorsque les éléments de la structure initiale sont cinématiquement actifs, la symétrie mécanique de la structure initiale est identique à la symétrie géométrique.

3) Lorsque certains éléments de la structure initiale sont passifs et les autres actifs, deux cas se présentent :
— Si les sous-structures actives ont même symétrie (structure homotactique) la symétrie mécanique est identique à la symétrie géométrique.
— Si les sous-structures ont des symétries différentes (structure hétérotactique) la symétrie mécanique coïncide avec la symétrie géométrique commune aux sous-structures actives.

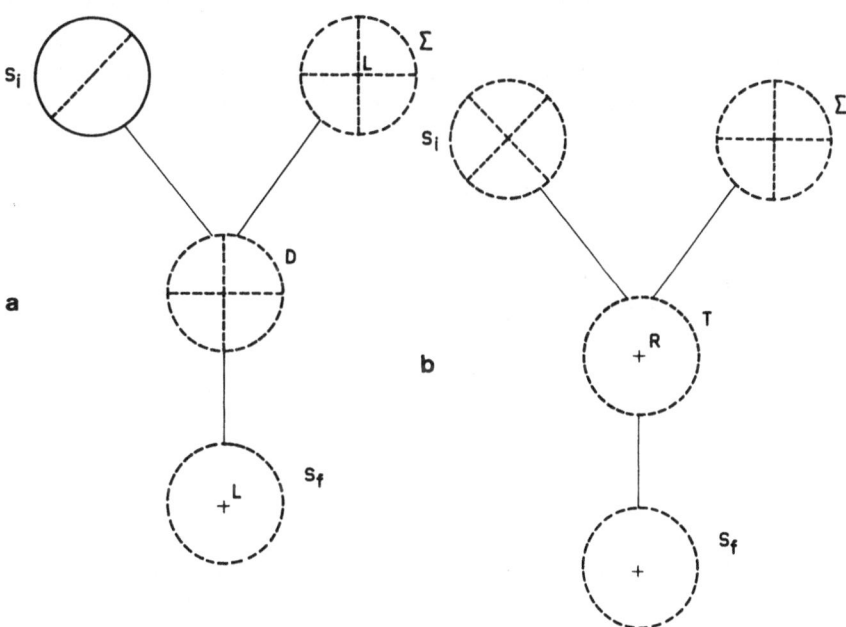

Figure 9. — *Structures tectoniques* (d'après Turner et Weiss [45]).

a : structure initiale S_i passive, axiale, tenseur contrainte Σ, orthorhombique, déformation moyenne D orthorhombique, structure finale S_f, monoclinique ; b : structure initiale S_i active, orthorhombique, tenseur contrainte Σ, orthorhombique, transformation T : déformation pure D et rotation R coaxiale à D, structure finale S_f monoclinique (une sous-structure imposée, une sous-structure héritée).

Envisageons trois exemples :

Exemple 1 :   Une structure initiale S_i passive (Fig. 9a) à symétrie axiale est soumise à tenseur contrainte orthorhombique, tel qu'une direction principale (L) de Σ soit contenue dans le plan isotrope de S_i. La déformation (moyenne) est de type orthorhombique (coaxiale à Σ), la structure finale est monoclinique d'axe L, commun à Σ et S_i.

Exemple 2 :   Cas du Calcaire de Pagny (Fig. 8a). La structure initiale S_i de type planaire, à stratification S_0 subhorizontale à symétrie de révolution a été soumise à des tenseurs tectoniques induisant deux familles de diaclases verticales orthogonales, la structure résultante est sensiblement orthorhombique.

Exemple 3 :   Une structure initiale S_i active (Fig. 9b) à symétrie orthorhombique est soumise à un tenseur contrainte orthorhombique, tel qu'un plan principal de Σ soit plan de symétrie de S_i. La transformation rotationnelle (moyenne) est de type monoclinique, l'axe de rotation R est coaxial à une direction principale de la déformation pure **D**. La structure finale est monoclinique, comportant une sous-structure imposée et une sous-structure héritée.

Les classes de symétrie des structures actuelles sont donc, comme celles des structures initiales, au nombre de cinq, mais d'un ordre généralement inférieur.

5.   Caractérisation de l'anisotropie structurale

L'anisotropie structurale étant définie géométriquement, son importance a été quantifiée par plusieurs processus :

1) Anisotropie morphologique – (Statistique d'orientation d'éléments) –(Fig. 10 et T1)– Le milieu est caractérisé par la proportion d'éléments rigides (par rapport à leur environnement) allongés ou plats, en fonction de leur orientation. Pour des gneiss [44] on utilise des éléments allongés de feldspath, dont les mesures d'orientation par rapport à la structure foliée, fournissent les diagrammes de répartition. Pour les argiles [7] on utilise l'intensité (en diffractométrie X) des raies de Kaolin, qui est proportionnelle au nombre de particules orientées parallèlement à la surface analysée, et l'on obtient les diagrammes de variation des pics I_θ/I_o des raies 001 et 002. Les résultats de ces mesures statistiques d'orientation sont reproduits sur des diagrammes de fréquences cumulées (Fig. 10).

2) Anisotropie cristalline – (Statistique d'orientation de l'axe optique d'une espèce minérale de la roche analysée).

On ne s'intéresse pas à la forme des grains, mais à leur orientation cristallographique par celle de leur axe optique. La répartition des axes optiques du quartz est fréquemment étudiée.

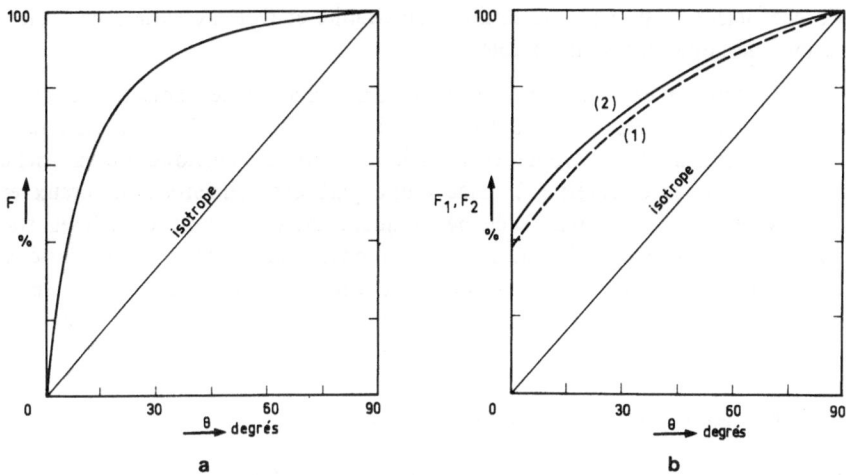

Figure 10. — *Orientation morphologique des grains.*
a : gneiss des Cammazes (d'après Sirieys et Debat [44]) ; b : Kaolin consolidé en laboratoire (d'après Boehler [7]) – (F, F_1, F_2 = fréquence cumulée, θ = orientation).

TABLE 1. Anisotropie d'Orientation Morphologique des Grains.

Référence Auteurs	Roche	θ degrés	0	15	30	45	60	75	90
[44]	Gneiss	F %		67	86,5	93,5	97	98	100
[7]	Argile	F_1 %	38,8	57,8	69,9	80,2	88,8	94	100
	consolidée	F_2 %	43,7	61,2	72,9	82,6	90,4	95,3	100

3) La déformation finie – Lorsque la dernière phase de déformation finie plastique est prépondérante, ayant éventuellement effacé les précédentes par réorientation des grains, c'est son intensité qui pourra servir d'élément quantitatif pour caractériser l'anisotropie structurale. Cette déformation peut être déterminée à l'aide de marqueurs tels que les éléments de l'anisotropie morphologique [44], des populations de fossiles ou la géométrie des plis.

4) L'indice de qualité micropétrographique K – On utilise [34], pour une roche, un coefficient tenant compte de la proportion de minéraux ayant une influence favorable sur son comportement et celle des minéraux altérés, favorisant le glissement, et microfissures ayant une influence défavorable sur le comportement mécanique, appelé indice de qualité micropétrographique.

Ce coefficient mesuré sur échantillons de Granite et Dolérite varie avec l'orientation selon une forme ellipsoïdale.

Le but de ces divers procédés de caractérisation de l'anisotropie structurale est sa corrélation avec l'anisotropie des caractéristiques mécaniques.

Finalement l'anisotropie structurale qui intègre l'histoire de la roche ou du massif se caractérise à l'aide d'une part des éléments structuraux et leur symétrie, d'autre part d'éléments quantitatifs résultant d'analyses statistiques de marqueurs. La succession des phases de déformation c'est-à-dire orientation, intensité et chemin de la déformation, quand il est accessible, sont des éléments essentiels dans l'étude des structures tectoniques et de leur anisotropie.

6. Anisotropie de dilatation thermique

La petite déformation de la roche résultant d'une élévation de la température est caractérisée par un tenseur du second ordre symétrique, c'est-à-dire par 6 paramètres : 3 d'orientation du trièdre principal et 3 d'intensité.

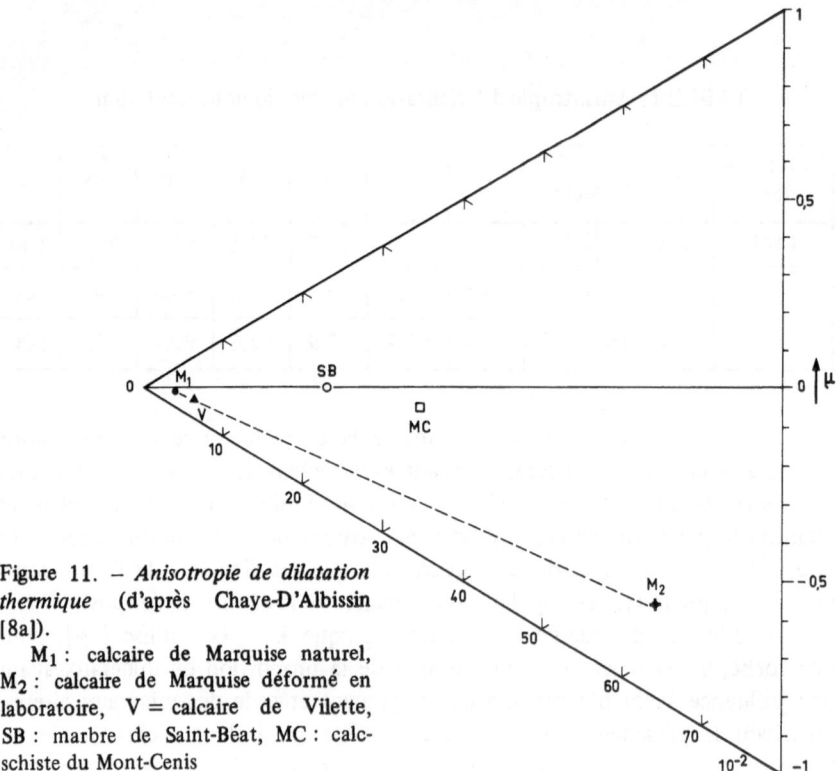

Figure 11. – *Anisotropie de dilatation thermique* (d'après Chaye-D'Albissin [8a]).

M$_1$: calcaire de Marquise naturel, M$_2$: calcaire de Marquise déformé en laboratoire, V = calcaire de Vilette, SB : marbre de Saint-Béat, MC : calcschiste du Mont-Cenis

Des mesures d'anisotropie de la dilatation thermique ont été effectuées [8a] sur des Calcaires naturels dans trois directions perpendiculaires, correspondant aux directions préférentielles structurales, utilisées dans le découpage des blocs en carrière d'extraction et sur des roches calcaires après déformation plastique en laboratoire (T2 et Fig. 11).

6.1. *Roches naturelles*

Les Calcaires de Marquise et de Vilette, avec des coefficients d'anisotropie de dilatation thermique q_1 (cf. Annexe) égaux respectivement à 3,87 % et 5,71 %, montrent une répartition sensiblement aléatoire des cristaux de calcite ; le Marbre de Saint-Béat et le Calcaire métamorphique du Mont-Cenis, avec des coefficients q_1 égaux respectivement à 19,9 et 30 %, font apparaître une orientation plus ou moins marquée.

6.2. *Roche après déformation plastique*

Les expériences ont été réalisées sur le Calcaire de Marquise, sensiblement isotrope initialement. La déformation plastique effectuée sous $p_c = 5$ kb et, dans la direction C, $\sigma_3 = 10$ kb, a été égale en moyenne à $\overline{e}_1 = \overline{e}_2 = 0,26$

TABLE 2. Anisotropie de dilatation thermique.

Référence Auteurs	Roche (Provenance)	α_m 10^{-6} degrés C^{-1}	q_1 %	μ
[8] a	Calcaire (Vilette)	18,27	5,71	− 0,36
	Marbre (St-Béat)	18,55	19,9	0,006
	Calcschiste (Mont-Cenis)	15,43	30	− 0,11
	Calcaire (Marquise) naturel	9,28	3,87	− 0,13
	Calcaire (Marquise) déformé en laboratoire	11,62	59,27	− 0,69

et, dans la direction C, $\bar{e}_3 = -0,52$ (valeurs moyennes en déformation logarithmique). Cette déformation de révolution a fait passer les 3 composantes des dilatations initialement égales à 8,99, 9,26 et $9,61 \times 10^{-6}$ par degré C, à : 9,06, 7,37 et, dans la direction C, $18,45 . 10^{-6}$ par degré C. La dilatation moyenne a varié de 9,28 à $11,62 . 10^{-6}$ par degré C, mais surtout est apparue une anisotropie caractérisée par un coefficient q_1 maintenant égal à 59 % et un coefficient μ de $-0,69$ montrant une symétrie proche de la symétrie de révolution, de même axe que celui de la déformation plastique. Ces résultats indiquent une orientation préférentielle des axes optiques des cristaux suivant une direction parallèle à celle de la compression et contraction majeure. La déformation plastique a réorienté les cristaux suivant une direction préférentielle.

7. Anisotropie élastique

Les roches, dans leur état actuel, ont un comportement sous charge qui vérifie de manière approchée (cf. § 10) les lois de l'Elasticité. C'est seulement pour la déformation initiale que l'on peut mettre en évidence une loi linéaire entre les tenseurs contrainte σ et petite déformation ε qui s'exprime par :

$$\epsilon_{ij} = s_{ijk\ell} \, \sigma_{k\ell} \tag{7a}$$

ou

$$\sigma_{ij} = c_{ijk\ell} \, \epsilon_{k\ell} \tag{7b}$$

Les tenseurs du 4e ordre s et c symétriques en i et j, en k et ℓ et en ij et kℓ sont appelés [29] tenseur élasticité et tenseur rigidité. De nombreuses expériences, en laboratoire et in situ, ont été réalisées sur des roches très variées pour caractériser leur anisotropie élastique. Les tests sur échantillons nécessitent un grand nombre d'essais et une interprétation statistique du fait de l'importante hétérogénéité des roches. Le nombre de constantes élastiques à mesurer dépend du type de structure.

7.1. *Structures pseudo-isotropes*

On ne peut envisager une stricte isotropie élastique des roches du fait même de leur mode de formation, mais parfois les lois élastiques isotropes :

$$\epsilon_{ik} = \frac{1 + \nu}{E} \, \sigma_{ik} - \frac{\nu}{E} \, \sigma_{mm} \, \delta_{ik} \tag{8a}$$

ou

$$\sigma_{ik} = \lambda \theta \delta_{ik} + 2\mu \epsilon_{ik} \tag{8b}$$

sont applicables, avec une bonne approximation, aussi qualifie-t-on ces roches de pseudo-isotropes, caractérisées par 2 constantes élastiques E et ν (ou λ et μ).

7.2. *Symétrie de révolution*

Le référentiel (x_i) utilisé est constitué par 2 axes orthogonaux quelconques XY dans le plan istrope S et Z normal à S. Le tenseur s est invariant dans une transmutation par symétrie par rapport au plan S et par rotation autour de Z. En adoptant la notation de Nye [29] telle que $s_{1111} \longrightarrow s_{11}$, $s_{1133} \longrightarrow s_{13}$, $s_{3333} \longrightarrow s_{33}$ et $4s_{1313} \longrightarrow s_{44}$, les composantes du tenseur élasticité s s'expriment dans (x_i) par :

$$
\begin{bmatrix}
s_{11} & s_{12} & s_{13} & 0 & 0 & 0 \\
 & s_{11} & s_{13} & 0 & 0 & 0 \\
 & & s_{33} & 0 & 0 & 0 \\
 & & & s_{44} & 0 & 0 \\
 & & & & s_{44} & 0 \\
\text{sym} & & & & & 2(s_{11} - s_{12})
\end{bmatrix}
\tag{9}
$$

s comporte 5 composantes indépendantes dont la détermination s'effectue à l'aide de deux types d'essais :

— Les essais coaxiaux, tels que σ et ε ont une direction principale confondue avec Z, permettent la détermination des 4 coefficients à indices 1, 2 et 3. Des compressions parallèles et normales à Z fournissent deux modules d'Young. $E_{33} = 1/s_{33}$ et $E_{11} = 1/s_{11}$ et deux coefficients de Poisson $\nu_{13} = -s_{13}/s_{33}$ et $\nu_{12} = -s_{12}/s_{11}$.

Les essais obliques, tels que des compressions monoaxiales, pour lesquelles Z n'est pas direction principale de σ et de ε, permettent d'obtenir le cinquième coefficient s_{44}, inverse de G_{13}, module de cisaillement dans XZ. Dans le référentiel (x_i') avec X_1' colinéaire à σ_1, les composantes du tenseur élastique s'expriment en fonction de ses composantes dans (x_i) par les relations de transmutation d'un tenseur du 4e ordre, la composante s_{1111}' notamment s'exprime par :

$$
s_{1111}' = a_{1m} a_{1n} a_{1p} a_{1q} s_{mnpq}
\tag{10}
$$

expression dans laquelle les a_{ik} sont les cosinus directeurs de (x_i') par rapport à (X_1). En posant $\theta = (X_1', Z)$ et en utilisant la notation précédente, on obtient :

$$
s_{11}' = s_{11} \sin^4 \theta + s_{33} \cos^4 \theta + (s_{44} + 2s_{13}) \sin^2 \theta \cos^2 \theta
\tag{10'}
$$

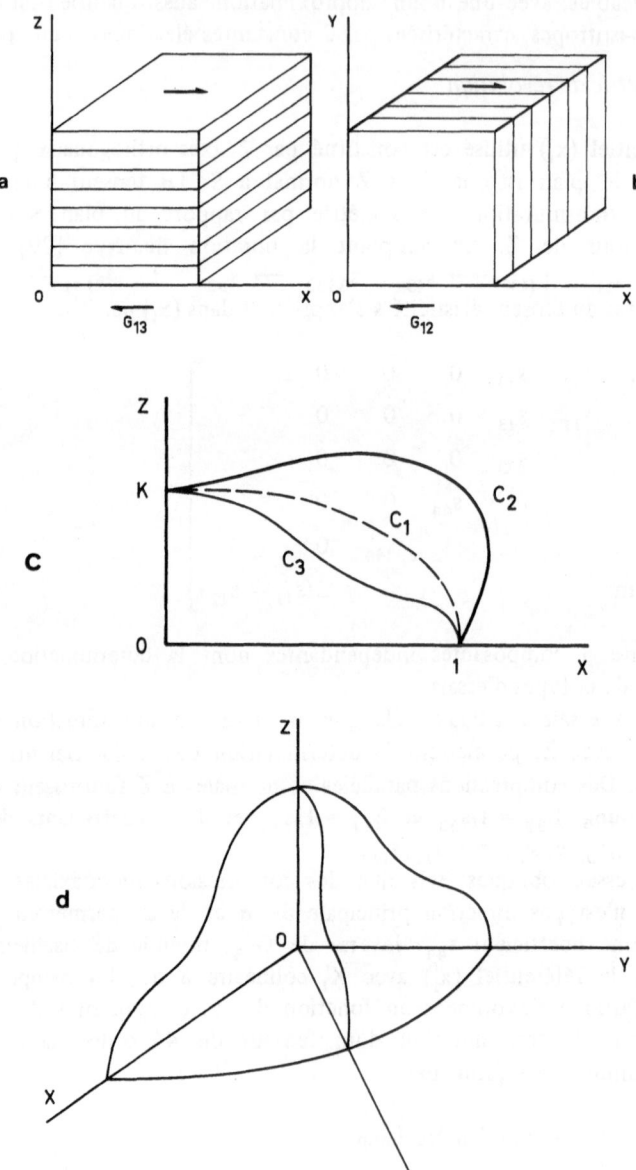

Figure 12. – *Anisotropie élastique.*
a : cisaillement structural de module G13 ; b : cisaillement astructural de module G12 ; c : courbes méridiennes des surfaces S_1 des modules élastiques ; d : surface S_1 des modules élastiques des roches schisteuses.

Une compression monoaxiale orientée de θ fournit le module $E_\theta = 1/s'_{11}$. Une seule mesure, théoriquement, suffirait pour déterminer G_{13}, en pratique on effectue plusieurs compressions monoaxiales sous différents angles θ et la dispersion des résultats des mesures, surabondantes, est traitée par la méthode des moindres carrés.

L'anisotropie élastique est visualisée par des surfaces représentatives S en reportant dans la direction de mesure une quantité fonction du module d'Young E_θ. Ces surfaces S étant de révolution autour de 0Z, on étudie leur méridienne. On utilise souvent la surface S_1 représentative du module d'Young E_θ pour laquelle le rayon vecteur est proportionnel au module. L'allure de S_1 est très influencée par le module G_{13}. En variables réduites obtenues en posant : $K = E_3/E_1$ ($K < 1$ pour une structure planaire, $K > 1$ pour une structure linéaire), la méridienne de S_1 a une forme caractérisée par le coefficient adimensionnel G_{13}/G_{12} (Fig. 12c). Pour les valeurs $K < 1$, des valeurs modérées de ce coefficient fournissent une méridienne régulière, à croissance continue de type C_1 (d'allure voisine d'une ellipse), des valeurs élevées donnent une courbe C_2 à maximum, des valeurs faibles donnent une méridienne C_3 à minimum.

On utilise parfois [29] la surface S_2 pour laquelle le rayon vecteur est proportionnel à s'_{11} c'est-à-dire à l'inverse du module.

Un troisième type de surface représentative S_3 peut être obtenu [42] en utilisant la racine quatrième du module E_θ, soit $E_\theta^{1/4}$, lorsque entre les coefficients élastiques on a la relation :

$$s_{44} + 2s_{13} = 2\sqrt{s_{11}s_{33}} \tag{11}$$

appelée relation d'anisotropie restreinte, qui ramène de 5 à 4 le nombre de constantes élastiques indépendantes. L'équation de transmutation (10') s'exprime alors par :

$$s'^{1/2}_{11} = s^{1/2}_{11} \sin^2\theta + s^{1/2}_{33} \cos^2\theta \tag{12}$$

le point image du module, dont le rayon vecteur est égal à $r = 1/s'^{1/4}_{11}$ décrit une ellipse (d'axes 0X et 0Z dont les demi-axes sont $E_1^{1/4}$ et $E_3^{1/4}$) qui est la méridienne de la surface S_3 définie par (12).

L'anisotropie des modules est caractérisée par deux paramètres : $K = E_3/E_1$, obtenu par les essais coaxiaux, ($K < 1$ avec les roches schisteuses et $K > 1$ avec les roches fibreuses), et un second paramètre tel que d [35], défini par :

$$E_3/G_{13} - 2(1 + \nu_{13}) = d \tag{13}$$

qui caractérise l'allure de la méridienne C de S_1 dont l'équation est alors :

$$E_\theta = \frac{E_3}{1 + d \sin^2 \theta \cos^2 \theta - (1 - K) \sin^4 \theta}$$

K donne l'excentricité et d la forme, avec ou sans extrêmum, de la courbe C ; d traduit l'influence du module de cisaillement G_{13} : d petit (G_{13} grand) fournit une courbe C_2 à maximum, d grand (G_{13} petit) fournit une courbe C_3 à minimum. L'anisotropie restreinte (11) s'exprime par : $d = 2(\sqrt{K} - 1)$, l'isotropie par : $K = 1, d = 0$.

Pour une orientation θ distincte de 0 et $\pi/2$, $\underline{\sigma}$ et $\underline{\varepsilon}$ ne sont pas coaxiaux, leur différence d'orientation, appelée déphasage, est une fonction de θ et des constantes élastiques en valeurs relatives, elle s'exprime en fonction de θ, K et d [35]. Selon les valeurs relatives de K et d, il peut, ou non, exister une direction de coaxialité (distincte de 0 et $\pi/2$) pour la valeur θ_0 telle que $\sin \theta_0 = \sqrt{d/2(d + 1 - K)}$.

TABLE 3. Anisotropie élastique des Roches Schisteuses. Méridiennes des surfaces des Modules.

Référence Auteurs	Roche (Provenance)	N°	K	d	ν_{12}/ν_{13}
[36]	Schiste (Portugal) (essais de laboratoire)	S 01 S 02 S 03	0,782 0,535 0,315	0,320 − 0,294 0,155	1,27 0,89 2
	Schiste Greywacke (Portugal) (essais in situ)	S 04 S 05 S 06	0,383 0,213 0,23	4,24 2,52	
[8] c	Schistes ardoisiers (Travassac)	S 07	0,75		
[8] b	Schistes ardoisiers (Fumay)	S 08	0,47		2,18
[41]	Schistes ardoisiers (Lacaune)	S 09	0,49		2,83

Pour les roches, les méridiennes C des surfaces S_1 sont généralement soit de type C_1, régulières, voisines d'une ellipse, soit de type C_3 à minimum. Une argile naturelle [7] est caractérisée par une courbe de type C_1 avec un coefficient K = 0,45. Avec les roches schisteuses, les méridiennes de S_1 sont des courbes de type C_3. Les coefficients K mesurés (T3) évoluent de 0,2 à 0,8, les rapports G_{12}/G_{13} compris entre 1,4 à 3,5 entraînent des coefficients d élevés. Ce résultat sur les courbes de type C_3 s'explique par la présence d'un feuilletage de la roche qui entraîne du fait du glissement "facile" dans S un coefficient G_{13} inférieur à G_{12} (Fig. 12a et 12b). Ce feuilletage des roches schisteuses résulte de déformations tectoniques très élevées. Les calculs de déformations sur les Gneiss [44] ont fourni des valeurs de contractions de l'ordre de 66 %, on peut même envisager des valeurs supérieures pour les schistes ardoisiers. C'est finalement ce feuilletage (analogie avec un empilement de très minces plaquettes) qui, entraînant un glissement facile sur les plans S, introduit un minimum dans la courbe des modules et donc une méridienne de type C_3.

7.3. *Symétrie orthorhombique*

Le tenseur élastique s'exprime dans le référentiel (x_i) constitué par les 3 plans de symétrie triorthogonaux par :

$$\begin{pmatrix} s_{11} & s_{12} & s_{13} & 0 & 0 & 0 \\ & s_{22} & s_{23} & 0 & 0 & 0 \\ & & s_{33} & 0 & 0 & 0 \\ & & & s_{44} & 0 & 0 \\ & & & & s_{55} & 0 \\ \text{sym} & & & & & s_{66} \end{pmatrix} \tag{15}$$

faisant intervenir 9 composantes indépendantes. Il est invariant dans une transmutation par symétrie par rapport aux 3 plans XY, YZ, ZX.

Une anisotropie apparaît dans le plan XY, lorsqu'elle est peu intense les structures rocheuses sont dites à faible orthorhombicité et assimilées à des structures de révolution. Les essais coaxiaux fournissent les 6 coefficients d'indices 1, 2 et 3, les trois autres étant obtenus par les essais obliques. Il convient de mesurer en outre, par rapport au cas de révolution ; $S_{22} = 1/E_2$, $S_{23} = -\nu_{23}/E_2$, $S_{55} = 1/G_{23}$, $S_{66} = 1/G_{12}$. Les 9 paramètres peuvent être obtenus par des compressions monoaxiales, les essais obliques étant caractérisés par deux angles d'orientation, les résultats étant également traités par la méthode des moindres carrés.

L'anisotropie est visualisée par les surfaces S_1 des modules et caractérisée, lorsque S_1 est un ellipsoïde, par les coefficients q et μ (cf. annexe) et lorsque S_1 est une surface plus complexe par q et μ de l'ellipsoïde circonscrit et deux paramètres supplémentaires.

Les résultats expérimentaux différent selon que l'on étudie des roches déformées naturellement telles que Granites et Dolérites et les roches feuilletées telles que les Schistes.

Pour les Granites et Dolérites [15] [32] [34] les surfaces S_1 des modules élastiques sont assimilables à des ellipsoïdes (T4 et Fig. 13). Les coefficients de variations, déduits de la méthode des moindres carrés, qui permettent de juger de la validité de l'assimilation à un ellipsoïde, ont des valeurs très acceptables. L'anisotropie, douce, est alors définie uniquement par les essais coaxiaux. Les modules moyens évoluent de 300 à 700 kb. L'intensité des déviateurs logarithmiques se situe entre 5 et 25.10^{-2} ce qui montre une anisotropie faible, c'est-à-dire le peu d'influence de la déformation tectonique (peut être même enregistre-t-on uniquement l'anisotropie de formation). L'orthorhombicité est caractérisée par le coefficient μ (si μ tend vers ± 1, on se rapproche de la symétrie de révolution). Sur les 13 roches analysées, dans la littérature consultée, 11 ont un coefficient μ négatif, c'est-à-dire une anisotropie élas-

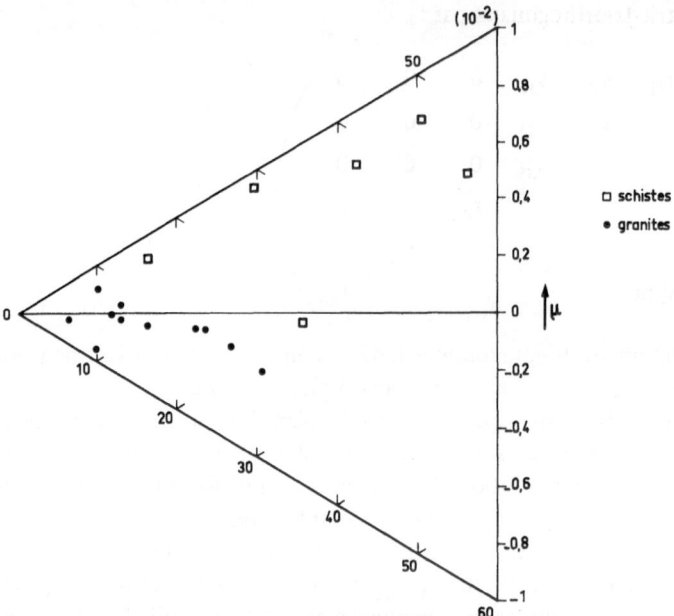

Figure 13. — *Coefficients d'anisotropie élastique.* Intensité q, et phase μ (• granites, Résultats d'après [15], [32] et [34] ; □ Schistes, Résultats d'après [25] et [33]).

TABLE 4. Coefficients d'anisotropie élastique.

Référence Auteurs	Roche (Provenance)	N°	$q\,(10^{-2})$	μ
[32]	Granite (Alvarenga)	G_1	23,5	− 0,25
		G_2	19,4	− 0,139
	Granite (Alto Lindoso)	G_3	11,68	− 0,10
		G_4	13,9	− 0,152
		G_5	11,8	0
	Granite (Vilarinho)	G_6	5,6	− 0,144
		G_7	9,1	− 0,72
[34]	Granite Gneissique	G_8	4,8	− 0,12
[15]	Granite (sous $p_c = 0,06$ Kb) (Barre)	G_9	20	− 0,13
	(Stanstead)	G_{10}	27	− 0,39
	(Laurentian)	G_{11}	9,2	0,51
[34]	Dolérite	D_1	4,8	− 0,12
		D_2	4,8	− 0,16
[33]	Schistes	S_1	51	0,52
		S_2	48	0,808
	ardoisiers	S_3	30,8	0,886
	Micaschistes à biotite	S_4	15,1	0,68
[25]	Phyllades (Revin)	S_5	39,2	0,746
	Schistes (Cheylas)	S_6	30,5	− 0,068

tique fibreuse. Ce résultat mériterait d'être relié à l'histoire de la formation de ces roches.

Pour les roches présentant une structure à foliation très prononcée, la surface S_1 des modules (comme observé dans le cas de révolution) ne peut pas être assimilée à un ellipsoïde, il convient [33] d'avoir recours à une quartique d'équation :

$$\frac{X^2}{\ell^2} + \frac{Y^2}{m^2} + \frac{Z^2}{n^2} + \frac{X^2 Z^2}{p^4} + \frac{Y^2 Z^2}{t^4} = 1 \qquad (16)$$

dans les axes de la structure. Dans le plan XY la courbe des modules est du type C_1 de forme elliptique (la relation (16) ne fait pas intervenir de terme en $X^2 Y^2$), mais dans un plan radial quelconque la courbe obtenue est de type C_3, avec minimum, traduisant le glissement facile sur les plans structuraux. L'ellipsoïde circonscrit à la quartique apporte des informations sur l'anisotropie élastique, ici plus violente, avec des coefficients q atteignant 50.10^{-2}. Les coefficients μ, évidemment positifs, sont supérieurs à 0.5. L'orthorhombicité est modérée, dans le plan de foliation XY le rapport axial de l'ellipse des modules qui suffit à caractériser l'anisotropie dans ce plan varie de 0,62 à 0,85. En plus des coefficients q et μ il convient de définir deux nouveaux coefficients, tels que p/m et t/n par exemple, qui caractérisent l'influence des coefficients S_{44} et S_{55}. Dans un plan radial la courbe des modules est du type C_3 et le rapport axial de son ellipse circonscrite atteint des valeurs beaucoup plus faibles, jusqu'à 0,2. Ces roches ont certainement subi des déformations tectoniques naturelles beaucoup plus intenses que les Granites et Dolérites précédents.

7.4. *Structure monoclinique*

Dans le référentiel (x_i), avec x_3 normal au plan de symétrie, le tenseur élastique s est invariant dans une transmutation par symétrie par rapport au plan $x_1 x_2$; il comporte 13 composantes indépendantes. Des mesures ont été effectuées [25] sur des schistes dont la structure linéaire comporte deux sous-structures, $a_1 Lc_1$ (schistosité) et $a_2 Lc_2$ (stratification), L étant une linéation d'intersection de la schistosité S et de la stratification S'_o. Les essais ont été réalisés dans les 3 directions triorthogonales : L et, dans le plan normal à L, les deux bissectrices de S et S'_o. Le coefficient q égal à 30.10^{-2} se situe dans le groupe des valeurs des roches schisteuses, le paramètre μ est légèrement négatif (ellipsoïde de type allongé). Sans doute des essais obliques auraient fourni une courbe de modules dans un plan normal à L de forme C'_3 à deux minimas.

Dans tous les cas de structures isotropes et anisotropes, les coefficients élastiques ne sont pas des constantes absolues, ils varient avec la déformation (cf. § 10) et avec la pression de confinement p_c. Par mécanisme de serrage des grains et de fermeture des fissures, on observe généralement un accroissement de la rigidité du milieu, les modules axiaux croissent avec p_c.

Pour les schistes [5] les modules de cisaillement G croissent avec p_c plus rapidement que les modules axiaux. Ces modules de cisaillement [24] passent par un maximum (un pic) vers 1 à 1,2 kb pour se stabiliser, après une légère décroissance sur un palier. Toutefois, l'anisotropie décroît lorsque p_c augmente. Cette croissance plus rapide des coefficients G [5] peut faire passer la courbe C des modules du type C_3 au type C_1. Cette variation de l'anisotropie [24] se stabiliserait pour p_c de l'ordre de 2 kb. Observons que ces résultats de laboratoire enregistrent les effets de la décompression anisotrope due à l'extraction de la roche de son milieu, la pression p_c pouvant dans certains cas la rapprocher partiellement de son état naturel.

En conclusion, l'anisotropie élastique des roches à structures planaires ou linéaires à symétrie orthorhombique ou axiale de type Granites et Dolérites, est relativement faible, et caractérisée par des surfaces S_1 ellipsoïdales à courbes axiales des modules de type C_1, les valeurs q sont relativement peu élevées. Pour les roches Schisteuses, l'anisotropie est généralement plus intense, les valeurs q sont plus importantes, les courbes axiales de modules du type C_3 à minimum du fait du glissement facile sur le feuilletage et les surfaces S_1 du 4e ordre. Les fortes déformations tectoniques sont responsables de cette intense anisotropie. Dans les deux cas l'anisotropie élastique diminue, jusqu'à un palier, quand la pression isotrope croît.

8. Plasticité

L'étude structurale a montré le rôle important de la plasticité sur l'anisotroqie des roches. Deux types d'anisotropie liée à la déformation plastique des roches apparaissent selon qu'elle s'effectue avec diminution de volume, sous l'effet d'une pression isotrope, ou à partir d'un seuil de contrainte, fonction de la pression et la température, de manière isovolume.

8.1. *Déformation plastique anisovolume*

Des roches tendres, poreuses, dont la densité est très inférieure à celle des grains constituants, soumises à une pression isotrope (Fig. 14) subissent une notable diminution de volume irréversible (fermeture des pores par glissement inter- ou intragranulaire). Ce mécanisme notamment observé sur des calcaires grossiers [17], des craies [13], des grès calcaires [31], une diatomite [3], fait apparaître un seuil (correspondant parfois à la destruction de la structure) à partir duquel la déformation croît plus rapidement avec la pression jusqu'à une troisième phase de raidissement général. L'anisotropie de ce mécanisme

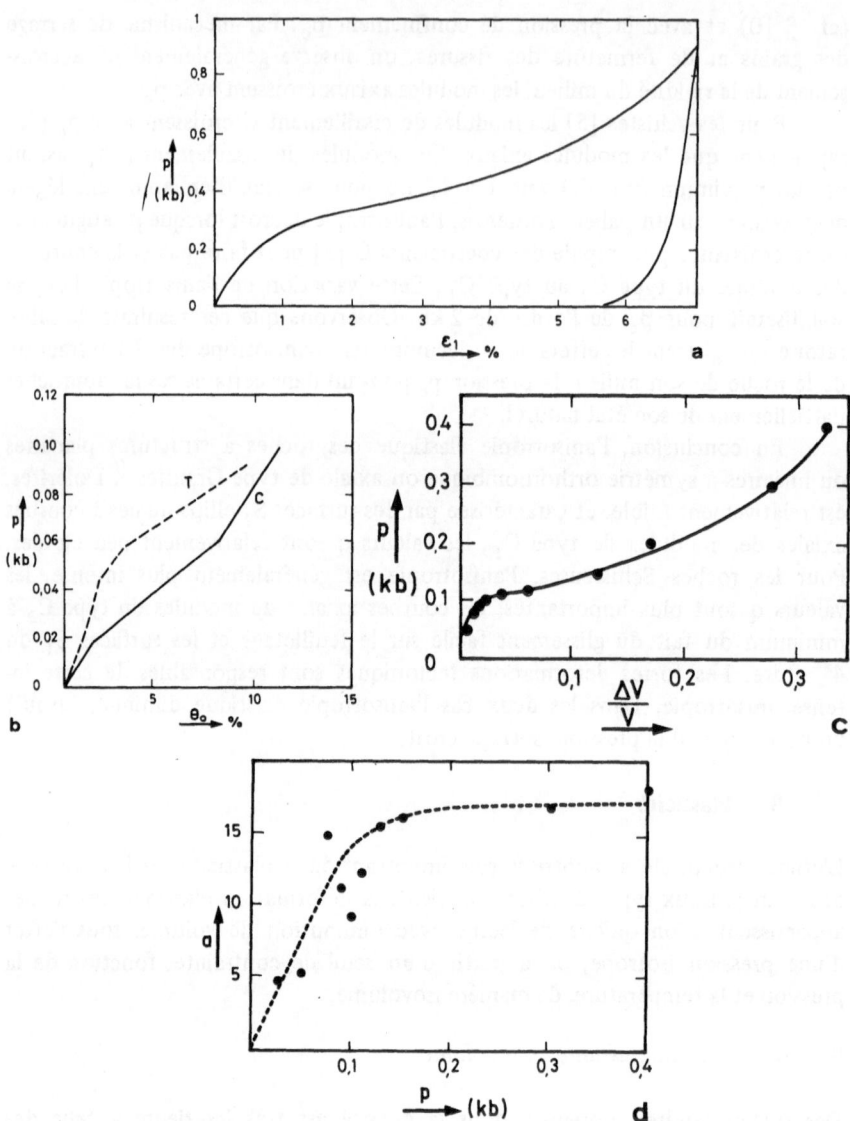

Figure 14. – *Anisotropie de déformation plastique sous pression isotrope.*
    a : craie de densité 1,63 (d'après Dayre, Dessenne et Wack [13] – (p = pression
isotrope, ϵ_1 = déformation axiale) ; b : calcarénite C de densité 1,4 et tuf T de densité 1,1
(d'après Pellegrino [31]) – (p = pression isotrope, θ_0 = variation de volume) ; c : diatomite
de densité 0,6 (d'après Allirot et Boehler [3]) – (p = pression isotrope, $\frac{\Delta V}{V}$ = variation de
volume) : d : même diatomite qu'en c (d'après Allirot [1]) – (p = pression isotrope,
a = ϵ_3/ϵ_1 = coefficient d'anisotropie).

a été étudiée sur la diatomite [1] ; la structure, de type planaire de révolution, est active ; la déformation comporte deux contractions sensiblement égales dans la stratification So, la troisième, normale à So, est beaucoup plus élevée, le coefficient d'anisotropie (Fig. 14d) a = ϵ_3/ϵ_1 croît et semble se stabiliser vers p_c = 0,4 kb à la valeur 17. Cette déformation plastique anisotrope a pour effet de modifier les propriétés anisotropes de cette roche (l'anisotropie élastique et l'anisotropie de résistance croissent avec p dans le domaine 0-60 bars) ; elle joue un rôle analogue à celui de la déformation tectonique naturelle.

8.2. *Plasticité isovolume*

Sous l'action conjuguée d'une pression isotrope et d'un déviateur de contrainte, à partir d'un seuil de contraintes, apparaît la déformation plastique isovolume avec les roches compactes, à faible porosité. Avec les roches poreuses, la variation de volume qui se manifeste lors de l'écoulement initial tend à s'annuler au cours de la déformation. Le seuil, ou limite d'écoulement, est caractérisé par une relation $f(\sigma_{ik}) = 0$.

Pour une roche pseudo-isotrope, f est indépendante de l'orientation de σ et assez bien représentée par une fonction du second ordre en σ_1, σ_2, σ_3 dont la représentation dans l'espace des contraintes E_3 a l'allure d'un paraboloïde de révolution, ou d'un cône pour certaines roches particulières. Dans l'écoulement plastique initial, la déformation isovolume vérifie les équations de linéarité de Saint-Venant entre la déformation infinitésimale et le déviateur de contrainte, elle induit une anisotropie de la roche (cf. § 4).

Pour une roche anisotrope, le critère d'écoulement f fait intervenir, en outre, l'orientation du tenseur par rapport à la structure ; il dépend alors, dans le cas général, de 6 paramètres. Les mécanismes de déformation plastique sont plus variés que ceux des roches isotropes, ils dépendent du caractère actif ou passif de la structure.

8.2.1. *Les seuils*

Les résultats expérimentaux sur la déformation plastique des roches anisotropes font apparaître deux groupes de roches quant à leur comportement plastique : celles à faible anisotropie mécanique telles que granites, dolérites, calcaires compacts, argiles consolidées, celles d'anisotropie plus violente, telles que schistes, schistes ardoisiers, calcaires dolomitiques.

a) Les granites étudiés [15], [32] ont des résistances en compression monoaxiale caractérisées par des surfaces s' assimilables à des ellipsoïdes. Leur anisotropie de seuil est donc définie par les coefficients q et μ (Fig. 15c et T5), les valeurs de q sont faibles, inférieures à 15 %, les coefficients μ sont variables. Ces ellipsoïdes sont évidemment de révolution pour les structures

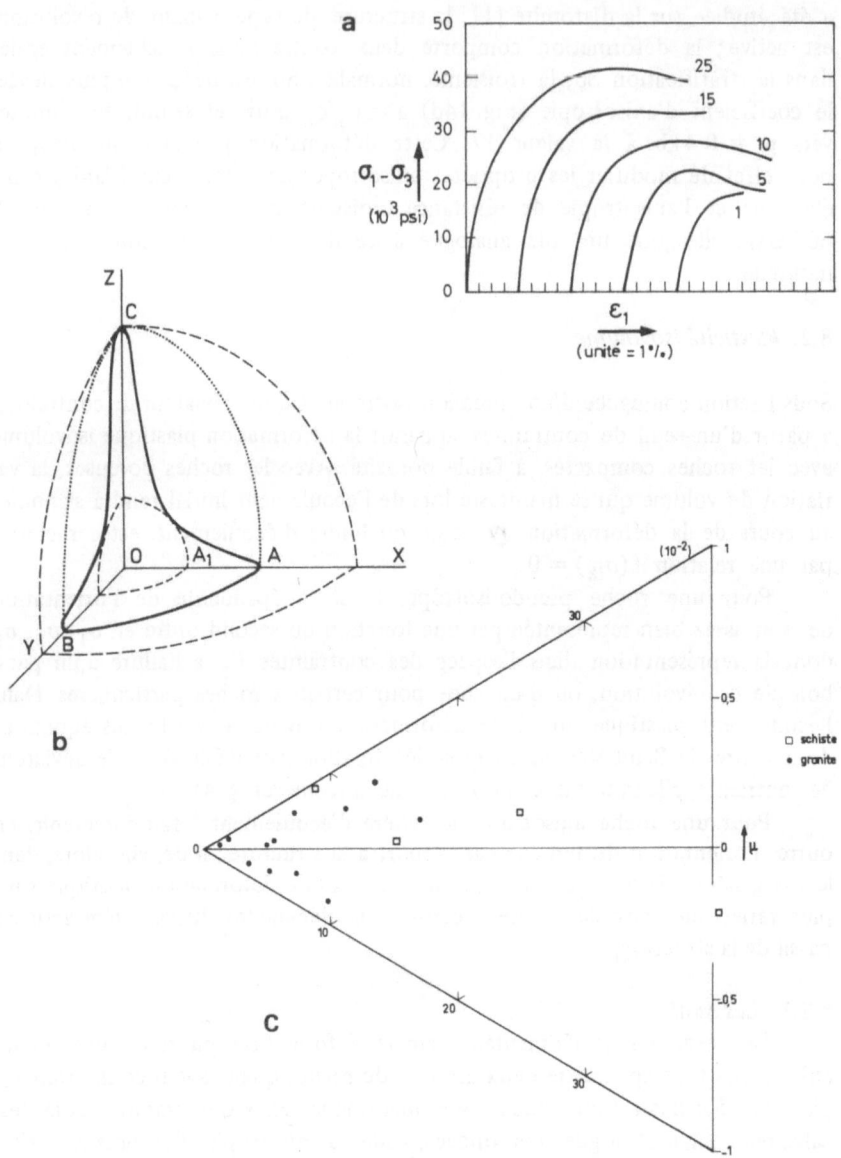

Figure 15. – *Anisotropie plastique des seuils.*
    a : courbes effort-déformation des schistes de Green River (d'après Mc Lamore et Gray [24]) sous différentes pressions de confinement, évoluant de 1 à 25 . 10^3 psi) ; b : surfaces limites de résistance du 4° ordre des roches schisteuses (d'après Peres-Rodriguez [33]) ; c : coefficients d'anisotropie q et μ des surfaces limites de résistance de Granites et de Schistes – • granites, □ schistes.

axiales. Toutefois, les axes extrémaux de ces ellipsoïdes sont inversés par rapport à ceux définissant l'anisotropie élastique. Des dolérites et des calcaires compacts conduisent à des conclusions analogues.

Pour une roche argileuse préalablement consolidée sous déformation monoaxiale (égale à $\epsilon_1 = 51\,\%$, sous $\sigma_1 = 78{,}4\,\mathrm{N/cm^2}$) on obtient [7] également un ellipsoïde sensiblement de révolution, dont l'anisotropie est caractérisée par $q = 8{,}8.10^{-2}$ et $\mu = -1$.

TABLE 5. Coefficients d'Anisotropie de Résistance

Référence Auteurs	Roche (Provenance)	N°	$q\,(10^{-2})$	μ
	Granite (Alvarenga)	G_1	7,65	0,68
		G_2	15,07	0,21
[32]	Granite (Alto Lindoso)	G_3	3,31	$-0{,}47$
		G_4	1,82	0,85
		G_5	10,1	0,45
	Granite (Vilarinho)	G_6	4,79	0,26
		G_7	8,94	$-0{,}71$
[34]	Granite Gneissique	G_8	12,72	0,65
[15]	Granite (Barre) (Stanstead) (Laurentian)	G_9	4,68	$-0{,}49$
		G_{10}	4,24	0,06
		G_{11}		
[8] b	Granite (Corbigny) (Senones)	G_{12}	6,37	$-0{,}43$
		G_{13}	1,2	0,22
[33]	Schistes ardoisiers	S_1	35,21	$-0{,}22$
		S_2	13,47	0,06
		S_3	21,62	0,17
	Micaschiste à biotite	S_4	8,42	0,98

b) Les roches à plans de foliation ont fait l'objet de nombreuses études. La plasticité des schistes [24] se développe d'autant mieux que la pression de confinement est élevée, les courbes effort-déformation (Fig. 15a) présentent une allure avec pic d'autant moins accusé que p_c est élevée, se rapprochant ainsi de la plasticité parfaite.

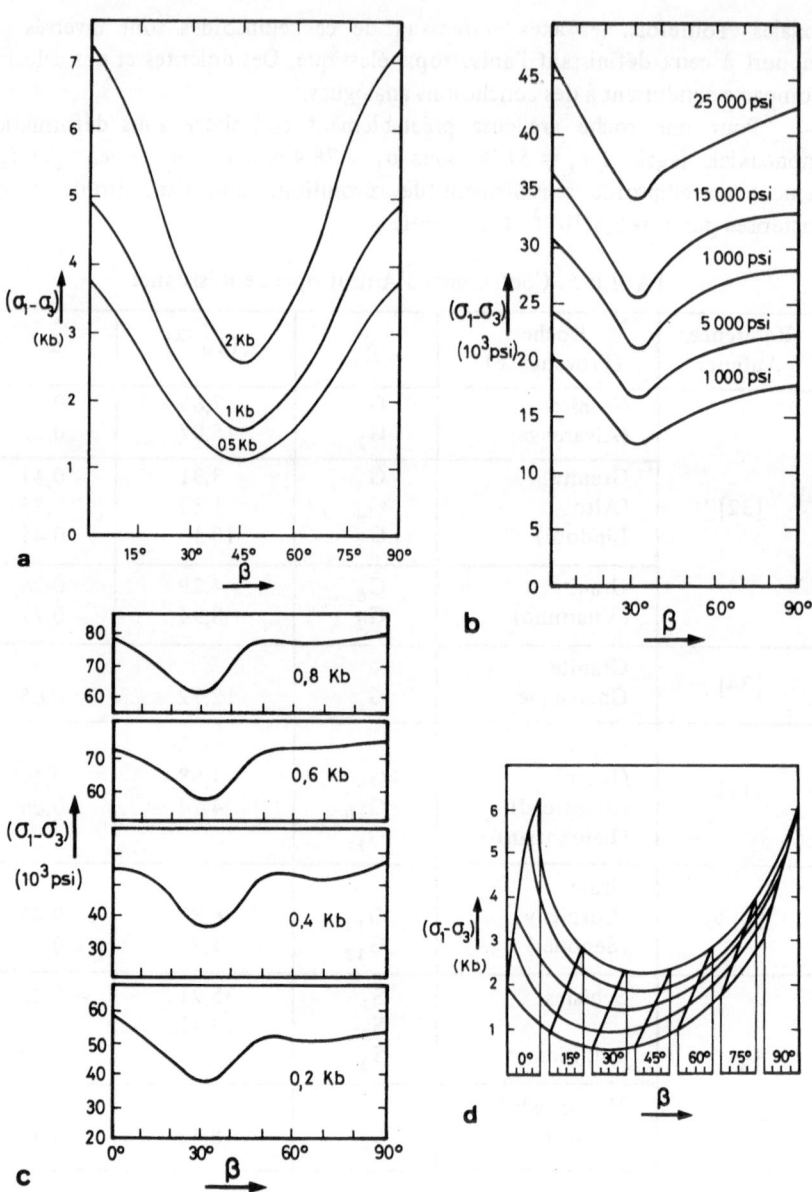

Figure 16. – *Anisotropie plastique des roches à structure planaire : Résistance en fonction de l'orientation.*

a : phyllite (d'après Donath [14]) ; b : schiste (d'après Mc Lamore et Gray [24]) ; c : calcaire dolomitique (d'après Mc Gill et Raney [23]) ; d : schistes ardoisiers (d'après Saint-Leu, Lérau et Sirieys [41]) .

L'anisotropie de résistance de Schistes (déjà analysés élastiquement) [33] est beaucoup plus intense (Fig. 15c et T5) que pour les Granites, Calcaires, Argile précédents, les rapports entre résistances extrémales pouvant atteindre les valeurs 4. Leur surface limite de résistance S′ (Fig. 15b) est représentée par une quartique, d'équation analogue à celle des modules élastiques (éq. 16), qui présente dans un plan normal à S une courbe de résistance avec un minimum très accusé et dans S une courbe régulière pouvant être assimilée à une ellipse. Ces courbes de résistance $(\sigma_1 - \sigma_3)$ en fonction de β (orientation de σ_1 avec la structure) à minimum (Fig. 16) ont également été observées sur des Phyllites [14], des Schistes [24] [27], des Calcaires dolomitiques [23], des Schistes ardoisiers [12], [5], [41], des Schistes cristallins [4]. Le minimum est fréquemment situé pour des valeurs β voisines de 30° ; par ailleurs l'ani-

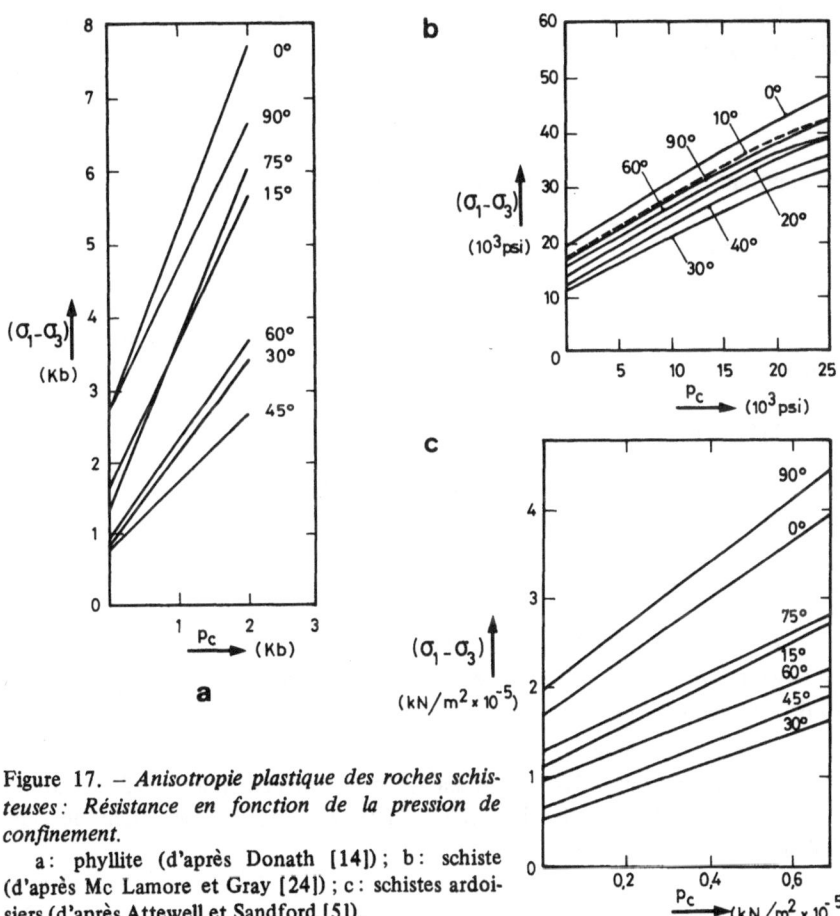

Figure 17. – *Anisotropie plastique des roches schisteuses: Résistance en fonction de la pression de confinement.*

a: phyllite (d'après Donath [14]) ; b: schiste (d'après Mc Lamore et Gray [24]) ; c: schistes ardoisiers (d'après Attewell et Sandford [5]).

sotropie, définie par le rapport $(\sigma_1 - \sigma_3)$ mini$/(\sigma_1 - \sigma_3)$ maxi décroît lorsque la pression de confinement croît. Cette anisotropie disparaîtrait, en extrapolant les résultats expérimentaux [24], pour $p_c = 5,3$ kb.

Pour une orientation β fixée la résistance (Fig. 17) varie linéairement avec σ_3 c'est-à-dire avec la pression de confinement, ce qui conduit à une théorie des critères d'écoulement de forme linéaire avec la pression isotrope.

Les seuils de déformation plastique, fonctions des contraintes principales et aussi de l'orientation des tenseurs contraintes, peuvent s'exprimer par des critères d'écoulement sous forme de loi phénoménologique. Olszak et Urbanowski [30] proposent un critère du type :

$$H_{ijk\ell}\, \sigma_{ij}\, \sigma_{k\ell} - 1 = 0 \tag{17}$$

pour lequel les symétries de $H_{ijk\ell}$ dépendent de celles de la structure. Boehler et Sawczuk [6] à partir de la transformation :

$$\overline{\sigma}_{ij} = A_{ijk\ell}\, \sigma_{k\ell}$$

et en notant $\overline{S}_{ij}$ le déviateur de $\overline{\sigma}_{ij}$, a et c deux constantes, proposent :

$$\overline{\sigma}_{ii} + a\, \sqrt{\overline{S}_{ij}\, \overline{S}_{ij}} = c \tag{18}$$

Cette anisotropie peut également être envisagée pour les roches feuilletées à partir d'une loi élémentaire reliant les composantes des vecteurs contraintes :

$$f_1(\sigma, \tau) = 0 \tag{19}$$

sur les surfaces de glissement. Pour les structures et les tenseurs contraintes de révolution, le critère s'exprime alors par une relation de la forme :

$$g(\sigma_1, \sigma_3, \beta) = 0 \tag{20}$$

Les données expérimentales sur les roches feuilletées sont susceptibles de deux modes d'interprétation, selon qu'on envisage une anisotropie de type continu ou discontinu.

1) Anisotropie continue

La déformation plastique apparaît par glissements irréversibles sur les surfaces de cisaillement, les lois élémentaires sur ces surfaces étant de type linéaire :

$$\tau = \tau_0 + \sigma \tan \Phi \tag{19a}$$

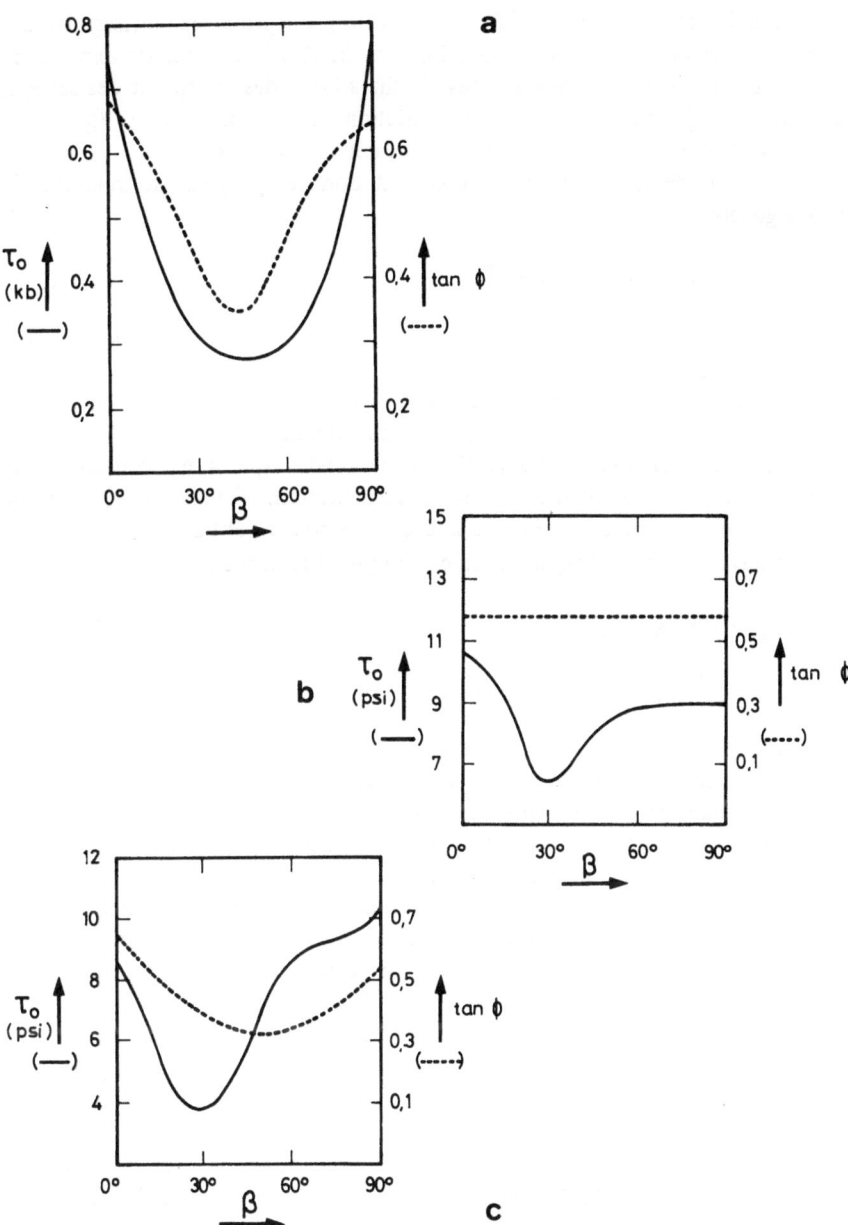

Figure 18. – *Anisotropie plastique continue. Variation de la cohésion* τ_0 *et de l'angle de frottement* Φ *avec l'orientation* β.

a: phyllite (d'après Donath [14]) ; b: schsite (d'après Mc Lamore et Gray [24]) ; c: schistes ardoisiers (d'après Mc Lamore et Gray [24]).

L'anisotropie de la loi élémentaire (19a) s'exprime en envisageant des constantes physiques τ_0 et Φ fonctions de β. Ainsi les résultats expérimentaux (Fig. 16a et b) obtenus sur des phyllites [14], des schistes et des schistes ardoisiers [24] conduisent-ils à des variations reproduites sur la figure 18. Parfois l'angle Φ peut être considéré comme constant (Fig. 18b).

Les lois de variation de τ_0 et Φ sont données [24] par des relations de forme générale :

$$\tau_0 = A - B\,[\cos 2(\varphi - \beta)]^m \tag{21a}$$

$$\tan \Phi = C - D[\cos 2(\varphi - \beta)]^n \tag{21b}$$

A, B, C, D, m, n, sont des constantes, β l'orientation de σ_1 par rapport à la structure, φ la valeur de β rendant τ_0 et/ou $\tan \Phi$ minimal. Jaeger [19] a proposé des relations (21) avec $m = 1$ et $D = 0$, c'est-à-dire que Φ est constant et la cohésion varie sinusoïdalement avec l'orientation. Donath [14] propose $m = n = 1$ c'est-à-dire des variations sinusoïdales pour τ_0 et $\tan \Phi$.

Dans tous les cas, l'équation du critère prend la forme :

$$\sigma_1 + H = K_p\,(\sigma_3 + H) \tag{20a}$$

où $K_p = (1 + \sin \Phi)/(1 - \sin \Phi)$ et $H = \tau_0 \cot \Phi$ varient avec β.

2) Anisotropie discontinue

Les résultats expérimentaux sur les seuils de résistance dans le cas de structures planaires peuvent être interprétés [42] par la théorie de l'anisotropie discontinue qui tient compte du type de surfaces de glissement irréversibles obtenues, structurales ou astructurales, donc du caractère actif ou passif de la structure. Pour certaines plages d'orientation dites plages astructurales, la structure est passive, les surfaces de cisaillement sont imposées par le tenseur σ, la résistance varie continûment avec β selon une loi représentée par les équations (20a) et (21) avec $m = n = 1$. Pour certaines plages d'orientation β, dites plages structurales, la structure est active, le glissement irréversible s'effectue selon des plans structuraux. Le critère, caractérisant le seuil, est alors constitué de deux lois physiques, relatives, chacune d'elles, au mécanisme de déformation irréversible apparaissant au seuil : La loi de type continu caractérisée par les équations (20) et (21) avec $m = n = 1$ et des rapports d'anisotropie B/A et D/C relativement faibles (anisotropie douce) et la loi caractérisant le glissement structural définie à partir d'une loi élémentaire de forme linéaire :

$$\tau = \overline{\tau}_0 + \sigma \tan \overline{\Phi} \tag{19b}$$

$\bar{\tau}_0$ et $\bar{\Phi}$ étant des constantes physiques caractéristiques du cisaillement structural, sur S. L'application d'un état de contrainte isotrope suggère que : $\bar{\tau}_0 \cot \bar{\Phi} = \tau_0 \cot \Phi$. Le critère correspondant (à cette loi élémentaire) prend la forme :

$$\sigma_1 + H = \bar{K}_p \, (\sigma_3 + H) \tag{20b}$$

à condition d'envisager un coefficient $\bar{K}_p$ (analogue au coefficient de pression passive du cas isotrope) dépendant de β, valable uniquement pour la plage de glissement structural, défini par :

$$\bar{K}_p = \frac{\sin(2\beta + \chi\bar{\Phi}) + \sin\bar{\Phi}}{\sin(2\beta + \chi\bar{\Phi}) - \sin\bar{\Phi}}$$

($\chi = 1$, cisaillement dextre, $\chi = -1$, cisaillement senestre).

Le critère global est donc caractérisé par l'ensemble de ces deux lois anisotropes (20a) et (20b). Souvent même on est conduit à négliger l'anisotropie continue dans la plage astructurale en regard de la violente anisotropie introduite par le glissement structural, c'est-à-dire à considérer B = D = 0. L'ensemble des deux lois (Fig. 19) comporte alors un critère isotrope et un critère anisotrope correspondant chacun à un mécanisme distinct. Dans la plage astructurale la résistance est constante d'où le nom de courbes à "plateaux" donné aux courbes de résistance du type de la figure 16c.

Dans les deux cas, d'anisotropie continue ou discontinue, on observe qu'une croissance de la pression isotrope entraîne une décroissance de l'anisotropie de seuil, définie comme le rapport entre résistance minimale et maximale.

8.2.2. *Mécanismes de la déformation plastique*

Les lois de la déformation plastique dépendent donc de la pression isotrope et de l'orientation β (Fig. 20), les mécanismes de déformation font, en outre, apparaître des instabilités et des hétérogénéités des champs de déformation.

a) Le glissement astructural s'effectue pour des structures anisotropes dont la surface des résistances est caractérisée par un ellipsoïde et également pour des structures planaires schisteuses pour des valeurs de β extérieures à la plage structurale. L'orientation des surfaces de glissement est imposée par le tenseur des contraintes (Fig. 20a).

b) Le glissement structural s'effectue pour des structures planaires (dont la surface S$'$ est caractérisée par une quartique) dans la plage structurale. Aux faibles pressions de confinement p_c un seul plan de cisaillement apparaît ; lorsque p_c croît, au contraire, apparaissent un grand nombre de plans de glissement, la déformation est alors du type glissement simple, elle s'apparente au

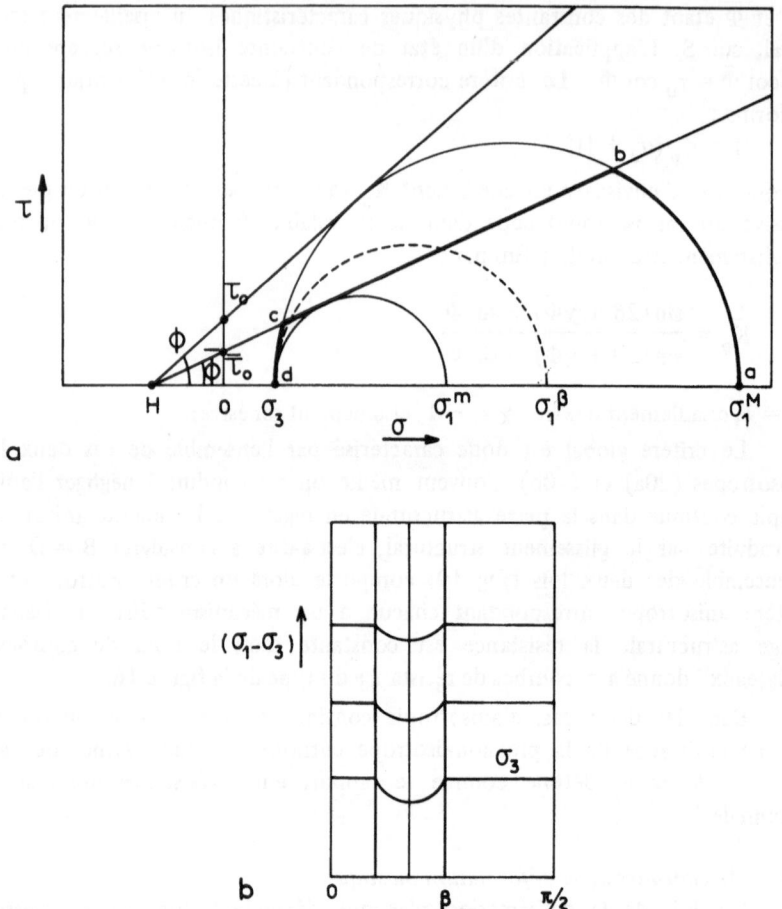

Figure 19. – *Anisotropie discontinue.*
    a : lois élémentaires de cisaillement, structural sur les plans S $(\bar{\tau}_0$ et $\Phi)$ et astructural sur un plan distinct de S $(\tau_0$ et $\Phi)$. La résistance varie entre σ_1^m et σ_1^M, le lieu des pôles des cercles de Mohr décrit a b c d lorsque β varie de 0 à $\pi/2$; b : résistance en fonction de β pour différentes valeurs de σ_3. Les plages structurales et astructurales.

glissement plastique des métaux [21]. Dans ce mécanisme, qui met en jeu une relation entre le tenseur contrainte σ et une transformation rotationnelle T, la structure tourne au cours de la déformation plastique (analogie avec le glissement cristallographique). Les tenseurs contraintes σ et déformation incrémentale ε ne sont plus nécessairement coaxiaux. Cobbold [10] met en évidence, dans le cas de la déformation plane, pour une structure orthorhombique, un tenseur d'anisotropie C_{mnij}, reliant σ et ε, caractérisé par deux valeurs principales N_1 et Q_1 telles que la non-coaxialité s'exprime par :

$$\tan 2\alpha = \frac{Q_1}{N_1} \tan 2\gamma \qquad (22)$$

(α, γ = orientations des tenseurs σ et ε).

c) Des phénomènes d'instabilité de la contraction, dans le cas du glissement structural, conduisent à des mécanismes de déformation plastique hétérogène tels que le plissement ou le pliage (kind-band). Le plissement est observé dans les milieux stratifiés, notamment les stratifiés hétérogènes (séries rythmiques) constitués de couches alternées de rigidités différentes, ainsi que dans les milieux ayant acquis une schistosité par déformation plastique tectonique antérieure. Le plissement (Fig. 20c) résultant d'un mécanisme de flambement, s'effectue en trois stades : déformation homogène, acquisition d'une longueur d'onde, enfin déformation homogène d'un milieu ondulé avec accroissement des amplitudes. Le pliage (Fig. 20d) se caractérise par une localisation de la déformation plastique, lorsque la rotation de la structure dans le glissement plastique est empêchée par des effets de parois (couches rigides des milieux naturels, ou plateaux de presses au laboratoire). Le rac-

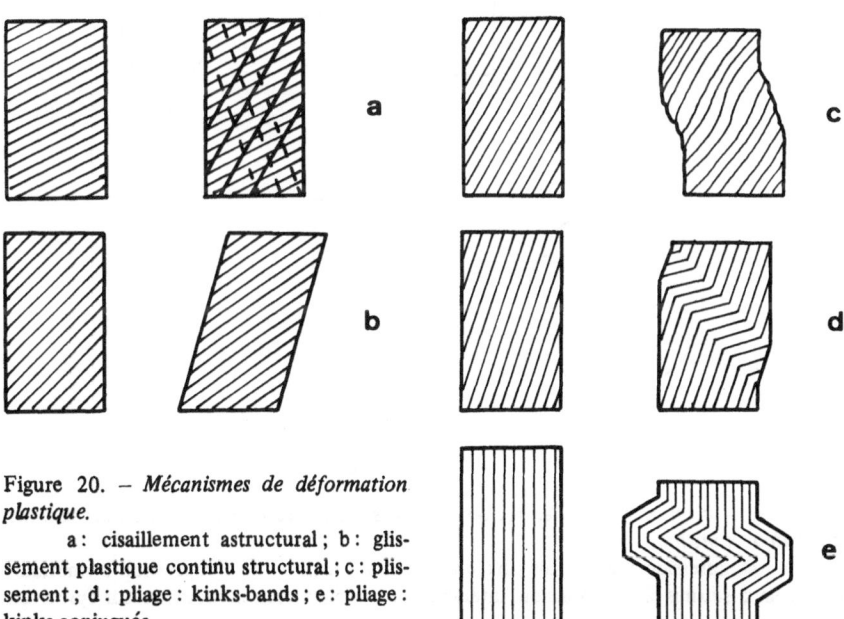

Figure 20. — *Mécanismes de déformation plastique.*

a : cisaillement astructural ; b : glissement plastique continu structural ; c : plissement ; d : pliage : kinks-bands ; e : pliage : kinks conjugués.

courcissement s'effectue par rotation locale de la structure, dans la kink-zone qui se développe au cours de la déformation finie. Pour $\beta = 0$ on peut obtenir (Fig. 20c) des kinks conjugués.

9.   La rupture fragile

A l'étage structural "faible pression-basse température", le comportement d'une roche devient fragile, il se manifeste par apparition de fissures intra- ou inter-granulaires qui au cours de la déformation inélastique (cf. §10) se relient et conduisent à des surfaces de rupture.

Pour un milieu pseudo-isotrope, à l'échelle macroscopique, les surfaces de rupture extensives sont globalement normales à l'extension principale majeure, elles s'orientent, en champ homogène, selon des surfaces isostatiques (Fig. 21). Une compression monoaxiale produit une rupture en colonnette (un degré de liberté pour la surface de rupture), en compressions biaxiales des

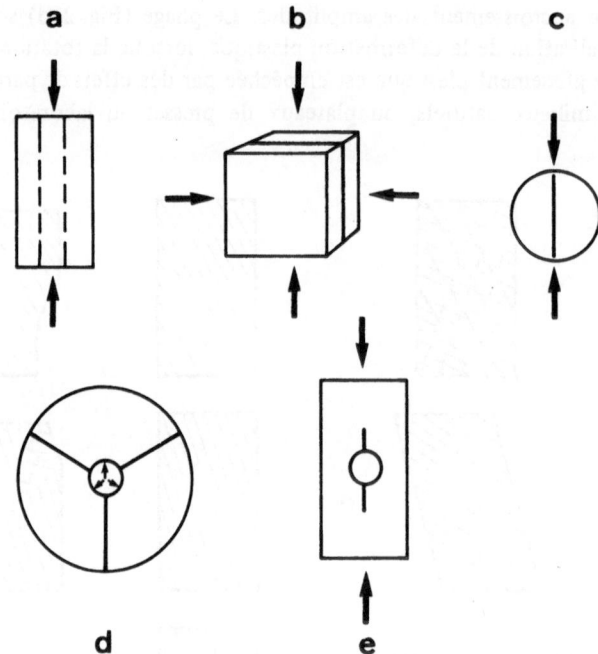

Figure 21. – *Rupture fragile des roches isotropes: ruptures extensives orientées suivant l'isostatique mineure.*

a : compression monoaxiale ; b : compression biaxiale ; c : compression diamétrale ; d : anneaux soumis à une pression intérieure ; e : compression monoaxiale sur éprouvette excavée.

plaques se fissurent parallèlement à leurs grandes faces, en traction indirecte, selon la technique de l'essai brésilien (compression diamétrale) la rupture est une surface axiale. Des anneaux de Granites soumis à une compression intérieure [42] se rompent suivant des plans radiaux (étoile à trois branches). Des éprouvettes excavées [43] se fissurent dans la direction normale à la fibre tendue de l'éprouvette. Jaeger [20] propose divers critères de rupture fragile, notamment le critère d'extension maximale : $\epsilon_1 \max = \epsilon_0$. Les ruptures fragiles de cisaillement sont orientées par l'angle μ (cf § 3.2) et se manifestent par l'apparition d'un plan unique de rupture. Ces cisaillements sont expliqués par une préfissuration isotrope.

En milieu anisotrope (Fig. 22) les mécanismes de rupture fragile dépendent du caractère actif ou passif de la structure. Les essais de compression (Fig. 22a et b) sur des roches à structure planaire, telles que des schistes ardoisiers, effectués à

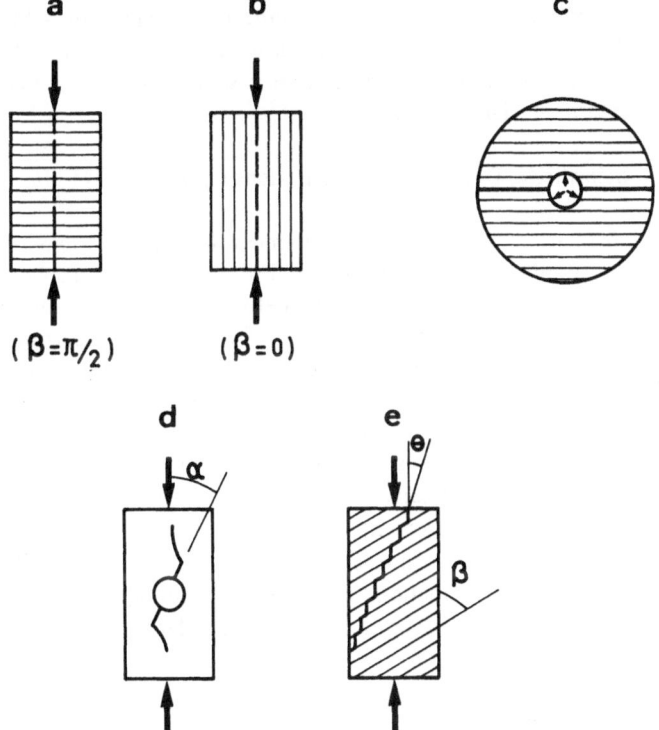

Figure 22. – *Rupture fragile des roches anisotropes.*
a : compression monoaxiale : rupture extensive astructurale ; b : compression monoaxiale : rupture extensive structurale ; c : anneaux, rupture extensive structurale ; d : compression monoaxiale sur éprouvette excavée préfissurée, bifurcation de la fissuration ; e : rupture mixte d'angle moyen $\theta < \beta$.

$\beta = 0$ et $90°$ et au voisinage de ces valeurs peuvent provoquer des ruptures extensives, structurales pour $\beta = 0$, astructurales pour $\beta = \pi/2$, elles font alors apparaître une nouvelle structure dans le milieu. Les essais de traction indirecte (Fig. 22c) sur anneaux provoquent une rupture orientée par la structure. Il convient de considérer pour ces milieux, un critère d'extension maximale structural de type $(\epsilon_y)\max = \epsilon'_0$, utilisé lorsque la rupture emprunte une surface structurale.

L'anisotropie d'une structure à fissuration initiale orientée, résultant d'une déformation fissurante antérieure, par exemple parallèle aux diaclases, peut influer sur la nouvelle fissuration (structure active). Une préfissuration dans une éprouvette excavée [43] réoriente la nouvelle fissure jusqu'à un angle de $20°$ environ (Fig. 22d). Ce mécanisme, connu sous le nom de "bifurcation de la fissuration" dépend de la densité et de la longueur moyenne de la fissuration initiale, il est reponsable des ruptures mixtes [12] telles que la surface de rupture est composée d'éléments structuraux et astructuraux. Finalement selon l'orientation de σ la nouvelle fissuration peut, soit emprunter l'ancienne (fissuration structurale) soit la réorienter (bifurcation) soit superposer une structure indépendante (structure initiale passive).

10. Inélasticité, Dilatance et Sollicitations Cycliques

La petite déformation n'est pas rigoureusement linéaire. Les expériences sur trois types de Granites [15] montrent que les modules initiaux évoluent, les modules tangents croissent avec la petite déformation axiale (sous l'effet de fermeture des pores et de serrage des grains). Par contre, l'anisotropie décroît lorsque la contrainte axiale passe de 0,06 à 1 kb, on observe (Fig. 23) une important diminution de q, ainsi qu'une variation de μ.

La structure cristalline des roches (mono ou polyminérales) entraine une hétérogénéité des déformations sous charges uniformes. Contraintes et déformations peuvent atteindre localement les seuils de rupture fragile pour certains minéraux composants favorablement orientés dans le milieu pseudo-isotrope. Ces mécanismes de fissuration intracristalline, éventuellement intercristalline, s'accompagnent d'une déformation, à dilatation volumique positive, qui se superpose à la déformation élastique du milieu. La déformation totale s'écrit donc :

$$\epsilon_{ik} = e_{ik} + \gamma_{ik} \tag{23}$$

en notant ε la déformation totale, e la déformation élastique, et γ la déformation inélastique, dilatante. Le premier invariant γ_{ii} de γ est positif, il caractérise la dilatance. Cette déformation inélastique comporte une partie réversible par fermeture des fissures qui viennent d'être créées, et une partie irréversible car

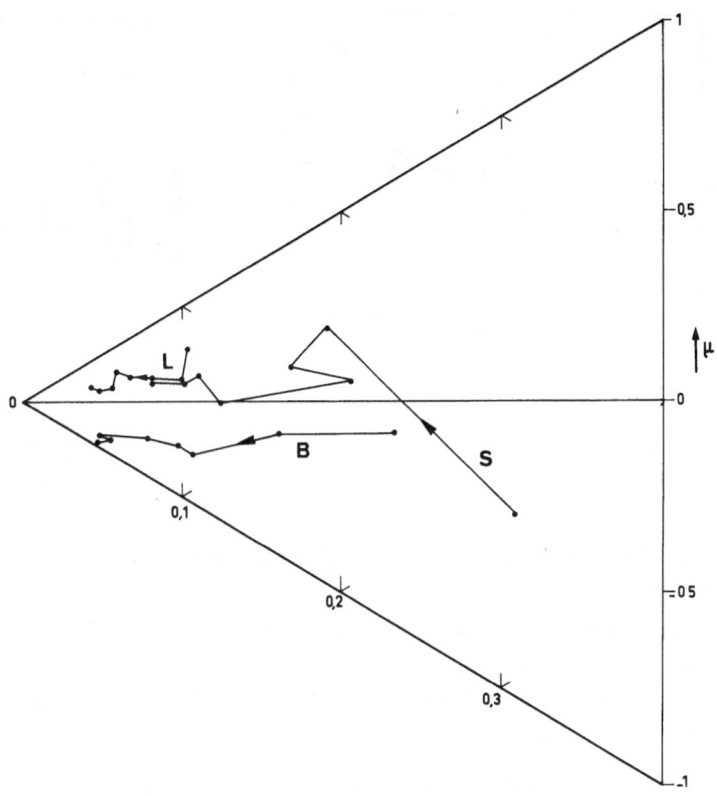

Figure 23. – *Inélasticité. Evolution de l'anisotropie élastique avec la déformation.* (d'après Douglas et Voight [15]) (S = granite de Stantead, B = granite de Barre, L = granite de Laurentian).

ces fissures refermées, à lèvres juxtaposées, entraînent une déformation permanente. La dilatance est donc la manifestation de l'apparition d'une structure de microfissuration associée à la structure initiale.

La dilatance apparaît après un seuil de contrainte, le seuil de dilatance (Fig. 24) caractérisé, pour une roche pseudo-isotrope, par une loi $f_1(\sigma_1, \sigma_2, \sigma_3) = 0$, représentée dans l'espace des contraintes [9] [40] par une surface D intérieure à la surface de rupture fragile F. Suivant la nature de la roche cette surface D est plus ou moins proche de F. En compression monoaxiale par exemple [38] le seul rapporté à la résistance à la rupture est de 30 % pour un Granite et nul pour un Marbre.

La loi de la dilatance est caractérisée par une relation entre l'évolution de la déformation dilatante et la déformation élastique. Nur [28] relie la défor-

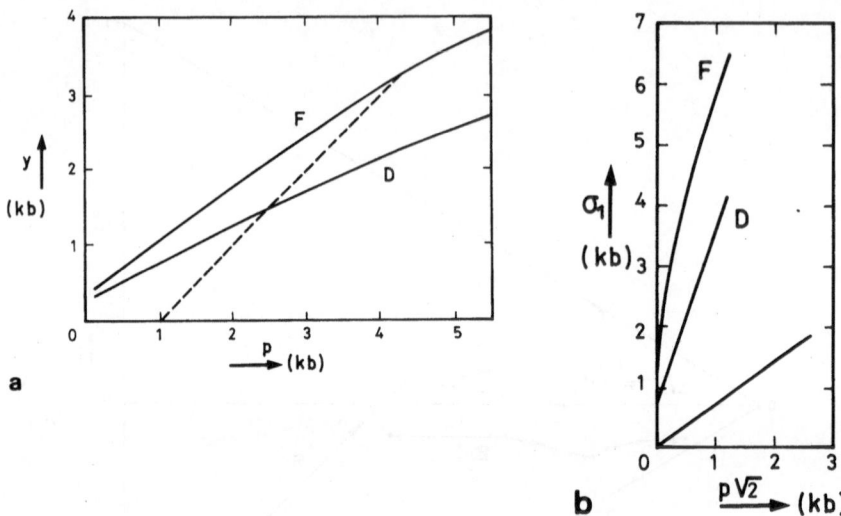

Figure 24. – *Seuil de dilatance.* (D = *surface limite de dilatance,* F = *surface limite de rupture fragile*).

a : granodiorite (d'après Cherry, Schock et Sweet [9]) ($p = \dfrac{\sigma_1 + \sigma_3}{2}$, $Y = \dfrac{3}{4}(J'_2)^{1/2}$,

J'_2 = second invariant du déviateur de σ_{ik}) ; b : granite du Sidobre (d'après Saint-Leu et Sirieys [40]) (p = pression de confinement, σ_1 = contrainte axiale).

mation dilatante au tenseur contrainte (c'est-à-dire au tenseur déformation élastique) par une relation, dans le cas biaxial, de la forme :

$$\gamma = (1 + K) \delta I_2^{n/2} \tag{24}$$

où $\gamma = \gamma_{ii}$, I_2 est le second invariant de σ_{ik}, δ et K des constantes. Cette équation se ramène dans le cas monoaxial à :

$$\gamma = \delta \tau^n \tag{24'}$$

l'exposant n dans (24) et (24′) étant égal à 2 pour la dilatance microfissurale. Les résultats expérimentaux sur un Granite et un Marbre [38] sont en accord avec cette théorie, les courbes de dilatance sont assimilables à des paraboles.

La déformation microfissurante γ comporte une partie déviatoire γ'_{ik} dont l'orientation dépend de l'orientation de σ_{ik} et éventuellement de la structure lorsqu'elle est active. Ce mécanisme de dilatance est donc un mécanisme anisotrope [18] [39], dont l'anisotropie évolue au cours de la déformation.

La stabilité des mécanismes de déformations irréversibles a été analysée expérimentalement sous chargements cycliques (Fig. 25) :

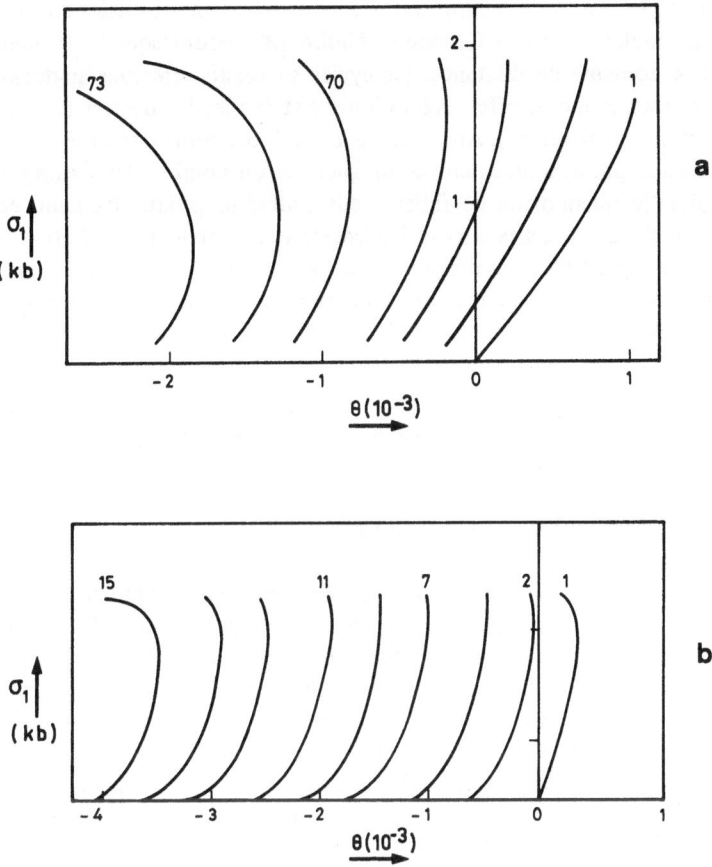

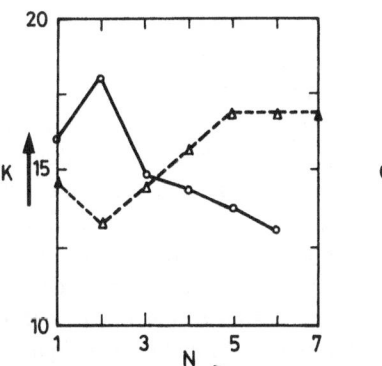

Figure 25. – *Sollicitations cycliques.*

a et b : d'après [39] : évolution des courbes de dilatance au cours de cycles de chargement – déchargement monoaxiaux – (θ = dilatation volumique, σ_1 = contrainte axiale) – a : granite, 73 cycles – b : marbre, 15 cycles ; c : évolution de l'anisotropie des déformations irréversibles sous cycles de pression isotrope, diatomite (d'après Allirot et Boehler [2]), 7 cycles.

Lors des essais de compression-décompression monoaxiales sur des roches microfissurables telles que Granite et Marbre [39] pour lesquelles le premier test atteint le domaine de dilatance, les cycles successifs effectués au-dessous d'un seuil ont tous le même effet, celui d'ouvrir et fermer les fissures (dit parfois de "respiration des fissures"). Au-dessus de ce seuil par contre, chaque cycle accroît la fissuration par accroissement de longueur et/ou nombre des fissures, la roche subit alors le phénomène de fatigue, elle s'affaiblit progressivement, en même temps que s'accroît la dilatance et l'anisotropie de la structure induite.

Pour des cycles sous pression isotrope effectués sur une roche poreuse, la Diatomite [2] : Sous une pression inférieure au seuil de modification de structure, la variation de volume (la contractance) se stabilise dès le 5° cycle, avec un coefficient d'anisotropie constant, voisin de 17. Sous pression supérieure au seuil, cette variation de volume croît avec le nombre de cycles (elle passe de 12 à 22,7 % du premier au sixième cycle), avec un coefficient d'anisotropie qui se stabilise vers 13,7.

11. Corrélations entre anisotropies

D'autres lois physiques anisotropes ont été étudiées, notamment la vitesse du son (pour les roches schisteuses les vitesses des ondes longitudinales sont plus élevées dans la direction de la schistosité que normalement à cette schistosité) et la susceptibilité magnétique. Il était alors naturel de comparer ces anisotropies, c'est-à-dire d'effectuer des corrélations entre les caractères d'anisotropie de différentes lois physiques. Deux cas de corrélations sont examinés ici, Elasticité-Résistance et Dilatation thermique-Susceptibilité magnétique.

Les anisotropies élastiques et de résistances sont caractérisées par les surfaces des modules et des résistances qui, pour les roches schisteuses, sont des quartiques présentant des minimums très accusés. La corrélation [34] a porté sur les orientations et les excentricités des ellipsoïdes circonscrits à ces quartiques. Ils ont un axe commun confondu avec la normale Z à la schistosité ; dans le plan de schistosité, par contre, ils présentent une non-coaxialité, les écarts pouvant atteindre des valeurs importantes. Mais surtout ces ellipsoïdes circonscrits ne sont pas de même nature, ils sont de type allongé selon Z pour la résistance et aplati pour l'élasticité.

La susceptibilité magnétique qui relie l'intensité d'aimantation à celle du champ magnétique, est caractérisée par un tenseur symétrique du second ordre noté ψ_{ik}. La corrélation entre anisotropie de susceptibilité magnétique et de dilatation thermique a été effectuée [11] sur des échantillons de roches calcitiques. Pour le Calcaire de Marquise (comme déjà noté en dilatation thermique) l'anisotropie de ψ_{ik}, définie par $a = (\psi_1 - \psi_3)/(\psi_1 + \psi_2 + \psi_3)/3$, croît avec la déformation plastique (elle varie de 0,5 à 2,4 % lorsqu'est effectuée une déformation plastique sous $p_c = 5$ kb et $\sigma_1 - \sigma_3 = 5$ kb). D'autre part, les essais effectués sur 7 roches calcitiques déformées naturellement montrent des écarts angulaires entre les directions principales de ψ et ε (les meilleures coïncidences étant comprises entre 2 et 24°). En outre, il n'y a identité entre même type d'ellipsoïde, allongé ou aplati, que dans 4 cas sur 7, sans qu'il y ait pour autant coïncidence entre axes majeurs ou mineurs. Le fait que les massifs d'où sont prélevés ces échantillons ont été affectés par plusieurs phases tectoniques pourrait être, selon les auteurs, un élément d'explication de ces écarts et inversions d'axes.

12. Conclusions

L'anisotropie mécanique des roches est étroitement liée à leurs structures qui résultent du mode de formation et des déformations tectoniques. Le caractère actif ou passif d'une structure est un élément essentiel de cette anisotropie, il conditionne les divers mécanismes de déformation plastique (le rejeu des failles qui provoque les séismes en est un exemple naturel).

Les anisotropies ont été quantifiées par des quantités adimensionnelles qui représentent les écarts réduits par rapport aux cas isotropes, elles ont permis de préciser l'ampleur des phénomènes et de comparer divers types de roches. On a observé notamment que les roches à structures feuilletées, ayant subi d'importantes déformations plastiques, présentent un cas plus caractéristique d'anisotropie. Des corrélations ont pu être effectuées entre anisotropies des différentes lois physiques, sous les aspects direction et intensité.

Actuellement, des recherches actives sur l'anisotropie mécanique des roches sont engagées notamment sur l'étude de la transition ductile-fragile, sur l'évolution de l'anisotropie avec la déformation plastique et enfin sur les lois de comportement, dites lois phénomènologiques, qui doivent rendre compte des divers mécanismes de déformation.

Annexe — Les coefficients d'anisotroqie

1) Lorsque l'anisotropie est caractérisée par un tenseur symétrique du second ordre a_{ik} (tel que ϵ_{ik} ou ψ_{ik}) on définit les coefficients d'anisotropie k et μ à l'aide des invariants du tenseur et de son déviateur, de composantes principales (a_1, a_2, a_3) et (a_1', a_2', a_3') :

$$I_1 = a_1 + a_2 + a_3$$

et $$J_2' = \frac{1}{6}[(a_1 - a_2)^2 + (a_2 - a_3)^2 + (a_3 - a_1)^2] = \frac{1}{2}(a_1'^2 + a_2'^2 + a_3'^2)$$

par les relations :

$$k = \frac{J_2'^{1/2}}{\dfrac{I_1}{3}} \tag{25a}$$

$$\mu = \frac{2\,a_2 - a_1 - a_3}{a_1 - a_3} = \frac{3\,a_2'}{a_1' - a_3'} \tag{25b}$$

k caractérise l'intensité de l'anisotropie et μ (paramètre de Lode) sa phase. Sur la figure 11 est reportée la quantité $q_1 = \dfrac{2}{\sqrt{3}}k$. Le coefficient k peut être aussi visualisé dans le plan de Mohr par un angle φ_a tel que

$$\sin \varphi_a = \frac{(a_1 - a_3)/2}{(a_1 + a_2 + a_3)/3}. \quad \text{Pour } \mu = 0, \quad k = \sin \varphi_a,$$

pour $\mu = \pm 1$, $k = \dfrac{2}{\sqrt{3}} \sin \varphi_a$.

Le taux d'anisotropie de susceptibilité [11] utilisé, égal à

$$(\psi_1 - \psi_3)/(\psi_1 + \psi_2 + \psi_3)/3, \quad \text{varie entre } k\sqrt{3} \text{ et } 2\,k.$$

2) Pour les Modules élastiques et les Résistances l'anisotropie est imagée par un ellipsoïde ou une quartique dont on étudie l'ellipsoïde circonscrit et caractérisée par les coefficients q et μ (coefficients d'anisotropie logarithmiques) définis par les équations (26).

X, Y, Z = demi-axes de l'ellipsoïde (avec $X \geqslant Y \geqslant Z$)

$\mathbf{x}$, y, z   = logarithmes décimaux de X, Y, Z.

t = (x + y + z)/3

x', y', z' = x − t, y − t, z − t (quantités adimensionnelles).

α, β tels que $3\alpha = x' - y'$ et $3\beta = y' - z'$ sont positifs.

$$q = 2(\alpha^2 + \beta^2 + \alpha\beta)^{1/2} \tag{26a}$$

$$\mu = (\beta - \alpha)/(\beta + \alpha) \tag{26b}$$

Ces coefficients q et μ, sans dimension, appelés intensité et phase de l'anisotropie, permettent de donner une image de l'anisotropie dans un diagramme ternaire, de comparer les roches et, pour une même roche, différentes lois anisotropes.

Les coefficients utilisés par les différents auteurs s'expriment en fonction de ces quantités (q, μ) ou (α, β) :

$$a_M' a_t \ [32]: \ \log a_M = 3(\alpha + \beta)$$

$$\log a_t = 2\alpha + \beta$$

$$R_a' \frac{x}{Z} \ [15]: \ R_a = a_t (1 - 1/a_M)$$

$$\frac{x}{z} = a_M$$

$$K[35]: K = \frac{Z}{X} = \frac{1}{a_M} \quad \text{dans le cas} \quad \mu = 1$$

REFERENCES

[1] ALLIROT D. — "Contribution à l'étude de l'anisotropie des déformations et de la rupture d'une roche stratifiée". *Thèse Spéc.*, Univ. Grenoble, 1976.

[2] ALLIROT D. et J.P. BOEHLER. — "Evolution de l'anisotropie d'une roche sous compression cyclique isotrope". *Bull. Acad. Pol. Sc.*, Sér. Sc. et Tech. XXIV, 9, (1976) : 405-409.

[3] ALLIROT D., J.P. BOEHLER et A. SAWCZUK. — "Irreversible déformations of an anisotropic rock under hydrostatic pressure". *Int. J. Rock. Min. & Geomech. Abstr.*, 14 (1977) : 77-83.

[4] AKAI K., K. YAMAMOTO et M. ARIOKA. — "Recherche expérimentale sur l'anisotropie des schistes cristallins". *C.R. 2° Cong. S.I.M.R.*, Belgrade, 3.26 (1970).

[5] ATTEWELL P.B. et M.R. SANDFORD. — "Intrinsic shear strength of a brittle. anisotropic rock – I. Experimental and mechanical interpretation". *Int. J. Rock Mech. Min. Sci. Geomech. Abstr.*, 11 (1974) : 423-430.

[6] BOEHLER J.P. et A. SAWCZUK. — "Equilibre limite des sols anisotropes". *Journal de Mécanique*, 9, 1 (1970) : 5-33.

[7] BOEHLER J.P. — "Contributions theoriques et expérimentales à l'étude des milieux plastiques anisotropes". *Thèse*, Univ. Grenoble, 1975.

[8] CHAYE D'ALBISSIN M. (a), VOUILLE G. et P. HABIB (b), M. DAYRE et P. SIRIEYS (c). — "Rapports d'études d'échantillons de Roches". *Action Concertée "Mécanique des Roches" D.G.R.S.T.*, B.R.G.M. 6-8 Rue Chasseloup Laubat 75 737 Paris cédex 15, 1965-67.

[9] CHERRY J.Th., R.N. SCHOCK and J. SWET. — "A theoretical model of the dilatant behavior of a brittle rock". *Pure and Applied Geophysics*, 113 (1975) : 183-196.

[10] COBBOLD P. – "Mechanical effects of anisotropy during large finite deformations". *Bull. Soc. Géol. Fr.*, (7) **XVIII**, 6 (1976) : 1 497-1 510.

[11] DALY L. et M. D'ALBISSIN. – "Corrélation entre les anisotropies de susceptibilité magnétique et de dilatation thermique des roches ; application en structurologie". *C.R. Acad. Sc.*, Paris, 267 (1968) : 473-476.

[12] DAYRE M. et P. SIRIEYS. – "Anisotropie des modules élastiques et des résistances à la rupture des roches métamorphiques". *C.R. Acad. Sc.*, Paris, 260 (1965) : 4 440-4 443.

[13] DAYRE M., J.L. DESSENNE et B. WACK. – "Variations locales et moyennes de la densité d'échantillons de craie soumis à l'essai triaxial". *C.R. Cong. S.I.M.R.*, Belgrade, 2-25 1970.

[14] DONATH F.A. – "Effects of cohesion and granularity on deformational behavior of anisotropic rock". *Geol. Soc. Amer. Memoir*, 135 (1972) : 95-128.

[15] DOUGLAS P.M. and B. VOIGHT. – "Anisotropy of granites : A reflection of microscopic fabric". *Geotechnique*, 19, 3 (1969) : 376-398.

[16] GERMAIN P. – *Cours de Mécanique des Milieux Continus. Tome 1. Théorie Générale*, Paris : Masson et Cie, 1973, pp. 1-417.

[17] GOGUEL J. – "Introduction à l'étude mécanique des déformations de l'écorce terrestre". *Mém. de la Carte Géol. de France*, 1948.

[18] HADLEY K. – "Azimuthal variation of dilatancy". *Jour. of Geophys. Research*, 80, 35 (1975) : 4 845-4 850.

[19] JAEGER J.C. – "Shear failure of anisotropic rocks". *Geol. Magazine*, 97 (1960) : 65-72.

[20] JAEGER J.C. – *Elasticity, Fracture and Flow with Engineering and Geological Applications.* Methuen & Co. Ltd, 1969.

[21] JAOUL B. – *Etude de la Plasticite et Application aux Métaux.* Paris : Dunod, 1965, pp. 1-600

[22] LAURENT D. et J.F. ALLARD. – "Rapports de prélevements" – *Action Concertée "Mécanique des Roches" D.G.R.S.T.* B.R.G.M. 6-8 rue Chasseloup Laubat 75 737 Paris cédex 15, 1965.

[23] Mc GILL G.E. et J.A. RANEY. – "Experimental study of faulting in an anisotropic, inhomogeneous, dolomitic limestone" *Geol. Soc. of Amer. Bull.*, 81, 10 (1970) : 2 949-2 958.

[24] Mc LAMORE R. et K.E. GRAY. – "The mechanical behavior of anistropic sedimentary rocks". *Transactions of the ASME* February 1967, pp. 62-73.

[25] MASURE Ph. – "Comportement mécanique des roches à anisotropie planaire discontinue" *C.R. 2° Cong. S.I.M.R.*, Belgrade, 1-27, 1970.

[26] MATTAUER M. – *Les Déformations des Matériaux de l'Ecorce Terrestre.* Paris : 1973, Herman, pp. 1-493.

[27] MELLO-MENDES F. – "About the anisotropy of uniaxial compressive strength in schistose rocks". *Symp. S.I.M.R.*, Nancy, 11-13, 1971.

[28] NUR J.F. – "A note on the constitutive law for dilatancy". *Pure and Applied Geophysics*, 113 (1975) : 197-206.

[29] NYE J.F. – *Propriétés Physiques des Cristaux.* Paris : Dunod, (1961), 1-344.

[30] OLSZAK W. et W. URBANOWSKI. – "Quelques problèmes fondamentaux relatifs à la théorie des milieux élastoplastiques anisotropes et non-homogènes". *Bull. Centre d'Etudes de Rech. et d'Essais Scien. Génie Civil*, X, Liège, 1960.

[31] PELLIGRINO A. – "Mechanical behaviour of soft rocks under high stresses". *Proc. 2e Congress I.S.R.M.*, Belgrade, 3-25, 1970.

[32] PERES-RODRIGUES F. – "Anisotropy of granites. Modulus of elasticity and ultimate strength ellipsoïds, joint systems, slopes attitudes, and their correlations". *Proc. 1er Congress I.S.R.M.*, Lisbon, I (1966): 721-731.

[33] PERES-RODRIGUES F. – "Anisotropy of rocks. Most probable surfaces of the ultimate stresses and of the moduli of elasticity". *Proc. 2^e Congress S.I.M.R.*, Belgrade, **1-20**, 1970.

[34] PERES-RODRIGUES F. et L. AIRES-BARROS. – "Anisotropy of Endogenetic Rocks. Correlation between micropetrographic index, ultimate strength and Modulus of Elasticity Ellipsoïds". *Proc. 2^e Congress S.I.M.R.*, Belgrade, 1-23, 1970.

[35] PINTO J.L. – "Stress and strain in an anisotropic-orthotropic body". *Proc. 1er Congress I.S.R.M.*, Lisbon, 7 (1966): 625-635.

[36] PINTO J.L. – "Deformability of schistous Rocks". *Proc. 2e Congress I.S.R.M.*, Belgrade, I, 2-30, 1970.

[37] PRICE N.J. – *Fault and joints development in brittle and semibrittle rocks.* Pergamon Press, (1966), 1-176.

[38] SAINT-LEU C. et P. SIRIEYS. – "Déformation des roches fragiles sous champs de contraintes homogènes et hétérogènes". *C.R. 2e Cong. S.I.M.R.*, Belgrade, 2-18. 1970.

[39] SAINT-LEU C. et P. SIRIEYS. – "La fatigue des roches". *Symp. S.I.M.R.*, Nancy, **11-18**, 1971.

[40] SAINT-LEU C. et P. SIRIEYS. – "Inélasticité et rupture du granite sous contraintes triaxiales". *C.R. Acad. Sci. Paris,* **276**, Série A (1973) : 817-820.

[41] SAINT-LEU C., LERAU J. et P. SIRIEYS. – "Mécanisme de rupture des schistes de Lacaune (Tarn). Influence de la pression istrope". *Bull. Soc. Fr. Miné. Cristall.*, 101 (1978): 437-442.

[42] SIRIEYS P. – "Contribution à l'étude des lois de comportement des structures rocheuses". *Thèse,* Univ. Grenoble, 1966.

[43] SIRIEYS P. et C. SAINT-LEU. – "Expériences relatives à l'influence du gradient des contraintes sur les lois de la fissuration des roches". *Revue de l'Industrie Minérale,* N° spécial Juillet 1969, 1-9.

[44] SIRIEYS P. et P. DEBAT. – "Les paléodéformations tectoniques des roches". *Problèmes de Rhéologie et de Mécanique des sols. PWN.* Varsovie: Editions Scientifiques de Pologne, (1977), 419-425.

[45] TURNER F.J. and L.E. WEISS. – *Structural Analysis of Metamorphic Tectonics.* New-York: Mc Graw-Hill, (1963), 1-545.

Abréviations utilisées dans la Bibliographie

Réf. 1. Thèse de Spécialité.

Réf. 2. Bulletin de l'Académie Polonaise des Sciences.

Réf. 3, 5. International Journal of Rock Mechanics and Mining Sciences & Geomechanics.

Réf. 4, 13, 25, 38. Rapport du 2e Congrès de la Société Internationale de Mécanique des Roches.

Réf. 10. Bulletin de la Société Géologique de France.

Réf. 11, 12, 40. Compte-Rendus de l'Académie des Sciences.

Réf. 14, 23. Geological Society of America Bulletin.

Réf. 17. Mémorial de la Carte Géologique de France. Imprimerie Nationale. Paris. 530 p.

Réf. 18. Journal of Geophysical Research.

Réf. 19. Geological Magazine.

Réf. 24. Transactions of the ASME. Journal of Engineering for Industry.

Réf. 27, 39. Fissuration des Roches. Symposium de la Société Internationale de Mécanique des Roches.

Réf. 30. Bulletin du Centre d'Etudes de Recherches et d'Essais scientifiques du Génie Civil.

Réf. 31, 33, 34, 36. Proceedings of the Second Congress of the International Society for Rock Mechanics.

Réf. 32, 35. Proceedings of the 1er Congress of the International Society for Rock Mechanics.

Réf. 41. Bulletin de la Société Française de Minéralogie et Cristallographie.

ABSTRACT

The mechanical anisotropy of rocks is connected to their structure, genetic and tectonic, acquired in the course of their history. Observations and experimental results studied in this paper refer, on the one hand, to the structural anisotropy genesis and, on the other hand, to the anisotropic characteristics of rocks in the present state. The studied laws of anisotropic behaviour refer to thermal strain, elasticity, elastic limit, plasticity, microcracking and failure. Comparisons on the anisotropy of several rocks are made by the aid of coefficients represented in ternary diagrams. Lastly, correlations between different laws of anisotropic behaviour of the same rock are examined.

Modélisation de l'Anisotropie par un Complexe Stratifié de Corps Isotropes. Application au Comportement au-delà de la Limite Élastique

J. Goguel

Bureau de Recherches Géologiques et Minières, Paris, France.

1. Introduction

Je me propose d'étudier un modèle de corps anisotrope, constitué par l'alternance de couches parallèles de corps isotropes différents, parfaitement adhérents le long de leurs surfaces de contact.

Il est clair que ce modèle ne constitue qu'un cas particulier, et que d'autres formes d'anisotropie mécanique existent. Il suffit d'évoquer la matière cristalline, avec ses propriétés vectorielles discontinues, ou le cas d'un agrégat cristallin, dans lequel les orientations des grains cristallins ne seraient pas distribuées d'une manière uniformément aléatoire, mais avec une fréquence supérieure à la moyenne autour de certaines orientations privilégiées. Aussi bien les coefficients élastiques, que les conditions de rupture, reflètent cette anisotropie, J'évoquerai simplement le cas de certains granites, où les cristaux de feldspath font un angle petit avec un certain plan, qui est macroscopiquement un plan de rupture privilégié ("feuille" des carriers), parce que la rupture suit les clivages des cristaux de feldspath.

2. Régime élastique

Revenons au modèle stratifié. Aussi bien les contraintes que les déformations peuvent se définir de deux manières : soit à une échelle fine, en distinguant ce qui se passe à l'intérieur de chaque strate (et que nous noterons avec l'indice correspondant, ϵ_{ij}^a ou σ_{ij}^a), soit globalement en définissant contrainte ou déformation pour un élément de volume, grand par rapport à l'épaisseur des strates. Les composantes moyennes de la contrainte ou de la déformation sont alors les moyennes des valeurs relatives aux différentes strates, pondérées dans le rapport des épaisseurs relatives. Nous nous bornerons à supposer que ces épaisseurs relatives, α, c'est-à-dire la proportion de chaque constituant dans l'épaisseur sur laquelle porte la moyenne, reste stable, sans avoir à préciser autrement les épaisseurs individuelles des strates.

Si les axes x0y sont parallèles au plan de stratification (ce que nous supposerons toujours par la suite), on voit immédiatement que les trois composantes de la contrainte qui s'exerce sur le plan de stratification, σ_{33}, σ_{13} et σ_{23} sont égales pour les diverses strates, et égales à leur moyenne. Il en est de même pour les composantes de la déformations du plan de stratification, ϵ_{11}, ϵ_{22} et ϵ_{12}.

Pour les autres composantes, nous aurons à distinguer les valeurs individuelles pour chaque assise et la valeur moyenne :

$$\sigma_{ij}^{m} \quad \text{ou} \quad \epsilon_{ij}^{m} = \alpha\epsilon_{ij}^{a} + \beta\epsilon_{ij}^{b} + \ldots = \Sigma\alpha\epsilon_{ij} \tag{1}$$

Connaissant les coefficients de Lamé, λ^{a} et μ^{a} pour chaque assise, on trouve les composantes de la matrice C_{ij} donnant les composantes de la contrainte en fonction de celles de la déformation ($\epsilon_{11}, \epsilon_{22}, \epsilon_{33}, \epsilon_{23}, \epsilon_{13}, \epsilon_{12}$). Inversement, on calcule les coefficients S_{ij} donnant la déformation en fonction de la contrainte.

$$C_{11} = C_{22} = \Sigma\alpha(\lambda + 2\mu) + \left(\Sigma\frac{\alpha\lambda}{\lambda + 2\mu}\right)^{2} \Big/ \Sigma\frac{\alpha}{\lambda + 2\mu} - \Sigma\frac{\alpha\lambda^{2}}{\lambda + 2\mu};$$

$$C_{33} = \frac{1}{\Sigma\dfrac{\alpha}{\lambda + 2\mu}}; \tag{2}$$

$$C_{23} = C_{13} = \Sigma\frac{\alpha\lambda}{\lambda + 2\mu}\Big/ \Sigma\frac{\alpha}{\lambda + 2\mu}; \quad C_{12} = C_{11} - C_{66};$$

$$C_{44} = C_{55} = 2\Big/\Sigma\frac{\alpha}{\mu}; \quad C_{66} = 2\,\Sigma\alpha\mu.$$

$$S_{11} = S_{22} = \frac{1}{4}\left(\frac{1}{\Sigma\alpha\mu} + 1\Big/\Sigma\frac{\alpha\mu(3\lambda + 2\mu)}{\lambda + 2\mu}\right);$$

$$S_{33} = \Sigma\frac{\alpha}{\lambda + 2\mu} + \left(\Sigma\frac{\alpha\lambda}{\lambda + 2\mu}\right)^{2}\Big/ \Sigma\frac{\alpha\mu(3\lambda + 2\mu)}{\lambda + 2\mu}; \tag{3}$$

$$S_{23} = S_{13} = -\Sigma\frac{\alpha\lambda}{\lambda + 2\mu}\Big/4\Sigma\frac{\alpha\mu(3\lambda + 2\mu)}{\lambda + 2\mu}; \quad S_{12} = \frac{1}{2}\,(S_{11} - S_{66});$$

$$S_{44} = S_{55} = \frac{1}{2}\,\Sigma\frac{\alpha}{\mu}; \quad S_{66} = \frac{1}{2\Sigma\alpha\mu}.$$

On notera que ces coefficients ne dépendent que de cinq paramètres indépendants, quel que soit le nombre de couches différentes.

Ceci peut d'ailleurs se démontrer directement, compte tenu de la symétrie du milieu. Toutefois, le milieu stratifié ne représente pas le cas le plus général d'une telle symétrie — on s'en rend compte en comparant les valeurs relatives des coefficients C_{44} et C_{66}, et on peut envisager également le cas de fibres (parallèles à Oz), de natures différentes, et dont on admettra que les sections peuvent être suffisamment aléatoires pour que la symétrie de révolution soit respectée. Les composantes identiques pour les différents constituants sont alors σ_{11}, σ_{22} et σ_{12} pour la contrainte, ϵ_{33}, ϵ_{23} et ϵ_{13} pour la déformation. Pour les autres, la valeur moyenne s'obtient par une pondération dans le rapport des sections et on calcule — par élimination— les coefficients C_{ij} et S_{ij} caractérisant le comportement global.

$$C_{11} = C_{22} = \frac{1}{\Sigma \dfrac{\alpha}{\mu}} + \frac{1}{\Sigma \dfrac{\alpha}{\lambda + \mu}} \;;$$

$$C_{33} = \Sigma \frac{\alpha \mu (3\lambda + 2\mu)}{\lambda + \mu} + \left(\Sigma \frac{\alpha\lambda}{\lambda + \mu} \right)^2 \Big/ \Sigma \frac{\alpha}{\lambda + \mu} \;; \qquad (4)$$

$$C_{23} = C_{13} = \Sigma \frac{\alpha\lambda}{\lambda + \mu} \Big/ \Sigma \frac{\alpha}{\lambda + \mu} \;;\; C_{12} = \frac{1}{\Sigma \dfrac{\alpha}{\lambda + \mu}} - \frac{1}{\Sigma \dfrac{\alpha}{\mu}} \;;$$

$$C_{44} = C_{55} = 2\,\Sigma\,\alpha\mu \;;\; C_{66} = \frac{2}{\Sigma \dfrac{\alpha}{\mu}} = C_{11} - C_{12}$$

$$S_{11} = S_{22} = \frac{1}{4} \left[\left(\Sigma \frac{\alpha\lambda}{\lambda + \mu} \right)^2 \Big/ \Sigma \frac{\alpha\mu (3\lambda + 2\mu)}{\lambda + \mu} + \Sigma \frac{\alpha(\lambda + 2\mu)}{\mu(\lambda + \mu)} \right] \;;$$

$$S_{33} = 1 \Big/ \Sigma \frac{\alpha\mu (3\lambda + 2\mu)}{\lambda + \mu} \;; \qquad (5)$$

$$S_{23} = S_{13} = -\frac{1}{2} \Sigma \frac{\alpha\lambda}{\lambda + \mu} \Big/ \Sigma \frac{\alpha\mu (3\lambda + 2\mu)}{\lambda + \mu} \;;\; S_{12} = S_{11} - S_{66} \;;$$

$$S_{44} = S_{55} = 1/2\,\Sigma\,\alpha\mu \;;\; S_{66} = \frac{1}{2} \Sigma \frac{\alpha}{\mu} \,,$$

Dans un cas comme dans l'autre, il n'y a aucune difficulté, connaissant les composantes moyennes de la contrainte, à calculer les contraintes locales dans chacune des strates, ou des fibres. Il est clair que la pression moyenne sera en général différente pour les différents constituants.

Une application essentielle de l'élasticité est la propagation des ondes sonores. Si la longueur d'onde est grande par rapport à l'épaisseur des strates, on trouvera trois types d'ondes, correspondant à des déplacements particulaires sensiblement longitudinaux (onde P), transversal dans un plan passant par Oz (SV), transversal et perpendiculaire à Oz (SH) dont les vitesses dépendront de l'angle de la surface d'onde avec Oz.

Si la longueur d'onde était de l'ordre de l'épaisseur des strates, ou plus petite, il faudrait analyser les reflexions et réfractions sur les surfaces d'accollement, qui se traduiraient globalement par une certaine diffusion et une absorption, voire des interférences.

Une autre application est le problème de Boussinesq : Comment se répartissent les contraintes sous une charge ponctuelle ? Il y a des raisons de penser qu'elles sont plus concentrées dans un milieu stratifié, et moins dans un milieu fibré.

3. La rupture

Il serait théoriquement possible d'aborder l'étude de la rupture à partir de la répartition élastique des contraintes. Empiriquement, on sait que la limite de rupture varie souvent dans le même sens que les coefficients élastiques, mais il n'y a pas de relation définie. On ne peut donc guère envisager une discussion des cas où la rupture apparaît d'abord dans le constituant le plus raide, ou dans celui où les coefficients d'élasticité sont les plus faibles.

Cette discussion n'aurait d'ailleur guère de sens, car il est rare que la distribution des contraintes, déterminée par les lois de l'élasticité à partir d'un état naturel sans contrainte, règne jusqu'à la rupture. Très souvent, un certain fluage sous charge modifie cette distribution.

4. Distribution des contraintes par relaxation

Pour les formations géologiques, la notion d'état naturel sans contrainte n'a aucun sens. Les roches se sont formées dans le champ de pesanteur, qui est responsable de l'essentiel des contraintes. Suivre l'évolution de celles-ci, depuis la formation de la roche, en passant par toutes les transformations qui ont pu se produire, devient vite inextricable.

Dans cette longue évolution, un rôle essentiel est joué par le fluage, qui permet un relâchement des contraintes (relaxation). Cette relaxation peut être plus ou moins rapide, et donc plus ou moins complète, par rapport aux conditions qui règnent actuellement, c'est-à-dire aux contraintes déterminées par le relief.

Suivre une relaxation partielle des contraintes, dans le cadre changeant d'un relief que l'érosion modifie, et sous l'action de forces internes, connues seulement par leurs effets, serait inextricable. Tout ce que l'on peut envisager est d'admettre une relaxation totale des contraintes, qui peut être suivie d'une phase d'érosion assez rapide pour que le comportement des roches soit élastique.

J'ai montré ailleurs (Goguel, 1942, 1943) qu'on peut étudier l'évolution des contraintes à la suite du fluage en donnant de celui-ci une définition assez large ; on admettra qu'il ne produit que des déformations infiniment petites, du même ordre que les déformations élastiques, si bien qu'il ne modifie pas la géométrie ; pour que la loi du fluage soit permanente, on doit admettre qu'il se fait à volume constant. On admettra que, pour un corps isotrope, le tenseur de déformation par fluage est semblable au tenseur de contrainte, et que le coefficient de proportionnalité φ est fonction seulement de C, le deuxième invariant du déviateur étant C^2.

Moyennant ces hypothèses, qui sont très peu restrictives, on démontre que, en posant

$$\Phi(C) = \int_0^c \varphi(C)\, C\, dC \tag{6}$$

l'intégrale $I = \iiint \Phi\, dV$ étendue à tout le volume V où un fluage est possible, ne peut que diminuer, les contraintes étant astreintes à rester en équilibre avec les forces extérieures. En particulier, si on peut trouver une distribution des contraintes rendant I minimum, on peut admettre que c'est la limite vers laquelle tendra le fluage.

Si, parmi les conditions aux limites, en figuraient de géométriques (par exemple : appui contre une surface rigide), il faudrait prendre comme inconnues les contraintes sur cette surface, et les déterminer par la condition de rendre I minimum.

Cette loi variationnelle s'applique aussi bien à un ensemble formé de roches différentes, à condition de prendre dans le volume occupé par chacune d'elles, la fonction Φ correspondante. Il n'y a donc aucune difficulté à l'appliquer au modèle de roche anisotrope par stratification que nous étudions ici.

Le lecteur s'étonnera peut être d'une contradiction apparente, entre ce régime de la relaxation, et celui de la déformation plastique, qui sera examiné par la suite ; précisons que notre calcul ne prétend pas décrire l'évolution des contraintes au cours de la relaxation — ce qu'on pourra essayer de faire dans le modèle de la déformation plastique — mais seulement indiquer une condition globale (diminution de I), d'où résulte l'existence d'un état limite pour la distribution des contraintes, dont on n'est pas sûr qu'il soit effectivement atteint. C'est cet état limite que nous allons étudier.

Nous nous bornerons à envisager un volume limité de cette roche, pour lequel nous supposons les contraintes moyennes déterminées (en réalité, la relaxation peut entraîner aussi une redistribution à grande échelle des ccontraintes), et nous allons examiner comment la contrainte se partage entre les strates alternantes.

Ce partage ne dépend pas des coefficients élastiques, mais uniquement du rapport des vitesses de fluage, $\varphi_a(C^a)$ et $\varphi_b(C^b)$ pour les valeurs de contraintes qui seront effectivement réalisées. En effet, dans le volume que nous considérons, et où nous supposons la contrainte moyenne déterminée et uniforme, le partage de cette contrainte entre les strates doit être tel que :

$$\Sigma\alpha\Phi^a \text{ minimum, ou } \Sigma\alpha\Delta\Phi^a = 0 , \tag{7}$$

Δ désignant une variation résultant d'une modification de distribution de la contrainte entre les strates. Mais :

$$\Delta\Phi^a = \varphi_a(C^a) \times C^a \times \Delta C^a \tag{8}$$

5.   La rupture après redistribution des contraintes par relaxation

S'il existe une couche relativement mince (α petit), proche de sa limite de rupture, où le fluage soit relativement rapide (φ_a grand), la distribution des contraintes sera telle que C^a y soit petit ; ceci entraîne que σ_{11}^a et σ_{22}^a soient proches de σ_{33} et que σ_{12}^a soit petit. Mais le cisaillement sur le plan de stratification σ_{13} et σ_{23} est imposé, et c'est lui qui déterminera la valeur de C^a. La rupture succédant à la relaxation des contraintes ne peut être qu'un cisaillement des lits plastiques, parallèlement à la stratification.

Si la contrainte globale est telle que cette composante de cisaillement soit faible ou nulle, elle se réparti entre les strates, et si β est proche de l'unité, la contrainte dans la couche b est voisine de la contrainte globale, et la rupture dépendra des propriétés de cette couche, sans que l'orientation de la contrainte ait beaucoup d'effet.

Globalement, on observera donc une rupture selon deux modes : soit par cisaillement des lits tendres, qui dépend du cisaillement sur le plan de stratification, donc de l'orientation de la contrainte, soit par rupture des bancs épais, pour une contrainte dépendant peu de l'orientation.

Supposons maintenant que les lits minces (α petit), soient constitués par la roche dure, au fluage le plus lent. La contrainte dans les lits épais (β proche de l'unité) est proche de la contrainte moyenne. Pour la distribution des contraintes résultant de la relaxation, C^a est beaucoup plus grand que C^b (proche de C^m). Cela signifie que σ^a_{11} et σ^a_{22} seront très différents de σ_{33}, et σ^a_{12} important.

Dans le cas particulier où la contrainte σ_{33} est supérieure à σ^m_{11} et à σ^m_{22}, σ^a_{11} et σ^a_{22} peuvent être très petits, voire négatifs, et correspondre à une traction, à laquelle la roche peut ne pas résister : on observe effectivement la rupture par traction des lits isolés résistants (c'est le boudinage).

Si au contraire σ_{33} est la plus faible composante, σ^a_{11} et σ^a_{22} peuvent être très grands, et ces lits résistants peuvent flamber — contrairement à notre hypothèse de travail selon laquelle la contrainte moyenne est distribuée d'une manière uniforme. C'est un point sur lequel nous reviendrons.

6. La pression moyenne et son influence sur l'eau d'imprégnation

Il est clair que la pression moyenne $(\sigma_{11} + \sigma_{22} + \sigma_{33})/3$ est différente dans les différentes strates, ce qui était d'ailleurs déjà le cas en régime élastique (également, nous le verrons, en régime de déformation plastique). Si les roches sont poreuses, et imprégnées d'une eau à une pression hydraulique localement uniforme, qui peut tenir certains éléments en solution (calcite, silice, éventuellement gypse, etc.), son comportement, et en particulier l'éventualité de cristallisation, dépend de la pression moyenne. Par conséquent, la manière dont ces cristallisations se répartissent dans les différents lits dépend de la contrainte. Par exemple, s'il existe des lits durs, relativement minces, perpendiculairement à la pression maximale σ_{33}, on a vu que σ_{11} et σ_{22} étaient très faibles et pouvaient même correspondre à des tractions. Des recristallisations secondaires se produiront facilement dans les fentes ouvertes par de telles tractions.

7. Relaxation suivie d'érosion

Lorsque la surface supérieure est horizontale, et si les densités des roches sont uniformes, la distribution limite des contraintes par relaxation est facile à trouver : elle correspond à une pression hydrostatique, fonction de la seule profondeur.

Supposons que cet état ait été atteint, et que se produise une érosion assez rapide pour que le fluage ne joue pas, mais que la modification des contraintes se fasse élastiquement.

On obtiendra les nouvelles contraintes en ajoutant :

1) une pression hydrostatique proportionnelle à la profondeur,

2) l'effet élastique d'une traction normale, s'exerçant sur la nouvelle surface d'érosion, et égale à la pression hydrostatique initiale à la même profondeur sous la surface primitive.

Cette traction s'exerçant sur un ensemble stratifié produira des contraintes inégales dans les strates alerternantes, et plus fortes dans celles où les coefficients d'élasticité sont les plus élevés. Au total, au voisinage de la nouvelle surface d'érosion, il existe des tractions dans les bancs rigides, tractions sous l'effet desquelles ils se fissurent, comme on le constate effectivement. Un tel modèle peut être très utile, dans nombre de problèmes techniques de travaux publics en montagne.

8.   Ecoulement plastique : la réfraction de la schistosité

Nous n'avons envisagé jusqu'ici qu'un fluage infiniment petit, ne se traduisant que par des modifications dans la distribution des contraintes.

Lorsqu'on atteint la limite élastique apparente, la déformation plastique se produit, et elle modifie la texture de la roche ; cette modification se traduit par l'apparition d'une schistosité, dont le plan est perpendiculaire à la direction de compression maximale. Dans ce plan, on discerne parfois une linéation, suivant la direction d'extension maximale. On peut admettre que cette déformation est lente, et que la contrainte dépasse de peu le seuil de plasticité ; il est naturel de caractériser celui-ci par la valeur du 2e invariant du déviateur (ou sa racine carré).

Si les strates qui alternent ne sont pas d'une nature complètement différente, elles peuvent atteindre simultanément ce régime de déformation plastique, mais avec des valeurs du seuil de plasticité différentes (pour fixer les idées, indiquons qu'on peut trouver des rapports atteignant 2 à 3). Les contraintes sont donc différentes dans les strates qui alternent. Outre les trois composantes communes, il faut tenir compte de ce que la déformation des plans de stratification est identique pour les diverses strates. Le tenseur de vitesse de déformation est, dans chaque strate, semblable au tenseur du déviateur, mais avec un coefficient de proportionnalité indéterminé (il dépend du faible écart entre le déviateur de la contrainte, et la valeur correspondant au seuil).

De ces hypothèses résulte que les directions principales sont différentes dans les différentes strates, et donc également les directions de schistosité.

Cette particularité, ou réfraction de la schistosité (*step of the cleavage*, en anglais) a été observée depuis longtemps. Une théorie simplifiée, à deux dimensions, en avait été donnée, conduisant au calcul du rapport des seuils de plasticité.

J'ai observé, dans l'Oisans, il y a une douzaine d'années, que les plans de schistosité des différentes strates pouvaient ne pas couper le plan de stratification suivant la même direction. Une théorie complète, à trois dimensions, était donc nécessaire, et on peut espérer tirer de l'observation des indications sur l'orientation de la contrainte et sa nature (rapport des composantes principales). La cohérence des résultats serait une confirmation des hypothèses mises en œuvre.

Il est évident que la contrainte ne pourra être déduite des observations d'angle que à un facteur inconnu près, et à une pression hydrostatique additive près. Il reste – pour la contrainte moyenne et pour chacun des milieux – quatre paramètres, trois pour l'orientation et le rapport des composantes principales du déviateur ("nature" de la contrainte). J'ai pu établir un ensemble de diagrammes qui permettent de déterminer ces grandeurs, à partir des observations de terrain. Le détail de ces calculs (qui sortent du thème de l'anisotropie) sera publié ailleurs (GOGUEL, 1982).

9. Instabilité d'une distribution uniforme

Il nous reste à examiner une question très importante, qui est celle de la stabilité d'une distribution uniforme.

Implicitement, dans tout calcul de mécanique des solides, on suppose que contrainte et déformation varient d'une manière continue, et se correspondent d'une manière univoque. Même pour une substance isotrope, ce n'est plus vrai au moment de la rupture. Au moins certaines formes de rupture peuvent s'interpréter en considérant la matière comme formée d'éléments, couplés en série-parallèle. Si les propriétés de certains éléments s'écartent aléatoirement de la valeur moyenne, la nature de la relation contrainte-déformation tend à limiter l'effet de ces fluctuations : si un élément est plus déformable, la contrainte se reporte sur ses voisins, et elle est plus faible pour lui.

Il n'en est plus de même si la dérivée de la relation contrainte-déformation change de signe : Le couplage des éléments de volume juxtaposés ne régularise plus la distribution, mais les irrégularités tendent à s'accentuer, jusqu'à dégénérer en rupture.

Un cas très instructif est celui de l'acier doux, où, à la limite élastique, une certaine déformation s'accompagne d'une chute de contrainte. Une distribution uniforme de la contrainte n'est plus stable, d'où l'individualisation des lignes, ou plutôt bandes, de Lüders (ou Hartmann), dont la forme résulte de l'accolement des domaines déformés ou non déformés, suivant le plan qui reste invariant dans cette déformation. Lorsque la déformation se poursuit, la contrainte augmente à nouveau, si bien que l'instabilité ne va pas jusqu'à la rupture.

Des instabilités de la déformation uniforme jouent un rôle très important pour des milieux stratifiés fortement contrastés. Sous l'action d'une contrainte uniforme, on peut toujours imaginer une déformation uniforme, contrainte et déformation se répartissant entre les strates.

Mais il peut également se produire une déformation plus complexe, non uniforme, dans laquelle les lits résistants se plissent, sans beaucoup se déformer dans leur plan tandis que les lits mous permettent un glissement relatif des précédents.

Nous allons aborder l'étude de ces instabilités, entraînant les plissements, pour les différentes lois de comportement que nous avons envisagées, à commencer par le régime élastique, pour lequel l'amorce de plissement est l'analogue d'un flambement ; il ne peut s'agir, bien entendu, que de l'amorce de ce phénomène, dont l'accentuation entraîne la sortie du régime élastique.

Le flambement peut tout aussi bien se produire — sinon plus facilement — à partir d'une distribution des contraintes qui ne résulte pas de ce jeu de l'élasticité (à partir d'un état naturel sans contrainte), mais d'une redistribution des contraintes déterminée par le fluage, dont nous avons vu qu'elle peut entraîner une compression des bancs durs, très supérieure à celle des bancs mous qui les encadrent.

Une fois le flambement déclenché, l'amplification de la déformation entraîne en général le dépassement de la limite élastique, et la suite de la déformation doit s'étudier avec un autre modèle. Nous étudierons celui de la plasticité, en le supposant valable pour les deux catégories de bancs, et nous verrons que des considérations énergétiques permettent de justifier l'hypothèse fondée sur l'observation — d'un plissement de la masse stratifiée, et d'en préciser certaines modalités. Mais il faudrait aussi envisager le cas où, à la suite du flambement, la sortie du domaine élastique correspondrait à une rupture. Si celle-ci se produisait dans les bancs mous, elle ne pourrait guère correspondre qu'à un cisaillement parallèle à la stratification, avec ensuite un frottement, jouant le même rôle que la valeur limite du cisaillement en régime plastique. Il faut cependant envisager le cas où, la pression perpendiculaire à la stratification devenant négative, la rupture conduirait à une ouverture des feuillets ; cela n'est guère à envisager dans les applications géologiques.

Si la rupture se produit dans les bancs durs, la compression longitudinale aura pour effet de les faire se redoubler, par pénétration dans les bancs mous. Le détail de tels phénomènes ne peut guère être analysé.

10.   Flambement en régime élastique

Le flambement peut apparaître en régime élastique, lorsque la compression maximale est dans le plan de stratification ; elle peut se décrire comme un flambement des bancs raides, auquel les couches molles qui les séparent ne peuvent s'opposer, et s'étudie de la même manière. L'amorce seule du flambement a lieu en régime élastique.

Considérons donc un milieu stratifié constitué d'un milieu a raide, en couches d'épaisseur e_a, alternant avec un milieu b mou, d'épaisseur e_b. On supposera, pour simplifier, les coefficients de Poisson égaux à 1/4, donc :

$$\lambda = \mu \quad ; \quad \mu^a \gg \mu^b \qquad (9)$$

En posant :

$$m = \frac{e_a\mu^a + e_b\mu^b}{e_a + e_b} \qquad p = (e_a + e_b)/\left(\frac{e_a}{\mu^a} + \frac{e_b}{\mu^b}\right) \qquad (10)$$

on trouve :

$$C_{11} = C_{22} = \frac{8}{3}\,m + \frac{1}{3}\,p\,; \quad C_{12} = \frac{2}{3}\,m + \frac{1}{3}\,p\,; \quad C_{33} = 3p\,;$$

$$C_{13} = C_{23} = p\,; \quad C_{44} = C_{55} = 2p\,; \quad C_{66} = 2m\,; \quad S_{11} = S_{22} = 2/5\,m\,;$$

$$S_{33} = \frac{1}{3p} + \frac{1}{15\,m}\,; \quad S_{12} = -1/20\,m = S_{13} = S_{23}\,; \quad S_{44} = S_{55} = 1/2\,p\,;$$

$$S_{66} = 1/2\,m \qquad (11)$$

Si les contraintes résultent d'une déformation élastique à partir d'un état naturel sans contraintes, on a :

$$\sigma_{11}^a = \frac{\mu^a}{m}\,\sigma_{11}^m + \frac{m - \mu^a}{3m}\,\sigma_{33} \qquad (12)$$

et comme $\mu^a > m > \mu^b$, σ_{11}^a est très supérieur à σ_{11}^m, et σ_{11}^b très inférieur.

Mais notre analyse est aussi valable si la distribution des contraintes résulte d'une adaptation par fluage, et nous avons vu qu'il peut en résulter des valeurs de σ_{11}^a très supérieures à σ_{11}^b, la "raideur" du milieu a tenant cette fois à la lenteur du fluage, par rapport au milieu b, et non plus à la valeur du module élastique. Nous avons tous présent à l'esprit des exemples où les deux choses vont de pair, mais on ne peut affirmer qu'il en sera toujours ainsi.

L'étude du flambement —que nous supposons semblable dans tous les bancs alternants— suit la démarche classique, mais il faut tenir compte de l'effet des intercallations molles sur les bancs durs, qui tend à s'y opposer (Fig. 1).

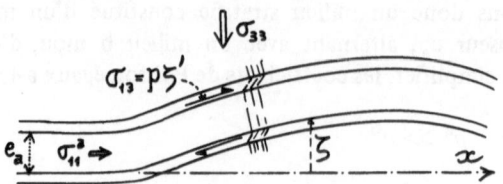

Figure 1. – *Calcul du flambement en régime élastique.*

Soit $\zeta(x)$ le profil pris par un banc quelconque, à partir de sa position initiale, au cours du flambement. Le moment de la compression longitudinale, $e_a \times \sigma_{11}^a$ par unité de largeur, au point x, est donné par $\zeta \times e_a \sigma_{11}^a$.

Globalement, le gauchissement correspondant à la déviation ζ' des bancs, correspond à une déformation $\epsilon_{13}^m = \frac{1}{2} \zeta'$. Mais cette déformation se répartit très inégalement entre les deux sortes de bancs, puisque :

$$\epsilon_{13}^a = \epsilon_{13}^m \times \frac{p}{\mu^a} \tag{13}$$

ϵ_{13}^a est beaucoup plus faible – dans le rapport de modules – que ϵ_{13}^b, ce qui justifie l'étude du flambement des bancs durs à l'approximation de Saint-Venant, c'est-à-dire en les traitant comme des poutres fléchies, dont les sections droites restent planes.

Néanmoins, les bancs mous exercent une composante de cisaillement sur les stratifications,

$$\sigma_{13} = 2p\epsilon_{13}^m = p\zeta' \tag{14}$$

qui se traduit, pour l'unité de largeur, et la longueur dx, de banc dur, par un couple $e_a \, p\zeta' dx$.

Enfin la courbure ζ'' du banc dur est liée au moment fléchissant $\mathfrak{M}$ par :

$$\mathfrak{M} = \frac{e_a^3}{4} \mu^a \zeta'' \tag{15}$$

Au total, en écrivant que ce moment fléchissant est égal à la somme des couples s'exerçant sur la partie de la poutre située d'un côté du point considéré, il vient

$$-\frac{e_a^3 \mu^a}{4} \zeta'' = \sigma_{11}^a \, e_a \, \zeta - e_a p \int^x \zeta' \, dx \tag{16}$$

ou

$$-\frac{e_a^3 \mu^a}{4} \zeta'' = e_a (\sigma_{11}^a - p) \zeta$$

équation qui comporte des solutions de la forme :

$$\zeta = \sin \left(x \times \sqrt{\frac{e_a^2 \mu^a}{4(\sigma_{11}^a - p)}} \right) \tag{17}$$

pour que le flambement soit possible, il faut que la longueur libre, entre deux encastrements, soit supérieure à la demi longueur d'onde,

$$\ell = \frac{\pi}{2} e_a \Big/ \sqrt{\frac{\sigma_{11}^a}{\mu^a} - \frac{p}{\mu^a}} \tag{18}$$

ce qui sera toujours possible, s'agissant d'un milieu indéfini, et nous apporte une information sur la forme que prendra le flambement.

Mais aussi, et surtout, il faut que la quantité sous le radical soit positive. Ici nous devons distinguer, dans la discussion, les deux cas, d'une distribution des contraintes due à l'élasticité, ou au fluage.

Dans le premier cas, il ne suffit pas que la contrainte σ_{11}^a soit suffisamment grande, il faut encore que l'hypothèse du régime élastique soit vérifiée. La condition

$$\sigma_{11}^a > p \tag{19}$$

compare une contrainte et un module, et on sait que le régime élastique n'est valable que pour des contraintes très inférieures au module. Posons :

$$\frac{\sigma_{11}^a}{\mu^a} < \frac{1}{R} \tag{20}$$

R étant grand (de l'ordre de 100, pour fixer les idées). Mais nous avons à comparer une contrainte dans les bancs les plus raides, et un module global, qui est une moyenne harmonique pondérée.

Il est toujours possible d'imaginer un modèle stratifié, pour lequel le rapport des modules soit de l'ordre de R, et pour lequel le flambement sera possible, quel que soit le rapport des épaisseurs.

Supposons maintenant que la structure fine du milieu stratifié soit indiscernable. Nous observons un milieu anisotrope, qui nous paraît homogène, et dont nous pouvons mesurer les coefficients élastiques, C_{ij} ou S_{ij} ou dans le cas simplifié que nous envisageons m et p (c'est-à-dire $C_{66} = 2m$, et $C_{44} = C_{55} = 2p$).

On calcule facilement que

$$\frac{m}{p} \geqslant 1 + \alpha(R - 1) \tag{21}$$

si m et p sont connus, on pourra toujours construire un modèle stratifié rendant compte de leurs valeurs, en prenant des bancs rigides suffisamment minces (α petit). On trouve ensuite facilement les modules à leur attribuer :

$$\frac{m}{p} = 1 + \alpha\beta \frac{\mu^a}{\mu^b} \left(1 - \frac{\mu^b}{\mu^a}\right)^2 \tag{22}$$

ce qui conduit à un rapport μ^a/μ^b voisin de R. Si la distribution des contraintes résulte du fluage, σ_{11}^a peut prendre une valeur élevée, si σ_{11}^m est supérieur à σ_{33} et si le milieu a ne permet qu'un fluage nul ou très lent, ceci indépendemment des valeurs des modules. Mais ceux-ci doivent être assez contrastés pour que σ_{11}^a soit supérieur à p, et les calculs ci-dessus restent valables.

Si le flambement —c'est-à-dire une instabilité de la distribution uniforme de la déformation, pour une contrainte uniforme— peut se produire dans le milieu stratifié, la même instabilité doit exister pour le milieu homogène, caractérisé par les mêmes coefficients élastiques, en l'espèce m et p. D'après la formule précédente, pour un rapport donné m/p, on peut toujours imaginer un milieu stratifié, qui en rende compte. Si l'on fixe un maximum pour le rapport μ^a/μ^b, on sera conduit à prendre une valeur suffisamment petite, pour α ou β.

La deuxième condition —fixant une borne supérieure 1/R à la déformation pour la limite élastique— qui nous imposait de prendre α petit, n'a pas de signification, lorsqu'il s'agit seulement de construire un modèle idéal pour représenter globalement le comportement d'un corps —tel un cristal— dont la structure réelle est toute différente, et pour lequel la sortie du domaine élastique s'exprime généralement par une rupture par clivage.

Pour certains micas (biotite, phlogopite), le rapport C_{66}/C_{44}, qui est l'équivalent de m/p, peut atteindre 13. On construirait facilement un modèle stratifié ayant un comportement global équivalent, avec des lits durs minces (α de l'ordre de 0,1), et des modules élastiques dans un rapport 100, qui donnerait lieu à flambement, pour une limite élastique des bancs durs qui ne serait que le centième de leur module.

Cela nous autorise-t-il à prévoir que le mica, comprimé parallèlement à son clivage, est susceptible de présenter une instabilité analogue à un flambement, tout en restant dans le domaine élastique ?

11. Application à la déformation des cristaux

Mais, quel que soit le modèle envisagé, modèle stratifié, ou cristal homogène, une telle instabilité se traduit par une augmentation brutale de la déformation, qui fait sortir du domaine élastique. Cela conduira, pour le cristal, à des ruptures suivant les plans de clivage, suivies de glissements des lamelles ainsi séparées : ce que l'on observe effectivement pour certains cristaux (micas, péridot), dans des roches déformées ; on y décrit des "kinks", mode de déformation défini pour des complexes stratifiés, et sur lequel nous allons revenir.

L'amorce, en régime élastique, d'une telle déformation, nous paraît donc pouvoir être attribuée à une instabilité de la déformation, pour une contrainte uniforme, dont les seules valeurs des constantes élastiques pour un milieu uniforme doivent pouvoir rendre compte, sans qu'il soit nécessaire d'imaginer pour le cristal l'équivalent de la structure stratifiée, que nous avons utilisée pour analyser une telle instabilité.

12. Plissement d'une masse stratifiée

Revenons au modèle stratifié, pour étudier son plissement, lorsque celui-ci atteint une grande amplitude, qu'il ait été amorcé par un flambement en régime élastique, ou à la suite d'une redistribution des contraintes par relaxation. Le plissement pourrait d'ailleurs aussi apparaître directement dans le régime plastique. Celui-ci est le plus commode pour étudier l'évolution du plissement ; on suppose que, pour chacun des constituants, la grandeur C (racine carré du 2e invariant du déviateur) atteint, ou dépasse de peu, une valeur caractéristique S, ou seuil de plasticité, et que le tenseur de déformation est semblable au tenseur qui mesure le déviateur.

Nous avons déjà utilisé le modèle de la plasticité dans un cas où le rapport des seuils de plasticité pouvait ne pas être très différent de 1 (pour fixer les idées, de 1/2 à 2), et où les différentes sortes de bancs se déformaient d'une manière analogue, avec apparition de schistosités, qui ne différaient que par leurs directions.

Mais ici, nous supposons qu'il y a un grand contraste entre les bancs — dont nous verrons plus loin comment on peut l'estimer. Il en résulte que les bancs durs s'incurvent, sans que leur épaisseur varie, et que les bancs mous subissent un cisaillement parallèle à la stratification, qui peut atteindre une amplitude notable.

L'observation montre l'extrême fréquence d'un tel type de déformation, à toutes les échelles, depuis celle des schistes cristallins, à lits sub-millimétriques, jusqu'aux formations sédimentaires dessinant des plis d'échelle kilométrique, ainsi que l'extrême variété de forme de ces plis.

Il nous apparaît très difficile de suivre la transition du régime élastique au régime plastique. Nous aborderons la question autrement. Si on connaît la forme des plis, répondant aux caractéristiques indiquées ci-dessus, et les seuils de plasticité caractéristiques de chacun des constituants, il est facile de calculer le travail total absorbée par les déformations.

De l'expression du travail pour l'unité de volume, en régime plastique :

$$W = 2\,SV \sqrt{\Sigma \epsilon_{12}'^2 - \Sigma \epsilon_{11}' \epsilon_{22}'} \qquad (23)$$

on tire facilement pour trois cas élémentaires :

Dans le cas d'un écrasement, dans le rapport a_1/a_0, en symétrie axiale :

$$\mathscr{C} = \sqrt{3}\,VS\,\mathrm{Log}\,(a_1/a_0) \qquad (24)$$

et à épaisseur constante :

$$\mathscr{C} = 2\,VS\,\mathrm{Log}\,(a_1/a_0) \qquad (25)$$

Dans le cas de la torsion d'un banc, d'épaisseur e, et de largeur ℓ, d'un angle α :

$$\mathscr{C} = \frac{1}{2}\,e^2\,\ell\,S\,\alpha \qquad (26)$$

expression qui ne dépend pas du rayon de courbure.

Pour un cisaillement :

$$\mathscr{C} = 2\,s\,dS \qquad (27)$$

où s est la surface affectée, d le déplacement. Cette expression ne dépend, ni de l'épaisseur, ni de la manière dont le cisaillement se répartit entre les différents lits. Il en résulte que le cisaillement le long d'une couche peut entraîner des variations d'épaisseur, sans travail supplémentaire.

Il est donc facile, pour toute forme de plis, de calculer le travail total absorbé, mais l'expression trouvée contiendra les seuils de plasticité des différentes sortes de roches, dont la valeur est généralement très mal connue. L'intérêt de cette estimation de l'énergie est qu'elle permet de comparer différentes formes de plis, a priori possibles.

Envisageons un complément de déformation de faible amplitude, partant du même état initial, entraînant la même déformation globale, mais qui peut se répartir de différentes manières entre les différents lits plissés. Il est évident que la déformation qui se produit réellement est celle pour laquelle le travail absorbé est le plus faible.

Envisageons maintenant une déformation de grande amplitude, à partir d'un même état initial (par exemple, stratification plane), et entraînant la même déformation globale. Dire que, de toutes les déformations possibles (à supposer qu'on puisse les dénombrer), celle qui se produit effectivement est celle qui absorbe le moindre travail n'est pas rigoureux, car on pourrait envisager une déformation qui demanderait, au départ, plus d'énergie qu'une autre, (et ne se déclencherait donc pas), bien que par la suite elle se poursuive facilement, au point d'absorber au total moins de travail. Cet énoncé constitue néanmoins une approximation commode.

On peut d'ailleurs l'utiliser de deux façons. Si on connaissait effectivement les seuils de plasticités des différentes roches, et si on pouvait dénombrer toutes les formes de plissements, on pourrait essayer de comprendre leur déterminisme.

Mais on peut également, observant un système de plis, chercher quels doivent être les rapports des seuils de plasticité, pour que ces plis se soient formés, plutôt que d'autres que nous pouvons imaginer. Cette deuxième démarche est souvent la plus féconde; elle nous permet, en particulier, d'estimer les rapports des seuils de plasticité de différents constituants, ou de leur fixer des bornes.

Suivant ce principe général, on présentera deux exemples. La forme de pli la plus simple possible est le "kink", dans lequel un panneau de largeur constante est incliné, entre deux blocs restés indéformés. On calcule facilement le travail absorbé, pour chacune des étapes successives de la déformation (Fig. 2)

$$\mathscr{C} = \ell E \left(S, e\alpha + S_2 \times 4L \, \text{tg} \, \frac{\alpha}{2} \right) \tag{28}$$

pour l'épaisseur E et la largeur ℓ.

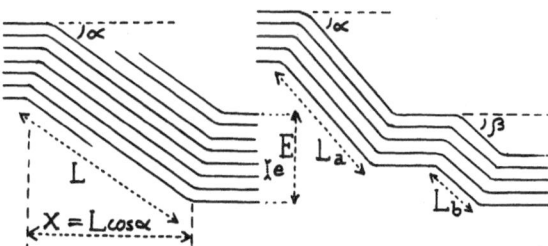

Figure 2. – *Schéma de Kink, pour le calcul du travail absorbé, et du rapport des seuils de plasticité de bancs et des joints, à partir des angles de deux "Kinks" voisins.*

La dérivée de cette expression, par rapport à la contraction résultant du jeu du kink est :

$$\frac{d\mathscr{C}}{dX} = \ell E \left(S_1 \frac{e}{L \sin \alpha} + S_2 \frac{2}{\cos^2 \dfrac{\alpha}{2} \sin \alpha} \right) \qquad (29)$$

Si on a des raisons de penser que deux kinks voisins ont atteint en même temps leur état final, c'est qu'ils absorbaient le même travail, pour la même déformation globale marginale. Si leurs largeurs diffèrent, ainsi que leurs angles, on peut de cette égalité déduire le rapport des seuils de plasticité des bancs et des joints.

$$\left(\frac{d\mathscr{C}}{dX} \right)_a = \left(\frac{d\mathscr{C}}{dX} \right)_b \qquad (30)$$

$$\frac{S_1}{S_2} = \left(\frac{2}{\cos^2 \dfrac{\alpha}{2} \sin \alpha} - \frac{2}{\cos^2 \dfrac{\beta}{2} \sin \beta} \right) \Big/ \left(\frac{e}{L_b \sin \beta} - \frac{e}{L_a \sin \alpha} \right) \qquad (31)$$

Nous avons trouvé, pour le Crétacé supérieur (calcaires très lités) des Alpes Maritimes, des rapports de 10 à 25.

Si on a affaire à des plis de forme très compliquée — tels qu'on peut les observer dans un flysch — on peut essayer de les caractériser statistiquement. Nous ferons le calcul pour un banc, d'épaisseur e, redressé d'un angle α entre deux charnières distantes de L, sans insister sur la manière dont on pourrait, statistiquement, calculer la moyenne pour un système complexe de plis (Fig. 3).

$$\mathscr{C} = \ell (S_1 e^2 \alpha + S_2 \times 4 L e \, \text{tg} \, \alpha) \qquad (32)$$

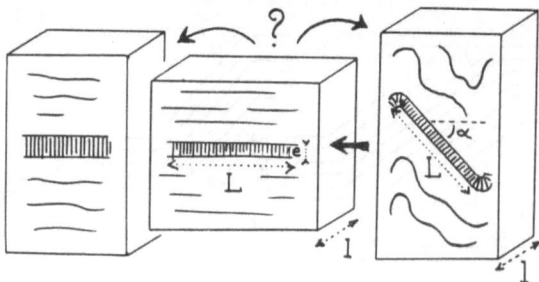

Figure 3. – *Comparaison du travail absorbé, pour une même déformation globale, par une compression uniforme et par plissement.*

alors que, dans l'hypothèse d'une compression uniforme, l'expression du travail serait :

$$\mathscr{C} = \ell\,L\,(e_a\,S_1 + e_b\,S_2)\,\text{Log}\cos\alpha \qquad (33)$$

On constate que, pourvu que $\dfrac{L}{e}$ soit grand, et les lits très contrastés (S_a/S_b petit), le travail absorbé est beaucoup plus faible, que ce qu'il serait pour une déformation uniforme. Celle-ci apparait donc comme instable, par rapport au plissement qui entraîne une déformation globale équivalente.

13. Conclusion

Telle est peut être la conclusion essentielle à retenir. L'anisotropie ne se traduit pas seulement par le maniement d'équations un peu plus compliquées que l'isotropie, mais elle peut mettre en défaut l'hypothèse, si répandue qu'elle en devient implicite, de la continuité et de la régularité de la relation tensorielle entre contrainte et déformation.

REFERENCES

GOGUEL J. – "Sur la modification des contraintes intérieures par relaxation". *C.R. Acad. Sci.,* Paris, 214 (1942) : 410-414.

GOGUEL J. – "Calcul des contraintes dans l'hypothèse de la relaxation". *C.R. Acad. Sci.,* Paris, 214 (1942) : 470-471.

GOGUEL J. – *Introduction à l'étude mécanique des déformations de l'écorce terrestre.* Mémoire Serv. Carte géol., 2e éd. 1948, chap. XIII.

GOGUEL J. – "Une interprétation mécanique de la réfraction de la schistosité" *Tectonophysics* 82 (1982) : 125-143.

ABSTRACT

A model for an anisotropic body is build by stratification of isotropic layers. The values of the global elastic parameters are computed. One can also consider a model build with parallel fibres. The same model can be used to compute the distribution of stress by relaxation, and after plastic deformation. An even distribution of strain, as an effect of uniform stress, may be unstable in the elastic field (buckling) and, much more, in the plastic field, where folding plays a major part.

Slip, Dilation and Closure of Joints in a Regularly Jointed Rock Mass

L.W. Morland

University of East Anglia, School of Mathematics and Physics, Norwich, U.K.

1. Introduction

Pre-fracture planes, or joints, are common features in natural rock masses, and have a significant effect on the overall mechanical response [3, 4], necessarily anisotropic. Major joints often occur as sets of approximately regularly spaced parallel planes, with a variety of orientations, which extend over great distances. Joint spacing ranges from centimetres to metres, and in many cases is small compared with the length scale of interest. It is then possible to construct a continuum model which describes the gross response of individual joints over a representative rock element with dimensions of several joint spacings [5]. Since joint thickness is much smaller than joint spacing, and misalignment of a block structure formed by intersecting joint sets commonly limits joint displacement magnitudes to joint thickness, the overall displacements are small compared with the representative element size. Thus strains associated with the continuous joint displacement fields are infinitesimal, and an additive strain decomposition is obtained for normal and tangential motions within each joint set, and for motions of different joint sets. In addition there is a strain contribution from the intact block material.

For elastic rock and elastic (reversible) joint response in closure and slip, the continuum model defines an anisotroqic elastic medium with moduli depending on the rock moduli, joint stiffnesses, and joint orientations [6]. Various weak and strong anisotropies arise as rock and joint moduli magnitudes differ in particular ways [7]. Allowing joint slip (irreversible) when a slip criterion on the joint tractions is satisfied leads to an anisotropic elastic-slip theory analogous to an elastic-plastic theory with non-associated flow rule [5]. The latter gives rise to non-interleaving bulk and shear plane wave speeds in elastic and slip regions, with consequent unsatisfactory wave interaction features [8]. For some joint orientations and slip conditions there are no real propagation speeds.

Goodman [1, 2] notes that joint closure under normal pressure is not recovered on unloading, and similarly that tangential slip under shear traction is largely irreversible. A conventional elastic joint theory then becomes invalid once unloading occurs. Here, closure and slip, with accompanying dilation, are described by time-independent rate laws during appropriate loading conditions, with no recovery during unloading. Since gross block structure or irregular joint planes inhibit continued slip, a slip-hardening factor is introduced. These laws for the response of a single joint are incorporated into a continuum model for the gross response [5]. A single joint set allows three distinct modes of joint motion : closure, no slip (c), slip and dilation, no closure (sd), slip and dilation, and closure (sdc), in addition to pure block deformation (b). For two joint sets the possible combinations of the single set motions plus block deformation allow ten distinct modes. This model is more complex than a two-mode elastic-plastic theory, though without the complication of a yield condition, but appears to be a minimal description of the joint motion effects.

The strain component histories arising from a cycle of simple shear stress on a single joint set are presented to illustrate the effects of joint orientation and relative closure, slip, and dilation factors, in the absence of block deformation.

2.   Joint closure, slip and dilation

Consider first a single joint in a rock element, and let σ_n be the normal traction relative to the initial seating pressure with U^n the normal displacement discontinuity across the joint, and τ the tangential traction in a fixed direction with U^s the tangential displacement discontinuity in the same direction. Figure 1 shows the qualitative joint responses presented by Goodman [1, 2]. In (a), which illustrates the response to a shear traction cycle at constant pressure, any peak strength and subsequent stress drop is neglected, leaving a monotonic loading curve. The unloading path neglects all recovery, and the reloading path shown with constant translation along the U^s axis illustrates a model in which $d\tau/dU^s$ depends only on τ independent of previous slip. Alternative history dependence may be used. It is also noted [1, 2] that the slip decreases as pressure increases or equivalently as the joint closes. While (a) relates to an unrestricted joint, within a block structure slip in a given direction will be increasingly inhibited as the net slip in that direction increases.

Following [5], introduce a slip hardening parameter by

$$q_\nu = \mathbf{U}^s . \boldsymbol{\nu}/\delta, \quad \tau = \tau\boldsymbol{\nu}, \tag{1}$$

where $\mathbf{U}^s$ is the displacement discontinuity in the joint plane, $\boldsymbol{\nu}$ is a unit vector defining the direction of the current shear traction $\boldsymbol{\tau}$, and δ is a joint thickness magnitude so that q_ν is of order unity. Similarly, a closure parameter of order unity is given by

$$q_n = -U^n/\delta, \quad \mathbf{U}^n = U^n \mathbf{n}, \tag{2}$$

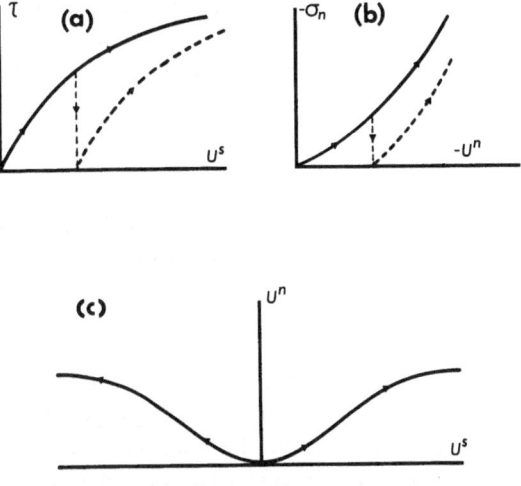

Figure 1. – *Joint motions: (a) slip, (b) closure, (c) dilation.*

where **n** is a unit normal vector to the joint plane. Thus, assuming that the current slip direction (velocity) is parallel to the traction, the simplest time-independent description of the above features is the seperable law

$$\dot{U}^s = |\dot{U}^s| \, \nu = g(\tau) \, h(q_\nu) \, f(q_n) \, \dot{\tau} H(\dot{\tau}) \nu, \tag{3}$$

where H is the Heaviside unit function. The superposed • denotes rate with respect to any increasing loading parameter, which may be time, and the slip velocity becomes zero when $\dot{\tau}.\nu = \dot{\tau} \leqslant 0$ as required by the irreversibility postulate. $g(\tau)$ is an increasing function, $h(q_\nu)$ a decreasing function, and $f(q_n)$ a decreasing function. Note that

$$\delta \dot{q}_\nu = |\dot{U}^s| + U^s . \dot{\nu}, \tag{4}$$

so that q_ν depends on the history of directions ν during slip, but for slip in a fixed joint plane direction ν, the increase in δq_ν is simply the increase in $|U^s|$. The restriction of slip by block rotation causing local closing of the joint [1, 2] is accounted for by the decreasing function $f(q_n)$.

Normal displacement decreases (closure) as pressure $- \sigma_n$ increases, at zero tangential traction, as shown in (b), and is not recovered as the pressure is removed. It is supposed that the slope $d\sigma_n/dU^n$ is an increasing function of the closure q_n, independent of the closure history, and independent of the shear traction. In addition, normal displacement increases (dilatancy) as slip takes place at constant pressure, shown in (c), and it is supposed that the two mechanisms are independent and additive. Dilatancy decreases as the pressure increases, or equivalently as q_n increases, and the slope dU^n/dU^s is zero at $q_\nu = 0$ and first increases then decreases as q_ν increases. The most simple law for the combined effects is

$$\dot{U}^n = \ell(q_n)\,\dot{\sigma}_n\,H(-\dot{\sigma}_n) + \alpha q_\nu\,k(q_\nu)\,r(q_n)\,|\dot{U}^s|, \tag{5}$$

where $\ell(q_n)$ is decreasing and $k(q_\nu)$, $r(q_n)$ are decreasing, both order unity, and α is a magnitude factor for dilation compared with slip.

3. Continuum model

It is supposed that the stress is approximately uniform over a representative element containing several joints with common properties for each set. The net infinitesimal displacement across the element due to the displacement discontinuities across the joints is equivalent to the change in a continuous displacement field, conveniently separated into a normal field $u^n(x)n$ and tangential field $u^s(x)$ for each joint set [5]. Then

$$\frac{\partial u^n(x)}{\partial n} = \frac{1}{d}\,\bar{U}^n(x), \qquad \frac{\partial u^s(x)}{\partial n} = \frac{1}{d}\,\bar{U}^s(x), \tag{6}$$

where $U^n(x)$, $U^s(x)$ are mean joint displacements over an element centred at x, determined by the relations $(1) - (5)$ in terms of mean stresses over the element. Focussing attention on a single joint set with normal n and unit orthogonal in-plane vectors s_1, s_2 as shown in Figure 2, the corresponding infinitesimal strain decomposition is

$$e = e^b + \tilde{e}^n + \tilde{e}^s, \tag{7}$$

where e^b is the strain in the intact block material and

$$\tilde{e}^n = \frac{\partial u^n}{\partial n}\,(n \otimes n), \quad \tilde{e}^s = \frac{1}{2}\,\frac{\partial u^s_\beta}{\partial n}\,(s_\beta \otimes n + n \otimes s_\beta), \quad \beta = 1, 2. \tag{8}$$

There are corresponding additive strains for each joint set defined by its own triad (s_1, s_2, n), in general with different laws $(3), (5)$.

Now $\dot{u}^s = |\dot{u}^s|\,\nu, \; = \dot{u}^s_\beta\,s_\beta$,   so

$$\dot{\tilde{e}}^n = \frac{\partial \dot{u}^n}{\partial n}\,(n \otimes n), \quad \dot{\tilde{e}}^s = \frac{1}{2}\,\left|\frac{\partial \dot{u}^s}{\partial n}\right|\,(\nu \otimes n + n \otimes \nu). \tag{9}$$

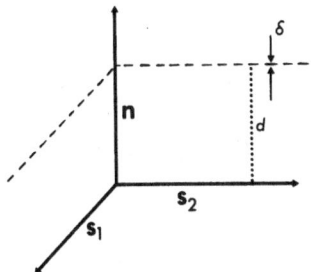

Figure 2. – *Joint set geometry.*

The relations (1) – (5), with interpretations (6), give

$$\left| \frac{\partial \dot{u}^s}{\partial n} \right| = \frac{1}{d} \, g(\tau) \, h(q_\nu) \, f(q_n) \, \dot{\tau} \, H(\dot{\tau}), \tag{10}$$

$$\frac{\partial \dot{u}^n}{\partial n} = \frac{1}{d} \, \ell(q_n) \, \dot{\sigma}_n \, H(-\dot{\sigma}_n) + \alpha \, q_\nu \, k(q_\nu) \, r(q_n) \left| \frac{\partial \dot{u}^s}{\partial n} \right|, \tag{11}$$

$$\dot{q}_n = -\frac{d}{\delta} \frac{\partial \dot{u}^n}{\partial n}, \quad \dot{q}_\nu = \frac{d}{\delta} \left\{ \left| \frac{\partial \dot{u}^s}{\partial n} \right| + \frac{\partial u^s}{\partial n} \cdot \dot{\nu} \right\}. \tag{12}$$

The block strain e^b is given by the intact rock properties, commonly an isotropic elastic law, but the laws (10), (11) are strictly anisotropic.

Four distinct modes of deformation are possible:

(b):   pure block deformation $\dot{\tau} \leqslant 0, \quad \dot{\sigma}_n \geqslant 0$;

(c):   joint closure $\dot{\tau} \leqslant 0, \quad \dot{\sigma}_n < 0$;

(sd):   slip and dilation $\dot{\tau} > 0, \quad \dot{\sigma}_n \geqslant 0$;

(sdc):   slip, dilation, and closure $\dot{\tau} > 0, \quad \dot{\sigma}_n < 0$;

(13)

with the latter three accompanied by block deformation in general. Each of the latter three modes is possible for each joint set, so for N joint sets there are $3^N + 1$ distinct modes, to be contrasted with a convention two-mode elastic-plastic theory. However, the mode criteria shown in (13) are simpler than a yield condition and plastic flow validity.

4.   Simple shear illustration

Consider an element with a single joint set given by

$$s = s_1 = (\sin \theta, -\cos \theta, 0), \quad s_2 = (0, 0, 1), \quad n = (\cos \theta, \sin \theta, 0)$$

$$0 \leqslant \theta < \pi, \quad (14)$$

in rectangular Cartesian axes $0x_1\, x_2\, x_3$ subject to a simple shear stress $\sigma_{12} = T$, other $\sigma_{ij} = 0$. The loading is a continuous two-stage cycle :

$$\text{I}: 0 \leqslant T \leqslant T_m \ \dot{T} > 0; \qquad \text{II}: T_m \geqslant T \geqslant 0, \ \ \dot{T} < 0; \qquad (15)$$

so convenient increasing loading parameters are T in stage I and $T_m - T$ in stage II. There is no motion in the $0x_3$ direction. In $0\,x_1\, x_2$,

$$(s \otimes n + n \otimes s) = \begin{pmatrix} \sin 2\theta & -\cos 2\theta \\ -\cos 2\theta & -\sin 2\theta \end{pmatrix}$$

$$(n \otimes n) = \begin{pmatrix} \cos^2 \theta & \dfrac{1}{2}\sin 2\theta \\ \dfrac{1}{2}\sin 2\theta & \sin^2 \theta \end{pmatrix}, \quad (16)$$

$$\tau_s = -T \cos 2\theta, \quad \sigma_n = T \sin 2\theta. \qquad (17)$$

It is supposed that the initial seating pressure is sufficient to prevent the actual normal joint stress becoming tensile. In this plane motion $v = ms$ $(m = \pm 1)$, $\dot{v} = 0$, $m\cos 2\theta \leqslant 0$,

$$\tau = m\,\tau_s = -mT \cos 2\theta, \qquad (18)$$

$$\dot{q}_n = -\frac{d}{\delta} \frac{\partial \dot{u}^n}{\partial n}, \quad \dot{q}_v = \frac{d}{\delta} \left| \frac{\partial \dot{u}^s}{\partial n} \right|. \qquad (19)$$

Thus, in stage I with loading rate defined by d/dT,

$$0 \leqslant \theta \leqslant \pi/2: \quad \dot{\tau} \geqslant 0, \quad \dot{\sigma}_n \geqslant 0, \quad \text{(sd)};$$

$$\pi/2 < \theta < \pi: \quad \dot{\tau} \geqslant 0, \quad \dot{\sigma}_n < 0, \quad \text{(sdc)}; \qquad (20)$$

while in stage II with loading rate defined by $- d/dT$,

$$0 \leqslant \theta \leqslant \pi/2: \quad \dot{\tau} \leqslant 0, \quad \dot{\sigma}_n \leqslant 0, \quad (c);$$

$$\pi/2 < \theta < \pi: \quad \dot{\tau} \leqslant 0, \quad \dot{\sigma}_n \geqslant 0, \quad (b);$$

$$\text{(21)}$$

with $\dot{\tau} \equiv 0$ for $\theta = \pi/4$ and $3\pi/4$, and $\dot{\sigma}_n \equiv 0$ for $\theta = 0$ and $\pi/2$. All four modes occur if values of θ in both first and second quadrants are considered. For values of θ in $0 \leqslant \theta \leqslant \pi/2$ symmetric about $\theta = \pi/4$ and in $\pi/2 < \theta < \pi$ symmetric about $3\pi/4$, $e_{11} \leftrightarrow e_{22}$ and e_{12} is the same. Examples are presented for the two values $\theta = \pi/\sigma$ and $\theta = \dfrac{5\pi}{\sigma}$, demonstrating modes (sd) $-$ (c) and (sdc) $-$ (b). The intact material is assumed rigid so that the unloading mode (b) leaves strain unchanged.

For illustration, the increasing function g and decreasing functions h, f, ℓ, k, r, are assumed to be exponentials, in qualitative agreement with the responses described earlier. Scaling stress and strain so that T and e are order unity, (10), (11), (17) $-$ (19), then give for stage I,

$$\frac{dq_n}{dT} = - p\lambda e^{-q_n} \sin 2\theta - \alpha q_\nu e^{-q_\nu - q_n} \frac{dq_\nu}{dT}, \quad p = \begin{cases} 0 & 0 \leqslant \theta \leqslant \pi/2 \\ 1 & \pi/2 < \theta < \pi \end{cases},$$

$$\frac{dq_\nu}{dT} = |\cos 2\theta| \, e^{T|\cos 2\theta| - q_\nu - q_n}, \quad \text{(22)}$$

and for stage II,

$$\frac{dq_\nu}{dT} = 0, \quad \frac{dq_n}{dT} = 0 \, (\pi/2 < \theta < \pi),$$

$$\frac{dq_n}{dT} = - \lambda e^{-q_n} \sin 2\theta \, (0 \leqslant \theta \leqslant \pi/2), \quad \text{(23)}$$

where λ measures the initial shear modulus relative to the initial closure modulus. Integrating $(22)_1$ from $T = q_n = q_\nu = 0$ gives

$$q_n = \ell n \, \{1 - p\lambda \sin 2\theta T - \alpha [1 - (1 + q_\nu) e^{-q_\nu}]\}, \quad \text{(24)}$$

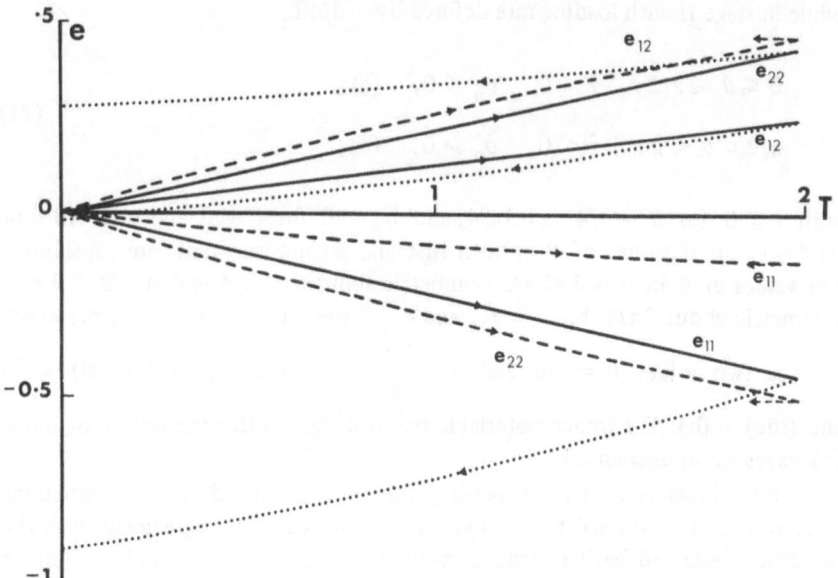

Figure 3. – *Strain histories for* $\alpha = 0$, $\lambda = 0.5$. $\theta = \pi/6$ (→, .. < ..), $\theta = 5\pi/6$ (---).

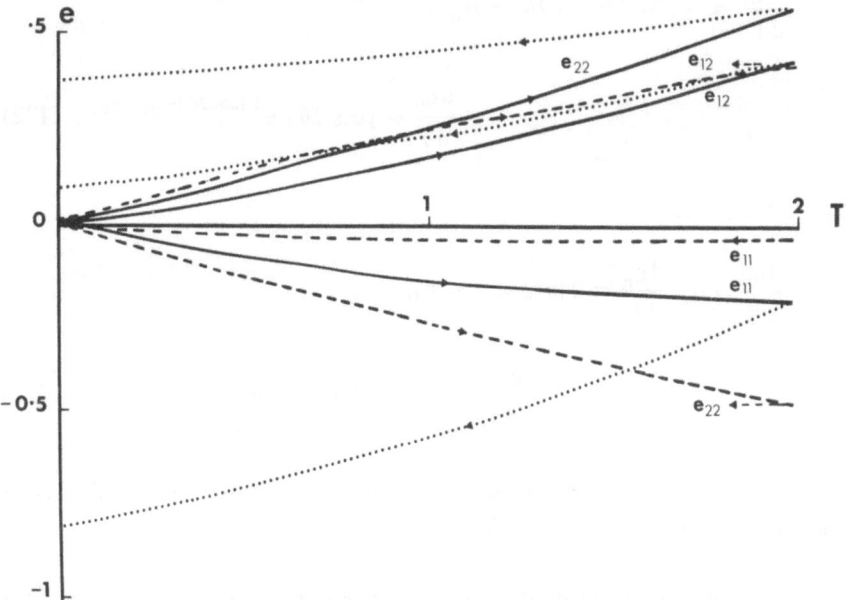

Figure 4. – *Strain histories for* $\alpha = 1$, $\lambda = 0.5$. $\theta = \pi/6$ (→, .. < ..), $\theta = 5\pi/6$ (---).

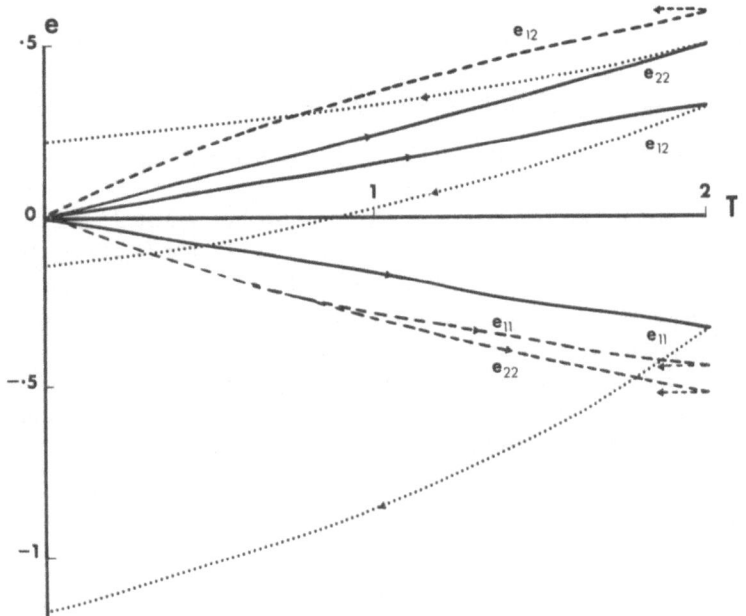

Figure 5. – *Strain histories for* $\alpha = 0.5$, $\lambda = 1$. $\theta = \pi/6$ (➤—, .. <..), $\theta = 5\,\pi/6$ (---).

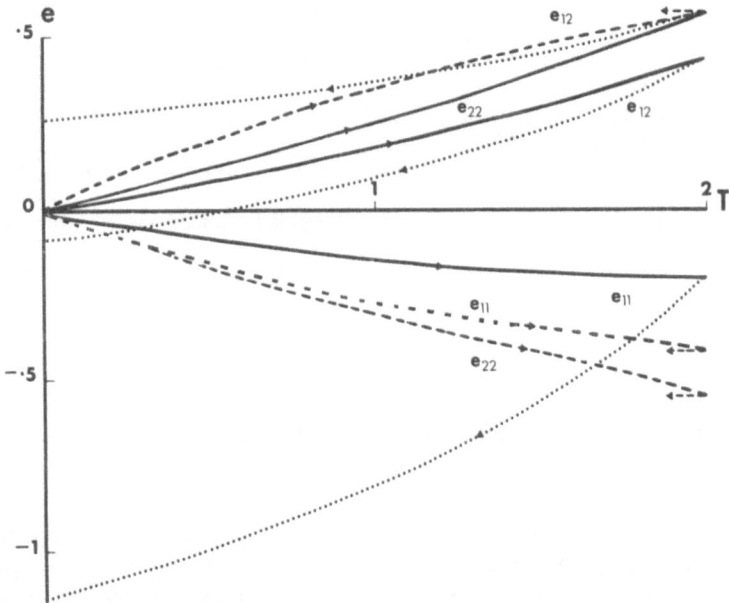

Figure 6. – *Strain histories for* $\alpha = 1$, $\lambda = 1$. $\theta = \pi/6$ (➤—, .. <..), $\theta = 5\,\pi/6$ (---).

then $(22)_2$ is solved numerically for $q_\nu(T)$. T can be expressed analytically in terms of q_ν for $p = 0$. If q_n^m, q_ν^m are the values at T_m, then in stage II,

$$q_\nu \equiv q_\nu^m, \quad q_n \equiv q_n^m \ (\pi/2 < \theta < \pi),$$

$$q_n = \ell n \ \{e^{q_n^m} + \lambda \sin 2\theta \ (T_m - T)\} \quad (0 \leqslant \theta \leqslant \pi/2). \tag{25}$$

The strain histories e_{11}, e_{12}, e_{22} can now be evaluated, and are shown in Figures 3-6 for $(\alpha, \lambda) = (0, \cdot 5)$, $(1, \cdot 5)$, $(\cdot 5, 1)$ and $(1,1)$ respectively to demonstrate the effects of dilatancy (α) and closure rate to shear rate (λ), with unloading from $T_m = 2$. For both values of θ the same qualitative patterns appear for each (α, λ) pair, and the main contrast is between the two values of θ (symmetric about $\pi/2$) with their different mode combinations. There is a marked difference from an elastic cycle because of the irreversible slip and closure response adopted in the model.

REFERENCES

[1] GOODMAN R.E. – "The mechanical properties of joints". *Proc. 3rd Cong. Int. Soc. Rock Mech.* Denver, (1974) : 127-140.

[2] GOODMAN R.E. – *Methods of Geological Engineering in Discontinuous Rocks.* St Paul, U.S.A. ; West Publishing, 1976.

[3] JAEGER J.C. – "Behaviour of closely jointed Rock", *Proc. 11th Symp. Rock Mech.* Berkeley, U.S.A. Port City Press : 1969, Chap. 4.

[4] JAEGER J.C. and N.G.W. COOK. – *Fundamentals of Rock Mechanics,* London : Methuen, 1969.

[5] MORLAND L.W. – "Continuum model of regularly jointed mediums" *J. Geophys. Res.,* 79, 2 (1974) : 357-362.

[6] MORLAND L.W. – "Elastic response of regularly jointed media" *Geophys. J.R. Astr. Soc.,* 37 (1974) : 435-446.

[7] MORLAND L.W. – "Elastic anisotropy of regularly jointed media", *Rock Mech,* 8 (1976) : 35-48.

[8] MORLAND L.W. – "Plane wave propagation in anisotropic jointed media" *Q.J. Mech. Appl. Math.,* 30, 1 (1977) : 1-21.

RESUME

(Glissement, ouverture et fermeture des joints dans une masse rocheuse à joints réguliers)

La présence d'un ou de plusieurs ensembles de joints parallèles régulièrement espacés dans une masse rocheuse provoque une anisotropie mécanique globale. Le glissement sur un plan de joint est généralement accompagné par une ouverture du joint ; la fermeture du joint apparaît sous une pression normale. De tels mouvements de joints sont hautement irréversibles et un modèle, négligeant entièrement toute réponse élastique, est proposé. Ce modèle est introduit dans une description continue pour des échelles de longueurs grandes par rapport à l'espacement des joints. Pour un seul ensemble de joints, on obtient quatre modes de déformations correspondant à : i) déformation des blocs pure, ii) glissement et ouverture, iii) glissement, ouverture et fermeture, iv) fermeture, en opposition avec la réponse à deux modes d'une théorie élasto-plastique. Pour deux ensembles, il y a dix modes. Un cycle en cisaillement pur illustre les effets pour un seul ensemble de joints.

Comportement Mécanique d'une Roche Anisotrope Thermiquement Fissurée

F. Homand-Étienne et R. Houpert
École Nationale Supérieure de Géologie, Nancy, France.

1. Introduction

Cette étude s'inscrit dans le cadre d'une Action Thématique Programmée financée par l'Institut National d'Astronomie et de Géophysique du Centre National de la Recherche Scientifique et concerne les transferts d'énergie thermique à travers l'écorce terrestre. L'objet de cette A.T.P. consiste en l'étude des propriétés géomécaniques de roches cristallines isotropes (granites) et anisotropes (gneiss) ainsi que de roches volcaniques du type basalte microfissurées par effet thermique. Le comportement mécanique des roches en fonction de la température a surtout été étudié jusqu'à présent pour des raisons d'ordre géologique, en combinant hautes températures et hautes pressions de confinement. Ces recherches avaient pour but l'étude du comportement plastique des matériaux rocheux. Il existe, par contre, très peu de recherches concernant les caractéristiques mécaniques des roches fissurées par effet thermique. La température provoque, dans les roches, une fissuration induite par l'anisotropie de dilatation thermique de ses principaux minéraux. L'intensité de la fissuration produite peut être caractérisée par la mesure de la célérité des ondes élastiques, encore appelée vitesse de propagation des ondes ultrasonores. L'étude du comportement mécanique de la roche préfissurée thermiquement est réalisée par l'examen de l'évolution des déformations axiales, latérales et volumiques, en fonction du degré de fissuration, d'éprouvettes soumises à des charges monoaxiales.

Les essais effectués dans le cadre du comportement des roches cristallines isotropes et anisotropes ont porté sur le granite de Remiremont (Vosges) et sur le gneiss de Saint-Evarzec (Finistère). Le granite de Remiremont à grains fins (0,5 à 1 mm) est formé de quartz (26 %), feldspaths (65 %) et de micas (9 %), il est à structure quasi isotrope (Fig. 1). Le gneiss de Saint-Evarzec est à grains très fins (< 0,1 mm), il est formé d'une alternance de très minces lits de biotite associée à un peu de sphène et de lits plus épais de quartz et feldspaths (Fig. 2). Localement, les feldspaths peuvent former des amas de quelques millimètres. Une linéation très nette de mica dans le plan de foliation permet de classer ce gneiss dans les roches à anisotropie planaire continue et à

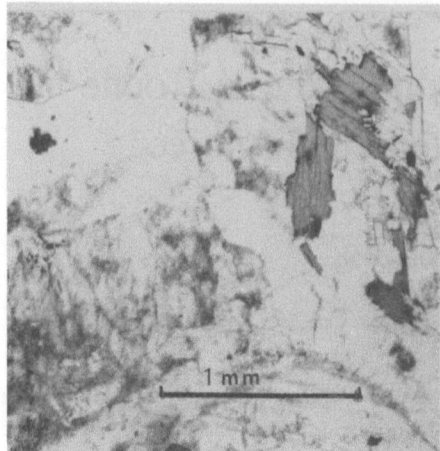

Figure 1. – *Granite de Remiremont.*

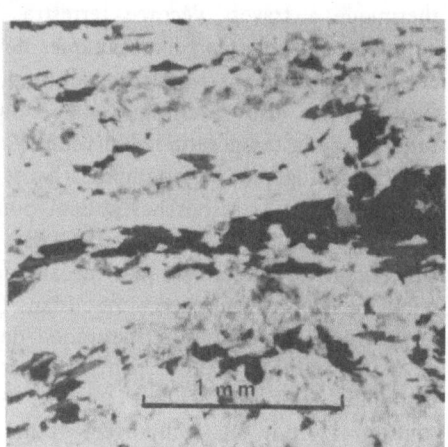

Figure 2. – *Gneiss de Saint-Evarzec.*

symétrie orthotrope. La composition minéralogique globale est voisine de celle du granite de Remiremont.

Les éprouvettes ont été portées à la température voulue à raison de 50°C/h jusqu'à 300°C et ensuite à 100°C/h (des expériences préliminaires ayant montré que cette augmentation de vitesse ne conduit pas à des modifications supplémentaires du matériau) et elles ont été maintenues à cette température pendant une durée de 5 h. Le retour à la température ambiante s'est effectué très lentement. Les températures auxquelles les éprouvettes ont été soumises sont respectivement 200, 400, 500, 600 et 700°C.

Les essais de compression ont été effectués à l'aide d'une machine d'essai asservie en contrôlant la vitesse de déplacement axial au moyen de capteurs placés entre les plateaux de la presse. Tous les essais ont été réalisés à vitesse de déformation axiale constante sur des éprouvettes cylindriques de 5 cm de diamètre et de 10 cm de hauteur. Trois jauges axiales et trois jauges latérales de 2 cm de longueur active, collées dans la partie centrale de l'éprouvette, permettent de mesurer les déformations axiales et latérales et, par suite, de calculer la déformation volumique.

La rupture fragile des roches en compression est un processus qui se développe progressivement avec l'augmentation de la charge appliquée et elle est accompagnée d'une émission de microbruits. Afin de détecter et d'enregistrer ces microbruits, un système comprenant un accéléromètre collé sur l'éprouvette et relié à un mesureur de vibration Brüel et Kjaer et à un enregistreur à réponse logarithmique, a été utilisé. Nous enregistrons ainsi les amplitudes crête à crête des microbruits dans les fréquences comprises entre 3 Hz et 15 kHz.

2.   Comportement mécanique du gneiss témoin

Les caractéristiques d'anisotropie du gneiss sont repérées par les directions orthogonales 1, 2 et 3, le plan (1, 2) étant le plan de foliation et la direction 1 celle de la linéation (Fig. 3). Les éprouvettes ont été prélevées dans le même bloc de gneiss en orientant leur axe, d'une part, suivant la direction 1 ($\alpha = 0°$), d'autre part, suivant la direction 3 ($\alpha = 90°$), ainsi que selon la direction $\alpha = 45°$ dans le plan 1-3 (Fig. 3 et 4). Les mesures de la vitesse

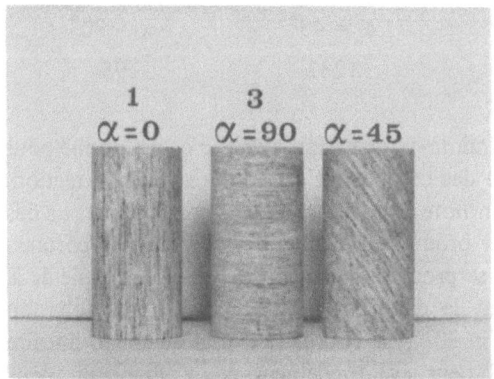

Figure 3. – *Orientation des éprouvettes de gneiss (hauteur: 10 cm).*

Figure 4. – *Structure du gneiss de Saint-Evarzec.*
a – *Trace du plan de foliation*
b – *Linéation.*

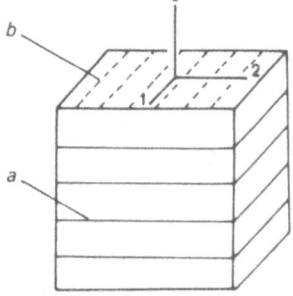

de propagation des ondes longitudinales mettent en évidence les variations suivantes (valeurs moyennes de plusieurs mesures) :

	1		3
	$\alpha = 0°$	$\alpha = 45°$	$\alpha = 90°$
V (m/s)	4978	4957	4390

On explique en général l'anisotropie de la vitesse de propagation des ondes par l'anisotropie de répartition de la fissuration [1], la vitesse la plus faible correspondant à la direction qui rencontre le plus d'obstacles, donc le plus de fissures situées dans un plan perpendiculaire à la direction de propagation. Dans ce cas, la résistance à la compression devrait être plus forte suivant cette direction. Si la présence d'éventuelles fissures est plus sensible dans les mesures de vitesse du son effectuées suivant la direction 3, ce n'est pas le seul phénomène qui joue pour expliquer cette chute de vitesse. Il faut faire intervenir également l'influence de l'orientation des micas. Les plaquettes de mica se présentent orientées perpendiculairement à la direction 3, la célérité des ondes étant plus faible dans le mica que dans le quartz et le feldspath, leur influence est donc plus grande dans cette direction, ce qui conduit à une diminution globale de V suivant la direction 3. Les valeurs de la résistance à la compression mettent en évidence une forte anisotropie de cette résistance :

	1		3
	$\alpha = 0°$	$\alpha = 45°$	$\alpha = 90°$
R_c (bar)	2290	1282	1954

Un exemple de courbe contrainte-déformation axiale $\sigma\text{-}\epsilon_a$ est donné pour chaque direction (Fig. 5). L'allure des courbes est différente selon la direction. Suivant la direction 1 ($\alpha = 0°$), on note des décrochements correspondant à des chutes d'écailles de matériau qui produisent d'ailleurs des bruits importants (Fig. 6). La ruine de l'éprouvette se produit finalement au niveau du pic de la courbe $\sigma\text{-}\epsilon_a$ par extension suivant la direction perpendiculaire à la foliation (Fig. 7). Dans la direction 3 ($\alpha = 90°$), la courbe est pratiquement linéaire jusqu'au voisinage du pic où il peut éventuellement se produire un petit écaillage ; la rupture se propage très brutalement provoquant la ruine de l'éprouvette. Ce phénomène est extrêmement rapide et on n'enregistre que quelques microbruits avant la désintégration de l'éprouvette (Fig. 6). Les éprouvettes taillées suivant $\alpha = 45°$ montrent un comportement identique à ceux des deux directions précédentes en ce qui concerne le début de la courbe $\sigma\text{-}\epsilon_a$. Ensuite, il se produit brusquement un décrochement correspondant à l'amorce d'un

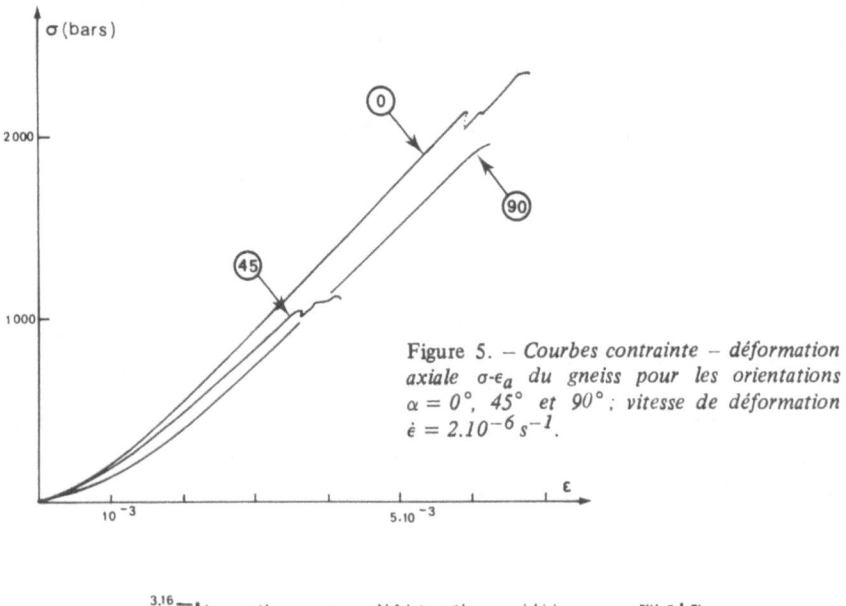

Figure 5. — *Courbes contrainte — déformation axiale* σ-ε$_a$ *du gneiss pour les orientations* α = 0°, 45° *et* 90° *; vitesse de déformation* ε̇ = 2.10^{-6} s^{-1}.

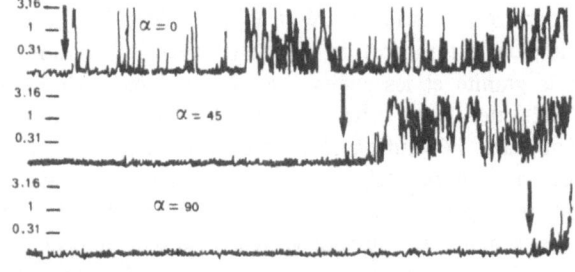

Figure 6. — *Diagramme d'enregistrement des amplitudes (m/s²) des microbruits émis pendant l'essai de compression du gneiss pour les orientations* α = 0°, 45° *et* 90°. *La position des flèches correspond à celle des flèches de la figure 5.*

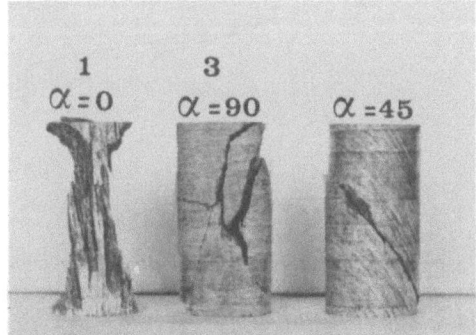

Figure 7. — *Ruptures caractéristiques des éprouvettes de gneiss (hauteur : 10 cm)*

cisaillement le long d'un   plan de foliation. Le frottement provoque de nombreux microbruits généralisés qui conduisent à un diagramme d'enregistrement (Fig. 6) différent des précédents.

3.   Fissuration induite par effet thermique

Une augmentation de température produit, d'une part, une variation différentielle des dimensions des minéraux dans les matériaux formés de cristaux à forte anisotropie de dilatation thermique (quartz et feldspath dans le gneiss et le granite) et, d'autre part, des défauts dans les cristaux. Ces dilatations différentielles des grains de la roche donnent lieu, après retour à la température ambiante, à la formation de microfissures. Cette microfissuration par effet thermique est essentiellement intergranulaire et provoque un déchaussement des grains [2], les microfissures se développant à partir d'un seuil de microfissuration thermique dont la valeur dépend de la nature (composition minéralogique) des roches.

Nous avons mis en évidence l'importance des fissures produites par traitement thermique dans le gneiss et le granite en représentant sur la figure 8 la variation du rapport V/V_0, V désignant la célérité des ondes élastiques longitudinales dans le matériau fissuré et V_0 la célérité dans le même matériau sain, en fonction de la température T. Les courbes de variation de V/V_0 sont linéaires pour le granite et les gneiss 1 et 3 jusqu'aux environs de 500°C et, ensuite, elles présentent une forte discontinuité qui doit marquer un changement important dans l'évolution des défauts produits par effet thermique. Des expériences antérieures effectuées sur un granite à grain plus gros [3] donnent une droite de pente plus forte et présentant la même discontinuité entre 500 et 600°C. Les minéraux, surtout le quartz, augmentent de volume avec la température. A 573°C se produit un changement de phase du quartz qui provoque une brusque augmentation de volume, ce qui accélère la microfis-

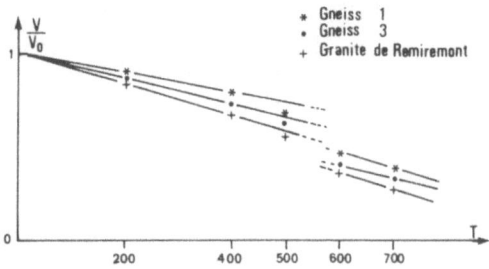

Figure 8. — *Variation de la célérité V des ondes élastiques longitudinales en fonction de la température T de préfissuration pour le granite de Remiremont et les gneiss 1 et 3 (V_0, célérité relative au matériau sain).*

suration. Une élévation de température provoque, en plus des fissures, des défauts en forme de pore dans les cristaux [4]. Les pentes des droites $V/V_0 - T$ correspondant au gneiss 1 et au gneiss 3 ne sont pas identiques. Le gneiss 3 semble plus influencé par le traitement thermique ; il doit s'agir là encore de l'influence, prépondérante suivant cette direction, de la biotite fortement transformée par l'action de la température. La droite relative au granite de Remiremont est caractérisée par une pente plus forte que celle du gneiss, elle correspond à une fissuration plus importante ; cette différence est liée à la dimension plus élevée des cristaux dans le cas du granite. Un matériau isotrope ayant des grains de taille équivalente à celle des minéraux du gneiss aurait une évolution intermédiaire entre gneiss 1 et gneiss 3.

4. Comportement mécanique en fonction de la température de préfissuration

Pour étudier le comportement mécanique du granite de Remiremont et du gneiss suivant les directions 1 et 3, nous avons réalisé des essais de compression à la vitesse de déformation constante $\dot{\varepsilon} = 2.10^{-6} \, \text{s}^{-1}$. Grâce à trois jauges collés parallèlement à l'axe de l'éprouvette et à trois jauges collées suivant la circonférence, on peut mesurer respectivement ε_a (déformation axiale) et ε_t (déformation transversale), calculer θ (variation de volume) $= \varepsilon_a + 2\varepsilon_t$ et ensuite tracer les courbes σ-ε_a, σ-ε_t et σ-θ. D'une façon générale, on distingue plusieurs domaines sur ces courbes (Fig. 9) :

1) La *phase de serrage* ou de tassement des fissures. La courbe ε_a a sa concavité tournée vers le haut. La durée de cette phase est fonction des fissures susceptibles de se fermer.

2) La *phase de déformation linéaire*. Les courbes ε_t et, par suite, θ sont rectilignes. Cette phase est limitée par le seuil de microfracturation.

3) La *phase de développement faible de la fracturation*. La courbe ε_t n'est plus linéaire, tandis que la courbe ε_a demeure linéaire.

4) La *phase de développement intense de la fracturation*. La courbe ε_a n'est plus linéaire ; ε_a et ε_t croissent rapidement.

Le point relatif au changement de sens de variation de θ correspond au début de la dilatance du matériau. Ce point peut se trouver soit en phase 3, soit en phase 4. Le diagramme σ-θ peut être caractérisé par le rapport de la surface limitée par l'axe σ et la courbe θ, à la surface qui serait occupée par le même matériau non dilatant (Fig. 9). Ce coefficient de dilatance, égal à 1 pour un matériau non dilatant, est de 0,74 pour le granite de Remiremont, de 0,86 pour le gneiss 1 et de 0,78 pour le gneiss 3.

Les figures 10 et 11 représentent, pour les gneiss 1 et 3, les courbes σ-ε_a relatives à différentes températures de microfissuration. Dans l'ensemble, ces

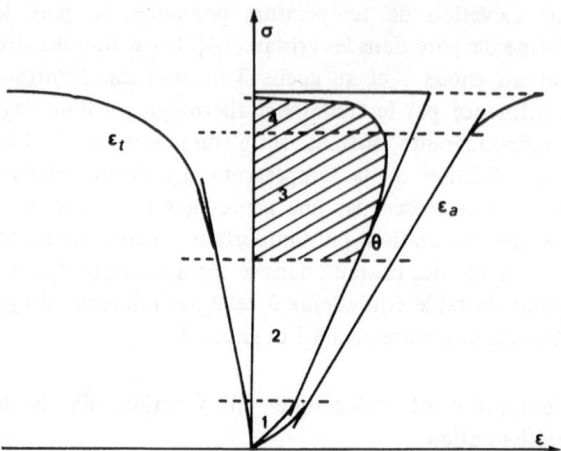

Figure 9. – *Courbes contrainte (σ) – déformation axiale (ϵ_a), latérale (ϵ_t) et volumique (θ) ; schéma théorique.*

courbes mettent en évidence l'importance de l'intensité de la fissuration sur la déformabilité du matériau ainsi que sur la résistance ultime. La figure 12 donne, de manière plus accentuée, le même type de comportement pour le granite de Remiremont. La déformation correspondant à la phase de serrage augmente avec T, tandis que le module de déformation (pente de la partie linéaire des courbes σ-ϵ_a) diminue avec T. Ces variations sont progressives jusqu'à 500°C

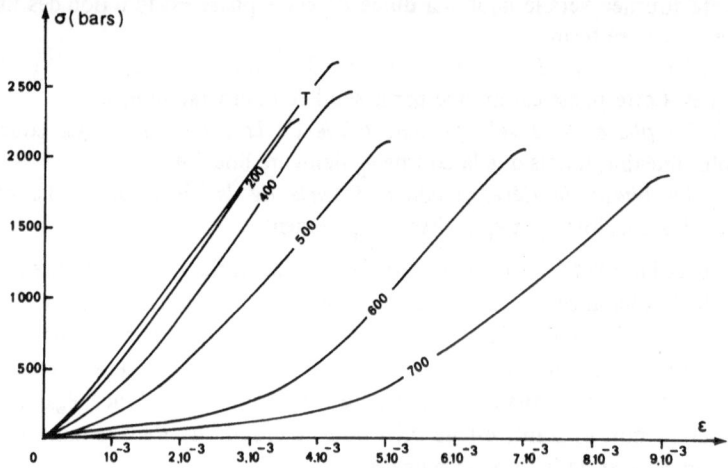

Figure 10. – *Courbes σ-ϵ_a du gneiss 1 pour différentes températures de préfissuration (T, courbe témoin) ; $\dot{\epsilon} = 2.10^{-6}\,s^{-1}$.*

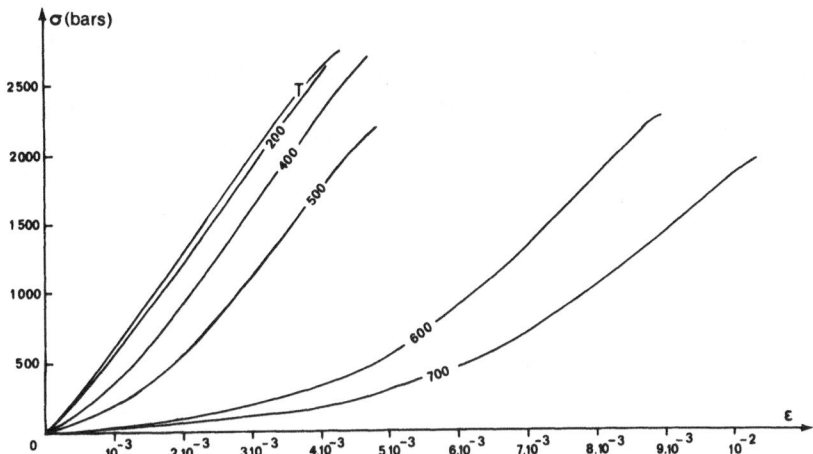

Figure 11. – *Courbes σ-ε_a du gneiss 3 pour différentes températures de préfissuration (T, courbe témoin)*; $\dot{\varepsilon} = 2.10^{-6} s^{-1}$.

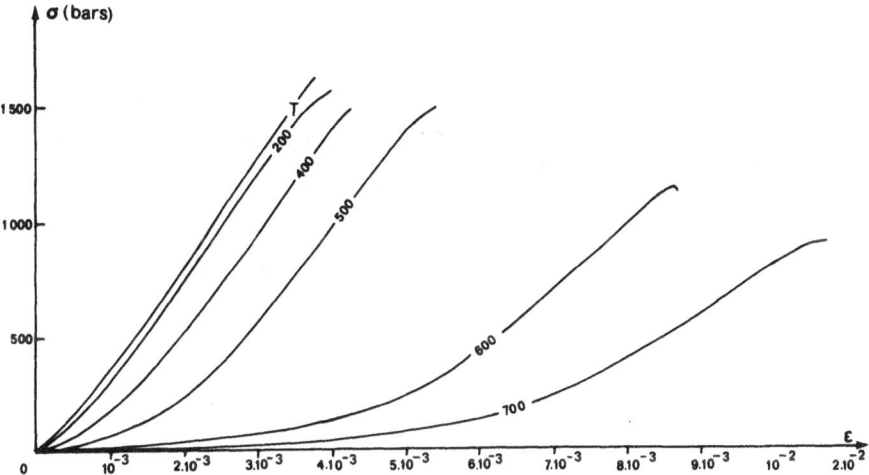

Figure 12. – *Courbes σ-ε_a du granite de Remiremont pour différentes températures de préfissuration (T, courbe témoin)*; $\dot{\varepsilon} = 10^{-6} s^{-1}$.

et, ensuite, elles s'accentent brusquement. Ce phénomène est à rapprocher de la discontinuité observée sur la courbe V/V_0 – T. Pour le gneiss 3, l'amplitude de la phase de serrage (par rapport à l'éprouvette témoin) augmente plus rapidement que dans le cas du gneiss 1. Ce comportement est à mettre en relation avec l'écrasement des fissures orientées perpendiculairement à la direction de compression dans le gneiss 3. Les éprouvettes taillées suivant cette

dernière direction présentent d'ailleurs des variations de hauteur, en fonction
du traitement thermique antérieur, plus importantes que celles de direction 1.
Il faut noter également une diminution du module de déformation plus sensible
pour le gneiss 3 que pour le gneiss 1. Quels que soient la direction de sollicitation et le degré de microfissuration, le gneiss présente un comportement
fragile et nous n'avons dans aucun cas pu obtenir la courbe σ-ε postérieure au

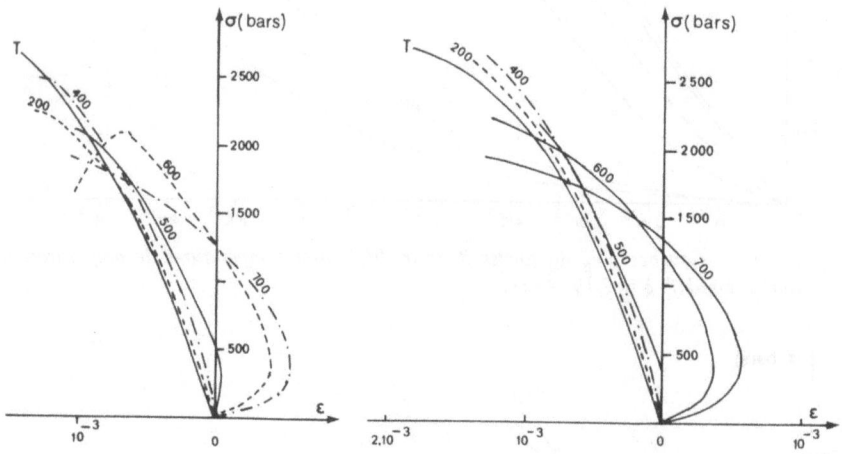

Figure 13. — *Courbes σ-ε$_t$ du gneiss
1 pour différentes températures de
préfissuration (T, courbe témoin);
ε̇ = 2.10^{-6}s^{-1}.*

Figure 14. — *Courbes σ-ε$_t$ du gneiss 3 pour différentes températures de préfissuration (T, courbe
témoin); ε̇ = 2.10^{-6}s^{-1}.*

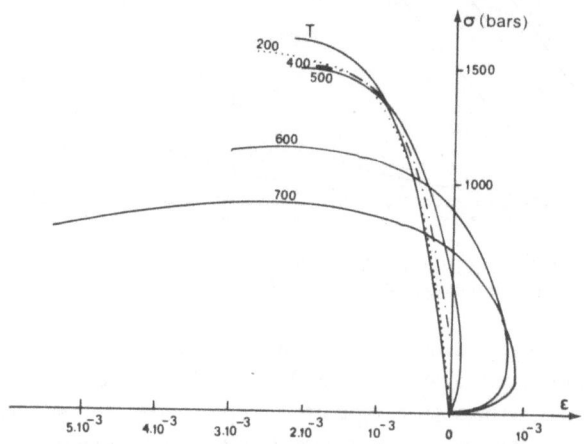

Figure 15. — *Courbes σ-ε$_t$ du granite de Remiremont pour différentes températures de
préfissuration (T. courbe témoin); ε̇ = 10^{-6}s^{-1}.*

pic. Par contre, le granite de Remiremont peut se contrôler à partir d'une température de préfissuration de 600°C.

Sur les figures 13, 14 et 15 sont portées les courbes $\sigma\text{-}\epsilon_t$ et sur les figures 16, 17 et 18, les courbes $\sigma\text{-}\theta$ respectivement pour les gneiss 1 et 3 et pour le granite de Remiremont. Ces courbes diffèrent rapidement du modèle théorique avec la température ; en effet, la partie linéaire de la courbe $\sigma\text{-}\theta$ disparaît

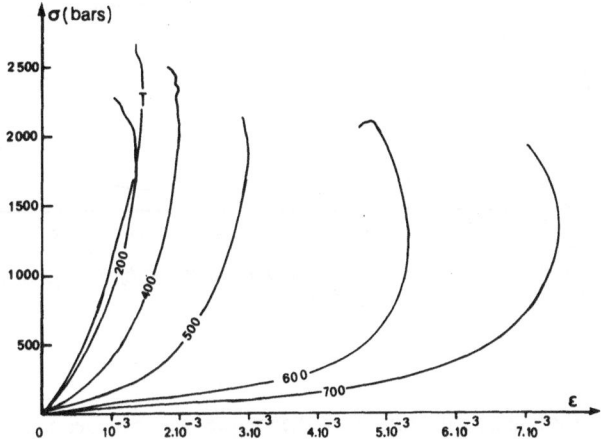

Figure 16. – *Courbes $\sigma\text{-}\theta$ du gneiss 1 pour différentes températures de préfissuration (T, courbe témoin) ; $\dot{\epsilon} = 2.10^{-6} s^{-1}$.*

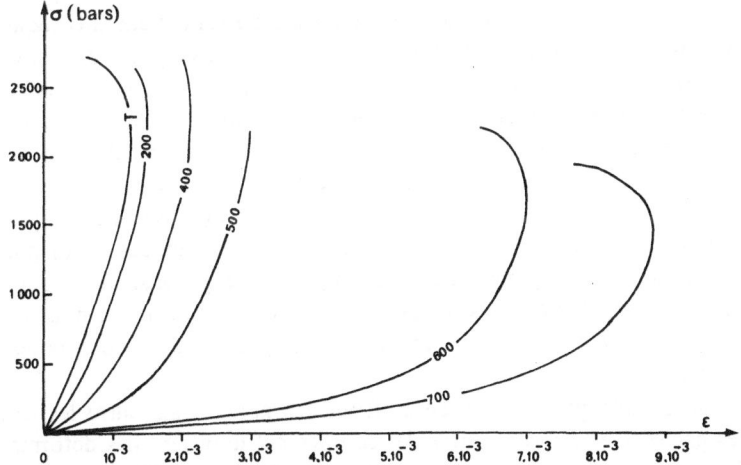

Figure 17. – *Courbes $\sigma\text{-}\theta$ du gneiss 3 pour différentes températures de préfissuration (T, courbe témoin) ; $\dot{\epsilon} = 2.10^{-6} s^{-1}$.*

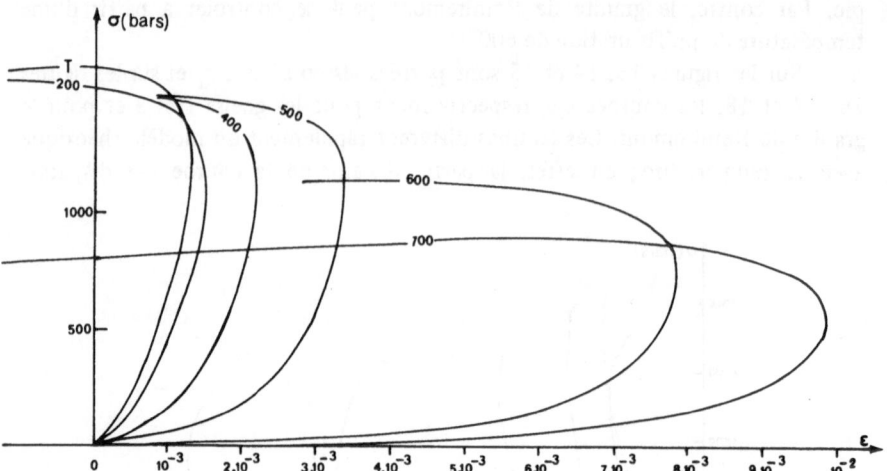

Figure 18. – *Courbes σ-θ du granite de Remiremont pour différentes températures de préfissuration (T, courbe témoin)*; $\dot{\epsilon} = 10^{-6}s^{-1}$.

très vite avec l'intensité du traitement thermique. Les courbes $\sigma\text{-}\epsilon_t$ ont (normalement), pour l'éprouvette témoin, une pente à l'origine négative. La valeur absolue de cette pente augmente avec la température T de préfissuration et elle devient positive à partir de $T = 400°C$ pour le granite de Remiremont et à partir de $T = 500°C$ pour les gneiss 1 et 3. Cette contraction globale du matériau, en début d'essai, évolue avec l'intensité du traitement thermique. On peut supposer qu'en plus de fissures, il y a formation de pores dus au départ d'inclusions fluides. Le matériau est très certainement fissuré et poreux. Les pores peuvent être assimilés à des sphères. L'analyse de la répartition des contraintes autour d'une sphère creuse placée dans un matériau élastique soumis à une compression uniaxiale, montre que la contrainte tangentielle (au voisinage immédiat de la sphère) située dans le plan équatorial est une compression [5]. L'existence de ces contraintes de compression autour des pores pourrait expliquer la contraction globale du matériau jusqu'à ce que le seuil de contrainte nécessaire à l'écrasement des pores soit atteint. Ce phénomène est plus sensible pour le granite de Remiremont, roche à grain plus gros, donc à pores plus importants.

Il ne semble pas y avoir de différence fondamentale entre les courbes $\sigma\text{-}\epsilon_t$ des gneiss 1 et 3; par contre, les courbes σ-θ montrent des déformations plus importantes pour le gneiss 3, reflétant ainsi les différences de déformation axiale.

5. Conclusion

Le gneiss faisant l'objet de cette étude est une roche à anisotropie planaire continue et à grain très fin. L'anisotropie est due à une alternance bien marquée de lits de biotite et de lits de quartz et de feldspaths. Suivant les deux directions principales d'anisotropie du matériau à l'état naturel, le comportement mécanique se différencie par la célérité des ondes élastiques, par le mode de rupture et la résistance ultime (en compression simple). Pour ces deux directions, on observe un comportement analogue en déformation axiale ; la méthode utilisée pour la détermination des déformations latérales ne permet pas de mettre en évidence une éventuelle différence en ce qui les concerne.

L'influence de l'effet thermique, mise en évidence au moyen de la célérité des ondes élastiques, est plus importante selon la direction 3, perpendiculaire au plan de foliation, que suivant la direction 1. Autrement dit, le nombre de fissures perpendiculaires à la direction 3 est supérieur à celui des fissures perpendiculaires à la direction 1. Les fissures thermiques se développent plus facilement entre les feuillets de mica, ce qui explique la formation d'un plus grand nombre de fissures parallèlement au plan de foliation. Il en découle une déformabilité axiale et volumique plus élevée du matériau préfissuré pour les éprouvettes orientées selon la direction 3, c'est-à-dire pour les éprouvettes perpendiculaires à la foliation.

REFERENCES

[1] HOUPERT R., F. HOMAND-ETIENNE et J.-P. TISOT. − "Mécanismes de propagation de la rupture en compression dans les roches cristallines". *Bull. Soc. Géol. France*, 7e sér., 18 (1976) : 1 583-1 589.

[2] PERAMI R. − "Formation des microfissures dans les roches sous l'effet de variations homogènes de températures". In *Fissuration des roches*, C.R. Symp. Internat. Méc. Roches, Nancy, 1971, vol. 1, comm. 1-6, 11 p.

[3] HOUPERT R. et F. HOMAND-ETIENNE. − "Influence de la température sur le comportement mécanique des roches"; *IVe Congr. Internat. Méc. Roches*, Montreux, 1979, à paraître.

[4] SPRUNT E.S. et W.F. BRACE. − "Direct observation of microcavities in crystalline rocks". *Internat. J. Rock Mech. Min. Sci.*, 11 (1974): 139-150.

[5] TIMOSHENKO S. et J.N. GOODIER. − *Théorie de l'élasticité*. Lib. Polytech. Ch. Béranger, Paris, 1961.

ABSTRACT

In the case of rocks composed of crystals, an increase in temperature produces differential changes in size and volume of the crystals. This results in microcracks whose intensity increases in proportion to the temperature. The present study deals with the mechanical behaviour under uniaxial compression (i.e., variation in axial and lateral deformation) of gneiss samples prefractured at temperatures ranging from 200°C-700°C. The intensity of the microcracks was determined by measuring elastic-wave velocities.

Session 8

Anisotropy of Consolidated Clays and of Materials with Internal Friction

Anisotropie des Argiles Consolidées et des Matériaux à Frottement Interne

Anisotropic Consolidation
of Initially Isotropic Soil

A. Cividini, G. Gatti and G. Gioda

Department of Structural Engineering,
Politecnico di Milano, Italy.

1. Introduction

It is recognized that almost all natural soil deposits show a certain amount of anisotropy that affects both the stress-strain relationships and the ultimate shear resistance [1, 2].

Among the causes of this behaviour, two seem to be of major importance [3], namely: a) the conditions under which the soil deposit was formed; b) the stress and strain histories which the deposit underwent after its formation. Both causes influence the relative position of the soil particles, their orientation and the way in which they are bonded together, or, in other words, the microscopic structure and fabric of the natural soil [4].

Under given conditions of deposition and/or stress and strain histories, the properties of soil fabric may vary with direction. As a consequence of the fabric anisotropy, an anisotropy of the soil mechanical characteristics shows up. Evidence of the correlation between the two anisotropies was given by various experimental studies (see, e.g. [5]). This correlation is particularly evident for soils having particles with plately shapes like clay or fine silt. In fact, it was observed in [6], [7] and [8] that isotropically consolidated specimens show random fabric and isotropic mechanical characteristics. On the other hand, anisotropically (or K_0) consolidated samples, in which most particles are oriented normally to the direction of the major principal stress, are usually characterized by anisotropy of the mechanical properties.

The present paper contains a numerical and experimental study of the anisotropy induced by the stress and strain histories in samples of saturated cohesive soil undergoing one-dimensional consolidation. The reorientation of the soil particles during the skeleton loading process, and the subsequent fabric modification, are assumed as the basic causes of this induced anisotropy.

In the numerical part of the study the soil is treated as a biphase medium and the consolidation problem is solved by means of a non-linear finite difference technique. The soil is assumed isotropic at the initial stage of consolidation, as each sample is completely remolded before testing it. During the consolidation process two different relationships are taken into account for the volumetric and deviatoric stress-strain behaviours. For the volumetric behaviour, the classical linear law between void ratio (or volumetric deformation) and logarithm of volumetric effective stress is assumed. As to the deviatoric behaviour, it is considered that, when increasing the shear deformation, a certain amount of particle reorientation is produced, which in turn causes a decrease of the instantaneous or tangent shear modulus. In a one-dimensional consolidation test no shear deformation develops in the horizontal plane, thus the vertical shear resistance becomes different from the horizontal one and the initially isotropic specimen becomes non-isotropic.

The soil skeleton behaviour is defined by means of a non-linear elastic incremental scheme; therefore such effects as the development of anisotropic shear strength are not accounted for.

The results of the numerical analysis are compared with those of laboratory oedometer tests while the soil parameters are derived from triaxial test data. On the basis of the numerical results, the variation of the horizontal and vertical tangent shear moduli in time is obtained.

The mere intention of this study is to show how the stress and strain histories may produce anisotropy in a sample of cohesive soil and no attempt is made to formulate a general law for the anisotropic behaviour of soils.

2.   Soil skeleton stress strain relationship

Consider a soil sample subjected to a compression triaxial test under drained conditions. Assume that the sample has been completely remolded before the test, and that it has been consolidated under a hydrostatic pressure. Thus, at the beginning of the shearing stage of the test, the soil has isotropic mechanical characteristics.

The loading process of the sample is characterized by two equal principal effective stresses in the horizontal directions $\sigma'_x = \sigma'_y$ and by the major principal effective stress σ'_z in the vertical direction. Assume that anisotropy of the soil fabric, and in turn anisotropy of the mechanical characteristics, is induced by the non-isotropic stress-strain history during the loading process. Since at every stage of the loading process σ'_x is equal to σ'_y, a cross anisotropy is produced, i.e. the soil mechanical characteristics are the same in every direction belonging to a horizontal plane and differ from those in the vertical direction. Because of this simple type of anisotropy, and due to the particular loading conditions, the horizontal deformation in the x direction is equal to that in the y direction at every stage of the test, and the principal stress directions coincide with those of principal strains.

Taking into account that $\sigma'_x = \sigma'_y$ and that $\epsilon_x = \epsilon_y$, the incremental stress-strain relationship for a soil skeleton element can be written in the form

$$\Delta\sigma' = E\,\Delta\epsilon, \tag{1}$$

where

$$\Delta\sigma' = \begin{Bmatrix} \Delta\sigma'_z \\ \Delta\sigma'_x \end{Bmatrix}; \quad E = \begin{bmatrix} e_{11} & e_{12} \\ e_{12}/2 & e_{22} \end{bmatrix}; \quad \Delta\epsilon = \begin{Bmatrix} \Delta\epsilon_z \\ \Delta\epsilon_x \end{Bmatrix}. \tag{2a, b, c}$$

In order to define the entries of the tangent constitutive matrix E it is convenient to express the sample stress and strain states by means of the quantities

$$\sigma'_{vol} = \frac{1}{3}(\sigma'_z + 2\sigma'_x), \tag{3a}$$

$$\tau = \frac{\sigma'_z - \sigma'_x}{2}, \tag{3b}$$

$$\epsilon_{vol} = \epsilon_z + 2\epsilon_x, \tag{3c}$$

$$\gamma = \epsilon_z - \epsilon_x, \tag{3d}$$

where σ'_{vol} is the volumetric effective stress; τ is the maximum shear stress; ϵ_{vol} is the volumetric deformation and γ is the maximum shear deformation.

The general incremental relationship between the quantities defined by Eqs. (3) can be expressed by the equation

$$\begin{Bmatrix} \Delta\sigma'_{vol} \\ \Delta\tau \end{Bmatrix} = \begin{bmatrix} a_{11} & a_{12} \\ a_{21} & a_{22} \end{bmatrix} \begin{Bmatrix} \Delta\epsilon_{vol} \\ \Delta\gamma \end{Bmatrix}, \tag{4}$$

where the coefficients a_{ij} depend upon the stress and strain histories.

By substituting Eqs. (3) into Eq. (4), the following relationships between the coefficients a_{ij} and e_{ij} are obtained:

$$e_{11} = a_{11} + \frac{2}{3}a_{22} + \frac{4}{3}a_{21}, \tag{5a}$$

$$e_{12} = 2a_{11} - \frac{2}{3} a_{22} + \frac{2}{3} a_{21}, \tag{5b}$$

$$e_{22} = 2a_{11} + \frac{1}{3} a_{22} - \frac{4}{3} a_{21}. \tag{5c}$$

In addition, it can be shown that

$$a_{12} = \frac{2}{3} a_{21}. \tag{6}$$

Eqs. (5) imply that the entries of the matrix **E** are functions of the coefficients a_{ij} which in turn can be determined on the basis of experimental data relating σ'_{vol} and τ to ϵ_{vol} and γ. Since only three out of the four coefficients a_{ij} are independent, for defining the matrix **E** it will be enough to establish the three following experimental relationships: σ'_{vol} vs. ϵ_{vol} ; τ vs. γ; τ vs. ϵ_{vol}.

3.   Triaxial test results

For defining the variation of the coefficients of the matrix **E** with the stress and strain histories, a series of drained compression triaxial tests was performed on samples of silty clay soil characterized by the following index properties:

Liquid limit $w_\varrho = 52\,\%$;   Plastic limit $w_p = 20\,\%$;   Plasticity index $I_p = 32\,\%$;
Clay fraction $= 42\,\%$;   Silt fraction $= 58\,\%$;   Activity index $= 0.76$.

Before testing, the soil was completely remolded at a water content higher than the liquid limit. A large sample (diameter about 12 cm and height about 20 cm) was prepared and consolidated under a hydrostatic pressure of $0.7\ \text{kg/cm}^2$. This first consolidation stage was performed in order to give to the remolded sample the resistance necessary for supporting its own weight while mantaining the original cylindrical shape. From the large sample six triaxial test specimens (3.9 cm × 7.8 cm) were trimmed and were consolidated in the triaxial cell under different all around pressures ranging from 1 to 8 kg/cm^2. In order to ensure an almost complete saturation of the samples they have been back pressurized before consolidation. A maximum back pressure value of 3 kg/cm^2 was applied in six to eight increments. In the last two increments a Skempton B coefficient of $0.95 \div 0.98$ was measured for all samples.

After consolidation, the drained triaxial tests were carried on by increasing the vertical load through small finite increments and by decreasing the cell pressure in such a way that the effective volumetric stress σ'_{vol} maintained a constant value, equal to the consolidation pressure, throught the test. In order to ensure a complete excess pore pressure dissipation before the application of a new load increment, the load increment rate was computed according to Gibson [9]. The excess pore pressure dissipation was controlled by connecting one sample base to a pore pressure measuring device. Drainage was allowed laterally and at the other sample base.

During the consolidation and shearing test stages the volume variation was evaluated by measuring the volume of water leaving the sample. A mechanical dial gauge measured the variation of sample height.

The results of the triaxial tests are summarized in Figure 1 to 4. In all the figures the dashed lines represent the numerical interpretation of the experimental data. Figures 1 and 2 refer to data obtained during the consolidation stage, while Figures 3 and 4 refer to data obtained in the shearing stage.

In Figure 1 the variation of the logarithm of the permeability coefficient K is reported versus the volumetric strain ϵ_{vol}. The data show a linear pattern, thus the following relationship can be established between K and ϵ_{vol} :

$$K = K^\circ \, 10^{\dfrac{\epsilon_{vol}^{\circ} - \epsilon_{vol}}{C_k}} \qquad (7)$$

In Eq. (7) K° and ϵ_{vol}° are the coordinates of a reference point and C_k is an experimental coefficient; for the soil under examination C_k has a value of 0.262.

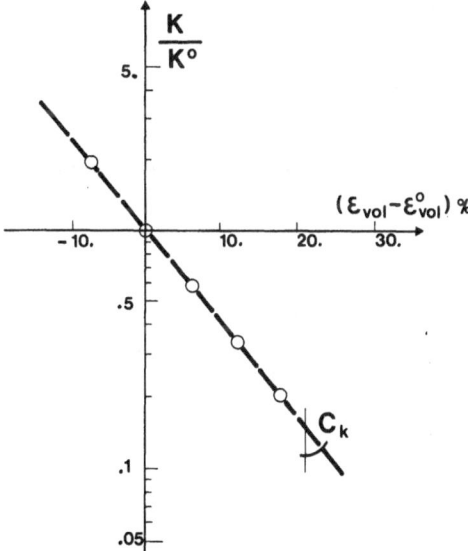

Figure 1. – *Relationship between the permeability coefficient K and the volumetric deformation* ϵ_{vol}.

Figure 2 shows the well known linear relationship between the logarithm of the volumetric effective stress σ'_{vol} and the volumetric strain ϵ_{vol}. The experimental data can be approximated by means of the law

$$\sigma'_{vol} = \sigma'^{o}_{vol} \, 10^{\dfrac{\epsilon_{vol} - \epsilon^{o}_{vol}}{C_\epsilon}} \tag{8}$$

where σ'^{o}_{vol} and ϵ^{o}_{vol} are the coordinates of a reference point and, in the present case, $C_c = 0.209$.

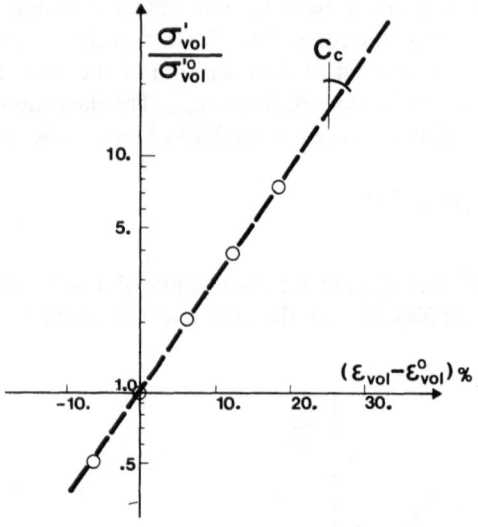

Figure 2. – *Relationship between the effective volumetric stress σ'_{vol} and the volumetric strain ϵ_{vol}.*

The relationship between the maximum shear stress τ (normalized with respect to the consolidation pressure $\sigma'_c = \sigma'_{vol}$) and the maximum shear deformation γ is reported in Figure 3. A reasonably good approximation of the experimental data can be obtained by means of the following exponential equation:

$$\frac{\tau}{\sigma'_c} = a_1 \, (\gamma \cdot 100) + a_2 \, (\gamma \cdot 100)^n. \tag{9}$$

The following values of the coefficients of Eq. (9) were computed by means of a best fitting least square algorithm: $a_1 = -0.026$, $a_2 = 0.172$; $n = 0.61$.

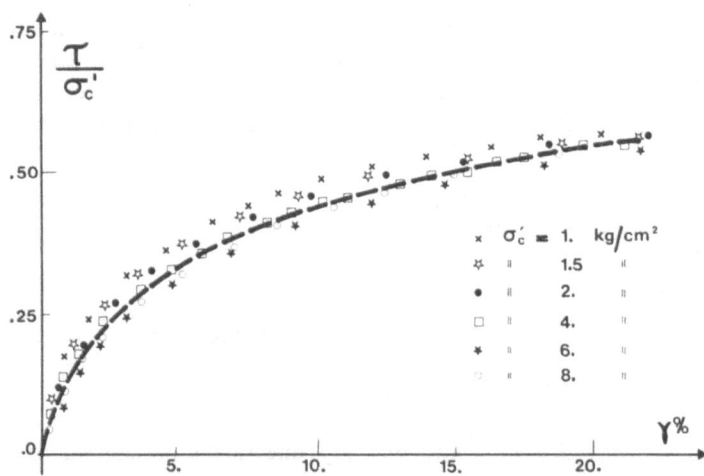

Figure 3. – *Relationship between the maximum shear stress* τ, *divided by the consolidation pressure* σ'_c, *and the maximum shear deformation* γ.

In Figure 4 the relationship between the volumetric strain and the dimensionless maximum shear stress is shown and approximated by means of the equation

$$\frac{\tau}{\sigma'_c} = a_3 + a_4 \, (\epsilon_{vol} \cdot 100)^{\frac{1}{2}} + a_5 \, (\epsilon_{vol} \cdot 100)^m, \tag{10}$$

where: $a_3 = 0.1$; $a_4 = 0.13$; $a_5 = 0.0003$; m = 3.45.

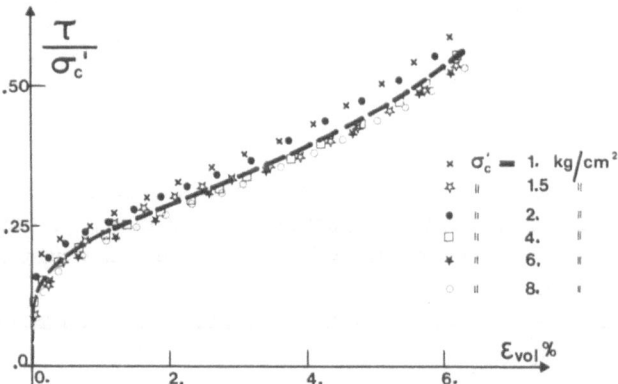

Figure 4. – *Relationship between the maximum shear stress* τ, *divided by the consolidation pressure* σ'_c, *and the volumetric strain* ϵ_{vol}.

Eqs. (8), (9) and (10) define the effective stress-strain behaviour of a soil element subjected to increasing volumetric and deviatoric stresses. In fact, deriving Eqs. (8) and (10) with respect to ϵ_{vol}, and Eq. (9) with respect to γ, the following expressions for the coefficients a_{ij} of Eq. (4) are obtained:

$$a_{11} = \sigma_{vol}^{\prime o} \frac{\ell n \, 10}{C_c} \, 10^{\frac{\epsilon_{vol} - \epsilon_{vol}^o}{C_c}} \tag{11a}$$

$$a_{21} = \sigma_c^{\prime} \, [50 \, a_4 \, (\epsilon_{vol} . 100)^{-\frac{1}{2}} + 100 \, a_5 \, m \, (\epsilon_{vol} . 100)^{m-1}], \tag{11b}$$

$$a_{22} = \sigma_c^{\prime} \, [100 \, a_1 + 100 \, a_2 \, n \, (\gamma . 100)^{n-1}]. \tag{11c}$$

Substituting Eqs. (11) into Eqs. (5), one finally obtains the coefficients of the constitutive matrix E.

4. Numerical approach

For checking whether the relationships obtained in the preceding section give a reasonable approximation of the soil behaviour, the displacement vs. time curve of a laboratory oedometer test have been compared with that obtained by a numerical analysis based on the above relationships. In addition to this comparison, the numerical results allow for defining the variation of vertical and horizontal tangent shear moduli during consolidation. In this section the procedure adopted for the numerical analysis is briefly outlined.

As well known, the behaviour of a soil element undergoing one-dimensional consolidation, neglecting the viscous effects, is governed by the equation (see e.g. [10])

$$\frac{1}{\gamma_w} \frac{\partial K_z}{\partial z} \frac{\partial u}{\partial z} + K_z \frac{\partial^2 u}{\partial z^2} = \frac{\partial \epsilon_z}{\partial t}, \tag{12}$$

where z is the direction in which both pore fluid flow and soil deformation ϵ_z take place; K_z is the coefficient of permeability; u is th pore pressure in excess of the steady state and hydrostatic values; γ_w is the unit weight of the pore fluid; t is the time.

Since $\epsilon_x = \epsilon_y = 0$, the following incremental stress-strain relationship can be derived from Eqs. (2):

$$\Delta \sigma_z^{\prime} = e_{11} \, \Delta \epsilon_z \, ; \quad \Delta \sigma_x^{\prime} = \frac{e_{12}}{2} \, \Delta \epsilon_z. \tag{13a, b}$$

Assume that the numerical integration of Eq. (12) is carried out by means of a step by step procedure in which the material parameters are assumed constant during each time step, while possible changes are allowed between a step and the next. On the basis of this assumption, Eq. (13a) can be derived with respect to time, assuming e_{11} to be constant along a generic time integration step,

$$\frac{\partial \sigma_z'}{\partial t} = e_{11} \frac{\partial \epsilon_z}{\partial t}. \tag{14}$$

Taking into account the relation between total and effective stresses (i.e. $\sigma = \sigma' + u$), one can rewrite Eq. (14) in the form

$$\frac{\partial \sigma_z'}{\partial t} - \frac{\partial u}{\partial t} = e_{11} \frac{\partial \epsilon_z}{\partial t}, \tag{15}$$

where σ_z is the total stress applied in the z direction.

Assume that the soil element loading process is subdivided into a series of instantaneous load increments separated by given time intervals. Along each interval σ_z remains constant, thus $\partial \sigma_z/\partial t$ vanishes and the following expression is obtained from Eq. (15):

$$\frac{\partial \epsilon_z}{\partial t} = -\frac{1}{e_{11}} \frac{\partial u}{\partial t}. \tag{16}$$

Substituting Eq. (16) into Eq. (12), one finally obtains the following expression:

$$\frac{1}{\gamma_w} \left(\frac{\partial K_z}{\partial z} \frac{\partial u}{\partial z} + K_z \frac{\partial^2 u}{\partial z^2} \right) = -\frac{1}{e_{11}} \frac{\partial u}{\partial t} \tag{17}$$

Eq. (17) can be easily integrated, adopting a finite difference discretization scheme both in the space and time domains.

Consider a soil sample limited at its upper and lower ends by pervious and impervious boundaries, respectively. The sample is discretized by means of n − 1 elements and n nodes at which the pore pressure u is defined; assume a linear variation of u in each element (Fig. 5). Adopting a central difference scheme for $\partial u/\partial z$ and $\partial^2 u/\partial z^2$ and a forward scheme for $\partial u/\partial t$, the following

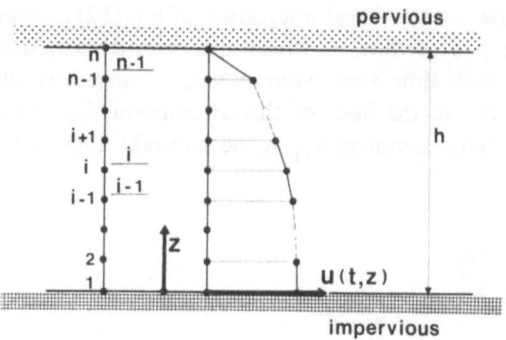

Figure 5. – *Finite difference scheme for the one-dimensional consolidation test.*

equation, relating for the generic node i, the pore pressure at time $t + \Delta t$ to that at time t, is obtained from Eq. (17)

$$u_i^{t+\Delta t} = u_i^t + \frac{\Delta t}{\gamma_w \, c_i^t} \left[\frac{(K_{z,i+1}^t - K_{z,i-1}^t)\,(u_{i+1}^t - u_{i-1}^t)}{(\Delta z_{(i)} + \Delta z_{(i-1)})^2} + \right.$$

$$\left. + \frac{2\,K_{z,i}^t}{\Delta z_{(i-1)} + \Delta z_{(i)}} \left(\frac{u_{i+1}^t - u_i^t}{\Delta z_{(i)}} - \frac{u_i^t - u_{i-1}^t}{\Delta z_{(i-1)}} \right) \right]. \qquad (18a)$$

Here: $c_i^t = \left(\dfrac{1}{e_{11}}\right)_i^t$; $\Delta z_{(i)} = z_{i+1} - z_i$; $\Delta z_{(i-1)} = z_i - z_{i-1}$.

For node 1, situated on the impervious boundary, the condition $(\partial u(z,t)/\partial z)_{z=0} = 0$ has to be imposed. Then Eq. (18a) simplifies to the following form:

$$u_1^{t+\Delta t} = u_1^t + \frac{2\,\Delta t}{\gamma_w \, c_1^t} \frac{K_{z,1}^t}{\Delta z_{(1)}^2} \, (u_2^t - u_1^t). \qquad (18b)$$

For node n, situated on the pervious boundary, the conditions $(u_n)_{t=0} = 0$ and $(\partial u(z,t)/\partial t)_{z=h} = 0$ have to be considered. Thus, the finite difference equation reduces to

$$u_n^t = u_n^{t+\Delta t} = 0. \qquad (18c)$$

The system of equations obtained by writing Eqs. (18) for each mesh node allows for the numerical integration of Eq. (17), when the initial distribution of pore pressure due to the external load increment is known. Note that, because of the simple nature of the problem at hand, the resulting system of equations is uncoupled; thus the solution can be carried out with an extremely limited computational effort.

In the present study, a constant initial pore pressure distribution equal to the increment of external load has been assumed throughout the sample, i.e. $\Delta u_i^{t=0} = \Delta \sigma_z$. Only for the node n, on the drainage boundary, $\Delta u_n^{t=0} = 0$ was assumed.

As usual when dealing with the numerical integration of consolidation equations, the values of the time increments and of the element size play a major role in the accuracy of the final results [11].

The choice of the element size has been made by considering that at the beginning of the consolidation process a discontinuity of the pore pressure distribution takes place at the sample drainage boundary. In fact, just after the load application, the pore pressure is different from zero at every sample point, except for the point on the drainage boundary where u = 0. Since the adopted model assumes linear variation of u within the elements, theoretically the height of the elements facing the drainage boundary should tend to zero in order to approximate the pore pressure discontinuity. In practice, an acceptable accuracy is achieved when the height of these elements has order of magnitude about one or two times smaller than that of the sample height.

As to the time increments, it can be observed that when their value is too large the computed pore pressure tends to oscillate both along the z direction and during time. With a rather empirical approach, in the present study the time increment values were decreased until the oscillations did not affect the first three digits of the pore pressure values.

Because of the very small value of the time increments adopted, the assumption that material properties are constant within each time step did not produce noticeable errors. In fact, only a negligible violation of the non-linear stress-strain and permeability-strain relationships was observed at the end of the numerical analysis.

Having computed the pore pressure distribution at a given time, we can evaluate the deformation increments by means of Eq. (16), rewritten in the form

$$\epsilon_{z,i}^{t+\Delta t} - \epsilon_{z,i}^{t} = -\left(\frac{1}{e_{11}}\right)_i^t (u_i^{t+\Delta t} - u_i^t), \tag{19}$$

and then the corresponding increments of effective stress can be obtained from Eqs. (13).

5. Oedometer versus numerical results

In order to bring out the influence of the stress and strain histories on the anisotropy of cohesive soils, a one-dimensional consolidation test was performed on a sample of the silty clay under examination. Before testing, the oedometer specimen was remolded at a water content ($w_i = 60\%$) higher than the liquid limit. This initial condition made it possible to consider the soil as isotropic at the beginning of consolidation. In order to avoid excessive deformation, due to the high initial water content and compressibility, a low value of the vertical pressure, $\sigma_z = 0.50$ kg/cm^2, was applied to the sample. In Figure 6 the dots represent the vertical deformation (normalized with respect to that at the end of primary consolidation) during time, obtained by the oedometer test.

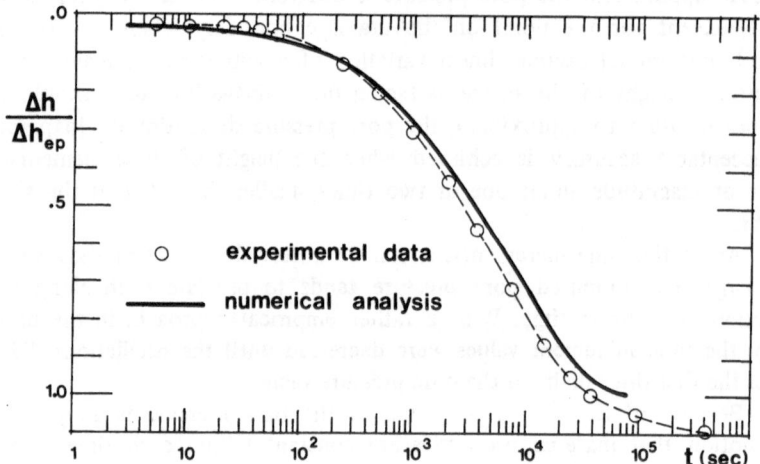

Figure 6. – *Total vertical deformation of the oedometer specimen vs. time, comparison between experimental and numerical data. (Δh_{ep} = variation of sample height at the end of primary consolidation).*

The laboratory test was simulated by means of a numerical analysis based on the finite difference procedure and on the non-linear stress-strain relationships described in the preceding sections. In the analysis the initial sample condition was defined by the effective stress state due to the soil's own weight. In Figure 6 the solid line represents the settlement versus time curve obtained by the numerical analysis.

A reasonable agreement was observed between experimental and numerical data. Their difference tends to increase towards the end of primary consolidation, due to the influence of secondary compression that was not accounted for in the finite difference calculations. On the basis of these results it can be concluded that the stress-strain relationships adopted for the soil skeleton approximate with acceptable accuracy the real soil behaviour.

Other diagrams obtained from the numerical analysis are shown in Figures 7, 8 and 9. In Figure 7 the ratio between vertical and horizontal effective stresses, σ'_z/σ'_x, is reported versus time, at two different locations in the soil sample. From the curves it appears that the ratio does not vary monotonously with time. This behaviour is a consequence of the stress-strain laws adopted in the analysis. According to these laws, the volumetric stiffness of the soil skeleton increases with increasing volumetric stress while two opposite effects influence the variation of the vertical tangent shear modulus G_z.

A first effect produces a decrease of G_z with increasing shear deformation (effect a); on the other hand an increase of G_z is produced by increasing volumetric effective stress (effect b). The diagram in Figure 8 shows that effect

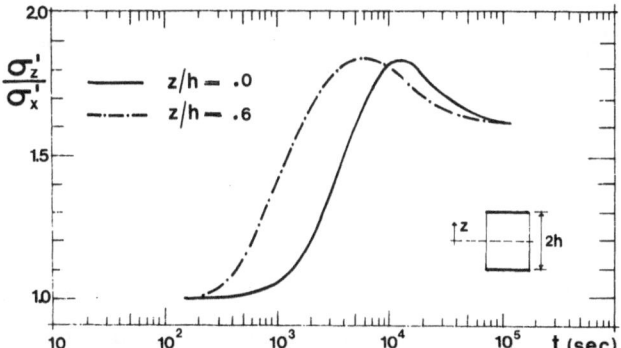

Figure 7. – *Ratio between vertical and horizontal effective stress vs. time.*

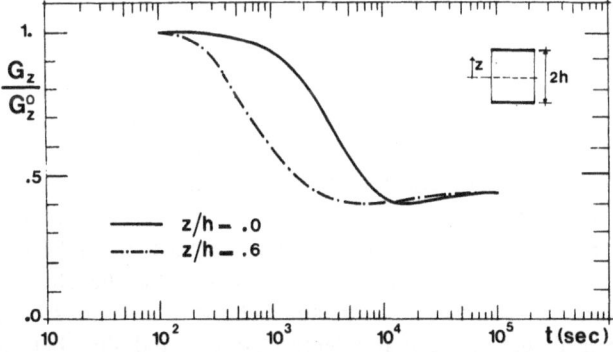

Figure 8. – *Variation of the vertical shear modulus during time.*

a prevails on effect *b* during the early stage of consolidation while, towards the end of consolidation, effect *b* is predominant. The simultaneous variations of the volumetric and deviatoric resistances, together with the soil dilatancy allowed for by Eq. (10), produce the σ'_z/σ'_x plot of Figure 7.

The increasing anisotropy of the soil during consolidation is illustrated in Figure 9, where the ratio between the vertical and the horizontal tangent shear moduli is reported versus time. As expected, the ratio decreases during consolidation. In fact, both G_z and G_x increase with increasing σ'_{vol}, while only G_z decreases with γ since the shear deformation is not present in the horizontal plane. Therefore, even though the soil is isotropic at the beginning of consolidation, the non-homogeneous distribution of shear deformations produces an anisotropy whose effects, in terms of tangent shear moduli, are shown in Figure 9.

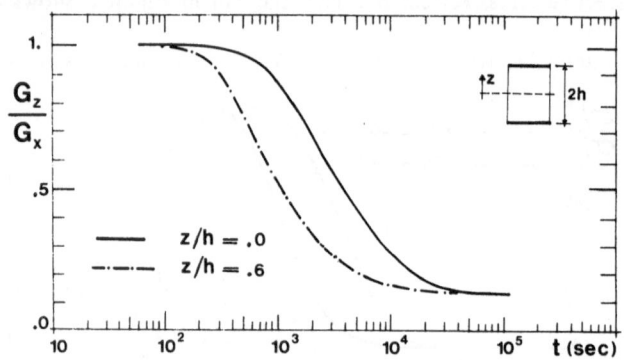

Figure 9. – *Ratio between vertical and horizontal shear moduli vs. time.*

6. Concluding remarks

A study has been presented of the anistropy induced by the stress and strain histories in samples of cohesive soil undergoing one-dimensional consolidation. The reorientation of the soil particles during the skeleton loading process is assumed as the basic cause of this induced anisotropy.

Simple stress-strain relationships, allowing for the effect of the deviatoric stresses on the volumetric deformation, are derived from drained triaxial test data. These relationships are applied to the numerical simulation of an oedometer test. On the basis of the numerical results it can be observed that the non-linear and non-isotropic soil skeleton behaviour influences both the ratio between vertical and horizontal effective stresses and that between vertical and horizontal shear moduli.

In particular, a pronounced increase of the horizontal shear modulus relatively to the vertical one shows up. This effect, in turn, causes the initially isotropic soil to become non-isotropic. The result seems to indicate that in most geotechnical problems concerning cohesive soils, even if normally consolidated, the initial anisotropy of the natural soil deposit should not be disregarded.

Both the numerical and the experimental research on this topic seem worth being persued, by considering the soil as a material whose anisotropic properties are not constant during time but strongly depend on the stress and strain history to which the soil is subjected.

Acknowledgements

This study is part of a research supported by the National (Italian) Research Council. The authors wish to thank Mr. L. Sevino and Mr. F. Iscandri for their foundamental contribution in performing the laboratory tests.

REFERENCES

[1] SAADA A.S. and Ou. CHIN-DER. – "Stress-strain relations and failure of anisotropic clays", *Proc. ASCE*, SM12 99 (1973): 1091.

[2] SANKARAN K.S. and R. BHASKARAN. – "Deformation and failure pattern in an anisotropic kaolinite clay". *Geotechnique*, XXIII (1973): 113.

[3] BASKARAN R. – "Anisotropy for finite element analysis". *Proc. 2nd Int. Conf. on Num. and Anal. Meth. in Geomechanics*, Blacksburg, USA, (1976): 345.

[4] LOH A.K. and R.T. HOLT. – "Directional variation of shear strength and fabric of Winnipeg upper brown clay". *Canadian Geotechnical J.*, 11 (1974): 430.

[5] MITCHELL J.K. – *"Fundamentals of Soil Behaviour"*. New York: J. Wiley & Son, 1976.

[6] KIRKPATRIC W.M. and I.A. RENNIE. – "Directional properties of consolidated kaolin". *Geotechnique*, XXII (1972): 166.

[7] MORGENSTERN N.R. and J.S. TCHALENKO. – "The optical determination of preferred orientation in clays and its application to the study of microstructure in consolidated kaolin". *Proc. of the Royal Society*, London, A300 (1967): 218.

[8] SKETCHLEY C.J. and P.L. BRANSBY. – "The behaviour of the overconsolidated clay in plane strain". *Proc. 8th Int. Conf. Soil Mech. Found. Eng.*, Moscow, (1973): 337.

[9] GIBSON R.E. and D.J. HENKEL. – "Influence of duration of tests at constant rate of strain on "drained" strength". *Geotechnique*, IV (1954): 6.

[10] ZARETSKII Y.K. – *"Theory of Consolidation"* (translated from Russian), Israel Program for Scientific Translation, Jérusalem, 1972.

[11] LEWIS R.W., G.K. ROBERTS and O.C. ZIENKIEWICZ. – "A non-linear flow and deformation analysis of consolidated problems". *Proc. 2nd Int. Conf. on Num. and Anal. Meth. in Geomechanics*, Blacksburg, USA, (1976): 1106.

RESUME

(Consolidation anisotrope des sols initialement isotropes)

Un modèle numérique pour l'anisotropie induite par l'histoire des contraintes et des déformations dans des échantillons de sols cohérents soumis à une consolidation mono-axiale, est développé à partir d'essais en laboratoire. L'essai de base est un essai triaxial draîné, qui permet d'obtenir une relation simple entre les contraintes et les déformations. Le sol est considéré comme étant isotrope au cours de la première phase de consolidation. Une réorientation des particules au cours du processus de chargement est supposée être à l'origine de l'anisotropie induite. Les résultats numériques montrent que les modules de cisaillement verticaux et horizontaux deviennent tout à fait différents au cours de la consolidation. Cette variation appréciable montre que dans la plupart des cas pratiques, l'anisotropie des couches argileuses naturelles ne devrait pas être négligée dans les calculs de la Géotechnique.

Description of Anisotropic Consolidation of Clays

St. Pietruszczak and Z. Mróz

Institute of Fundamental Technological Research, Warsaw, Poland.

1. Introduction

One of the major problems in soil mechanics is proper understanding and description of anisotropy induced by one-dimensional consolidation. It is well-known that specimens of homogeneous clay, trimmed at different directions to the major consolidating stress, exhibit different failure stresses in undrained triaxial compression and different stiffness moduli prior to failure. Such anisotropy can be attributed to preferred orientation of clay particles during consolidation process (textural anisotropy) or to residual stress resulting from inhomogeneity of deformation on the microscale (residual stress anisotropy).

This problem was investigated by numerous authors, cf. [2, 5, 7, 8, 13, 14, 15, 21]. Though preferred orientation has been observed in some natural clays [13], this was not detected in others [15, 21]. Undrained triaxial tests reported by several authors [7, 8, 14, 21] indicate that the undrained strength is highest in specimens trimmed along the line of major preconsolidation stress but the failure envelope in terms of effective stress does not vary significantly with the direction of specimens. The differences in undrained strength result mostly from different stress paths and different evolution of pore water pressure prior to failure. Similarly, the stiffness moduli in the elasto-plastic domain depend essentially on the relative orientation of axes of consolidating stresses and of subsequent testing stresses.

The aim of this paper is to modify the anisotropic hardening model discussed in detail by Mróz, Norris and Zienkiewicz [16, 17] and apply it to simulate quantitatively some deformation processes of K_0-consolidated clays. The major assumption is that textural anisotropy resulting from preferred orientation is neglected and only residual stress anistropy is accounted for. This anistropy is described by a modified kinematic hardening model for which the translation and expansion or contraction of the yield surface depend on its relative position with respect to the consolidation surface characterizing the material prestressing. In Section 2 the formulation of our model will be presented for the case of triaxial testing and in Section 3 some model predictions will be discussed and compared with available experimental data. In Section 4 a general stress state will be considered and the constitutive relations will be derived which constitute generalization of the relations discussed in Section 2.

2. Model description for the case of triaxial test

In this section we shall restrict our analysis to the case of a triaxial test, that is when two effective principal stresses are equal, $\sigma_2 = \sigma_3$. If the compressive stresses and strains are regarded as negative, the stress and strain states are

$$p = -\frac{1}{3}(\sigma_1 + 2\sigma_2), \quad q = \sigma_2 - \sigma_1,$$

$$\epsilon_v = -(\epsilon_1 + 2\epsilon_2), \quad \epsilon_q = \frac{2}{3}(\epsilon_2 - \epsilon_1). \tag{1}$$

The total stresses are $p^t = p + p_w$, $q^t = q$, where p_w denotes the pore water pressure.

Similarly as in [16] the development of our model is based on the following assumptions:

1) The degree of consolidation of soil is represented by the consolidation surface $F = 0$ which depends on varying material density or porosity.
2) The yield surface $f_0 = 0$ encloses the elastic domain in the stress space and translates within the domain enclosed by the consolidation surface $F = 0$.
3) The hardening modulus of the material varies along the stress path from an infinite or very large value on the yield surface to a prescribed value on the consolidation surface. This variation may be described by introducing a set of nesting surfaces and formulating an interpolation rule which relates the hardening modulus to the diameters of the instantaneous active loading and consolidation surfaces [16]. An alternative interpolation rule relates the hardening modulus to the distance between the yield and the consolidation surfaces without the use of nesting surfaces, similarly as in models developed by Krieg [10], and Dafalias and Popov [6] for metals. In this paper, we shall discuss a two-surface description of the field of hardening moduli. An alternative description using an infinite number of nesting surfaces produced practically identical results and will be discussed elsewhere.
4) The usual associated flow rule is assumed and viscous effects are neglected.

   The present formulation differs from classical isotropic or kinematic hardening rules since it enables one to incorporate two basic phenomena: the effect of initial consolidation and the subsequent elastic-plastic response for stress states below the consolidation level.

The consolidation surface in [16] was assumed in the form

$$F(p, q, \eta^p) = (p - c)^2 + \frac{q^2}{n^2} - a^2(\eta^p) = 0, \tag{2}$$

where

$$c = a\frac{n}{m} \qquad m = \tan \omega \qquad n = \tan \zeta.$$

This equation in the p, q-plane corresponds to a one-parameter family of ellipses whose centre and semiaxes change with the irreversible density η^p or void ratio e^p but the ratio of semiaxes remains constant, Figure 1a. The critical state lines OB and OD are inclined at the angle ω to the p-axis. When n = m, the consolid-

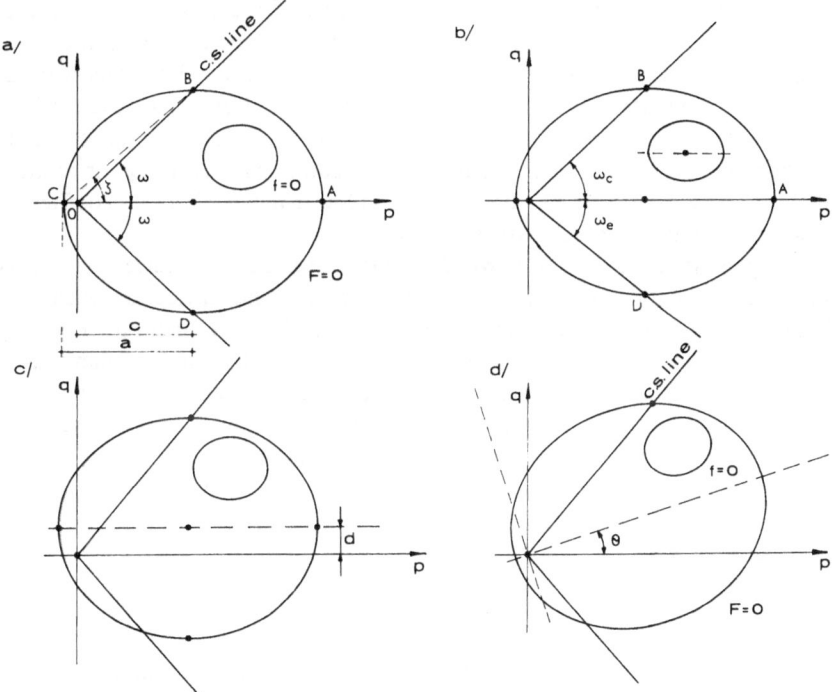

Figure 1. – *Consolidation and yield surfaces in the p,q-plane*
a) elliptical locus
b) consolidation and yield locus composed of two semi-ellipses $\omega_c \neq \omega_e$
c) translated and d) rotated consolidation surface

ation surface passes through the origin. Let us note that this surface was assumed as the yield surface by Roscoe and Burland [22] for study of deformation of clays in the "wet" state corresponding to the angular domain BOD. Here however, we present a model valid in the whole stress space.

In writing (2) it is assumed that the consolidation surface may expand, contract or translate along the p-axis, and that it represents the isotropic properties of the material. A modification of (2) can be proposed by accounting for asymmetry of this surface with respect to the p-axis. Assuming the critical state lines to be inclined at angles ω_c and ω_e to the p-axis in compression and extension domains and taking these angles from the Coulomb yield condition, we have

$$m_c = \frac{q_B}{c} = \tan \omega_c = \frac{6 \sin \varphi}{3 - \sin \varphi}, \quad m_e = \frac{q_D}{c} = \tan \omega_e = -\frac{6 \sin \varphi}{3 + \sin \varphi}, \quad (3)$$

where φ is the angle of internal friction at failure. Figure 1b presents the consolidation surface composed of two semiellipses (2) with different values of n_e and n_c, where $n_c/m_c = n_e/m_e$. These two forms of the consolidation surface were studied in [17] and both isotropic and anisotropic consolidation was considered. It was concluded that K_0-consolidation and rebound curves can not be well simulated by this model and further modifications as to the rule of translation of the consolidation surface are to be introduced or its form is to be changed. It was suggested that introducing additional translation or rotation of this surface, Figure 1c, d, should result in lower values of the K_0 coefficient, which correspond to actual experimental data.

In the present work, however, we shall not pursue this idea since preliminary numerical tests did not provide satisfactory results and difficulties existed in fitting K_0 and critical states line to experimental data. We shall rather modify the form of both consolidation and yield surfaces and assume, similarly as in [17] that the consolidation surface may translate only along the p-axis, thus representing the isotropic properties of the material. Figure 2 presents the modified form of the consolidation surface. It is composed of two portions of rotated ellipses intersecting at 0 and tangential to each other at A on the p-axis. The critical state lines form different angles ω_c and ω_e with the p-axis. Starting from a general equation

$$F = Ap^2 + 2Bpq + Cq^2 + 2Dp + 2Eq + F = 0, \qquad (4)$$

we shall determine the parameters A, B, C, D, E, F from the conditions

$$p = 0 \rightarrow q = 0, \quad p = 2a \rightarrow \frac{\partial F}{\partial q} = 0, \quad p = c \rightarrow \frac{\partial F}{\partial p} = 0,$$

$$p = c \rightarrow q = n_c c. \qquad (5)$$

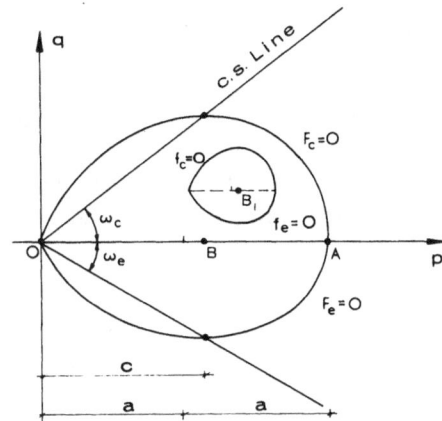

Figure 2. — *Consolidation and yield surfaces in the p,q-plane formed by two rotated ellipses.*

These conditions imply the following from of the upper portion

$$F_c = p^2 + \frac{2(\xi - 1)}{n_c} pq + \frac{(1 - 2\xi)^2}{n_c^2} q^2 - 2ap - \frac{4(\xi - 1)}{n_c} aq = 0 \; ;$$

$$(q > 0), \qquad (6)$$

where

$$\xi = \frac{a}{c} \; ; \quad n_c = \operatorname{tg} \omega_c$$

For $q < 0$, the lower portion is described by a similar equation satisfying the condition $p = c \rightarrow q = - n_e c$, where $n_e = \tan \omega_e \neq n_c$. We obtain

$$F_e = p^2 + \frac{2(1 - \xi)}{n_e} pq + \frac{(1 - 2\xi)^2}{n_e^2} q^2 - 2ap - 4\frac{(1 - \xi)}{n_e} aq = 0 \; ;$$

$$(q < 0). \qquad (7)$$

Let us note for $\xi = 1$, that is for $a = c$, the principal axis of each ellipse coincides with the p-axis. Then

$$F = (p - a)^2 + \frac{q^2}{n^2} - a^2 = 0 \; , \qquad (8)$$

and we obtain the case discussed in [17]. In general, $\frac{1}{2} < \xi \leqslant 1$ and ξ is an additional material parameter which will be identified from K_0-consolidation tests. Let us assume that n_e, n_c, ξ are constant and $a = a(\eta^p)$ or $a = a(e^p)$.

Since

$$d\eta = \eta\, d\epsilon_v = \eta\, d\epsilon_v^e + \eta\, d\epsilon_v^p,$$

$$de = -(1 + e)\, d\epsilon_v = -(1 + e)\, d\epsilon_v^e - (1 + e)\, d\epsilon_v^p,$$

$$(9)$$

where η and e denote respectively the relative bulk density and void ratio, we can relate the irreversible increments of η or e to the plastic volume change, thus

$$de^p = -(1 + e)\, d\epsilon_v^p \quad , \quad d\eta^p = \eta\, d\epsilon_v^p$$

and   (10)

$$(1 + e)\, \eta = 1.$$

Consider first the consolidation process for which the stress point lies on the consolidation surface which then can be used as the yield surface and generate the plastic strain increments

$$d\epsilon_v^p = \frac{\left(\dfrac{\partial F}{\partial p}\, dp + \dfrac{\partial F}{\partial q}\, dq\right)}{(1 + e)\dfrac{\partial F}{\partial e^p}\dfrac{\partial F}{\partial p}}\frac{\partial F}{\partial p} \quad ; \quad d\epsilon_q^p = \frac{\left(\dfrac{\partial F}{\partial p}\, dp + \dfrac{\partial F}{\partial q}\, dq\right)}{(1 + e)\dfrac{\partial F}{\partial e^p}\dfrac{\partial F}{\partial p}}\frac{\partial F}{\partial q}. \quad (11)$$

For the portion $F_c = 0$, we have

$$\frac{\partial F_c}{\partial p} = 2\left(p + \frac{(\xi - 1)}{n_c}\, q - a\right) \quad ;$$

$$\frac{\partial F_c}{\partial q} = 2\left(\frac{(1 - 2\xi)^2}{n_c^2}\, q + \frac{(\xi - 1)}{n_c}\, p - \frac{2(\xi - 1)}{n_c}\, a\right), \quad (12)$$

and the hardening modulus represented by the denominator of (11) is given by

$$H = (1 + e)\frac{\partial F_c}{\partial e^p}\frac{\partial F_c}{\partial p}$$

$$= -4(1 + e)\frac{\partial a}{\partial e^p}\left(p + \frac{2(\xi - 1)}{n_c}\, q\right)\left(p + \frac{(\xi - 1)}{n_c}\, q - a\right) \quad (13)$$

For the portion $F_e = 0$ the formulae (13) and (12) should be modified replacing $\xi - 1$ by $1 - \xi$ and n_c by n_e.

When the consolidation process is terminated, the domain enclosed by the surface $F = 0$ is not elastic since the yield surface $f = 0$ encloses a much smaller domain. It is assumed that the yield surface is similar to the consolidation surface and is composed of portions of two ellipses, thus

$$f_c = (p - \alpha_p)^2 + \frac{2(\xi - 1)}{n_c} (q - \alpha_q)(p - \alpha_p)$$

$$+ \frac{(1 - 2\xi)^2}{n_c^2} (q - \alpha_q^2) + 2a_0 (p - \alpha_p) \frac{1 - \xi}{\xi}$$

$$+ \frac{2(\xi - 1)(1 - 2\xi)}{n_c \xi} a_0 (q - \alpha_q)$$

$$+ a_0^2 \frac{1 - 2\xi}{\xi^2} = 0 ; \qquad (q - \alpha_q > 0) \qquad (14a)$$

$$f_e = (p - \alpha_p)^2 + \frac{2(1 - \xi)}{n_e} (q - \alpha_q)(p - \alpha_p)$$

$$+ \frac{(1 - 2\xi)^2}{n_e^2} (q - \alpha_q)^2 + 2a_0 (p - \alpha_p) \frac{1 - \xi}{\xi} \qquad (14b)$$

$$+ \frac{2(1 - \xi)(1 - 2\xi)}{n_e \xi} a_0 (q - \alpha_q) + a_0^2 \frac{1 - 2\xi}{\xi^2} = 0 ; \qquad (q - \alpha < 0) ,$$

where α_p, α_q defines the position of the point B_1, see Fig. 2. The parameters ξ, n_c, n_e are the same as those for the consolidation surface, but $a_0 = a_0 (e^p) < a$. The flow rule now takes the form

$$de_v^p = \frac{1}{H} \frac{\partial f}{\partial p} \left(\frac{\partial f}{\partial p} dp + \frac{\partial f}{\partial q} dq \right) ;$$

$$(15)$$

$$de_q^p = \frac{1}{H} \frac{\partial f}{\partial q} \left(\frac{\partial f}{\partial p} dp + \frac{\partial f}{\partial q} dq \right) .$$

The parameters n_c, n_e can be determined from (3), whereas the function $a = a(e^p)$ is obtained from the isotropic consolidation test. Using the well-known exponential relationship, we can write

$$p = p_i \exp \left(\frac{e_i - e}{\lambda} \right) , \qquad p = p_0 \exp \left(\frac{e_0^e - e^e}{k} \right) , \qquad a = a_i \exp \left(\frac{e_i^p - e^p}{\lambda - k} \right) ,$$

$$(16)$$

where p_i, e_i are initial values of p and e in the consolidation process and p_0, e_0 are initial values of the unloading process ; k and λ are material parameters. Further, let us assume that the ratio $a_0/a = \gamma$ remains constant, that is, the yield surface shrinks or expands proportionally to the variation of the size of the consolidation surface.

We complete our description of the model by specifying the translational rule of the yield surface and the rule of variation of the hardening modulus. We shall assume that the instantaneous motion of the stress point P occurs along the line connecting P with the associated point R on the consolidation surface. Since

$$d\sigma_P = d\alpha_P + (\sigma_P - \alpha_P)\frac{da_0}{a_0} \quad d\sigma_R = d\alpha_R + (\sigma_R - \alpha_R)\frac{da}{a}, \quad (17)$$

where α_P and α_R denote the centre positions of the yield and the consolidation surfaces, and

$$\frac{\sigma_P - \alpha_P}{a_0} = \frac{\sigma_R - \alpha_R}{a}, \quad (18)$$

the translation of the yield surface is governed by the relation

$$d\alpha_P = d\alpha_R + d\mu\beta + (\sigma_P - \alpha_P)\frac{da - da_0}{a_0}, \quad (19)$$

where

$$\beta = \sigma_R - \sigma_P = \frac{1}{a_0}[(a - a_0)\sigma_P - (a\alpha_P - a_0\,\alpha_R)]. \quad (20)$$

In the p , q-plane, the relations (19) take the form

$$d\alpha_P = dc + d\mu\left[c + \frac{a}{a_0}(p - \alpha_p) - p\right] + \frac{da - da_0}{a_0}(p - \alpha_p), \quad (21)$$

$$d\alpha_q = d\mu\left[\frac{a}{a_0}(q - \alpha_q) - q\right] + \frac{da - da_0}{a_0}(q - \alpha_q),$$

where the multiplier $d\mu$ can be determined from the consistency condition

$$\frac{\partial f}{\partial p}dp + \frac{\partial f}{\partial q}dq - \frac{\partial f}{\partial p}d\alpha_p - \frac{\partial f}{\partial q}d\alpha_q + \frac{\partial f}{\partial a_0}da_0 = 0. \quad (22)$$

Substituting (21) into (22), we obtain

$$
d\mu = \frac{\left(\dfrac{\partial f}{\partial p}dp + \dfrac{\partial f}{\partial q}dq\right) + \dfrac{\partial f}{\partial a_0}da_0 - \dfrac{\partial f}{\partial p}\left[dc + \dfrac{da - da_0}{a_0}(p - \alpha_p)\right] - \dfrac{\partial f}{\partial q}\dfrac{da - da_0}{a_0}(q - \alpha_q)}{\dfrac{\partial f}{\partial p}\left[c + \dfrac{a}{a_0}(p - \alpha_p) - p\right] + \dfrac{\partial f}{\partial q}\left[\dfrac{a}{a_0}(q - \alpha_q) - q\right]}
\tag{23}
$$

Let us define the hardening modulus K_p as

$$
K_p = \frac{H_p}{\left(\dfrac{\partial f}{\partial \sigma}\right)^T \dfrac{\partial f}{\partial \sigma}} = \frac{H_p}{\left[\left(\dfrac{\partial f}{\partial p}\right)^2 + \left(\dfrac{\partial f}{\partial q}\right)^2\right]}
\tag{24}
$$

It is assumed that K_p varies from its initial value on the yield surface to the value K_{PR} on the consolidation surface defined by (13), where K_{PR} is related to H by (24).

Let us assume the following interpolation rule for K_p

$$
K_P = K_{PR} + K_{po}\left(\frac{\delta}{\delta^0}\right)^\gamma ,
\tag{25}
$$

where K_{po} denotes the initial value of K_p on the yield surface; δ is the actual distance of the two surfaces, $\delta = PR$, and δ^0 denotes the initial maximal distance attained during the previous loading history. It is assumed that $K_{po} \gg K_{PR}$, hence for $\delta = \delta^0$ one has $K_p \approx K_{po}$, whereas for $\delta = 0$ one has $K_p = K_{PR}$. When the stress path after reaching the consolidation surface continues to proceed in the exterior of this surface, we associate the flow rule with the surface $F = 0$ and equations (11), (13) apply.

3. Study of material response for "triaxial" stress states

In this section, we shall discuss some typical deformation paths occuring in edometric and triaxial tests, when $\sigma_2 = \sigma_3$.

3.1. K_0-consolidation process

Let $\epsilon_2 = \epsilon_3 = 0$ and ϵ_1 corresponds to non-vanishing principal strain. Since $\epsilon_1 = -\dfrac{1}{3}(\epsilon_v + 3\epsilon_q)$ and $\epsilon_2 = -\dfrac{1}{3}\left(\epsilon_v - \dfrac{3}{2}\epsilon_q\right)$, the uniaxial deformation is characterized by the relationship

$$
d\epsilon_q - \frac{2}{3}d\epsilon_v = 0.
\tag{26}
$$

If we assume that the elastic strain increments are defined by the relations

$$d\epsilon_v^e = \frac{dp}{K} \quad , \quad d\epsilon_q^e = \frac{dq}{3G}$$
(27)

where $K = (1 + e)\, p/k$, then equation (26) implies that

$$\frac{dq}{dp} = \frac{\dfrac{H_p}{K} + \left(\dfrac{\partial f}{\partial p}\right)^2 - \dfrac{3}{2}\dfrac{\partial f}{\partial p}\dfrac{\partial f}{\partial q}}{\dfrac{H_p}{2G} + \dfrac{3}{2}\left(\dfrac{\partial f}{\partial q}\right)^2 - \dfrac{\partial f}{\partial p}\dfrac{\partial f}{\partial q}}.$$
(28)

When the stress state corresponds to the consolidation surface, the derivatives $\partial f/\partial p$ and $\partial f/\partial q$ should be replaced by $\partial F/\partial p$ and $\partial F/\partial q$. For a rigid-plastic behaviour, we then obtain

$$\frac{\partial F}{\partial q} - \frac{2}{3}\frac{\partial F}{\partial p} = 0$$
(29)

and

$$s_{K_0} = \frac{q}{p}$$

$$= \frac{2n^2}{[[4(\xi - 1)(3\xi - n) + 3]^2 + 4n(1 - 2\xi)^2(n + 3\xi - 3) - 4n(\xi - 1)^2]^{1/2} + 4(\xi - 1)(3\xi - n) + 3}$$
(30)

moreover

$$K_0 = \frac{\sigma_2}{\sigma_1} = \frac{3p - 2q}{3p + 2q} = \frac{3 - s_{K_0}}{3 + 2s_{K_0}}.$$
(31)

Thus for a rigid-plastic material, the consolidation path will be a straight line in the p,q-plane. For $\xi = 1$ we obtain the relation

$$s_{K_0} = \frac{q}{p} = \frac{2n^2}{(9 + 4n^2)^{1/2} + 3}$$
(32)

already obtained in [17].

Equation (30) can be used to determine the parameter ξ by comparing (30) with the experimental K_0-line in the p,q-plane.

Figures 3, 5 show predicted paths in the σ_1, σ_2-plane for K_0-loading and unloading programs. The material parameters for the Weald clay were assumed similar as in [17]: $\varphi = 26°$; $\lambda = 0,091$; $k = 0,030$; $e_0 = 0,57$; $a/a_0 = 5,0$; $G/K = 0,55$; $K = \dfrac{(1 + e)\,p}{k}$.

Figure 3 presents the K_0 consolidation paths for four values of ξ. It is seen that for $\xi = 0,85$ the present model can well simulate the consolidation path obtained experimentally by Skempton and Sowa in [24]. For $\xi = 1$, we arrive at the case of the yield condition studied in [17]. Figure 4 shows the predicted depedence of K_0 on the angle of internal friction φ for four values of ξ. It is seen that collected experimental data lie between the lines corresponding to $\xi = 0,85$ and $\xi = 0,70$.

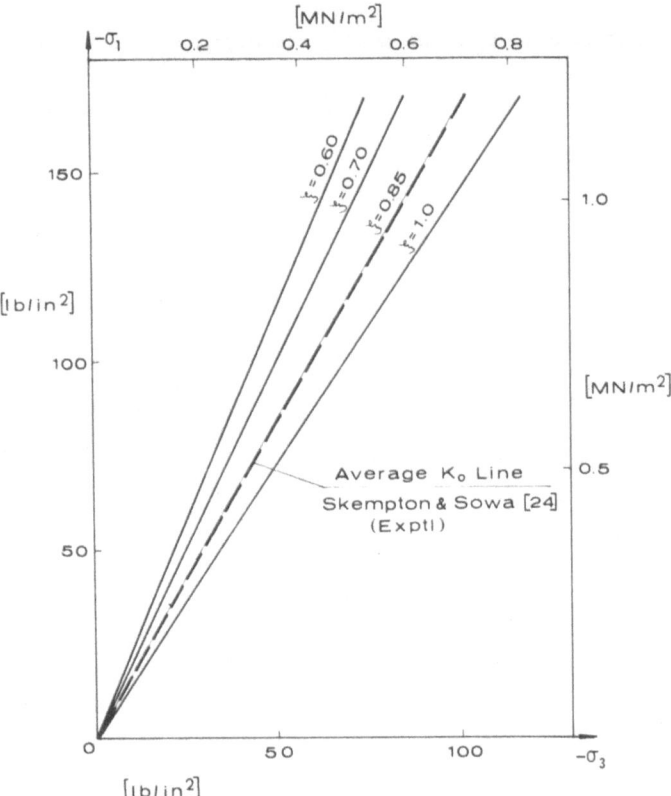

Figure 3. – *Comparison of experimental K_0-consolidation of Weald clay with numerical analyses for different values of ξ.*

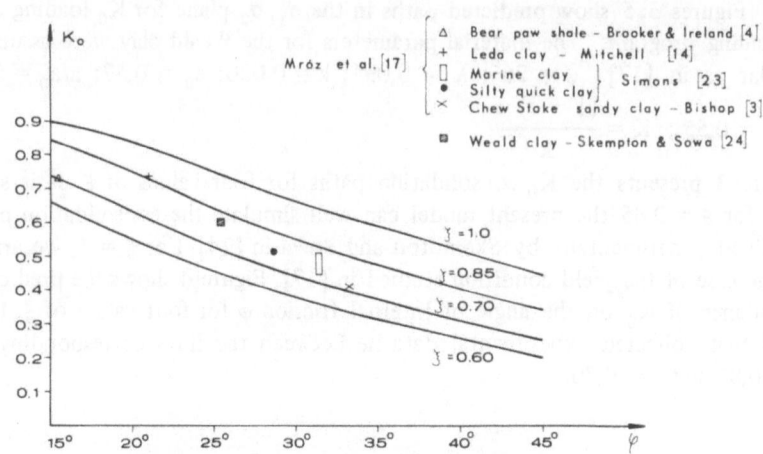

Figure 4. – *Depedence of K_0 on the angle of internal friction for a rigid-plastic material.*

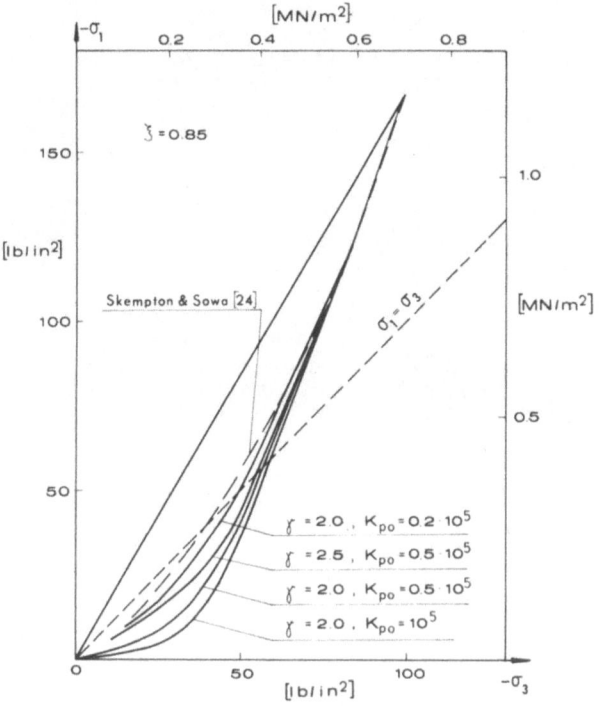

Figure 5. – *K_0-unloading paths for different values of γ and K_{po}.*

Figure 5 shows the K_0-loading and rebound curves for different values of the exponent γ occuring in the hardening rule (25) and for a selected value of $\xi = 0.85$. It is seen that assuming $\gamma = 2.0$ and $K_{po} = 0.2 \cdot 10^5$, we can simulate sufficiently accurately the rebound curve obtained in [24]. However, it turns out that for small values of p the rule (25) implies very large reverse strains. To avoid this inconvenience, a modified hardening rule was proposed for $p \leqslant p_0$, whereas for $p \geqslant p_0$ the rule (25) applies. Here p_0 denotes an experimentally determined value of transition pressure. Assume that

$$K_p = K_{PR} [K_{po} + C (p_0 - p)] (\delta/\delta^0)^\gamma \; ; \quad p \leqslant p_0 , \tag{33}$$

where : $C = \alpha \cdot K_{po}$; $0.1 \leqslant \alpha \leqslant 0.5$. For $p = p_0$, the formulae (33) and (25) provide the same value of K_p. Figure 6 shows the modified rebound curves for different values of C. The undrained compression path is also shown in Figure 6

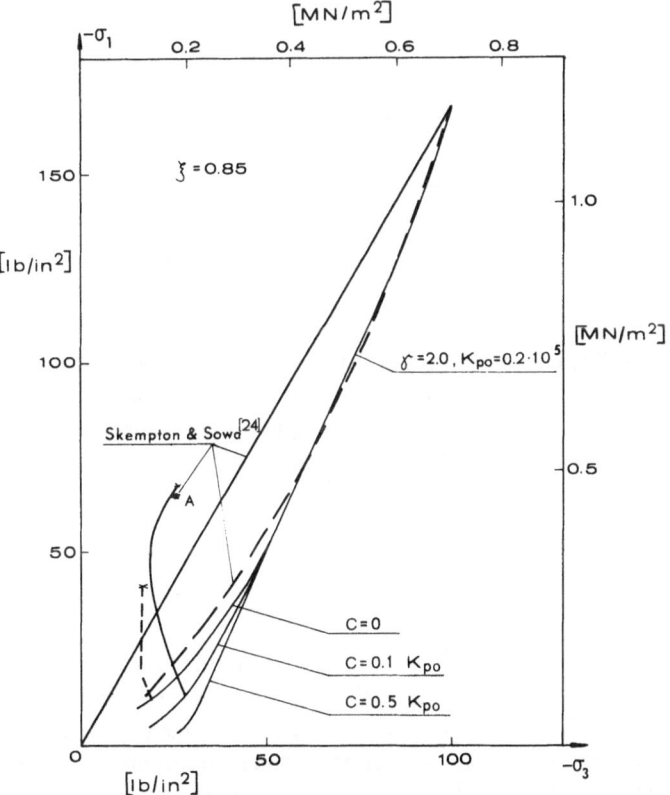

Figure 6. – *Modified rebound curves for different values of C.*

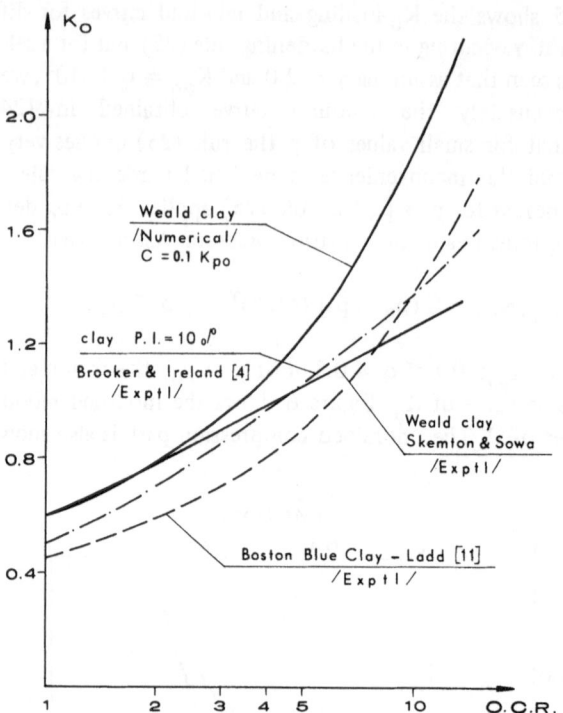

Figure 7. – *Depedence of K_0 on the overconsolidation ratio.*

for overconsolidated clay. It turns out that modifying the K_0 rebound curves for low values of pressure, we improve significantly the subsequent undrained paths. Therefore the modification of the hardening rule seems expedient.

Figure 7 illustrates the depedence of K_0 on the overconsolidation ratio (O.C.R.) defined as

$$O.C.R. = \frac{\sigma_{vm}}{\sigma_{vc}}, \tag{34}$$

where σ_{vm} is the maximum value of σ_1 reached in the loading process and σ_{vc} is the actual value of σ_1 during K_0-unloading. It is seen that the model overestimates the K_0 values for higher values of O.C.R, as compared to experimental data of [24] for Weald clay. Therefore, further improvement of the model is required. On the other hand, the data for two other clays seem to fit better the calculated curve.

The numerical tests carried out so far enable us to specify the best values of material parameters: $\xi = 0.85$, $\gamma = 2.0$, $K_{po} = 0.2 \cdot 10^5$, $p_0 = 50$ lb/in^2, $C = 0.1 K_{po}$.

3.2. *Material response after anisotropic K_0-consolidation*

Let us now discuss the undrained compression and extension tests following the initial K_0-consolidation. If we assume both the pore water and skeleton as incompressible, then the undrained loading becomes an isochronic deformation process, that is

$$d\epsilon_v = d\epsilon_v^e + d\epsilon_v^p = 0 \tag{35}$$

Using the flow rule and the elastic incremental relations (27), we get that the undrained stress path is given by the equations:

$$\frac{dq}{dp} = \frac{-\dfrac{\partial f}{\partial p}\dfrac{\partial f}{\partial q}}{\left(\dfrac{\partial f}{\partial p}\right)^2 + \dfrac{H_p}{K}} = -\frac{1}{a_w} \tag{36}$$

and

$$dp_w = dp^t + a_w \, dq.$$

When the stress state corresponds to the consolidation surface, we have

$$\frac{dq}{dp} = -\frac{\dfrac{\partial F}{\partial p}\dfrac{\partial F}{\partial q}}{\left(\dfrac{\partial F}{\partial p}\right)^2 + \dfrac{H}{K}}, \tag{37}$$

where H and $\partial F/\partial p$, $\partial F/\partial q$ are determined from (13) and (12).

Figure 8 presents the undrained stress paths in the plane of effective stresses σ_1, σ_2. The predicted and experimental paths [24] are similar. Moreover, the failure points 1, 2, 3, 4 for the same overconsolidation ratios are accurately predicted by the present model. Figure 9 shows the functions C_u/σ_{vc} versus O.C.R. predicted and obtained experimentally for several clays. It is seen that the agreement with the data of [24] is very good and the respective curves for other clays indicate a similar shape.

Finally, Figure 10 shows the $q - \epsilon_q$ response curves corresponding to compression - extension undrained paths issuing from C and D in Figure 8. It is interesting to note that not only failure stresses are different for compression and extension tests, but also *initial stiffness* moduli differ significantly, i.e. the

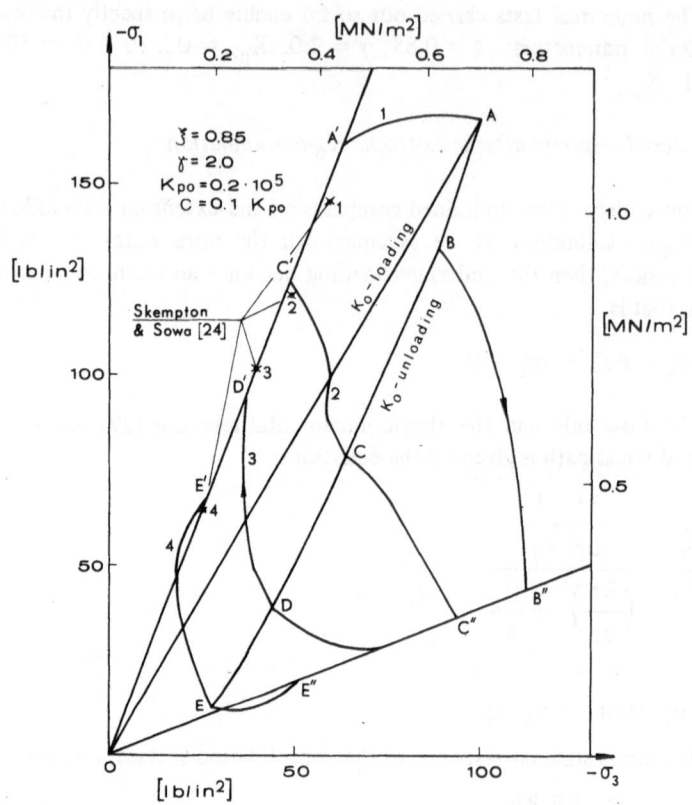

Figure 8. − K_0 *consolidation-unloading followed by undrained axial loading and unloading.*

tangent modulus in the compression test is greater than that in extention. This problem was recently discussed by James and Cairncross [5] and by Mitchell [15] who indicated that directional variation of stiffness within the elasto-plastic domain is one of the essential features of clay anisotropy.

We have thus presented a sufficiently complete study of model predictions for different loading histories. It is demonstrated that K_0-loading, unloding and undrained compression − extension paths can be simulated with sufficient accuracy. Moreover, the standard $K_0 = f(O.C.R.)$ and $C_u/\sigma_{vc} = f(O.C.R.)$ curves for clays are also simulated with fair accuracy, though further refinements are obviously needed in the case of particular clays. In general, a considerable improvement was reached with respect to the previous version of the model discussed in [17].

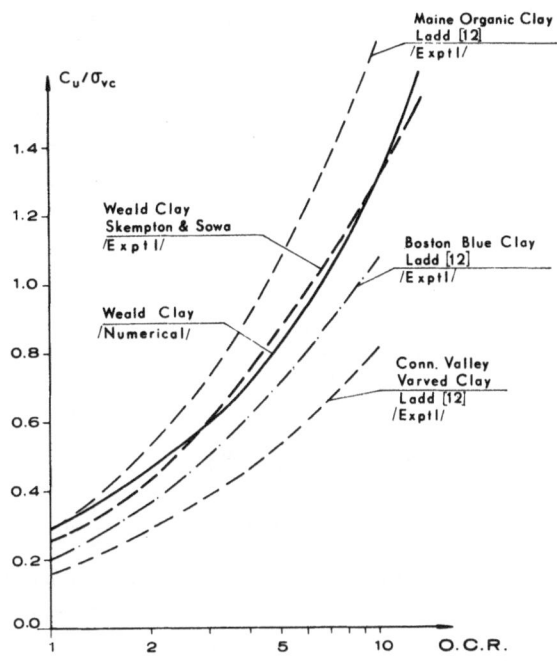

Figure 9. – *Effect of overconsolidation on undrained shear strength* (c_u).

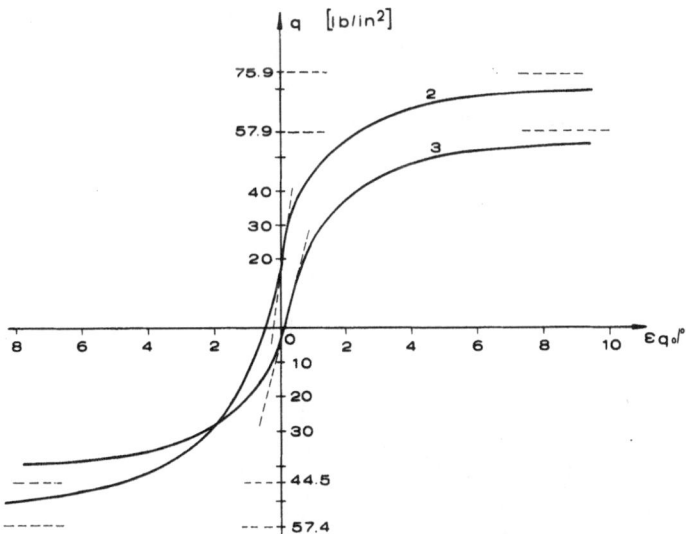

Figure 10. – *Shear stress versus deviatoric strain for stress paths (2) and (3) in Figure 8.*

4. Constitutive relations for the general stress state

So far, we have considered only the "triaxial" stress states for which the two principal stresses are equal. However, the applicability of the model can be extended considerably by formulating the constitutive relations for the general stress state. It is the aim of this section to provide such a generalization.

Consider the three invariants

$$
J_m = \frac{1}{3}(\sigma_{ii} - \alpha_{ii}) ,
$$

$$
\bar{\sigma} = \left[\frac{1}{2}(s_{ij} - \bar{\alpha}_{ij})(s_{ij} - \bar{\alpha}_{ij}) \right]^{1/2}
\tag{38}
$$

$$
J_3 = \frac{1}{3}(s_{ij} - \bar{\alpha}_{ij})(s_{ki} - \bar{\alpha}_{ki})(s_{kj} - \bar{\alpha}_{kj})
$$

of the "translated" stress $\sigma_{ij} - \alpha_{ij}$. Here $s_{ij} = \sigma_{ij} - \frac{1}{3}\sigma_{kk}\delta_{ij}$ is the stress deviator and $\alpha_{ii}, \bar{\alpha}_{ij}$ denote the spherical and deviator components of the translation tensor α_{ij} of the yield surface. Further, let us introduce the angle measure of the third invariant

$$
\theta = \frac{1}{3} \arcsin\left(-\frac{3\sqrt{3}}{2} \frac{J_3}{\bar{\sigma}^3} \right) \quad ; \quad -\pi/6 \leqslant \theta \leqslant \pi/6 ,
\tag{39}
$$

where the angle θ can be found in the octahedral plane π: $\sigma_1 + \sigma_2 + \sigma_3 = $ const. For the "triaxial" stress state, $\sigma_2 = \sigma_3$, we have

$$
p - \alpha_p = -J_m = -\sigma_m + \frac{1}{3}\alpha ,
$$

$$
q - \alpha_q = \sqrt{3}\,\bar{\sigma}_+ ,
\tag{40}
$$

where $\sigma_m = \frac{1}{3}\sigma_{ii}$, $\alpha = \alpha_{ii}$ and $\bar{\sigma}_+$ denotes the value of $\bar{\sigma}$ for $\sigma_2 = \sigma_3, \theta = \pi/6$. Generally, the principal stresses can be expressed in terms of $J_m, \bar{\sigma}$ and θ:

$$
\left\{ \begin{array}{c} \sigma_1 \\[6pt] \sigma_2 \\[6pt] \sigma_3 \end{array} \right\} = \frac{2}{\sqrt{3}}\bar{\sigma} \left\{ \begin{array}{c} \sin\left(\theta + \frac{2}{3}\pi\right) \\[6pt] \sin\theta \\[6pt] \sin\left(\theta + \frac{4}{3}\pi\right) \end{array} \right\} + \sigma_m .
\tag{41}
$$

Assume that the yield curve in the π plane is expressed as

$$\bar{\sigma} = \bar{\sigma}_+ \, g(\theta), \tag{42}$$

where

$$g(\theta) = \frac{2\,k}{(1 + k) - (1 - k)\sin 3\theta} \quad ; \quad k = \frac{3 - \sin\varphi}{3 + \sin\varphi}, \tag{43}$$

so that $g(\pi/6) = 1$. This relation follows from earlier developments proposed by Zienkiewicz and Pande [25], Gudehus [9] and Argyris [1]. The yield condition (14) in the p,q-plane can be rewritten as follows

$$
f = \left(\sigma_m - \frac{1}{3}\alpha\right)^2 - \frac{2\sqrt{3}\,(\xi - 1)}{n_c}\left(\bar{\sigma}_m - \frac{1}{3}\alpha\right)\bar{\sigma}_+
$$

$$
+ 3\,\frac{(1 - 2\xi)^2}{n_c^2}\,\bar{\sigma}_+^2 - 2a_0\,\frac{1 - \xi}{\xi}\left(\sigma_m - \frac{1}{3}\alpha\right) + \tag{44}
$$

$$
+ \frac{2\sqrt{3}\,(\xi - 1)\,(1 - 2\xi)}{n_c\,\xi}\,a_0\,\bar{\sigma}_+ + a_0^2\,\frac{1 - 2\xi}{\xi^2} = 0,
$$

and since for $\theta = \pi/6$, we have $n_c = 6\sin\varphi/(3 - \sin\varphi)$, we obtain

$$
f = 3\sin^2\varphi\left(\sigma_m - \frac{\alpha}{3}\right)^2 - 6\sin^2\varphi\,\frac{1 - \xi}{\xi}\,a_0\left(\sigma_m - \frac{\alpha}{3}\right) -
$$

$$
- \sqrt{3}\,(\xi - 1)\sin\varphi\,(3 - \sin\varphi\sin 3\theta)\cdot
$$

$$
\left[\left(\sigma_m - \frac{\alpha}{3}\right) - \frac{1 - 2\xi}{\xi}\,a_0\right]\bar{\sigma} + \left[\frac{1}{2}(1 - 2\xi)(3 - \sin\varphi\sin 3\theta)\right]^2\bar{\sigma}^2 \tag{45}
$$

$$
+ 3a_0^2\sin^2\varphi\,\frac{1 - 2\xi}{\xi^2} = 0.
$$

Equation (45) is geometrically represented by a surface with similar cross-sectional shapes in all π-planes, Figure 11. The center of this surface is determined by the tensor α_{ij}. The gradient tensor of this surface is given by

$$
\frac{\partial f}{\partial \sigma} = c_1\left(\frac{\partial \sigma_m}{\partial \sigma}\right) + \left(c_2' + \frac{3\sqrt{3}\,J_3}{2\bar{\sigma}^4\cos 3\theta}\,c_3'\right)\left(\frac{\partial \bar{\sigma}}{\partial \sigma}\right)
$$

$$
+ \left(-c_3'\frac{\sqrt{3}}{2\bar{\sigma}^3\cos 3\theta}\right)\left(\frac{\partial J_3}{\partial \sigma}\right) \tag{46}
$$

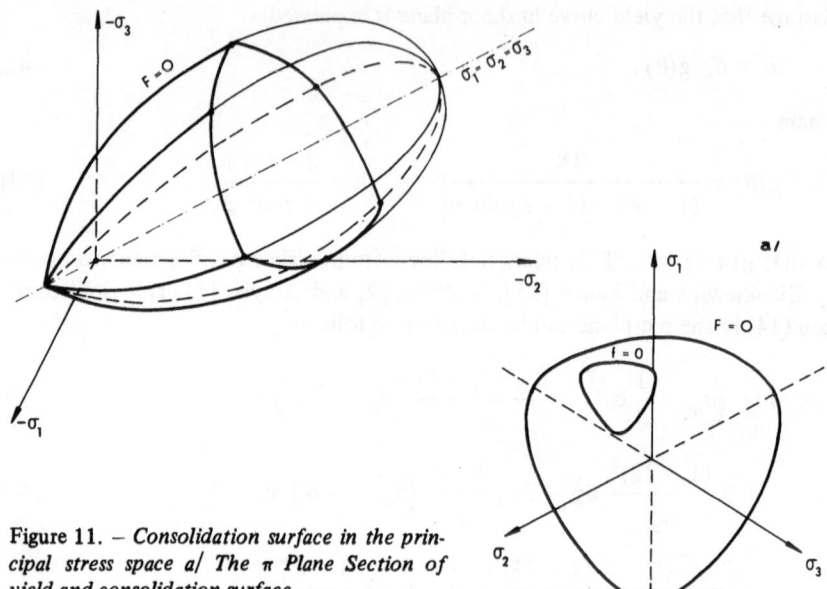

Figure 11. – *Consolidation surface in the principal stress space a/ The π Plane Section of yield and consolidation surface.*

where

$$c_1 = \frac{\partial f}{\partial \sigma_m} \quad ; \quad c_2' = \frac{\partial f}{\partial \bar{\sigma}} \quad ; \quad c_3' = \frac{\partial f}{\partial \theta}. \tag{47}$$

Further, from (45), we have

$$\frac{\partial f}{\partial \sigma_m} = 6 \sin^2 \varphi \left(\sigma_m - \frac{\alpha}{3} \right) - 6 \sin^2 \varphi \frac{1 - \xi}{\xi} a_0 -$$

$$- \sqrt{3} (\xi - 1) \sin \varphi (3 - \sin \varphi \sin 3\theta) \bar{\sigma}$$

$$\frac{\partial f}{\partial \bar{\sigma}} = \frac{1}{2} (1 - 2\xi)^2 (3 - \sin \varphi \sin 3\theta)^2 \bar{\sigma}$$

$$- \sqrt{3} (\xi - 1) \sin \varphi (3 - \sin \varphi \sin 3\theta) \left(\sigma_m - \frac{\alpha}{3} - \frac{1 - 2\xi}{\xi} a_0 \right) \tag{48}$$

$$\frac{\partial f}{\partial \theta} = 3 \bar{\sigma} \cos 3\theta \left[\sqrt{3} (\xi - 1) \sin^2 \varphi \left(\sigma_m - \frac{\alpha}{3} - \frac{1 - 2\xi}{\xi} a_0 \right) - \right.$$

$$\left. - \frac{1}{2} \bar{\sigma} (1 - 2\xi)^2 (3 - \sin \varphi \sin 3\theta) \sin \varphi \right]$$

and the gradient tensor $\partial f/\partial \boldsymbol{\sigma}$ is finally expressed as

$$\frac{\partial f}{\partial \boldsymbol{\sigma}} = c_1 \frac{\partial \sigma_m}{\partial \boldsymbol{\sigma}} + c_2 \frac{\partial \bar{\sigma}}{\partial \boldsymbol{\sigma}} + c_3 \frac{\partial J_3}{\partial \boldsymbol{\sigma}}, \tag{49}$$

where

$$c_1 = 6 \sin^2 \varphi \left(\sigma_m - \frac{\alpha}{3} - \frac{1 - \xi}{\xi} a_0 \right)$$
$$- \sqrt{3} (\xi - 1) \sin \varphi (3 - \sin \varphi \sin 3\theta) \bar{\sigma}$$

$$c_2 = (2 \sin 3\theta \sin \varphi + 3) \left[\frac{1}{2} (1 - 2\xi)^2 (3 - \sin \varphi \sin 3\theta) \bar{\sigma} \right.$$
$$\left. - \sqrt{3} (\xi - 1) \sin \varphi \left(\sigma_m - \frac{\alpha}{3} - \frac{1 - 2\xi}{\xi} a_0 \right) \right] \tag{50}$$

$$c_3 = -\frac{2\sqrt{3}}{2\bar{\sigma}^2} \left[\sqrt{3}(\xi - 1) \sin^2 \varphi \left(\sigma_m - \frac{\alpha}{3} - \frac{1 - 2\xi}{\xi} a_0 \right) \right.$$
$$\left. - \frac{1}{2} \bar{\sigma} (1 - 2\xi)^2 (3 - \sin \varphi \sin 3\theta) \sin \varphi \right]$$

and

$$\frac{\partial \sigma_m}{\partial \boldsymbol{\sigma}} = \frac{1}{3} \begin{Bmatrix} 1 \\ 1 \\ 1 \\ 0 \\ 0 \\ 0 \end{Bmatrix} \quad ; \quad \frac{\partial \bar{\sigma}}{\partial \boldsymbol{\sigma}} = \frac{1}{2\bar{\sigma}} \begin{Bmatrix} s_x - \bar{\alpha}_x \\ s_y - \bar{\alpha}_y \\ s_z - \bar{\alpha}_z \\ 2(\tau_{yz} - \alpha_{yz}) \\ 2(\tau_{xz} - \alpha_{xz}) \\ 2(\tau_{xy} - \alpha_{xy}) \end{Bmatrix} ,$$

$$\tag{51}$$

$$\frac{\partial J_3}{\partial \boldsymbol{\sigma}} = \begin{Bmatrix} (s_y - \bar{\alpha}_y)(s_z - \bar{\alpha}_z) - (\tau_{zy} - \alpha_{zy})^2 \\ (s_x - \bar{\alpha}_x)(s_z - \bar{\alpha}_z) - (\tau_{xz} - \alpha_{xz})^2 \\ (s_x - \bar{\alpha}_x)(s_y - \bar{\alpha}_y) - (\tau_{xy} - \alpha_{xy})^2 \\ 2\{(\tau_{xz} - \alpha_{xz})(\tau_{xy} - \alpha_{xy}) - (s_x - \alpha_x)(\tau_{yz} - \alpha_{yz})\} \\ 2\{(\tau_{yz} - \alpha_{yz})(\tau_{xy} - \alpha_{xy}) - (s_y - \alpha_y)(\tau_{xz} - \alpha_{xz})\} \\ 2\{(\tau_{yz} - \alpha_{yz})(\tau_{xz} - \alpha_{xz}) - (s_z - \alpha_z)(\tau_{xy} - \alpha_{xy})\} \end{Bmatrix} + \frac{1}{3}\bar{\sigma}^2 \begin{Bmatrix} 1 \\ 1 \\ 1 \\ 0 \\ 0 \\ 0 \end{Bmatrix} .$$

Setting $\bar{\alpha}_{ij} = 0$, $\alpha_{ij} = -3\dfrac{a}{\xi}$, $a_0 = a$ we obtain from (45) the equation of the consolidation surface $F = 0$. Now, the invariants (38) become

$$\sigma_m = \frac{1}{3}\sigma_{ii} \;,\; \bar{\sigma} = \left(\frac{1}{2}s_{ij}\,s_{ij}\right)^{1/2} \;,\; J_3 = \frac{1}{3}s_{ij}\,s_{ki}\,s_{kj} \;,$$

$$\theta = \frac{1}{3}\arcsin\left(-\frac{3\sqrt{3}}{2}\frac{J_3}{\bar{\sigma}^3}\right) \qquad (52)$$

and the consolidation surface is expressed as

$$F = 3\sin^2\varphi\,(\sigma_m + a)^2 - \sqrt{3}\,(\xi - 1)\sin\varphi\,(3 - \sin\varphi\sin 3\theta)\,(\sigma_m + 2a)\,\bar{\sigma}$$

$$+ \left[\frac{1}{2}(1 - 2\xi)\,(3 - \sin\varphi\sin 3\theta)\right]^2\,\bar{\sigma}^2 - 3a^2\sin^2\varphi = 0, \qquad (53)$$

with the gradient tensor

$$\frac{\partial F}{\partial\sigma} = c_1\frac{\partial\sigma_m}{\partial\sigma} + c_2\frac{\partial\bar{\sigma}}{\partial\sigma} + c_3\frac{\partial J_3}{\partial\sigma} \qquad (54)$$

and

$$c_1 = 6\sin^2\varphi\,(\sigma_m + a) - \sqrt{3}\,(\xi - 1)\sin\varphi\,(3 - \sin\varphi\sin 3\theta)\,\bar{\sigma}$$

$$c_2 = (2\sin\varphi\sin 3\theta + 3)\left[-\sqrt{3}\,(\xi - 1)\sin\varphi\,(\sigma_m + 2a) + \right.$$

$$\left. + \frac{1}{2}(1 - 2\xi)^2\,(3 - \sin\varphi\sin 3\theta)\,\bar{\sigma}\right].$$

$$c_3 = -\frac{3\sqrt{3}}{2\bar{\sigma}^2}\left[\sqrt{3}\,(\xi - 1)\sin^2\varphi\,(\sigma_m + 2a) - \right.$$

$$\left. - \frac{1}{2}(1 - 2\xi)^2\,(3 - \sin\varphi\sin 3\theta)\sin\varphi\bar{\sigma}\right] \qquad (55)$$

The hardening modulus for states corresponding to the consolidation surface $F = 0$ can be obtained by starting from the flow rule

$$d\varepsilon^p = \frac{1}{H}\frac{\partial F}{\partial\sigma}\left(\frac{\partial F}{\partial\sigma}\right)^T d\sigma \qquad (56)$$

and the consistency condition

$$\left(\frac{\partial F}{\partial\sigma}\right)^T d\sigma + \frac{\partial F}{\partial e^p}\,de^p = 0, \qquad (57)$$

where $de^p = (1 + e)\, \mathrm{tr}\,(d\boldsymbol{\varepsilon}^p)$. From (56), (57) we obtain

$$H = -(1 + e) \frac{\partial F}{\partial e^p}\, \mathrm{tr}\left(\frac{\partial F}{\partial \boldsymbol{\sigma}}\right), \tag{58}$$

where

$$\frac{\partial F}{\partial e^p} = 2 \frac{\partial a}{\partial e^p}\, [3 \sin^2 \varphi\, \sigma_m - \sqrt{3}\,(\xi - 1)\sin\varphi\,(3 - \sin\varphi\sin 3\theta)\,\overline{\sigma}] \,;$$

$$\mathrm{tr}\left(\frac{\partial F}{\partial \boldsymbol{\sigma}}\right) = c_1 . \tag{59}$$

The translation rule is given by

$$d\boldsymbol{\alpha} = d\boldsymbol{\alpha}_R + d\mu\boldsymbol{\beta} + (\boldsymbol{\sigma}_p - \boldsymbol{\alpha})\,\frac{da - da_0}{a_0}, \tag{60}$$

where

$$d\mu = \frac{\left(\dfrac{\partial f}{\partial \boldsymbol{\sigma}}\right)^T d\boldsymbol{\sigma} + \dfrac{\partial f}{\partial a_0}\,da_0 - \left(\dfrac{\partial f}{\partial \boldsymbol{\sigma}}\right)^T \left[d\boldsymbol{\alpha}_R + (\boldsymbol{\sigma}_p - \boldsymbol{\alpha})\,\dfrac{da - da_0}{a_0}\right]}{\left(\dfrac{\partial f}{\partial \boldsymbol{\sigma}}\right)^T \boldsymbol{\beta}} \tag{61}$$

and

$$\boldsymbol{\alpha}_R = \{-a/\xi,\, -a/\xi,\, -a/\xi,\, 0,\, 0,\, 0\}^T . \tag{62}$$

The other relations remain identical to those in Section 2. The transition to the case $\sigma_2 = \sigma_3$ can be now easily obtained. In fact, the invariants $\overline{\sigma}, J_3, \theta$ become

$$\overline{\sigma} = \frac{1}{\sqrt{3}}[(\sigma_2 - \alpha_2) - (\sigma_1 - \alpha_1)] \quad , \quad J_3 = -\frac{2}{27}[(\sigma_2 - \alpha_2) - (\sigma_1 - \alpha_1)]^3, \tag{63}$$

$$\theta = \pi/6 \quad \text{for} \quad [(\sigma_2 - \sigma_1) - (\alpha_2 - \alpha_1)] \geqslant 0 \quad \text{or} \quad q - \alpha_q > 0,$$

$$\theta = -\pi/6 \quad \text{for} \quad [(\sigma_2 - \sigma_1) - (\alpha_2 - \alpha_1)] < 0 \quad \text{or} \quad q - \alpha_q < 0. \tag{64}$$

5. Concluding remarks

The present study supplements the previous work [16, 17] on the modelling of elastic-plastic response under monotonic and cyclic loading. The basic assumption is that the preconsolidated soil behaves as an isotropic material for stress states exceeding the preconsolidation level measured by the surface $F = 0$, whereas anisotropic effects occur for stress states below the consolidation level ($F < 0$). It was demonstrated that K_0-loading, unloading and undrained paths can be simulated with sufficient accuracy. On the other hand, for some clays, where the textural anisotropy becomes an essential factor, it may turn out necessary to allow for translation and/or rotation of the consolidation surface. This however, would involve additional material functions and imply further complexity of the model.

REFERENCES

[1] ARGYRIS J.H., G. FAUST, J. SZIMMAT, E.P. WARNKE and K.J. WILLIAM. – "Recent developements in the finite element analysis of PCRV". *2nd Int. Conf. SMIRT*, Berlin, 3 (1973).

[2] BARDEN L. – "Stress and displacements in a cross-anisotropic soil". *Géotechnique*, 13, (1963): 198-210.

[3] BISHOP A.W. – "Test requirements for measuring the coefficient of earth pressure at rest". *Proc. Conf. on Earth Pressure Problems, Brussels*, 1 (1958): 2-14.

[4] BROOKER E.W. and H.O. IRELAND. – "Earth pressures at rest related to stress history" *Can. Geotechn. J.*, 2 (1965): 60-83.

[5] CAIRNCROSS A.M. and R.G. JAMES. – "Anisotropy in over-consolidated soil", *Géotechnique*, 27 (1977): 31-36.

[6] DAFALIAS Y.F., and E.P. POPOV. – "A model of non-linearly hardening materials for complex loadings" *Acta Mechanica*, 21 (1975): 173-192.

[7] DUNCAN I.M. and H.B. SEED. – "Anisotropy and stress reorientation in clay". *Proc. Am. Soc. Civ. Engrs*, 92, 5. (1966): 21-50.

[8] DUNCAN I.M. and H.B. SEED. – "Strength variations along failure surfaces in clay" *Proc. Am. Soc. Civ. Engrs*, 92, 6 (1966): 81-104.

[9] GUDEHUS G. – "Elastoplastische Stoffgleichungen für trockenen Sand" *Ingenieur Archiv*, 42 (1973).

[10] KRIEG R.D. – "A practical two-surface plasticity theory" *J. Appl. Mech.* ser., E42 (1975): 641-646.

[11] LADD C.C. – "Stress-strain behaviour of anisotropically consolidated clays during undrained shear". *Proc. 6th. Int. Conf. S.M. and F.E., Montreal*, 1 (1965): 282-286.

[12] LADD C.C. – "Strength parameters and stress-strain behaviour of saturated clays". *Soils Publ. M.I.T.* (1971): 278-281.

[13] LAMBE T.W. – "The engineering behaviour of compated clay". *Proc. Am. Soc. Civ. Engrs*, 84, 2, (1958): 16-55.

[14] MITCHELL R.J. – "On the yielding and mechanical strength of Leda clays". *Can. Geotechn. Journ.*, 7 (1970): 297-312.

[15] MITCHELL R.J. – "Some deviations from isotropy in a lightly overconsolidated clay". *Géotechnique*, 22 (1972): 459-467.

[16] MROZ Z., V.A. NORRIS and O.C. ZIENKIEWICZ. – "An anisotropic hardening model for soils and its application to cyclic loading". *Int. J. Num. Anal. Meth in Geomech.*, 2 (1978): 203-221.

[17] MROZ Z., V.A. NORRIS and O.C. ZIENKIEWICZ. – "Application of an anisotropic hardening model in the analysis of the elasto-plastic deformation of soils". *Géotechnique*, **29** (1979).

[18] MROZ Z. – "On the description of anisotropic workhardening". *J. Mech. Phys. Sol.*, **15** (1967): 163-175.

[19] POULOS H.G. – "Settlement of isolated foundation". *Res. report no R265, Civil Engineering Laboratories*, The University of Sydney, Sydney, 1975.

[20] PREVOST J.H. – "Mathematical modelling of monotonic and cyclic undrained clay behaviour". *Int. J. Num. Anal. Meth. Geom.* **1** (1977): 195-216.

[21] QUIGLEY R.M. and C.D. THOMSON. – "The fabric of anisotropically consolidated marine clay". *Can. Geotech. Journ.* **3**, (1966): 61-73.

[22] ROSCOE K.H. and J.B. BURLAND. – "On the generalized stress-strain behaviour of "wet" clay". In *Engineering Plasticity*, J. Heyman and F.A. Leckie, Eds., Cambridge Univ. Press (1968): 535-609.

[23] SIMONS N. – "Discussion on coefficient of earth pressure at rest. *Proc. Conf. on Earth Pressure Problems"*, Brussels, **3** (1958): 50-53.

[24] SKEMPTON A.W. and V.A. SOWA. – "The behaviour of saturated clays during sampling and testing". *Géotechnique*, **13** (1963): 269-290.

[25] ZIENKIEWICZ O.C. and G.N. PANDE. – "Some useful forms of isotropic yield surfaces for soil and rock mechanics". *University College of Swansea Report*, C/R/248/75, 1975.

RESUME

Une description de la consolidation anisotrope des argiles

Un modèle d'écrouissage combiné isotrope-cinématique pour les sols est discuté en utilisant le concept de la surface de consolidation et de la surface d'écoulement pour définir l'état du matériau. La consolidation Ko, la décharge et les chemins de contrainte non-drainés sont étudiés et simulés par le modèle. Une formulation générale des lois de comportement est présentée dans la dernière partie.

DISCUSSION

COMMENT BY Y. F. DAFALIAS ; Dr. Pietruszczak presented a very interesting adaptation of a two surface plasticity model to soil mechanics taking into account anisotropy as an extension of a previous work by Mróz, Norris and Zienkiewicz (1978). Since it has not been explicitly stated, I would like to mention that the above work by Mróz *et al.* is based on the concept of the "Bounding Surface" in Plasticity enclosing the yield surface, as

originally proposed by Dafalias and Popov (9th U.S. National Congress
of Applied Mechanics, *Proceedings*, Boulder, Colorado, June 1974). This
concept has been further elaborated within the framework of plastic
internal variables in a series of papers by Dafalias and Popov (Acta Mechanica (1975), Mechanics Research Communications (1976), J. Applied
Mechanics (1976), Nuclear Engin. and Design (1977)), and independently
proposed in a more restrictive form by Krieg (J. Applied Mechanics
(1975)). Finally, I would like to mention that an even simpler model in
terms of a Bounding Surface only without a yield surface, has been adap-
ted by critical state soil cyclic plasticity by Dafalias (7th Canadian Conference of Applied Mechanics, *Proceedings*, May 1979). In conclusion, it
appears that the concept of the Bounding Surface (called the Consolidation Surface in the present paper) can treat successfully complex aspects
of material behavior including cyclic response and anistropy effects.

REPLY BY St. PIETRUSZCZAK : The model proposed by Mróz et al [16] follows
from a multisurface hardening rule for metals developed by Mróz [18] for
both a discrete and an infinite set of loading surfaces. In the present paper
we use a simplified interpolation rule which indeed is similar to that developed by Krieg [10] and Dafalias and Popov [6]. In the paper only a short
recapitulation of the theory discussed by Mróz *et al.* is given (with the
references to [16]). The aim of this study is to modify that concept and
apply it to simulate some deformation processes. However, even in this
paper, the Reader can find the references to Dafalias' and Popov's works
as well as to Krieg's.

Thank you for the more general review of your papers given in the
comment.

A Generalized Failure Condition
for Orthotropic Solids

R. Nova and G. Sacchi

*Istituto di Scienza e Tecnica delle Costruzioni,
Politecnico di Milano, Milano, Italy.*

1. Introduction

Every engineering material exhibits anisotropic behaviour to a certain extent. In fact, even if the material's inherent structure would have been initially perfectly isotropic, some degree of anisotropy would be induced by a super-imposed non-isotropic state of strain. Nevertheless for many materials, such as steel, the influence of anisotropy is negligible and is therefore usually disregarded. For many others, on the contrary, inherent and induced anisotropy greatly influence their mechanical behaviour. To the latter class of materials belong, for instance, many types of rocks, varved clays and in practice every soil that underwent a consolidation process with a preferential direction. In fact, it has been demonstrated, see e.g. Mitchell [7], that clay particles tend to become oriented perpendicularly to the major principal stress direction during one-dimensional consolidation. The preferred orientation of clay particles causes strength and deformability of clay to vary with direction. Also, the way of transportation during deposition may result in anisotropic fabric and consequently in anisotropic behaviour. Even the behaviour of granular materials depends on the angle between the principal axes of stress and on the deposition orientation, as demonstrated by Arthur and Menzies [1]. For this class of materials analyses based on the assumption of purely isotropic behaviour may be often misleading.

In this paper the influence of anisotropy on the strength of materials enjoying cohesion and friction, and then mainly soils and rocks, is investigated. Strength is by no means the sole property of a geologic material affected by anisotropy, since the latter also widely influences the stress-strain characteristics and the mode of failure, as thoroughly discussed by Mitchell [8], but these latter aspects will eventually be considered in a further paper.

The analysis will be restricted to orthotropic materials. This makes the problem of the derivation of the failure condition much easier and does not limit the practical applicability of the theory, since the way in which soils and rocks are formed suggests that these materials enjoy that property.

It will be shown that the proposed failure condition reduces to either the Tresca or the Coulomb failure criteria in special cases for isotropic materials. The results obtained from the theory will be compared with test data, available from the literature, in conventional triaxial loading conditions.

In the following, the term stress means effective stress in the usual sense of Soil Mechanics, if not otherwise specified.

2. The failure condition

Consider a material element of infinitesimal dimensions and a reference frame m, n, t centred in the element and a plane passing through it whose normal is m. Let m^i (i = 1, 2, 3) be its directon cosines with respect to a reference frame of axes x^i that will be specified later. We shall assume that the scalar shear component σ_{mn} on that plane cannot exceed a certain value, in general dependent on the direction of m, so that

$$\sigma_{mn} \leqslant S(m). \tag{1}$$

Failure is attained when the equality sign holds. We shall postulate that the shear strength S(m) is due to the concurrence of two scalar terms f_{mm} and c_{mm} which vary with the orientation as the normal components of a double tensor do.

The latter can be physically linked to the cohesion of the material. The former is defined as

$$f_{mm} = \phi_{mmrs}\, \sigma^{rs}, \quad (r, s = m, n, t), \tag{2}$$

where ϕ is a quadruple tensor that must be intended as a simple array of numbers, see also Boehler and Sawczuk [4]. The tensor ϕ can be physically linked to the internal friction of the material. It will be assumed that $\phi_{mnrs} = \phi_{nmrs} = \phi_{mnsr}$. Therefore one has

$$\sigma_{mn} \leqslant \phi_{mmrs}\, \sigma^{rs} + c_{mm}. \tag{3}$$

It is possible to formulate a proper failure criterion starting from (3) by the same path of reasoning as proposed by Capurso and Sacchi [5]. In fact, consider the reference frame x^i and express the terms σ_{mn}, f_{mm} and c_{mm} with reference to x^i. One has

$$\sigma_{mn} = \sigma_{ij}\, m^i n^j, -$$

$$f_{mm} = f_{ik}\, m^i m^k, \quad (i, k, \ell = 1, 2, 3), \tag{4}$$

$$c_{mm} = c_{ik}\, m^i m^k,$$

so that condition (3) becomes

$$\sigma_{ij} m^i n^j \leqslant \{f_{ik} + c_{ik}\} m^i m^k. \tag{5}$$

The directon cosine n^j can be expressed in terms of the other two through the Ricci alternating tensor $e^i{}_k{}^\ell$,

$$n^j = t_\varrho \, e^\ell{}_k{}^j m^k. \tag{6}$$

From (3) and (6) one has

$$\{f_{ik} + c_{ik} - \sigma_{ij} t_\varrho \, e^\ell{}_k{}^j\} m^i m^k \geqslant 0. \tag{7}$$

This expression is a positive semidefinite quadratic form. Thus one must have

$$\det [f_{ik} + c_{ik} - \sigma_{ij} t_\varrho \, e^\ell{}_k{}^j] = 0, \tag{8}$$

i.e. the determinant of the *symmetric* part of the matrix of the quadratic form must be non-negative. In particular, if condition (3) holds with the equality sign, so does condition (8). Therefore Eq. (8), explicitly written, gives the required expression of the failure criterion.

To start with, set

$$t_\varrho \, e^\ell{}_k{}^j = b^j_k. \tag{9}$$

It is easy to see that $b_{kk} = 0$, $b_{kj} = -b_{jk}$, $b_{12} = t_3$, $b_{13} = -t_2$, $b_{23} = t_1$. Therefore the tensor

$$p_{ik} = \sigma_{ij} b^j_k \tag{10}$$

has the components

$$p_{11} = \sigma_{12} t_3 - \sigma_{13} t_2 \,;\, p_{12} = -\sigma_{11} t_3 + \sigma_{13} t_1 \,;\, p_{13} = \sigma_{11} t_2 - \sigma_{12} t_1,$$

$$p_{21} = \sigma_{22} t_3 - \sigma_{23} t_2 \,;\, p_{22} = -\sigma_{21} t_3 + \sigma_{23} t_1 \,;\, p_{23} = \sigma_{21} t_2 - \sigma_{22} t_1,$$

$$p_{31} = \sigma_{32} t_3 - \sigma_{33} t_2 \,;\, p_{32} = -\sigma_{31} t_3 + \sigma_{33} t_1 \,;\, p_{33} = \sigma_{31} t_2 - \sigma_{32} t_1. \tag{11}$$

The only non-symmetric term in the matrix of Eq. (8) is the last one To get the symmetric part of that matrix we then have to put only

$$\hat{p}_{ik} = \frac{1}{2}(p_{ik} + p_{ki}) \,;\, \hat{f}_{ik} = f_{ik} \,;\, \hat{c}_{ik} = c_{ik}. \tag{12}$$

The determinant D,

$$D \equiv \det [f_{ik} + c_{ik} - \hat{p}_{ik}], \tag{13}$$

is in general a rather involved function of the stress tensor components σ_{ik}. If we assume as reference frame that of the axes of principal stresses, the expression for D is much simpler. In fact, one has

$$\hat{p}_{11} = \hat{p}_{22} = \hat{p}_{33} = 0 \, ; \, \hat{p}_{12} = -\frac{1}{2} \, T_{12} \, t_3 \, ; \, \hat{p}_{13} = \frac{1}{2} \, T_{13} \, t_2 \, ;$$

$$\hat{p}_{23} = \frac{1}{2} \, T_{32} \, t_1 , \tag{14}$$

having defined

$$T_{ik} = \sigma_{ii} - \sigma_{kk}, \qquad \text{(no summation over i or k).} \tag{15}$$

Therefore at failure

$$D \equiv (f_{11} + c_{11}) \, [(f_{22} + c_{22}) \, (f_{33} + c_{33}) - (f_{23} + c_{23} - \hat{p}_{23})^2] -$$

$$- (f_{12} + c_{12} - \hat{p}_{12}) \, [(f_{12} + c_{12} - \hat{p}_{12}) \, (f_{33} + c_{33}) -$$

$$- (f_{23} + c_{23} - \hat{p}_{23}) \, (f_{13} + c_{13} - \hat{p}_{13})] + (f_{13} + c_{13} - \hat{p}_{13}) \cdot \tag{16}$$

$$\cdot \, [(f_{12} + c_{12} - \hat{p}_{12}) \, (f_{23} + c_{23} - \hat{p}_{23}) - (f_{22} + c_{22})$$

$$\cdot \, (f_{13} + c_{13} - \hat{p}_{13})] = 0.$$

Because of (9) the direction cosines t_i are still present in Eq. (16). To get rid of them, one can make use of the theory of the envelopes.

$$\partial D/\partial t_1 \equiv T_{32} \, [(f_{11} + c_{11}) \, (f_{23} + c_{23} - \hat{p}_{23}) - (f_{12} + c_{12} - \hat{p}_{12}) \cdot$$

$$\cdot \, (f_{13} + c_{13} - \hat{p}_{13})] = 0. \tag{17}$$

$$\partial D/\partial t_2 \equiv T_{13} \, [(f_{12} + c_{12} - \hat{p}_{12}) \, (f_{23} + c_{23} - \hat{p}_{23}) - (f_{22} + c_{22}) \cdot$$

$$\cdot \, (f_{13} + c_{13} - \hat{p}_{13})] = 0. \tag{18}$$

$$\partial D/\partial t_3 \equiv T_{21} \, [(f_{12} + c_{12} - \hat{p}_{12}) \, (f_{33} + c_{33}) - (f_{23} + c_{23} - \hat{p}_{23}) \cdot$$

$$\cdot \, (f_{13} + c_{13} - \hat{p}_{13})] = 0. \tag{19}$$

Substituting (17) and (19) in Eq. (10) one sees easily that

$$(f_{11} + c_{11}) (f_{33} + c_{33}) - (f_{13} + c_{13} - \hat{p}_{13})^2 = 0 \qquad (20)$$

which implies

$$\hat{p}_{13} = f_{13} + c_{13} \mp [(f_{11} + c_{11}) (f_{33} + c_{33})]^{\frac{1}{2}} . \qquad (21)$$

Squaring each member of (21) and taking into account that

$$t_1^2 + t_2^2 + t_3^2 = 1, \qquad (22)$$

one gets

$$\{[(f_{11} + c_{11}) (f_{33} + c_{33})]^{\frac{1}{2}} \pm (f_{13} + c_{13})\}^2 \, t_1^2 + [\{[(f_{11} + c_{11}).$$
$$\cdot (f_{33} + c_{33})]^{\frac{1}{2}} \pm (f_{13} + c_{13})\}^2 - (1/2 \, T_{13})^2] \, t_2^2 +$$
$$+ \{[(f_{11} + c_{11}) (f_{33} + c_{33})]^{\frac{1}{2}} \pm (f_{13} + c_{13})\}^2 \, t_3^2 = 0. \qquad (23)$$

Eq. (23) is again a quadratic form. The associated matrix is diagonal and its determinant must be nil. This occurs when

$$\{[(f_{11} + c_{11}) (f_{33} + c_{33})]^{\frac{1}{2}} \pm (f_{13} + c_{13})\}^2 - (1/2 \, T_{13})^2 = 0. \qquad (24)$$

The least value of $|T_{13}/2|$ for which Eq. (24) is fulfilled, and then failure attained, is therefore

$$|T_{13}/2| = |[(f_{11} + c_{11}) (f_{33} + c_{33})]^{\frac{1}{2}} - |f_{13} + c_{13}||. \qquad (25)$$

Following the same path of reasoning and substituting Eqs. (17) in (18) and then (18), (19) in (10), one gets

$$|T_{12}/2| = |[(f_{11} + c_{11}) (f_{22} + c_{22})]^{\frac{1}{2}} - |f_{12} + c_{12}||, \qquad (26)$$

$$|T_{23}/2| = |[(f_{22} + c_{22}) (f_{33} + c_{33})]^{\frac{1}{2}} - |f_{23} + c_{23}||, \qquad (27)$$

Failure is attained when either one of the conditions (25), (26), (27) is fulfilled. Conditions (25), (26), (27) state that failure is attained when the

largest shear stress reaches a value that is dependent on the inclination of
the material element with respect to the principal axes of stress.

The strength parameters f_{ij}, c_{ij}, appropriate for each considered direction,
can be derived once some material constants have been experimentally determined. In fact, it is necessary to know the principal cohesive strengths c_I, c_{II},
c_{III} to uniquely define the double tensor c. Moreover, as shown by Boehler
and Sawczuk [4], the number of independent components of the tensor ϕ to
be experimentally determined reduce to six non-zero terms

$$\phi_{mmmm} , \quad \phi_{mnmn} , \quad (m, n = 1, 2, 3) , \tag{28}$$

since one has

$$\phi_{mnmn} = \phi_{nmnm} . \tag{29}$$

Since the principal axes of strength do not coincide in general with
those of stress, the usual tensor transformations must be employed. Therefore the relevant strength parameters are given by

$$c_{ij} = c_{mm} \, \partial x_m/\partial x_i \, \partial x_m/\partial x_j , \tag{30}$$

$$\phi_{ijkk} = \phi_{mnmn} \, \partial x_m/\partial x_i \, \partial x_n/\partial x_j \, \partial x_m/\partial x_k \, \partial x_n/\partial x_k , \tag{31}$$

$$f_{ij} = \phi_{ijkk} \, \sigma_{kk} . \tag{32}$$

3. Specialization to the isotropic case

To check the validity of the proposed theory, let us first consider the special
case when the material is isotropic. To start with, assume that the material
is purely cohesive. Then, if δ_{ij} is the Kronecker delta,

$$c_{ij} = c\delta_{ij} \quad ; \quad f_{ij} = 0 . \tag{33}$$

Assume that Eq. (25) is fulfilled. Then we have

$$|T_{13}/2| = c \tag{34}$$

which coincides with the Tresca criterion.

Consider now a purely granular material such that

$$c_{ij} = 0 \quad ; \quad f_{ij} = \phi_{ijkk} \, \sigma_{kk} , \tag{35}$$

where

$$\phi_{ijk\ell} = \lambda \delta_{ij} \delta_{k\ell} + \mu(\delta_{ik} \delta_{j\ell} + \delta_{i\ell} \delta_{jk}) \tag{36}$$

which is the most general expression for an isotropic tensor enjoying the assumed symmetries. Since the frictional strength of an isotropic solid is directly proportional to the normal component of the effective stress on the considered plane, it is easy to show that λ must be nil. This result has been also obtained by Boehler and Sawczuk [4] with similar considerations. Therefore, from Eq. (25) one has

$$|T_{13}/2| = 2\mu \sqrt{\sigma_{11}\sigma_{33}} \tag{37}$$

which is an unusual way of writing the Coulomb criterion where

$$2\mu = \tan \phi \tag{38}$$

and ϕ is the angle of internal friction. In fact, the Coulomb failure condition is usually written as

$$\tau = \sigma \tan \phi . \tag{39}$$

This implies that, at failure,

$$\sigma_{33} = \frac{1 - \sin \phi}{1 + \sin \phi} \sigma_{11} . \tag{40}$$

Therefore

$$1/4(\sigma_{11} - \sigma_{33})^2 = 1/4 \sigma_{11}^2 \left(1 - \frac{1 - \sin \phi}{1 + \sin \phi}\right)^2 = \frac{\sin^2 \phi}{1 - \sin^2 \phi} \sigma_{11}\sigma_{33}$$

$$= \sigma_{11}\sigma_{33} \tan^2 \phi \tag{41}$$

which implies Eq. (37).

Thus the Coulomb criterion can be also formulated as stating that the maximum shear stress cannot exceed a quantity proportional to the geometric mean of the maximum and minimum principal stresses,

$$|\tau_{max}| \leqslant \tan \phi \sqrt{\sigma_{11}\sigma_{33}} . \tag{42}$$

Finally, if the material enjoys both cohesion and friction, Eq. (25) becomes

$$|T_{13}/2| = \sqrt{(c + 2\mu\sigma_{11})(c + 2\mu\sigma_{33})} \tag{43}$$

which is again the correct formulation of the Coulomb criterion as can be easily verified substituting $(\sigma_i + c/2\mu)$ for σ_i in Eq. (42).

4.   Transversely isotropic materials

For a generic orthotropic material with friction and cohesion there are nine independent constants to be determined to characterize its strength. These constants can be found in principle by suitably designed experimental procedures, but their practical determination appears complex, because of the difficult mathematical expression of the failure criterion. To make things easier, let us consider only materials which enjoy a polar symmetry with respect to an axis, or transversely isotropic materials. This limitation does not restrict the practical interest of what follows since most soils and rocks are stratified and enjoy this property. Soils like varved clays or rocks as diatomite are examples of this class of materials.

The determination of the experimental constants can be performed with a standard triaxial apparatus in which the principal axes of stress are the vertical axis x_1 and any two orthogonal axes x_2, x_3 in the horizontal plane. Therefore, also the state of stress enjoys a polar symmetry with respect to the vertical axis.

Assume that x_t is the axis of symmetry of the strength and that x_r and x_s are any two orthogonal axes in the plane whose normal is x_t; for the assumed symmetry $x_s = x_2$.

Let us consider separately the case of purely cohesive and purely frictional materials.

a) Purely cohesive material

Assume that the principal components of the strength tensor are

$$c_r = c_s = \alpha c_t , \quad \alpha \geqslant 1 . \tag{44}$$

Note that by definition c_r is the cohesion on the plane whose normal is x_r. It is easy to see that failure is attained when Eq. (25) holds. If θ is the angle between x_r and x_1 one has

$$c_{11} = c_r \cos^2 \theta + c_t \sin^2 \theta ,$$

$$c_{33} = c_t \cos^2 \theta + c_r \sin^2 \theta , \tag{45}$$

$$c_{13} = \frac{1}{2} (c_r - c_t) \sin 2\theta .$$

Therefore Eq. (25) becomes

$$\frac{1}{2} (\sigma_{11} - \sigma_{33}) = c_t \{ [(\alpha \cos^2 \theta + \sin^2 \theta)(\alpha \sin^2 \theta + \cos^2 \theta)]^{\frac{1}{2}}$$

$$- \frac{\alpha - 1}{2} \sin^2 \theta \} . \tag{46}$$

In Figure 1 the polar diagram of strength is plotted for various values of the anisotropy coefficient α. The shape of this diagram is prominently dependent on α. In fact, it passes from a circle for $\alpha = 1$, to a convex curve for $1 \leqslant \alpha \leqslant 2$, to a straight line for $\alpha = 2$ up to a concave curve for $\alpha \geqslant 2$. What is most interesting is that for every value of α the least strength is attained when $\phi = 45°$, and it coincides with the least possible strength c_t. On the contrary, the largest strength c_r can never be reached. The maximum available strength is

$$\frac{1}{2}(\sigma_{11} - \sigma_{33}) = \sqrt{c_r c_t} = c_t \sqrt{\alpha} \tag{47}$$

and is reached when $\theta = 0°$ or $\theta = 90°$. The diagram is symmetric with respect to the axis inclined at $45°$.

The physical meaning of the results obtained can be deduced from Figure 2 where some samples of a stratified material with different inclinations of the stratification with respect to the principal axes of stress are schematized. In fact, assuming that the plane of the stratifications is that of least strength, it is obvious that if the least strength is reached for $\theta = 45°$, failure occurs independently of the strength at different inclinations. It is also evident that the largest available strength is reached when $\theta = 0°$ or $\theta = 90°$ and that the polar diagram of strength should be symmetric with respect to the axis of least strength.

b) Purely frictional material
Following Boehler and Sawczuk [4], in an orthotropic tensor written with reference to the principal axes of orthotropy the only independent

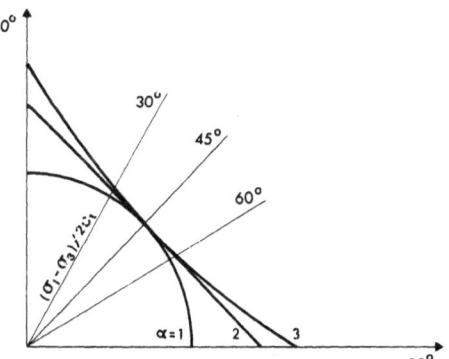

Figure 1. – *Polar diagram of strength of a purely cohesive transversely isotropic material.*

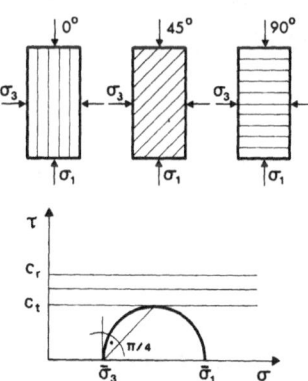

Figure 2. – *Strength of specimens with different trimming directions for a purely cohesive material.*

parameters are ϕ_{rrrr}, ϕ_{ssss}, ϕ_{tttt}, ϕ_{rtrt}, ϕ_{stst}, ϕ_{rsrs}. Because of the assumed transverse isotropy, a rotation of the reference axes in the isotropic plane does not affect the components of the tensor ϕ. It is possible to see, by means of the usual tensor transformation rules, that $\phi_{rrrr} = \phi_{ssss} = 2\phi_{rsrs}$, $\phi_{rtrt} = \phi_{stst}$, so that the tensor ϕ becomes

$$
\phi = \begin{bmatrix}
2\mu_r & & & & & & & \\
 & 2\mu_r & & & & & & \\
 & & 2\mu_t & & \mathbf{0} & & & \\
 & & & \mu_r\mu_r & & & & \\
 & & & & \mu_r\mu_r & & & \\
 & \mathbf{0} & & & & \mu_{rt}\mu_{rt} & & \\
 & & & & & & \mu_{rt}\mu_{rt} & \\
 & & & & & & & \mu_{rt}\mu_{rt} \\
 & & & & & & & \mu_{rt}\mu_{rt}
\end{bmatrix}
\tag{48}
$$

with only three independent constants μ_r, μ_t and μ_{rt} to be determined. Since the failure criterion is expressed in terms of principal stresses, one has to transform the tensor (48) to express it in terms of x_1, x_2, x_3. It can be easily seen that Eq. (25) holds at failure. Then, since

$$
f_{11} = \phi_{1111}\sigma_{11} + \phi_{1122}\sigma_{22} + \phi_{1133}\sigma_{33} ,
$$
$$
f_{33} = \phi_{3311}\sigma_{11} + \phi_{3322}\sigma_{22} + \phi_{3333}\sigma_{33} ,
\tag{49}
$$
$$
f_{13} = \phi_{1311}\sigma_{11} + \phi_{1322}\sigma_{22} + \phi_{1333}\sigma_{33} ,
$$

only the nine ϕ_{ikrs} terms appearing in Eqs. (49) are relevant. Since

$$
\partial x_s/\partial x_1 = \partial x_s/\partial x_3 = \partial x_r/\partial x_2 = \partial x_t/\partial x_2 = 0 ,
\tag{50}
$$

it is easy to see after some calculation that

$$
\phi_{1122} = \phi_{3322} = \phi_{1322} = 0 ,
\tag{51}
$$

and that

$$
\phi_{1111} = 2\mu_r(\partial x_r/\partial x_1)^4 + 2\mu_t(\partial x_t/\partial x_1)^4 + 4\mu_{rt}(\partial x_r/\partial x_1\,\partial x_t/\partial x_1)^2 ,
\tag{52}
$$
$$
\phi_{3333} = 2\mu_r(\partial x_r/\partial x_3)^4 + 2\mu_t(\partial x_t/\partial x_3)^4 + 4\mu_{rt}(\partial x_r/\partial x_3\,\partial x_t/\partial x_3)^2 ,
\tag{53}
$$

$$\phi_{1133} = 2\mu_r(\partial x_r/\partial x_1 \, \partial x_r/\partial x_3)^2 + 2\mu_t(\partial x_t/\partial x_1 \, \partial x_t/\partial x_3)^2$$
$$+ 4\mu_{rt}(\partial x_r/\partial x_1 \, \partial x_t/\partial x_1 \, \partial x_r/\partial x_3 \, \partial x_t/\partial x_3), \tag{54}$$

$$\phi_{1311} = 2\mu_r(\partial x_r/\partial x_1)^3 \, (\partial x_r/\partial x_3) + 2\mu_t(\partial x_t/\partial x_1)^3 \, (\partial x_t/\partial x_3)$$
$$+ 2\mu_{rt}(\partial x_r/\partial x_1)\,(\partial x_t/\partial x_1)\,(\partial x_r/\partial x_1 \, \partial x_t/\partial x_3$$
$$+ \partial x_r/\partial x_3 \, \partial x_t/\partial x_1), \tag{55}$$

$$\phi_{1333} = 2\mu_r(\partial x_r/\partial x_1)\,(\partial x_r/\partial x_3)^3 + 2\mu_t(\partial x_t/\partial x_1)\,(\partial x_t/\partial x_3)^3$$
$$+ 2\mu_{rt}(\partial x_r/\partial x_3)\,(\partial x_t/\partial x_3)\,(\partial x_r/\partial x_1 \, \partial x_t/\partial x_3$$
$$+ \partial x_r/\partial x_3 \, \partial x_t/\partial x_1). \tag{56}$$

Defining again θ as the angle between x_r and x_1, one has

$$x_r = x_1 \cos\theta - x_3 \sin\theta ,$$
$$x_t = x_1 \sin\theta + x_3 \cos\theta , \tag{57}$$

which implies

$$\partial x_r/\partial x_1 = \cos\theta \quad ; \quad \partial x_r/\partial x_3 = -\sin\theta \quad ; \quad \partial x_t/\partial x_1 = \sin\theta \quad ;$$
$$\partial x_t/\partial x_3 = \cos\theta . \tag{58}$$

Substituting (51)-(56) in (49) and taking account of (58), one gets

$$f_{11} = 2\mu_r \cos^2\theta \,(\sigma_{11} \cos^2\theta + \sigma_{33} \sin^2\theta) + 2\mu_t \sin^2\theta \,(\sigma_{11} \sin^2\theta$$
$$+ \sigma_{33} \cos^2\theta) + 4\mu_{rt} \cos^2\theta \sin^2\theta \,(\sigma_{11} - \sigma_{33}), \tag{59}$$

$$f_{33} = 2\mu_r \sin^2\theta \,(\sigma_{11} \cos^2\theta + \sigma_{33} \sin^2\theta) + 2\mu_t \cos^2\theta \,(\sigma_{11} \sin^2\theta$$
$$+ \sigma_{33} \cos^2\theta) - 4\mu_{rt} \cos^2\theta \sin^2\theta \,(\sigma_{11} - \sigma_{33}),$$

$$f_{13} = -2\mu_r \cos\theta \sin\theta \,(\sigma_{11} \cos^2\theta + \sigma_{33} \sin^2\theta)$$
$$+ 2\mu_t \cos\theta \sin\theta \,(\sigma_{11} \sin^2\theta + \sigma_{33} \cos^2\theta) + 2\mu_{rt} \sin\theta \cos\theta$$
$$(\cos^2\theta - \sin^2\theta)\,(\sigma_{11} - \sigma_{33}).$$

For practical applications it is often convenient to express Eq. (49) in terms of stresses referred to the principal axes of orthotropy. Considering that

$$\sigma_r = \sigma_{11} \cos^2\theta + \sigma_{33} \sin^2\theta ,$$
$$\sigma_t = \sigma_{11} \sin^2\theta + \sigma_{33} \cos^2\theta , \tag{60}$$
$$\tau_{rt} = \frac{1}{2}\,(\sigma_{11} - \sigma_{33}) \sin 2\theta ,$$

one gets

$$f_{11} = 2\mu_r \sigma_r \cos^2\theta + 2\mu_t \sigma_t \sin^2\theta + 4\mu_{rt}\tau_{rt}\sin\theta\cos\theta ,$$

$$f_{33} = 2\mu_r \sigma_r \sin^2\theta + 2\mu_t \sigma_t \cos^2\theta - 4\mu_{rt}\tau_{rt}\sin\theta\cos\theta , \qquad (61)$$

$$f_{13} = (-\mu_r \sigma_r + \mu_t \sigma_t)\sin 2\theta + 2\mu_{rt}\tau_{rt}\cos 2\theta .$$

The experimental constants can be found with a triaxial test. In fact, in terms of principal stresses, taking account of Eqs. (59), one has

$$\frac{1}{2}(R-1) = [(2\mu_r \cos^2\theta \, (R\cos^2\theta + \sin^2\theta) + 2\mu_t \sin^2\theta \, (R\sin^2\theta + \cos^2\theta) +$$

$$+ 4\mu_{rt}\cos^2\theta \sin^2\theta \, (R-1))\,(2\mu_r \sin^2\theta \, (R\cos^2\theta + \sin^2\theta)$$

$$+ 2\mu_t \cos^2\theta \, (R\sin^2\theta + \cos^2\theta) - 4\mu_{rt}\cos^2\theta \sin^2\theta \, (R-1))]^{\frac{1}{2}}$$

$$- \sin 2\theta \, | -\mu_r \, (R\cos^2\theta + \sin^2\theta) + \mu_t \, (R\sin^2\theta + \cos^2\theta) +$$

$$+ \mu_{rt} \, (R-1)\cos 2\theta \, | , \qquad (62)$$

where $R = \sigma_{11}/\sigma_{33}$. Eq. (62) is in implicit form. It can be solved for R, by giving a trial value to R in the r.h.s. and calculating R in the l.h.s. until convergence is reached. It is easy to see that if $\theta = 0°$ or $\theta = 90°$ one has

$$\frac{R-1}{2\sqrt{R}} = 2\sqrt{\mu_r \mu_t}, \qquad (63)$$

Since

$$\frac{R-1}{2\sqrt{R}} \equiv \tan\phi, \qquad (64)$$

where ϕ is the angle between the σ axis and the tangent to the Mohr circle at failure passing through the origin in the Mohr plane, one may define

$$\beta = \mu_r/\mu_t ; \qquad \beta \geqslant 1 \qquad (65)$$

so that Eq. (63) becomes

$$\tan\phi = \sqrt{\beta} \quad 2\mu_t, \qquad (66)$$

i.e. the overall angle of friction, when $\theta = 0°$ or $90°$, is proportional to the least available angle of friction and the proportionality constant is the square root of the orthotropy ratio.

The only parameter still to be determined is μ_{rt}. It is possible to show that if $\mu_{rt} \neq \mu_t$ for some values of θ, the strength decreases with increasing β, which is physically untenable. In fact, it does not seem reasonable that if the largest strength is increased and the least is kept constant, the overall strength could decrease. On the other hand, if $\mu_{rt} = \mu_t$, it is possible to show that the least strength is given by

$$\tan \phi = 2\mu_t \tag{67}$$

for any value of β and that this occurs when

$$\theta = \pi/4 - \frac{1}{2} \tan^{-1} 2\mu_t \equiv \hat{\theta}, \tag{68}$$

as shown in Figure 3. This result is reasonable. In fact, it is clear, from Figure 4 that if the least strength is reached on a plane inclined at $\hat{\theta}$ to the axis of the largest principal stress, failure occurs independently of the strength on planes at other inclinations. Thus in the following we shall assume

$$\mu_r = \beta\mu \; ; \; \mu_t = \mu_{rt} = \mu. \tag{69}$$

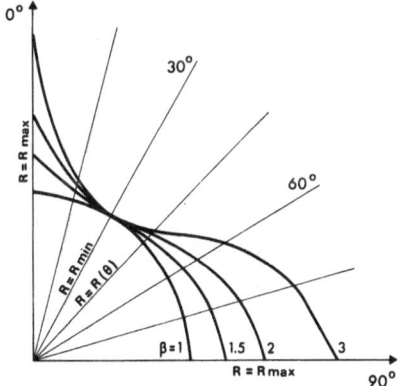

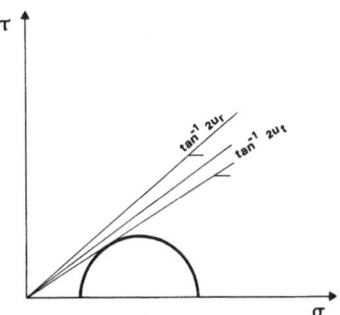

Figure 3. – *Polar diagram of strength for a purely frictional transversely isotropic material* – $\tan^{-1} 2\mu_t = 30°$.

Figure 4. – *Least strength for a purely frictional material.*

Hence, Eq. (62) can be rewritten as

$$\frac{R-1}{4} = [(\beta \cos^2\theta \ (R\cos^2\theta + \sin^2\theta) + \sin^2\theta \ (R + (R-1)\cos^2\theta))$$

$$\cdot \ (\beta \sin^2\theta \ (R\cos^2\theta + \sin^2\theta) + \cos^2\theta (1 - (R-1)\sin^2\theta))]^{\frac{1}{2}}$$

$$- (R\cos^2\theta + \sin^2\theta)(\beta - 1)\sin 2\theta. \tag{70}$$

c) Material with cohesion and friction

It will be assumed that the principal axes of strength for friction and cohesion coincide. In general, however,

$$c_r/c_t = \alpha, \qquad \mu_r/\mu_t = \beta. \tag{71}$$

From Eqs. (25), (45) and (71) we then have

$$\frac{R-1}{4\mu} = [\{(\alpha\cos^2\theta + \sin^2\theta)\rho + \beta\cos^2\theta \ (R\cos^2\theta + \sin^2\theta) +$$

$$+ \sin^2\theta \ (R + (R-1)\cos^2\theta)\} \{(\cos^2\theta + \alpha\sin^2\theta)\rho +$$

$$+ \beta\sin^2\theta \ (R\cos^2\theta + \sin^2\theta) + 2\cos^2\theta \ (1 - (R-1)\sin^2\theta)\}]^{\frac{1}{2}} -$$

$$- \sin 2\theta \ |(R\cos^2\theta + \sin^2\theta)\frac{\beta - 1}{2} + \frac{\alpha - 1}{2} \ \rho\,|, \tag{72}$$

where ρ has been defined as

$$\rho = c_t/(2\mu_t \sigma_3). \tag{73}$$

A parametric study has been performed and the following conclusions have been drawn:

1) whatever the values of α and β, the least overall strength is given by

$$R_{min} = N\phi_t + \rho(N\phi_t - 1); \quad N\phi_t = \tan^2(\pi/4 - \hat\theta). \tag{74}$$

This value is reached when

$$\theta = \hat\theta.$$

2) ρ has little influence on the shape of the polar diagram of strength. It affects mainly the magnitude of R.

3) if $\alpha = \beta$, then $R(0°) = R(90°)$,

    if $\alpha > \beta$, then $R(0°) < R(90°)$,

    if $\alpha < \beta$, then $R(0°) > R(90°)$.

4) $\theta = 0°$ implies $(R - 1)/4\mu = \sqrt{(\alpha\rho + \beta R)(\rho + 1)}$;

    $\theta = 90°$ implies $(R - 1)/4\mu = \sqrt{(\rho + R)(\alpha\rho + \beta)}$

In Figure 5 there are shown typical polar diagrams of strength for various α, β and ρ.

5. Comparison with experimental data and the determination of mechanical constants

At first glance the results obtained for purely cohesive materials may seem to contradict actual experimental results published in the literature. In fact, it is well known that the strength of a saturated clay specimen in undrained

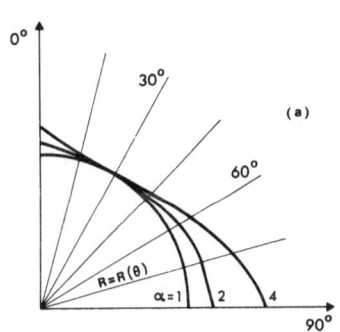

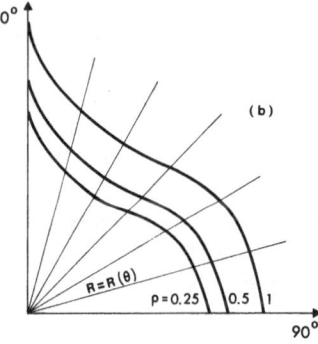

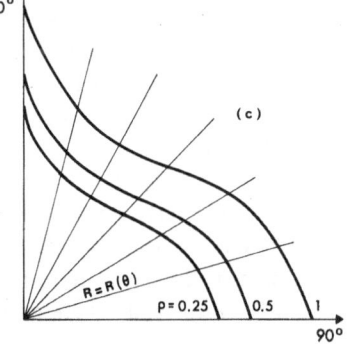

Figure 5. – *Polar diagrams of strength for materials with friction and cohesion with different values of the material constants*

a) $\beta = 1$, $\tan^{-1} 2\mu_t = 30°$, $\rho = 0.5$

b) $\alpha = 1$, $\beta = 2$, $\tan^{-1} 2\mu_t = 30°$

c) $\alpha = \beta = 2$, $\tan^{-1} 2\mu_t = 30°$.

compression, usually considered a purely cohesive material in terms of total stresses, is dependent on the sampling orientation. The various functions proposed to take account of this fact, mainly on empirical grounds, are rather different from the one derived here. In particular, they do not exhibit symmetry about the axis of least overall strength. The experimental points can be well interpolated in a polar diagram of strength by an ellipse or by more complex trigonometric functions (Bishop [2]).

It must be noted, however, that since the material is anisotropic even before failure, the effective stress path to failure will in general depend on the sampling orientation with respect to the principal axes of strength. Therefore even if the true failure locus, i.e. the failure locus in terms of effective stresses, would be independent of the sampling orientation, the strength in terms of total stresses would be anisotropic, as shown in Figure 6. Indeed this is very much what actually occurs, since, following Bishop, Webb and Lewin [3] and Duncan and Seed [6], who took account of the values of the pore pressures at failure, the anisotropic undrained strength of clay is much more due to different pore pressures at failure than to a large variation in effective strength parameters. Without knowing the value of the pore pressures at failure it is not possible to know which are the effective stresses and thus to make any comparison with the theory presented. Clay can be considered an anisotropic purely cohesive material in terms of total stresses, but it must be emphasized that its strength depends on the stress path followed and on its deformability properties prior to failure. In terms of effective stresses, clay is not a purely cohesive material but a material with friction and cohesion, if heavily over-consolidated, and a purely frictional material, if normally consolidated or lightly overconsolidated. Therefore the theory proposed for purely cohesive materials cannot be applied to clays but could be used for solids without internal friction, such as steel.

In principle for this kind of materials, if the principal directions of strength are known, the material constants can be easily found by performing two tests for $\theta = 45°$ and $\theta = 0°$. From the former it is possible to find c_t whilst α can be derived from the latter.

More important for practical geotechnical purposes is the problem concerning purely frictional materials. The constants to be determined are β and μ but one has also to find the position of the principal axes of strength, wich is not a priori obvious. By the results of Section 4b, the least overall strength is reached when $\theta = \hat{\theta}$. Knowing which is the orientation of the least strength, it is possible to find next the principal axes of strength. The parameters μ and β can be determined from Eqs. (64), (66), respectively.

Consider the tests performed by Arthur and Menzies [1] on rounded Leighton Buzzard sand. These authors prepared dry samples at almost the same

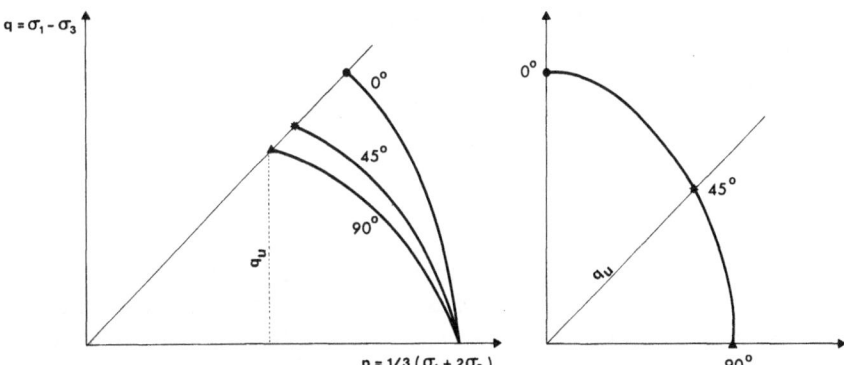

Figure 6. – *Anisotropic undrained strength of a material with isotropic failure condition, but anisotropic constitutive law prior to failure.*

porosity (34.1-34.2 %) in a special cubical triaxial cell. Samples were prepared in a tilting mould to give different directions of sample deposition with respect to the sample axes and applied principal stress directions. Their experimental results are shown by the dots in Figure 7. Note that the principal stress ratio R at failure is plotted for various angles of tilt ψ and not, as in Figure 3, for various angles θ.

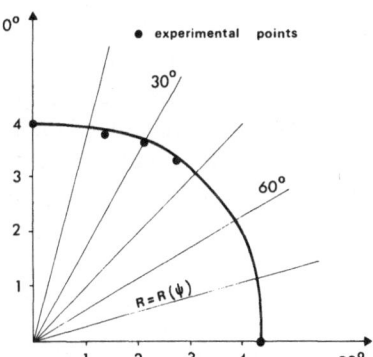

Figure 7. – *Comparison between calculated and experimental data for Leighton Buzzard sand-data after Arthur and Menzies (1)* $\beta = 1.25$, $\tan^{-1} 2\mu_t = 36\overset{\circ}{.}58$.

From the least value of the overall strength one has $\mu = 371$ which corresponds to a minimum angle of friction $\phi_t = 36\overset{\circ}{.}58$. The anisotropy factor has been estimated to be $\beta = 1.25$. Since the least value of strength, corresponding in this case to $\hat{\theta} = 26\overset{\circ}{.}71$, is reached at $\psi = 0$, one has

$$\psi = \theta - 26\overset{\circ}{.}71. \tag{75}$$

The calculated results well agree with experimental data. It may be noted that the anisotropy parameter β is low and the ratio of the largest to the least strength is only 1.12. This confirms that the conclusions of Duncan and Seed [6] are valid also for sand. By similar considerations, it is possible to find μ, ρ, ψ, α and β for a material enjoying cohesion and friction.

6. Conclusions

This paper presents a compact derivation of a general failure condition that can be considered as the extension to anisotropic solids of the Coulomb criterion. It is assumed that the strength varies as a double tensor does. This limits the validity of the theory presented to the class of orthotropic materials, to which most soils and rocks belong. The failure condition can be reduced to a quadratic form that must be positive semidefinite. The condition of positive semidefiniteness of the associated matrix yields the desired failure criterion. Specialization to the isotropic case gives the Coulomb criterion. Failure criteria for purely cohesive and purely frictional media and materials enjoying both cohesion and friction are derived next. Where possible, a comparison with experimental findings is made ; calculated and experimental results are shown to be in good agreement. The experimental findings discussed in this paper confirm the conclusions drawn by Duncan and Seed that in soils anisotropy affects more the behaviour up to failure than the effective strength parameters. Nevertheless, it should be noted that a small variation of the strength parameters, for high values of the internal angle of friction, may result, in much larger differences in failure loads. Disgregarding then the anisotropy of strength may lead to profoundly erroneous predictions of safety factors.

To verify the validity of the theory proposed, it is necessary to perform a series of tests on anisotropic materials, at least under conventional triaxial conditions. For the time being, it is possible to conclude that the theory presented gives a reasonable interpretation of the anisotropy of strength, at least for transversely isotropic materials. Also, for these materials, the angle θ appearing in the expressions for the failure conditions, can be easily eliminated by means of the Mohr circle of stress, so that the failure conditions can be written in terms of stresses only. The failure criteria can be easily taken into account in any computer routine for elastic perfectly plastic analysis and used for practical purposes.

Acknowledgments

The authors acknowledge the financial support of the Italian Research Council (C.N R.).

REFERENCES

[1] ARTHUR J., B.K. MENZIES. – "Inherent anisotropy in a sand", *Geotechnique,* **22** (1972): 115-128.

[2] BISHOP A.N. – "The strength of soils as engineering materials". 6th Rankine Lecture, *Geotechnique,* **16** (1966): 86-130.

[3] BISHOP A.W., D.L. WEBB and P.I. LEWIN. – "Undisturbed samples of London clay from the Ashford common shaft: Strength-effective stress relationships". *Geotechnique* **15**, 1 (1965): 1-31.

[4] BOEHLER J.P. and A. SAWCZUK. – "Equilibre limite des sols anisotropes". *Journal de Mécanique,* **9**, 1 (1970): 5-33.

[5] CAPURSO M. and G. SACCHI. – "Una condizione di plasticità per solidi anisotropi", *Techn. Rep. No. 512, Ist. di Scienza a Tecnica delle Costruzioni del Politecnico di Milano,* 1970.

[6] DUNCAN J.M. and H.B. SEED. – "Anisotropy and stress reorientation in clay". *Journal of Soil Mechanics and Found. Eng., ASCE,* **92**, 5 (1966): 21-50.

[7] MITCHELL J.K. – "The fabric of natural clays and its relation to engineering properties". *Proc. Highway Res. Bd.* Washington, D.C., **35**, (1956): 693-713.

[8] MITCHELL J.K. – *Fundamentals of Soil Behaviour.* Wiley, 1976, p. 422.

RESUME

(Un critère de rupture généralisé pour les solides orthotropes)

Dans ce mémoire, nous présentons une condition de rupture pour les solides orthotropes. On montre que cette condition peut être formulée à l'aide d'une forme quadratique non négative et telle que les conditions classiques de Tresca et de Coulomb peuvent être considérées comme des cas particuliers. On examine les différents diagrammes polaires de la résistance en fonction du coefficient d'orthotropie dans les conditions de charges triaxiales conventionnelles. On considère des matériaux purement cohérents et des matériaux à frottement interne pur. Le cas du matériau cohérent et à frottement interne est également étudié. On compare ensuite les résultats théoriques avec des données expérimentales disponibles dans la littérature et on propose une méthode de détermination expérimentale des paramètres d'orthotropie.

Cisaillement Positif, Négatif et Neutre

Z. Sobotka

Institut de Mécanique Théorique et Appliquée, Prague, Tchécoslovaquie.

1. Introduction

Sur la base des résultats de recherches théoriques et expérimentales, l'auteur a introduit la notion du cisaillement positif et négatif comme des types nouveaux de contraintes qui représentent dans un certain sens une analogie avec la compression et la traction dans le domaine des contraintes normales. Mais les différences entre le cisaillement positif et négatif dépendent d'une manière essentielle des angles entre les directions principales du cisaillement et les directions caractéristiques de la structure des matériaux. Il existe une direction neutre, pour laquelle les différences entre le cisaillement positif et négatif disparaissent. Dans cette direction, nous avons le cisaillement neutre ou le cisaillement dans le sens classique.

Par la notion du cisaillement positif et négatif, une signification physique a pu être donnée au sens des contraintes de cisaillement. L'existence de deux types opposés de contraintes de cisaillement est lié à l'anisotropie, à la configuration de la structure, aux différences entre les caractéristiques mécaniques correspondantes à la traction et à la compression, à la non-homogénéité, à la porosité, à l'orientation des réseaux de microfissures, à la stratification, à la superposition des contraintes de cisaillement avec des contraintes normales et aux rugosités orientées sur les surfaces de glissement.

Le cisaillement positif, qui est caractérisé par des valeurs plus élevées pour les modules de déformation et les contraintes limites, est accompagné par un accroissement du volume, tandis que le volume diminue au cours du cisaillement négatif. Contrairement à la dilatance, qui est un effet du second ordre, les variations de volume provoquées par le cisaillement positif et négatif sont du premier ordre, c'est pourquoi elles sont généralement plus élevées.

Dans certains cas, les variations de volume provoquées par le cisaillement positif et négatif sont de signe opposé à celles dues à la dilatance. Si la non-homogénéité et l'anisotropie du matériau sont relativement faibles et si la dilatance est très forte, les variations de volume dues à la dilatance peuvent prédominer sur celles provoquées par le cisaillement orienté.

Les effets du cisaillement positif et négatif entraînent également un comportement dissymétrique dans l'état limite, qui correspond à une forme dissymétrique des courbes intrinsèques de Mohr.

2.   Courbe de cisaillement

La résistance au cisaillement des matériaux anisotropes dépend de l'orientation de la direction de la contrainte principale du cisaillement par rapport aux directions caractéristiques de la structure.

Les valeurs limites des contraintes de cisaillement, qui agissent en un point dans les différentes directions sont définies par la courbe de cisaillement indiquant les variations de la résistance aux cisaillement suivant l'orientation du matériau.

La forme la plus simple pour une telle courbe de cisaillement est l'ellipse représentée sur la figure 1a pour les matériaux anisotropes.

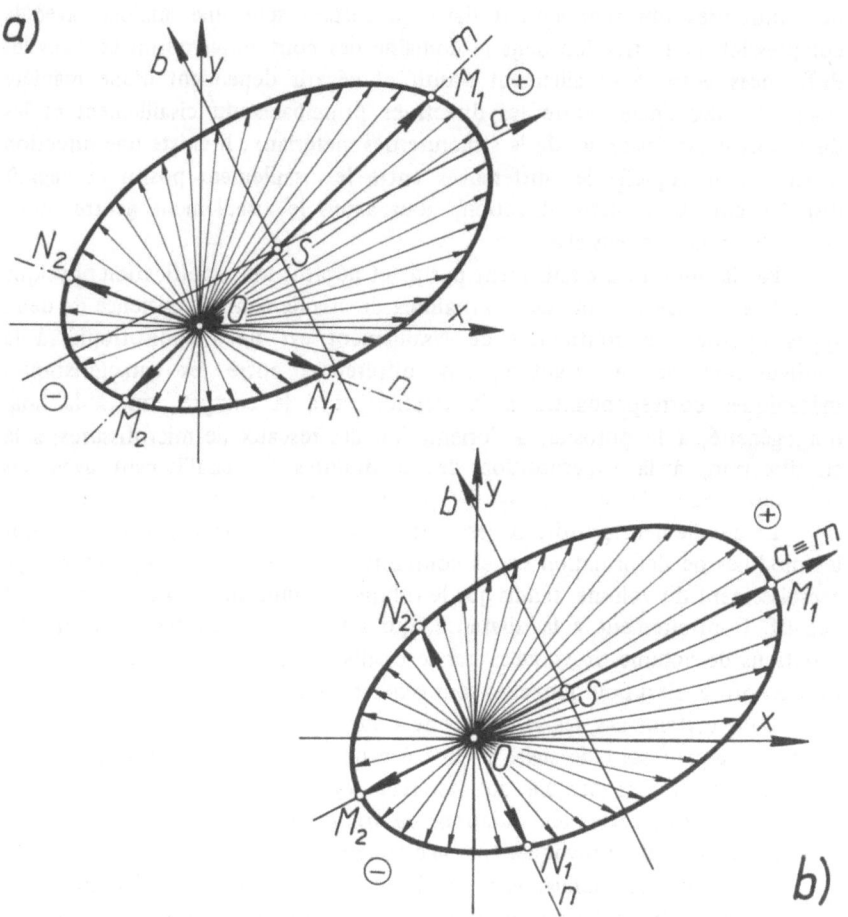

Figure 1. — *Courbes de cisaillement pour les matériaux anisotropes et orthotropes.*

La droite $n \equiv N_1 N_2$ indique la direction neutre, pour laquelle nous avons le cisaillement neutre, indépendant du sens de la contrainte. Cette droite divise le plan des contraintes de cisaillement en deux régions: la région du cisaillement positif et celle du cisaillement négatif. Dans la direction de cette droite, il n'y a pas de différences entre les cisaillements dans les deux directions opposées.

Dans la direction conjuguée définie par la droite $m \equiv M_1 M_2$, la différence entre les contraintes limites en cisaillement positif et négatif atteint sa valeur maximale. La courbe elliptique sur la figure 1b représente les variations des résistances au cisaillement en un point d'un milieu orthotrope.

Le centre S de cette ellipse est situé sur la droite m, support de l'axe principal. La droite n, qui correspond au cisaillement neutre, est dans ce cas perpendiculaire à la droite m, qui indique la direction pour laquelle la différence entre les valeurs limites des contraintes de cisaillement positives et négatives est minimale.

Dans le cas général d'un corps anisotroqe, la courbe de cisaillement peut avoir une forme plus complexe. Elle peut être composée non seulement de secteurs convexes mais également de secteurs concaves.

3. Enveloppes Dissymétriques des Cercles de Mohr

Le cisaillement positif et négatif provoque une dissymétrie pour les enveloppes des cercles de Mohr. L'auteur explique cette dissymétrie par des relations entre les courbes intrinsèques dans le plan des contraintes principales et l'enveloppe linéaire dissymétrique des cercles de Mohr pour les matériaux anisotropes cohérents avec un angle de frottement interne constant.

Les courbes intrinsèques présentées sur la figure 2 correspondent à une orientation donnée des directions principales de l'état de contrainte par rapport aux axes de symétrie matérielle du milieu.

Les droites limites a et b sur la figure 2a coupent les axes des contraintes principales σ_1 et σ_2 aux points A, B, C et D, qui délimitent les segments représentant les contraintes limites σ_{C1} et σ_{C2} en compression simple et les contraintes limites σ_{T1} et σ_{T2} en traction simple.

En introduisant dans l'équation de la droite limite a de la figure 2a

$$\frac{\sigma_1}{\sigma_{C1}} - \frac{\sigma_2}{\sigma_{T2}} = 1 \tag{1}$$

la contrainte limite en cisaillement pur positif

$$\tau_{PO} = \sigma_1 = -\sigma_2, \tag{2}$$

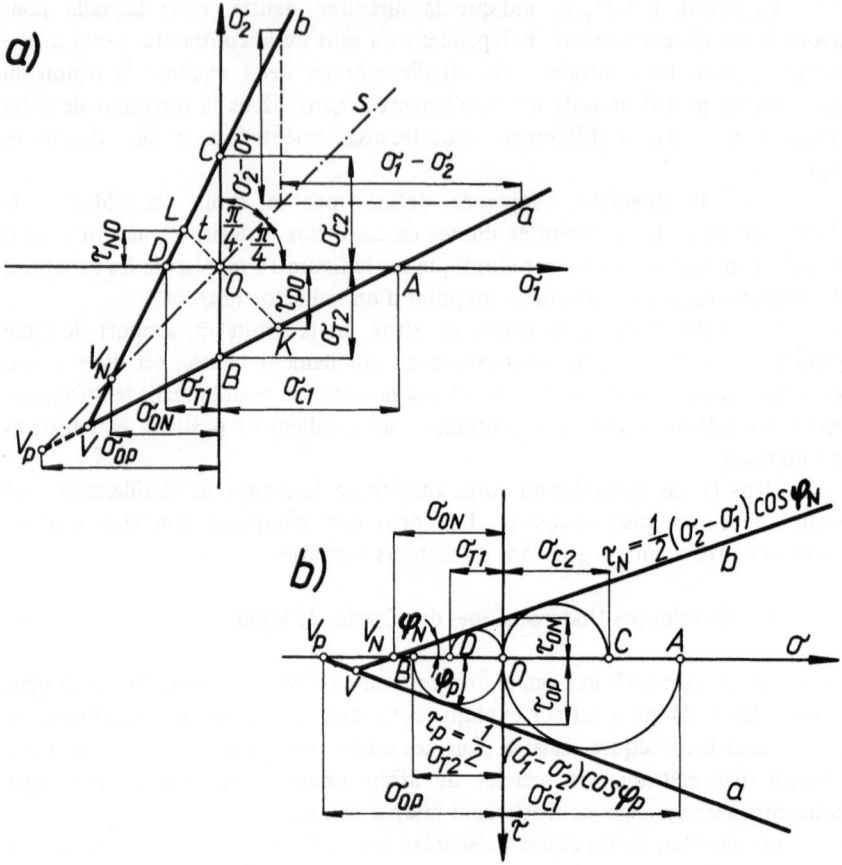

Figure 2. — a) Droites limites dissymétriques dans le plan des contraintes principales;
b) enveloppe limite dissymétrique des cercles de Mohr.

nous obtenons :

$$\tau_{PO} = \frac{\sigma_{C1} \, \sigma_{T2}}{\sigma_{C1} + \sigma_{T2}} \tag{3}$$

En substituant d'une manière analogue la contrainte limite en cisaillement pur négatif

$$\tau_{NO} = -\sigma_1 = \sigma_2 \tag{4}$$

dans l'équation de la droite limite b :

$$-\frac{\sigma_1}{\sigma_{T1}} + \frac{\sigma_2}{\sigma_{C2}} = 1,$$ (5)

nous obtenons

$$\tau_{NO} = \frac{\sigma_{T1}\,\sigma_{C2}}{\sigma_{C2} + \sigma_{T1}}.$$ (6)

La contrainte limite en cisaillement positif et négatif est définie par les coordonnées des points K et L, qui sont situés sur l'axe t, deuxième bissectrice des axes σ_1 et σ_2.

En posant $\sigma_1 = \sigma_2$, nous obtenons à partir des équations (1) et (5) les coordonnées des points V_P et V_N, qui sont les intersections des droites limites a et b avec l'axe s et qui définissent les contraintes limites en traction isotrope :

$$\sigma_{OP} = -\frac{\sigma_{C1}\,\sigma_{T2}}{\sigma_{C1} - \sigma_{T2}},$$ (7)

$$\sigma_{ON} = -\frac{\sigma_{T1}\,\sigma_{C2}}{\sigma_{C2} - \sigma_{T1}}.$$ (8)

Les deux points V_N et V_P ne peuvent pas représenter simultanément tous les deux l'état limite. En effet, les états de contrainte correspondants étant des états isotropes, les contraintes de cisaillement sont nulles sur toutes les facettes et on ne peut plus considérer les différences entre cisaillement positif et négatif. Si le milieu entre en plasticité sous la contrainte isotrope σ_{ON}, il ne peut pas, dans le même état, atteindre la contrainte isotrope σ_{OP}. L'axe σ représente alors un lieu singulier correspondant à des états limites différents.

Les coordonnées du sommet V, qui est le point d'intersection des droites limites a et b, sont déduites des équations (1) et (5) :

$$\sigma_{O1} = -\frac{\sigma_{C1}\,\sigma_{T1}\,(\sigma_{C2} + \sigma_{T2})}{\sigma_{C1}\,\sigma_{C2} - \sigma_{T1}\,\sigma_{T2}},$$ (9)

$$\sigma_{O2} = -\frac{\sigma_{C2}\,\sigma_{T2}\,(\sigma_{C1} + \sigma_{T1})}{\sigma_{C1}\,\sigma_{C2} - \sigma_{T1}\,\sigma_{T2}}.$$ (10)

Les branches droites a et b de l'enveloppe limite de Mohr représentées sur la figure 2b sont tangentes aux demi-cercles de Mohr ayant comme diamètres σ_{C1}, σ_{T2}, σ_{C2} et σ_{T1}, qui sont les contraintes limites en compression simple et en traction simple dans les directions des contraintes principales σ_1 et σ_2.

La droite a de l'enveloppe limite de Mohr forme, avec l'axe des contraintes normales σ dans le diagramme 2b, l'angle de frottement interne positif φ_P, tandis que la droite b forme avec l'axe σ l'angle de frottement interne négatif φ_N. Ces angles sont définis par les expressions suivantes

$$\sin \varphi_P = \frac{\sigma_{C1}}{\sigma_{C1} - 2\sigma_{OP}} = \frac{\sigma_{T2}}{\sigma_{T2} - 2\sigma_{OP}}, \tag{11}$$

$$\sin \varphi_N = \frac{\sigma_{C2}}{\sigma_{C2} - 2\sigma_{ON}} = \frac{\sigma_{T1}}{\sigma_{T1} - 2\sigma_{ON}}. \tag{12}$$

En reportant dans les relations précédentes les expressions (7) et (8), nous obtenons :

$$\sin \varphi_P = \frac{\sigma_{C1} - \sigma_{T2}}{\sigma_{C1} + \sigma_{T2}}, \tag{13}$$

$$\sin \varphi_N = \frac{\sigma_{C2} - \sigma_{T1}}{\sigma_{C2} + \sigma_{T1}}. \tag{14}$$

En comparant les expressions (13) et (14), nous obtenons la condition d'égalité des deux angles φ_P et φ_N :

$$\sigma_{C1}\, \sigma_{T1} = \sigma_{C2}\, \sigma_{T2} \tag{15}$$

ou

$$\frac{\sigma_{C1}}{\sigma_{C2}} = \frac{\sigma_{T2}}{\sigma_{T1}}. \tag{16}$$

Cette condition est satisfaite en particulier par les matériaux isotropes, qui sont caractérisés par les égalités $\sigma_{C1} = \sigma_{C2}$ et $\sigma_{T1} = \sigma_{T2}$.

Les équations des droites limites de l'enveloppe des cercles de Mohr, qui définissent les contraintes de cisaillement sur les surfaces de glissement, ont la forme suivante :

$$\tau_P = \sigma \sin \varphi_P \cos \varphi_P + \tau_{OP}, \tag{17}$$

$$\tau_N = \sigma \sin \varphi_N \cos \varphi_N + \tau_{ON}, \tag{18}$$

où σ est la contrainte moyenne.

Les relations entre les contraintes de cisaillement limites sur les surfaces de glissement et les contraintes de cisaillement maximales sont représentées par les formules

$$\tau_P = \frac{1}{2}(\sigma_1 - \sigma_2)\cos\varphi_P, \tag{17a}$$

$$\tau_N = \frac{1}{2}(\sigma_2 - \sigma_1)\cos\varphi_N. \tag{18a}$$

Les droites limites a et b délimitent sur l'axe des contraintes de cisaillement les segments qui représentent la cohésion positive et négative :

$$\tau_{OP} = |\sigma_{OP}|\,\mathrm{tg}\,\varphi_P = \frac{1}{2}\sqrt{\sigma_{C1}\,\sigma_{T2}}, \tag{19}$$

$$\tau_{ON} = |\sigma_{ON}|\,\mathrm{tg}\,\varphi_N = \frac{1}{2}\sqrt{\sigma_{T1}\,\sigma_{C2}}. \tag{20}$$

Ces deux contraintes sont différentes des contraintes limites τ_{PO} et τ_{NO}, en cisaillement pur positif et négatif, définies par les coordonnées des points K et L sur le diagramme de la figure 2a.

Les relations entre les contraintes de cisaillement τ_{PO} et τ_{OP}, ainsi que celles entre τ_{NO} et τ_{ON} ont la forme suivante :

$$\tau_{PO} = \tau_{OP}\cos\varphi_P,$$

$$\tau_{NO} = \tau_{ON}\cos\varphi_N.$$

Les enveloppes limites des cercles de Mohr sont les sections droites de la surface gauche représentée sur la figure 3. L'axe τ_n est situé dans le plan dans lequel les enveloppes de cercles de Mohr sont symétriques et qui correspond au cisaillement neutre. L'axe τ_m définit avec l'axe σ le plan dans lequel les enveloppes dissymétriques correspondent aux différences maximales entre le cisaillement positif et négatif.

La figure 4 représente les courbes limites dissymétriques pour les matériaux ayant, dans les directions principales, les mêmes contraintes limites en compression σ_C et en traction σ_T, la même cohésion normale σ_O, mais des contraintes limites différentes en cisaillement positif et négatif. L'auteur a

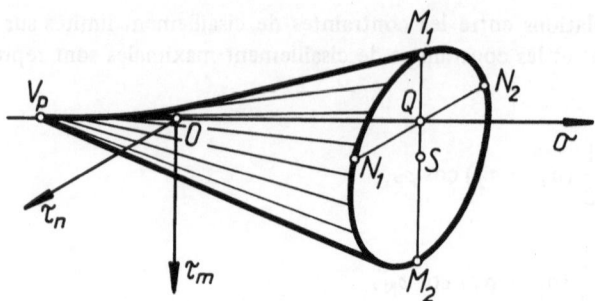

Figure 3. – *Surface gauche dont les sections droites représentent les droites limites des enveloppes des cercles de Mohr.*

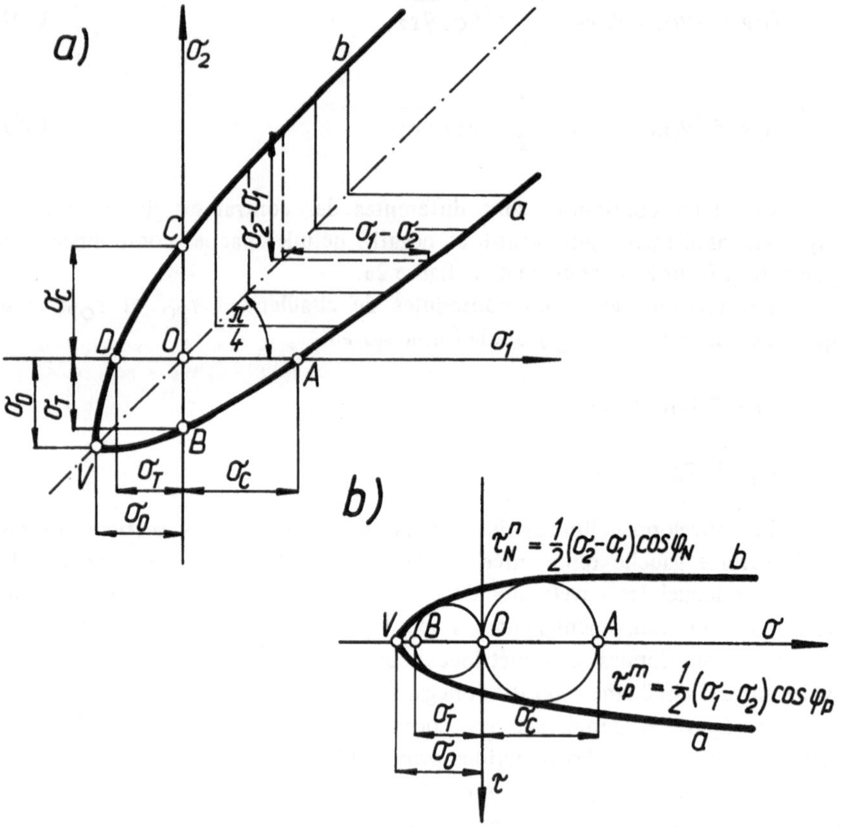

Figure 4. – *Courbes limites pour les matériaux quasi-isotropes.*

appelé de tels matériaux quasi-isotropes. Les branches dissymétriques de l'enveloppe de Mohr peuvent être définies par les lois aux puissances :

$$\tau_P^m = \frac{1}{2}(\sigma_1 - \sigma_2)\cos\varphi_P, \tag{21}$$

$$\tau_N^n = \frac{1}{2}(\sigma_2 - \sigma_1)\cos\varphi_N. \tag{22}$$

4. Enveloppes de Mohr fermées

Pour de nombreux matériaux anisotropes, les critères de plasticité et de résistance dans l'état plan sont représentées sur le diagramme des contraintes principales par des courbes fermées.

La figure 5a représente, pour les matériaux anisotroqes, le critère des contraintes normales et des contraintes de cisaillement maximales, qui correspond au critère de Tresca pour les corps isotropes. L'anisotropie provoque une irrégularité de l'hexagone limite.

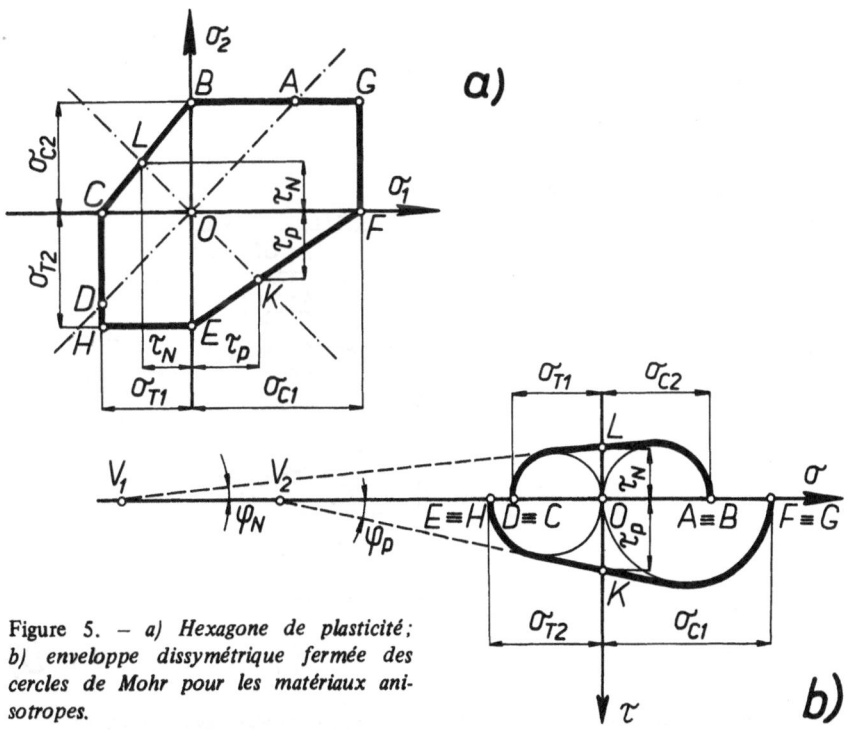

Figure 5. — *a) Hexagone de plasticité; b) enveloppe dissymétrique fermée des cercles de Mohr pour les matériaux anisotropes.*

Ce critère est représenté d'une autre façon sur la figure 5b par l'enveloppe fermée dissymétrique des cercles de Mohr. Cette courbe est l'enveloppe de Mohr pour des états de contraintes planes. La branche du cisaillement positif EKF est composée par un segment de la droite inclinée passant par le point K sur l'axe τ, par des arcs de deux demi-cercles, dont les diamètres OF et OH représentent les contraintes limites σ_{C1}, σ_{T2}, et par les deux points A et D. La branche du cisaillement négatif DLA est composée par un segment de la droite inclinée passant par le point L sur l'axe τ et par des arcs de demi-cercles, dont les diamètres OB et OC représentent les contraintes limites σ_{C2} et σ_{T1}.

La figure 6 représente un critère d'écoulement du deuxième degré pour les matériaux anisotropes. Dans le diagramme 6a des contraintes principales, nous obtenons une ellipse occupant une position quelconque. Cette ellipse correspond à l'ellipse de Mises pour les corps isotropes.

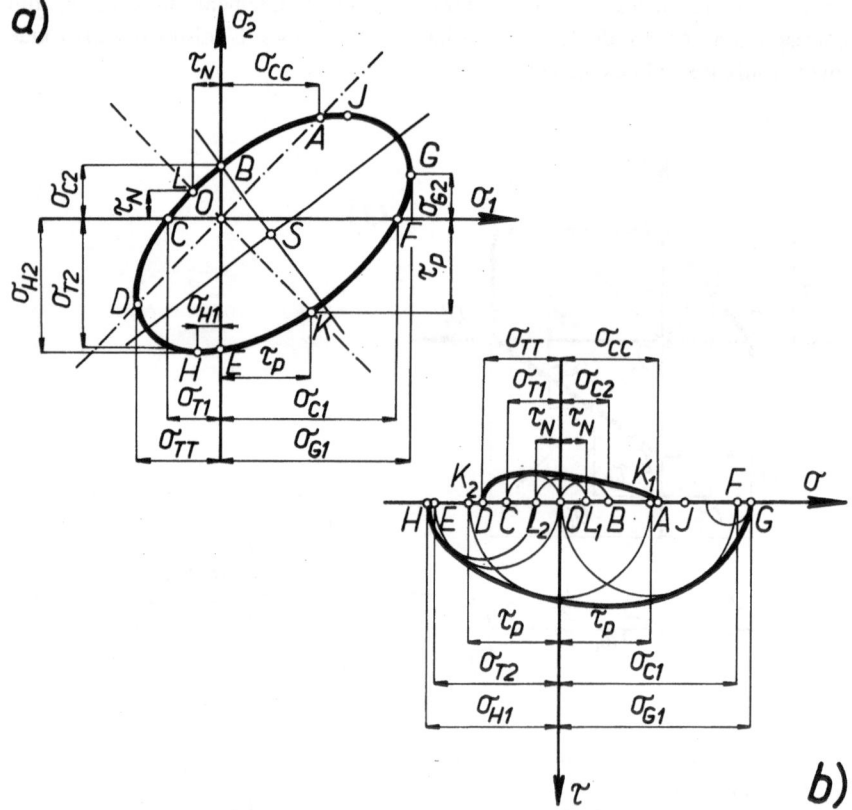

Figure 6. – *Courbes limites du deuxième degré pour les matériaux anisotropes.*

L'enveloppe correspondante des cercles de Mohr pour des états de contrainte planes est représentée sur la figure 6b. Elle est composée par deux branches curvilignes dissymétriques. La branche positive enveloppe les quatre demi-cercles de Mohr ayant comme diamètres σ_{C1}, σ_{T2}, $\sigma_{G1} - \sigma_{G2}$ et $\sigma_{H2} - \sigma_{H1}$. La branche négative est constituée par la courbe qui enveloppe les demi-cercles ayant comme diamètres σ_{C2} et σ_{T1}. On remarque que les demi-cercles correspondant aux arcs GJ et DH, délimités sur l'ellipse d'écoulement par les points à tangentes verticales et horizontales, sont inscrits les uns dans les autres et n'admettent pas d'enveloppe réelle.

Cette propriété n'est pas spécifique pour un milieu anisotrope, puisqu'elle apparaît également dans le critère de Von Misès pour les milieux isotropes.

5. Lois de Comportement

L'auteur a établi les lois de comportement pour les déformations inélastiques des matériaux anisotropes en introduisant le tenseur matériel transformé des contraintes, donné par

$$\omega_{ij} = (a_{ijk\ell} - \chi_{ij} a_{k\ell mn} \delta_{mn}) \sigma_{k\ell}, \tag{23}$$

où $\sigma_{k\ell}$ est le tenseur actuel des contraintes,
$a_{k\ell mn}$ est le tenseur d'anisotropie du quatrième ordre,
δ_{mn} est le symbole de Kronecker et où

$$\chi_{ij} = \left\| \begin{matrix} \chi_{11} & \chi_{12} & \chi_{13} \\ \chi_{21} & \chi_{22} & \chi_{23} \\ \chi_{31} & \chi_{32} & \chi_{33} \end{matrix} \right\| \tag{24}$$

est le tenseur matériel qui exprime le degré de compressibilité des matériaux au cours de la déformation inélastique.

Le tenseur matériel transformé des contraintes a les mêmes axes principaux que le tenseur des déformations. C'est pourquoi, le tenseur des déformations peut être exprimé par une fonction tensorielle isotrope du tenseur matériel transformé des contraintes :

$$\epsilon_{ij} = f_{ij}(\omega_{k\ell}). \tag{25}$$

En développant cette fonction selon les lois de l'algèbre tensorielle, nous obtenons la formule suivante :

$$\epsilon_{ij} = \Phi_0 \delta_{ij} + \Phi_1 \omega_{ij} + \Phi_2 \omega_{i\lambda} \omega_{\lambda j}, \tag{26}$$

où Φ_K sont des fonctions scalaires des invariants.

Pour déterminer ces fonctions, l'auteur a décomposé l'équation (26) en deux relations :

$$^1\epsilon_{ij} = {}^1\Phi_0 \delta_{ij} + \Phi_1 \omega_{ij} , \tag{27}$$

$$^2\epsilon_{ij} = {}^2\Phi_0 \delta_{ij} + \Phi_2 \omega_{i\lambda} \omega_{\lambda j} . \tag{28}$$

En multipliant l'équation (27) successivement par δ_{ij} et $^1\epsilon_{ij} = {}^1\Phi_0 \delta_{ij} + \Phi_1 \omega_{ij}$, nous obtenons

$$^1\epsilon_{ij} \delta_{ij} = 3 \, {}^1\Phi_0 + \Phi_1 \omega_{ij} \delta_{ij} , \tag{29}$$

$$^1\epsilon_{ij} \, {}^1\epsilon_{ij} = 3 \, {}^1\Phi_0^2 + 2 \, {}^1\Phi_0 \Phi_1 \omega_{ij} \delta_{ij} + \Phi_1^2 \omega_{ij} \omega_{ij} , \tag{30}$$

d'où

$$^1\Phi_0 = \frac{1}{3} \left(I_{1\epsilon} - I_\omega \sqrt{\frac{3 \, II_{1\epsilon} - I_{1\epsilon}^2}{3 \, II_\omega - I_\omega^2}} \right) = \epsilon_M - \frac{3 \, {}^1\epsilon_I}{2 \, \omega_I} \, \omega_M , \tag{31}$$

$$\Phi_1 = \sqrt{\frac{3 \, II_{1\epsilon} - I_{1\epsilon}^2}{3 \, II_\omega - I_\omega^2}} = \frac{3 \, {}^1\epsilon_I}{2 \, \omega_I} , \tag{32}$$

où

$$I_{1\epsilon} = {}^1\epsilon_{ij} \delta_{ij} , \quad II_{1\epsilon} = {}^1\epsilon_{ij} \, {}^1\epsilon_{ij} , \quad I_\omega = \omega_{ij} \delta_{ij} , \quad II_\omega = \omega_{ij} \omega_{ij} , \tag{33}$$

sont les invariants des déformations et des contraintes transformées,

$$^1\epsilon_M = \frac{1}{3} ({}^1\epsilon_{11} + {}^1\epsilon_{22} + {}^1\epsilon_{33}) \tag{34}$$

est la déformation partielle moyenne,

$$^1\epsilon_I = \frac{2}{3} \sqrt{{}^1\epsilon_{11}^2 + {}^1\epsilon_{22}^2 + {}^1\epsilon_{33}^2 - {}^1\epsilon_{11} \, {}^1\epsilon_{22} - {}^1\epsilon_{22} \, {}^1\epsilon_{33} - {}^1\epsilon_{33} \, {}^1\epsilon_{11} + 3({}^1\epsilon_{12}^2 + {}^1\epsilon_{23}^2 + {}^1\epsilon_{31}^2)} \tag{35}$$

est l'intensité du tenseur des premières déformations partielles,

$$\omega_M = \frac{1}{3} (\omega_{11} + \omega_{22} + \omega_{33}) \tag{36}$$

est la contrainte matérielle transformée moyenne et

$$\omega_I = \sqrt{\omega_{11}^2 + \omega_{22}^2 + \omega_{33}^2 - \omega_{11} \omega_{22} - \omega_{22} \omega_{33} - \omega_{33} \omega_{11} + 3(\omega_{12}^2 + \omega_{23}^2 + \omega_{31}^2)} \tag{37}$$

est l'intensité du tenseur matériel transformé des contraintes.

En multipliant l'équation (28) successivement par δ_{ij} et $^2\epsilon_{ij} = {}^2\Phi_0 \delta_{ij} + \Phi_2 \omega_{i\lambda} \omega_{\lambda j}$, nous obtenons le système des deux équations :

$$^2\epsilon_{ij} \delta_{ij} = 3\,^2\Phi_0 + \Phi_2 \,\omega_{ij}\,\omega_{ij}\,, \tag{38}$$

$$^2\epsilon_{ij}\,^2\epsilon_{ij} = 3\,^2\Phi_0^2 + 2\,^2\Phi_0 \Phi_2 \,\omega_{ij}\,\omega_{ij} + \Phi_2^2\,\omega_{ij}\,\omega_{j\lambda}\,\omega_{\lambda\mu}\,\omega_{\mu i}\,, \tag{39}$$

pour déterminer les fonctions scalaires

$$^2\Phi_0 = \frac{1}{3}\left(I_{2\epsilon} - II_\omega \sqrt{\frac{3\,II_{2\epsilon} - I_{2\epsilon}^2}{3\,IV_\omega - II_\omega^2}}\right) = {}^2\epsilon_M - \frac{9\,^2\epsilon_I}{4\,\omega_{II}^2}\,\omega_s^2\,, \tag{40}$$

$$\Phi_2 = \sqrt{\frac{3\,II_{2\epsilon} - I_{2\epsilon}^2}{3\,IV_\omega - II_\omega^2}} = \frac{9\,^2\epsilon_I}{4\,\omega_{II}^2}\,, \tag{41}$$

où $^2\epsilon_M$ est la deuxième déformation partielle moyenne, $^2\epsilon_I$ est l'intensité du tenseur des deuxièmes déformations partielles,

$$\omega_S = \sqrt{\frac{II_\omega}{3}} = \frac{1}{\sqrt{3}}\sqrt{\omega_{11}^2 + \omega_{22}^2 + \omega_{33}^2 + 2(\omega_{12}^2 + \omega_{23}^2 + \omega_{31}^2)} \tag{42}$$

est la contrainte matérielle transformée moyenne du second ordre et

$$\omega_{II} = \sqrt{\frac{2}{3}}\,\sqrt[4]{\left\{\frac{1}{2}\,[(\omega_{11}^2 - \omega_{22}^2)^2 + (\omega_{22}^2 - \omega_{33}^2)^2 + (\omega_{33}^2 - \omega_{11}^2)^2] + \right.}$$

$$+ (\omega_{12}^2 + \omega_{23}^2 + \omega_{31}^2)^2 + 2\omega_{12}^2\,[2(\omega_{11} + \omega_{22})^2 - \omega_{11}\,\omega_{22} - \omega_{33}^2] +$$

$$+ 2\omega_{23}^2\,[2(\omega_{22} + \omega_{33})^2 - \omega_{22}\,\omega_{33} - \omega_{11}^2] + 2\omega_{31}^2\,[2(\omega_{33} + \omega_{11})^2 -$$

$$\left. - \omega_{33}\,\omega_{11} - \omega_{22}^2] + 12\,\omega_{12}\,\omega_{23}\,\omega_{31}\,(\omega_{11} + \omega_{22} + \omega_{33})\right\} \tag{43}$$

est l'intensité du tenseur matériel transformé du second ordre et

$$IV_\omega = \omega_{ij}\,\omega_{j\lambda}\,\omega_{\lambda\mu}\,\omega_{\mu i}\,. \tag{44}$$

En introduisant les expressions (31), (32), (40) et (41) dans l'équation (26), nous obtenons :

$$\epsilon_{ij} - \epsilon_M\,\delta_{ij} = \frac{3\,^1\epsilon_I}{2\,\omega_I}\,(\omega_{ij} - \omega_M\,\delta_{ij}) + \frac{9\,^2\epsilon_I}{4\,\omega_{II}^2}\,(\omega_{i\lambda}\,\omega_{\lambda j} - \omega_S^2\,\delta_{ij}) \tag{45}$$

Cette relation peut être écrite sous la forme

$$\epsilon_{ij} - \epsilon_M \, \delta_{ij} = \frac{1}{2} \, \widetilde{J} \, (\omega_I) \, (\omega_{ij} - \omega_M \, \delta_{ij}) + \frac{1}{4} \, \widetilde{M} \, (\omega_{II}) \, (\omega_{i\lambda} \, \omega_{\lambda j} - \omega_S^2 \, \delta_{ij}),$$

(46)

où $\widetilde{J} \, (\omega_I) = \dfrac{3 \, {}^1\epsilon_I}{\omega_I}$, $\widetilde{M}(\omega_{II}) = \dfrac{9 \, {}^2\epsilon_I}{\omega_{II}^2}$ sont les fonctions de déformation.

6. Critère d'écoulement

L'auteur a établi un critère d'écoulement pour les matériaux anisotropes à partir de la relation entre le tenseur des vitesses de déformation et le tenseur matériel transformé des contraintes

$$\dot{\epsilon}_{ij} = f_{ij} \, (\omega_{k\ell})$$

(47)

qui peut être développée sous la forme

$$\dot{\epsilon}_{ij} = \phi_0 \, \delta_{ij} + \phi_1 \, \omega_{ij} + \phi_2 \, \omega_{i\lambda} \, \omega_{\lambda j}.$$

(48)

En limitant nos considérations aux effets du premier ordre, nous ne retenons que deux termes dans la formule (48). En multiplicant cette formule réduite par ω_{ij}, nous obtenons la relation invariante

$$\omega_{ij} \, \epsilon_{ij} = \phi_0 \, \omega_{ij} \delta_{ij} + \phi_1 \, \omega_{ij} \, \omega_{ij}.$$

(49)

L'auteur s'est placé dans l'hypothèse des quotients constants à la limite d'écoulement, qui est exprimée par

$$H_0 = \frac{\phi_0}{\omega_{ij} \, \dot{\epsilon}_{ij}} , \qquad H_1 = \frac{\phi_1}{\omega_{ij} \, \dot{\epsilon}_{ij}}$$

(50)

En divisant l'équation (49) par $\omega_{ij} \, \dot{\epsilon}_{ij}$ et en y introduisant les expressions (50), nous obtenons la forme générale du critère de plasticité

$$H_0 \, \omega_{ij} \delta_{ij} + H_1 \, \omega_{ij} \, \omega_{ij} = 1.$$

(51)

En utilisant la relation (23), nous pouvons écrire l'équation (51) sous la forme suivante

$$P_{ijk\ell} \, \sigma_{ij} \, \sigma_{k\ell} + Q_{ij} \, \sigma_{ij} = 1$$

(52)

En considérant les symétries matérielles, nous avons dans la relation (51) 27 produits des constantes H_0 et H_1, des coefficients d'anisotropie $a_{ijk\ell}$ et des composantes χ_{ij} du tenseur matériel, c'est-à-dire 27 constantes qui peuvent être déterminées par 27 limites de plasticité particulières. La relation (52) contient 21 coefficients $P_{ijk\ell}$ et six coefficients Q_{ij}, donc également 27 constantes.

Le critère d'écoulement, exprimé en fonction de limites de plasticité particulières, admet la forme suivante :

$$
\frac{\sigma_{ij}\,\sigma_{ij}}{\sigma_{Pij}\,\sigma_{Nij}} - \left(\frac{1}{\sigma_{Pij}\,\sigma_{Nij}} + \frac{1}{\sigma_{Pk\ell}\,\sigma_{Nk\ell}} - \frac{1}{\sigma_{PPijk\ell}^2} - \frac{\sigma_{Pij} - \sigma_{Nij}}{\sigma_{Pij}\,\sigma_{Nij}\,\sigma_{PPijk\ell}} - \right.
$$

$$
\left. - \frac{\sigma_{Pk\ell} - \sigma_{Nk\ell}}{\sigma_{Pk\ell}\,\sigma_{Nk\ell}\,\sigma_{PPijk\ell}} \right) \sigma_{ij}\,\sigma_{k\ell} - \frac{\sigma_{Pij} - \sigma_{Nij}}{\sigma_{Pij}\,\sigma_{Nij}}\,\sigma_{ij} = 1, \qquad (53)
$$

où σ_{Pij} sont les limites particulières d'écoulement en contraintes monoaxiales positives, σ_{Nij} celles en contraintes monoaxiales négatives et $\sigma_{PPijk\ell}$ celles en contraintes biaxiales isotropes positives.

En limitant les considérations à l'état plan, nous pouvons exprimer ce critère en fonction des composantes du tenseur des contraintes dans les coordonnées cartésiennes, sous la forme suivante :

$$
\frac{\sigma_x^2}{\sigma_{Cx}\,\sigma_{Tx}} + \frac{\sigma_y^2}{\sigma_{Cy}\,\sigma_{Ty}} + \frac{\sigma_z^2}{\sigma_{Cz}\,\sigma_{Tz}} -
$$

$$
- \left(\frac{1}{\sigma_{Cx}\,\sigma_{Tx}} + \frac{1}{\sigma_{Cy}\,\sigma_{Ty}} - \frac{1}{\sigma_{CCxy}^2} - \frac{\sigma_{Cx} - \sigma_{Tx}}{\sigma_{Cx}\,\sigma_{Tx}\,\sigma_{CCxy}} - \frac{\sigma_{Cy} - \sigma_{Ty}}{\sigma_{Cy}\,\sigma_{Ty}\,\sigma_{CCxy}} \right) \sigma_x\,\sigma_y -
$$

$$
- \left(\frac{1}{\sigma_{Cx}\,\sigma_{Tx}} + \frac{1}{\tau_{Pxy}\,\tau_{Nxy}} - \frac{1}{\sigma_{CPxxy}^2} - \frac{\sigma_{Cx} - \sigma_{Tx}}{\sigma_{Cx}\,\sigma_{Tx}\,\sigma_{CPxxy}} - \frac{\tau_{Pxy} - \tau_{Nxy}}{\tau_{Pxy}\,\tau_{Nxy}\,\sigma_{CPxxy}} \right) \sigma_x\,\tau_{xy} -
$$

$$
- \left(\frac{1}{\sigma_{Cy}\,\sigma_{Ty}} + \frac{1}{\tau_{Pxy}\,\tau_{Nxy}} - \frac{1}{\sigma_{CPyxy}^2} - \frac{\sigma_{Cy} - \sigma_{Ty}}{\sigma_{Cy}\,\sigma_{Ty}\,\sigma_{CPyxy}} - \frac{\tau_{Pxy} - \tau_{Nxy}}{\tau_{Pxy}\,\tau_{Nxy}\,\sigma_{CPyxy}} \right) \sigma_y\,\tau_{xy} -
$$

$$
- \frac{\sigma_{Cx} - \sigma_{Tx}}{\sigma_{Cx}\,\sigma_{Tx}}\,\sigma_x - \frac{\sigma_{Cy} - \sigma_{Ty}}{\sigma_{Cy}\,\sigma_{Ty}}\,\sigma_y - \frac{\tau_{Pxy} - \tau_{Nxy}}{\tau_{Pxy}\,\tau_{Nxy}}\,\tau_{xy} = 1, \qquad (54)
$$

où σ_{Cx} et σ_{Cy} sont les limites d'écoulement en compression simples, σ_{Tx} et σ_{Ty} celles en traction simple, σ_{CCxy} la limite d'écoulement en compression

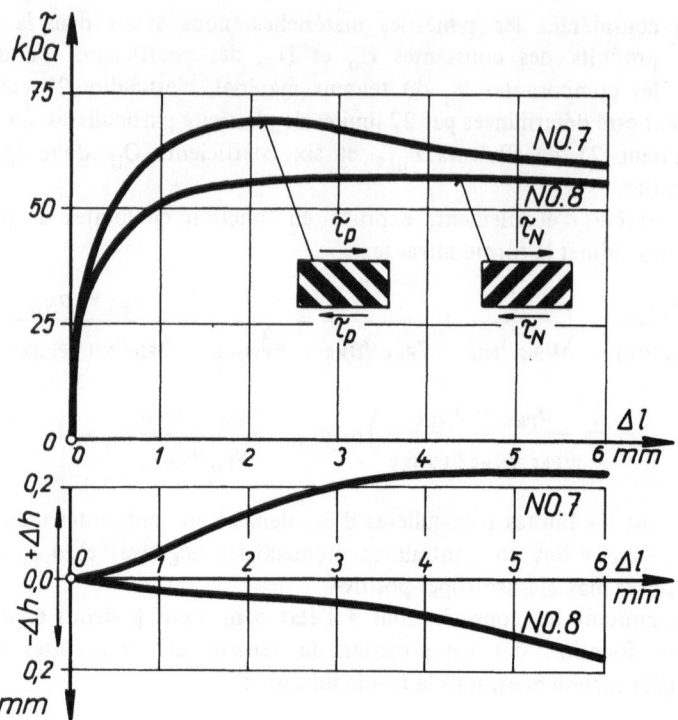

Figure 7. – *Diagramme des déformations des éprouvettes à couches inclinées d'un angle de 45°, testées en cisaillement positif et négatif.*

biaxiale isotrope, τ_{Pxy} celle en cisaillement positif, τ_{Nxy} celle en cisaillement négatif, σ_{CPxxy} et σ_{CPyxy} celles pour un état de contraintes composé par une compression et un cisaillement positif ayant des valeurs égales.

Si les directions principales des contraintes coïncident avec celles de l'orthotropie, nous obtenons pour les corps orthotropes le critère suivant :

$$\frac{\sigma_1^2}{\sigma_{\mathrm{C1}}\,\sigma_{\mathrm{T1}}} + \frac{\sigma_2^2}{\sigma_{\mathrm{C2}}\,\sigma_{\mathrm{T2}}} - \left(\frac{1}{\sigma_{\mathrm{C1}}\,\sigma_{\mathrm{T1}}} + \frac{1}{\sigma_{\mathrm{C2}}\,\sigma_{\mathrm{T2}}} - \frac{1}{\sigma_{\mathrm{CC12}}^2} - \frac{\sigma_{\mathrm{C1}} - \sigma_{\mathrm{T1}}}{\sigma_{\mathrm{C1}}\,\sigma_{\mathrm{T1}}\,\sigma_{\mathrm{CC12}}}\right.$$

$$\left. - \frac{\sigma_{\mathrm{C2}} - \sigma_{\mathrm{T2}}}{\sigma_{\mathrm{C2}}\,\sigma_{\mathrm{T2}}\,\sigma_{\mathrm{CC12}}}\right) \sigma_1\,\sigma_2 - \frac{\sigma_{\mathrm{C1}} - \sigma_{\mathrm{T1}}}{\sigma_{\mathrm{C1}}\,\sigma_{\mathrm{T1}}}\,\sigma_1 - \frac{\sigma_{\mathrm{C2}} - \sigma_{\mathrm{T2}}}{\sigma_{\mathrm{C2}}\,\sigma_{\mathrm{T2}}}\,\sigma_2 = 1,$$

$$(55)$$

qui est représenté par l'ellipse de la figure 6a.

7. Vérification Expérimentale

L'existence des cisaillements positif et négatif a été vérifiée par des essais sur des sols stratifiés, des matières plastiques, du bois, du zinc et un alliage d'aluminium.

La figure 7 représente les courbes contraintes-déplacements et les courbes exprimant en fonction du déplacement $\Delta \ell$, les variations de la hauteur Δh des éprouvettes constituées par des couches de sable et d'argile inclinées sous une contrainte normale de 212 kPa et testées ensuite en cisaillement positif et négàtif sous une contrainte normale de 112 kPa. La courbe N° 7 du cisaillement positif présente un pic, alors que la courbe N° 8, correspondant au cisaillement négatif, ne présente pas ce phénomène. La figure 8 représente les résultats des essais sur des éprouvettes à couches inclinées de 30° et qui ont été testées en cisaillement positif et négatif sous une contrainte normale de 212 kPa.

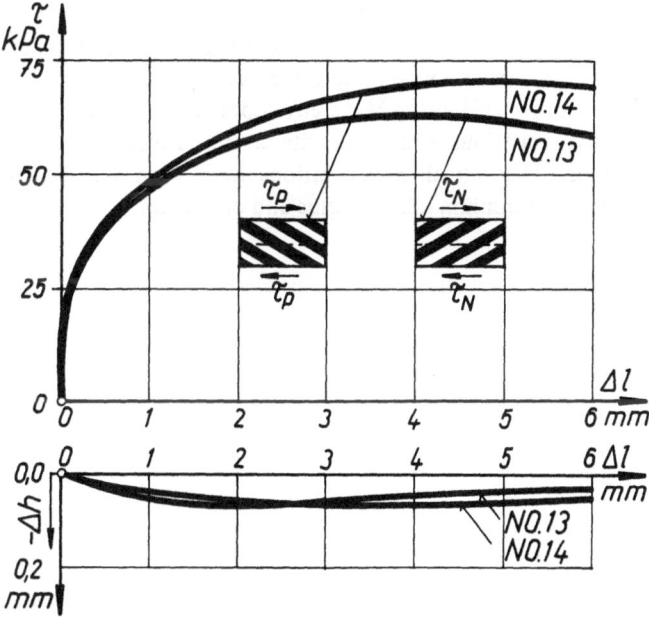

Figure 8. – *Diagramme des déformations des éprouvettes à couches inclinées d'un angle de 30°.*

REFERENCES

[1] BOEHLER J.P. – *"Contributions théoriques et expérimentales à l'étude des milieux plastiques anisotropes"*. Thèse présentée à l'Université Scientifique et Médicale de Grenoble. Grenoble, 1975.

[2] MANDEL J. – *"Introduction à la Mécanique des Milieux Continus Déformables"*. Varsovie : PWN-Editions Scientifiques de Pologne, 1974.

[3] SOBOTKA Z. – "Positive and negative shear in anisotropic soils". *Istanbul Conference on Soil Mechanics and Foundation Engineering"*, 2. Istanbul: Istanbul Teknik Universitesi, Istanbul Fakültesi Matbasi, (1975): 67-74.

[4] SOBOTKA Z. – "Positive and negative shear causing volume changes in anisotropic soils". *Proceedings of the 5th Budapest Conferences on Soil Mechanics and Foundation Engineering*. Budapest: Academiai Kiadó, (1976): 167-181.

ABSTRACT

The results of theoretical and experimental investigations have led the author to the concept of positive and negative shears which are, in some manner, analogous to the compression and tension at the normal stressing. The existence of two opposite kinds of shear is caused by anisotropy, structural configuration and non-homogeneity of materials. Differences between the positive and negative shear involve the asymmetry of the deformation as well as asymmetrical forms of limiting Mohr envelopes and of the slip-lines network.

Constitutive equations and yield criteria are derived on the basis of tensorial expansions. Theoretical considerations have been verified by tests of layered soils, of an aluminium alloy, of zinc and of plastics.

Session 9

Vibrations, Waves' Propagation and Induced Anisotropy

Vibrations, Propagation des Ondes et Anisotropie Induite

On the Problem of Oscillations
of the Anisotropic Electroconductive Plates
in the Magnetic Fields

S.A. Ambartsumyan

Yerevan State University, Yerevan, USSR.

1. Introduction

Proceeding from the general principles of the theory of anisotropic plates which considers transverse shear [1] and from the fundamental hypotheses of magneto-elasticity of thin bodies [2], a general vibration theory of orthotropic plates in a transverse magnetic field is developed.

Let us take a thin elastic orthotropic plate of constant thickness $h_0 = 2h$ and finite electroconductivity $\sigma(\sigma_1, \sigma_2, \sigma_3)$, oscillating in an external magnetic field with constant intensity $H_0(O, O, H_{03})$, normal to the middle plane of the plate.

The plate is considered in a Cartesian system of coordinates x_i, so that the middle plane coincides with the coordinate plane $x_1 O x_2$. The coordinate lines x_1 and x_2 are assumed to coincide with the principal directions of elasticity.

The problem of magnetostatics in an unperturbed case is assumed to be solved.

The dielectric constants for the outside and the inside of the plate are assumed to be $\epsilon^{(e)}$ and ϵ, respectively, and the magnetic permeability coefficients μ and $\mu^{(e)}$ for both the inside region (vacuum) and the plate are taken equal to one.

Further on, we shall use the linearized magnetoelastic equations [2].

2. Initial hypotheses and general equations

The initial hypotheses on which the present theory is based are :

a) the hypothesis of the magnetoelasticity of thin bodies [2] in accordance with which it is assumed that approximately

$$e_1 = \varphi(x_1, x_2, t) \quad , \quad e_2 = \psi(x_1, x_2, t) \quad , \quad h_3 = f(x_1, x_2, t), \quad (2.1)$$

where $h(h_1, h_2, h_3)$, $e(e_1, e_2, e_3)$ are the inducted electromagnetic field components, and φ, ψ, f are sought functions;

b) the hypothesis of the exacted bending theory of plates [1], according to which

$$u_1 = u(x_1, x_2, t) - x_3 \frac{\partial w}{\partial x_1} + a_{55} J_0 \Phi(x_1, x_2, t),$$

$$u_2 = v(x_1, x_2, t) - x_3 \frac{\partial w}{\partial x_2} + a_{44} J_0 \Psi(x_1, x_2, t), \qquad (2.2)$$

$$u_3 = w(x_1, x_2, t),$$

where u, v, w are the sought displacements of the plate middle surface, Φ, Ψ are the sought functions characterizing sharp deformations of the plate. J_0 is the given function of x_3 characterizing the shear value changes through the plate thickness, Particularly, for J_0 it is assumed [1] that

$$J_0 = \frac{x_3}{2}\left(h^2 - \frac{x_3^2}{2}\right). \qquad (2.3)$$

$a_{55} = \dfrac{1}{G_{13}}$, $a_{44} = \dfrac{1}{G_{23}}$ are the elasticity coefficients, G_{13} and G_{23} are the shear moduli.

From (2.1), (2.2) for the linearized electrodynamic equations we get [2]

$$\frac{\partial h_1}{\partial x_3} = \frac{\partial f}{\partial x_1} + \frac{4\pi\sigma_2}{c}\left[\psi - \frac{1}{c}\left(B_{03}\frac{\partial u}{\partial t} - \right.\right.$$
$$\left.\left. - x_3 B_{03}\frac{\partial^2 w}{\partial x_1 \partial t} + a_{55} J_0 B_{03}\frac{\partial\Phi}{\partial t}\right)\right], \quad (2.4)$$

$$\frac{\partial h_2}{\partial x_3} = \frac{\partial f}{\partial x_2} - \frac{4\pi\sigma_1}{c}\left[\varphi + \frac{1}{c}\left(B_{03}\frac{\partial v}{\partial t} - \right.\right.$$
$$\left.\left. - x_3 B_{03}\frac{\partial^2 w}{\partial x_2 \partial t} + a_{44} J_0 B_{03}\frac{\partial\Psi}{\partial t}\right)\right],$$

$$\frac{4\pi\sigma_3}{c} e_3 = \frac{\partial h_2}{\partial x_1} - \frac{\partial h_1}{\partial x_2},$$

$$\frac{\partial\psi}{\partial x_1} - \frac{\partial\varphi}{\partial x_2} = -\frac{1}{c}\frac{\partial f}{\partial t},$$

where c is the electrodynamic constant numerically equal to the light velocity in vacuum, and B_{03} is the initial magnetic field intensity.

Considering the equations (2.4) we can see that in the first two equations there appear terms representing a new type of interactions [3] between the magnetic field and the shear deformations:

$$4 \pi \sigma_2 \, a_{55} \, J_0 \, \frac{B_{03}}{c} \, \frac{\partial \Phi}{\partial t} \quad , \quad 4 \pi \sigma_1 \, a_{44} \, J_0 \, \frac{B_{03}}{c} \, \frac{\partial \Psi}{\partial t} \quad ,$$

with anisotropic coefficients $\sigma_2 a_{55}$ and $\sigma_1 a_{44}$.

Integrating the first two equations (2.4) with respect to x_3 in the range from 0 to x_3 and taking into account the surface conditions

$$h_i = h_i^+ \quad \text{when} \quad x_3 = h , \quad$$
$$h_i = h_i^- \quad \text{when} \quad x_3 = -h , \tag{2.5}$$

we obtain for h_i

$$h_1 = \frac{h_1^+ + h_1^-}{2} + x_3 \left(\frac{\partial f}{\partial x_1} + \frac{4 \pi \sigma_2}{c} \psi \right) -$$

$$- \frac{4 \pi \sigma_2}{c^2} \left[a_2 \frac{\partial u}{\partial t} - d_3 \frac{\partial^2 w}{\partial x_1 \partial t} + c_3 a_{55} \frac{\partial \Phi}{\partial t} \right] ,$$

$$h_2 = \frac{h_2^+ + h_2^-}{2} + x_3 \left(\frac{\partial f}{\partial x_2} + \frac{4 \pi \sigma_1}{c} \varphi \right) -$$

$$- \frac{4 \pi \sigma_1}{c^2} \left[a_3 \frac{\partial v}{\partial t} - d_3 \frac{\partial^2 w}{\partial x_2 \partial t} + c_3 a_{44} \frac{\partial \Psi}{\partial t} \right] ,$$

where

$$a_3 = B_{03} x_3 \quad , \quad d_3 = B_{03} \frac{1}{2} (x_3^2 - h^2) ,$$

$$c_3 = B_{03} \frac{1}{4} \left(x_3^2 h - \frac{x_3^4}{6} - \frac{5}{6} h^4 \right) . \tag{2.7}$$

Then from the third equation (2.4), according to (2.6), for the third component of the excited electric field we obtain

$$
\begin{aligned}
e_3 = \frac{c}{4\pi\sigma_3} & \left[\frac{1}{2}\left(\frac{\partial h_2^+}{\partial x_1} + \frac{\partial h_2^-}{\partial x_1}\right) - \frac{1}{2}\left(\frac{\partial h_1^+}{\partial x_2} + \frac{\partial h_1^-}{\partial x_2}\right) \right] - \\
& - \frac{\sigma_1}{\sigma_3}\left[x_3 \frac{\partial\varphi}{\partial x_1} + \frac{a_3}{c}\frac{\partial^2 v}{\partial x_1 \partial t} - \frac{d_3}{c}\frac{\partial^3 w}{\partial x_1 \partial x_2 \partial t} + \right. \\
& \left. + \frac{c_3}{c} a_{44} \frac{\partial^2 \Psi}{\partial x_1 \partial t} \right] + \frac{\sigma_2}{\sigma_3}\left[x_3 \frac{\partial\psi}{\partial x_2} + \frac{a_3}{c}\frac{\partial^2 u}{\partial x_2 \partial t} - \right. \\
& \left. - \frac{d_3}{c}\frac{\partial^3 w}{\partial x_1 \partial x_2 \partial t} + \frac{c_3}{c} a_{55} \frac{\partial^2 \Phi}{\partial x_2 \partial t} \right].
\end{aligned}
$$

Thus, assuming the magnetoelasticity hypothesis [2] and the plate exacted theory hypotheses [1] we derive all the components of the excited electromagnetic field in the plate. The components are given by eight sought functions u, v, w, Φ, Ψ, φ, ψ, f and by the values of the inducted magnetic field components on the plate surface h_1 and h_2.

3.   Magnetoelastic equation system for thin anisotropic plates

The equations of motion are [1, 2] :

$$
\frac{\partial\sigma_{11}}{\partial x_1} + \frac{\partial\sigma_{12}}{\partial x_2} + \frac{\partial\sigma_{13}}{\partial x_3} + \rho K_1 = \rho \frac{\partial^2 u_1}{\partial t^2} ,
$$

$$
\frac{\partial\sigma_{22}}{\partial x_2} + \frac{\partial\sigma_{12}}{\partial x_1} + \frac{\partial\sigma_{23}}{\partial x_3} + \rho K_2 = \rho \frac{\partial^2 u_2}{\partial t^2} , \qquad (3.1)
$$

$$
\frac{\partial\sigma_{33}}{\partial x_3} + \frac{\partial\sigma_{32}}{\partial x_2} + \frac{\partial\sigma_{13}}{\partial x_1} + \rho K_3 = \rho \frac{\partial^2 u_3}{\partial t^2} ,
$$

where for the components of the "cargo" term we have

$$
\rho K_1 = \frac{\sigma_2}{c}\left[B_{03}\,\psi - \frac{1}{c}\frac{\partial u_1}{\partial t} B_{03}^2 \right],
$$

$$
\qquad (3.2)
$$

$$
\rho K_2 = \frac{\sigma_1}{c}\left[-B_{03}\,\varphi - \frac{1}{c}\frac{\partial u_2}{\partial t} B_{03}^2 \right] , \quad \rho K_3 = 0 ,
$$

and ρ is the material density.

For stresses in the plate we have the known formulas of the exact theory [1]

$$\sigma_{11} = B_{11} \frac{\partial u}{\partial x_1} + B_{12} \frac{\partial v}{\partial x_2} - x_3 \left(B_{11} \frac{\partial^2 w}{\partial x_1^2} + \right.$$

$$\left. + B_{12} \frac{\partial^2 w}{\partial x_2^2} \right) + J_0 \left(B_{11} a_{55} \frac{\partial \Phi}{\partial x_1} + B_{12} a_{44} \frac{\partial \Psi}{\partial x_2} \right),$$

$$\sigma_{22} = B_{22} \frac{\partial v}{\partial x_2} + B_{12} \frac{\partial u}{\partial x_1} - x_3 \left(B_{22} \frac{\partial^2 w}{\partial x_2^2} + \right.$$

$$\left. + B_{12} \frac{\partial^2 w}{\partial x_1^2} \right) + J_0 \left(B_{22} a_{44} \frac{\partial \Psi}{\partial x_2} + B_{12} a_{55} \frac{\partial \Phi}{\partial x_1} \right),$$

$$\sigma_{12} = B_{66} \left(\frac{\partial u}{\partial x_2} + \frac{\partial v}{\partial x_1} \right) - 2 x_3 B_{66} \frac{\partial^2 w}{\partial x_1 \partial x_2} +$$

$$+ J_0 B_{66} \left(a_{55} \frac{\partial \Phi}{\partial x_2} + a_{44} \frac{\partial \Psi}{\partial x_1} \right),$$

(3.3)

$$\sigma_{13} = f_0 (x_3) \Phi \quad, \quad \sigma_{23} = f_0 (x_3) \Psi, \tag{3.4}$$

where

$$B_{11} = \frac{E_1}{1 - \nu_1 \nu_2} , \quad B_{22} = \frac{E_2}{1 - \nu_1 \nu_2} , \quad B_{12} = \frac{\nu_2 E_1}{1 - \nu_1 \nu_2} =$$

$$= \frac{\nu_1 E_2}{1 - \nu_1 \nu_2} , \quad B_{66} = G_{12} , \quad a_{44} = \frac{1}{G_{23}} a_{55} = \frac{1}{G_{13}} ,$$

(3.5)

$$f_0 (x_3) = \frac{1}{2} (h^2 - x_3^2), \tag{3.6}$$

E_i are the moduli of elasticity, G_{ik} are the shear moduli, and ν_i are the Poissons coefficients.

Considering (2.2), (3.2)-(3.4), we note that both the plate stress component and the "cargo" terms are written in terms of the sought values which are functions of only two coordinates x_1, x_2 and the time t. This means that now we can use the plate motion equations reduced to the middle plane.

Now we perform the integration of each equilibrium equation in (3.1) with respect to x_3 from $x_3 = -h$ to $x_3 = h$, and next we multiply the first two equations of (3.1) by x_3. Then, integrating this result with respect to x_3 within the same limits, according to (2.2), (3.2), (3.3), we obtain the following five equations of the plate motion, averaged over the plate thickness

$$B_{11} \frac{\partial^2 u}{\partial x_1^2} + B_{66} \frac{\partial^2 u}{\partial x_2^2} + (B_{12} + B_{66}) \frac{\partial^2 v}{\partial x_1 \partial x_2} +$$

$$+ \frac{\sigma_2}{c} \left[B_{03} \psi - \frac{1}{c} B_{03}^2 \frac{\partial u}{\partial t} \right] = \rho \frac{\partial^2 u}{\partial t^2} , \quad (3.7)$$

$$B_{22} \frac{\partial^2 v}{\partial x_2^2} + B_{66} \frac{\partial^2 v}{\partial x_1^2} + (B_{12} + B_{66}) \frac{\partial^2 u}{\partial x_1 \partial x_2} -$$

$$- \frac{\sigma_1}{c} \left(B_{03} \varphi + \frac{1}{c} B_{03}^2 \frac{\partial v}{\partial t} \right) = \rho \frac{\partial^2 v}{\partial t^2} , \quad (3.8)$$

$$\frac{2}{3} h^3 \left(\frac{\partial \Phi}{\partial x_1} + \frac{\partial \Psi}{\partial x_2} \right) = 2\rho h \frac{\partial^2 w}{\partial t^2} , \quad (3.9)$$

$$B_{11} \frac{\partial^3 w}{\partial x_1^3} + (B_{12} + 2B_{66}) \frac{\partial^3 w}{\partial x_1 \partial x_2^2} - \frac{2h^2}{5} \left[a_{55} \left(B_{11} \frac{\partial^2 \Phi}{\partial x_1^2} + \right. \right.$$

$$\left. + B_{66} \frac{\partial^2 \Phi}{\partial x_2^2} \right) + a_{44} (B_{12} + B_{66}) \frac{\partial^2 \Psi}{\partial x_1 \partial x_2} \right] + \Phi -$$

$$- \frac{\sigma_2 B_{03}^2}{c^2} \left[\frac{\partial^2 w}{\partial x_1 \partial t} - \frac{2h^2}{5} a_{55} \frac{\partial \Phi}{\partial t} \right] = \rho \frac{\partial^3 w}{\partial x_1 \partial t^2} -$$

$$- \frac{2h^2}{5} \rho a_{55} \frac{\partial^2 \Phi}{\partial t^2} , \quad (3.10)$$

$$B_{22} \frac{\partial^3 w}{\partial x_2^3} + (B_{12} + 2B_{66}) \frac{\partial^3 w}{\partial x_1^2 \partial x_2} - \frac{2h^2}{5} \left[a_{44} \left(B_{22} \frac{\partial^2 \Psi}{\partial x_2^2} + \right. \right.$$

$$\left. + B_{66} \frac{\partial^2 \Psi}{\partial x_1^2} \right) + a_{55} (B_{12} + B_{66}) \frac{\partial^2 \Phi}{\partial x_1 \partial x_2} \right] + \Psi -$$

$$- \frac{\sigma_1 B_{03}^2}{c^2} \left[\frac{\partial^2 w}{\partial x_2 \partial t} - \frac{2h^2}{5} a_{44} \frac{\partial \Psi}{\partial t} \right] = \rho \frac{\partial^3 w}{\partial x_2 \partial t^2} -$$

$$- \frac{2h^2}{5} \rho a_{44} \frac{\partial^2 \Psi}{\partial t^2} . \quad (3.11)$$

The equations of electrodynamics averaged over the plate thickness are to be added to the system (3.7)-(3.11). Integrating the first two equations (2.4) with respect to x_3 in the range from $x_3 = -h$, to $x_3 = h$, and taking into account the conditions on the planes (2.5) and adding the fourth equation (2.4), we finally obtain

$$\frac{\partial \psi}{\partial x_1} - \frac{\partial \varphi}{\partial x_2} + \frac{1}{c}\frac{\partial f}{\partial t} = 0 , \tag{3.12}$$

$$\frac{\partial f}{\partial x_1} + \frac{4\pi\sigma_2}{c}\left(\psi - \frac{B_{03}}{c}\frac{\partial u}{\partial t}\right) = \frac{h_1^+ - h_1^-}{2h} , \tag{3.13}$$

$$\frac{\partial f}{\partial x_2} - \frac{4\pi\sigma_1}{c}\left(\varphi + \frac{B_{03}}{c}\frac{\partial v}{\partial t}\right) = \frac{h_2^+ - h_2^-}{2h} . \tag{3.14}$$

Thus, we have a complete system of magnetoelastic equations for the elastic thin anisotropic plates with finite electroconductivity.

It is evident that in a general case these equations should be considered together with the equations of electrodynamics in the outside region.

The obtained complete system of differential equations allows us to consider the wave processes in the plate in a magnetic field.

However, in the present paper the wave processes will not be considered. Instead, we shall be interested in the plate oscillations with allowance for the transversal shears and anisotropic electroconductivity.

4. The problem of transversal oscillations of rectangular orthotropic plate

Let us consider the problem of transversal oscillations of a symply-supported (along the entire contour) rectangular (axb) orthotropic plate with isotropic and orthotropic electroconductivity in a transversal magnetic field having intensity $B_{03} = B = $ const. For the considered case we have the following initial equations

$$\frac{h^2}{3}\left(\frac{\partial \Phi}{\partial x_1} + \frac{\partial \Psi}{\partial x_2}\right) = \rho \frac{\partial^2 w}{\partial t^2} ,$$

$$B_{11}\frac{\partial^3 w}{\partial x_1^3} + (B_{12} + 2B_{66})\frac{\partial^3 w}{\partial x_1 \partial x_2^2} - \frac{2h^2}{5}\left[a_{55}\left(B_{11}\frac{\partial^2 \Phi}{\partial x_1^2} + B_{66}\frac{\partial^2 \Phi}{\partial x_{22}^2}\right) + \right.$$

$$\left. + a_{44}(B_{12} + B_{66})\frac{\partial^2 \Psi}{\partial x_1 \partial x_2}\right] + \Phi - \frac{\sigma_2 B^2}{c^2}\left(\frac{\partial^2 w}{\partial x_1 \partial t} - \frac{2h^2}{5}a_{55}\frac{\partial \Phi}{\partial t}\right) =$$

$$\rho \frac{\partial^3 w}{\partial x_1 \partial t^2} - \frac{2h^2}{5}\rho a_{55}\frac{\partial^2 \Phi}{\partial t^2} , \tag{4.1}$$

$$B_{22} \frac{\partial^3 w}{\partial x_2^3} + (B_{12} + 2B_{66}) \frac{\partial^3 w}{\partial x_1^2 \, \partial x_2} - \frac{2h^2}{5} \left[a_{44} \, B_{22} \frac{\partial^2 \Psi}{\partial x_2^2} + \right.$$

$$\left. + \left(B_{66} \frac{\partial^2 \Psi}{\partial x_1^2} \right) + a_{55} (B_{12} + B_{66}) \frac{\partial^2 \Phi}{\partial x_1 \, \partial x_2} \right] + \Psi -$$

$$- \frac{\sigma_1 B^2}{c^2} \left(\frac{\partial^2 w}{\partial x_2 \, \partial t} - \frac{2h^2}{5} a_{44} \frac{\partial \Psi}{\partial t} \right) = \rho \frac{\partial^3 w}{\partial x_2 \, \partial t^2} -$$

$$- \frac{2h^2}{5} \rho \, a_{44} \frac{\partial^2 \Psi}{\partial t^2} \, ,$$

and the boundary conditions

at $x_1 = 0$, $x_1 = a$: $w = 0$, $M_1 = 0$, $\Psi = 0$,

at $x_2 = 0$, $x_2 = b$: $w = 0$, $M_2 = 0$, $\Phi = 0$.

(4.2)

The solution of the system (3.15) is sought in the form

$$w = A_{mn} \, e^{\Omega_{mnt}} \sin \frac{m\pi x_1}{a} \sin \frac{n\pi x_2}{B} ,$$

$$\Phi = B_{mn} \, e^{\Omega_{mnt}} \cos \frac{m\pi x_1}{a} \sin \frac{n\pi x_2}{B} ,$$ (4.3)

$$\Psi = C_{mn} \, e^{\Omega_{mnt}} \sin \frac{m\pi x_1}{a} \cos \frac{n\pi x_2}{B} ,$$

which satisfies all the conditions of the simple supports along the plate contour. Here A_{mn}, B_{mn}, C_{mn} are constants, m and n are integer numbers.

Substituting the expressions for w, Φ, Ψ, from (4.3) into (4.1), we obtain the following system of algebraic equations with respect to the unknown constants A_{mn}, B_{mn}, C_{mn}:

$$\rho \Omega_{mn}^2 \, A_{mn} + \frac{h^2}{3} \lambda_m B_{mn} + \frac{h^2}{3} \lambda_n C_{mn} = 0 ,$$

$$\left[\rho \lambda_m \Omega_{mn}^2 + B_{11} \lambda_m^3 + (B_{12} + 2B_{66}) \lambda_m \lambda_n^2 + \frac{\sigma_2 B^2}{c^2} \lambda_m \Omega_{mn} \right] A_{mn} -$$

$$-\left[1 + a_{55}\,\frac{2h^2}{5}\,(B_{11}\,\lambda_m^2 + B_{66}\,\lambda_n^2) + \frac{\sigma_2\,B^2}{c^2}\,\frac{2h^2}{5}\,a_{55}\,\Omega_{mn} + \right.$$

$$\left. + \frac{2h^2}{5}\,\rho\,a_{55}\,\Omega_{mn}^2\right]B_{mn} - a_{44}\,\frac{2h^2}{5}\,(B_{12} + B_{66})\,\lambda_m\lambda_n\,C_{mn} = 0\,,$$

$$\left[\rho\lambda_n\Omega_{mn}^2 + B_{22}\,\lambda_n^3 + (B_{12} + 2B_{66})\,\lambda_m^2\,\lambda_n + \frac{\sigma_1\,B^2}{c^2}\,\lambda_n\,\Omega_{mn}\right]A_{mn} -$$

$$-\left[1 + a_{44}\,\frac{2h^2}{5}\,(B_{22}\,\lambda_n^2 + B_{66}\,\lambda_m^2) + \frac{\sigma_1\,B^2}{c^2}\,\frac{2h^2}{5}\,a_{44}\,\Omega_{mn} + \right.$$

$$\left. + \frac{2h^2}{5}\,\rho\,a_{44}\,\Omega_{mn}^2\right]C_{mn} - a_{55}\,\frac{2h^2}{5}\,(B_{12} + B_{66})\,\lambda_m\lambda_n\lambda_{mn} = 0\,,$$

$$(4.4)$$

where

$$\lambda_m = \frac{m\pi}{a}\quad,\quad \lambda_n = \frac{n\pi}{B}\,. \tag{4.5}$$

In order to have non-trivial solutions of the system (4.4), the determinant is set equal to zero. Thus, we obtain the necessary characteristic equation to determine Ω_{mn}, which shows that the frequencies ($J_m\,\Omega_{mn}$) and the damping coefficients (Re Ω_{mn}) of the natural vibrations of the plate depend on the anisotropic parameters of the plate (b_{ik}, a_{ik}, σ_i).

Let us consider the problem of an infinitely long plate ($b = \infty$, $\lambda_n = 0$, $C_{mn} = 0$). In this case, to determine Ω_m we obtain from the equation system the characteristic equation

$$\frac{2\rho h^2\,\omega_m^2}{5}\,a_{55}\,\bar{\Omega}_m^4 + \frac{12\rho\,\omega_m^2}{5\lambda_m^2}\,a_{55}\,\beta\bar{\Omega}_m^3 + \left[1 + \frac{h^2\,\lambda^2}{3} + \right.$$

$$\left. + \frac{2h^2\,\lambda_m^2\,a_{55}\,B_{11}}{5}\right]\bar{\Omega}_m^2 + 2\beta\bar{\Omega}_m + 1 = 0\,, \tag{4.6}$$

where

$$\bar{\Omega}_m = \frac{\Omega_m}{\omega_m}\quad,\quad \omega_m^2 = \frac{h^2\,\lambda_m^4\,B_{11}}{3\rho}\quad,\quad \beta = \frac{\sigma_2\,h^2\,\lambda_m^2\,B^2}{6c^2\,\rho\omega_0} \tag{4.7}$$

and ω_m are the frequencies of the plate's natural vibration in the absence of the absence of the magnetic field and electroconductivity of the plate.

Inspecting the characteristic equation (4.6), we note that the relative oscillation frequencies $I_m \bar{\Omega}_m$ and the relative damping coefficient $R_e \bar{\Omega}_m$ do not depend on the transversal coefficient of electroconductivity σ_1 (along the tx_1 lines).

Considering futher the equation (4.6), we can see that here the magneto-elastic interaction is represented by two terms. The first term is describing the classic interaction $(2\beta\bar{\Omega}_m)$ and the second term $\left(\dfrac{12\rho\omega_m^2 \, a_{55}\beta}{5\lambda_m^2} \bar{\Omega}_m^3 \right)$ is describing a new type of interaction with parameters a_{55} and σ_2 [3].

Here the new type of interaction essentially changes the picture of oscillation. At first the oscillation picture resembles the classic form, i.e. up to a definite value of β ($\beta \approx 1$) we observe oscillations with damping, and their frequency decreases, reaching zero while the magnetic field intensity B (or the electroconductivity coefficient σ_2) increases. Then we have a region where perturbations consist of damping without oscillations. The further increase of B or σ_2, i.e. β, leads to a quite new phenomenon, namely a new oscillation process appears with an essentially high frequency and the same damping coefficients which have a tendency to decrease with the increase of the magnetic field intensity or electroconductivity coefficients.

The dependence of the roots of the characteristic equation on various values of the parameter β is considered in [3].

Let us consider the problem, neglecting the terms arising when the inertia of rotation is taken into account (the last terms in the second and third equations of the system (4.1)). In particular, this is so if we return to the hypothesis of non-deformable normals. Then from (4.4) we obtain

$$\rho\Omega_{mn}^2 \, A_{mn} + \frac{h^3}{3} \lambda_m \, B_{mn} + \frac{h^2}{3} \lambda_n \, C_{mn} = 0 \, ,$$

$$\left[\rho\lambda_m \, \Omega_{mn}^2 + B_{11} \lambda_m^3 + (B_{12} + 2B_{66}) \lambda_m \lambda_n^2 + \right.$$

$$\left. + \frac{\sigma_2 \, B^2}{c^2} \lambda_m \, \Omega_{mn} \right] = B_{mn} \, , \tag{4.8}$$

$$\rho\lambda_n \Omega_{mn}^2 + B_{22} \lambda_n^3 + (B_{12} + 2B_{66}) \lambda_m^2 \lambda_n + \frac{\sigma_1 \, B^2}{c^2} \lambda_n \, \Omega_{mn} = C_{mn} \, ,$$

and the following characteristic equation

$$\bar{\Omega}_{mn}^2 + 2\beta_1 \, \bar{\Omega}_{mn} + 1 = 0 \, , \tag{4.9}$$

where

$$\overline{\Omega}_{mn} = \frac{\Omega_{mn}}{\omega_{mn}} \ , \qquad \beta = \frac{h^2}{3\rho\,\omega_0} \ \frac{B^2}{c^2} \ \sigma_2 \ (\lambda_m^2 + \alpha\lambda_n^2) \ ,$$

$$\omega_{mn}^2 = \frac{h^2}{3\beta} \ [B_{11} \ \lambda_m^4 + 2 \ (B_{12} + 2B_{66}) \lambda_m^2 \lambda_n^2 + B_{22} \ \lambda_n^4] \ , \qquad (4.10)$$

and ω_{mn} are the natural vibration frequencies of the plate in the absence of the magnetic field.

$\alpha = \dfrac{\sigma_1}{\sigma_2}$ is the degree of the electroconductivity's anisotropy.

Considering the characteristic equation (3.23), we can see that with the increase of the electroconductivity's anistropy, i.e. α or β_1), the oscillation frequency ($J_m \ \overline{\Omega}$) decreases and reaches zero when the parameter β_1 equals 1. The futher increase of β_1 leads to the perturbation damping without oscillations. Here the oscillation picture resembles the classic form.

REFERENCES

[1] AMBARTSUMYAN S.A. – *The Theory of Anisotropic Plates*. Nauka, 1967 (in Russian).
[2] AMBARTSUMYAN S.A., G.E. Bagdassarian and M.V. BELUBEKIAN. – *Magneto-elasticity of Thin Shells and Plates*. Nauka, 1977 (in Russian).
[3] AMBARTSUMYAN S.A. – On the problem of oscillations of the electroconductive plates in the transverse magnetic Field". *Izvestia AN SSSR* MTT, 3 (1979): 164-173. (in Russian).

RESUME

(Sur le problème des oscillations des plaques électroconductrices anistropes dans un champ magnétique)

Une théorie générale des vibrations des plaques orthotropes dans un champ magnétique transversal est développée à partir des principes généraux de la théorie des plaques anisotropes prenant en compte le cisaillement transversal, et des hypothèses fondamentales de la magnétoélasticité des corps minces. Plusieurs cas de vibrations transversales des plaques sont discutés. Il est montré par une analyse comparative que les théories améliorées conduisent à de nouvelles formes pour les vibrations des plaques. Lorsque l'intensité du champ magnétique croît, la fréquence des vibrations "naturelles" diffère de plus en plus des fréquences correspondantes déduites de la théorie classique de la magnétoélasticité.

On Deformational Anisotropy

A.N. Guz'

Institute of Mechanics, The Ukrainian Academy of Sciences, Kiev, USSR

1. Introduction

This report deals with some problems of deformational anisotropy in a non-linear compressible prestrained body. Within recent years significant results have been obtained in this field in some countries. However it is difficult to analyse them completely.

In [1] an attempt was made to analyse some investigations on the linearized theory which are familiar to the author. In the present report a separate aspect of the problem is discussed, i.e. some results obtained in the Institute of Mechanics of the Academy of Sciences of the Ukrainian SSR in Kiev are presented. These results are described extensively in [2, 5] as well as in the references cited there.

We consider the following three states of an elastic body: the natural state (without stresses and strains); the initial state, and the distrubed stressed-strained state. Values characterizing the disturbed state are given in the form of a sum of values characterizing the initial state and the disturbed one.

The disturbances are assumed to take considerably lesser values in comparison with respective values in the initial state. To specify the initial state the theory of large (finite) initial strains and two forms of the theory of small initial strains will be used. Values related to the initial state will be denoted by the index 'zero', while for the disturbances no additional index will be used. An elastic body in its natural state is supposed to be isotropic and to possess an elastic potential, represented by a twice continuously-differentiable function of algebraic invariants of the Green strain tensor.

In [2, 5] the case of an anisotropic body in its natural state is also discussed. We use the Lagrangian coordinates $x_i \equiv x^i$ which coincide in the natural state with the Cartesian coordinates. All values are related to the basis in the initial reference state. Linearization relative to disturbances will be carried out and subsequent items will be discussed within the linearized theory. We shall give the main results for compressible elastic bodies, the results for incompressible elastic bodies may be similar. The following summation rule will be used: with indices occuring two or more times only on one side of an equality we shall summarize from 1 to 3, unless stated otherwise.

2. General considerations

Linearized relations may be represented as follows.
Equation of motion:

$$(\omega_{im\alpha\beta} u_{\alpha,\beta})_{,i} - \rho \ddot{u}_m = 0 \; ; \tag{2.1}$$

boundary conditions in stresses on a surface region S_1:

$$Q_m |_{S_1} = P_m \; ; \quad Q_m \equiv N_i \, \omega_{im\alpha\beta} u_{\alpha,\beta} \; ; \tag{2.2}$$

and boundary conditions in displacements in a surface region S_2:

$$u_m |_{S_2} = 0 \; . \tag{2.3}$$

In these relations
u_m = component of the displacement vector,
P_m = component of the vector of the right-hand boundary stresses,
N_m = component of the unit vector normal to the body surface in the natural state,
$\omega_{im\alpha\beta}$ = component of a fourth rank tensor, specified by the elastic potential in the initial state.
In the theory of large (finite) initial strains we obtain

$$\omega_{im\alpha\beta} = (\delta_{mn} + u^0_{m,n}) (\delta_{\alpha j} + u^0_{\alpha,j}) A_{inj\beta} \Phi^0 + \delta_{\alpha m} S^0_{i\beta} \; ;$$

$$S^0_{i\beta} = \Sigma_{i\beta} \Phi^0 \; ; \quad 2\epsilon^0_{ij} = u^0_{i,j} + u^0_{j,i} + u^0_{i,k} u^0_{k,j} \; ; \quad \Phi^0 = \Phi^0(A^0_1, A^0_2, A^0_3) \; . \tag{2.4}$$

In the first version of the theory of small initial strains, the elongations and displacements are small as compared with unity, and thus can be neglected. In that case expressions (2.4) remain valid, but the geometrical changes of the body under initial strain should be neglected. Besides, we assume

$$S^0_{i\beta} \approx \sigma^0_{i\beta} \; , \tag{2.5}$$

where $\sigma^0_{i\beta}$ is the true stress.
In the second version of small initial strains theory, in addition to the first version, it is assumed that the precritical state can be determined on the

basis of the geometrically-linear theory. In that case the following expressions are obtained:

$$\omega_{im\alpha\beta} = A_{im\alpha\beta} \Phi^0 + \delta_{\alpha m} \sigma^0_{i\beta} ; \quad \sigma^0_{i\beta} = \Sigma_{i\beta} \Phi^0 ;$$

$$2 \epsilon^0_{ij} = u^0_{i,j} + u^0_{j,i} ; \quad \Phi^0 = \Phi^0(A^0_1, A^0_2, A^0_3) . \tag{2.6}$$

In (2.4)-(2.6) the following notation regarding the initial state is introduced:

ϵ^0_{ij} = component of the Green strain tensor,

Φ^0 = elastic potential,

A^0_i = algebraic invariant of the Green strain tensor.

Besides, the following differential operators are used:

$$A_{inj\beta} = \Sigma_{in} \Sigma_{j\beta} + (\delta_{i\beta} \delta_{jn} + \delta_{ij} \delta_{\beta n}) \frac{\partial}{\partial A^0_2} + \frac{3}{2} (\delta_{i\beta} \epsilon^0_{jn} + \delta_{ij} \epsilon^0_{n\beta})$$

$$+ \delta_{nj} \epsilon^0_{i\beta} + \delta_{n\beta} \epsilon^0_{ij}) \frac{\partial}{\partial A^0_3} ; \tag{2.7}$$

$$\Sigma_{i\beta} = \delta_{i\beta} \frac{\partial}{\partial A^0_1} + 2 \epsilon^0_{i\beta} \frac{\partial}{\partial A^0_2} + 3 \epsilon^0_{ik} \epsilon^0_{k\beta} \frac{\partial}{\partial A^0_3} .$$

With the Green strain tensor the above problems remain general. As a matter of fact, invariants of any system of invariants of any strain tensor could be represented by algebraic invariants of the Green strain tensor. Let us consider now the case of a uniform initial state characterized by the following values for a body which is isotropic in its natural state:

$$u^0_m = \delta_{im} (\lambda_i - 1) x_i ; \quad \epsilon^0_{ij} = \delta_{ij} \epsilon^0_{jj} ; \quad S^0_{ij} = \delta_{ij} S^0_{jj} ,$$

$$\sigma^0_{ij} = \delta_{ij} \sigma^0_{jj} ; \quad \lambda_i, \epsilon^0_{ij}, S^0_{ij}, \sigma^0_{ij} = \text{const.} \tag{2.8}$$

In this case, the theory of large (finite) initial strains yields

$$\omega_{im\alpha\beta} = \lambda_m \lambda_\alpha [\delta_{im} \delta_{\alpha\beta} a_{i\beta} + (1 - \delta_{im}) (\delta_{i\alpha} \delta_{m\beta} + \delta_{i\beta} \delta_{m\alpha}) \mu_{im}]$$

$$+ \delta_{m\alpha} \delta_{i\beta} S^0_{\beta\beta} ; \quad \omega_{im\alpha\beta} = \text{const.} \tag{2.9}$$

In (2.9) the following notation is used:

$$a_{i\beta} = A_{i\beta} \Phi^0 ; \quad \mu_{im} = B_{im} \Phi^0 ; \quad S^0_{\beta\beta} = \Sigma_{\beta\beta} \Phi^0 \tag{2.10}$$

In (2.10) we introduced the following differential operators:

$$A_{i\beta} = \Sigma_{ii}\Sigma_{\beta\beta} + 2\delta_{i\beta}B_{ii} \; ; \quad B_{im} = \frac{\partial}{\partial A_2^0} + \frac{3}{2}\,(\epsilon_{ii}^0 + \epsilon_{mm}^0)\,\frac{\partial}{\partial A_3^0} \; ;$$

$$\Sigma_{ii} = \frac{\partial}{\partial A_1^0} + 2\epsilon_{ii}^0\,\frac{\partial}{\partial A_2^0} + 3\epsilon_{ii}^{02}\,\frac{\partial}{\partial A_3^0} \; .$$

$$(2.11)$$

In case of the first and the second versions of the small initial strains theory, the following transformations should be made in (2.9)-(2.11).

Equations (2.1)-(2.3) resemble the equations of the linear theory of elasticity for an anisotropic non-homogeneous body with the elastic moduli tensor ω. So the problem of possible analogies between linear and linearized problems arises. It has been proved that in the general case the following di-symmetries hold:

$$\omega_{im\alpha\beta} \neq \omega_{mi\alpha\beta} \; ; \quad \omega_{im\alpha\beta} \neq \omega_{im\beta\alpha} \; ; \quad \omega_{im\alpha\beta} \neq \omega_{\alpha\beta im} \; . \qquad (2.12)$$

Hence, in the general case there is no such analogy. It should also be noted here that the tensor ω meets the following conditions of symmetry:

$$\omega_{im\alpha\beta} \equiv \omega_{\beta\alpha mi} \qquad (2.13)$$

On the basis of (2.13), the self-conjugation of respective problems can be demonstrated. Moreover, these conditions give rise to general variational principles of the Hu-Washizu type. These principles have been formulated for dynamic boundary problems as well as for dynamic problems with initial conditions. Similar principles are formulated for static problems as well.

For particular cases, one demonstrates that the following equations for existence of similar linear and linearized problems are necessary and sufficient:

$$\omega_{im\alpha\beta} = \omega_{mi\alpha\beta} \; ; \quad \omega_{im\alpha\beta} = \omega_{im\beta\alpha} \; . \qquad (2.14)$$

In the particular case of a uniform initial state (2.8), conditions (2.14) yield:

$$e_{ijk}\,e_{k\alpha\beta}[(1 - \delta_{ij})\,\lambda_\alpha\,\mu_{\alpha\beta}(\lambda_\beta - \lambda_\alpha) - S_{\beta\beta}^0] = 0 \; ;$$

$$e_{ijk}\,e_{k\alpha\beta}[(1 - \delta_{ij})\,(\lambda_\alpha\,\lambda_\beta - \lambda_i\,\lambda_j)\,\mu_{\alpha\beta} - S_{ii}^0] = 0 \; .$$

$$(2.15)$$

For the second variation of small initial deformations theory, conditions (2.15) have the following form.

$$e_{ijk} e_{k\alpha\beta} \sigma^0_{\beta\beta} = 0 ; \quad e_{ijk} e_{k\alpha\beta} \sigma^0_{ii} = 0 .$$ (2.16)

In (2.15) and (2.16) e_{ijk} denotes the components of the alternator tensor of third rank.

If follows from (2.15) and (2.16) that even in the uniform initial state expressed by (2.8), in the general case the linear and linearized problems are not similar. It is worth noting that even under omnidirectional uniform compression, the mentioned analogy exists rather under the 'tracking' load than under the 'dead' one. There are two types of problems when the necessary and sufficient conditions of such analogy are satisfied. These are anti-plane problems and problems of plane strained body under axial load. Our results on the analogies are independent of the basis, as can be easily proved by direct verification.

3. Wave propagation

From now on we shall consider the uniform initial state in the form (2.8), and we shall discuss problems of wave propagation in bodies with deformational anisotropy. Let us consider a propagation of a plane wave, the phase surface of which is determined by the phase normal n. Equation (2.1) is solved by

$$u_\alpha = \hat{u}_\alpha \exp [i(k \cdot r - \Omega \tau)] ; \quad k = kn ; \quad n \cdot n = 1 .$$ (3.1)

Then the characteristic equation is obtained in the form

$$\det \| \Lambda_{m\alpha} - C^2 \delta_{m\alpha} \| = 0 ; \quad C^2 = \left(\frac{\Omega}{k} \right)^2 .$$ (3.2)

For dimensional changes of the initially strained body or in the absence of such changes, the values of $\Lambda_{m\alpha}$ are, respectively,

$$\Lambda_{m\alpha} = \rho^{-1} \omega_{im\alpha\beta} n_i \lambda_i n_\beta \lambda_\beta ; \quad \Lambda_{m\alpha} = \rho^{-1} \omega_{im\alpha\beta} n_i n_\beta$$ (3.3)

According to the condition of symmetry (2.13) it follows from (3.3) that the matrix $\| \Lambda_{m\alpha} \|$ is symmetrical:

$$\Lambda_{m\alpha} = \Lambda_{\alpha m}$$ (3.4)

Now let us consider the case when the linearized problem has a unique solution (R. Hill, *J. Mech. and Phys. Solids,* Vol. 5, N° 4, 1957). The condition of uniqueness is satisfied when, in our notation,

$$H_1 > 0, \quad H_1 = \omega_{im\alpha\beta} \, \zeta_{mi} \zeta_{\alpha\beta}, \tag{3.5}$$

for all ζ_{ij} not simultaneously equal to zero. Consider the following relation for all ξ_i not simultaneously equal to zero :

$$H_2 = \Lambda_{m\alpha} \xi_m \, \xi_\alpha \tag{3.5'}$$

From (3.3), (3.5) and (3.5)' it follows that

$$H_2 = H_1 \rho^{-1} > 0, \quad \zeta_{mi} = \xi_m \lambda_i n_i, \quad \zeta_{\alpha\beta} = \xi_\alpha \lambda_\beta n_\beta \tag{3,6}$$

(3.4)-(3.6) imply that the matrix $\|\Lambda_{m\alpha}\|$ is symmetrical and the corresponding quadratic form is positively determined. Consequently we have 3 positive eigenvalues C_1^2, C_2^2, C_3^2. We conclude that in the uniform initial state of a non-linear elastic isotropic body with an arbitrary form of elastic potential, 3 isonormal waves propagate in an arbitrary direction of the phase normal. A similar situation exists in the linear theory of elasticity of an anisotropic body. The results cited make it possible to propose an experimental procedure for measurement of isonormal waves propagation velocity with different wave normals, to determine the values of the tensor ω of the form (2.9). In particular, in [6] the second version of the small initial strains theory is discussed, for which the components of the tensor ω may be represented in the form

$$\omega_{im\alpha\beta} = \delta_{im} \delta_{\alpha\beta} a_{i\beta} + (1 - \delta_{im}) (\delta_{i\alpha} \delta_{m\beta} + \delta_{i\beta} \delta_{m\alpha}) \mu_{im} +$$
$$+ \delta_{\alpha m} \delta_{i\beta} \sigma^\circ_{\beta\beta}. \tag{3.7}$$

In [6], propagation velocities of isonormal waves with wave normals are discussed in the form

$$\{1, 0, 0\}; \quad \{0, 1, 0\}; \quad \{0, 0, 1\}; \quad \{2^{-\frac{1}{2}}, 2^{-\frac{1}{2}}, 0\};$$

$$\{2^{-\frac{1}{2}}, 0, 2^{-\frac{1}{2}}\}; \quad \{0, 2^{-\frac{1}{2}}, 2^{-\frac{1}{2}}\}. \tag{3.8}$$

As a result, 12 relations are derived to determine the constants (3.7), using the corresponding velocities. In the general case, (3.7) contains 15 constants.

So in every specific case, 3 additional conditions are required. In particular, in the case

$$\sigma^{\circ}_{33} = 0 \qquad (3.9)$$

we have [6] the additional conditions

$$\omega_{1313} = \omega_{3113} \; ; \; \omega_{2323} = \omega_{3223} \; ; \; \omega_{1212} = \omega_{3113} \; +$$

$$+ \; \omega_{1221} - \omega_{1331} \qquad (3.10)$$

Thus we conclude that in some cases the results of experimental studies for the determination of values of linearized relations of elasticity may be used without fixing the form of the elastic potential.

4. The structure of elastic potential

Now let us consider the problem of the structure of the elastic potential $\Phi^{\circ} = \Phi^{\circ} (A^{\circ}_1, A^{\circ}_2, A^{\circ}_3)$ which allows to explain the experimentally observed regularities of wave qropagation in bodies under initial stresses. The body in its natural state is assumed to be isotropic. In some cases, this means to assume for particular materials that the body is quasi-isotropic in its natural state [2, 4]. Results of experiments on plexiglass, steel 09G2S and steel EI-702 were analysed. It was conclusively stated that the elastic potential should depend on 3 invariants,

$$\Phi^{\circ} = \Phi^{\circ} (A^{\circ}_1, A^{\circ}_2, A^{\circ}_3) \; ; \; \Phi^{\circ} \neq \Phi^{\circ} (A^{\circ}_1, A^{\circ}_2). \qquad (4.1)$$

This is an interesting case when the dependence on the third invariant has to be accepted as an explanation of experimental results. Such a situation is quite uncommon in the mechanics of strained bodies. The obtained result explains why, in the analytical treatment of experimental results on wave propagation in initially stressed bodies, the simplest form of elastic potential depending on 3 invariants is employed ; this is the potential of the Murnaghan type :

$$\Phi^{\circ} = \frac{1}{2} \lambda A^{\circ 2}_1 + \mu A^{\circ}_2 + \frac{a}{3} A^{\circ 3}_1 + b A^{\circ}_1 A^{\circ}_2 + \frac{c}{3} A^{\circ}_3. \qquad (4.2)$$

The third order constants a, b and c are easily determined by an ultrasonic technique. It should be noted that in differently formulated problems (the large initial strains theory and two versions of the small initial strains theory) different numerical values of third order constants are obtained. Nevertheless the regularities of wave propagation in all formulations of problems can be explained.

5.   Non-destructive method of determination of the biaxial stressed state

We consider a plane sheet $|x_1| \leqslant h$ with initial stresses induced in it. We shall excite the waves propagating along the axis Ox_1. The body may be a shell of a large radius. We restrict the analysis to relatively rigid materials satisfying the inequalities

$$\frac{\overset{\circ}{\sigma}_{11}}{\mu} < 1 \; ; \quad \frac{\overset{\circ}{\sigma}_{22}}{\mu} < 1, \tag{5.1}$$

where μ is a shear modulus. In all relations the values of the order indicated by (5.1) are retained. Let us apply the above theory. For the considered case, we obtain the result

$$\overset{\circ}{\sigma}_{33} - \overset{\circ}{\sigma}_{22} \doteq A \left(\frac{C_{Sx_3} - C_{SO}}{C_{SO}} - \frac{C_{Sx_2} - C_{SO}}{C_{SO}} \right) ;$$

$$\overset{\circ}{\sigma}_{33} + \overset{\circ}{\sigma}_{22} = B \left(\frac{C_{Sx_3} - C_{SO}}{C_{SO}} + \frac{C_{Sx_2} - C_{SO}}{C_{SO}} \right) . \tag{5.2}$$

In (5.2), C_{SO} is the velocity of the shear wave in the body in its natural state (in the unstressed body),

C_{Sx_3} is the velocity of the shear wave in the body with initial stresses (with the wave propagating in the direction of the Ox_1 axis and polarized in the $x_1 Ox_3$ plane),

C_{Sx_2} is the velocity of the shear wave in the body with initial stresses (the wave is propagating in the direction of the Ox_1 axis and polarized in the $x_1 Ox_2$ plane),

A, B are constants expressed by constants of the second and the third orders.

The constants A and B can be determined experimentally. Expressions (5.2) relate the difference of main stresses to the difference of relative velocity changes, and the sum of main stresses to the sum of relative velocity changes. We shall refer to relations (5.2) as to the relations of acoustoelasticity. Relations of acoustoelasticity are similar to relations of photoelasticity, being more general in character. In fact, in photoelasticity we have only one relation of the type as the first relation of (5.2), hence two main stresses cannot be determined from this one relation. In acoustoelasticity, two relations (5.2) exist which makes it possible to determine two main stresses without destruction and addtional procedures, while in photoelasticity, additional procedures are used. In [2, 4] the relations of acoustoelasticity for quasi-isotropic materials are derived.

The effect of defomational anisotropy under elastic strains on the changes in wave propagation velocity is small, it is of the order of tenths and hundredths of a percent. Measuring instruments are thus required to measure velocity changes to a mentioned precision. A special instrument was constructed to obtain the results given below. Its operating frequency is 5MHz, to a precision of the order of a thousandth of a percent. The instrument is described in detail in the references cited [2, 4]. With a view to specify the possibilities and precision of stress measurements, using the described acoustical technique, the following experiment was carried out. A round solid disk was compressed along its diameter by concentrated forces. According to the theoretical solution, in the centre of the disk arises a stressed state approximating a biaxial uniform state. For various materials (disk specimens) the calculated (theoretical) and measured (experimental) results were compared. The results of the comparison are given in the table.

The tabulated findings show that the proposed non-destructive ultrasonic technique for determination of biaxial stresses has proved to be precise and effective. This technique was used for the determination of uniaxial and biaxial stresses which occur during the welding of structure elements. Measurements were made in laboratory conditions, as well as on full-scale structures.

TABLE

Materials	Stresses			
	measured		calculated	
	$\overset{\circ}{\sigma}_{22}$	$\overset{\circ}{\sigma}_{33}$	$\overset{\circ}{\sigma}_{22}$	$\overset{\circ}{\sigma}_{33}$
Plexiglass	0,196	$-$ 0,53	0,182	$-$ 0,55
Steel 09G2S	3,02	$-$ 9,25	3,27	$-$ 9,80
Steel 15 X	4,27	$-$ 13,68	4,5	$-$ 13,3
Steel 45G17VZ	6,73	$-$ 16,30	5,79	$-$ 17,37
Steel ST-3 sp	3,56	$-$ 10,26	3,26	$-$ 9,78
Steel ST-20	4,75	$-$ 13,8	4,55	$-$ 13,65
Alloy D16	4,05	$-$ 7,90	3,00	$-$ 9,00
AMG 3V	3,4	$-$ 7,01	2,55	$-$ 7,65
Alloy 1516	2,91	$-$ 10,4	3,54	$-$ 10,6
High-strength steel	6,7	$-$ 19,1	6,4	$-$ 19

The above results and some other results, obtained in the institute of Mechanics of the Academy of Sciences of the Ukrainian SSR, are cited in references [1-6].

REFERENCES

[1] BABICH S., A. GUZ., A. ZHUK. − "Elastic waves in bodies with initial stresses". *Prikladnay Mekhanika*, 15, 4 : (1979): 3-23. (in Russian).

[2] GUZ A., Ph MAHORT., O. GUSHCHA., V. LEBEDEV. − *Foundations of Ultrasound Non-Destructive Technique for Determination of Stresses in Solid Bodies* Kiev : Naukova dumka Publishing House, 1974, p. 108. (in Russian).

[3] GUZ A., A. ZHUK., Ph. MAHORT. − *Waves in a Layer with Initial Stresses*. Kiev: Naukova dumka Publishing House, 1976, p. 104 (in Russian).

[4] GUZ A., Ph. MAHORT., O. GUSHCHA. − *Introduction to Acoustoelasticity*. Kiev : Naukova dumka Publishing House, 1977, p. 152. (in Russian).

[5] GUZ A. − "On linearized theory of elastic wave propagation in bodies with initial stresses". *Prikladnaya mekhanika*, 14, 4, (1978) : 3-32 (in Russian).

[6] GUZ A. − "On determination of elastic constants in linearized theory of elasticity". *Dopovidi AN UkrSSR* Series A, 1, (1975) : (in Russian).

RESUME

(Sur l'anisotropie en déformation)

Nous donnons une brève discussion des résultats sur l'anisotropie en déformation, obtenus à l'Institut de Mécanique de l'Académie des Sciences de la RSS d'Ukraine (Kiev, URSS). Ces résultats comportent : quelques problèmes généraux de la théorie de l'élasticité tridimensionnelle linéarisée en déformations initiales petites et finies ; une étude sur les régularités de la propagation des ondes dans des corps ayant subi des déformations initiales ; la vérification de la possibilité d'une détermination expérimentale des coefficients des relations linéarisées ; les bases montrant la nécessité de tenir compte de la dépendance du potentiel élastique par rapport au troisième invariant du tenseur déformation ; des discussions sur les fondements de la méthode non-destructive par ultra-sons pour la détermination des états de contrainte biaxiaux.

Anisotropie Induite par les Fissures dans les Roches sous Contrainte : Modules Élastiques Effectifs et Atténuation

M. Piau

Institut de Mécanique, Grenoble, France.

1. Introduction

La présence de nombreuses fissures à faibles charges ainsi que le développement de fissures dans des directions privilégiées sous fortes contraintes sont des phénomènes bien connus dans les roches. Lorsqu'on veut étudier le comportement de tels matériaux sous des charges compressives croissantes, par exemple, on a besoin de moyens non-destructifs pour suivre l'évolution du champ des fissures dans l'échantillon. On propose donc d'utiliser la variation de la vitesse de propagation d'une onde et de son atténuation due aux fissures présentes dans le matériau à chaque instant.

Beaucoup de roches polycristallines, où les dimensions des microfissures sont commensurables avec celles des cristaux, peuvent être considérées comme homogènes et isotropes d'un point de vue macroscopique, tout au moins au début du programme de charge, et sont souvent étudiées comme telles. On dispose alors pour caractériser le champ de fissures de deux modules élastiques isotropes modifiés et des atténuations des ondes de compression (P) et de cisaillement (S) correspondantes. Parmi les articles qui traitent ce cas des fissures réparties uniformément dans toutes les directions, fissures convexes ou fissures infiniment longues, on peut citer les références [3, 5, 9, 10, 14-17].

Mais quand les conditions de charge deviennent plus sévères, après la fermeture de certaines microfissures initiales, les fissures se développent dans des directions privilégiées, de préférence parallèles aux directions principales du tenseur des contraintes. On est alors conduit à déterminer le tenseur d'élasticité anisotrope effectif associé à une distribution particulière de fissures.

Ce modèle de milieu anisotrope est étudié ici pour des matériaux formés d'une matrice dont la réponse est supposée linéaire élastique, contenant une concentration suffisamment faible de fissures planes, de dimensions telles que les conditions de Rayleigh soient satisfaites pour l'onde plane excitatrice ; aucun contact n'a lieu entre les lèvres de la fissure.

Dans le chapitre 2, les fissures planes sont supposées circulaires, et l'onde harmonique plane est diffractée par une distribution diluée de fissures toutes parallèles dont le plan fait un angle donné avec la normale à l'onde. La forme asymptotique de l'onde permet d'obtenir une estimation des modules équivalents associés : les composantes du tenseur d'élasticité orthotrope de révolution effectif sont déduites de la partie réelle de l'onde et correspondent aux vitesses effectives des ondes P et S ; l'atténuation par unité de longueur à la traversée du milieu fissuré est donnée par la partie imaginaire de l'onde diffractée. On est amené à introduire une porosité effective associée aux fissures ϵ. Les composantes du tenseur d'élasticité sont modifiées par des termes correctifs fonctions de ϵ et du coefficient de Poisson. L'atténuation dépend en plus, pour les longueurs d'onde considérées, du nombre d'onde adimensionnel associé au rayon de la fissure.

Le tenseur d'anisotropie ainsi déterminé sert de base à ceux qu'on peut associer à toute distribution particulière de fissures. Le chapitre 3 considère un autre cas de symétrie de révolution où les fissures circulaires sont telles que leurs plans aient tous une direction commune. Les modules effectifs obtenus sont comparés à ceux du chapitre précédent, et aussi à leur valeur pour des milieux contenant des fissures orientées dans toutes les directions. Ces derniers milieux peuvent être représentés par un milieu isotrope équivalent ; si les fissures sont elliptiques, mais orientées dans toutes les directions, les coefficients de ϵ ont le même ordre de grandeur, à condition de bien choisir la porosité effective associée, et les résultats seront valables par extension pour des fissures planes de frontière convexe quelconque. On le montre au chapitre 4.

Les chapitres 2 et 3 supposaient les fissures sèches, avec des conditions aux limites sur les lèvres du type libre. Le chapitre 5 traite le cas de fissures remplies d'un liquide non-visqueux, dont le module de compressibilité K_f n'est plus négligeable par rapport à celui de la matrice solide, pour l'épaisseur relative de fissure considérée. L'anisotropie induite par les fissures est alors bien moindre.

Il est intéressant, et c'est l'objet du chapitre 6, de comparer les valeurs obtenues pour les fissures circulaires à celles qu'on peut déduire des études bi-dimensionnelles sur les fissures infiniment longues de type Griffith. Les termes correctifs intervenant dans le module font apparaître une porosité effective bi-dimensionnelle et l'atténuation des ondes est proportionnelle au cube du nombre d'onde adimensionnel, alors que c'est la puissance quatrième de ce nombre qui intervient pour les fissures circulaires (on retrouve la différence qui existe entre la diffraction par des cylindres et par des sphères).

En général, on n'a accès expérimentalement aux fissures dans un échantillon que par les traces qu'elles laissent sur des plans de coupe. Il n'est pourtant pas indifférent de les modéliser par des fissures circulaires ou par des fissures de Griffith puisque les résultats obtenus avec ces deux modèles sont loin d'être équivalents, surtout lorsque le développement des fissures dans des directions privilégiées induit une anisotropie notable.

2. Fissures circulaires parallèles

On suppose que n_0 fissures circulaires par unité de volume, de plan normal à l'axe z, sont réparties dans un milieu infini élastique de constantes de Lamé λ et μ. Les fissures sont suffisamment espacées pour qu'on puisse, au premier ordre, négliger les diffractions multiples entre obstacles. Elles sont sèches, c'est-à-dire que sur les lèvres les conditions de bord libre sont satisfaites :

$$\sigma_{zz} = \sigma_{zr} = \sigma_{z\Phi} = 0 \tag{1}$$

où r, Φ, z sont les coordonnées cylindriques. Une onde harmonique plane de la forme $\exp ik_i (\sin\theta_0 x + \cos\theta_0 z)$ est incidente sur ces fissures en faisant un angle θ_0 avec l'axe z. Le nombre d'onde k_i, $i = 1, 2$, selon qu'il s'agit d'une onde de compression (onde P) ou d'une onde de cisaillement (onde S), est relié à la pulsation ω par

$$k_i = \frac{\omega}{V_i} \qquad V_1 = \left(\frac{\lambda + 2\mu}{\rho}\right)^{1/2} \qquad V_2 = \left(\frac{\mu}{\rho}\right)^{1/2} \tag{2}$$

Pour une dépendance du temps en $\exp(-i\omega t)$, on résout trois équations aux dérivées partielles pour les trois potentiels scalaires de Helmotz. La solution obtenue [9, 10] est développée pour des nombres d'onde adimensionnels suffisamment petits, soit $k_i a \ll 1$, où a est le rayon de la fissure, donc lorsque l'onde incidente satisfait les conditions de Rayleigh par rapport à l'obstacle.

On en déduit [14] la forme asymptotique de l'amplitude diffractée à grande distance R d'une fissure le long d'une direction d'angles polaires θ, Φ, sous la forme :

$$u_R = u_0 g(\theta, \Phi) \exp \frac{ik_1 R}{R}$$

$$u_\theta = u_0 h_1(\theta, \Phi) \exp \frac{ik_2 R}{R} \tag{3}$$

$$u_\Phi = u_0 h_2(\theta, \Phi) \exp \frac{ik_2 R}{R}$$

où u_0 est l'amplitude du déplacement associé à l'onde incidente. La vitesse effective de l'onde V_i' et son atténuation α_L par unité de longueur à la traversée d'un milieu élastique contenant n_0 fissures de rayon a par unité de volume sont alors données par [2]

$$\frac{\omega}{V_i} = k_i \left\{ 1 + \frac{2\pi n_0 a^3}{(k_i a)^2} \, \mathcal{R}e \, f(\theta_0, 0) \right\} \tag{4}$$

$$\alpha_L = k_i \left\{ \frac{2\pi n_0 a^3}{(k_i a)} \, \mathcal{I}m \, f(\theta_0, 0) \right\} \tag{5}$$

les termes d'ordre $(k_i a)^2$ étant négligés. Pour une onde P incidente $f(\theta, \Phi) = g(\theta, \Phi)$; pour une onde S, si **b** est la direction de polarisation de l'onde incidente, $\mathbf{h} = (h_1, h_2)$ l'amplitude transversale de l'onde diffractée, $f(\theta_0, 0) = \mathbf{b} \cdot \mathbf{h}(\theta_0, 0)$.

Les calculs étant explicités en [14, 16], on donne ici les principaux résultats obtenus pour des fissures parallèles. Une onde de cisaillement incidente se sépare alors en deux composantes qui se propagent chacune avec sa vitesse propre : une composante parallèle au plan des fissures (SH) qui garde une direction constante, et une composante (SV) contenue dans le plan normal aux fissures qui passe par la direction d'incidence. L'onde SV subit à la traversée du milieu anisotrope un couplage avec l'onde P mais reste principalement transversale. Les vitesses et les atténuations obtenues pour les 3 ondes de volume (qP – qSV – SH) sont alors, au premier ordre :

$$\rho(V_P)^2 = (\lambda + 2\mu) \left\{ 1 - \frac{4\pi}{3} n_0 a^3 [a_1 + (a_2 - a_1) \sin^2 \theta_0 \right.$$
$$\left. + \sigma^2 g \sin^2 \theta_0 \cos^2 \theta_0] \right\}$$

$$\rho(V_{SV})^2 = \mu \left\{ 1 - \frac{4\pi}{3} n_0 a^3 (b_1 - g \sin^2 \theta_0 \cos^2 \theta_0) \right\} \tag{6}$$

$$\rho(V_{SH})^2 = \mu \left\{ 1 - \frac{4\pi}{3} n_0 a^3 b_1 (1 - \sin^2 \theta_0) \right\}$$

$$\alpha_p = \frac{2\pi}{9} n_0 k_1^4 a^6 \{ f_1 + 4\sigma^2 (h_1 - f_1) \sin^2 \theta_0 - 4\sigma^2 (h_1 - \sigma^2 f_1) \sin^4 \theta_0 \}$$

$$\alpha_{SV} = \frac{2\pi}{9} n_0 k_1^4 a^6 \sigma^{-1} \{ h_1 (1 - 2\cos^2 \theta_0)^2 + 4\sigma^2 f_1 \sin^2 \theta_0 \cos^2 \theta_0 \} \tag{7}$$

$$\alpha_{SH} = \frac{2\pi}{9} n_0 k_1^4 a^6 \sigma^{-1} h_1 \cos^2 \theta_0$$

$$a_1 = [\pi\sigma^2(1-\sigma^2)]^{-1} \qquad a_2 = (1-2\sigma^2)^2\,a_1$$

$$b_1 = 4[\pi(3-2\sigma^2)]^{-1} \qquad g = 4(b_1 - \sigma^2 a_1) \tag{8}$$

$$f_1 = \frac{4}{3\pi^2(1-\sigma^2)^2\,\sigma^5}\left(\frac{2}{5} + \frac{3\sigma}{4} - 2\sigma^3 + \frac{8\sigma^5}{5}\right)$$

$$h_1 = \frac{4}{3\pi^2(3-2\sigma^2)^2}\,\frac{8}{5\sigma^3}\,(3+2\sigma^5) \tag{9}$$

Le rapport des nombres d'onde $\sigma = k_1/k_2$, qui varie entre $1/\sqrt{2}$ et 0 est associé au coefficient de Poisson ν par $\nu = (1-2\sigma^2)/2(1-\sigma^2)$. Lorsque les fissures considérées ne sont pas identiques, a^3 et a^6 sont remplacés par leurs moyennes sur les valeurs prises par les rayons de fissures.

Pour les fissures planes circulaires de rayon a et de demi-ouverture c, si $\varphi = c/a$ est le rapport d'aspect ou l'épaisseur relative d'une fissure, $v_c = (4/3)\pi a^2 c$ son volume, l'effet des fissures sur les vitesses d'onde est donc décrit par une porosité effective $\epsilon = n_0 v_c/\varphi$ et non par la porosité réelle qui en diffère d'un ou plusieurs ordres de grandeur.

On peut aussi faire apparaître ϵ dans l'atténuation qui dépend alors du produit $\epsilon k_i^4\,\bar{a}^3$, où $\bar{a}^3$ est le rapport des moyennes $\langle a^6 \rangle/\langle a^3 \rangle$ ($\geqslant \langle a^3 \rangle$) [14]).

Les équations (6) qui donnent les vitesses d'onde effectives montrent qu'on peut leur associer un tenseur d'élasticité orthotrope de révolution $C_{ijk\ell}$, d'axe de symétrie z ; les modules élastiques sont des modules dynamiques pour des nombres d'onde adimensionnels petits. On a cinq composantes distinctes

$$A = C_{1111} = C_{2222} = (\lambda + 2\mu)(1 - \epsilon a_2)$$

$$C = C_{3333} = (\lambda + 2\mu)(1 - \epsilon a_1)$$

$$F = C_{1133} = C_{2233} = \lambda(1 - \epsilon a_1) \tag{10}$$

$$L = C_{1313} = C_{2323} = \mu(1 - \epsilon b_1)$$

$$N = C_{1212} = \mu$$

On a négligé les diffractions multiples entre fissures donc les termes d'ordres ϵ^2 et supérieurs. Les coefficients des termes en ϵ^2 qui apparaîtraient, d'une part en prenant l'inverse de (4), d'autre part en prenant en compte les diffractions entre deux fissures, étant d'ordre 1 ou inférieurs (on peut le prévoir d'après les calculs faits en [4] pour des sphères), la limite de validité des équations (6) et des modules élastiques linéaires (10) peut être estimée comme $\epsilon < 10^{-1}$. Cette linéarité des modules par rapport à ϵ permet de combiner de façon simple les tenseurs anisotropes relatifs à des distributions de fissures biplanaires ou triplanaires [16].

Par souci de symétrie et pour tenir compte de l'atténuation des ondes dans la description du milieu, il est possible aussi de généraliser les milieux isotropes presque élastiques de [18]. On introduit, pour la sollicitation sinusoïdale considérée, un milieu anisotrope presque-élastique équivalent : les 5 modules élastiques ont alors une forme complexe $M + iM^*$, avec $\Theta = M^*/M$ $(\ll 1)$, et l'atténuation est reliée à cet angle de phase Θ par $\alpha_L = k_i \Theta/2$. Les parties imaginaires des modules, désignées par des lettres étoilées, sont telles que :

$$A^* = A(k_1 a)^3 \frac{f_1}{3} (1 - 2\sigma^2)^2 \epsilon$$

$$C^* = C(k_1 a)^3 \frac{f_1}{3} \epsilon$$

$$F^* = F(k_1 a)^3 \frac{f_1}{3} (1 - 2\sigma^2) \epsilon \qquad\qquad (11)$$

$$L^* = L(k_1 a)^3 \frac{h_1}{3} \epsilon$$

$$N^* = 0$$

Toutefois, ce modèle n'a pas la maniabilité souhaitée puisque les angles de phase dépendent du nombre d'onde adimensionnel, et ne peuvent donc être considérés comme des grandeurs caractéristiques du milieu dans une gamme de fréquences données.

3.   Distribution axisymétrique de fissures

Sous l'effet de charges compressives croissantes, on sait que les fissures tendent à se développer parallèlement aux directions principales du tenseur des contraintes [8], tant que le mécanisme de rupture fragile est prédominant.

On considère ici un cas particulier de distribution de fissures qu'on peut s'attendre à voir apparaître dans des cylindres en compression axiale. Les fissures sont réparties uniformément dans les différents plans méridiens, de telle sorte que pour chacune d'elles un diamètre soit parallèle à l'axe z. On intègre alors les expressions des vitesses d'onde par rapport à l'angle que fait chaque fissure avec un méridien de référence, et on obtient un nouveau tenseur d'élasticité orthotrope de révolution de composantes

$$A = (\lambda + 2\mu)(1 - \epsilon a_r) \qquad N = \mu(1 - \epsilon b_r)$$

$$C = (\lambda + 2\mu)(1 - \epsilon a_2) \qquad L = \mu\left(1 - \epsilon \frac{b_1}{2}\right) \qquad (12)$$

$$F = \lambda - \epsilon(\lambda + 2\mu)\{a_1(1 - \sigma^2 + \sigma^4) - \sigma^2 b_r\}$$

$$a_r = a_1(1 - \sigma^2)^2 + \sigma^2 b_r$$

$$b_r = \frac{1}{2}(b_1 + \sigma^2 a_1) \qquad (13)$$

Les coefficients de ϵ dans les modules effectifs sont des fonctions de σ (ou du coefficient de Poisson ν). Leur variation dans l'intervalle $\sqrt{2} \leqslant \sigma^{-1} \leqslant \sqrt{6}$ ou $0 \leqslant \nu \leqslant 0,4$ est donnée figures 1a-1b en même temps que les variations de a_1, a_2, b_1 obtenus au chapitre 2. On trace aussi les courbes donnant les coefficients qu'on obtient pour des milieux contenant des fissures orientées dans toutes les directions et qui peuvent donc être remplacés par un milieu équivalent isotrope. On a, dans ce dernier cas, les deux modules élastiques habituels modifiés par des termes correctifs

$$(\lambda + 2\mu)' = (\lambda + 2\mu)\{1 - \epsilon a_m(\sigma)\}$$

$$\mu' = \mu\{1 - \epsilon b_m(\sigma)\} \qquad (14)$$

$$a_m(\sigma) = \frac{4}{\pi}\left\{ \frac{1 - \dfrac{8\sigma^2}{3} + \dfrac{32\sigma^4}{15}}{4\sigma^2(1 - \sigma^2)} + \frac{8\sigma^2}{15(3 - 2\sigma^2)} \right\} \qquad (15)$$

$$b_m(\sigma) = \frac{4}{\pi}\left\{ \frac{1}{15(1 - \sigma^2)} + \frac{2}{5(3 - 2\sigma^2)} \right\}$$

Des formules analogues aux équations (14), avec des coefficients $a_s(\sigma)$ et $b_s(\sigma)$ sont valables pour des milieux contenant n_0 cavités sphériques de rayon a par unité de volume, et dont la porosité effective ϵ est égale à la porosité réelle $(4\pi/3)\,n_0\,a^3$. On a :

$$a_s(\sigma) = \frac{3}{4\sigma^2} - 1 + \frac{20}{9 - 4\sigma^2}$$

$$b_s(\sigma) = \frac{15}{9 - 4\sigma^2}$$

(16)

On constate (Fig. 1) que l'effet des fissures planes circulaires est du même ordre de grandeur, à un coefficient de réduction près, que celui de cavités sphériques ayant même porosité effective.

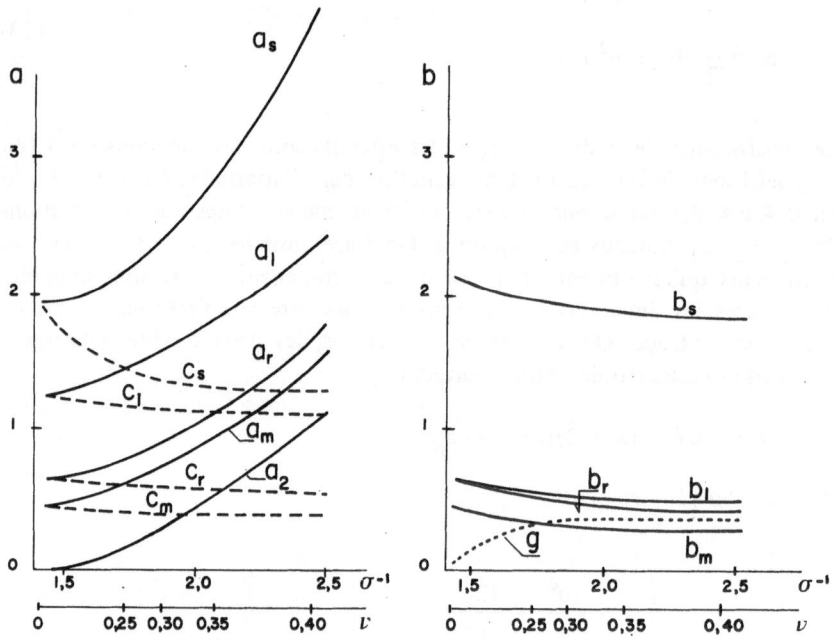

Figure 1. – *Coefficients de la porosité effective ϵ dans l'expression des modules effectifs d'un milieu élastique fissuré.* Fissures sèches : a_i pour les modules de compression, b_i pour les modules de cisaillement, c_i pour les modules d'Young, les indices i se rapportant aux diverses distributions de fissures considérées.

On peut admettre en première approximation, avec une erreur inférieure à 10 % que $a_m \cong (2/\pi)\,(a_s - 4/3)$.

Une autre façon de rendre compte du comportement des milieux fissurés est de déterminer le tenseur $M_{ijk\ell}$, inverse de $C_{ijk\ell}$ et qui donne les déformations à partir des contraintes. Si E est le module d'Young, ν étant le coefficient de Poisson, $G = E/2(1 + \nu) = \mu$ le module de torsion, les tenseurs orthotropes de révolution associés aux deux distributions de fissures considérées sont :

— pour des fissures parallèles

$$[M_{ijk\ell}] = \frac{1}{E}
\begin{bmatrix}
1 & -\nu & -\nu & & & \\
-\nu & 1 & -\nu & & O & \\
-\nu & -\nu & 1+\epsilon c_1 & & & \\
& & & 2(1+\nu)(1+\epsilon b_1) & 0 & 0 \\
& O & & 0 & 2(1+\nu)(1+\epsilon b_1) & 0 \\
& & & 0 & 0 & 2(1+\nu)
\end{bmatrix}
\qquad (17)$$

— pour des fissures réparties dans des plans méridiens

$$[M_{ijk\ell}] = \frac{1}{E}
\begin{bmatrix}
1+\epsilon c_r & -\nu + \epsilon \frac{c_1}{2} & -\nu & & & \\
-\nu + \epsilon \frac{c_1}{2} & 1+\epsilon c_r & -\nu & & O & \\
-\nu & -\nu & 1 & & & \\
& & & 2(1+\nu)\left(1+\epsilon \frac{b_1}{2}\right) & 0 & 0 \\
& O & & 0 & 2(1+\nu)\left(1+\epsilon \frac{b_1}{2}\right) & 0 \\
& & & 0 & 0 & 2(1+\nu)(1+\epsilon b_r)
\end{bmatrix}$$

$$(18)$$

où b_1, b_r sont définis comme en (8), (13) et

$$c_1 = \frac{3-4\sigma^2}{1-\sigma^2}\,\sigma^2 a_1 \qquad c_r = \frac{3-4\sigma^2}{1-\sigma^2}\left(\frac{3\sigma^2 a_1 + b_1}{4}\right) \qquad (19)$$

Les composantes $M_{ijk\ell}$ sont obtenues par inversion des relations d'élasticité anisotrope associées à $C_{ijk\ell}$. Elles coïncident avec le développement au premier ordre des calculs de [13]. On les complète par les valeurs c_m et c_s des coefficients de ϵ apparaissant dans le module d'Young effectif des milieux contenant respectivement des fissures circulaires réparties dans toutes les directions et des cavités sphériques de même porosité effective ϵ.

$$c_m = \frac{3-4\sigma^2}{1-\sigma^2}\left(\frac{\sigma^2 a_1}{5} + \frac{2b_1}{15}\right)$$

$$c_s = \frac{3}{4(1-\sigma^2)} - \frac{3-4\sigma^2}{1-\sigma^2}\left(\frac{b_s}{3}\right) \qquad (20)$$

4.   Cas des fissures elliptiques

La notion de porosité effective, introduite au chapitre 2 pour des fissures circulaires, peut être avantageusement étendue aux fissures elliptiques.

4.1. *Fissures elliptiques orientées dans toutes les directions [3]*

Le module de rigidité à la compression ne dépend pas explicitement du rapport des longueurs d'axes de l'ellipse b/a si on choisit un paramètre représentatif de la densité de fissures proportionnel à $n_0 A^2/P$, où A et P sont respectivement l'aire et le périmètre de l'ellipse. On peut définir pour des fissures elliptiques une porosité effective

$$\epsilon = \frac{8}{3} n_0 \frac{A^2}{P} \tag{21}$$

dont la définition de ϵ donnée pour les fissures circulaires est un cas particulier (ϵ diffère du paramètre choisi en [3] par un facteur $4\pi/3$). La valeur effective du module de rigidité à la compression $\bar{K}$ est alors reliée à sa valeur K dans la matrice élastique par :

$$\frac{\bar{K}}{K} = 1 - \frac{3 - 4\sigma^2}{3} \epsilon a_1 = 1 - \frac{3 - 4\sigma^2}{3} \epsilon \left(a_m - \frac{4}{3} \sigma^2 b_m \right) \tag{22}$$

Le module d'Young $\bar{E}$ et le module de cisaillement $\bar{G} (= \bar{\mu})$ sont donnés par

$$\frac{\bar{E}}{E} = 1 - \frac{3 - 4\sigma^2}{1 - \sigma^2} \epsilon \left(\frac{\sigma^2 a_1}{5} + \frac{t}{15} \right) \tag{23}$$

$$\frac{\bar{G}}{G} = 1 - \epsilon \left(\frac{4\sigma^2 a_1}{15} + \frac{t}{5} \right) \tag{24}$$

$t(b/a, \sigma^2)$ varie de $2 b_1$ pour un cercle à $(3 - 2\sigma^2)/\pi(1 - \sigma^2)$ pour des ellipses infiniment longues (b/a $\to$ 0), soit une variation relative maximum de 6 % dans l'intervalle de variation de σ considéré figure 1.

4.2. *Fissures elliptiques dans des plans parallèles*

Elles peuvent être obtenues comme un cas particulier des inclusions en forme d'ellipsoïdes étudiées en [6]. Quand on passe de la fissure circulaire à l'ellipse, les modules C_{1111} et C_{2222} deviennent différents, ainsi que C_{1313} et C_{2323}. Toutefois, cette différence est surtout appréciable (d'après les figures 1 et 2 de [6]) pour les valeurs de c/a suffisantes (supérieures à 0,1 pour b/a = 1/3, par exemple). Le paramètre de densité d'inclusions qui apparaît dans [6] est $n_0 v_c a^2/bc$, si a, b, c, sont les longueurs des demi-axes de l'ellipsoïde, ce qui revient à remplacer chaque inclusion par la sphère circonscrite. On constate alors, pour des fissures sèches d'épaisseur relative c/a = 0,01 et un coefficient de Poisson $\nu = 1/4$ (ou $\sigma^2 = 1/3$), quand b/a passe de 1 à 1/3, une chute du coefficient de ce paramètre de 1,49 à 0,23 dans le module C_{3333}. Mais cette différence n'est pas significative. En effet, il est préférable, pour les fissures, de choisir comme paramètre la généralisation de la porosité effective proposée au paragraphe 4-a. Cela revient à remplacer la fissure par un ellipsoïde de même ellipse de base, et d'une hauteur intermédiaire entre les axes a et b, soit $2A/P = \pi b/2E(k)$, où E(k) est l'intégrale elliptique complète de seconde espèce, et $k = (1 - (b^2/a^2))^{1/2}$. Le nouveau rapport d'aspect de la fissure est $\varphi = c/(\pi b/2E(k))$, et le coefficient de ϵ passe seulement à 1,47.

Cet exemple montre que la porosité effective permet aussi de décrire une distribution de fissures elliptiques parallèles, en réduisant à quelques pour cents l'influence du rapport b/a sur les termes donnant la variation des modules effectifs.

Par extension, avec la même définition de la porosité effective, on peut espérer avoir une bonne estimation des variations des modules effectifs pour toute distribution de fissures planes de forme convexe.

5. Fissures circulaires saturées

Les fissures considérées dans les chapitres 2 et 3 étaient sèches, c'est-à-dire remplies d'un gaz de compressibilité négligeable par rapport à celle de la matrice solide élastique K. Pour des fissures remplies de liquide non-visqueux, on distingue deux cas selon la compressibilité du liquide.

5.1. *Liquide incompressible*

Les conditions aux limites du type libre sur les lèvres sont remplacées par :
– la continuité des contraintes normales σ_{zz}.
– la nullité des contraintes tangentielles $\sigma_{zr} = \sigma_{z\Phi} = 0$.
Les résultats se déduisent des précédents en remplaçant a_1 par 0 [11] partout où il apparaît, le coefficient b_1 n'étant pas modifié. Des développements analogues donnent les nouvelles valeurs, affectées d'un tilde, des coefficients de ϵ respectivement pour des fissures réparties dans des plans méridiens et dans toutes les directions

$$\widetilde{a}_r = \frac{\sigma^2 b_1}{2} \qquad \widetilde{b}_r = \frac{b_1}{2} \qquad \widetilde{a}_m = \frac{8}{15}\sigma^2 b_1 \qquad \widetilde{b}_m = \frac{2}{5} b_1 \qquad (25)$$

Les valeurs de ces coefficients restent faibles (Fig. 2).

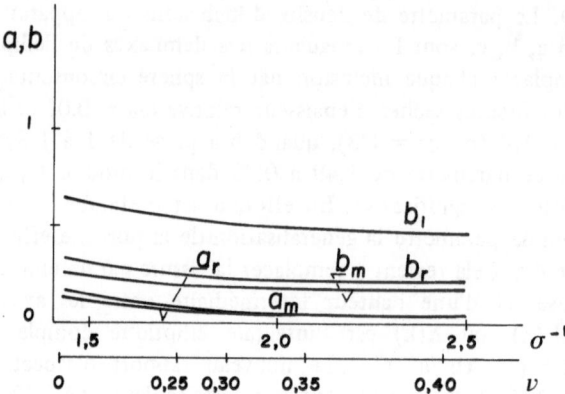

Figure 2. – *Coefficients a_i et b_i, définis Figure 1, pour des fissures remplies d'un liquide non visqueux incompressible.*

L'anisotropie induite par des fissures remplies d'un fluide incompressible modifie surtout les composantes du cisaillement, et est beaucoup moins intense que dans le cas de fissures sèches.

5.2. *Compressibilité intermédiaire*

En fait l'influence de la compressibilité K_f du liquide contenu dans une fissure est prise en compte par un paramètre caractéristique

$$w = \frac{a}{c} \frac{K_f}{K} \tag{26}$$

introduit en [3] par des considérations d'énrgie. Il apparaît comme terme prédominant, lorsque c/a tend vers 0, dans l'expression de l'amplitude diffractée par une inclusion sphéroïdale (avec $\mu_f = 0$) développée sur les fonctions d'onde sphériques pour des nombres d'onde adimensionnels petits [7, 16, 19].

Si $K_f/K \ll 1$, il peut très bien se faire que w soit $0(1)$ ou plus grand, si bien que la présence d'un liquide dans des fissures minces joue un rôle plus important que dans des cavités sphériques. On a alors une estimation des coefficients de ϵ dans les termes correctifs des modules en remplaçant a_1 et a_2 par $a_1 D$ et $a_2 D$ où

$$D = \left\{ 1 + \frac{4}{3\pi} \frac{3 - 4\sigma^2}{4\sigma^2 (1 - \sigma^2)} w \right\}^{-1} \tag{27}$$

Les cas des fissures sèches et de liquides incompressibles sont obtenus comme limites respectivement quand w tend vers 0 et l'infini, et encadrent les résultats intermédiaires. Un exemple des valeurs numériques obtenues pour un matériau élastique contenant des fissures remplies, soit d'un gaz de compressibilité K_g, soit d'un liquide de compressibilité K_ϱ, tels que $\lambda = \mu = 4\,K_\varrho = 4\,000\,K_g$, est donné dans la Table. Pour chaque valeur de c/a, la colonne de gauche correspond au gaz, celle de droite au liquide.

TABLE 1

	$\dfrac{c}{a} = 1$		$\dfrac{c}{a} = 0,1$		$\dfrac{c}{a} = 0,01$	
D	1,00	0,84	1,00	0,46	0,99	0,08
a_1	2,12	1,78	1,43	0,66	1,42	0,11
a_2	2,12	1,78	0,16	0,07	0,16	0,01
b_1	1,96	1,96	0,55	0,55	0,55	0,55
a_r	2,12	1,78	0,81	0,42	0,80	0,15
b_r	1,96	1,96	0,52	0,38	0,51	0,29
a_m	2,12	1,78	0,60	0,33	0,59	0,14
b_m	1,96	1,96	0,35	0,28	0,34	0,23
f_1	9,34	5,98	2,69	0,57	2,64	0,02
f_r	–	–	1,45	0,39	1,42	0,11
f_m	–	–	1.05	0,31	1,03	0,12
h_1	12,09	9,79	0,64	0,64	0,64	0,64

Les développements menés pour les modules élastiques peuvent être conduits de la même façon pour les atténuations. Des valeurs intermédiaires f_r, f_m, ..., sont aussi obtenues pour les coefficients d'atténuation dus à des distributions de fissures réparties dans des plans méridiens et dans toutes les directions. Pour les fissures saturées, les termes en f_1 sont multipliés par D^2, les termes en h_1 sont inchangés. On peut se reporter en [16] pour les détails des calculs.

6. Fissures de Griffith

Lorsque les deux dimensions caractéristiques d'une fissure convexe ne sont plus du même ordre, le problème devient à peu près bidimensionnel, pour des fissures parallèles, et on se ramène à une fissure infiniment longue dite

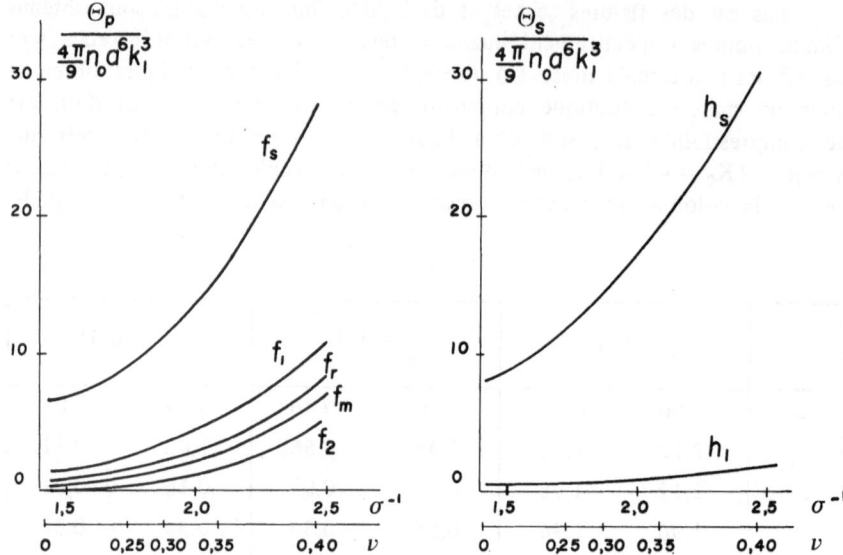

Figure 3. – *Angles de phase adimensionnels associés aux atténuations des ondes de compression (f$_i$) et de cisaillement (h$_i$) pour divers systèmes de fissures sèches.*

fissure de Griffith. Les modules effectifs calculés pour ce modèle le sont en général pour des fissures dont la direction varie dans le plan ; ils sont fonctions d'une porosité effective bidimensionnelle $(N/S) \pi b^2$ où b est la demi-longueur de fissure et (N/S) le nombre de fissures par unité de surface.

Si ces fissures rectangulaires infiniment longues sont réparties dans toutes les directions de l'espace, on peut leur associer une porosité tridimensionnelle effective définie comme au chapitre 4, et qui vaut $\epsilon = (32/3) \, n_0 ab^2$. Les résultats obtenus sont alors très voisins de ceux qu'on trouverait pour des fissures elliptiques infiniment longues de même rapport A^2/P [3], car les coefficients de ϵ sont dans un rapport $3 \pi^2/32$ très voisin de 1.

Mais lorsque ces fissures de Griffith se développent dans des directions privilégiées, l'anisotropie induite n'est pas la même que celle due à des fissures convexes.

La diffraction des ondes scalaires et des ondes de compression par une fissure de Griffith a été étudiée [1, 12]. Les expressions de l'amplitude diffractée à grande distance de la fissure et l'analogue bidimensionnel des relations (4) (5) [2] permettent d'obtenir le module de compression associé à une onde P incidente, et le module de cisaillement pour une onde SH d'amplitude portée par la grande dimension de la fissure. La section droite de diffraction Q calculée en [12] pour une onde P en incidence normale permet par ailleurs de donner l'atténuation de cette onde

$$\frac{\overline{\lambda + 2\mu}}{\lambda + 2\mu} = 1 - \frac{N}{S} \frac{\pi b^2}{\sigma^2 (1 - \sigma^2)} \{1 - 4\sigma^2(1 - \sigma^2) \sin^2\theta - 8\sigma^4 \sin^2\theta \cos^2\theta\} \quad (28)$$

$$\frac{\overline{\mu}_{SH}}{\mu} = 1 - \frac{N}{S} \pi b^2 \cos^2\theta \quad (29)$$

$$\alpha_P(\theta = 0) = \frac{N}{S} \pi k_1^3 b^4 \frac{(1 - \sigma^2)^2 + \dfrac{\sigma^4}{32}}{8\sigma^4(1 - \sigma^2)^2} \quad (30)$$

Le plan (x, z) étant normal à la grande dimension de la fissure, les coefficients d'élasticité C_{1111} et C_{3333} sont modifiés, les termes correctifs étant dans le rapport $(1 - 2\sigma^2)^2$ comme pour les fissures circulaires, mais C_{2222} reste inchangé comme C_{1212}. Introduisant la porosité effective tridimensionnelle dans les équations (28) et (29), on trouve les nouvelles valeurs des coefficients a_1 et b_1 : $\overline{a}_1 = 3\pi^2 a_1/16$, $\overline{b}_1 = 3\pi/16$. Il y a donc peu de changement pour les modules de cisaillement, la variation des modules de compression est multipliée par $(3\pi^2/16)$ pour A et C, et nulle pour $B = C_{2222}$. Ces différences s'estompent quand les fissures sont réparties dans toutes les directions, mais lorsque les fissures se développent dans des directions privilégiées, il reste important de savoir si le milieu sera décrit convenablement.

Le choix est encore plus crucial pour l'atténuation ; l'équation (30) montre qu'on retrouve entre fissures circulaires et fissures infiniment longues la différence qui existe entre l'atténuation par des sphères et par des cylindres circulaires, puisqu'on passe de la puissance quatrième du nombre d'onde adimensionnel au cube de ce nombre.

REFERENCES

[1] ANG D.D. and L. KNOPOFF. – "Diffraction of vector elastic waves by a finite crack". *Proceedings of the National Academy of Sciences,* 52 (1964): 1075-1081.
[2] BARRAT P.J. and W.D. COLLINS. – "The scattering cross section of an obstacle in an elastic solid for plane harmonic waves". *Proceedings of the Cambridge Philosophical Society,* 61 (1965): 969-981.
[3] BUDIANSKY B. and R.J. O'CONNELL. – "Elastic moduli of a cracked solid". *International Journal of Solids and Structures,* 12 (1976): 81-97.
[4] CHATTERJEE A.K., A.K. MAL and L. KNOPOFF. – "Elastic moduli of two-component systems". *Journal of Geophysical Research,* 83 (1978): 1785-1792.
[5] DATTA S.K. – "Propagation of SH-waves through a fiber-reinforced composite-elliptic cylindrical fibers". *Journal of Applied Mechanics,* 42 (1975): 165-170.
[6] DATTA S.K. – "A self-consistent approach to multiple scattering by elastic ellipsoidal inclusions". *Journal of Applied Mechanics,* 44 (1977): 657-662.

[7] DATTA S.K. – "Diffraction of plane elastic waves by ellipsoidal inclusions". *Journal of the Acoustical Society of America*, 61 (1977): 1432-1437.

[8] DRAGON A. and Z. MROZ. – "A continuum model for plastic-brittle behaviour of rock and concrete". *International Journal of Engineering Science*, 17 (1979): 121-137.

[9] GARBIN H.D. and L. KNOPOFF. – "The compressional modulus of a material permeated by a random distribution of circular cracks". *Quarterly of Applied Mathematics*, 30 (1973): 453-464.

[10] GARBIN H.D. and L. KNOPOFF. – "The shear modulus of a material permeated by a random distribution of free circular cracks". *Quarterly of Applied Mathematics*, 33 (1975): 296-300.

[11] GARBIN H.D. and L. KNOPOFF. – "Elastic moduli of a medium with liquid-filled cracks". *Quarterly of Applied Mathematics*, 33 (1975): 301-303.

[12] HARUMI K. – "Scattering of plane waves by a cavity ribbon in a solid". *Journal of Applied Physics*, 33 (1962): 3588-3593.

[13] HOENIG A. – "Elastic moduli of a non-randomly cracked body". *International Journal of Solids and Structures*, 15 (1979): 137-154.

[14] PIAU M. – "Attenuation of a plane compressional wave by a random distribution of thin circular cracks". *International Journal of Engineering Science*, 17 (1979): 151-167.

[15] PIAU M. – "Crack scattering in polycrystalline media". *Letters in Applied and Engineering Science*, 16 (1978): 565-570.

[16] PIAU M. – "Crack-induced anisotropy and scattering in stressed rocks : effective elastic moduli and attenuation". *International Journal of Engineering Science*, 18 (1980): 549-568.

[17] WALSH J.B. – "The effect of cracks on the uniaxial compression of rocks". *Journal of Geophysical Research*, 70 (1965): 399-411.

[18] WHITE J.E. – *Seismic Waves: Radiation, Transmission and Attenuation*. International Series in the Earth Sciences: Mac Graw-Hill Book Company, 1965.

[19] WU T.T. – "The effect of inclusion shape on the elastic moduli of a two-phase material". *International Journal of Solids and Structures*, 2 (1966): 1-8.

ABSTRACT

An effective anisotropic elasticity tensor is proposed to describe the near-elastic dynamic response of cracked stressed rocks. For loosely distributed thin circular cracks, a relationship is developed between two crack distributions: parallel and axisymetrical, and the associated effective transversely isotropic elasticity tensor; the corresponding compressional and shear wave attenuations are determined for small adimensional wave numbers. An effective porosity is defined, which makes the results useful for elliptical and even for arbitrary convex plane crack shapes. Saturated cracks, filled with non viscous liquids, and bidimensional or Griffith cracks are also discussed.

Electroacoustic Waves in Elastic Ferroelectrics

G.A. Maugin and J. Pouget

Université Paris VI, Paris, France.

1. Introduction

Anisotropy is the realm of crystalline structures. This is even more so in structures which exhibit electro-mechanical couplings such as piezoelectricity in ferroelectrics. The purpose of this work is to establish first a complete, exact, nonlinear theory of elastic ferroelectrics, and then to determine the essential features of harmonic wave propagation in such structures.

Ferroelectricity is characterized by the existence of a large static electrical polarization even in the absence of an applied electric field. The presence of this initial polarization field induces an anisotroqy in the elastic, electric, and coupled properties of the material in the same manner as an initial stress induces an anisotropy in the optical behavior of certain transparent materials in photoelasticity. To account for this anisotropy effect, the generalized 'kinematics' of the problem is envisaged in three steps. First, an elastically released, polarization-free configuration is considered as a reference configuration K_R. The material which is supposed to be nonlinear elastic and a nonlinear dielectric (because of ferroelectricity), is assumed to be isotropic with respect to.this ideally defined (not necessarily existing) configuration. Then a static, rigid-body initial configuration K_i is considered, which corresponds to the real initial state of the ferroelectric (locally nonvanishing finite, electrically polarized). Finally, corresponding to the current configuration K_t, small disturbances (in view of the applications to signal processing) in strains and electromagnetic fields are superimposed. As a result of the introduction of the initial configuration, the propagation of these disturbances takes place in an anisotropic body (the ideal symmetry has been broken), whose symmetry group is that of transverse isotropy about the locally nonvanishing initial electrical polarization. Because of the fully dynamical features of the treatment (albeit only Galilean invariant), the approach allows us to exhibit all branches, coupled or uncoupled, of elementary oscillations such as quasi-electromagnetic branches, elastic branches which are converted in polariton branches at certain critical wavenumbers, etc. A birefringence of transverse polaritons and an acoustic Faraday effect (rotation of the polarization plane) are placed in evidence. The allied dissipative processes, viscosity and dielectric relaxation, are introduced, and their effects studied, along the same lines of reasoning. The practical results are illustrated with several qualitative and quantitative dispersion curves (in ferroelectric BaTiO$_3$). This research leans heavily on previous investigations [1] which led to a theory of dielectrics more intricate than Voigt's theory. Strong similarities are offered to the problem of coupled magnetoelastic waves in elastic ferromagnets [5]. The same methodology can be applied to more complicated cases such as those of elastic antiferroelectrics and ferrielectrics.

2. General equations

Following a now well-established method, it can be shown (cf. [1, 2]) that all field equations and thermodynamical equations, apart from Maxwell's equations, which are needed in this work can be deduced from three basic principles written in global form for a material body occupying the volume D_t with boundary ∂D_t, equiped with the unit outward normal n, and closure $\bar{D}_t$ in the configuration K_t. We have:

a) The principle of virtual power expressed by

$$P^*_{(a)}(D_t) = P^*_{(i)}(D_t) + P^*_{(v)}(D_t) + P^*_{(c)}(\partial D_t) ; \qquad (2.1)$$

b) The first principle of thermodynamics

$$\dot{K}(D_t) + \dot{E}(D_t) + \dot{U}^{em}(D_t) = P_{(d)}(\bar{D}_t) + \dot{Q}_h(\bar{D}_t) ; \qquad (2.2)$$

c) The second principle of thermodynamics

$$\dot{N}(D_t) - \mathfrak{N}(\bar{D}_t) \geqslant 0 . \qquad (2.3)$$

Here

$$P^*_{(i)}(D_t) = - \int_{D_t} p^*_{(i)} \, dv, \, p^*_{(i)} = \sigma_{ij} D^*_{ij} - \rho \, {}^{L}E \cdot \hat{\pi}^* + {}^{L}E_{ij} \hat{\Pi}^*_{ij} , \qquad (2.4)$$

$$P^*_{(a)}(D_t) = \int_{D_t} \rho \, [\dot{U} \cdot U^* + d\ddot{\pi} \cdot (\dot{\pi})^*] \, dv, \qquad (2.5)$$

$$P^*_{(v)}(D_t) = \int_{D_t} [(f + f^{em}) \cdot U^* + \rho \, \mathcal{E} \cdot (\dot{\pi})^*] \, dv, \qquad (2.6)$$

$$P^*_{(c)}(\partial D_t) = \int_{\partial D_t} [(T + T^{em}) \cdot U^* + \rho \, \tilde{P} \cdot (\dot{\pi})^*] \, da, \qquad (2.7)$$

$$K(D_t) = \int_{D_t} \frac{1}{2} \rho (U^2 + d\dot{\pi}^2)\, dv,$$

$$E(D_t) = \int_{D_t} \rho\, e\, dv, \quad N(D_t) = \int_{D_t} \rho\, \eta\, dv, \tag{2.8}$$

$$U^{em}(D_t) = \int_{D_t} \frac{1}{2} (E^2 + B^2)\, dv, \tag{2.9}$$

$$P_{(d)}(\overline{D}_t) = \int_{D_t} f \cdot U\, dv + \int_{\partial D_t} (T \cdot U + \rho\, \widetilde{P} \cdot \dot{\pi})\, da, \tag{2.10}$$

$$\dot{Q}_h(\overline{D}_t) \equiv \int_{D_t} \rho h\, dv - \int_{\partial D_t} q \cdot n\, da,$$

$$\mathfrak{N}(\overline{D}_t) = \int_{D_t} \rho\, \sigma\, dv - \int_{\partial D_t} \Phi \cdot n\, da, \tag{2.11}$$

and

$$\left.\begin{aligned}
& U_i \equiv \partial \mathfrak{X}_i / \partial t \,\big|_{X_K}, \quad x_i = \mathfrak{X}_i(X_K, t) : K_R \xrightarrow{\ t\ } K_t, \\[4pt]
& D_{ij} \equiv U_{(i,j)}, \quad \Omega_{ij} \equiv U_{[i,j]}, \quad (\dot{\ }) \equiv d/dt\,(-) \equiv \partial/\partial t + U \cdot \nabla, \\[4pt]
& \hat{\pi}_i \equiv \dot{\pi}_i - \Omega_{ij}\, \pi_j, \quad \hat{\Pi}_{ij} \equiv (\dot{\pi}_i)_{,j} - \Omega_{ik}\, \pi_{k,j},
\end{aligned}\right\} \tag{2.12}$$

$$f^{em} = \frac{1}{c} P^* \times B + (P \cdot \nabla)\, \mathcal{E}, \quad \mathcal{E} \equiv E + \frac{1}{c} U \times B, \tag{2.13}$$

$$P^* = P - (P \cdot \nabla) U + P(\nabla \cdot U), \quad \pi \equiv P/\rho.$$

In these equations the right upper asterisks in (2.1) and (2.4)-(2.7) indicate virtual fields. $P_{(a)}$ is the total power of acceleration or inertia forces (d is the so-called polarization inertia). $P_{(i)}$ is the total power of internal forces, written as a linear form on a set of generalized objective velocities. $\sigma_{ij} = \sigma_{ji}$ is the intrinsic stress tensor, ^LE is the local electric field, and $^LE_{ij}$ accounts for polarization gradients. $P_{(v)}$ is the total power of external volume forces, and $P_{(c)}$ the total power of cohesion, or surface, forces. In this formulation, Maxwellian electromagnetic fields are assumed to contribute to external forces of the volume and surface types (this is a deliberate choice). f^{em} is the form taken by the volume ponderomotive force in nonmagnetizable, nonrelativistically moving, dielectrics; T^{em} is the associated surface ponderomotive force ; cf. [3]. However, the total power of prescribed volume and surface forces, $P_{(d)}$, does not include the effects of Maxwellian electromagnetic fields. $\widetilde{P}$ is a prescribed surface polarization density. $E, B, D, H, P = D - E$ are the usual electromagnetic fields in a fixed (laboratory) Galilean frame. $\mathcal{E}, \mathcal{B}, \mathcal{D}, \mathcal{H}, \mathcal{P}$ are the same fields in a frame moving with the matter element. For a nonmagnetizable dielectric we have the energetic identity [3]

$$\dot{U}^{em}(D_t) = -\int_{D_t} (f^{em} \cdot U + \rho \mathcal{E} \cdot \dot{\pi}) \, dv$$

$$-\int_{\partial D_t} [c(\mathcal{E} \times \mathcal{H}) \cdot n + T^{em} \cdot U] \, da. \tag{2.14}$$

Furthermore, if θ is the thermodynamical temperature and we assume that the entropy source σ and the entropy flux Φ are such that

$$\sigma = h/\theta, \quad \Phi = \theta^{-1}(q - c \mathcal{E} \times \mathcal{H}) \equiv \theta^{-1} \hat{q} \tag{2.15}$$

then eqs. (2.1)-(2.3) yield the following results :

$$\rho \dot{U} = \text{div } t + f + f^{em} \text{ in } D_t, \quad T_i + T_i^{em} = t_{ij} n_j \text{ on } D_t, \tag{2.16}$$

$$\mathcal{E}_i + {}^L E_i + \rho^{-1} \, {}^L E_{ij,j} = d\dot{\pi}_i \text{ in } D_t, \quad \rho^{-1} \, {}^L E_{ij} n_j = \widetilde{P}_i \text{ on } D_t \tag{2.17}$$

with

$$t_{ij} = \sigma_{ij} + t_{[ij]}^{INT}, \quad t_{ij}^{INT} \equiv \rho \, {}^L E_i \pi_j - {}^L E_{ip} \pi_{j,p} \tag{2.18}$$

and, on account of (2.14) and (2.1) written for real fields,

$$\dot{E}(D_t) + P_{(i)}(D_t) = \dot{Q}_h(\overline{D}_t) + \dot{Q}^{em}(D_t), \quad \dot{Q}^{em}(D_t) \equiv \int_{D_t} (c\mathcal{E} \times \mathcal{H}) \, da \tag{2.19}$$

or, locally,

$$\rho \dot{e} = p_{(i)} - \nabla \cdot \hat{q} + \rho h \tag{2.20}$$

while, with $\psi \equiv e - \theta\eta$ and on account of (2.15), (2.3) one has the Clausius-Duhem inequality

$$-\rho(\dot{\psi} + \eta\dot{\theta}) + p_{(i)} - \theta^{-1} \hat{q} \cdot \nabla\theta \geq 0. \tag{2.21}$$

In addition we recall the continuity equation and the form taken by Maxwell equations in nonmagnetizable dielectrics (in D_t) :

$$\dot{\rho} + \rho \nabla \cdot U = 0 \tag{2.22}$$

and

$$\nabla \times \mathbf{E} + \frac{1}{c} \, (\partial \mathbf{B}/\partial t) = \mathbf{0}, \nabla \cdot \mathbf{B} = 0,$$

$$(2.23)$$

$$\nabla \times \mathbf{H} - \frac{1}{c} (\partial \mathbf{D}/\partial t) = \mathbf{0}, \, \nabla \cdot \mathbf{D} = 0, \, \mathbf{D} = \mathbf{E} + \mathbf{P}, \, \mathbf{H} = \mathbf{B} + \frac{1}{c} \, \mathbf{U} \times \mathbf{P}.$$

The local balance of moment of momentum is built in eq. (2.18). It proves convenient to introduce a symmetric stress tensor $^{E}t_{ij}$ and a convected-time derivative $\hat{\pi}_{ij}$ such that

$$^{E}t_{ij} \equiv \sigma_{ij} - \rho \, ^{L}E_{(i}\pi_{j)} + \, ^{L}E_{(i|k|}\pi_{j),k},$$

$$\hat{\pi}_{ij} \equiv (\pi_{i,j})^* = (\dot{\overline{\pi_{i,j}}}) + \pi_{i,k} \, U_{k,j} - U_{i,k}\pi_{k,j}.$$

$$(2.24)$$

Then

$$t_{ij} = \, ^{E}t_{ij} + t_{ij}^{INT}, \, t_{[ij]} = t_{[ij]}^{INT},$$

$$(2.25)$$

and (2.21) reads

$$- J^{-1} \, (\dot{\Sigma} + \rho_0 \, \eta\dot{\theta}) + \, ^{E}t_{ij} \, D_{ij} - \, ^{L}\mathbf{E} \cdot \mathbf{P}^* + \, ^{L}E_{ij}\hat{\pi}_{ij} - \theta^{-1}\hat{q} \cdot \nabla\theta \geqslant 0,$$

$$(2.26)$$

$$\Sigma \equiv \rho_0 \psi, \, \rho_0 = \rho(K_R), \, J = \rho_0/\rho.$$

$$(2.27)$$

Following a routine procedure, one shows that the thermodynamically reversible contribution to the 'elastic' stress tensor ^{E}t, the local electric field, the tensor ^{L}E and the entropy density η are derived from the frame-independent energy per unit of undeformed volume [4],

$$\left.\begin{array}{l} \Sigma = \hat{\Sigma} \, (E_{KL}, \Pi_K, \Pi_{KL}, \theta), \\[2mm] E_{KL} \equiv \frac{1}{2} \, (\mathscr{X}_{i,K}\mathscr{X}_{i,L} - \delta_{KL}), \, \Pi_K \equiv JX_{K,i}P_i, \, \Pi_{KL} \equiv X_{K,i}\pi_{i,L}, \end{array}\right\} \quad (2.28)$$

while the dissipative parts in ^{E}t and ^{L}E, and $\hat{q}$, are provided by complementary laws which are linear forms in D_{ij}, P_i^* and $\theta_{,j}$, respectively (no dissipative process being associated with the tensor ^{L}E, [4]). We note that, as a consequence of the decomposition (2.25), both viscosity and dielectric-relaxation processes contribute to the dissipative component of the Cauchy stress tensor.

3.   Linearization and symmetry breaking

Let us assume that there exists a rigid-body, static, uniformly polarized solution of eqs. (2.16)-(2.17) and (2.22)-(2.23) such that

$$S_0 = \{x_i = \delta_{iK} X_K, {}_0P \neq 0, {}_0B \neq 0, {}_0\rho, \theta_0 \ll \theta_C = \text{Ferroelectric}$$
$$\text{Curie temperature}$$

(3.1)

It can be shown that such a solution exists and is stable if the body has an ellipsoidal shape (or a degenerate form of this) and if there are applied a spatially uniform magnetic field and electric fields and adequate mechanical surface loadings on ∂D_{t_0} (with $f = 0$ in D_{t_0}). The solution (3.1) to which there correspond initial fields $({}_0^E t, {}_0^L E, {}_0^L E)$ may be considered, via the constitutive equations as defining a one-domain ferroelectric state corresponding to an initial configuration $K_i \neq K_R$. Let us pull back to K_R the equations (2.16)-(2.17) and (2.22)-(2.23), effecting a socalled Lagrangian variation of these equations about K_i. If $(u, p, e, b, \tilde{\theta} \equiv \theta - \theta_0)$ is the set of small dynamical fields which describe the dynamical solution S_t^p of the pulled back equations which remains in the neighborhood of S_0, we obtain the following perturbation equations:

$$_0\rho \ddot{u} = \text{div } \tau + \frac{1}{c} ({}_0\rho \dot{p} \times {}_0B) + ({}_0P \cdot \nabla) e , \quad \left[(\dot{-}) \equiv \frac{\partial}{\partial t} (-) \right] , \quad (3.2)$$

$$e_i + \frac{1}{c} (\dot{u} \times {}_0B)_i + {}^L e_i + {}_0\rho^{-1} \, {}^L e_{ij,j} = d\ddot{p}_i , \quad (3.3)$$

$$\nabla \times e - \frac{1}{c} (\dot{e} + {}_0\rho \dot{p}) = \frac{1}{c} ({}_0P \cdot \nabla) \dot{u} , \quad \nabla \times e + \frac{1}{c} \dot{b} = 0 , \quad (3.4)$$

$$\nabla \cdot b = 0 , \quad \nabla \cdot (e + {}_0\rho p) = \nabla \cdot [({}_0P \cdot \nabla) u] ,$$

where

$$\tau_{ij} = {}_0C_{ijpq} u_{p,q} + {}_0\epsilon_{ijk} ({}_0\rho p_k - {}_0P_m u_{k,m}) + {}_0^E t_{(i|k|} u_{j),k} + {}^L e_i \, {}_0P_j$$
$$+ {}_0^L E_i ({}_0\rho p_j - {}_0P_k u_{j,k}) - {}_0^L E_{ip} p_{j,p} + {}_0\nu_{ij} \tilde{\theta} + {}_0\eta_{ijpq} \dot{u}_{p,q}$$
$$- {}_0C_{ik} ({}_0\rho \dot{p}_k - {}_0P_m \dot{u}_{k,m}) {}_0P_j ,$$

$$^L e_i = - {}_0^L E_j u_{j,i} - {}_0\epsilon_{pqi} u_{p,q} - {}_0\chi_{ij}^{-1} ({}_0\rho p_j - {}_0P_m u_{j,m}) + {}_0a_i \tilde{\theta}$$
$$- {}_0C_{ij} ({}_0\rho \dot{p}_j - {}_0P_m \dot{u}_{j,m}) , \quad (3.5)$$

$$^L e_{ij} = -\,^L_0 E_{m\,j}\, u_{m,i} + {}_0\delta_{ijpq}\, P_{q,p}\,,$$

$$\hat{q}_i = -\,K_{ij}\,\widetilde{\theta}_{,j}\,.$$

All tensor coefficients introduced possess obvious symmetries and, apart from $_0\eta_{ijpq}$ and $_0C_{ij}$, are defined as second-order partial derivatives of Σ taken at S_0, [4]. If the body behaves isotropically with respect to K_R, then Σ depends on 36 scalar invariants built from the arguments of Σ, and it is shown that the elasticity tensor $_0C_{ijpq}$, the piezoelectricity tensor $_0\epsilon_{ijk}$ the initial stress $^E_0 t_{ij}$, the initial fields $^L_0 E_i$ and $^L_0 E_{ij}$, the reciprocal electrical susceptibility $_0\chi_{ij}^{-1}$, the pyroelectricity vector $_0 a_i$ and the thermoelasticity tensor $_0\nu_{ij}$, which are defined by second-order derivatives of Σ with respect to the invariants possess hexagonal symmetry (i.e., transverse isotropy with respect to $_0\mathbf{P}$) so that the model obtained applies to the treatment of materials of the type of $BaTiO_3$ in their ferroelectric phase. The initial polarization $_0\mathbf{P}$ has broken the ideal symmetry (isotropy) or, in other words, we witness a phenomenon of anisotropy inducement by an initial field (a phenomenon similar to what occurs in photoelasticity and, more generally, in crystal optics). Let us introduce the following notational device: we shall denote by $L[\mathbf{A},\ldots,\mathbf{C}\,|\mathbf{D}]$ a linear operator in $\mathbf{A},\ldots,\mathbf{C}$ with parameters $\mathbf{D}$. Then eqs. (3.2), (3.3) and (3.4) can be symbolically written as

$$\ddot{\mathbf{u}} = L_{(u)}[\nabla\nabla\mathbf{u}\,,\nabla p,\,\nabla\nabla\underline{p},\;\dot{\underline{p}},\;\nabla\underline{e}\;;\;\nabla\nabla\dot{\underline{u}},\,\nabla\dot{\underline{p}}\,|\,S_0]\,, \tag{3.6}$$

$$\ddot{\mathbf{p}} = L_{(p)}[\nabla\mathbf{u},\,\dot{\mathbf{u}},\,p,\,\nabla\nabla p,\;\underline{e}\,,\;\nabla\nabla\underline{u};\;\nabla\dot{\underline{u}},\,\underline{\dot{p}}\,|\,S_0]\,, \tag{3.7}$$

$$\ddot{\mathbf{e}} = L_{(e)}[\nabla\nabla\nabla\mathbf{u},\,\nabla\ddot{\mathbf{u}},\,\nabla\nabla p,\,\ddot{\mathbf{p}},\,\nabla\nabla e\,|\,S_0] \tag{3.8}$$

for isothermal processes ($\widetilde{\theta} = 0$). Equation (3.8) follows from eqs. $(3.4)_{1,2,4}$ which combine to yield the equations ($\mathrm{tr} = \mathrm{trace}$)

$$\square e = -\,\square\cdot({}_0\rho\,\mathbf{p} - ({}_0\mathbf{P}\cdot\nabla)\mathbf{u})\,,$$

$$(\square)_{ij} \equiv \delta_{ij}\frac{1}{c^2}\frac{\partial^2}{\partial t^2} - \nabla_i\nabla_j,\quad \square = \mathrm{tr}\,\square - \frac{2}{c^2}\frac{\partial^2}{\partial t^2} \tag{3.9}$$

where $\square$ is the usual d'Alembertian (in rectangular coordinates); the remaining equation $\nabla\cdot\mathbf{b} = 0$ indicates the transversality of electromagnetic perturbations. Equations (3.6)-(3.8) exhibit the different wave modes that can occur in the model (elastic modes, polarization, or polariton, modes and electromagnetic modes). The nonhomogeneity of the polynomials of differentiation in these equations shows that most wave modes will be dispersive, the variables to the right of semi-colons indicating damping. The linear contributions in $\dot{\mathbf{p}}$ and $\dot{\mathbf{u}}$

in eqs. (3.6) and (3.7), respectively, correspond to a so-called magnetoelastic dragging. This effect disappears with $_0B = 0$. Henceforth we consider a solution S_t^p such that $_0B = 0$ and $\tilde{\theta} = 0$.

4. Wave-propagation modes

We consider plane waves of the form

$$(u, p, e) = (\tilde{u}, \tilde{p}, \tilde{e}) \exp \left[i\omega \left(t - \frac{n}{c} s \cdot r \right) \right], \tag{4.1}$$

where ω is the circular frequency, s is a unit vector in the direction of propagation, n is the refractive index, $k = \omega n/c$ is the wave number, and $q = ck/\omega$ is the dimensionless reciprocal phase velocity. The wave length $\lambda = k^{-1}$ is such that $\lambda \ll L$, where L is a characteristic linear dimension of the body, so that in fact, neglecting boundary effects, we consider bulk waves and study their various propagation modes and their dispersion and damping. The qualitative study of the dispersion in absence of damping (ω and k real) is made in terms of four small dimensionless parameters (c = velocity of light in vacuum)

$$\epsilon_L \equiv c_L^2/c^2, \quad \epsilon_T = c_T^2/c^2, \quad \epsilon_P = {}_0P^2/{}_0\rho c^2, \tag{4.2}$$

and ϵ_{GP} which give a measure of, respectively, the importance of longitudinal and transverse elastic waves (when these two concepts can be defined) as compared to optical phenomena, the importance of the initial polarization (typical ferroelectric effect), and the influence of polarization gradients (another ferroelectric effect). These parameters are of the order of $10^{-11} - 10^{-10}$ (for the first two), $10^{-12} - 10^{-11}$ and $10^{-22} - 10^{-23}$, respectively. The general study of wave propagation in a direction at any angle θ_k to $_0P$ and the complete discussion of the special cases $\theta_k = 0$ and $\theta_k = \pi/2$ are to be found in the original works [4]. Here we simply present the most relevant qualitative and quantitative results for an orthogonal setting of the bias field, i.e., when $_0P \cdot s = 0$ or $\theta_k = \pi/2$. First, consider the effects of $_0P$ on elastic wave propagation. The first effect is a change in the slope of the otherwise nondispersive elastic branch $\omega_E(k) = c_E k$, which is initially replaced by $\omega_E(k) = \hat{c}_E k$ with $\hat{c}_E^2 = c_E^2[1 + (\epsilon_P/\epsilon_E)]$, where $\epsilon_E = c_E^2/c^2$. The second effect is to render that branch slightly dispersive, i.e., to yield a frequency spectrum of the type $\omega_E(k) = \hat{c}_E k + \epsilon_P f(k)$, where f(k) is a slowly varying function of k such that f(k = 0) = 0 and $f(k \to \infty) \to 0$. A third effect results from the existence of a new type of elementary excitations provided by eq. (3.7) and the coupling with it. In absence of couplings with the displacement field and the electromagnetic field, (3.7) yields two dispersive branches (of the hyperbolic type), a so-called high optical branch and a low optical branch, or solft mode of polaritons. This soft mode couples with the electromagnetic wave to give rise to an optical splitting and a dispersion of the electromagnetic wave. Finally, the same soft mode couples with an elastic-wave branch giving rise, at the point of acoustic-polariton resonance (ω_0, k_0), to a repulsion of the branches, a phenomenon arising from piezoelectric couplings and which, in spite of the smallness of the parameters involved, has a drastic effect. This effect is quite similar to the magnon-phonon coupling arising in elastic ferromagnets (See, e.g., [5]). The situation is illustrated in Figure 1 for the case $_0\mathbf{P} \cdot \mathbf{s} = 0$ in absence of damping. Curves are drawn in dimensionless units, the frequencies being rendered dimensionless by Ω^*, the starting frequency of the quasi-electromagnetic branch, and wavenumbers by $q = ck/\Omega^*$, i.e., all phenomena are essentially normalized as functions of the optical ones. The branch I of transverse phonons (elastic excitations) is slightly dispersive but otherwise uncoupled. Branch II is an optical longitudinal branch. Branch V is a quasi-electromagnetic branch which couples (band splitting) with the acoustic-polariton branch III. The latter starts, with increasing q, as a quasi-electromagnetic branch, then becomes a polariton branch to end, after passing through the cross-over region defined by the resonance (Ω_0, q_0), as a longitudinal acoustic branch. Finally, branch IV starts as a longitudinal acoustic branch for small q's and becomes, after passing through the cross-over region (Ω_0, q_0) a pure polariton branch. The latter branch, as well as branch II, is practically level in the macroscopic region of the spectrum, so that the group velocity $V_g = (\partial\omega/\partial k)_{IV}$ diminishes continuously throughout the cross-over region with increasing k, and practically tends toward zero, an effect which has been detected in ferroelectrics by means of the Raman and Brillouin spectroscopies, cf. [6]. Similarly, the slope of the asymptote to branch V is very steep, so that in fact we can replace Figure 1 by the schematic Figure 2.

In Figure 3 there is shown a magnified view of the repulsion region, computed numerically for ferroelectric $BaTiO_3$ ($\theta_0 = 52°C$). In the case of propagation at an angle $0 < \theta_k < \pi/2$ to $_0\mathbf{P}$, the low polariton branch may couple with both longitudinal and transverse acoustic branches, so that there are two cross-over regions (Ω_0, q_0) and $(\Omega \simeq \Omega_0, q_1)$, cf. Figure 4. This demonstrates the possibility, at static spatially weakly nonuniform bias electric field, to convert phonons into polaritons, and at slowly time-varying, spatially uniform, bias electric field, to convert polaritons into phonons. It has been verified that for the values of material coefficients at our disposal, Ω_0 is situated in the hypersound region and that q_0 corresponds practically to the middle of the Brillouin zone of lattice theory, so that the phenomenological approach considered in this work is feasible.

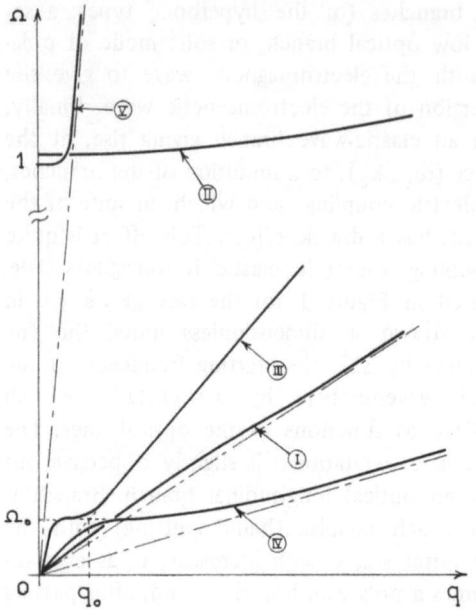

Figure 1

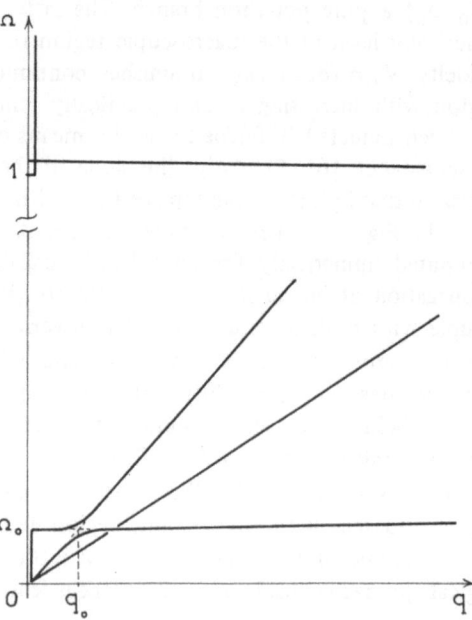

Figure 2

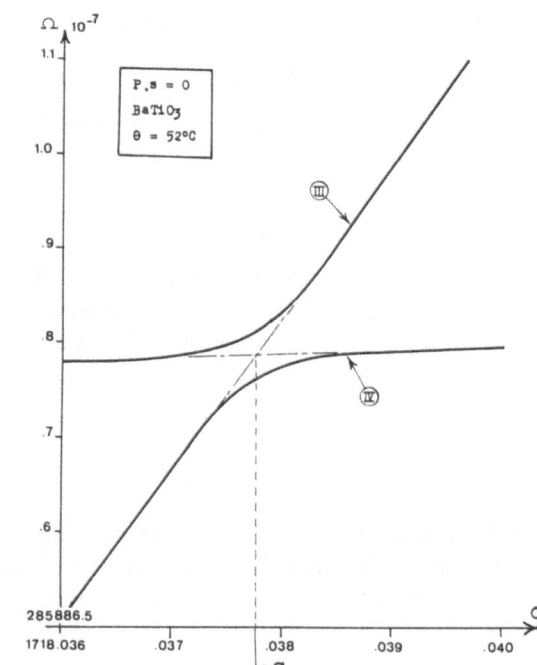

Figure 3

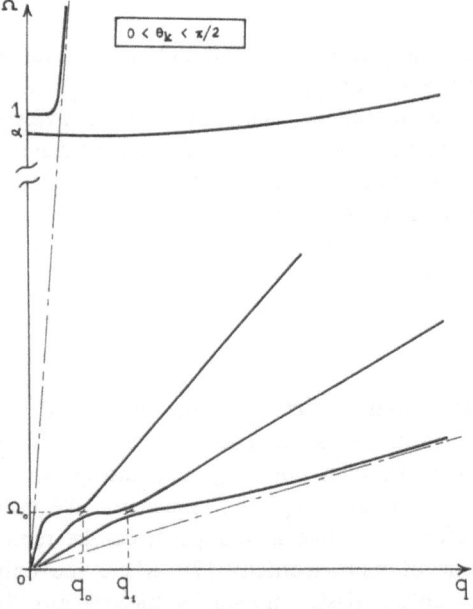

Figure 4

5. Birefringence and acoustical activity

Transverse polaritons present a phenomenon of birefringence since there are two polariton modes with refractive indices n_I and n_{II}, polarized along $_0P$ and $s \times {_0P}$, respectively (for $\theta_k = \pi/2$). It can be shown that

$$\Delta n = n_I - n_{II} = [B_1 + B_2 + B_3(q)] {_0P^2} , \tag{5.1}$$

where B_1 accounts for electric susceptibility, B_2, which is negligible, accounts for polarization gradients, and B_3, accounting for piezoelectric couplings, blows up at the resonance point. As to transverse acoustic waves, they are shown to possess the mechanism of acoustical activity, such that the two phase velocities satisfy the relation

$$v_I^{-2} - v_{II}^{-2} = [A_1 + A_2(q)] {_0P^2} , \tag{5.2}$$

where A_1 accounts for elastic anisotropy (due to $_0\underline{P}$) and A_2 includes a resonance effect. The latter, however, cannot be realistically reached, as the frequencies involved are far outside the acoustic range. The exact expressions of the coefficients introduced in eqs. (5.1) and (5.2) can be found in the original works [4].

6. Damping

The tensorial coefficients $_0\eta_{ijpq}$ and $_0C_{ij}$ of viscosity and dielectric relaxation can now a forciori be taken with the uniaxial symmetry induced by $_0P$. With ω now complex in eq. (4.1), the corresponding analytical study is involved and yields a (reciprocal relaxation time vs. real part of the frequency) diagram of the type drawn in Figure 5. Here branch II is the electromagnetic branch, I is essentially the elastic branch and II is essentially the low polariton branch. Again, a typical repulsion effect occurs at $(\tau_0^{-1}, \sigma_0 \simeq \Omega_0)$. A numerical evaluation of this repulsion effect for damping in $BaTiO_3$ is given in Figure 6.

7. Conclusion

In the above-enunciated results, the essential facts are those that $d \neq 0$ and $_0P \neq 0$, and that $_0P$ induces an anisotropy, so that there indeed exists a piezoelectric coupling. If we let $d \to 0$, then both high and low polariton branches rise to escape from the acoustical range of frequencies. We note that none of the mode-coupling effects exhibited in this paper can be reproduced with the usual Voigt scheme of piezoelectricity [7], where the only effect in anisotropic bodies is to replace elastic moduli by piezoelectrically

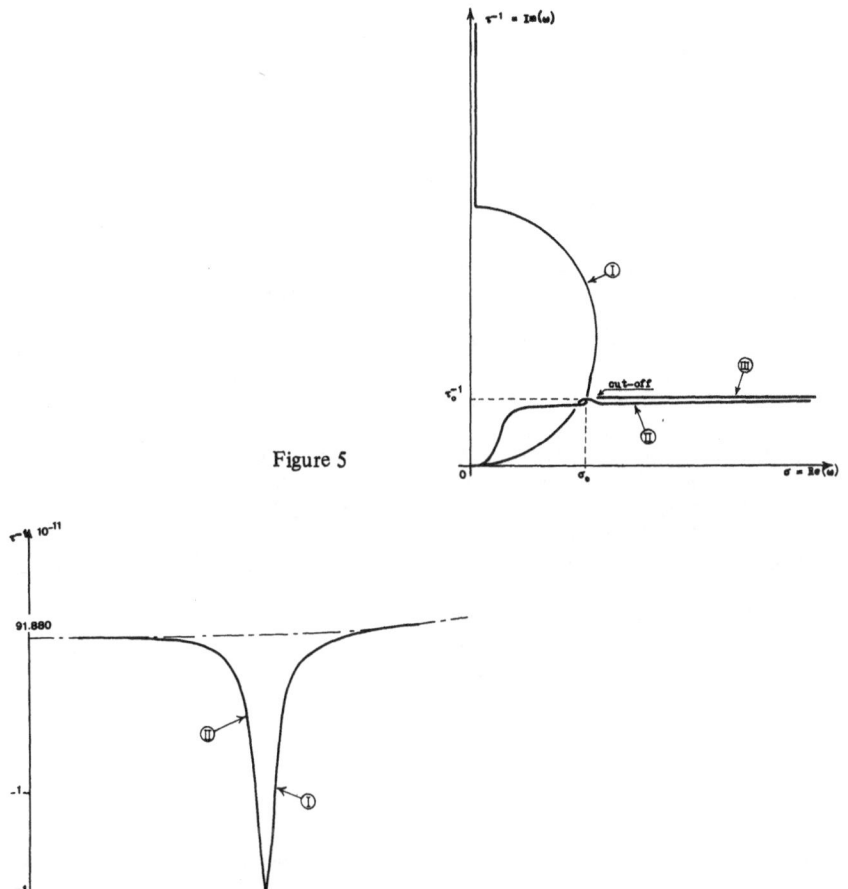

Figure 5

Figure 6

stiffened moduli, which can only lead to a change in the elastic speeds. Another peculiarity of the present model is that a linearization about a finite state $(_0\mathbf{P})$ of polarization requires starting from a fully nonlinear theory in order to guarantee a correct linearization procedure. One consequence of this is that not only the strain per se is involved in the linearized equations, but also the infinitesimal rotations $u_{[i,j]}$ (which do not appear in Voigt's theory). The above model can also be used to study simple wave propagation problems in presence of material boundaries, e.g., the excitation of electroacoustic elastic material sandwiched between two electrodes, as well as the half-space problem leading to the consideration of surface electroacoustic waves. Then we have couplings between Rayleigh (elastic) waves and Bleustein-Gulayev (electric) surface modes*.

REFERENCES

[1] MAUGIN G.A. – *Arch. Mechanics* (Poland) **28** (1976): 679-692, *ibid*, **29**, 143-151, 251-258.
COLLET B. – *Thesis* (3e cycle), Université de Paris VI, February, 1976.
COLLET B. and G.A. MAUGIN. – *C.R. Acad. Sci. Paris*, **279B** (1974): 379-382, 439-442.

[2] MAUGIN G.A. – Review Article in *Acta Mechanica*, **35** (1980) : 1-70.

[3] MAUGIN G.A. and A.C. ERINGEN. – *J. Mécanique*, **16** (1977): pp. 101-147.

[4] POUGET J. – *Thesis* (3e cycle), Université de Paris VI, April 1979, and paper in preparation (with G.A. Maugin).

[5] AKHIEZER A.I., *et al.*– *Spin Waves*, Amsterdam North-Holland, 1968.
MAUGIN G.A. – *Int. J. Engng. Sci.*, **17** (1979) : 1073-1108.

[6] FLEURY P.A. and P.D. LAZAY. – *Phys. Rev. Lett.*, **36**, (1971): 1331-1334.
DVORAK V. – *Phys. Rev.*, **167**, 525-528, Scott J.F., *Rev. Mod. Phys.*, **46**, 83-123, Pitte E. 1974: *Phys. Rev.*, **B1**, 924-930, Reese R.L. *et al.*, (1970), *Phys. Rev.*, **B7**, (1973): 4105.
BRODY E. and H.Z. CUMMINS. – *Phys. Rev. Lett.*, **21**, (1968): 1263-1266.
RIMAI L. *et al.* – *Phys. Rev.*, **168**, (1968): 623-630.

[7] AULD B.A. – *Acoustic Fields and Waves in Solids*, New York: J. Wiley, 1973.
DIEULESAINT E. and D. ROYER. – *Ondes élastiques dans les solides-application au traitement du signal*. Paris: Masson, 1974.

RESUME

(Ondes électroacoustiques dans les diélectriques élastiques ferroélectriques)

Etude des ondes électroacoustiques couplées, dispersives et atténuées dans les diélectriques élastiques ferroélectriques dans lesquels la symétrie hexagonale est induite par le champ de polarisation électrique initial (symétrie brisée).

---

* *Note added in proof.* – This problem has been treated in the meantime, cf. J. POUGET and G. A. MAUGIN, *J. Acoust. Soc. Amer.*, **69** (1981): 1304-1325.

Microfissuration du Béton
et Propagation d'Ondes Ultrasonores

Y. Bamberger, J.J. Marigo
Laboratoire Central des Ponts et Chaussées, Paris, France.

G. Cannard
Laboratoire Régional des Ponts et Chaussées, Bron, France.

1. Introduction

Le but de l'étude est la mise au point d'une méthode d'auscultation dynamique non destructive du béton qui permettrait à terme de contrôler les ouvrages in situ. Il s'agit donc de déduire l'état du matériau à partir de son comportement dynamique et plus précisément d'obtenir des informations sur les états de contraintes et de fissurations grâce aux mesures des célérités de propagation des ondes.

Dans un premier temps, on décrit et analyse les résultats expérimentaux, puis on construit un modèle théorique susceptible d'expliquer qualitativement et quantitativement les phénomènes.

2. Résultats expérimentaux

2.1. *Dispositif expérimental et notations* [2]

On effectue des cycles de chargement en compression uniaxiale d'une éprouvette en béton, figure 1 ; ces cycles ont une amplitude croissante, figure 2, jusqu'à rupture de l'éprouvette. A chaque palier de charge on mesure le temps de traversée d'ondes ultrasonores longitudinales et transversales se propageant les unes parallèlement, les autres perpendiculairement à la contrainte : ces ondes sont émises par des palpeurs placés sur les faces de l'éprouvette.

Eprouvette : Les dimensions de l'éprouvette (15 x 15 x 50 cm) ont été choisies pour qu'on puisse considérer le matériau comme statistiquement homogène (puisqu'il contient des graviers de diamètre pouvant atteindre 25 mm).

Chargement : L'éprouvette est placée verticalement entre les plateaux d'une presse suivant sa plus grande dimension. Les résultats expérimentaux

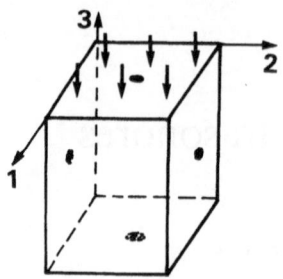

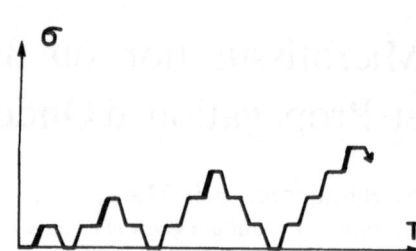

Figure 1. – *Dispositif ; définition du repère;*
• palpeurs ; ↓ compression simple σ.

Figure 2. – *Histoire du chargement* (σ , T) :
—phases de déchargement-rechargement ;
—phases de surchargement.

décrits plus loin conduisent à distinguer dans l'histoire du chargement σ en fonction du temps (représenté sur la figure 2) trois phases :

— Les phases de surchargement où l'on dépasse la valeur maximum notée σ_m de la contrainte antérieurement atteinte : elles sont indiquées en trait fort sur la figure ;
— les phases de rechargement où la contrainte augmente sans dépasser la valeur σ_m ;
— les phases de déchargement où la contrainte diminue.

Les paliers de charge ont la durée nécessaire aux mesures des temps de propagation (cinq à dix minutes suivant leurs nombres).

Types d'ondes : On considère que vis-à-vis des ondes ultrasonores le béton sous compression uniaxiale donnée est un matériau élastique, homogène et isotrope de révolution (autour de la direction de la contrainte) — ce que l'expérience confirme. La théorie des ondes d'accélération dans un tel matériau montre que des ondes longitudinales ou transversales peuvent se propager suivant des directions parallèles ou perpendiculaires (notées respectivement 3,1 et 2) à la direction de chargement. On est conduit à distinguer cinq types d'ondes suivant leur vecteur de propagation $\vec{n}$ et leur vecteur polarisation $\vec{\delta}$, de célérité $V_{n\delta}$:

1. longitudinale parallèle : $V_{L/\!/}$ ou V_{33} ;
2. transversales parallèles : $V_{T/\!/}$ ou $V_{31} = V_{32}$;
3. longitudinales perpendiculaires : $V_{L\perp}$ ou $V_{11} = V_{22}$;
4. transversales perpendiculaires parallèles : $V_{T\perp/\!/}$ ou $V_{13} = V_{23}$;
5. transversales perpendiculaires perpendiculaires: $V_{T\perp\perp}$ ou $V_{12} = V_{21}$.

Palpeurs : Les émissions et réceptions de signaux se font à l'aide de céramiques piézoélectriques polarisées disposées par couples sur deux faces opposées. Les mesures sont effectuées par transparence, l'un des palpeurs servant d'émetteur, l'autre de récepteur. Une colle cyanoacrylique — qui assure une transmission satisfaisante des signaux et résiste aux déformations différentielles du béton et des céramiques sous charge — maintient les palpeurs latéraux. Pour des ondes se propageant parallèlement à la direction de compression, les céramiques sont collées sur les faces supérieures et inférieures puis noyées dans une épaisseur de mortier destinée à transmettre la charge au béton.

Grandeurs mesurées : On ne s'intéresse qu'aux temps de traversée des signaux, en fait aux variations de ces temps au cours de l'histoire du chargement : on ne mesure pas les amplitudes et on ne cherche pas de fonction de transfert. Les phénomènes de dispersion et de dissipation ne sont donc pas étudiés.

Chaîne de mesures : Les célérités de propagation dans le béton sont de l'ordre de 4 500 m/s pour les ondes longitudinales, 2 400 m/s pour les transversales. Les temps de traversée varient donc entre 30 μs (pour V_{11}) et 250 μs (pour V_{31}). Les variations de ces temps ne dépassent pas 10 % soit quelques centaines de nanosecondes ou quelques microsecondes.

Pour obtenir une résolution de l'ordre de quelques nanosecondes, une chaîne de mesures relativement complexe est nécessaire. Les émetteurs sont excités les uns après les autres à l'aide d'un générateur d'impulsions de quelques μs de largeur, de fréquence centrée autour de 50 Khz. Après filtrage, amplification et moyenne, on obtient un signal de sortie stable : le pointage est effectué sur le "début" du signal — ce qui n'est pas aisé pour les ondes transversales, celles-ci étant toujours précédées d'une onde longitudinale (qui, heureusement, est en général de plus faible amplitude) en raison des réflexions, ... qui ont lieu au cours de la traversée de l'onde.

2.2. *Résultats expérimentaux*

Les essais font apparaître que les temps de traversée des signaux des différents types d'ondes t_{ij} ne dépendent que des deux paramètres de chargement σ_m et σ définis précédemment. Autrement dit, à charge maximale passée σ_m donnée, les temps ne dépendent que de la contrainte actuelle σ : les phases de rechargement-déchargement sont réversibles. Par contre, les phases de surchargement sont irréversibles.

Les distances parcourues par les différents types d'ondes n'étant pas égales, il est plus commode d'utiliser les vitesses V_{ij} que les temps t_{ij} et plus précisément les rapports

$$V_{ij}(\sigma_m, \sigma) / V_{ij}(0, 0)$$

qui sont au nombre de cinq et qu'on appellera, comme en mécanique des roches, indices de qualité, car, comme on le verra, ils caractérisent l'état de fissuration du béton. L'allure des courbes, où la contrainte est rapportée à la contrainte de rupture R_c (Fig. 3) appelle les commentaires suivants :

Courbes de chargement-déchargement

$$\sigma \rightarrow V_{ij}(\sigma_m, \sigma) / V_{ij}(\sigma_m, 0)$$

Lorsqu'on recharge ($\sigma\nearrow$), toutes les célérités augmentent, plus ou moins suivant le type d'onde. Pour $\sigma_m = 0,75\ R_c$, on note les augmentations suivantes entre $\sigma = 0$ et $\sigma = \sigma_m$: environ 10 % pour V_{33}, 2 % pour V_{13} et V_{31}, 1 % pour V_{11}, quasiment nulle pour V_{12}. Par ailleurs, pour un type d'onde, les courbes en cause sont sensiblement parallèles.

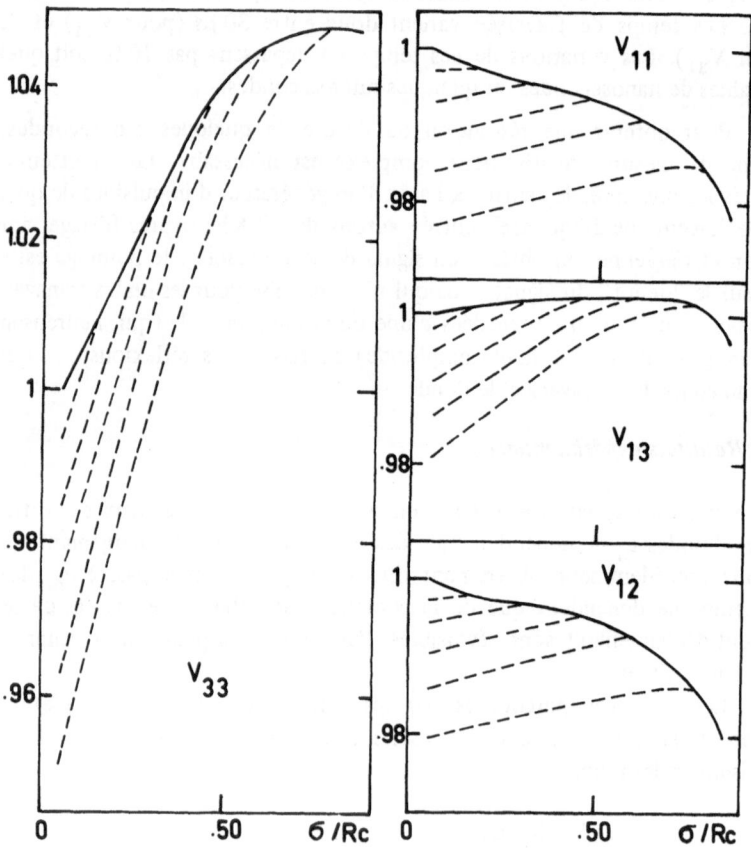

Figure 3. – *Courbes expérimentales* : $V_{ij}(\sigma_m, \sigma)/V_{ij}(0, 0)$.
—— phases de surchargement : σ_m croît.
- - - phases (réversibles) de déchargement-rechargement : σ_m constant.

Courbes d'état déchargé

$$\sigma_m \rightarrow V_{ij}(\sigma_m, 0) / V_{ij}(0, 0)$$

Lorsque σ_m croît, jusqu'à 0,75 R_c environ, toutes les célérités diminuent, et ce de manière sensiblement linéaire ; la variation est la plus forte pour V_{33} (5 % de diminution relative pour $\sigma = 0,75$ R_c).

Courbes de surchargement

Leur sens de variation dépend de l'onde :

— pour V_{33}, une augmentation très forte jusqu'à 0,5 R_c, puis plus faible jusqu'à 0,75 R_c ; enfin une diminution rapide au-delà ;
— pour V_{11}, une diminution faible jusqu'à 0,75 R_c, puis brutale au-delà ;

Observations complémentaires

— Pour tester l'homogénéité globale des éprouvettes, on a placé des palpeurs à plusieurs niveaux sur les faces latérales : les résultats sont similaires et concordent remarquablement pour V_{11}.
— On vérifie que V_{13} et V_{31} sont égales, ce qui est en concordance avec la théorie des ondes en milieu élastique isotrope de révolution ; on vérifie en outre que V_{11} et V_{22}, V_{12} et V_{21}, V_{13} et V_{23} ont des valeurs voisines.
— Des expériences d'émission acoustique ont été effectuées : il n'y a pratiquement pas d'émission dans les phases de rechargement-déchargement ; on en observe seulement durant les phases de surchargement.

2.3. Interprétation des résultats

La théorie de l'élasticité linéaire ne permet pas de rendre compte des phénomènes précédents, tant en raison des irréversibilités que de l'ordre de grandeur des variations de vitesse observées — qu'elle prévoit de l'ordre de σ/E c'est-à-dire au plus 10^{-3} et non 10^{-2} ou 10^{-1} comme pour V_{33}.

La prise en compte de la microfissuration du béton et de son évolution permet par contre une interprétation cohérente :

a) les phases de rechargement-déchargement correspondent à une "respiration" des microfissures qui se ferment et s'ouvrent, et ceci dès la première montée en charge puisqu'il est connu qu'il existe des microfissures "de naissance" dues à la dessication — l'augmentation initiale des vitesses serait d'ailleurs difficilement explicable sinon.

b) les phases de surchargement correspondent à une extension ou une création de microfissures.

Compte tenu de l'ordre de grandeur des variations de déformations et de contraintes induites par les ondes ultrasonores, il est naturel de supposer

c) qu'une fissure fermée n'influence pas le passage des ondes.

Les hypothèses a, b et c sont cohérentes avec les résultats décrits en 2 : réversibilité des courbes de rechargement, augmentation des vitesses par rechargement, émission acoustique non nulle uniquement au surchargement, diminution des vitesses à contrainte nulle lorsque σ_m croit.... En particulier, les variations de pente observées lorsque σ passe par σ_m correspondent à la superposition du serrage d'une partie des microfissures et de l'extension de certaines microfissures.

Par ailleurs, puisque $V_{33}(\sigma_m,0)$ décroît plus vite que $V_{11}(\sigma_m,0)$ lorsque σ_m augmente, la fissuration se développe de manière anisotrope et, semble-t-il, plutôt suivant les directions voisines de la perpendiculaire au chargement, au moins jusqu'à 0,75 R_c.

Les inflexions observées à partir d'environ 0,75 R_c correspondent à un accroissement plus rapide de la microfissuration : les microfissures, qui existent dès l'origine au contact agrégats-pâte et qui progressent en restant à leur interface jusqu'à 0,75 R_c environ pénètrent dans la pâte au-delà puis se rejoignent jusqu'à créer les fissures de rupture, visibles en surface.

3.   Modèle de matériau microfissuré [1]

3.1. *Idées directrices*

On assimile le béton à un ensemble de cellules cubiques égales contenant chacune une fissure ; les cubes sont composés du même matériau $\mathfrak{M}$ linéairement élastique, homogène et isotrope (on néglige ici les différences pâte-agrégats), figure 4 ; les fissures sont des ellipsoïdes aplatis (pour faciliter les calculs analytiques) de même centre que le cube associé, figure 5. Leurs dimensions et orientations sont supposées quelconques à cela près que le grand axe est supposé petit par rapport à l'arête du cube de sorte que les interactions des fissures soient négligeables.

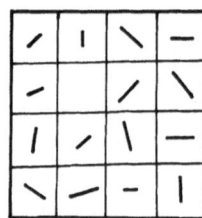

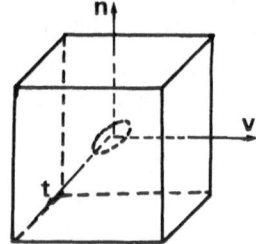

Figure 4. – *Modélisation du milieu fissuré.*

Figure 5. – *Cellule microfissurée élémentaire et axes principaux de la fissure ellipsoïdale* $(\vec{t}, \vec{v}, \vec{n})$.

L'hypothèse de non-interaction des fissures permet d'étudier une cellule isolément. On montre alors que la cellule cubique formée du matériau $\mathcal{M}$ fissurée a même énergie de déformation qu'une cellule homogène formée d'un matériau $\mathcal{M}^\star$ (dont les modules sont dits modules effectifs du matériau fissuré) lorsqu'elle est soumise à des contraintes constantes sur ses faces.

La densité de probabilité des paramètres de fissuration étant supposé connue, on en déduit par moyenne les grandeurs cherchées : supposant qu'une onde traversant l'éprouvette rencontre "toutes" les cellules possibles (milieu statistiquement homogène), on obtient en particulier les célérités moyennes.

L'approche par les modules effectifs employée ici se trouve notamment dans les travaux de Hashin [3] et Budiansky et O'Connell [6] dans le cas particulier d'une fissuration isotrope.

3.2. *Homogénéisation d'une cellule*

Position du problème et notations. La cellule cubique Ω de coté ℓ contient une fissure ellipsoïdale de grand axe a $(\ll \ell)$, de petit axe b, d'épaisseur e $(\ll$ a) ; le repère orthonormé principal de l'ellipsoïde $(0, \vec{t}, \vec{v}, \vec{n})$ est supposé parallèle aux cotés du cube ce qui ne restreint pas la généralité du problème puisque a $\ll \ell$. On note Ω_m le domaine occupé par le matériau $\mathcal{M}$ constitutif de la cellule, Ω_f celui occupé par la fissure.

On suppose la cellule Ω en équilibre sous l'action de charges uniformes $\vec{T}^\nu_\infty$ exercées sur les faces du cube (ν = t, v ou n). Le champ de contraintes $\underset{\sim}{\tau}$ à l'équilibre est solution des équations

$$
\left\{
\begin{array}{ll}
\text{div } \tau = 0, \quad \text{dans } \Omega_m : & (1) \\[2mm]
\tau \cdot \vec{\nu} = \vec{T}^\nu_\infty, \text{ sur la face } \Gamma^\nu \text{ (normale extérieure } \vec{\nu}) & (2) \\
\qquad\qquad\qquad (i, j \text{ et } \nu \text{ égaux à t, v ou n)}, & \\[2mm]
\tau \cdot \vec{n} = 0 , \text{ sur le bord } \Gamma_f \text{ de } \Omega_f & (3)
\end{array}
\right.
$$

et est lié au champ de déformations e par la loi de Hooke

$$
\mathbf{e} = \mathbf{s} \cdot \tau \quad \text{ou} \quad \tau = \mathbf{c} \cdot \mathbf{e} \tag{4}
$$

(c tenseur de rigidité, s tenseur de souplesse).

Dans le cas isotrope examiné ici s et c sont caractérisés par E et ν et λ et μ.

En raison de la fissure, les champs e et τ ne sont pas uniformes. On se propose de chercher s'il existe un matériau élastique $\mathcal{M}^\star$ homogène tel qu'un cube de coté ℓ de ce matériau ait le même potentiel élastique que la cellule Ω, autrement dit s'il existe un tenseur de souplesse $s^{\star}$ tel que, si τ_0 est le champ de contraintes constant associé aux conditions $\overrightarrow{T}_{\infty}$ sur les faces, on a

$$\forall T_{\infty}, \quad \int_{\Omega_m} \tau \cdot s \cdot \tau = \int_{\Omega} \tau_0 \cdot s^{\star} \cdot \tau_0. \tag{5}$$

Enoncé du résultat. Le matériau. $\mathcal{M}^{\star}$ existe ; son tenseur de souplesse $s^{\star}$ défini par ses composantes dans le repère associé aux faces de la fissure est donné par les expressions

$$s_{ijk\ell}^{\star} = s_{ijk\ell} + \frac{1}{\mu} K_p (\delta_{ni} \delta_{pj})^S (\delta_{nk} \delta_{p\ell})^S, \quad [\text{sommation sur p}], \tag{6}$$

où l'exposant S désigne le symétrisé du tenseur entre parenthèses par rapport au 2e et au 4e indices et où les constantes K_n, K_t et K_v sont

$$\left\{ \begin{array}{l} K_n = \dfrac{4}{3} \pi \dfrac{1 - \nu}{E(k)} \dfrac{ab^2}{\ell^3}, \\[3mm] K_t = \dfrac{4}{3} \pi \dfrac{(1 - \nu) k^2}{(k^2 - \nu) E(k) + \nu k'^2 K(k)} \dfrac{ab^2}{\ell^3}, \\[3mm] K_v = \dfrac{4}{3} \pi \dfrac{(1 - \nu) k^2}{(k^2 + \nu k'^2) E(k) - \nu k'^2 K(k)} \dfrac{ab^2}{\ell^3}, \end{array} \right. \tag{7}$$

avec $k^2 = 1 - b^2/a^2$, $k'^2 = 1 - k^2$, $K(k)$ et $E(k)$ intégrales elliptiques complètes de première et de deuxième espèce[1].

Démonstration : Pour trouver s^1 et établir (5), on remarque que l'on a

$$\int_{\Omega_m} \tau \cdot e = \int_{\Gamma^{i+}} \tau_0^{ij} (u_j^+ - u_j^-) \quad \text{avec} \left\{ \begin{array}{l} \Gamma^{i+} \text{ face du cube } x_i = + \ell/2, \\[2mm] u_j^+ \text{ } j^{\text{ième}} \text{ composante du dépla-} \\ \text{cement u sur cette face,} \\[2mm] u_j^- \text{déplacement correspondant} \\ \text{sur la face opposée,} \end{array} \right.$$

soit encore, puisque τ_0 est constant,

$$\int_{\Omega_m} \tau \cdot e = \tau_0^{ij} \int_{\Gamma^{i+}} (u_j^+ - u_j^-) = \tau_0^{ij} \int_{\Omega_m} e_{ij} + \tau_0^{ni} \int_{\Gamma_f} [u_i],$$

---

[1]. Pour une fissure circulaire (k = 0) il faut utiliser d'autres intégrales pour K_t et K_v car les expressions précédentes deviennent singulières.

en notant $[u_i]$ la discontinuité de u_i à travers la fissure. Le problème consiste donc à exprimer la déformation moyenne

$$\langle e_{ij}\rangle = \left\{ \int_{\Gamma^{i+}} (u_j^+ - u_j^-)\right\}^S = \frac{1}{|\Omega_m|}\left[\int_{\Omega_m} e_{ij} + \int_{\Gamma_f} ([u_i]\,n_j)^S \right],$$
$$(|\Omega_m| \# \ell^3) \quad (8)$$

en fonction de $\boldsymbol{\tau}_0$. A cet effet, on va faire travailler les contraintes $\boldsymbol{\sigma}_0$ d'un état de contraintes uniformes dans $\langle e\rangle$ et utiliser le théorème de Maxwell-Betti. Cet état de contraintes uniformes est caractérisé par

$$\begin{cases} \operatorname{div} \boldsymbol{\sigma} = 0 & (9) \\[2mm] \boldsymbol{\sigma}\cdot\vec{\nu} = \boldsymbol{\sigma}_0\cdot\vec{\nu} \ \text{ sur } \ \Gamma^{\nu} & (10) \\[2mm] \boldsymbol{\sigma}\cdot\vec{n} = \boldsymbol{\sigma}_0\cdot\vec{n} \ \text{ sur } \ \Gamma_f\,. & (11) \end{cases}$$

La déformation $\boldsymbol{\varepsilon}_0 = s\cdot\boldsymbol{\sigma}_0$ est uniforme et le déplacement $\vec{u}_0$ linéaire par rapport aux coordonnées

$$u_0^i = \epsilon_0^{ij}\, x_j\,. \quad (12)$$

Le théorème de Maxwell-Betti s'écrit

$$\int_{\Omega_m} \boldsymbol{\tau}\cdot\boldsymbol{\varepsilon}_0 = \int_{\Omega_m} \boldsymbol{\tau}\cdot s\cdot\boldsymbol{\sigma}_0 = \int_{\Omega_m} e\cdot\boldsymbol{\sigma}_0\,. \quad (13)$$

Pour faire apparaître $\langle e\rangle$, on transforme l'intégrale de droite. Il vient

$$\int_{\Omega_m} e\cdot\boldsymbol{\sigma}_0 = \int_{\Gamma^{\nu+}} \vec{\sigma}_0^{\nu}\cdot[\vec{u}^+ - \vec{u}^-] - \int_{\Gamma_f} \vec{\sigma}_0^n\cdot[\vec{u}]\,,$$

(le signe $-$ vient de ce que la fissure est intérieure), soit, puisque $\vec{\sigma}_0^{\nu}$ est constant et compte tenu de (8)

$$\int_{\Omega_m} e\cdot\boldsymbol{\sigma}_0 = \ell^3\,\boldsymbol{\sigma}_0\cdot\langle e\rangle - \int_{\Gamma_f} \vec{\sigma}_0^n\cdot[\vec{u}]\,. \quad (14)$$

Par ailleurs, on a, compte-tenu notamment de (12)

$$\int_{\Omega_m} \boldsymbol{\tau}\cdot\boldsymbol{\varepsilon}_0 = \int_{\Gamma^{\nu}} \vec{T}_{\infty}^{\nu}\cdot\vec{u}_0 = \int_{\Gamma^{\nu+}} \vec{T}_{\infty}^{\nu}\cdot(\vec{u}_0^+ - \vec{u}_0^-)$$

$$= \tau_0^{\nu i}\,\epsilon_{ij}^0 \int_{\Gamma^{\nu+}} (x_j^+ - x_j^-) = \ell^3\,\boldsymbol{\tau}_0\cdot\boldsymbol{\varepsilon}_0\,. \quad (15)$$

Les égalités (13), (14) et (15) entraînent

$$\sigma_0 \cdot \langle e \rangle = \varepsilon_0 \cdot \tau_0 + \frac{1}{\ell^3} \sigma_0^{ni} \int_{\Gamma_f} [u^i]. \tag{16}$$

De cette égalité vraie pour tout σ_0, on déduit la valeur $\langle e \rangle$ dès que l'on connaît celle de $[u]$: celle-ci est calculée comme si la fissure était placée dans un milieu *infini* — cas où l'on connaît une expression analytique exacte (Kassir et Sih, [4]. Il vient

$$[u^i] = c_i \tau_0^{ni} \left(1 - \frac{x^2}{a^2} - \frac{y^2}{b^2}\right)^{\frac{1}{2}} \text{ (sans sommation sur i).}$$

Après calcul (16) se met sous la forme

$$\sigma_0^{ij} \langle e_{ij} \rangle = \sigma_0^{ij} s_{ijk\ell} t_0^{k\ell} + \sigma_0^{ij} \cdot \frac{K_\ell}{2\mu} (\delta_{ni} \delta_{\ell j} + \delta_{nj} \delta_{\ell i}) t_0^{n\ell},$$

d'où le résultat cherché

$$\langle e_{ij} \rangle = \left[s_{ijk\ell} + \frac{K_p}{\mu} (\delta_{ni} \delta_{pj})^S (\delta_{nk} \delta_{p\ell})^S \right] t_0^{k\ell}, \tag{17}$$

(sommation sur k, ℓ et p).

Commentaires. Des calculs sans difficulté mais qu'on ne détaillera pas ici faute de place montrent que

- la fissure assouplit le matériau : les coefficients d'élasticité (en particulier celui de compressibilité) sont diminués ;
- le matériau possède six coefficients indépendants ;
- dans le repère propre de la fissure, cisaillements et efforts normaux sont découplés.

3.3. *Propagation des ondes*

On suppose les longueurs d'onde λ nettement supérieures à la grande dimension a des fissures ce qui est le cas pour le béton ($\lambda \sim 10$ cm, a ~ 2–10 mm) : on considère alors les fissures comme des affaiblissements du matériau et on admet que le matériau microfissuré se comporte vis-à-vis des ondes comme un matériau élastique plus souple, défini à l'aide des cellules cubiques définies précédemment.

Célérités locales. On étudie ici la propagation des ondes dans le milieu $\mathfrak{M}^\star$ associé à une cellule, figure 6. On note $(\vec{e}_1, \vec{e}_2, \vec{e}_3)$ le repère fixe, $(\vec{t}, \vec{v}, \vec{n})$ le repère principal de la fissure, qui est ici d'orientation quelconque, caractérisé par les angles θ entre $\vec{e}_1$ et $\vec{n}'$ (projection de $\vec{n}$ sur le plan $(1, 2)$), ϕ entre $\vec{n}'$ et n, ψ entre $\vec{t}'$ (projection de $\vec{t}$ sur le plan $(1, 2)$ et t. On appelle densité de fissuration et l'on note ϵ le nombre sans dimension $\dfrac{2}{3} \cdot \dfrac{\pi^2}{E(k)} \cdot \dfrac{ab^2}{\ell^3}$.

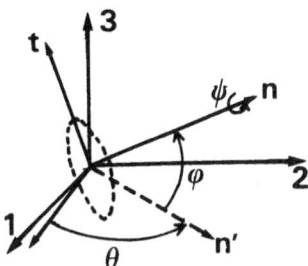

Figure 6. – *Repère attaché aux axes de la fissure* $(\vec{t}, \vec{v}, \vec{n})$ *et repère fixe* $(\vec{e}_1, \vec{e}_2, \vec{e}_3)$.

Les calculs, sans difficulté, montre que, comme $\epsilon \ll 1$, il existe des ondes "quasi-longitudinales" et "quasi-transversales". En particulier, on obtient (en ne gardant que les termes du premier ordre en ϵ)

$$v_{33} = \frac{c_{33} - c_{33}^\star}{c_{33}} = \frac{2}{\pi} \{[\nu + (1 - 2\nu) \sin^2 \phi]^2 +$$

$$+ (1 - 2\nu)^2 \sin^2 \phi \cos^2 \phi \, [k_t \sin^2 \psi + k_v \cos^2 \psi]\} \frac{\epsilon}{1 - 2\nu} \qquad (18)$$

$$v_{11} = \frac{c_{11} - c_{11}^\star}{c_{11}} = \frac{2}{\pi} \{[\nu + (1 - 2\nu) \cos^2 \theta \cos^2 \phi]^2 +$$

$$+ (1 - 2\nu)^2 \cos^2 \theta \cos^2 \phi \, [k_t \, t_1^2 + k_v \, v_1^2]\} \frac{\epsilon}{1 - 2\nu} \qquad (19)$$

où $K_\varrho = \dfrac{2}{\pi} k_\varrho \dfrac{\epsilon}{1 - \nu}$ cf. [5]). La présence de la fissure ralentit toujours d'onde.

Célérités moyennes dans le milieu aléatoire. Comme annoncé, on les définit comme les moyennes des célérités sur l'ensemble des cellules. La fissuration étant faible, on considère que les raies sont rectilignes et les moyennes harmoniques

$$\frac{1}{\langle c \rangle} = \frac{1}{\|AB\|} \int_A^B \frac{ds}{c(s)}$$

peuvent être remplacées par des moyennes arithmétiques. En introduisant une densité de probabilité p et en supposant le milieu statistiquement homogène, il vient

$$\langle c \rangle = \int c(\epsilon, k, \theta, \phi, \psi) \, p(\epsilon, k, \theta, \phi, \psi) \, d\epsilon \, dk \, d\theta \, d\phi \, d\psi \qquad (20)$$

avec $\epsilon \in [0, + \infty[$, $k \in [0, 1]$, $\theta \in [0, 2\pi]$, $\phi \in [0, \pi/2]$, $\psi \in [0, \pi]$

et $\int p = 1$.

Cas particuliers
- pour un milieu statistiquement isotrope contenant des fissures circulaires de même rayon (ϵ uniforme, $k = 0$, $p(\theta, \phi, \psi) = \cos \phi/2\pi^2$), on retrouve les résultats de Budiansky et O'Connell [6].
- pour un milieu où toutes les fissures ont la même inclinaison ϕ sur la verticale, on observe que v_{33} dépend fortement de ϕ, figure 7 ; pour v_{11} la dépendance est moins marquée.

4. Critère de fermeture

4.1. *Définition*

L'interprétation des courbes expérimentales, et plus particulièrement des courbes de rechargement, nécessite l'introduction d'un critère de fermeture des fissures, i.e. d'un seuil de contrainte au-delà duquel les lèvres sont en contact.

On dira donc qu'une fissure ouverte dans un état libre de charge se ferme sous une compression σ lorsque l'écartement des lèvres s'annule.

4.2. *Formulation*

Le critère va être formulé pour une fissure en forme d'ellipsoïde aplati : grand axe a, petit axe b, demi-épaisseur (ouverture) c, avec $c \ll a$. On appelle n la normale unitaire au "plan de fissure". Cette fissure est placée dans un milieu élastique homogène isotrope, infini et chargé uniformément à l'infini suivant

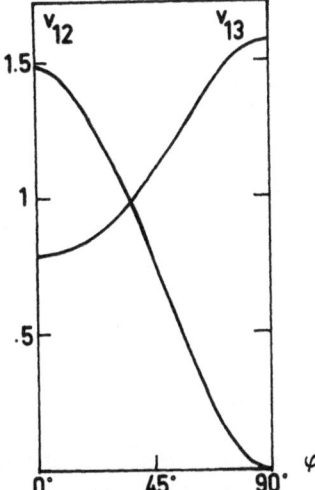 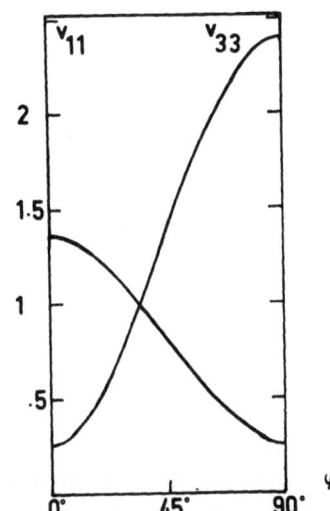

Figure 7. – *Influence d'une anisotropie de fissuration sur les célérités* c_{ij} *des ondes ; cas d'un milieu isotrope de révolution où toutes les fissures sont inclinées de* φ *par rapport à la direction 3 :* $C_{ij}(\varphi) = C_{ij}[1 - V_{ij}(\varphi)]$, C_{ij} = célérité dans le milieu non fissuré.
Courbes théoriques représentant, à un facteur multiplicatif près, la diminution relative $V_{ij}(\varphi)$ *des célérités due à la fissuration dans un milieu de coefficient de Poisson* $\nu = 0,25$:
a) ondes transversales
b) ondes longitudinales.

un processus de chargement croissant $\lambda \sigma^{ij}$. Dans son état libre de charge ($\lambda = 0$), le demi-écartement initial des lèvres est noté $c_0 > 0$.

Pour un état de chargement λ, le demi-écartement c_λ est relié à c_0 et à la discontinuité des déplacements normaux $[\![u_n]\!]$ des lèvres de la fissure par la relation :

$$c_\lambda = c_0 + \frac{1}{2}[\![u_n]\!].\tag{21}$$

L'expression de $[\![u_n]\!]$ en un point (x, y) de la surface fissurée ne fait intervenir que la contrainte normale au plan de fissure $\lambda \sigma^{nn}$, cf. [5], et est obtenue à l'aide de la solution du champ de déplacement autour d'une cavité ellipsoïdale, cf. [8] :

$$[\![u_n]\!] = \frac{4(1-\nu^2)}{E(k)} \cdot \frac{\lambda \sigma^{nn}}{E} \cdot \left[1 - \frac{x^2}{a^2} - \frac{y^2}{b^2}\right]^{\frac{1}{2}},\tag{22}$$

où

k est le facteur d'ellipticité, $k = (1 - b^2/a^2)^{\frac{1}{2}}$

E (k) l'intégrale elliptique complète de deuxième espèce, et

E, ν le module d'Young et coefficient de Poisson de la matrice.

En reportant (22) dans (21), il vient :

$$c_\lambda (x, y) = c_0 \left[1 + \frac{\lambda \sigma^{nn}}{\alpha \tau} \right] \cdot \left\{ 1 - \frac{x^2}{a^2} - \frac{y^2}{b^2} \right\}^{\frac{1}{2}}, \tag{23}$$

avec $\alpha = c_0/b$ appelé facteur de forme de la fissure, et $\tau = \dfrac{E(k)}{2(1 - \nu^2)} E$, qui a la dimension d'une contrainte, dépend à la fois de l'élasticité de la matrice et de l'ellipticité de la fissure.

La fissure reste ouverte tant que $c_\lambda (x, y)$ est positif; lorsqu'on augmente la charge, λ croissant et $\sigma_{nn} < 0$ en compression, le terme entre crochets dans (23) décroît et s'annule pour un changement critique λ_c tel que :

$$\sigma_c^{nn} + \alpha \tau = 0.$$

Alors l'écartement en tout point (x, y) de la surface fissurée est nul, les lèvres sont entièrement en contact.

D'où le critère de fermeture suivant :

– Une fissure elliptique de facteur de forme α, d'ellipticité k, placée dans un milieu élastique homogène et isotrope, de constantes E et ν, soumis à un processus de chargement croissant :

   • se ferme lorsque la contrainte normale de compression au plan de fissure atteint la valeur critique $\sigma_c^{nn} = - \alpha \tau$;

   • et elle est dite ouverte tant que $\sigma^{nn} + \alpha \tau > 0$.

4.3. *Cas d'une compression simple*

Lorsque le solide est soumis à une compression simple σ parallèle à l'axe 3, la contrainte de fermeture dépend de l'inclinaison ϕ de la normale n avec le plan horizontal (e_1, e_2) : (Fig. 8),

$$\sigma^{nn} = \sigma \sin^2 \phi$$

et la fissure (k, α) est fermée si :

$$\sigma \sin^2 \phi + \alpha \tau (k) \leqslant 0.$$

Remarque : Ce critère s'apparente à celui introduit par Siegfried et Simmons [7] pour l'interprétation et l'utilisation de l'essai de compressibilité en mécanique des roches.

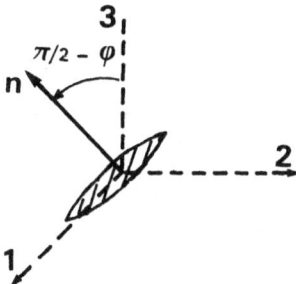

Figure 8. – *Orientation de la normale $\vec{n}$ au plan de fissure.*

5. Application au béton

5.1. *Introduction du chargement*

Le modèle théorique présenté dans le § 2 permet de déduire les célérités de propagation des ondes dans un milieu fissuré dont on se donne a priori la densité de probabilité p des paramètres de fissuration ξ. Lorsqu'on charge le matériau, la fissuration évolue et p va dépendre également du chargement.

Les résultats expérimentaux nous conduisent à distinguer deux types de fissuration, la fissuration réelle et la fissuration apparente :

— On appelle fissuration réelle du matériau dans un état de chargement l'ensemble des fissures, tant ouvertes que fermées ;

— On appelle fissuration apparente dans un état de chargement, l'ensemble des fissures ouvertes, i.e. celles dont l'écartement des lèvres est non nul.

La fissuration réelle n'évolue que lorsqu'on surcharge le matériau, autrement dit sa densité de probabilité ne dépend que de σ_m :

$$p_{réel} = P(\xi ; \sigma_m).$$

La fissuration apparente dépend en plus de la contrainte actuelle σ, on la définit par sa densité de probabilité $Pa(\xi ; \sigma_m , \sigma)$.

L'utilisation du critère de fermeture précédent permet de relier la fissuration apparente à la fissuration réelle :

$$Pa(\xi ; \sigma_m , \sigma) = \begin{cases} P(\xi ; \sigma_m), & \text{si } \xi \in \mathcal{E}(\sigma) \\ 0, & \text{si } \xi \notin \mathcal{E}(\sigma), \end{cases} \tag{24}$$

avec $\mathcal{E}(\sigma) = \{\xi / \sigma \sin^2 \phi + \alpha \tau(k) > 0\}$.

Dans le cas de fissures circulaires (k = 0), les paramètres de fissuration se réduisent à $\xi = (\epsilon, \alpha, u)$, où $u = \sin^2 \phi$ est le seul paramètre d'orientation pour un matériau statistiquement isotrope de révolution. Le paramètre τ devient alors une constante du matériau.

En introduisant les variables adimensionnelles t et t_m :

$$t = \frac{\sigma}{\tau} \quad \text{et} \quad t_m = \frac{\sigma_m}{\tau}, \quad t_m \leqslant t \leqslant 0,$$

les expressions des célérités de propagation dans un milieu isotrope de révolution soumis au chargement (t_m, t) s'écrivent :

$$V_{ij}(t_m, t) = V_{ij}(0, 0) \times \{1 - \mathcal{V}_{ij}(t_m, t)\}, \quad (1 \leqslant i, j \leqslant 3) \qquad (25)$$

avec

$$\mathcal{V}_{ij}(t_m, t) = \int_{\mathscr{E}(t)} \epsilon\, Q_{ij}(u)\, P(\epsilon, \alpha, u; t_m)\, d\xi, \qquad (26)$$

$$\mathscr{E}(t) = \{(\epsilon, \alpha, u) / \alpha + ut > 0\} \subset \mathscr{E}_0,$$

$$\mathscr{E}_0 = \{(\epsilon, \alpha, u) / \epsilon \in [0, \infty], \; \alpha \in [0, 1], \; u \in [0, 1]\},$$

et $Q_{ij}(u)$ des polynômes du deuxième degré en u dont les coefficients ne dépendent que du coefficient de Poisson ν de la matrice.

5.2. *Evolution de la fissuration du béton*

Hypothèse

En introduisant les valeurs expérimentales de $\mathcal{V}_{ij}(t_m, t)$ dans (26), il s'agit d'inverser les expressions pour obtenir des informations sur l'évolution de la fissuration du matériau avec le chargement par l'intermédiaire de sa densité de probabilité $P(\xi; t_m)$.

Le problème est traité dans un cas simple, cf. [5], en se donnant a priori P sous la forme :

$$P(\epsilon, \alpha, u; t_m) = p(\epsilon; t_m)\, r(\alpha; t_m)\, q(u). \qquad (27)$$

Autrement dit, on suppose en particulier que l'allongement des fissures se fait indépendamment de leur orientation et que le chargement ne tend pas à privilégier une direction de fissuration qui reste orientée comme la fissuration initiale.

Au vu des courbes expérimentales, ces hypothèses paraissent raisonnables tout au moins durant la première phase de chargement : $\sigma_m < 0,75$ Rc.

Etude de la phase de déchargement

On se limite ici à l'étude de la phase de déchargement (t_m, $t = 0$). En appelant $\mathcal{V}_{ij}^d (t_m) \equiv \mathcal{V}_{ij} (t_m, 0)$ et en reportant (27) dans (26), il vient :

$$\mathcal{V}_{ij}^d (t_m) = Q_{ij} \{\langle \epsilon (t_m) \rangle - \langle \epsilon (0) \rangle\}, \tag{28}$$

où $\langle \epsilon (t_m) \rangle$ et $\langle \epsilon (0) \rangle$ représentant respectivement les densités de fissurations moyennes du matériau lors des surcharges t_m et 0, et

$$Q_{ij} = \int_0^1 Q_{ij} (u)\, q(u)\, du. \tag{29}$$

Comme $Q_{ij} (u)$ sont des polynômes du 2e degré en u, les coefficients Q_{ij} font intervenir les moments d'ordre 1 et 2, q_1 et q_2, de $q(u)$:

$$Q_{ij} = Q_{ij}^0 + Q_{ij}^1 q_1 + Q_{ij}^2 q_2.$$

Les courbes expérimentales de déchargement sont sensiblement linéaires, cf. figure 9 : $\mathcal{V}_{ij}^d (t_m) = k_{ij}^d \cdot t_m$.

Comme les densités de fissuration sont indépendantes du type d'onde (i, j), l'équation (28) n'est satisfaite que si :

$$Q_{ij} = Q \cdot k_{ij}^d, \tag{30}$$

qui constitue un système linéaire de 4 équations à 3 inconnues Q, q_1 et q_2 qui n'admet une solution que si les k_{ij}^d vérifient une condition de compatibilité du type :

$$a_{ij} (\nu)\, k_{ij}^d = 0. \tag{31}$$

Pour $\nu = 0,30$, les coefficients a_{ij} prennent respectivement les valeurs :

$$a_{12} = 1, \quad a_{13} = 0,1566, \quad a_{11} = -0,5795, \quad a_{33} = 0,0538.$$

Si la condition de compatibilité (31) est satisfaite, 3 des 4 équations de (30) fournissent Q, q_1 et q_2, dont on déduit $\langle \epsilon (t_m) \rangle$:

$$\langle \epsilon (t_m) \rangle = \langle \epsilon (0) \rangle + \frac{t_m}{Q}.$$

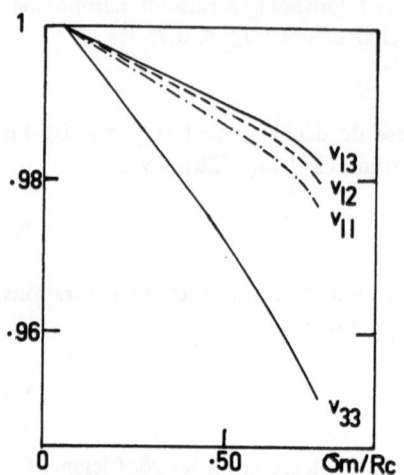

Figure 9. – *Courbes expérimentales des états déchargés* : $V_{ij}(\sigma_m)$ à $\sigma = 0$.

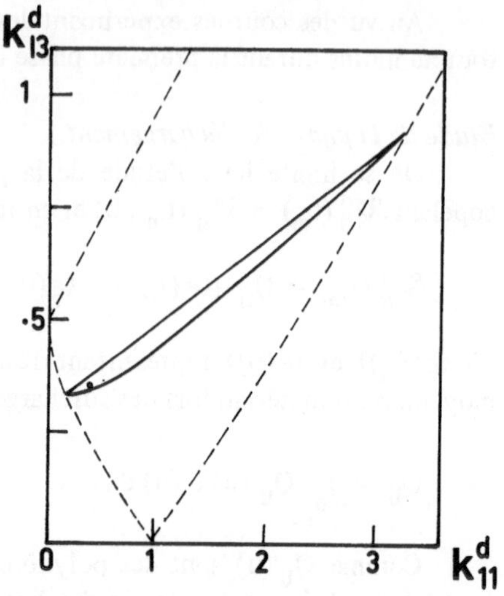

Figure 10. – *Domaines d'admissibilité des pentes des courbes expérimentales d'états déchargés.*

—— domaine d'admissibilité pour $\nu = 0.30$;
--- domaine d'admissibilité pour $\nu \in [0, 1/2]$;
●   point expérimental

Autrement dit, les informations recueillies sur l'évolution de la fissuration sont :

— la variation de la densité de fissuration moyenne avec t_m,
— l'orientation moyenne de la fissuration : $q_1 = \langle u \rangle$,
— l'écart-type de l'orientation par rapport à $\langle u \rangle$ grâce à q_2.

On peut également s'assurer au préalable de l'applicabilité du modèle, car, outre la relation de compatibilité, les coefficients k_{ij}^d doivent appartenir à un domaine d'admissibilité provenant des inégalités auxquelles doivent satisfaire q_1 et q_2

$$q_1^2 \leqslant q_2 \leqslant q_1.$$

Pour ν fixé, ce domaine d'admissibilité est délimité dans le plan $(k_{11}^d, k_{13}^d)^1$ par une droite (D_ν) et une branche d'hyperbole (H_ν). Ce domaine est représenté dans la figure 10 pour $\nu = 0,30$, ainsi que le point expérimental relatif à l'éprouvette présentée précédemment.

---

1 On peut fixer arbitrairement $k_{33}^d = 1$, et on suppose que k_{12}^d vérifie (31).

En pratique, compte tenu des approximations du modèle et de l'imprécision des mesures, on ne peut espérer obtenir une estimation significative de q_2, mais on conservera néanmoins le domaine d'admissibilité.

6. Conclusion

La méthode dynamique expérimentale mise en œuvre paraît plus fine que les méthodes statiques, puisqu'elle permet de mesurer avec précision des modifications, même faibles, de la structure du matériau. Elle fait apparaître deux comportements antagonistes de la fissuration sous compression : création de nouvelles surfaces fissurées dans les phases de surchargement et serrage des fissures existantes ; de plus, elle semble mettre en défaut les hypothèses de phase élastique du matériau.

Le modèle théorique introduit permet de rendre compte qualitativement des phénomènes. Il fournit, de plus, des informations quantitatives sur la fissuration, telles la densité moyenne de fissuration et l'orientation moyenne des fissures.

Les résultats expérimentaux sont toutefois en nombre insuffisants pour pouvoir en déduire des lois empiriques du développement de la microfissuration du béton. L'approche théorique présentée ne constitue qu'un outil de mesure de la fissuration qu'il va s'avérer nécessaire de compléter par un modèle spécifique de loi de comportement ou d'évolution de la fissuration avec le chargement.

REFERENCES

[1] BAMBERGER Y. et J.J. MARIGO. – "Propagation des ondes en milieu fissuré", *C.R. Acad. Sc.,* Paris, t. **289** (19 Nov. 1979), 215-218.
[2] CANNARD G. – Rapports internes des Laboratoires des Ponts et Chaussées. 1977-1978-1979.
[3] HASHIN Z. – "Theory of composite materials". *5th Symposium on Structural Mechanics,* Philadelphia, 1967.
[4] KASSIR M.K. and G.C. SIH. – "Three dimensional stress distribution around an elliptical crack under arbitrary loadings". *J. Appl. Mech. Trans. ASME,* **33,** 3, 1966.
[5] MARIGO J.J. – "Propagation des ondes ultrasonores et microfissuration du béton". Thèse de 3e cycle, Université Pierre et Marie Curie, Paris, Janvier 1980.
[6] O'CONNELL R.J. and B. BUDIANSKY. – "Seismic velocities in dry and saturated cracked solids". *Journal of Geophysical Research,* **79,** 1974.
[7] SIEGFRIED R. and G. SIMMONS. – "Characterization of oriented cracks with differential strain analysis". *Journal of Geophysical Research,* **83,** 1978.
[8] SIH G.C. and H. LIEBOWITZ. – "Theory of cracks in three dimensions". *Fracture II,* Academic Press, (1968): 131-166.

ABSTRACT

We observe experimentally some variations of ultrasonic waves velocities in an uniaxially compressed concrete, which are attributed to microcracks. We give a propagation theoretical model in a randomly microcracked elastic medium. The model shows a strong anisotropy when all cracks have the same orientation. We initiate a closure criterion and apply this model to a transversely isotropic concrete.

Session 10

Damage and Creep

Endommagement et Fluage

Le Concept de Contrainte Effective Appliqué à l'Élasticité et à la Viscoplasticité en Présence d'un Endommagement Anisotrope

J. L. Chaboche

Office National d'Études et de Recherches Aérospatiales,
(ONERA), Châtillon, France.

1. Cadre général

1.1. *Notion classique d'endommagement*

Depuis les travaux de Kachanov [19] et Rabotnov [32] le mécanicien des milieux continus dispose d'un nouveau type de variable pour décrire l'état du matériau : l'endommagement de rupture qui indique le degré de détérioration de la matière et sa perte de résistance. Contrairement aux variables de déformation qui décrivent des mouvements (réversibles ou non) l'endommagement est lié à la création et à la propagation de microdécohésions (microfissures, microvides) qui diminuent la résistance du matériau et conduisent finalement à l'amorçage d'une fissure macroscopique.

Bien que cela ne soit pas fondamental pour notre propos, précisons tout de suite ce que le mécanicien appelle fissure macroscopique : c'est la fissure prépondérante, celle qui va continuer à se propager au détriment des autres, et qui a déjà une taille suffisante pour traverser quelques grains de matière et présenter une géométrie moyenne suffisamment bien définie : à ce stade, la Mécanique du Milieu Continu Endommagé ne s'applique plus et l'on fait appel à la Mécanique de Rupture.

Cette variable endommagement décrit globalement la présence des micro-défauts : dans le cas unidimensionnel, on introduit classiquement une variable scalaire D, nulle dans les conditions initiales du matériau vierge, atteignant une valeur caractéristique D_c à la rupture (dans la plupart des applications, on considère que $D_c \simeq 1$).

L'intérêt de l'approche introduite par Kachanov et Rabotnov est de permettre de décrire l'influence de l'endommagement sur la résistance à la déformation, ce qui fournit un moyen indirect de mesure de cette variable

[25, 4]. Ceci se fait au travers de la définition d'une contrainte effective : c'est la contrainte $\widetilde{\sigma}$ qu'il faut appliquer à l'élément de volume vierge pour obtenir la même déformation ϵ que celle provoquée par la contrainte σ appliquée à l'élément de volume endommagé (Fig. 1).

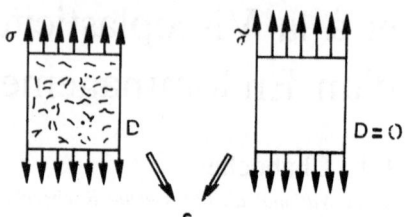

Figure 1. – *Le concept de contrainte effective.*

De façon classique, on suppose la relation suivante qui revient à considérer D comme une perte de section résistante effective :

$$\widetilde{\sigma} = \sigma/(1 - D) \tag{1}$$

Ainsi dans le cas particulier du comportement élastique linéaire, la mesure du module d'élasticité $\widetilde{E}$ du matériau endommagé fournit une mesure de D par comparaison avec la valeur E du module d'élasticité du matériau vierge [26] :

$$\widetilde{E} = E(1 - D)$$

Dans d'autres types de comportement, en viscoplasticité par exemple, il suffit de remplacer la contrainte σ par $\widetilde{\sigma}$ dans la loi d'écoulement pour obtenir une description assez précise de l'effet accélérateur du processus de déformation, dû à l'augmentation de l'endommagement.

1.2. *L'anisotropie de l'endommagement*

Les notions unidimensionnelles d'endommagement et de contrainte effective se généralisent facilement au cas tridimensionnel tant que l'endommagement est isotrope : la présence des microdéfauts se traduit par une perte de résistance identique quelle que soit la direction considérée dans l'espace des contraintes et la variable scalaire D est suffisante pour développer des théories satisfaisantes : on se reportera entre autres aux travaux de Rabotnov [33] Martin et Leckie [21] Hayhurst [13], Lemaître et Chaboche [25, 22, 3] sur ce sujet.

Malheureusement, le comportement réel est beaucoup moins simple et de nombreux travaux expérimentaux démontrent l'existence d'une anisotropie de l'endommagement, aussi bien pour ce qui concerne sa structure micro-scopique [17, 18] que pour les effets macroscopiques : ainsi on observe généra-lement que les microdécohésions se multiplient et se propagent dans des directions plus ou moins perpendiculaires à la direction de la contrainte principale maximale (en cas de chargement proportionnel bien sûr). Ceci rejoint évidemment les observations faites en Mécanique de la Rupture pour des fissures macroscopiques.

Pour traduire ce phénomène Rabotnov [34], Leckie, Hayhurst et Martin [27, 14], Goel [12] introduisent le tenseur de contrainte effectif de la forme suivante dans le système principal des contraintes :

$$\tilde{\sigma}_1 = \sigma_1/(1 - D) \quad \tilde{\sigma}_2 = \sigma_2 \quad \tilde{\sigma}_3 = \sigma_3 \tag{2}$$

où $\sigma_1, \sigma_2, \sigma_3$ sont les contraintes principales dans l'ordre décroissant. Une telle théorie avec une variable D restant scalaire est utilisable tant que l'on ne considère qu'un chargement proportionnel. Dans le cas contraire, il faut envisager l'introduction d'un tenseur endommagement : deux sortes de théorie ont été proposées jusqu'à présent :

— la première est basée sur un tenseur endommagement **D** du second ordre et dans laquelle le tenseur des contraintes effectives s'exprime sous la forme générale :

$$\tilde{\sigma} = \frac{1}{2} [(1 - D)^{-1} \cdot \sigma + \sigma \cdot (1 - D)^{-1}] \tag{3}$$

Cette approche due à Murakami et Ohno [28] est très intéressante par sa généralité et sa simplicité pour l'étude du matériau viscoplastique mais elle ne permet pas de traiter le cas du matériau élastique, comme nous le verrons plus loin.

— la seconde utilise la loi d'élasticité du matériau endommagé pour définir en même temps le tenseur endommagement et le tenseur des contraintes effectives. Dans ce cadre, plusieurs formulations peuvent être proposées suivant que l'on utilise une équivalence en déformation, en contrainte, ou en énergie : ce dernier cas est celui étudié par Sidoroff et Cordebois [8]. Nous nous intéressons ci-dessous au cas d'une équivalence en déformation.

1.3. *Lignes générales de l'approche proposée*

Commençons par deux remarques :

l'anisotropie de l'endommagement peut se mesurer ou s'identifier de deux manières différentes pour un même état actuel endommagé : soit en mesurant la perte de résistance à la déformation dans diverses directions de l'espace des contraintes, dans le cadre du comportement élastique par exemple, soit en effectuant divers essais jusqu'à rupture (essais de fluage en traction dans des directions différentes) à partir d'un même état d'endommagement.

— l'anisotropie n'est pas une caractéristique de l'endommagement lui-même mais plutôt de sa loi d'évolution : on peut ainsi imaginer un chargement non simple produisant un état endommagé quasi isotrope (**D** sphérique) sur un matériau dans lequel la loi d'évolution est anisotrope.

La théorie que nous proposons consiste à introduire l'anisotropie dans la loi d'évolution de **D**, conformément à la remarque ci-dessus.

La loi est choisie de façon à rendre compte du comportement élastique du matériau endommagé par un essai en chargement simple (proportionnel) : elle est particularisée en étudiant deux cas limites : — celui de l'endommagement isotrope pour lequel les défauts sont répartis en direction de façon statistiquement isotrope, — et un cas idéal d'anisotropie élastique provoquée par des réseaux de microfissures toutes parallèles. Une hypothèse simplificatrice est faite pour que la non linéarité des phénomènes d'endommagement puisse être traduite par la loi d'évolution d'un paramètre scalaire D.

On verra que le moyen le plus simple de satisfaire à ces conditions est d'utiliser un tenseur d'endommagement du quatrième ordre mais non symétrique et que l'anisotropie de sa loi d'évolution peut être traduite par trois coefficients seulement, dépendant du matériau et éventuellement de la température. Un exemple concret sera étudié dans le cas particulier d'une loi d'endommagement de fluage.

2. Le matériau élastique soumis à un endommagement anisotrope

2.1. Généralités

Cet aspect de la théorie étant abordé de façon plus détaillée par Sidoroff et Cordebois [8] on se contente ici d'apporter les compléments nécessaires à l'approche particulière qui est envisagée. Tout ce qui suit est évidemment écrit dans le cadre des petites transformations.

La façon la plus générale d'introduire des modifications anisotropes du comportement élastique est d'introduire un opérateur d'ordre huit [5] : en effet le comportement élastique est lui-même caractérisé par un opérateur d'ordre quatre Λ permettant de passer du tenseur des déformations élastiques au tenseur des contraintes :

$$\sigma = \Lambda : \varepsilon_e \tag{4}$$

Le comportement élastique du matériau endommagé est décrit par un tenseur d'élasticité $\widetilde{\mathbf{\Lambda}}(\mathbf{D})$:

$$\boldsymbol{\sigma} = \widetilde{\mathbf{\Lambda}}(\mathbf{D}) : \boldsymbol{\varepsilon}_e \tag{5}$$

$\mathbf{D}$ représente formellement un opérateur d'ordre huit puisqu'il assure la transformation de $\mathbf{\Lambda}$ en $\widetilde{\mathbf{\Lambda}}$.

Le tenseur de contrainte effectif s'exprime alors facilement. En suivant le concept de l'équivalence en déformation, c'est celui qu'il faut appliquer au matériau vierge pour obtenir le même tenseur de déformation élastique que celui produit par le tenseur $\boldsymbol{\sigma}$ appliqué au matériau endommagé (la déformation élastique est bien sûr mesurée par rapport à un même état de référence ; l'état relaché) :

$$\widetilde{\boldsymbol{\sigma}} = \mathbf{\Lambda} : \boldsymbol{\varepsilon}_e \tag{6}$$

La combinaison des relations (5) et (6) donne :

$$\widetilde{\boldsymbol{\sigma}} = \mathbf{M}(\mathbf{D}) : \boldsymbol{\sigma} = \mathbf{\Lambda} : \mathbf{\Lambda}^{-1}(\mathbf{D}) : \boldsymbol{\sigma} \tag{7}$$

La manipulation de l'opérateur $\mathbf{D}$ d'ordre huit dans le cas d'un chargement complexe pose de difficiles problèmes et il est donc souhaitable de se ramener à des opérateurs d'ordre inférieur. Dans ce but, Murakami [28] remplace la relation (7) définissant la contrainte effective par la relation (3) indiquée plus haut dans laquelle $\mathbf{D}$ est un tenseur d'ordre deux : malheureusement une équivalence en déformation ne peut plus s'appliquer car elle conduit à un tenseur d'élasticité $\widetilde{\mathbf{\Lambda}}(\mathbf{D})$ non symétrique ($\widetilde{\Lambda}_{1122} \neq \widetilde{\Lambda}_{2211}$) condition cependant imposée par le principe de Onsager par exemple : en se plaçant dans le système d'axes principaux de $\mathbf{D}$ on vérifie facilement que cette condition de symétrie n'est vérifiée que si $\mathbf{D}$ est sphérique (ou isotrope : $D_1 = D_2 = D_3$).

En utilisant une équivalence en énergie de déformation élastique Sidoroff et Cordebois [8] proposent eux aussi l'utilisation d'un tenseur d'endommagement d'ordre deux qui permet d'exprimer la contrainte effective et le tenseur d'élasticité par :

$$\widetilde{\boldsymbol{\sigma}} = (1 - \mathbf{D})^{-1/2} \cdot \boldsymbol{\sigma} \cdot (1 - \mathbf{D})^{-1/2}$$

$$\widetilde{\mathbf{\Lambda}}(\mathbf{D}) = (1 - \mathbf{D})^{1/2} \otimes (1 - \mathbf{D})^{1/2} \otimes (1 - \mathbf{D})^{1/2} \otimes (1 - \mathbf{D})^{1/2} :: \mathbf{\Lambda} \tag{8}$$

où $\otimes$ désigne le produit tensoriel et $::$ le produit tensoriel contracté sur quatre indices. Cette forme de $\widetilde{\mathbf{\Lambda}}$ présente l'inconvénient de ne pouvoir être identifiée avec une anisotropie élastique quelconque et en particulier avec le cas idéal que nous utiliserons ci-dessous.

2.2. *Tenseur d'endommagement d'ordre quatre*

C'est une formulation avec un tenseur d'ordre quatre non symétrique qui nous semble fournir le meilleur compromis entre un souci légitime de simplicité et un désir non moins légitime de décrire le mieux possible les phénomènes réels : les possibilités phénoménologiques restent très larges avec un nombre restreint de coefficients caractéristiques pour définir l'évolution des trente six composantes du tenseur (six composantes pour le tenseur d'ordre deux, deux cent trente et un pour le tenseur d'ordre huit) dans le cas le plus général.

Remarquons que l'introduction d'un tel tenseur se justifie par les théories modernes d'homogénéisation qui montrent qu'un endommagement orthotrope[1] modifie les composantes du tenseur de rigidité élastique Λ par une relation du type [10] [11] [30] :

$$\Lambda_{ijrs} = \frac{1}{V} \left[\int_{V*} \Lambda_{ijrs}\, dv - \int_{V*} \Lambda_{ijk\ell}\, b_{k\ell rs}\, dv \right]$$

où V est le volume apparent de matière, V* le volume solide (c'est-à-dire le volume apparent diminué du volume des vides), $b_{k\ell rs}$ une famille de champs de déformation, solutions élémentaires du problème d'élasticité sur la maille élémentaire définissant la structure à homogénéiser.

Si la matière solide est homogène on a :

$$\widetilde{\Lambda}_{ijrs} = \frac{V*}{V} \Lambda_{ijrs} - \Lambda_{ijk\ell} \frac{1}{V} \int_{V*} b_{k\ell rs}\, dv$$

qui peut encore se mettre sous la forme :

$$\widetilde{\Lambda}_{ijrs} = \Lambda_{ijk\ell} \left[\delta_{k\ell rs} - \left(1 - \frac{V*}{V}\right) \delta_{k\ell rs} - \frac{1}{V} \int_{V*} b_{k\ell rs}\, dv \right]$$

On voit ainsi apparaître un tenseur endommagement d'ordre quatre, de composantes :

$$D_{k\ell rs} = \left(1 - \frac{V*}{V}\right) \delta_{k\ell rs} + \frac{1}{V} \int_{V*} b_{k\ell rs}\, dv$$

---

[1] La géométrie des défauts est périodique et comporte plusieurs plans de symétrie.

Ce cas orthotrope est intéressant puisqu'il implique une géométrie de défauts comportant plusieurs plans de symétrie, ce qui correspond bien à l'idée du développement des microfissures dans des plans perpendiculaires a la contrainte principale maximale.

L'obtention des champs $b_{k\ell rs}$ dans un cas quelconque n'est pas possible. Par contre des travaux récents [9, 16] portant sur le comportement élastique d'un matériau contenant divers arrangements de fissures planes et parallèles (Fig. 2) peuvent être utilisés pour obtenir la forme du tenseur endommagement dans ce cas idéal orthotrope.

Dans le cas d'un milieu bidimensionnel les auteurs montrent que la matrice des complaisances élastiques apparentes a la forme suivante (c'est une matrice 6x6 correspondant aux six composantes σ_{11}, σ_{22}, σ_{33}, σ_{23}, σ_{31}, σ_{12} dans cet ordre) lorsque le matériau sans fissure est isotrope de module d'Young E et coefficient de Poisson ν [9] :

$$\tilde{\Lambda}^{-1} = \begin{bmatrix} \dfrac{1}{E_1} & -\dfrac{\nu_{21}}{E_2} & -\dfrac{\nu_{31}}{E_3} & & & \\ -\dfrac{\nu_{12}}{E_1} & \dfrac{1}{E_2} & -\dfrac{\nu_{32}}{E_3} & & & \\ -\dfrac{\nu_{13}}{E_1} & -\dfrac{\nu_{23}}{E_2} & \dfrac{1}{E_3} & & & \\ & & & \dfrac{1}{2\mu_{23}} & & \\ & & & & \dfrac{1}{2\mu_{31}} & \\ & & & & & \dfrac{1}{2\mu_{12}} \end{bmatrix} = \dfrac{1}{E} \begin{bmatrix} 1+Q_1 & -\nu & -\nu & & & \\ -\nu & 1 & -\nu & & & \\ -\nu & -\nu & 1 & & & \\ & & & 1+\nu & & \\ & & & & (1+\nu)(1+Q_5) & \\ & & & & & (1+\nu)(1+Q_6) \end{bmatrix}$$

$$(9)$$

où Q_1, Q_5, Q_6 dépendent de la surface, de l'espacement et de l'arrangement des fissures. La matrice des raideurs s'obtient facilement par inversion :

$$\tilde{\Lambda} = \dfrac{E}{(1+\nu)(1-2\nu)} \begin{bmatrix} (1-\nu)(1-D_1) & \nu(1-D_1) & \nu(1-D_1) & & & \\ \nu(1-D_1) & (1-\nu)(1-D_2) & \nu(1-D_{23}) & & & \\ \nu(1-D_1) & \nu(1-D_{23}) & (1-\nu)(1-D_2) & & & \\ & & & 1-2\nu & & \\ & & & & (1-2\nu)(1-D_5) & \\ & & & & & (1-2\nu)(1-D_6) \end{bmatrix}$$

$$(10)$$

où l'on a posé :

$$D_1 = \dfrac{(1-\nu)\,Q_1}{1-\nu-2\nu^2+(1-\nu)\,Q_1} \qquad D_2 = \left(\dfrac{\nu}{1-\nu}\right)^2 D_1$$

$$D_{23} = \dfrac{\nu}{1-\nu}\,D_1 \qquad D_5 = \dfrac{Q_5}{1+Q_5} \qquad D_6 = \dfrac{Q_6}{1+Q_6}$$

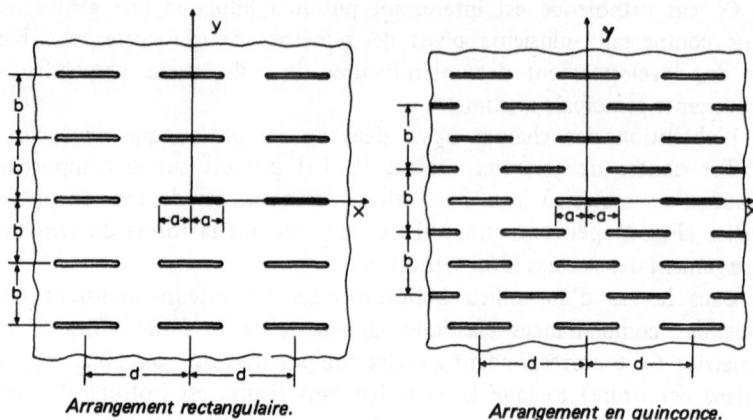

Arrangement rectangulaire.   Arrangement en quinconce.

Figure 2. – *Arrangements de fissures permettant l'établissement d'une solution quasi analytique.*

Comme on le voit, trois variables endommagement seulement (D_1, D_5, D_6) sont nécessaires pour décrire ce cas particulier. Le tenseur d'endommagement est celui qui permet de passer de Λ à $\widetilde{\Lambda}$ par :

$$\widetilde{\Lambda}(D) = (1 - D) : \Lambda \tag{11}$$

où 1 est l'opérateur unité d'ordre quatre. Il s'obtient facilement par :

$$D = 1 - \widetilde{\Lambda} : \Lambda^{-1} \tag{12}$$

c'est-à-dire, dans le cas présent :

$$D = \begin{bmatrix} D_1 & 0 & 0 & & & \\ \dfrac{\nu}{1-\nu} D_1 & 0 & 0 & & & \\ \dfrac{\nu}{1-\nu} D_1 & 0 & 0 & & & \\ & & & 0 & & \\ & & & & D_5 & \\ & & & & & D_6 \end{bmatrix} \tag{13}$$

Le tenseur des contraintes effectives est donné alors par (7) et (11)

$$\widetilde{\sigma} = (1 - D)^{-1} : \sigma \tag{14}$$

où la matrice représentative de l'opérateur $(1 - D)^{-1}$ s'exprime :

$$(1-D)^{-1} = \begin{bmatrix} \dfrac{1}{1-D_1} & 0 & 0 & & & \\[2mm] \dfrac{\nu}{1-\nu}\dfrac{D_1}{1-D_1} & 1 & 0 & & & \\[2mm] \dfrac{\nu}{1-\nu}\dfrac{D_1}{1-D_1} & 0 & 1 & & & \\[2mm] & & & 1 & & \\[2mm] & & & & \dfrac{1}{1-D_5} & \\[2mm] & & & & & \dfrac{1}{1-D_6} \end{bmatrix}$$

L'état d'endommagement décrit par la matrice (13) correspond à une anisotropie idéale créée en état de chargement simple dans le système d'axes principaux avec $\sigma_1 > \sigma_2 > \sigma_3$. On vérifie facilement que les axes principaux du tenseur des contraintes effectives sont les mêmes et que $\widetilde{\sigma}_1 > \widetilde{\sigma}_2 > \widetilde{\sigma}_3$. Remarquons que les variables d'endommagement D_5 et D_6 n'interviennent qu'en cas de chargement non simple et que la contrainte effective diffère du cas anisotrope pur (2) introduit par Leckie et ses collaborateurs [27, 14, 12].

L'intérêt de la particularisation (13) pour le tenseur d'endommagement dans le cas de chargement simple réside dans la possibilité d'identification des rapports d'anisotropie à partir des résultats de l'analyse élastique du cas idéal. Dans le cas tridimensionnel, on a $D_5 = D_6$ [16] et la relation (9) donne :

$$\begin{cases} \dfrac{E_1}{E} = \dfrac{\nu_{12}}{\nu} = \dfrac{\nu_{13}}{\nu} = \dfrac{(1-\nu)(1-D_1)}{1-\nu-2\nu^2 D_1} & \quad \dfrac{\mu_{31}}{\mu} = \dfrac{\mu_{12}}{\mu} = 1 - D_6 \\[3mm] \nu_{21} = \nu_{31} = \nu_{23} = \nu_{32} = \nu & \quad \mu_{23} = \mu \end{cases} \tag{15}$$

Nous supposerons par la suite que $D_6 = aD_1$. Les résultats de l'analyse élastique montrent que $a \# 0,5$ dans le domaine $0 < D_1 < 0,5$ qui intéresse la plus grande partie de la vie de l'élément de volume.

3. Loi d'évolution de l'endommagement

3.1. *Forme générale proposée*

La loi d'evolution du tenseur **D** peut avoir une forme très complexe faisant apparaître d'une part, l'anisotropie de l'endommagement, d'autre part la forte non linéarité des phénomènes physiques liés à la nucléation et la progression des microfissures. De façon à disposer d'une formulation suffisamment simple pour être utilisable de façon pratique et identifiable par quelques essais particuliers seulement, tout en fournissant une modélisation assez fine des phénomènes d'anisotropie, nous nous limitons au cadre fourni par les trois hypothèses qui suivent :

1) L'anisotropie de la loi d'évolution est indépendante de l'instant considéré au cours de la vie de l'élément de volume et du type de sollicitation : elle dépend seulement de l'orientation de la direction de contrainte principale maximale (comme dans [28]) et se met sous la forme :

$$\dot{\mathbf{D}} = \mathbf{Q}(\widetilde{\sigma})\,\dot{D} \tag{16}$$

où **Q** est un tenseur d'ordre quatre qui dépend du matériau, éventuellement de la température, et des directions principales du tenseur de contrainte effectif. Toute la non linéarité des phénomènes est contenue dans la loi d'évolution du paramètre scalaire D qui est représentatif de la quantité de microdéfauts par unité de volume, indépendamment de leur orientation.

2) Dans le système d'axes principaux de $\widetilde{\sigma}$ (avec $\widetilde{\sigma}_1 > \widetilde{\sigma}_2 > \widetilde{\sigma}_3$) le tenseur **Q** ne dépend plus que du matériau et de la température. La loi s'exprime alors :

$$\dot{\mathbf{D}} = \mathbf{Q}^*\,\dot{D} \tag{17}$$

3) La forme du tenseur de répartition de l'endommagement **Q*** est particularisée à partir des résultats de l'analyse élastique faite en 2-2 :

$$\mathbf{Q}^* = \gamma\,\mathbf{1} + (1 - \gamma)\,\mathbf{\Gamma} \tag{18}$$

où **1** représente le tenseur unité du quatrième ordre (de composantes $\Delta_{ijk\ell} = \delta_{ik}\,\delta_{j\ell}$, δ étant le symbole de Kroenecker) et **Γ** désigne le tenseur du quatrième ordre représenté dans le système principal par :

$$\Gamma = \begin{bmatrix} 1 & & 0 & 0 & & \\ \nu/1-\nu & & 0 & 0 & & \\ \nu/1-\nu & & 0 & 0 & & \\ & & & & 0 & \\ & & & & & a \\ & & & & & & a \end{bmatrix} \tag{19}$$

Il est facile de vérifier que $\gamma = 1$ correspond au cas parfaitement isotrope et qu'avec $\gamma = 0$ on retrouve exactement le cas idéal d'élasticité anisotrope examiné précédemment. Cette forme de loi, assez générale puisque permettant une combinaison linéaire quelconque du cas isotrope et du cas anisotrope idéal ne nécessite la détermination que des deux coefficients γ et a en plus de la loi d'évolution du dommage scalaire D. La présence des termes $\nu/1-\nu$ permet dans tous les cas d'obtenir un tenseur d'élasticité $\widetilde{\Lambda}$ symétrique (pour un matériau élastique isotrope initialement) de la forme (10).

Dans le cas d'un chargement non proportionnel c'est la loi (16) qui doit être employée. En pratique on obtient la matrice représentative de **Q** dans le système de référence :

$$Q_{ijk\ell} = V_{ijrs} \, Q^*_{rspq} \, \overline{V}_{pqk\ell} \tag{20}$$

où V s'exprime en fonction de la matrice de passage du système principal au système de référence

$$V_{ijrs} = P_{ir} \, P_{js} \tag{21}$$

En fait, les relations (20) et (21) ne s'appliquent que pour chaque indice variant de 1 à 3 ce qui fait que l'on utilise des matrices 9 x 9 au lieu des matrices 6 x 6 qui ont été indiquées ci-dessus : cela ne pose cependant aucun problème de principe.

Le critère de rupture peut n'être défini que sur la variable scalaire D puisque, si a $<$ 1 (mais on a vu que a était proche de 0,5) toute composante de **D** est inférieure à D quels que soient les chargements subis. Nous poserons donc comme valeur limite définissant l'amorçage d'une fissure macroscopique ou la rupture de l'élément de volume : $D = D_c \simeq 1$.

3.2. *Loi d'endommagement de fluage*

La loi d'évolution de la variable scalaire D est maintenant établie dans le cas de l'endommagement par fluage qui se produit à température élevée sous des

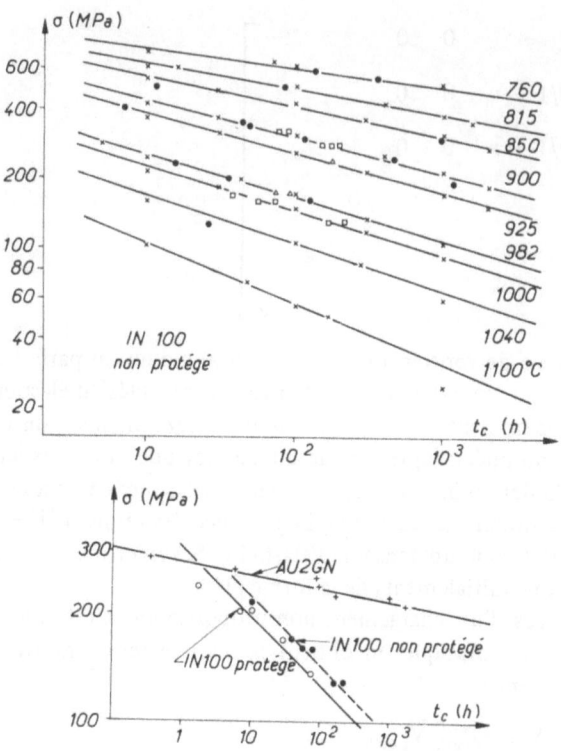

Figure 3. − *Temps à rupture en fluage ; loi de Rabotnov- Kachanov.*

sollicitations constantes ou lentement variables. Dans ce domaine, les lois d'endommagement sont généralement basées sur les travaux initiaux de Kachanov et Rabotnov. Dans le cas de la traction, la loi d'endommagement de Rabotnov s'écrit [33] :

$$\dot{D} = \left[\frac{\sigma}{A(1-D)} \right]^r (1-D)^{-k} \qquad (22)$$

Pour une contrainte constante, le temps à rupture dans l'essai de fluage s'exprime après intégration entre $D = 0$ et $D = 1$:

$$t_\sigma = \frac{1}{k+1} \left(\frac{\sigma}{A} \right)^{-r} \qquad (23)$$

Cette relation s'applique dans un assez large domaine pour de nombreux matériaux, en particulier pour l'alliage réfractaire IN 100 comme le montre la figure 3 : les essais de fluage jusqu'à rupture permettent la détermination des coefficients r et A. Le coefficient k, lié à la rapidité d'évolution de l'endommagement, peut être déterminé par les courbes de fluage tertiaire [5].

En fait, des études expérimentales montrent que k dépend légèrement de la contrainte [24, 25], mais nous nous limitons ici à la généralisation tridimensionnelle de la loi de Rabotnov (22) (le cas particulier où k = 0 correspond à la théorie initiale de Kachanov [19]).

Les travaux de Leckie et ses collaborateurs [27, 14, 12] puis de Murakami et Ohno [28] montrent comment généraliser la loi de Rabotnov en tridimensionnel : on utilise une fonction invariante $\chi(\widetilde{\sigma})$ du tenseur de contrainte effectif et du tenseur d'endommagement **D**. D'après les observations de Johnson et ses collaborateurs [18] sur des éprouvettes sous chargement multiaxial, il y a deux types essentiels de critères de rupture : l'un est caractérisé par la contrainte principale maximale de traction $\widetilde{\sigma}_1$ (comme le cuivre), l'autre par la contrainte de cisaillement octahédrale (comme l'aluminium) :

$$J(\widetilde{\sigma}) = \left[\frac{1}{2} \left[(\widetilde{\sigma}_1 - \widetilde{\sigma}_2)^2 + (\widetilde{\sigma}_2 - \widetilde{\sigma}_3)^2 + (\widetilde{\sigma}_3 - \widetilde{\sigma}_1)^2 \right] \right]^{1/2} \qquad (24)$$

Ceci n'implique pas nécessairement que le premier cas corresponde à un endommagement prépondérant dans la direction associée à $\widetilde{\sigma}_1$, ni que le second corresponde à un endommagement isotrope, des contre-exemples ayant été observés sur divers matériaux [18, 15]. En utilisant le critère de rupture proposé par Hayhurst [13] qui donne aussi une influence de la contrainte hydrostatique on pose :

$$\chi(\widetilde{\sigma}) = \alpha \widetilde{\sigma}_1 + \beta \, T_r(\widetilde{\sigma}) + (1 - \alpha - \beta) \, J(\widetilde{\sigma}) \qquad 0 < \alpha + \beta < 1 \qquad (25)$$

La formulation proposée dans les références [27, 14, 12] consiste dans le cas anisotrope, à choisir le tenseur de contrainte effectif donné par la relation (2) dans le système principal. La variable D est scalaire et la loi d'évolution généralise la loi de Rabotnov :

$$\dot{D} = \left\langle \frac{\chi(\sigma)}{A} \right\rangle^r (1 - D)^{-k} \qquad (26)$$

où $<u>$ est nul si $u \leqslant 0$, égal à u si $u > 0$.

Dans le cas de la traction on retrouve exactement la relation (22) quelles que soient les valeurs de α et β. On remarquera que l'endommagement en compression uniaxiale peut soit être nul ($\alpha + \beta \geqslant 0,5$) soit égal à celui de la traction ($\alpha = \beta = 0$), soit prendre des valeurs intermédiaires. Signalons que la loi (26), valable dans le cas du chargement simple, a été généralisée à un cas quelconque par Murakami à l'aide du concept de contrainte effective (3) dont nous avons parlé plus haut.

Dans le cadre de la théorie que nous proposons, qui prend en compte l'endommagement en élasticité, deux modifications mineures sont nécessaires :

— on considère que la loi d'évolution ne dépend de **D** que par la contrainte effective mais dépend aussi de la variable scalaire D.
— pour retrouver la loi de Rabotnov (22) dans le cas particulier de la traction simple on remplace $\chi(\widetilde{\sigma})$ par la fonction :

$$\chi(\widetilde{\sigma}, D) = \alpha \widetilde{\sigma}_1 + \beta \, T_r \, (\widetilde{\sigma}) + \frac{(1 - \alpha - \beta) \, J(\widetilde{\sigma})}{1 - \dfrac{\nu}{1 - \nu} \dfrac{(1 - \gamma) \, D}{1 - \gamma D}} \tag{27}$$

Compte tenu de ces choix la loi d'évolution proposée s'exprime dans le cas général par :

$$\dot{\mathbf{D}} = Q(\widetilde{\sigma}) \, \dot{D} \qquad \dot{D} = \left\langle \frac{\chi(\widetilde{\sigma}, D)}{A} \right\rangle^r (1 - D)^{-k} \tag{28}$$

$Q(\widetilde{\sigma})$ étant donné par (18) et (19) dans le système des axes principaux par (18), (20) et (21) dans un cas quelconque.

Cette formulation englobe les deux cas limites (isotrope et anisotrope) et les cas intermédiaires, elle s'applique dans le cas d'un chargement non proportionnel quelconque, elle permet de prendre en compte l'effet de l'endommagement sur le comportement élastique et viscoplastique (voir ci-dessous). En plus des coefficients α, β, r, k, A propres à la loi d'évolution de D, seuls deux coefficients supplémentaires γ et a sont nécessaires pour traduire les effets d'anisotropie.

Bien sûr ces avantages se font au prix d'un nombre de variables plus élevé (la variable scalaire D et les trente six composantes de **D** dans le cas le plus général) que dans la théorie de Murakami qui ne fait intervenir que les six composantes d'un tenseur symétrique d'ordre 2.

3.3. *Loi de comportement viscoplastique*

On sait traduire assez finement de nombreux phénomènes de viscoplasticité, en chargement monotone ou cyclique, au moyen de variables internes d'écrouissage et dans un cadre thermodynamique précis [3, 15] : fluage primaire et secondaire, essai de relaxation, effets Bauschinger, adoucissement ou durcissement cyclique, courbes d'écrouissage cyclique, effets de restauration, effets d'histoire de la température [11], effets d'histoire du trajet de déformation [6]. Nous nous limitons ici au cas particulier d'une loi de fluage secondaire (loi du type Norton, sans effet d'écrouissage), en examinant les conséquences de l'apparition de l'endommagement.

On suppose l'existence d'un pseudo-potentiel de dissipation obéissant à un critère de type Von Mises (déformation plastique insovolumique) :

$$\Omega = \frac{K}{N + 1} \left[\frac{J(\widetilde{\sigma})}{K} \right]^{N+1} \tag{29}$$

où $J(\widetilde{\sigma})$ s'exprime par (24) ou ce qui revient au même :

$$J(\widetilde{\sigma}) = \left[\frac{3}{2} \widetilde{\sigma}'_{ij} \widetilde{\sigma}'_{ij} \right]^{1/2}$$

$\widetilde{\sigma}'_{ij}$ désignant les composantes du déviateur de $\widetilde{\sigma}$. Suivant l'hypothèse de dissipativité normale la vitesse de déformation plastique s'écrit :

$$\dot{\varepsilon}_P = \frac{\partial \Omega}{\partial \sigma} = \frac{3}{2} \left[\frac{J(\widetilde{\sigma})}{K} \right]^N \frac{(1-D)^{-1} : \widetilde{\sigma}'}{J(\widetilde{\sigma})} \tag{30}$$

ou encore :

$$\dot{\varepsilon}_{P_{ij}} = \frac{3}{2} \left[\frac{J(\widetilde{\sigma})}{K} \right]^N \frac{M_{ijk\ell} \, \widetilde{\sigma}'_{k\ell}}{J(\widetilde{\sigma})}$$

$M_{ijk\ell}$ représentant la matrice représentative du tenseur d'ordre quatre $(1 - D)^{-1}$.

Dans le cas du matériau non endommagé ($D = 0$) on retrouve la loi de fluage secondaire proposée par Odqvist et Hult [29]. La présence de l'endommagement se traduit par le phénomène de fluage tertiaire : pour une loi d'endommagement isotrope ($\gamma = 1$) on obtient :

$$\dot{\varepsilon}_{P_{ij}} = \frac{3/2}{(1-D)^{N+1}} \left[\frac{J(\sigma)}{K} \right]^N \frac{\sigma'_{ij}}{J(\sigma)} \tag{31}$$

qui diffère de la formulation utilisée dans les références [27] et [14] par la présence de l'exposant $N + 1$ au lieu de N pour le facteur d'endommagement : ceci est dû au fait que l'hypothèse de dissipativité normale doit s'appliquer dans l'espace des contraintes (et non des contraintes effectives) pour être

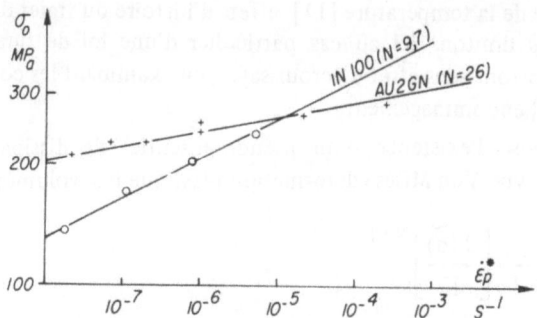

Figure 4. – *Loi de Norton de l'IN 100 à 1 000° C et l'AU2GN à 180° C.*

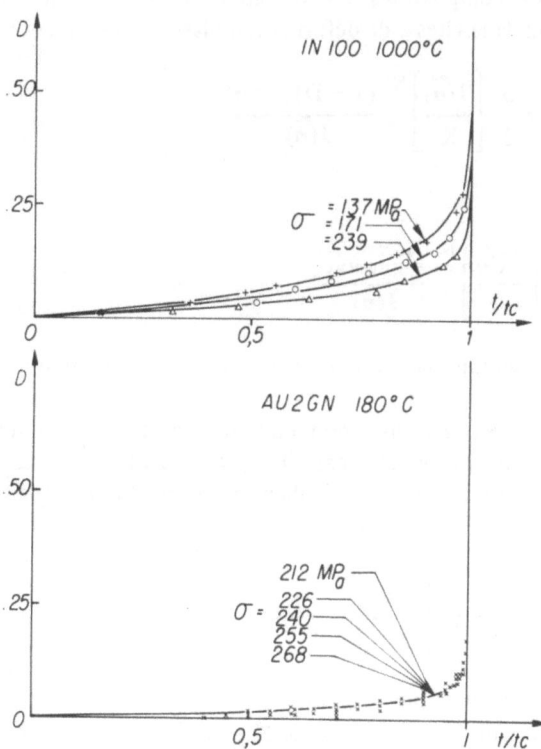

Figure 5. – *Evolution de l'endommagement de fluage pour l'IN 100 et l'AU2GN.*

cohérente avec le cadre thermodynamique et vérifier en particulier le second principe, même lorsque l'endommagement est anisotrope [3, 5].

Une telle loi de comportement viscoplastique (30) permet de décrire le fluage tertiaire d'un matériau à endommagement isotrope ou anisotrope : elle permet inversement une mesure indirecte de l'endommagement dans la phase de fluage tertiaire comme cela a été pratiqué pour plusieurs matériaux avec l'hypothèse d'isotropie [25]. Les figures 4 et 5 indiquent les résultats de l'IN 100 à 1 000°C et de l'AU2GN-T6 à 180°C pour lesquels l'exposant N est déterminé par quelques essais de fluage jusqu'à rupture, la mesure du rapport des vitesses de fluage secondaire et de fluage tertiaire donnant une mesure de D. En effet, de la relation (31) appliquée en traction on tire facilement :

$$D = 1 - \left(\frac{\dot{\epsilon}_p^*}{\dot{\epsilon}_p}\right)^{1/N+1}$$

où $\dot{\epsilon}_p^*$ est la vitesse de fluage secondaire (lorsque D # 0).

Pour l'IN 100 on observe une dépendance entre les courbes d'évolution de D en fonction du pourcentage de vie t/t_c et la contrainte appliquée ; ceci devrait se traduire par une dépendance entre l'exposant k de la loi d'endommagement de fluage et la contrainte, comme cela a été proposé dans [25] et [5].

4. Endommagement de fluage en traction-torsion

4.1. *Torsion constante*

Dans ce cas on est en chargement simple et le système principal du tenseur de contrainte effectif $\underset{\sim}{\tilde{\sigma}}$ est fixe. On obtient alors facilement le tenseur **D** à un instant quelconque pour lequel la mesure scalaire D est connue :

$$\mathbf{D} = \begin{bmatrix} D & 0 & 0 & & & \\ \dfrac{\nu(1-\gamma)}{1-\nu}D & \gamma D & 0 & & & \\ \dfrac{\nu(1-\gamma)}{1-\nu}D & 0 & \gamma D & & & \\ & & & \gamma D & & \\ & & & & [\gamma+(1-\gamma)a]D & \\ & & & & & [\gamma+(1-\gamma)a]D \end{bmatrix} \quad (32)$$

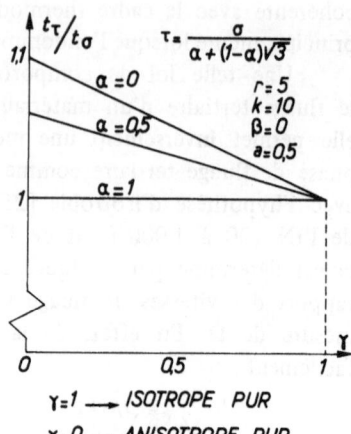

Figure 6. – *Influence du coefficient α et du degré d'anisotropie γ sur le temps à rupture en torsion.*

Les composantes principales de $\widetilde{\sigma}$ s'expriment en fonction de la contrainte de torsion :

$$\widetilde{\sigma}_1 = \frac{\tau}{1-D}\;;\;\widetilde{\sigma}_2 = \frac{\nu}{1-\nu}\,\frac{(1-\gamma)\,D}{1-\gamma\,D}\,\frac{\tau}{1-D}\;;\;\widetilde{\sigma}_3 = \widetilde{\sigma}_2 - \frac{\tau}{1-\gamma\,D}\quad(33)$$

Dans le cas isotrope ($\gamma = 1$) on retrouve évidemment

$$\sigma_1 = -\sigma_2 = \frac{\tau}{1-D}\qquad\text{et}\qquad \sigma_3 = 0,$$

et la loi d'endommagement (27) (28) s'intègre facilement. Dans le cas contraire l'intégration n'est pas analytique et l'on a recours à une méthode numérique. Un programme basé sur la méthode de Runge-Kutta et un calcul automatique du pas d'intégration [35] a donc été élaboré : il permet d'intégrer la loi d'endommagement générale pour un chargement quelconque comprenant plusieurs niveaux de contrainte (y compris en chargement non simple). Le calcul d'un essai de fluage à deux niveaux prend moins d'une seconde de temps CPU sur CDC 7600. Ce programme a donc été appliqué systématiquement et la figure 6 donne par exemple l'influence des coefficients α et γ sur le rapport entre temps à rupture en torsion et en traction lorsque $\tau = \sigma/[\alpha + (1-\alpha)\sqrt{3}]$. Dans le cas d'un critère de rupture influencé par la contrainte octahédrale ($\alpha \neq 1$), ce rapport dépend du degré d'anisotropie comme dans la théorie classique [14].

La figure 7 indique l'évolution des caractéristiques d'élasticité en fonction du temps pour trois exemples avec $\gamma = 0$ et a = 0,5. Elle est comparée avec

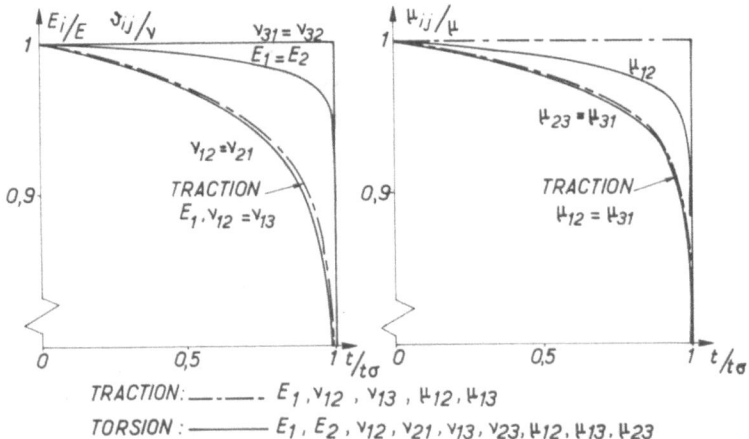

Figure 7. — *Influence de l'anisotropie de l'endommagement sur l'évolution des caractéristiques d'élasticité au cours du fluage en traction ou en torsion*
$\nu = 0,33$, $k + r = 15$, $\alpha = 1$, $\beta = 0$, $\gamma = 0$, $a = 0,5$.

l'évolution obtenue en fluage en traction uniaxiale par les relations (15) et l'intégration de la loi (22) qui donne :

$$D = 1 - \left(1 - \frac{t}{t_\sigma}\right)^{1/k+r+1}$$

On remarquera les évolutions très différentes par rapport au cas de la traction pour les modules E_1, E_2 pour les coefficients de Poisson $\nu_{13}, \nu_{23}, \nu_{21}$, pour les modules de cisaillement μ_{12}, μ_{23}.

4.2. *Chargement en traction suivi d'une torsion*

C'est l'exemple traité par Murakami et Ohno [28] pour étudier l'influence de l'anisotropie : essai sous la contrainte de traction σ jusqu'à l'instant t*, poursuivi en torsion sous la contrainte $\tau = \sigma$ jusqu'à rupture. La comparaison porte sur le rapport du temps total à rupture au temps à rupture en torsion t_R/t_τ. Dans le cas traité par Murakami ($\alpha = 1, \beta = 0$) les temps à rupture en traction constante et en torsion constante sont égaux et les courbes qu'il prévoit sont indiquées en pointillé sur la figure 8 (cas particulier où $k = 0$ et $r = 1$ ou 5). Dans notre théorie des courbes analogues sont obtenues pour le matériau à anisotropie idéale ($\gamma = 0$, $a = 0,5$). Par contre en faisant varier le degré d'anisotropie on observe des variations sensibles des résultats comme l'indique la figure 9 : en particulier, une valeur nulle pour le coefficient a donné des courbes similaires

à celles obtenues par la théorie de Kachanov [20] (pour r = 5 et r = 1). De plus,
l'effet de l'anisotropie est diminué lorsque le coefficient k est non nul (courbe
r = 5, k = 10, γ = 0, a = 0,5). La figure 10 montre l'influence du coefficient a
lorsque $\alpha = 0$, c'est-à-dire lorsque le temps à rupture en torsion est différent
du temps à rupture en traction (pour $\tau = \sigma/\sqrt{3}$), sauf dans le cas isotrope.

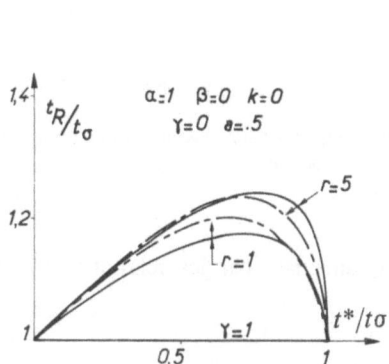 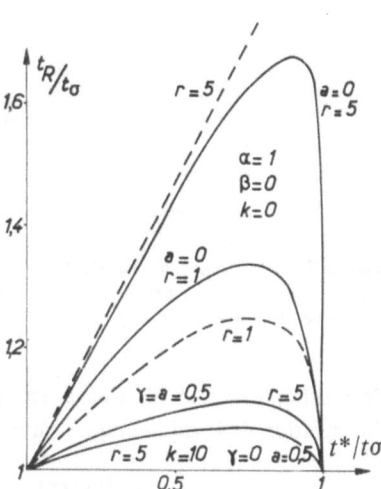

Figure 8. – *Influence du temps en traction
sur le temps total à rupture (—— théorie proposée ; — · —— · théorie de Murakami [28]).*

Figure 9. – *Influence du degré d'anisotropie
et de l'exposant k sur la rupture en traction-
torsion (—— théorie proposée, - - - - - - théorie de Kachanov [20]).*

Tous ces exemples montrent la souplesse et les possibilités de la théorie
proposée puisqu'elle permet un éventail de prévisions assez large en traction-
torsion tout en rendant possible la prise en compte de l'effet de l'endommagement sur les caractéristiques d'élasticité. Il reste maintenant à appliquer cette
théorie sur un matériau réel en réalisant des expériences de fluage en traction,
en torsion et en traction suivie de torsion (ou l'inverse) de façon à déterminer
le critère de rupture (valeur de α) et le degré d'anisotropie (valeur de γ).

Les essais de fluage en traction donnent aussi les valeurs des coefficients
r, k, A ; par ailleurs, des mesures de module de cisaillement au cours des
essais de torsion devraient permettre de vérifier la validité de la théorie (la
procédure inverse peut être utilisée, à savoir détermination de γ et a par
des mesures des caractéristiques d'élasticité du matériau endommagé et
vérification par les essais de rupture en fluage sous traction puis torsion).

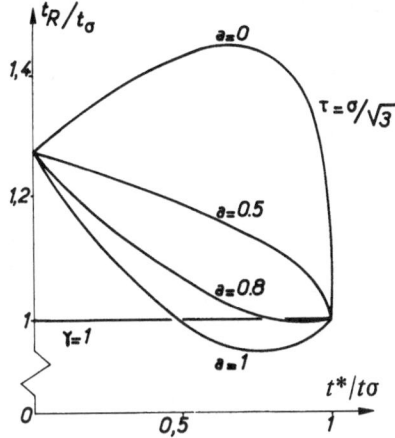

Figure 10. – *Influence du coefficient a sur la rupture en traction-torsion (α = β = 0, r = 5, k = 0, γ = 0).*

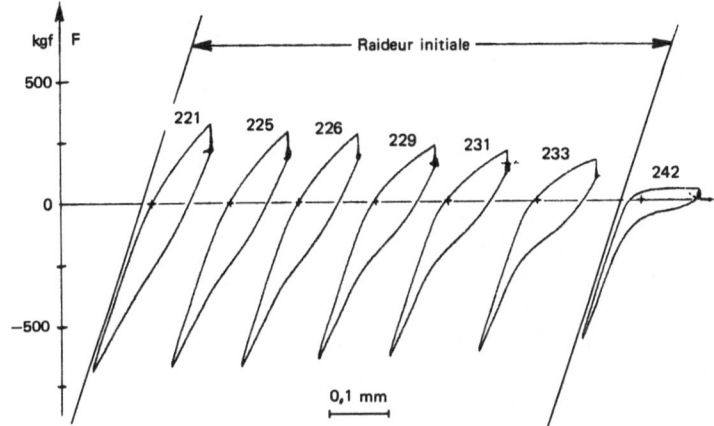

Figure 11. – *Evolution des derniers cycles d'un essai à allongement imposé avec période de maintien en traction : IN 100 à 1 000°C ± 0,1 mm, t_m = 300s).*

5. Conclusion

La théorie d'endommagement anisotrope développée ici s'appuie sur l'analyse élastique de cas idéaux de matériau endommagé. Elle permet de traduire par deux coefficients seulement (γ et a) le degré d'anisotropie du matériau endommagé et le concept utilisé pour définir le tenseur de contrainte effectif est compatible avec la description de l'effet de l'endommagement anisotrope sur le comportement élastique. La loi d'endommagement proposée scinde les difficultés : la non-linéarité des phénomènes d'évolution de l'endommagement est rejetée sur une seule variable scalaire, l'anisotropie étant introduite par un tenseur de répartition ne dépendant que du matériau et de la direction de contrainte principale maximale.

Cette théorie, développée ici dans le cas de l'endommagement de fluage, peut s'appliquer sans difficulté particulière pour d'autres types d'endommagement (en plasticité [26, 22] ou en fatigue [25, 2]), et même dans des cas de couplage, par exemple à température élevée pour des chargements cycliques lents dans lesquels les endommagements de fatigue et de fluage se cumulent et interagissent [25, 7, 31, 23]. Dans le cas des chargements cycliques cependant, une difficulté supplémentaire surgit lorsque l'on s'intéresse à l'influence de l'endommagement sur le comportement à la déformation, en particulier sur les caractéristiques d'élasticité : dans les derniers cycles avant la rupture, lorsque plusieurs microfissures existent dans l'éprouvette, il se produit en compression un phénomène de fermeture de ces microfissures qui implique une différence entre les caractéristiques d'élasticité en traction et en compression : on a ainsi besoin d'un comportement élastique non linéaire gouverné par un "seuil de fermeture" : ce seuil en contrainte est négatif mais d'autant plus proche de zéro que l'endommagement est important comme le montrent les cycles de la figure 11 pour un essai à allongement imposé. La prise en compte de ce phénomène complexe par introduction d'une variable interne supplémentaire constitue un sujet intéressant.

La théorie proposée pour traduire l'endommagement anisotrope des métaux cherche à généraliser le plus simplement possible des théories développées antérieurement en traction-compression ou dans le cas tridimensionnel isotrope. En effet, deux coefficients supplémentaires seulement sont introduits pour décrire l'effet d'anisotropie. On conçoit bien alors la nécessité de vérifications expérimentales de cette formulation, avec par exemple des essais particuliers de traction-torsion combinées, comme nous en avons discuté au paragraphe 4.

REFERENCES

[1] CAILLETAUD G. and J.L. CHABOCHE. – "Macroscopic description of the microstructural changes induces by varying temperature: example of IN 100 cyclic behaviour". *Conf. ICM3*, Cambridge, 1979.
[2] CHABOCHE J.L. – "Une loi différentielle d'endommagement de fatigue avec cumulation non linéaire". *Revue Française de Mécanique*, 50-51, 1974.
[3] CHABOCHE J.L. – "Sur l'utilisation des variables d'état interne pour la description du comportement viscoplastique et de la rupture par endommagement". *Symp. Franco-Polonais de Rhéologie et Mécanique*, Cracovie, 1977.

[4] CHABOCHE J.L. – "Méthodes de calcul en fatigue à chaud", *Ecole d'été sur la Fatigue, Sherbrooke* (Québec), 1978.

[5] CHABOCHE J.L. – "Description thermodynamique et phénoménologique de la viscoplasticité cyclique avec endommagement". Thèse univ. Paris VI et publication ONERA n° 1978-3.

[6] CHABOCHE J.L., K. DANG VAN and G. CORDIER. – "Modelization of the strain memory effect on the cyclic hardening of 316 Stainless steel". *SMIRT 5 Conference*, Berlin, 1979.

[7] CHABOCHE J.L., H. POLICELLA and S. SAVALLE. – "Application of the Continuous Damage approach to the prediction of high temperature low-cycle fatigue". Conf. sur les alliages à Haute Température pour Turbines à Gaz, Liège, 1978.

[8] CORDEBOIS J.P. and F. SIDOROFF. – "Anisotropie élastique induite par endommagement". Colloque Euromech 115, *Comportement Mécanique des solides anisotropes*. Grenoble, 1979.

[9] DELAMETER W.R. and G. HERMANN. – "Weakening of elastic solids by doubly-periodic arrays of cracks". *In Topics in Applied Continuum Mechanics*, Springer Verlag, 1974.

[10] DUVAUT G. – "Analyse fonctionnelle. Mécanique des milieux continus. Homogénéisation". *In Theoretical and Applied Mechanics*, North-Holland, 1976.

[11] ENGRAND D. – "Homogénéisation des propriétés thermoélastiques statiques des milieux à structure périodique". *La Recherche Aérospatiale*, 5 (1978): 283-286.

[12] GOEL R.P. – "On the creep rupture of a tube and a sphere". *J. Applied Mech. Trans. ASME*, (1975): 625-628.

[13] HAYHURST D.R. – "Creep rupture under multi-axial state of stress". *J. Mech. Phys. Solids*, 20 5 (1972) ; 381-390.

[14] HAYHURST D.R. and F.A. LECKIE. – "The effect of creep constitutive and damage relationships upon the rupture time of a solid circular torsion bar" *J. Mech. Phys. Solids*, 21, 6 (1973) : 431-446.

[15] HAYHURST D.R. and B. STORAKERS. – "Creep rupture of the Andrade shear disk", *Proc. Royal Soc., London*, 349 (1976), 369-382.

[16] HOENIG A. – "Elastic moduli of a non randomly cracked body" *Int. J. Solids Struct.*, 15, 2 (1979), 137-154.

[17] JOHNSON A.E., J. HENDERSON and V.D. MATHUR. – "Combined stress creep fracture of a commercial copper at 250°C", *Engineer* London, 202 (1956) : 261-265 et 299-301.

[18] JOHNSON A.E., J. HENDERSON, B. KHAN. – *Complex stress. Creep, Relaxation and Fracture of Metallic Alloys*. National engeng. Laboratory, H.M.S.O., London, 1962.

[19] KACHANOV L.M. – "Time of the rupture process under creep conditions". *Izv. Akad. Nauk.*, SSR OTd Tekh. Nauk, n° 8, 1958.

[20] KACHANOV L.M. – *Foundations of Fracture Mechanics* Nauka, Moscow, 1974.

[21] LECKIE F.A. and D.R. HAYHURST. – "Creep rupture of structures" *Proc. Royal Soc., London*, 340 (1974), 323-347.

[22] LEMAITRE J. – Theorie mécanique de l'endommagement isotrope appliquée à la fatigue des métaux. Séminaire "Matériaux et structures sous chargement cycliques", Palaiseau, 28-29 Sept. 1978.

[23] LEMAITRE J. – "Damage modelling for prediction of plastic or creep fatigue failure in structures", *SMIRT 5 Conference*, Berlin, 1979.

[24] LEMAITRE J. and J.L. CHABOCHE. – "A non linear model of creep-fatigue damage cumulation and interaction". dans Jan Hult-Springer, Ed. *Mechanics of Visco-Plastic Media and Bodies*. 1975, pp. 291-301.

[25] LEMAITRE J. and J.L. CHABOCHE. – "Aspect phénoménologique de la rupture par endommagement" *J. Mécanique Appliquée,* **2,** 3 (1978) : 317-365.
[26] LEMAITRE J. and J. DUFAILLY. – "Modélisation et identification de l'endommagement plastique des métaux. 3e Congrès Français de Mécanique, Grenoble, 1977.
[27] MARTIN J.B. and F.A. LECKIE. – "On the creep rupture of structures" *J. Mech. Phys. Solids,* **20,** 4 (1972) : 223-238.
[28] MURAKAMI S. and N. OHNO. – "A constitutive equation of creep damage in polycrystalline metals", *IUTAM Coll. Euromech 111,* "Constitutive Modelling in Inelasticity" Marienbad, Tchécoslovaquie, 1978.
[29] ODQVIST F.K.J. and J. HULT. – *Kriechfestigkeit Metallisher Werkstoffe.* Berlin, Springer Verlag : 1962.
[30] OHAYON F. – "Homogénéisation par développements asymptotiques mixtes. Calcul des vibrations de milieux élastiques à structure périodique" *La Recherche Aérospatiale,* **2** (1979) : 121-130.
[31] PLUMTREE A. and J. LEMAITRE. – "Application of damage concepts to predict creep-fatigue failures". *ASME. Pressure Vessels and Piping Conf.,* Montreal, June 1978.
[32] RABOTNOV Y.N. – "Creep rupture" *12th Int. Congress of Applied Mechanics,* Stanford, 1968.
[33] RABOTNOV Y.N. – *Creep Problems in Structural Members,* North Holland Publishing Co, 1969.
[34] RABOTNOV Y.N. – "Advances in creep design". In Smith-Nicholson, eds, *Applied Science,* London : 1971.
[35] SAVALLE S. and J.P. CULIE. – "Méthodes de calcul associées aux lois de comportement cyclique et d'endommagement" *La Recherche Aérospatiale,* **5** (1978) : 263-278.

ABSTRACT

The concept of effective stress applied to elasticity and to viscoplasticity in présence of anisotropic damage.

For some years various articles have shown the possibility of applying continuum mechanics to the model representation of the evolution of variable damage, initially introduced by Kachanov. Of interest here are the complex problems posed by the anisotropy of this damage, an anisotropy which affects both the elastic behaviour and the viscoplastic behaviour, and also rupture.

It is shown how a damage tensor of the fourth order can be introduced through the concept of effective stress in order to take into consideration the effect of the damage on the elastic behaviour, on the basis of results furnished by homogenisation techniques. A three-dimensional law of anisotropy is proposed for creep damage, introducing a minimum number of characteristic coefficients. The possibilities of this law are studied for the case of the traction-torsion of thin tubes and are compared with previous formulations.

Damage Induced Elastic Anisotropy

J.P. Cordebois
Laboratoire de Mécanique et Technologie,
ENSET, Cachan, France.

F. Sidoroff
Laboratoire de Mécanique des Solides,
École Centrale de Lyon, Ecully, France.

1. Introduction

The concept of damage has been introduced by Rabotnov and Kachanov in 1958 for the description of creep rupture. Since then this concept has been used to account for various types of rupture occuring in metals and other kinds of solids: creep rupture, fatigue, plastic rupture. For the one-dimensional case a survey of the basic ideas as well as some recent results can be found in [1]. The three-dimensional case has been far less investigated and many conceptual problems remain to be solved. Most of these problems are related to the anisotropy of the damage and of the resulting properties, for instance the elastic response. This paper is devoted to an approach to these problems: a model will be presented for the description of the damage concept in the three-dimensional case and its consequences, as regards the induced elastic anisotropy, will be investigated in order to allow the evaluation of this model by comparison with experimental results or numerical simulation.

A crude but significant insight into the physical nature of damage can be gained by viewing the damage processe as the generation and growth of micro-defects (micro-holes or micro-cracks) within an initially perfect material. The material remains the same but its macroscopic properties change with its microscopic geometry. The damage variable is a macroscopic measure of the microscopic geometrical deterioration of the material. Thus the elastic response, or more generally any mechanical property, of the damaged material will depend on the true elastic response of the material and on the microscopic deterioration, i.e. on the damage.

In the following we shall consider an elastic-plastic material and we shall limit ourselves to the influence of damage on the elastic response of the material. There are two reasons for this. First, most of the problems mentioned above are already present within this limited framework. Secon, this is the only part of the theory which can be directly tested against experiment or numerical simulation. Indeed, even in the one-dimensional case, the variation of the elastic modulus is the simplest way for identification of the damage, [2].

2. Three dimensional damage

2.1. *The thermodynamic framework*

In this section a brief outline of the model for three-dimensional damage will be presented. For a more complete presentation and for discission of the underlying assumptions and motivations, the reader is referred to [3]. In subsection 2.1 the general thermodynamic framework will be presented while the model itself will be developed in the following subsections.

We consider an elastic-plastic material and define the plastic strain as the residual strain after unloading. Therefore the specific energy and enthalpy for the damaged material are

$$W(\varepsilon^e, d, \alpha) = \frac{1}{2} \varepsilon^e : \Lambda(d) [\varepsilon^e] + \phi(\alpha),$$

$$V(\sigma, d, \alpha) = \sigma : \varepsilon^e - W = \frac{1}{2} \sigma : A(d) [\sigma] - \phi(\alpha),$$

where d denotes the damage variable, α the plastic internal variables (hardening parameter for instance), $A(d)$ and $\Lambda(d) = A(d)^{-1}$ the damaged elasticity tensors. The corresponding relations for the virgin (undamaged) material are

$$W_0(\varepsilon^e, \alpha) = \frac{1}{2} \varepsilon^e : \Lambda_0 [\varepsilon^e] + \phi(\alpha),$$

$$V_0(\sigma, \alpha) = \frac{1}{2} \sigma : A_0 [\sigma] - \phi(\alpha), \tag{2}$$

where, assuming the virgin material to be isotropic,

$$A_0[\sigma] = \frac{1 + \nu}{E} \sigma - \frac{\nu}{E} (\text{tr } \sigma) \mathbf{1}. \tag{3}$$

The constitutive equation is then obtained as

$$\varepsilon^e = \frac{\partial V}{\partial \sigma} = A(d) [\sigma], \quad \sigma = \frac{\partial W}{\partial \varepsilon^e} = \Lambda(d) [\varepsilon^e], \tag{4}$$

while the dissipation is given by

$$\Phi = \sigma : \dot{\varepsilon}^p + G\,\dot{d} - A\,\dot{\alpha} \geq 0, \tag{5}$$

$$G = \frac{\partial V}{\partial d} = \frac{1}{2}\,\sigma : A'(d)\,[\sigma]. \tag{6}$$

The thermodynamic force associated to the damage, G, is analogous to the Griffith's energy release rate in fracture mechanics.

2.2. *The general model*

In the one-dimensional case, damage is introduced in the constitutive equations through the concept of effective stress. Extending this to the three-dimensional case, we define the effective stress tensor $\tilde{\sigma}$ by a linear transformation law acting upon the actual stress tensor σ,

$$\tilde{\sigma} = M(d)\,[\sigma] \tag{7}$$

where $M(d)$ is a linear transformation of the space S of all symmetric tensors into itself (i.e. a fourth order tensor $M_{ijk\ell}$ such that $M_{ijk\ell} = M_{jik\ell} = M_{ij\ell k}$). Since (7) is a generalization of the one-dimensional $\tilde{\sigma} = \sigma/(1-D)$, the transformation $M(d)$ depends non-linearly upon the damage. The damaged constitutive equation is then obtained by an energy identification

$$V(\sigma, d, \alpha) = V_0(\tilde{\sigma}, \alpha), \tag{8}$$

damage being introduced by replacing the actual stress by the effective stress in the energy function; see [3] for details. The damaged elastic tensor is then derived as

$$A(d) = M^T(d)\,A_0\,M(d) \tag{9}$$

and the constitutive equation (4) can be interpreted as the undamaged elastic law relating the effective stress to an effective elastic strina tensor $\tilde{\varepsilon}^e$ defined by

$$\tilde{\varepsilon}^e = (M^T)^{-1}(d)\,[\varepsilon^e]\,, \quad \tilde{\varepsilon}^e = A_0[\tilde{\sigma}]. \tag{10}$$

It can further be shown that this model involves no restriction upon the damaged elastic tensor $A(d)$, since any symmetric ($A_{ijk\ell} = A_{k\ell ij}$) positive defined $A(d)$ can be written in the form (9) with a proper choice of $M(d)$ which can even be assumed to be symmetric, as will be the case later.

2.3. *Specialization*

Two choices remain to be made in order to specialize the theory; they concern
1) the nature of the damage variable d, 2) the special form of the transformation (7) or (10). Si we want to describe an anotropic damage, the damage
variable must be of tensorial character. The simplest case is a
symmetric second-order tensor **D**. This implies some limitations which will
be discussed later and the possibility of other choices [4]. However, in view
of practical applications, this seems a reasonable assumption. The transformation law (7) or (10) is supposed to be a generalization of the one-dimensional
law $\tilde{\sigma} = \sigma/(1 - D)$. In the case where the stress tensor and the damage tensor
have the same eigendirections, this extends naturally to

$$\tilde{\sigma} = \sigma(1 - D)^{-1}, \quad \tilde{\sigma}_i = \frac{\sigma_i}{1 - D_i}. \tag{11}$$

To obtain an appropriate transformation law, we have to symmetrise
this expression which can be achieved in at least three different ways, lading
to the following transformation laws:

A. $\tilde{\sigma} = (1 - D)^{-1/2} \cdot \sigma \cdot (1 - D)^{-1/2}$,

B. $\tilde{\sigma} = \dfrac{1}{2} [(1 - D)^{-1} \cdot \sigma + \sigma \cdot (1 - D)^{-1}]$,   (12)

C. $\tilde{\sigma} = \dfrac{1}{2} [(1 - D) \cdot \tilde{\sigma} + \tilde{\sigma} \cdot (1 - D)]$.

The corresponding constitutive equations are obtained from (9) and
(3) as

A. $\varepsilon = \dfrac{1 + \nu}{E} (1 - D)^{-1} \cdot \sigma \cdot (1 - D)^{-1} - \dfrac{\nu}{E} \mathrm{tr}[\sigma \cdot (1 - D)^{-1}]$
$$(1 - D)^{-1}$$

$\varepsilon = \dfrac{1 + \nu}{4E} \{(1 - D)^{-2} \cdot \sigma + 2(1 - D)^{-1} \cdot \sigma \cdot (1 - D)^{-1}$

$$\tag{13}$$

$+ \sigma \cdot (1 - D)^{-2}\} - \dfrac{\nu}{E} \mathrm{tr}[\sigma(1 - D)^{-1}] (1 - D)^{-1}$

C. $\varepsilon = \dfrac{4(1 + \nu)}{E} x^2(1 - D)[\sigma] - \dfrac{\nu}{E} \mathrm{tr}[\sigma \cdot (1 - D)^{-1}] (1 - D)^{-1}$

corresponding to (12), A, B or C respectively, and where x is the linear operator such that $\mathbf{X} = x(\mathbf{A})[\mathbf{H}]$ is the unique solution of the equation

$$\mathbf{A} \cdot \mathbf{X} + \mathbf{X} \cdot \mathbf{A} = \mathbf{H} \tag{14}$$

(cf. [3], appendix E).

The remaining part of this paper will be devoted to an analysis of the induced elastic anisotropy and of the experimental and numerical programs developed to test the validity of this model.

3. Induced elastic anisotropy

3.1. *Interpretation of the transformation laws*

Since damage is described by a symmetric tensor $\mathbf{D}$, it is convenient to work in its principal axes, i.e. in the coordinate system in which $\mathbf{D}$ is diagonal; we shall denote by D_1, D_2 and D_3 the 'principal damages', i.e. the eigenvalues of $\mathbf{D}$. With this choice of axes, the transformation laws (12) A, B or C can be given a phenomenological interpretation. For the normal stress σ_{11}, (12) gives in all cases

$$\tilde{\sigma}_{11} = \frac{\sigma_{11}}{1 - D_1} = \frac{\sigma_{11}}{w_1} \;, \quad w_1 = 1 - D_1 \tag{15}$$

which can be viewed as expressing some kind of 'weakening' of the material in the direction x_1. For the shear stress σ_{12}, the transformation laws (12) A, B or C can be written as

$$\tilde{\sigma}_{12} = \frac{\sigma_{12}}{w_{12}} \;, \tag{16}$$

where the shear-weakening coefficient w_{12}, given by

A. $\quad w_{12} = w_1 w_2 = (1 - D_1)(1 - D_2)$,

B. $\quad w_{12} = \frac{2 w_1 w_2}{w_1 + w_2} = 2 \left[\frac{1}{(1 - D_1)} + \frac{1}{(1 - D_2)} \right]^{-1}$ \qquad (17)

C. $\quad w_{12} = \frac{1}{2}[w_1 + w_2] = \frac{1}{2}[(1 - D_1) + (1 - D_2)]$

is the geometrical (in case A), inverse arithmetic (B) or arithmetic (C) mean of the weakening coefficients w_1, w_2 valid for the normal stresses.

3.2. *Damaged elastic response*

It is clear that in our model, $\widetilde{\boldsymbol{\varepsilon}}$ is a linear isotropic function of $\widetilde{\boldsymbol{\sigma}}$, and therefore $\boldsymbol{\varepsilon}$ is an isotropic function of $\boldsymbol{\sigma}$ and $\mathbf{D}$, linear in $\boldsymbol{\sigma}$. The damaged elastic response, i.e. the linear relationship between $\boldsymbol{\varepsilon}$ and $\boldsymbol{\sigma}$ at fixed damage $\mathbf{D}$ is then orthotropic with respect to the principal axis of $\mathbf{D}$. Accordingly, with respect to the principal axes of $\mathbf{D}$, this linear relationship can be written as

$$
\begin{bmatrix} \epsilon_{11}^e \\ \epsilon_{22}^e \\ \epsilon_{33}^e \\ \epsilon_{12}^e \\ \epsilon_{23}^e \\ \epsilon_{31}^e \end{bmatrix} = \begin{bmatrix} A_1 & B_{12} & B_{13} & 0 & 0 & 0 \\ B_{12} & A_2 & B_{23} & 0 & 0 & 0 \\ B_{13} & B_{23} & A_3 & 0 & 0 & 0 \\ 0 & 0 & 0 & C_{23} & 0 & 0 \\ 0 & 0 & 0 & 0 & C_{31} & 0 \\ 0 & 0 & 0 & 0 & 0 & C_{12} \end{bmatrix} \begin{bmatrix} \sigma_{11} \\ \sigma_{22} \\ \sigma_{33} \\ \sigma_{12} \\ \sigma_{23} \\ \sigma_{31} \end{bmatrix}. \tag{18}
$$

Starting from (13), we obtain for the coefficients A and B the values

$$
A_1 = \frac{1}{E}\frac{1}{(1-D_1)^2} = \frac{1}{E}\frac{1}{w_1^2}
$$

$$
B_{12} = -\frac{\nu}{E}\frac{1}{(1-D_1)(1-D_2)} = -\frac{\nu}{E}\frac{1}{w_1 w_2} \tag{19}
$$

which hold true in the three cases A, B, C while we obtain for C

$$
C_{12} = \frac{1+\nu}{E}\left(\frac{1}{w_{12}}\right)^2, \tag{20}
$$

where the weakening coefficients w_{12} are defined by (17), depending on the choice A, B or C.

This analysis shows that our approach imposes certain limitations on the damaged elastic response:

1) It must be orthotropic; the directions of orthotropy being those of the principal axes of $\mathbf{D}$. This limitation is a consequence of choosing a symmetric second-order tensor for the damage.

2) By (19) and (20), it depends only on three constants D_1, D_2 and D_3 (E and ν are known and fixed) while a general orthotropic law depends on 9 coefficients.

Whether these limitations are important or not can only be answered on experimental grounds, and the remaining part of this paper will be devoted to the experimental and numerical programs now being developed to test this model and to identify the damage tensor in a given state.

4. Experimental verifications

4.1. *Uniaxial damage*

In the process of damage growth due to plastic deformations, significant modifications of the elastic properties occur only for large deformations. The major experimental difficulty is to create a significant homogeneous damage and at the same time, to be able to measure the anisotropic elastic properties. From a practical point of view, it is possible to create a uniaxial state of damage only by a loading in pure traction. Due to the symmetry of the loading, it is quite clear that the damage tensor and the damaged elastic response will show a transverse isotropy around the direction of traction Ox_1,

$$
\mathbf{D} = \begin{bmatrix} D_1 & 0 & 0 \\ 0 & D_2 & 0 \\ 0 & 0 & D_2 \end{bmatrix}
\tag{21}
$$

By elastic unloading we are able to measure the apparent modulus $\tilde{E}$ and the apparent contraction coefficient $\tilde{\nu}$ (Fig. 1). From (18) and (19) we obtain

$$
\tilde{E} = E(1 - D_1)^2 \ , \quad \tilde{\nu} = \nu \, \frac{1 - D_1}{1 - D_2} \ ,
\tag{22}
$$

so that, in a given state, the damage tensor can be determined from measurements of $\tilde{E}$ and $\tilde{\nu}$.

The ratio D_2/D_1 characterizes the degree of anisotropy of damage (isotropic damage is obtained for $D_1 = D_2$ while a perfectly uniaxial damage would be $D_2 = 0$). It may be reasonable to assume that this ratio δ is a material constant for a given material. Starting from (22), we get Figure 2 showing the relation between $\tilde{\nu}$ and $\tilde{E}$. These curves can be compared to experimental results, and thus the material constant δ can be determined. Once this is done, damage will depend only on $\tilde{E}$.

The first experimental results are described in subsection 4.4.

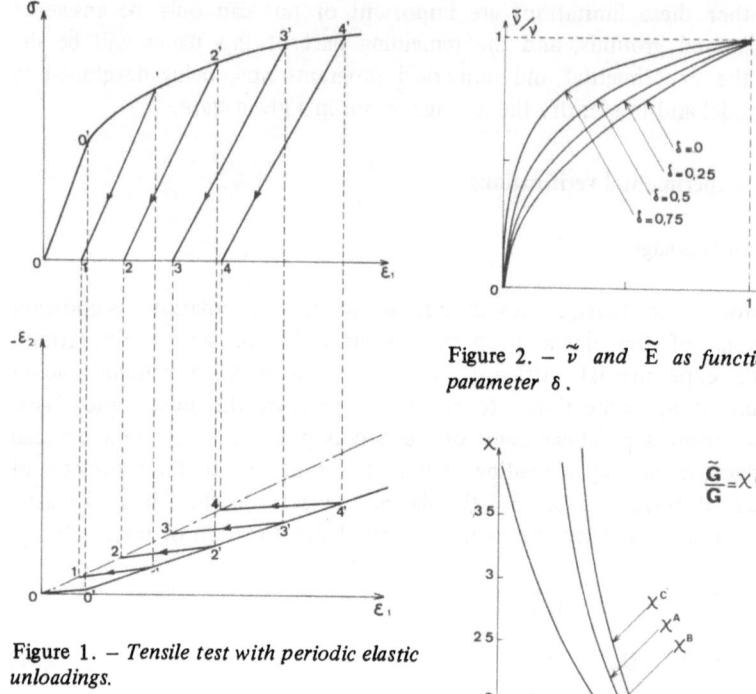

Figure 2. – $\tilde{\nu}$ and $\tilde{E}$ as functions of the parameter δ.

Figure 1. – *Tensile test with periodic elastic unloadings.*

$$\frac{\tilde{G}}{G} = \chi(\alpha) \frac{\tilde{E}}{E}$$

Figure 3. – *Dependence of* $\tilde{G}$ *on* $\tilde{E}$ *and* $\tilde{\nu}$.

4.2. *Apparent shear modulus*

The traction test alone leads to the identification of the damage tensor. However it doesn't allow any verification of the model. This verification requires other kinds of experiments for measuring other elastic properties.

A first kind of experiment can be performed on a traction-torsion machine: after a loading in traction, which generates an uniaxial damage state as described above, the specimen is unloaded and then loaded in torsion, which gives an apparent shear modulus $\tilde{G}$. This apparent shear modulus can be computed from $\tilde{E}$, $\tilde{\nu}$, E, ν and G, and the results are expressed in Figure 3 giving $(\tilde{G}/G)/(\tilde{E}/E)$ as a function of $\tilde{\nu}/\nu$, according to whether A, B or C is chosen in (12).

A comparison of these results with experiments will allow testing the theory and choosing between A, B or C.

4.3. *Apparent directional modulus*

Another kind of experiment consists in performing traction tests on different specimens obtained from a plate, previously subjected to a damage-generating traction, and with different orientation θ with respect to the direction of traction. For each specimen we obtain an apparent directional modulus $\tilde{E}_\theta$ which can be computed from $\tilde{E}, \tilde{\nu}, E, \nu$. Some results are presented in Figure 4 where we have ploted $\tilde{E}_\theta/E$ as a function of θ for $\nu = 0.3$ and for two values of $\tilde{\nu}/\nu$: 0.8 (when the three curves corresponding to A, B or C cannot be distinguished) and 0.5. In Figure 5 we have presented the curves giving $\tilde{E}_{45°}/E$ as a function of $\tilde{E}/E$ for different values of δ.

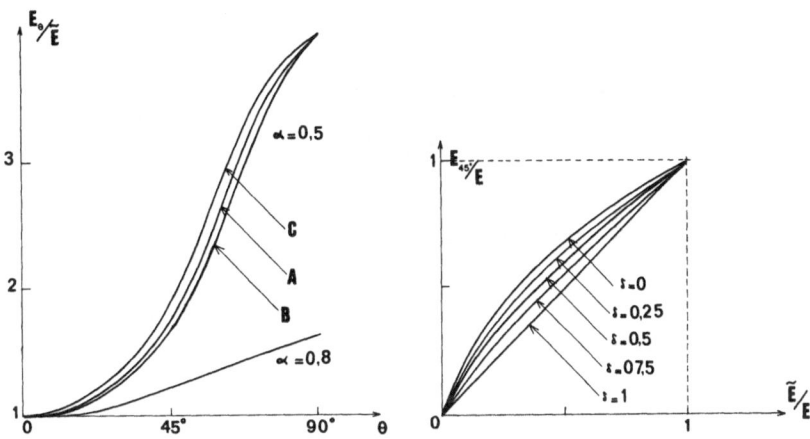

Figure 4. – E_θ *as a function of* θ.   Figure 5. – $\tilde{E}_{45°}$ *as a function of* $\tilde{E}$ *and* δ.

4.4. *Experimental results*

The experimental results obtained so far concern only measurements of $\tilde{E}$ and $\tilde{\nu}$, see section 4.1.

In the case of AU4GT4, a material of average ductility, Figures 6, 7 give the variations of $\tilde{E}$ and $\tilde{\nu}$ as functions of the plastic deformation ϵ_p. Even though the plastic deformation at rupture is not very large (35 %, we can obtain variations of the elastic modulus up to 20 %. Evolution of the longitudinal damage, while D_2 remains very small, is shown in Figure 8. Accordingly, the material constant δ is very small, but this seems to be specific to this material.

We also have performed experiments on CU_{a1}, a very ductile material exhibiting very important variations of $\tilde{E}$, Figure 9.

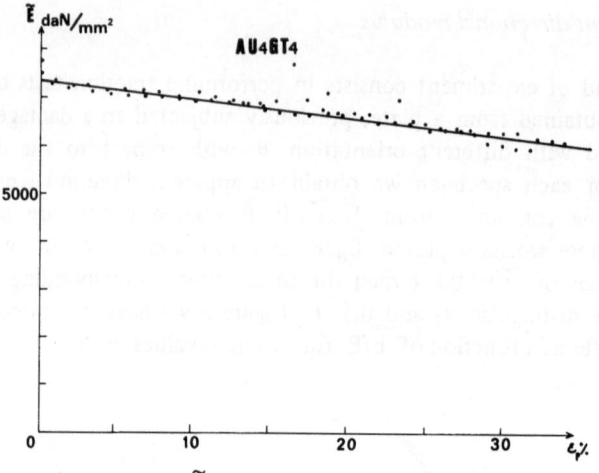

Figure 6. — $\tilde{E}$ *as a function of* ϵ_p *for AU4GT4.*

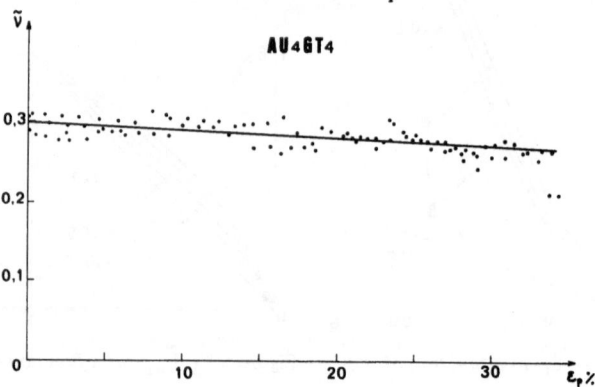

Figure 7. — $\tilde{\nu}$ *as a function of* ϵ_p *for AU4GT4.*

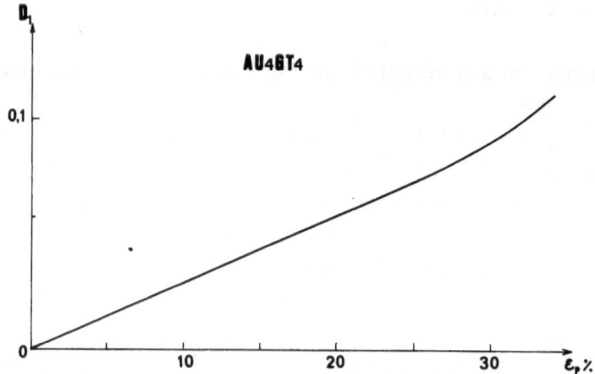

Figure 8. — *The longitudinal damage parameter* D_1 *as a function of* ϵ_p *for AU4GT4.*

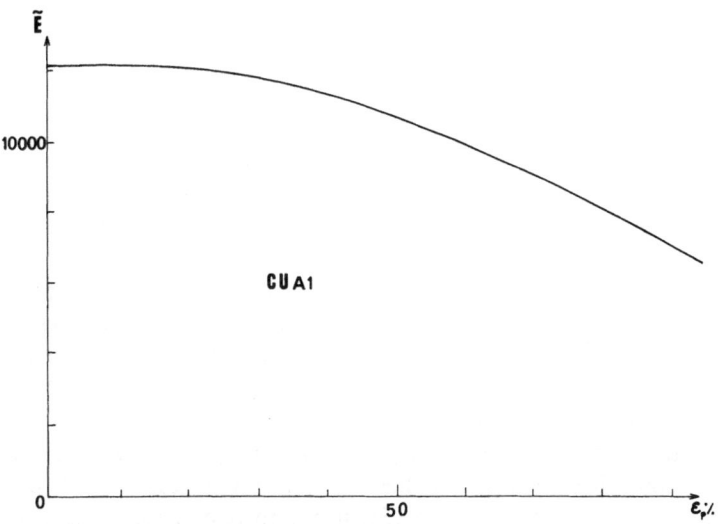

Figure 9. – $\tilde{E}$ *as function of* ϵ_p *for CUA1.*

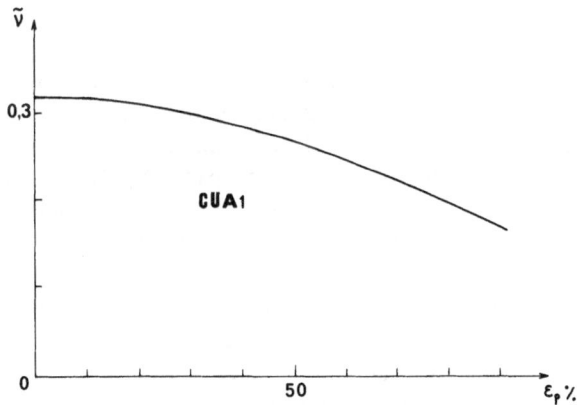

Figure 10. – $\tilde{\nu}$ *as a function of* ϵ_p *for CUA1.*

For all materials we obtain $\tilde{\nu} < \nu$, which by (22) shows that $D_2 < D_1$. This corresponds to a damage which is more prominent in the longitudinal direction than in the transverse direction. This result seems reasonable.

5. Numerical simulation

5.1. *Damage simulation*

Experimental investigations are limited by the actual damage process: testing the elastic properties of a damaged material requires the homogeneous creation of this damage and therefore of large plastic deformations on a sufficiently large scale. This is a strong limitation both on the 'shape' of the damage and on its 'magnitude'. Practically, we are limited to uniaxial damage. For instance, the orthotropy of the damaged elastic response, which is a consequence of our approach as mentioned in sub-section 3.2, cannot be tested experimentally since this would require the generation of a potentially non-orthotropic damage, i.e. a non-orthotropic damage-generating history (!).

Therefore it is quite interesting to be able to generate *artificially* a given, perfectly controlled, damage. In the two-dimensional case, this is physically possible by perforation of a plate [5]. In the three-dimensional case however, this is not so easy and it is more convenient to use a numerical simulation: a damaged material is simulated by a material with a given distribution of micro-holes. The damaged elastic response can then be computed by solving appropriate boundary value problems for the perforated material.

5.2. *Homogenisation*

The appropriate mathematical technique for this problem is 'homogenisation'. Generally speaking, homogenisation is the process of obtaining the global homogeneous behaviour of a locally non homogeneous material with inhomogeneities periodically distributed in space [6]. In the case of linear elasticity this problem is completely solved and the homogenised elastic law resulting from a given damage, i.e. a given repartition of micro-holes, can be computed by solving a simple boundary value problem [7].

Strictly speaking, this result holds only for elastic materials with a periodic repartition of holes, while we are dealing here with an elastic plastic material with a random distribution of holes. Plasticity has no influence on the result because it is well known that the global rigidity of a structure does not depend on previous plastic deformations. However, the assumption of a periodic distribution of holes may seem to be a strong limitation of our simulation. Nevertheless, it seems reasonable to hope that the results are not essentially affected the precise distribution of holes.

A finite-element program is now being developed to solve the remaining boundary value problem and to obtain the damaged elastic law for a given damage. Comparison of this law with (18) will lead to an estimation of the value of our model. In particular, it will be possible to answer the following questions:

— How is the damage tensor **D** related to the actual damage, i.e. to the geometry of the micro-defects ?
— Is the limitation of an orthotropic damaged elastic law a strong one ? (since it is quite easy to generate a non-orthotropic damage).
— More generally, are the limitations imposed by our model, for instance, a cubic damage resulting in a isotropic damaged elastic law, not too strong ?

Comparison of these results with measurements of the density variations and of other physical quantities will also provide information about the microscopic nature of damage, for instance, about the shape of the micro-defects (micro-voids or micro-cracks ?).

6. Conclusion

Much experimental and numerical work remains to be done along the lines described above. However, in view of the experimental difficulties mentioned in Section 4, it is doubtful that we shall ever be able to distinguish between our model and more refined ones. In fact, our model is the simplest one taking into account isotropic damage. In particular, even though options A, B and C in (12) can, in principle, be distinguished experimentally (cf. Section 4, Fig. 3 and 4), this is not possible in actual experiments because of the covered range of damage. Therefore we shall choose the simplest one, A, in future developments of the theory.

In order to obtain a constitutive theory including damage for application to metal forming process, two fundamental problems have to be solved:

1) Extension of the theory to plasticity, with the difficult problem of deciding how damage affects hardening;

2) Assumption of the evolution equations for damage, with the help of thermodynamic restrictions imposed by the second law.

These problems are now being investigated.

REFERENCES

[1] LEMAITRE J., J.L. CHABOCHE. — "Aspect phénoménologique de la rupture par endommagement". *Journal de Mécanique Appliquée*, **2, 3** (1978): 317-365.
[2] LEMAITRE J., J. DUFAILLY. — "Modélisation et identification de l'endommagement plastique des métaux". 3e Congrès français de mécanique, Grenoble, Septembre 1977.
[3] CORDEBOIS J.P., F. SIDOROFF. — "Endommagement tridimensionnel en élasticité". A paraître.

[4] CHABOCHE J.L.. – "Description thermodynamique et phénoménologique de la visco-plasticité cyclique avec écrouissage. Thèse, Université P. et M. Curie, 1978.

[5] CORDEBOIS J.P. – *Comportement et résistance des milieux métalliques multi-perforés*, Thèse de 3e Cycle, Université P. et M. Curie.

[6] DUVAUT G. – "Analyse fonctionnelle et mécanique des milieux continus. Application à l'étude des matériaux composites élastiques à structure périodique. Homogénéisation". In *Theoretical and Applied Mechanics, I.U.T.A.M. Congress Delft*. W.T. Koiter, ed., Amsterdam: North Holland, 1976, pp. 119-132.

[7] LENE F. – "Comportement macroscopique de matériaux élastiques comportant des inclusions rigides ou des trous répartis périodiquement", *C.R.A.S.*, Paris, série A, 286 (1978): 75-78.

RESUME

Anisotropie élastique induite par endommagement

On présente un modèle thermodynamique d'endommagement tridimensionnel et on étudie l'anisotropie élastique qui en résulte. On discute ensuite la confrontation de la thétorie proposée avec l'expérience ou la simulation numérique.

Creep Characterization of Eutectic Composites

A.F. Johnson

National Physical Laboratory, Teddington, England.

1. Introduction

Directionally-solidified eutectic alloys are metallic composite materials with excellent high temperature creep resistance which are being developed for gas turbine blades. These materials consist of a metallic or inter-metallic matrix reinforced by strong inter-metallic or carbide fibres. By cooling the molten alloy in an intense uniaxial temperature gradient, it is possible to precipitate fibres which align to within a few degrees of the solidification direction. Much of the recent research on eutectic composites (see for example [5]) has concentrated on understanding and controlling the solidification process and on the development of materials with improved high temperature performance. However, before these materials can be used confidently in applications, further work is needed to provide design information and to develop relevant design methods. At the high temperatures encountered inside gas turbines, eutectic composites exhibit excellent creep resistance in the fibre direction ; but for the same stress, the creep rate may be as much as three orders of magnitude higher in directions transverse to the fibres. Thus the materials are highly anisotropic in their creep properties and it is this aspect of behaviour which can be critical in design and which concerns us here. We model directionally-solidified eutectic composites as homogeneous, transversely isotropic metallic materials and consider the steady state creep properties, which are important in the assessment of creep and are readily incorporated into design rules (see for example [2]).

The steady-state or secondary creep of isotropic metals is adequately modelled for many applications by the Bailey-Norton law (see for example [2, 3]) which connects the strain-rate and applied stress by a power-law relation. In an earlier paper [4], this expression was generalized to triaxial stress states in a transversley-isotropic metal and this theory is summarized in Section 2. The method adopted is to express the constitutive equation for the strain rate tensor in terms of a scalar dissipation potential, which for isotropic materials is a function of the second principal invariant of the deviatoric stress.

The extension to anisotropic materials is made by allowing the dissipation potential to depend on additional stress invariants, relevant to the appropriate symmetry class of the material. It is then found that three material parameters, in addition to the power law index, are necessary for characterizing the secondary creep properties of a transversely isotropic metallic material. In Section 3 we consider the response predicted by the theory to uniaxial tensile loading at an angle to the fibres and to combined tension-torsion loading along the fibre axis. The predictions of the theory are compared in Section 4 with experimental data obtained at NPL from creep tests on a Co $-$ Cr $-$ C eutectic alloy. The data enable the parameters of the theory to be determined and provide a check on the validity of the proposed constitutive equation. Finally the influence of combined tension-torsion loading on the creep behaviour of a eutectic composite is discussed and its significance for design assessed.

2.   Secondary creep in anisotropic metals

Secondary creep in isotropic metals under uniaxial loading may be adequately modelled for many applications by the Bailey-Norton creep law

$$\dot{e} = A\,\sigma^n, \tag{1}$$

(see, for example, [2, 3]). Here e is the creep strain and $\dot{e}$ the strain rate which result when a uniaxial stress σ is applied to the material. In the secondary stage of creep considered here, the strain increases linearly with time thus (1) refers to a steady state and A, n are material constants, independent of time. In general they will depend on temperature, although this is not considered explicitly. Experimental evidence on creep of metals suggests that $n \geqslant 1$, which we assume here.

For three-dimensional deformations we adopt the referential or material description, so that referred to a rectangular co-ordinate system (x_1, x_2, x_3) fixed in the undeformed body, σ_{ij} and e_{ij} $(i, j = 1, 2, 3)$ denote respectively the stress components measured per unit area in the undeformed body and the components of the finite material strain tensor. In the initial development of the theory we use direct tensor notation. The deviatoric stress tensor s is defined by

$$s = \sigma - \frac{1}{3}\,(\mathrm{tr}\,\sigma)\,I \tag{2}$$

and we note that $\mathrm{tr}\,s = s_{ii} = 0$. Extensions of (1) to triaxial loading of isotropic metals are discussed by Hult [3], who gives the constitutive equation

$$\dot{e} = \frac{3}{2}\,A\,(\bar{\sigma})^{n-1}\,s, \tag{3}$$

where

$$\bar{\sigma} = \left(\frac{3}{2} \, \mathrm{tr} \, s^2\right)^{\frac{1}{2}},$$ (4)

is the effective stress. Equation (3) satisfies the requirements of objectivity and reduces to (1) for uniaxial loading. It also defines a strain rate tensor $\dot{e}$ which, through its dependence on s is independent of a superposed hydrostatic stress, in accord with creep data on metals. We show in [4] that the constitutive equation (3) can be written in the alternative form

$$\dot{e} = \frac{\partial \Phi}{\partial \sigma},$$ (5)

where the scalar Φ is defined by

$$\Phi = A(\bar{\sigma})^{n+1}/(n + 1).$$ (6)

·We refer to Φ as the dissipation potential. In the absence of heat conduction, it can be shown that Φ must satisfy the thermodynamic restriction $\Phi \geqslant 0$.

As a general framework for characterizing the secondary creep of metals, we assume a basic constitutive equation in the form (5) in which Φ is a non-negative scalar function of the deviatoric stress s. For an isotropic material, Φ must depend on the principal invariants of s. We denote the principal invariants of a second-order tensor A by I_A, II_A, III_A, where

$$I_A = \mathrm{tr} \, A, \quad II_A = \tfrac{1}{2} \, [(\mathrm{tr} \, A)^2 - \mathrm{tr} \, A^2], \quad III_A = \det A,$$

and since $\mathrm{tr} \, s = 0$, it follows that

$$I_s = 0, \quad II_s = -\tfrac{1}{2} \, \mathrm{tr} \, s^2, \quad III_s = \det s.$$ (7)

Thus for isotropic materials, the dissipation potential has the general form

$$\Phi = \Phi(II_s, III_s)$$

and the supporting experimental evidence for equations (3)-(6) suggests that Φ is a power function of $-II_s$ and independent of III_s.

For anisotropic metals, we assume that the constitutive equation for secondary creep may again be expressed in the form (5), with the dissipation potential Φ now dependent on the invariants of s appropriate to the symmetry class of the material. Let the fibre direction in a transversely-isotropic composite material be denoted by a unit vector **d** and introduce the tensor product **I***, defined by

$$I_{ij}^* = d_i d_j. \tag{8}$$

It can then be shown [1], [4] that Φ satisfies the invariance requirements for a transversely-isotropic material if and only if it is expressible as a function of the five scalar invariants

$$I_s, \ II_s, \ III_s, \ tr(I^* s), \ tr(s I^* s).$$

Rather than consider general forms for Φ, we seek the simplest extension of (6) to the transversely isotropic case. Since (6) is independent of the third order invariant III_s, this suggests we allow Φ to be a power-law function of combinations of invariants which are second order in s. We therefore assume

$$\Phi = \frac{1}{2m} U^m, \tag{9}$$

where

$$U = \tfrac{1}{2} \alpha \, tr \, s^2 + \tfrac{1}{2} \beta \, (tr \, I^* \, s^2 + \tfrac{1}{2} \gamma \, tr \, (s \, I^* \, s), \tag{10}$$

and $\alpha, \beta, \gamma, \ m$ are material constants.

Making use of the results:

$$\frac{\partial \Phi}{\partial \sigma} = \frac{\partial \Phi}{\partial s} - \frac{1}{3} tr \left(\frac{\partial \Phi}{\partial s} \right) I, \ \frac{\partial}{\partial s} (tr \, s^2) = 2 \, s,$$

$$\frac{\partial}{\partial s} tr \, (I^* \, s) = I^*, \ \frac{\partial}{\partial s} tr \, (s \, I^* \, s) = I^* s + s I^*,$$

equations (5), (9) and (10) yield the constitutive equation

$$\dot{e} = \tfrac{1}{2} \, U^{m-1} \left\{ \alpha s + \beta \, tr \, (I^* \, s) \, I^* + \tfrac{1}{2} \, \gamma \, (I^* s + s I^*) \right.$$

$$\left. - \tfrac{1}{3} \, (\beta + \gamma) \, tr \, (I^* \, s) \, I \right\}, \tag{11}$$

which is the proposed equation for secondary creep in directionally-solidified eutectic composites. Creep behaviour is characterized in the theory by three material parameters α, β, γ and an index m. It can be shown that the thermodynamic restriction $\Phi \geqslant 0$ is satisfied for all stress states provided that these parameters satisfy the conditions $\alpha, \beta, \gamma \geqslant 0$, $m \geqslant 1$. We note that on setting $\beta = \gamma = 0$, (11) reduces to the corresponding equation for isotropic materials (3) with $n = 2m - 1$ and $A = \left(\frac{1}{3} \alpha \right)^m$.

We observe that the creep index m is independent of the loading direction, which is in accord with creep tests on a directionally solidified eutectic performed by Miles and McLean [6] and discussed in Section 4. A more complex theory than that given here is required when this assumption is not valid. From the discussion given above this could be derived by adding further invariant functions to (9) giving terms with additional indices. These generalizations are not considered in the present paper, but may be necessary for modelling creep data on certain materials.

3. Applications of the theory

Before considering the predictions of the theory in particular loading situations, we derive an explicit form for the constitutive equation (11). We choose Cartesian axes in the undeformed material so that the fibre direction is aligned with the x_3-axis and introduce an independent set of non-negative material constants λ, μ, ν, defined by

$$\lambda = \frac{1}{3} \alpha + \frac{2}{9} (\beta + \gamma), \quad \mu = \frac{1}{3} \alpha + \frac{1}{18} (\beta + \gamma), \quad \nu = \alpha + \frac{1}{2} \gamma, \qquad (12)$$

which are more meaningful physically than α, β, γ as we shall show later. On setting $d_i = \delta_{i3}$, $I_{ij}^* = \delta_{i3} \delta_{j3}$, where δ_{ij} is the Kronecker delta, and replacing s by the stress σ, equations (10) and (11) may be written in component form

$$\dot{e}_{11} = U^{m-1} \left\{ \mu \, \sigma_{11} - \frac{1}{2} (2\mu - \lambda) \, \sigma_{22} - \frac{1}{2} \lambda \, \sigma_{33} \right\},$$

$$\dot{e}_{22} = U^{m-1} \left\{ -\frac{1}{2} (2\mu - \lambda) \, \sigma_{11} + \mu \, \sigma_{22} - \frac{1}{2} \lambda \, \sigma_{33} \right\},$$

$$\dot{e}_{33} = U^{m-1} \left\{ -\frac{1}{2} \lambda (\sigma_{11} + \sigma_{22}) + \lambda \, \sigma_{33} \right\},$$

$$\dot{e}_{23} = \frac{1}{2} U^{m-1} \nu \sigma_{23},$$

$$\dot{e}_{13} = \frac{1}{2} U^{m-1} \nu \sigma_{13},$$

$$\dot{e}_{12} = \frac{1}{2} U^{m-1} (4\mu - \lambda) \sigma_{12}, \tag{13}$$

where U is given by

$$U = \frac{1}{2} (2\mu - \lambda)(\sigma_{11} - \sigma_{22})^2 + \frac{1}{2} \lambda [(\sigma_{22} - \sigma_{33})^2 + (\sigma_{33} - \sigma_{11})^2]$$

$$+ \nu(\sigma_{23}^2 + \sigma_{13}^2) + (4\mu - \lambda)\sigma_{12}^2. \tag{14}$$

We note from (12) that for an isotropic material $\lambda = \mu = \frac{1}{3} \nu$.

3.1. *Uniaxial loading*

A physical interpretation of the material constants λ, μ, ν follows from considering the response predicted by equations (13) to some simple uniaxial stress states. For a uniaxial tensile creep test in the fibre direction

$$\sigma_{33} = \sigma_L, \quad \sigma_{ij} = 0, \quad i, j \neq 3,$$

the theory gives

$$\dot{e}_{33} = \lambda^m \sigma_L^{2m-1}.$$

For a second uniaxial tensile creep test in a direction transverse to the fibres,

$$\sigma_{11} = \sigma_T, \quad \sigma_{ij} = 0, \quad i, j \neq 1,$$

$$\dot{e}_{11} = \mu^m \sigma_T^{2m-1}.$$

For a shear creep test parallel to the fibres

$$\sigma_{13} = \sigma_S, \quad \sigma_{ij} = 0, \quad i \neq 1, \quad j \neq 3,$$

$$\dot{e}_{13} = \frac{1}{2} \nu^m \sigma_S^{2m-1}.$$

Thus λ^m, μ^m may be determined from uniaxial tensile creep tests along and transverse to the fibre direction and $\frac{1}{2}\nu^m$ from a longitudinal shear creep test. Furthermore these three uniaxial loading tests completely characterize the secondary creep response. It follows that when analysing the influence of composite structure on the secondary creep properties it is only necessary to consider these three loading conditions. Since λ is dominated by the fibre tensile properties and μ, ν by the matrix tensile and shear properties respectively, the understanding of the influence of fibre and matrix properties on the composite material is simplified.

In order to compare the predictions of the theory with the creep tests performed by Miles and McLean [6], we consider the response of a transversely isotropic material subjected to uniaxial tensile loading in directions inclined at an angle θ to the fibre direction. For an applied stress in the 1-3 plane at an angle θ to the 3-axis the non-zero stress components are

$$\sigma_{11} = \sigma \sin^2 \theta, \quad \sigma_{13} = \sigma \sin \theta \cos \theta, \quad \sigma_{33} = \sigma \cos^2 \theta \tag{15}$$

and

$$U = [\lambda \cos^4 \theta + (\nu - \lambda) \sin^2\theta \cos^2\theta + \mu \sin^4\theta] \sigma^2. \tag{16}$$

The strain rate $\dot{e}$ in the loading direction is given by the transformation

$$\dot{e} = \dot{e}_{11} \sin^2 \theta + 2 \dot{e}_{13} \sin \theta \cos \theta + \dot{e}_{33} \cos^2\theta,$$

whence from (13) and (15) with $n = 2m - 1$,

$$\dot{e} = F^{\frac{1}{2}(n+1)} \sigma^n, \tag{17}$$

where the function $F(\theta)$ is given by

$$F(\theta) = \lambda \cos^4\theta + (\nu - \lambda) \sin^2\theta \cos^2\theta + \mu \sin^4\theta. \tag{18}$$

We see that the response predicted by the theory to this "off-axis" uniaxial loading is in the form of the Bailey-Norton law (1), with the parameter A now being given as a specific function of loading angle θ and the material constants λ, μ, ν. We note that for isotropic materials when $\mu = \lambda$, $\nu = 3\lambda$ then $F(\theta) = \lambda$, which is independent of the loading direction as it should be in this case. It follows from (17) and (18) that tensile creep tests at any three distinct values of θ will enable the material constants to be determined, and a check on the theory is obtained from further creep tests at other values of θ.

3.2. *Combined tension-torsion loading*

A convenient experimental method of examining biaxial stress states, which is also amenable to simple analysis, is to subject a thin-walled tube to both tension and torsion. Consider a thin-walled tube of radius a, wall thickness h, with $h \ll a$, composed of a transversely-isotropic metallic composite in which the fibres are aligned along the tube axis. In keeping with the notation used above, we choose co-ordinates (x_1, x_2, x_3) with origin and x_3 — coordinate measured along the tube axis, as shown schematically in Figure 1. The tube is subjected to an axial tensile load T combined with a twisting moment M about the tube axis. For a thin-walled tube the wall through-thickness stresses may be neglected and the stress field is biaxial consisting of a tensile stress σ and shear stress τ, where

$$\sigma = T/2\pi a h, \quad \tau = M/2\pi a^2 h.$$

At a point (x_1, x_2, x_3) in the tube wall, the nonzero components of the stress tensor σ are given by

$$\sigma_{13} = \sigma_{31} = -\tau x_2/a, \quad \sigma_{23} = \sigma_{32} = \tau x_1/a, \quad \sigma_{33} = \sigma. \tag{19}$$

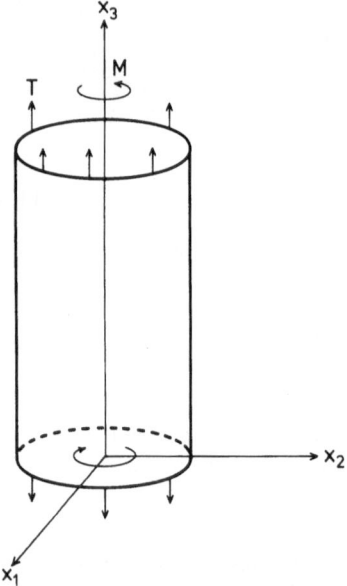

Figure 1. — *Schematic diagram of thin-walled tube.*

It follows from (14) that

$$U = \lambda\sigma^2 + \nu(x_1^2 + x_2^2)\,\tau^2/a^2$$
$$= \lambda\sigma^2 + \nu\tau^2, \tag{20}$$

since we may assume for a thin-walled tube that $x_1^2 + x_2^2 = a^2$. The creep rate along the tube axis $\dot{e}_{33}$ is given from (13) and (20) by

$$\dot{e}_{33} = \lambda(\lambda\sigma^2 + \nu\tau^2)^{m-1}\,\sigma,$$

which may be written in the alternative form

$$\dot{e}_{33} = \lambda^m\,\sigma^{2m-1}\left[1 + \frac{\nu}{\lambda}\left(\frac{\tau}{\sigma}\right)^2\right]^{m-1}. \tag{21}$$

Expressed in this way, the significance of a superimposed shear stress on the axial creep rate in the fibre direction can be assessed.

We note that the solution given above can be generalized to apply to the combined tension and torsion of a thick-walled tube or a solid cylinder by using the methods of Hult [3]. This extension is not considered here since the simpler solution (21) is sufficient to illustrate the effects of combined loading on the axial creep properties of a transversely-isotropic material.

4. Comparison of theory with creep test data

Miles and McLean [6] have carried out uniaxial tensile creep tests at NPL on the unidirectionally solidified eutectic C73 provided by Brown Boveri and Co., Baden. The material has composition by weight Co 56.9 %, Cr 40.7 %, C 2.4 % and consists of a (Co, Cr) matrix reinforced by faceted $(Co, Cr)_7\,C_3$ fibres. The tests were performed to rupture under constant load in air at 825 °C on specimens cut at angles of $0°, 10°, 15°, 20°, 40°, 60°, 90°$ to the fibre direction. In the secondary creep region it was found that a plot of log $\dot{e}$ against log σ for different loading angles θ gave a set of almost parallel straight lines, see [6, Fig. 5]. These data are in agreement with the form of equation (17) and it was concluded that, to a reasonable approximation, the power law index n was independent of loading angle θ, and a value n = 5.5 ± 1 was obtained for this material.

Miles and McLean [6, Table 2] also determined the stress level σ to give a constant creep rate of $\dot{e} = 2.5 \times 10^{-4}$/hour for each loading angle θ and their experimental results are reproduced in Table 1. These data enable the parameters λ, μ, ν of the theory to be determined and the predicted form of $F(\theta)$ in equation (18) to be checked. We use the experimental data from Table 1 at $\theta = 0°, 40°, 90°$, with n = 5.5, in equations (17) and (18) to give the calculated values

$$\lambda = 0.556, \quad \mu = 5.85, \quad \nu = 11.2 \tag{22}$$

TABLE 1. Measured (σ_{exp}) and calculated (σ_{th}) values of stress to give constant creep rate of 2.5 × 10^{-4}/hour. (Experimental results from [6]).

$\theta°$	σ_{exp} (MPa)	$\sigma_{th.}$ (MPa)
0	313	313
10	266	243
15	226	200
20	178	166
40	101	101
60	81	82
90	78	78

where the units are $(\text{hours})^{-2/(n+1)} (\text{MPa})^{-2n/(n+1)}$. These three parameters and n characterize the secondary creep behaviour and we may now calculate $F(\theta)$ from equation (18) and thus determine the influence of loading direction on the creep properties. Figure 2 shows the calculated function $F(\theta)$ and, as a test for the validity of the theory, compares it with experimentally measured values of $(\dot{\epsilon}/\sigma^n)^{2/(n+1)}$ taken from Table 1 with n = 5.5. There is fair agreement between the theoretical curve and the experimental data. We note that the agreement at $\theta = 0°, 40°, 90°$ follows from the choice of these angles for determining λ, μ, ν.

As a further test of the theory, we may employ the theoretically determined $F(\theta)$ in Figure 2 and equation (17) to calculate the stress level at various loading angles θ which yield a particular steady creep rate. Choosing a creep rate of 2.5 × 10^{-4}/hour, we compare in Table 1 stress values calculated in this way with the measured values given in [6]. Again there is reasonable agreement between theory and experiment. We note that the theoretical calculations are very sensitive to the value chosen for n, which did show some scatter about the mean value 5.5.

Since $F(\theta)$ is constant for isotropic materials, Figure 2 shows how strongly anisotropic are the creep properties. For the eutectic alloy studied $F(90)/F(0) = 10.5$ at the test temperature of 825 °C. As the creep rate for a fixed stress is proportional to $F^{3.25}$, it follows that the creep rate increases rapidly for loads inclined to the fibres and, transverse to the fibres, it is

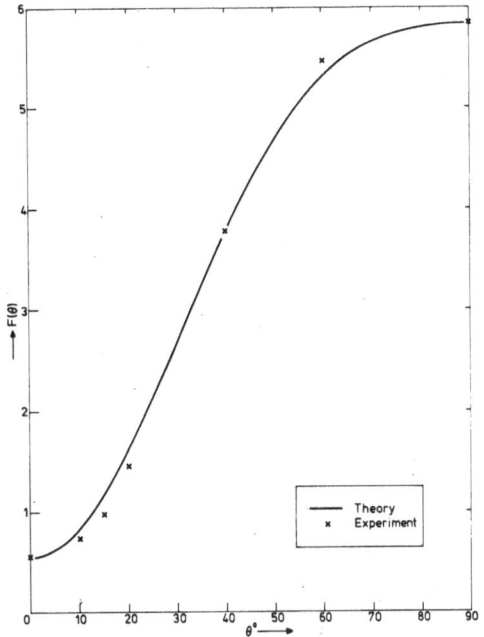

Figure 2. – *Comparison of experimental data with theoretical values of* F(θ) *calculated from equation (18).*

2.08×10^3 times greater than in the fibre direction. This highlights the design problems which may arise when using eutectic composites, since small misalignments of fibres from the main load direction, or the presence of secondary transverse stresses could lead to excessive creep behaviour.

It is also of interest to use equation (21) and the materials data (22) to assess the significance of superposed shear stresses on creep behaviour along the fibres. This type of loading has some relevance to the design of composite turbine blades where the main loads are uniaxial along the blade and fibre axis, but which may be subjected to secondary shear or torsional loads. The term $\lambda^m \sigma^{2m-1}$ in equation (21) is the uniaxial creep rate in the fibre direction and, since $m \geqslant 1$, it follows that the presence of a torsional stress τ will always increase the uniaxial tensile creep rate. This is shown in Figure 3 where $\dot{e}_{33}/\lambda^m \sigma^{2m-1}$ is plotted as a function of τ/σ using the materials data (22). For this eutectic composite the figure indicates a rapid increase in the axial creep rate as the shear stress increases from zero. Thus the tensile creep rate along the fibres is increased by a factor of 10 for relatively small superposed torsional loads for which $\tau/\sigma = M/aT = 0.3$. We see

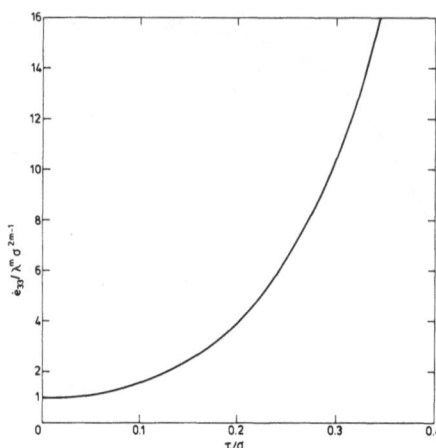

Figure 3. – *The influence of torsion on tensile creep properties as predicted by equation (21).*

that if the excellent high temperature creep resistance of eutectic composites is to be fully exploited then great care will be needed in component design.

5. Concluding remarks

The main theoretical results of the paper are equations (13) and (14) which provide a constitutive equation for secondary creep of a transversely isotropic metal. They were derived as the simplest generalization to transversely isotropic materials of the Bailey-Norton law for secondary creep in isotropic metals. The creep behaviour is found to be characterized by three material constants λ, μ, ν and a power law index m or n. In the theory it is assumed, on the basis of experimental evidence, that the index is independent of loading direction in the material. Uniaxial tensile creep tests on a directionally solidified eutectic composite were found to be modelled reasonably well by the theory, although further tests involving biaxial and triaxial loading are necessary to establish its more general validity.

The theory enables creep experiments on eutectic composites to be analysed; it provides a basis for design calculations on these materials and defines the materials data needed by the designer. Experimental data on a Co-Cr-C eutectic composite obtained at NPL shows the highly anisotropic creep properties of these materials, with excellent creep resistance along the fibres but with creep rates up to three orders of magnitude higher transverse to the fibres. The predictions of the theory, when uniaxial loads are applied at an angle to the fibres or when combined tension and torsion loads are applied, indicates the design problems which arise in using these highly anisotropic materials. It is clear that the successful application of eutectics to components such as turbine blades will require more emphasis on stress analysis and detailed design methods. However, the results of the idealized analyses given here suggest that some improvement in the matrix creep resistance is needed if best use is to be made of the excellent fibre creep properties.

Acknowledgements

The writer wishes to thank his colleagues Dr. L N McCartney, Dr. M McLean and Dr. D E Miles for valuable discussions during the course of this work.

REFERENCES

[1] ERICKSEN J.L. and R.S. RIVLIN. – "Large elastic deformations of homogeneous anisotropic materials". *J. Rat. Mech. Anal.*, 3 (1954): 281-301.

[2] HOFF N.J. – "Rules and methods of stress and stability calculations in the presence of creep". *J. of Appl. Mech.*, 45 (1978): 669-675.

[3] HULT J. – *Creep in Engineering Structures.* Waltham, MA: Blaisdell, 1966.

[4] JOHNSON A.F. – "Creep characterization of transversely-isotropic metallic materials". *J. Mech. Phys. Solids,* 25 (1977): 117-126.

[5] McLEAN M. and F. SCHUBERT. – "Mechanical properties of directionally solidified super-alloys and eutectics". *Proceedings of Liege Conference on "High temperature alloys for gas turbines".* Applied Science, London, (1978): 423-458.

[6] MILES D.E. and M. McLEAN. – "Anisotropic creep behaviour of (Co, Cr) (Co, Cr), C_3 eutectic composite". *Metal Science*, 11 (1977): 563-570.

RESUME

Caractérisation du fluage des composites eutectiques.

Une équation de comportement, basée sur une généralisation pour les matériaux orthotropes de révolution de la loi de Bailey-Norton, est proposée pour la modélisation du fluage secondaire des alliages eutectiques à solidification directionnelle. Nous montrons, qu'en plus de l'indice de la loi de puissance, trois paramètres matériels sont nécessaires pour la caractérisation du comportement en fluage secondaire ; la détermination expérimentale de ces paramètres est discutée. La théorie est comparée avec des données expérimentales du fluage d'un alliage eutectique Co-Cr-C.

DISCUSSION

QUESTION BY R.G.C. ARRIDGE : What was the aspect ratio of the particles in
the eutectic ?

REPLY BY A.F. JOHNSON : The fiber aspect can be controlled during the fabrication process. The Co-Cr-C eutectic alloy tested at N.P.L. had fibers about
100 μm long with diameter about 1 μm.

Relations entre la Vitesse du Fluage de la Glace Polycristalline et la Texture Cristalline

H. Le Gac, J. Meyssonnier et P. Duval

Laboratoire de Glaciologie - CNRS, Grenoble, France.

1. Introduction

Le comportement plastique de la glace polycristalline a principalement été étudié sur des glaces isotropes. Cependant, pour des déformations supérieures à quelques %, la testure initiale (orientation statistique des axes optiques) est modifiée soit par les processus de recristallisation soit par la déformation elle-même [3]. Le changement de texture induit le plus souvent des variations de la vitesse de déformation [5]. L'écoulement des glaciers parait lui-même fortement influencé par la texture de la glace [14]. Ainsi les vitesses mesurées en surface en Antarctique ne peuvent être expliquées qu'à partir de modèles où l'influence de la texture sur la viscosité est prise en compte [14].

Dans ce travail, nous nous proposons de décrire d'abord le comportement plastique du monocristal, de déterminer dans des cas particuliers les variations de la viscosité de la glace polycristalline avec la texture et finalement de tenter d'expliquer les vitesses mesurées sur la calotte polaire près de Byrd Station (Antarctique).

2. Déformation plastique du monocristal de glace

La glace des glaciers de notre planète est cristallisée dans le système hexagonal. Elle se déforme plastiquement surtout par glissement sur les plans de base $\langle 0001 \rangle$. L'axe optique ou axe "C" est perpendiculaire aux plans de base. La déformation résulte du mouvement de dislocations dont le vecteur de Burgers est $\frac{a}{3} \langle 11\bar{2}0 \rangle$ [8]. D'autres types de dislocations ont été observés, par exemple, des boucles de dislocations prismatiques [16]. La dissociation des dislocations basales, même si elle n'a pu être observée expérimentalement du fait du manque de résolution des appareils utilisés, semble possible [18].

Plusieurs courbes contrainte-déformation sont montrées sur les figures 1 et 2. Pour les cristaux orientés pour le glissement basal, les courbes présentent d'importants pics de contrainte et aucun indice d'écrouissage (Fig. 1). Dans le cas où le cristal est orienté pour éliminer le glissement basal, les courbes montrent l'effet de l'écrouissage et la contrainte est environ 20 fois plus

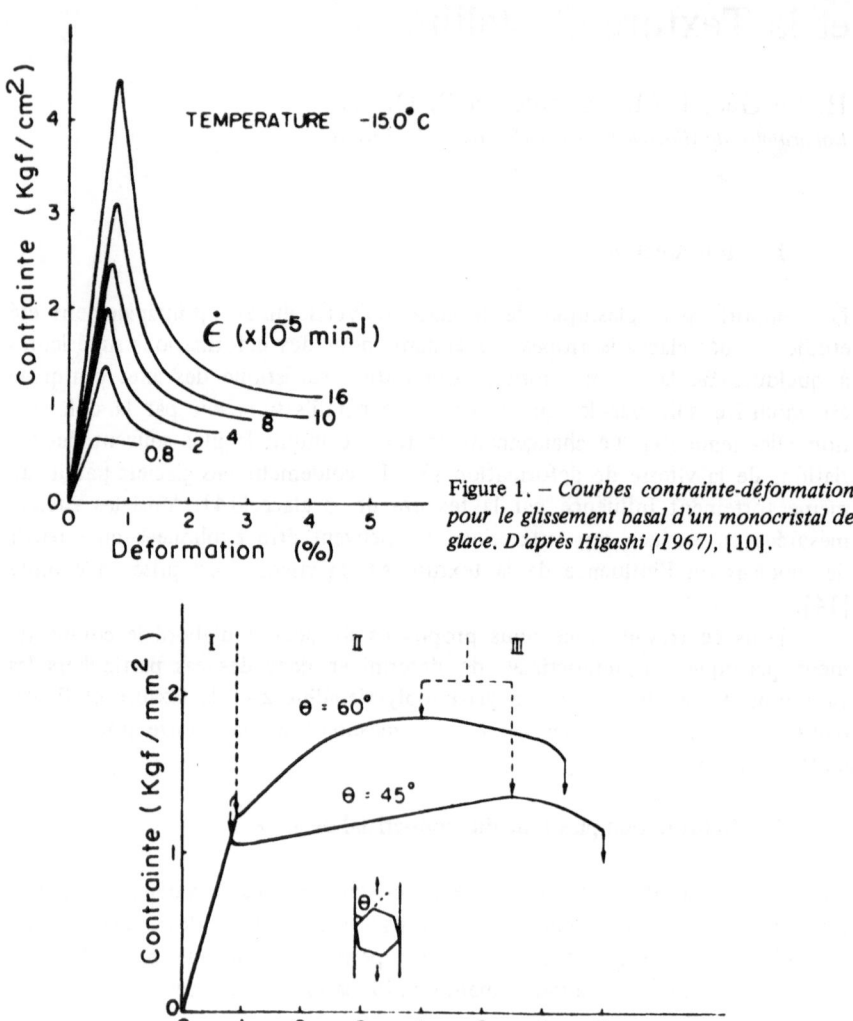

Figure 1. – *Courbes contrainte-déformation pour le glissement basal d'un monocristal de glace. D'après Higashi (1967), [10].*

Figure 2. – *Courbes contrainte-déformation pour le glissement non basal pour différentes orientations de l'axe "a". L'axe de traction est dans le plan de base. Vitesse de déformation : 310^{-6} sec^{-1}. D'après Higashi et al (1964), [11].*

élevée que dans le cas où le cristal est orienté pour le glissement basal [10]. Pour des essais à contrainte imposée (fluage), le rapport des vitesses de déformation est supérieur à 1 000 pour les deux orientations extrêmes. Aussi la déformation des monocristaux de glace est-elle contrôlée principalement par l'intensité de la cission dans le plan de base ⟨0001⟩.

3. Fluage de la glace polycristalline

La déformation plastique de la glace polycristalline a principalement été étudiée par des tests de fluage [7, 17, 2]. A l'échelle macroscopique, comme les métaux, la glace polycristalline présente d'abord un fluage transitoire qui diminue avec le temps, ensuite un fluage stationnaire (ou fluage secondaire). Près du point de fusion et pour des déformations supérieures à quelques %, la vitesse du fluage augmente et un nouveau fluage stationnaire est obtenu (fluage tertiaire).

Pour le fluage secondaire de glaces polycristallines isotropes, les vitesses de déformation $\dot{\epsilon}_{ij}$ sont liées au déviateur des contraintes τ'_{ij} suivant la relation de Bailey-Norton :

$$\dot{\epsilon}_{ij} = \frac{\tau'_{ij}}{2\eta} = \frac{B}{2} \, \tau^{n-1} \, \tau'_{ij} \tag{1}$$

où η est la viscosité, B un paramètre qui dépend principalement de la température et τ la cission efficace définie par : $\tau^2 = \frac{1}{2} \Sigma (\tau'_{ij})^2$. L'équation (1) a été vérifiée par des expériences de fluage en torsion-compression [2]. La valeur de n est de l'ordre de 3 pour des cissions efficaces comprises entre 0,5 et 5 bars. Les courbes de fluage des glaces polycristallines et des monocristaux orientés pour éviter le glissement basal sont similaires et les vitesses de fluage sont du même ordre de grandeur [1]. Pour ces glaces, la déformation n'est pas contrôlée par la vitesse de déplacement des dislocations mais plutôt par des processus de restauration [2].

L'influence de la texture sur la vitesse du fluage a été étudiée sur des glaces polycristallines naturelles. L'échantillon en forme de cylindre était soumis soit à une torsion soit à une compression selon l'axe [2]. La température était fixée à $-7,2\,°C$. La texture initiale de la glace étudiée est donnée sur la figure 3. La plus grande partie des axes optiques est suivant l'axe de l'échantillon. Cette glace provient d'un glacier polaire (Antarctique). La taille des cristaux était de l'ordre de 6 mm. Sur la figure 4 sont représentées la courbes de fluage de cette glace en torsion et en compression. Une courbe de fluage d'une glace polycristalline isotrope est représentée sur cette même figure pour comparaison. Le rapport des vitesses minimum de torsion et de

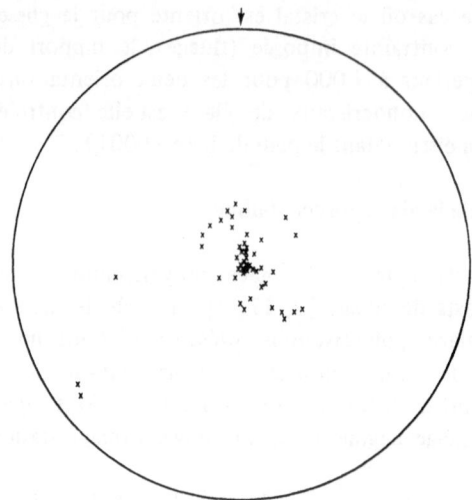

Figure 3. – *Orientation initiale des axes C d'une glace polycristalline, en provenance de l'Antarctique, déformée en torsion et compression (cf. Fig. 4).*

compression en fin d'expérience était de l'ordre de 40. Les valeurs respectives de B calculées à partir de l'équation (1) avec $n = 3$ sont de 0,20 et $0,005\ bar^{-3}\ an^{-1}$. Des variations beaucoup plus faibles (rapport des vitesses inférieur à 10) étaient trouvées par Lile [12] à partir d'expériences du même

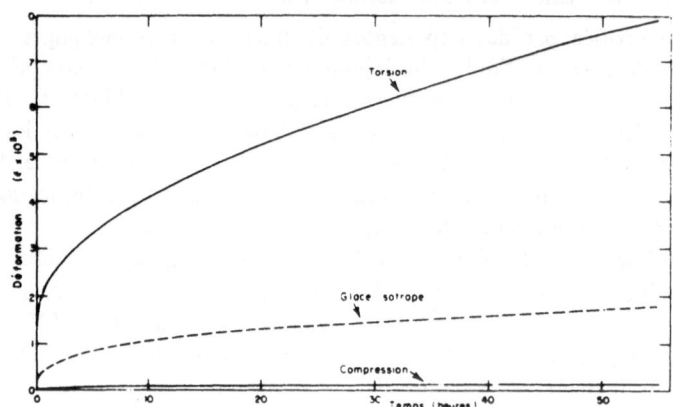

Figure 4. – *Courbes de fluage en torsion et compression d'un échantillon de glace polycristalline présentant une texture à un maximum suivant l'axe de torsion et de compression. La courbe en pointillés a été calculée à partir des résultats obtenus pour les glaces dont l'orientation des axes C était aléatoire. Température : $-7,2°C$. Cission efficace $\tau = 1,35\ bar$.*

type. Ce type d'essais ne permet pas de déterminer les trois coefficients de viscosité caractéristiques du comportement anisotrope de la glace polycristalline. Des essais en torsion-compression sont en cours pour déterminer ces trois paramètres nécessaires pour caractériser le fluage permanent.

4. Texture des glaces de glaciers

De nombreuses études pétrographiques de glaces naturelles ont pu être faites grâce aux différents carottages effectués ces dernières années aussi bien dans les glaciers polaires que dans les glaciers alpins. La glace des glaciers alpins est dans sa plus grande partie au point de fusion. Aussi la glace recristallise-t-elle facilement pendant la déformation. Les textures observées sont principalement induites par la recristallisation dynamique [5]. Les dislocations produites pendant la déformation forment des sous-joints. De nouveaux cristaux sont créés à partir de sous-grains fortement désorientés. Les cristaux orientés pour minimiser l'énergie de la déformation plastique grossissent aux dépens des cristaux présentant une énergie de déformation plus importante. Les textures formées sont le plus souvent à plusieurs maximums (3 à 5) [19, 20]. Une texture typique de la glace des glaciers tempérés est montrée sur la figure 5. L'influence de ces textures à plusieurs maximums sur la vitesse de déformation est faible [5].

La température de la glace des glaciers polaires est le plus souvent inférieure à $-10\,°C$. Elle peut atteindre $-55\,°C$ dans les parties centrales de la

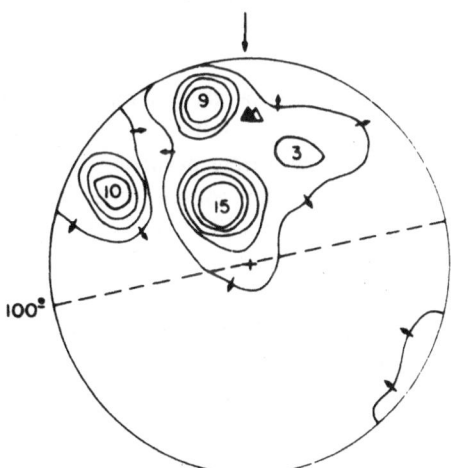

Figure 5. – *Texture typique d'une glace de glacier tempéré – Vallée Blance. Profondeur :* *171 m. Contours à 0, 3, 6, 9 et 12 % des points dans 1 % de l'aire – 88 cristaux.*

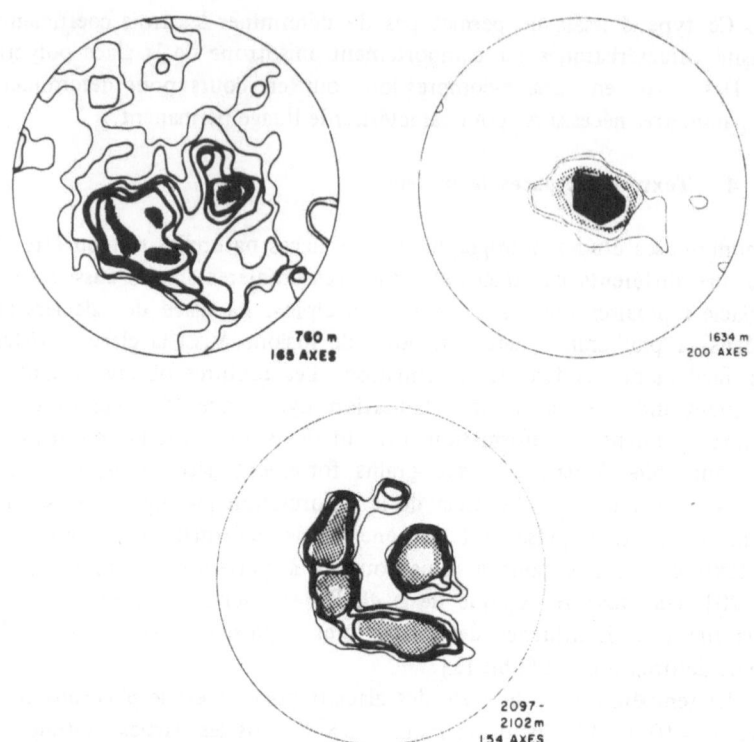

Figure 6. – *Textures de la glace de Byrd (Antarctique) à différentes profondeurs. D'après Gow et Williamson (C.R.R.E.L.), [9].*

calotte Antarctique. Cependant, de la glace tempérée semble être localisée dans les parties les plus profondes de la calotte [9].

La texture de la glace froide diffère beaucoup de celle de la glace tempérée. Des textures à 1 seul maximum, comme celle de la glace étudiée dans ce travail, sont fréquemment observées [9, 13]. Gow and Williamson [9] ont étudié les textures des glaces du carottage de Byrd (Antarctique) (épaisseur du glacier : 2 160 m). Trois de ces textures sont représentées sur la figure 6. Elles caractérisent l'évolution observée avec la profondeur. En effet aucune orientation préférentielle n'apparait jusqu'à 100 m de profondeur. Au-dessous de 100 m, les axes optiques s'orientent progressivement suivant la direction verticale. Entre 1 200 et 1 300 m, la texture à un seul maximum apparaît et persiste jusqu'à 1 800 m. Au-dessous de 1 800 m, la texture à plusieurs maximums, du même type que celle apparaissant dans les glaciers tempérés, est observée. La formation de cette texture est probablement liée aux températures proches du point de fusion mesurées entre 1 800 m et le lit rocheux [9]. En effet la recristallisation dynamique ne semble intervenir, comme processus de restauration ou d'adoucissement de la glace, que pour les températures les plus élevées.

5. Essais d'interprétation des vitesses mesurées sur un carottage en tenant compte de la variation de la viscosité avec la texture

Un profil de vitesse a été mesuré in situ à Byrd Station, Antarctique par Garfield et Ueda [6].

Nous avons supposé que dans cette région l'écoulement de la glace était assimilable à l'écoulement bidimensionnel d'un nappe d'épaisseur uniforme, inclinée d'un angle α par rapport à l'horizontale, soumise à l'action de la pesanteur et adhérant au lit rocheux. La densité de la glace a été supposée constante avec la profondeur.

La vitesse dans le sens x de l'écoulement est notée u. Elle ne dépend que de la profondeur z mesurée perpendiculairement à la surface de la nappe. Ces hypothèses ne sont probablement pas valables pour les couches supérieures du glacier. Mais, la température restant très basse jusqu'à 1 000 m de profondeur (Fig. 7), la déformation de la glace doit rester très faible dans les couches proches de la surface.

Notant σ la cission s'exerçant dans la couche et p la pression isotrope, les équations de l'équilibre se réduisent à :

$$\frac{d\sigma}{dz} + \rho g \sin \alpha = 0$$

$$\frac{dp}{dz} + \rho g \cos \alpha = 0$$

d'où l'on tire, la surface de la nappe étant libre :

$$\sigma(z) = - \rho g z \sin \alpha$$

$$p(z) = - \rho g z \cos \alpha.$$

D'après la relation (1) liant les vitesses de déformation au déviateur des contraintes :

$$\frac{du}{dz} = B |\sigma^{n-1}| \sigma. \tag{2}$$

Ayant supposé que l'état de déformation de la glace en profondeur est un cisaillement simple dont l'axe peut être confondu avec la verticale, nous

pouvons obtenir un ordre de grandeur de la valeur du paramètre B de l'équation (1) à partir des résultats de la figure 4 et des textures données par Gow and
Williamson [9]. La valeur de référence de B a été prise égale à 0,02 bar^{-3} an^{-1}
à la température de $-7\,^{\circ}$C, valeur obtenue avec des glaces isotropes en fluage
secondaire [4]. Il a été alors, tenu compte de la variation de B avec la température grâce au profil de température donné par Gow and Williamson [9]. Nous
avons adopté deux valeurs pour l'énergie d'activation du fluage : 15 kcal/mole
pour $T \langle -7\,^{\circ}$C et 25 kcal/mole pour $T \rangle -7\,^{\circ}$C [15]. Enfin, pour tenir compte
de l'effet de la texture sur la vitesse de déformation, la valeur de B était multipliée par 10 entre 1 300 et 1 800 m. Au-delà de ces profondeurs la glace était
supposée isotrope.

L'intégration de (2) pour une variation quelconque de B avec la profondeur est possible numériquement. Compte tenu des hypothèses grossières
faites sur l'écoulement ainsi que de l'incertitude sur la valeur de l'inclinaison
de la nappe, nous avons approché la courbe B(z) obtenue ci-dessus par une
ligne polygonale représentée sur la figure 7. (Sur cette figure on donne la
variation de la température de la glace avec la profondeur).

Dans chaque couche pour laquelle la variation de B avec z est linéaire,
l'intégration exacte de (2) est possible.

Dans la couche i, comprise entre les profondeurs z_{i-1} et z_i $(z_i > z_{i-1})$
nous posons :

$$B(z) = \mu_i z + \nu_i$$

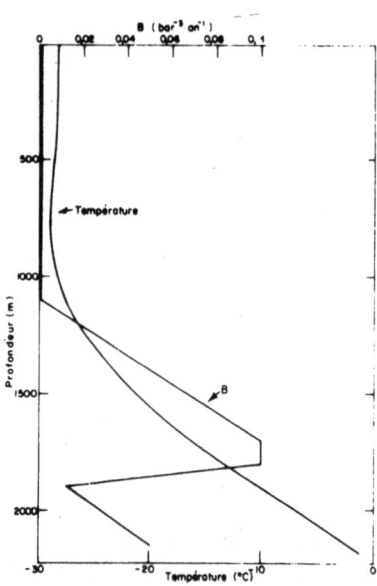

Figure 7. — *Variations de la température et
du paramètre B de l'équation (1) avec la
profondeur (Byrd, Antarctique).*

et (2) donne :

$$u(z) = u(z_i) + (\rho g \sin \alpha)^n \left[\frac{\mu_i(z_i^{n+2} - z^{n+2})}{n+2} + \frac{v_i(z_i^{n+1} - z^{n+1})}{n+1} \right]$$

Nous avons tracé sur la figure 8 le profil de vitesse obtenu avec $n = 3$, pour les valeurs de B adoptées, en estimant l'angle $\alpha = 0,18°$ et avec $\rho g = 0,0911$ bar/m.

La forme de la courbe obtenue est semblable au profil publié par Whillans [21], mais la vitesse calculée en surface est inférieure à la vitesse observée (12,7 . an–1). Cette différence peut s'expliquer par une valeur trop faible de α ou par une valeur trop faible du facteur d'amplification dû à la texture. Pour obtenir la vitesse observée, il faudrait prendre soit $\alpha = 0,23°$ soit un facteur d'amplification de 25 (au lieu de 10).

Notons qu'en supposant la glace isotrope, avec $n = 3$ et $B = 0,02$ $\text{bar}^{-3} \text{an}^{-1}$ à $-7°C$, la vitesse calculée en surface est de 1,7 m/an.

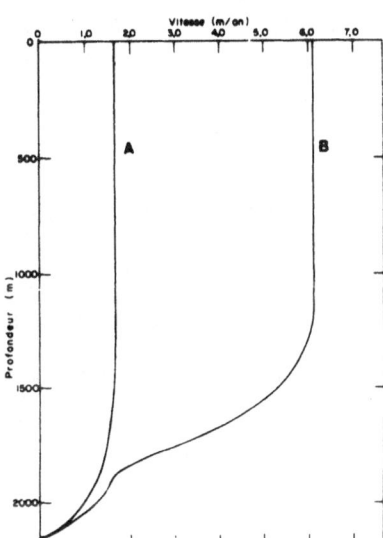

Figure 8. – *Variations de la vitesse horizontale avec la profondeur (Byrd, Antarctique) (Profilts théoriques). A – Viscosité indépendante de la texture.*

6. Conclusion

Ces résultats ont permis de montrer la forte influence de la texture sur la vitesse du fluage de la glace polycristalline. Comme montré dans ce papier, l'anisotropie plastique devra être prise en compte pour calculer l'écoulement des calottes glaciaires.

Les trois paramètres nécessaires pour caractériser l'anisotropie de la glace devront être déterminés en laboratoire à partir de divers types d'essais. La connaissance de ces paramètres permettra de calculer l'écoulement des calottes glaciaires pour des situations plus complexes que celles traitées dans ce papier.

REFERENCES

[1] BUTKOVITCH T.R. and J.K. LANDAUER. – "The flow law for ice." *USA SIPRE Research Report*, **56**, 1959.

[2] DUVAL P. – "Lois du fluage transitoire ou permanent de la glace polycristalline pour divers états de contrainte". *Ann. Geophys.*, **32**, 4 (1976): 335-350.

[3] DUVAL P. – "Creep and recrystallization of polycrystalline ice." *Bull. Mineral.*, **102** (1979): 80-85.

[4] DUVAL P. and H. LE GAC. – "Does the permanent creep rate of polycrystalline ice increase with crystal size." *Journal of Glaciology*, **25**, 91 (1980).

[5] DUVAL P. – "Creep and fabrics of polycrystalline ice under shear and compression." *Journal of Glaciology*, **27**, 95 (1981).

[6] GARFIELD D.E. and H.T. UEDA. – "Resurvey of the Byrd Station, Antarctica, drill hole." *Journal of Glaciology*, **17**, 75 (1976): 29-34.

[7] GLEN J.W. – "The creep of polycrystalline ice." *Proc. of the Royal Society*, Ser. A, **228**, 1175 (1955): 519-538.

[8] GLEN J.W. – "The mechanics of ice." *USA CRREL, Monograph II*, C2 b (1975): 1-43.

[9] GOW A.J. and T. WILLIAMSON. – Rheological implications of the internal struc-drilling at Byrd Station." *USA, CRREL* Report 76-35 (1976): 1-25.

[10] HIGASHI A. – "Mechanisms of plastic deformation in ice single cyrstals." *Physics of snow and ice, Conference on physics of snow and ice*, Oura, 1967.

[11] HIGASHI A., S. KOINUMA and S. MAE. – "Plastic yielding in ice single crystals." *JJAP*, **3** (1964): 612.

[12] LILE R.C. – "The effect of anisotropy on the creep of polycrystalline ice." *Journal of Glaciology*, **21**, 85 (1978): 475-483.

[13] LORIUS C. et M. VALLON. – "Etude structurographique d'un glacier antarctique." *C.R. Acad. Sci.*, Paris, Série D., **265** (1967): 315-318.

[14] LLIBOUTRY L. – "Analytical models for the flow of cold ice sheets." *Journal of Glaciology* (sous presse).

[15] MELLOR M. and R. TESTA. – "Effect of temperature on the creep of ice." *Journal of Glaciology*, **8**, 52 (1969): 131-145.

[16] OGURO M. and A. HIGASHI. – "Concentric dislocation loops with [0001]. Burgers vectors in ice single cyrstals doped with NH_3." *Phil. Mag.*, **24** (1971): 713-718.

[17] STEINEMANN S. – "Experimentalle Untersuchungen zur Plastizitât von Eis." *Beiträge zur Geologie der Schweiz, Hydrology*, 1958.

[18] TYSON W.R. – "Elastic strain energy of dislocations in ice." *Canadian Journal of Physics*, **49** (1971): 2181-2186.

[19] VALLON M. – "Contribution à l'étude de la Mer de Glace." *Thèse de Doctorat d'Etat*, Université de Grenoble, 1967.

[20] VALLON M., J.R. PETIT and B. FABRE. – "Study of an ice core to the bedrock from an alpine glacier." *Journal of Glaciology*, 17, 75 (1976): 13-27.

[21] WHILLANS I.M. – "The equation of continuity and its application to the ice sheet near Byrd Station, Antarctica." *Journal of Glaciology*, 18, 80 (1977): 359-371.

ABSTRACT

Ice single crystals undergo plastic deformation by slipping on the basal planes $\langle 0001 \rangle$. Strain rates of crystals orientated for basal glide are of 3 orders of magnitude higher than those of crystals orientated for hard glide.

Creep tests in torsion and compression were performed on natural ices with strong fabrics. Creep rates were compared to the ones of randomly orientated polycrystalline ices. Ice fabrics of temperate and cold glaciers were described.

Measured velocities near Byrd Station (Antarctica) were approached by considering the two-dimensional flow of the glacier supposed of uniform thickness and by taking into account the influence of fabrics on the ice viscosity.

Session 11

*Experimental Investigations and Interpretation
of Mechanical Tests*

*Recherches Expérimentales et Interprétation
des Essais Mécanique*

On a Continuum Approach to Plastic Anisotropy of Perforated Materials

A. Litewka
Technical University, Poznań.

A. Sawczuk
Institute of Basic Engineering Research, Warsaw.

1. Introduction

The overall response of perforated metal sheets to mechanical agencies depends on the hole arrangement specified by the penetration pattern and on the magnitude of holes characterized by the ligament efficiency. In general this response is direction dependent, thus specific to materials with oriented internal structure. A regularly perforated isotropic material behaves under stress as an equivalent anisotropic solid, thus as a homogeneous material of oriented internal structure. This results in direction dependent material constants such as the Young moduli, the yield stress and alike. The mechanical anisotropy of equivalent materials is not the same in the elastic and plastic ranges. Therefore different methods of homogenisations are needed for establishing the above equivalence.

In order to evaluate the effective elastic constants of a perforated elastic material, the homogenisation usually compares the elastic energy and deformability of respective isotropic and equivalent materials. Slot and O'Donnell [22] as well as Grigolyuk and Filshtinskii [8] evaluated effective elastic constants.

In the plastic range a homogenisation procedure is intended to give the yield condition and the flow rule of an equivalent material. Usually the attention is focused on evaluation of the yield locus. Porowski and O'Donnell [6, 16] employed the methods of limit analysis to derive bounds to the yield surfaces for perforated materials with square and triangular penetration patterns. Winnicki, Kwieciński and Kleiber attempted to derive the yield surface for perforated materials with triangular penetration pattern [24]. The finite element technique and the Huber-Mises yield condition were used for this purpose. In these homogenisation techniques as well as in applications of the derived anisotropic criteria to plate problems by Sawczuk, O'Donnell and

Porowski [19] the plastic potential flow rule was assumed to hold both for
the original and for the equivalent material.

It appears worthwhile to study the problem of plastic homogenisation
both as regards the yield condition and the plastic flow law. The perforated
materials often are deformed so that the volume changes cannot be neglected and the flow law should account for this fact. The theoretical basis allowing for a consistent analysis of the plastic behaviour of perforated materials
is furnished by the tensor function representations. The basic notions of
tensor functions polynomial representations due to Rivlin and Ericksen [17]
proved useful in studying the fluid and solid materials behaviour, cf. Rivlin
[18]. To investigate the plastic flow of isotropic materials polynomial representations were used by Thomas [23] and by Sawczuk and Stutz [20]. It has
been shown in [20] that both the yield condition and the flow law, but not
necessarily that of the potential rule, can be obtained via such an approach
in a consistent manner. Boehler [1] and Boehler and Sawczuk [4] applied
the tensor functions to studying anisotropic materials at plasticity and rupture.
Boehler [2] developed further the method of tensor functions representations,
considering non-polynomial representations.

As the behaviour of perforated materials is not isotropic and an appropriate account has to be made when deriving the flow law and the yield condition on the perforated material symmetries, the tensor function approach
to the question is adopted in the paper when outlining the theoretical setting.
To allow for the anisotropic response of equivalent materials, the notion of
perforation tensor is introduced. Its components will be related to the penetration pattern and to the perforation density. The present note concerns
mostly the experimental motivation for further studies on specifying a suitable
representation.

To establish the overall plastic anisotropy tensor, experiments are needed
on the plastic response of a perforated material. Experiments performed on
perforated tubes of a square penetration pattern are presented in order to
furnish the experimental data necessary when specifying the perforation
tensor, the yield condition and the flow law for perforated materials. Attention is focused in the paper on experiments regarding yielding and deformability. The theoretical aspects are outlined. A theoretical study will be given
elsewhere, employing further tests including triangular penetration patterns
as well.

2. Experimental setting

Experiments on perforated sheets should allow to record the overall response
at different stress biaxiality ratios for a given penetration pattern. The influence
of the orientation of the overall stress principal directions with respect to the

privileged directions of the considered penetration pattern should be taken into account.

Studies on plastic anisotropy of perforated sheets with square and triangular penetration patterns, subjected to uniaxial loading, were initiated by Litewka and Rogalska [12, 13], whereas Cordebois [5] considered a square penetration pattern. When the direction of loading does not coincide with a material symmetry axis, the known off-axis effect occurs and the principal directions of the overall stress and strain do not coincide. The actual stress state in perforated sheets in the case of a uniaxial off-axis loading is disturbed with the increasing plastic strain so that the influence of shear cannot be neglected at larger deformations [13]. The deviations are negligibly small for overall plastic strains lesser then 0,5 per cent. It means that for the onset of yielding the off-axis effect has practically no influence in perforated materials at least for the large ligament efficiences.

The experiments reported here concern perforated materials with a square penetration pattern, subjected to a plane stress. A unit cell, and the notation employed are shown in Figure 1. The angle α specifies the inclination of the overall stress directions to the direction of one of the symmetry lines of the perforation pattern. The tests were carried out on tubular specimens of mild steel St3S subjected to an axial load, internal pressure and torsion. The yield stress of the unperforated material in uniaxial tension of the tube was $\sigma_0 = 236 \frac{MN}{m^2}$. The specimen had the outer diameter $d = 46$ mm and the wall thickness $h = 0.7, 1.0$ and 2.0 mm. Details of the experimental technique employed can be found in [14].

The tests concerned the inclinations $\alpha = 0°$, $\pi/8$ and $\pi/4$, Figure 2, and the ligament efficiency $\mu = \frac{h}{P} = 0.3, 0.5, 0.7$ and 1.0. The material tested showed a slight anisotropy since the yield stress in the circumferential

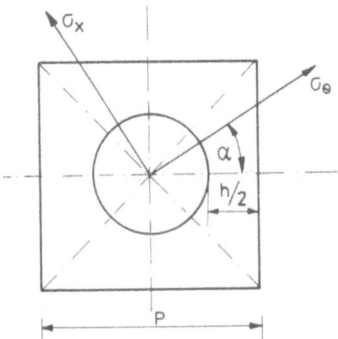

Figure 1. – *Unit cell of the perforated material.*

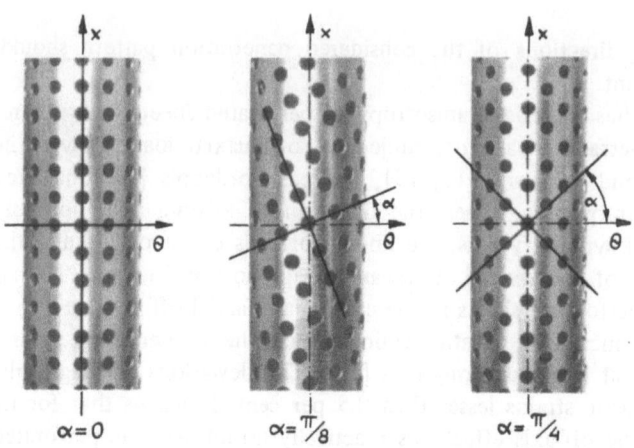

Figure 2. — *Tubular specimens with various orientations of the holes array.*

direction of the tube did amount to about 0.90 σ_0. The tests were made for seven biaxiality ratios.

3. Recorded yield loci

The recorded yield stress concerning the overall behaviour at the permanent overall equivalent strains $\epsilon_i = 0.1$ percent is presented in dimensionless form in Figure 3. The respective diagrams concern different ligament efficiencies μ. In each case the multiaxial state is referred to the actual yield stress in uniaxial tension in order to exhibit the perforation induced anisotropy.

It should be noticed first of all that the material tested showed a slight anisotropy, as for the unperforated tube, $\mu = 1.0$, the recorded yield locus differs from the Huber-Mises ellipse. The yield locus cannot be approximated by an ellipse, according to the Hill criterion [9], as the yield points in tension and in compression are not equal.

The recorded yield points for different angles of inclination of the overall stress directions to the preferred orientations of the original perforated material indicate that the overall behaviour is anisotropic. When the perforation efficiency decreased to $\mu = 0.3$, thus at increasing diameter of holes, the resulting curves formed more elanced figures when the direction of stress and material structure differed more. In Table 1 the length of the stress vectors is compared with the respective results concerning the Huber-Mises ellipse. The changes of the recorded yield point and the ligament efficiency are not the same and therefore the overal anisotropy depends not only on the penetration pattern but also on another geometrical parameter related presumably to the area of the plastically deformed material.

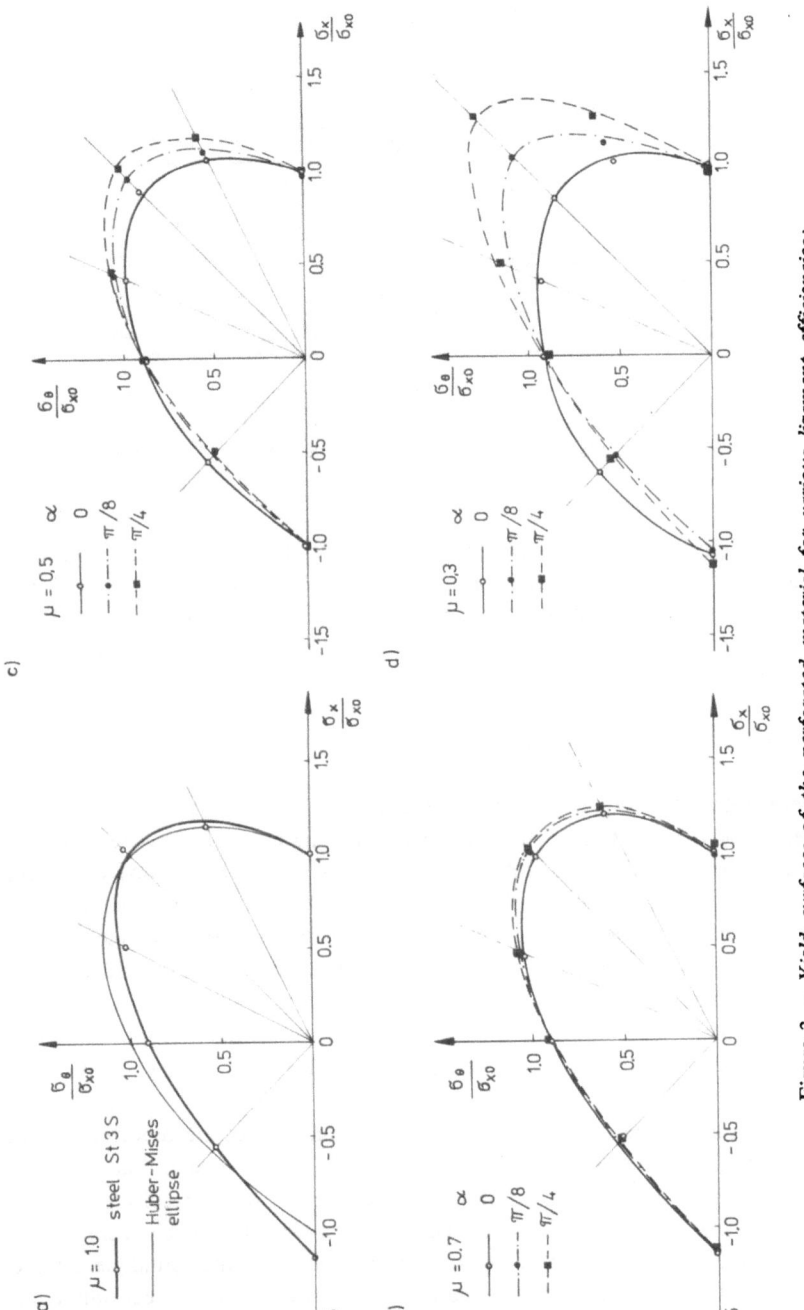

Figure 3. – *Yield surfaces of the perforated material for various ligament efficiencies:*
a) $\mu = 1.0$, b) $\mu = 0.7$, c) $\mu = 0.5$, d) $\mu = 0.3$.

TABLE 1

$\sigma_x/\sigma_\bullet$	Stress vector σ/σ_0^*					Stress vector ratio $\dfrac{\text{experimental}}{\text{isotropic}}$			
	Huber Mises ellipse	Experimental results							
μ		1.0	0.7	0.5	0.3	1.0	0.7	0.5	0.3
∞	1.0	1.0	1.0	1.0	1.0	1.0	1.0	1.0	1.0
2.0	1.258	1.29	1.34	1.24	1.12	1.02	1.07	0.99	0.89
1.0	1.414	1.46	1.38	1.35	1.20	1.03	0.97	0.95	0.85
0.5	1.258	1.14	—	—	—	0.91	—	—	—
0.430	1.229	—	1.13	1.15	0.99	—	0.92	0.93	0.81
0	1.0	0.90	0.90	0.90	0.91	0.90	0.90	0.91	0.90
−1.0	0.816	0.78	0.71	0.77	0.84	0.95	0.87	0.94	1.03
−∞	1.0	1.15	1.14	0.99	1.03	1.15	1.14	0.99	1.03

In Figure 4 the equivalent overall stress is traced versus the equivalent overall strain for various loading paths. The results concern the ligament efficiency $\mu = 0.3$. It is clear that a conventional yield stress should be employed to specify transition from the elastic to the plastic range. The material exhibits hardening.

4.   Flow law and volume changes

Experimentally obtained strains were measured, using mechanical extensometers. The overall strains of the diameter and the base length were recorded. The directions of the strain rate vector were found by the method developed by Miastkowski and Szczepiński [15]. The method consists in making plots of the circumferential and longitudinal strains versus the equivalent stress σ_i. The slope of the tangent to each curve gives the respective values of the strain increments measured at the actually considered yield stress. The obtained vectors are plotted in Figure 5 for two orientations of the perforation pattern with respect to the principal overall stress directions, namely $\alpha = 0$ and $\alpha = \pi/4$.

In Table 2 the inclinations of the strain rate vector are reported for the loading path and the ligament efficiences considered. It is seen that the strain rate vector has a different orientation in comparison with that resulting from the plastic potential flow law for isotropic materials. However, it is yet impossible to state whether the strain rate vector is perpendicular to the actual yield curve, as no analytic expression for the yield curve is known. In this respect only quantitative information can be obtained from Figure 5.

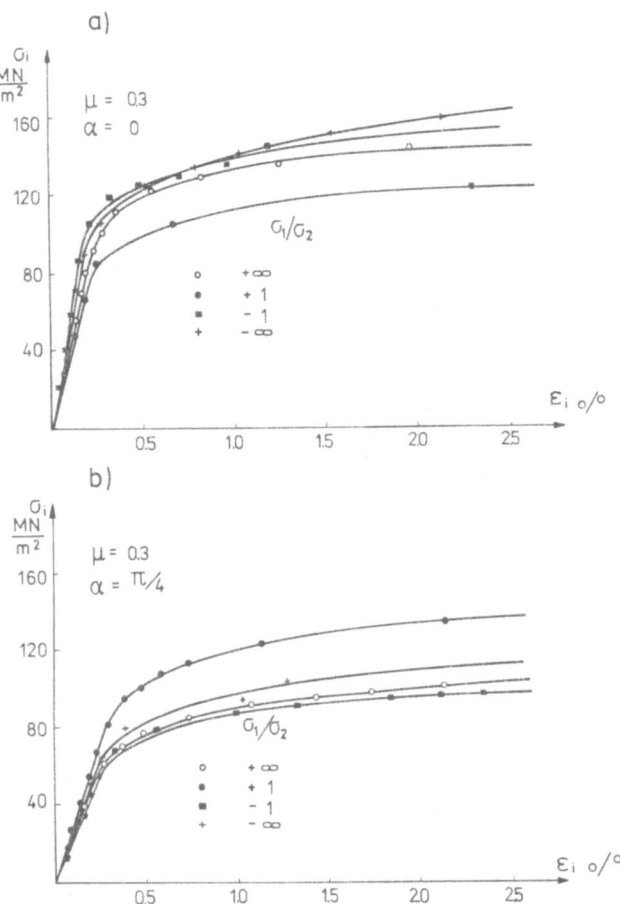

Figure 4. – *Equivalent stress versus equivalent strain for various loading paths:* a) $\alpha = 0$, b) $\alpha = \pi/4$.

The second important subject in the plasticity of perforated materials are the permanent overall plastic volume changes of the specimen. These changes are due to the changing geometry of the holes. It is assumed at present that the overal plastic strain in the thickness direction of the perforated plate is negligibly small. In Figure 6 the permanent volume change is related to a dimensionless hydrostatic pressure $s = \mathrm{tr}\ \sigma/Y_0$, where Y_0 represents, for the given loading path, the tr σ at the considered yielding.

It is seen from Figure 6 that for a given value of the ligament efficiency, the volume change depends on the angle α specifying the perforation orien-

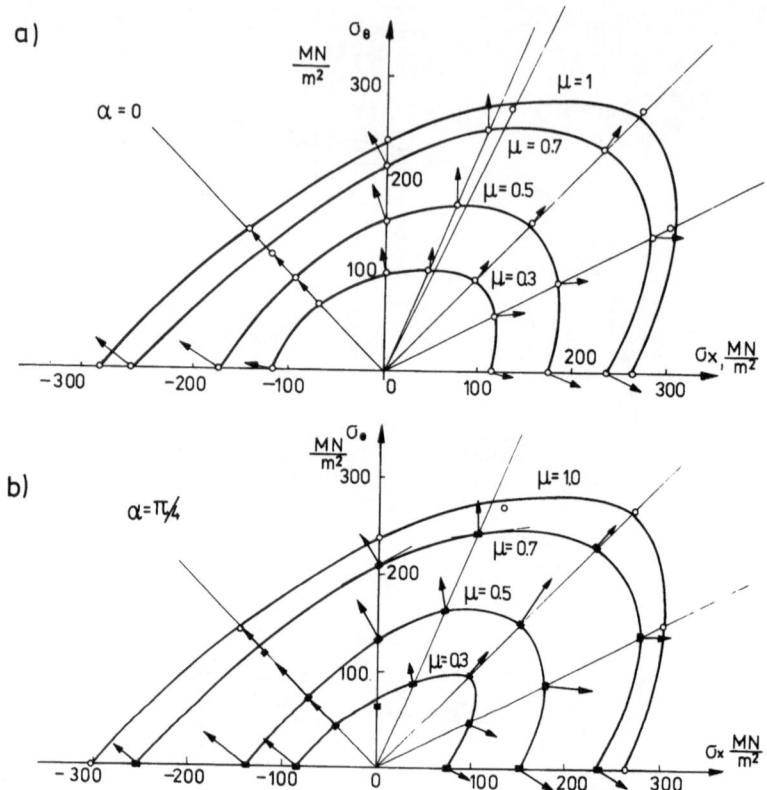

Figure 5. – *Permanent strain increment vectors:* a) $\alpha = 0$, b) $\alpha = \pi/4$.

TABLE 2

Stress ratio σ_x/σ_θ	arc tg$(d\varepsilon_\theta^p/d\varepsilon_x^p)$						
	$\alpha=0$			$\alpha=\pi/4$			Isotropy
	μ 0.3	0.5	0.7	0.3	0.5	0.7	
∞	-0.05π	-0.12π	-0.15π	-0.12π	-0.16π	-0.17π	-0.15π
2.0	0.04π	0.01π	0	-0.16π	-0.04π	0	0
1.0	0.29π	0.30π	0.29π	0.28π	0.31π	—	0.25π
0.430	0.46π	0.50π	0.48π	0.57π	0.54π	0.51π	0.50π
0	0.55π	0.62π	0.65π	0.62π	0.66π	0.67π	0.65π
-1.0	0.75π	0.75π	0.75π	0.75π	0.75π	0.75π	0.75π
$-\infty$	0.96π	0.83π	0.80π	0.79π	0.80π	0.83π	0.85π

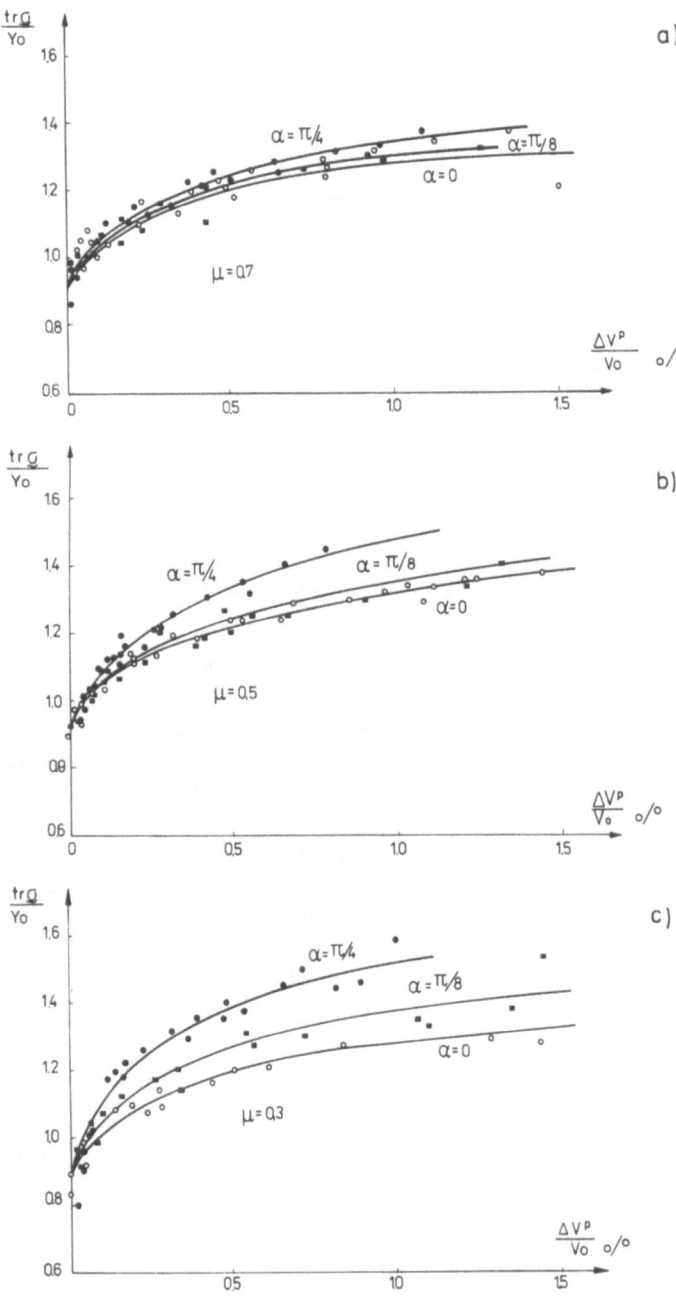

Figure 6. – *Permanent volume deformation versus dimensionless hydrostatic pressure:* a) $\mu = 0.7$, b) $\mu = 0.5$, c) $\mu = 0.3$.

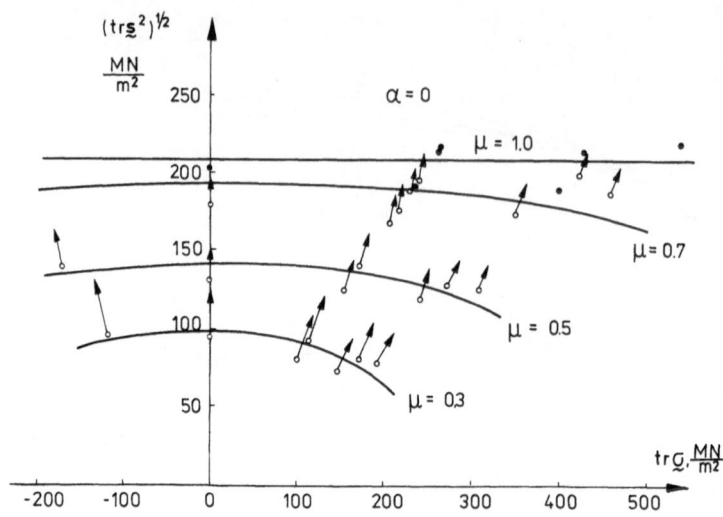

Figure 7. — *Yield surfaces in the coordinate system* $(\mathrm{tr}\, S^2)^{1/2}$, $\mathrm{tr}\, \underset{\sim}{\sigma}$.

tation with respect to the principal direction of overall stresses. It can be concluded that the yield locus probably depends on the hydrostatic pressure. In Figure 7 the obtained plastic strain rate vectors are plotted in the plane specified by the first two basic invariants of the stress. The experimental points are scattered, as concerns the stresses at yield. In the appropriate space of stress and perforation tensor invariants these points belong to the actual yield surface, whereas on the plane $(\sqrt{\mathrm{tr}\, S^2}, \mathrm{tr}\, \sigma)$ they are only projections. Had the equivalent material been isotropic and exhibiting no overall volume change, the stress points would have been on a surface which would be independent of the perforation, except in the case of material porosity where a sensitivity to hydrostatic pressure at yielding is also observed. For comparison sake the plastic potential flow law was used to draw the respective yield curves on the plane $(\sqrt{\mathrm{tr}\, S^2}, \mathrm{tr}\, \sigma)$. Evidently, curves are obtained similar to those for a porous solid. The purpose of giving Figure 7 here was to demonstrate the permanent overall volume changes which occur in perforated materials, and also that the yield condition is probably sensitive to hydrostatic pressure.

5. Theoretical setting

We assume that the material perforation, and thus the induced directional properties of an equivalent material, are described by a perforation tensor **P**. For example in the case of plasticity of porous solids, usually an isotropic second order tensor is introduced to specify the porosity and to introduce the tr σ into the yield condition. A porosity tensor for rectilinear cylindrical pores was derived by Kubik [11]. Similarly, the continuum theory of material damage at brittle fracture employs the concept of the crack density tensor as defined by Kachanov [10] and Dragon [7].

When an appropriate tensor associated with the penetration pattern and with the ligament efficiency is given, the tensor function representations lead to some conclusions regarding the general form of the flow law and of the yield criterion for an equivalent material.

The most appropriate way would be to account for the material symmetries by the number of vectors specifying the privileged directions associated with the perforation pattern. Representations involving the respective number of vectors and the stress tensor would follow next. Such an approach was successfully used by Boehler [2] and Boehler and Raclin [3] for certain cases of anisotropy. There is, however a too large number of invariants usually involved to permit a wieldy expression involving few tensor generators G_i and scalar functions α_i, expressing the stress tensor T in the form

$$T = \alpha_i\, G_i\,. \tag{1}$$

The tensor independent variables in the proposed approach are the strain rate tensor D and the tensor parameter P specifying the perforation. The perforation tensor is an array of material constants of the equivalent material. These constants are determined experimentally or by an appropriate homogenisation procedure. In the case of isotropy the perforation tensor is

$$P = f(\mu)\, I\,, \tag{2}$$

where $f(\mu)$ is a scalar function of the ligament efficiency and I stands for a unit tensor. In the isotropic case, all the yield curves in Figure 3 would not depend on α and would represent an appropriately diminished yield locus corresponding to $\mu = 1$, the $f(\mu)$ specifying the law of reduction.

In Figure 8 there is shown the experimentally obtained relation between the ligament efficiency and the equivalent stress at the equivalent plastic strain $\epsilon_i^P = 0.1$. It is seen that the effective equivalent stress depends on the orientation of the perforation pattern. In a somewhat different form, this conclusion is illustrated in Figure 9. It appears justified to assume that the perforation tensor has the form analogous to (2), except that in the case of anisotropy the unit tensor is replaced by an appropriate array of numbers related to the penetration pattern. A detailed study of this array will be given elsewhere on the basis of the tests described, as well as others concerning a triangular penetration pattern [21].

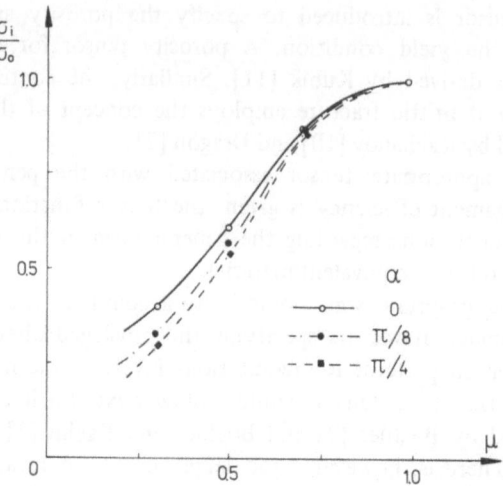

Figure 8. – *Dimensionless equivalent stress at yielding versus ligament efficiency.*

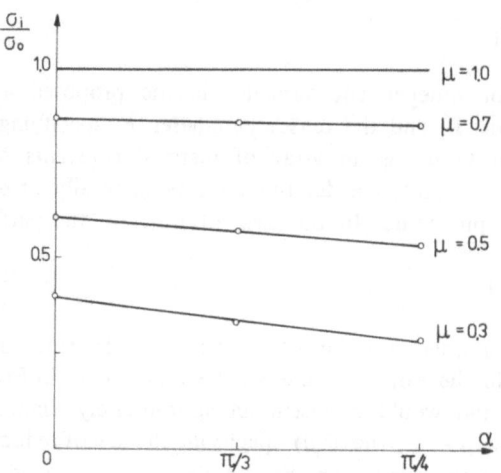

Figure 9. – *Dimensionless equivalent stress at yielding versus angle* α.

The overall stress for an equivalent material has to be looked for in the form

$$T = F(D, P),\tag{3}$$

where D denotes the plastic strain rate tensor. The generators and the scalar multipliers appearing in the representation (1) are known for the function (3) involving two independent tensor variables. To describe the perfectly plastic behaviour, the representation (3) has to satisfy the requirement of homogeneity of order zero with respect to tr $\mathbf{D}$ as discussed in [20]. Eventually the flow law is obtained as well as the most general form of the yield condition. It appears reasonable to look for the simplest possible form of the yield condition involving only certain stress and mixed stress-anisotropy invariants,

$$f(\text{tr } \mathbf{T}, \text{tr } \mathbf{S}^2, \text{tr } \mathbf{PS}) = 0, \tag{4}$$

where $\mathbf{S}$ denotes the deviatoric part of the stress tensor. A form, quadratic in stress and constituting the sum of invariants, would give a yield condition for a porous material.

It can be concluded from Figure 8 that anisotropy in the considered case of perforation is but slightly pronounced. To certain extent a similar conclusion could be drawn already from Figures 3 and 5.

6. Final remark

The paper concerns a homogenisation of perfectly plastic perforated materials. Experiments undertaken on perforated tubes with a square penetration pattern are described and discussed. They were intended to give an indication regarding tensor functions suitable to specify the flow law and the yield condition. The notion of perforation tensor is introduced in order to account for the anisotropic behaviour of an equivalent homogeneous material replacing the perforated one. Experiments on tubes with a triangular penetration pattern and eventually a homogenisation procedure leading to the perforation tensor will be reported separately.

REFERENCES

[1] BOEHLER J.P. – "A simple derivation of representations for non-polynomial constitutive equations in some cases of anisotropy." *ZAMM*, 59, 4 (1979): 157-167.

[2] BOEHLER J.P. – "Lois de comportement anisotrope des milieux continus", *J. Mécanique*, 17, 2 (1978): 153-190.

[3] BOEHLER J.P.. and J. RACLIN. – "Représentations irreducibles des fonctions tensorielles non-polynomiales de deux tenseurs symétriques dans quelques cas d'anisotropie." *Arch. Mech. Stos.*, 29, 3 (1977): 431-444.

[4] BOEHLER J.P. and A. SAWCZUK. – "On yielding of oriented solids." *Acta Mechanica*, 27 (1977): 185-206.

[5] CORDEBOIS J.P. – *Comportement et résistance des milieux métalliques multiperforés*. Thèse, Université Pierre et Marie Curie, Paris, 1976.

[6] O'DONNEL W.J. and J. POROWSKI. – "Yield surface for perforated materials." *Trans. ASME, J. Appl. Mech.*, 40, 1 (1973): 263-270.

[7] DRAGON A. – "On phenomenological description of rock-like materials with account for kinetics of brittle fracture." *Arch. Mech. Stos.*, 28, 1 (1976): 13-30.

[8] GRIGOLYUK E.I. and L.A. FILSHTINSKII. – *Perforated Plates and Shells.* Moscow: 1970 (in Russian).

[9] HILL R. – "A theory of the yielding and plastic flow of anisotropic metals". *Proc. Roy. Soc.*, A193 (1948): 281-297.

[10] KACHANOV M.L. – *Continuum model of medium with cracks,* Brown Univ. Rep. 16 Porous Media Series, September 1978.

[11] KUBIK I. – "Permeability tensor and porosity of material with rectilinear channels." *Bull. Acad. Pol. Sci.* Ser. Tech., 27 (1979), 445-453.

[12] LITEWKA A. and E. ROGALSKA. – "Experimental study of plastic properties of perforated materials under uniaxial tension." *8th Conf. on Experimental Investigation in Mechanics of Solids,* Warsaw, 2 (1978): 179-190 (in Polish).

[13] LITEWKA A. and E. ROGALSKA. – "Stress state in perforated materials for off-axis tension." To be published.

[14] LITEWKA A. and E. ROGALSKA. – "Plastic flow of perforated materials with square penetration pattern." *Trans. 5th Int. Conf. SMIRT.* Berlin, 1979.

[15] MIASTKOWSKI J. and W. SZCZEPINSKI. – "An experimental study of yield surfaces of prestrained brass." *Mech. Teoret. Stos.*, 3, (1965): 55-66 (in Polish).

[16] POROWSKI J. and W.J. O'DONNELL. – "Plastic strength of perforated plates with square penetration patterns." *Trans. ASME, J. Press. Vessel Techn.*, 97, 3 (1975): 146-154.

[17] RIVLIN R.S. and J.L. ERICKSEN. – "Stress-deformation relations for isotropic materials." *J. Rat. Mech. An.*, 4 (1955): 323-425.

[18] RIVLIN R.S. – "The fundamental equations of nonlinear continuum mechanics." In: S.I. Pai, ed., *Dynamic of Fluids and Plasmas.* New York: Academic Press, (1966): 83-126.

[19] SAWCZUK A., W.J. O'DONNELL and J. POROWSKI. – "Plastic analysis of perforated plates for orthotropic yield criteria." *Int. J. Mech. Sci.*, 17 (1975): 411-417.

[20] SAWCZUK A. and P. STUTZ. – "On formulation of stress-strain relation for soils at failure." *ZAMP*, 19 (1968): 769-778.

[21] LITEWKA A. and A. SAWCZUK. – "A yield criterion for perforated sheets", *Ing. Archiv*, 50 (1981) : 242-250.

[22] SLOT T. and W.J. O'DONNELL. – "Effective elastic constants for thick perforated plates with square and triangular penetration patterns." *Trans. ASME, J. Engng Ind.*, 93, 4 (1971): 935-942.

[23] THOMAS T.Y. – *Flow and Fracture of Solids.* New York: Academic Press, 1961.

[24] WINNICKI L., M. KWIECINSKI and M. KLEIBER. – "Numerical limit analysis of perforated plates." *Int. J. Num. Meth. Engng.*, 11 (1977): 553-561.

RESUME

Sur une approche par un continu de l'anisotropie plastique des matériaux perforés

Dans ce travail, nous présentons une étude expérimentale sur la réponse plastique globale des matériaux perforés à réseau carré, soumis à des états de contrainte plans. Les surfaces d'écoulement effectives ont été déterminées expérimentalement pour plusieurs valeurs de l'efficacité de la partie utile. On remarque que l'anisotropie plastique des matériaux perforés est une fonction de cette efficacité. Pour la description du comportement plastique du matériau équivalent, les représentations des fonctions tensorielles ont été utilisées, en prenant en compte les symétries spécifiques du matériau.

Apparatus Effects in the Determination of Strength Variation in Anisotropic Rock

F. A. Donath[1]
University of Illinois, Urbana, Illinois, U.S.A.

K.W. Schuler and J.R. Tillerson
Sandia Laboratories, Albuquerque, New-Mexico, U.S.A.

1. Observed and theoretical strength variation

Experimental studies of strength variation in anisotropic rock have demonstrated that compressive strength can vary significantly depending upon the orientation of the anisotropy with respect to the direction of maximum compression, e.g. [1, 3, 14]. The presence of anisotropy can greatly influence the geometry and the mode of deformation in the rock as well as its strength, [5]. Of the several theories presently available in the literature to explain this strength variation, the continuously variable shear strength theory of Jaeger [13] has been shown by Donath [3, 7] and McLamore and Gray [14] to explain as well as any the failure of rocks characterized by pervasive planar anisotropy.

Jaeger [13] derived his criterion for failure in anisotropic rock by assuming the Coulomb relationship

$$\tau = \tau_0 + \sigma \tan \Phi \tag{1}$$

to hold, the internal friction ($\tan \Phi$) to remain constant, and the cohesive strength (τ_0) to vary continuously according to the relation

$$\tau_0 = a - b \cos 2(\alpha - \beta), \tag{2}$$

where β is the inclination of the anisotropy to the direction of maximum compression, a and b are constants, and α is the orientation of β for which τ_0 is a minimum. Because, as shown in [3], the internal friction as well as cohesive strength can vary with inclination of the anisotropy, a more general form of the failure criterion is given by

$$\sigma_1 - \sigma_3 = \frac{2(\tau_0 + \sigma_3 \tan \Phi)}{\sec \Phi - \tan \Phi} \tag{3}$$

---

[1] Present affiliation: CGS, Urbana, Illinois 61801.

where

$$\tan \Phi = c - d \cos 2(\alpha - \beta) \tag{4}$$

and τ_0 varies as in (2). Thus, both τ_0 and $\tan \Phi$ are made to vary systematically with β, [7, 14].

Use of the failure criterion represented by (3) requires the empirical determination of the variation of cohesive strength and of internal friction from experimental data. These values are normally determined by plotting the Mohr circles for failure at several confining pressures on a τ vs. σ diagram and drawing a linear envelope tangent to the circle [11]. The straight line thus obtained satisfies Equation (1) and is commonly referred to as the Coulomb-Mohr criterion. The slope of the line ($\tan \Phi$) represents the coefficient of internal friction and its intercept on the τ axis represents the cohesive strength. It should be noted that the Coulomb-Mohr envelope may not be linear; indeed, it will always be concave toward the σ axis if the normal stress is sufficiently high. At lower pressures, however, the envelope for most common rocks can be represented by a straight line.

Because of the difficulty of drawing a single line tangent to several circles of differing diameters, it is usually preferable to obtain the Mohr envelope by transformation of a plot such as maximum differential stress vs. confining pressure, [15]. Figure 1 shows the least-squares fits to the experimental data for several orientations of phyllite deformed at 50, 100, and 200 MPa confining pressure. The slopes and intercepts of these lines provided the data plotted in Figure 2. Equations of the general types (2) and (4) which best satisfy the data, as determined by a least-squares fit, are indicated in the figure. These equations permit solutions to be obtained for Equation (3) for the different orientations of anisotropy (cleavage) at the several pressures tested. The resulting theoretical curves of strength variation are superposed on the experimental data in Figure 3. Excellent agreement exists between the observed and predicted values for most orientations, although the rock is seen to be slightly stronger in the 0-degree orientation and somewhate weaker in the 90-degree orientation than predicted by the theoretical curves.

Similar results have been obtained for other anisotropic rocks, and certain effects of anisotropy on the strength and deformational characteristics of Martinsburg slate can be found elsewhere, [3, 4, 5, 7]. In connection with the cited studies on slate, the experimentally determined values of strength were found to be highly sensitive to differences in test apparatus and procedure, as well as to differences in material properties that affect the strength parameters (especially cohesive strength) [7]. Some typical results are shown in Figure 4.

The data presented in Figure 4 are all from triaxial compression tests run at room temperature, dry, at a strain rate of about 3.0×10^{-4} per second,

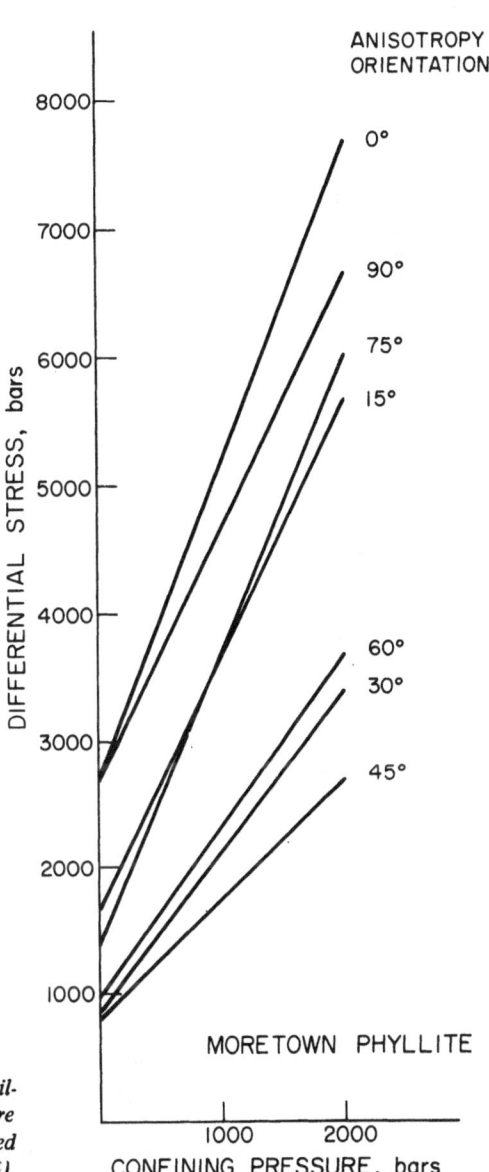

Figure 1. – *Differential stress at failure (σ₁ – σ₃) vs. confining pressure (σ₃), Moretown phyllite (reprinted from GSA Memoir 135, 1972, [7]).*

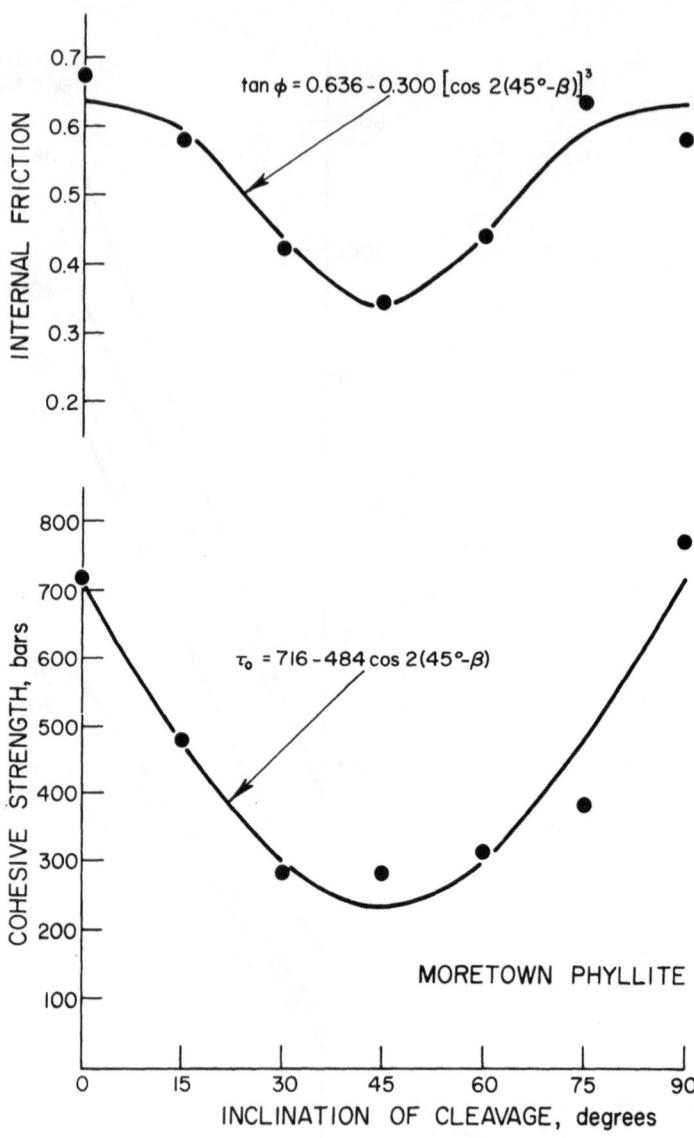

Figure 2. – *Variation of the coefficient of internal friction and cohesive strength in Moretown phyllite as a function of anisotropy orientation (reprinted from GSA Memoir 135, 1972, [7]).*

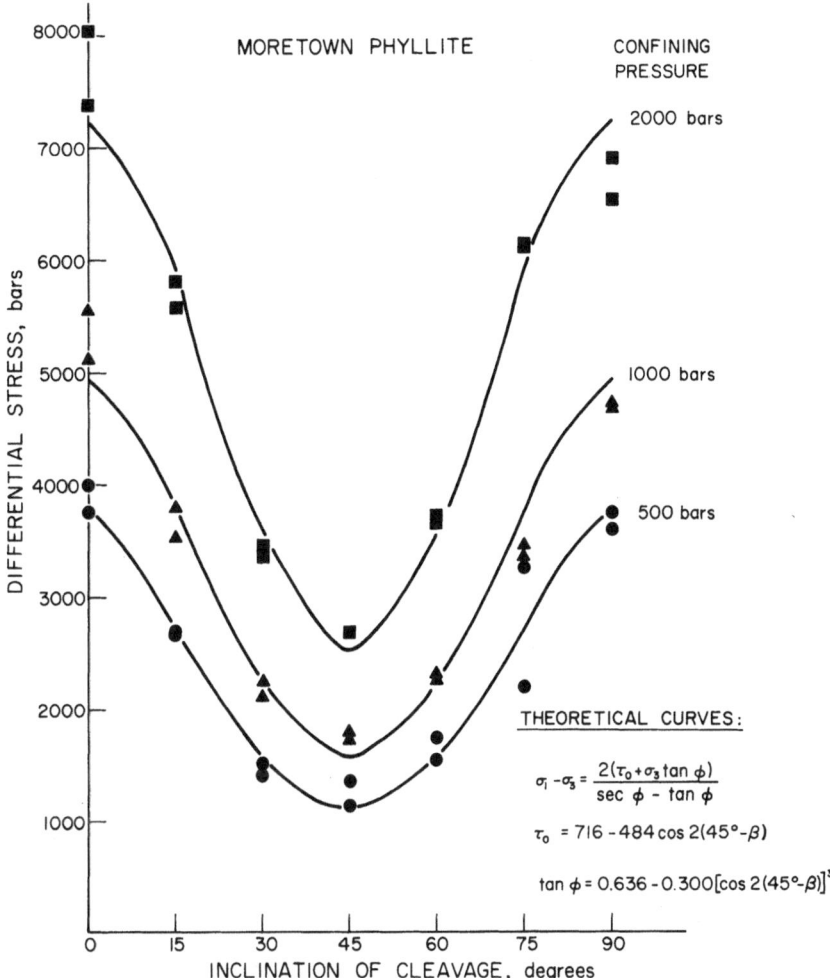

DIFFERENTIAL STRESS, bars

MORETOWN PHYLLITE

CONFINING PRESSURE

2000 bars

1000 bars

500 bars

THEORETICAL CURVES:

$$\sigma_1 - \sigma_3 = \frac{2(\tau_0 + \sigma_3 \tan \phi)}{\sec \phi - \tan \phi}$$

$$\tau_0 = 716 - 484 \cos 2(45° - \beta)$$

$$\tan \phi = 0.636 - 0.300[\cos 2(45° - \beta)]^3$$

INCLINATION OF CLEAVAGE, degrees

Figure 3. – *Observed strength variation in Moretown phyllite (symbols) as a function of anisotropy orientation, compared with curves of theoretical strength variation (reprinted from GSA Memoir 135, 1972, [7]).*

and at a confining pressure of 200 MPa. Right circular cylinders of slate 1/2-inch diameter by 1-inch length, were prepared under identical, closely controlled conditions from cores taken from a single block of slate at 0°, 15°, 30°, 45°, 60°, 75°, and 90° to the plane of anisotropy. Two different pressure vessels were used. Both were of conventional design, but one utilized a Newhall controlled-clearance packing for the piston seal whereas the other incorporated a Bridgman unsupported-area packing for that seal. A description of the

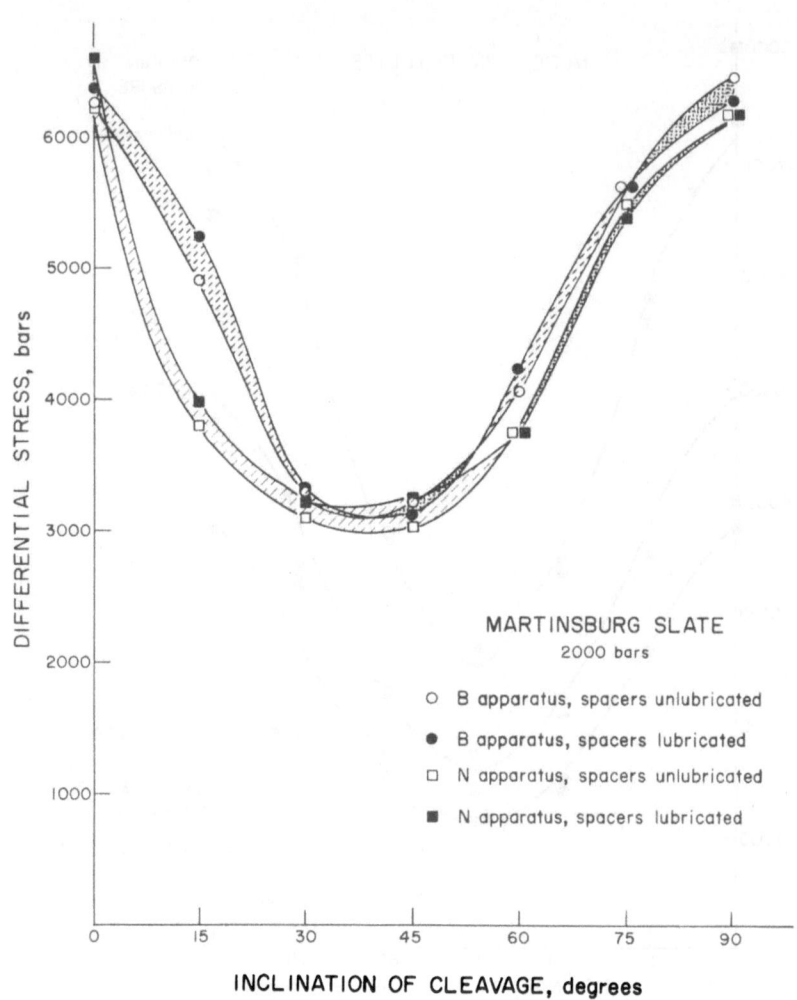

INCLINATION OF CLEAVAGE, degrees

Figure 4. – *Observed strength variation in Block 12 of Martinsburg slate, related to different apparatus with and without specimen end lubrication. All data are for tests run at 200 MPa confining pressure.*

controlled-clearance packing can be found in [9], and a schematic drawing of the vessel utilizing the unsupported-area packing is shown in [6]. The principal difference between the two apparatuses, referred to as the N and B apparatuses, respectively, is the rigidity with which the piston is held. In both units the piston passes through a gland to which it has a lapped tolerance, but the cylindrical surface of contact is 1/2-inch long in the N apparatus and 4 inches long in the B apparatus. Thus, the piston is held much more rigidly in the B apparatus. The different apparatuses used to obtain the results, along with the presence or absence of lubricant on the surfaces of the spacers that separate the specimen from the piston and anvil, respectively, are indicated in Figure 4. Some aspects of the observed apparatus effects were discussed in [7] and will be reviewed after presenting the results of a finite element analysis of triaxial compression of anisotropic rock.

2. Calculated stress nonuniformities

In interpreting test results such as those in Figure 4 it is generally assumed that the stress distribution in the sample is uniform. Moreover, investigators normally assume that the axial stress in triaxial compression tests is a principal stress, although it clearly cannot be if shear stress exists on the ends of the specimen. If the axial stress is not a principal stress, then measured values of axial stress will, of course, be lower than the true strength of the rock under the imposed test conditions. For isotropic rock, analysis has shown that end effects caused by constraint of radial expansion at the end loading plates are insignificant in the center region of the test sample and that the assumption of a uniform stress distribution is reasonable for this region in samples having a length-to-diameter ratio of 2 : 1 [2]. When anisotropic rock is loaded, stress nonuniformity can result as a consequence of the anisotropy, in addition to the usual end effects. This is caused by the tendency for the anisotropic sample to shear as it is compressed if the anisotropy is inclined to the loading axis. End constraints inhibit this shearing and results in a net lateral shear stress being applied to the ends of the sample. The normal stress distribution on the sample ends must therefore become skewed to preserve equilibrium [16].

To analyze the possible effects of stress nonuniformity on the strength of Martinsburg slate, we assumed the rock to be a linearly elastic, transversely isotropic medium and ran the finite-element computer program BMINES [12] in the two-dimensional mode to determine the magnitude and nature of this stress nonuniformity. Although actually a three-dimensional problem, previous results had indicated that a plane strain approximation can illustrate the effects. The material constants determined for Martinsburg slate from ultrasonic wave velocity measurements are listed in Table 1 along with the values used in the analysis, assuming transverse isotropy for this rock.

The two-dimensional idealization of the test specimen and attached steel end plates is shown in the upper right corner of Figure 5. The element spacing in the slate specimen was selected to provide element controls located on lines oriented at $35°$ to the vertical axis. Four node isoparametric quadrilateral elements were employed, and bilinear displacement functions were used within the elements. The mesh consisted of 533 elements surrounded by 588 nodal

TABLE 1. Elastic coefficients and moduli for Martinsburg slate, MPa $\times 10^4$

Determined	Used in Analysis		
$C_{11} = 11.11$	$C_{11} = C_{22} = 10.8$	$E_{11} = E_{22} = 9.15$	
$C_{22} = 10.37$			
$C_{33} = 7.16$	$C_{33} = \quad = 7.2$	$E_{33} = \quad 5.91$	
$C_{44} = 2.40$	$C_{44} = C_{55} = 2.4$		
$C_{55} = 2.37$			
$C_{66} = 3.89$	$C_{66} = \dfrac{C_{11} - C_{12}}{2} = 3.8$		
$C_{12} = 3.21$	$C_{12} = \quad 3.2$	$G_{12} = \quad 3.80$	
$C_{13} = 3.17$	$C_{13} = C_{23} = 3.0$	$G_{13} = G_{23} = 2.40$	
$C_{23} = 2.89$		$\nu_{12} = \quad 0.204$	
		$\nu_{13} = \quad 0.331$	

points, giving 1 130 total degrees of freedom for the system after the boundary conditions were applied.

Both stress and displacement constraints were used as boundary conditions. Nodes at the bottom end were not allowed to displace vertically, and axial pressure was applied uniformly to the upper steel plate. Rollers were simulated three-fourths of the distance from the ends of each steel end-plate. This restricted the lateral movement of those boundary nodes, and ensured that the loading ram would vertically displace the sample. An inconsistency in the field near the steel-slate interface would have resulted if the lateral movement of the end-plates were completely restrained in the presence of lateral loads. Lateral confining pressure loads were applied at nodes below the rollers in the simulation of multiaxial loading. No slip between the steel end plates and the slate specimen was allowed to occur; this condition results in the greatest amount of stress nonuniformity.

The stress nonuniformity produced at the interface (AB) between the upper end plate and slate during uniaxial loading is plotted in Figure 5 for eight orientations of the anisotropy. The stress distribution is seen to be uniform only for the 0 and 90-degree orientations, and is similar to that calculated by other investigators for an isotropic cylinder whose end faces are restrained from free radial motion. The maximum nonuniformity in the stress distribution occurs when the anisotropy is oriented at 35 degrees to the central axis of the specimen. It is apparent from a comparison of the plots that the stress distribution skews rapidly as the anisotropy is inclined from 0 to

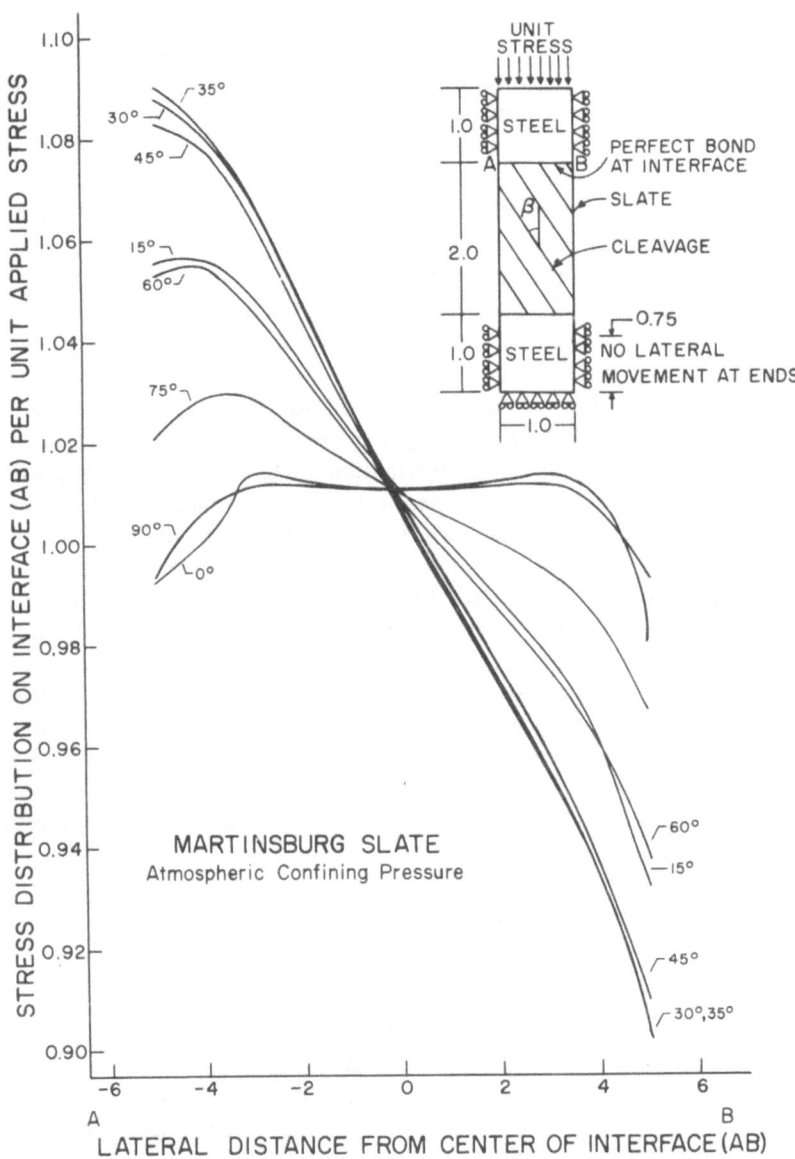

Figure 5. – *Stress nonuniformities produced at the interface (AB) between the upper end plate and the specimen during uniaxial loading; no confining pressure is applied. A two-dimensional idealization of the test specimen and steel end plates is shown at upper right.*

35 degrees. In contrast, the skewness diminishes rapidly for inclinations greater than 45 degrees; it has dropped off markedly for the 75-degree orientation.

An understanding of why the 30° to 35° orientations result in the worst nonuniformities can be gained by considering the stress-strain relationships governing plane-strain deformations of transversely isotropic materials. Of particular interest is the compliance coefficient S_{16} which determines the shear strain-normal stress coupling. Calculations performed for various representative sets of coefficients have shown that the maximum value of S_{16}, and hence shear strain-normal stress coupling, occurs at approximately 30 degrees, [16]. Thus, the maximum amount of stress nonuniformity appears to be strongly related to the manner in which S_{16} transforms.

The calculations and analysis have established that the worst non-uniformities occur for slate samples cored at about 35 degrees to the bedding planes, and we therefore simulated triaxial testing only for this orientation. Because failure invariably occurs along the cleavage for this orientation, it is of interest to examine the stress distribution along the cleavage plane which passes through the center of the specimen. All of the numerical calculations show that both the shear and normal stresses on this plane have maximum values at the extremities of the plane and minima at the center of the specimen. Plots of the variation in vertical stress along this cleavage plane (SS') for different values of applied axial stress are shown in Figure 6 for a simulated test at 200 MPa confining pressure. It should be noted that the 500 MPa axial stress with 200 MPa confining pressure provides 300 MPa differential stress which, as seen from Figure 4, is approximately that required to cause failure in this orientation of Martinsburg slate. The nonuniform stress distribution on the cleavage strongly suggests that failure might initiate in the high stress regions in the extemities of the plane and progress catastrophically inward to the center of the plane.

3. Discussion

Although the coefficient of internal friction and the cohesive strength are somewhat unsatisfying physically, they are, nevertheless, useful strength parameters by which to compare the deformational properties of different rocks or the effects of various factors on the behavior of a single rock, [10]. Because the internal friction and cohesive strength must be empirically determined from the experimental strength determinations, values for these parameters are highly sensitive to boundary conditions imposed by apparatus. The analysis of stress nonuniformities has demonstrated that the degree of nonuniformity increases as the piston is more highly constrained laterally, that it is most severe for inclinations of the anisotropy near 30 degrees, and that the variation of stress

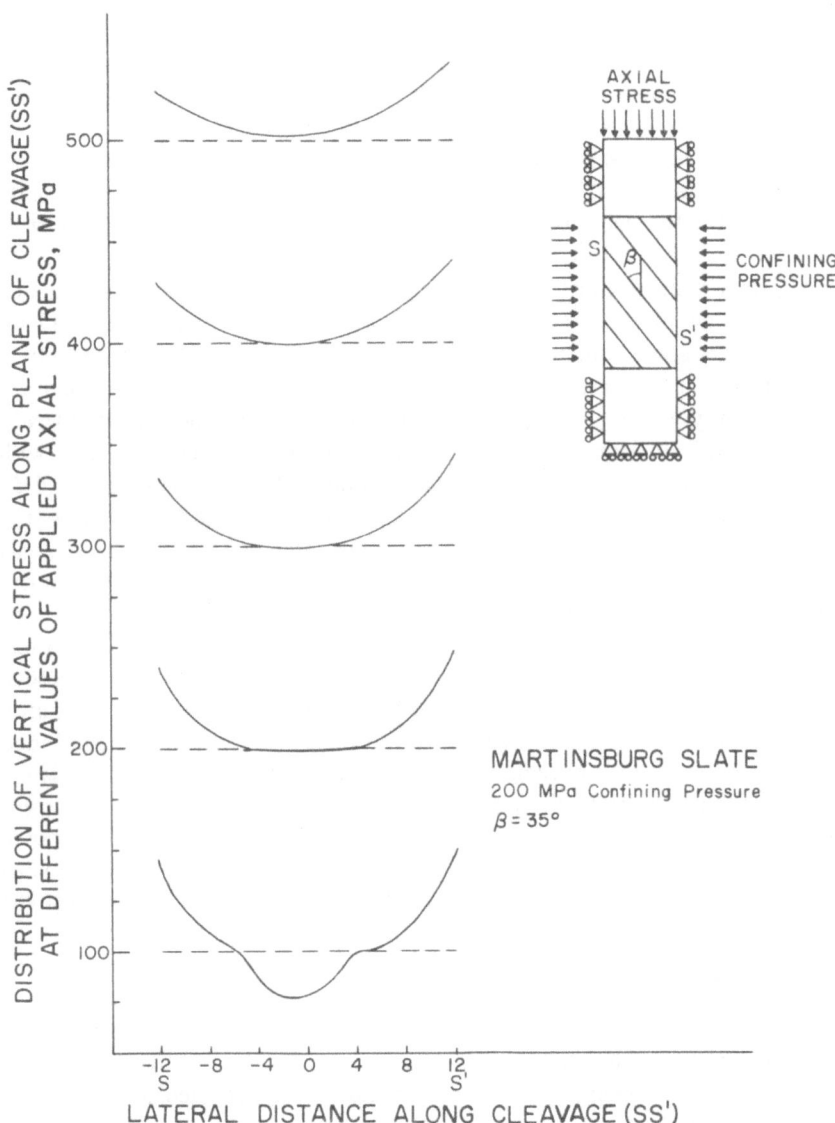

Figure 6. – *Plots of the distributions of vertical stress along the plane of cleavage (SS')*
for different values of applied axial stress in a simulated triaxial test at 200 MPa confining
pressure.

along the anisotropy is characterized by maxima at the extremities. As shown by the simulated triaxial testing (Fig. 6), the nominal vertical (axial) stress is realized only at the center of the specimen; it is significantly higher at the margins of the test sample. Thus, values measured by the experimental investigator at failure will be lower than those which actually exist in the specimen and cause the failure. The measured values, when plotted on a $\sigma_1 - \sigma_3$ vs. σ_3 or similar diagram, will therefore yield unrepresentative values for the strength parameters of the material being tested. The results will be strictly valid only for the behavior of a single orientation under the identical testing conditions.

Stress nonuniformity of the type described can be reduced or eliminated by lubricating the interface between the specimen and end plates[1] and by removing the lateral constraint on the end plates. This has been confirmed by a calculation that simulated a specimen with thin lubricated end caps; the stress nonuniformity virtually disappeared, [17]. Thus, the calculations predict that a specimen with lubricated end caps appears to be stronger than one with unlubricated end caps, since the stress nonuniformities are minimized and do not act as stress risers that precipitate failure at lower applied loads. Indeed, the experimental results shown in Figure 4 for lubricated and unlubricated triaxial tests of slate confirm this prediction; strength values for lubricated specimens are normally higher than for unlubricated specimens.

Other apparatus-related effects on strength determinations in anisotropic rock are seen in the data of Figure 4. These are the differences between results from tests using the B and N apparatuses, respectively, which reflect significantly different boundary conditions. Although the apparatus configurations and their influence on stress distributions in the test sample have not yet been simulated, it appears that significant shear stress can develop at the ends of test specimens during laboratory testing of anisotropic rock, [7]. This has also been noted during displacement on faults in pre-faulted specimens under triaxial testing conditions, [8]. A simplified analysis of the results shown in Figure 4 for the 15-degree orientation suggests that the direction of maximum principal stress might have been rotated from the direction of loading by as much as 30°, [7]. Consequently, the measured differential stress at failure in such instances can be moderately to appreciably less than the true strength of the rock at the conditions being evaluated.

It is clear from experimental results and theoretical calculations that the stress distributions leading to failure in anisotropic rock can be significantly affected by the anisotropy orientation, lubrication of specimen end surfaces,

---

[1] We recognize that lubrication of the end pieces might introduce other effects in the stress distribution depending on the specific rock type and conditions of testing.

and constraints on the loading piston of the test apparatus. Therefore, in addition to consideration of the selection and preparation of test samples, which affect the variation of *material* properties, the determination of truly representative values of strength parameters for anisotropic rock requires careful evaluation by the investigator of apparatus-related factors.

Acknowledgement

The authors wish to thank J.J. Sweeney for providing the elastic coefficients for Martinsburg slate.

REFERENCES

[1] ATTEWELL P.B. and M.R. SANDFORD. – "Intrinsic shear strength of a brittle, anisotropic rock". *Int. J. Rock Mech. Min. Sci. and Geomech. Abstr.*, 11 (1974): 423-451.

[2] BALIGH M.M. – "Numerical study of uniaxial and triaxial rock compression tests". Systems Science and Software Report SSS-R-73-1658, 1973.

[3] DONATH F.A. – "Experimental study of shear failure in anisotropic rocks". *Geol. Soc. America Bull.*, 72 (1961): 985-990.

[4] DONATH F.A. – "Strength Variation and deformational behavior in anisotropic rock". In *State of Stress in the Earth's Crust* (W. R. Judd, editor), 1964, p. 281-297.

[5] DONATH F.A. – "The development of kink bands in brittle anisotropic rock". In L.H. Larsen, ed., *Igneous and Metamorphic Geology*. Geol. Soc. America Memoir 115 (1968): 453-493.

[6] DONATH F.A. – "Some information squeezed out of rock". *Am. Scientist*, 58 (1970): 54-72.

[7] DONATH F.A. – "Effects of cohesion and granularity on deformational behavior of anisotropic rock". In B.R. Doe and D.K. Smith, eds., *Studies in Mineralogy and Precambrian Geology*. Geol. Soc. America Memoir 135 (1972): 95-128.

[8] DONATH F.A., L.S. FRUTH, Jr. and W.A. OLSSON. – "Experimental study of frictional properties of faults". *Proc. 14th Symposium on Rock Mechanics*, Am. Soc. Civil Engineers, (1973): 189-222.

[9] HANDIN J. – "An application of high pressure in geophysics: Experimental rock deformation". *Am. Soc. Mech. Engineers Trans.*, 75 (1953): 315-324.

[10] HANDIN J. – "On the Coulomb-Mohr failure criterion". *J. Geophys. Res.*, 74 (1969): 5343-5348.

[11] HUBBERT M.K. – "Mechanical basis for certain familiar geologic structures". *Geol. Soc. America Bull.*, 62 (1951): 355-372.

[12] ISENBERG J. – "Analytic modeling of rock-structure interaction". Vols. 1 and 2, Final Tech. Rpt., U.S. Bureau Mines Contract H0220035, April 1973.

[13] JAEGER J.C. – "Shear failure of anisotropic rocks". *Geol. Mag.*, 97 (1960): 65-72.

[14] McLAMORE R. and K.E. GRAY. – "The mechanical behavior of anisotropic sedimentary rocks". *Am. Soc. Mech. Engineers Trans.* Ser. B, 1967, pp. 62-76.

[15] RALEIGH C.B. and M.S. PATERSON. – "Experimental deformation of serpentinite and it's tectonic implications". *J. Geophys. Res.*, 70 (1965) : 3965-3985.

[16] SCHULER K.W. and J.R. TILLERSON. – "A preliminary analysis of stress nonuniformities in triaxial tests of bedded rock". Sandia Laboratories Rpt. SAND 76-0526,
1976.

[17] SCHULER K.W. and J.R. TILLERSON. – "Finite-element analysis of oriented
triaxial tests". In D.E. Munson, ed. *In Situ Oil Shale Bed Preparation Study*. Sandia
Laboratories Rpt. SAND 78-1950, (1978): 87-107.

RESUME

Effets d'appareil dans la détermination de la variation des résistances des roches
anisotropes.

Des résultats expérimentaux et des calculs théoriques de la variation des résistances des roches anisotropes sont passés en revue et discutés avec les résultats
d'une analyse en éléments finis des effets d'appareil. Des distributions de
contraintes fortement non-uniformes peuvent être imposées sur les échantillons d'essais par les appareils ; on conclue que les distributions de contraintes
conduisant à la rupture des roches anisotropes sont particulièrement affectées
par l'orientation et la lubrification des bases des échantillons, ainsi que par les
conditions imposées au piston de charge de l'appareil d'essais. Négliger ces
facteurs peut résulter en la détermination de valeurs non représentatives pour
les paramètres de résistance du matériau testé.

The Experimental Behavior of Anisotropic Clay

A. S. Saada and L.P. Shook

Case Western Reserve University, Cleveland, Ohio, U.S.A.

1. Forward

The mechanical behavior of saturated clay soils has been intensively investigated in the last 25 years within a framework valid only for isotropic materials. It is only relatively recently that researchers have started thinking in terms of determining the mechanical properties of clays taking into account their natural structural and mechanical anisotropy. It is a well established fact that deposition followed by one dimensional consolidation arranges the clay particles and results in bonds such that the material acquires the property of cross anisotropy with the axis of symmetry along the direction of consolidation.

In this paper the methods of testing anisotropic clay are briefly discussed. The static behavior of normally consolidated and overconsolidated clays is examined under a variety of stress paths. The dynamic behavior under cyclic loading is studied for both small and large strains. The fabric is found to have a pronounced effect on both static and dynamic response.

2. Equipment and testing procedures

2.1. *Introduction*

The directional properties of cross-anisotropic materials have generally been studied by subjecting specimens which have been cut at different inclinations to the axis of symmetry to direct compression or extension tests. Under such conditions, initially circular cross sections of specimens do not remain circular and initially square cross sections do not remain square. In addition, the prevention of the rotation and shearing strains by the end platens during an unconfined compression or a triaxial compression test on inclined specimens results in shearing stresses and couples of substantial magnitude.

Figure 1 shows schematically what will subsequently be called a vertical, an inclined and a horizontal specimen. Figure 2 indicates the forces and moments that act on inclined specimens in the triaxial cell when it is subjected to an axial compression [6]. Figure 3 illustrates the expected deform-

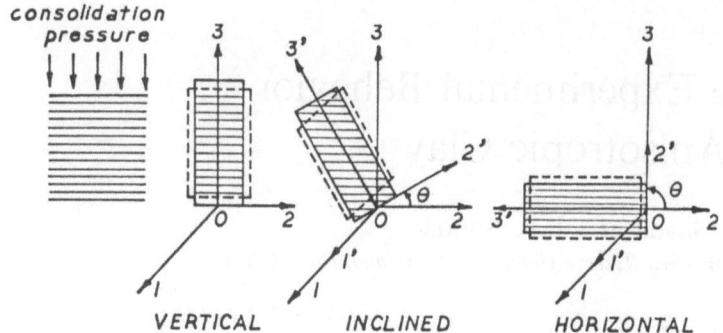

Figure 1. – *Systems of Reference Axes and Types of Test Specimens.*

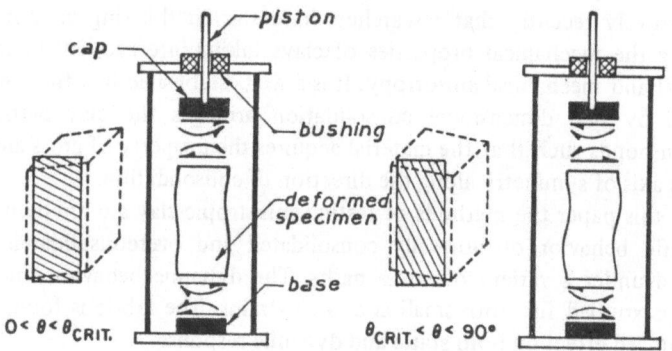

Figure 2. – *End Effects and Deformed Shapes of Inclined Test Specimens.*

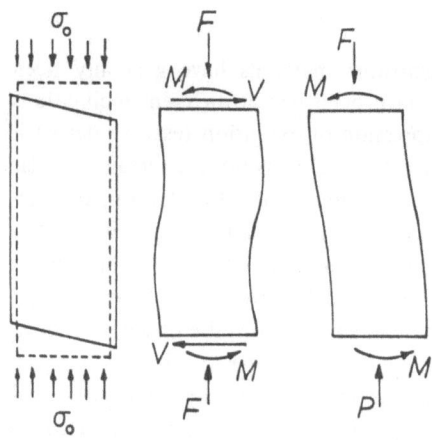

Figure 3. – *Exaggerated Deformation of Inclined Specimen :*
a) under a Uniform State of Stress;
b) when the end Platens are Horizontal Rough and Rigid. c) when the End Platens are Horizontal Smooth and Rigid.

ation of an inclined specimen tested in compression a) under uniform stress, b) between rough rigid plates and c) between smooth rigid plates. In an effort to prove that a cross-anisotropic clay will indeed react as in Figure 3, a series of unconfined compression tests were conducted on square specimens [25 mm x 25 mm x 76 mm] cut out of a block of a one-dimensionally consolidated slurry of Kaolinite [10]. The inclination of the specimens varied between 0° and 90° with the axis of symmetry. For each inclination two identical specimens were tested, one between greased smooth plates and one between plates covered with coarse sanding cloth. As expected the inclined

Figure 4. – *Compression of Inclined Specimens Between Rough and Smooth Platens.*

specimens took different shapes according to their inclination. Figure 4 shows photographs of these tests; the labels give the inclination and the type of end platens, smooth (S) or rough (F) used in the test. The forces and moments acting at the ends of the specimen are not self equilibrating. If the ends are not free to deform, a uniform state of stress will not exist in the material.

2.2. Testing Cross-Anisotropic Clays

In view of the previous discussion it is seen that the use of the standard triaxial cell will give reliable results only when a vertical specimen is tested under axial loads. For a horizontal specimen, although the initially circular cross section becomes an ellipse, no bending effects are generated at the ends and the results should be reliable enough.

To avoid all the extraneous end effects that develop when inclined specimens are used, one can keep the specimens vertical and incline the principal stresses. If an inclination is produced by a loading system which gives uniform and axially symmetric stresses, no tendencies for out of plane shears will be generated. One way to accomplish this is to combine the effects of axial and torsional stresses on thin hollow vertical specimens. For the results to be meaningful, the inclination of the principal stresses must be constant during each test. Since this inclination is given by the well known formula (Fig. 5)

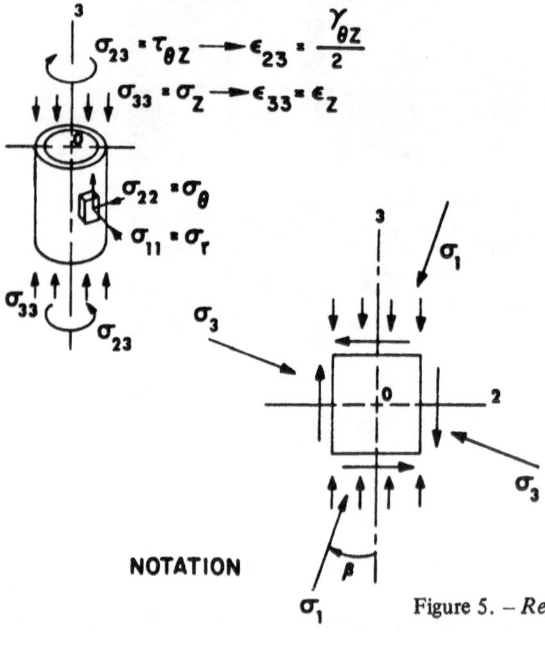

Figure 5. – Reference Axes and Stress Systems.

Tan $2\beta = 2\sigma_{23}/(\sigma_{33} - \sigma_{22})$ each experiment must be conducted in a way such that $\sigma_{23}/(\sigma_{33} - \sigma_{22})$ does not change during the shearing process. As shown in Figure 5, if the hollow cylinder is within a cell and subjected to a hydrostatic pressure, the same pressure will act inside and outside the cylinder and this pressure will be equal to the intermediate principal stress. By changing the ratio in the previous formula from test to test, a complete spectrum of inclinations of principal stresses on the axis of symmetry can be covered without creating the ends effects associated with the testing of inclined specimens.

The only end effects remaining under the conditions above are the ones due to the prevention of uniform radial expansion or contraction during axial compression or extension. The resulting system of radial shearing forces is self equilibrating and its influence vanishes away from the ends. Using the equations of the theory of thin elastic shells, Wright, Gilbert and Saada [14] deduced that the length of hollow cylinders should be equal to or larger than $5.44 \sqrt{r_0^2 - r_i^2}$ where r_0 and r_i are the outer and inner radii. In a hollow cylinder subjected to torque, the shearing stress varies with the radius. The assumption of a uniform distribution leads to an error which decreases as the ratio $n = r_i/r_0$ increases. It has been estimated that a value of $n \geqslant 0.65$ would lead to a very acceptable approximation.

2.3. Testing Equipment

There are two types of equipment in use at Case Western Reserve University to study the bavior of anisotropic clay; one strictly dynamic namely the resonant column and the other both static and dynamic namely the Saada Pneumatic Analog Computer.

The Resonant Column used at Case is the fixed-free Drnevich apparatus. Using two independent circuits it has the capability of inducing both axial and torsional vibrations in a specimen under a spherical state of stress. It was modified to accomodate thin, hollow, circular cylinders and equipped with a central piston rod through the base plate. The rod is used to apply axial loads prior to the shaking of the specimen and can be actuated by a pneumatic servomechanism to produce K_0 consolidation condition. Two accelerometers mounted on the top cap monitor the axial and torsional motions. This apparatus is used to determine Young's modulus and the shear modulus of the soil and it can develop at resonance strains as high as 10^{-3} for relatively soft clays.

The function of the Saada Pneumatic Analog Computer is to drive two actuators, one applying axial stresses and the other torsional stresses to a specimen inside a cell; as well as to vary the cell pressure such that a given relation among the stresses applied to a soil specimen is continuously satisfied. The machine can apply combinations of axial and torsional stresses at any ratio to either a hollow or a solid cylinder. The ratio can be kept constant or can be

varied during a given test. This allows one to incline the major and the minor
principal stresses at will. If desired, the pressure in the cell (which is equal to
both the internal and external pressures acting on the hollow cylinder) can also
be changed simultaneously so that the intermediate principal stress can maintain
a given ratio with the two others. The changes in cross section of the specimen
are continuously accounted for by the computer. The above operations can
also be conducted in a cyclic way at speeds as high as 4 cycles per minute.
This testing device can generate practically any stress path and induce any
inclination of the principal stresses on the axis of symmetry.

   Attached to each of the devices above is a K_0 consolidation servome-
mechanism, which during the consolidation phase of any specimen, measures
the water expelled and imposes an axial displacement such that the cross section
of the sample remains constant.

2.4. *Sample Preparation*

For undisturbed clay the standard shelby samples (2.8 in O.D.) were placed
in a mould and a 2 in diameter core cut with a piano wire following a template
at both ends [3, 5]. As for laboratory prepared specimens, they were obtained
through a one-dimensional consolidation of slurries in large consolidometers.
2.8 in. diameter samples were then cut from the resulting blocks and cored as
previously mentioned. All specimens were placed in the cell and if needed
additional K_0 consolidation induced. Such K_0 consolidation in the cell added
substantially to the degree of orientation of the particles obtained during the
initial densification of the slurries.

3.   Static behavior of anisotropic clay

3.1. *Introduction*

A very extensive testing program was conducted at Case Institute of Technology
between 1967 and 1977 on anisotropic clays, both natural and laboratory
prepared. All the tests were conducted on thin long hollow cylinders first K_0
consolidated in the cell then rebounded to the cell pressure. This resulted in an
axial overconsolidation ratio generally close to 2 and a lateral one equal to unity.
To obtain highly overconsolidated specimens the cell pressure was additionally
reduced to the desired amount.

   Various clays at different water contents were tested under as many as 13
stress paths in which the mean stress was either allowed to change or kept
constant [7, 8]. In all cases, because of the way SPAC operates, a condition of
proportional stressing was present in the specimen.

3.2. *Normally consolidated and Slightly Overconsolidated Clay*

Following a pattern established by Dorn [2], Saada and Zamani [7] suggested that, under static testing, the stress-strain relations for cross anisotroqic normally consolidated clay could be written in the form $\dot{\epsilon}_{ij} = \dot{S}_{ijk\varrho} \, \sigma_{k\varrho}$ with $\dot{S}_{ijk\varrho} = \alpha_{ijk\varrho} (\dot{\epsilon}_{eq}/\sigma_{eq})$ and $\epsilon_{ij} = 1/2 \, (\partial u_i/\partial x_j + \partial u_j/\partial x_i)$; u_1, u_2, u, are the three components of the displacements; $\dot{\epsilon}_{ij}$ are the instantaneous strain rates; σ_{ij} are the instantaneous components of the stress; $\dot{S}_{ijk\varrho}$ are variable coefficients having dimensions of strain rate over stress and depending on the mechanical history of the material; $\alpha_{ijk\varrho}$ are constants of anisotropy; $\dot{\epsilon}_{eq}$ is an equivalent strain rate; σ_{eq} is an equivalent stress. The equivalent stress and increment of equivalent strain were defined by assuming that the increase in the deviator stress is a function of the work done per unit volume and that the increment in the specific work dW is always given by $\sigma_{eq} \, d\epsilon_{eq}$. The constants of anisotropy were obtained through a series of simple tests and a variety of stress paths gave results supporting the proposed stress-strain relations. Later, Saada and Ou [8] found that clays K_0 consolidated in the triaxial cell had a high degree of anisotropy and that their behavior in extension and compression differed quite substantially (Fig. 6). The material was found to be weaker in ultimate strength but much more ductile in extension than in compression. Two different sets of $\alpha_{ijk\varrho}$ were sought, to be used depending on whether the sample was extended or compressed along the axis of symmetry. Experimental data obtained in an extensive program supported the theory above [8].

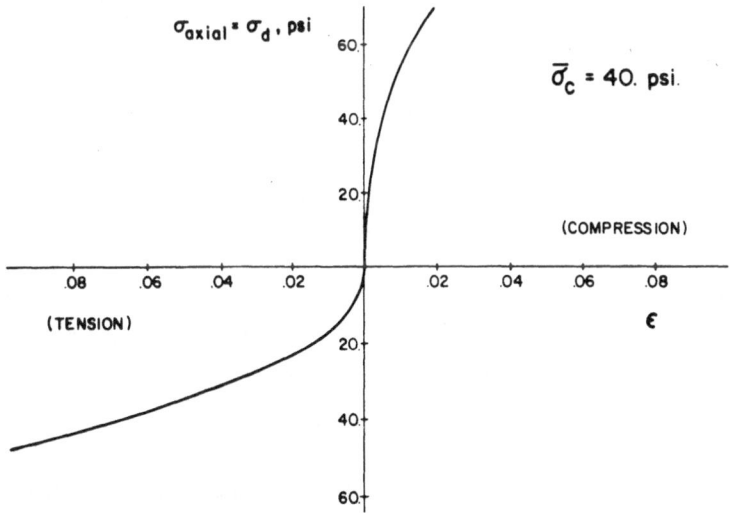

Figure 6. – *Typical Tension and Compression Behavior of a K_0 Consolidated Clay.*

Saada and Bianchini [9] showed that K_0 consolidated clays led to different effective stress paths to failure for different inclinations of the principal stresses on the axis of symmetry. For example, Figure 7 shows the total and effective stress paths for a Kaolinite clay one dimensionally consolidated at a cell pressure of 80 psi. The notation and the details of those tests can be found in Reference [9]. If the ends of the effective stress paths are joined to the origin of a (p, q) system of axes with straight lines they define angles of friction Φ terms of effective stresses varying between $26°$ and $45°$. The classical Mohr-Coulomb failure criterion obviously does not apply to K_0 consolidated clays.

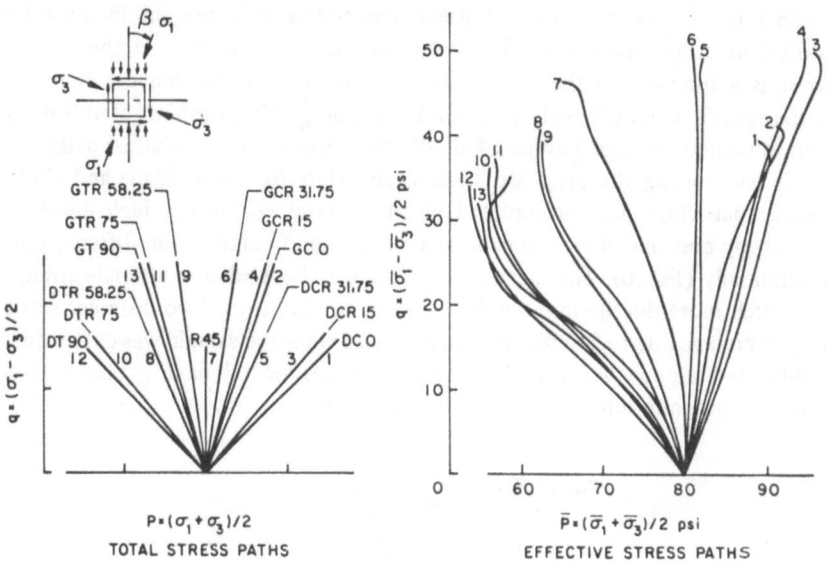

Figure 7. – *Total and Effective Stress Paths for a K_0 Consolidated Kaolinite.*

A failure criterion due to Goldenblat and Kopnov was found to be quite satisfactory for the states of stress present in hollow cylinders subjected to axial and torsional stresses [8]. Its drawback is that it requires too many parameters to characterize the failure of the material. It does not seem possible, however, to escape this situation because of the difference in behavior in extension and compression.

3.3. *Overconsolidated Clay*

Overconsolidation seems to result in a material which, while still being more brittle in compression than in extension, fails at a stress that is much higher in extension: Overconsolidation of a material with cross anisotropy seems to reverse what is to be expected from clays under stress. The pore water pressure increases slightly in compression, but then quickly decreases. Both extension and torsion result in significant decreases in the pore water pressure even for relatively small degrees of oversconsolidation. Fig. 8 shows typical examples of compression, tension and torsion for a cell pressure of 10 psi and an overconsolidation ratio of 2.

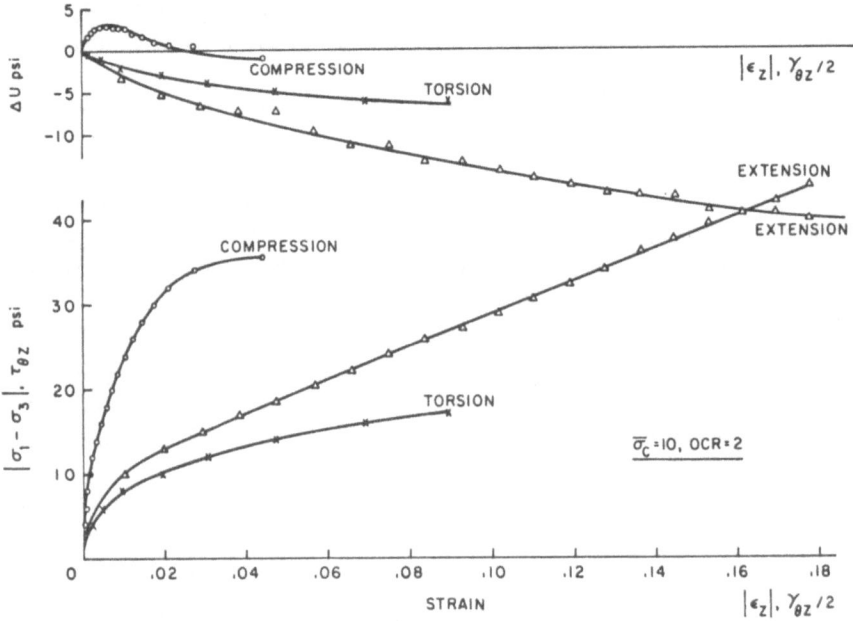

Figure 8. – *Test Results for an Anisotropic Overconsolidated Clay.*

4. Dynamic behavior of anisotropic clay

4.1. *Introduction*

The response of clay to dynamic loading was studied under small strains in the resonant column and large strains in SPAC as previously mentioned. Anisotropic specimens were prepared as described above through K_0 consolidation of slurries. Isotropic specimens were prepared by hand mixing and kneading of the clay followed by consolidation under hydrostatic stress. It is interesting to note that the blocks of clay removed from the large one dimensional consolidometers (refered to earlier) at 48 per cent water content were much stiffer than the isotropic clay at 39 to 42 per cent water content. The system of stresses applied during the preparation phase of the specimen has a very significant effect on particle orientation and/or bonds.

4.2. *Normally Consolidated and Slightly Overconsolidated Clay at Small Strain and High Frequency*

Moduli were studied under both longitudinal and torsional modes of vibration. The specimens were placed in resonance in a fixed-free configuration and strains and natural frequencies were measured. These allow one to compute the Young's Moduli and the shear moduli for isotropic clays. Actually, these moduli are secant moduli. For isotropic clays the assumption of constant volume was very well supported by the result [11]; E was found nearly equal to 3G. For anisotropic clays E, E', G and G' (Fig. 9) were found to decrease as the strain increases. Notice that the water content versus the effective consolidation pressure of isotropic and anisotropic clays is quite different (Fig. 10). Figure 11 shows the variations of the Young's and the shear moduli with strain for various values of the effective consolidation pressure.

The damping ratios were measured by the magnification factor method and by the logarithmic decrement method. Figure 12 shows damping at small and large strain. Measurements showed that damping increases with strain, but that no particular difference could be found between isotropic and anisotroqic clays.

Hysteresis loop fitting using Rambert-Osgood's and Masing's criteria was found unsatisfactory in predicting both modulii and damping [1]. For example

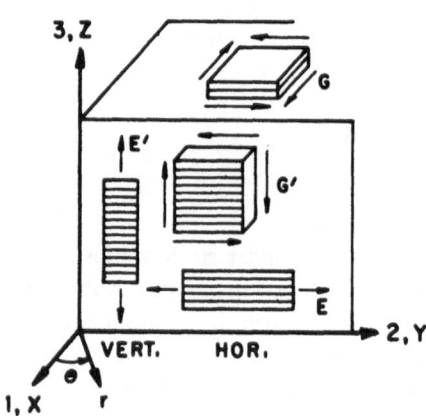

Figure 9. – *System of Coordinates and Moduli.*

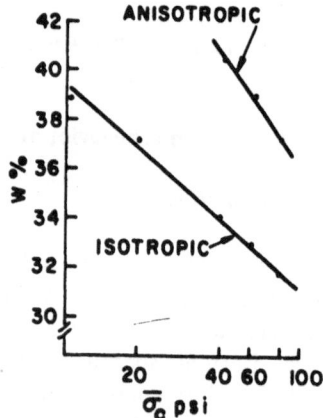

Figure 10. – *Water Content Versus Consolidation Pressure.*

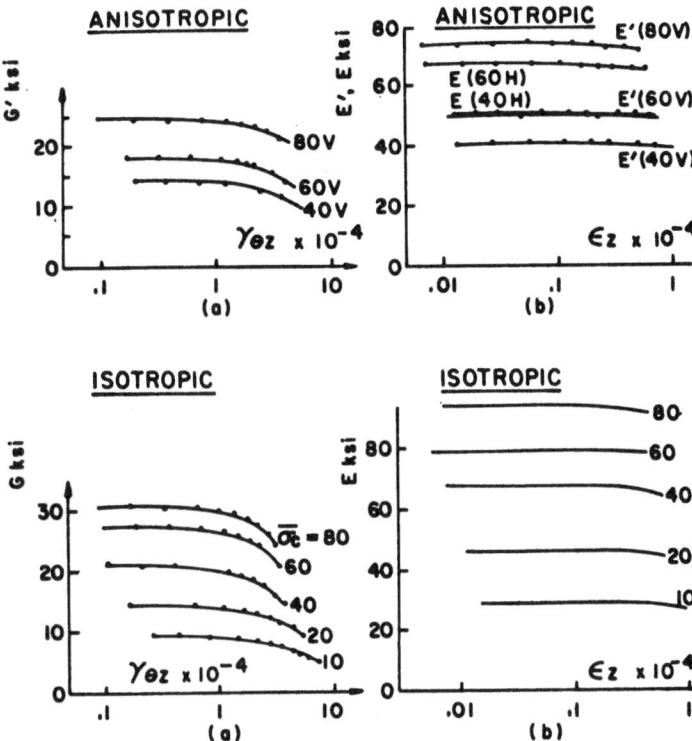

Figure 11. – *Variation of Modulii With Strain.*

experimentally obtained damping was found to be considerably smaller than the damping computed using a backbone curve derived from secant moduli. Inversely, the measured damping leads to a prediction of modulii that are considerably larger than the corresponding measured valus (Fig. 13).

4.3. *Normally Consolidated and Slightly Overconsolidated Clay at Large Strain and Low Frequency*

Cross-anisotropic specimens as well as isotropic specimens were subjected to cyclic loadings at approximatley 2 cycles per minute. The cyclic stress patterns involved one directional as well as bi-directional loading in compression, extension and torsion. Figure 14 shows the envelopes of the axial and shearing strains and those of the pore water pressures versus the number of cycles of one directional cyclic loading. The difference in behavior between the anisotropic and the isotropic specimens is dramatically illustrated by the compression and extension tests. The brittleness and ductility characteristics

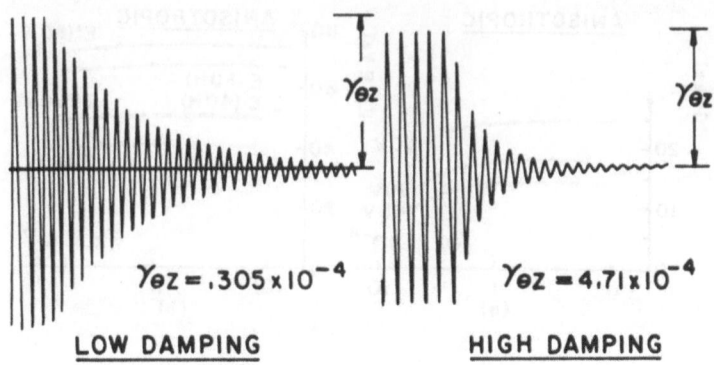

Figure 12. – *Decay Curves for Small and Large Strain.*

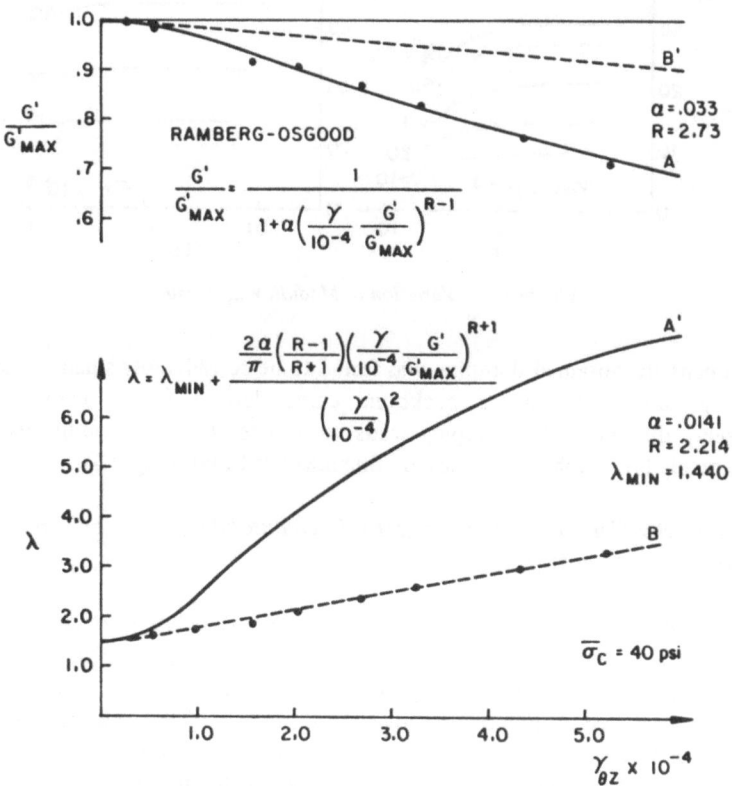

Figure 13. – *Measured and Predicted G'/G'_{max} and λ using the Ramberg-Osgood-Masing Model.*

indicated by the static behavior (Fig. 6) are very noticeable in spite of the enormous remolding that takes place at high strain. The contention that high cyclic straining erases the memory of clays is certainly not true for this clay. In the torsional mode the effects of anisotropy are the least noticeable. In spite of the fact that the stress level applied was as high as 85 percent of the static failure load some specimens did not fail, supporting Sangrey's observations [12] of a threshold stress level below which failure does not occur.

Figure 15 shows the envelopes of the axial and shearing strains and those of the pore water pressure, versus the number of cycles for bi-directional loading. Because of the high ductility of the anisotropic clay in extension and its brittleness in compression, the strain build up is seriously biased towards the tension side to the extent that the first cycle both envelopes fell on the tension side of the reference axis. The isotropic clay reacted in a totally different way. The reference axis was crossed at every cycle but with a larger build up of strain on the tension side. The isotropic samples failed after a comparatively smaller number of cycles that the anisotropic ones.

As expected the pore pressure build up was much higher and faster for the bi-directional than it was for the one directional loading; and catastrophic failure was observed to take place after a relatively small number of cycles.

Hysteresis loops were used to compute the moduli and the damping. Obvious difficulties arise where a material behaves differently in extension and compression. There is no symmetry with respect to the origin and a Ramberg-Osgood representation yields different coefficients in extension and compression. The use of the one dimensional kinematic hardening model (which can be made to coincide with the Masing hypothesis for materials having the same behavior in tension and compression) leads to hysteresis loops that will move away from the origin along the strain axis each time the level of the cycling stress is increased. If cycling is conducted between two fixed levels of strain the loop will move along the stress axis each time this level is changed. If there is degradation of the backbone curve, additional displacement and rotation of the loops will take place. Figure 16 illustrates the motion of the loop for a material whose behavior is different in tension and compression and which is cycled between constant levels of stress, but with no degradation.

Figure 17 shows the hysteresis loops obtained when a Kaolinite K_0 consolidated clay was axially and cyclically loaded for two different levels of stress. One notices that even for very small strains, the loops move and rotate showing the effects of both differences in tension and compression behavior and degradation. Figure 18 shows the hysteresis loops obtained in torsional loading. Because of the cross anisotropy the direction of the rotation is immaterial. The loop moves very little along the strain axis and degradation results in a rotation of the loops about the origin. A brief examination of Figure

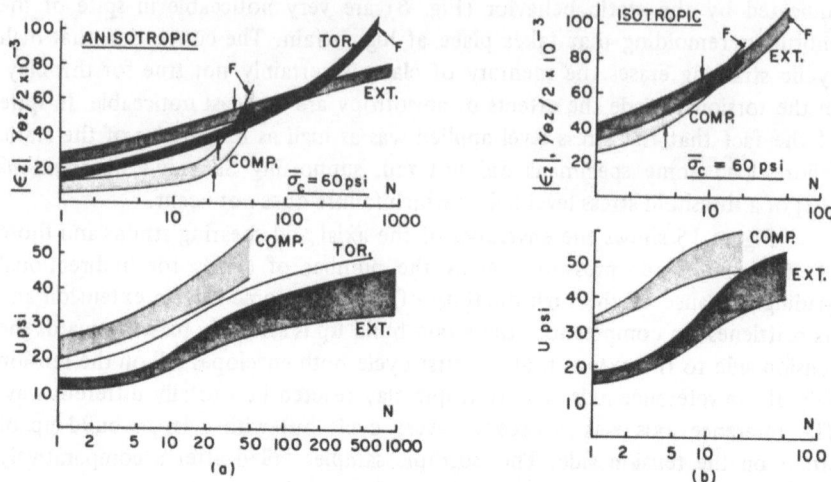

Figure 14. – *Strain Versus Number of Cycles for One Directional Loading on Isotropic and Anisotropic Clays.*

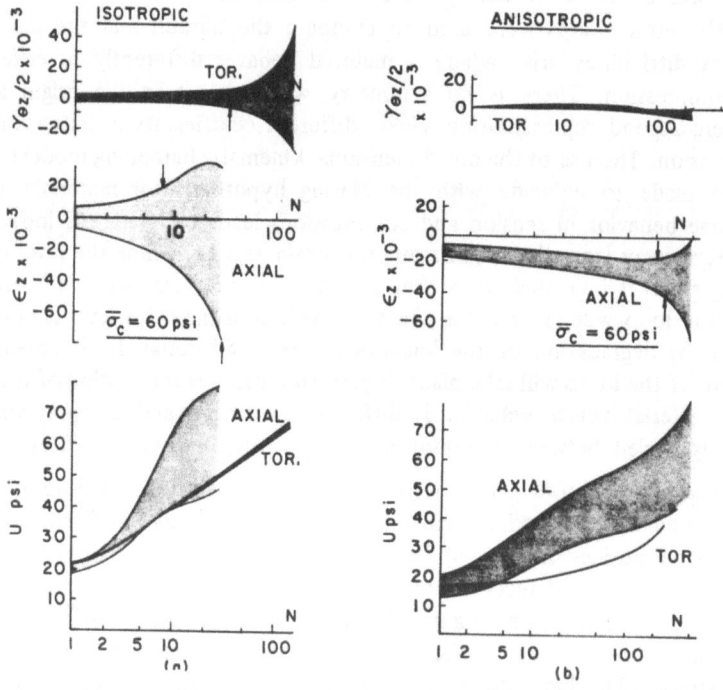

Figure 15. – *Strain Versus Number of Cycles for Bi-directional Loadings on Isotropic and Anisotropic Clays.*

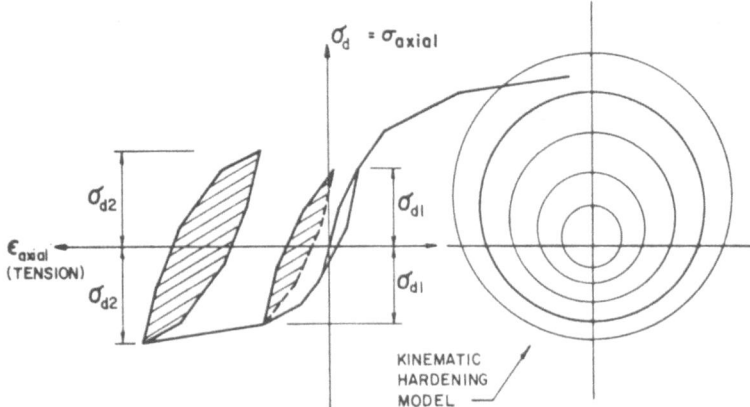

Figure 16. — *Kinematic Hardening Model and Motion of Loops when Tension and Compression are different.*

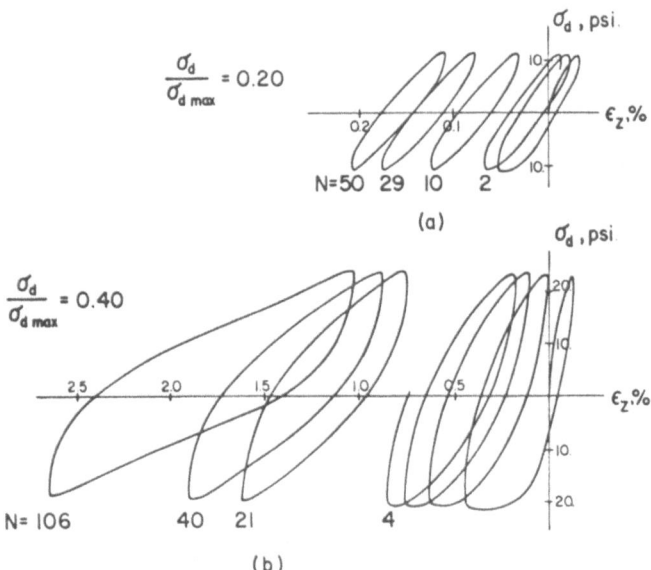

Figure 17. — *Axial Cyclic Loading of a K_0 Consolidated Clay.*

17 and 18 indicates that whatever interparticle mechanism is responsible for mechanical degradation in the axial direction it is quite different from the one operating in the shearing direction along the fabric's preferential orientation.

The relation between the number of cycles N and the degradation index δ [4] is not easy to get. While the secant moduli E' and G' are in theory

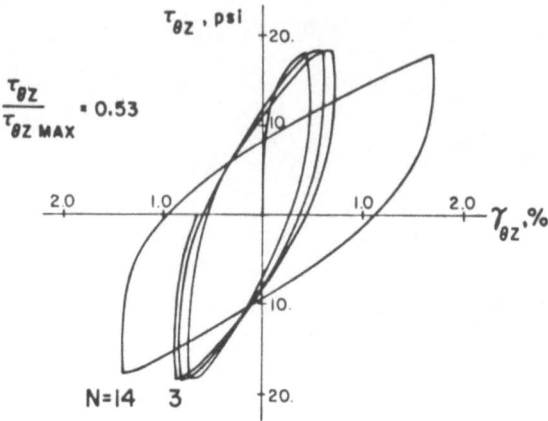

Figure 18. – *Torsional Shear Cyclic Loading of a K_0 Consolidated Clay.*

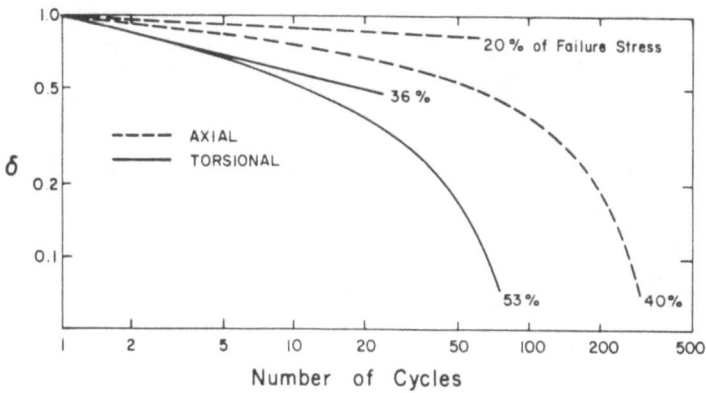

Figure 19. – *Degradation Indices G'/G'_1 and E'/E'_1 Versus Number of Cycles for a K_0 Consolidated Clay.*

obtained by joining the ends of the loops, one has to wait for the loops to stabilize and stop moving along the strain axis before those ends are defined. Thus the first cycle relative to which degradation is measured has a secant modulus that, at best is approximately determined. It was found by extrapolation of the values obtained from higher numbered cycles on a Log G' or Log E' versus Log N diagram. Figure 19 shows the relation between Log δ and Log N in the axial and torsional modes. These relations are only linear for very small strains. Here too, hysteresis loop fittings using Ramberg-Osgood's and Masing's criteria led to results somewhat similar to those mentioned for small strain behavior; in other words unsatisfactory (19). Damping energy computed as the area of the hysteresis loop was found to increase with the strain and consequently with the number of cycles. On the other hand specific damping which involves dividing this area by a multiple of the maximum potential energy did not indicate a clear trend when plotted with respect to the number of cycles (13).

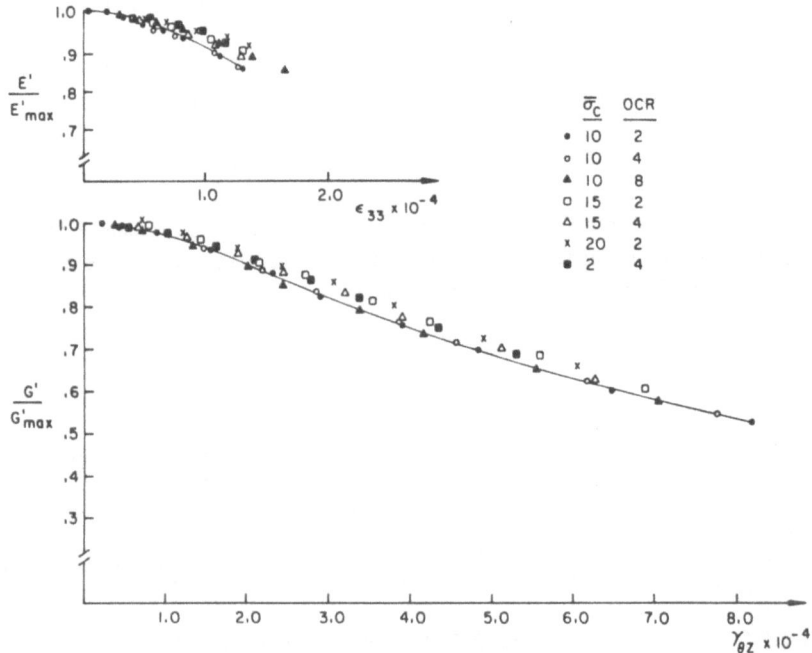

Figure 20. – *Normalized Moduli Versus Strain for an Overconsolidated Anisotropic Kaolinite Clay.*

4.4. Overconsolidated Clay

Resonant column tests as well as slow cyclic loading tests were conducted on clays with overconsolidation ratios of 2, 4 and 8 and final effective mean stresses of 10, 15 and 20 PSI. In the resonant column it was found that if the shear and Young's moduli were normalized with respect to their maximum value a simple curve would be obtained between G'/G'_{max} and the strain (Fig. 20). Damping follows a trend similar to the one established for normally consolidated clays.

Slow cyclic loading tests presented a trend faily consistent with normally consolidated clay. Again, the damping predicted by the Ramberg-Osgood-Masing model was unsatisfactory when compared with the measured one.

5. Conclusion

Our present understanding of the behavior of clays under static loadings while extensive, has not yet allowed us to find general constitutive equations that can be used in the solution of many common problems in foundation engineering. More laboratory investigations involving combined states of stress are necessary before any of the present theories can be accepted; since all of these theories have, at best, been supported by axially symmetric tests in the standard triaxial cell.

In the dynamics area clays have held second place to sands in the minds of investigators because of the paramount importance of the liquefaction phenomena. However, with the spread of offshore structures research in the response of clays to cyclic loading has substantially incrased and many design criteria are being formulated.

The research described in the present paper has been sponsored by the National Science Foundation over the last seven to eight years. The authors wish to express their gratitude for this support.

REFERENCES

[1] BIANCHINI G.F. – "Effects of anisotropy and strain on the dynamic properties of clay soils" Ph.D. Thesis, Case Western Reserve University, 1981.

[2] DORN J.E. – "Stress-strain relations for anisotropic plastic flow" *Journal of Applied Physics,* 20, 15-20.

[3] HVORSLEV M.J. and R.I. KAUFMAN. – "Torsion shear apparatus and testing procedures" *Bulletin No. 38,* USAE Waterways Experiment Station, Vicksburg, Miss.

[4] IDRISS I.M.. DOBRY R. and R.D. SINGH. – "Nonlinear behavior of soft clays during cyclic loading". *Journal of the Geotechnical Engineering Division,* ASCE, 104, GT12, Proc. Paper 14265, Dec. 1978, pp. 1427-1447.

[5] SAADA A.S. and A.K. BAAH. – "Deformation and failure of a cross anisotropic clay under combined stresses" *Proceedings,* Third Panamerican Conference on Soil Mechanics and Foundation Engineering, Caracas, Venezuela, 1, 1967.

[6] SAADA A.S. – "Testing of anisotropic clay soilds" *Journal of the Soil Mechanics and Foundation Division* ASCE, 96, SM5, Proc. Paper 7502, Sept., 1970, 1847-1852.

[7] SAADA A.S. and K.K. ZAMANI. – "The mechanical behavior of cross anisotropic clay", *Proceedings,* Seventh International Conference on Soil Mechanics and Foundation Engineering, Vol. 1, Mexico, (1969) : 351-359.

[8] SAADA A.S. and C.D. OU. – "Stain-stress relations and failure of anisotropic clays" *Journal of the Soil Mechanics and Foundations Division,* ASCE, Vol. 99, N. SM12, Proc. Paper 10225, Dec., (1973) : 1091-1111.

[9] SAADA A.S. and G.F. BIANCHINI. – "Strength of one dimensionally consolidated clays", *Journal of the Geotechnical Engineering Division,* ASCE, Vol. 101, No. GT11, Nov. (1975): 1151-1164.

[10] SAADA A.S. and G.F. BIANCHINI. – "Closure of strength of one dimensionally consolidated clays", *Journal of the Geotechnical Engineering Division,* ASCE, 103., GT 6, June 1977.

[11] SAADA A.S., G.F. BIANCHIN and L.P. SHOOK. – "The dynamic response of normally consolidated anisotropic clay". *Proceedings* of the 1978 ASCE Specialty Conference on Earthquake Engineering and Soil Dynamics, Pasadena, Cal.

[12] SANGREY D.A., D.J. HENKEL and M.I. ESRIG. – "The effective stress response of a saturated clay to repeated loading" *Canadian Geotechnical Journal,* 6, 3, 241-252.

[13] SHOOK L.P. – "Behavior of clays subjected to slow cyclic loading" M.SC. Thesis, in preparation.

[14] WRIGHT D.K., P.A. GILBERT and A.S. SAADA. – "Shear devices for determining dynamic soil properties". *Proceeding* of the 1978 ASCE Specialty Conference on Earthquake Engineering and Soil Dynamics, Pasadena, Cal.

RESUME

Comportement expérimental des argiles anisotropes

Dans cette communication, on discute les méthodes utilisées dans les essais sur les argiles anisotropes. Un appareil permettant une rotation contrôlée des contraintes principales par rapport à l'axe de symétrie est décrit. Le comportement statique et dynamique des argiles consolidées sous k_0 et soumises à de petites et à de grandes déformations est étudié. On trouve que les modules d'élasticité dynamiques, ainsi que l'amortissement, dépendent de la structure du matériau. Cette structure n'est pas complètement détruite, malgré la dégradation qui a lieu pendant les chargements cycliques, même après de très grandes déformations. On montre que les modèles mathématiques actuellement utilisés ne sont pas satisfaisants.

Saint-Venant's Principle
in Anisotropic Elasticity Theory

C. O. Horgan
Michigan State University, East Lansing, Michigan, U.S.A.

1. Introduction

The simplification arising in elasticity theory from consideration of resultant boundary conditions instead of mathematically exact pointwise conditions has been the key to widespread application of the subject. We cite, for example, the theories for strength of materials, plates and shells having such relaxed boundary conditions as cornerstones of their development. The justification for consideration of load resultants is usually based on some form of Saint-Venant's principle characterizing the boundary layer behavior involved. Hence there has been continued interest in investigation of the issues underlying this 'principle', both from a mathematical and physical point of view.

In this paper, we summarize the results of various investigations, both theoretical and experimental, concerned with analysis of Saint-Venant's principle in the context of anisotropic elasticity theory. In particular, we draw attention to the fact that the routine invocation of Saint-Venant's principle in the solution of problems involving highly anisotropic or composite materials is not justified in general. In view of the widespread use of some form of Saint-Venant's principle in diverse areas of solid mechanics, the ramifications of this are many. One of the more serious may be the implications for accurate measurement of the mechanical properties of such materials (at least for static testing).

Some time ago, the author considered the effect of anisotropy on the decay of stresses with distance from the boundary of an elastic solid subject to self-equilibrated loads [10-12]. Using widely applicable techniques involving energy-decay inequalities of the type developed by Knowles [13] and Toupin [20], lower bounds (in terms of the elastic constants) were obtained for the rate of exponential decay of stresses giving rise to upper bounds for 'characteristic decay lengths'. In [10, 11] the plane problem was treated, while the torsionless axisymmetric problem for a circular cylinder was considered in [12]. In particular, the case of transverse isotropy relevant to fiber-reinforced composites was examined in [11]. For such highly anisotropic materials, a characteristic decay length of order $b(E/G)^{1/2}$ was predicted, where b is the maximum dimension perpendicular to the fibers and E, G are the longitudinal Young's modulus and shear modulus respectively. When E/G is large, therefore, very slow stress decay is anticipated with end effects being transmitted a considerable distance from the loaded ends. Similar results were obtained in [12] for the torsionless axisymmetric problem.

Recently we have considered the plane problem from a different point of view [5]. For an anisotropic rectangular strip under self-equlibrating end loads, the *exact* stress decay is analyzed using analogues of the Fadle-Papkovich eigenfunctions. (See Section 4 here). For highly anisotropic transversely isotropic materials slow decay rates were obtained confirming the lower bound estimates obtained in [10, 11]. In particular, for the case of a graphite/epoxy composite, the exact decay rate is about four times smaller than that for an isotropic material (see Table 1 here). A similar analysis was carried out in [6] for sandwich strips, composed of homogeneous isotropic layers perfectly bonded at the interfaces. For the case of a relatively soft inner core, we obtain a slow decay rate confirming earlier photoelasticity studies of Alwar [1] on the same effect. (See Section 6 here).

The 'extended end effects' of concern here have been investigated experimentally by other authors [2, 3, 8]. Initial attention to the question at hand was drawn in the course of conducting torsion pendulum tests for measurements of the longitudinal shear modulus for a highly anisotropic polymeric microcomposite [8]. It was found [8] that the calculated values of the shear modulus varied with the specimen aspect ratio (length/width ratio), uniform results being obtained only for samples with aspect ratios of about 100. This is, of course, in complete contrast with results of testing procedures for isotropic materials. For the block copolymer microcomposite used in [8], E/G is about 280 and so characteristic decay lengths of the order of 16-17 sample widths might be expected on the basis of the theoretical predictions. Further tests are described in [2, 3]. In particular, the results of a simple tension test for the measurement of the longitudinal Young's modulus for a highly drawn polyethylene are described in [2, 3]. For this material, $(E/G)^{1/2} =$ 15.1 [3] and so the theoretical results would suggest that specimens with aspect ratios of at least 30 would be required for satisfactory measurements. This was found out to be the case [2, 3].

As is discussed in [2, 3, 8], this failure of the classic application of Saint-Venant's principle raises some serious questions for practical stress analysis, particularly with regard to current testing practice for measurement of elastic constants. Thus, if end effects cannot be neglected, one alternative is to obtain exact elasticity solutions for the particular end conditions employed. This is extremely difficult in general. Otherwise sample aspect ratios have to be chosen sufficiently large such that end effects can be ignored. This may not always be possible, in view of the fact that many materials may present themselves only in minute 'whisker' form.

We note that various facets of such 'failures' of classic applications of Saint-Venant's principle in highly anisotropic elasticity theory have been encountered by other authors. Thus in the context of idealized theories for fiber-reinforced composites, Everstime and Pipkin [7] have observed a 'stress channelling' phenomenon. A similar 'load diffusion' is discussed in [15]. The fundamental 'skin effect' treated by Biot in [4] is of a similar nature. Many investigations in composite material mechanics are concerned with the effects of end constraints in mechanical testing (see e.g. Pagano and Halpin [16]). An interesting extended boundary-layer has been observed by Wan [21] in the problem of bending of a pretwisted strip.

2. Plane Deformation for Transversely Isotropic Materials

The stress-strain relations governing the in-plane stresses and strains in plane strain of an homogeneous anisotropic elastic solid, transversely isotropic about the x-axis, are given by [10, 11].

$$
\left.
\begin{aligned}
e_{xx} &= \beta_{11}\,\sigma_x + \beta_{12}\,\sigma_y \\
e_{yy} &= \beta_{21}\,\sigma_x + \beta_{22}\,\sigma_y \\
2e_{xy} &= \beta_{66}\,\tau_{xy},
\end{aligned}
\right\}
\tag{2.1}
$$

where the elastic constants $\beta_{pq} = \beta_{qp}$ (p, q = 1, 2, 6) may be written in terms of the usual engineering constants as

$$
\beta_{11} = \frac{1}{E_L}\,(1 - \nu_{LT}^2\,E_T/E_L), \quad \beta_{12} = -\nu_{LT}(1 + \nu_{TT})/E_L,
\tag{2.2}
$$

$$
\beta_{22} = (1 - \nu_{TT}^2)/E_T, \quad \beta_{66} = 1/G_{LT}.
$$

Here L denotes the direction parallel to the x-axis, T the transverse direction and ν, E, G denote Poisson's ratio, Young's modulus and shear modulus respectively. The strain-energy density is assumed to be positive definite and so the elastic constants are such that

$$
\beta_{11} > 0, \quad \beta_{11}\beta_{22} - \beta_{12}^2 > 0, \quad \beta_{66} > 0.
\tag{2.3}
$$

We are concerned with the traction boundary-value problem of linear elasticity for plane domains in the x-y Cartesian coordinate plane. Introducing the Airy stress function $\Phi(x, y)$ in the usual way by

$$\sigma_x = \Phi_{yy}, \quad \sigma_y = \Phi_{xx}, \quad \tau_{xy} = -\Phi_{xy}, \tag{2.4}$$

where the subscript notation on Φ denotes partial differentiation, we obtain the governing elliptic partial differential equation [10, 11]

$$\beta_{22}\, \Phi_{xxxx} + (2\beta_{12} + \beta_{66})\, \Phi_{xxyy} + \beta_{11}\, \Phi_{yyyy} = 0 . \tag{2.5}$$

For the special case of an isotropic material $\beta_{11} = \beta_{22} = (1 - \nu)/2G$, $\beta_{66} = 1/G$, $\beta_{12} = -\nu/2G$ and (2.5) reduces to the familiar biharmonic equation.

3.   Saint-Venant's Principle for Plane Deformation of Highly Anisotropic Materials: General Discussion

Our interest in this work is to assess the effect of anisotropy on the exponential decay of stresses inherent in Saint-Venant's principle. In particular, we are chiefly concerned with the case of transversely isotropic materials with a high degree of anisotropy in the axial direction. Such transversely isotropic materials have had wide usage as models for fiber-reinforced composites. (See e.g. [9, 19]). For highly anisotropic materials, we assume that

$$E_T/E_L \ll 1, \quad G_{LT}/E_L \ll 1, \quad E_T/G_{LT} \simeq 1 . \tag{3.1}$$

Methods involving energy-decay inequalities were employed by the author in [10, 11] to investigate Saint-Venant's principle in plane elasticity for a wide class of anisotropic materials in a general bounded domain. Lower bounds (in terms of the elastic constants) were obtained for the rate of exponential decay of stresses (with distance from self-equilibrated boundaries) giving rise to upper bounds for "characteristic decay lengths". Thus the typical results obtained in [10, 11] for a representative stress component τ are of the form

$$|\tau| \leqslant K\, e^{-kx}, \quad x \geqslant 0 , \tag{3.2}$$

where K is a constant and the decay rate k is characterized explicitly in terms of the elastic constants and cross-sectional dimensions of the body. For highly anisotropic transversely isotropic materials satisfying (3.1), it is shown in [11] that

$$k = 0\left[\frac{(G_{LT}/E_L)^{1/2}}{b}\right], \tag{3.3}$$

where b is the maximum cross-sectional dimension. When E_L/G_{LT} is large, therefore, very slow stress decay is anticipated with end effects being transmitted a considerable distance from the loaded ends. For example, for the case of a graphite/epoxy composite, this ratio has the value 33.3 (see Table 1 here) compared with a range of values of between 2 and 3 for the isotropic case. Thus for the problem of a rectangular strip loaded only at the short ends, the usual engineering approximation of neglecting Saint-Venant end effects at distances of about one width from the ends is justified in the isotropic case (see e.g. Fig. 1 here). However, for the graphite/epoxy composite, such approximations are justified only at distances about four times larger. Of course, for even more highly anisotropic materials the effects are more pronounced.

The slow exponential decay rate (3.3) predicted by the analyses of [10, 11] might also be anticipated from another viewpoint. Introducing the notation

$$\epsilon_t^2 = G_{LT}/E_L , \tag{3.4}[1]$$

and using the conditions (3.1) characterizing a highly anisotropic material, the differential equation (2.5) can be written in this case as

$$A\Phi_{xxxx} + B\Phi_{xxyy} + \epsilon_t^2 \Phi_{yyyy} = 0, \qquad A, B \text{ constants.} \tag{3.5}$$

In the limit as $\epsilon_t \to 0$, the equation (3.5) ceases to be elliptic — in fact, the equation becomes parabolic with the straight lines y = const being characteristic curves. Thus, in this limit, (3.3) implies the complete breakdown of a Saint-Venant principle — the "end effects" are transmitted without attenuation along the fibers (characteristics). A similar stress "channelling" phenomenon was observed in [7] in the context of idealized theories for fiber-reinforced composites.

4.   Illustrative Examples for Rectangular Strips

To provide explicit illustration of the extended end effects of concern here, we investigate the problem of a rectangular strip, of half width unity, traction free at the lateral sides y = ± 1, and subject to self-equilibrated load conditions at the ends x = ± ℓ. (See Fig. 1). Eigenfunction expansion techniques may be used to assess the *exact* exponential decay rate and to solve specific boundary value problems [5]. Solutions of (2.5) are sought in the form

$$\Phi = \sum_k (a_k \cosh \gamma_k x + b_k \sinh \gamma_k x) F_k(y) , \tag{4.1}$$

---

1 The small parameter ϵ_t was introduced by Everstine and Pipkin [7] in their idealized theory involving the constraint of inextensibility in the fiber (x-) direction.

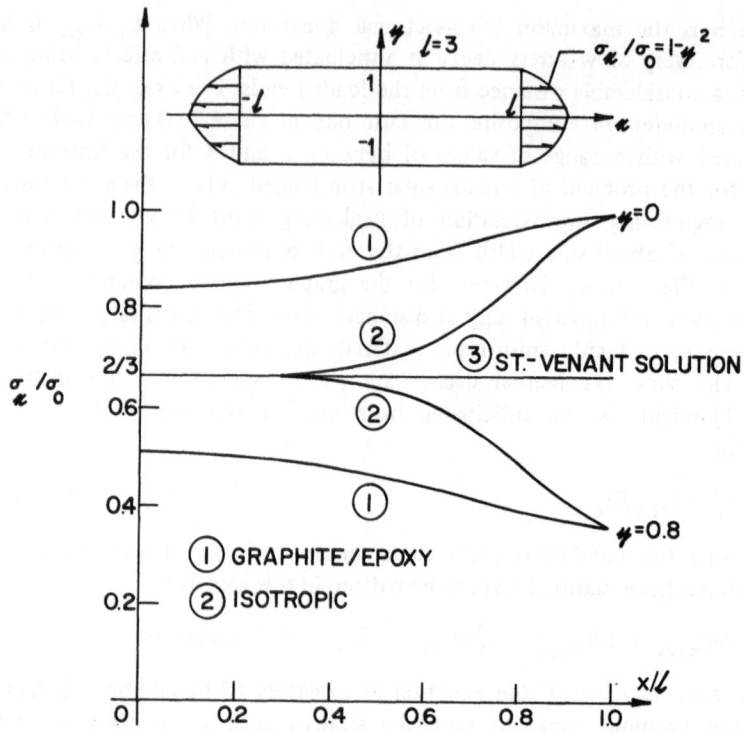

Figure 1. – (after Choi and Horgan [5]). Comparison of the decay of σ_x between isotropic and graphite/epoxy materials for extension problem.

where a_k, b_k are constants to be determined from the end conditions, and $F(y)$ are analogues of the well-known Fadle-Papkovich eigenfunctions. Thus F, γ are the eigenfunctions and eigenvalues (complex) of the problem

$$\beta_{11} F^{iv}(y) + (2\beta_{12} + \beta_{66}) \gamma^2 F''(y) + \beta_{22} \gamma^4 F(y) = 0 , \tag{4.2}$$

$$F(\pm 1) = 0 , \quad F'(\pm 1) = 0 . \tag{4.3}$$

From (4.2), (4.3) it is readily verified that

$$\int_{-1}^{1} F'(y) \, dy = 0 , \quad \int_{-1}^{1} F''(y) \, dy = 0 , \quad \int_{-1}^{1} yF''(y) \, dy = 0 , \tag{4.4}$$

and so from (4.1) and (2.4) it follows that

$$\int_{-1}^{1} \tau_{xy}(x, y) \, dy = 0 , \quad \int_{-1}^{1} \sigma_x(x, y) \, dy = 0 , \quad \int_{-1}^{1} y\sigma_x(x, y) \, dy = 0 , \tag{4.5}$$

valid for each x in the strip. Thus the stresses associated with F are self-equilibrating and so the representation given by (4.1), (2.4) provides a convenient framework for investigating the decay of end effects underlying Saint-Venant's principle. The dominant exponential decay rate is governed by the smallest real part of the eigenvalues γ. It is shown in [5] that the eigenfunctions F satisfy the orthogonality property

$$\int_{-1}^{1} (\beta_{11} F_k'' F_s'' - \beta_{22} \gamma_k^2 \gamma_s^2 F_k F_s) \, dy = 0, \quad (k \neq s) \tag{4.6}$$

which can be used to determine the constants a_k, b_k of (4.1) for prescribed end conditions.

The analysis of (4.2), (4.3) is conveniently broken down into three cases. [5]

Case A $\quad (2\beta_{12} + \beta_{66})^2 - 4\beta_{11}\beta_{22} > 0$:

The eigenfunctions may be separated into even and odd functions corresponding to symmetric and anti-symmetric deformations respectively. Thus the even eigenfunctions are given by

$$F_e(y) = \frac{\cos \gamma q_1 y}{\cos \gamma q_1} - \frac{\cos \gamma q_2 y}{\cos \gamma q_2}, \tag{4.7}$$

where γ is given as the root of the transcendental equation

$$q_1 \tan \gamma q_1 - q_2 \tan \gamma q_2 = 0, \tag{4.8}$$

and the odd eigenfunctions by (4.7) with cos replaced by sin and (4.8) with tan replaced by cot. Here

$$2q_{1,2} = \left\{ \frac{2\beta_{12} + \beta_{66}}{\beta_{11}} + 2\left(\frac{\beta_{22}}{\beta_{11}}\right)^{1/2} \right\}^{1/2} \mp \left\{ \frac{2\beta_{12} + \beta_{66}}{\beta_{11}} - 2\left(\frac{\beta_{22}}{\beta_{11}}\right)^{1/2} \right\}^{1/2}. \tag{4.9}$$

Case B $\quad (2\beta_{12} + \beta_{66})^2 - 4\beta_{11}\beta_{22} < 0$:

The even eigenfunctions are given by

$$F_e(y) = \frac{\cosh \gamma p y \cos \gamma q y}{\cos \gamma p \cos \gamma q} - \frac{\sinh \gamma p y \sin \gamma q y}{\sinh \gamma p \sin \gamma q}, \tag{4.10}$$

where

$$\frac{\sinh 2\gamma p}{p} + \frac{\sin 2\gamma q}{q} = 0, \tag{4.11}$$

while the odd eigenfunctions are given by (4.10) with cos replaced by sin, sin replaced by cos and the plus sign in (4.11) replaced by a minus sign. Here

$$2p = \left\{ 2\left(\frac{\beta_{22}}{\beta_{11}}\right)^{1/2} - \frac{(2\beta_{12} + \beta_{66})}{\beta_{11}} \right\}^{1/2},$$

$$2q = \left\{ 2\left(\frac{\beta_{22}}{\beta_{11}}\right)^{1/2} + \frac{(2\beta_{12} + \beta_{66})}{\beta_{11}} \right\}^{1/2}. \tag{4.12}$$

Case C $(2\beta_{12} + \beta_{66})^2 - 4\beta_{11}\beta_{22} = 0$:   In this case, we have

$$F_e(y) = \frac{\cos \gamma\, ry}{\cos \gamma\, r} - \frac{y \sin \gamma\, ry}{\sin \gamma\, r}, \qquad \sin 2\gamma r + 2\gamma r = 0, \tag{4.13}$$

$$F_0(y) = \frac{\sin \gamma\, ry}{\sin \gamma\, r} - \frac{y \cos \gamma\, ry}{\cos \gamma\, r}, \qquad \sin 2\gamma r - 2\gamma r = 0, \tag{4.14}$$

where

$$r = \left(\frac{2\beta_{12} + \beta_{66}}{2\beta_{11}}\right)^{1/2} = \left(\frac{\beta_{22}}{\beta_{11}}\right)^{1/4}. \tag{4.15}$$

It is readily verified that the case of an isotropic material is a subcase of Case C in which $r = 1$ and then (4.13), (4.14) characterize the Fadle-Papkovich eigenfunctions.

In Table 1, we have tabulated the elastic constants for the transversely isotropic materials magnesium and a highly anisotropic graphite/epoxy composite. It is readily verified that Case A governs these two materials. The eigenvalues with smallest real part are also given, computed from equation (4.8). For comparison purposes, we have included the well-known isotropic result[2], computed from (4.13) with $r = 1$. In the table, we also show the lower-bound estimates for the real part given by the energy method [11]. The decay rate for the graphite/epoxy composite is almost four times slower than that for

---

2 Note that the stresses are independent of the elastic constants in the isotropic case.

TABLE 1. Plane Elasticity.

Material	Elastic Constants $\begin{bmatrix} 10^{-8} \text{ m}^2/\text{kN} \\ 10^6 \text{ kPa} \end{bmatrix}$	Eigenvalue with smallest real part (Computed from (4.8))	Lower bound for Re (γ) (from eq. (3.3) of [11])
Magnesium (from [18])	$\beta_{11} = 1.892, \beta_{12} = -0.606$ $\beta_{22} = 1.968, \beta_{66} = 5.95$ $E_T = 44.754, E_L = 50.3$ $G_{LT} = 16.794, \nu_{TT} = .344,$ $\nu_{LT} = .227$	$2.1884 + i\,0.9425$	0.6548
Graphite/Epoxy (from [17])	$\beta_{11} = 0.722, \beta_{12} = 0.227$ $\beta_{22} = 13.6, \beta_{66} = 24.2$ $E_T = 6.833, E_L = 137.65$ $G_{LT} = 4.13, \nu_{TT} = .25, \nu_{LT} = .25$	0.5640 (real root)	0.1726
Isotropic	Stresses independent of elastic constants	$2.1062 + i\,1.1254$	0.7149

the isotropic material. It is seen that the energy approach gives rise to an accurate prediction of the ratio of decay rates between the graphite/epoxy and isotropic materials.

On using the conditions (3.1) characterizing a highly anisotropic material, an asymptotic estimate for the decay rate may be obtained [5]. It can be readily verified that, under the conditions (3.1), Case A holds. Using (2.2) and (3.1), it follows from (4.9) that

$$q_1 \simeq (G_{LT}/E_T)^{1/2} \simeq 1, \quad q_2 \simeq (E_L/G_{LT})^{1/2} \gg 1. \tag{4.16}$$

Under these conditions, it follows that the transcendental equation (4.8) has a smallest real root. Using appropriate Taylor expansions in (4.8), it can be shown that

$$\gamma \simeq \pi q_2/(q_2^2 - q_1^2) \cong \pi/q_2 \simeq \pi(G_{LT}/E_L)^{1/2}. \tag{4.17}$$

Thus, for the rectangular strip of half-width unity, (4.17) provides an approximate formula for the decay rate γ. For a strip of width b, we thereby obtain the approximate result

$$k \simeq \frac{2\pi}{b} (G_{LT}/E_L)^{1/2} \tag{4.18}$$

for the decay rate k in (3.2). One anticipates that (4.18) might also be used for a general simply-connected plane domain which is 'close' to rectangular, where b is now the maximum lateral dimension. It should be pointed out that the constant inherent in the order estimate (3.3) has been ignored in the interpretation of the work [11] by the authors of [2, 3, 8]. The result (4.18) suggests a representative value of 2π for this constant. For the graphite/epoxy material, (4.17) yields the result $\gamma \simeq .5442$ which is in excellent agreement with the exact value $\gamma = .5640$ given in Table 1.

Finally, we consider briefly the problem of extension of a finite strip composed of the graphite/epoxy composite and compare results with the case of an isotropic material [5]. For uniform tension of $2/3\sigma_0$ at both ends, we have the Saint-Venant solution

$$\sigma_x = 2/3\sigma_0, \quad \sigma_y = 0, \quad \tau_{xy} = 0. \tag{4.19}$$

Consider the statically equivalent load conditions shown in Figure 1. By subtraction, we obtain a boundary-value problem involving self-equilibrating load conditions, which are symmetric about the x-axis. Using an eigenfunction expansion technique involving the even eigenfunctions, the resulting problem was solved numerically [5] for both the graphite/epoxy and isotropic materials (with Poisson's ratio equal to zero). In Figure 1, the decay of σ_x with distance from the end $x = \ell$ is illustrated for both materials. In the latter case, the normal stress distribution reaches that of the Saint-Venant solution at approximately one width from the end. For the highly anisotropic material, the contrasting slow decay is evident — the characteristic decay length is about four times larger.

5. Torsionless Axisymmetric Problem for Circular Cylinders

Similar results to those described above for plane problems have also been obtained for axisymmetric problems. In [12] the torsionless axisymmetric problem for a circular cylinder, transversely isotropic about the axial direction, under the action of self-equilibrated end loads is considered. Using energy-decay inequalities, a result of the form (3.2) is obtained for the exponential decay of stresses and compared with previous results obtained in the isotropic case [14]. The exact decay rate is given by the root of a transcendental equation involving Bessel functions (see equation (9.5) of [12].) For highly anisotropic materials, with $\epsilon \equiv E_T/E_L \ll 1$, the decay rate k predicted in [12] is of the form

$$k = 0\left(\frac{\epsilon^{1/2}}{c}\right) \quad \text{as} \quad \epsilon \to 0, \tag{5.1}$$

where c is the radius of the cylinder. An asymptotic analysis of the transcendental equation for the exact decay rate also yields the result (5.1). Thus, in the present context, we establish again the slow attenuation of end effects for highly anisotropic materials. The order estimate (5.1) was confirmed experimentally in [2, 3, 8]. In Table 2 below, we provide numerical results

TABLE 2. Torsionless Axisymmetric Problem.

Material	(Exact Decay Rate) xc (from eqn. (9.5) of [12])	Lower bound for ck (from eqn. (7.24) of [12])
Magnesium	2.87	1.45
Graphite/epoxy	0.69 (real root)	0.46
Isotropic ($\nu = 1/4$)	2.70	1.43

for the exact decay rates and lower bounds obtained in [12]. Again we see that the decay rate for the graphite/epoxy material (where $\epsilon = 1/20$) is about four times slower than that for the isotropic case.

6. Sandwich Strips

Finally, we discuss briefly the case of plane deformation of a sandwich strip, composed of identical homogeneous isotropic face materials occupying two layers of equal thickness enclosing a dissimilar homogeneous isotropic core. For the case when the Young's modulus of the core is small compared with that for the face layers, it might be expected that the decay of Saint-Venant end effects should be much slower than that for a single strip. Such a result was demonstrated experimentally by Alwar [1] using photoelasticity techniques.

Recently a theoretical analysis of this issue was carried out in [6], through investigation of the exact decay rates characterized as eigenvalues (cf. Section 4 here). An exponential decay result of the form (3.2) was established. Using the notation $f = (4C_f)/(4C_f + 2C_c)$ for the volume fraction of face material (where $2C_f$, $2C_c$ denote the width of face layer and core layer respectively), and E_f, E_c, ν_f, ν_c for the respective Young's moduli and Poisson ratios, it is shown in [6] that when $E_f/E_c \gg 1$,

$$k \text{ (width)} \sim 2 \left[\frac{2 (f^3 - 3f + 3) (1 - \nu_f^2) E_c}{f^3 (1 - f) (1 + \nu_c) E_f} \right]^{1/2} , \tag{6.1}$$

yielding slow exponential decay for the case of a relatively soft inner core. If we assume that $E_f/E_c = 3600$, $f = 0.8$ as in [1], and let $\nu_f = \nu_c = 0.3$, then from (6.1) we obtain the estimate

$$k \text{ (width)} \sim 0.14 , \tag{6.2}$$

which is about thirty times smaller than the value for a homogeneous strip. Thus the neglect of Saint-Venant end effects is justified in this case only at a distance of about thirty widths from the ends. The photoelasticity studies of Alwar [1] for sandwich beams under concentrated loads lead to similar conclusions.

Acknowledgement

This work was supported by the U.S. National Science Foundation under Grants ENG 75-13643 and ENG 78-26071.

REFERENCES

[1] ALWAR R.S. – "Experimental verification of Saint-Venant's Principle in a sandwich beam". *AIAA J.*, 8 (1970): 160-162.

[2] ARRIDGE R.G.C., P.J. BARHAM, C.J. FARRELL and A. KELLER. – "The importance of end effects in the measurement of moduli of highly anisotropic materials". *J. of Materials Science*, 11 (1976): 788-790.

[3] ARRIDGE R.G.C. and M.J. FOLKES. – "Effect of sample geometry on the measurement of mechanical properties of anisotropic materials". *Polymer*, 17 (1976): 495-500.

[4] BIOT M.A. "Fundamental skin effect in anisotropic solid mechanics". *Int. J. Solids Structures*, 2 (1966): 645-663.

[5] CHOI I. and C.O. HORGAN. – "Saint-Venant's Principle and end effects in anisotropic elasticity". *J. Applied Mechanics* (Trans. ASME), 44 (1977): 424-430.

[6] CHOI I. and C.O. HORGAN. – "Saint-Venant end effects for plane deformation of sandwich strips". *Int. J. Solids Structures*, 14 (1978): 187-195.

[7] EVERSTINE G.C. and A.C. PIPKIN. – "Stress channelling in transversely isotropic elastic composites". *J. Appl. Math. Phys.* (ZAMP), 22 (1971): 825-834.

[8] FOLKES M.J. and R.G.C. ARRIDGE. – "The measurement of shear modulus in highly anisotropic materials: The validity of St. Venant's Principle". *J. Physics D: Appl. Phys.*, 8 (1975): 1053-1064.

[9] HASHIN Z. – "Theory of Composite Materials", in *Mechanics of Composite Materials*, p. 201 (Proc. 5th Symp. on Naval Structural Mechanics, 1967, edited by F.W. Wendt, H. Liebowitz and N. Perrone) Pergamon Press, 1970.

[10] HORGAN C.O. – "On Saint-Venant's Principle in plane anisotropic elasticity". *J. of Elasticity*, 2 (1972): 169-180.

[11] HORGAN C.O. – "Some remarks on Saint-Venant's Principle for transversely isotropic composites". *J. of Elasticity*, 2 (1972): 335-339.

[12] HORGAN C.O. "The axisymmetric end problem for transversely isotropic circular cylinders". *Int. J. Solids Structures*, 10 (1974): 837-852.

[13] KNOWLES J.K. – "On Saint-Venant's Principle in the two-dimensional linear theory of elasticity". *Archive for Rational Mechanics and Analysis*, 21 (1966): 1-22.

[14] KNOWLES J.K. and C.O. HORGAN. – "On the exponential decay of stresses in circular elastic cylinders subject to axisymmetric self-equilibrated end loads". *Int. J. Solids Structures*, 5 (1969): 33-50.

[15] MANSFIELD E.H. and D.R. BEST. – "The concept of load diffusion length in fibre reinforced composites". *Aeronautical research council current paper*, 1338. Her Majesty's Stationery Office, London, 1976.

[16] PAGANO N.J. and J.C. HALPIN. – 'Influence of end constraint in the testing of anisotropic bodies". *J. of Composite Materials*, 2 (1968): 18-31.

[17] PAGANO N.J. and J.M. WHITNEY. – "Geometric design of composite cylindrical characterization specimens". *J. of Composite Materials*, 4 (1970): 360-378.

[18] SEITZ F. and T.A. READ. – "Theory of the Plastic Properties of Solids I". *J. Appl. Phys.*, 12 (1941): 100-118.

[19] SPENCER A.J.M. – *"Deformations of Fibre-Reinforced Materials"*, Oxford University Press, 1972.

[20] TOUPIN R.A. – "Saint-Venant's Principle". *Archive for Rational Mechanics and Analysis*, 18 (1965): 83-96.

[21] WAN F.Y.M. – "An eigenvalue problem for a semi-infinite pretwisted strip". *Studies in Appl. Math.*, 54 (1975): 351-358.

RÉSUMÉ

Le principe de Saint-Venant dans la théorie de l'élasticité anisotrope.

Dans cet article, nous présentons un sommaire des recherches théoriques et expérimentales sur l'analyse du principe de Saint-Venant dans la théorie de l'élasticité anisotrope, en soulignant les matériaux fortement anisotropes ou composites. Nous montrons que, pour de tels matériaux, l'application courante du principe de Saint-Venant n'est pas valable en général. Des illustrations sont données pour les problèmes plans des bandes anisotropes et des bandes sandwich, ainsi que pour les problèmes axisymétriques. Les résultats ont des implications pour la mesure précise des propriétés des matériaux anisotropes.

DISCUSSION

QUESTION BY I. MÜLLER: Could you please comment on the physical interpretation of the fact that the rate of decay of the inhomogeneity in the stress field is so much different in an isotropic and in an anisotropic material ?

REPLY BY C.O. HORGAN: Yes. Consider, for example, the highly anisotropic transversely-isotropic case discussed earlier. In the limit as the anisotropic small parameter $\epsilon_t \to 0$, we have the case of *rigid* fibers. Physically, then, conditions at one end of a fiber are transmitted without decay along the fiber. Mathematically, one is considering the idealized constraint of *inextensibility* in the fiber direction. As pointed out in my lecture, the governing equation then ceases to be elliptic. In fact the equation becomes *parabolic,* with the fibers acting as a parallel family of characteristics. Boundary data is then transmitted without attenuation along the characteristics. The fact that the governing equations in inextensible elasticity are *parabolic* seems not to have been pointed out in the literature.

COMMENT BY R.G.C. ARRIDGE: I should like to elaborate on the experimental results obtained in our laboratory, as mentioned in Horgan's lecture. The importance of the effects discussed in the paper has been demonstrated by measurements made at the H.H. Wills Physics Laboratory, Bristol, U.K. by the discusser and his collaborators. In work by M.J. Folkes and the discusser the compliance S_{44} of the copolymer "Kraton" was measured by a torsion pendulum. Kraton is a copolymer consisting of 20 % by volume polystyrene embedded as rods of 150 Å diameter, with hexagonal spatial symmetry, in a matrix of polybutadiene. It is therefore a 'molecular' fibre composite in which E, the longitudinal Young's modulus is very much greater than the longitudinal shear modulus G.

In the experiments of Arridge and Folkes the value of S_{44} was found to depend strongly upon the aspect ratio (length/diameter) of the sample (Fig. A). This effect was explained (Ref. [1]) by considering the sample as made of a central portion and two end portions, the stiffness of the end regions being greater than that of the central region. Constraints at the grips were considered, following Horgan, to propagate for a distance $\ell = c \sqrt{E/G} \, d$, where d is the maximum lateral dimension of the sample and c a constant. Satisfactory agreement between ℓ, as determined experimentally and by calculation from a knowledge of E and G was obtained.

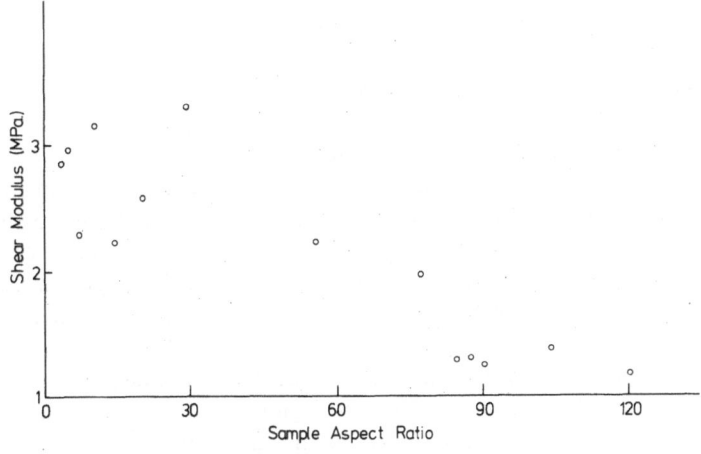

Figure A.

In the second example (Ref. [2]) the material used was linear polyethylene (Rigidex 50) drawn to a draw ratio of 28 by Barham, following the method of Capaccio and Ward. Young's modulus at 0.1 % strain was determined using the grip separation (corrected for machine effects) to measure the mean strain. In Figure B it is shown that the apparent modulus E increases with aspect ratio approaching a constant value only at aspect ratios of about 100.

In such highly drawn polyethylene it is known from X-ray diffraction that chain orientation is virtually perfect and therefore that the measured elastic moduli should approach the theoretical crystal moduli. The longitudinal Young's modulus for a crystal of polyethylene should

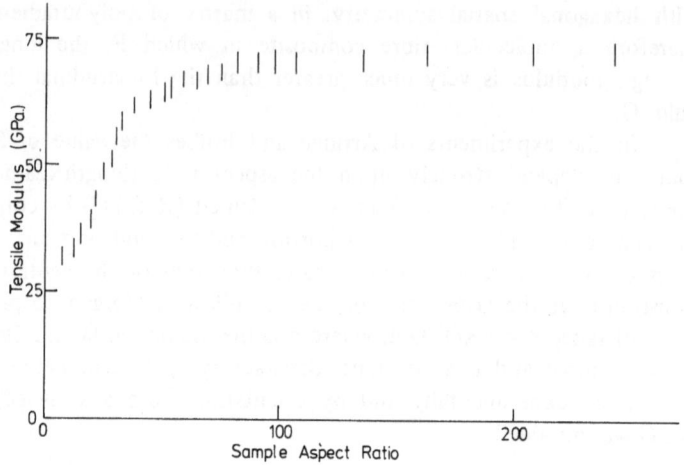

Figure B.

theoretically lie between the values of about 250 GPa and 320 GPa, while the shear modulus is about 1 GPa. The ratio $\sqrt{E/G}$ for this material is therefore very large and stress channelling along the outer surface of the sample may be expected, because of the mode of gripping. This stress channelling means that uniform stress across the fibre cross section will not be obtained until the aspect ratio exceeds a value of about 32.

The measured Young's modulus for a sample gripped in the usual way will therefore increase with aspect ratio in just the way that is found experimentally.

The importance of the modified St. Venant principle in the determination of the elastic constants of highly anisotropic materials cannot be over-emphasised. This is particularly true for the currently studied ultra-oriented polymer fibres such as polyethylene, polypropylene, Kevlar and graphite.

[1] FOLKES M.J. and R.G.C. ARRIDGE. – *J. Phys. D.*, **8** (1975): 1053-1064.
[2] ARRIDGE R.G.C., P.J. BARHAM, C.J. FARRELL, A. KELLER. – *J. Materials Sci.*, **11** (1976): 788-790.

REPLY BY C.O. HORGAN: It was a great pleasure for me to finally meet Mr. Arridge at this meeting. We have corresponded by mail on these issues. The author was gratified to find experimental confirmation of the earlier theoretical predictions. My subsequent work has been largely motivated by these experimental results. We hope to plan a joint effort in this area in the near future.

Interprétation d'Essais de Traction, de Flexion Trois Points et de Dilatométrie de Plaques Stratifiées Epoxy/Carbone

J.L. Seichepine, A. Vautrin et D. Guitard
Laboratoire d'Énergétique et de Mécanique Théorique et Appliquée, Nancy, France.

LISTE DES SYMBOLES

(x, y, z) : Repère cartésien lié à une plaque stratifiée.

(a, b, z) : Repère cartésien lié à une monocouche.

$\phi^{(p)}$: Angle des directions x et a.

$e^{(p)}$: Epaisseur d'une couche.

$z^{(p)}$: Cote du plan moyen d'une couche par rapport au plan moyen de la plaque.

$I^{(p)} = \dfrac{(e^{(p)})^3}{12}$

$S_{ijkl}^{(p)}$: Composantes du tenseur des complaisances élastiques d'une couche,

$\alpha_{ij}^{(p)}$: Coefficient de dilatation thermique d'une couche.

ϵ_{ij} : Composantes du tenseur des déformations linéarisé.

σ_{ij} : Composantes du tenseur des contraintes.

ΔT : Champ de température uniforme.

$Q_{ijkl}^{(p)}$: Composantes du tenseur des rigidités élastiques d'une couche.

u_x^0, u_y^0, w : Composantes du déplacement d'un point du plan moyen.

Pour l'ensemble des grandeurs définies ci-dessous, $(i, j, k, l) \in (1, 2)^4$.

$\epsilon_{ij}^0 = \dfrac{1}{2}(u_{i,j}^0 + u_{j,i}^0)$: Déformations de membrane.

$K_{ij} = -w_{,ij}$: Courbures.

$N_{ij} = \int_{-h/2}^{h/2} \sigma_{ij}\, dz$: Efforts de membrane.

$M_{ij} = \int_{-h/2}^{h/2} \sigma_{ij}\, z\,dz$: Moments de flexion-torsion.

$A_{ijkl} = \sum_{p=1}^{n} Q_{ijkl}^{(p)}\, e^{(p)}$: Rigidités de membrane.

$D_{ijkl} = \sum_{p=1}^{n} Q_{ijkl}^{(p)}\, (z^{(p)^2} e^{(p)} + I^{(p)})$: Rigidités de flexion-torsion.

$B_{ijkl} = \sum_{p=1}^{n} Q_{ijkl}^{(p)}\, z^{(p)}\, e^{(p)}$: Termes de couplage mécanique.

$E_{ij} = \sum_{p=1}^{n} Q_{ijkl}^{(p)}\, \alpha_{kl}^{(p)}\, e^{(p)}$: Termes traduisant la contribution thermique aux contraintes de membrane.

$F_{ij} = \sum_{p=1}^{n} Q_{ijkl}^{(p)}\, \alpha_{kl}^{(p)}\, e^{(p)}\, z^{(p)}$: Termes traduisant la contribution thermique aux moments de flexion-torsion.

h   : Epaisseur des plaques.

$\overline{A}_{ijkl}$: Souplesse de membrane.

$\overline{D}_{ijkl}$: Souplesse de flexion-torsion.

$R_{ij} = A_{ijkl}\, E_{kl}$: Coefficient de dilatation thermique relatif à la plaque.

$(1, 2, z)$: Repère cartésien lié à la plaque,

θ : Angle des directions 1 et x.

J_{1111}, J_{2211} : Complaisances élastiques mesurées en traction,

L   : Longueur entre appuis au cours de l'essai de flexion 3 points,

ℓ : largeur de l'éprouvette

F   : Force appliquée au centre de l'éprouvette lors de l'essai de flexion 3 points.

y   : Flèche produite par la force F.

E   : Rigidité à la flexion.

L   : Longueur de l'éprouvette de dilatométrie.

1. Introduction

Le comportement thermoélastique de plaques stratifiées équilibrées à fibres croisées est prévu à partir :

— du comportement thermoélastique du composite à fibres unidirectionnelles.

— des paramètres géométriques d'arrangement des couches, en s'appuyant sur la théorie classique des plaques stratifiées.

Après un bref rappel théorique, le modèle, adapté aux matériaux étudiés (Epoxy/Carbone), conduit à la définition des souplesses de membrane et de flexion-torsion, ainsi qu'au tenseur des dilatations thermiques.

L'identification de ces paramètres est réalisée à l'aide des mesures en traction et dilatométrie effectuées sur le composite à fibres unidirectionnelles constituant les couches du stratifié. Compte tenu de l'interprétation des résultats expérimentaux obtenus en traction, flexion trois points et dilatométrie sur le matériau à fibres croisées à 45° et 90°, on apporte une vérification expérimentale des prévisions du modèle.

2. Le modèle théorique du comportement thermoélastique des stratifiés

Les plaques planes de stratifiés à fibres croisés sont constituées de n couches à fibres unidirectionnelles supposées macroscopiquement homogènes, thermoélastiquement orthotropes, parfaitement soudées et d'orientations relatives connues.

2.1. *Les données*

a) Un système de coordonnées cartésiennes (x, y, z) est choisi pour repérer un point courant de la plaque, z étant la cote par rapport au plan moyen.

b) Chaque couche de rang (p) admet un repère principal d'orthotropie associé aux coordonnées cartésiennes (a, b, z) ; soit $\phi^{(p)}$ l'angle formé par la direction **a** et la direction x liée à la plaque.

c) L'épaisseur $e^{(p)}$ et la cote $z^{(p)}$ du plan moyen de couche par rapport au plan moyen de la plaque sont, pour chaque couche, des données du problème. On notera $I^{(p)} = \dfrac{(e^{(p)})^3}{12}$.

d) On suppose connus dans le référentiel d'orthotropie de couche (a, b, z) les *seules* composantes du tenseurs des complaisances élastiques notées $S_{ijkl}^{(p)}$ et les *seuls* coefficients de dilatations thermiques notés $\alpha_{ij}^{(p)}$, nécessaires pour définir l'état des déformations dans le plan de couche, résultant d'un état de contrainte supposé plan et d'un champ de température supposé *uniforme.*

$$\epsilon_{ij} = S^{(p)}_{ijkl}\ \sigma_{kl} + \Delta T\ \alpha^{(p)}_{ij} \tag{1}$$

avec $(i,j,k,l) \in (a,b)^4$

En procédant au changement de base, correspondant à une *rotation* $\phi^{(p)}$ autour de **z** suivie d'une *inversion,* on accède à l'expression des contraintes planes, dans le repère lié à la plaque.

$$\sigma_{ij} = Q^{(p)}_{ijkl}\ (\epsilon_{kl} - \Delta T\ \alpha^{(p)}_{kl}) \tag{2}$$

avec $(i,j,k,l) \in (x,y)^4$

Remarque : l'ensemble de définition des indices permet tout au long du texte de savoir si l'on travaille dans le repère de couche, de plaque, ou d'échantillon.

2.2. *Loi de comportement thermoélastique du stratifié*

Dans la suite on se place dans le cadre des hypothèses de la théorie classique des plaques stratifiées [3, 10] et on admet que le champ de température ΔT est uniforme [5].

La loi de comportement du stratifié s'écrit alors :

$$N_{ij} = A_{ijkl}\ \epsilon^0_{kl} + B_{ijkl}\ K_{kl} - E_{ij}\ \Delta T$$
$$M_{ij} = B_{ijkl}\ \epsilon^0_{kl} + D_{ijkl}\ K_{kl} - F_{ij}\ \Delta T \tag{3}$$

avec $(i,j,k,l) \in (x,y)^4$.

Les relations de définition des grandeurs utilisées sont regroupées dans la liste des symboles.

3. Les matériaux

3.1. *Les composants*

Les plaques planes stratifiées d'épaisseur $h = 1,5$ mm sont obtenues par polymérisation de douze couches ($n = 12$) isoépaisseur d'un préimprégné à fibres unidirectionnelles.

Une résine Epoxy NARMCO 5208 est associée à des fibres "haute résistance" de carbone TORAYCA type T 300 B 6000 conditionnées en mèches non torsadées de 6000 filaments de 7μ de diamètre [8].

Les trois types de composites testés correspondent à des angles de croisement des fibres $2\phi = 0$; 45 et 90 degrés obtenus en orientant les 12 couches :

$$\phi^{(p)} = \phi^{(13-p)} \qquad \text{avec} \quad p \in (1, \ldots, 6)$$

$$\phi^{(p)} = -\phi^{(p+1)} \qquad \text{avec} \quad p \in (1, \ldots, 5) \tag{4}$$

L'ensemble des couches sont donc groupées en deux familles définies par : $\phi^{(1)} = -\phi^{(2)}$.

3.2. *Conséquences sur la loi de comportement thermoélastique*

Les stratifiés sont dits *équilibrés* [6]. Du fait de la disparition des termes de couplage mécanique ($B_{ijkl} = 0$) et des contributions thermiques aux flexions torsions ($F_{ij} = 0$) les relations (3) se réduisent à :

$$N_{ij} = A_{ijkl} \, \epsilon^0_{kl} - E_{ij} \, \Delta T$$

$$M_{ij} = D_{ijkl} \, K_{kl} \tag{5}$$

avec :

$$A_{ijkl} = h \left[\frac{1}{2} Q^{(1)}_{ijkl} + \frac{1}{2} Q^{(2)}_{ijkl} \right] \tag{6}$$

$$D_{ijkl} = \frac{h^3}{12} \left[\frac{5}{8} Q^{(1)}_{ijkl} + \frac{3}{8} Q^{(2)}_{ijkl} \right] \tag{7}$$

$$E_{ij} = h \left[\frac{1}{2} Q^{(1)}_{ijkl} \, \alpha^{(1)}_{kl} + \frac{1}{2} Q^{(2)}_{ijkl} \, \alpha^{(2)}_{kl} \right] \tag{8}$$

avec $(i, j, k, l) \in (x, y)^4$.

Remarque 1 : χ Les axes x et y du référentiel de la plaque sont des axes de symétrie pour les rigidités de membrane.

χ Cette propriété n'est plus vraie en ce qui concerne les rigidités de flexion-torsion du fait que les couches externes (p = 1 ; 12) sont alors privilégiées (les coefficients $\frac{5}{8}$ et $\frac{3}{8}$ sont liés au nombre n = 12 des couches).

Pour des raisons pratiques, on introduit les souplesses de membrane $\overline{A}_{ijkl}$ et de flexion-torsion $\overline{D}_{ijkl}$ telles que :

$$\overline{A}_{ijkl} \, A_{klpq} = \delta_{ip} \, \delta_{jq}$$

$$\overline{D}_{ijkl} \, D_{klpq} = \delta_{ip} \, \delta_{jq} \tag{9}$$

et la loi de comportement s'écrit :

$$\epsilon_{ij}^0 = \bar{A}_{ijkl}\, N_{kl} + R_{ij}\, \Delta T$$
$$K_{ij} = \bar{D}_{ijkl}\, M_{kl} \tag{10}$$

pour $(i, j, k, l) \in (x, y)^4$

et avec :

$$R_{ij} = \bar{A}_{ijkl}\, E_{kl} \tag{11}$$

Remarque 2 : Dans le cas particulier d'un angle de croisement des fibres nul ($2\phi = 0$ degré) les référentiels de couche et de plaque sont confondus et il vient :

$$\epsilon_{ij}^0 = \frac{1}{h}\, S_{ijkl}^{(p)}\, N_{kl} + \alpha_{ij}^{(p)}\, \Delta T$$

$$K_{ij} = \frac{12}{h^3}\, S_{ijkl}^{(p)}\, M_{kl}$$

Les axes x et y étant dans ce cas particulier, axes de symétrie pour les souplesses de flexion-torsion.

4.   Aspect expérimental

Les essais mécaniques mis en œuvre sont de trois sortes :

 — Essais de traction ;
 — Essais de flexion trois points norme ASTM n° D790 ;
 — Essais dilatométriques.

4.1. *Les éprouvettes*

Les éprouvettes sont des bandes de 10 mm de large prélevées dans les plaques de stratifiés. Le référentiel lié à chaque éprouvette noté $(1, 2, z)$ (où 1 est le grand axe) se déduit du référentiel de plaque (x, y, z) par une rotation connue d'angle θ autour de z.

On dispose pour chaque stratifié de lots d'éprouvettes d'orientation θ fixée :

TABLEAU 1. Eprouvettes d'essais.

2ϕ : angle de croisement en degrés	θ : orientation des éprouvettes en degrés
$2\phi = 0$	0 ; 45 ; 90
$2\phi = 45$	0 ; 15 ; 22,5 ; 30 ; 45 ; 60 ; 75 ; 90
$2\phi = 90$	0 ; 15 ; 30 ; 45

Chaque lot comprend :
   2 éprouvettes de traction (10 mm x 250 mm)
   3 éprouvettes de flexion (10 mm x 70 mm)
   1 éprouvette de dilatation (10 mm x50 mm)

4.2. *Les mesures et grandeurs déduites*

   a) En traction : on mesure l'effort axial N_{11}, les déformations ϵ_{11}^0 longitudinales et ϵ_{22}^0 transversales dont on déduit :

$$J_{1111} = h \frac{\epsilon_{11}^0}{N_{11}} \quad ; \quad J_{2211} = h \frac{\epsilon_{22}^0}{N_{11}}$$

Les grandeurs sont interprétées comme les valeurs effectives des composantes du tenseur des souplesses de membrane exprimées dans le référentiel $(1, 2, z)$ lié à l'éprouvette soit :

$$J_{1111} = h \overline{A}_{1111} \quad ; \quad J_{2211} = h \overline{A}_{2211} \tag{14}$$

   b) en flexion : de la mesure de la flèche y résultant d'une force F imposée, compte-tenu de la largeur 1 de l'éprouvette et de la distance L entre appuis, on déduit une constante élastique E par une formule classique de résistance des matériaux

$$E = \frac{L^3}{4 \, lh^3} \frac{F}{y} \tag{15}$$

Cette grandeur est interprétée comme la mesure de l'une des composantes du tenseur des souplesses en flexion-torsion exprimé dans le référentiel de l'éprouvette :

$$\frac{1}{E} = \frac{h^3}{12} \overline{D}_{1111} \tag{16}$$

c) en dilatométrie : la mesure de l'élongation longitudinale de l'éprouvette due à une variation uniforme ΔT permet de chiffrer la composante R_{11} du tenseur des coefficients de dilatation thermique plan.

$$R_{11} = \frac{\Delta L}{L \Delta T} \tag{17}$$

5. Comparaison des prévisions et des mesures

5.1. Comportement du monocouche

Le comportement thermoélastique plan du monocouche est déduit des seuls essais de traction et de dilatométrie effectués sur le stratifié à couches parallèles ($2\phi = 0$ degré) au moyen des éprouvettes orientées suivantes $\theta = 0$; 45 ; 90 degrés, soit, en tenant compte de la remarque 2 § 3.2 :

$$\begin{vmatrix} S_{aaaa} & S_{aabb} & S_{aaab} \\ S_{bbaa} & S_{bbbb} & S_{bbab} \\ S_{abaa} & S_{abbb} & S_{abab} \end{vmatrix} = \begin{vmatrix} 6,9 & -0,6 & 0 \\ -2,2 & 90,8 & 0 \\ 0 & 0 & 35,8 \end{vmatrix} \quad \text{en } (10^{12} \text{ Pa})^{-1} \tag{18}$$

et

$$\begin{vmatrix} \alpha_{aa} & \alpha_{ab} \\ \alpha_{ba} & \alpha_{bb} \end{vmatrix} = \begin{vmatrix} 0,30 & 0 \\ 0 & 26,7 \end{vmatrix} \quad \text{en } \mu\epsilon/^{\circ}C \tag{19}$$

Les résultats (18) et (19) sont considérés comme des données pour l'exploitation du modèle défini § 2 et 3.

Remarques :

La non symétrie $S_{aabb} \neq S_{bbaa}$ observée expérimentalement sera conservée tel quel, bien que contradictoire à l'hypothèse d'existence d'un potentiel élastique forme quadratique définie positive des composantes du tenseur des contraintes.

Les termes nuls en (18) et (19) résultent de l'hypothèse d'orthotropie qui n'a pas fait l'objet d'une confrontation expérimentale.

5.2. Calcul du comportement des stratifiés

La démarche suivie pour calculer le comportement des stratifiés à fibres croisées se décompose en 5 étapes :

1) Inversion du tenseur complaisance du monocouche qui permet le passage de S_{ijkl} à Q_{ijkl} dans le référentiel de couche $(i,j,k,l) \in (a,b)^4$.

2) Pour un croisement des fibres 2ϕ, en posant $\phi^{(1)} = \phi$; $\phi^{(2)} = -\phi$ on procède aux changements de base donnant :

$$Q_{ijkl}^{(1)}, Q_{ijkl}^{(2)} \quad \text{avec } (i, j, k, l) \in (x, y)^4$$

dans le référentiel de plaque.

3) Compte tenu des relations (6), (7), (8) les grandeurs suivantes sont calculées :

$$\frac{1}{h} A_{ijkl} \quad ; \quad \frac{12}{h^3} D_{ijkl} \quad ; \quad \frac{1}{h} E_{ij}$$

4) Par inversion des rigidités on accède aux souplesses de membrane et de flexion-torsion

$$h \cdot \overline{A}_{ijkl} \quad ; \quad \frac{h^3}{12} \overline{D}_{ijkl}$$

5) Les coefficients de dilatation thermique des plaques se déduisent des résultats des étapes 3) et 4) soit

$$R_{ij} = \overline{A}_{ijkl} E_{kl}$$

Ces résultats calculés acquis dans le référentiel de plaque (x, y, z) relatifs à $2\phi = 45$ et 90 degrés sont indiqués dans le Tableau n° 2.

5.3. *Prévision des grandeurs déduites expérimentalement*

Les grandeurs A_{1111}, A_{2211}, D_{1111} et R_{11} exprimées dans le référentiel $(1, 2, z)$ lié à chaque éprouvette sont déterminées à partir des composantes thermoélastiques calculées au paragraphe précédent dans le référentiel de plaque (x, y, z). On procède à un changement de base, rotation d'angle θ autour de z. Soit :

$$\overline{A}_{1111} = \overline{A}_{xxxx} C^4 + \overline{A}_{yyyy} S^4 + (\overline{A}_{xxyy} + \overline{A}_{yyxx} + 4\overline{A}_{xyxy}) C^2 S^2 \quad (20)$$

$$\overline{A}_{2211} = \overline{A}_{yyxx} C^4 + \overline{A}_{xxyy} S^4 + (\overline{A}_{xxxx} + \overline{A}_{yyyy} - 4\overline{A}_{xyxy}) C^2 S^2 \quad (21)$$

$$\overline{D}_{1111} = \overline{D}_{xxxx} C^4 + \overline{D}_{yyyy} S^4 + (\overline{D}_{xxyy} + \overline{D}_{yyxx} + 4\overline{D}_{xyxy}) C^2 S^2 +$$
$$[(\overline{D}_{xxxy} + \overline{D}_{xyxx}) C^2 + (\overline{D}_{yyxy} + \overline{D}_{xyyy}) S^2] 2CS \quad (22)$$

$$R_{11} = R_{xx} C^2 + R_{yy} S^2$$

avec $C = \cos \theta$; $S = \sin \theta$.

TABLEAU 2. Composantes thermoélastiques du comportement des stratifiés dans le rère de plaque x, y, z

Angle de croisement (degré)	Souplesses de membrane $(10^{12}\ Pa)^{-1}$ $\begin{vmatrix}\bar{A}_{xxxx} & \bar{A}_{xxyy} & \bar{A}_{xxxy}\\ \bar{A}_{yyxx} & \bar{A}_{yyyy} & \bar{A}_{yyxy}\\ \bar{A}_{xyxx} & \bar{A}_{xyyy} & \bar{A}_{xyxy}\end{vmatrix}$ Calcul	Mesure en traction	Souplesses de flexion-torsion $(10^{12}\ Pa)^{-1}$ $\begin{vmatrix}\bar{D}_{xxxx} & \bar{D}_{xxyy} & \bar{D}_{xxxy}\\ \bar{D}_{yyxx} & \bar{D}_{yyyy} & \bar{D}_{yyxy}\\ \bar{D}_{xyxx} & \bar{D}_{xyyy} & \bar{D}_{xyxy}\end{vmatrix}$ Calcul	Mesure en flexion	Coefficient de dilatation thermique $(10^6\ ^\circ C)^{-1}$ $\begin{vmatrix}R_{xx} & R_{xy}\\ R_{xy} & R_{yy}\end{vmatrix}$ Calcul	Mesure dilatométrie
$2\phi = 0$		6,86 −0,62 — −2,17 90,8 — — — 35,8	6,86 −0,6 0 −2,17 90,8 0 0 0 35,8	7,48 — — — 95,1 — — — —		0,29 26,7
$2\phi = 45$	11,1 −12,4 0 −13,7 80,9 0 0 0 11,1	11,8 −13,5 — 14,2 79,0 — — — 11,0	11,4 −12,2 1,94 −13,5 81,0 0,92 1,89 0,87 11,5	13,4 — — — 88,8 — — — —	−1,59 0 0 19,8	−2,58 — — −20,0
$2\phi = 90$	42,0 −29,7 0 −29,7 42,0 0 0 0 6,55	47,5 — — −33,8 — — — — 6,32	42,3 −29,4 1,46 −29,4 42,3 1,46 1,41 1,41 6,87	45,2 — — — — — — — —	2,66 0 0 2,66	2,27 — — —

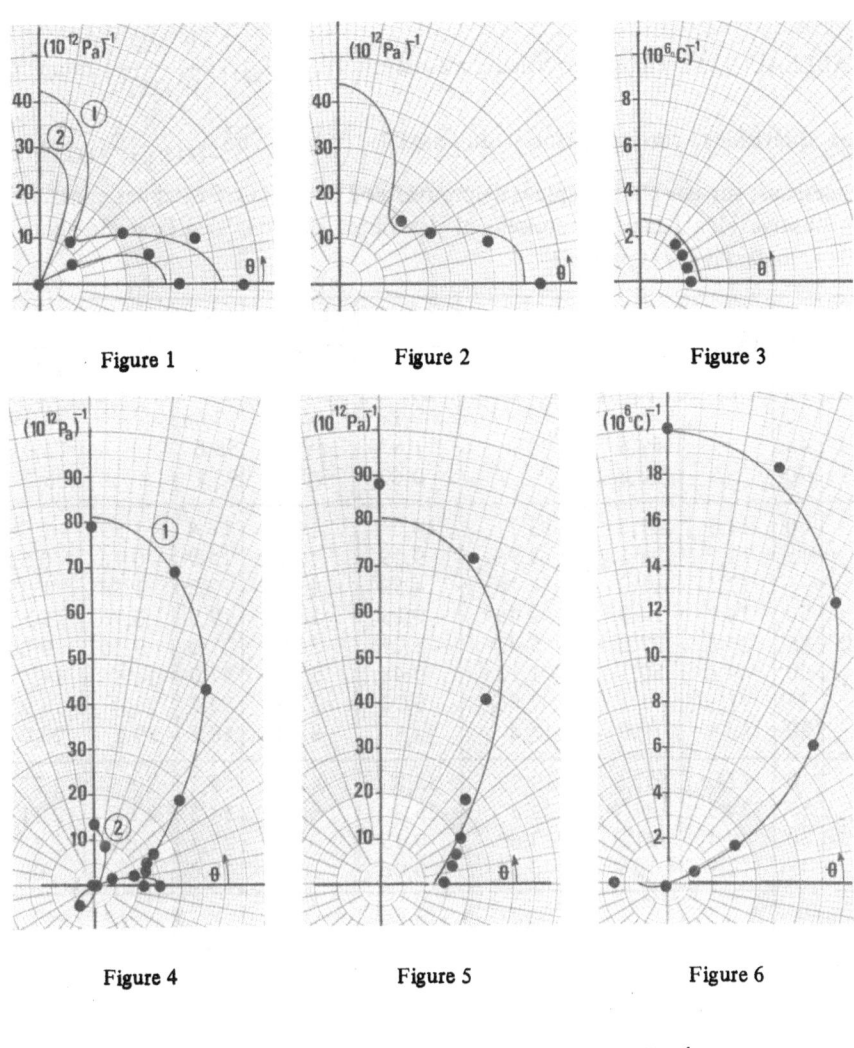

Figure 1   Figure 2   Figure 3

Figure 4   Figure 5   Figure 6

Figure 1. $2\Phi = 90$ degrés ;
    courbe 1 $h\,\bar{A}_{1111}$,   ● J_{1111}
    courbe 2 $-h\,\bar{A}_{2211}$,   ● $-J_{2211}$

Figure 2. $2\Phi = 90$ degrés ;
    courbe $\dfrac{h3}{12}\,\bar{D}_{1111}$,   ● $\dfrac{1}{E}$

Figure 3. $2\Phi = 90$ degrés ;
    courbe R_{11},   ● $\dfrac{\Delta L}{L\Delta T}$

Figure 4. $2\Phi = 45$ degrés ;
    courbe 1 $h\,\bar{A}_{1111}$,   ● J_{1111}
    courbe 2 $-h\,\bar{A}_{2211}$,   ● $-J_{2211}$

Figure 5. $2\Phi = 45$ degrés ;
    courbe $\dfrac{h3}{12}\,\bar{D}_{1111}$,   ● $\dfrac{1}{E}$

Figure 6. $2\Phi = 45$ degrés ;
    courbe R_{11},   ● $\dfrac{\Delta L}{L\Delta T}$

TABLEAU 3. Ensemble des valeurs mesurées $\left(J_{1111}, J_{2211}, \frac{1}{E}, \frac{\Delta L}{L\Delta T}\right)$ comparées aux prévisions correspondantes du modèle $\left(h\bar{A}_{1111}, h\bar{A}_{2211}, \frac{h^3}{12}\bar{D}_{1111}, R_{11}\right)$. Ci-dessous, sur des photographies d'éprouvettes à $\theta = 0$ degré rompues en traction, on observe bien l'angle de croisement des fibres ($2\phi = 45$ degrés sur la photo 1, $2\phi = 90$ degrés sur la photo 2).

2ϕ	θ	$h\bar{A}_{1111}$	J_{1111}	$-h\bar{A}_{2211}$	$-J_{2211}$	$\frac{h^3}{12}\bar{D}_{1111}$	$\frac{1}{E}$	R_{11}	$\frac{\Delta L}{L\Delta T}$
0	0		6,86		2,17	6,86	7,48		0,29
	45		59,8		11,8	59,9	53,6		11,4
	90		90,8		0,62	90,8	95,1		26,7
45	0	11,1	11,8	13,7	14,2	11,4	13,4	− 1,59	− 2,58
	15	11,2	11,1	8,98	9,19	13,4	15,9	− 0,16	− 0,51
		12,1	11,6	4,26	4,50	15,1	17,2	1,54	1,25
	30	14,7	14,3	− 0,48	− 0,15	18,2	19,9	3,76	3,53
	45	27,6	27,2	− 5,42	− 5,27	31,0	27,0	9,10	9,04
	60	49,6	50,1	− 1,13	− 0,48	52,1	47,0	14,5	14,4
	75	71,6	71,1	7,86	9,23	72,8	73,2	18,4	18,9
	90	80,9	79,0	12,4	13,5	81,0	88,8	19,8	20,0
90	0	42,0	47,5	29,7	33,8	42,3	45,2	2,66	2,27
	15	34,5	37,4	22,3	25,3	36,5	34,8	2,66	2,11
	30	20,1	21,3	7,69	9,15	23,1	22,4	2,66	2,15
	45	12,7	12,3	0,36	0,36	16,2	18,0	2,66	2,09
degré	$(10^{12}\ Pa)^{-1}$							$(10^6\ °C)^{-1}$	

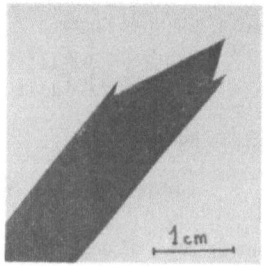

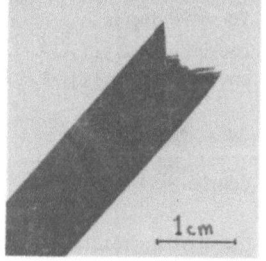

Les résultats expérimentaux définis par les relations (14), (16) et (17) sont comparés aux prévisions données par les relations (20), (21), (22) et (23) et présentés dans le Tableau n° 3 et sur les Figures n° 1 à 6.

6.   Discussion et conclusion

La première remarque est de constater que les écarts relatifs entre une grandeur élastique calculée et la mesure de cette même grandeur sont dans le plus mauvais cas de 6 % par un angle de croisement de fibres $2\phi = 45$ degrés et de 12 % pour $2\phi = 90$ degrés.

Lorsque l'on souligne que la comparaison "calcul-mesure" implique l'intégration des incertitudes portant sur 35 valeurs d'origine expérimentale pour les grandeurs élastiques, et de 40 valeurs pour les coefficients de dilatation thermique, on conçoit que les écarts constatés ci-dessus restent dans la plage d'incertitude expérimentale.

Nous concluons : le comportement thermoélastique de plaques stratifiées, équilibrées, d'époxy/carbone à fibres croisées est prévu de façon satisfaisante à partir des caractéristiques du composite à fibres unidirectionnelles en utilisant la théorie classique des plaques stratifiées.

La qualité de la vérification expérimentale présentée ici résulte de certains choix techniques, de l'utilisation de certaines méthodes propres développées par les auteurs [9], et au respect de conseils que donne la littérature quant à la réalisation et à l'interprétation de tests mécaniques pratiqués sur des matériaux renforcés par fibres.

Le choix de stratifiés équilibrés, en éliminant les possibilités de couplage membrane, flexion-torsion [6] est un élément simplificateur.

Pour les essais de traction :

L'utilisation d'éprouvettes de traction ayant un facteur de forme (longueur/largeur) égal à 25 permet de négliger les effets d'extrémité [1] et de minimiser l'influence de la rigidité à la rotation des machoires, particulièrement dans le cas d'éprouvettes taillées en dehors des axes de symétrie des plaques [4] [7].

Les mesures des élongations longitudinales et transverses effectuées au moyen de jauges résistives, placées au centre des éprouvettes semble la technique extensométrique la mieux adaptée [11].

Le stratifié $2\phi = 90$ degrés a été testé par un essai de traction à vitesse imposée. La méthode de traction modulée [9] a été appliquée au stratifié $2\phi = 45$ degrés. La concordance meilleure (6 %) observée entre calcul et mesure est considérée comme un indice supplémentaire de l'intérêt technique de cette méthode.

Pour les essais en flexion :

En s'appuyant sur les travaux de J.M. Whitney [12], compte tenu du facteur de forme des éprouvettes de flexion $\dfrac{L}{\ell} = S$ (où L est la distance entre appuis) et du nombre des couches (n = 12) constituant le composite, l'interprétation des mesures données au paragraphe 4.2.b est raisonnable.

L'observation attentive des mesures confirme la non-symétrie des rigidités de flexion par rapport aux axes (x , y , z) du référentiel de plaque.

Pour les essais en dilatométrie :

L'anisotropie du tenseur des dilatations d'origine thermique est fortement sensible à l'angle de croisement des fibres.

Dans le cas du stratifié à $2\phi = 90$ degrés, l'équivalence des deux axes de référence de la plaque conduit à une isotropie dans son plan. Ce résultat théorique est nettement confirmé par l'expérience (cf. Fig. n° 3).

Pour le stratifié à $2\phi = 45$ degrés, le coefficient de dilatation thermique longitudinal, fortement positif suivant la bissectrice du grand angle des fibres ($\theta = 90$ degrés) devient sensiblement négatif suivant l'axe de la plaque. Cette particularité, prévue par le modèle, observée par ailleurs [2], est confirmée ici expérimentalement.

Remerciements

Les auteurs remercient le Centre de Recherches du Bouchet (Société Nationale des Poudres et Explosifs) et tout particulièrement Madame Gourdin et Monsieur Dumas qui apportèrent à ce travail une contribution capitale en fournissant le matériau étudié.

BIBLIOGRAPHIE

[1] CHOI Y. et C.O. HORGAN. – "Saint Venant's Principle and end effects in anisotropic elasticity". *Transactions of the ASME* (1977) 424.

[2] CLOUET M. et M. DUMAS. – "Etude dilatométrique de matériaux composites Résine époxy-fibres de carbone". *Note technique n° 2014 du CRB* (SNPE), 1974.

[3] Engineering Sciences Data Unit, "Stiffnesses of laminated flat plates", *Item Number 75002*, 1975.

[4] PAGANO N.J. and J.C. HALPIN. – "Influence of end constraint in the testing of anisotropic bodies". *J. Composite Materials*, **2**, 1, (1968) : 18.

[5] PAGANO N.J. – "Exact module of anisotropic laminates". *Composite Materials*, **2**, (1974) : 23-44.

[6] REISSNER E. and Y. STAVSKY. – "Bending and stretching of certain types of heterogeneous aelotropic elastic plates". *Transactions of the ASME*. (1961): 402-408.

[7] RIZZO R.R. – "More on the influence of end constraints on off axis tensile tests". *J. Composite Materials*, **3**, 4 (1969) : 202.

[8] SEICHEPINE J.L. – Comportement des matériaux composites Carbone/Epoxy à fibres croisées". *Rapport de stage ENSEM,* 1978.

[9] VAUTRIN A. et D. GUITARD. – "Caractérisation du comportement inélastique des solides par des essais en traction modulée". *Dans Problème de rhéologie et de mécanique des sols.* p. 187-196. Académie Polonaise des Sciences, Varsovie, 1977.

[10] VERCHERY G. – *Théorie classique des plaques stratifiées.* Cours : ENSEM, 1977.

[11] VIDOUSE F. – "Détermination des constantes techniques des stratifiées dans les directions obliques". *Rapport PL4,* Novembre 1973, CRIF Bruxelles.

[12] WHITNEY J.M. – "Bending-extensional coupling in laminated plates under transverse loading". *J. Composite Materials,* 3, 1, (1969) : 20.

ABSTRACT

Traction, three points flexure and dilatometric test results obtained on Carbon/Epoxy angle ply composites (45°-90°) are discussed. The thermoelastic behaviour prediction of those materials is based on the laminated flat plates classical theory and takes in account, the geometrical design of the multilayers, seven experimental parameters measured on the unidirectional fiber reinforced material. Predicted and experimental results are in good agreement, even for off axis measurements.

DISCUSSION

QUESTIONS POSEES PAR R.G.C. ARRIDGE : 1) Was the figure of 32 for the ratio length/thickness in three-point bending, recommanded by ASTM, determined experimentally on a composite or on an isotropic specimen ? If the latter, then I think experiments should be made at higher values of the ratio.

2) How were the elastic constants for the unidirectional composite combinded to form the constants for the cross ply laminates ?

RÉPONSES DE D. GUITARD : 1) Le facteur de forme longueur/épaisseur = 32, que nous avons utilisé, est celui recommandé par la norme ASTM n° D 790, applicable aux échantillons isotropes. Nous n'avons pas, jusqu'à présent, cherché à optimiser ce paramètre dans le cas des matériaux composites. Je pense en effet que des valeurs plus élevées de ce facteur de forme seraient mieux adaptées.

2) Les caractéristiques des stratifiés à couches croisées ont été évaluées à partir des constantes élastiques de chaque couche en admettant des états de contraintes planes.

QUESTION POSEE PAR F. SIDOROFF : Ces dernières années on a vu apparaître dans la littérature des modèles de composites renforcés par fibres présentant des modules différents en traction et compression, cet effet étant attribué au fait que les fibres contribuent à la résistance en traction mais non à la résistance à la compression. Avez-vous observé expérimentalement de tels effets soit en traction-compression, soit en flexion ? Références ?

REPONSE DE D. GUITARD : Ne pratiquant pas d'essais en compression uniaxiale, nous ne sommes pas en mesure d'apporter une conclusion expérimentale à la question de savoir si des matériaux renforcés par fibres présentent une rigidité à la traction supérieure à la rigidité à la compression. Dans le cas d'essais en flexion, menés sur des échantillons renforcés par fibres parallèles au grand axe de l'éprouvette, nous avons observé sur epoxy/kevlar [1] et ici sur epoxy-carbone [2] que les valeurs mesurées de la souplesse à la flexion sont supérieures aux valeurs prédites par le calcul à partir des souplesses de traction. Une explication, *parmi d'autres*, serait en effet d'admettre une souplesse des couches comprimées supérieure à la souplesse des couches tendues.

[1] MOREAUX C. — "Etude du comportement mécanique des matériaux composites à fibre Kevlar", Mémoire de D.E.A. : Institut National Polytechnique de Lorraine, Nancy, octobre 1977.

[2] SEICHEPINE. — "Comportement des matériaux composites carbone/résine à fibres croisées", Mémoire de D.E.A. : Institut National Polytechnique de Lorraine, Nancy, octobre 1978.

Session 12

Problems of Civil Engineering
Problèmes de Génie Civil

Engineering Problems Caused by the Anisotropic Behaviour of the Rock Mass

P. Egger

Ecole Polytechnique Fédérale de Lausanne, Lausanne, Switzerland.

1. Introduction

Contrarily to the structural engineer who works with materials which meet well defined specifications, the rock engineer deals with a material given by the nature and which excells by its heterogeneity and complexity. In most cases, the rock mass is not a continuous medium, but it exhibits one or several sets of discontinuities such as bedding planes, joints or schistosity planes. Frequently, these weakness planes confer to the rock mass a 'quasi-homogeneous' transverse isotropy provided the considered rock volume be large enough to contain a sufficient number of unit elements.

In the following, a series of typical engineering problems are described which are caused by the effects of deformational and strength anisotropy of the rock mass.

2. Problems encountered in underground works

2.1. *Layered and schistous rocks*

The major part of tunnels, particularly those with small overburden, are situated in bedded sedimentary rocks. These rocks are generally composed by alternating, more or less deformable strata exhibiting quite different strengths. The interfaces between the layers are often very smooth or covered by e.g. a clay film and may show extremely low shear strengths.

Tunnels at great depth cross frequently zones which has undergone metamorphic and tectonic processes and where the rock mass shows a clearly defined orientation of weakness planes.

The mechanical behaviour of these types of rock masses may be described with a good approximation by a 'quasi-homogeneous' transverse isotropy (Sonntag [12], Fig. 1).

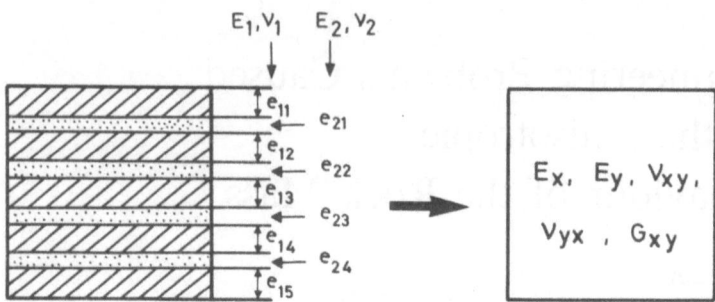

Figure 1. – *Simulation of stratified rock by a 'quasi-homogeneous' transversely isotropic medium.*

It is obvious that the response of these rocks to a given stress field depends strongly on the angle between the principal stress axes and the axis of anisotropy. As long as the rock behaves elastically, this relation can be expressed by the direction curve of Youngs' modulus (Lekhnitskii [8]). But in tunnelling rare are the cases where the rock mass remains entirely in the elastic state; local failures and even the plastification of a continuous zone around the excavation can generally not be prevented both for technical and economical reasons.

For engineering purposes, the effects of the strength anisotropy resulting from parallel weakness planes are generally more pronounced and cause more trouble than those of deformational anisotropy. This is particularly true in the case of sound rock (high strength of rock matrix) crossed by smooth discontinuity planes (small angle of joint friction, no cohesion). Depending on the angle between the principal stresses and the axis of anisotropy, failure may occur at very low values of the maximum possible principal stress ratio by sliding along the discontinuities (John [6], Fig. 2).

Near the wall of unlined tunnels, the major principal stress is tangent to the excavation and the minor principal stress is zero. When there is no cohesion along the discontinuities, non-zero stresses exist only at that part of the circumference where the radius vector and the weakness planes either form an angle inferior or equal to the friction angle or are perpendicular to each other.

The stress distribution around a circular tunnel in an isostatic stress field ($\lambda_0 = 1$) is thus not uniform any more (Fig. 3), and two characteristic phenomena are to be observed:

First, the stresses concentrate parallel and more or less perpendicularly to the discontinuities. Between these directions, there appear nearly unstressed zones. Rock blocks situated in the unstressed zones are unstable; they can easily fall out, e.g. at the blasting of the next rounds and create overbreak.

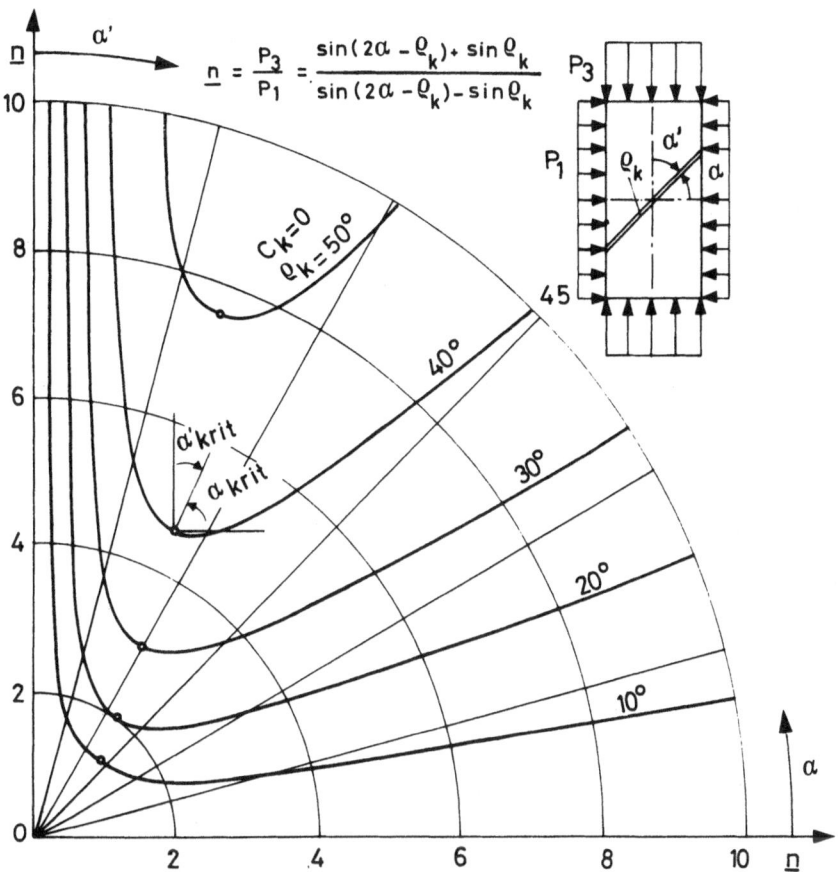

Figure 2. – *Maximum principal stress ratio* $n = \dfrac{p_3}{p_1}$ *vs. angle* α *between principal stresses and weakness plane at limit equilibrium* (ρ_k = *angle of friction on weakness plane*).

The second, even more dangerous phenomenon occurs along and near the radius vector which is perpendicular to the discontinuities. There, a deep zone appears where the radial stresses remain very small. The rock is stressed nearly uniaxially and since the circumferential stresses are high, the rock strata tend to foliate and to buckle; the consequences are important rock break and high deformations.

The example given in Figure 4 shows the results of a comparative numerical analysis for the cases of respectively isotropic and anisotropic, continuous and horizontally fissured rock. For $\lambda_0 = 1$, the small values of the radial

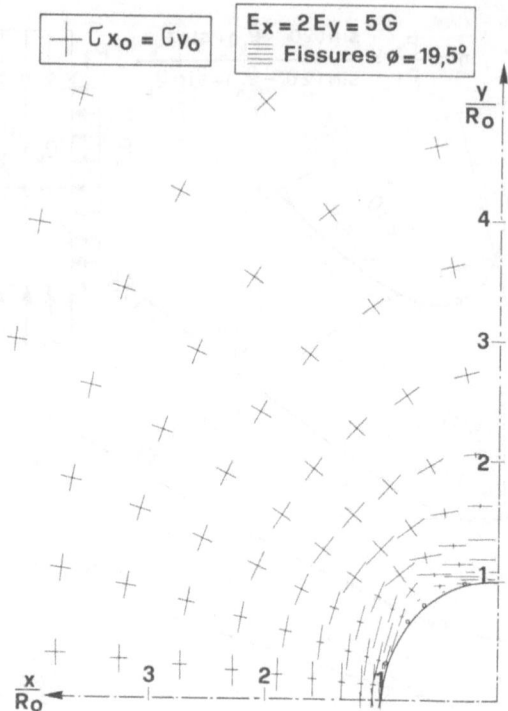

Figure 3. – *Stress distribution around a circular tunnel in horizontally fissured rock; isostatic primary stresses* ($\lambda_0 = \sigma_{x_0} : \sigma_{y_0} = 1$).

stress σ_r at the crown (point B) up to a depth of about one tunnel radius are clearly evidenced for the two cases of fissured rock.

Further, the radial displacement of the crown is 1,7 (isotropic, fissured) to 2,9 (anisotropic, fissured) times larger than that of the abutment. For values $\lambda_0 < 1$, this phenomenon is even more pronounced.

It must be noted that the analysis was performed by means of a special finite element programme (Malina [9]) taking into account elastic anisotropy and strength anisotropy due to one (or more) weakness directions; but it is unable to simulate buckling and opening of joints and therefore underestimates the displacements.

The theoretical results were confirmed by observations made at several tunnels driven through schistous rocks, such as the Arlberg tunnel in Austria and Fréjus tunnel in France. This latter was driven through lustred schists parallel to their direction, dipping West up to about 40° (Fig. 5a). Crosswise arranged convergence measurements (Fig. 5b) showed e.g. in the zone from

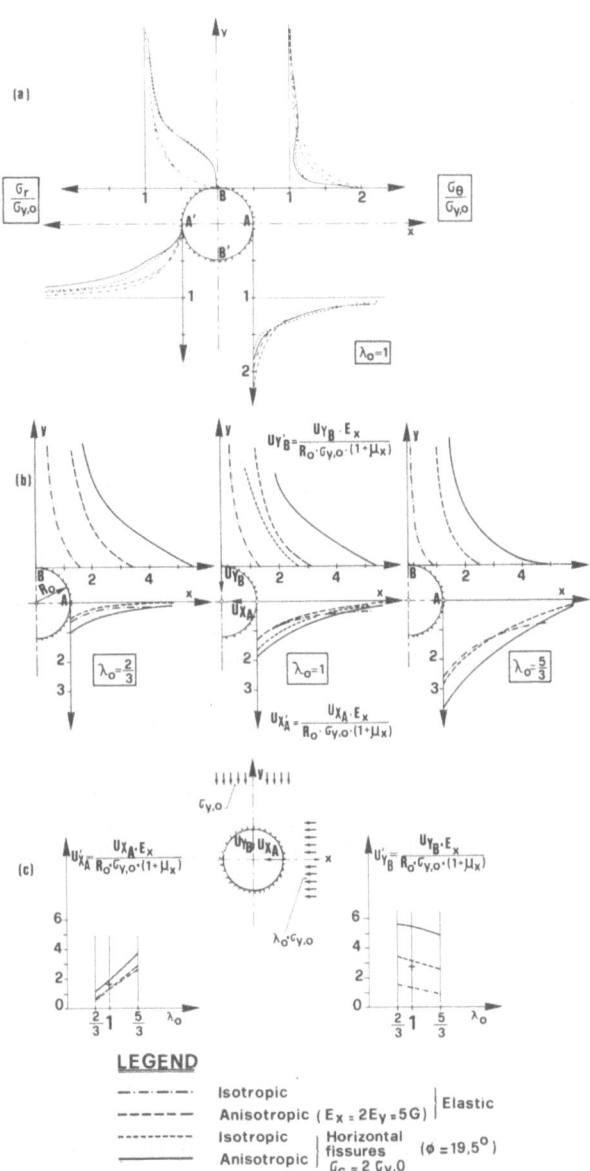

Figure 4. – *Stresses and displacements around a circular tunnel in isotropic and aniso-tropic, continuous and horizontally fissured rock (a) radial and circumferential stresses along the vertical and horizontal axes (b) radial displacements along the vertical and horizontal axes for different values of* $\lambda_0 = \sigma_{x_0} : \sigma_{y_0}$ *(c) radial displacements of the tunnel abutment and crown vs. the* λ_0*-value.*

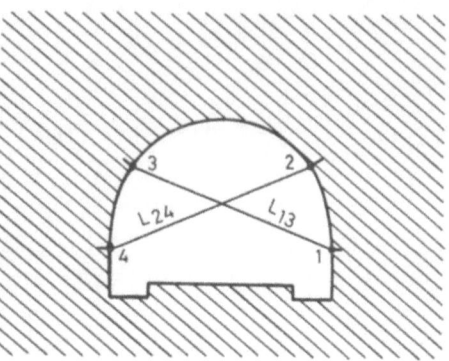

Figure 5. – *Fréjus tunnel situated in lustred schists (a) view of the tunnel face (b) layout of convergence measurements* $(L_{13}, L_{24}...$ *monitored distances).*

chainage 1872 to 1998 m, displacement ratios $\Delta L_{24} : \Delta L_{13} = 2,1$ to 5,0 (average 3,0), the absolute values of the major convergence reaching 30 to 40 cm. It is obvious that deformations of this order of magnitude cause serious troubles because of heavy local rock destruction and by the need of reprofiling the deformed tunnel section.

Increased deformations perpendicularly to the weakness planes are also observed at near-surface tunnels in stratified rocks. At e.g. the Grigny tunnel (Egger [1]), important vertical tunnel convergences were accompanied by horizontal divergences; at ground level, the settlements were characterized by a concentration to a considerably narrower zone than in an isotropic medium (Fig. 6).

2.2. *Swelling rocks*

Amongst the rock types exhibiting anisotropic behaviour, the ill-reputed swelling rocks take a particular place. Their anisotropy may be caused by different reasons: preferred orientation of swelling clay minerals in homogeneous rocks or appearance of swelling minerals in stratified rocks forming thin strata or just a coating of the bedding planes. The latter case is frequently observed in sedimentary rocks of the Secondary Age (Triassic, Jurassic) where swelling clay minerals (e.g. Montmorillonite, Corrensite) as well as anhydrite may cause most serious troubles to underground works.

The swelling phenomena are generally characterized by a heave of the bottom which may reach important values and lead to severe damage of the lining. For example in the Kappelesberg double track railway tunnel (SW-

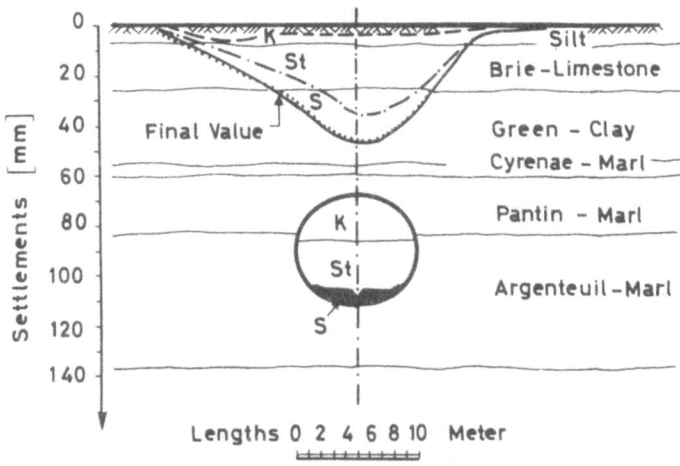

Figure 6. − *Settlement trough above Grigny tunnel situated in Tertiary deposits.*

Germany) which crosses claystones and sulfatic rocks, the bottom heaves required 19 successive excavations for levelling the tracks since its construction in 1880 (Krause and Wurm [7], Fig. 7). The total amount of heave was evaluated to 4,7 m.

In order to avoid these repeated bottom excavations, tunnels in swelling rock are generally provided with invert arches. Care must be taken for a sufficiently strong design because high contact pressures may develop. At the Belchen Tunnel (Grob [5]) in Switzerland, e.g., contact pressures up to 300 N/cm^2 were measured which destroyed the original invert and required important strengthening measures.

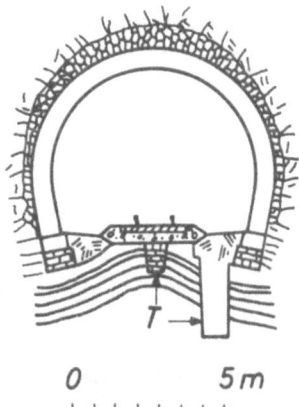

Figure 7. − *Kappelesberg tunnel − Investigation trenches (T) showing deformed strata at the bottom of the tunnel.*

3. Problems encountered in surface works

3.1. *Excavations and rock slopes*

At both natural and artificial rock slopes, a main feature of 'quasi-homogeneous' anisotropic rock masses is the modification of the failure mode with respect to a continuous isotropic medium. Whereas in the latter case the failure surface is curved (Fig. 8a), plane sliding is generally observed in stratified or schistous rocks when the layers are cut by the excavation (Fig. 8b). In many cases, the shear strength along the weakness planes is low and causes serious troubles to excavation works or endangers the stability of natural slopes. For instance, in the molasse hills near Lausanne, frequent rock slides or creep occur along thin clay or coal seams which cover rather competent sandstone banks and exhibit friction angles of only 9 to 10° at some places.

A typical example for the risk of plane sliding can be seen at the abutments of Ridracoli dam (Italy) which is now under construction. This dam site is situated in a regularly dipping flysch formation (Fig. 9) consisting of sandstone and marl beds with intercalation of several laminated clayey seams.

Laboratory and large in-situ tests permitted to evaluate the friction angle of these seams to about 13°, whereas their dip angle is 27°. The foundations of the dam required excavations up to 20 meters depth which cut the foot of a certain number of layers. In order to prevent their sliding, important stabilizing works had to be undertaken: e.g. at the right abutment shown in Figure 9, a part of the potential sliding mass was removed and the remainder was secured by approximatively 500 prestressed 180 ton anchors of some 40 m average length.

When the weakness planes are dipping rather steeply towards the mountain (Fig. 8c), so called toppling may occur: the rock layers bend, joints open up and tension cracks appear which increase the deformability of the rock mass. This phenomenon is frequently observed at the slopes of steep Alpine valleys where erosion or the melting of glaciers took away lateral constraint.

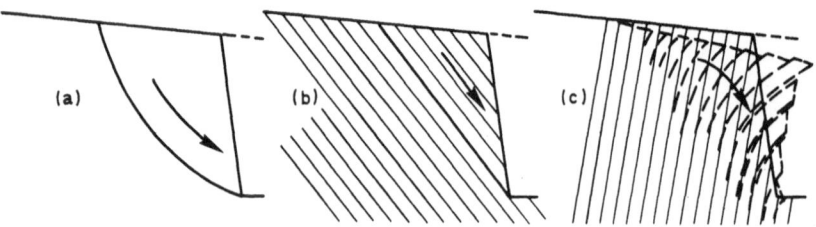

Figure 8. – *Typical slope failure modes (a) in continuous isotropic rock and in stratified rock dipping (b) downhill, (c) uphill.*

Figure 9. – *Ridracoli dam – Excavations for the right abutment; note the anchor heads at the exposed flysch beds.*

In certain cases where there is a second set of weakness planes parallel to the slope, toppling may be combined with buckling of the superficial rock layers (Fürlinger [3], Fig. 10): This latter phenomenon occurs mainly at the toe of the slope and accelerates toppling at the higher parts.

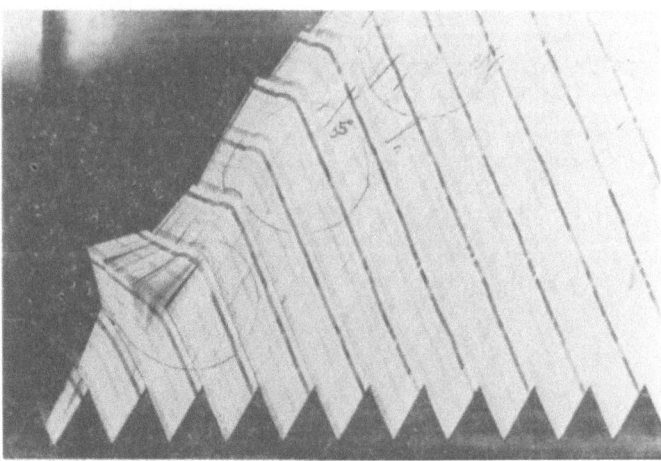

Figure 10. – *Combined toppling and buckling of a slope with schistosity parallel to the surface.*

3.2. *Foundations*

In the cases where loads are to be transmitted into stratified or schistous rocks, problems may arise as well from their deformational as from their strength anisotropy. Deformational anisotropy has particularly to be taken into account in hyperstatic systems where the rock-structure-interaction must not be neglected. A typical example is an arch dam where the stress distribution in the dam and in the abutments depends strongly from the deformations at the contact surface and where magnitude and direction of the external loads change considerably with the reservoir level. The problem of rock-structure-interaction exists certainly also for isotropic rocks, but it is more complicated in anisotropic ones because there the principal axes of deformations do not necessarily coincide with the principal stress directions. To avoid the difficulties caused by a deformational anisotropy of the rock mass, foundations should be designed as isostatic systems whenever possible.

The effects of strength anisotropy of the rock mass on the stress distribution below a foundation are similar to what can be observed in underground works. Experimental (Gaziev [4]) and numerical (Malina [9], Fig. 11) investigations show clearly that the stress trajectories tend either to follow the direction of weakness planes or to cross them more or less perpendicularly. The stresses spread out much less than in the case of an isotropic continuous medium and remain surprisingly concentrated down to important depths.

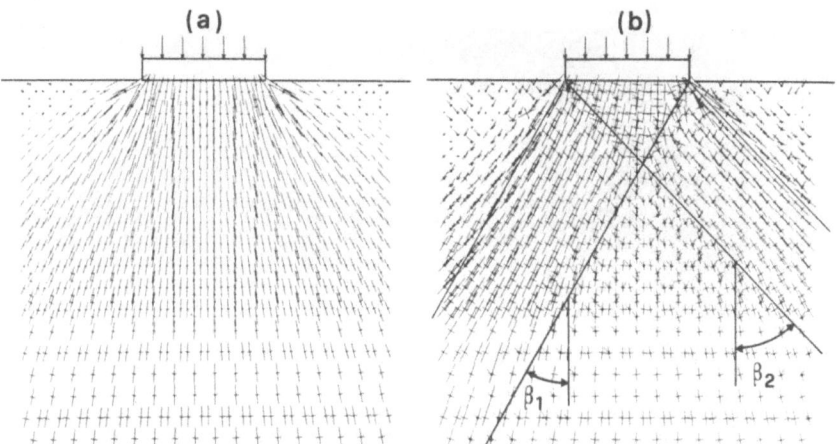

Figure 11. – *Stresses below a foundation on (a) isotropic continuous rock, (b) rock with two sets of discontinuities (angles to the vertical* $\beta_1 = 30°$, $\beta_2 = -45°$; *angle of friction along joints* $\phi_j = 15°$).

As was shown by Sonntag [11] and Maury [10], this holds also in the case of horizontal discontinuities and vertical loads (Fig. 12).

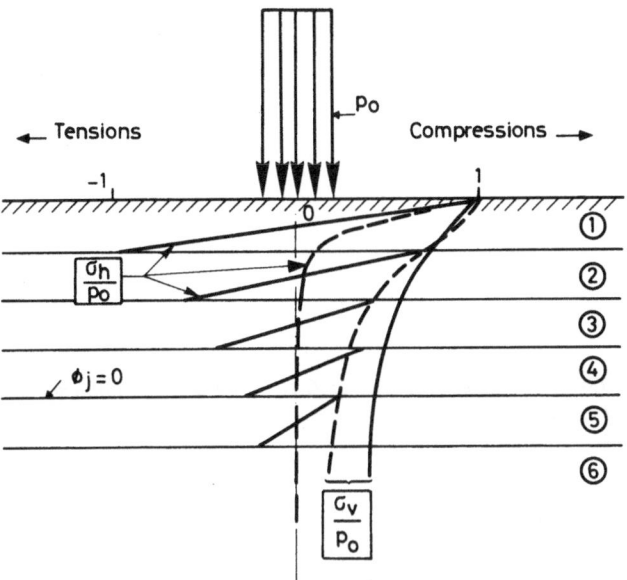

Figure 12. – *Stresses in the centerline below a foundation of a horizontally layered* (——) *and continuous* (---) *elastic medium.*

These theoretical findings were confirmed by stress measurements done at a test drive for the Grigny tunnel (Egger [2], Fig. 13). When an earthfill of limited extension was deposed at the ground level, 20 m above a 2 meter diameter test drive, the contact pressure at its crown increased suddenly by 90 % (!) of its weight.

A second effect of strength anisotropy due to horizontal weakness planes is the appearance of high bending stresses in the rock (see Fig. 12) proper to entail tensile failure in the upper layers. Particular attention must be paid in the case of moving loads, because they may completely dislocate the rock near the surface.

4. Testing and monitoring

In the previous chapters, it was pointed out that anisotropic behaviour of the rock mass is generally caused either by the existence of regularly arranged weakness surfaces such as bedding or schistosity planes or by alternating competent and weak rock strata. To evaluate the mechanical characteristics of

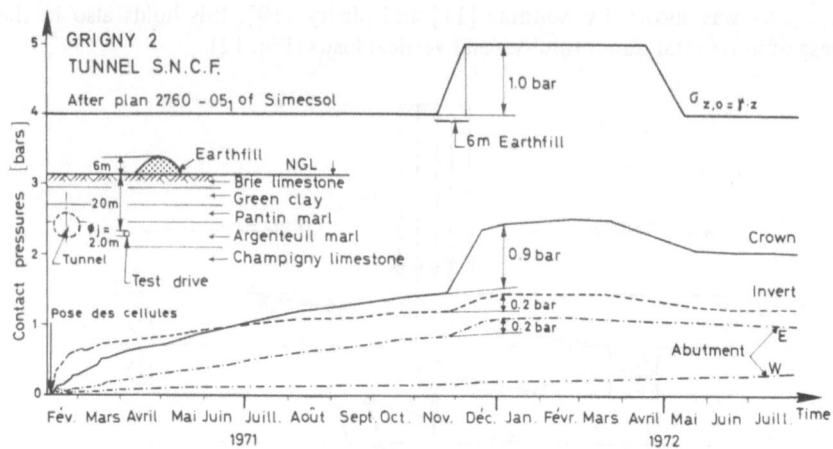

Figure 13. – *Contact pressures vs. time at Grigny test drive – influence of a 6 m high earthfill at the ground level.*

those 'quasi-homogeneous' rock masses, representative samples must be statistically homogeneous. This means that for layered rocks, the characteristic length of the sample or of the zone which is interested by an in-situ test must be at least five to six times the average thickness of the strata.

This condition is in most cases prohibitive for conventional laboratory testing on NX-cores or so. As long as the thickness of the layers or the joint spacing does not exceed 5 to 10 cm, special large scale laboratory tests can still be performed. For example, the servo-controlled triaxial cell Triroc (Fig. 14) – now under construction – will handle samples of 30 cm diameter or square length and of 60 cm height.

But for rocks with a larger joint spacing or when the wavelength of the joint roughnesses exceeds some 20 to 30 cm, only in-situ tests, although rather expensive, are feasible. The Figure 15 (Egger [2]) gives an indication about the rock volume which is interested by different test methods and may help to define the best suited method for a given case.

The experience shows that despite the high importance for a realistic design the credits for pre-construction testing are generally very limited, and the information about the rock mass behaviour remains rather poor. In order to avoid inestimable risks for the safety and the cost of the works, a close monitoring of displacements and stresses during the construction period and afterwards is indispensable. But monitoring is only useful if the results are interpreted immediately and if provisions for a rapid intervention on the site are made.

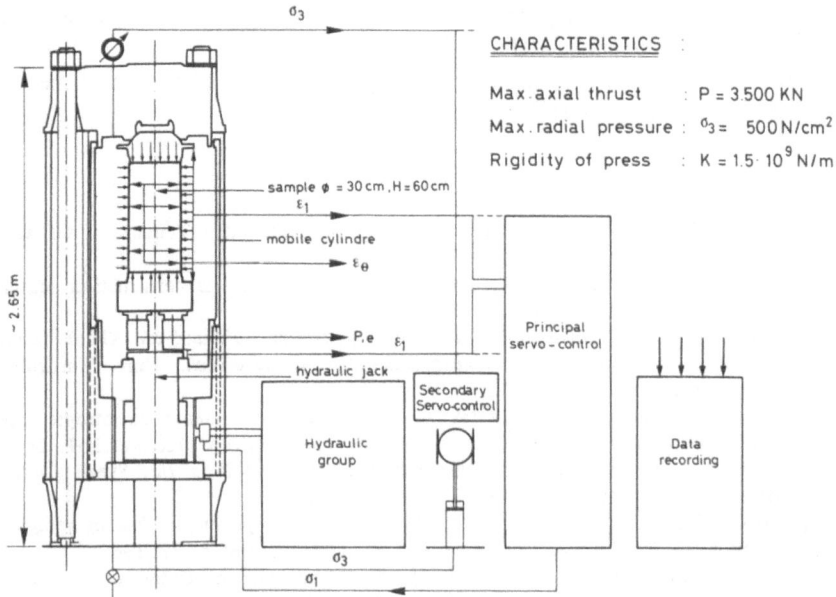

Figure 14. — *Layout and characteristics of the triaxial cell TRIROC for large samples.*

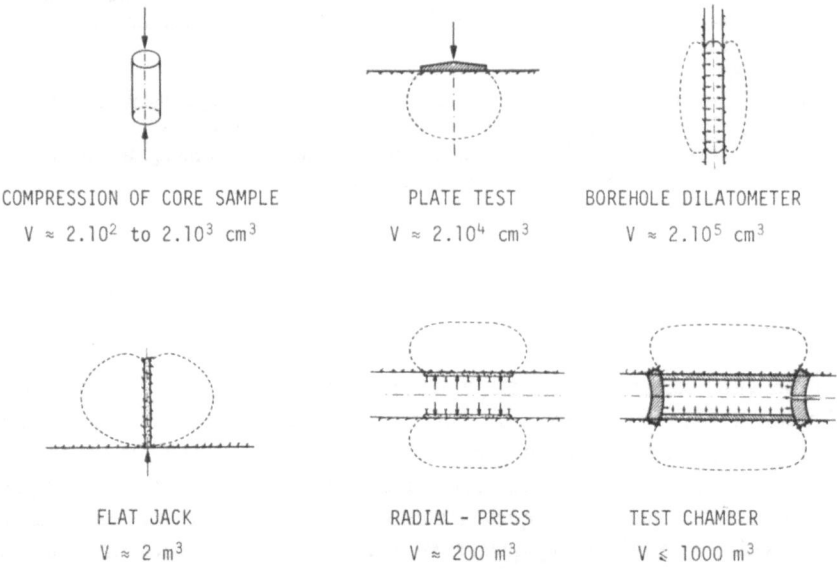

COMPRESSION OF CORE SAMPLE
$V \approx 2.10^2$ to 2.10^3 cm^3

PLATE TEST
$V \approx 2.10^4$ cm^3

BOREHOLE DILATOMETER
$V \approx 2.10^5$ cm^3

FLAT JACK
$V \approx 2$ m^3

RADIAL - PRESS
$V \approx 200$ m^3

TEST CHAMBER
$V \leqslant 1000$ m^3

Figure 15. — *Deformation tests — rock volume interested by different testing methods.*

The validity of this general rule is confirmed by the success of e.g. the New Austrian Tunnelling Method; it holds in particular for the rock types considered in this paper because of their reputedly high dispersion of mechanical characteristics at short distances.

REFERENCES

[1] EGGER P. – "Erfahrungen beim Bau eines seichtliegenden Tunnels in tertiären Mergeln". *Rock Mech. Suppl.*, 4 (1975): 41-54.

[2] EGGER P. – "Le rôle des mesures et auscultations dans la construction souterraine". *Doc. SIA*, 12 (1976): 33-43.

[3] FÜRLINGER W. – *Talzuschübe in schiefrig-plattigem Fels und die Klärung ihres Bewegungsmechanismus im gefügeäquivalenten Modellversuch*. Univ. Karlsruhe: SFB 77, Jahresbericht 1971, 1972.

[4] GAZIEV E.G. and S.A. ERLIKHMAN. – "Stresses and strains in anisotropic rock foundation (model studies)". Nancy: *Proc. Int. Symp. Rock Fracture*: II-1, 1971.

[5] GROB H. – "Schwelldruck im Belchentunnel". Luzern: *Proc. Int. Symp. Underground Openings* (1972): 99-119.

[6] JOHN K.-W. – *Festigkeit und Verformbarkeit von druckfesten, regelmässig gefügten Diskontinuen*. Univ. Karlsruhe: Veröff. Inst. Bodenm. und Felsm. 37, 1969.

[7] KRAUSE H. and F. WURM. – "Geologische Grundlagen und Untersuchungen zum Problem der Sohlhebungen in Keupertunneln Baden-Württembergs". *Sohlhebungen beim Tunnelbau im Gipskeuper*. Stuttgart: Min. f. WMV, 1975.

[8] LEKHNITSKII S.G. – *Theory of Elasticity of an Anisotropic Elastic Body*. San Francisco: Holden-Day, 1963.

[9] MALINA H. – *Berechnungen von Spannungsumlagerungen in Fels und Boden mit Hilfe der Elementenmethode*. Univ. Karlsruhe: Veröff. Inst. Bodenm. und Felsm. 40, 1969.

[10] MAURY V. – *Mécanique des milieux stratifiés*. Paris: Dunod, 1970.

[11] SONNTAG G. – "Die in Schichten gleicher Dicke reibungsfrei geschichtete Halbebene mit periodisch verteilter Randbelastung". *Forsch. Ing. Wes.*, 23 (1957): H. 1/2.

[12] SONNTAG G. – "Einfluss der Anisotropie auf die Beanspruchung des Gebirges in der Umgebung von Stollen". *Bauing.* (1958): 8-9.

RESUME

Problèmes d'ingéniérie provoqués par le comportement anisotrope des massifs rocheux

La présente communication traite des principaux problèmes rencontrés par l'ingénieur dans les roches anisotropes. L'anisotropie "quasi-homogène" des roches stratifiées ou des schistes, par exemple, est plus fréquente et, de ce fait, plus pertinente pour l'ingénieur que l'anisotropie homogène de, par exemple, certaines roches métamorphiques. L'anisotropie du massif rocheux est la cause d'une série de problèmes typiques, aussi bien dans les travaux souterrains, qu'en surface. Ces problèmes sont décrits et des recommandations concernant le choix des méthodes d'essais et des auscultations appropriées sont données.

Theory of Explosions in Anisotropic Media

J.K. Dienes

University of California, Los Alamos Scientific Laboratory, Los Alamos, New Mexico, U.S.A.

1. Introduction

Vast deposits of oil shale in the western United States could be exploited to solve our energy problem if an economical means of processing it were available. One of the most promising approaches is the modified in-situ method, in which a portion of the oil shale is mined out, and a large bed of rubble is prepared with explosives. Then, combustion supported by forced air causes the organic matter to separate from the rock. A portion of it supports the flame, and the remainder flows to the bottom of the retort, whence it is pumped out.

Conversion of the competent shale to rubble with explosives is complicated by the variability of the shale, which involves natural voids, joints, and faults, and is randomly stratified. Still, we think that many systematic features of the process can be investigated by developing a theoretical model and studying its response to explosives by computer simulation. We are particularly interested in the effects of anisotropy, and the possibility that anisotropy can influence the propagation of shock waves and the subsequent fracture processes. From the theoretical point of view, it is convenient to separate anisotropic effects into three classes. In the elastic regime, the moduli vary with orientation by roughly a factor of 2, and it is relatively straightforward to determine them by acoustic methods. In the plastic regime the flow stress also varies by a factor of about two, as determined by triaxial tests. Finally, fractures can propagate under certain combinations of principal stress, and the tensile fracture strength appears to vary by a factor between two and five with orientation. More important than these effects, though, is the possibility that the entire mechanism of wave propagation and fracture is modified by the anisotropy and new phenomena appear. To investigate this possibility we have concentrated our first efforts on the effects of spherical charges, since any absence of point symmetry can then be attributed unambiguously to anisotropic effects. Theoretical studies have predicted, and experiments confirm, that enhanced fracture appears in the neighborhood of a cone that makes an angle of about 45° with the bedding planes. Though in general agreement, many details of this enhanced fracture process are different in the theory (in its rudimentary form) and in experiments. What is of most interest, however, is that this kind of fracture does not exist in isotropic materials, and is a clear consequence of the angular dependence of wave propagation.

In this paper we will show how the associated flow law of plasticity can be used to construct a constitutive law for an ideally elastic-plastic material, and how the law can be inverted to provide stress rate as a function of strain rate and stress. This is an essential step in performing numerical calculations. The resulting constitutive law is used in the YAQUI computer program to compute explosions in a transversely isotropic material, a calculation that requires a two-dimensional (i.e., axisymmetric) code. The calculations show that the cavity remains essentially spherical, but that tensile hoop stresses develop in the neighborhood of the $45°$ cone through the center of symmetry. An explanation for this phenomenon in terms of wave propagation phenomena will be given.

2. Anisotropic plasticity

If we assume a plastic potential, then it is possible to construct a constitutive relation describing plastic flow by means of the associated flow rule. It is shown in this section that the equations can be solved explicitly for the stress rate, even in the case of anisotropic materials. The importance of material rotation is greater for anisotropic than for isotropic materials, and for that reason the analysis begins with a discussion of the kinematics of rotation. Following this we construct a potential involving a 4-index plasticity tensor which is a generalization of the scalar yield strength that appears in isotropic plasticity. The flow law involves a Lagrangian multiplier, λ, which we determine as a function of the stress, strain rate and material properties. Frequently, papers discussing plasticity provide expressions for λ which involve the stress rate. This is not adequate for numerical work, since the stress rate is what we need to find, and for this reason it turns out that the calculation indicated here is somewhat more involved than in the usual treatments. It will be shown that there is no change in plastic volume, as previously indicated by Hill [5]. Finally, we indicate how the stress can be separated into an isotropic part, which dominates the behavior at high pressures, and an anisotropic part which dominates the low pressure behavior.

To account for material rotation in the constitutive law, we relate the stress in space axes, σ, to the stress in material axes, $\bar{\sigma}$, through the rotation matrix $\mathbf{R}$ by means of the equation

$$\sigma = \mathbf{R}\,\bar{\sigma}\,\mathbf{R}^{\mathrm{T}} . \tag{1}$$

as discussed by Dienes [4]. A similar relation transforms the strain rate matrix, **D**, into material axes, i.e.,

$$\bar{\mathbf{D}} = \mathbf{R}^T \mathbf{DR} .\tag{2}$$

Since the constitutive law which we will develop expresses stress rate in terms of stress and strain rate, it is necessary to consider how the stress rate is affected by rotation. Physically, it is clear that the state of material stress is not affected by material rotation, but as the material rotates the components of stress in fixed space axes will vary. If we define the rate of material rotation, **Ω**, by

$$\mathbf{\Omega} = \dot{\mathbf{R}}\mathbf{R}^T\tag{3}$$

then the result of differentiating (1) can be written

$$\dot{\sigma} = \mathbf{\Omega}\bar{\sigma} + \dot{\bar{\sigma}} - \bar{\sigma}\mathbf{\Omega}.\tag{4}$$

The quantity

$$\hat{\sigma} = \dot{\sigma} - \mathbf{\Omega}\bar{\sigma} + \bar{\sigma}\mathbf{\Omega}\tag{5}$$

is the Jaumann-Noll stress rate if **Ω** represents the vorticity. It is shown in Ref. 4 that the quantity **Ω** given in (3) is approximated by the vorticity for small deformations, and hence we may put

$$\mathbf{\Omega} = (\omega_{ij})\tag{6}$$

where

$$\omega_{ij} = \frac{1}{2}\left(u_{i,j} - u_{j,i}\right) .\tag{7}$$

We assume that a plastic potential exists having the form

$$f = \frac{1}{2}\, b_{ijk\ell}\, \sigma_{ij}\, \sigma_{k\ell}\tag{8}$$

for anisotropic materials. By comparison with the expression

$$2f = F(\sigma_y - \sigma_z)^2 + G(\sigma_z - \sigma_x)^2 + H(\sigma_x - \sigma_y)^2 + 2L\,\tau_{yz}^2$$
$$+ 2M\,\tau_{zx}^2 + 2N\tau_{xy}^2\tag{9}$$

given by Hill we may deduce the terms of the $b_{ijk\varrho}$ tensor. It is convenient, however, to reduce the number of indices by defining the stresses as a nine-element vector, Σ_K, with the values of K being given by the matrix

$$
\begin{array}{cccc}
j\backslash i & 1 & 2 & 3 \\
1 & \begin{pmatrix} 1 & 6 & 4 \\ 2 & 7 & 2 & 8 \\ 3 & 5 & 9 & 3 \end{pmatrix} .
\end{array}
\tag{10}
$$

Then the plastic potential takes the form

$$
f = \frac{1}{2} B_{KL} \Sigma_K \Sigma_L .
\tag{11}
$$

The elements of the matrix are given by

$$
B = \begin{pmatrix}
G+H & -H & -G & 0 & 0 & 0 & 0 & 0 & 0 \\
-H & H+F & -F & 0 & 0 & 0 & 0 & 0 & 0 \\
-G & -F & F+G & 0 & 0 & 0 & 0 & 0 & 0 \\
0 & 0 & 0 & M & 0 & 0 & 0 & 0 & 0 \\
0 & 0 & 0 & 0 & M & 0 & 0 & 0 & 0 \\
0 & 0 & 0 & 0 & 0 & N & 0 & 0 & 0 \\
0 & 0 & 0 & 0 & 0 & 0 & N & 0 & 0 \\
0 & 0 & 0 & 0 & 0 & 0 & 0 & L & 0 \\
0 & 0 & 0 & 0 & 0 & 0 & 0 & 0 & L
\end{pmatrix}
\tag{12}
$$

It is shown by Hill that for transversely isotropic materials

$$
F = G ,
\tag{13}
$$

$$
L = M ,
\tag{14}
$$

and

$$
N = F + 2H .
\tag{15}
$$

This is the case of interest in our numerical calculations, which will assume axially symmetric flow. If it is assumed that the plastic strain rates can be derived from the plastic potential, f, then

$$
D_K^p = \lambda \frac{\partial f}{\partial \Sigma_K} = \lambda B_{KL} \Sigma_L .
\tag{16}
$$

Also, the elastic strain rate for an anisotropic material is given by

$$D_K^e = Q_{KL} \hat{\Sigma}_L .$$ (17)

The total strain rate

$$D_K = D_K^p + D_K^e$$ (18)

is then given by

$$D_K = \lambda B_{KL} \Sigma_L + Q_{KL} \hat{\Sigma}_L .$$ (19)

If we define

$$C = Q^{-1}$$ (20)

and express the relation for D_K in matrix notation as

$$D = \lambda B \Sigma + Q \hat{\Sigma}$$ (21)

then we can solve for $\hat{\Sigma}$

$$\hat{\Sigma} = C(D - \lambda B \Sigma) .$$ (22)

We may write the stress rate as

$$\hat{\Sigma}_K = \dot{\Sigma}_K + T_K$$ (23)

where

$$t_{ij} = \sigma_{im} \omega_{mj} + \sigma_{mj} \omega_{mi}$$ (24)

and the T_K vector is related to the t_{ij} matrix in the same manner as previously indicated for the stresses and strain rates. Then (22) reduces, in matrix notation, to

$$\dot{\Sigma} = C(D - \lambda B \Sigma) - T.$$ (25)

To determine λ we note that when the yield condition

$$\frac{1}{2} B_{KL} \Sigma_K \Sigma_L = 1$$ (26)

is differentiated we obtain the relation

$$B_{KL} \Sigma_K \dot{\Sigma}_L = 0 \tag{27}$$

involving $\dot{\Sigma}_L$. In combination with (25) we can obtain an expression for the Lagrangian multiplier

$$\lambda = \frac{G_{LM} \Sigma_L D_M - \gamma}{H_{LN} \Sigma_L \Sigma_N} \tag{28}$$

where

$$H_{LN} = B_{KL} B_{MN} C_{KM}, \tag{29}$$

$$G_{LM} = B_{KL} C_{KM}, \tag{30}$$

and

$$\gamma = B_{KL} \Sigma_L T_K. \tag{31}$$

The rotational contribution, γ, can be expressed as the sum of two terms

$$\gamma_1 = b_{ijkm} \sigma_{i\varrho} \sigma_{km} \omega_{\varrho j} \tag{32}$$

and

$$\gamma_2 = b_{ijkm} \sigma_{\varrho j} \sigma_{km} \omega_{\varrho i}. \tag{33}$$

In the case of current interest the rotation is about a fixed axis as the result of axial symmetry of the problem, and only the terms ω_{13} and ω_{31} do not vanish. Then, if write

$$\omega_{13} = \Omega \tag{34}$$

we find

$$\gamma_1 = \gamma_2 = [(H + 2F - L) \Sigma_1 - (H - F) \Sigma_2 + (L - 3F) \Sigma_3] \Sigma_4 \Omega \tag{35}$$

and, hence

$$\gamma = 2\gamma_1. \tag{36}$$

For isotropic materials,

$$L = M = N = 3F = 3G = 3H \tag{37}$$

as discussed by Hill and, consequently,

$$\gamma = 0. \tag{38}$$

To show that the rate of change of plastic volume is zero we write

$$\dot{\theta}^P = d_{ii}^P = \lambda b_{iik\ell} T_{k\ell}. \tag{39}$$

In the reduced index notation this can be expressed as

$$\dot{\theta}^P = \lambda(B_{1J} + B_{2J} + B_{3J}) \Sigma_J. \tag{40}$$

However, for any J, the sum of the first three terms in a row of the B matrix given as (12) is zero, and consequently

$$\dot{\theta}^P = 0. \tag{41}$$

This demonstrates that the theory does not allow for permanent changes in volume.

At high pressure the stress is essentially that associated with the Mie-Gruneisen equation of state

$$p = p(\rho, I) \tag{42}$$

where ρ is the material density and I the specific internal energy. At intermediate pressures the stress can be expressed as

$$\sigma_{ij} = \sigma_{ij}^L - (p - k\mu) \delta_{ij} \tag{43}$$

where σ_{ij}^L denotes the low pressure component of stress determined in the preceding pages and $p - k\mu$ is the excess of the pressure over the linear approximation. Otherwise expressed,

$$\text{stress} = \begin{pmatrix} \text{low pressure} \\ \text{component of} \\ \text{stress} \end{pmatrix} + \begin{pmatrix} \text{high pressure} \\ \text{component of} \\ \text{stress} \end{pmatrix} - \begin{pmatrix} \text{common} \\ \text{part} \\ -k\mu \delta_{ij} \end{pmatrix}.$$

Here, the compression is expressed as

$$\mu = 1 - \rho_0/\rho. \quad . \tag{44}$$

3. Anisotropy Fragmentation

Lagrangian codes determine the movements in a continuum by tracking elements of mass with the momentum equation

$$\dot{\rho u_i} = \sigma_{ij,j}. \tag{45}$$

Though Yaqui [6] is an Arbitrary Lagrangian-Eulerian Code, for oil-shale calculations it can be made to function as a Lagrangian code by a special choice of the mesh-moving equations. We resist discussing the finite-difference method here, focussing on the physical interpretation of the numerical results.

The effects of a spherical charge of explosive can be approximated by representing the charge as a sphere of uniform, polytropic gas with index 3. Though this neglects the transit time of the detonation wave initiated at the center, the average value of the pressure at the interface turns out to be approximated very well by this simplification. The cavity remains spherical to within a few percent when the best estimates of oil-shale elastic and plastic properties are made. The elastic parameters for 2 g/cm^3 material are [2], in GPa,

$$C_{11} = 24.5, \ C_{33} = 15.1, \ C_{44} = 5.1, \ C_{66} = 8.0, \ C_{12} = 8.5, \ C_{13} = 6.2 \tag{46}$$

and the variation of strength with bedding angle is given graphically in Figure 1. It shows that the plasticity theory used to model the triaxial test results of McLamore and Gray [9] gives a minimum strength at 45°, where the test results indicate the minimum at 60°. It is not clear whether the discrepancy represents a defect of the theory or of the test method, which tends to cause failures across the diagonal of the specimen. Since the test specimens have an aspect ratio of 2, when the bedding angle is 60° (see Fig. 1), the diagonal coincides with bedding plane, and this may contribute to the experimental minimum at 60°. The high pressure constitutive law takes its form from the empirical relation [7].

$$u_S = C + Su_p \tag{47}$$

between shock velocity and particle velocity. This, together with the constitive law defined above, defines the theoretical behavior of the shale.

The effect of bedding on material strength that is illustrated in Figure 1 can be understood by considering a stack of plates of elastic-plastic material

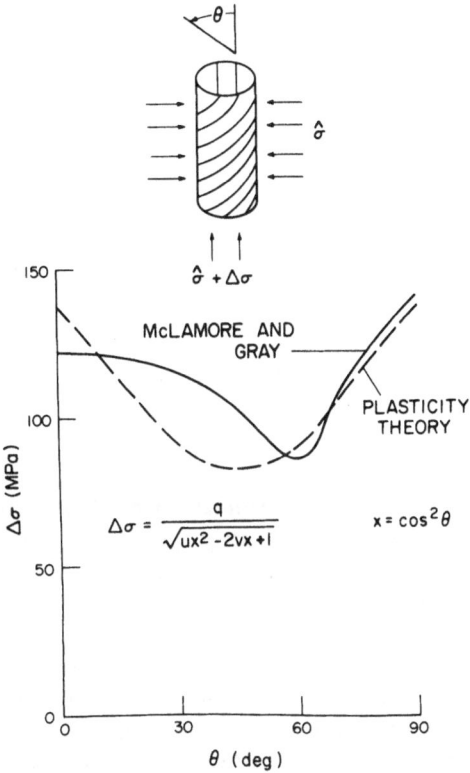

Figure 1. – *Comparison of theoretical and measured strength-vs-orientation curves using data of McLamore and Gray, Ref.* [9].

separated by a lubricating material of lower strength. When stressed in compression by loads in either the vertical or horizontal direction the stack has the same strength as the plate material, but when a core with an axis at 45° to the vertical is loaded along the axis of the core the strength is reduced because of slip along the bedding planes.

The result of a spherical explosion is illustrated in Figure 2, which shows contours of the hoop stress σ_θ. This principle stress is associated with the directions normal to planes through the axis of symmetry of the problem. Tensile values of σ_θ tend to cause failures analogous to the separation of orange sections. In an explosion in an isotropic medium the stress contours are spheres and the stresses are all compressive out to very large radii. The effect of anisotropy is to cause a region of tensile hoop stress centered at about 5 charge radii from the center of the explosive. The tensile stresses exceed a kilobar over a significant region, and half a kilobar over a very large volume.

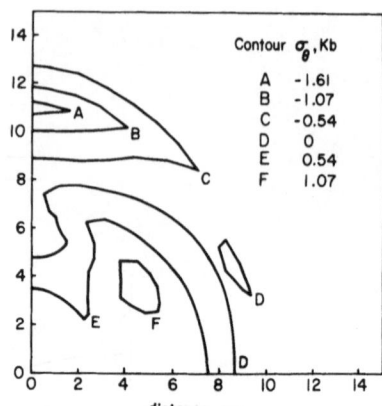

Figure 2. – *Contours of constant hoop stress $\sigma_{\theta\theta}$ at 43 μs resulting from a spherical detonation in anisotropic oil shale.*

In a spherical explosion in horizontally-bedded material the upper portion of the wave front (which is approximately spherical) is attenuated at a rate that depends on the vertical strength, and the horizontal wave front attenuates at essentially the same rate, since it enters material that appears to have the same strength. Analytic solutions for spherical waves have been discussed by Luntz [8], Chadwick and Morland [3] and Blake and Dienes [1], but in realistic calculations material nonlinearities in the oil shale and the complexities of the pressure history in the explosive make analytic solutions intractable. Intuitively, one expects the wave to attenuate more rapidly where the strength is higher, and this is borne out by the calculation in spherical geometry illustrated in Figure 3, where the stress profiles for spherical waves in isotropic shales having shear strengths of 0.05 and 0.0913 GPa are compared. Since the wave moving along a 45° cone is an anisotropic shale encounters material of apparantly lower strength than the horizontal and vertical waves, it attenuates more slowly. Consequently, the hoop stress in the vertical plane is greatest on the 45° cone, and it causes material to move away from the cone. Though this effect is relatively small in terms of stress gradients, it is large enough to cause a small reduction in residual density, leaving a residual tensile stress that exceeds a kilobar in certain regions. We have estimated that the volume of the toroidal tensile region is about 60 times the volume of the charge. The formation of this tensile region suggests a new method of fracturing shales.

Acknowledgment

This work was supported by the Department of Energy, Division of Oil, Gas, Shale and In Situ Technology.

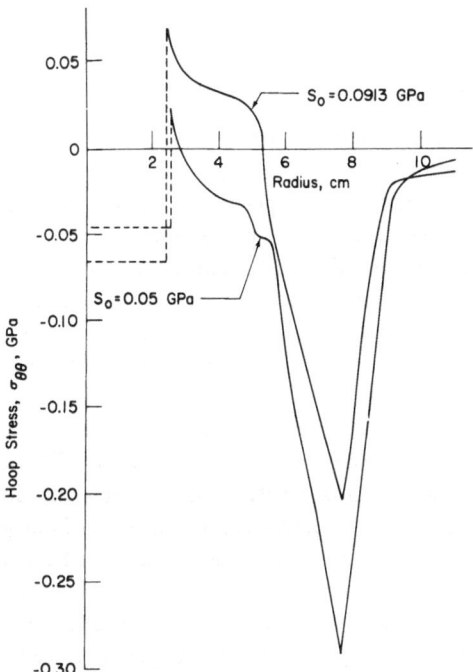

Figure 3. — *Profiles of spherical waves in isotropic shales with shear strengths (S_Q) of 0.05 and 0.0913 GPa.*

REFERENCES

[1] BLAKE T.R. and J.K. DIENES. — "On Viscosity and the inelastic nature of waves in geological media" *Bull. Seis. Soc.*, **66**, 1976.

[2] CARTER W.J. *et al.* — *"Explosively produced fracture of oil Shale"* Los Alamos Scientific Laboratory report LA-6817-PR, Sept. 1977.

[3] CHADWICK P. and L.W. MORLAND. — "The starting problem for spherical elastic-plastic waves of small amplitude" *J. Mech. Phys. Solids*, **17**, 1969.

[4] DIENES J.K. — "On the analysis of rotation and stress rate in deforming bodies" Acta Mechanica, **32**, (1979): 217.

[5] HILL R. — *Plasticity,* Oxford, Clarendon Press, 1950.

[6] HIRT C.W., AMSDEN A.A. and J.L. COOK. — "An arbitrary lagrangian-eulerian computing method for all flow speeds" *J. Comp. Phys.*, **14**, (1974): 227-253.

[7] KINSLOW R. — *Hypervelocity Impact Phenomena.* Academic Press, New-York ; 1968.

[8] LUNTZ Y.L. — "The propagation of spherical waves in an elastic-plastic medium" *Prikl. Mat. Mekh.*, **13**, 1949.

[9] McLAMORE R. and K.E. GRAY. — "The mechanical behavior of anisotropic sedimentary rocks" *J. of Engineering for Industry*, **89**, 1967.

RESUME

Théorie des explosions dans les milieux anitropes

Lorsqu'un procédé pratique pour le raffinage des schistes bitumineux aura été
découvert, il deviendra possible d'extraire de grandes quantités de kérogène,
un hydrocarbure peu différent du pefrole. Nous étudions la mécanique de la
fracturation des schistes bitumineux par explosifs, en particulier les effets dûs
a l'anisotropie. Dans ce travail, nous discutons, du point de vue théorique, des
effets des explosions sphériques dans des matériaux orthotropes de révolution :
c'est un problème axisymétrique. La loi constitutive a été construite de telle
sorte que pour les hautes pressions l'équation d'état (isotrope) de Mie-Gruneisen
prédomine, tandis que pour les basses pressions les équations se comportent
comme dans une théorie pour des matériaux élasto-plastiques. Pour les basses
pressions, la théorie est basée sur le potentiel de Hill pour les matériaux
plastiques anisotropes. Le taux de déformation est donc donné par la superposition des taux élastiques et plastiques. Comme la loi constitutive doit servir
dans un programme pour ordinateur, il est nécessaire de l'inverser pour obtenir
les taux de contraintes comme fonction des contraintes et des taux de
déformation. La loi constitutive a été introduite dans le programme Yaqui,
du type Euler-Lagrange. Les calculs montrent que la cavité reste sphérique avec
une précision de quelques pour cent, tandis qu'il apparaît de grandes déformations au voisinage d'un cône d'angle au sommet de 45° et d'axe normal aux
couches. D'abord, l'onde de choc est à peu près sphérique, mais plus tard
l'effet de l'anisotropie devient important et des tractions apparaissent au
voisinage du cône. On explique ce phénomène par une moindre résistance dans
les directions orientées à 45° par rapport aux couches.

Axisymmetric Loads on a Semi-Infinite Transversely Isotropic Body

M. Dahan

Laboratoire de Mécanique des Solides, Ecole Polytechnique, Palaiseau, France.

1. Introduction

In this paper, we resolve the problem of finding the stresses and the displacements produced inside a transversely isotropic halfspace when various loads are applied on its free surface ; these loads are symmetrical about the axis of the medium.

Two classes of contact problems are considered :

Problem 1 : The boundary conditions are given in normal stresses (σ_{zz}) or in shearing stresses (σ_{rz}) on the surface.

Problem 2 : The boundary conditions are given in vertical displacements (w) or in radial shear displacements (u_r).

In Section 2, concerned with Problem 1, we determine the stress and displacement distribution in terms of an intermediate function p^H which depends on the loading.

In Section 3, we reduce Problem 2 to Problem 1, by determining the surface stress distributions which correspond to the boundary conditions in displacements. Thus, we obtain the function p^H and the solution of Problem 2.

In the next section, the punch has the equation $f(r) = \sum\limits_{n=1}^{\infty} c_n r^n$, and the results corresponding to this case are deduced from the general formula. In the last section we show the influence of anisotropy.

It is found that the exact stress and displacement distributions can be obtained in closed form [1, 2] for the main special cases of loading.

2. Problem 1

Let us consider a homogeneous transversely isotropic half-space whose surface lies in the horizontal (r, θ)-plane and whose axis of elastic symmetry (z) is vertical. For this material, there are five elastic coefficients a_{ij}.

The stress-strain relationships associated with axi-symmetric loading are

$$\epsilon_{rr} = a_{11}\,\sigma_{rr} + a_{12}\,\sigma_{\theta\theta} + a_{13}\,\sigma_{zz},$$

$$\epsilon_{\theta\theta} = a_{12}\,\sigma_{rr} + a_{11}\,\sigma_{\theta\theta} + a_{13}\,\sigma_{zz},$$

$$\epsilon_{zz} = a_{13}\,\sigma_{rr} + a_{13}\,\sigma_{\theta\theta} + a_{33}\,\sigma_{zz}, \tag{1}$$

$$\gamma_{rz} = a_{44}\,\sigma_{rz},$$

where the strain components are defined as

$$\epsilon_{rr} = \frac{\partial u_r}{\partial r}\ ,\quad \epsilon_{\theta\theta} = \frac{u_r}{r},$$

$$\epsilon_{zz} = \frac{\partial w}{\partial z}\ ,\quad \gamma_{rz} = \frac{\partial u_r}{\partial z} + \frac{\partial w}{\partial r}. \tag{2}$$

The equations of equilibrium are

$$\frac{\partial \sigma_{rr}}{\partial r} + \frac{\partial \sigma_{rz}}{\partial z} + \frac{\sigma_{rr} - \sigma_{\theta\theta}}{r} = 0,$$

$$\frac{\partial \sigma_{rz}}{\partial r} + \frac{\partial \sigma_{zz}}{\partial z} + \frac{\sigma_{rz}}{r} = 0, \tag{3}$$

and the equations of compatibility transformed by relations (1) are

$$(a_{11} - a_{12})(\sigma_{rr} - \sigma_{\theta\theta}) - r\frac{\partial}{\partial r}(a_{12}\,\sigma_{rr} + a_{11}\,\sigma_{\theta\theta} + a_{13}\,\sigma_{zz}) = 0,$$

$$\frac{\partial^2}{\partial z^2}(a_{11}\,\sigma_{rr} + a_{12}\,\sigma_{\theta\theta} + a_{13}\,\sigma_{zz}) + \frac{\partial^2}{\partial r^2}(a_{13}\,\sigma_{rr} + a_{13}\,\sigma_{\theta\theta} + a_{33}\,\sigma_{zz})$$

$$- a_{44}\frac{\partial^2 \sigma_{rz}}{\partial r \partial z} = 0, \tag{4}$$

We take the following boundary conditions:
a) for r or z infinite,

$$\sigma_{rr} = \sigma_{\theta\theta} = \sigma_{zz} = \sigma_{rz} = 0; \tag{5}$$

b) on the horizontal surface (z = 0), we distinguish between the two cases considered in the introduction, according to the boundary conditions prescribed in

i) the normal stresses σ_{zz} :

$$\sigma_{rz}(r, 0) = 0, \tag{6}$$

$$\sigma_{zz}(r, 0) = -p(r), \tag{7}$$

ii) the shearing stresses σ_{rz} :

$$\sigma_{zz}(r, 0) = 0, \tag{6'}$$

$$\sigma_{rz}(r, 0) = -p(r), \tag{7'}$$

where p(r) is the stress distribution on the area of contact.

Both cases will be solved simultaneously.
If we introduce an auxiliary function φ such that [4]

$$\sigma_{rr} = -\frac{\partial}{\partial z}\left(\frac{\partial^2\varphi}{\partial r^2} + \frac{b}{r}\frac{\partial\varphi}{\partial r} + a\frac{\partial^2\varphi}{\partial z^2}\right),$$

$$\sigma_{\theta\theta} = -\frac{\partial}{\partial z}\left(b\frac{\partial^2\varphi}{\partial r^2} + \frac{1}{r}\frac{\partial\varphi}{\partial r} + a\frac{\partial^2\varphi}{\partial z^2}\right),$$

$$\sigma_{zz} = \frac{\partial}{\partial z}\left(c\frac{\partial^2\varphi}{\partial r^2} + \frac{c}{r}\frac{\partial\varphi}{\partial r} + d\frac{\partial^2\varphi}{\partial z^2}\right), \tag{8}$$

$$\sigma_{rz} = \frac{\partial}{\partial r}\left(\frac{\partial^2\varphi}{\partial r^2} + \frac{1}{r}\frac{\partial\varphi}{\partial r} + a\frac{\partial^2\varphi}{\partial z^2}\right),$$

where the coefficients a, b, c, d are defined as

$$a = a_{13}(a_{11} - a_{12})/(a_{11}a_{33} - a_{13}^2),$$

$$b = [a_{13}(a_{13} + a_{44}) - a_{12}a_{33}]/(a_{11}a_{33} - a_{13}^2),$$

$$c = [a_{13}(a_{11} - a_{12}) + a_{11}a_{44}]/(a_{11}a_{33} - a_{13}^2), \tag{9}$$

$$d = (a_{11}^2 - a_{12}^2)/(a_{11}a_{33} - a_{13}^2),$$

then the corresponding values of the displacements are:

$$u_r = -(1-b)(a_{11}-a_{12})\frac{\partial^2 \varphi}{\partial r \, \partial z},$$

$$u_\theta = 0,$$ (10)

$$w = a_{44}\left(\frac{\partial^2 \varphi}{\partial r^2} + \frac{1}{r}\frac{\partial \varphi}{\partial r}\right) + (a_{33} \, d - 2\,a_{13}\, a)\frac{\partial^2 \varphi}{\partial z^2}.$$

As described in [1, 2], the solution of the problem is obtained if the auxiliary function is taken in the form

$$\varphi(r,z) = \frac{1}{f(s_1-s_2)}\int_0^\infty (p_2 \, e^{-s_1 mz} - p_1 \, e^{-s_2 mz})\, p^H(m)\, J_0(mr)\,\frac{dm}{m^2}$$
(11)

$$\varphi(r,z) = -\frac{1}{f(s_1-s_2)}\int_0^\infty (s_2\, g_2\, e^{-s_1 mz} - s_1\, g_1\, e^{-s_2 mz})\, p^H(m)$$

$$J_0(mr)\,\frac{dm}{m^2} \quad (11')$$

with the notations

$$\left.\begin{array}{l} s_1 \\ s_2 \end{array}\right\} = [(a+c\pm\sqrt{(a+c)^2-4\,d})/2\,d]^{1/2},$$

$$\left.\begin{array}{l} p_i = 1 - a s_i^2, \\[4pt] g_i = c - d s_i^2, \\[4pt] q_i = s_i^2\,(a_{33}\,d - 2a_{13}\,a) - a_{44}, \end{array}\right\} \quad i = 1, 2$$

$$e = (a_{11}-a_{12})(1-b),$$

$$f = (d-ac)/\sqrt{d},$$

$$\lambda = (1-b)/f,$$

$$\mu = (b-1)(a+\sqrt{d})/(ac-d),$$

$$q = \lambda(a_{11}-a_{12})(s_1+s_2),$$

(12)

and where p^H is the function obtained by the Hankel transformation of the loading p. We have

$$p^H(m) = \int_0^\infty r J_0(mr) p(r) \, dr \,, \tag{13}$$

$$p^H(m) = \int_0^\infty r J_1(mr) p(r) \, dr \,. \tag{13'}$$

Let us introduce the following notation:

$$A_i = \int_0^\infty e^{-ms_i z} J_0(mr) p^H(m) m \, dm \,,$$

$$B_i = \frac{1}{r} \int_0^\infty e^{-ms_i z} J_0(mr) p^H(m) \, dm \,,$$

$$\qquad\qquad\qquad\qquad i = 1, 2 \tag{14}$$

$$C_i = \int_0^\infty e^{-ms_i z} J_1(mr) p^H(m) m \, dm \,,$$

$$D_i = \frac{1}{r} \int_0^\infty e^{-ms_i z} J_1(mr) p^H(m) \, dm \,.$$

Then, if we substitute expression (11) or (11') into formulae (8) and (10), the expressions for stress and displacement are given by:

$$\sigma_{rr} = [\lambda(s_1 p_2 D_1 - s_2 p_1 D_2) - (s_1 A_1 - s_2 A_2)/\sqrt{d}]/(s_1 - s_2) \,,$$

$$\sigma_{\theta\theta} = \{-\lambda(s_1 p_2 D_1 - s_2 p_1 D_2)$$
$$\qquad - [s_1 p_2 (b - as_1^2) A_1 - s_2 p_1 (b - as_2^2) A_2]/f\}/(s_1 - s_2) \,,$$

$$\sigma_{zz} = (s_2 A_1 - s_1 A_2)/(s_1 - s_2) \,, \tag{15}$$

$$\sigma_{rz} = (C_1 - C_2)/(s_1 - s_2)\sqrt{d} \,,$$

$$u_r = -\lambda r(s_1 p_2 D_1 - s_2 p_1 D_2)(a_{11} - a_{12})/(s_1 - s_2) \,,$$

$$w = r(q_1 p_2 B_1 - q_2 p_1 B_2)/f(s_1 - s_2) \,,$$

$$\sigma_{rr} = [\lambda(g_2 D_1 - g_1 D_2)/\sqrt{d} - (s_1^2 A_1 - s_2^2 A_2)]/(s_1 - s_2) \,,$$

$$\sigma_{\theta\theta} = \{-\lambda(g_2 D_1 - g_1 D_2)/\sqrt{d}$$
$$\qquad - [g_2 (b - as_1^2) A_1 - g_1 (b - as_2^2) A_2]/f\sqrt{d}\}/(s_1 - s_2) \,,$$

$$\sigma_{zz} = (A_1 - A_2)/(s_1 - s_2) \, , \tag{15'}$$

$$\sigma_{rz} = (s_1 C_1 - s_2 C_2)/(s_1 - s_2) \, ,$$

$$u_r = - \lambda(a_{11} - a_{12}) r(g_2 D_1 - g_1 D_2)/(s_1 - s_2) \sqrt{d} \, ,$$

$$w = r(q_1 s_2 g_2 B_1 - q_2 s_1 g_1 B_2)/f(s_1 - s_2) \, .$$

Distribution along special lines.

i) *Horizontal surface.* For $z = 0$, the formulae for stresses and displacements reduce to the relations

$$\sigma_{rr} = - \frac{1}{\sqrt{d}} L + \mu N \, ,$$

$$\sigma_{\theta\theta} = \left(\mu - \frac{1}{\sqrt{d}} \right) L - \mu N \, ,$$

$$\sigma_{zz} = - p(r) \, , \tag{16}$$

$$u_r = - \mu(a_{11} - a_{12}) r N \, ,$$

$$w = q r M \, ,$$

$$\sigma_{rr} = (s_1 + s_2)(- L + \lambda \sqrt{d} N) \, ,$$

$$\sigma_{\theta\theta} = (s_1 + s_2) [(\lambda \sqrt{d} - 1) L - \lambda \sqrt{d} N] \, ,$$

$$\sigma_{zz} = 0 \, ,$$

$$\sigma_{rz} = - p(r) \, , \tag{16'}$$

$$u_r = - q \sqrt{d} r N \, ,$$

$$w = (q_1 s_2 g_2 - q_2 s_1 g_1) r M/f(s_1 - s_2) \, ,$$

with the following notation:

$$L = \int_0^\infty J_0(mr) p^H(m) m \, dm \, ,$$

$$M = \frac{1}{r} \int_0^\infty J_0(mr) p^H(m) \, dm \, , \tag{17}$$

$$N = \frac{1}{r} \int_0^\infty J_1(mr) p^H(m) \, dm \, .$$

ii) *Axis of loading.* If in the expressions (15) or (15'), r approaches zero, we obtain

$$\sigma_{rr} = \sigma_{\theta\theta} = -\left[s_1 p_2 \left(\frac{1}{2} + \frac{b}{2} - as_1^2 \right) R_1 \right.$$

$$\left. - s_2 p_1 \left(\frac{1}{2} + \frac{b}{2} - as_2^2 \right) R_2 \right] / f(s_1 - s_2),$$

$$\sigma_{zz} = (s_2 R_1 - s_1 R_2)/(s_1 - s_2), \tag{18}$$

$$\sigma_{rz} = 0,$$

$$u_r = 0,$$

$$w = z(q_1 p_2 S_1 - q_2 p_1 S_2)/f(s_1 - s_2),$$

$$\sigma_{rr} = \sigma_{\theta\theta} = -\left[g_2 \left(\frac{1}{2} + \frac{b}{2} - as_1^2 \right) R_1 \right.$$

$$\left. - g_1 \left(\frac{1}{2} + \frac{b}{2} - as_2^2 \right) R_2 \right] / f\sqrt{d}(s_1 - s_2),$$

$$\sigma_{zz} = (R_1 - R_2)/(s_1 - s_2), \tag{18'}$$

$$\sigma_{rz} = 0,$$

$$u_r = 0,$$

$$w = z(q_1 s_2 g_2 S_1 - q_2 s_1 g_1 S_2)/f(s_1 - s_2),$$

where

$$R_i = \int_0^\infty e^{-ms_i z} \, p^H(m) \, m \, dm,$$

$$\hspace{5cm} i = 1, 2 \tag{19}$$

$$S_i = \frac{1}{z} \int_0^\infty e^{-ms_i z} \, p^H(m) \, dm.$$

The integrals (14), (17), (19) are known in terms of loading through the function p^H.

3. Problem 2

When the boundary conditions are given in vertical displacement, the problem corresponds to the penetration of a rigid punch into the half-space [1, 6]. For the points inside the contact area, we can write

$$w(r, 0) = w_0 - f(r) \qquad r \leqslant r_0 \tag{20}$$

where the function f is prescribed by the fact that, if the tip is the origin, the punch has the equation $z = f(r)$, so that $f(0) = 0$; r_0 is the (as yet unspecified) radius of the circle of contact and w_0 is the depth to which the tip of the punch penetrates into the elastic half-space.

For the problem in radial displacement, we have simply

$$u_r(r, 0) = f(r), \qquad r \leqslant r_0. \tag{20'}$$

In this case, we must determine the function p introduced in (7) or (7'). From equations (16), (16'), we obtain the boundary values

$$p(r) = \int_0^\infty m \, J_0(mr) \, p^H(m) \, dm = 0, \qquad r > r_0, \tag{21}$$

$$p(r) = \int_0^\infty m \, J_1(mr) \, p^H(m) \, dm = 0, \qquad r > r_0, \tag{21'}$$

$$w(r, 0) = q \int_0^\infty m J_0(mr) \, p^H(m) \, dm = w_0 - f(r), \qquad r \leqslant r_0, \tag{22}$$

$$u_r(r, 0) = -q\sqrt{d} \int_0^\infty J_1(mr) \, p^H(m) \, dm = f(r), \qquad r \leqslant r_0. \tag{22'}$$

If we represent p^H by a formula of the type

$$p^H(m) = \int_0^{r_0} \chi(t) \cos(mt) \, dt, \tag{23}$$

$$p^H(m) = \int_0^{r_0} \chi(t) \sin(mt) \, dt, \tag{23'}$$

we find that equation (21) or (21') is automatically satisfied. Then, equation (22) or (22') is equivalent to the Abel integral equation:

$$q \int_0^r \chi(t) \, (r^2 - t^2)^{-1/2} \, dt = w_0 - f(r), \tag{24}$$

$$q\sqrt{d} \int_0^r \chi(t) \, t(r^2 - t^2)^{-1/2} \, dt = -r \, f(r). \tag{24'}$$

The solution of this equation is known to be

$$\chi(t) = \frac{2}{q\pi} \left[w_0 - \frac{d}{dt} \int_0^t r f(r) (t^2 - r^2)^{-1/2} \, dr \right],$$ (25)

$$\chi(t) = \frac{-2}{q\sqrt{d}\pi t} \cdot \frac{d}{dt} \int_0^t r^2 f(r) (t^2 - r^2)^{-1/2} \, dr .$$ (25')

An integration by parts yields

$$\chi(t) = \frac{2}{q\pi} \left[w_0 - t \int_0^t f'(r) (t^2 - r^2)^{-1/2} \, dr \right],$$ (26)

$$\chi(t) = \frac{-2}{\sqrt{d}q\pi} \int_0^t [r f'(r) + f(r)] (t^2 - r^2)^{-1/2} \, dr .$$ (26')

The function f depends on the loading. From relations (23) and (21) we obtain successively the functions p^H and p; we have reduced Problem 2 to Problem 1, whose solution is given by the formulae (15) or (15'), for any point of the medium.

We now determine the radius r_0 of the area of contact when the boundary conditions are given in vertical displacements. We know the depth w_0 of penetration of the tip of the punch. From the physical condition that for a punch of a smooth profile, the normal component of stress σ_{zz} must remain finite along the circle of contact, we deduce the relation

$$w_0 = r_0 \int_0^{r_0} f'(r) (r_0^2 - r^2)^{-1/2} \, dr ,$$ (27)

by writing $\chi(r_0) = 0$.

It is also interesting to know the form of the displacement on the horizontal surface for the points outside the contact area. We have

$$w(r, 0) = q \int_0^{r_0} \chi(t) (r^2 - t^2)^{-1/2} \, dt , \quad r \geq r_0 ,$$ (28)

$$u_r(r, 0) = - q \sqrt{d} \int_0^{r_0} \chi(t) (r^2 - t^2)^{-1/2} \frac{t \, dt}{r} , \quad r \geq r_0 .$$ (28')

4. Application

For the isotropic half-space, Segedin [5] has considered the case in which the profile of the punch is given by

$$f(r) = \sum_{n=1}^{\infty} c_n r^n .$$

(29)

We assume this relation for the imposed displacement w or u_r. If we substitute this expression into equations (26), (26'), we find that the function χ is given by the formula

$$\chi(t) = \frac{2}{q\pi}\left(w_0 - \sum_{n=1}^{\infty} n c_n a_{n-1} t^n\right),$$

(30)

$$\chi(t) = \frac{-2}{q\sqrt{d\pi}} \sum_{n=1}^{\infty} (n+1) c_n a_n t^n ,$$

(30')

with the recurrence relation

$$a_0 = \frac{\pi}{2} ,$$

$$a_n a_{n-1} = \frac{\pi}{2n} .$$

(31)

From equations (23), (23') we obtain the function p^H:

$$p^H(m) = \frac{2}{q\pi m^2} \sum_{n=1}^{\infty} n^2 c_n a_{n-1} i_{n-1}(m) ,$$

(32)

$$p^H(m) = \frac{-2}{q\sqrt{d\pi}m} \sum_{n=1}^{\infty} (n+1) c_n a_n i_n(m) .$$

(32')

The integrals $i_n(m)$ can be determined by the recurrence relation

$$i_n(m) = r_0^n \left(\frac{n}{m r_0} \sin(m r_0) - \cos(m r_0)\right) - \frac{n(n-1)}{m^2} i_{n-2}(m)$$

(33)

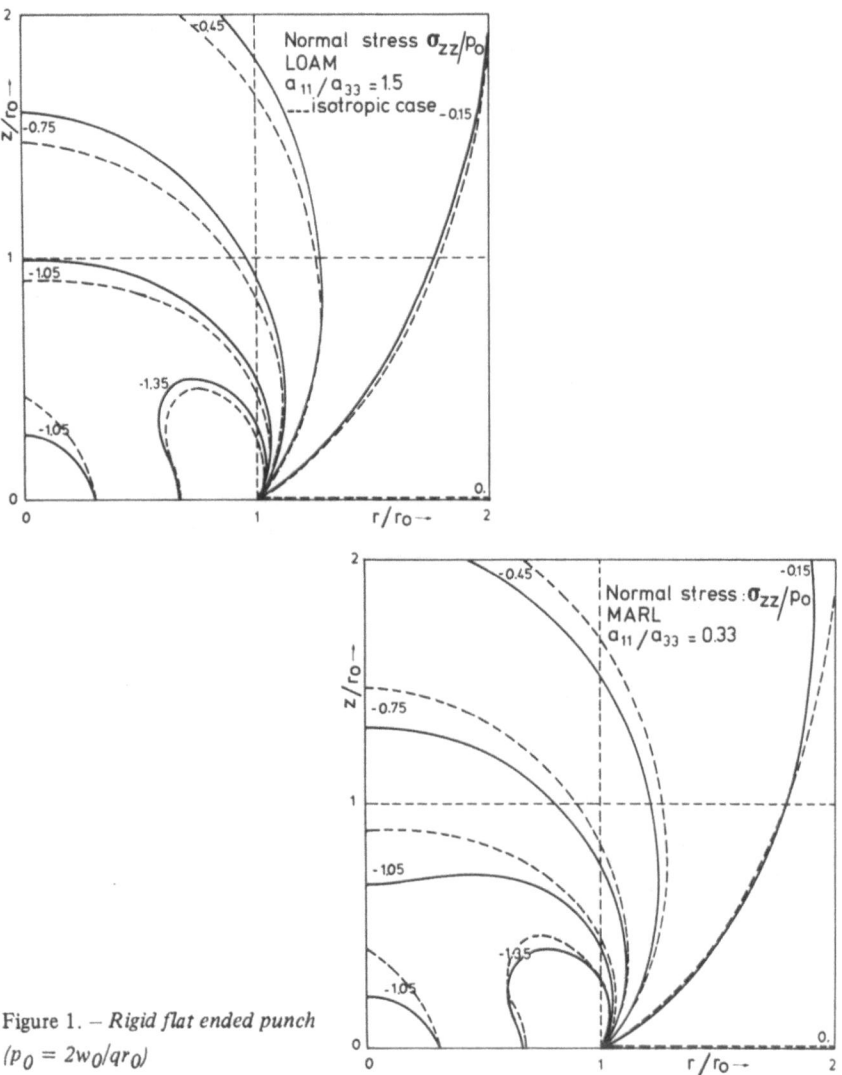

Figure 1. — *Rigid flat ended punch*
$(p_0 = 2w_0/qr_0)$

with

$$i_0(m) = 1 - \cos(m\, r_0),$$

$$i_1(m)\, r_0 \left(\frac{\sin(m\, r_0)}{m\, r_0} - \cos(m\, r_0) \right). \tag{34}$$

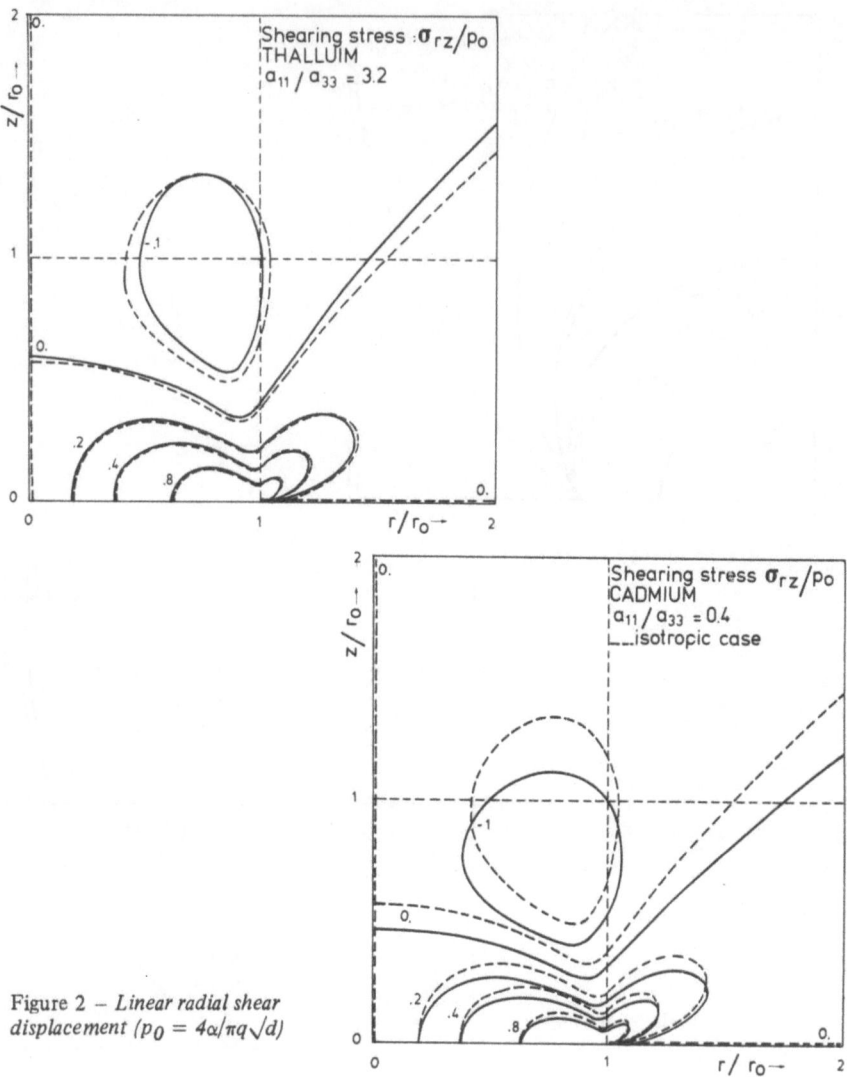

Figure 2 – *Linear radial shear displacement* $(p_0 = 4\alpha/\pi q\sqrt{d})$

Similarly, we can find the distribution of pressure under the loading $p(r)$ and the deformed shape of the free surface $w(r, 0)$ or $u_r(r, 0)$.

5. Influence of anisotropy

From the results $(26, 26')$ and $(28, 28')$, we deduce that the surface displacement is identical for a transversely isotropic half-space and an isotropic half-space; the associated stress components, under the loading, are identical to within a multiplying factor. The coefficient q must be changed by the modulus $2(1 - \nu^2)/E$ of isotropic case.

To show the influence of the anisotropy, we have plotted the stresses when the imposed displacement has the form $w = w_0$ (rigid flat-ended punch), Figure 1, and $u_r = \alpha r$ (linear radial shear displacement), Figure 2. The curves can be obtained by a simple calculation from formulas (15) and (15') or from Gerard Harrison's results [3].

When the inequality

$$a_{11}/a_{33} < 1 \tag{35}$$

is satisfied, the iso-stress curves are situated under the isotropic ones. These curves are situated above the isotropic curves when (35) is not satisfied.

The present solution can be reduced to the results for isotropic materials [6] by setting $a_{11} = a_{33} = E$, $a_{12} = a_{13} = -\dfrac{\nu}{E}$ and $a_{44} = 2\dfrac{(1 + \nu)}{E}$, where E and ν are Young's modulus and Poisson's ratio of the elastic solid.

REFERENCES

[1] DAHAN M. – "Poinçons axisymétriques rigides sur un massif semi-infini transversalement isotrope". *J. de Mécanique Appliquée,* 3 (1979) : 379-386.

[2] DAHAN M. – "Cisaillement et Déplacement radial imposés à la surface d'un massif transversalement isotrope". *Transactions of the CSME,* 5 (1979) : 181-186.

[3] GERRARD C.M. and W.J. HARRISON. – "Circular loads applied to a cross anisotropic half-space". Tech. Paper 8, *Dvn Applied Geomechanics C.S.I.R.O.,* Australia, 1970.

[4] LEKHNITSKI S.G. – *Theory of elasticity of an anisotropic elastic body.* San Francisco : Holden Day, 1963.

[5] SEGEDIN C.M. – "The relation between load and penetration for a spherical punch". *Mathematika,* 4 (1957) : 151-161.

[6] SNEDDON I.N. – "The relation between load and penetration in the axisymmetric Boussinesq problem for a punch of arbitrary profile". *Int. J. Engng. Sc.,* 3 (1965) : 47-57.

RESUME.

Chargements axisymétriques d'un massif à isotropie transversale.

La distribution des contraintes et des déplacements à l'intérieur d'un massif semi-infini transversalement isotrope soumis à une pression normale ou à un cisaillement axisymétrique est déterminée. On calcule ensuite la répartition des contraintes sur le cercle de contact pour un déplacement, vertical ou radial, imposé à la surface. On se ramène ainsi au problème précédent. Des résultats numériques sont présentés sous forme de courbes, qui montrent l'influence de l'anisotropie.

COMMENT BY C. O. HORGAN : I have a question and a comment. Have you looked at cases of severe anisotropy where the ratio of Young's moduli (E_T/E_L) may be as low as 1/20 ? Such cases are of current considerable interest for fiber-reinforced composites (see e.g. the references contained in my paper at this Colloquium) and may be also relevant in rock mechanics. From the results you have shown, it is not clear that you have provided an effective assessment of the influence of anisotropy, surely this is the chief objective of a study such as yours.

My comment concerns your use of the Lekhnitski stress function, which satisfies a 4th order elliptic partial differential equation, the stresses being given by *third* derivatives of this function. Perhaps you may be interested in an alternative formulation involving two stress functions satisfying a coupled system of second-order elliptic partial differential equations. The stresses now involve *second* derivatives of the relevant functions. Such a representation was used in [1] and proved convenient in analyzing the effect of anisotropy. I draw your attention to this approach as it may be worth investigating for your problem.

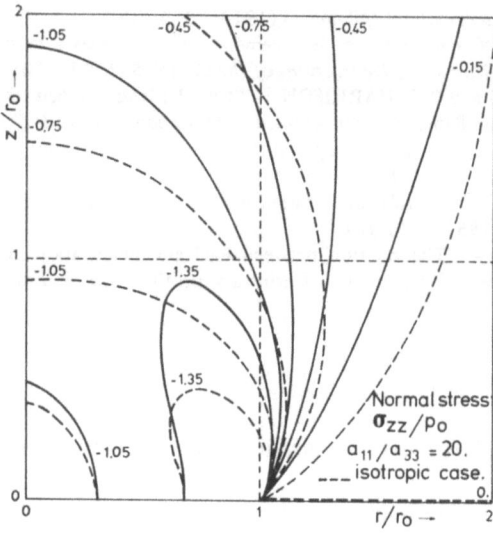

Figure A. – *Variation of the normal stress for a highly anisotropic material.*

[1] C.O. HORGAN. – "The Axisymmetric End Problem for Transversely Isotropic Circular Cylinders". *Int. J. Solids Structures,* 10 (1974): 837-852.

REPLY BY M. DAHAN: For the material where the ratio of Young's moduli (E_1/E_2) is very small, we have the same results as for loam.

However, the divergence with the isotropic case is proportional to the ratio of the anisotropy. For the severe anisotropy, $E_1/E_2 = 0.05$, I give above the plotting of the stress σ_{zz} when the imposed normal displacement is obtained by a flat-ended punch.

Thank you very much for your comment; I am interested in the method in the paper you mentioned above.